evolution

SECOND EDITION

evolution

SECOND EDITION

DOUGLAS J. FUTUYMA

Stony Brook University

Chapter 20, "Evolution of Genes and Genomes"
by Scott V. Edwards, Harvard University

Chapter 21, "Evolution and Development"
by John R. True, Stony Brook University

SINAUER ASSOCIATES, INC. • *Publishers*
Sunderland, Massachusetts U.S.A.

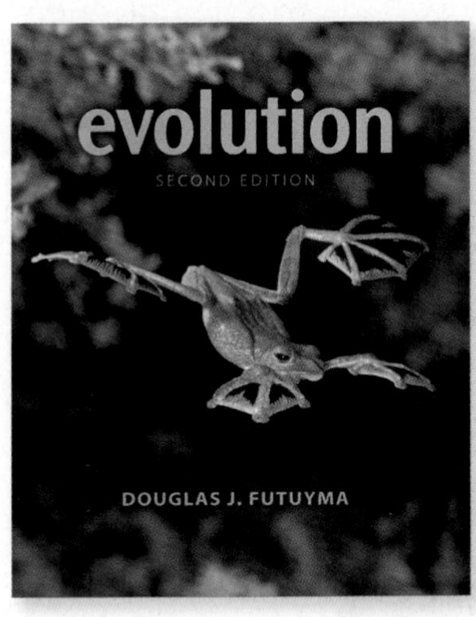

On the Cover

Wallace's flying frog (*Rhacophorus nigropalmatus*), an inhabitant of the rain forest canopy in southeastern Asia, glides from tree to tree with the aid of its toe webbing, which is much more extensive than in most other tree frog species. Such modification of ancestral characteristics to serve new functions is a common theme in evolution, and is often associated with adaptation to new ways of life. This species was named in 1869 by the naturalist Alfred Russel Wallace, who conceived of evolution by natural selection independently of Darwin. (Photograph © Stephen Dalton/Minden Pictures, Inc.)

evolution SECOND EDITION

Sinauer Associates, Inc., 23 Plumtree Road, Sunderland, MA 01375 USA

FAX: 413-549-4300

Email: publish@sinauer.com, orders@sinauer.com

Website: www.sinauer.com

Sources of the scientists' photographs appearing in Chapter 1 are gratefully acknowledged:
C. Darwin and A. R. Wallace courtesy of The American Philosophical Library
R. A. Fisher courtesy of Joan Fisher Box
J. B. S. Haldane courtesy of Dr. K. Patau
S. Wright courtesy of Doris Marie Provine
E. Mayr courtesy of Harvard News Service and E. Mayr
G. L. Stebbins, G. G. Simpson, and Th. Dobzhansky courtesy of G. L. Stebbins
M. Kimura courtesy of William Provine

Library of Congress Cataloging-in-Publication Data

Futuyma, Douglas J., 1942-
Evolution / Douglas J. Futuyma. — 2nd ed.
p. cm.
Includes index.
ISBN 978-0-87893-223-8 (hardcover)
1. Evolution (Biology) I. Title.
QH366.2.F87 2009
576.8—dc22 2009010142

Fourth printing, October 2011
Printed in the U.S.A.

To my teachers

Larry Slobodkin, Dick Lewontin, Bill Brown

Brief Contents

Contents

CHAPTER 5

A History of Life on Earth 101

CHAPTER 6

The Geography of Evolution 133

CHAPTER 7

The Evolution of Biodiversity 161

CHAPTER 8

The Origin of Genetic Variation 187

CHAPTER 9

Variation 215

CHAPTER 13

Phenotypic Evolution 337

CHAPTER 14

The Evolution of Life Histories 369

CHAPTER 15

Sex and Reproductive Success 387

CHAPTER 16

Conflict and Cooperation 413

Preface

The year 2009 marks the bicentennial anniversary of the birth of Charles Darwin and the sesquicentennial anniversary of the publication of *On the Origin of Species*. It is pleasing to release the second edition of *Evolution* in this year, when the accomplishments and future of evolutionary biology are being celebrated by symposia and by scientific and popular books and articles throughout the world. Evolution has clearly established itself as the most important principle in biology, the single scientific theory that unites the phenomena of life at every level from macromolecule to ecosystem.

During the last such celebratory year, 1959, the theoretical foundations of evolution, the joining of Darwin's insights to modern genetics, had been established, but the "molecular revolution" of biology had only just started, and as yet had had little impact on evolutionary research. Since 1959, evolutionary science has been transformed by the knowledge and techniques of molecular biology, as well as by advances in theory and in computation, and every subject area in evolution has seen immense and rapid advance. Phylogenetic relationships among organisms are understood with greater confidence and on a far greater scale; the roles of natural selection and genetic drift in the evolution of genes and phenotypes are increasingly clarified. The evolution of whole classes of organismal features, such as behavior, life histories, adaptations of species to one another, and DNA sequences, had hardly been addressed 50 years ago, but are well understood today. Our knowledge of human evolution has been transformed.

Evolutionary research today continues to illuminate these subjects, and is now exploring such frontiers as the evolution of genomes, metabolomes, complex systems such as cells and brains, and the developmental processes that translate DNA sequences into the physical structures of organisms. The diverse applications of evolutionary biology to human affairs, ranging from medicine to environmental change, have attracted increased attention and research. Scientists working in molecular, genomic, cell, and developmental biology increasingly recognize that the interpretive framework, the principles, and the analytical methods of evolutionary biology are critically important for their fields.

For these reasons, I suggest, the principles of evolution should be a foundation of the scientific education of everyone—future biologist, doctor, farmer, citizen—especially as opposition to evolution, and to the principles of sound science, spreads. I hope this book will help, however slightly, to advance the unification of biology and the understanding of science.

As in the first edition, the book begins with phylogeny and evolutionary history, continues with the many facets of evolutionary process, and returns to macroevolution, approached as a synthesis of evolutionary process and pattern. Human evolution is treated throughout the text rather than as a separate topic, an approach that reveals the social and personal relevance of the various topics in evolutionary science and points to the commonality of humans with the rest of the biological world. The final chapter, as before, treats the increasingly indispensable topics of the evidence for evolution, the nature of science, and the failings of creationism, ending on a positive note with a brief survey of some of the social applications of evolutionary biology. The most conspicuous difference between this and the previous edition is the reorganization of material on life histories, sex and sexual selection, and cooperation into three rather than two chapters, but every chapter has been updated and in some cases differs substantially from the first edition.

It is increasingly difficult to keep abreast of progress in any one field of evolutionary studies, and nearly impossible to follow—or to understand fully—all of them. So I am once again immensely grateful to Scott Edwards (Harvard University) and John True (Stony Brook University) for agreeing to continue in this venture, by contributing revised chapters on the rapidly moving subjects of genome evolution and evolutionary developmental biology.

I am grateful to Michael Foote, Bruce Lyon and Donald Prothero for indispensable review and advice on several chapters, and to Rodrigo Cogni for help with obtaining figures. Steve Dudgeon, Doug Erwin, John Relethford, Joel Sachs, and Mark Wilson provided invaluable feedback on the first edition. Among the many colleagues who provided information or answered questions as I worked on the new edition are Richard Bambach, L. Jardón Barbolla, Michael Bell, Daniel Bolnick, Carlos Bustamante, Jerry Coyne, Joel Cracraft, Daniel Dykhuizen, Walter Eanes, Catherine Graham, Joseph Graves, David Jablonski, Andrew Knoll, David Krause, Jeffrey Levinton, Maureen O'Leary, Massimo Pigliucci, David Pfennig, Joshua Rest, A. Vásquez-Lobo, and John Wiens. Scott Edwards also thanks Allan Drummond, Kevin Peterson, and Andrew Shedlock for their comments on Chapter 20.

I have learned from many others, besides those listed, during the preparation of this edition, and I apologize to those whom I have inadvertently omitted. Of course it cannot be assumed that my colleagues agree with all my interpretations or emphases, and any faults or errors are mine alone. I am very fortunate to enjoy the continued intellectual stimulation and support of the faculty and graduate students of one of the best Departments of Ecology and Evolution in the world, for which I am most grateful.

Every page of this book bears witness in its design, esthetic values, and accuracy to the extraordinary dedication and competence of the Sinauer team, including Elizabeth Pierson, David McIntyre, Jefferson Johnson, Chris Small, Carol Wigg, and, of course, Andy Sinauer. To all, my warmest thanks.

DOUGLAS J. FUTUYMA
STONY BROOK, NEW YORK
MARCH 2009

To the Student

The great geneticist François Jacob, who won the Nobel Prize in Biology and Medicine for discovering mechanisms by which gene activity is regulated, wrote in 1973 that "there are many generalizations in biology, but precious few theories. Among these, the theory of evolution is by far the most important, because it draws together from the most varied sources a mass of observations which would otherwise remain isolated; it unites all the disciplines concerned with living beings; it establishes order among the extraordinary variety of organisms and closely binds them to the rest of the earth; in short, it provides a causal explanation of the living world and its heterogeneity."

Jacob did not himself do research on evolution, but like most thoughtful biologists, he recognized its pivotal importance in the biological sciences. Today molecular biologists, developmental biologists, and genome biologists, as well as ecologists, behaviorists, anthropologists, and many psychologists, share Jacob's view. Evolution provides an indispensable framework for understanding phenomena ranging from the structure and size of genomes to some features of human behavior. Moreover, evolutionary biology is increasingly recognized for its usefulness; in fields as disparate as public health, agriculture, and computer science, the concepts, methods, and data of evolutionary biology make indispensable contributions to both basic and applied research. Everyone should know something about evolution, and for anyone who envisions a career based in the life sciences—whether as physician or as biological researcher—an understanding of evolution is indispensable. As James Watson, co-discoverer of the structure of DNA, wrote, "today, the theory of evolution is an accepted fact for everyone but a fundamentalist minority."

The core of evolutionary biology consists of describing and analyzing the history of evolution and of analyzing its causes and mechanisms. The scope of evolutionary biology is far greater than any other field of biological science, because all organisms, and all their characteristics, are products of a history of evolutionary change. Because of this enormous scope, courses in evolution generally do not emphasize the details of the evolution of particular groups of organisms—the amount of information would be simply overwhelming. Rather, evolution courses emphasize the general principles of evolution, the hypotheses about the causes of evolutionary change that apply to most or all organisms, and the major patterns of change that have characterized many different groups. In this book, concepts are illustrated with examples drawn from research on a great variety of organisms, but it is less important to know the details of these examples than to understand how the data obtained in those studies bear on a hypothesis.

Determining the causes and patterns of evolution can be difficult, in part because we often are attempting to understand how and why something happened in the past. In this way, evolutionary biology differs from most other biological subjects, which deal with current characteristics of organisms. However, evolutionary biology and other biological disciplines share the fact that we often must make inferences about invisible processes or objects. We do not see past evolutionary changes in action; but neither can we actually see DNA replication, nor can we see the hormones that we know regulate growth and reproduction. Rather, we make inferences about these things by (1) posing informed hypotheses about what they are and how they work, then (2) generating predictions (making deductions) from these hypotheses about data that we can actually obtain, and finally (3) judging the validity of each hypothesis by the match between our observations and what we expect to see if the hypothesis were true. This is the "hypothetico-deductive method," of which Darwin was one of the first successful exponents, and which is widely and powerfully used throughout science.

In science, a group of interrelated hypotheses that have been well supported by such tests, and which together explain a wide range of phenomena or observations, is called a *theory*. "Theory" in this sense—the sense in which Jacob and Watson used the word—does not mean mere speculation. "Theory" is a term of honor, reserved for principles such as quantum theory, atomic theory, or cell theory, all of which are well supported and provide a broad framework of explanation. The emphasis in your course in evolutionary biology, then, will probably be, first, on learning the theory (the principles of evolutionary change that together explain a vast variety of observations about organisms); and second, on learning how to test evolutionary hypotheses with data: the hypothetico-deductive method applied to questions about what has happened, and how it has happened—whether it be the history of corn from its wild state to mass cultivation, or the evolution of complex societies in some insect species, or the spread of human immunodeficiency virus (HIV).

For many students, the emphasis in studying evolution may differ from what they have experienced in other biology courses. I suggest that you pay special attention to chapters or passages where the fundamental principles and methods are introduced, be sure you understand them before moving on, and reread these passages after you have gone through one or more later chapters. (You may want to revisit Chapters 2, 9, 10, and 12–13 in particular.) Be sure to emphasize understanding, not memorization, and test your understanding, using the questions at the end of each chapter or in this book's associated Web site, or that your instructor assigns.

The material in this book builds cumulatively. Almost every concept, principle, or major technical term introduced in any chapter is used again in later chapters. You will need to understand the early chapters as thoroughly for your final exam as for a midterm exam. Evolutionary biology is a unified whole; just as carbohydrate metabolism and amino acid synthesis cannot be divorced in biochemistry, so it is for topics as seemingly different as the phylogeny of species and the theory of genetic drift.

Finally, I cannot emphasize too strongly that in every field of science, the unknown greatly exceeds the known. Thousands of research papers on evolutionary topics are published each year, and many of them raise new questions even as they attempt to answer old ones. No one, least of all a scientist, should be afraid to say "I don't know" or "I'm not sure"; those refrains will sound fairly often in this book. To recognize where our knowledge and understanding are uncertain or lacking is to see where more research may be warranted, or where exciting new research trails might be blazed. I hope that some readers will find evolution so rich a subject, so intellectually challenging, so fertile in insights, and so deep in its implications that they will adopt evolutionary biology as a career. But all readers, I hope, will find in evolutionary biology the thrill of understanding, the excitement of finding both answers and intriguing new questions about the living world, including ourselves. *Felix qui potuit rerum cognoscere causas,* wrote Virgil: happy is the person who can learn the causes of things.

Media and Supplements

to accompany **evolution** SECOND EDITION

eBOOK (www.coursesmart.com; ISBN 978-0-87893-361-7)

Evolution, Second Edition is available as an eBook via CourseSmart. This affordable electronic version of the textbook can be purchased online at a substantial discount off the price of the printed book, as either an online or downloadable ebook. Both versions include features such as highlighting, full-text search, and a notes tool.

FOR INSTRUCTORS (Available to qualified adopters)

Instructor's Resource Library (ISBN 978-0-87893-360-0)

The *Evolution,* Second Edition Instructor's Resource Library includes a variety of resources to help instructors in developing their courses and delivering their lectures. The Library includes:

- *Textbook Figures and Tables:* All of the figures (including photographs) and tables from the textbook are provided as JPEGs (both high- and low-resolution), reformatted and relabeled for optimal readability.
- *PowerPoint Presentations:* For each chapter, all figures and tables are provided in a ready-to-use PowerPoint presentation, making it easy to quickly insert figures into lecture presentations.
- *Answers to the textbook end-of-chapter Problems and Discussion Topics.*
- *Companion Website Online Quiz questions.*
- *Companion Website Data Analysis and Simulation Exercises,* for use in class.

Online Quizzing

Available via the Companion Website, instructors have access to a set of online quizzes that can be used either as assigned homework or as a self-study tool. Instructors choose how they want the quizzes to be used by their students, and can quickly see student results via the quiz administration site. (Instructor registration is required in order for students to be able to access the quizzes. Instructors can request registration via the Companion Website.)

FOR STUDENTS

Companion Website (www.sinauer.com/evolution)

New for the Second Edition, the *Evolution* Companion Website features review and study tools to help students master the material presented in the textbook. Access to the site is free of charge, and requires no passcode. (Instructor registration is required in order for students to be able to access the quizzes.) The site includes:

- *Chapter Outlines and Summaries:* Concise overviews of the important topics covered in each chapter.
- *Data Analysis Exercises:* Inquiry-based problems designed to sharpen the student's ability to reason as a scientist, drawing on data from real experiments and published papers.
- *Simulation Exercises:* Interactive modules that allow students to explore many of the dynamic processes of evolution and answer questions based on the results they observe.

- *Online Quizzes:* Quizzes that cover all the major concepts introduced in each chapter. These quizzes are assignable by the instructor.
- *Flashcards & Key Terms:* Easy-to-use activities that help students learn all the key terminology introduced in each chapter.
- The complete *Glossary.*

CHAPTER 1

Evolutionary Biology

A plethora of adaptations. The cryptic color pattern, hypodermic-like fangs, fast-acting venom, and high-speed reflexes of an African puff adder (*Bitis arietans*) are exquisite adaptations. The only scientific explanation of adaptations is the theory of evolution by natural selection. (Photo © Tony Allen/OSF/Photolibrary.com.)

Biologists, looking broadly at living things, are stirred to ask thousands of questions. Why do whales have lungs, and why do snakes lack legs? Why do some insects attack only a single species of plant, while others eat many different species? Why do salamanders have more than ten times as much DNA as humans? Why are there 19,000 species of orchids and more than 300,000 species of beetles? Why do people differ in facial features, skin color, attitudes, behavior? What accounts for the astonishing variety of organisms? What accounts for their intricate, multitudinous adaptations?

In one of the most breathtaking ideas in the history of science, Charles Darwin proposed that *"all the organic beings which have ever lived on this earth have descended from some one primordial form."* From this idea, it follows that every characteristic of every species—the fangs of a puff adder, the number and sequence of its genes, the catalytic abilities of its enzymes, the structure of its cells and organs, its physiological tolerances and nutritional requirements, its life span and reproductive system, its capacity for behavior—is the outcome of an evolutionary history. The evolutionary perspective illuminates every subject in biology, from molecular biology to ecology. Indeed, *evolution is the unifying theory of biology*. "Nothing in biology makes sense," said the geneticist Theodosius Dobzhansky, "except in the light of evolution."

What Is Evolution?

The word "evolution" comes from the Latin *evolvere*, "to unfold or unroll"—to reveal or manifest hidden potentialities. Today "evolution" has come to mean, simply, "change." It is sometimes used to describe changes in individual objects such as stars. **Biological** (or **organic**) **evolution**, however, is *change in the properties of groups of organisms over the course of generations*. The development, or ONTOGENY, of an individual organism is not considered evolution: individual organisms do not evolve. Groups of organisms, which we may call **populations**, undergo *descent with modification*. Populations may become subdivided, so that several populations are derived from a *common ancestral population*. If different changes transpire in the several populations, the populations *diverge*.

The changes in populations that are considered evolutionary are those that are passed via the genetic material from one generation to the next. Over the course of many generations, many changes may accumulate. Thus, biological evolution may be slight or substantial: it embraces everything from slight changes in the proportions of different forms of a gene within a population to the alterations that led from the earliest organism to corals, grasshoppers, tomatoes, and humans.

No more dramatic example of evolution by natural selection can be imagined than that of today's crisis in antibiotic resistance. Before the 1940s, most people in hospital wards did not have cancer or heart disease. They suffered from tuberculosis, pneumonia, meningitis, typhoid fever, syphilis, and many other kinds of bacterial infection—and they had little hope of being cured (Figure 1.1). Infectious bacterial diseases condemned millions of people in the developed countries to early death. Populations in developing countries bore not only these burdens, but also diseases such as malaria and cholera, even more heavily than they do today.

By the 1960s, the medical situation had changed dramatically. The discovery of antibiotic drugs and subsequent advances in their synthesis led to the conquest of most bacterial diseases, at least in developed countries. The sexual revolution in the 1970s was encouraged by the confidence that sexually transmitted diseases such as gonorrhea and syphilis were merely a temporary inconvenience that penicillin could cure. In 1969, the Surgeon General of the United States proclaimed that it was time to "close the book on infectious diseases."

He was wrong. Today we confront not only new infectious diseases such as AIDS, but also a resurgence of old diseases such as tuberculosis. The same bacteria are back, but now

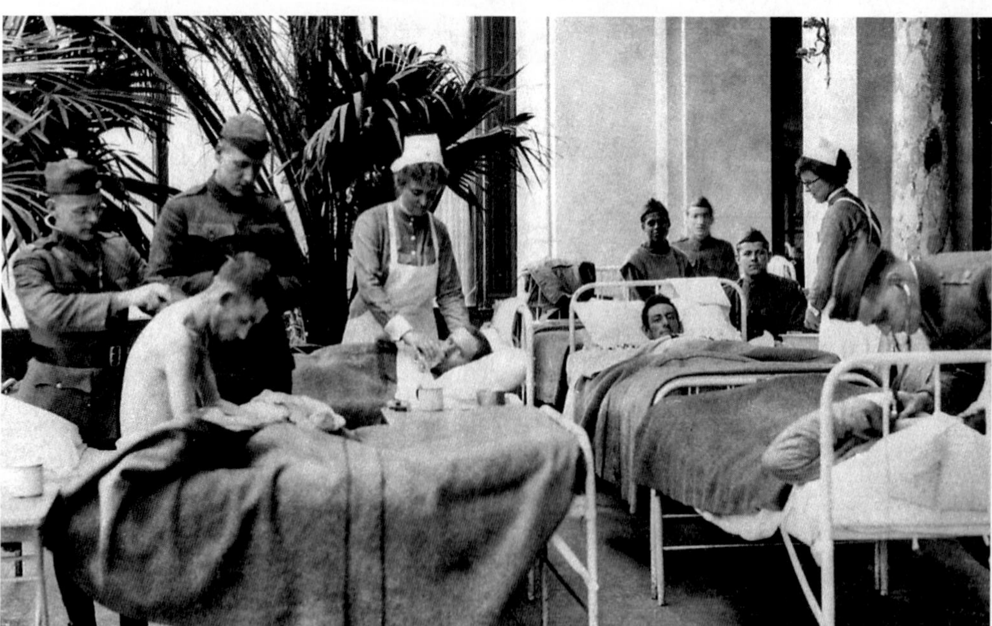

Figure 1.1 A tuberculosis ward at a U.S. Army base hospital in France during World War I. Until recently, it was thought that antibiotics, which came into widespread use after World War II, had conquered this devastating bacterial disease. (Photo courtesy of the National Library of Medicine.)

they are resistant to penicillin, ampicillin, erythromycin, vancomycin, fluoroquinolones—all the weapons that were supposed to have vanquished them. Almost every hospital in the world treats casualties in this battle against changing opponents, and unintentionally may make those opponents stronger. In this and many other ways, humans are instigating an explosion of evolutionary change (Palumbi 2001).

For example, *Staphylococcus aureus*, a bacterium that causes many infections in surgical patients, has become resistant to a vast array of antibiotics, starting with penicillin and working its way through many others. Almost the only effective remedy that remains against *S. aureus* is vancomycin, but it too is becoming less effective. Likewise, drug-resistant strains of *Neisseria gonorrheae*, the bacterium that causes gonorrhea, have steadily increased in abundance; many strains of the pneumonia and cholera bacteria are highly resistant to antibiotics; throughout the tropics, the microorganism that causes malaria is now resistant to chloroquine, and is becoming resistant to other drugs as well. HIV, the human immunodeficiency virus that causes AIDS, rapidly became resistant to AZT and other drugs that at first were effective. People now survive with HIV by taking a combination of three drugs, but there is great concern about how long it will be until virus populations evolve resistance to these as well.

As the use of antibiotics increases, so does the incidence of bacteria that are resistant to those antibiotics; thus any gains made are almost as quickly lost (Figure 1.2). Why is this happening? Do the drugs cause drug-resistant mutations in the bacteria's genes? Do the mutations occur even without exposure to drugs—that is, are they present in unexposed bacterial populations? How many mutations result in resistance to a drug? How often do they occur? Do the mutations spread from one bacterium to another? Are they spread only among bacteria or viruses of the same species, or can they pass between different species? How is the growth of the organism's population affected by such mutations? Can the evolution of resistance be prevented by using lower doses of drugs? Higher doses? Combinations of different drugs? Can an individual recover from infection by drug-resistant organisms by faithfully following a physician's prescription, or will this work only if everyone else is just as conscientious?

The principles and methods of evolutionary biology have provided some answers to these questions, and to many others that affect society. Evolutionary biologists and other scientists trained in evolutionary principles have traced the transfer of HIV to humans from chimpanzees and mangabey monkeys (Gao et al. 1999; Korber et al. 2000). They have studied the evolution of insecticide resistance in disease-carrying and crop-destroying insects. They have helped to devise methods of nonchemical pest control and have laid the foundations for transferring genetic resistance to diseases and insects from wild plants to crop plants. Evolutionary principles and knowledge are being used in biotechnology to design new drugs and other useful products. In computer science and artificial intelligence, "evolutionary computation" uses principles taken directly from evolutionary the-

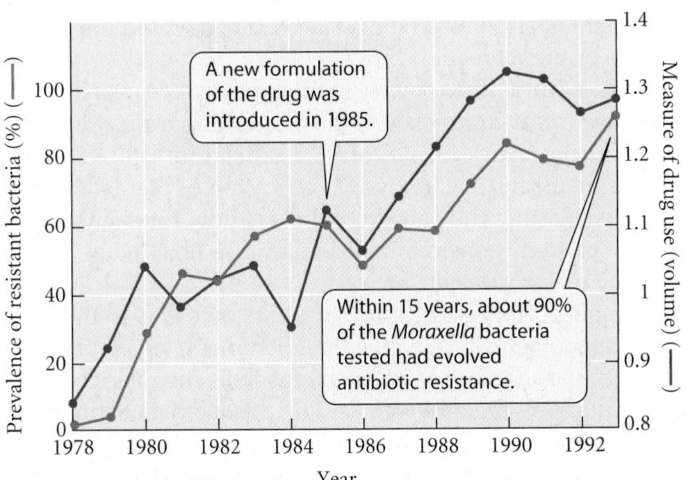

Figure 1.2 Evolution of drug resistance. The increase in the consumption of penicillin-like antibiotics in a community in Finland between 1978 and 1993 was matched by a dramatic increase in the percentage of resistant isolates of the bacterium *Moraxella catarrhalisis* from middle-ear infections in young children. (After Levin and Anderson 1999.)

ory to solve mathematically intractable practical problems, such as constructing complex timetables or processing radar data.

The importance of evolutionary biology goes far beyond its practical uses, however. An evolutionary framework provides answers to many questions about ourselves. How do we account for human variation—the fact that almost everyone is genetically and phenotypically unique? Are there human races, and if so, how do they differ, and how and when did they develop? What accounts for behavioral differences between men and women? How did exquisitely complex, useful features such as our hands and our eyes come to exist? What about apparently useless or even potentially harmful characteristics such as our wisdom teeth and appendix? Why do we age, undergo senescence, and eventually die? Why are medical researchers able to use monkeys, mice, and even fruit flies and yeasts as models for processes in the human body? Moreover, as soon as Darwin published *The Origin of Species*, the evolutionary perspective was perceived to bear on longstanding questions in philosophy. If humans, with all their mental and emotional complexity, originated by natural processes, where do ethics and moral precepts find a foundation and origin? What, if anything, does evolution imply about the meaning and purpose of life?

Before Darwin

Darwin's theory of biological evolution is one of the most revolutionary ideas in Western thought, perhaps rivaled only by Newton's theory of physics. It profoundly challenged the prevailing world view, which had originated largely with Plato and Aristotle. Foremost in Plato's philosophy was his concept of the *eidos*, the "form" or "idea," a transcendent ideal form imperfectly imitated by its earthly representations. For example, the reality—the "essence"—of the true equilateral triangle is only imperfectly captured by the triangles we draw or construct. Likewise, horses (or any other species) have an immutable essence, but each individual horse has imperfections. In this philosophy of **essentialism**, variation is accidental imperfection.

Plato's philosophy of essentialism became incorporated into Western philosophy largely through Aristotle, who developed Plato's concept of immutable essences into the notion that species have fixed properties. Later, Christians interpreted the biblical account of Genesis literally and concluded that each species had been created individually by God in the same form it has today. (This belief is known as "special creation.") Christian theologians and philosophers elaborated on Platonic and Aristotelian philosophy, arguing that since existence is good and God's benevolence is complete, He must have bestowed existence on every creature of which he could conceive. Because order is superior to disorder, God's creation must follow a plan: specifically, a gradation from inanimate objects and barely animate forms of life, through plants and invertebrates, up through ever "higher" forms of life. Humankind, which is both physical and spiritual in nature, formed the link between animals and angels. This "Great Chain of Being," or *scala naturae* (the scale, or ladder, of nature), must be permanent and unchanging, since change would imply that there had been imperfection in the original creation (Lovejoy 1936).

As late as the eighteenth century, natural history was justified partly because it might reveal the plan of creation so that we might appreciate God's wisdom. Carolus Linnaeus (1707–1778), who established the framework of modern classification in his *Systema Naturae* (1735), won worldwide fame for his exhaustive classification of plants and animals, undertaken in the hope of discovering the pattern of the creation. Linnaeus classified "related" species into genera, "related" genera into orders, and so on. To him, "relatedness" meant propinquity in the Creator's design.

Belief in the literal truth of the biblical story of creation started to give way in the eighteenth century, when a philosophical movement called the Enlightenment, largely inspired by Newton's explanations of physical phenomena, adopted reason as the major basis of authority. The foundations for evolutionary thought were laid by astronomers, who developed theories of the origin of stars and planets, and by geologists, who amassed evidence that the Earth had undergone profound changes, that it had been populated by many creatures now extinct, and that it was very old. The geologists James Hutton and Charles Lyell expounded the principle of **uniformitarianism**, holding that the same processes operated in the past as in the present, and that the observations of geology should therefore be explained by causes that we can now observe. Darwin was greatly influenced by Lyell's teachings, and he adopted uniformitarianism in his thinking about evolution.

JEAN-BAPTISTE PIERRE ANTOINE DE MONET, CHEVALIER DE LAMARCK

In the eighteenth century, several French philosophers and naturalists suggested that species had arisen by natural causes. The most significant pre-Darwinian evolutionary hypothesis, representing the culmination of eighteenth-century evolutionary thought, was proposed by the Chevalier de Lamarck in his *Philosophie Zoologique* (1809). Lamarck proposed that each species originated individually by spontaneous generation from nonliving matter, starting at the bottom of the chain of being. A "nervous fluid" acts within each species, he said, causing it to progress up the chain. Species originated at different times, so we now see a hierarchy of species because they differ in age (Figure 1.3A).

Lamarck argued that species differ from one another because they have different needs, and so use certain of their organs and appendages more than others. The more strongly exercised organs attract more of the "nervous fluid," which enlarges them, just as muscles become strengthened by work. Lamarck, like most people at the time, believed that such alterations, acquired during an individual's lifetime, are inherited—a principle called **inheritance of acquired characteristics**. In the most famous example of Lamarck's theory, giraffes originally had short necks, but stretched their necks to reach foliage above them. Hence their necks were lengthened; longer necks were inherited, and over the course of generations, their necks got longer and longer. This could happen to any and all giraffes, so the entire species could have acquired longer necks because it was composed of individual organisms that changed during their lifetimes (see Figure 1.4, top panel).

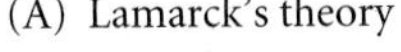

(A) Lamarck's theory

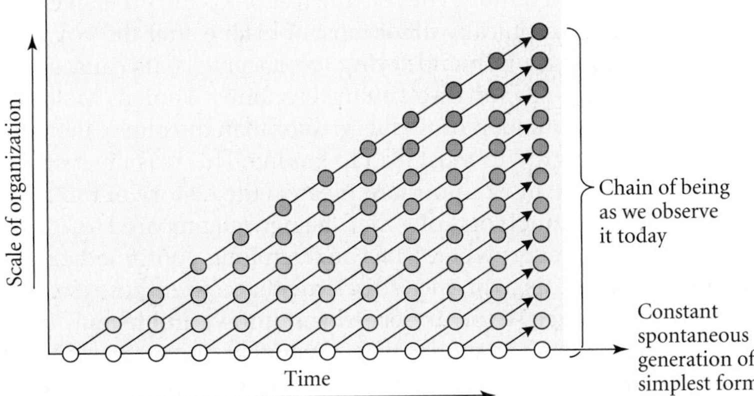

(B) Darwin's theory

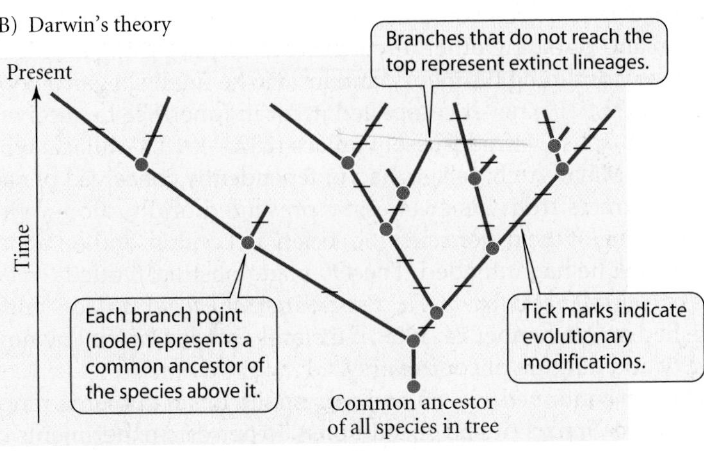

Figure 1.3 (A) Lamarck's theory of organic progression. Over time, species originate by spontaneous generation, and each evolves up the scale of organization, establishing a *scala naturae*, or chain of being, that ranges from newly originated simple forms of life to older, more complex forms. In Lamarck's scheme, species have not originated from common ancestors. (B) Darwin's theory of descent with modification, represented by a phylogenetic tree. Lineages (species) descend from common ancestors, undergoing various modifications in the course of time. Some (such as the leftmost lineages) may undergo less modification from the ancestral condition than others (the rightmost lineages). (A after Bowler 1989.)

CHARLES ROBERT DARWIN

Lamarck's ideas had little impact during his lifetime, partly because they were criticized by respected zoologists, and partly because after the French Revolution, ideas issuing from France were considered suspect in most other countries. Lamarck's ideas of how evolution works were wrong, but he deserves credit for being the first to advance a coherent theory of evolution.

Charles Darwin

Charles Robert Darwin (February 12, 1809–April 19, 1882) was the son of an English physician. He briefly studied medicine at Edinburgh, then turned to studying for a career in the clergy at Cambridge University. He apparently believed in the literal truth of the Bible as a young man. He was passionately interested in natural history and became a companion of the natural scientists on the faculty. In 1831, at the age of 22, his life was forever changed when he was invited to serve as a naturalist and captain's gentleman companion on the H.M.S. *Beagle*, a ship the British navy was sending to chart the waters of South America.

The voyage of the *Beagle* lasted from December 27, 1831 to October 2, 1836. The ship spent several years traveling along the coast of South America, where Darwin observed the natural history of the Brazilian rain forest and the Argentine pampas, then stopped in the Galápagos Islands, which lie on the Equator off the coast of Ecuador. In the course of the voyage, Darwin became an accomplished naturalist, collected specimens, made innumerable geological and biological observations, and conceived a new (and correct) theory of the formation of coral atolls. Soon after Darwin returned, the ornithologist John Gould pointed out that Darwin's specimens of mockingbirds from the Galápagos Islands were so different from one island to another that they represented different species. Darwin then recalled that the giant tortoises, too, differed from one island to the next. These facts, and the similarities between fossil and living mammals that he had found in South America, triggered his conviction that different species had evolved from common ancestors.

Darwin's comfortable finances enabled him to devote the rest of his life exclusively to his biological work (although he was chronically ill for most of his life after the voyage). He set about amassing evidence of evolution and trying to conceive of its causes. On September 28, 1838, at the age of 29, he read an essay by the economist Thomas Malthus, who argued that the rate of human population growth is greater than the rate of increase in the food supply, so that unchecked growth must lead to famine. This was the inspiration for Darwin's great idea, one of the most important ideas in the history of thought: **natural selection**. Darwin wrote in his autobiography that "being well prepared to appreciate the struggle for existence which everywhere goes on from long-continued observation of the habits of animals and plants, it at once struck me that under these circumstances favourable variations would tend to be preserved and unfavourable ones to be destroyed." In other words, if individuals of a species with superior features survived and reproduced more successfully than individuals with inferior features, and if these differences were inherited, the average character of the species would be altered.

ALFRED RUSSEL WALLACE

Mindful of how controversial the subject would be, Darwin then spent 20 years amassing evidence about evolution and pursuing other researches before publishing his ideas. In 1844, he wrote a private essay outlining his theory, and in 1856 he finally began a book he intended to call *Natural Selection*. He never completed it, for in June 1858 he received a manuscript from a young naturalist, Alfred Russel Wallace (1823–1913). Wallace, who was collecting specimens in the Malay Archipelago, had independently conceived of natural selection. Darwin had extracts from his 1844 essay presented orally, along with Wallace's manuscript, at a meeting of the major scientific society in London, and set about writing an "abstract" of the book he had intended. The 490-page "abstract," titled *On the Origin of Species by Means of Natural Selection, or The Preservation of Favoured Races in the Struggle for Life*, was published on November 24, 1859; it instantly made Darwin, by now 50 years old, both a celebrity and a figure of controversy.

For the rest of his life, Darwin continued to read and correspond on an immense range of subjects, to revise *The Origin of Species* (it had six editions), to perform experiments of all sorts (especially on plants), and to publish many more articles and books, of which *The Descent of Man* is the most renowned. Darwin's books reveal an irrepressibly inquisitive man, fascinated with all of biology, creative in devising hypotheses and in bringing evidence to bear upon them, and profoundly aware that every fact of biology, no matter how seemingly trivial, must fit into a coherent, unified understanding of the world.

Darwin's Evolutionary Theory

The Origin of Species has two major theses. The first is Darwin's theory of **descent with modification**. It holds that all species, living and extinct, have descended, without interruption, from one or a few original forms of life (Figure 1.3B). Species that diverge from a common ancestor were at first very similar, but accumulated differences over great spans of time, so that they may have come to differ radically from one another. Darwin's conception of the course of evolution is profoundly different from Lamarck's, in which the concept of common ancestry plays almost no role.

The second theme of *The Origin of Species* is Darwin's theory of the causal agents of evolutionary change. This was his theory of natural selection: "if variations useful to any organic being ever occur, assuredly individuals thus characterised will have the best chance of being preserved in the struggle for life; and from the strong principle of inheritance, these will tend to produce offspring similarly characterised. This principle of preservation, or the survival of the fittest, I have called natural selection." This theory is a VARIATIONAL THEORY of change, differing profoundly from Lamarck's TRANSFORMATIONAL THEORY, in which individual organisms change (Figure 1.4).

What is often called "Darwin's theory of evolution" actually includes five theories (Mayr 1982a):

1. *Evolution as such* is the simple proposition that the characteristics of lineages of organisms change over time. This idea was not original with Darwin, but it was Darwin who so convincingly marshaled the evidence for evolution that most biologists soon accepted that it has indeed occurred.

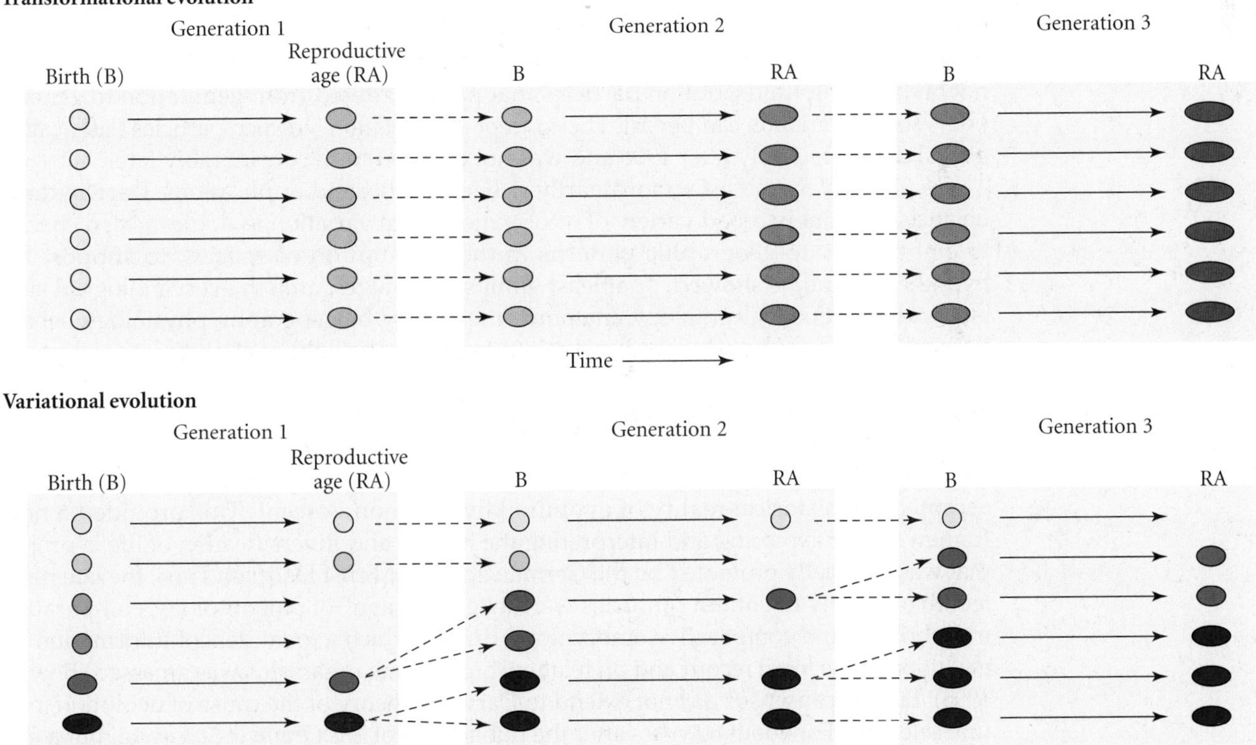

Figure 1.4 A diagrammatic contrast between transformational and variational theories of evolutionary change, shown across three generations. Within each generation, individuals are represented earlier and later in their lives. The individuals in the left column in each generation are the offspring of those in the right column of the preceding generation. In transformational evolution, individuals are altered during their lifetimes, and their progeny are born with these alterations. In variational evolution, hereditarily different forms at the beginning of the history are not transformed, but instead differ in survival and reproductive rate, so that their proportions change from one generation to another.

2. *Common descent* is a radically different view of evolution than the scheme Lamarck had proposed (see Figure 1.3). Darwin was the first to argue that species had diverged from common ancestors and that all of life could be portrayed as one great family tree.
3. *Gradualism* is Darwin's proposition that the differences between even radically different organisms have evolved incrementally, by small steps through intermediate forms. The alternative hypothesis is that large differences evolve by leaps, or SALTATIONS, without intermediates.
4. *Populational change* is Darwin's thesis that evolution occurs by changes in the *proportions* of individuals within a population that have different inherited characteristics (see Figure 1.4, bottom panel). This concept was a completely original idea that contrasts both with the sudden origin of new species by saltation and with Lamarck's account of evolutionary change by transformation of individuals.
5. *Natural selection* was Darwin's brilliant hypothesis, independently conceived by Wallace, that changes in the proportions of different types of individuals are caused by differences in their ability to survive and reproduce—and that such changes result in the evolution of **adaptations**, features that appear "designed" to fit organisms to their environment. *The concept of natural selection revolutionized not only biology, but Western thought as a whole.*

Darwin proposed that the various descendants of a common ancestor evolve different features because they are adaptive under different "conditions of life"—different habitats or habits. Moreover, the pressure of competition favors the use of different foods or habitats by different species. He believed that no matter how extensively a species has diverged from its ancestor, new hereditary variations continue to arise, so that given enough time, there is no evident limit to the amount of divergence that can occur.

Where, though, do these hereditary variations come from? This was the great gap in Darwin's theory, and he never filled it. The problem was serious, because according to the prevailing belief in BLENDING INHERITANCE, variation should decrease, not increase. Because offspring are often intermediate between their parents in features such as color or size, it was widely believed that characteristics are inherited like fluids, such as different colors of paint. Blending white and black paints produces gray, but mixing two gray paints doesn't yield black or white: variation decreases. Darwin never knew that Gregor Mendel had solved the problem in a paper that was published in 1865, but not widely noticed until 1900. Mendel's theory of PARTICULATE INHERITANCE proposed that inheritance is based not on blending fluids, but on particles that pass unaltered from generation to generation—so that variation can persist. The concept of "mutation" in such particles (later called genes) developed only after 1900 and was not clarified until considerably later.

The Origin of Species is extraordinarily rich in insights and implications. Darwin drew on an astonishingly broad variety of information, from variation in domesticated species to embryology to geographic patterns in the distribution of species, to support his hypotheses. And he showed, or at least glimpsed, how research in every biological subject—classification, paleontology, anatomy, embryology, biogeography, physiology, behavior, ecology—could be advanced and reinterpreted in the light of evolution.

Evolutionary Theories after Darwin

Although *The Origin of Species* raised enormous controversy, by the 1870s most scientists accepted the historical reality of evolution by common descent. This provided a new framework for exploring and interpreting the history and diversification of life, a project that was especially promoted by the German zoologist Ernst Haeckel. Thus, the late nineteenth and early twentieth centuries were a "golden age" of paleontology, comparative morphology, and comparative embryology, during which a great deal of information on evolution in the fossil record and on relationships among organisms was amassed (Bowler 1996). But the consensus did not extend to Darwin's theory of the cause of evolution, natural selection. For about 60 years after the publication of *The Origin of Species*, all but a few faithful Darwinians rejected natural selection, and numerous theories were proposed in its stead. These theories included neo-Lamarckian, orthogenetic, and mutationist theories (Bowler 1989).

NEO-LAMARCKISM includes several theories based on the old idea of inheritance of modifications acquired during an organism's lifetime. Such modifications might have been due, for example, to the direct effect of the environment on development (as in plants that develop thicker leaves if grown in a hot, dry environment). In a famous experiment, the German biologist August Weismann (1834–1914) cut off the tails of mice for many generations and showed that this mutilation had no effect on the tail length of their descendants. Extensive subsequent research has provided no evidence that specific hereditary changes can be induced by environmental conditions under which they would be advantageous.

Theories of ORTHOGENESIS, or "straight-line evolution," held that the variation that arises is directed toward fixed goals, so that a species evolves in a predetermined direction without the aid of natural selection. Some paleontologists held that such trends need not be adaptive and could even drive species toward extinction. None of the proponents of orthogenesis ever proposed a mechanism for it.

RONALD A. FISHER

MUTATIONIST theories were advanced by some geneticists who observed that discretely different new phenotypes can arise by a process of mutation. They supposed that such mutant forms constituted new species, and thus believed that natural selection was not necessary to account for the origin of species. Mutationist ideas were advanced by Hugo de Vries, one of the biologists who "discovered" Mendel's neglected paper in 1900, and by Thomas Hunt Morgan, the founder of *Drosophila* genetics. The last influential mutationist was Richard Goldschmidt (1940), an accomplished geneticist who nevertheless erroneously argued that evolutionary change within species is entirely different in kind from the origin of new species and higher taxa. These, he said, originate by sudden, drastic changes that reorganize the whole genome. Although most such reorganizations would be deleterious, a few "hopeful monsters" would be the progenitors of new groups.

The Evolutionary Synthesis

These anti-Darwinian ideas were refuted in the 1930s and 1940s by the **evolutionary synthesis**, or **modern synthesis**, forged from the contributions of geneticists, systematists, and paleontologists who reconciled Darwin's theory with the facts of genetics (Mayr and Provine 1980; Smocovitis 1996). Ronald A. Fisher and John B. S. Haldane in England and Sewall Wright in the United States developed a mathematical theory of population genetics, which showed that mutation and natural selection *together* cause adaptive evolution: mutation is not an alternative to natural selection, but is rather its raw material. The study of genetic variation and change in natural populations was pioneered in Russia by Sergei Chetverikov and continued by Theodosius Dobzhansky, who moved from Russia to the United States. In his influential book *Genetics and the Origin of Species* (1937), Dobzhansky conveyed the ideas of the population geneticists to other biologists, thus influencing their appreciation of the genetic basis of evolution.

J. B. S. HALDANE

Other major contributors to the synthesis included the zoologists Ernst Mayr, in *Systematics and the Origin of Species* (1942), and Bernhard Rensch, in *Evolution Above the Species Level* (1959); the botanist G. Ledyard Stebbins, in *Variation and Evolution in Plants* (1950); and the paleontologist George Gaylord Simpson, in *Tempo and Mode in Evolution* (1944) and its successor, *The Major Features of Evolution* (1953). These authors argued persuasively that mutation, recombination, natural selection, and other processes operating *within species* (which Dobzhansky termed **microevolution**) account for the *origin of new species and for the major, long-term features of evolution* (termed **macroevolution**).

Fundamental principles of evolution

The principal claims of the evolutionary synthesis are the foundations of modern evolutionary biology. Although some of these principles have been extended, clarified, or modified since the 1940s, most evolutionary biologists today accept them as fundamentally valid. These, then, are the fundamental principles of evolution, to be discussed at length throughout this book.

SEWALL WRIGHT

ERNST MAYR

G. LEDYARD STEBBINS, GEORGE GAYLORD SIMPSON, AND THEODOSIUS DOBZHANSKY

1. *The phenotype* (observed characteristic) *is different from the genotype* (the set of genes in an individual's DNA); phenotypic differences among individual organisms may be due partly to genetic differences and partly to direct effects of the environment.
2. Environmental effects on an individual's phenotype do not affect the genes passed on to its offspring. In other words, *acquired characteristics are not inherited.*
3. Hereditary variations are based on particles—*genes*—that *retain their identity as they pass through the generations; they do not blend* with other genes. This is true of both discretely varying traits (e.g., brown vs. blue eyes) and continuously varying traits (e.g., body size, intensity of pigmentation). Genetic variation in continuously varying traits is based on several or many discrete, particulate genes, each of which affects the trait slightly (*polygenic inheritance*).
4. Genes mutate, usually at a fairly low rate, to equally stable alternative forms, known as *alleles*. The phenotypic effect of such mutations can range from undetectable to very great. The variation that arises by mutation is amplified by recombination among alleles at different loci.
5. *Evolutionary change is a populational process*: it entails, in its most basic form, a change in the relative abundances (proportions, or *frequencies*) of individual organisms with different genotypes (hence, often, with different phenotypes) within a population. One genotype may gradually replace other genotypes over the course of generations. Replacement may occur in only certain populations, or in all the populations that make up a species.
6. The rate of mutation is too low for mutation by itself to shift a population from one genotype to another. Instead, the change in genotype proportions within a population can occur by either of two principal processes: random fluctuations in proportions (*genetic drift*), or nonrandom changes due to the superior survival and/or reproduction of some genotypes compared with others (i.e., natural selection). Natural selection and random genetic drift can operate simultaneously.
7. Even a slight intensity of natural selection can (under certain circumstances) bring about substantial evolutionary change in a realistic amount of time. *Natural selection can account for both slight and great differences among species*, as well as for the earliest stages of evolution of new traits. Adaptations are traits that have been shaped by natural selection.
8. Natural selection can alter populations beyond the original range of variation by increasing the frequency of alleles that, by recombination with other genes that affect the same trait, give rise to new phenotypes.

9. Mutations can accumulate in natural populations. Hence populations have *genetic variation*, and so can often evolve rapidly when environmental conditions change.
10. Populations of a species in different geographic regions differ in characteristics that have a genetic basis. These differences are often adaptive, and thus are the consequence of natural selection.
11. The differences between different species, and between different populations of the same species, are often based on differences at several or many genes, many of which have a small phenotypic effect. This pattern supports the hypothesis that the differences between species evolve by rather small steps.
12. Species are not defined simply by phenotypic differences. Rather, different species of sexually reproducing organisms represent distinct "gene pools"; that is, species are groups of interbreeding or potentially interbreeding individuals that do not exchange genes with other such groups.
13. Speciation is the origin of two or more species from a single common ancestor. Speciation usually occurs by the genetic differentiation of geographically segregated populations. Because of the geographic segregation, interbreeding does not prevent incipient genetic differences from developing.
14. Among living organisms, there are many gradations in phenotypic characteristics among species assigned to the same genus, to different genera, and to different families or other higher taxa. Such observations provide evidence that higher taxa arise by the prolonged, sequential accumulation of small differences, rather than by the sudden mutational origin of drastically new "types."
15. All organisms form a great "tree of life," or *phylogeny*, that has developed by the branching of common ancestors into diverse lineages, chiefly through speciation. All forms of life appear to have descended from a single common ancestor in the remote past.
16. The fossil record includes many gaps among quite different kinds of organisms. Such gaps may be explained by the incompleteness of the fossil record. But the fossil record also includes examples of gradations from apparently ancestral organisms to quite different descendants. These data support the hypothesis that the evolution of large differences proceeds incrementally. Hence the principles that explain the evolution of populations and species may be extrapolated to the evolution of higher taxa.

Evolutionary Biology since the Synthesis

Since the evolutionary synthesis, a great deal of research has elaborated and tested its basic principles, and these principles have withstood the tests. But progress in both evolutionary studies and other fields of biology has required some modifications and many extensions of the basic principles of the evolutionary synthesis, and has spurred additional theory to account for biological phenomena that were unknown in the 1940s. Since Watson and Crick established the structure of DNA in 1953, advances in genetics and molecular biology, and in molecular and information technology, have revolutionized the study of evolution. Molecular biology has provided tools for studying a vast number of evolutionary topics, such as mutation, genetic variation, species differences, development, and the phylogenetic history of life, in greater depth and detail than ever before, and new mathematical theory has been developed to provide evolutionary interpretations of newly discovered molecular and genetic phenomena.

As molecular and computational technology has become more sophisticated and available, new fields of evolutionary study have developed. Among these fields is MOLECULAR EVOLUTION (analyses of the processes and history of change in genes), in which the NEUTRAL THEORY OF MOLECULAR EVOLUTION has been particularly important. This hypothesis, developed especially by Motoo Kimura (1924–1994), holds that most of the evolution of DNA sequences occurs by genetic drift rather than by natural selection, but it provides a foundation for detecting effects of natural selection on DNA sequences. Because entire genomes can now be sequenced, molecular evolutionary studies have expanded into

MOTOO KIMURA

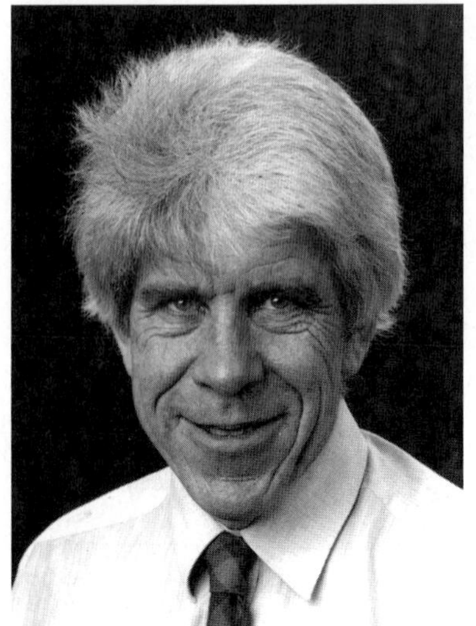

WILLIAM D. HAMILTON

EVOLUTIONARY GENOMICS, which is concerned with variation and evolution in multiple genes or even entire genomes. EVOLUTIONARY DEVELOPMENTAL BIOLOGY is an exciting field devoted to understanding how the evolution of developmental processes underlies the evolution of morphological features at all levels, from cells to whole organisms.

The advances in these fields, though, are complemented by vigorous research, new discoveries, and new ideas about longstanding topics in evolutionary biology, such as the evolution of adaptations and of new species. Since the mid 1960s, evolutionary theory has expanded into areas such as ecology, animal behavior, and reproductive biology, and detailed theories that explain the evolution of particular kinds of characteristics such as life span, ecological distribution, and social behavior were pioneered by the evolutionary theoreticians William Hamilton and John Maynard Smith in England and George Williams in the United States. The study of macroevolution has been renewed by provocative interpretations of the fossil record and by new methods for studying phylogenetic relationships. Research in evolutionary biology is progressing more rapidly than ever before. This is an exciting time to learn about evolution—or to be an evolutionary biologist.

Philosophical Issues

Thousands of pages have been written about the philosophical and social implications of evolution. Darwin argued that every characteristic of a species can vary, and can be altered radically, given enough time. Thus he rejected the essentialism that Western philosophy had inherited from Plato and Aristotle and put variation in its place. Darwin also helped to replace a static conception of the world—one virtually identical to the Creator's perfect creation—with a world of ceaseless change. It was Darwin who extended to living things, including the human species, the principle that change, not stasis, is the natural order.

Above all, Darwin's theory of random, purposeless variation acted on by blind, purposeless natural selection provided a revolutionary new kind of answer to almost all questions that begin with "Why?" Before Darwin, both philosophers and people in general answered "Why?" questions by citing purpose. Since only an intelligent mind, with the capacity for forethought, can have purpose, questions such as "Why do plants have flowers?" or "Why are there apple trees?"—or diseases, or earthquakes—were answered by imagining the possible purpose that God could have had in creating them. This kind of explanation was made completely superfluous by Darwin's theory of natural selection. The adaptations of organisms—long cited as the most conspicuous evidence of intelligent design in the universe—could be explained by purely mechanistic causes. For evolutionary biologists, the pink color of a magnolia's flower has a *function*, but not a *purpose*. It was not designed in order to propagate the species, much less to delight us with its beauty, but instead came into existence because magnolias with brightly colored flowers reproduced more prolifically than magnolias with less brightly colored flowers. The unsettling implication of this pure-

Figure 1.5 In 1999, Ernst Mayr, George C. Williams, and John Maynard Smith received the Crafoord Prize in Biological Sciences from the Royal Swedish Academy of Sciences in honor of their "pioneering contributions to broadening, deepening and refining our understanding of biological evolution and related phenomena." (Photo © Royal Swedish Academy of Sciences.)

ly material explanation is that, except in the case of human behavior, we need not invoke, nor can we find any evidence for, any design, goal, or purpose anywhere in the natural world.

It must be emphasized that all of science has come to adopt the way of thought that Darwin applied to biology. Astronomers do not seek the purpose of comets or supernovas, nor chemists the purpose of hydrogen bonds. The concept of purpose plays no part in scientific explanation.

Ethics, Religion, and Evolution

In the world of science, the reality of evolution has not been in doubt for more than a hundred years, but evolution remains an exceedingly controversial subject in the United States and a few other countries. The **creationist movement** opposes the teaching of evolution in public schools, or at least demands "equal time" for creationist beliefs. Such opposition arises from the fear that evolutionary science denies the existence of God, and consequently, that it denies any basis for rules of moral or ethical conduct.

Our knowledge of the history and mechanisms of evolution is certainly incompatible with a *literal* reading of the creation stories in the Bible's Book of Genesis—as it is incompatible with hundreds of other creation myths people have devised. A literal reading of some passages in the Bible is also incompatible with physics, geology, and other natural sciences. But does evolutionary biology deny the existence of a supernatural being or a human soul? No, because science, including evolutionary biology, is silent on such questions. By its very nature, science can entertain and investigate only hypotheses about material causes that operate with at least probabilistic regularity. It cannot test hypotheses about supernatural beings or their intervention in natural events.

Evolutionary biology has provided natural, material causes for the diversification and adaptation of species, just as the physical sciences did when they explained earthquakes and eclipses. The steady expansion of the sciences, to be sure, has left less and less to be explained by the existence of a supernatural creator, but science can neither deny nor affirm such a being. Indeed, some evolutionary biologists are devoutly religious, and many nonscientists, including many priests, ministers, and rabbis, hold both religious beliefs and belief in evolution (see Chapter 23).

Wherever ethical and moral principles are to be found, it is probably not in science, and surely not in evolutionary biology. Opponents of evolution have charged that evolution by natural selection justifies the principle that "might makes right," and certainly more than one dictator or imperialist has invoked the "law" of natural selection to justify atrocities. But evolutionary theory cannot provide any such precept for behavior. Like any other science, it describes how the world *is*, not how it *should be*. The supposition that what is "natural" is "good" is called by philosophers the NATURALISTIC FALLACY.

Various animals have evolved behaviors that we give names such as cooperation, monogamy, competition, infanticide, and the like. Whether or not these behaviors *ought* to be, and whether or not they are moral, is not a scientific question. The natural world is amoral—it lacks morality altogether. Despite this, the concepts of natural selection and evolutionary progress were taken as a "law of nature" by which Marx justified class struggle, by which the Social Darwinists of the late eighteenth and early nineteenth centuries justified economic competition and imperialism, and by which the biologist Julian Huxley justified humanitarianism (Hofstadter 1955; Paradis and Williams 1989). All these ideas are philosophically indefensible instances of the naturalistic fallacy. Infanticide by lions and langur monkeys does not justify infanticide in humans, and evolution provides no basis for human ethics.

Evolution as Fact and Theory

Is evolution a fact, a theory, or a hypothesis? Biologists often speak of the "theory of evolution," but they usually mean by that something quite different from what most nonscientists understand by that phrase.

In science, a **hypothesis** is an informed conjecture or statement of what might be true. Most philosophers (and scientists) hold that we do not know anything with absolute certainty. What we call facts are hypotheses that have acquired so much supporting evidence that we act as if they were true. A hypothesis may be poorly supported at first, but it can gain support, to the point that it is effectively a fact. For Copernicus, the revolution of the Earth around the Sun was a hypothesis with modest support; for us, it is a hypothesis with such strong support that we consider it a fact.

In everyday use, a "theory" refers to an unsupported speculation. Like many words, however, this term has a different meaning in science. A **scientific theory** is a mature, coherent body of interconnected statements, based on reasoning and evidence, that explain some aspect of nature—usually many aspects. Or, to quote the *Oxford English Dictionary*, a theory is "a scheme or system of ideas and statements held as an explanation or account of a group of facts or phenomena; a hypothesis that has been confirmed or established by observation or experiment, and is propounded or accepted as accounting for the known facts; a statement of what are known to be the general laws, principles, or causes of something known or observed." Thus atomic theory, quantum theory, and the theory of plate tectonics are elaborate schemes of interconnected ideas, strongly supported by evidence, that account for a great variety of phenomena.

Given these definitions, evolution is a fact. But *the fact of evolution is explained by evolutionary theory.*

In *The Origin of Species,* Darwin propounded two major hypotheses: that organisms have descended, with modification, from common ancestors; and that the chief cause of modification is natural selection acting on hereditary variation. Darwin provided abundant evidence for descent with modification, and hundreds of thousands of observations from paleontology, geographic distributions of species, comparative anatomy, embryology, genetics, biochemistry, and molecular biology have confirmed this hypothesis since Darwin's time. Thus the hypothesis of descent with modification from common ancestors has long had the status of a scientific fact.

The explanation of *how* modification occurs and *how* ancestors give rise to diverse descendants constitutes the theory of evolution. We now know that Darwin's hypothesis of natural selection on hereditary variation was correct, but we also know that there are more causes of evolution than Darwin realized, and that natural selection and hereditary variation themselves are more complex than he imagined. A body of ideas about the causes of evolution, including mutation, recombination, gene flow, isolation, random genetic drift, the many forms of natural selection, and other factors, constitute our current theory of evolution, or "evolutionary theory." Like all theories in science, it is a work in progress, for we do not yet know the causes of all of evolution, or all the biological phenomena that evolutionary biology will have to explain. Indeed, some details may turn out to be wrong. But the main tenets of the theory, as far as it goes, are so well supported that most biologists confidently accept evolutionary theory as the foundation of the science of life.

GO TO THE

Companion Web Site at
www.sinauer.com/evolution
for quizzes, data analysis and simulation exercises, and other study aids.

Summary

1. Evolution is the unifying theory of the biological sciences. Evolutionary biology aims to discover the history of life and the causes of the diversity and characteristics of organisms.
2. Darwin's evolutionary theory, published in *The Origin of Species* in 1859, consisted of two major hypotheses: first, that all organisms have descended, with modification, from common ancestral forms of life, and second, that a chief agent of modification is natural selection.
3. Darwin's hypothesis that all species have descended with modification from common ancestors is supported by so much evidence that it has become as well established a fact as any in biology. His theory of natural selection as the chief cause of evolution was not broadly supported until the "evolutionary synthesis" that occurred in the 1930s and 1940s.
4. The evolutionary theory developed during and since the evolutionary synthesis consists of a body of principles that explain evolutionary change. Among these principles are (a) that genetic variation in phenotypic characters arises by random mutation and recombination; (b) that

changes in the proportions of alleles and genotypes within a population may result in replacement of genotypes over generations; (c) that such changes in the proportions of genotypes may occur either by random fluctuations (genetic drift) or by nonrandom, consistent differences among genotypes in survival or reproduction rates (natural selection); and (d) that as a result of different histories of genetic drift and natural selection, populations of a species may diverge and become reproductively isolated species.

5. Evolutionary biology makes important contributions to other biological disciplines and to social concerns in areas such as medicine, agriculture, computer science, and our understanding of ourselves.
6. The implications of Darwin's theory, which revolutionized Western thought, include the ideas that change rather than stasis is the natural order; that biological phenomena, including those seemingly designed, can be explained by purely material causes rather than by divine creation; and that no evidence for purpose or goals can be found in the living world, other than in human actions.
7. Like other sciences, evolutionary biology cannot be used to justify beliefs about ethics or morality. Nor can it prove or disprove theological issues such as the existence of a deity. Many people hold that, although evolution is incompatible with a literal interpretation of some passages in the Bible, it is compatible with religious belief.

Terms and Concepts

adaptation
creationist movement
descent with modification
essentialism
evolution (biological evolution; organic evolution)
evolutionary synthesis (= modern synthesis)
hypothesis
inheritance of acquired characteristics
macroevolution
microevolution
natural selection
population
scientific theory
uniformitarianism

Suggestions for Further Reading

The readings at the end of each chapter include major works that provide a comprehensive treatment and an entry into the professional literature. The references cited within each chapter also serve this important function.

No one should fail to read at least part of Darwin's *On the Origin of Species by Means of Natural Selection, or The Preservation of Favoured Races in the Struggle for Life*, in either the first edition (1859) or the sixth edition (1872), in which Darwin deleted "on" from the title. After some adjustment to Victorian prose, you should be enthralled by the craft, detail, completeness, and insight in Darwin's arguments. It is an astonishing book.

Among biographies of Darwin, *The Reluctant Mr. Darwin: An Intimate Portrait of Charles Darwin and the Making of His Theory of Evolution* by David Quammen (W. W. Norton, New York, 2006) is enthralling and is a "must." The best scholarly biographies of Darwin include Janet Browne's superb two-volume work, *Charles Darwin: Voyaging* and *Charles Darwin: The Power of Place* (Knopf, New York, 1995 and 2002, respectively); and *Darwin*, by A. Desmond and J. Moore (Warner Books, New York, 1991), which emphasizes the role played by the religious, philosophical, and intellectual climate of nineteenth-century England on the development of his scientific theories.

Important works on the history of evolutionary biology include P. J. Bowler, *Evolution: The History of an Idea* (University of California Press, Berkeley, 1989); E. Mayr, *The Growth of Biological Thought: Diversity, Evolution, and Inheritance* (Harvard University Press, Cambridge, MA, 1982), a detailed, comprehensive history of systematics, evolutionary biology, and genetics that bears the personal stamp of one of the major figures in the evolutionary synthesis; and E. Mayr and W. B. Provine (eds.), *The Evolutionary Synthesis: Perspectives on the Unification of Biology* (Harvard University Press, Cambridge, MA, 1980), which contains essays by historians and biologists, including some of the major contributors to the synthesis.

Recent books that expose the fallacies of creationism and explain the nature of science and of evolutionary biology include R. T. Pennock, *Tower of Babel: The Evidence against the New Creationism* (MIT Press, Cambridge, MA, 1999); B. J. Alters and S. M. Alters, *Defending Evolution: A Guide to the Creation/Evolution Controversy* (Jones and Bartlett, Sudbury, MA, 2001); M. Pigliucci, *Denying Evolution: Creationism, Scientism, and the Nature of Science* (Sinauer Associates, Sunderland, MA, 2002); and E. C. Scott, *Evolution versus Creationism: An Introduction*, second edition (University of California Press, Berkeley, 2009). J. A. Coyne describes the evidence for evolution in *Why Evolution Is True* (Viking, New York, 2009).

Web Sites

Several excellent Web sites provide good introductions to evolution; most of them also include material on teaching evolution and on creationism.

"Understanding Evolution" (http://evolution.berkeley.edu) is an excellent site, developed by the Museum of Paleontology at the University of California, Berkeley.

"Evolution" (www.pbs.org/wgbh/evolution) is based on a superb series of WGBH/NOVA programs on the Public Broadcasting System.

The National Academy of Sciences of the U.S.A., the members of which are leaders in science, has a site devoted to evolution (www.nationalacademies.org/evolution) and has published an excellent 70-page booklet, *Science, Evolution, and Creationism* (2008), that can be accessed for free through the Web site or can be purchased at low cost (order at www.nap.edu).

Problems and Discussion Topics

1. How does evolution unify the biological sciences? What other principles might do so?
2. Analyze and evaluate Ralph Waldo Emerson's couplet,

   Striving to be man, the worm
   Mounts through all the spires of form.

   What pre-Darwinian concepts does it express? What fault in it will a Darwinian find?
3. Some scientists vigorously rejected Darwin's ideas when *On the Origin of Species* was published. Richard Owen, perhaps the most respected biologist in England, wrote (among many other objections): "Are all the recognised organic forms of the present date, so differentiated, so complex, so superior to conceivable primordial simplicity of form and structure, as to testify to the effects of Natural Selection continuously operating through untold time? Unquestionably not. The most numerous living beings ... are precisely those which offer such simplicity of form and structure, as best agrees...with that ideal prototype from which...vegetable and animal life might have diverged." How might Darwin, or you, argue against Owen's logic?
4. In the same spirit as the preceding question, find and analyze other arguments made by Owen or other critics of Darwin such as François Pictet, Adam Sedgwick, St. George Jackson Mivart, or Louis Agassiz. Some of their essays are reprinted in *Darwin and His Critics* by D. L. Hull (1973). Many of their criticisms are exactly those raised by modern-day creationists.
5. Discuss how a creationist versus an evolutionary biologist might explain some human characteristics, and the implications of their differences. Sample characteristics: eyes; wisdom teeth; individually unique friction ridges (fingerprints); five digits rather than some other number; susceptibility to infections; fever when infected; variation in sexual orientation; limited life span.
6. In February 2001, two research groups published preliminary sequences of the entire human genome. If humans, along with all other forms of life, have evolved from a common ancestor, what evidence of this would be expected in the human genome? In what ways might the history and processes of evolution help us interpret and make sense of genome sequence data?
7. How might the evolution of antibiotic resistance in pathogenic bacteria be slowed down or prevented? What might you need to know in order to achieve this aim?
8. Based on sources available in a good library, discuss how the "Darwinian revolution" affected one of the following fields: philosophy, literature, psychology, or anthropology.
9. Should both evolution and creationism be taught as alternative theories in science classes?

CHAPTER 2

The Tree of Life: Classification and Phylogeny

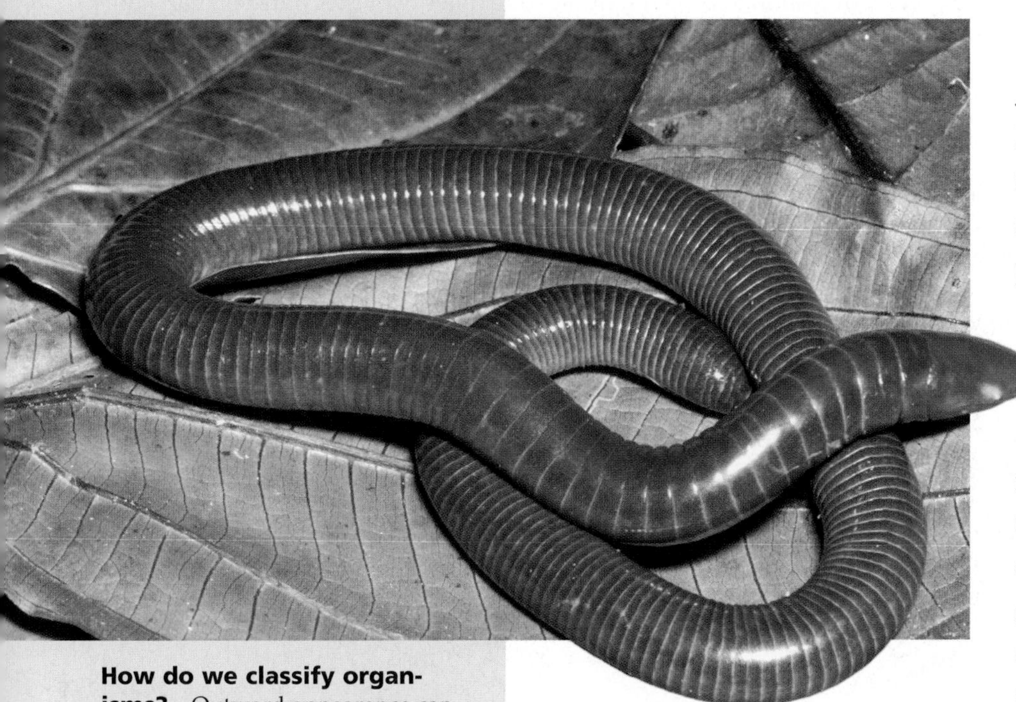

How do we classify organisms? Outward appearance can be deceiving. This legless, burrowing, segmented animal looks like an earthworm—but it possesses a vertebral column, and thus might be taken for a snake. It lacks scales, however, and its life cycle and reproductive physiology are distinct from those of snakes. In fact, this caecilian (*Gymnophis multiplicata*) is an amphibian, a distant relative of frogs and salamanders. (Photo © Michael Fogden/OSF/Photolibrary.com.)

About 2000 million years ago, a bacterium, not unlike the *Escherichia coli* bacteria in our intestines, took up residence within the cell of another bacteria-like organism. This partnership flourished, since each partner evidently provided biochemical services to the other. The "host" in this partnership developed a modern nucleus, chromosomes, and mitotic spindle, while the "guest" bacterium within it evolved into the mitochondrion. This ancestral eukaryote gave rise to diverse one-celled descendants. Some of those descendants subsequently became multicellular when the cells they produced by mitosis remained together and evolved mechanisms of regulating gene expression that enabled groups of cells to form different tissues and organs. One such lineage became the progenitor of green plants, another of the fungi and animals.

Between about 1000 million and 600 million years ago, a single animal species gave rise to two species that became the progenitors of two quite astonishingly different groups of animals: one evolved into the starfishes, sea urchins, and other echinoderms, and the other into the chordates, including the vertebrates. Most of the species derived from the earliest vertebrate are fishes, but one of these species would prove to be the ancestor of the tetrapod (four-legged) vertebrates. About 150 million years after the first land-dwelling tetrapod evolved, some of its descendants stood on the brink of mammalhood. After another 125 million years or so, the mammals had diversified into many groups, including the first primates, adapted to life in trees. Some primates became small, some evolved prehensile tails, and one became the ancestor of large, tailless apes. About 14 million years ago, one such ape gave rise to the Asian orangutan on the one hand and an African descendant on the other. The African descendant split into the gorilla and another species. About 6 to 8 million years ago, that species, in turn, divided into one lineage that became today's chimpanzees and into another lineage that underwent rapid evolution of its posture, feet, hands, and brain: our own quite recent ancestor.

This is an astonishing story, but it is our best current understanding of some of the high points in our history, in which, metaphorically speaking, the human species developed as one twig in a gigantic tree, the great Tree of Life. (We will look at this history in more detail in Chapters 5 and 7.) With the passage of time, a species is likely to "branch"—to give rise to two species that evolve different modifications of some of their features. Those species branch in turn, and their descendants may be altered further still. By this process of branching and modification, repeated innumerable times over the course of many millions of years, many millions of kinds of organisms have evolved from a single ancestral organism at the very base, or root, of the tree (Figure 2.1).

Evolutionary biologists have developed methods of "reconstructing" or "assembling" the tree of life—of estimating the **phylogeny**, or genealogical relationships, among organisms (i.e., which species share a recent **common ancestor**, which share distant ancestors, and which share even more remote ancestors). The resulting portrayal of relationships not only is fascinating in itself (have you ever thought of yourself as related to a starfish, a butterfly, a mushroom?) but is also an important foundation for understanding many aspects of evolutionary history, such as the pathways by which various characteristics have evolved.

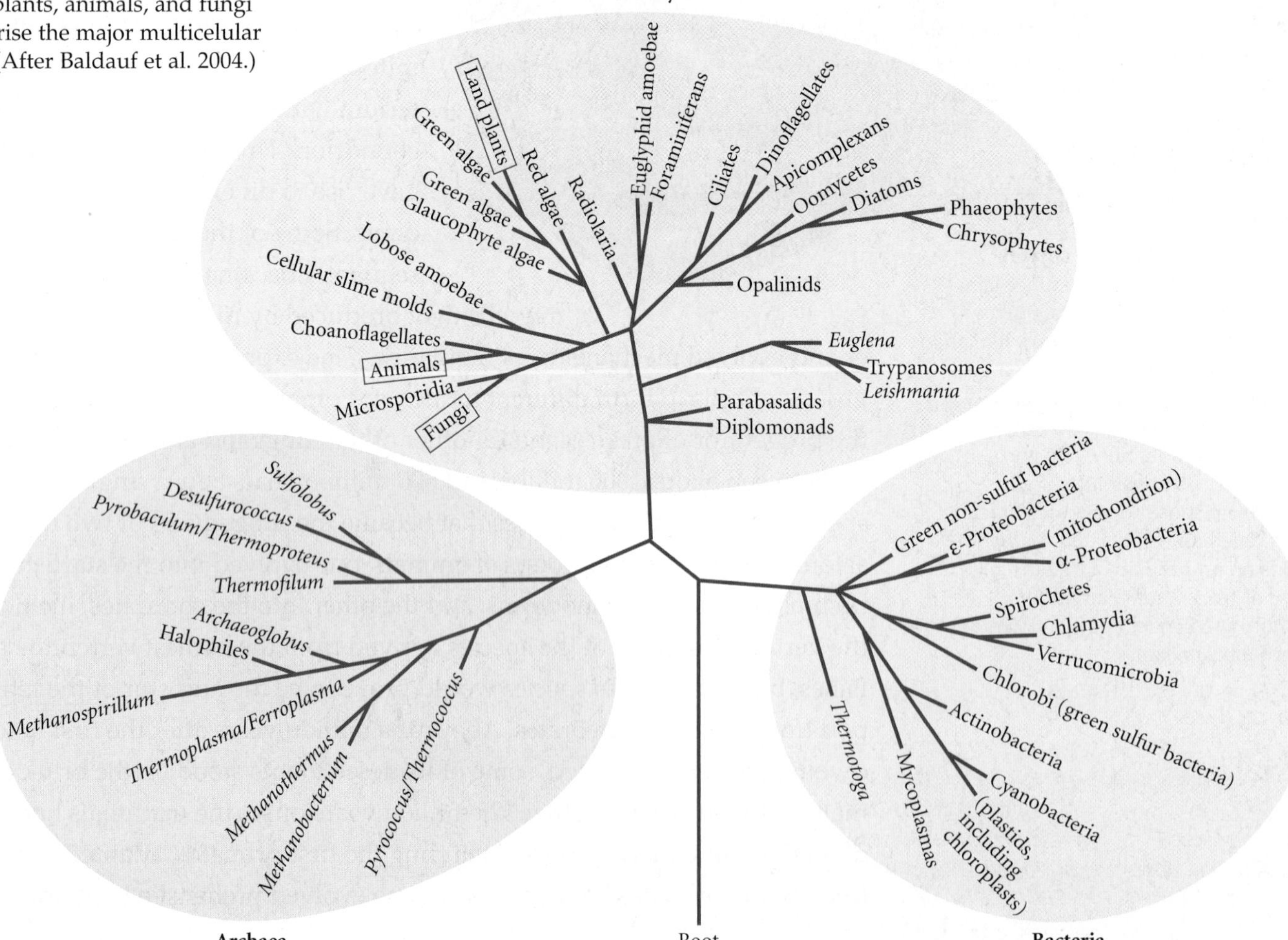

Figure 2.1 The Tree of Life. This estimate of the relationships among some of the major branches is based mostly on DNA sequences, especially those of genes encoding ribosomal RNA. Of the three empires of life, Archaea and Eucarya appear to have the most recent common ancestor. The majority of taxa in the tree are unicellular; the land plants, animals, and fungi comprise the major multicelular taxa. (After Baldauf et al. 2004.)

We cannot directly observe evolutionary history, so we must infer it using deductive logic, like Sherlock Holmes reconstructing the history of a crime. A few points in our brief sketch of human origins, such as the timing of certain events, have been learned from the fossil record. However, most of this history has been determined from studying not fossils, but living organisms. In this chapter, we will become acquainted with some of the methods by which we can infer phylogenetic relationships, and we will see how our understanding of those relationships is reflected in the classification of organisms. In the following chapter, we will examine some common evolutionary patterns that these approaches have helped to elucidate.

Classification

Phylogenetic analysis—the study of relationships among species—has historically been closely associated with the classification and naming of organisms (known as TAXONOMY). Both are among the tasks of the field of SYSTEMATICS.

In the early 1700s, European naturalists believed that God must have created species according to some ordered scheme, as we saw in Chapter 1. It was therefore a work of devotion to discover "the plan of creation" by cataloging the works of the Creator and discovering a "natural," true, classification. The scheme of classification that was adopted then, and is still used today, was developed by the Swedish botanist Carolus Linnaeus (1707–1778). Linnaeus introduced BINOMIAL NOMENCLATURE, a system of two-part names consisting of a genus name and a specific epithet (such as *Homo sapiens*). He proposed a system of grouping species in a HIERARCHICAL CLASSIFICATION of groups nested within larger groups (such as genera nested within families) (Box 2A). The levels of classification, such as kingdom, phylum, class, order, family, genus, and species, are referred to as **taxonomic categories**, whereas a particular group of organisms assigned to a categorical rank is a **taxon** (plural: **taxa**). Thus the rhesus monkey is placed in the genus *Macaca*, in the family Cercopithecidae, in the order Primates; *Macaca*, Cercopithecidae, and Primates are taxa that exemplify the taxonomic categories genus, family, and order, respectively. Several "intermediate" taxonomic categories, such as SUPERFAMILY and SUBSPECIES, are sometimes used in addition to the more familiar and universal ones. In assigning species to **higher taxa** (those above the species level), Linnaeus used features that he imagined represented propinquity in God's creative scheme. For example, he defined the order Primates by the features "four parallel upper front [incisor] teeth; two pectoral nipples," and on this basis included bats among the Primates. But without an evolutionary framework, naturalists had no objective basis for classifying mammals by their teeth rather than by, say, their color or size. Indeed, because we now use an evolutionary framework, bats are not combined in the same order with primates today.

Classification took on an entirely different significance after *On the Origin of Species* was published in 1859. In the Galápagos Islands, as we saw in Chapter 1, Darwin had observed that different islands harbored similar, but nevertheless distinguishable, mockingbirds. He came to suspect that the different forms of mockingbirds had descended from a single ancestor, acquiring slight differences in the course of descent. But this thought, logically extended, suggested that that ancestor itself had been modified from an ancestor further back in time, which could have given rise to yet other descendants—various South American mockingbirds, for instance. By this logic, some remote ancestor might have been the progenitor of all species of birds, and a still more remote ancestor the progenitor of all vertebrates.

Darwin proposed, then, that on occasion, an ancestral species may split into two descendant species, which at first are very similar to each other, but which **diverge** (become more different) in the course of time. Each of those species, in turn, may divide and diverge to yield two descendant species, and the process may be repeated again and again throughout the immensely long history of life. Darwin therefore gave meaning to the notion of "closely related" species: they are descended from a relatively recent common ancestor, whereas more distantly related species are descended from a more remote (i.e., further back in time) common ancestor. The features that species hold in common, such as the vertebrae that all vertebrates possess, were not independently bestowed on each of those species by

BOX 2A Taxonomic Practice and Nomenclature

Standardized names for organisms are essential for communication among scientists. To ensure that names are standardized, taxonomy has developed rules of procedure.

Most species are named by taxonomists who are experts on that particular group of organisms. A new species may be one that has never been seen before (e.g., an organism dredged from the deep sea), but many unnamed species are sitting in museum collections, awaiting description. Moreover, a single species often proves, on closer study, to be two or more very similar species. A taxonomist who undertakes a REVISION—a comprehensive analysis—of a group frequently names new species. A species name has legal standing if it is published in a journal, or even in a privately produced publication that is publicly available.

The name of a species consists of its genus and its specific epithet; both are Latin or latinized words. These words are always *italicized* (or underlined), and the genus is always capitalized. In entomology and certain other fields, it is customary to include the name of the AUTHOR (the person who conferred the specific epithet); for example, the corn rootworm beetle *Diabrotica virgifera* LeConte.

Numerous rules govern the construction of species names (e.g., genus and specific epithet must agree in gender: *Rattus norvegicus*, not *Rattus norvegica*, for the brown rat). It is recommended that the name have meaning [e.g., *Vermivora* ("worm eater") *chrysoptera* ("golden-winged") for the golden-winged warbler; *Rana warschewitschii*, "Warschewitsch's frog"], but it need not. Often a taxonomist will honor another person by naming a species after him or her.

The first rule of nomenclature is that no two species of animals, or of plants, can bear the same name. (It is permissible, however, for the same name to be applied to both a plant and an animal genus; for example, *Alsophila* is the name of both a fern genus and a moth genus.) The second is the rule of PRIORITY: the valid name of a taxon is the oldest available name that has been applied to it. Thus it sometimes happens that two authors independently describe the same species under different names; in this case, the valid name is the older one, and the younger name is a junior SYNONYM. Conversely, it may turn out that two or more species have masqueraded under one name; in this case, the name is applied to the species that the author used in his or her description. To prevent the obvious ambiguity that could arise in this way, it has become standard practice for the author to designate a single specimen (the TYPE SPECIMEN, or HOLOTYPE) as the "name-bearer" so that later workers can determine which of several similar species rightfully bears the name. The holotype, usually accompanied by other specimens (PARATYPES) that exemplify the range of variation, is deposited and carefully preserved in a museum or herbarium.

In revising a genus, a taxonomist usually introduces changes in the taxonomy. Some examples of such changes are:

- Species that were placed by previous authors in different genera may be brought together in the same genus because they are shown to be closely related. These species retain their specific epithets, but if they are shifted to a different genus, the author's name is written in parentheses.
- A species may be removed from a genus and placed in a different genus because it is determined not to be closely related to the other members.
- Forms originally described as different species may prove to be the same species, and be synonymized.
- New species may be described.

The rules for naming higher taxa are not all as strict as those for species and genera. In zoology (and increasingly in botany), names of subfamilies, families, and sometimes orders are formed from the stem of the type genus (the first genus described). Most family names of plants end in -aceae. In zoology, subfamily names end in -inae and family names in -idae. Thus *Mus* (from the Latin *mus, muris*, "mouse"), the genus of the house mouse, is the type genus of the family Muridae and the subfamily Murinae; *Rosa* (rose) is the type genus of the family Rosaceae. Endings for categories above family are standardized in some, but not all, groups (e.g., -formes for orders of birds, such as Passeriformes, the perching birds that include *Passer*, the house sparrows). Names of taxa above the genus level are not italicized but are always capitalized. Adjectives or colloquial nouns formed from these names are not capitalized: thus we may refer to murids or to murid rodents, without capitalization.

a Creator, but were inherited from the ancestral species in which the features first evolved. With breathtaking daring, Darwin ventured that all species of organisms, in all their amazing diversity, had descended by endless repetition of such events, through long ages, from perhaps only one common ancestor—the universal ancestor of all life.

In Darwin's words, all species, extant and extinct, form a great "Tree of Life," or **phylogenetic tree**, in which closely adjacent twigs represent living species derived only recently from their common ancestors, whereas twigs on different branches represent species derived from more ancient common ancestors (Figure 2.2). He expressed this metaphor in some of his most poetic language:

> The affinities of all the beings of the same class have sometimes been represented by a great tree. I believe this simile largely speaks the truth. The green and

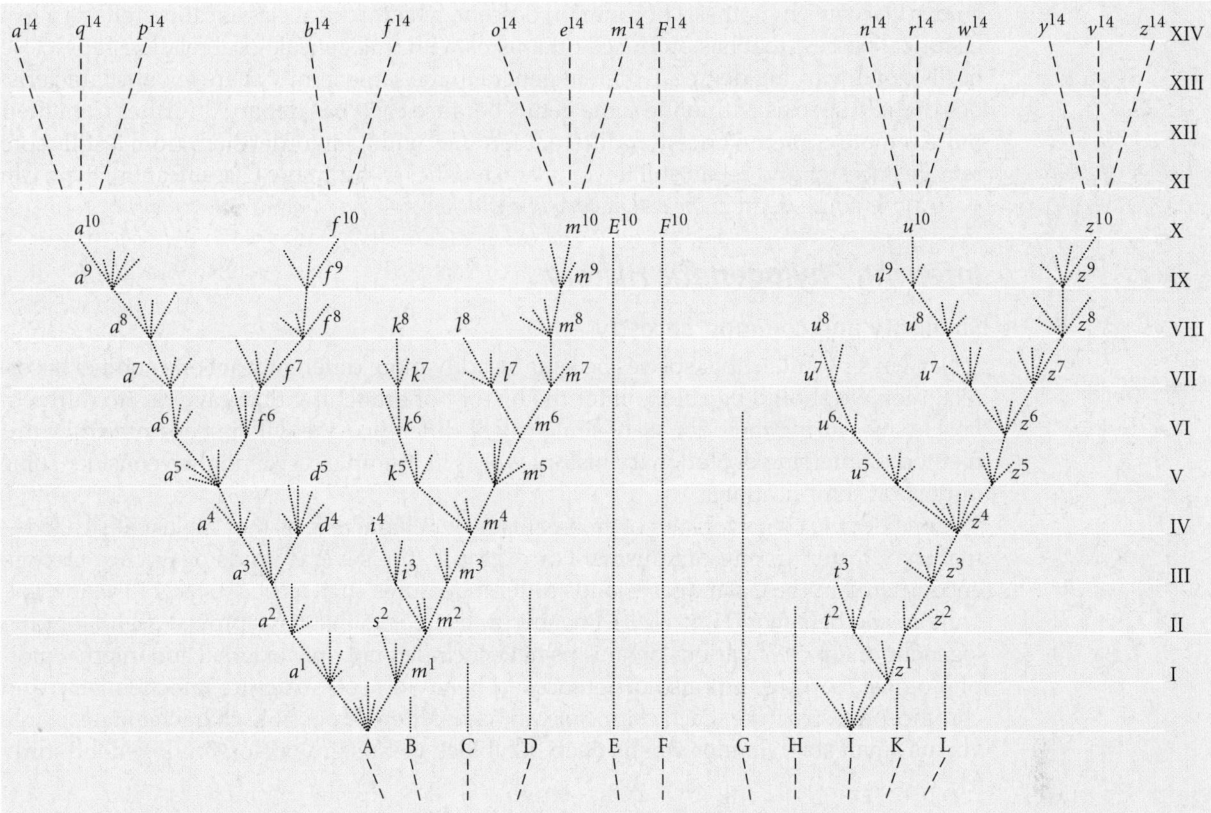

Figure 2.2 Darwin's representation of hypothetical phylogenetic relationships, showing how lineages diverge from common ancestors and give rise to both extinct and extant species. Time intervals (between Roman numerals) represent thousands of generations. Darwin omitted the details of branching for intervals X through XIV. Extant species (at time XIV) can be traced to ancestors A, F, and I; all other original lineages have become extinct. Distance along the horizontal axis represents degree of divergence (as, for example, in body form). Darwin recognized that rates of evolution vary greatly, showing this by different angles in the diagram; for instance, the lineage from ancestor F has survived essentially unchanged. (From Darwin 1859.)

budding twigs may represent existing species; and those produced during former years may represent the long succession of extinct species. At each period of growth all the growing twigs have tried to branch out on all sides, and to overtop and kill the surrounding twigs and branches, in the same manner as species and groups of species have at all times overmastered other species in the great battle for life. The limbs divided into great branches, and these into lesser and lesser branches, were themselves once, when the tree was young, budding twigs; and this connection of the former and present buds by ramifying branches may well represent the classification of all extinct and living species in groups subordinate to groups. Of the many twigs which flourished when the tree was a mere bush, only two or three, now grown into great branches, yet survive and bear the other branches; so with the species which lived during long-past geological periods, very few have left living and modified descendants. From the very first growth of the tree, many a limb and branch has decayed and dropped off; and these fallen branches of various sizes may represent those whole orders, families, and genera which have now no living representatives, and which are known to us only in a fossil state. As we here and there see a thin, straggling branch springing from a fork low down in a tree, and which by some chance has been favoured and is still alive on its summit, so we occasionally see an animal like the Ornithorhynchus or Lepidosiren,* which in some small degree connects by its affinities two large branches of life, and which has apparently been saved from fatal competition by having inhabited a protected station. As buds give rise by growth to fresh buds, and these, if vigorous, branch out and overtop on all sides many a feebler branch, so by generation I believe it has been with the great Tree of Life, which fills with its dead and broken branches the crust of the earth, and covers the surface with its ever-branching and beautiful ramifications.

---

*_Ornithorhynchus_, the duck-billed platypus, is a primitive, egg-laying mammal. _Lepidosiren_ is a genus of living lungfishes, a group that is closely related to the ancestor of the tetrapod (four-legged) vertebrates, and which is known from ancient fossils.

Under Darwin's hypothesis of common descent, a hierarchical classification reflects a real historical process that has produced organisms with true genealogical relationships, close or distant in varying degree. Different genera in the same family share fewer characteristics than do species within the same genus because each has departed further from their more remote common ancestor; different families within an order stem from a still more remote ancestor and retain still fewer characteristics in common. Classification, then, can portray, to some degree, *the real history of evolution*.

Inferring Phylogenetic History

Similarity and common ancestry

If Darwin's postulate that species become steadily more different from one another is correct, then we should be able to infer the history of branching that gave rise to different taxa by measuring their degree of similarity or difference. We will first examine how this method of inferring evolutionary history works in a simple case, and then consider some important complications.

Consider the characteristics of an organism—or **characters**, as they are usually called—that may differ among organisms. For example, the several kinds of finches Darwin encountered in the Galápagos Islands differ in features such as body size, bill shape, coloration, and behavior (Figure 2.3). Phenotypic characters that have proved useful for phylogenetic analyses of various organisms have included not only external and internal morphological features, but also differences in behavior, cell structure, biochemistry, and chromosome structure. Each character can have different possible **character states**: large versus small size, slender versus deep bill, black versus gray color. Phylogenetic study

Figure 2.3 The finch species that inhabit the Galápagos Islands—often referred to as "Darwin's finches"— differ in size, bill form, coloration, and behavior, as seen in these examples. (A) The sharp-beaked ground finch (*Geospiza difficilis*). (B) The large ground finch (*Geospiza magnirostris*). (C) The woodpecker finch (*Cactospiza pallida*). (D) The warbler finch (*Certhidea olivacea*). (A © M. Hyett/VIREO; B courtesy of B. Rosemary Grant; C © Mary Plage/OSF/Photolibrary.com; D © Celia Mannings/Alamy.)

(A)

(B)

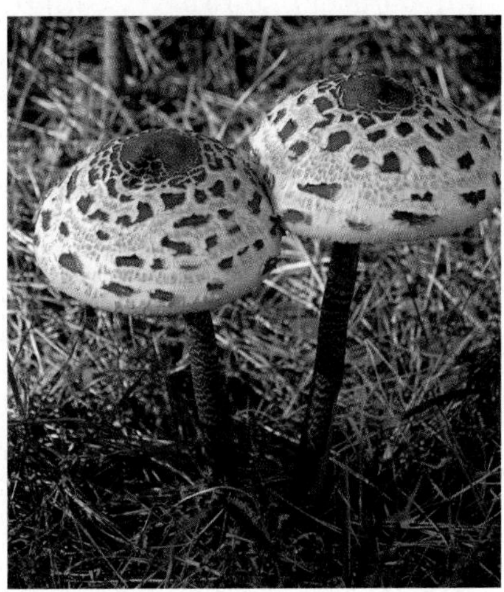

Figure 2.4 Phylogenetic analysis, based on DNA sequences, has revealed the relationships of formerly puzzling organisms. (A) *Rafflesia arnoldii* (left), a parasite with the largest flowers of any plant, is now known to be related to a group of families that includes the spurges such as *Euphorbia sikkimensis* (right). (B) The fungus cultivated underground by leafcutter ants (left) has been identified as a member of the Lepiotaceae, a family that includes mushrooms such as *Macrolepiota procera* (right). (A © A&J Visage / Alamy and Holmes Garden Photos / Alamy; B courtesy of U. Mueller, Texas / Austin and © J. González-Zarraonandia / ShutterStock).

has been revolutionized, and placed on a far firmer basis, by DNA sequencing, which can reveal variation at hundreds or even thousands of base pair positions. At each such position ("site"), the identity of the nucleotide base (A, T, C, or G) may be considered a character. (We note only one base, such as A, rather than the pair A–T, because specifying the complementary base adds no information.) DNA sequences often provide more data than morphological or other phenotypic features, and have enabled systematists to find the evolutionary relationships of many organisms that could not formerly be classified because they have so few variable phenotypic features (Figure 2.4). Moreover, the four possible states of each character in a DNA sequence are much less ambiguous than many phenotypic features (there can be gradations of size or coloration).

As a first step, let us look at a group of four species (1–4) with ten variable characters of interest (a–j). Our task is to determine which of the species are derived from recent, and which from more ancient, common ancestors. That is, we wish to arrange them into a phylogenetic tree such as that in Figure 2.5A, in which each branch point (NODE) represents the common ancestor of the two lineages. For simplicity, let us suppose that each character can have two states, labeled 0 and 1, and that 0 is the **ancestral** state, found in the common ancestor (ancestor 1) of the group. State 1 is a **derived** state—that is, a state that has evolved from the ancestral state. For example, the nucleotide base adenine (A, the ances-

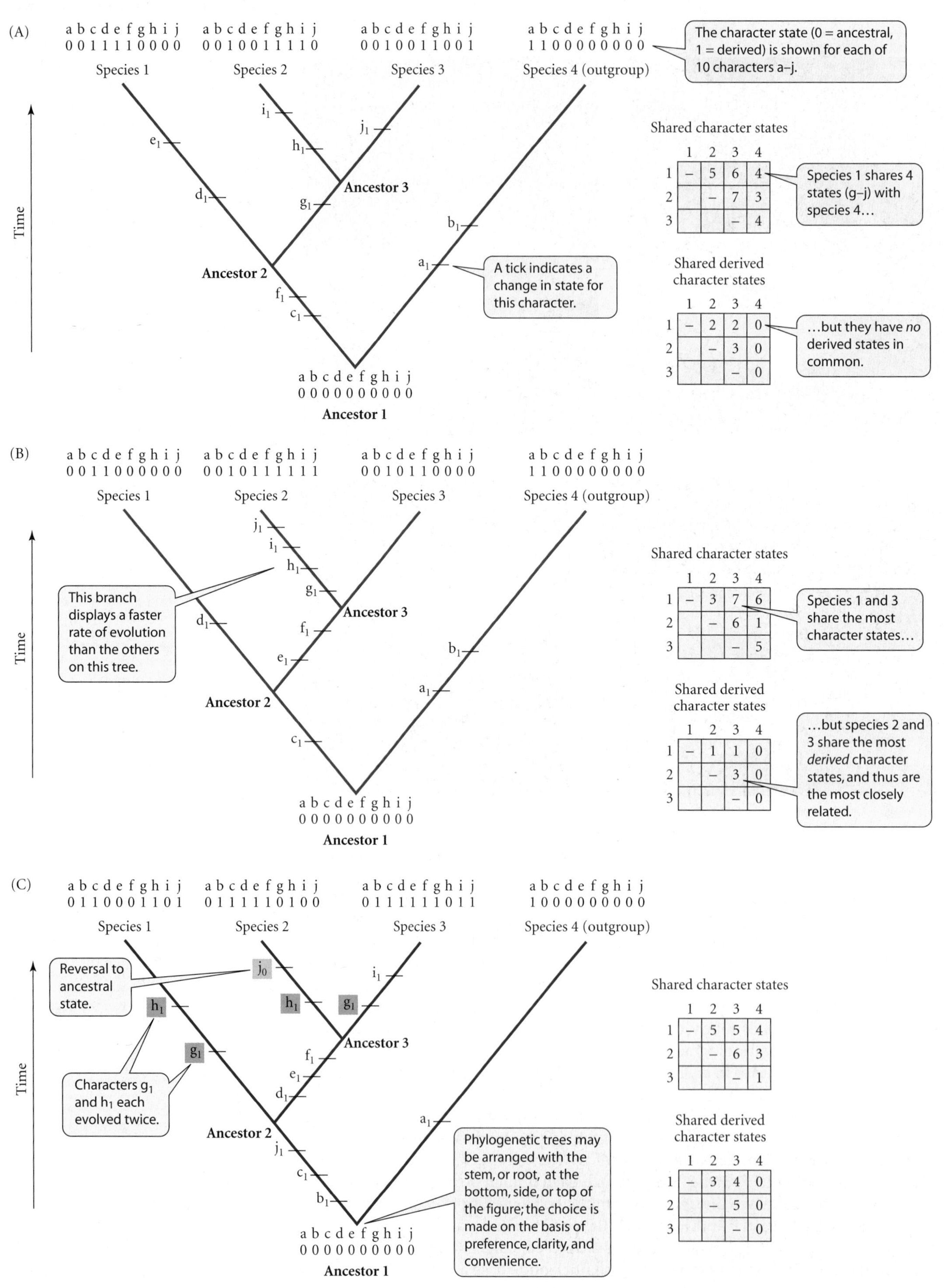

(A) Shared character states

	1	2	3	4
1	–	5	6	4
2		–	7	3
3			–	4

(A) Shared derived character states

	1	2	3	4
1	–	2	2	0
2		–	3	0
3			–	0

(B) Shared character states

	1	2	3	4
1	–	3	7	6
2		–	6	1
3			–	5

(B) Shared derived character states

	1	2	3	4
1	–	1	1	0
2		–	3	0
3			–	0

(C) Shared character states

	1	2	3	4
1	–	5	5	4
2		–	6	3
3			–	1

(C) Shared derived character states

	1	2	3	4
1	–	3	4	0
2		–	5	0
3			–	0

◀ **Figure 2.5** An example of phylogenetic analysis applied to three data sets. In each tree, species 1–4 are shown at the top of the tree and their common ancestor at the bottom. The character state (0 or 1) for each of ten characters (a–j) is shown for ancestor 1 and the extant species. The labeled tick marks along the branches show evolutionary changes in character states. To the right of each tree, the upper matrix shows the total number of character states shared between each pair of species, and the lower matrix shows the number of shared derived character states, as implied by state changes along the branches. (A) A tree with relatively constant evolutionary rates and no homoplasy. (B) The rate of evolution is faster in one branch (ancestor 3 to species 2) than in other branches. (C) Three characters are homoplasious: character states g_1 and h_1 both independently evolved twice, and character j underwent evolutionary reversal, from j_1 back to the ancestral state j_0. In all these cases, we assume that species 4 is an "outgroup"—i.e., species 4 is more distantly related to the other three species than they are to one another.

tral state) might be replaced by cytosine (C, the derived state), or the ancestral state red eyes by the derived state yellow eyes, during the evolution of one or more descendant taxa. The Greek-derived adjectives PLESIOMORPHIC and APOMORPHIC are often used for "ancestral" and "derived," respectively. We use such data on character states to infer the phylogenetic relationships among the species.

Figure 2.5A portrays a hypothetical phylogeny in which four species descend from ancestor 1. Each evolutionary change, such as evolution from character state a_0 to character state a_1, is indicated by a tick mark along the branch in which it occurs. We refer to the set of species derived from any one common ancestor as a **monophyletic group**, also called a **clade**. Thus Figure 2.5A shows three monophyletic groups: species 2 + 3, species 1 + 2 + 3, and species 1 + 2 + 3 + 4. Suppose that Figure 2.5A does indeed represent the true phylogeny of the four species; can we infer, or estimate, this phylogeny from the data?

We may calculate the similarity of each pair of descendant species as the number of character states they share. Species 1 and 2, for example, both have character states a_0, b_0, c_1, f_1, and j_0, as tallied in the matrix of shared character states in Figure 2.5A. In this example, species 2 and 3 are most similar, and evolved from the most recent common ancestor (ancestor 3). Species 1 is more similar to species 2 and species 3, with which it shares the common ancestor (ancestor 2), than it is to species 4, with which the entire group (species 1–3) shares the most remote common ancestor (ancestor 1). In this example, the degree of similarity is a reliable index of recency of common ancestry, and it enables us to infer the monophyletic groups—that is, the phylogeny of these species.

In this hypothetical case, we suppose that we know which character states are ancestral and which are derived. When we measured similarity, we counted both the shared characters that did not evolve during the ancestry of any two species (e.g., the ancestral state a_0 shared by species 1 and 2) and the shared characters that did evolve (c_1 in this instance). If we count only the shared *derived* character states—those that did evolve—we obtain another matrix, in which, again, species 2 and 3 are most similar, and species 1 is more similar to them than to species 4. Shared derived character states are sometimes called **synapomorphies**.

Complications in inferring phylogeny

In Figure 2.5A, the number of character state changes is roughly the same from the ancestor of the group (ancestor 1) to each descendant species. That is, the rate of evolution is about equal among the lineages. This need not be the case, however. In Figure 2.5B, we assume that the rate of evolution between ancestor 3 and species 2 is greater than elsewhere in the phylogeny. Perhaps this difference represents more base pair substitutions* in a DNA sequence in species 2 than in the others. The matrix of shared character states now indicates that species 1 and 3 are most similar. We might therefore be misled into

*A **base pair substitution** is the replacement of one nucleotide base pair (e.g., A–T) by another (e.g., G–C) in an entire population or species. Such substitutions sometimes, but not always, change the amino acid specified by the genetic code, as explained in detail in Chapter 8.

thinking that they are the most closely related species, although they are not. (Remember, *degree of relationship means relative recency of common ancestry, not similarity*.) In this case, similarity is not an adequate indicator of relationship. But the number of shared *derived* character states (3) accurately indicates that the closest relatives are species 2 and 3: it accurately indicates phylogeny. Why the difference?

Species 1 and 3 share not only one derived character state (c_1), but also six ancestral character states (a_0, b_0, g_0, h_0, i_0, j_0). Because species 2 underwent four evolutionary changes (to g_1, h_1, i_1, and j_1) that species 3 does not share, it is less similar to species 3 than species 1 is, even though it is more closely related. But species 2 shares more derived character states with species 3 (c_1, e_1, f_1) than it does with species 1 or 4: these are the features that truly indicate their close relationship. Taxa may be similar because they share ancestral character states or derived character states, but *only the derived character states that are shared among taxa* (i.e., synapomorphies) *indicate monophyletic groups* and enable us to infer the phylogeny successfully. By the same token, derived character states that are restricted to a single lineage, sometimes called **autapomorphies** (such as state j_1 in species 2), do not provide any evidence about its relationship to other lineages.

In the previous examples, each character changed only once across the whole phylogeny. Hence all the taxa sharing a character state inherited it without change from their common ancestor. Such a character state is said to be **homologous** in all the taxa that share it. Again, this need not be the case. A character state is **homoplasious** if it has independently evolved two or more times, and so does not have a unique origin. The taxa that share such a character have not all inherited it from their common ancestor. Figure 2.5C shows three homoplasious characters. State g_1 has independently evolved from g_0 in species 1 and 3, and h_1 has independently evolved in species 1 and 2. These two homoplasies are examples of **convergent evolution**, the independent origin of a derived character state in two or more taxa. Conversely, character j has evolved from j_0 to j_1 in the evolution of ancestor 2, and then has undergone **evolutionary reversal** to j_0 in species 2.

Homoplasious characters—those that undergo convergent evolution or evolutionary reversal—provide misleading evidence about phylogeny. In Figure 2.5C, characters g and j would both lead us to mistake species 1 and 3 for the closest relatives; character h erroneously suggests that species 1 and 2 are a monophyletic group. Thus shared derived character states are valid evidence of monophyletic groups only if they are *uniquely* derived.

Many systematists trace the modern practice of inferring phylogenetic relationships to the German entomologist Willi Hennig (1966). Hennig pointed out that taxa may be similar because they share (1) uniquely derived character states, (2) ancestral character states, or (3) homoplasious character states, and that only the similarity due to uniquely derived character states provides evidence for the clades, the nested monophyletic groups, that make up a phylogenetic tree. Thus, for example, we are confident that the tetrapod limb, the amnion, and the feather each evolved only once. We conclude that all tetrapods (vertebrates with four limbs rather than fins) form a single clade (Figure 2.6). Within the tetrapods, the amniotes* form a monophyletic group. Within the amniotes, all feather-bearing animals (birds) are, again, a single monophyletic group. However, this does not mean that all vertebrates without feathers form a single branch, and indeed they do not. The lack of feathers is an ancestral character state that does not provide evidence that featherless animals are

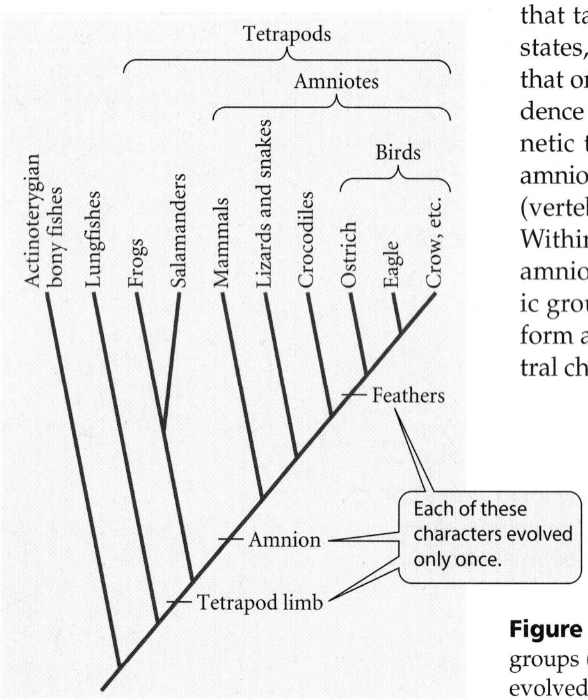

Figure 2.6 A phylogeny of some groups of vertebrates, showing monophyletic groups (tetrapods, amniotes, birds) whose members share derived character states that evolved only once.

*AMNIOTES are those vertebrates—reptiles, birds, and mammals—characterized by a major adaptation for life on land: the amniotic egg, with its tough shell, protective embryonic membranes (chorion and amnion), and a membranous sac (allantois) for storing embryonic waste products.

all more closely related to one another than to birds. (Fishes, lizards, and frogs all lack feathers, but so, after all, do all invertebrates.)

The method of maximum parsimony

Hennig's principle, that uniquely derived character states define monophyletic groups, poses two difficulties: First, how can we tell which state of a character is derived? Second, how can we tell whether it is uniquely derived or homoplasious? For our hypothetical phylogenies in Figure 2.5, we were free to dictate that state 0 was ancestral and that g_1 evolved twice. In real life, we are not given such information—we have to determine it somehow. It might be supposed that the fossil record would tell us what the ancestor's characteristics were, and sometimes it is indeed very helpful. But as we will see, the relationships among fossils and living species have to be interpreted, just like those among living species alone. Besides, the great majority of organisms have a very incomplete fossil record, as we will see in Chapter 4.

Some of the methods devised to deal with these problems depend on the concept of **parsimony**. Parsimony is the principle, dating at least from the fourteenth century, that the simplest explanation, requiring the fewest undocumented assumptions, should be preferred over more complicated hypotheses that require more assumptions for which evidence is lacking. The method based on parsimony is among the simplest methods of phylogenetic analysis. More elaborate methods that are widely used but are too complex to describe here relax the parsimony assumption. For many data sets, these methods yield the same tree topology as parsimony; only in extreme situations will different phylogenetic methods yield different trees (Box 2B).

In phylogenetic analysis, the principle of parsimony suggests that among the various phylogenetic trees that can be imagined for a group of taxa, *the best estimate of the true phylogeny is the one that requires us to postulate the fewest evolutionary changes*. Here is a highly contrived example. Suppose, for the sake of argument, that someone postulates that birds and humans form a monophyletic group because they are bipedal, and that all the other creatures we call mammals form another monophyletic group (Figure 2.7A). This phylogeny would require us to postulate (on the basis of no evidence) that many features shared by humans and other mammals (e.g., hair, milk glands, three middle-ear bones) each evolved twice. In contrast, if we postulate that humans are descended from the same ancestor as other mammals, then each of these features evolved once, and only bipedal-

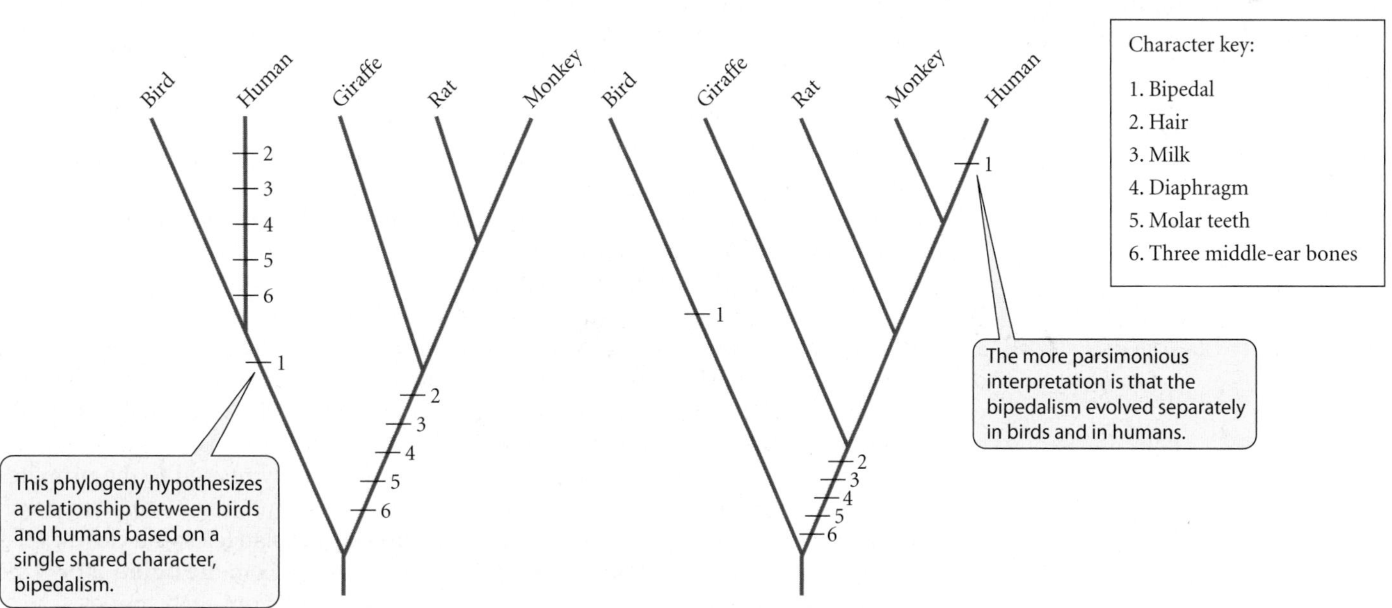

Figure 2.7 Two possible hypotheses for the phylogenetic relationships of humans. (A) A hypothetical phylogeny postulating a close relationship between humans and birds, based on the shared bipedal (two-legged) posture and locomotion. (B) The accepted phylogeny, in which humans are most closely related to other mammals. Tick marks show the changes in several characters that are implied by each phylogeny. Characters 2–6 are considered uniquely derived synapomorphies of mammals. The accepted phylogeny requires fewer evolutionary changes than the phylogeny in (A), and is therefore a more parsimonious hypothesis.

BOX 2B More Phylogenetic Methods

Many methods have been proposed for inferring phylogenies, and their strengths and drawbacks have been extensively argued and analyzed (Felsenstein 2004). Some are *algorithmic* methods that calculate a single tree from the data using a specific and repeatable protocol. Among these, the NEIGHBOR-JOINING METHOD, which does not assume equal rates of DNA sequence evolution among lineages, is the most frequently used. Algorithmic methods have the advantage of yielding the same result from the same data, and therefore being perfectly repeatable by different researchers or computer runs. They are also computationally fast, and are in some cases the only practical approach with the very large data sets that are becoming common. Many practitioners, however, prefer *tree-searching* methods that examine a large sample of the huge number of trees that are possible for even a few taxa. These methods use various search routines to maximize the chance of finding the shortest trees that are compatible with the data. Unlike algorithmic methods, the course of a given tree search may differ (sometimes substantially) from run to run, and thus tree-searching methods run the risk of arriving at different trees for the same data set if the search is not thorough enough.

Some tree-searching programs, such as those based on MAXIMUM PARSIMONY, save many of the trees that are examined, so that the shortest tree found can be compared with those that are nearly as short. We are then interested in knowing if the shortest tree (or particular groupings within that tree) is reliable, or if it differs from other short trees only by chance. This statistical problem is often addressed by a procedure called BOOTSTRAPPING, in which many random subsamples of the characters (e.g., nucleotide sites) are used for repeated phylogenetic analyses. Our confidence in the reliability of a particular grouping is greater if this grouping is consistently found by using these different data sets (bootstrap samples). A group of three or more taxa whose relationships cannot be confidently resolved is often represented by a node with three or more branches—a POLYTOMY. A CONSENSUS TREE is one that portrays both the relationships in which we can be confident and the polytomies that represent "unresolved" relationships.

Two powerful and widely used tree-searching methods are the MAXIMUM LIKELIHOOD (ML) and BAYESIAN methods. They are too complicated to explain in detail here, but both use a model of the evolution of the data (usually DNA sequences). For example, the model might assume that all nucleotide substitutions are equally likely and that a constant substitution rate can be estimated from the data (the "one-parameter model"), or it might assume that different kinds of substitutions occur at different characteristic rates ("Kimura's two-parameter model"). Given the model and a possible tree with branch lengths (measured in substitutions per site), the ML method calculates the likelihood of observing the data. The best estimate of the phylogeny is the tree that maximizes this likelihood.

The more recently developed, and increasingly popular, Bayesian method differs from the ML method by maximizing the probability of observing a particular tree, given the model and the data. (Note the seemingly subtle, but nevertheless very important, difference.) Unlike the ML method, the Bayesian method provides and calculates the probabilities of a set of trees so that they can be compared and summarized, often as a consensus tree (Huelsenbeck et al. 2001).

Phylogenetic analyses based on any one gene yield a tree of alleles—a gene tree—that could in principle differ from the species tree of species because of processes such as incomplete lineage sorting (see Figure 2.20). Yet newer Bayesian methods are able to estimate the species tree directly, even when the topologies and branch lengths vary from gene to gene (e.g., Liu and Pearl 2007; Edwards 2009). However, neither maximum parsimony, maximum likelihood, nor Bayesian methods are guaranteed to find the shortest tree or the tree that maximizes the data, because of the very large number of trees that must be searched. Thus there is a sometimes substantial chance that the best tree is not discovered.

ism evolved twice (Figure 2.7B). The "extra" evolutionary changes we must postulate in a proposed phylogeny are homoplasious changes: those that we must suppose occurred more than once. Thus parsimony holds that *the best phylogenetic hypothesis is the one that requires us to postulate the fewest homoplasious changes.*

Suppose we wish to determine the relationships among species 1, 2, and 3 (Figure 2.8). We are quite sure that they form a monophyletic group relative to taxa 4 and 5, which we are confident are successively more distantly related. These distantly related taxa are called **outgroups**, relative to the **ingroup**, the monophyletic set of species whose relationships we wish to infer. (In our example, species 1, 2, and 3 might be species of primates, 4 might be a rodent, and 5 might be a marsupial such as a kangaroo. From extensive prior evidence, we are confident that rodents and marsupials are more distantly related to the primates than the various primates are to one another.) As Figure 2.8 shows, there are three possible branching relationships (trees 1–3) among the three ingroup species: the closest relatives might be species 1 and 2, 1 and 3, or 2 and 3. The data matrix in the figure shows the nucleotide bases at seven sites in a homologous DNA sequence from each species.

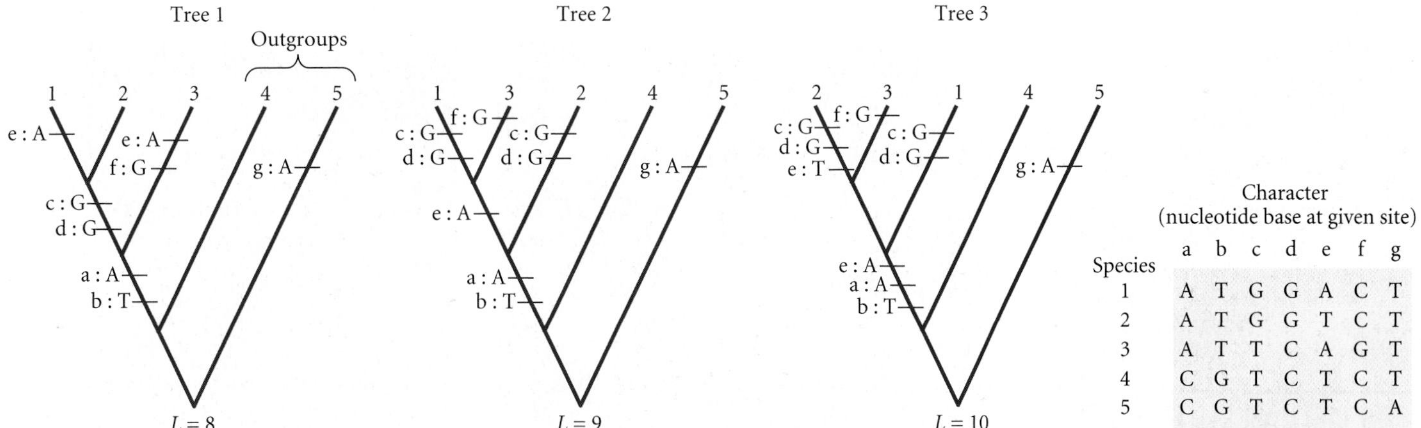

Species	Character (nucleotide base at given site)						
	a	b	c	d	e	f	g
1	A	T	G	G	A	C	T
2	A	T	G	G	T	C	T
3	A	T	T	C	A	G	T
4	C	G	T	C	T	C	T
5	C	G	T	C	T	C	A

Figure 2.8 Inferring a phylogeny by the method of maximum parsimony. The matrix gives the character states (nucleotide bases) for seven sites (a–g) in a DNA sequence. On each of the three trees representing possible relationships among species 1, 2, and 3, the locations of changes to derived character states are shown. When the lengths (*L*) of the three trees are compared, we see that tree 1 is the shortest (*L* = 8); that is, it requires us to postulate the fewest character changes.

Our task is to determine which tree structure implies the smallest number of evolutionary changes in the characters. We go about this by plotting, on each tree, the position at which each character must have changed, minimizing the number of state changes. Consider tree 1, and examine first the variation in character *a* among species. The simplest explanation for the fact that species 1, 2, and 3 share state A and species 4 and 5 share state C is that C was replaced by A in the common ancestor of the ingroup; that single change divides the tree into species with A and species with C. We infer that the change was from C to A, rather than vice versa, because if A had been the ancestral character state, two independent changes from A to C in the evolution of species 4 and 5 would have to be postulated. This pattern would not be parsimonious.

Following the same procedure for each character individually, we find that tree 1 requires us to plot evolutionary change in characters c and d in the ancestor of species 1 and 2, providing evidence that they are **sister groups** (groups derived from a common ancestor that is not shared with any other groups). Character e must be convergent, evolving twice from T to A; if this tree is the true branching history, there is no way in which it could have changed only once and yet have the same state in species 1 and 3. Characters f and g are autapomorphies, changing only in the individual species 3 and 5, respectively. They have exactly the same positions in the other possible phylogenetic trees and carry no information about branching sequence.

The same procedure is then followed for every other possible phylogeny. (This very tedious procedure is carried out by computer programs in real phylogenetic analyses, which typically involve more species and many more characters.) The topology of tree 2 implies that character e is not homoplasious, but evolved from T to A in the common ancestor of species 1 and 3, which are sister taxa in this tree. However, both characters c and d must have evolved convergently in this case. Tree 3 likewise implies that characters c and d evolved convergently, and that character e underwent evolutionary reversal from T to A and back to T.

If, having completed this procedure for all three possible branching trees, we count the number of character state changes (the LENGTH of each tree), we find that tree 1 is shortest, requiring the fewest character state changes. Moreover, more characters support the monophyly of species 1 and 2 (tree 1) than of 1 and 3 or of 2 and 3. According to the parsimony criterion, tree 1 is our *best estimate* of the history of branching (the "true tree"). In any real case, of course, we would want the difference in length between the most parsimonious tree and other possible trees to be much greater before we would have confidence in our estimate.

An example of phylogenetic analysis

In traditional classifications, the primate superfamily Hominoidea consists of three families: the gibbons (Hylobatidae), the human family (Hominidae), and the great apes (Pongidae). The great apes are the orangutan (*Pongo pygmaeus*) in southeastern Asia; the

(A) Gibbon

(B) Orangutan

(D) Chimpanzee

(C) Gorilla

Figure 2.9 Members of the primate superfamily Hominoidea. (A) White-handed gibbon (*Hylobates lar*), family Hylobatidae. (B) Orangutan (*Pongo pygmaeus*). (C) Lowland gorilla (*Gorilla gorilla*). (D) Common chimpanzee (*Pan troglodytes*). Two other living hominoid species, bonobo (*Pan paniscus*) and humans (*Homo sapiens*), are not shown. (A © Steve Bloom/Alamy Images; B © Radovan Spurny/ShutterStock; C © Lars Christensen/istockphoto.com; D © Gerry Ellis/DigitalVision.)

gorilla (*Gorilla gorilla*) in Africa; and two species of chimpanzees (*Pan*), also in Africa. From anatomical evidence, it has long been accepted that the Hominoidea is a monophyletic group, and that the Hylobatidae are more distantly related to the other species than those species are to one another.

Physically, *Pongo*, *Pan*, and *Gorilla* appear more like one another than like *Homo* (Figure 2.9). Hence the traditional view was that Pongidae is monophyletic and that *Homo* branched off first. However, molecular data have definitively shown that humans and chimpanzees are each other's closest relatives (Ruvolo 1997). Among the many DNA analyses leading to this conclusion is a study by Morris Goodman and his coworkers (Goodman et al. 1989; Bailey et al. 1991), who sequenced more than 10,000 base pairs of a segment of DNA that includes a hemoglobin PSEUDOGENE—a nonfunctional DNA sequence derived early in primate evolution by duplication of a hemoglobin gene (see Chapter 8). The outgroups used were the New World spider monkey, *Ateles*, a distant relative of the Hominoidea, and the Old World rhesus monkey, *Macaca*, which belongs to a more closely related family (Cercopithecidae).

On the whole, the hominoids, especially *Homo* and *Pan*, have very similar sequences (Table 2.1), but they do differ by some

Table 2.1 ***Divergence between nucleotide sequences of the ψη-globin pseudogene among orangutan (Pongo), gorilla (Gorilla), chimpanzee (Pan), and human (Homo)[a]***

	Gorilla	Pan	Homo
Pongo	3.39	3.42	3.30
Gorilla		1.82	1.69
Pan			1.56
Homo			0.38

Source: Data from Bailey et al. 1991.

[a]The percentage of divergence between two human sequences is given in the lower right cell. Divergences between *Homo* and other species are calculated using the average of these two sequences. Values are not corrected for multiple substitutions.

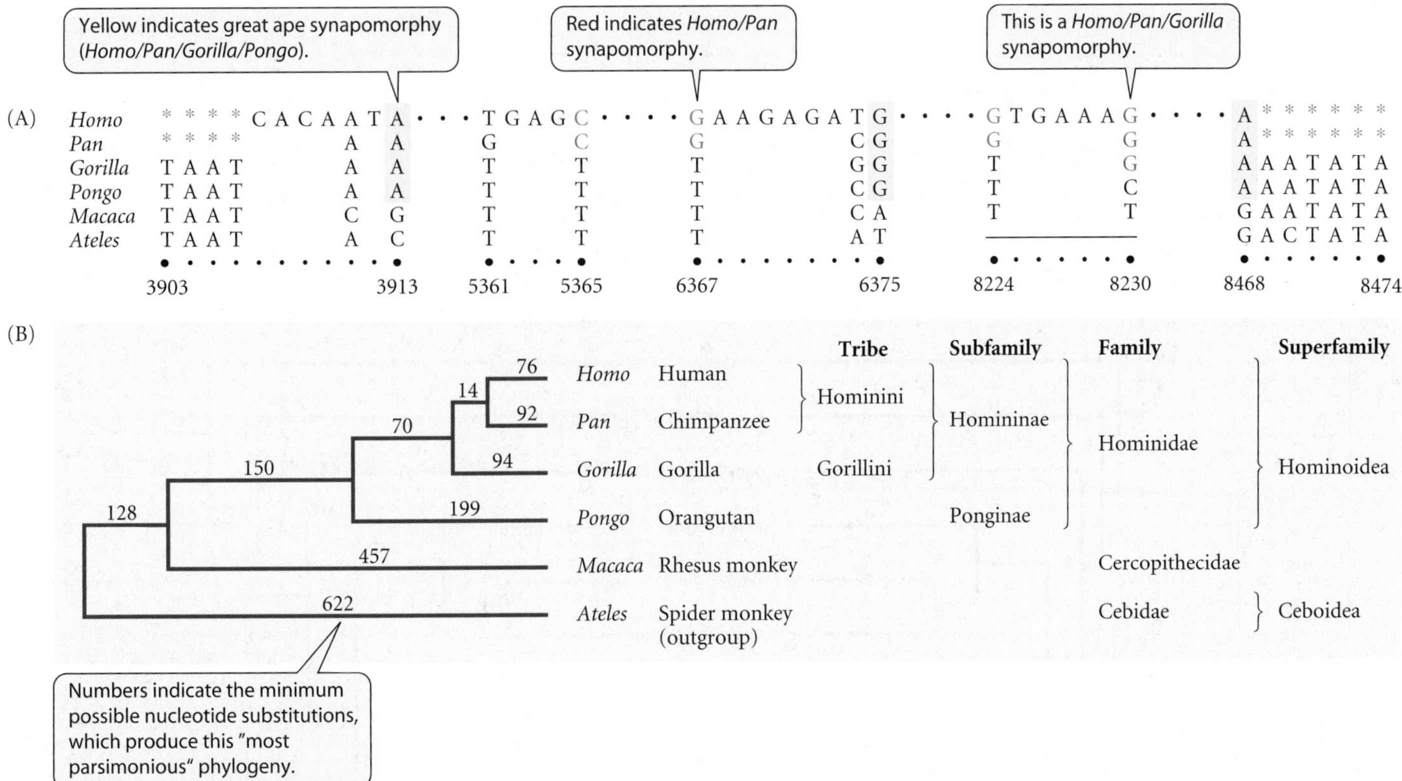

Figure 2.10 Evidence for phylogenetic relationships among primates, based on the ψη-globin pseudogene. (A) Portions of the sequence in six primates. *Macaca*, an Old World monkey, and *Ateles*, a New World monkey, are successively more distantly related outgroups with reference to the Hominoidea. Sequences are identical except as indicated. Using *Ateles* and *Macaca* as outgroups, positions 3913, 6375, and 8468 exemplify synapomorphies of the other four genera, and position 8230 provides a synapomorphy of *Gorilla*, *Pan*, and *Homo*. Synapomorphies of *Pan* and *Homo* include base pair substitutions at positions 5365, 6367, and 8224, and deletions at 3903–3906 and 8469–8474 (red asterisks). Autapomorphies (unshared derived states) include 3911 and 3913 (*Macaca*), 8230 (*Pongo*), 6374 (*Gorilla*), 5361 (*Pan*), and 6374 (*Homo*). (B) The most parsimonious phylogeny based on the ψη-globin sequence, using *Ateles* as an outgroup. The minimal number of changes is indicated along each branch. A tree that split up the *Homo-Pan-Gorilla* group would be 65 steps longer, and one that split *Homo* and *Pan* would be 8 steps longer. The figure includes one of several proposed classifications for humans and apes (Delson et al. 2000). (A after Goodman et al. 1989; B after Shoshani et al. 1996.)

base pair substitutions and short insertions and deletions of bases (Figure 2.10). Some sites, such as number 8230 in Figure 2.10A, distinguish the orangutan from the chimpanzee, gorilla, and human. Fourteen synapomorphies unite chimpanzee and human as sister groups, including deletions of short sequences such as sites 8468–8474. In contrast, only three sites (none of which is shown in this figure) support the hypothesis that chimpanzee and gorilla are the closest relatives.

The phylogeny in Figure 2.10B, with chimpanzee and human as closest relatives and gorilla as sister group to this pair, is eight steps (evolutionary changes) shorter than a tree that separates chimpanzee from human. Many other molecular data sets support this conclusion. Such phylogenetic studies have led taxonomists to alter the classification of apes and humans (Figure 2.10B), as is explained in Chapter 3.

Evaluating phylogenetic hypotheses

How can we judge the validity of our phylogenetic inferences? The estimate of a phylogenetic tree that we obtain from a particular set of data is a phylogenetic *hypothesis* that is *provisionally* accepted (as is any scientific statement). Additional data may lead us to modify or abandon the hypothesis, or may lend further support to it. In phylogenetic studies, as is true throughout science, *the chief way of confirming a phylogenetic hypothesis is to see if it agrees with multiple, independent sources of data*. Morphological features and DNA sequences, for example, evolve largely independently of each other and thus provide independent phylogenetic information (see below and Chapter 20). With some exceptions,

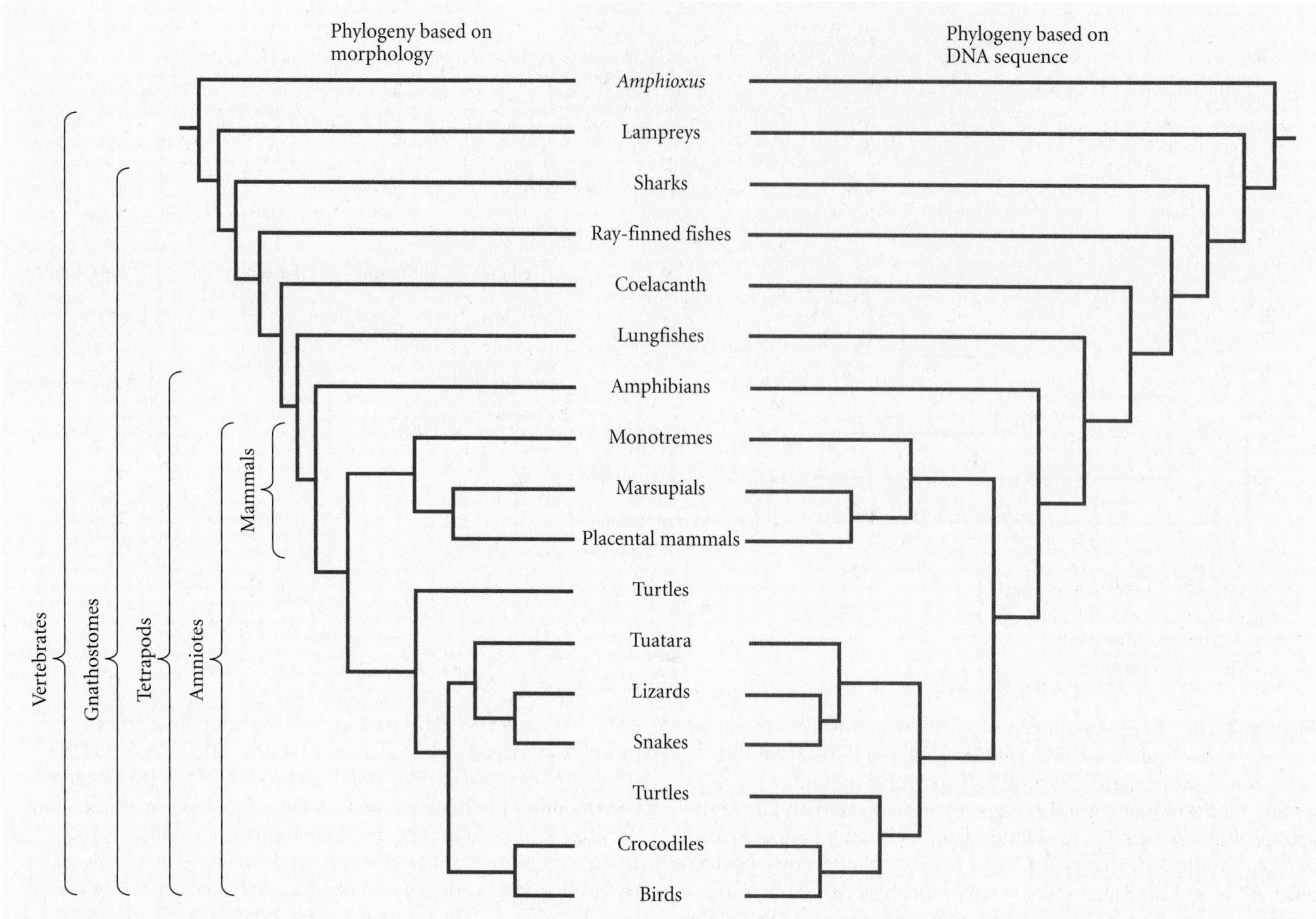

Figure 2.11 Relationships among major groups of living vertebrates, as estimated from morphological characters (left) and from DNA sequences (right). On the whole, these two sources of information provide similar estimates of the phylogeny. The relationships of turtles are unclear from morphological data. Among these taxa, the two sources of data conflict only with respect to the relationships of the turtles. (After Meyer and Zardoya 2003; van Rheede et al. 2006.)

these two kinds of data usually yield similar estimates of phylogeny. For instance, the phylogenetic relationships among higher taxa of vertebrates based on DNA sequences are the same, with few exceptions, as those inferred from morphological features (Figure 2.11). In the same vein, the genome sequence of sea urchins, determined in 2006 (Sodergren et al. 2006), confirmed that among invertebrate phyla, echinoderms are the closest relatives of the chordates (including humans)—a conclusion reached in the 1890s by comparisons of embryonic development.

We can also test the validity of phylogenetic methods by applying them to phylogenies that are absolutely known. Many investigators have simulated evolution on a computer, allowing computer-generated lineages to branch and their characters to change according to various models of the evolutionary process. (For example, characters might change at random at different average rates, or one of two species derived from a common ancestor might evolve faster than the other.) The investigators then see whether or not the various phylogenetic methods can use the final characteristics of the simulated lineages to yield an accurate history of their branching. Perhaps more convincing to the skeptic are a few studies in which phylogenetic methods have been applied to experimental populations of real organisms that have been split into separate lineages by investigators (creating artificial branching events, or **nodes**) and allowed to evolve. In one such study, David Hillis and coworkers (1992; Cunningham et al. 1998) successively subdivided lineages of T7 bacteriophage and exposed them to a mutation-causing chemical, which caused them to accumulate DNA sequence differences rapidly over the course of about 300 generations. The investigators then scored the eight resulting lineages (Figure 2.12) for sequence differences and performed a phylogenetic analysis of the data. For this many populations, there are 135,135 possible BIFURCAT-

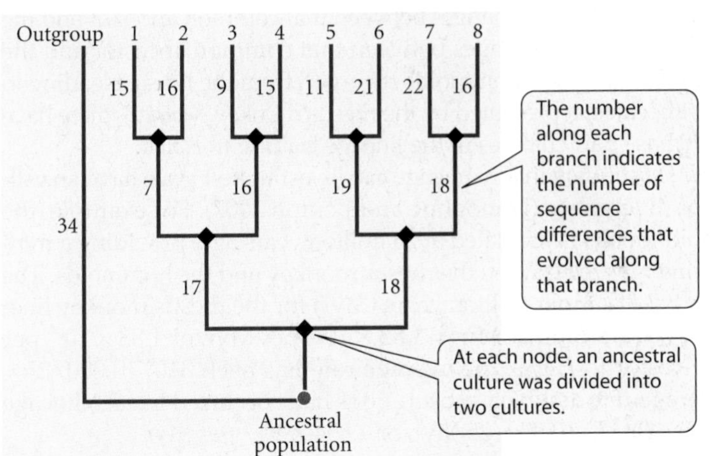

Figure 2.12 The true phylogeny of the experimental population of T7 bacteriophage studied by Hillis et al. The estimate of phylogeny, based on sequence differences among populations at the end of the experiment, was exactly the same as the true phylogeny. (After Hillis et al. 1992.)

ING trees (in which each lineage branches into two others), but the phylogenetic analysis the investigators used correctly found the one true tree.

Molecular Clocks

If evolution were only divergent (i.e., if there were no homoplasy), and if all lineages evolved at the same, constant rate, then the number of differences between two species would be a straightforward index of the time since they diverged from their common ancestor (Figure 2.13). In that case, we could determine the phylogeny—i.e., the relative order of branching—simply by the degree of difference between pairs of taxa (as in Figure 2.5A).

Early in the history of molecular phylogenetic studies, the data suggested that DNA sequences may indeed evolve and diverge at a constant rate (which is certainly not true of morphological features). This concept has been dubbed the **molecular clock** (Zuckerkandl and Pauling 1965). To the extent that a precise molecular evolutionary clock exists, it can provide a simple way of estimating phylogeny. It can also enable us to *estimate the absolute time* since different taxa diverged—if we can determine how fast it is "ticking." (Bear in mind that the phylogenetic trees we have considered so far portray only the branching sequences of taxa, and therefore their *relative* times of divergence, rather than absolute times.)

In order to estimate the rate of molecular evolution, we calculate the number of differences (e.g., in base pairs) that have accrued among pairs of species since their common ancestor by parsimoniously plotting on our estimated phylogeny where each change took place (as in Figures 2.5 and 2.8). For example, Figure 2.10B shows 76 changes between *Homo* and its com-

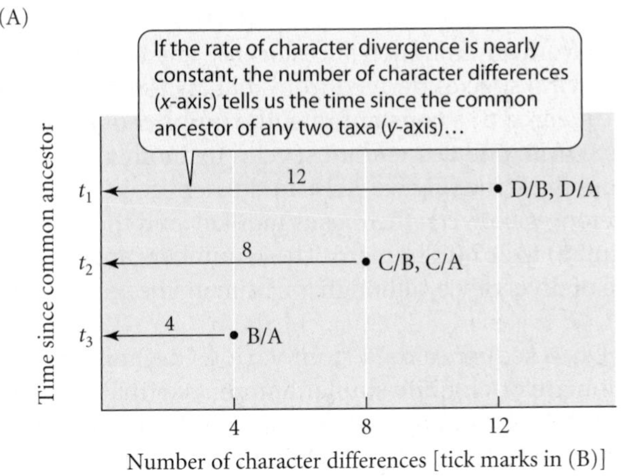

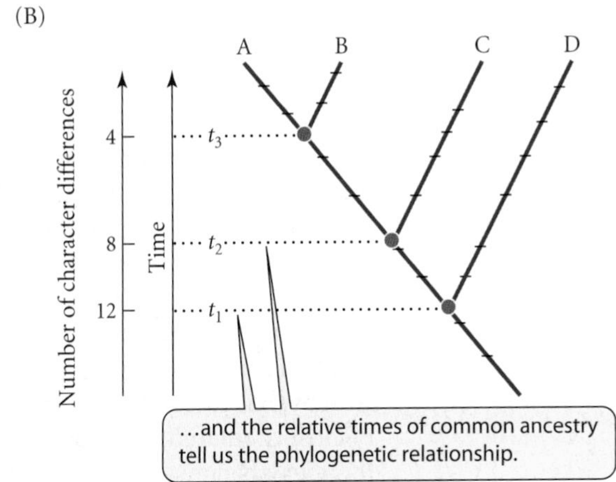

Figure 2.13 (A) If divergence occurred at a nearly constant rate, the relative times of divergence of lineages could be determined from the overall differences (or similarities) between taxa, and the phylogeny of the taxa could then be estimated. (B) A hypothetical phylogeny in which evolution occurs at a nearly constant rate, and which therefore could be estimated from the differences among taxa. Each tick mark represents the evolution of a new character state.

mon ancestor with *Pan* (e.g., at site 6374), 14 changes between that common ancestor and the *Gorilla* branch (e.g., at site 5365), and 70 changes between that common ancestor and the *Pongo* branch (e.g., at site 8230). Since these hominoids diverged from the lineage leading to Old World monkeys (Cercopithecidae, represented by the rhesus monkey, *Macaca*), there have been 76 + 14 + 70 + 150 = 310 base pair changes in the lineage leading to *Homo*.

The *average rate of base pair substitution* in any lineage can be estimated if we have an estimate of the absolute time of divergence (Donoghue and Benton 2007). For example, the oldest fossils of cercopithecoid monkeys are dated at 25 million years ago, providing a minimal estimate of time since divergence between the rhesus monkey and the hominoids. The number of substitutions per base pair per million years (Myr) for the rhesus monkey lineage is 457/10,000 base pairs sequenced/25 Myr = 1.83×10^{-3} per Myr, or 1.83×10^{-9} per year. From the common ancestor to *Homo*, the average rate has been 310/10,000/25 = 1.24×10^{-3} per Myr. The average rate at which substitutions have occurred in each lineage is therefore $r = (457 + 310)/2 = 383.5/10{,}000/25$ Myr, or 1.534×10^{-3} per Myr.

In this way, information from the fossil record on the absolute time of divergence of certain taxa may be used to calibrate a molecular clock—to determine its rate—and to estimate the divergence times of other taxa that have not left a good fossil record. For example, suppose the proportion of base pairs that differ between the ψη-globin pseudogene sequences of two primate species is 0.0256. Assuming a molecular clock,

$$D = 2rt$$

where D is the proportion of base pairs that differ between the two sequences, r is the rate of divergence per base pair per Myr, t is the time (in Myr) since the species' common ancestor, and the factor 2 represents the two diverging lineages. If $D = 0.0256$ and $r = 0.001534$, as estimated from the data in Figure 2.10, then $t = D/2r = 8.3$, and 8.3 Myr is our best estimate of when the two species diverged from their common ancestor.

The absolute rate of sequence evolution can sometimes be calibrated not only from fossils, but also from geological events of known age. For example, the major islands of the Hawaiian archipelago have arisen as volcanoes in sequence, and range from 0.43 to 5.1 Myr old (Figure 2.14A). Robert Fleischer and colleagues (1998) used these dates to estimate the rate of sequence evolution in a mitochondrial gene of a group of birds, the Hawaiian honeycreepers, and a nuclear gene of several of the fruit flies (*Drosophila* spp.) that are restricted to these islands. For example, the Kauai and Oahu creepers have had 3.7 Myr to diverge, assuming their common ancestor already occupied Kauai when Oahu arose 3.7 Myr ago, and colonized Oahu soon after its origin. When several such points are plotted, based on the phylogenies and sequence differences among several species of either birds or fruit flies, they show a remarkably strong correlation between sequence difference (sometimes called the GENETIC DISTANCE) and the amount of time the species have been physically separated (Figure 2.14B,C).

The rate of sequence divergence is not always as nearly constant as in this example, however (Mindell and Thacker 1996; Arbogast et al. 2002). There are several ways of determining whether sequence evolution conforms to a molecular clock, even without information on divergence time. One such method is the RELATIVE RATE TEST (Wilson et al. 1977). We know that the time that has elapsed from any common ancestor (i.e., any branch point on a phylogenetic tree) to each of the living species derived from that ancestor is *exactly the same*. Therefore, if lineages have diverged at a constant rate, the number of changes along all paths of the phylogenetic tree from one descendant species to another through their common ancestor should be about the same (Figure 2.15). In the hominoid example (see Figure 2.10B), the number of differences between the rhesus monkey and the various hominoids ranges from 806 (to orangutan) to 767 (to human). These numbers are so close that they indicate a fairly constant rate of divergence, although the human lineage appears to have slowed down somewhat.

The relative rate test, applied to DNA sequence data from various organisms, has shown that rates of sequence evolution are often quite similar among taxa that are fairly closely related, but may differ considerably among distantly related taxa. Moreover, there exists considerable intrinsic variation in the rates of mutation and other processes

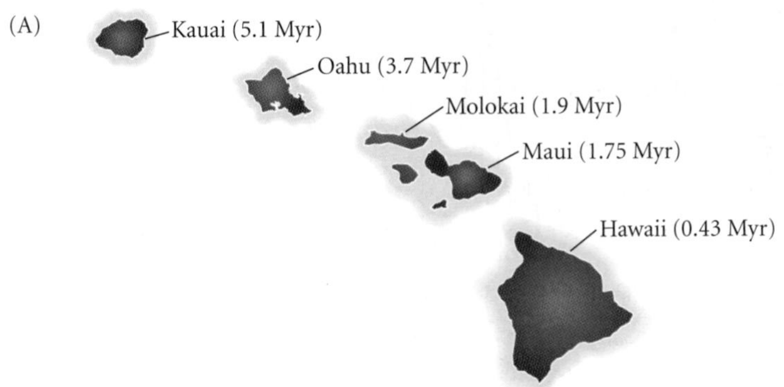

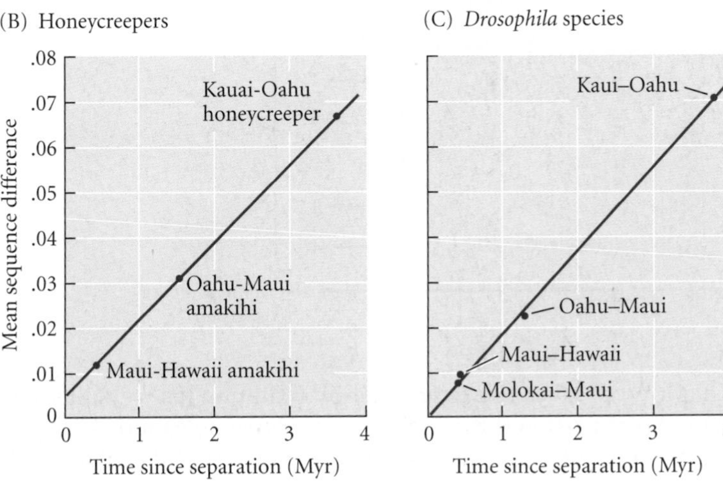

Figure 2.14 Calibration of molecular clocks in Hawaiian organisms. (A) The major islands, showing their approximate ages, from oldest (Kauai) to youngest (Hawaii). (B) In Hawaiian honeycreepers, the DNA sequence of the mitochondrial cytochrome *b* gene has diverged at an effectively constant rate of 0.016/Myr, as determined from the slope of the line that best fits the relationship between sequence divergence of pairs of species and the time since separation of the islands they inhabit. (C) A similar pattern is found among closely related species of *Drosophila*, in which the sequence of the *Yp1* gene has diverged at an average rate of 0.019/Myr. (After Fleischer et al. 1998.)

that affect the rate of sequence evolution, so there always exists an appreciable "estimation error" around even the best estimate of divergence time from DNA sequences. Many methods have been developed to identify and correct for variations in rates of sequence evolution and other sources of error (Arbogast et al. 2002; Welch and Bromham 2005). Molecular clocks are seldom used to estimate phylogenetic relationships among taxa, but they are often useful for estimating approximate dates of divergence, and we will encounter many examples of their use later in this book.

Gene Trees

So far, we have been concerned with inferring phylogenetic trees of species. Because genes replicate, different copies of a gene, within a single species or more than one species, have a history of descent from common ancestral genes, just as species do. Using the same principles that we use for inferring phylogenetic relationships among species, we can infer the historical relationships among variant DNA sequences of a gene (**haplotypes**), and construct a phylogeny of genes, often called a **gene tree** or a **gene genealogy** (Figure 2.16).

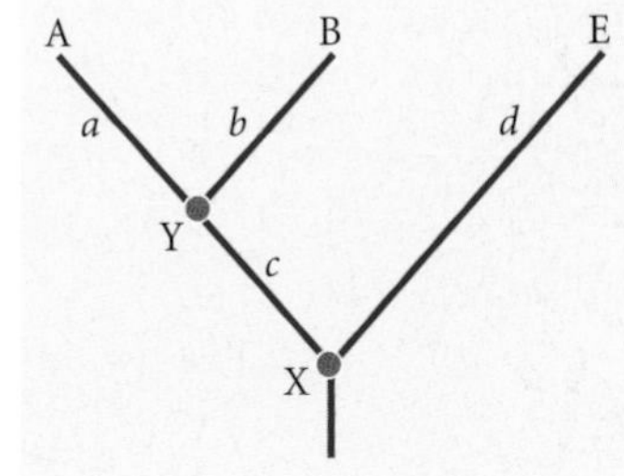

Figure 2.15 The relative rate test for constancy of the rate of molecular divergence. Sequences are obtained for living species A and B and for outgroup species E. Y and X represent ancestral species. Lowercase italic letters represent the number of character differences (e.g., nucleotide changes) along each branch. The genetic distance between A and E is $D_{AE} = a + c + d$. That between B and E is $D_{BE} = b + c + d$. If the rate of nucleotide substitution is constant, then $a = b$, so $D_{AE} = D_{BE}$. If rate constancy holds throughout the tree, the distance between any pair of species that have species X as a common ancestor will equal that between any other such pair of species.

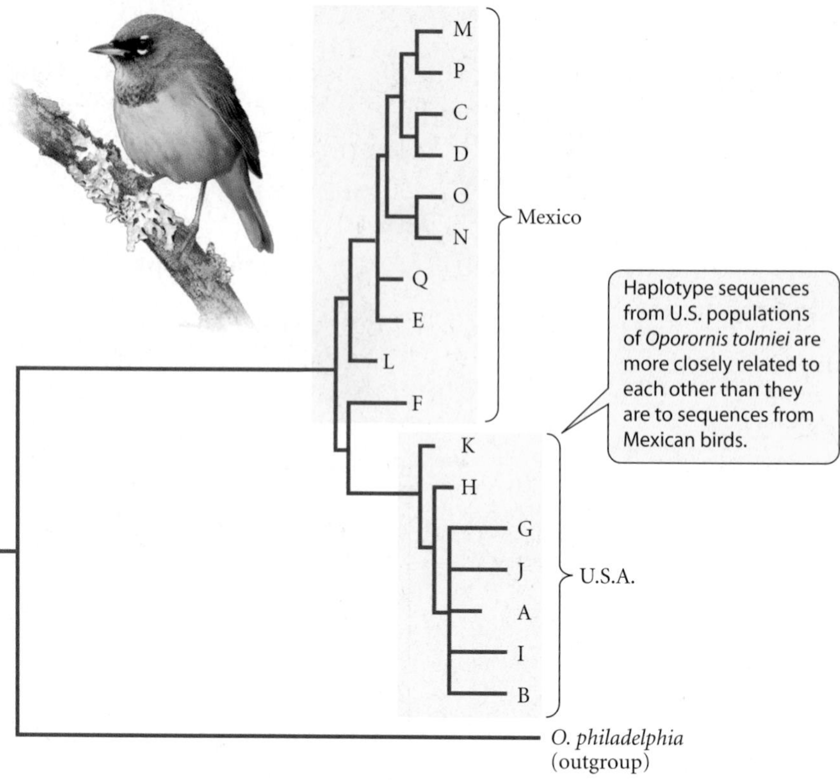

Figure 2.16 A gene tree, showing the relationships among the sequences of haplotypes A–Q of the mitochondrial cytochrome *b* gene in MacGillivray's warbler (*Oporornis tolmiei*), using a sequence from the mourning warbler (*O. philadelphia*) as an outgroup. (After Milá et al. 2000; Photo © Tim Zurowski/Photolibrary.com.)

One DNA sequence (haplotype) gives rise to another by mutation (as we will see in Chapter 8). The simplest kind of mutation is the replacement of one nucleotide base pair by another at a single site. Consider a hypothetical series of haplotypes that evolve in this way. For illustrative purposes, we'll hypothesize seven haplotypes, arising by a series of base pair mutations:

	Sites				
Haplotype	1 2 3	4 5 6	7 8 9	10 11 12	13 14 15
3	A T A	C T A	**T** A T	G T T	G C C
2	A T A	C T A	**C** A **T**	G T T	G C C
1	A T A	C T A	C A **C**	G T T	**G** C C
4	A T A	C T A	C A C	G T T	**A** C **C**
5	A T A	C T **A**	C A C	G T T	A C **T**
6	A T **A**	C T **G**	C A C	G T T	A C T
7	A T **G**	C T G	C A C	G T T	A C T

Thus haplotype 3 arises from haplotype 2 by substitution of T for C at position 7 (or vice versa). The new haplotype (3) arises by alteration of only one of the many copies of the gene that are carried by different individuals in a population, so its ancestor (haplotype 2) continues to exist, at least for a while.

Now imagine that we find all seven haplotypes in a sample of organisms. We can correctly arrange them in an UNROOTED GENE TREE by assuming that the sequences that differ least have the closest ancestor-descendant relationship; that is, that the most closely related haplotypes are connected to each other by the smallest possible number of mutations. Haplotype 1 is placed between haplotypes 2 and 4 because a greater number of mutations would be required in order to connect 1 to any other haplotype. In this way, we find the most parsimonious unrooted tree (which doesn't look very treelike):

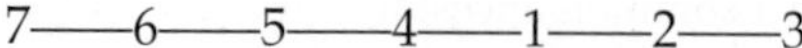

At this point, the tree is "unrooted," meaning that we do not know what the direction of evolution was (i.e., which sequence is closest to the common ancestor of them all). Suppose, moreover, that we find only haplotypes 2 through 7 in our sample. We would infer that haplotype 1 exists, or existed in the past, because it is the necessary step between haplotypes 2 and 4, which differ by two mutations (at positions 9 and 13). It is very unlikely that haplotype 4 arose directly from haplotype 2 (or vice versa), because of the improbability that two mutations would occur in the same individual gene. Therefore, we would postulate haplotype 1 as a hypothetical intermediate, or ancestor.

How can we infer the direction of evolution? Suppose haplotypes 4, 5, 6, and 7 are found in species A, and haplotypes 2 and 3 are found in closely related species B. We already know from the unrooted tree that haplotypes 2 and 3 are close relatives, as are haplotypes 4–7. The most parsimonious inference is that the common ancestor of the two species had a haplotype—such as haplotype 1—that gave rise to these two groups of haplotypes in the two species. This ancestral sequence allows us to ROOT the tree:

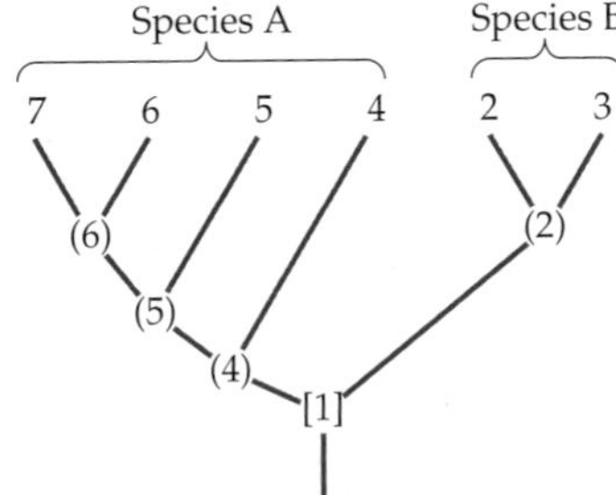

Since haplotype 1 is the postulated ancestral haplotype, we can conclude that the evolutionary sequence of haplotypes was $4 \rightarrow 5 \rightarrow 6 \rightarrow 7$ in species 1, and $2 \rightarrow 3$ in species 2. This conclusion remains the same if we should find haplotype 1 in either (or both) of the species.

Difficulties in Phylogenetic Analysis

In practice, it can be quite difficult to resolve relationships among taxa. We will mention some of the problems here, although the methods developed to solve them are outside the scope of this book. Nevertheless, it is very important to understand the evolutionary processes that can make phylogenetic analysis difficult; we will encounter examples of these situations quite often.

- *Scoring characters can be hard.* Several problems can arise in obtaining basic data for phylogenetic analysis. For example, anatomical characters are important in systematics, and they are the only data typically available for extinct organisms. Deciding whether or not organisms have the same character state often requires extensive knowledge of anatomical details and is not an easy or trivial task. Another problem lies in deciding how many independent characters there are. If some mammals have, on each side of each jaw, two incisors, one canine, three premolars, and four molars, and if other mammals (e.g., anteaters) have no teeth at all, does this represent a single character difference (loss of teeth), four differences (loss of four kinds of teeth), or ten differences? A similar problem can arise in DNA sequence data if changes at several sites are likely to evolve in concert, to preserve function. For instance, the structure of the ribosomal RNA molecule includes short sequences whose bases must pair to form "stems," so changes in those sequences are not independent. Furthermore, DNA sequences must be aligned so that homologous sites can be compared. The more different the sequences, the harder they are to align, especially if there have been insertions, deletions, or inversions of portions of the sequence.
- *Homoplasy is very common.* For this reason, a data set may yield several different phylogenetic estimates that are equally good, or almost so. It is unwise to rely on a particular phylogenetic estimate if other phylogenetic hypotheses imply just a few extra evolutionary changes. Instead, one should try to get more data (i.e., on other characters).

Figure 2.17 The proportion of base pairs in the DNA sequences of the mitochondrial gene *COI* that differ between pairs of vertebrate species, plotted against the time since their most recent common ancestor, based on the fossil record. Sequence differences evolve most rapidly at third-base positions and most slowly at second-base positions within codons (as we will see in Chapter 8). Divergence at third-base positions increases rapidly at first, then levels off as a result of multiple substitutions at the same sites. Thus these positions provide no phylogenetic information for taxa that diverged more than about 75 million years ago (Mya). The species used for this analysis include two cypriniform fish species, one frog, one bird, one marsupial, two rodents, two seals, two baleen whales, and *Homo*. (After Mindell and Thacker 1996.)

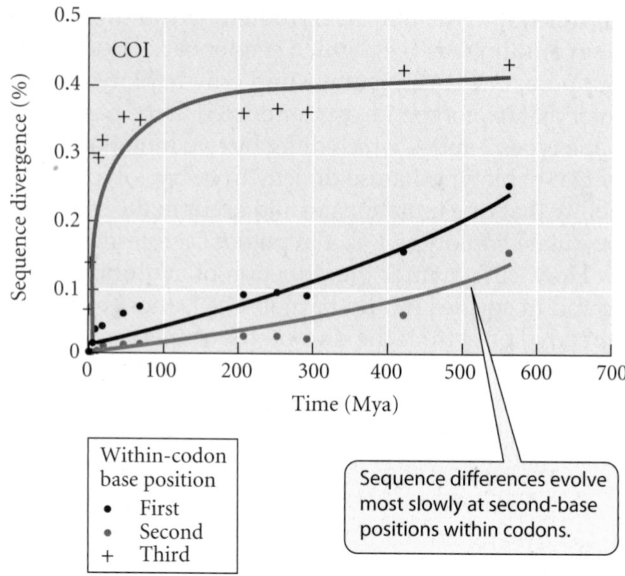

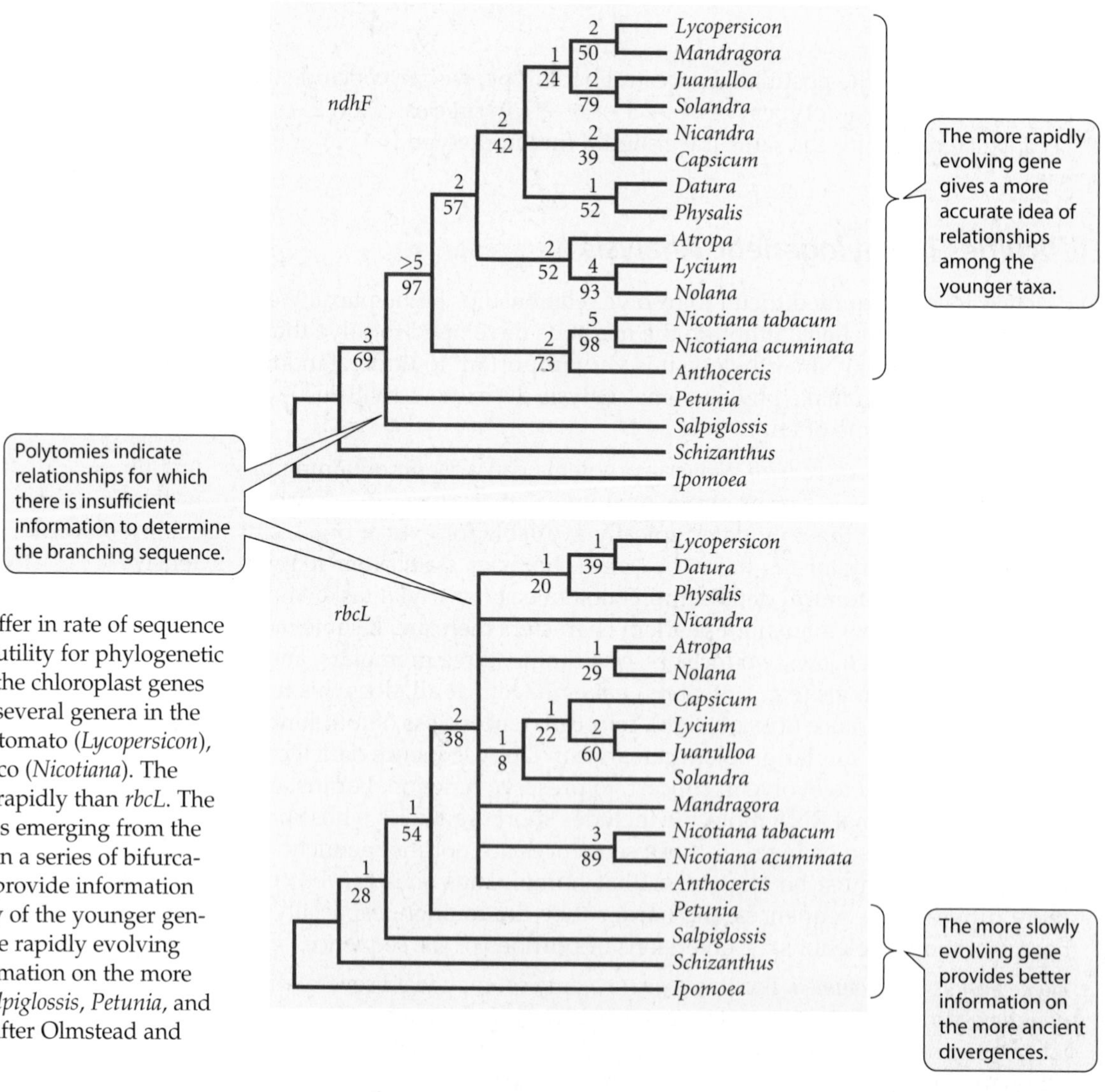

ndhF

Lycopersicon
Mandragora
Juanulloa
Solandra
Nicandra
Capsicum
Datura
Physalis
Atropa
Lycium
Nolana
Nicotiana tabacum
Nicotiana acuminata
Anthocercis
Petunia
Salpiglossis
Schizanthus
Ipomoea

The more rapidly evolving gene gives a more accurate idea of relationships among the younger taxa.

Polytomies indicate relationships for which there is insufficient information to determine the branching sequence.

rbcL

Lycopersicon
Datura
Physalis
Nicandra
Atropa
Nolana
Capsicum
Lycium
Juanulloa
Solandra
Mandragora
Nicotiana tabacum
Nicotiana acuminata
Anthocercis
Petunia
Salpiglossis
Schizanthus
Ipomoea

The more slowly evolving gene provides better information on the more ancient divergences.

Figure 2.18 Genes that differ in rate of sequence evolution may differ in their utility for phylogenetic analysis. Estimates based on the chloroplast genes *ndhF* and *rbcL* are shown for several genera in the family Solanaceae, including tomato (*Lycopersicon*), pepper (*Capsicum*), and tobacco (*Nicotiana*). The *ndhF* gene has evolved more rapidly than *rbcL*. The polytomies (multiple branches emerging from the same branch point, rather than a series of bifurcations) show that *rbcL* fails to provide information on relationships among many of the younger genera. However, unlike the more rapidly evolving gene, *rbcL* does provide information on the more ancient divergence among *Salpiglossis*, *Petunia*, and the sister group of *Petunia*. (After Olmstead and Sweere 1994.)

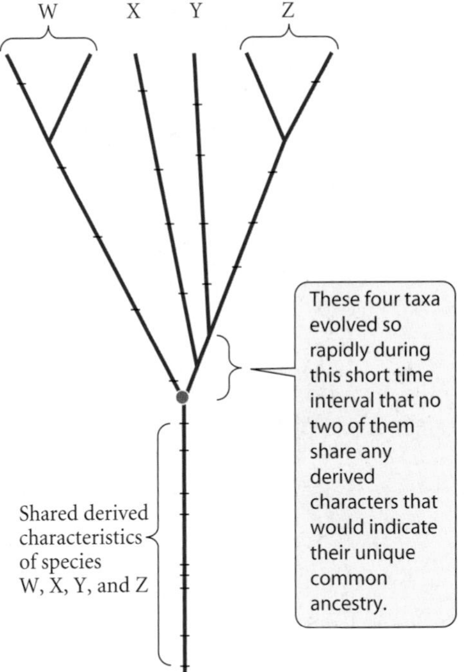

W
X
Y
Z
These four taxa evolved so rapidly during this short time interval that no two of them share any derived characters that would indicate their unique common ancestry.
Shared derived characteristics of species W, X, Y, and Z

Figure 2.19 Rapid evolutionary radiation. The relationships among species W, X, Y, and Z cannot be determined because they all diverged within such a short interval of time that few or no derived character states evolved in the common ancestors of any subset of these taxa.

- *The process of evolution often erases the traces of prior evolutionary history.* If the taxa under study diverged long ago or have evolved very rapidly, many of their characteristics will have diverged so greatly that homologous characters may be difficult to discern. For instance, teeth provide important features for determining relationships among many mammals, but they cannot be used to assess the relationships among toothless anteaters. In DNA sequences, multiple substitutions may occur at a site over the course of time, erasing earlier synapomorphies. Mutation at a site from A to C provides a shared derived character only as long as no further changes occur at that site; if in a descendant taxon, C is replaced by G, or by reversal to A, the evidence of common ancestry is lost. Furthermore, as time passes, it becomes more likely that the same substitution will occur in parallel in different lineages. Therefore, convergent evolution and successive substitutions ("multiple hits") cause the amount of sequence divergence between taxa eventually to level off, at a plateau that is attained sooner in rapidly evolving than in slowly evolving sequences (Figure 2.17). For this reason, rapidly evolving DNA sequences are useful for phylogenetic analyses of taxa that have diverged recently (Figure 2.18), whereas slowly evolving sequences are required to assess relationships among taxa that diverged in the distant past.
- *Some lineages diverge so rapidly that there is little opportunity for the ancestors of each monophyletic group to evolve distinctive synapomorphies* (Figure 2.19). Such "bursts" of divergence of many lineages during a short period, termed EVOLUTIONARY RADIATIONS, are often called ADAPTIVE RADIATIONS because the lineages have often acquired different adaptations. Many families of songbirds, for example, appear to have diversified within a short time, and so their relationships are not well understood.
- *An accurately estimated gene tree may imply the wrong species phylogeny.* This is a challenging concept, but it has some important implications (see Chapters 11 and 17). Suppose that two successive branching events occur, giving rise first to species A and then to species B and C (Figure 2.20). Now suppose that the ancestral species that gave rise to A, B, and C carries two different haplotypes of the gene we are studying (is POLYMORPHIC). Perhaps, just by chance, haplotype 1 becomes FIXED in species A (that is, the other haplotype is lost from that species by genetic drift; we will see how this can happen in Chapter 10), and haplotype 2 is fixed in the common ancestor of B and C, and then diverges by further mutations into haplotypes 2a in species B and 2b in species C. Then the gene tree will reflect the species phylogeny (Figure 2.20A). But if such sorting of haplotypes into the diverging species does not occur, and the common ancestor of B and C is also polymorphic (i.e., if there is INCOMPLETE SORTING of gene lineages), then the haplotypes may become fixed, by chance, in the three species in such a way that the most closely related species do not inherit the same haplotype (Figure 2.20B,C). A phylogeny based on these haplotypes will therefore imply incorrect relationships among the species. Figure 2.21 provides a real example, in which sequences of six independently segregating genes were taken from several individuals of each of four species of grasshoppers (Carstens and Knowles 2007). At each gene locus, the relationships among haplotypes differ, in many different ways, from the best estimate of the species phylogeny, based on all the available data. Each species has inherited several gene lineages from its common ancestor with other species.
- *Relationships among closely related species may be obscured by low levels of interbreeding between species* (HYBRIDIZATION; see the next section). This is another reason why gene trees may not faithfully reflect the phylogeny of the species from which the genes are

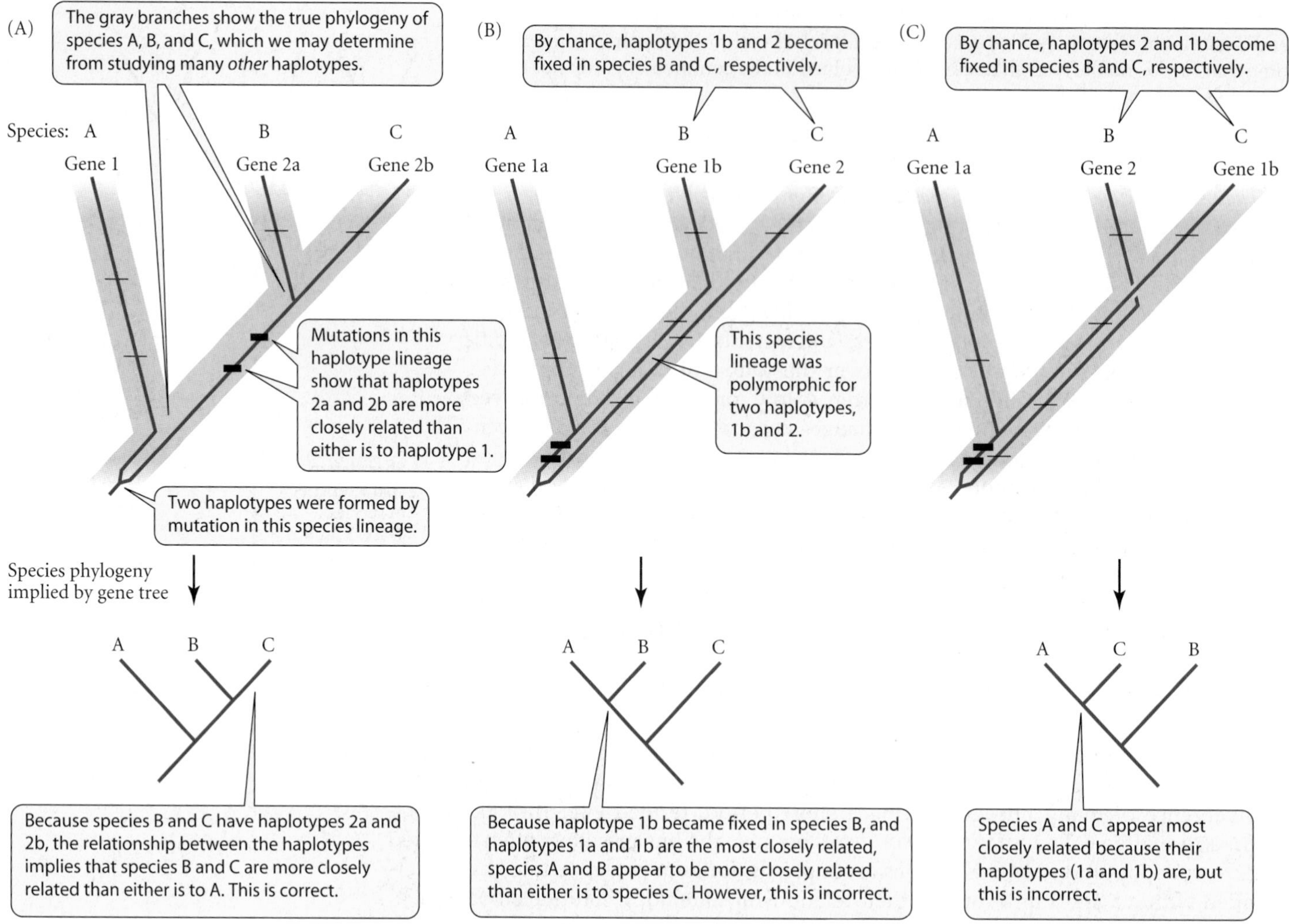

Figure 2.20 A gene tree (red lines) may or may not reflect the phylogeny of the species from which the genes are sampled (outer envelopes). Species A, B, and C have arisen by two successive branching events. (A) The true phylogeny, which we might infer from other data. Haplotypes 2a and 2b form a monophyletic group, as indicated by two shared base pair substitutions (thick tick marks), and the species phylogeny implied by the gene tree matches the actual phylogeny of the species. Thin tick marks show other mutations that distinguish the three haplotypes. (B) Haplotypes 1a and 1b form a monophyletic group, and falsely imply that species A and B are sister species. This is because the common ancestor of species B and C was polymorphic for two gene lineages (1b and 2). The gene lineages that became fixed in the sister species B and C are not sister lineages. (C) The common ancestor of species B and C is again polymorphic for gene lineages 2 and 1b, but the fixation of these genes is the reverse of diagram B. The gene tree falsely implies that species A and C are most closely related. (After Maddison 1995.)

sampled. It is often difficult to distinguish the effects of hybridization and incomplete lineage sorting, both of which are most likely to occur between recently diverged species or populations.

Methods are being developed to reliably infer phylogenies from DNA sequences despite these problems (Maddison and Knowles 2006; Whitfield and Lockhart 2007). The greater the number of independent genes used, the more reliable the phylogeny is likely to be. Phylogenetic studies of some taxa, such as primates, now employ dozens or hundreds of gene sequences that have been deposited in the database GenBank, and some studies use the sequence of entire genomes. For example, the orders of mammals form four major

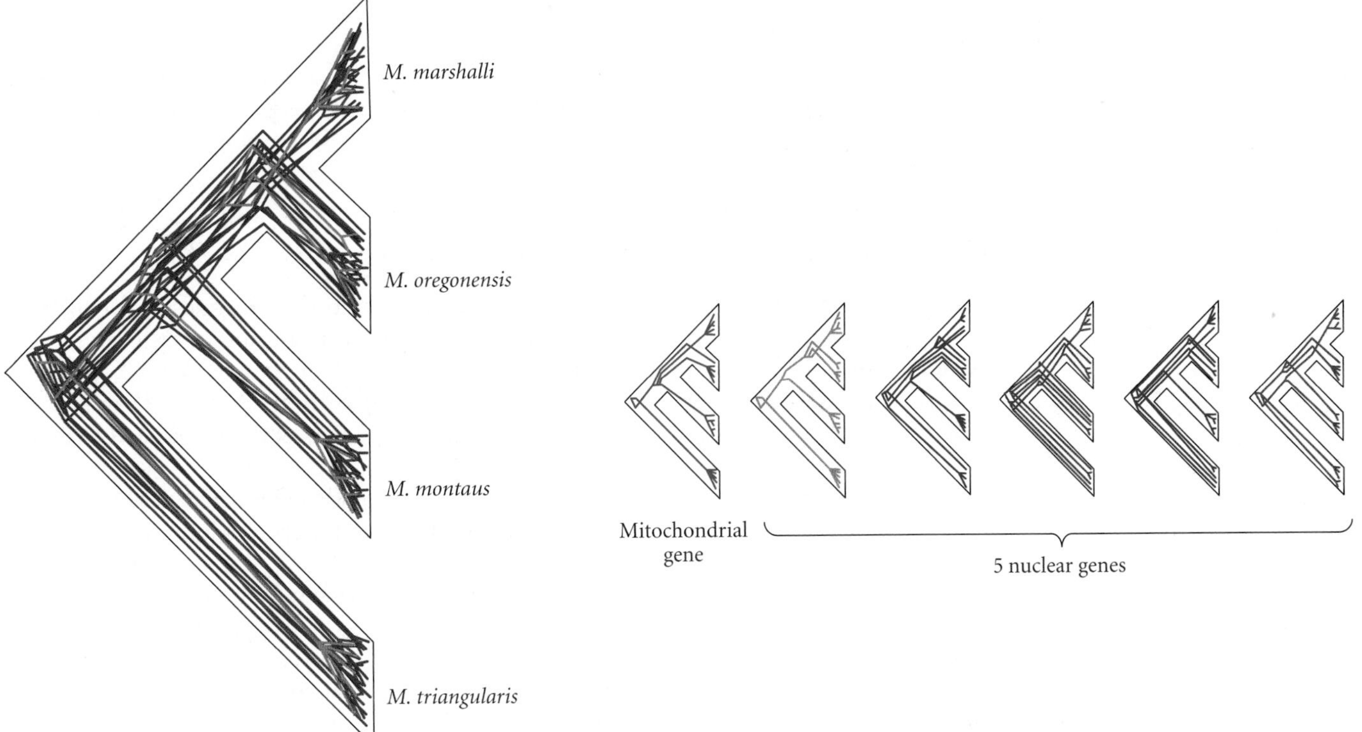

Figure 2.21 An estimate of the phylogeny of four species of grasshoppers (*Melanoplus*), each sampled for multiple sequences of each of six genes. Trees for the individual genes (at right) show relationships among haplotypes that vary in many different ways from the best estimate of the species phylogeny (black outer lines), indicating that each of these four species inherited several gene lineages from its common ancestor with the other species. The diagram at left shows all six gene trees nested within the species phylogeny. (From Carstens and Knowles 2007, courtesy of L. L. Knowles.)

clades, but the branches connecting these major groups are short, indicating that these lineages diverged during a relatively brief time. In 2007, two research groups independently analyzed relationships among representative species from each clade, for which most of the genome has been sequenced. One group (Wildman et al. 2007) used 1,698 gene sequences, summing to 1,443,825 base pairs; the other (Hallström et al. 2007) used 1,961 gene sequences (1,569,141 base pairs). Both groups obtained the same strongly supported phylogeny (Figure 2.22).

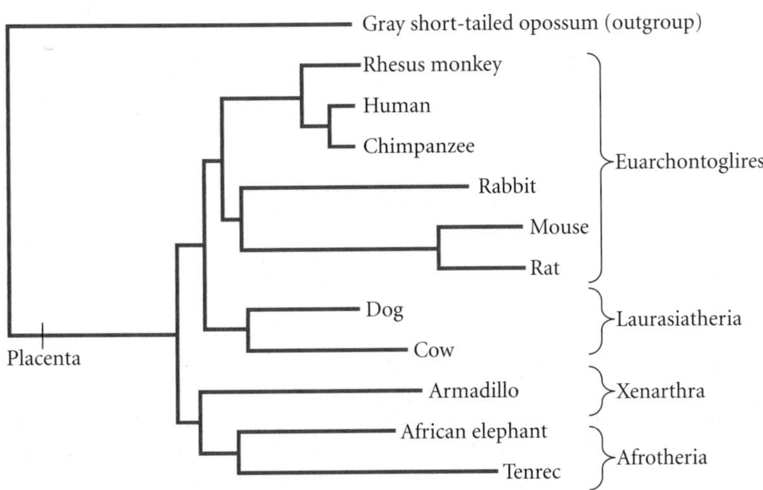

Figure 2.22 Relationships among 11 species of eutherian (placental) mammals that represent four major clades. This phylogeny, based on 1,698 gene sequences, shows that Xenarthra are the sister group of Afrotheria, rather than Laurasiatheria + Euarchontoglires, as some investigators had postulated. Hallström et al. obtained the same tree, based on a somewhat different set of 1,961 genes. (After Wildman et al. 2007.)

(A) Phylogeny based on sequence of gene 1

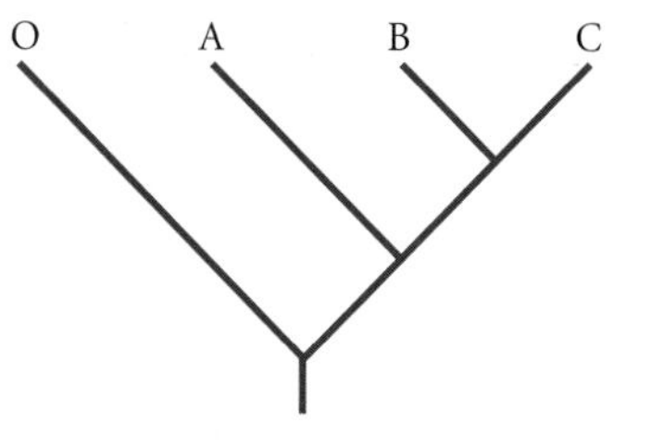

(B) Phylogeny based on sequence of gene 2

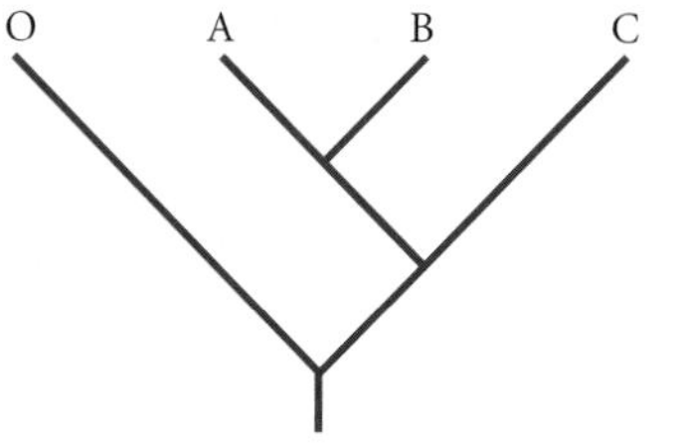

(C) Reticulate phylogeny

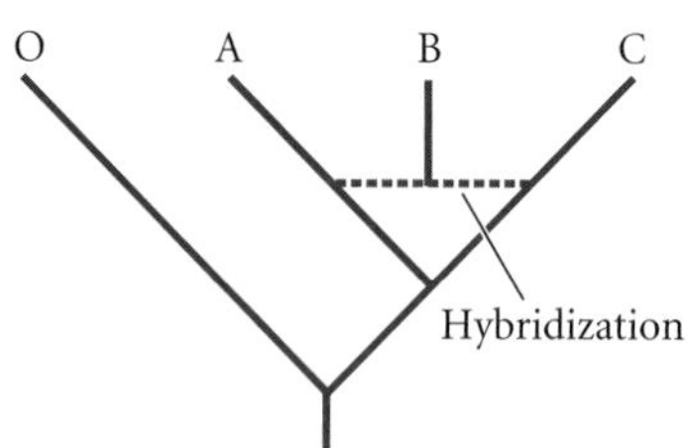

Figure 2.23 Hybridization and reticulate evolution. (A, B) Phylogenies inferred for species A, B, C, and an outgroup (O), based on the sequences of two different genes. (C) The incongruence suggests that species B has arisen by hybridization between species A and C, and that the history of this taxon can thus best be portrayed as reticulate (netlike) evolution. (After Sang and Zhong 2000.)

Hybridization and Horizontal Gene Transfer

When hybridization (interbreeding) occurs, it usually just transfers genes between populations or species. But hybridization between two ancestral species has also given rise to entirely new species, quite often in plants, and occasionally in animals (see Chapter 18). In such cases, part of the phylogeny will be RETICULATED (netlike) rather than strictly branching, and some genes in the hybrid population will be most closely related to genes in each of two other species lineages (Figure 2.23).

In contrast to hybridization, which may involve much of the genome, **horizontal** (or **lateral**) **gene transfer** usually incorporates just a few genes of one species into the genome of another species. (These terms contrast with the normal "vertical" transfer of genes from parent to offspring.) For example, a certain "virogene" has been found only in one group of closely related species of cats and in Old World monkeys. This gene would imply that these cats are more closely related to monkeys than they are to other cats, so it is clearly incongruent with the phylogeny indicated by other genes (Figure 2.24). The gene may have been transferred from monkeys to cats by a virus. Although most genes studied in the parasitic plant *Rafflesia* (see Figure 2.4A) show that it is related to spurges and other members of the Malpighiales, a mitochondrial gene was closely related to that of its host, a member of the grape family, implying that the parasite acquired this gene from its host (Davis and Wurdack 2004). Lateral gene transfer is very important in the evolution of bacteria, in which genes are transferred among distantly related species by phage, by uptake of naked DNA released from dead cells, and by conjugation (Ochman et al. 2000). Phylogenetic incongruence and several other lines of evidence have shown that lateral transfer has provided various bacteria with genes for antibiotic resistance, for the ability to invade hosts and cause disease, and for adaptation to extreme environments such as hot springs.

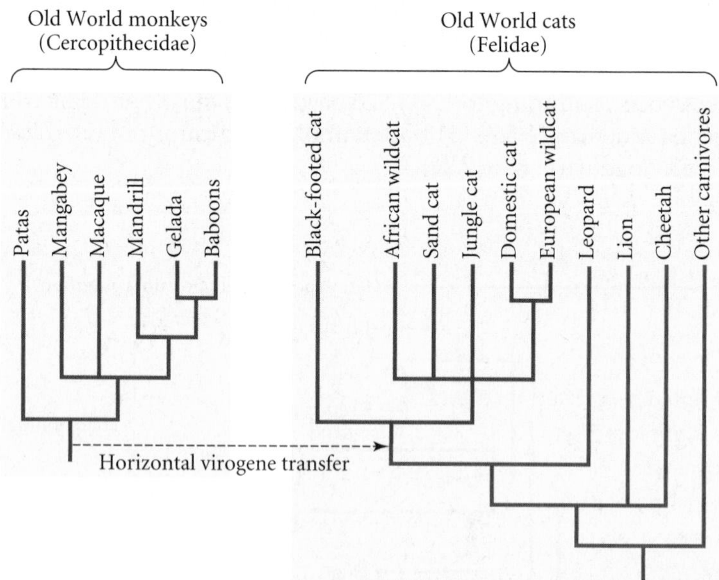

Figure 2.24 Phylogenies of some Old World monkeys and cats (Felidae). Both the monkeys and one group of cats (blue branches) have a virogene with such similar sequence that horizontal transfer from the ancestor of these monkeys to the ancestor of the small cats is the most reasonable interpretation. (After Li and Graur 1991.)

Summary

1. A phylogeny is the history by which species or other taxa have successively originated from common ancestors. It may be depicted by a phylogenetic tree, in which each branch point represents the division of an ancestral species into two lineages. Closely related species have more recent common ancestors than distantly related species. A group of species descended from a particular common ancestor is a monophyletic group; a phylogenetic tree portrays nested sets of monophyletic groups. Phylogenetic trees, estimated from the characteristics of the taxa, represent evolutionary relationships and provide a framework for analyzing many aspects of evolution.
2. Overall similarity among organisms is not the best indicator of phylogenetic relationships. Two species may be more similar to each other than to a third because they retain ancestral character states (whereas the third has diverged), because they independently evolved similar character states (homoplasy), or because they share derived character states that evolved in their common ancestor. Only unique shared derived character states are evidence of phylogenetic relationship. Thus a monophyletic group is marked by uniquely derived character states shared by the group's members.
3. Phylogenetic relationships can be obscured by different rates of evolution among lineages and by homoplasy. Several methods are used to estimate phylogenies in the face of these misleading features. The method of maximum parsimony is frequently used, according to which the best estimate of the phylogeny is the tree that requires one to postulate the smallest number of evolutionary changes to account for the differences among species. Some other methods are sometimes more reliable.
4. A phylogenetic tree is a statement about evolutionary relationships, and like all scientific statements, it is a hypothesis. We gain confidence in the validity of the hypothesis when new data, such as different characters, support it. Uncertainties remain about the phylogenetic relationships among many taxa, but there are also many well-supported phylogenies.
5. Molecular data are increasingly used to infer phylogenies. The rate of evolution of DNA sequences can be shown in some cases to be fairly constant (providing a "molecular clock"), such that sequences in different lineages diverge at a roughly constant rate. The absolute rate of sequence evolution can sometimes be calibrated if fossils of some lineages are known. This rate, in turn, can be used to estimate the absolute age of some evolutionary events, such as the origin of other taxa.
6. Evolutionary processes may make it difficult to infer phylogenetic relationships. For example, shared derived character states may be erased by subsequent evolution, as when successive base pair substitutions occur at the same site in a DNA sequence. If multiple lineages originate from a common ancestor within a short time, the relationships among them may be unresolved because there was not enough time for derived character states to evolve between successive branching events. An accurately estimated phylogeny of genes (DNA sequences) from different species may differ from the phylogeny of the species themselves. With enough data, these difficulties can usually be surmounted.
7. A phylogeny must consider more than just the branching pattern if some species have originated by hybridization between different ancestral species, or if there has been horizontal (lateral) transfer of some genes among different lineages. Such phenomena may be suspected if different genes imply different branching phylogenies.

Terms and Concepts

ancestral character state
autapomorphy
character
character state
clade
common ancestor
convergent evolution (convergence)
derived character state
divergence (divergent evolution)
evolutionary reversal
gene tree (gene genealogy)
haplotype
higher taxon
homology (homologous)
homoplasy (homoplasious)
horizontal (lateral) gene transfer
ingroup
molecular clock
monophyletic group (monophyly)
node
outgroup
parsimony (maximum parsimony)

phylogenetic tree
phylogeny
sister group
substitution (base pair or amino acid)
synapomorphy
taxon (plural: taxa)
taxonomic category

Suggestions for Further Reading

An introduction to phylogenetic methods using molecular data is given by B. G. Hall in *Phylogenetic Trees Made Easy: A How-to Manual for Molecular Biologists* (third edition, Sinauer Associates, Sunderland, MA, 2007), which includes instructions for several widely used tree-construction software programs. For deep coverage of phylogenetic analysis by a leader in the field, see J. Felsenstein, *Inferring Phylogenies* (Sinauer Associates, Sunderland, MA, 2004).

A package of computer programs for estimating phylogenies from data, and especially for estimating the evolutionary history of characters from their pattern of variation among taxa, is W. P. Maddison and D. R. Maddison's *MacClade 4* (Sinauer Associates, Sunderland, MA, 2000).

Problems and Discussion Topics

1. Suppose we are sure, because of numerous characters, that species 1, 2, and 3 are more closely related to one another than to species 4 (an outgroup). We sequence a gene and find ten nucleotide sites that differ among the four species. The nucleotide bases at these sites are:
   (species 1) GCTGATGAGT; (species 2) ATCAATGAGT;
   (species 3) GTTGCAACGT; (species 4) GTCAATGACA.
   Estimate the phylogeny of these taxa by plotting the changes on each of the three possible phylogenies for species 1, 2, and 3 and determining which tree requires the fewest evolutionary changes.
2. There is evidence that many of the differences in DNA sequence among species are not adaptive. Other differences among species, both in DNA and in morphology, are adaptive (as we will see in Chapters 12 and 20). Do adaptive and nonadaptive variations differ in their utility for phylogenetic inference? Can you think of ways in which knowledge of a character's adaptive function would influence your judgment of whether or not it provides evidence for relationships among taxa?
3. It is possible for two different genes to imply different phylogenetic relationships among a group of species. What are the possible reasons for this? If there is only one true history of formation of these species, what might we do in order to determine which (if either) gene accurately portrays it? Is it possible for both phylogenetic trees to be accurate even if there has been only one history of species divergence?
4. Explain why rapidly evolving DNA sequences are useful for determining relationships only among taxa that evolved from quite recent common ancestors, and why slowly evolving sequences are useful only for resolving relationships among taxa that diverged much longer ago.
5. Given the principle stated in question 4, suppose you determine the nucleotide sequence of a particular gene from each of several species. (Perhaps, for the sake of argument, these include horse, sheep, giraffe, kangaroo, and human.) How could you tell if this gene has evolved at a rate (not too fast or slow) that would make it useful for estimating the relationships among these animals?
6. Suppose the nucleotide sequence in the gene you used in question 5 has evolved much faster in some lineages (say, horse and human) than in others. Could that affect the estimate of the phylogeny? How? Is there any way you could tell if there was indeed a big difference in the rate of sequence evolution among lineages?
7. What should a biologist do if she or he finds that different methods of analyzing the same data (say, the maximum parsimony method and the maximum likelihood method) provide different estimates of the relationships among certain taxa? What should she do if the different analytical methods give the same estimate, but the estimate differs depending on which of two different genes has been sequenced? (Your answers do not depend on knowing how maximum likelihood works.)
8. Do the quandaries described in the previous question ever occur? Choose a group of organisms that interests you, find recent phylogenetic studies of this group, and see whether such problems have been encountered. (You may use keywords, such as "phylogeny" and "[taxon name, e.g., deer]," in any of several literature search engines that your instructor can point you to.)

CHAPTER 3

Patterns of Evolution

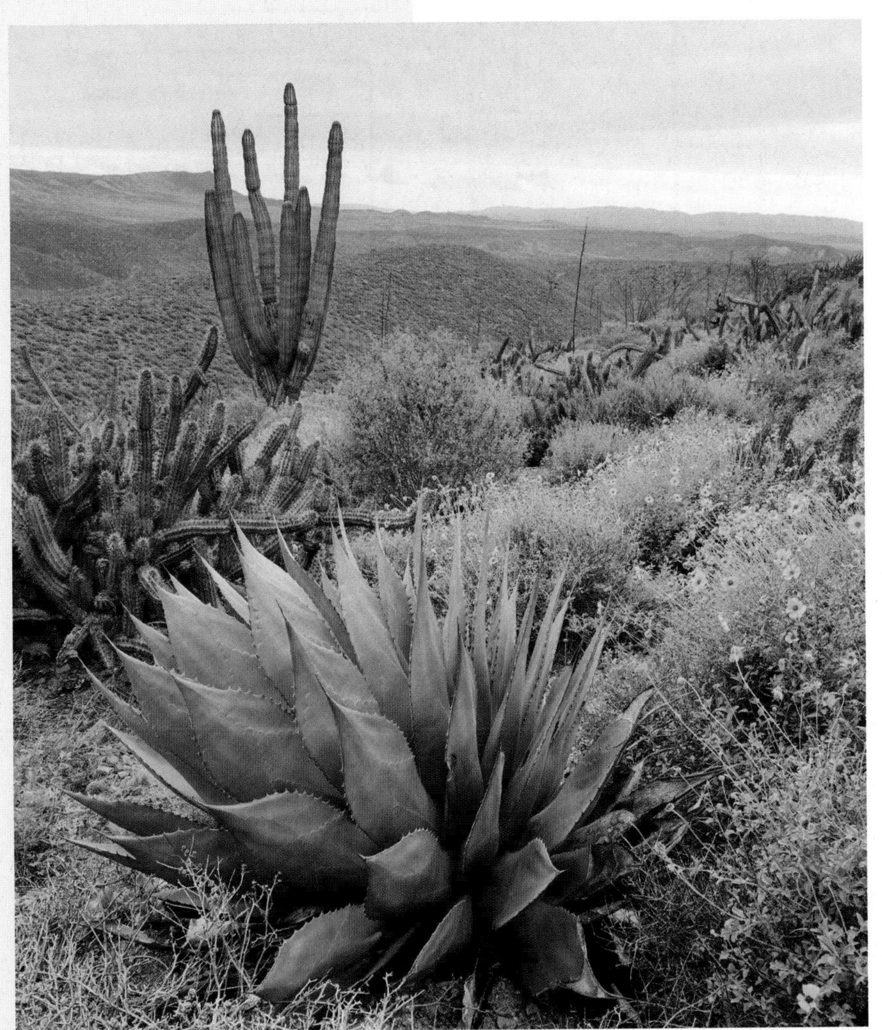

Diverse adaptations to a dry environment. The adaptations of plants in a Mexican desert include fleshy, leafless stems that store water; dense spines that reduce heat load by reflecting light; thick, succulent leaves; widely spreading roots that catch rare rainwater; and brief seasonal flowering. Each of these adaptations has evolved independently in many distantly related plants. (Photo © D. M. Schrader/ShutterStock.)

In order to classify organisms, systematists compare their characteristics. These comparisons provide a foundation for a great deal of what we know about evolution. Especially when coupled with the methods of phylogenetic analysis introduced in the previous chapter, they are an indispensable basis for inferring the history of evolution of various groups of organisms, and such histories are often interesting in their own right. The phylogenetic relationships among various birds, plants, or fungi are fascinating to a person who has developed an interest in and knowledge of these organisms—and tracing our own origins back through the last 2 billion years (Figure 3.1) cannot fail to stir our imagination.

Phylogenies provide more than just the branching relationships among taxa, however. They enable us to infer, with considerable confidence, the history of changes in the characteristics of organisms, even in the absence of a fossil record (Pagel 1999). In fact, they are the only way of inferring the history of evolution of features that leave little or no fossil trace, such as DNA sequences, biochemical pathways, and behaviors. Phylogenetic and systematic studies enable us to describe past changes in genes, genomes, biochemical and physiological features, development and morphology, and life histories and behavior, as well as associated changes in geographic distribution, habitat associations, and ecological interactions among different species. Figure 3.1 shows the sequence by which our own ancestors acquired

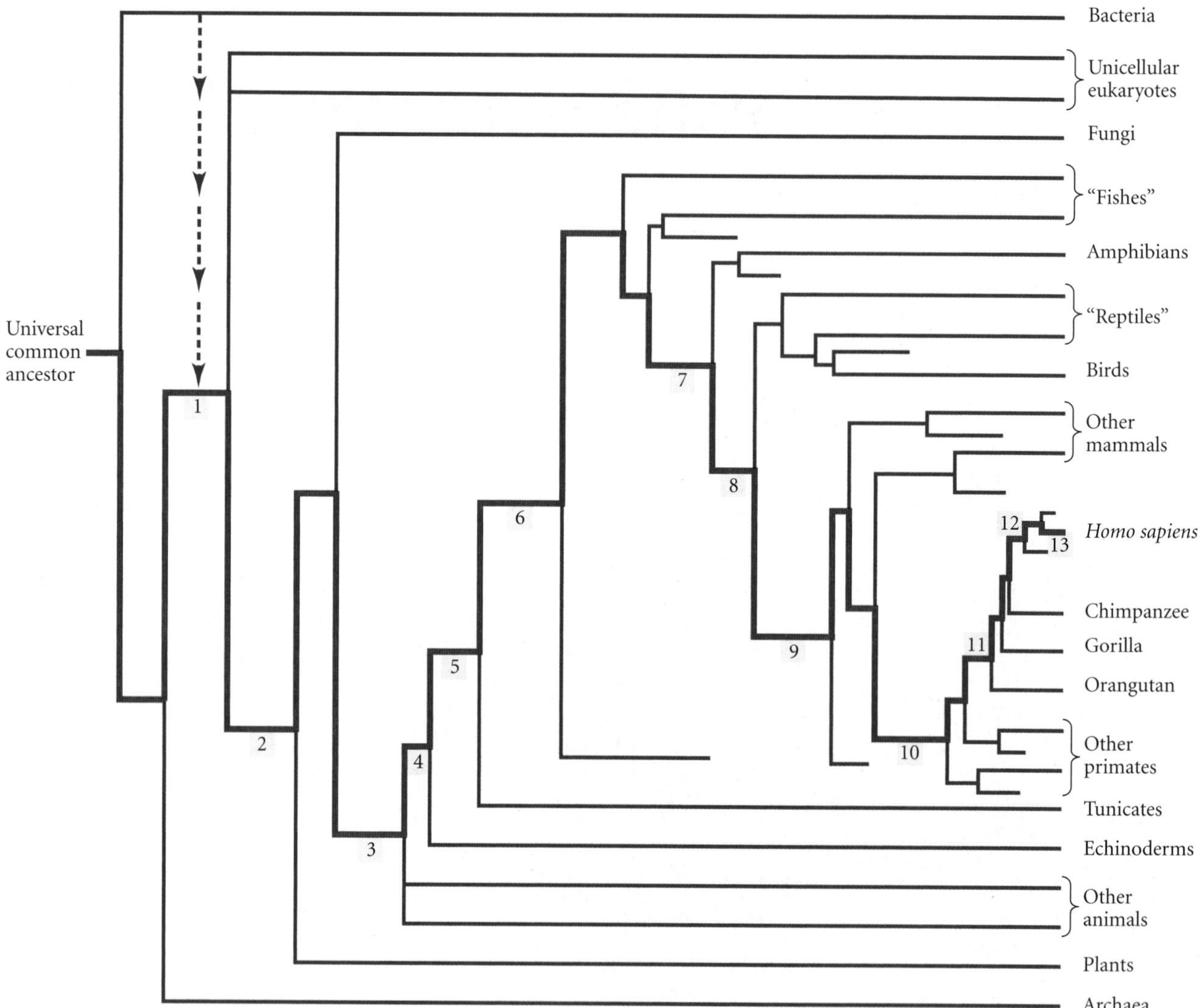

1. Origin of eukaryotes: a symbiotic bacterium becomes the mitochondrion
2. Multicellularity: cell and tissue differentiation
3. Animals: internal digestive cavity; muscles
4. Deuterostomes: embryonic blastopore develops into anus
5. Chordates: notochord; dorsal nerve cord
6. Vertebrates: bony skeleton
7. Tetrapods: limbs
8. Amniotes: amniotic egg; other water-conserving features
9. Mammals: unique jaw joint; bones of middle ear; milk
10. Primates: binocular vision; arboreality
11. Anthropoid apes: loss of tail
12. Hominins: bipedalism
13. *Homo sapiens* spreads from Africa

Figure 3.1 Tracing the path (in red) of evolution to *Homo sapiens* from the universal ancestor of all life, in the context of the tree of life. Some of the major events are shown here as character changes mapped onto the phylogenetic tree. Such evolutionary histories are often intrinsically fascinating.

important features, such as a skeleton, amnion, middle-ear bones, binocular vision, and bipedal locomotion. From our inferences of such changes in many kinds of organisms, it has been possible to find common themes, or **patterns of evolution**.

Because organisms are so diverse, there are few universal "laws" of biology of the kind that are known in physics (Mayr 2004). However, we can make generalizations about what kinds of evolutionary changes have been prevalent—and developing such general statements is one of the chief tasks of science. Furthermore, general patterns of change are the most important phenomena for evolutionary biology to explain. For instance, we know that the size of genomes—the amount of DNA—in various organisms varies greatly. The search for an explanation of genome size might become more interesting (and perhaps easier) if we determined in which clades genome size increased and in which it decreased, rather than knowing only that the genome of one species is larger than that of another.

Phylogenetic and comparative studies provide insights into almost every aspect of evolution and help us understand evolutionary processes (Futuyma 2004). This chapter describes a few of the most important patterns that have emerged from a long history of systematic and phylogenetic analyses, concentrating on morphological and genomic characteristics. In later chapters, we will encounter phylogenetic insights into many other topics.

Evolutionary History and Classification

Darwin's hypothesis of descent with modification from common ancestors provided a scientific foundation for classifying organisms because it explained similarities among species as the result of a real evolutionary history. Many systematists therefore adopted the position that classification should reflect evolution. In almost any discussion of long-term patterns of evolution and biological diversity, it becomes necessary to use classification, since we must refer to named taxa such as Chordata, Vertebrata, Mammalia, or *Homo*. What, then, does it mean for classification to reflect evolution? And does it?

Evolution has two major features: the *branching* of a lineage into two or more descendant lines, called **cladogenesis** (from the Greek *clados*, "branch"), and evolutionary *change* of various characteristics in each of the descendants, called **anagenesis** (from the Greek *ana*, "up," referring to directional change). Some evolutionary changes, such as the evolution of the bird wing from a dinosaur forelimb, are particularly striking and adaptively important. Many traditional classifications were constructed so as to convey both cladogenesis and anagenesis. For example, birds were placed in a different class (Aves) from other amniote vertebrates because of their wings and other adaptations for flight, and humans were placed in a different family (Hominidae) from other primates because of our erect posture, large brain, and egocentricity. However, it can be difficult to convey both cladogenesis and anagenesis in a classification. In the hominoid phylogeny in Figure 2.10B, for example, it is obvious that placing humans in a separate family from the great apes reflects the great divergence in humans' brain size and some other features. But it obscures the fact that humans are more closely related to chimpanzees than gorillas are.

On the basis of phylogenetic relationships, a taxon (a named group of organisms) may be monophyletic, polyphyletic, or paraphyletic (Figure 3.2). A **monophyletic** taxon, as noted previously, is the set of all known descendants from a single common ancestor. For example, the birds (Aves), the beetles (Coleoptera), and the flowering plants (Angiospermae) are believed to be monophyletic groups. A **polyphyletic** taxon includes unrelated lineages that are more closely related to species that are placed in other taxa; modern taxonomists do not consider polyphyletic taxa acceptable. For instance, a taxon that included whales with fishes would be polyphyletic, since whales stem from ancestors (e.g., the earliest tetrapod, the earliest mammal) that fishes do not share. A **paraphyletic** taxon is a group that is monophyletic except that some descendants of the common ancestor have been placed in other taxa.

Paraphyletic taxa usually lack species that have been placed in another taxon in order to emphasize their distinctive adaptations. For example, the traditional family Pongidae consists of orangutans, gorillas, and chimpanzees; the family is paraphyletic, because humans—the closest relatives of the chimpanzees—have been placed not in the Pongidae,

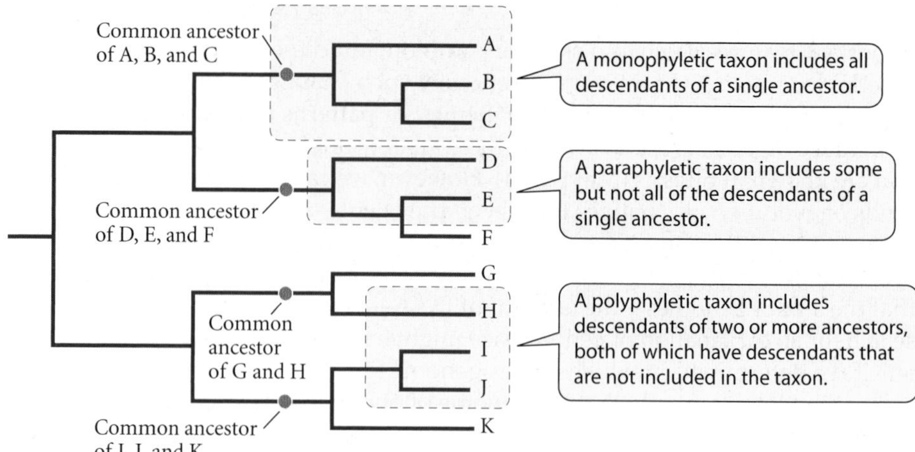
Common ancestor of A, B, and C
A
B
C
A monophyletic taxon includes all descendants of a single ancestor.
D
E
F
A paraphyletic taxon includes some but not all of the descendants of a single ancestor.
Common ancestor of D, E, and F
G
H
I
J
K
Common ancestor of G and H
A polyphyletic taxon includes descendants of two or more ancestors, both of which have descendants that are not included in the taxon.
Common ancestor of I, J, and K

Figure 3.2 Monophyletic, paraphyletic, and polyphyletic groups. Monophyletic groups are preferred by most systematists, who consider paraphyletic and polyphyletic groupings inappropriate in modern taxonomy.

but in the Hominidae. And if birds, with their distinctive adaptations for flight, are placed in the class Aves while their relatives, such as dinosaurs and crocodiles, are placed in the class Reptilia, then "Reptilia" is paraphyletic. It is sometimes useful to refer to traditional groups, such as "Reptiles," but to enclose their names in quotation marks in order to indicate that they are not used in modern, formal classification.

Increasingly, systematists have adopted a philosophy of classification urged by Willi Hennig, who argued that all taxa should be monophyletic and thus reflect common ancestry. Hennig and his followers argued for the abolition of paraphyletic taxa such as "Reptilia." Hennig greatly influenced systematics, first, by proposing a *method* for discovering the real branching pattern of evolutionary history (see Figure 2.5), and second, by articulating an *opinion* on criteria for classification. These two very different proposals have together come to be known as **cladistics**. Phylogenetic trees are sometimes referred to as **cladograms**, and monophyletic groups are called **clades**.

Even if we have an accurate (true) estimate of a phylogeny, certain aspects of classification are still difficult. For example, we often find that an extinct group of species early in evolutionary time (which may be termed a **stem group**, such as the dinosaur order Theropoda) has given rise to a later group with distinctive derived characters (referred to as a **crown group**, such as the class Aves, or birds; see Figure 4.9). Any name for the stem group that excludes the crown group will designate a paraphyletic taxon—an unsatisfactory element in classification.

Furthermore, decisions as to whether the members of a monophyletic group should be SPLIT into several taxa or LUMPED into one taxon may be quite arbitrary. For instance, given the phylogeny shown in Figure 2.10, we might form monophyletic families either by placing the orangutan in one family (Pongidae) and the gorilla, chimpanzee, and human in another family (Hominidae), or by placing all of them in a single family (Hominidae). The latter classification (shown in Figure 2.10B) has been widely adopted; within the single family of humans and great apes, the subfamily Homininae includes the African apes and humans, and the tribe Hominini designates a clade consisting of *Pan* (the chimpanzee and the bonobo), *Homo sapiens*, and extinct human relatives.

Inferring the History of Character Evolution

One of the most important uses of phylogenetic information is to reconstruct the history of evolutionary change in interesting characteristics by "mapping" character states on the phylogeny and inferring the state in each common ancestor, using the principle of parsimony (see Chapter 2). That is, we assign to ancestors those character states that require us to postulate the fewest homoplasious evolutionary changes for which we lack independent evidence. This method enables us to infer when (i.e., on which branch or segment of a phylogeny) changes in characters occurred, and thus to trace their history.

Humans, for example, have nonopposable first (great) toes, while the orangutan, gorilla, and chimpanzee have opposable first toes (like our thumbs). In Figure 3.3, we consid-

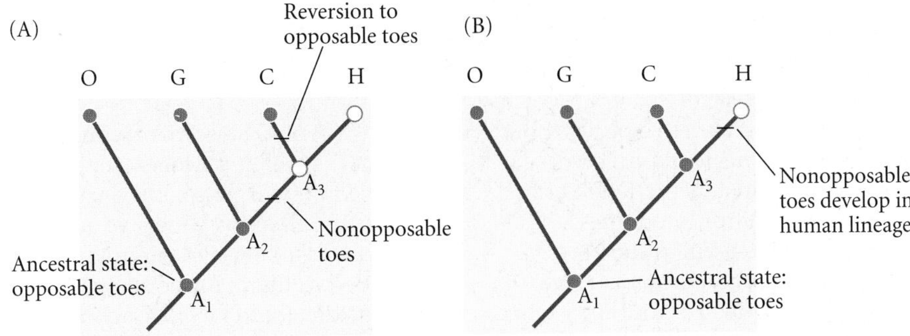

Figure 3.3 Two possible histories of change of a character (opposable versus nonopposable toes) in the Hominoidea (O, orangutan; G, gorilla; C, chimpanzee; H, human). (A) If nonopposable toes (open circles) are hypothesized for A_3, the common ancestor of chimpanzee and human, two state changes (tick marks) must be postulated. (B) If opposable toes are hypothesized for A_3, only one change need be postulated. It is therefore most parsimonious to conclude that humans evolved from an ancestor with opposable toes.

er two possible evolutionary histories. The common ancestors are labeled A_1, A_2, and A_3 in order of increasing recency. If we assume (Figure 3.3A) that A_1 and A_2 had opposable toes and that A_3, the immediate common ancestor of chimpanzees and humans, had nonopposable toes, we would have to postulate two changes, with the chimpanzee reverting to the ancestral state. If, however, we assume (Figure 3.3B) that A_3, like A_1 and A_2, had opposable toes, we need to infer only one evolutionary change; namely, the shift to nonopposable toes in the human lineage. This is the more parsimonious hypothesis, so our best estimate is that the common ancestor of humans and chimpanzees had opposable toes.

Figure 3.4 illustrates a more complex and important application of this approach. The host species that a parasite, such as a virus, inhabits can be considered a character of the virus. Based on their nucleotide sequences, strains of the human immunodeficiency virus (HIV) that causes AIDS are classified into two major groups, HIV-1 and HIV-2, that differ in disease etiology. When AIDS surfaced in the 1980s, the question of where it had come from was a major issue, not least because of fundamentalist religious leaders who shouted that it was God's punishment of homosexual men. Science, of course, provides more useful and testable hypotheses, and so biologists soon showed, by nucleic acid sequencing, that HIV is related to lentiviruses carried by cats and other mammals. Phylogenetic analysis of viruses from various primates provided strong evidence that HIV-1 has evolved at least twice from the virus found in chimpanzees, and that HIV-2 has been acquired from the sooty mangabey monkey. Infection of humans probably occurred by butchering these animals, which are eaten in parts of Africa.

In this book, we will encounter many examples of evolutionary inferences from phylogenetic trees. They are fundamentally important in the study of molecular evolution, and they have even been used to infer the amino acid sequences and functions of ancestral proteins, which have then been experimentally synthesized (Thornton 2004). For example, by comparing DNA sequences of the visual pigment protein rhodopsin in diverse vetebrates, Belinda Chang and colleagues (2002) inferred the most probable amino acid sequence of rhodopsin in the common ancestor of living archosaurs: alligators and birds. This reconstruction provides our best estimate of the protein in dinosaurs, from which birds evolved. The research team synthesized the postulated rhodopsin, expressed it in a mammalian cell line in tissue culture, and found that the protein performed its proper function, and had its greatest light absorbtion slightly red-shifted, compared with living archosaurs.

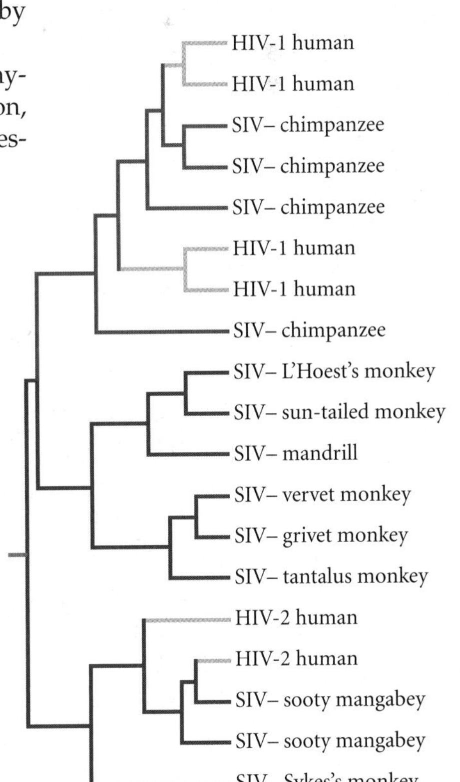

Figure 3.4 A phylogeny of strains of human immunodeficiency virus (HIV-1 and HIV-2) and simian immunodeficiency viruses (SIV) from chimpanzees and various African monkeys. The internal branches are colored to show the most parsimonious interpretation of the host species of ancestral virus lineages (green, monkeys; red, chimpanzee; gold, human). For example, strains of HIV-1 are most closely related to different strains of chimpanzee SIV, and are most parsimoniously interpreted as having evolved from the chimpanzee virus. Strains of HIV-1 fall into several groups that apparently originated as separate transfers from chimpanzees to humans. HIV-2 probably originated by at least one transfer from sooty mangabeys. (After Hahn et al. 2000.)

BOX 3A Evidence for Evolution

Systematists have always classified organisms by comparing characteristics among them. The early systematists had amassed an immense body of comparative information even before *The Origin of Species* was published. Their data suddenly made sense in light of Darwin's theory of descent from common ancestors; indeed, Darwin drew on much of this information as evidence for his contention that evolution has occurred. Since Darwin's time, the amount of comparative information has increased greatly, and today includes data not only from the traditional realms of morphology and embryology, but also from cell biology, biochemistry, and molecular biology.

All of this information is consistent with Darwin's hypothesis that living organisms have descended from common ancestors. Indeed, innumerable biological observations are hard to reconcile with the alternative hypothesis, that species have been individually created by a supernatural being, unless that being is credited with arbitrariness, whimsy, or a devious intent to make organisms *look* as if they have evolved. From the comparative data amassed by systematists, we can identify several patterns that confirm the historical reality of evolution, and which make sense only if evolution has occurred.

1. **The hierarchical organization of life.** Before Linnaeus, there had been many attempts to classify species, but those early systems just didn't work. One author, for instance, tried to classify species in complicated five-parted categories—but organisms simply don't come in groups of five. But organisms do fall "naturally" into the hierarchical system of groups-within-groups that Linnaeus described. A historical process of branching and divergence will yield objects that can be hierarchically ordered, but few other processes will do so. For instance, languages can be classified in a hierarchical manner, but elements and minerals cannot.
2. **Homology.** Similarity of structure despite differences in function follows from the hypothesis that the characteristics of organisms have been modified from the characteristics of their ancestors, but it is hard to reconcile with the hypothesis of intelligent design. Design does not require that the same bony elements form the frame of the hands of primates, the digging forelimbs of moles, the wings of bats, birds, and pterodactyls, and the flippers of whales and penguins (see Figure 3.5). Modification of pre-existing structures, not design, explains why the stings of wasps and bees are modified ovipositors, and why only females possess them. All proteins are composed of "left-handed" (L) amino acids, even though the "right-handed" (D) optical isomers would work just as well if proteins were composed only of those. But once the ancestors of all living things adopted L amino acids, their descendants were committed to them; introducing D amino acids would be as disadvantageous as driving on the right in Great Britain or on the left in the United States. Likewise, the nearly universal, arbitrary genetic code (see Chapter 8) makes sense only as a consequence of common ancestry.
3. **Embryological similarities.** Homologous characters include some features that appear during development, but would be unnecessary if the development of an organism were not a modification of its ancestors' ontogeny. For example, tooth primordia appear and then are lost in the jaws of fetal anteaters, and some terrestrial frogs and salamanders pass through a larval stage within the egg, with the features of typically aquatic larvae, but hatch ready for life on land. Early in development, human embryos briefly display branchial pouches similar to the gill slits of fish embryos.
4. **Vestigial characters.** The adaptations of organisms have long been, and still are, cited by creationists as evidence of the Creator's wise beneficence, but no such claim can be made for the features, displayed by almost every species, that served a function in the

Some Patterns of Evolutionary Change Inferred from Systematics

Many important patterns and principles of evolution have been based on systematic studies over the course of the last century or more, and have been clarified by modern phylogenetic studies. Indeed, systematic studies of living organisms have provided a vast amount of evidence for the reality of evolution (Box 3A). In this section, we will illustrate some patterns of evolution, drawing both on a long tradition of comparative morphology and on contemporary studies at the molecular level.

Most features of organisms have been modified from pre-existing features

One of the most important principles of evolution is that *the features of organisms almost always evolve from pre-existing features of their ancestors*; they do not arise de novo, from nothing. The wings of birds, bats, and pterodactyls are modified forelimbs (Figure 3.5); they do not arise from the shoulders (as in angels), presumably because the ancestors of these animals had no shoulder structures that could be modified for flight. In Chapter 4, we will

BOX 3A (*Continued*)

species' ancestors, but do so no longer. Cave-dwelling fishes and other animals display eyes in every stage of degeneration. Flightless beetles retain rudimentary wings, concealed in some species beneath fused wing covers that would not permit the wings to be spread even if there were reason to do so. In *The Descent of Man*, Darwin listed a dozen vestigial features in the human body, some of which occur only as uncommon variations. They included the appendix, the coccyx (four fused tail vertebrae), rudimentary muscles that enable some people to move their ears or scalp, and the posterior molars, or wisdom teeth, that fail to erupt, or do so aberrantly, in many people. At the molecular level, every eukaryote's genome contains numerous nonfunctional DNA sequences, including pseudogenes: silent, nontranscribed sequences that retain some similarity to the functional genes from which they have been derived (see Chapter 19).

5. **Convergence.** There are many examples, such as the eyes of vertebrates and cephalopod molluscs, in which functionally similar features actually differ profoundly in structure (see Figure 3.6). Such differences are expected if structures are modified from features that differ in different ancestors, but are inconsistent with the notion that an omnipotent Creator, who should be able to adhere to an optimal design, provided them. Likewise, evolutionary history is a logical explanation (and creation is not) for cases in which different organisms use very different structures for the same function, such as the various modified structures that enable vines to climb (see Figure 3.13).
6. **Suboptimal design.** The "accidents" of evolutionary history explain many features that no intelligent engineer would be expected to design. For example, the paths followed by food and air cross in the pharynx of terrestrial vertebrates, including humans, so that we risk choking on food. The human eye has a "blind spot," which you can find at about 45° to the right or left of your line of sight. It is caused by the functionally nonsensical arrangement of the axons of the retinal cells, which run forward into the eye and then converge into the optic nerve, which interrupts the retina by extending back through it toward the brain (see Figure 3.6).
7. **Geographic distributions.** The study of systematics includes the geographic distributions of species and higher taxa. This subject, known as biogeography, is treated in Chapter 6. Suffice it to say that the distributions of many taxa make no sense unless they have arisen from common ancestors. For example, many taxa, such as marsupials, are distributed across the southern continents, which is easily understood if they arose from common ancestors that were distributed across the single southern land mass that began to fragment in the Mesozoic.
8. **Intermediate forms.** The hypothesis of evolution by successive small changes predicts the innumerable cases in which characteristics vary by degrees among species and higher taxa. Among living species of birds, we see gradations in beaks; among snakes, some retain a vestige of a pelvic girdle and others have lost it altogether. At the molecular level, the difference, in DNA sequences for the same protein ranges from almost none among very closely related species through increasing degrees of difference as we compare more remotely related taxa.

For each of these lines of evidence, hundreds or thousands of examples could be cited from studies of living species. Even if there were no fossil record, the evidence from living species would be more than sufficient to demonstrate the historical reality of evolution: all organisms have descended, with modification, from common ancestors. We can be even more confident than Darwin, and assert that all organisms that we know of are descended from a single original form of life.

see that the middle-ear bones of mammals evolved from jaw bones of reptiles. In Chapter 20, we will encounter examples of proteins that have been modified from ancestral proteins and have new functions. In other words, related organisms have **homologous characters**, which have been inherited (and sometimes modified) from an equivalent organ in the common ancestor. Homologous characters generally have similar genetic and developmental underpinnings, although these foundations sometimes have undergone substantial divergence among species.

A *character* may be homologous among species (e.g., "toes"), but a given *character state* may not be (e.g., a certain number of toes). The five-toed state is homologous in humans and crocodiles (as far as we know, both have an unbroken history of pentadactyly as far back as their common ancestor), but the three-toed state in guinea pigs and rhinoceroses is not homologous, for these animals have evolved this condition independently from five-toed ancestors.

A character (or character state) is *defined* as homologous in two species if it has been derived from their common ancestor, but *diagnosing* **homology**—that is, determining

Figure 3.5 The forelimb skeletons of some tetrapod vertebrates. Compared with the "ground plan," as seen in the early amphibian, bones have been lost or fused (e.g., horse, bird), or modified in relative size and shape. Modifications for swimming evolved in the porpoise, and for flight in the bird, bat, and pterodactyl. All the bones shown are homologous among these organisms, except for the sesamoid bones (S) in the pterodactyl; these bones have a different developmental origin from the rest of the limb skeleton. (After Futuyma 1995.)

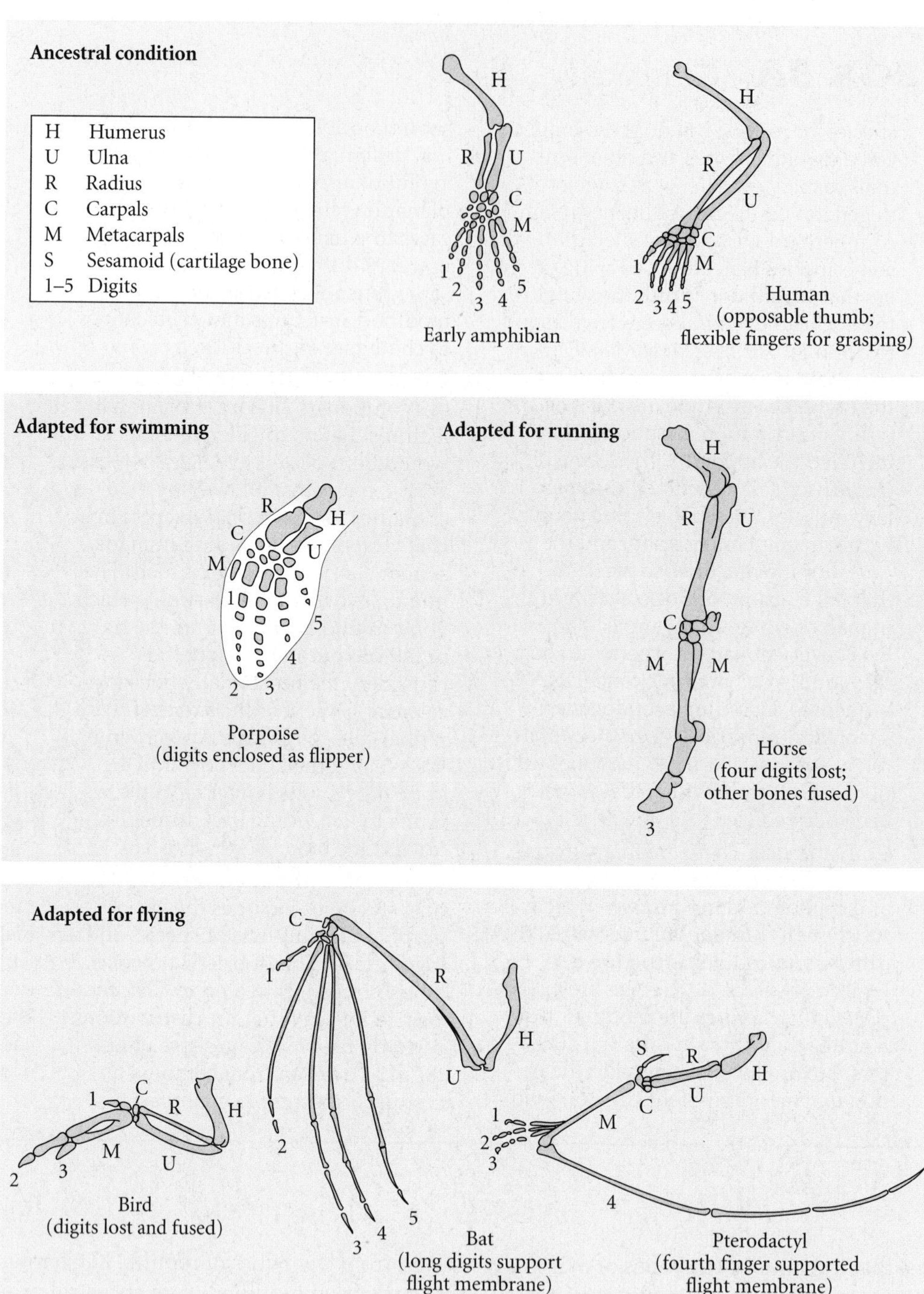

whether or not characters of two species are homologous—can be difficult. The most common criteria for *hypothesizing* homology of anatomical characters are correspondence of *position* relative to other parts of the body and correspondence of *structure* (the parts of which a complex feature is composed). Correspondence of shape or of function are not useful criteria for homology (consider the forelimbs of a horse and an eagle). Embryological studies often are important for hypothesizing homology. For example, the structural correspondence between the hindlimbs of birds and crocodiles is more evident in the embryo than in the adult because many of the bird's bones become fused as development proceeds. However, the most important criterion for judging whether a character is homologous among species is to see if its distribution on a phylogenetic tree (based on other characters) indicates continuity of inheritance from their common ancestor.

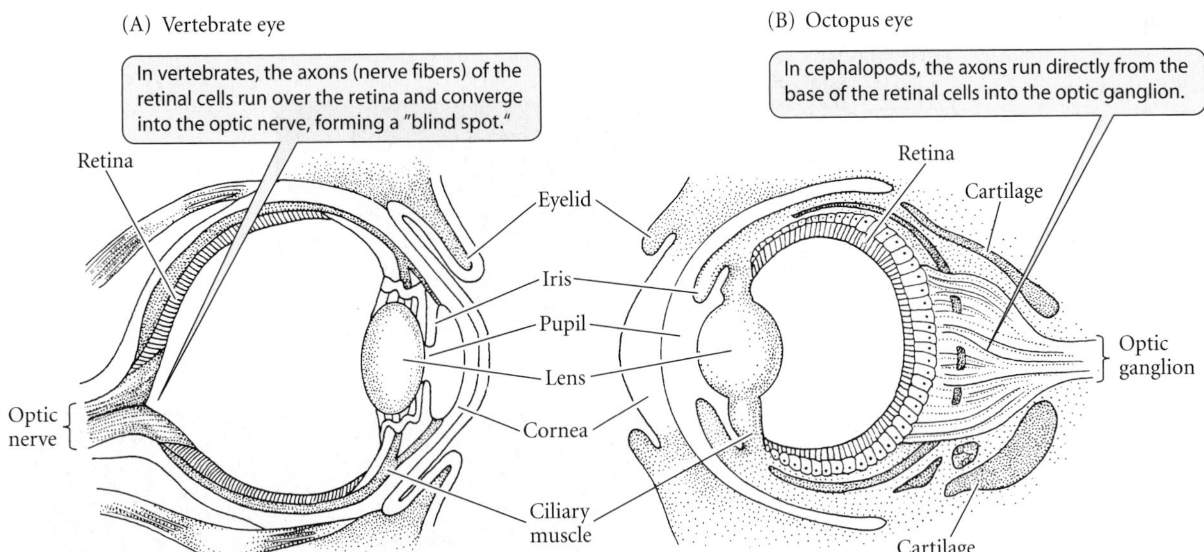
(A) Vertebrate eye
In vertebrates, the axons (nerve fibers) of the retinal cells run over the retina and converge into the optic nerve, forming a "blind spot."
Retina
Optic nerve
Eyelid
Iris
Pupil
Lens
Cornea
Ciliary muscle
(B) Octopus eye
In cephalopods, the axons run directly from the base of the retinal cells into the optic ganglion.
Retina
Cartilage
Optic ganglion
Cartilage

Figure 3.6 The eyes of (A) a vertebrate and (B) a cephalopod mollusc are an extraordinary example of convergent evolution. Despite their many similarities, note the several differences, including interruption of the retina by the optic nerve in the vertebrate, but not in the cephalopod. In vertebrates, the axons (nerve fibers) of the retinal cells run over the surface of the retina and converge into the optic nerve, forming a "blind spot." In cephalopods, the axons run directly from the base of the retinal cells into the optic ganglion. (From Brusca and Brusca 1990.)

Homoplasy is common

We noted in Chapter 2 that **homoplasy**—the independent evolution of a character or character state in different taxa—includes convergent evolution and evolutionary reversal. Some authors distinguish convergent evolution from parallel evolution, but others feel that this is not a meaningful distinction (Arendt and Reznick 2008). In **convergent evolution (convergence)**, superficially similar features are formed by different developmental pathways. The eyes of both vertebrates and cephalopod molluscs (such as squids and octopuses) have a lens and a retina, but their many profound differences indicate that they evolved independently: for example, the axons of the retinal cells arise from the cell bases in cephalopods, but from the cell apices in vertebrates (Figure 3.6).

Parallel evolution (parallelism), by contrast, is thought to involve similar developmental modifications that evolved independently (often in closely related organisms, because they are likely to have similar developmental mechanisms to begin with). For example, in several diverse crustacean lineages, from one to as many as three pairs of feeding structures called maxillipeds develop not on the head segments that bear the animals' mouthparts, but on their anteriormost thoracic segments, which ancestrally bear legs. Michael Averoff and Nipam Patel (1997) studied differences in the expression of two "master" regulatory genes (Hox genes, which will be discussed in detail in Chapter 21). The two genes they studied, called *Ultrabithorax* (*Ubx*) and *abdominal A* (*abdA*), determine how each segment develops in all arthropods, and are expressed in leg-bearing thoracic segments, but not in mouthparts-bearing head segments. Averoff and Patel found that the evolutionary transformation of crustacean legs into maxillipeds corresponds exactly with a loss of expression of *Ubx* and *abdA* in the thoracic segment(s) involved (Figure 3.7), and that this has happened in parallel in the several lineages.

There is increasing evidence that the same gene may contribute to similar evolutionary changes in distantly related organisms. For example, mutations in the melanocortin-1-receptor (*Mc1r*) gene cause the difference between dark versus pale populations of both the pocket mouse *Perognathus intermedius* and the beach mouse *Peromyscus polionotus*, which belong to distantly related families, and appear to underlie dark versus pale forms

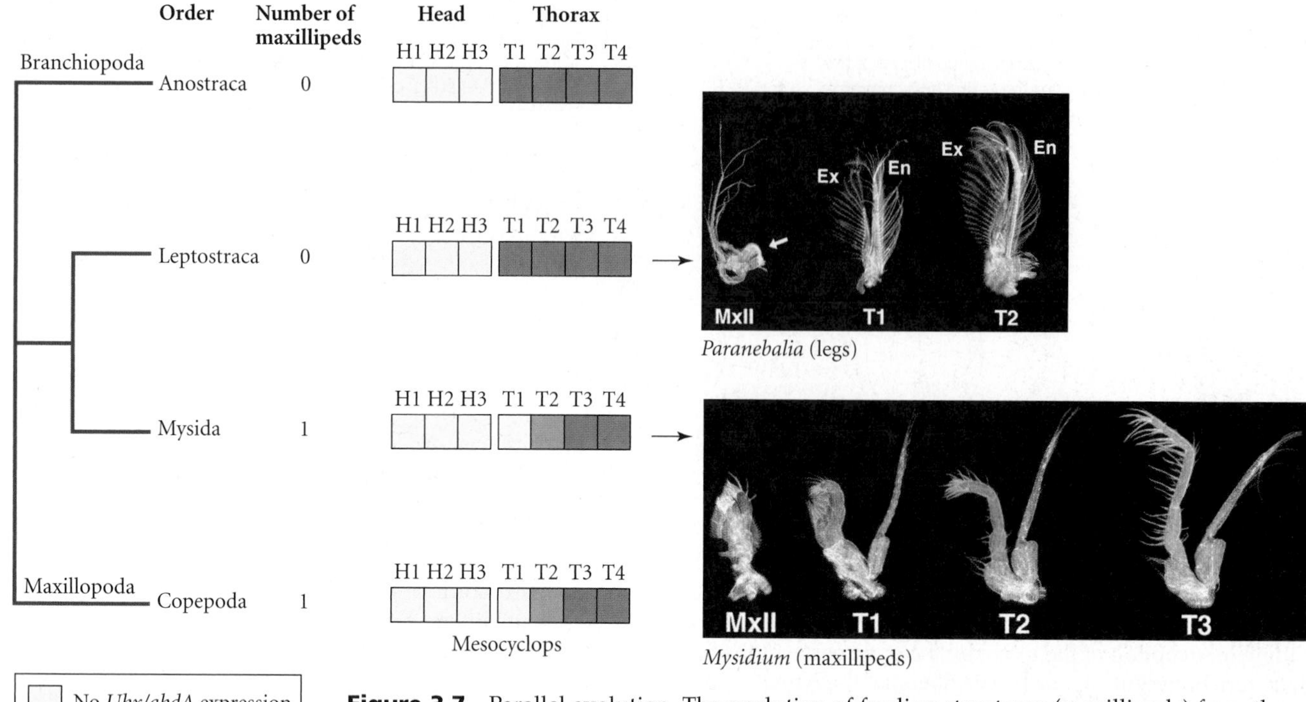

Figure 3.7 Parallel evolution. The evolution of feeding structures (maxillipeds) from thoracic legs in crustaceans is marked by parallel reduction or loss of expression of the genes *Ubx* and *abdA* in the same thoracic segments. The top photo (*Paranebalia*, order Leptostraca) shows the ancestral condition: there is a mouthpart (the maxilla, MxII) on head segment H3, and legs (each with two branches, En and Ex) on thoracic segments T1 and T2. In *Mysidium* (order Mysida), in the bottom photo, thoracic segment T3 has a normal leg (En branch in yellow), but the appendages on T2 and (especially) T1 show maxilla-like modifications of the En branch (green and red, respectively). The copepod *Mesocyclops* (not pictured) has similar modifications of gene expression and morphology. (After Averoff and Patel 1997.)

of several lizards and birds (Figure 3.8). However, some dark populations of both species of mice owe their coloration not to a mutation of this gene, but to other genes that have not yet been identified. Thus, even within a single species, similar phenotypes may have either similar or different genetic and developmental foundations (Nachman et al. 2003; Hoekstra et al. 2006; Arendt and Reznick 2007).

Evolutionary reversals constitute a return from an "advanced," or derived, character state to a more "primitive," or ancestral, state (Porter and Crandall 2003). For example, winged insects evolved from wingless ancestors, but many lineages of insects have lost the wings in the course of subsequent evolution. It has long been thought that complex characteristics, once lost, are unlikely to be regained, a principle known as DOLLOS'S LAW. However, recent phylogenetic and experimental evidence casts some doubt on this principle. Birds have not had teeth for at least 70 million years, but tooth primordia develop in embryos of chickens with the ta^2 mutation (which also results in death before hatching) (Harris et al. 2006). The ancestral life history pattern of salamanders includes an aquatic larval stage, with gills, whereas adults lack gills and have other distinctive features. The aquatic larval stage has been lost in the evolution of the wholly terrestrial plethodontine salamanders, but phylogenetic analysis showed that it has been regained in a lineage of dusky salamanders (*Desmognathus*; Figure 3.9).

Homoplasious features are often (but not always) adaptations by different lineages to similar environmental conditions. In fact, a correlation between a particular homoplasious character in different groups and a feature of those organisms' environment or niche is often the best initial evidence of the feature's adaptive significance. For example, a long, thin beak has evolved independently in at least six different lineages of nectar-feeding

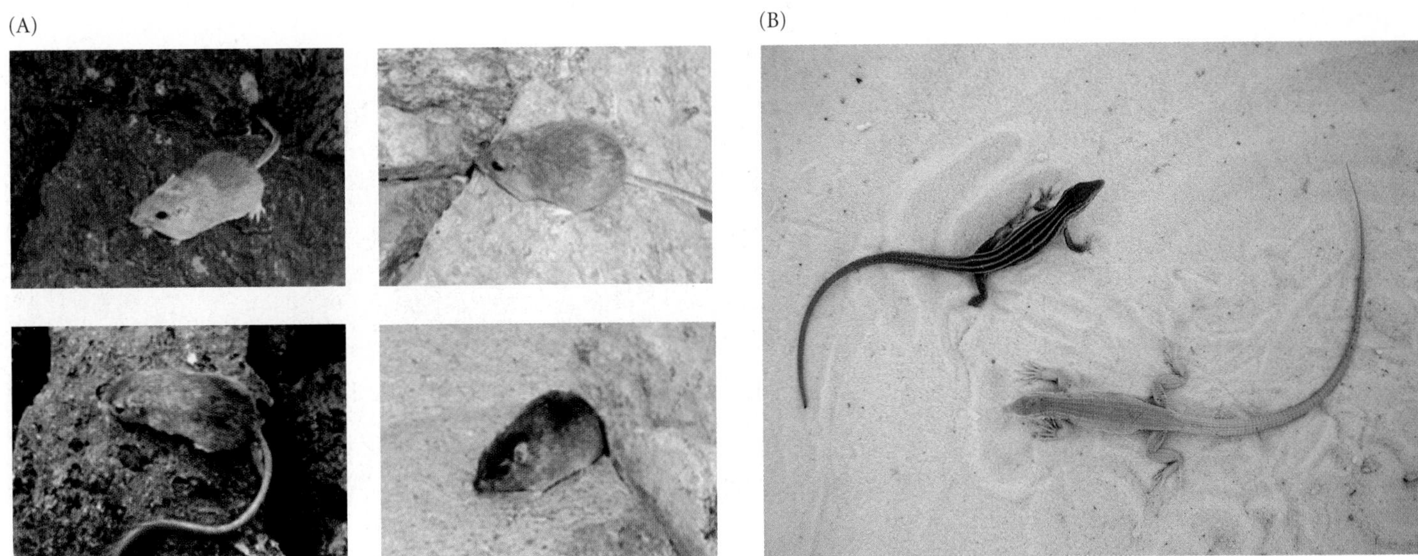

Figure 3.8 Convergent (parallel) evolution based on mutations of the same gene, *Mc1r*. (A) Dark and pale populations of the pocket mouse *Perognathus intermedius* occupy dark rocky outcrops (left) or pale desert soils (right). The difference is believed to reduce visibility to predators, as seen in the lower two photos. (B) A similar difference, also probably due to mutations of *Mc1r*, is seen among populations of the whiptail lizard *Aspidoscelis inornata* that are associated with different desert soils. (A courtesy of Michael Nachman; B courtesy of Erica Bree Rosenblum.)

birds. Such a beak enables these birds to reach nectar in the bottoms of the long, tubular flowers in which they often feed (Figure 3.10).

MIMICRY, a condition in which features of one species have specifically evolved to resemble those of another species, provides especially interesting examples of convergent evolution. Two of the most common kinds of mimicry are BATESIAN MIMICRY and MÜLLERIAN MIMICRY, named after the nineteenth-century naturalists who first described the phenomena. Both of these are defenses against predators. A Batesian mimic is a palatable or innocuous animal that has evolved resemblance to an unpalatable or dangerous animal (the *model*); for instance, many harmless flies have bright yellow and black color patterns like those of wasps, and some palatable butterfly species closely resemble other species that are toxic (see Figure 9.1B). Predators that learn, from unpleasant experience, to avoid the model also will tend to avoid attacking the mimic. In Müllerian mimicry, two or more distasteful or dangerous species have evolved similar characteristics, and a predator that associates the features of one with an unpleasant experience avoids attacking both species. There are many examples of mimicry "rings" that involve several species of distantly related distasteful butterflies (Müllerian mimics; see Figure 19.25) as well as palatable butterflies and moths (Batesian mimics) that share the

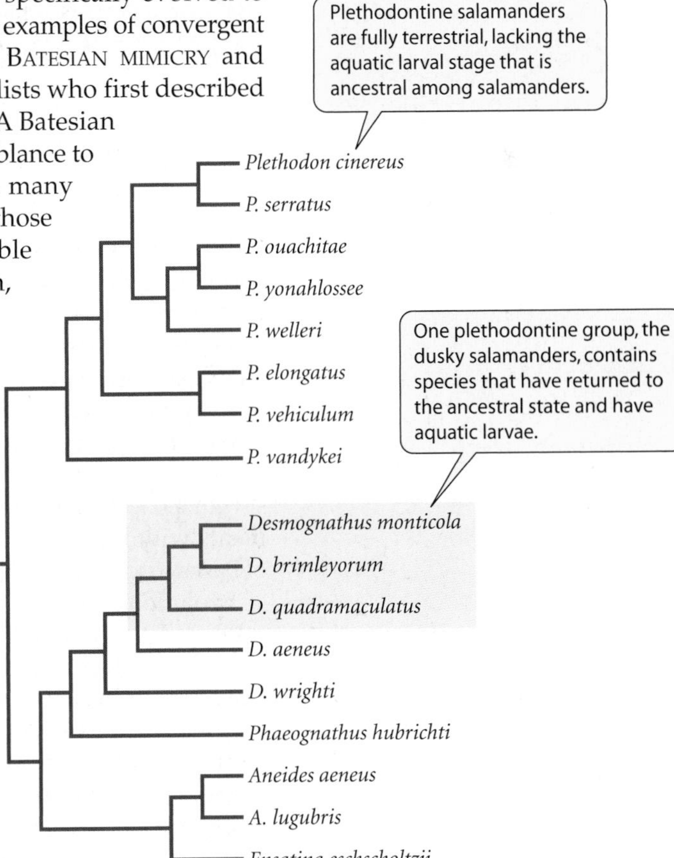

Figure 3.9 Phylogeny, based on DNA sequences, of part of the salamander family Plethodontidae, showing that species of *Desmognathus* with aquatic larvae (red) are nested within a large group of taxa that lack the larval stage (blue). Most outgroups (i.e., other salamander families) possess the larval stage, which is an ancestral feature of salamanders generally. (After Chippindale et al. 2004.)

Figure 3.10 Four bird groups in which similar bill shape has evolved independently as an adaptation for feeding on nectar. (A) A South American honeycreeper, family Thraupidae. This is the purple honeycreeper (*Cyanerpes caeruleus*). (B) Iiwi (*Vestiaria coccinea*), one of the many Hawaiian honeycreepers, family Fringillidae. The Hawaiian honeycreepers are a well-studied example of adaptive radiation. (C) Hummingbirds, family Trochilidae. This violet sabrewing (*Campylopterus hemileucurus*) is from Costa Rica. (D) Sunbirds, family Nectariniidae. The red-chested sunbird (*Nectarinia pulchella*) is native to the Great Lakes region of Africa. (Photos: A © fotolincs/Alamy Images; B, Photo Resource Hawaii/Alamy Images; C, Anthony Mercieca/Photo Researchers, Inc.; D © Gregory G. Dimijian/Photo Researchers, Inc.)

same color pattern. Some species represent intermediate conditions between the Batesian and Müllerian extremes (Mallet and Joron 1999).

Rates of character evolution differ

Different characters evolve at different rates, as is evident from the simple observation that any two species differ in some features, but not in others. Some characters, often called **conservative characters**, are retained with little or no change over long periods among the many descendants of an ancestor. For example, humans retain the pentadactyl limb that first evolved in early amphibians (see Figure 3.5); all amphibians and reptiles have two aortic arches, and all mammals have only the left arch. Body size, in contrast, evolves rapidly; within orders of mammals, it often varies at least a hundredfold. As we saw in Chapter 2, the rate of DNA sequence evolution varies among genes, among segments within genes, and among the three positions within amino acid–coding triplets of bases (codons).

Evolution of different characters at different rates within a lineage is called **mosaic evolution** (Figure 3.11). It is one of the most important principles of evolution, for it says that a species evolves not as a whole, but piecemeal: many of its features evolve quasi-independently. (There are important exceptions; for example, features that function together may evolve in concert.) This principle largely justifies the theory of evolutionary mechanisms, in which we analyze evolution not in terms of whole organisms, but in terms of changes in individual features, or even individual genes underlying such features.

Every species is a mosaic of plesiomorphic (ancestral, or "primitive") and apomorphic (derived, or "advanced") characteristics. For example, the amphibian lineage leading to

Heliocidaris tuberculata (indirect development)

Adult

Larva

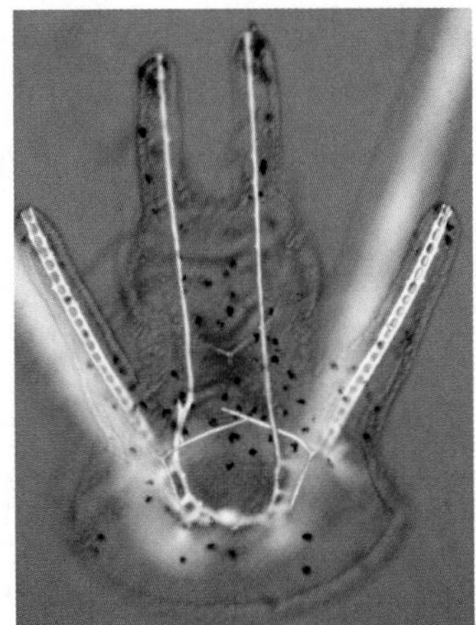

Heliocidaris erythrogramma (direct development)

Adult

Larva

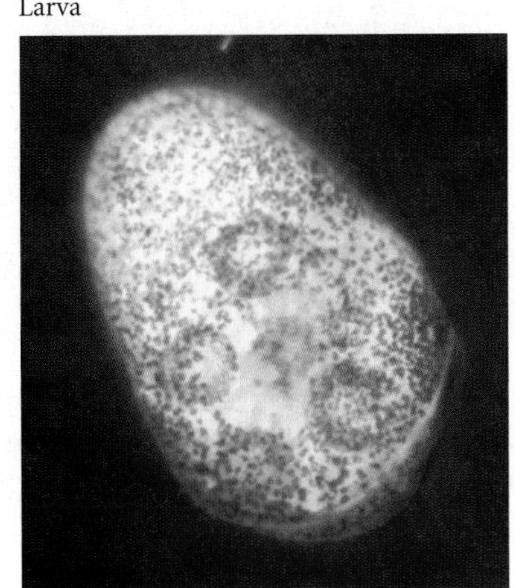

Figure 3.11 An example of mosaic evolution. Adults of the closely related sea urchins *Heliocidaris tuberculata* and *H. erythrogramma* are very similar and can even be hybridized in the laboratory; hence most adult features have evolved slowly or not at all. In contrast, larval characteristics have evolved rapidly. *H. tuberculata* has a pluteus larva that is typical of most sea urchins, whereas *H. erythrogramma* has evolved a nonfeeding, direct-developing larval form. (Adult photos © David Harasti; larvae photos courtesy of Rudolf A. Raff.)

frogs split from the lineage leading to mammals before the mammalian orders diversified, so in terms of *order of branching*, frogs are an older branch than cows and humans. In that sense, frogs might be assumed to be more primitive. But relative to early Paleozoic amphibians, frogs have both ancestral features (e.g., five toes on the hind foot, multiple bones in the lower jaw) and features (such as lack of teeth in the lower jaw) that are more advanced than those of most mammals, in that they have changed further from the ancestral state. Moreover, numerous differences among frog species have evolved in the recent past. For example, one genus has direct development without a tadpole stage, and another gives birth to live young. Humans likewise have both "primitive" characters (five digits on hands and feet, teeth in the lower jaw) and some that are "advanced" compared with frogs (e.g., a single lower jawbone). *Because of mosaic evolution, it is inaccurate or even wrong to consider one living species more "advanced" than another.*

Evolution is often gradual

Darwin argued that evolution proceeds by small successive changes (GRADUALISM) rather than by large "leaps" (SALTATIONS). Whether or not evolution is always gradual is unknown, and the issue is much debated (see Chapter 21). Many higher taxa that diverged in the distant past (e.g., the animal phyla; many orders of insects and of mammals) are very different and are not bridged by intermediate forms, either among living species or in the fossil record. However, the fossil record does document intermediates in the evolution of some higher taxa, as we will see in Chapter 4.

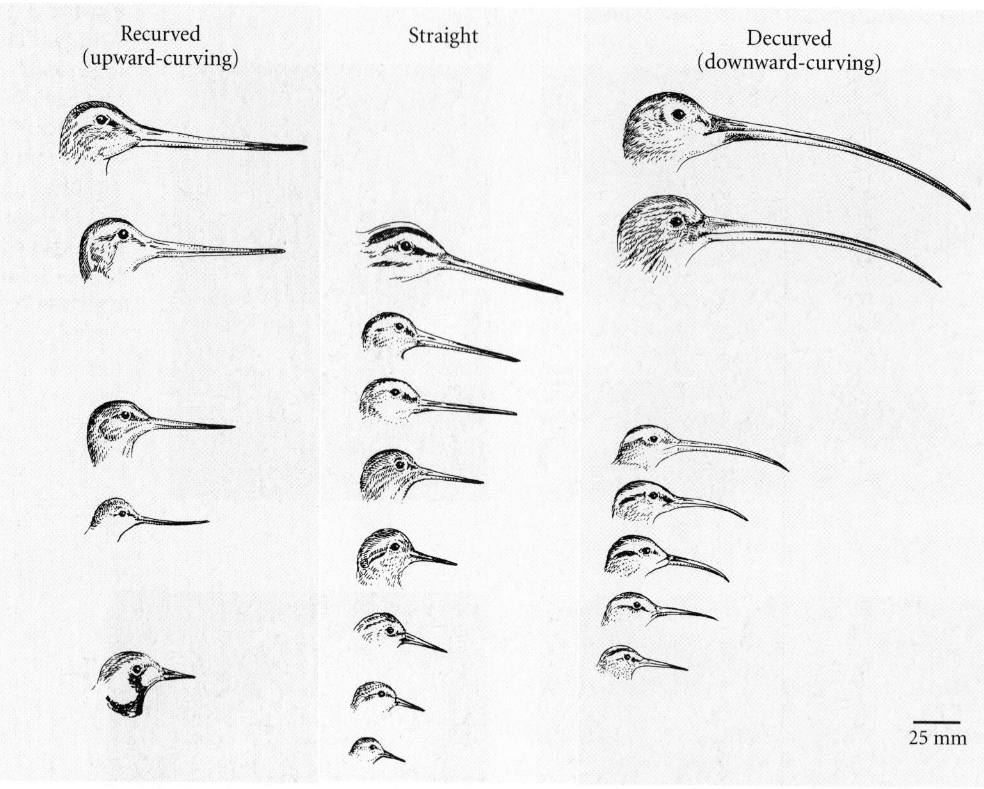

Figure 3.12 Variation in the shape and length of the bill among sandpipers (Scolopacidae). The three vertical series are drawn to scale and are spaced to match the differences in bill length, which range from 18 mm (bottom center) to 166 mm (upper right). Note the gradations in both curvature and length. The phylogenetic relationships among these species are not well resolved, but the variation shows how very different bills could have evolved through small changes. (After Hayman et al. 1986.)

Gradations among living species are commonplace and provide support for gradual evolution. For example, the length and shape of the bill differs greatly among species of sandpipers, but the most extreme forms are bridged by species with intermediate bills (Figure 3.12).

Change in form is often correlated with change in function

One of the reasons a homologous character may differ so greatly among taxa is that its form may have evolved as its function has changed. The sting of a wasp or bee, for example, is a modification of the ovipositor that other members of the Hymenoptera use to insert eggs into plant or arthropod hosts (that is why only female wasps and bees sting). In the many groups of plants that have independently evolved a vinelike climbing habit, the structures that have been modified into climbing organs include roots, leaves, leaflets, stipules, and inflorescences (Figure 3.13).

Similarity among species changes throughout ontogeny

Species are often more similar as embryos than as adults. Karl Ernst von Baer noted in 1828 that the features common to a more inclusive taxon (such as the subphylum Vertebrata) often appear in ontogeny (development) before the specific characters of lower-level taxa (such as orders or families). This generalization is now known as VON BAER'S LAW. Probably the most widely known example is the similarity of many tetrapod vertebrate embryos, all of which display pharyngeal clefts (gill slits), a notochord, segmentation, and paddlelike limb buds before the features typical of their class or order become apparent (Figure 3.14).

One of Darwin's most enthusiastic supporters, the German biologist Ernst Haeckel, reinterpreted such patterns to mean that "ontogeny recapitulates phylogeny"; that is, that the development of the individual organism (ONTOGENY) repeats the evolutionary history of the adult forms of its ancestors. Haeckel thus supposed that by studying embryology, one could read a species' phylogenetic history, and therefore infer directly phylogenetic relationships among organisms. By the end of the nineteenth century, it was already

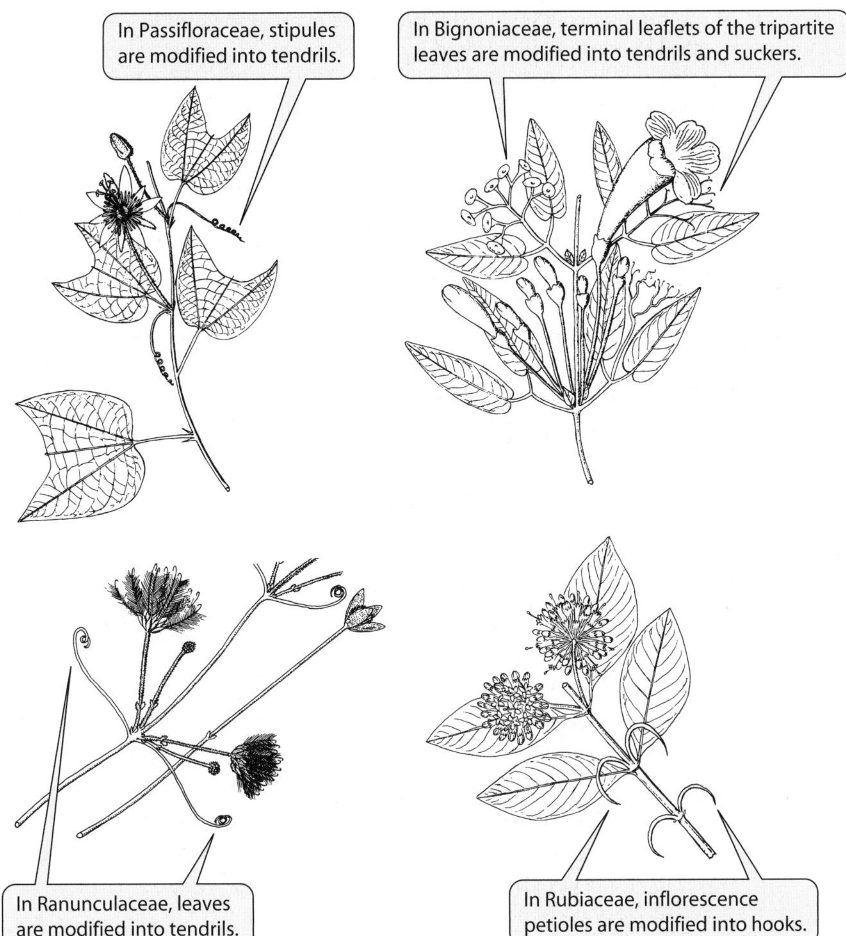

In Passifloraceae, stipules are modified into tendrils.
In Bignoniaceae, terminal leaflets of the tripartite leaves are modified into tendrils and suckers.
In Ranunculaceae, leaves are modified into tendrils.
In Rubiaceae, inflorescence petioles are modified into hooks.

Figure 3.13 Structures modified for climbing in vines from different plant groups show that structures can become modified for new functions, and that different evolutionary paths to the same functional end may be followed in different groups. (After Hutchinson 1969.)

clear that Haeckel's dictum (known as the "biogenetic law") rather seldom holds (Gould 1977). The pharyngeal clefts and branchial arches of embryonic mammals and "reptiles" never acquire the form typical of adult fishes. Moreover, various features develop at different rates, relative to one another, in descendants than in their ancestors, and embryos and juvenile stages have stage-specific adaptations of their own (such as the amnion in the embryo). Thus the biogenetic law is certainly not an infallible guide to phylogenetic history. However, embryological similarities provided Darwin with some of his most important evidence of evolution, and they continue to shed important light on how characteristics have been transformed during evolution.

Development underlies some common patterns of morphological evolution

Until a few decades ago, classification and phylogenetic studies relied chiefly on analyses of morphological characters, including their change during embryonic development. In the course of their work, systematists and comparative morphologists documented many common patterns of evolution, and expressed these patterns in terms of underlying developmental changes. Some patterns of developmental change in morphology are individualization, heterochrony, allometry, heterotopy, and changes in complexity (Rensch 1959; Müller 1990; Raff 1996; Wagner 1996). Today, one of the most active research areas concerns the genetic and developmental basis of such evolutionary changes,

Figure 3.14 Micrographs show the similarities—and differences—among several vertebrate embryos at different stages of development. Each begins (at top) with a similar basic structure, although they acquire this structure at different stages and sizes. As the embryos develop, they become less and less alike. (Adapted from Richardson et al. 1998; photo courtesy of M. Richardson and R. O'Rahilly.)

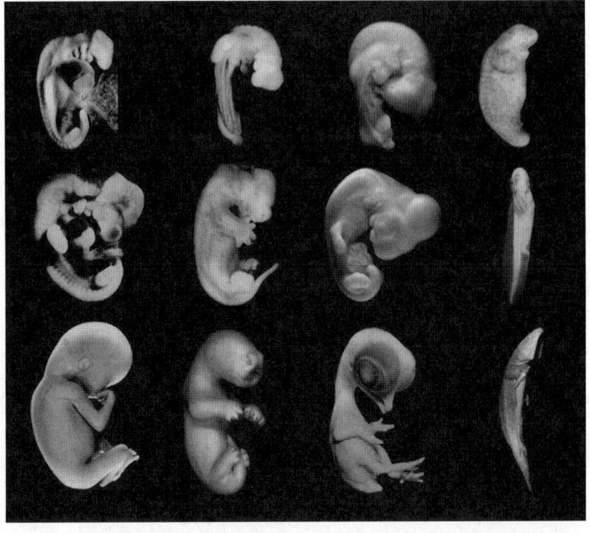

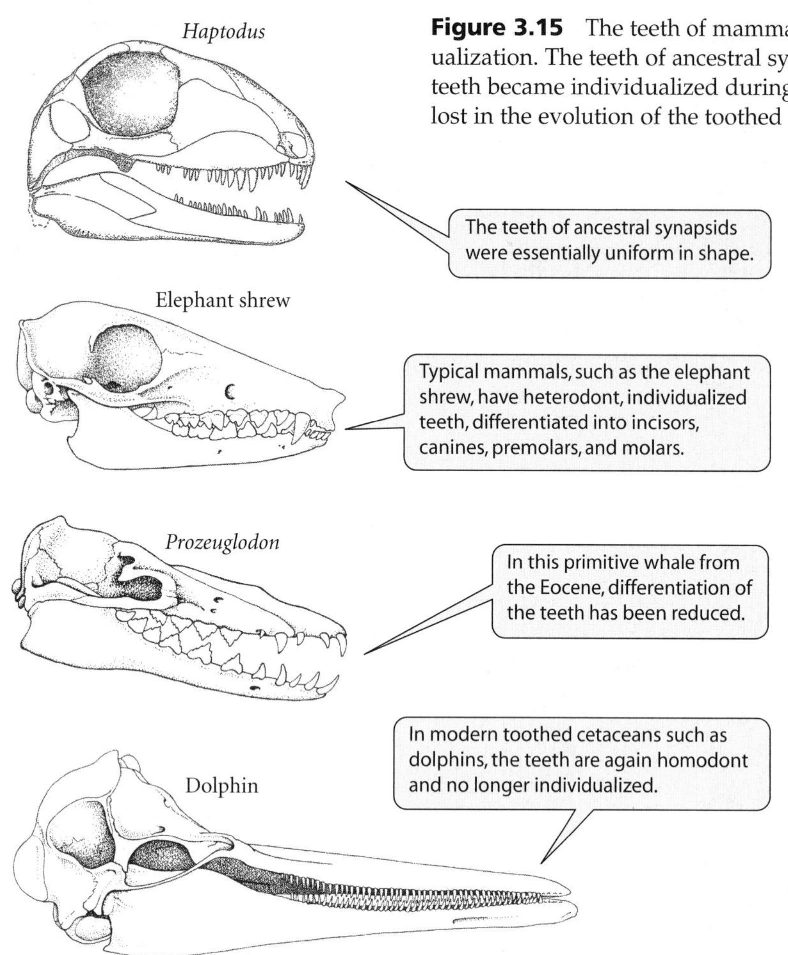

Haptodus
The teeth of ancestral synapsids were essentially uniform in shape.
Elephant shrew
Typical mammals, such as the elephant shrew, have heterodont, individualized teeth, differentiated into incisors, canines, premolars, and molars.
Prozeuglodon
In this primitive whale from the Eocene, differentiation of the teeth has been reduced.
In modern toothed cetaceans such as dolphins, the teeth are again homodont and no longer individualized.
Dolphin

Figure 3.15 The teeth of mammals provide an example of the acquisition and loss of individualization. The teeth of ancestral synapsids, illustrated by the Permian *Haptodus*, are uniform; teeth became individualized during the evolution of mammals. Distinct tooth identity was later lost in the evolution of the toothed whales. (A after Romer 1966; B–D after Vaughan 1986.)

such as the evolution of crustacean appendages portrayed in Figure 3.7 (see also Chapter 21).

INDIVIDUALIZATION. The bodies of many organisms consist of MODULES—distinct units that have distinct genetic specifications, developmental patterns, locations, and interactions with other modules (Raff 1996). Some modules (e.g., leaves of many plants, teeth of many fishes) lack distinct individual identities and may be considered aspects of a single character. Such structures are termed SERIALLY HOMOLOGOUS if they are arrayed along the body axis (as vertebrae are) and HOMONYMOUS if they are not. An important evolutionary phenomenon is the acquisition of distinct identities by such units, called **individualization** (Müller and Wagner 1996; Wagner 1996), which in turn is an important basis for mosaic evolution. For instance, the teeth of most reptiles are uniform, but they became individualized (differentiated into incisors, canines, premolars, and molars) during the evolution of mammals. Distinct tooth identity was later lost during the evolution of the toothed whales (Figure 3.15).

HETEROCHRONY. **Heterochrony** (Gould 1977; McKinney and McNamara 1991) is broadly defined as *an evolutionary change in the timing or rate of developmental events*. Many phenotypic changes appear to be based on such changes in timing, but several other developmental mechanisms can produce similar changes (Raff 1996).

Relatively *global* heterochronic changes, affecting many characters simultaneously, are illustrated by cases in which the time of development of most SOMATIC features (those other than the gonads and related reproductive structures) is altered relative to the time of maturation of the gonads (i.e., initiation of reproduction). The axolotl, for example, is a salamander that does not undergo metamorphosis, as most salamanders do, but instead reproduces while retaining most of its larval (juvenile) characteristics (Figure 3.16). Such evolution of a more "juvenilized" morphology of the reproductive adult is called **paedomorphosis** (from the Greek *paedos*, "child," and *morphos*, "form"). Paedomorphosis can be caused by reducing the growth rate of somatic characters (an evolutionary process called **neoteny**) or by cessation of growth at an earlier age (a process called **progenesis**).

Figure 3.16 Paedomorphosis in salamanders. (A) The tiger salamander (*Ambystoma tigrinum*), like most salamanders, undergoes metamorphosis from an aquatic larva (left; note the presence of gills) to a terrestrial adult (right). (B) The adult axolotl (*Ambystoma mexicanum*), with gills and tail fin, resembles the larva of its terrestrial relative. The axolotl remains aquatic throughout its life span. These individuals are a white laboratory strain. (A © R. Nistri and C. Mattison/Alamy; B © Arco Images GmbH/Alamy.)

(A)

(B)

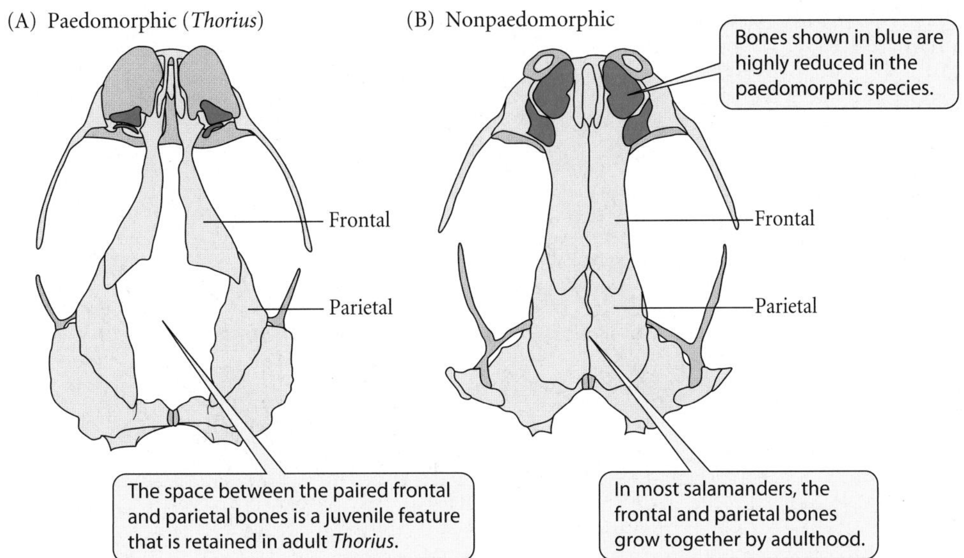
(A) Paedomorphic (Thorius)
(B) Nonpaedomorphic
Bones shown in blue are highly reduced in the paedomorphic species.
Frontal
Parietal
Frontal
Parietal
The space between the paired frontal and parietal bones is a juvenile feature that is retained in adult Thorius.
In most salamanders, the frontal and parietal bones grow together by adulthood.

Figure 3.17 Comparison of the skulls of the progenetic dwarf salamander *Thorius* and a typical nonprogenetic relative, *Pseudoeurycea*. The skull of adult *Thorius* has a number of juvenile features. (After Hanken 1984.)

The axolotl is a neotenic salamander species that reaches the same size as its metamorphosing relatives, whereas tiny salamanders in the genus *Thorius* are progenetic: they have many features characteristic of the juveniles of larger species of salamanders, as if their development had been abbreviated (Figure 3.17).

In contrast to paedomorphosis, evolution of delayed maturity may result in reproduction at a larger size, associated with the extended development of "hyperadult" features. Such an evolutionary change is called **peramorphosis**. The large size of the human brain, for example, has been ascribed to humans' extended period of growth (McNamara 1997).

ALLOMETRY. **Allometric growth**, or **allometry**, refers to the *differential rate of growth of different parts or dimensions of an organism* during its ontogeny. For example, during human postnatal growth, the head grows at a slower rate than the body as a whole, and the legs grow at a higher rate. Changes in rates of allometric growth of individual characters—that is, "local" heterochrony—appear to have played an extremely important role in evolution. For example, many evolutionary changes can be described as if local heterochronies had altered the *shape* of one or more characters: an increased rate of elongation of the digits "accounts for" the shape of a bat's wing compared with the forelimbs of other mammals (see Figure 3.5); an elephant's tusks are incisor teeth that have grown much faster than the other teeth.

Allometric growth is often described by the equation

$$y = bx^a$$

where y and x are two measurements, such as the height and width of a tooth or the size of the head and the body. (In many studies, x is a measure of body size, such as weight, because many structures change disproportionately with overall size.) The ALLOMETRIC COEFFICIENT, a, describes their relative growth rates. If $a = 1$, growth is ISOMETRIC, meaning that the two structures or dimensions increase at the same rate, and shape does not change. If y increases faster than x, as for human leg length relative to body size or weight, $a > 1$ (positive allometry); if it increases relatively slowly, as for human head size, $a < 1$ (negative allometry; Figure 3.18). These curvilinear relationships between y and x often appear more linear if transformed to logarithms, yielding the equation for a straight line: $\log y = \log b + a \log x$. For example, antler size in deer increases allometrically with body mass. The largest species of deer, the extinct Irish elk (*Megaceros giganteus*), had much larger antlers, relative to body mass, than those of any other deer (Figure 3.19).

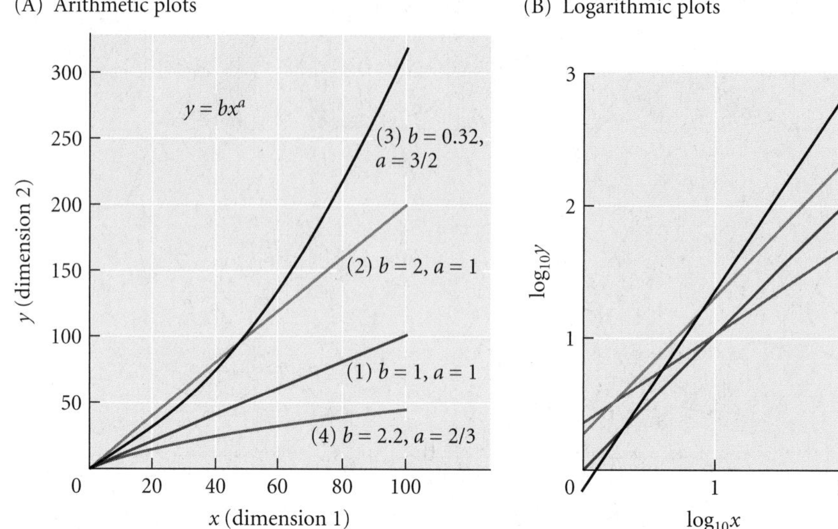

Figure 3.18 Hypothetical curves showing various allometric growth relationships between two body measurements, y and x, according to the equation $y = bx^a$. (A) Arithmetic plots. Curves 1 and 2 show isometric growth ($a = 1$), in which y is a constant multiple (b) of x. Curves 3 and 4 show positive ($a > 1$) and negative ($a < 1$) allometry, respectively. (B) Logarithmic plots of the same curves have a linear form. The slope differences depend on a. Curves 1 and 2 have slopes equal to 1.

HETEROTOPY. **Heterotopy** is an evolutionary change in the position within an organism at which a phenotypic character is expressed. Studies of the distribution of gene products have revealed many heterotopic differences among species in sites of gene expression. For instance, certain species of deep-sea squids have organs on the body that house light-emitting bacteria. The light is diffracted through small lenses made up of the same two proteins that compose the lenses of the squids' eyes (Raff 1996).

Heterotopic differences among species are very common in plants. For example, the major photosynthetic organs of most plants are the leaves, but photosynthesis is carried out in the stem in cacti and many other plants that occupy dry environments. In many unrelated species of lianas (woody vines), roots grow along the aerial stem (Figure 3.20). In some lianas these roots serve as holdfasts, while in others they grow down to the soil from the canopy high above.

The bones of vertebrates provide many examples of heterotopy. For example, many phylogenetically new bones have arisen as SESAMOIDS—bones that develop in tendons or other connective tissues subject to stress (Müller 1990). Many dinosaurs had ossified ten-

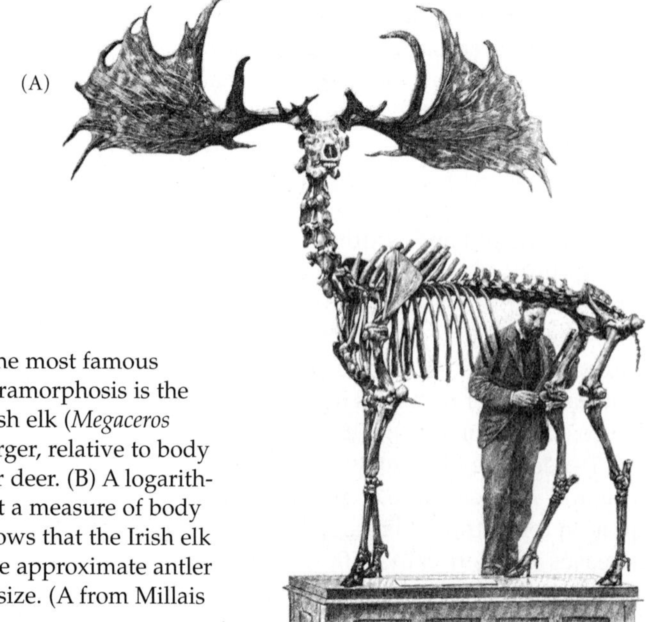

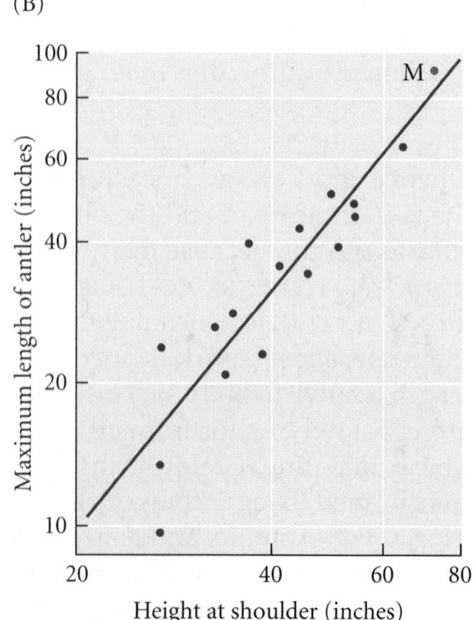

Figure 3.19 (A) Perhaps the most famous example of allometry and peramorphosis is the largest of deer, the extinct Irish elk (*Megaceros giganteus*). Its antlers were larger, relative to body mass, than those of any other deer. (B) A logarithmic plot of antler size against a measure of body size for 20 species of deer shows that the Irish elk (red point marked M) had the approximate antler size predicted from its body size. (A from Millais 1897; B after Gould 1974.)

Figure 3.20 Plants of the genus *Philodendron*, such as this Jamaican climber, are lianas. Many lianas, which include several genera besides *Philodendron*, have evolved exposed roots that grow from an aerial stem. (Photo © Acevedo Melanie/photolibrary.com.)

dons in the tail, and the giant panda (*Ailuropoda melanoleuca*) is famous for having a "thumb" that is not a true jointed digit, but a single sesamoid (see Figure 22.11).

INCREASES AND DECREASES IN COMPLEXITY. The earliest organisms must have had very few genes and must have been very simple in form. Both phylogenetic and paleontological studies show that there have been great increases in complexity during the history of life, exemplified by the origin of eukaryotes, then of multicellular organisms, and later of the elaborate tissue organization of plants and animals (Chapter 5). Given our impression of overall increases in complexity during evolution, it can be surprising to learn that simplification of morphology—reduction and loss of structures—is one of the most common trends within clades (Adamowicz and Purvis 2006). The primitive "ground plan" of flowers, for example, includes numerous sepals, petals, stamens, and carpels, but the number of one or more of these elements has been reduced in many lineages of flowering plants, and many clades have lost petals and/or sepals altogether. The number of digits in tetrapod vertebrates has been reduced many times (consider the single toe of horses), but has increased only once (in extinct ichthyosaurs). Early lobe-finned fishes had far more skull bones than their amniote descendants (Figure 3.21). Many such changes can be ascribed to increased functional efficiency.

Phylogenetic Analysis Documents Evolutionary Trends

The term **evolutionary trend** can refer to repeated changes of a character in the same direction, either within a single lineage or, often, in many lineages independently. For example, Justen Whittall and Scott Hodges (2007) documented both kinds of trends in a phylogenetic analysis of columbines (*Aquilegia*; Figure 3.22). During the diversification of columbine species, there have been repeated shifts from pollination by short-tongued bumblebees to longer-tongued hummingbirds, and from hummingbirds to still longer-tongued hawkmoths. In concert with these repeated shifts, the plants have consistently evolved longer nectar spurs, tubular extensions of the petals from which the pollinator extracts nectar. Whittall and Hodges suggest that consistent shifts to longer-tongued, rather than shorter-tongued, pollinators occurs because short-tongued animals cannot reach the nectar in long spurs and avoid such flowers.

Figure 3.21 An example of reduction and loss of structures during evolution. The number of bones in the skull is higher in early lobe-finned fishes (such as the Devonian *Eusthenopteron*), from which amniotes were derived, than in early amniotes (such as the Permian *Milleretta*). Among the later amniotes are placental mammals, such as the domestic dog (*Canis*), in which the skull has fewer elements still. The reduction in the number of bones in the lower jaw is particularly notable. (After Romer 1966.)

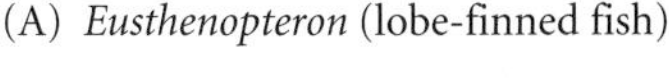
(A) *Eusthenopteron* (lobe-finned fish)

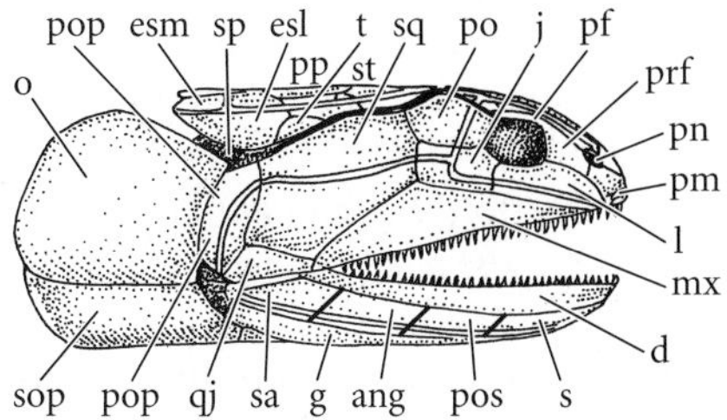

(B) *Milleretta* (early amniote)

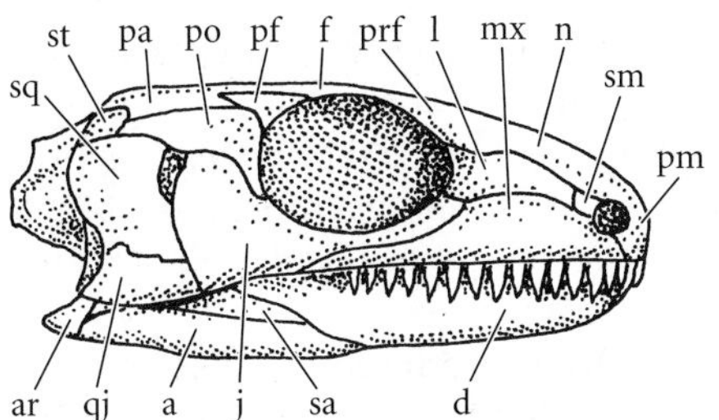

(C) *Canis* (modern mammal)

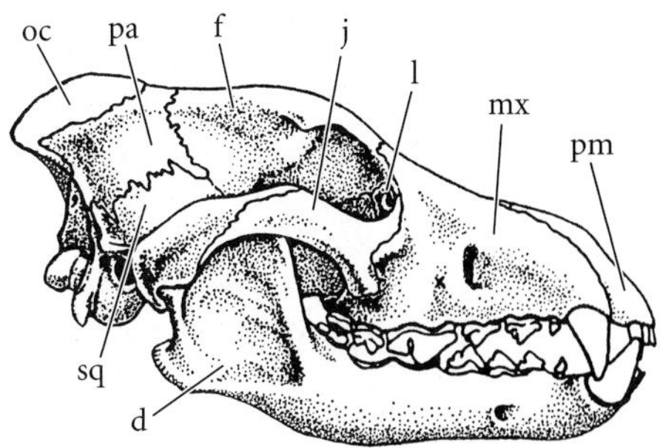

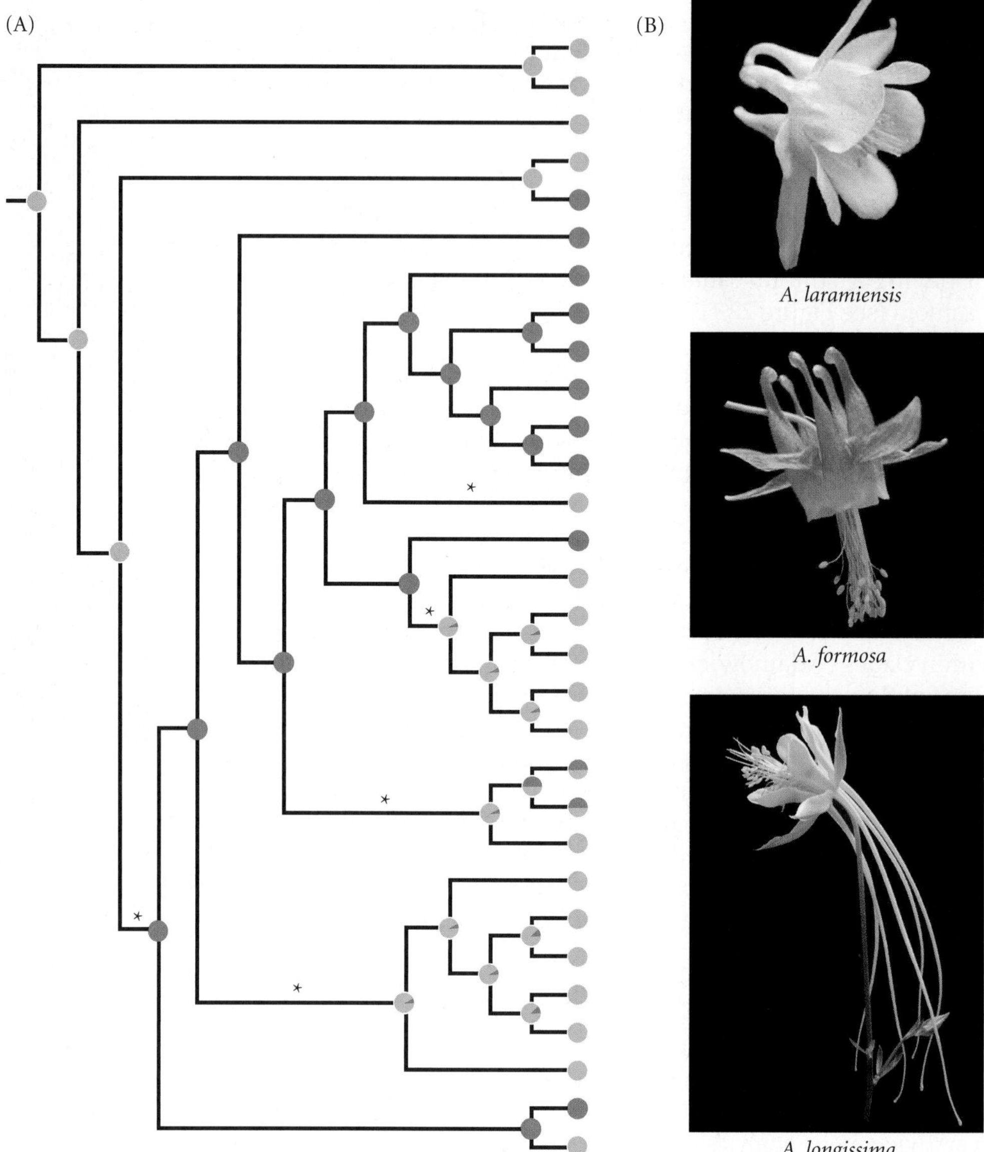

Figure 3.22 (A) Phylogenetic analysis of species of columbines (*Aquilegia*) shows shifts from pollination by bees (blue circles) to pollination by hummingbirds (red) and then by hawkmoths (lime). These shifts are associated with changes in several flower characteristics, including increasing length of nectar spurs, as illustrated by the three species pictured. The circles at the tips of the phylogenetic tree are different species; the colors of circles at internal nodes show the inferred pollination habit of ancestors; and the asterisks mark shifts between pollination modes. (B) Representative species of (top to bottom) bee-, hummingbird-, and hawkmoth-pollinated columbines. (After Whittall and Hodges 2007; photographs courtesy of Justen Whittall.)

Among flowering plants, many groups independently display trends from low to high chromosome number and DNA content, from many to few flower parts (e.g., stamens or carpels), from separate to fused flower parts, from radial to bilateral symmetry of the flower, from animal to wind pollination, and from woody to herbaceous structure. We will analyze the kinds and causes of evolutionary trends in Chapter 22.

Many Clades Display Adaptive Radiation

Evolutionary radiation, as we saw in the previous chapter, is divergent evolution of numerous related lineages within a relatively short time. In most cases, the lineages are modified for different ways of life, and the evolutionary radiation may be called an **adaptive radiation** (Schluter 2000). The characteristics of the members of an evolutionary radiation usually do not show a directional trend in any one direction. *Evolutionary radiation, rather than sustained, directional evolutionary trends, is probably the most common pattern of long-term evolution.*

Several adaptive radiations have been extensively studied and are cited in many evolutionary contexts. The most famous example is the adaptive radiation of Darwin's finches in the Galápagos archipelago. These finches, which are all descended from a single ances-

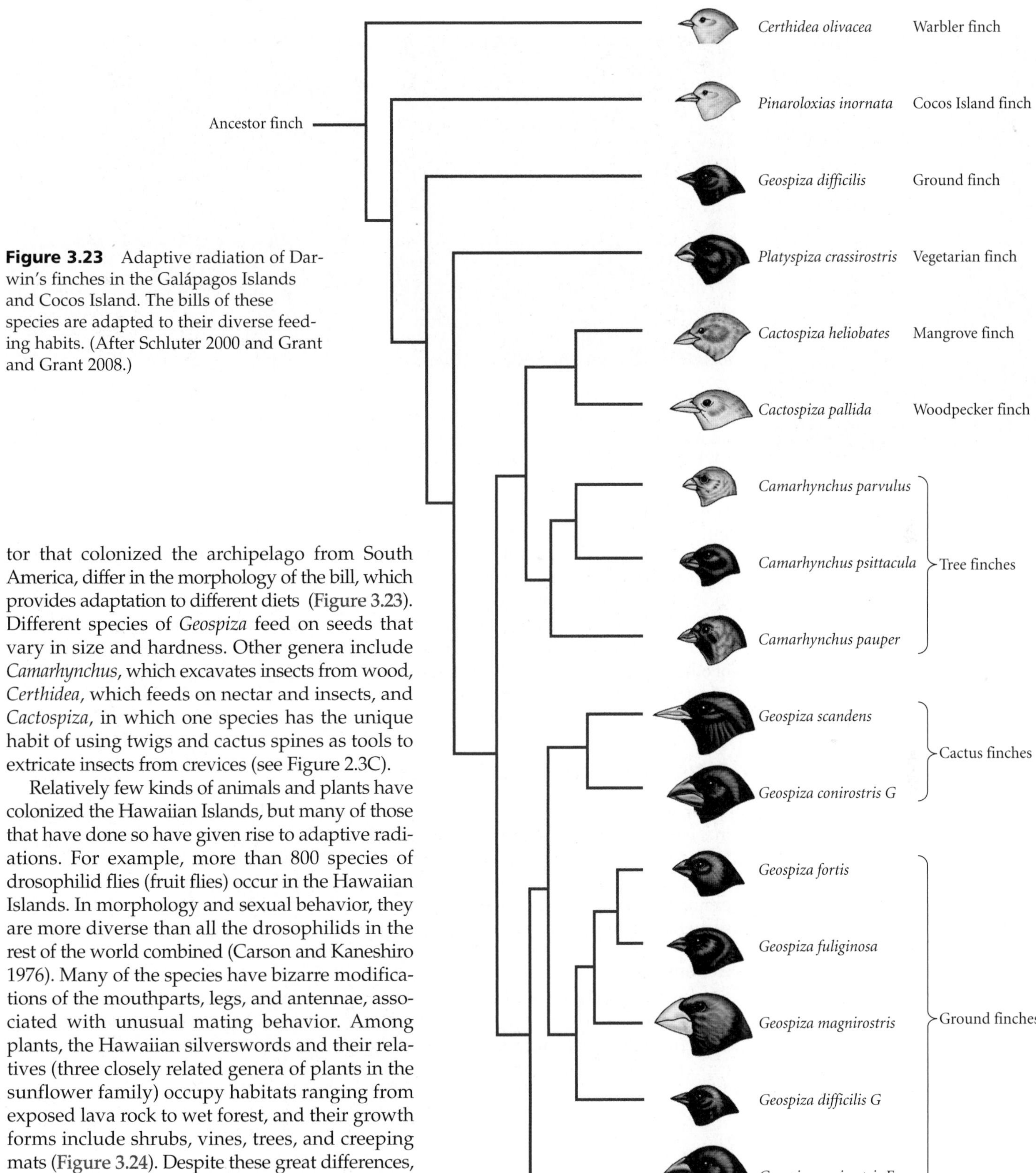

Figure 3.23 Adaptive radiation of Darwin's finches in the Galápagos Islands and Cocos Island. The bills of these species are adapted to their diverse feeding habits. (After Schluter 2000 and Grant and Grant 2008.)

tor that colonized the archipelago from South America, differ in the morphology of the bill, which provides adaptation to different diets (Figure 3.23). Different species of *Geospiza* feed on seeds that vary in size and hardness. Other genera include *Camarhynchus*, which excavates insects from wood, *Certhidea*, which feeds on nectar and insects, and *Cactospiza*, in which one species has the unique habit of using twigs and cactus spines as tools to extricate insects from crevices (see Figure 2.3C).

Relatively few kinds of animals and plants have colonized the Hawaiian Islands, but many of those that have done so have given rise to adaptive radiations. For example, more than 800 species of drosophilid flies (fruit flies) occur in the Hawaiian Islands. In morphology and sexual behavior, they are more diverse than all the drosophilids in the rest of the world combined (Carson and Kaneshiro 1976). Many of the species have bizarre modifications of the mouthparts, legs, and antennae, associated with unusual mating behavior. Among plants, the Hawaiian silverswords and their relatives (three closely related genera of plants in the sunflower family) occupy habitats ranging from exposed lava rock to wet forest, and their growth forms include shrubs, vines, trees, and creeping mats (Figure 3.24). Despite these great differences, most silverswords can produce fertile hybrids when crossed (Carlquist et al. 2003).

The cichlid fishes in the Great Lakes of eastern Africa have undergone some of the most spectacular adaptive radiations (Kornfield and Smith 2000; Kocher 2004). Lake Victoria has more than

(A) *Argyroxiphium sandwicense*

(B) *Wilkesia hobdyi*

(C) *Dubautia menziesii*

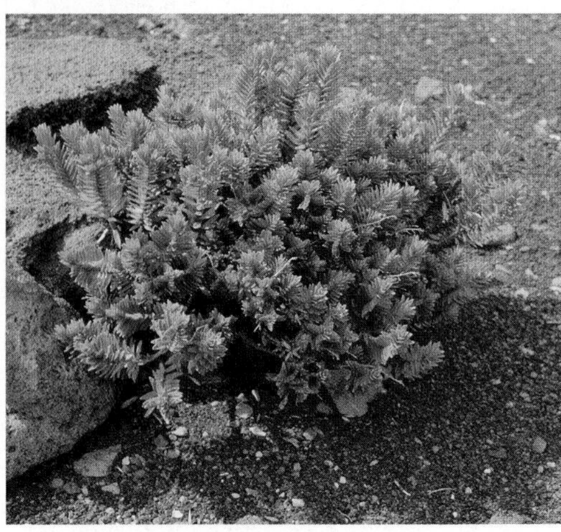

Figure 3.24 Some members of the Hawaiian silversword alliance with different growth forms. (A) *Argyroxiphium sandwicense*, a rosette plant that lacks a stem except when flowering (as it is here). (B) *Wilkesia hobdyi*, a stemmed rosette plant. (C) *Dubautia menziesii*, a small shrub. (A, Elizabeth N. Orians; B, Gerald D. Carr; C © Noble Procter/Photo Researchers, Inc.)

200 species, Lake Tanganyika at least 140, and Lake Malawi more than 500—perhaps as many as 1000. The species within each lake vary greatly in coloration, body form, and the form of the teeth and jaws (Figure 3.25), and they are correspondingly diverse in feeding habits. There are specialized feeders on insects, detritus, rock-encrusting algae, phytoplankton, zooplankton, molluscs, baby fishes, and large fishes. Some species feed on the scales of other fishes, and one has the gruesome habit of plucking out other fishes' eyes. The teeth of certain closely related species differ more than among some families of fishes. The species in each lake are a monophyletic group and have diversified very rapidly.

Patterns in Genes and Genomes

So far, we have seen how phylogenetic analysis can reveal patterns of historical evolution in the phenotypic characteristics of organisms, such as their morphology. Exactly the same methods are now used to probe the evolution of DNA sequences, such as genes, and of the structure of genomes—the full DNA complement of an organism. Here are just a few examples of evolutionary patterns at the molecular level; more instances will arise in subsequent chapters.

Genome size

Genome sizes are frequently measured in picograms (pg) of DNA; 1 pg is roughly equivalent to 1 Gb (1 gigabase, or 1 billion base pairs) of actual sequence. As data on genome sizes from hundreds of organisms were compared, a curious pattern emerged. It was expected that physiologically and behaviorally complex organisms, such as mammals,

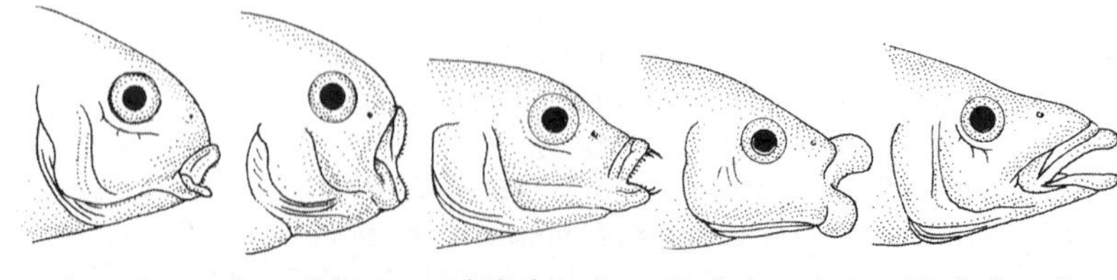

Figure 3.25 A sample of the diverse head shapes among the Cichlidae of the African Great Lakes. The differences in morphology are associated with differences in diet and mode of feeding. (After Fryer and Iles 1972.)

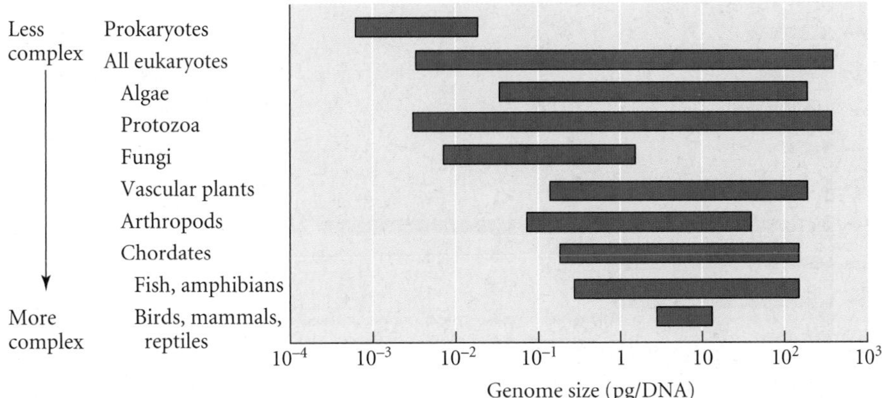

Figure 3.26 Genome size variation. The bars indicate the range of genome size for particular clades. Taxa are arranged from top to bottom in order of supposedly increasing organismal complexity. Within eukaryotes, there is little relationship between genome size and organismal complexity. This "C-value paradox" may be a result of great variation among lineages in the amount of repetitive (noncoding) DNA . (1 pg of DNA is approximately equivalent to 1 billion base pairs.) (After Gregory 2001.)

would have more complex, and therefore larger, genomes than simpler organisms. On a very broad scale, this holds true (Figure 3.26). Viral and bacterial genomes are tiny compared with those of eukaryotes, particularly those of mammals, amphibians, and some plants. But major groups of eukaryotes, such as vertebrates and flowering plants, seem not to differ as had originally been expected. For example, the haploid genome of the pufferfish is about 0.5 Gb, and human and mouse genomes are both about 3 Gb—but salamanders have enormous haploid genome sizes of up to 50 Gb. Moreover, genome size varies more than tenfold among species of salamanders, and about twofold even between species in the same genus. Similarly, flowering plants have a range of genome sizes spanning three orders of magnitude, from 10^8 to 10^{11} base pairs, entirely encompassing the range seen across all vertebrates. The amount of DNA per genome was called the C-value, and so lack of correspondence between genome size and phenotypic complexity in eukaryotes was dubbed the C-VALUE PARADOX.

This seeming paradox was resolved in the 1960s by Roy Britten, Mary Lou Pardue, and others, who showed that eukaryotic genomes usually contain a great deal of noninformative, highly repetitive DNA that varies greatly in amount among species. Many of the highly repetitive sequences are thought not to be useful for the organism. Today, entire genomes are being sequenced and protein-coding genes are identified, so that it is possible to compare the number of functional genes among species, instead of simply amounts of DNA. The phylogenetic transition from unicellular to multicellular eukaryotes seems to have been marked by considerable increase in gene number, but there is surprisingly little variation among animals and plants, and it is unsettling to find that rice appears to have more than twice as many genes as mammals (Figure 3.27).

Comparative studies are beginning to shed some light on the evolution of genome size. For example, endosymbiotic microorganisms, parasites or mutualists that live within eukaryotic host organisms, generally have smaller genomes than their free-living relatives. Nancy Moran and colleagues have extensively studied the evolution of reduced genomes in *Buchnera*, a bacterial clade that has been an intracellular symbiont of aphids for about 200 Myr (Moran 2003; van Ham et al. 2003; see Figure 19.2). Genome comparisons have revealed that the common ancestor of *Buchnera* species lost over 2000 genes, compared with its relative *E. coli*. Because *Buchnera*'s host aphids provide many essential nutrients and metabolic functions for the symbionts (see Figure 16.22), natural selection for retaining many genes has been relaxed.

Duplicated genes and genomes

Gene duplication is a mutational event (Chapter 8) by which a new gene (say β) arises as a copy of a preexisting gene (α), so that a single gene locus in an ancestor is represented by two loci in the descendant. These two genes will undergo different evolutionary changes in sequence, and so can be distinguished. If two species (1 and 2) inherit the duplicated pair (α , β) from their common ancestor, the relationships among the genes represent two forms of homology, and so warrant different terms: **orthology** and **paralo-**

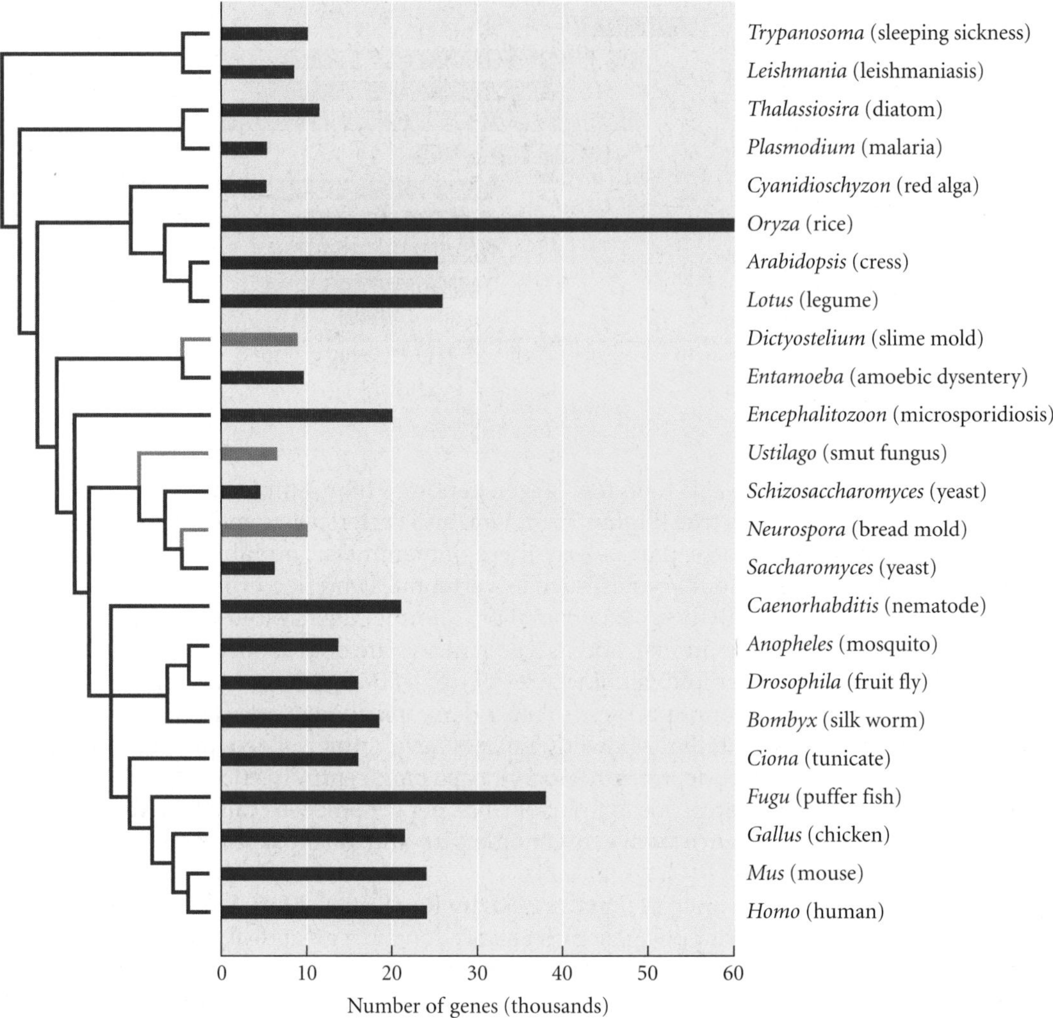

Figure 3.27 Number of genes estimated for some eukaryotes with fully sequenced genomes, arranged on a provisional phylogeny. Multicellular organisms with tissue organization (plants and animals, represented by blue branches) have more genes than single-celled organisms (red branches) or multicellular organisms that lack pronounced tissue organization (some fungi and slime molds; green branches). (Data from Lynch 2007.)

gy. The genes that originated from a common ancestral gene duplication are paralogous, whereas the genes that diverged from a common ancestral gene by phylogenetic splitting at the organismal level are orthologous (Figure 3.28). The phylogenetic relationships among the orthologous and paralogous genes in two or more species can be determined by standard phylogenetic methods.

This process may occur repeatedly over evolutionary time, generating a gene family. In the human genome, for example, the twelve members of the globin gene family include genes that encode myoglobin and several α and β hemoglobin chains (Figure 3.29). The origin of myoglobin and hemoglobin by duplication of an ancestral globin gene occurred during the ancestry of the vertebrates, all of which have both genes. The α and β hemoglobins arose by gene duplication in the ancestor of jawed vertebrates, all of which have a functional hemoglobin composed of both α and β chains, whereas the jawless verte-

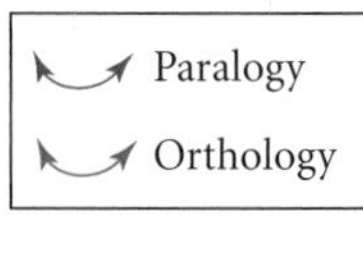

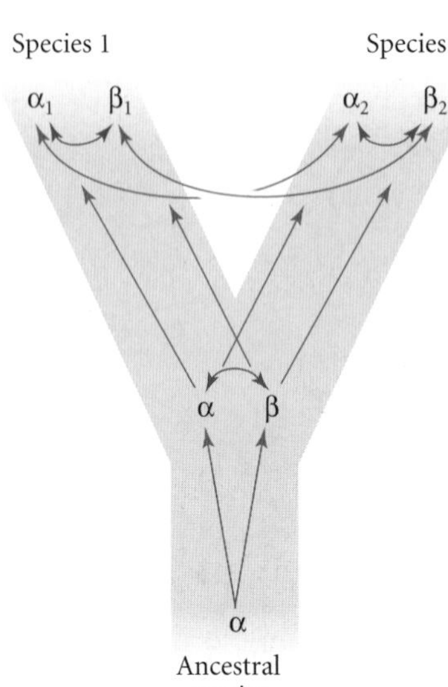

Figure 3.28 Orthology and paralogy in gene families. When an ancestral gene (α) undergoes duplication, the two resulting genes (α + β) have a paralogous relationship to one another (blue arrow). A speciation event after duplication results in divergence of the ancestral set of two paralogous genes. Within the genomes of the two diverged species, the α + β still have a paralogous relationship (blue arrows). However, the copies of α in species 1 and species 2 are orthologous (red arrows), because the two genes are related to one another via speciation, not duplication. Likewise, the copies of β in species 1 and 2 are orthologous.

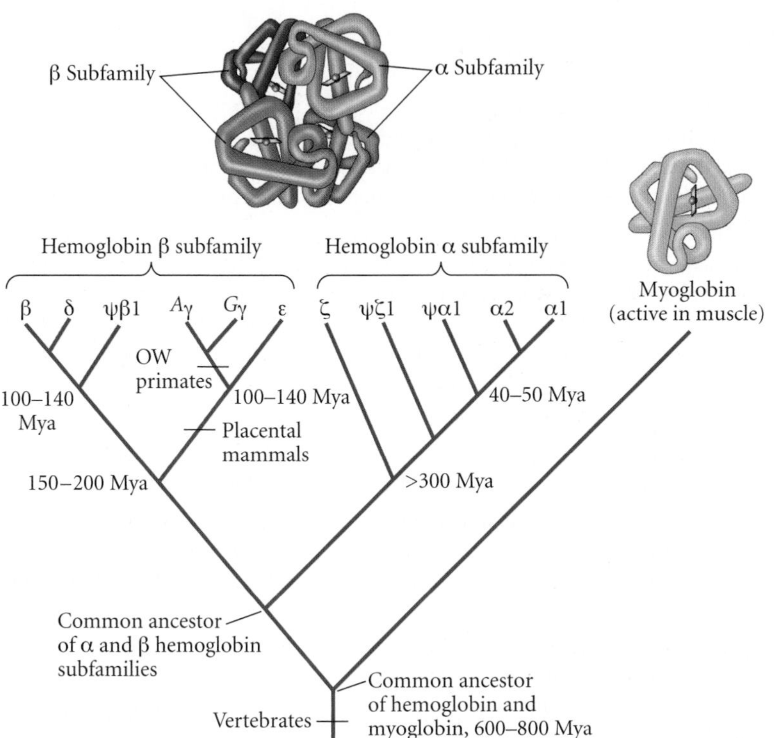

Figure 3.29 The phylogeny of genes in the globin family in the human genome. Myoglobin consists of a single protein unit, whereas mammalian hemoglobins consist of four subunits, two each from the α and β subfamilies. Each branch point on the tree denotes a gene duplication event; some of these events are marked with estimates of when the duplication occurred. The origin of hemoglobin and myoglobin from a common ancestral gene occurred in the ancestor of all vertebrates, but the α and β hemoglobin subfamilies originated by duplication in an ancestor of the jawed vertebrates. The duplication of the β hemoglobin into two genes occurred in the ancestor of placental mammals, since the $A_\gamma/G_\gamma/\varepsilon$ genes are lacking in monotremes and marsupials. In some instances, one of the pair of genes formed by duplication became a nonfunctional pseudogene, symbolized by ψ. (After Li 1997 and Hartwell et al. 2000.)

brates (e.g., lamprey) have only a single hemoglobin chain. The origin of the other globin genes can similarly be traced, based on their sequences and phylogenetic distribution.

Many genes are duplicated as parts of large chromosomal blocks (paralogous regions, or "paralogons") that often contain hundreds of genes. Indeed, **polyploidy**, whereby the entire nuclear genome is duplicated, occurs commonly in plants and occasionally in animals. The entire set of Hox genes, which play a major role in the embryonic development of animals, is represented by one set of genes in nonvertebrate chordates, four sets in most vertebrates, and seven sets in teleost fishes (Figure 3.30). Some other groups of genes show similar patterns. At this time, it is unclear whether the entire genome, or just parts of it, underwent two successive duplications in the ancestor of jawed vertebrates, but it does appear likely that the teleosts had a polyploid ancestor (Panopoulou and Poustka 2005).

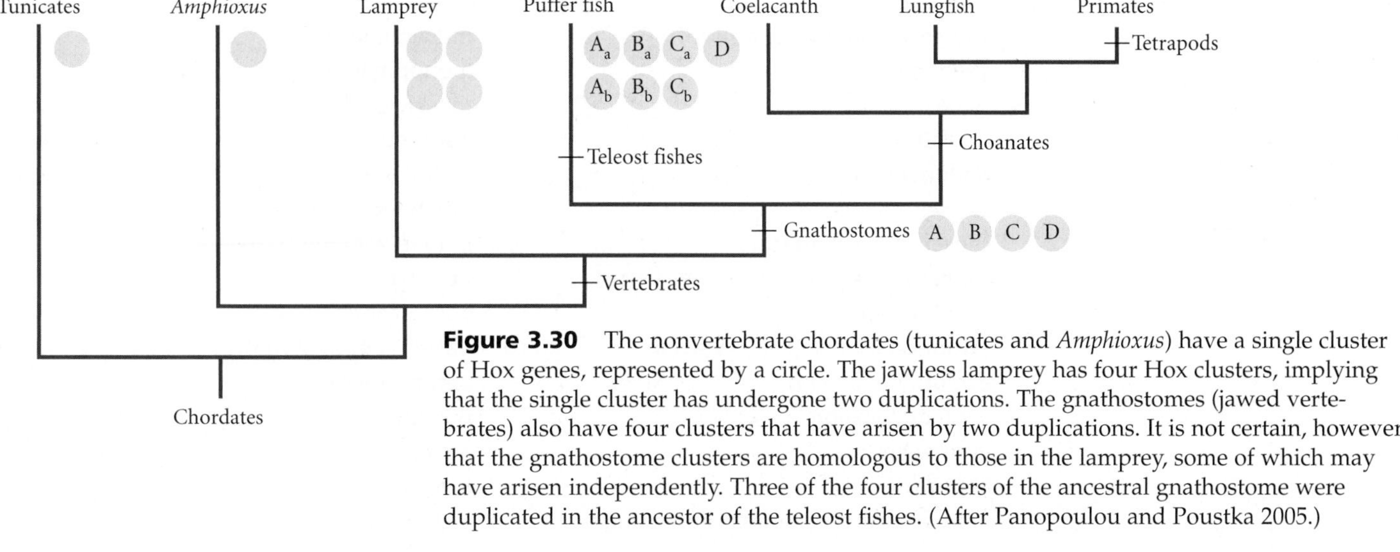

Figure 3.30 The nonvertebrate chordates (tunicates and *Amphioxus*) have a single cluster of Hox genes, represented by a circle. The jawless lamprey has four Hox clusters, implying that the single cluster has undergone two duplications. The gnathostomes (jawed vertebrates) also have four clusters that have arisen by two duplications. It is not certain, however, that the gnathostome clusters are homologous to those in the lamprey, some of which may have arisen independently. Three of the four clusters of the ancestral gnathostome were duplicated in the ancestor of the teleost fishes. (After Panopoulou and Poustka 2005.)

Summary

1. Most traditional classifications of organisms have been devised to reflect both branching (phylogeny) and phenotypic divergence. Many contemporary systematists adopt the "cladistic position" that classification should explicitly reflect phylogenetic relationships, and that all higher taxa should be monophyletic.
2. Phylogenetic analyses have many uses in addition to describing the branching history of life. An important one is inferring the history of evolution of interesting characteristics by parsimoniously "mapping" changes in character states onto a phylogeny that has been derived from other data. Such systematic studies have yielded information on common patterns and principles of character evolution.
3. New features almost always evolve from pre-existing characters.
4. Homoplasy, which is common in evolution, is often a result of similar adaptations in different lineages. It includes convergent evolution, parallel evolution, and reversal.
5. For the most part, different characters evolve piecemeal, at different rates. Conservative characters are those that are retained with little or no change over long time periods; other characters may evolve rapidly and vary widely across a single lineage. This phenomenon is called mosaic evolution.
6. Differences among related species illustrate that large differences often evolve gradually, by small steps. Although this pattern is common, it may not be universal.
7. Changes in structure are often associated with change in a character's function. Different structures may be modified to serve similar functions in different lineages.
8. The evolution of morphological features involves changes in their development. Such evolutionary changes include individualization of repeated structures, alterations in the timing (heterochrony) or site (heterotopy) of developmental events, and increases and decreases in structural complexity. Heterochrony can result in changes in the shape of features, often because of allometric growth.
9. Trends in the evolution of a character may be documented by phylogenetic analysis. Evolutionary trends may occur within a single lineage or repeatedly among different lineages.
10. In an adaptive radiation, numerous related lineages arise in a relatively short time and evolve in many different directions as they adapt to different habitats or ways of life. Radiation is perhaps the most common pattern of long-term evolution.
11. Phylogenetic and comparative studies also reveal many aspects of the evolution of genes and genomes, such as changes in the amount of DNA, the number of genes, and the origin of new genes by gene duplication.

Terms and Concepts

adaptive radiation
allometry (allometric growth)
anagenesis
clade
cladistics
cladogenesis
cladogram
conservative characters
convergent evolution (convergence)
crown group
evolutionary reversal (reversal)
evolutionary trend
gene duplication
heterochrony
heterotopy
homology (homologous characters)
homoplasy
individualization
monophyletic
mosaic evolution
neoteny
orthology
paedomorphosis
parallel evolution (parallelism)
paralogy
paraphyletic
patterns of evolution
peramorphosis
polyphyletic
polyploidy
progenesis
stem group

Suggestions for Further Reading

Most books and other sources cover only some of the many subjects discussed in this chapter.

A broad perspective on the uses of phylogenies in studying evolution and ecology is provided by D. R. Brooks and D. A. McLennan in *The Nature of Diversity: An Evolutionary Voyage of Discovery* (University of Chicago Press, 2002) and, briefly, by D. J. Futuyma, "The fruit of the tree of life: Insights into evolution and ecology" [(pp. 25–39 in *Assembling the Tree of Life*, J. Cracraft and M. J. Donoghue (eds.), Oxford University Press, New York, 2004)].

The early history of interpretation of morphological evolution in a developmental context was reviewed by Stephen Jay Gould in *Ontogeny and Phylogeny* (Harvard University Press, Cambridge, MA, 1977). More contemporary perspectives are provided by R. A. Raff in *The Shape of Life: Genes, Development, and the Evolution of Animal Form* (University of Chicago Press, 1996). G. P. Wagner addressed the nature of homology in "The biological homology concept" (*Annual Review of Ecology and Systematics* 20: 51–69, 1989).

Many aspects of the evolution of genes and genomes are treated from a phylogenetic perspective by W.-H. Li in *Molecular Evolution* (Sinauer, Sunderland, MA, 1997). S. B. Carroll treats many of this chapter's topics in *The Making of the Fittest: DNA and the Ultimate Forensic Record of Evolution* (W. W. Norton, New York, 2006).

Problems and Discussion Topics

1. In Problem 1 in Chapter 2, you were asked to estimate the phylogeny of species 1, 2, and 3, based on the following DNA sequences:

   (1) GCTGATGAGT; (2) ATCAATGAGT; (3) GTTGCAACGT; (4) GTCAATGACA

   Species 4 is an outgroup. Suppose the species are primates that differ in mating system: species 1 and 2 are monogamous, and species 3 and 4 are polygamous. We also happen to know that another polygamous species, 5, is more distantly related to species 1–4 than those species are to one another. Given your best estimate of the phylogenetic history, what has been the probable history of evolution of the mating system?

2. In the absence of a fossil record, how might phylogenetic analysis tell us whether the rate of diversification (increase in number of species) has differed among evolutionary lineages?

3. The antlers of the extinct Irish elk (Figure 3.19) were so large that some paleontologists proposed that the species became extinct because the antlers simply became too unwieldy and prevented the elk from escaping predators. Other biologists wondered why such big antlers evolved; some suggested that males with larger antlers could win fights with other males and gain greater access to mates. Stephen Jay Gould (1974) calculated that, given the allometric relationship between antler size and body size among species of deer, the Irish elk's antlers were exactly as large as we would expect in such a large deer. If antler size is a result of a growth correlation with body size, is it necessary to postulate an adaptive advantage for the Irish elk's large antlers in order to account for their great size? And would such a correlation imply that natural selection could not prevent the antlers from becoming so large that they caused the species' extinction?

4. The previous question illustrates questions about evolutionary processes that arise when patterns of evolution are described. Specify some questions about evolutionary processes that might be raised by each of the following patterns: (a) modification of different structures to achieve similar adaptive functions (e.g., Figure 3.13); (b) evolutionary trends (e.g., Figure 3.21); (c) evolutionary reversal (e.g., salamander life history, Figure 3.9); and (d) reduction or loss of structures or genes during evolution (e.g., Figure 3.21).

5. The first analyses of the human genome offered many insights based on an evolutionary, phylogenetic perspective. After reading these analyses (International Human Genome Sequencing Consortium 2001; Venter et al. 2001) or associated commentaries, describe five questions about or insights into the human genome that could be addressed with the help of phylogenetic analyses.

6. Early in this chapter, the claim was made that phylogenetic information is the basis for describing patterns of evolution, yet some examples of such patterns were presented without phylogenetic trees. Consider the following examples and discuss what phylogenetic evidence or inference was left unstated: (a) The fusion of leg bones during embryonic development of birds is a derived trait, not an ancestral trait, relative to the unfused condition in crocodiles. (b) Pentadactyly (five digits) is homologous in humans and crocodiles. (c) The sting of a wasp is a derived from an ovipositor, but modified in both structure and function. (d) Teeth became indi-

vidualized in the evolution of mammals, but this pattern was reversed in the whales. (e) Paedomorphosis is a derived (not ancestral) condition in salamanders.

7. The C-value paradox was a paradox only because many biologists assumed that amphibians are more complex than fishes, "reptiles" more complex than amphibians, and mammals more complex than "reptiles." Might the "paradox" disappear if this assumption is false? How might you objectively determine whether or not this assumption is valid?
8. Why is it difficult to tell if the variation in the number of sets of Hox genes among vertebrates is a result of repeated duplication of the entire genome or only parts of the genome? What processes can you imagine that would make this distinction difficult? What kinds of data might help to determine which has occurred?
9. Why would it be surprising if Dollo's law were violated, and complex characteristics that have been lost during evolution (such as birds' teeth) were regained much later?

CHAPTER 4

Evolution in the Fossil Record

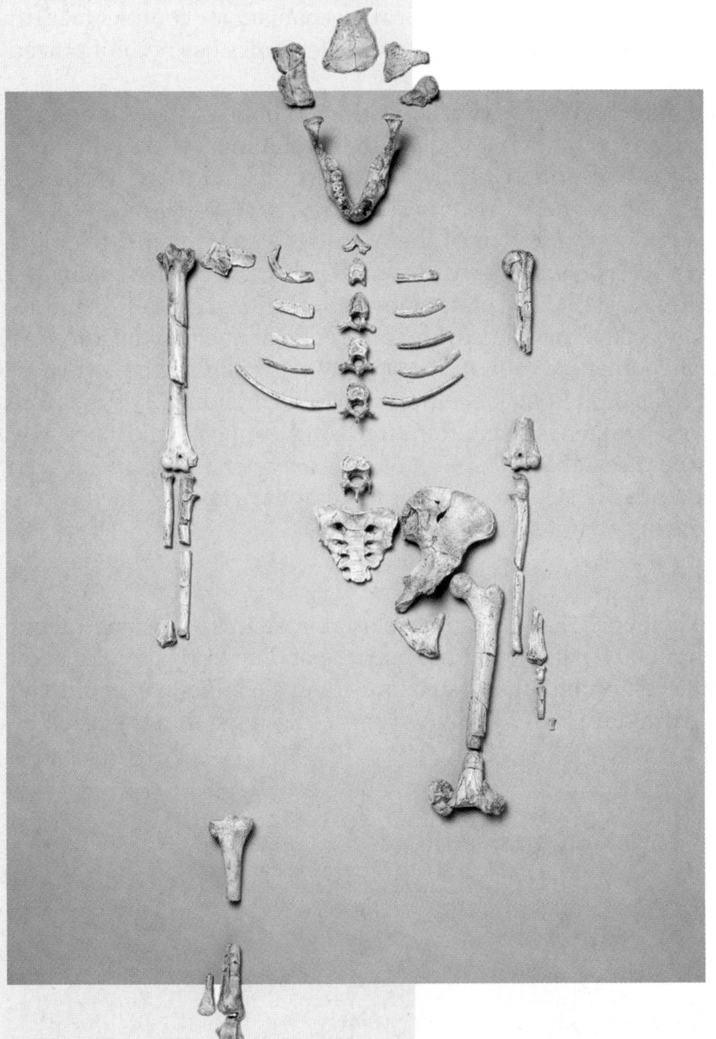

Skeletal remains of the Pliocene hominin *Australopithecus afarensis*. This famous specimen, nicknamed "Lucy," is unusually complete. Studying fossils such as Lucy has enabled scientists to document the evolutionary history of the human lineage. (Photo © The Natural History Museum, London.)

Although some of the history of evolution can be inferred from living organisms, the most direct evidence of that history is found in the fossil record. Fossils tell us of the existence of innumerable creatures that have left no living descendants; of great episodes of extinction and diversification; and of the movements of continents and organisms that explain their present distributions. Only from this record can we obtain an absolute time scale for evolutionary events, as well as evidence of the environmental conditions in which they transpired. The fossil record also provides many details of which we would otherwise have no knowledge; for instance, although we can infer some of the evolutionary changes in the human lineage from comparisons with other primates, research on fossils verifies some such inferences and reveals information—such as the sequence of different anatomical changes—that comparisons with other living species cannot provide.

The fossil record provides evidence on two particularly important themes: phenotypic transformations in particular lineages, and changes in biological diversity over time. The first of these themes is the chief subject of this chapter.

Some Geological Fundamentals

Rocks at the Earth's surface originate as molten material (magma) that is extruded from deep within the Earth. Some of this extrusion occurs via volcanoes, but much rock originates as new crust forms at mid-oceanic ridges (see Figure 4.1). Rock so formed is called IGNEOUS (Latin, "from fire") rock. SEDIMENTARY (Latin, "settle," "sink") rock is formed by the deposition and solidification of sediments, which are usually formed either by the breakdown of older rocks or by precipitation of minerals from water. Under high temperature and pressure, both igneous and sedimentary rocks are altered, forming METAMORPHIC (Greek, "changed form") rocks.

Most fossils are found in sedimentary rocks; they are never found in igneous rocks, and they are usually altered beyond recognition in metamorphic rocks. A few fossils are found in other situations; for example, insects are found in amber (fossilized plant resin), and some mammoths and other species have been found frozen in permafrost.

Plate tectonics

Alfred Wegener first broached the idea of continental drift in 1915, but it was not until the 1960s that both definitive evidence and a theoretical mechanism for continental drift convinced most geologists of its reality. The theory of **plate tectonics** has revolutionized geology.

The lithosphere, the solid outer layer of the Earth bearing both the continents and the crust below the oceans, consists of eight major and a number of minor PLATES that move over the denser, more plastic asthenosphere below. The heat of the Earth's core sets up convection cells within the asthenosphere. At certain regions, such as the mid-oceanic ridge that runs longitudinally down the floor of the Atlantic Ocean, magma from the asthenosphere rises to the surface, cools, and spreads out to form new crust, pushing the existing plates to either side (Figure 4.1). The plates move at velocities of 5 to 10 centimeters per year. Where two plates come together, the leading edge of one may be forced to plunge under the other (a phenomenon known as SUBDUCTION), rejoining the asthenosphere. The pressure of these collisions is a major cause of mountain building. When a plate moves over a "hot spot" where magma is rising from the asthenosphere, volcanoes may be born, or a continent may be rifted apart. The Great Lakes of eastern Africa lie in such a rift valley; the Hawaiian Islands are a chain of volcanoes that have been formed by the movement of the Pacific plate over a hot spot (see Chapter 6).

Geological time

Astronomers have amassed evidence that the universe originated in a "big bang" about 14 billion years ago (1 billion = 1,000 million) and has been expanding from a central point since then. The Earth and the rest of the solar system are about 4.6 billion years old, but the oldest known rocks on Earth are about 3.8 billion years old. Living things existed by

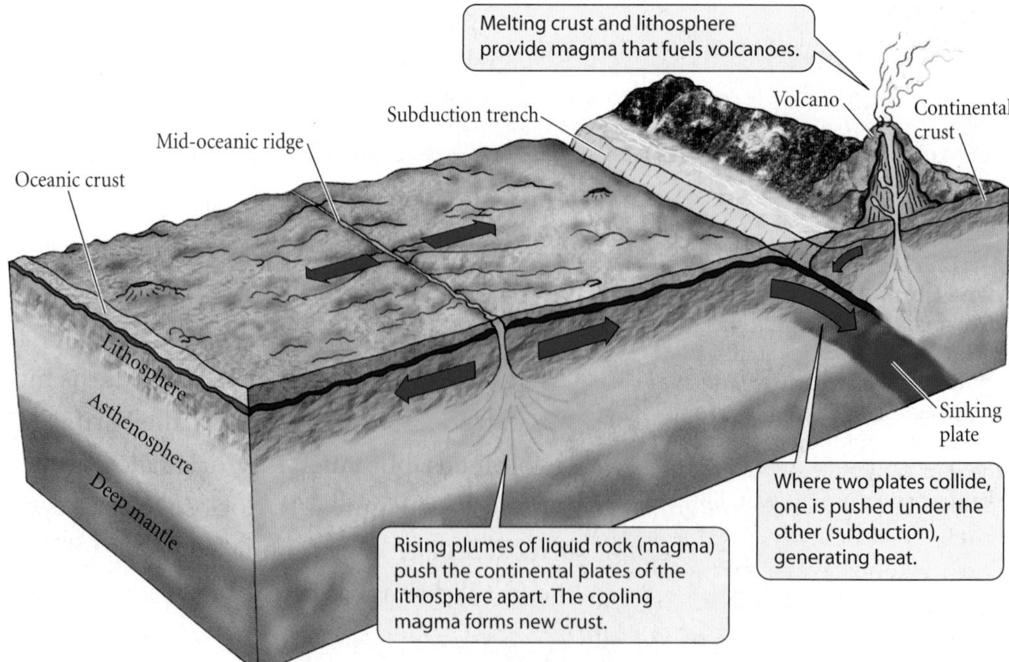

Figure 4.1 Plate tectonic processes. At a mid-oceanic ridge, rising magma creates new lithosphere and pushes the existing plates to either side. When moving lithospheric plates meet, one plunges under the other at a subduction trench, frequently causing earthquakes and mountain building. Heat generated by subduction melts the lithosphere, causing volcanic activity.

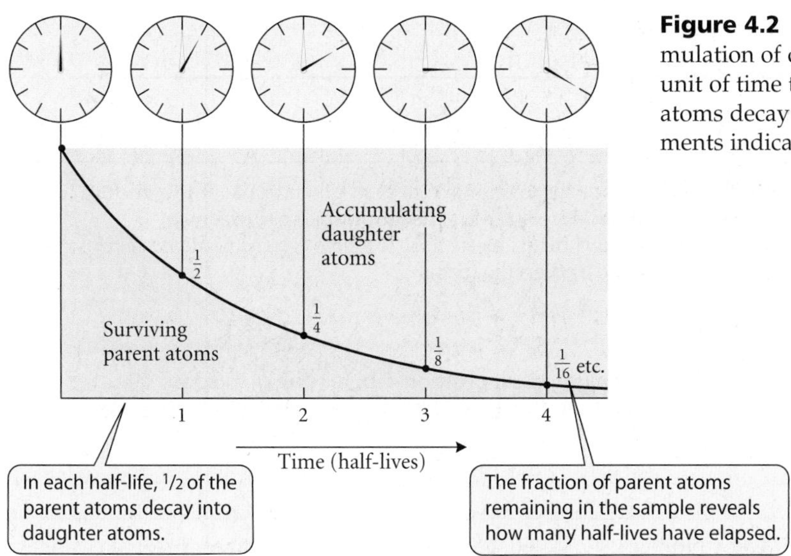

Figure 4.2 Radiometric dating. The loss of parent atoms and the accumulation of daughter atoms occurs at a constant rate. In each half-life—a unit of time that is specific to each element—half the remaining parent atoms decay into daughter atoms. The relative numbers of the two elements indicate how many half-lives have elapsed. (After Eicher 1976.)

about 3.5 billion years ago; the first fossil evidence of animal life is about 800 million (0.8 billion) years old (see Chapter 5).

Such time spans are hard for us to comprehend. As an analogy, if Earth's age is represented by one year, the first life appears in late March, the first marine animals make their debut in late October, the dinosaurs become extinct and the mammals begin to diversify on December 26, the human and chimpanzee lineages diverge at about 13 hours before midnight on December 31, and the Common Era, dating from the birth of Christ, is about 13 seconds before midnight.

"Absolute" ages of geological events can often be determined by **radiometric dating**, which measures the decay of certain radioactive elements in minerals that form in igneous rock. The decay of radioactive parent atoms into stable daughter atoms (e.g., uranium-235 to lead-207) occurs at a constant rate, such that each element has a specific half-life. The half-life of U-235, for example, is about 0.7 billion years, meaning that in each 0.7 billion-year period, half the U-235 atoms present at the beginning of the period will decay into Pb-207. The ratio of parent to daughter atoms in a rock sample thus provides an estimate of the rock's age (Figure 4.2). Only igneous rocks can be dated radiometrically, so a fossil-bearing sedimentary rock must be dated by bracketing it between younger and older igneous formations.

Long before radioactivity was discovered—indeed, before Darwin's time—geologists had established the relative ages (i.e., earlier vs. later) of sedimentary rock formations, based on the principle that younger sediments are deposited on top of older ones. Layers of sediment deposited at different times are called **strata**. Different strata have different characteristics, and they often contain distinctive fossils of species that persisted for a short time and are thus the signatures of the age in which they lived. Using such evidence, geologists can match contemporaneous strata in different localities. In many locations, sediment deposition has not been continuous, and sedimentary rocks have eroded; thus any one area usually has a very intermittent geological record, and some time intervals are well represented at only a few localities on Earth. In general, the older the geological age, the less well it is represented in the fossil record, because erosion and metamorphism have had more opportunity to take their toll.

The geological time scale

Most of the eras and periods of the **geological time scale** (Table 4.1) were named and ordered before Darwin's time, by geologists who did not entertain the idea of evolution. These geological eras and periods were distinguished, and are still most readily recognized in practice, by distinctive fossil taxa. Great changes in faunal composition—the result of mass extinction events—mark many of the boundaries between them. The

TABLE 4.1 *The geological time scale*

Era	*Period (abbreviation)*	*Epoch*	*Millions of years from start to present*	*Major events*
CENOZOIC	Quaternary (Q)	Recent (Holocene)	0.01	Continents in modern positions; repeated glaciations and lowering of sea level; shifts of geographic distributions; extinctions of large mammals and birds; evolution of *Homo erectus* to *Homo sapiens*; rise of agriculture and civilizations
		Pleistocene	1.8	
	Tertiary (T)	Pliocene	5.3	Continents nearing modern positions; increasingly cool, dry climate; radiation of mammals, birds, snakes, angiosperms, pollinating insects, teleost fishes
		Miocene	23.0	
		Oligocene	33.9	
		Eocene	55.8	
		Paleocene	65.5	
MESOZOIC	Cretaceous (K)		145	Most continents separated; continued radiation of dinosaurs; increasing diversity of angiosperms, mammals, birds; mass extinction at end of period, including last ammonoids and dinosaurs
	Jurassic (J)		200	Continents separating; diverse dinosaurs and other "reptiles"; first birds; archaic mammals; "gymnosperms" dominant; evolution of angiosperms; ammonoid radiation; "Mesozoic marine revolution"
	Triassic (Tr)		251	Continents begin to separate; marine diversity increases; "gymnosperms" become dominant; diversification of "reptiles," including first dinosaurs; first mammals
PALEOZOIC	Permian (P)		299	Continents aggregated into Pangaea; glaciations; low sea level; increasing "advanced" fishes; diverse orders of insects; amphibians decline; "reptiles," including mammal-like forms, diversify; major mass extinctions, especially of marine life, at end of period
	Carboniferous (C)		359	Gondwana and small northern continents form; extensive forests of early vascular plants, especially lycopsids, sphenopsids, ferns; early orders of winged insects; diverse amphibians; first "reptiles"
	Devonian (D)		416	Diversification of bony fishes; trilobites diverse; origin of ammonoids, amphibians, insects, ferns, seed plants; mass extinction late in period
	Silurian (S)		444	Diversification of agnathans; origin of jawed fishes (acanthodians, placoderms, Osteichthyes); earliest terrestrial vascular plants, arthropods
	Ordovician (O)		488	Diversification of echinoderms, other invertebrate phyla, agnathan vertebrates; mass extinction at end of period
	Cambrian (€)		542	Marine animals diversify: first appearance of most animal phyla and many classes within relatively short interval; earliest agnathan vertebrates; diverse algae
PROTEROZOIC			2500	Earliest eukaryotes (ca. 1900–1700 Mya); origin of eukaryotic kingdoms; trace fossils of animals (ca. 1000 Mya); multicellular animals from ca. 640 Mya, including possible Cnidaria, Annelida, Arthropoda
ARCHEAN			Lower limit not defined	Origin of life in remote past; first fossil evidence at ca. 3500 Mya; diversification of prokaryotes (bacteria); photosynthesis generates oxygen, replacing earlier oxygen-poor atmosphere; evolution of aerobic respiration

Source: Gradstein, et al. 2004.

absolute time of these boundaries is only approximate, and is subject to slight revision as more information accumulates. Phanerozoic time (marked by the first appearance of diverse animals) is divided into three **eras**, each of which is divided into **periods**. We will frequently refer to these divisions, and to the **epochs** into which the Cenozoic periods are divided. *Every student of evolution should memorize the sequence of the eras and periods, as well as a few key dates*, such as the beginning of the Paleozoic era (and the Cambrian period, 542 million years ago, or 542 Mya), the Mesozoic era (and Triassic period, 251 Mya), the Cenozoic era (and Tertiary period, 65.5 Mya), and the Pleistocene epoch (1.8 Mya).

The Fossil Record

Some short parts of the fossil record, in certain localities, provide detailed evolutionary histories, and some groups of organisms, such as abundant planktonic protists with hard shells, have left an exceptionally good record. In some respects, such as the temporal distribution of many higher taxa (e.g., phyla and classes), the fossil record is adequate to provide a reasonably good portrait (Benton et al. 2000). In other respects, the fossil record is extremely incomplete (Jablonski et al. 1986). Consequently, the origins of many taxa have not been well documented. We *know* that the fossil record is very incomplete because continuing exploration constantly yields new discoveries; for instance, a huge new dinosaur, perhaps the most massive terrestrial animal in Europe and one of the largest animals known, was discovered in Spain in 2006 (Royo-Torres et al. 2006). Most of the discoveries that have documented the origin of birds from dinosaurs have been made in Chinese deposits within the last few years.

The incompleteness of the fossil records has several causes. First, many kinds of organisms rarely become fossilized because they are delicate, or lack hard parts, or occupy environments—such as humid forests—where decay is rapid. Second, because sediments generally form in any given locality very episodically, they typically contain only a small fraction of the species that inhabited the region over time. Third, if fossils are to be found, the fossil-bearing sediments must first become solidified into rock; the rock must persist for millions of years without being eroded, metamorphosed, or subducted; and the rock must then be exposed and accessible to paleontologists. Finally, the evolutionary changes of interest may not have occurred at the few localities that have strata from the right time; a species that evolved new characteristics elsewhere may appear in a local record fully formed, after having migrated into the area. Paleontologists agree that the approximately 250,000 described fossil species represent much less than 1 percent of the species that lived in the past.

Two of the most basic questions we may ask of the fossil record are, first, whether it provides evidence for evolutionary change—for descent with modification—and second, whether or not evolution is gradual, as Darwin proposed. The answer to the first question is an unequivocal "yes," as the following examples show. Regarding the second question, the fossil record provides many examples of gradual change, but we must admit the possibility that some characteristics have evolved by large, discontinuous changes.

Evolutionary changes within species

A detailed, continuous sedimentary record can be expected only over short geological time spans. Some such records have preserved histories of change within single evolving lineages—single species. Change in a characteristic within a single, unbranching lineage is often called **anagenesis**. In many such cases, characters tend to change gradually (Levinton 2001). Such changes have been particularly well documented in planktonic foraminiferans, abundant unicellular protists with shells of calcium carbonate that drop to the ocean floor in huge numbers. For example, Michal Kucera and Björn Malmgren (1998) found that in *Contusotruncana*, the shell became increasingly cone-shaped over the course of about 3 million years (Figure 4.3).

In another detailed study, Michael Bell and his colleagues (1985) studied Miocene fossils of a stickleback fish, *Gasterosteus doryssus*, in strata that were laid down annually for 110,000 years. They took samples from layers that were on average about 5000 years

Figure 4.3 Gradual evolution of shell shape in the foraminiferan *Contusotruncana*. (A) Changes in mean conicity, from older (at bottom) to younger (at top) samples over the course of 3.5 Myr. The circles represent the mean of a sample from a specific time interval, and the horizontal bars indicate variation among individuals within samples. Red and blue circles represent localities in the North and South Atlantic seafloor, respectively. Note that the change (anagenetic trend) began after more than 2.5 Myr of constant shape (stasis). The ancestral, flatter form is named *C. fornicata*, and the later, more conical phenotype is named *C. contusa*. (B) Variation among individual specimens at each time interval. Note the increase in variation (left/right axis) after the anagenetic trend began. (After Kucera and Malmgren 1998.)

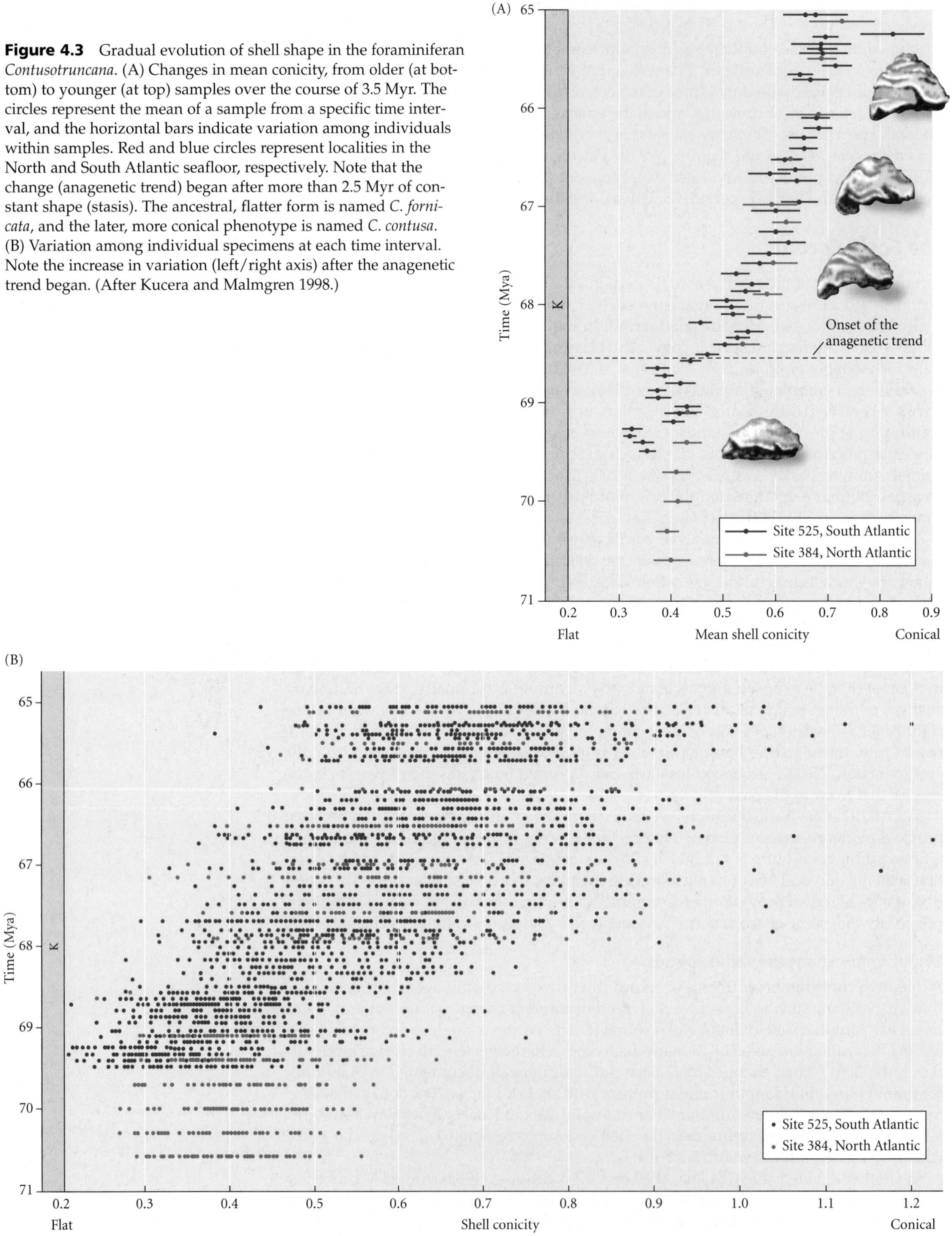

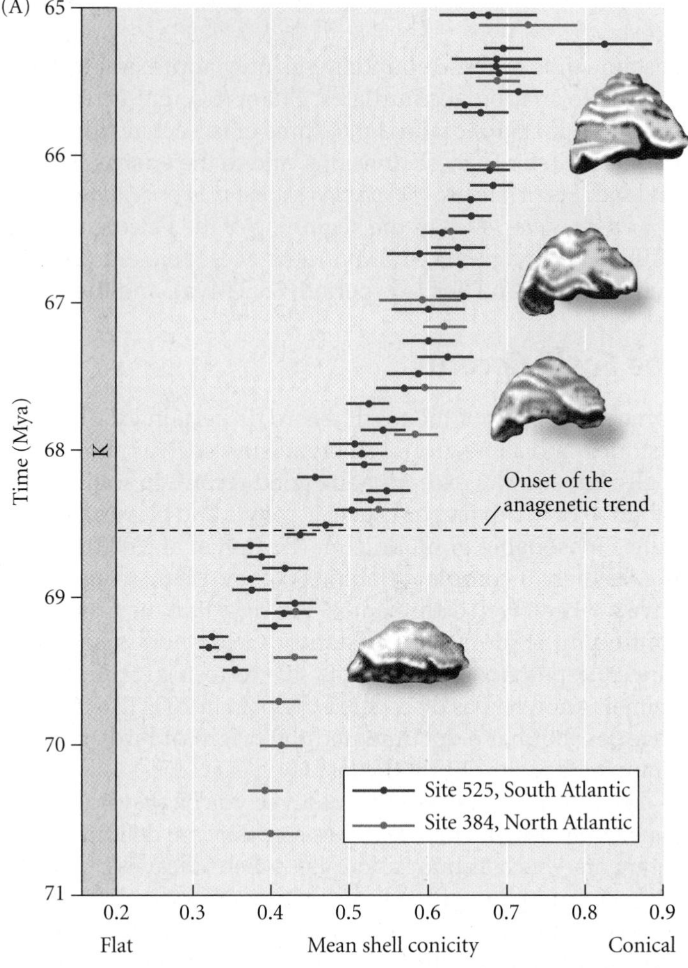

(B)

65
66
67
68
69
70
71
Time (Mya)
K
• Site 525, South Atlantic
• Site 384, North Atlantic
0.2 0.3 0.4 0.5 0.6 0.7 0.8 0.9 1.0 1.1 1.2
Flat
Shell conicity
Conical

(A)

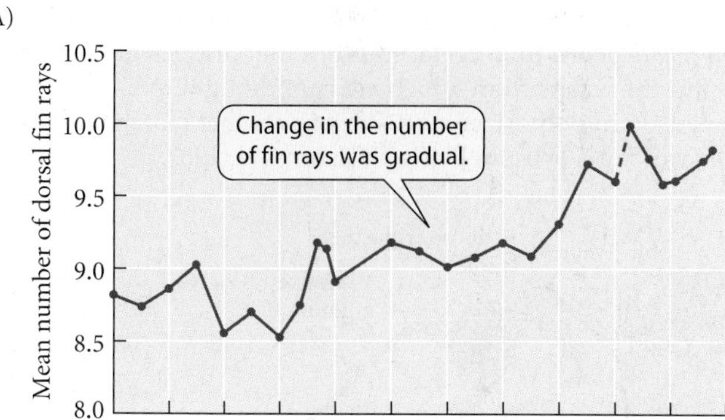

(B)

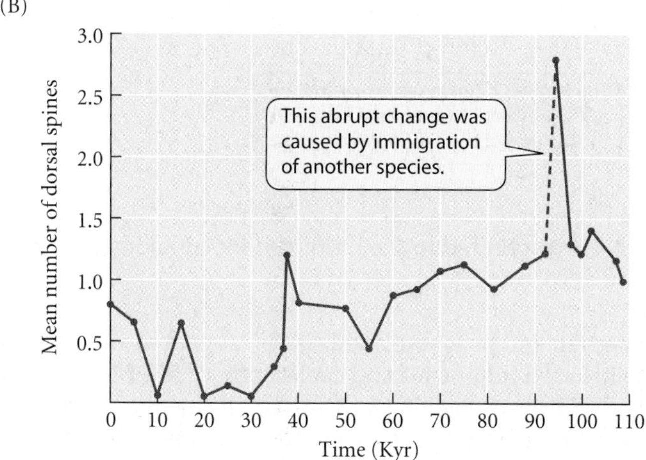

Figure 4.4 Changes in the mean values of characters in fossil sticklebacks, *Gasterosteus doryssus* (photo below). In a study of five characters, three characters, including the number of dorsal fin rays (A), showed independent, gradual change; dorsal spine number (B) was one of two characters displaying abrupt changes in value. Values are given for intervals of 5000 years, over a span of 110,000 years. (After Bell et al. 1985; photo courtesy of M. Bell.)

Gasterosteus doryssus

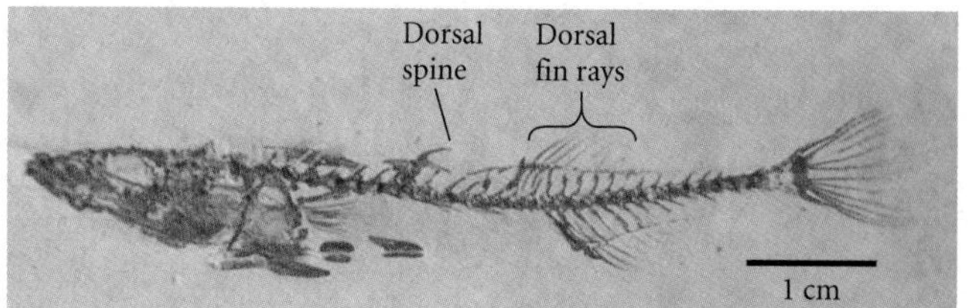

apart. Three of the characters they studied changed more or less independently and gradually (Figure 4.4A). Two of these features also changed abruptly at one point in time. This shift was caused by the extinction of the local population and immigration of another population—apparently a different species—in which these features were different (Figure 4.4B). An interval of 5000 years between samples is a long time by normal human standards, but is unusually fine-scaled by the standards of geological time. The gradual evolution that Bell and colleagues documented would not have been evident if the samples had been spaced hundreds of thousands of years apart, as is more typical of paleontological data.

Origins of higher taxa

In this section, we present several examples of macroevolutionary change—the origin of higher taxa over long periods of geological time. These examples illustrate several important principles, such as homology, change of structure with function, and mosaic evolution, that were introduced in Chapter 2. Moreover, they give the lie to claims by antievolutionists that the fossil record fails to document macroevolution. Anyone educated in biology should be able to counter such charges with examples such as these.

Recall from Chapter 3 that it is often possible to infer the ancestral characteristics of a taxon from phylogenetic comparisons among living species; we are quite sure, for example, that humans descended from primates that had a tail and opposable toes (Figure 3.3). Consequently, we can often recognize a fossil as a transitional stage in the origin of a taxon because even before it is found, some of its critical features are predictable. For instance, E. O. Wilson, F. M. Carpenter, and W. L. Brown hypothesized what features the ancestor of ants should have had, based on comparisons between primitive species of living ants and related families of wasps. Several years after Wilson and colleagues published their hypothesis, Cretaceous ants preserved in amber—older than any previously known

(A) *Sphecomyrma freyi*

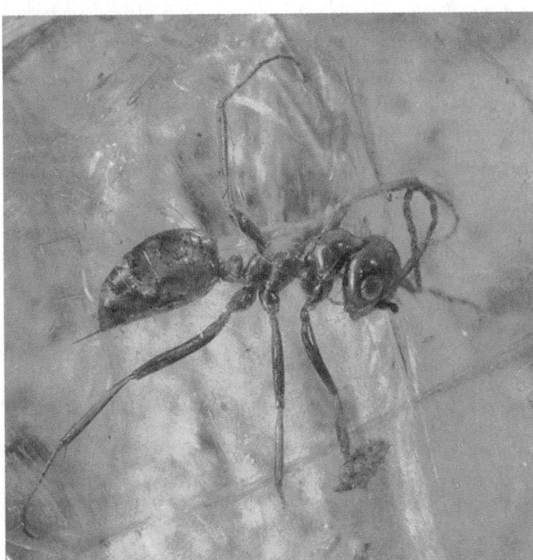

Figure 4.5 A fossil can help confirm an evolutionary hypothesis. (A) Fossilized in amber, this *Sphecomyrma freyi* is a mid-Cretaceous ant that bridges the gap between modern ants and the wasps from which ants are thought to have arisen. (B) Features of the previously hypothesized ancestor of ants that matched those of *Sphecomyrma*. (A courtesy of E. O. Wilson; B after Wilson et al. 1967.)

(B) Hypothesized ancestor

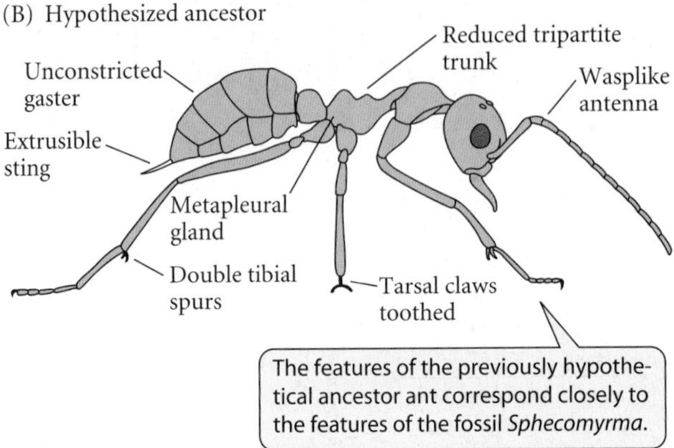

fossil ants—were found, and they corresponded to the predicted morphology in almost all their features (Figure 4.5).

THE ORIGIN OF TETRAPODS. Sarcopterygii, or lobe-finned fishes, appeared in the early Devonian, about 408 Mya. They included lungfishes and coelacanths, a few of which are still alive, as well as extinct groups such as osteolepiforms that had distinctive tooth structure and skull bones (Figure 4.6A). Osteolepiforms had a tail fin and fleshy paired fins, with a central axis of several large bones to which lateral bones and slender, jointed rays (radials) articulated. The head could not be flexed relative to the body. The first definitive tetrapods, such as *Ichthyostega* from the very late Devonian of Greenland, had the same tail fin and distinctive teeth and skull, but the gill cover bones at the rear of the skull had been lost and the head could now be moved on a more flexible neck. Most importantly, they had larger pectoral and pelvic girdles and fully developed tetrapod limbs that bore more than five digits (unlike almost all later tetrapod vertebrates; Clack 2002).

Clearly, ichthyostegids show a mosaic of sarcopterygian and tetrapod features. Until recently, only a few fossils, such as *Panderichthys*, provided evidence of intermediate steps in the transition from fin to limb (see Figure 4.6A). Hoping to fill in more of this evolutionary sequence, Neil Shubin and colleagues explored likely Devonian deposits in northern Canada, and found just what they sought: a rich fossil deposit of a new "tetrapodomorph" sarcopterygian that they named *Tiktaalik roseae* (Daeschler et al. 2006; Shubin et al. 2006). Like ichthyostegids, *Tiktaalik* had a flat head and elongate snout, and lacked gill cover bones so that the head was mobile; it had overlapping ribs, providing support for the body that a partially terrestrial animal requires (Figure 4.6B). Most importantly, the pectoral (shoulder) girdle and fins of *Tiktaalik* are intermediate between the sarcopterygian and tetrapod conditions. The humerus, ulna, radius, and wrist bones are clearly homologous to those of early tetrapods and reveal a critical feature: the limb could be flexed at the elbow and wrist (Figure 4.6C). All the anatomical details of girdle, limbs, and ribs show that *Tiktaalik* could hold its body off the ground—it could do push-ups. Features of the skull suggest that its mode of breathing was intermediate between that of lungfishes and terrestrial tetrapods. *Tiktaalik* probably crawled on the water bottom and at the water's edge; it is a transitional form that makes the distinction between fish and the earliest tetrapods difficult to draw.

THE ORIGIN OF BIRDS. In recent years, it has become clear that birds are dinosaurs. Not long ago, birds (placed until recently in the class Aves) were defined by their feathers. But because of the many extraordinary fossils that have recently been discovered in China,

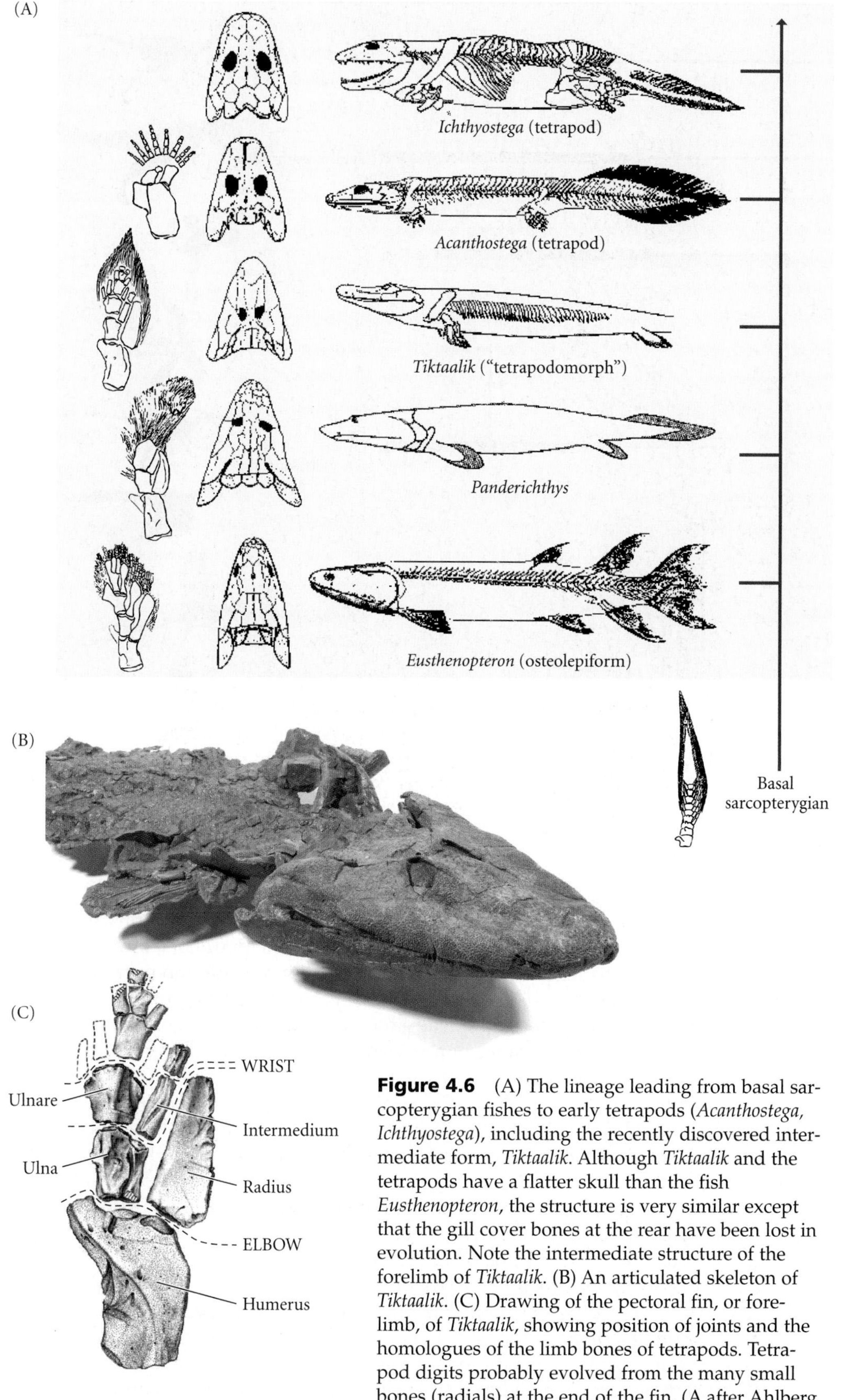

Figure 4.6 (A) The lineage leading from basal sarcopterygian fishes to early tetrapods (*Acanthostega, Ichthyostega*), including the recently discovered intermediate form, *Tiktaalik*. Although *Tiktaalik* and the tetrapods have a flatter skull than the fish *Eusthenopteron*, the structure is very similar except that the gill cover bones at the rear have been lost in evolution. Note the intermediate structure of the forelimb of *Tiktaalik*. (B) An articulated skeleton of *Tiktaalik*. (C) Drawing of the pectoral fin, or forelimb, of *Tiktaalik*, showing position of joints and the homologues of the limb bones of tetrapods. Tetrapod digits probably evolved from the many small bones (radials) at the end of the fin. (A after Ahlberg and Clack 2006 and Shubin et al. 2006; B © Ted Daeschler/VIREO; C after Shubin et al. 2006.)

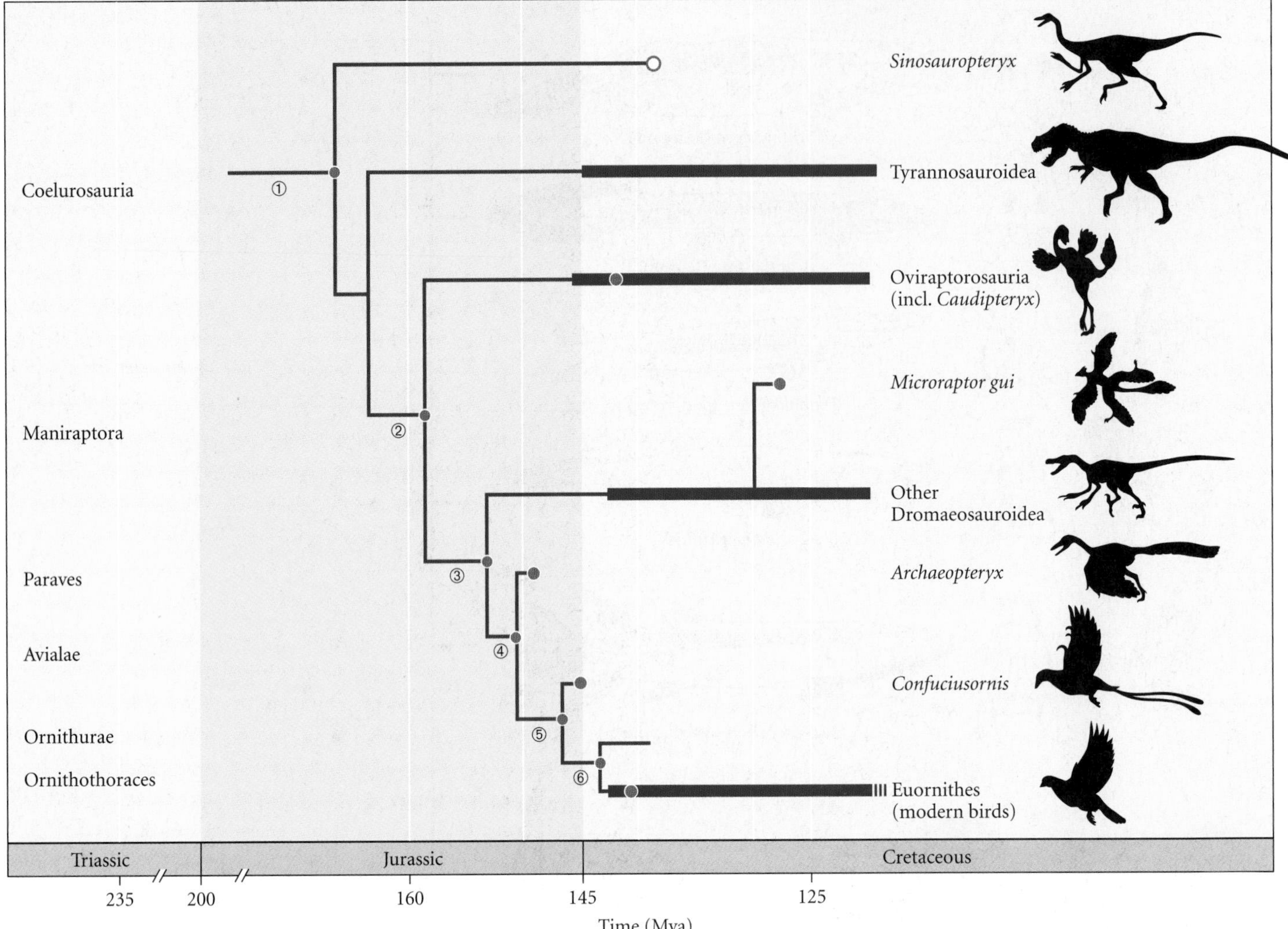

Figure 4.7 A phylogeny of some groups of theropod dinosaurs, showing the hypothesized relationships of *Archaeopteryx* and later birds and the origin of some derived characters (indicated by numbers and names of successively derived clades). Hollow long bones, three-fingered hand, and elevated first toe are features of all theropods, including more basal lineages than shown here. (1) Expanded breastbone, filamentous integumentary feathers in some Coelurosauria; (2) vaned feathers in some Maniraptora; (3) sickle-shaped claw on foot in Paraves, which include the four-winged dromaeosaur *M. gui*; (4) opposable hind (first) toe in Avialae, including *Archaeopteryx*; (5) short tail with fused vertebrae in more "advanced" birds, Ornithurae; (6) breast keel and eventual loss of teeth and finger claws in "true" birds, Euornithes. Heavier lines indicate the existence of an abundant fossil record. (After Sereno 1999; Chiappe and Dyke 2002; Norell and Xu 2005.)

the distinction between birds and dinosaurs has now become arbitrary (Chiappe and Dyke 2002; Xu et al. 2003; Norell and Xu 2005; Chiappe 2007). Birds are clearly theropod dinosaurs, the group that includes *Tyrannosaurus* and members of the Dromaeosauridae (e.g., *Velociraptor* and *Deinonychus*). Dromaeosauridae is the family most closely related to the Avialae—the group that includes birds and *Archaeopteryx*, one of the most famous non-missing links of all time (Figure 4.7).

Archaeopteryx (see Figure 4.9D), discovered in 1860 in Upper Jurassic strata in Germany, has only a few of the many modifications of the skeleton of modern birds that accommodate the flying habit (Figure 4.8A,B). Except for feathers that enabled it to fly, most of its features—such as teeth, fully developed hands with long, clawed fingers, elongate tail, and leg structure—closely resemble those of other theropod dinosaurs, many of which were similarly small in size (Figure 4.8C). It was to be expected, then, that feathers, which had been considered the defining feature of birds, would be found in some other theropods.

Since 1996, an astonishing variety of feathered dinosaurs have been found in China (Figure 4.9). These include the compsognathid *Sinosauropteryx* and the dromaeosaurid *Sinornithosaurus*, with filamentous feathers coating the body; the oviraptorosaur *Caudipteryx*, with long, broad feathers on the hand and tail; and an extraordinary four-winged dromaeosaurid, *Microraptor gui*, that had flight feathers on all four limbs. The distinctive synapomorphy of *Archaeopteryx* and other Avialae is no longer feathers, but merely their opposable hind toe! Some of the features of

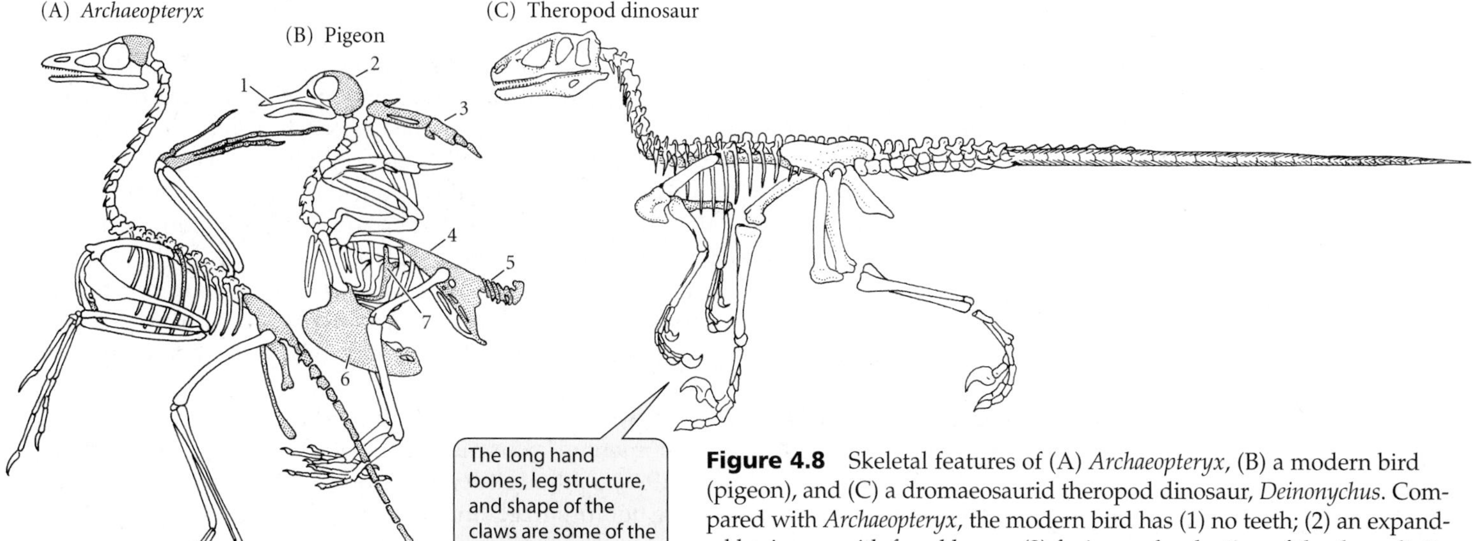

Figure 4.8 Skeletal features of (A) *Archaeopteryx*, (B) a modern bird (pigeon), and (C) a dromaeosaurid theropod dinosaur, *Deinonychus*. Compared with *Archaeopteryx*, the modern bird has (1) no teeth; (2) an expanded braincase with fused bones; (3) fusion and reduction of the three digits of the hand; (4) fusion of the pelvic bones and several vertebrae into a single structure; (5) fewer tail vertebrae, several of which are fused; (6) a greatly enlarged, keeled sternum (breastbone); and (7) horizontal processes that strengthen the rib cage. Theropod dinosaurs share many features with *Archaeopteryx*, the structure of the hand and leg being perhaps most evident in this figure. (A, B after Colbert 1980; C after Ostrom 1976.)

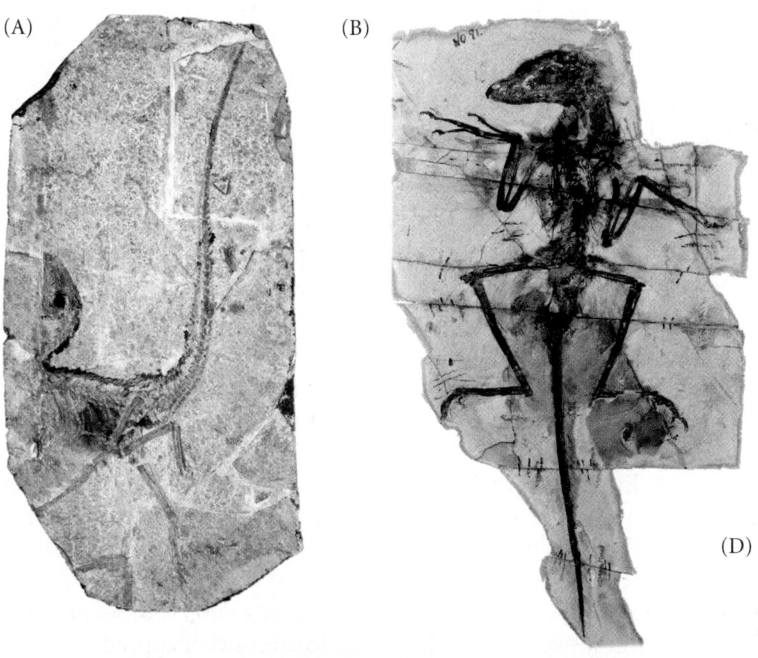

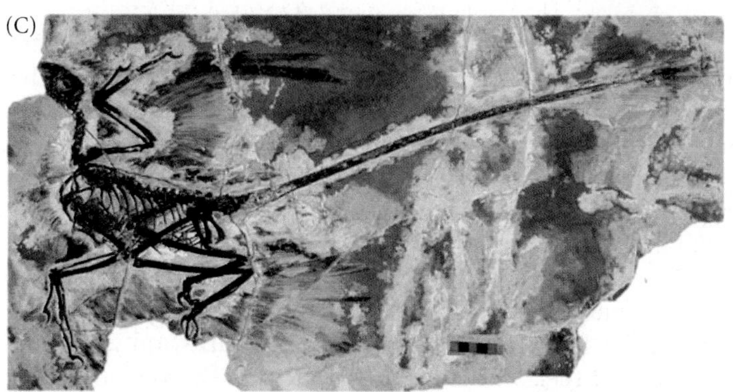

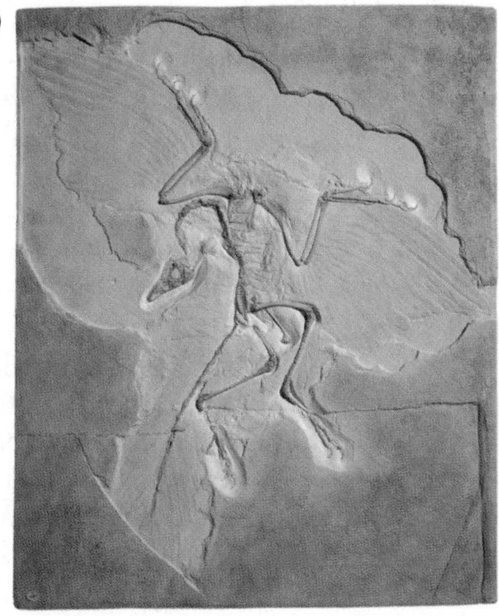

Figure 4.9 Feathered dinosaurs. (A) *Sinosauropteryx*, a compsognathid with filamentous feathers on the body (about 125 Mya). (B) *Sinornithosaurus*, a dromaeosaurid with short, filamentous feathers on the body and longer filaments on the arms, visible as shadowy rays (125 Mya). (C) *Microraptor gui*, a dromaeosaurid with long feathers at the back of both the front and hindlimbs (120–110 Mya). (D) A well-preserved specimen of *Archaeopteryx lithographica*, an early example of a bird, showing the wing feathers and the long tail with feathers on both sides. (A, B © Mick Ellison, American Museum of Natural History; C from Xu et al. 2003; D © The Natural History Museum, London.)

modern birds, such as hollow limb bones, evolved in theropods long before *Archaeopteryx*, and other characters, such as fusion of the tail vertebrae, evolved later (Figure 4.7). Later still, one subgroup of Avialae evolved such features as the keeled breastbone, loss of teeth, and loss of claws on the hand that typify the 10,000 species of living dinosaurs.

We are not sure of the adaptive function of hollow bones in theropods or the body feathers of *Sinosauropteryx*—though perhaps the feathers provided insulation. Certainly, however, these features, in modified form, later became useful in flight. (Hollow bones lighten the load, and some feathers became modified as airfoils.) These features provide an example of evolutionary change of function and of "preadaptation," possession of a feature that fortuitously plays a different, useful role at a later time (see Chapter 11).

THE ORIGIN OF MAMMALS. The origin of mammals from the earliest amniotes (see Figure 2.6) is one of the most fully documented examples of the evolution of a major taxon (Kemp 1982; Sidor and Hopson 1998). Although some features of living mammals, such as hair and mammary glands, are not fossilized, mammals have diagnostic skeletal features, and it is important to learn these in order to understand how mammals evolved. We will describe only changes in the skull and jaw, but the evolution of the postcranial skeleton has also been well described.

The lower jaw consists of several bones in reptiles, but only a single bone (the dentary) in mammals. The primary (and in all except the earliest mammals, the exclusive) jaw articulation is between the dentary and the squamosal bones, rather than between the articular and the quadrate bones, as in other tetrapods. Early amniotes have a single bone (the stapes, or stirrup) that transmits sound, whereas mammals have three such bones (hammer, anvil, stirrup) in the middle ear. Mammals' teeth are differentiated into incisors, canines, and multicusped (multipointed) premolars and molars, whereas most other tetrapods have uniform, single-cusped teeth. Other features that distinguish most mammals from other amniotes include an enlarged braincase, a large space (temporal fenestra) behind the eye socket, and a secondary palate that separates the breathing passage from the mouth cavity.

Soon after the first amniotes originated in the Carboniferous, they gave rise to the Synapsida, a group characterized by an opening (temporal fenestra) behind the eye socket, which probably provided space for enlarged jaw muscles to expand into when contracted (Figure 4.10A). The temporal fenestra became progressively enlarged in later synapsids (Figure 4.10B–D).

Permian synapsids in the order Therapsida (Figure 4.10B) had large canine teeth, and the center of the palate was recessed, suggesting that the breathing passage was partially separated from the mouth cavity. The hind legs were held rather vertically, more like a mammal than a reptile.

Cynodont therapsids, which lived from the late Permian to the late Triassic, represent several steps in the approach toward mammals. The rear of the skull was compressed, giving it a doglike appearance (Figure 4.10C,D), the dentary was enlarged relative to the other bones of the lower jaw, the cheek teeth had a row of several cusps, and a bony shelf formed a secondary palate that was incomplete in some cynodonts and complete in others. The quadrate was smaller and looser than in previous forms and occupied a socket in the squamosal.

In the advanced cynodonts of the middle and late Triassic (Figure 4.10E), the cheek teeth had not only a linear row of cusps, but also a cusp on the inner side of the tooth. This innovation begins a history of complex cusp patterns in the cheek teeth of mammals, which are modified in different lineages for chewing different kinds of food. The cynodont lower jaw had not only the old articular/quadrate articulation with the skull, but also an articulation between the dentary and the squamosal, marking a critical transition between the ancestral condition and the mammalian state.

In *Morganucodon* of the late Triassic and very early Jurassic (Figure 4.10F), the teeth are typical of mammals. *Morganucodon* has both a weak articular/quadrate hinge and a fully developed mammalian articulation between the dentary and the squamosal. The articular and quadrate bones are sunk into the ear region, and together with the stapes, they

(A) Synapsid (*Haptodus*)

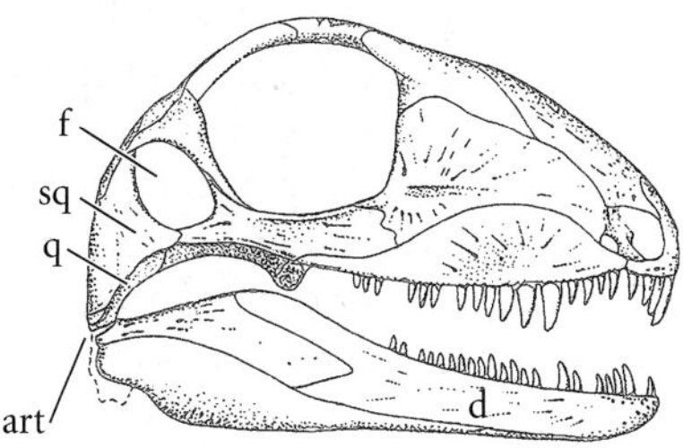

(B) Therapsid (*Biarmosuchus*)

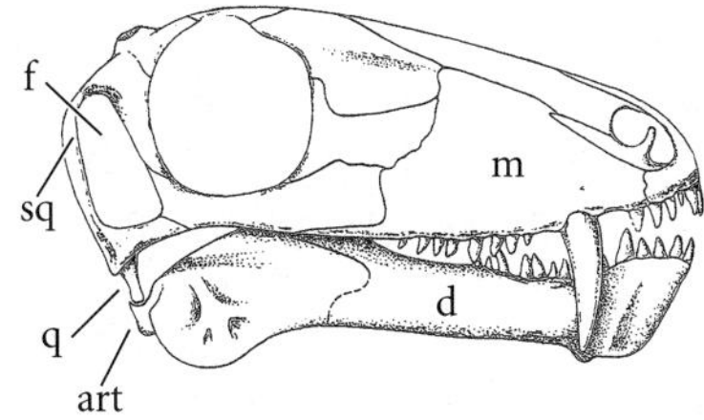

(C) Early cynodont (*Procynosuchus*)

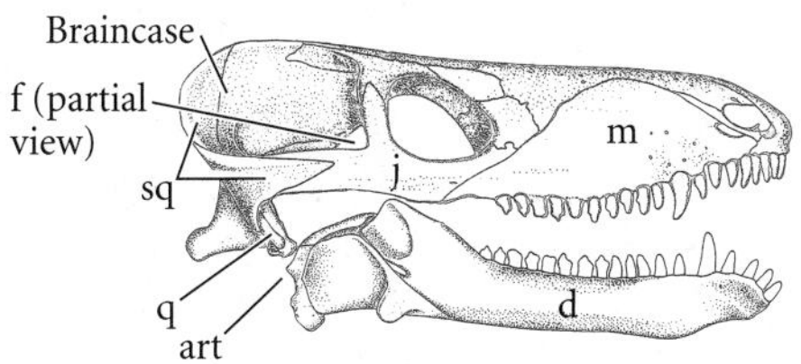

(D) Cynodont (*Thrinaxodon*)

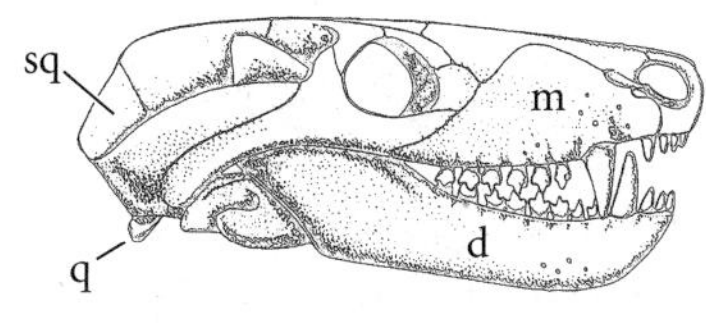

(E) Advanced cynodont (*Probainognathus*)

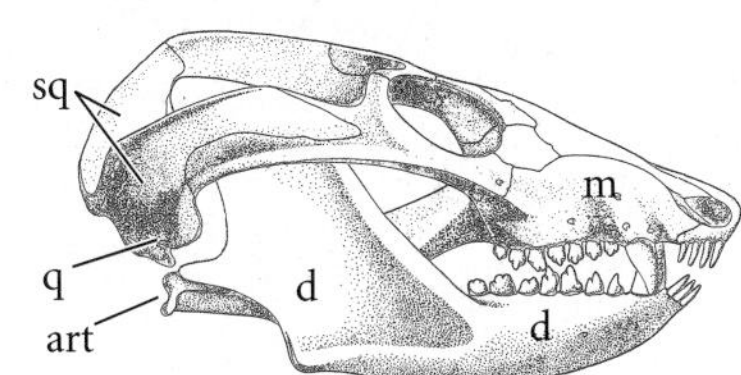

(F) *Morganucodon*

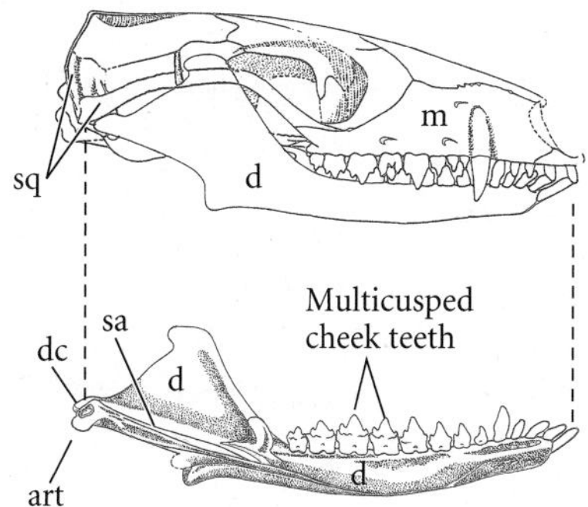

Figure 4.10 Skulls of some stages in evolution from early synapsids to early mammals. (A) A pelycosaur, *Haptodus*. Note temporal fenestra (f), multiple bones in lower jaw, single-cusped teeth, and articular/quadrate (art/q) jaw joint. (B) An early therapsid, *Biarmosuchus*. Note enlarged temporal fenestra. (C) An early cynodont, *Procynosuchus*. The side of the braincase is now vertically oriented, separated by a large temporal fenestra from a lateral arch formed by the jugal (j) and squamosal (sq). Note the enlarged dentary (d). (D) A later cynodont, *Thrinaxodon*. Note multiple cusps on rear teeth, large upper and lower canine teeth, and greatly enlarged dentary with a vertical extension to which powerful jaw muscles were attached. (E) An advanced cynodont, *Probainognathus*. The cheek teeth had multiple cusps, and the lower jaw articulated with the squamosal by the articular (art), and posterior process of the dentary (d). (F) *Morganucodon* was almost a mammal. Note multicusped cheek teeth (including inner cusps) and double articulation of lower jaw, including articulation of a dentary condyle (dc) with the squamosal (sq). (After Futuyma 1995; based on Carroll 1988 and various sources.)

closely approach the condition in modern mammals, in which these bones transmit sound to the inner ear.

Hadrocodium, a recently described fossil from the early Jurassic, carries the trend from *Morganucodon* to the very brink of mammalhood (Luo et al. 2001). This tiny animal is very similar to *Morganucodon*, but the articular and quadrate bones are fully separated from the jaw joint and fully lodged in the middle ear, and the lower jaw consists entirely of the dentary.

This description touches on only the highlights of a complex history. For example, several of the changes in the lower jaw and middle ear occurred independently in different cynodont lineages (Luo 2007). Nonetheless, the fossil record shows that most mammalian characters (e.g., posture, tooth differentiation, skull changes associated with jaw musculature, secondary palate, reduction of the elements that became the small bones of the middle ear) evolved gradually. Evolution was mosaic, with different characters "advancing" at different rates. No new bones evolved; in fact, many bones have been lost in the transition to modern mammals (Sidor 2001), and all the bones that persist are modified from those of the stem amniotes (and in turn, from those of amphibians and even lobe-finned fishes). Some major changes in the form of structures are associated with changes in their function. The most striking example is the articular and quadrate bones, which serve for jaw articulation in all other tetrapods, but became the sound-transmitting middle-ear bones of mammals. Because the evolution of mammals from synapsids, over the course of more than 130 Myr, has been gradual, there is no cutoff point for recognizing mammals: the definition of "Mammalia" is arbitrary.

THE ORIGIN OF CETACEA. Whales and dolphins, traditionally distinguished as the order Cetacea, evolved from terrestrial ancestors. Among living mammals, their closest relatives, based on molecular phylogenetic analyses, are hippopotamuses (Gatesy et al. 1999). Thus cetaceans fall within the order Artiodactyla—even-toed hoofed mammals—along with camels, pigs, and ruminants such as cattle and antelopes. The group as a whole is now often called Cetartiodactyla.

Compared with basal mammals, living cetaceans are greatly modified, owing to their adaptations for aquatic life. All share a uniquely shaped tympanic bone that encloses the ear, one of several modifications for hearing in water; a nasal opening far back on top of the skull; stiff elbow, wrist, and finger joints, all enclosed in a paddlelike flipper; a rudimentary pelvis (sometimes associated with a hindlimb rudiment) that is disconnected from the vertebral column; and a lack of the fused, differentiated sacral (lower back) vertebrae that land mammals have. Toothed whales have a large cavity (foramen) in the lower jaw (mandible) that contains a sound-transmitting pad of fat.

Philip Gingerich, J. G. M. Thewissen, and colleagues have discovered many Eocene fossils, mostly in Pakistan, that document many details of the evolution of cetaceans from about 50 to 35 Mya, including features such as the transition to aquatic locomotion (Figure 4.11) and underwater hearing (Figure 4.12; Gingerich et al. 2001; Thewissen and Bajpai 2001; Thewissen and Williams 2002). The closest known relatives of cetaceans appear to

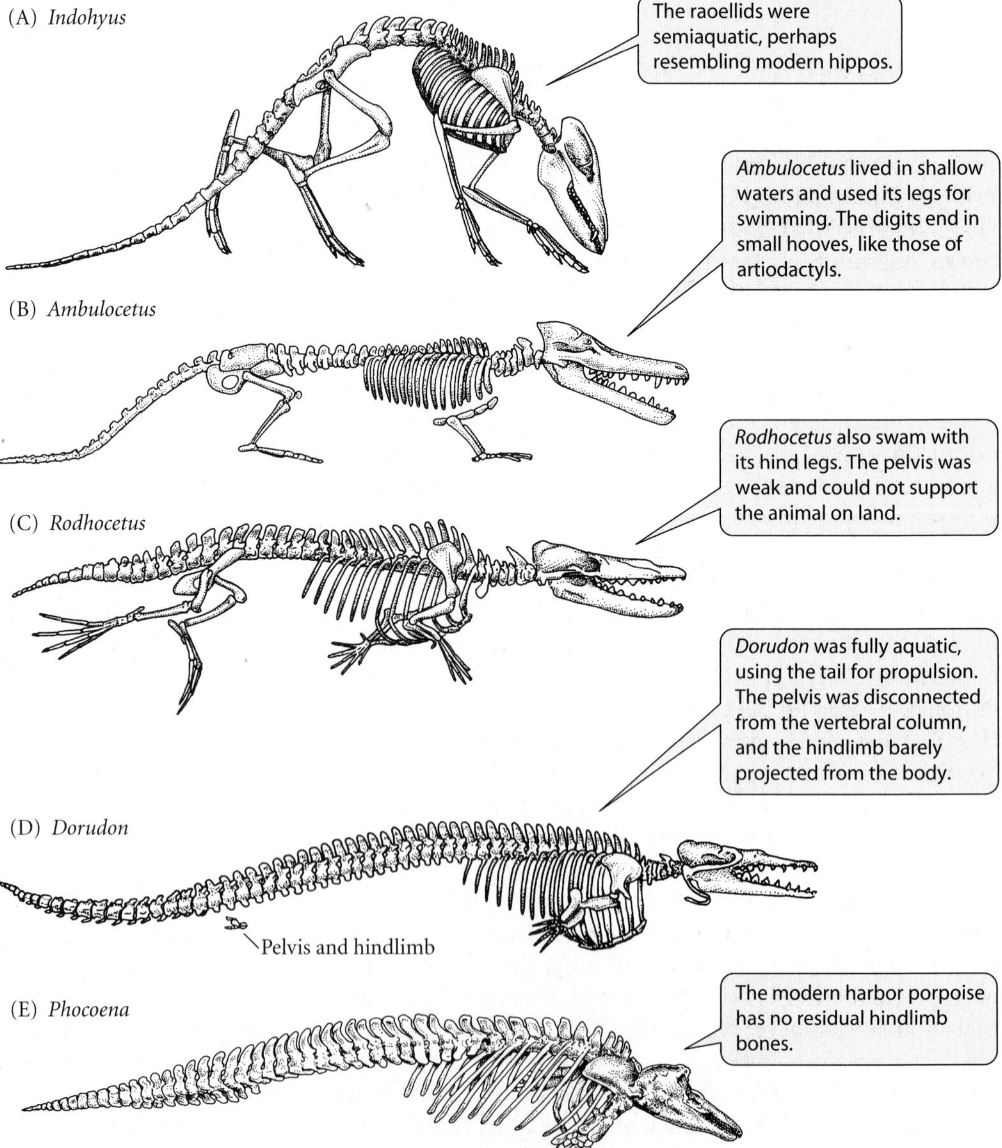
(A) *Indohyus*
The raoellids were semiaquatic, perhaps resembling modern hippos.
Ambulocetus lived in shallow waters and used its legs for swimming. The digits end in small hooves, like those of artiodactyls.
(B) *Ambulocetus*
Rodhocetus also swam with its hind legs. The pelvis was weak and could not support the animal on land.
(C) *Rodhocetus*
Dorudon was fully aquatic, using the tail for propulsion. The pelvis was disconnected from the vertebral column, and the hindlimb barely projected from the body.
(D) *Dorudon*
Pelvis and hindlimb
The modern harbor porpoise has no residual hindlimb bones.
(E) *Phocoena*

Figure 4.11 Reconstruction of stages in the evolution of cetaceans from terrestrial artiodactyl ancestors. (A) Eocene raoellids, perhaps the sister group of the Cetacea, were terrestrial but show some evidence of semiaquatic life. (B) The amphibious *Ambulocetus*. (C) The middle-Eocene protocetid *Rodhocetus* had the distinctive ankle bones of artiodactyls but had numerous cetacean characters. (D) *Dorudon*, of the middle to late Eocene, had most of the features of modern cetaceans, although its nonfunctional pelvis and hindlimb were larger. (E) A modern toothed whale, the harbor porpoise, *Phocoena phocoena*. The nostrils, forming a blowhole, are far back on the top of the head, accounting for the peculiar shape of the skull. (A after Thewissen et al. 2007; B–D after Gingerich 2003 and de Muizon 2001; E art by Nancy Haver.)

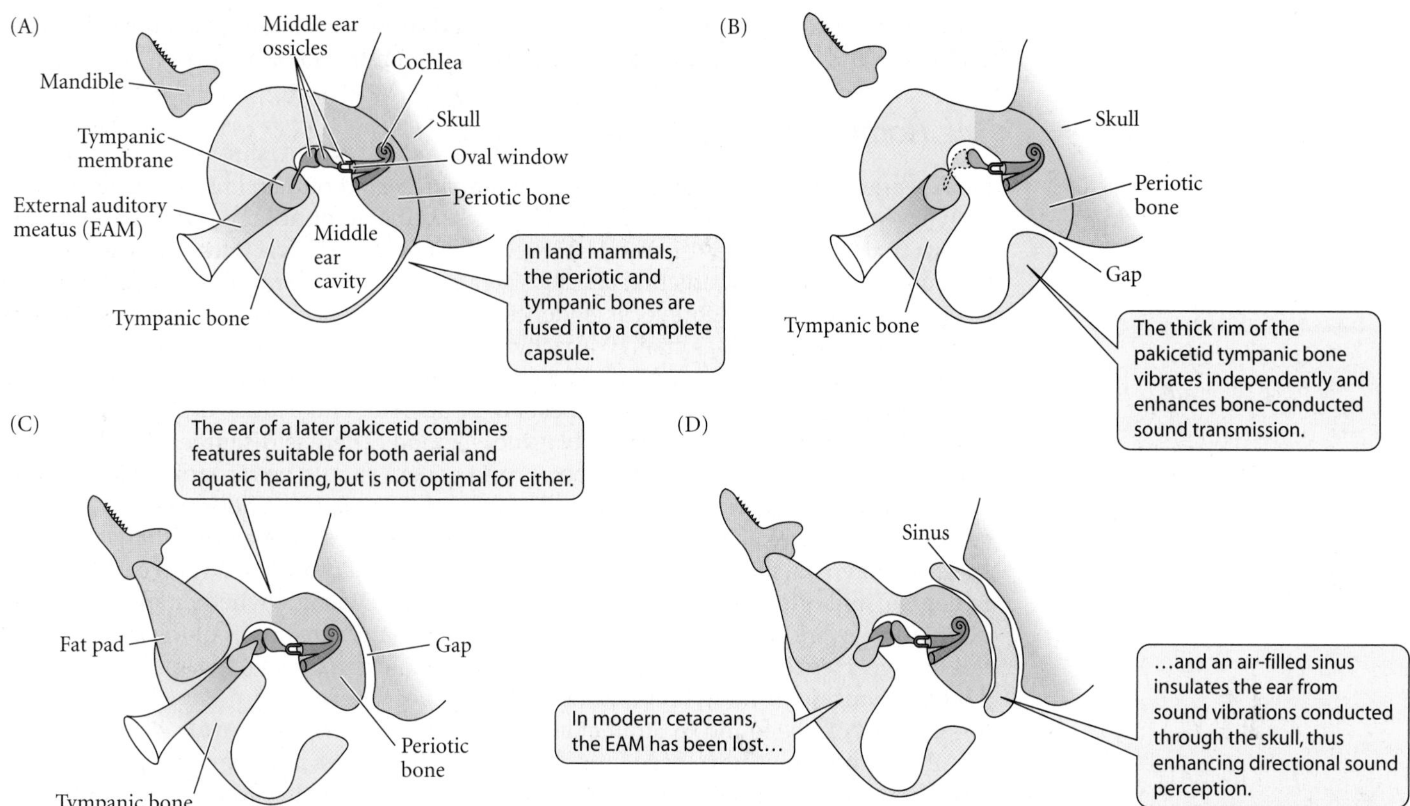

Figure 4.12 Stages in the evolution of the cetacean ear, adapted for directional hearing in water. Schematic diagrams are of the left ear, viewed from the rear, with the lower jawbone (the mandible) projecting forward. (A) In a land mammal, the tympanic and periotic bones form a complete capsule around the middle-ear cavity. This capsule is connected to the rest of the skull. Sound entering through the air-filled ear canal, or external auditory meatus, vibrates the tympanic membrane; vibrations are transmitted by the three ear ossicles (maleus, incus, and stapes) to the oval window, a membrane of the inner ear's cochlea. (B) In pakicetids, the earliest known cetaceans, the tympanic bone is separated from the periotic bone, and has a thick rim—a distinctive feature of cetaceans and raoellids. (C) In the later protocetids, a fat pad conducts sound vibrations from the lower jaw; relative to the tympanic membrane (which is oval, like that of modern cetaceans), the auditory ossicles are more massive, and thus better suited for conducting low frequencies. (D) The final steps, seen in modern toothed whales, are loss of the external auditory meatus, and formation of an air-filled sinus. (After Nummela et al. 2004.)

be the Raoellidae, small Eocene artiodactyls that share the distinctive cetacean tympanic bone and certain features of the teeth, and evidently were semiaquatic like hippos (Figure 4.11A). The oldest known cetacean, *Pakicetus* (53–48 Mya), was a terrestrial or semiaquatic animal with the distinctive cetacean tympanic bone (Figure 4.12B). The slightly younger *Ambulocetus* (48–47 Mya) was adapted for life in shallow coastal waters. It had short hind legs but large feet, with separate digits that bore small hooves. The mandibular foramen was larger than in pakicetids, starting a steady increase in size. *Ambulocetus* was a predator that had long jaws and teeth with somewhat reduced cusps. In the protocetids (e.g., *Rodhocetus*; 49–39 Mya), the fusion between sacral vertebrae was reduced, tooth form was simpler, and the nasal opening was farther back from the tip of the snout. Protocetids had an artiodactyl ankle bone and small hooves at the tips of the toes, but the pelvis and hindlimbs of some of the aquatic protocetids were too small and weak to bear their weight. The enlarged mandibular foramen shows that sound was transmitted by a fat pad from the lower jaw to the ear (Figure 4.12C). Complete adaptation to life in water is shown by the dorudontine basilosaurids, of about 35 Mya, in which the teeth were even simpler, the nostrils were farther back, and the front limbs were flipperlike, with an almost inflexible wrist and elbow. The pelvis and hindlimbs were completely nonfunctional. The small pelvis was disconnected from the vertebral column, and the hind feet and legs barely protruded from the surface of the body. Dorudontines probably had a horizontal tail fin (fluke), and were, indeed, a small step away from modern cetaceans.

The Hominin Fossil Record

DNA sequence differences imply that the chimpanzee and human lineages diverged 5 to 6 Mya (see Chapter 2). No chimpanzee fossils have been found, but hominin fossils are known from as far back as the late Miocene, between 6 and 7 Mya. (Here, we use the term "hominin" for the hominids that are the sister group of the chimpanzees.) The hominin fossil record provides unequivocal evidence of general, more or less unidirectional trends in many characters, such as cranial capacity, a measure of brain size (Figure 4.13). It shows beyond any doubt that modern humans evolved through many intermediate steps from ancestors that in most anatomical respects were apelike. Thus, much of the broad sweep of human evolution has been superbly documented. There is disagreement, however, about how many distinct hominin species and genera should be recognized (e.g., Wood and Collard 1999), in part because fossil specimens are too few, and too widely separated in time and space, to characterize variation within species compared with that between species. Moreover, the differences among various hominins are quantitative (differences in degree) and often rather slight, so it is difficult to determine whether specific earlier populations were ancestral to later ones, or were collateral relatives of the actual ancestral lineage. Hence, even if the overall pattern of evolution is clear, the specific phylogenetic relationships among hominin taxa may not be. (Useful summaries of the hominin fossil record can be found in Strait et al. 1997 and Cela-Conde and Ayala 2007.)

All early hominins have been found in Africa. The earliest hominin fossil, a skull from Chad in western Africa named *Sahelanthropus tchadensis*, is believed to date from the late Miocene, between 6 and 7 Mya. In many respects, such as its small brain, it is the most primitive known hominin, but it also has derived features, such as small canine teeth and a rather flat face, that resemble those of much later hominins. It was almost certainly bipedal (walked on two legs).

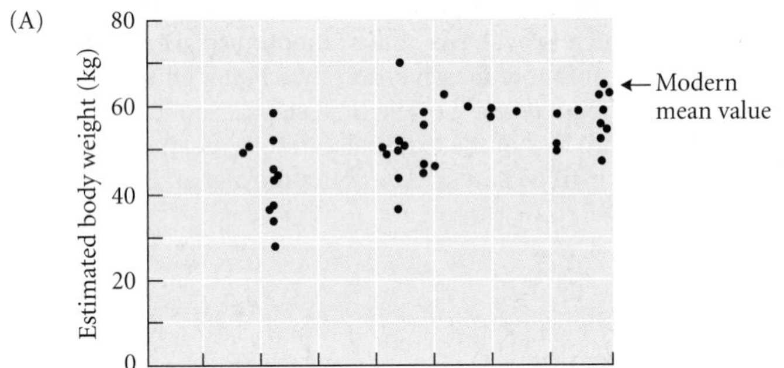

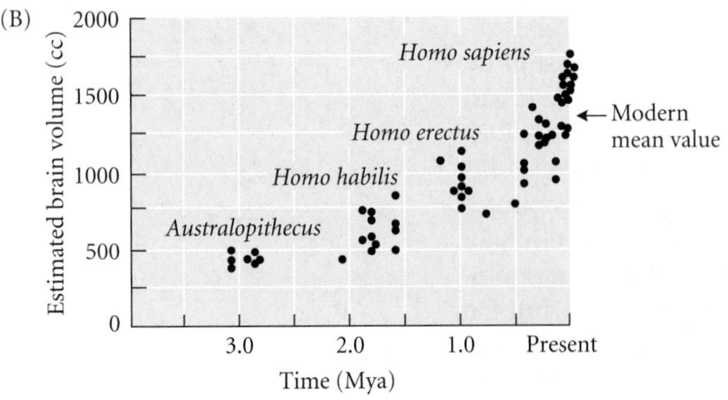

Figure 4.13 Estimated body weights (A) and brain volumes (B) of fossil hominins. There has been a steady, fairly gradual increase in brain volume, even though body size has not increased very much in the last 2 million years. The arrows indicate modern averages. (After Jones et al. 1992.)

Somewhat later hominins include several recently described forms, such as *Orrorin tugenensis* (ca. 6 Mya), *Ardipithecus kadabba* (5.8–5.2 Mya), *Ardipithecus ramidus* (4.4 Mya), and *Australopithecus anamensis* (4.2–3.9 Mya), that are represented by skull fragments, jaws, teeth, and leg and arm bones, as well as a 3.5 Myr old skull named *Kenyanthropus platyops*. The most extensive and informative early fossil material, also about 3.5 Myr old, has been named *Australopithecus afarensis* (see the photograph at the beginning of this chapter).

That these forms are not far removed from the common ancestor of humans and chimpanzees is indicated by their many "primitive" or ancestral features. They had a lower face projecting far beyond the eyes (less so in *platyops*), relatively large canine teeth, long arms relative to the legs, a small brain (about 400 cc), and curved bones in the fingers and toes (which imply that they climbed trees). However, the structure of the pelvis and hindlimb clearly shows that *anamensis* and *afarensis* were bipedal. In fact, "fossilized" footprint traces have been found in rock formed from volcanic ash near an *afarensis* site in Tanzania.

Following *afarensis* in the late Pliocene and early Pleistocene, the number of hominin species and the relationships among them have not yet been resolved (Figure 4.14). Most authorities agree that hominin species were quite diverse at this time. A lineage of "robust" australopithecines (*Paranthropus*), of which three species have been named, had large molars and premo-

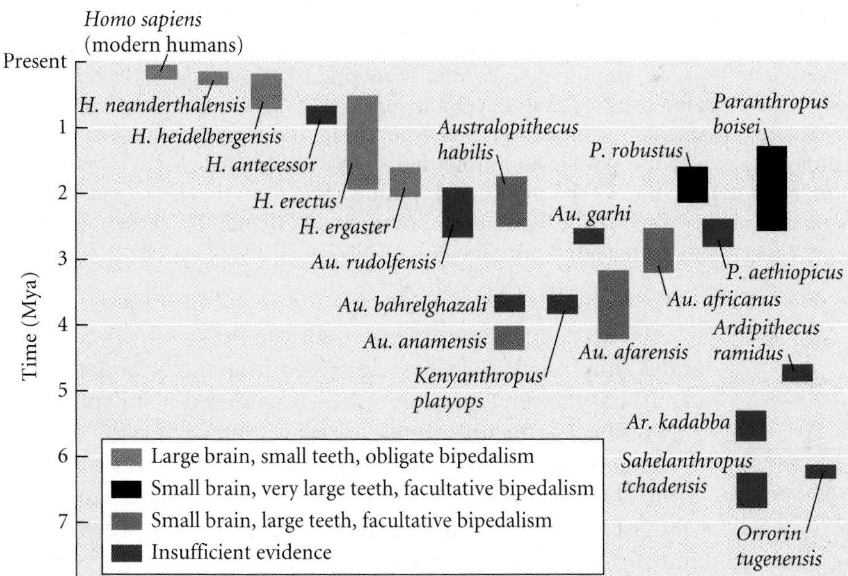

Figure 4.14 The approximate temporal extent of named hominin taxa in the fossil record. The time span indicates either the range of dated fossils or the range of estimated dates for taxa known from few specimens. (After Wood 2002.)

lars and other features adapted for powerful chewing; they probably fed on tubers and hard plant material. The robust australopithecines may have made stone tools, the oldest of which are 2.6 to 2.3 Myr old. However, they became extinct without having contributed to the ancestry of modern humans. A more slender form was *Australopithecus africanus*, which is generally thought to have descended from *A. afarensis*, but had a greater cranial capacity (about 450 cc), indicating a larger brain (Figure 4.15B,C). Some of the derived characters of *africanus* resemble those of robust australopithecines, so it may not be in the line of direct ancestry of modern humans.

The earliest fossils that are usually assigned to the genus *Homo* range from about 1.9 to 1.5 Mya (late Pliocene and early Pleistocene). Originally called *Homo habilis*, they are variable enough to be assigned by some authors to two or even three species, *H. habilis*, *H. ergaster*, and *H. rudolfensis*. *Homo habilis* (in the broad sense) is the epitome of a rediscovered missing link (Figure 4.15D), and might better be assigned to *Australopithecus* (Wood and Collard 1999). The oldest specimens are very similar to *Australopithecus africanus*, and the younger ones grade into the later form *Homo erectus*. Compared with *Australopithecus*, *Homo habilis* more nearly resembles modern humans in its greater cranial capacity (610 to nearly 800 cc), flatter face, and shorter tooth row. Although the limbs retain apelike proportions that suggest an ability to climb, the structure of the leg and foot indicates that its bipedal locomotion was more nearly human than that of the australopithecines. *Homo habilis* is associated with stone tools (referred to as Olduwan technology) and with animal bones that bear cut marks and other signs of hominin activity (Potts 1988).

Later hominin fossils, from about 1.6 Mya to about 200,000 years ago (200 Kya), are referred to as *Homo erectus*. Most authorities think that *habilis*, *erectus*, and *sapiens* are a single evolutionary lineage. In most respects, *erectus* from the middle Pleistocene onward has fairly modern human features in its skull, in its postcranial anatomy, and in indications of its behavior. The skull is rounded, the face is less projected than in earlier forms, the teeth are smaller, and the cranial capacity is larger, averaging about 1000 cc and evidently increasing over time (Figure 4.15E). At least 1 Mya (perhaps as far back as 1.7 Mya), *erectus* spread from Africa into Asia, extending eastward to China and Java. Throughout its range, *erectus* is associated with stone tools, termed the Acheulian culture, that are more diverse and sophisticated than the Olduwan tools of *H. habilis*. The use of fire was widespread by half a million years ago.

Starting about 400 or 300 Kya, *Homo erectus* grades into forms with more *sapiens*-like features, with mean cranial capacity increasing from about 1175 cc at 200 Kya to its modern mean of 1400 cc. The best known of these populations are the Neanderthals of Europe

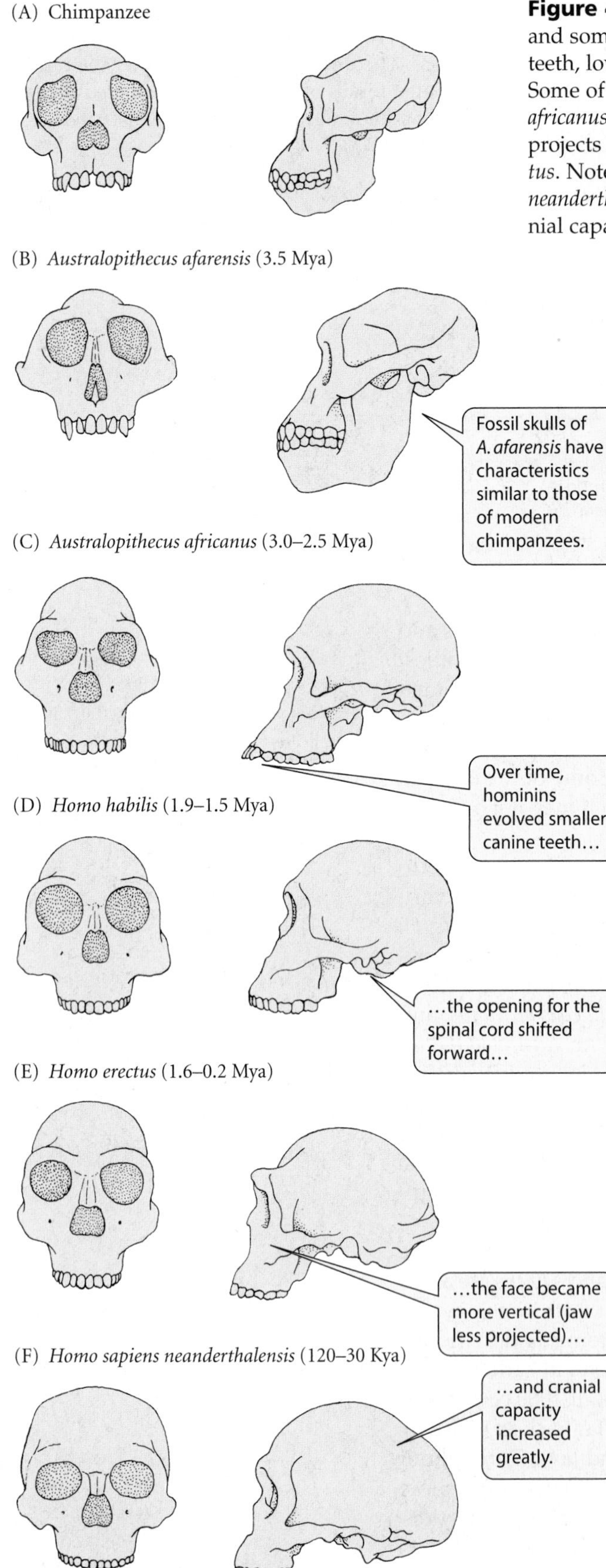
(A) Chimpanzee
(B) *Australopithecus afarensis* (3.5 Mya)
Fossil skulls of *A. afarensis* have characteristics similar to those of modern chimpanzees.
(C) *Australopithecus africanus* (3.0–2.5 Mya)
Over time, hominins evolved smaller canine teeth…
(D) *Homo habilis* (1.9–1.5 Mya)
…the opening for the spinal cord shifted forward…
(E) *Homo erectus* (1.6–0.2 Mya)
…the face became more vertical (jaw less projected)…
(F) *Homo sapiens neanderthalensis* (120–30 Kya)
…and cranial capacity increased greatly.

Figure 4.15 Frontal and lateral reconstructions of the skulls of a chimpanzee and some fossil hominins. (A) *Pan troglodytes*, the chimpanzee. Note large canine teeth, low forehead, prominent face, and brow ridge. (B) *Australopithecus afarensis*. Some of the same features as in the chimpanzee are evident. (C) *Australopithecus africanus* has smaller canines and a higher forehead. (D) *Homo habilis*. The face projects less, and the skull is more rounded, than in earlier forms. (E) *Homo erectus*. Note the still more vertical face and rounded forehead. (F) *Homo* (*sapiens*) *neanderthalensis*. The rear of the skull is more rounded than in *H. erectus*, and cranial capacity is greater. (A, B after Jones et al. 1992; C–F after Howell 1978.)

and southwestern Asia, first described from the Neander Valley (Thal) of western Germany. Once considered a subspecies of *Homo sapiens*, Neanderthals are now regarded as a distinct species, *Homo neanderthalensis*, partly because DNA extracted from Neanderthal bone shows substantial differences from DNA of *H. sapiens*. Neanderthals had dense bones, thick skulls, and projecting brows (Figure 4.15F); contrary to the popular image of a stooping brute, Neanderthals walked fully erect, had brains as large as or even larger than ours (up to 1500 cc), had a fairly elaborate culture that included a variety of stone tools (Mousterian culture), and probably practiced ritualized burial of the dead. Their remains extend from about 120 to 30 Kya.

Modern *sapiens*, anatomically virtually indistinguishable from today's humans, appeared earlier in Africa (ca. 170 Kya) than elsewhere. Modern humans overlapped with Neanderthals in the Middle East for much of the Neanderthals' history, but abruptly replaced them in Europe about 40 Kya. By 12 Kya, and possibly earlier, modern humans had spread from northeastern Asia across the Bering Land Bridge to northwestern North America, and thence rapidly throughout the Americas. (In addition to "standard" *H. sapiens*, enigmatic fossil material of a meter-tall "dwarf" hominin has been described from the Indonesian island Flores, dated at 18 Kya. There is great controversy about whether this represents a genetically distinct population or a pathological condition in standard *H. sapiens*.)

"Upper Paleolithic" culture emerged about 40 Kya. The earliest of several successive cultural "styles" in Europe, the Aurignacian, is marked by stone tools more varied and sophisticated than those of the Mousterian culture. Moreover, culture became more than simply utilitarian: art, self-adornment, and possible mythical or religious beliefs are increasingly evident from about 35 Kya onward. Agriculture, which resulted in an enormously increased human population density and began the human transformation of the Earth, is about 11,000 years old. There is, at least at present, no way of knowing which (if any) of these cultural advances were associated with genetic changes in the capacity for reason, imagination, and awareness, but they are not paralleled by any increase in brain size or other anatomical changes.

Throughout hominin evolution, different hominin features evolved at different rates ("mosaic" evolution). On average, brain size (cranial capacity) increased throughout hominin history, although not at a constant rate, and there were progressive changes, from *afarensis* to *africanus* to *erectus* to *sapiens*, in many other features, such as the teeth, face, pelvis, hands, and feet. The very fuzziness of the taxonomic distinctions among the named forms attests to the mosaic and gradual evolution of hominin features. Although many issues remain unresolved, the most important point is fully documented: modern humans evolved from an apelike ancestor.

Why these changes occurred—what advantages they may have provided—is the subject of much speculation but only slight evidence (Lovejoy 1981; Fedigan 1986). What evidence exists is indirect, consisting mostly of inferences from studies of other primates, contemporary cultures, and anatomy and artifacts.

The erect posture and bipedal locomotion are the first major documented changes toward the human condition. A plausible hypothesis is that bipedalism freed the arms for carrying food back to the social unit, especially to an individual's mate and offspring. Food sharing occurs in chimpanzees, which have a complex social structure that includes matrilocal family groups and "friendships." Chimpanzees also make and use a variety of simple tools, such as stone and wooden hammers they use to crack nuts, and twigs fashioned to "fish" termites out of their nests. The advantages gained by using a greater variety of tools may have selected for greater intelligence and brain size. However, many authors, beginning with Darwin, have emphasized that social interactions, such as learning how to provide parental care, forming cooperative liaisons with other group members, detecting cheaters in social exchanges, and competing for resources within and among groups, would place a selective premium on intelligence, learning, and communication—thus selecting for greater intelligence and a larger brain.

Phylogeny and the Fossil Record

In inferring phylogenetic relationships among living taxa, we conclude that certain taxa share more recent common ancestors than others. If such statements are correct, then there should be some correspondence between the relative times of origin of taxa, as inferred from phylogenetic analysis, and their relative times of appearance in the fossil record. We can expect this correspondence to be imperfect because of the great imperfection of the fossil record; for example, a group that originated in the distant past might be recovered only from recent deposits. Moreover, although a lineage may have branched off early, it may not have acquired its diagnostic characters until much later. The synapsid clade, for example, did not acquire the diagnostic characters of mammals until long after it had diverged from other reptiles. In many taxa, nevertheless, there is a strong overall correspondence between phylogenetic branching order and order of appearance in the fossil record (Norell and Novacek 1992; Benton and Hitchin 1997). Just by phylogenetic analysis of living species, we infer that the common ancestors of the different orders of mammals, of mammals and "reptiles," of these groups and amphibians, and of all tetrapods and sarcopterygian fishes are sequentially older. The sequence in which these groups appear in the fossil record matches the phylogeny. Figure 4.16 shows the correlation

Figure 4.16 Correlations between clade rank, the relative order of branching from the base of a phylogenetic tree, and age rank, the relative order of first appearance of the clades in the fossil record. (A) Illustration of how age rank (left) and clade rank (right) are determined for four hypothetical clades, A–D. (B) Correlation between clade rank and age rank for 13 clades in the horse family, Equidae. (C) Same, for 12 clades of dinosaurs. S is the Spearman rank correlation coefficient which ranges from 0 (no correlation) to 1.0 for perfectly correlated variables. (After Norell and Novacek 1992.)

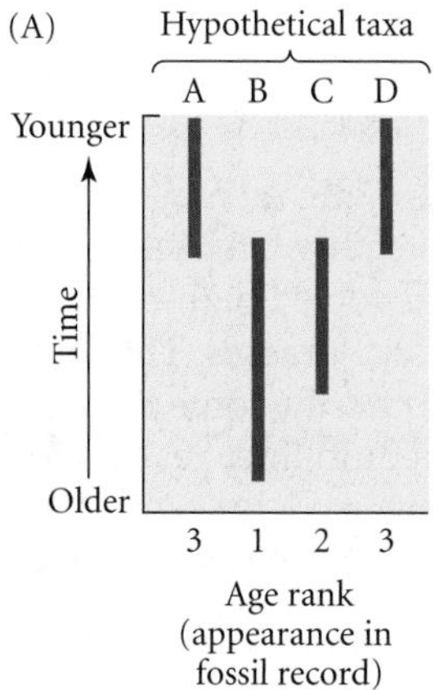

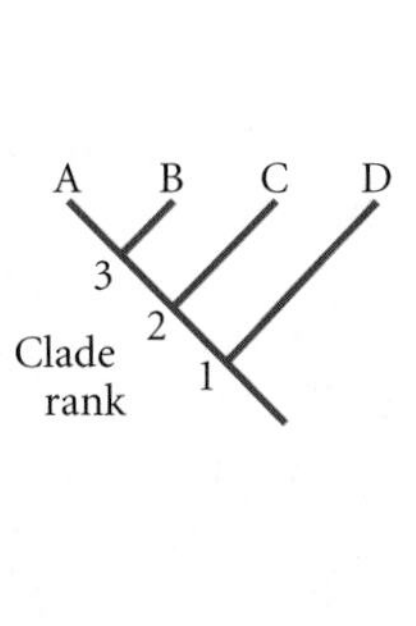

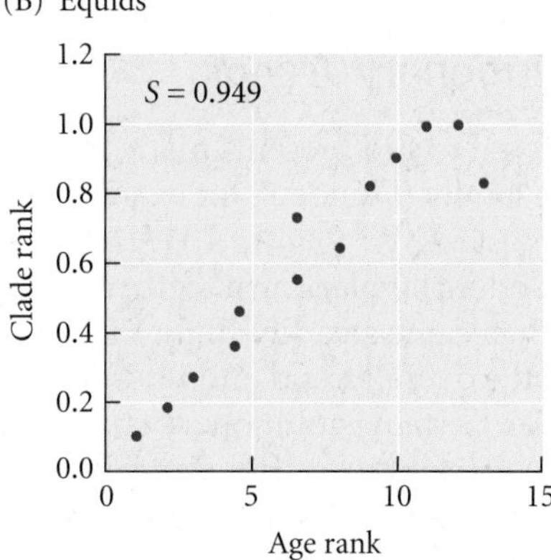

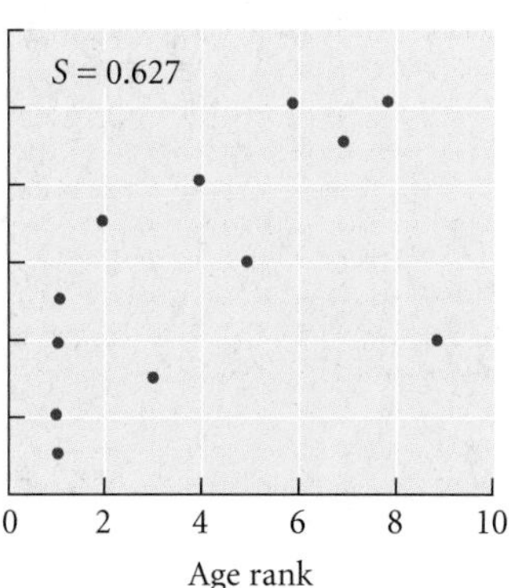

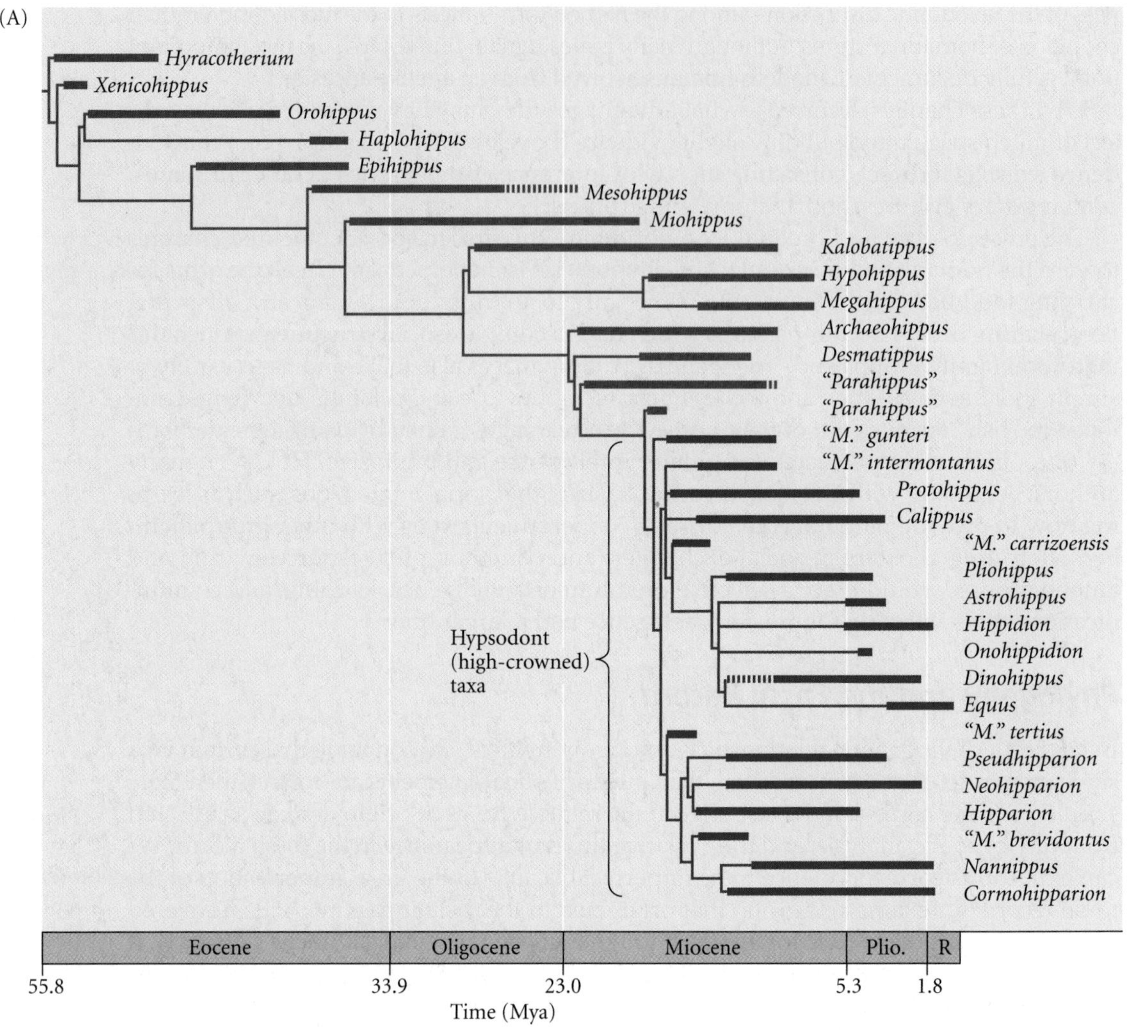

between phylogenetic branching order (the clade rank) and the sequence of appearance of the branches (the age rank) in the fossil record of two sample lineages.

Phylogenetic relationships can often be clarified by information from extinct species (Donoghue et al. 1989). Some characters may have been so highly modified that it is difficult to trace their evolutionary transformations, or even to determine their homology among living organisms. Fossils may provide the crucial missing information. For example, some authors had postulated that mammals and birds are sister taxa, but the fossil record of mammal-like reptiles and dinosaur-like birds showed that this hypothesis was wrong.

Evolutionary Trends

The fossil record presents many instances of **evolutionary trends**. Among members of the horse family (Equidae), for example, average body size increased steadily over the course of almost 50 Myr (Figure 4.17A,B). In the mid-Miocene, one lineage evolved high-crowned (hypsodont) molariform teeth, an adaptation for feeding on abrasive grasses. The tooth crown became steadily higher in this lineage (Figure 4.17C). Commonly, some lineages buck the overall trend and undergo reversal; for example, several equid lineages became smaller. Certain evolutionary changes appear never to have been reversed, however. Ever since *Morganucodon*, for example, mammals have had a single lower jawbone and have never reverted to the multiple bones of their ancestors.

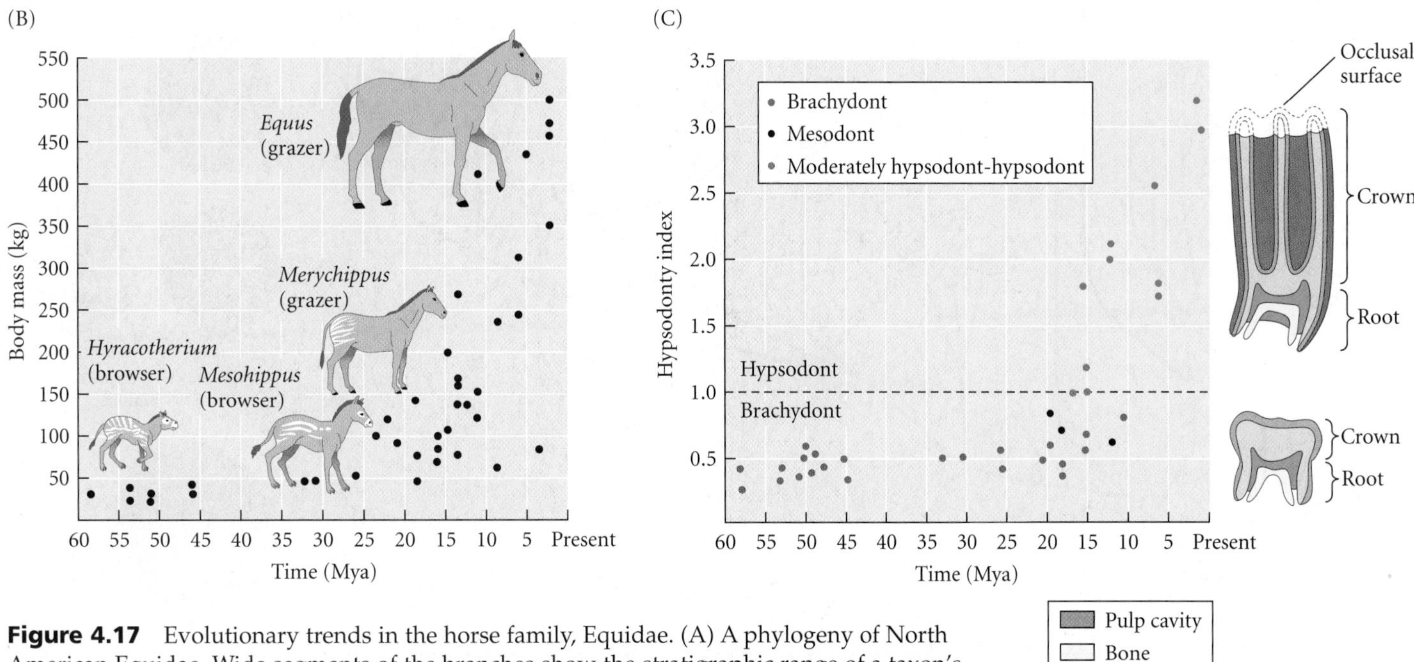

Figure 4.17 Evolutionary trends in the horse family, Equidae. (A) A phylogeny of North American Equidae. Wide segments of the branches show the stratigraphic range of a taxon's fossils. Hypsodont (high-crowned) taxa are shown in dark blue. (B) Estimated body masses of 40 species, plotted against geological time. Although some small species occurred throughout the history of the family, the average body size increased over time. (C) Crown height (hypsodonty index) of the cheek teeth of North American equids, showing the trend within the hypsodont lineage (blue circles). Vertical sections show the difference between low-crowned and high-crowned cheek teeth. See Figure 4.20 for a similar trend in a vole. (A, C after Strömberg 2006; B after MacFadden 1986.)

Many taxa in the fossil record display parallel trends. For example, the horse family is only one of many animal clades in which average body size has increased—a generalization called **Cope's rule**. Multiple lineages often evolve through similar stages, called **grades**. For example, balanomorph barnacles, which first appeared in the Cretaceous, are enclosed by a cone-shaped skeleton made up of a number of slightly overlapping plates. Ancestrally, there were eight plates, but as the balanomorphs diversified in the Cenozoic, the proportion of genera with eight plates steadily declined, and the proportion with fewer plates increased. Several lineages independently evolved through six-plate, four-plate, and even one-plate grades (Figure 4.18). Shells with fewer plates, and hence fewer lines of vulnerability between them, may provide greater protection against predatory snails (Palmer 1982).

Punctuated Equilibria

Although we have described paleontological examples of gradual transitions through intermediate states, these kinds of transitions are by no means universally found in the fossil record. Intermediate stages in the evolution of many higher taxa are not known, and many closely related species in the fossil record are separated by smaller but nonetheless distinct gaps. Most paleontologists have followed Darwin in ascribing these gaps to the great incompleteness of the fossil record. In 1972, Niles Eldredge and Stephen Jay Gould proposed a more complicated, and much more controversial, explanation, which they called **punctuated equilibria**. Their hypothesis applies to the abrupt appearance of closely related species, not to higher taxa.

"Punctuated equilibria" refers to both a *pattern* of change in the fossil record and a *hypothesis* about evolutionary processes. A common pattern, Eldredge and Gould said, is

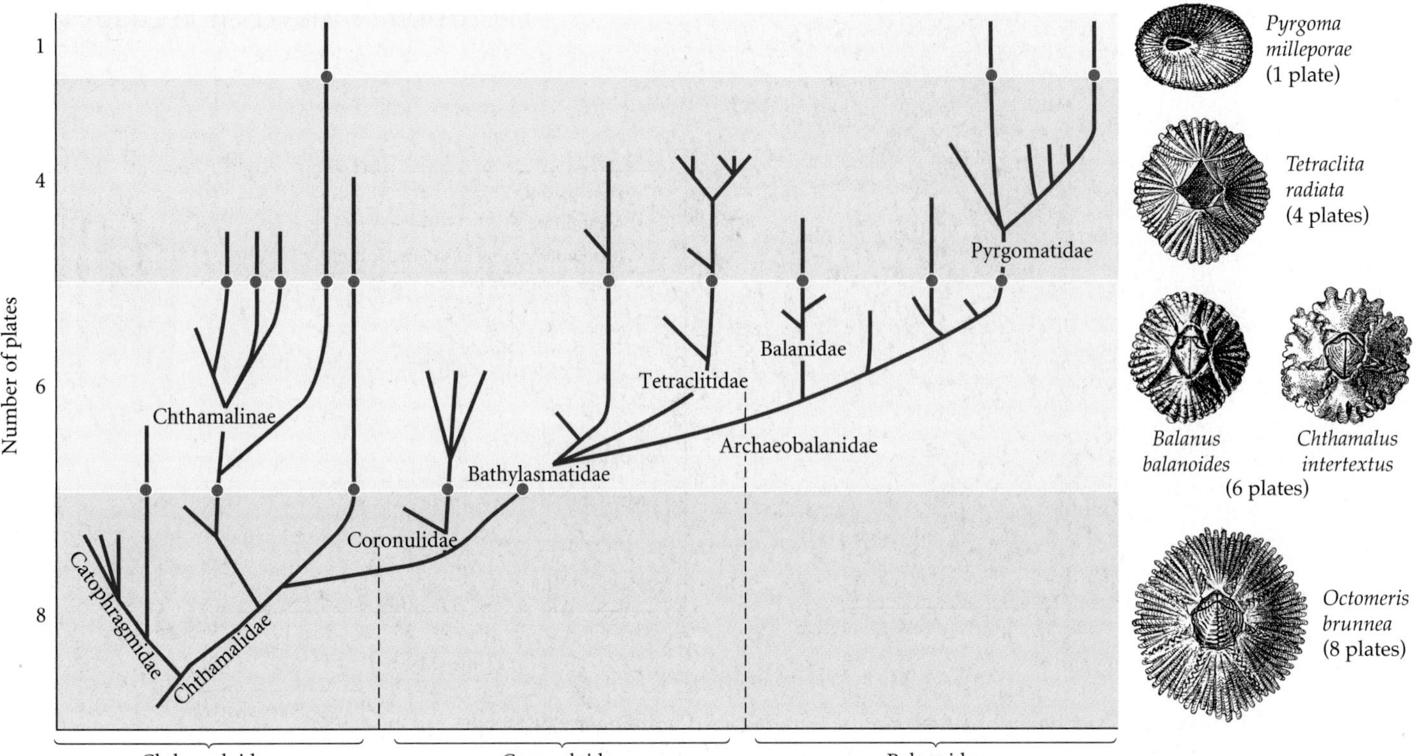

Figure 4.18 A parallel trend. The phylogeny of balanomorph barnacles shows the reduction in the number of shell plates that occurred during the Cenozoic in several independent lineages. The vertical axis is not time, but grade of organization (plate number). The drawings are from an extensive monograph on barnacles by Charles Darwin. (After Palmer 1982; drawings by G. Sowerby, from Darwin 1854.)

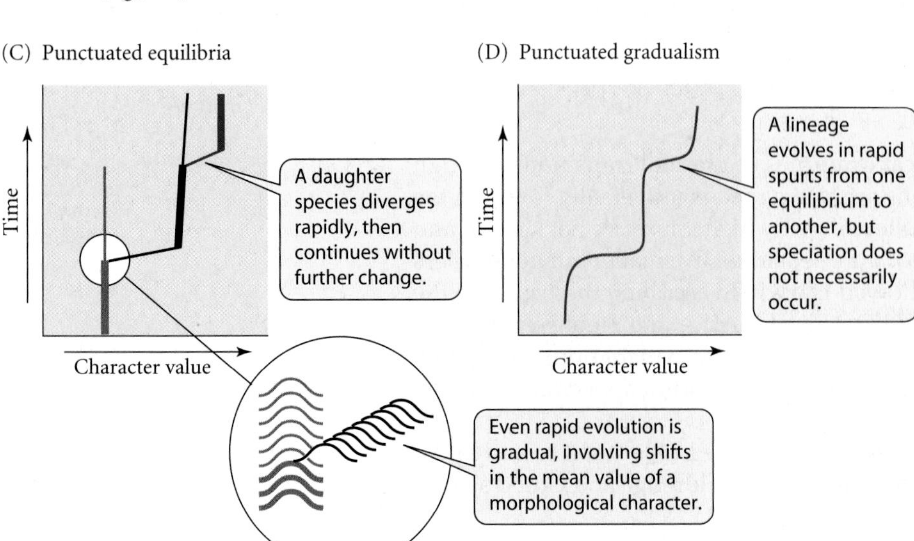

Figure 4.19 Three models of evolution, as applied to a hypothetical set of fossils. (A) Hypothetical values for a character in fossils recorded from different time periods. These data might correspond to any of the models shown in panels B–D. (B) The traditional "phyletic gradualism" model. (C) The "punctuated equilibria" model of Eldredge and Gould, in which morphological change occurs in new species. Morphological evolution, although rapid, is still gradual, as shown in the inset. (D) The "punctuated gradualism," or "punctuated anagenesis," model of Malmgren et al. (1983).

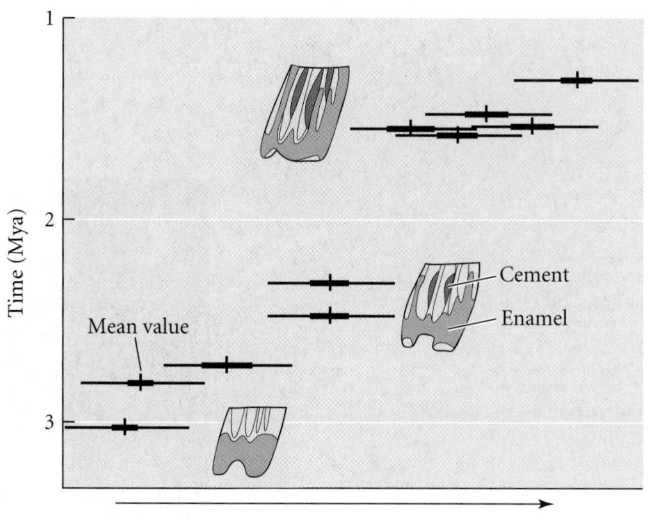

Figure 4.20 Phyletic gradualism: change in a molar of the grass-feeding vole *Mimomys*. Grass wears down the molar surface, so it is advantageous to have a high (hypsodont) tooth with enamel (pink) and cement (brown) forming grinding ridges at the tooth's surface. An index of change in these several features shows a gradual increase over more than 1.5 Myr. Horizontal bars show variation around the mean, which is indicated by vertical marks. Enamel, cement, and tooth height all increased. (After Chaline and Laurin 1986.)

one of long periods in which species exhibit little or no detectable phenotypic change, interrupted by rapid shifts from one such "equilibrium" state to another; that is, **stasis** that is "punctuated" by rapid change (Figure 4.19A). They contrasted this pattern with what they called **phyletic gradualism**, the traditional notion of slow, incremental change (Figure 4.19B).

The fossil record offers examples of both gradual and punctuated patterns. A particularly well documented instance of phyletic gradualism has been described for changes in the molar teeth of a lineage of grass-feeding voles (*Mimomys occitanus*) in the late Pliocene and Pleistocene (Figure 4.20). Several molar characters of these rodents changed directionally throughout Europe, indicating that gene flow among populations enabled the entire species to respond as a whole to selection for increased tooth height (Chaline and Laurin 1986). In contrast, *Metrarabdotos*, a Miocene genus of bryozoans, or "moss animals," clearly shows a pattern of punctuated equilibria (Figure 4.21; Cheetham 1987). Species persisted with little change for several million years, while new species appeared abruptly, without evident intermediates.

The hypothesis that Eldredge and Gould introduced is that characters evolve primarily in concert with true speciation—that is, the branching of an ancestral species into two species (Figure 4.19C). They based their hypothesis on a model, known as "founder-effect speciation" or "peripatric speciation," proposed by Ernst Mayr in 1954, that we will consider in Chapter 17. The thrust of that model is that new species appear suddenly in the fossil record because they evolved in small populations separated from the ancestral species and then, fully formed, migrated into the region where the fossil samples were taken. The evolutionary change they underwent may have been gradual, but it was rapid and took place "off stage."

This hypothesis would not have been so controversial if Eldredge and Gould had not further proposed that, except in populations that are undergoing speciation, morphological characters generally *cannot* evolve, due to internal genetic "constraints." This proposition is contradicted by considerable evidence from populations of living species (see Chapters 9 and 13), and Eldredge and Gould's hypothesis that evo-

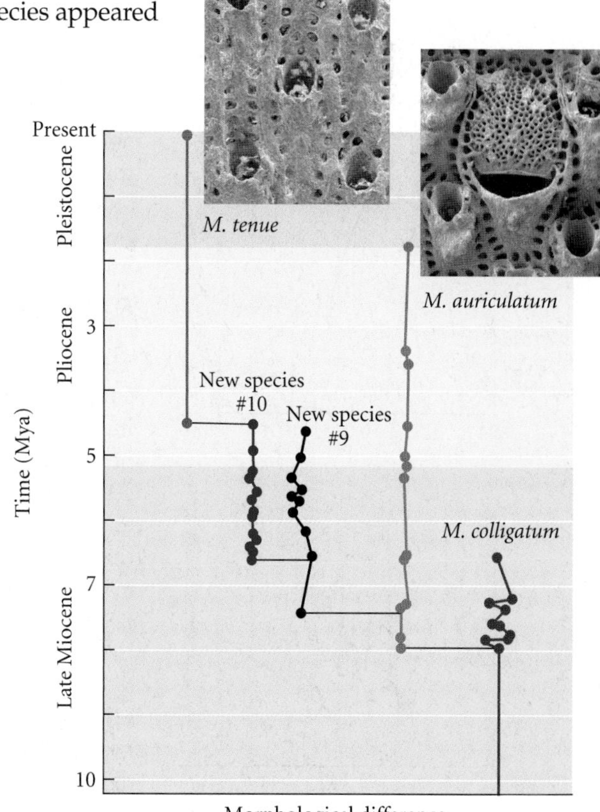

Figure 4.21 Punctuated equilibria: the phylogeny and temporal distribution of a lineage of bryozoans (*Metrarabdotos*). The horizontal distance between points represents the amount of morphological difference between samples. The general pattern is one of abrupt shifts to new, rather stable morphologies. (After Cheetham 1987; photos courtesy of A. Cheetham.)

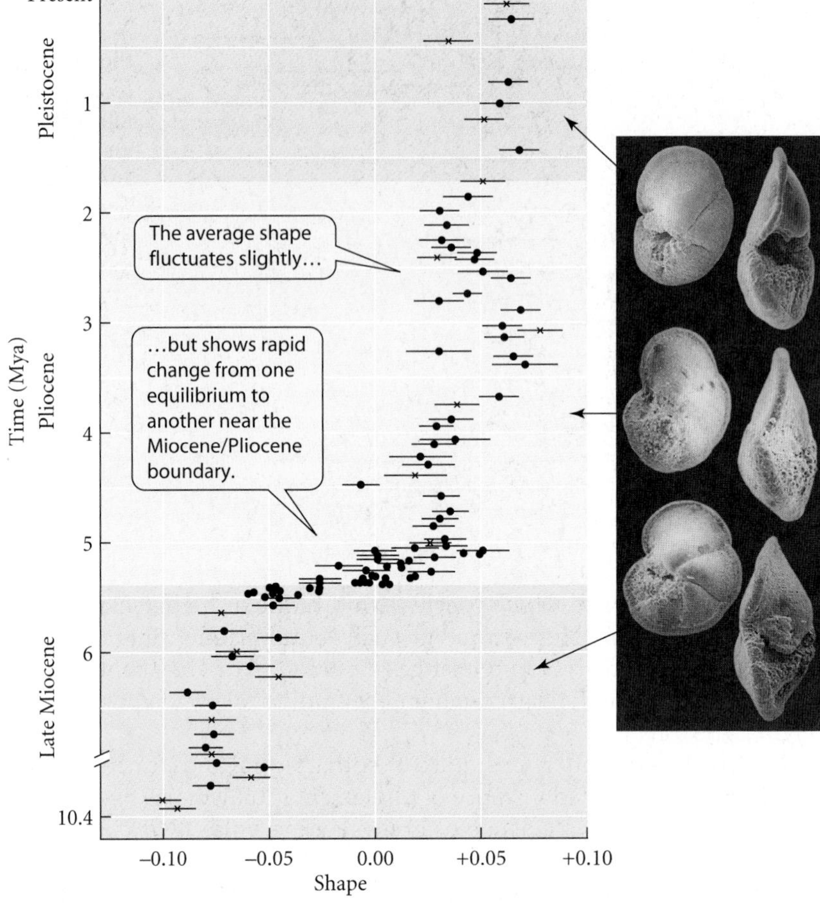

Figure 4.22 Punctuated anagenesis: evolution of shell shape in the foraminiferan lineage *Globorotalia*. Side and edge views of specimens from the late Pleistocene (top), early Pliocene (middle), and late Miocene (bottom) appear at the right. In the graph, a mathematical index of shape is plotted against the ages of fossil samples. There is no evidence that speciation occurred in this lineage. (After Malmgren et al. 1983; photos courtesy of B. Malmgren.)

lutionary change requires speciation is not widely accepted. Furthermore, the fossil record shows that characters may evolve between long-stable states in populations that do not undergo speciation (Figure 4.19D). This pattern, which has been called **punctuated gradualism** or **punctuated anagenesis**, is illustrated by the detailed fossil record of abundant shell-bearing protists called foraminiferans (Figure 4.22). Gene Hunt (2007) analyzed data for more than 250 phenotypic traits in fossil lineages and concluded that most fit a model of either stasis, or random fluctuations that result in a gradual net change over time (Figure 4.23). Consistent directional change was seldom recorded, perhaps because it often occurs too quickly to be preserved in a coarse fossil record.

Rates of Evolution

The rate of evolutionary change varies greatly among characters, among evolving lineages, and within the same lineage over time. Although one may describe the change in, say, the height of a horse's tooth in millimeters (mm) per million years, an increase of 1 mm represents far greater change if the original tooth was 5 mm high than if it was 50 mm high. Therefore, evolutionary rates are usually described in terms of proportional, rather than absolute, change by using the logarithm of the measurement rather than the original scale of measurement. J. B. S. Haldane, one of the pioneers of the evolutionary synthesis, proposed a unit of measurement of evolutionary rate that he called the DARWIN, which he defined as a change by a factor of 2.718 (the base of natural loga-

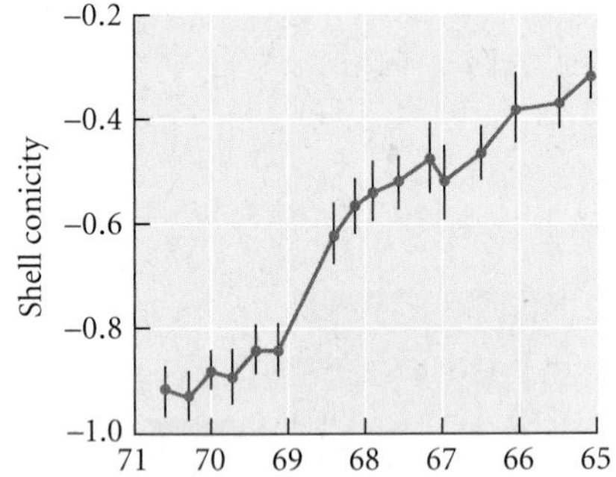

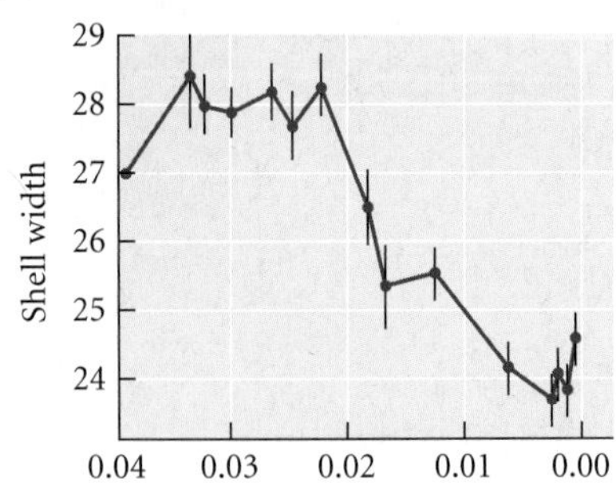

(C) Stasis

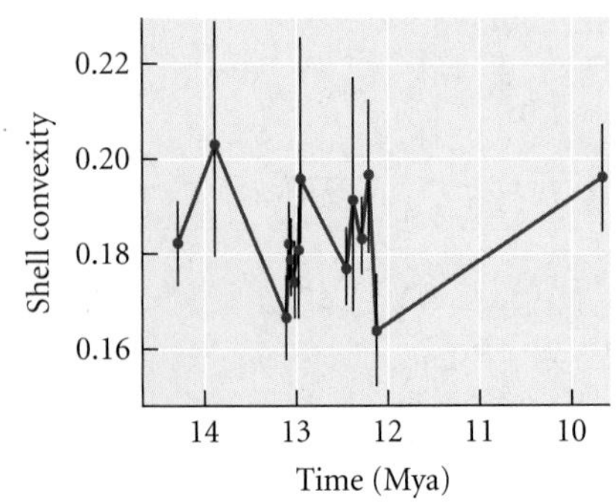

Figure 4.23 Examples of traits that fit three models of evolution in the fossil record. (A) Directional evolution: steady change in shell conicity in the foraminiferan *Contusotruncana* (cf. Figure 4.3). (B) A random walk: random fluctuations in shell width of the land snail *Mandarina*. (C) Stasis in shell shape in a scallop, *Chesapecten nefrens*. (After Hunt 2007.)

(A)

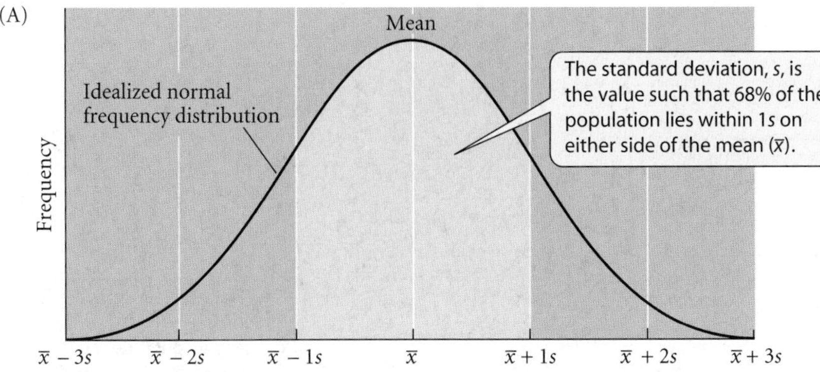
Mean
Idealized normal frequency distribution
The standard deviation, s, is the value such that 68% of the population lies within 1s on either side of the mean ($\bar{x}$).
Frequency
$\bar{x} - 3s$
$\bar{x} - 2s$
$\bar{x} - 1s$
$\bar{x}$
$\bar{x} + 1s$
$\bar{x} + 2s$
$\bar{x} + 3s$

(B)

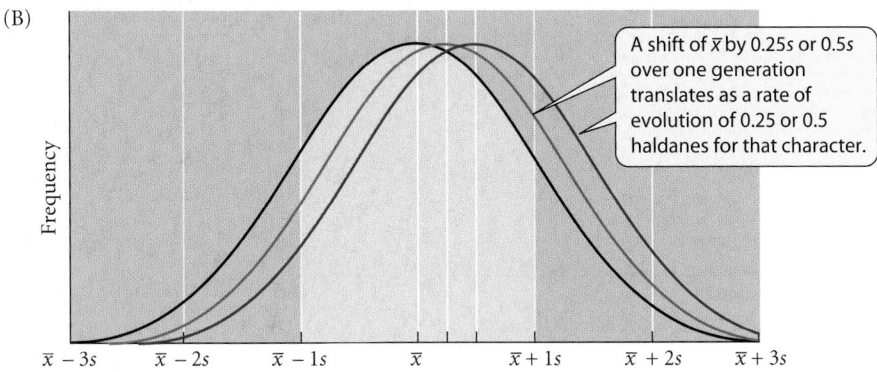
A shift of $\bar{x}$ by 0.25s or 0.5s over one generation translates as a rate of evolution of 0.25 or 0.5 haldanes for that character.
Frequency
$\bar{x} - 3s$
$\bar{x} - 2s$
$\bar{x} - 1s$
$\bar{x}$
$\bar{x} + 1s$
$\bar{x} + 2s$
$\bar{x} + 3s$

Figure 4.24 (A) In an idealized normal frequency distribution, the standard deviation (*s*) is the value at which 68% of the population lies within 1*s* above and 1*s* below the mean ($\bar{x}$). 95% of the population lies within 2*s*, and 99.7% lies within 3*s*, on either side of the mean. (B) Given the standard deviation of the "original" frequency distribution (black curve), the distribution shown as a red line lies 0.25*s*, and the distribution shown as a blue line lies 0.5*s*, away from the original mean. If these shifts in the mean occurred over just one generation, the rate of evolution of the mean character value would be 0.25 haldanes or 0.5 haldanes, respectively.

rithms) per million years. The darwin, however, has some drawbacks, and researchers now tend to measure evolutionary rate as the number of standard deviations by which a character mean changes per generation: a unit that paleontologist Philip Gingerich (1993) dubbed the HALDANE (Hendry and Kinnison 1999). The standard deviation is a measure of the amount of variation within a population (Figure 4.24; see also Box 9B.)

Most morphological characters display very low rates of evolution in fossil lineages, reflecting the pattern of stasis to which Eldredge and Gould called attention. Usually, the feature is not absolutely constant, but instead fluctuates (see Figure 4.23C, as well as the stickleback data in Figure 4.4). For example, even though the size of the first molar of the early Eocene horse *Hyracotherium grangeri* fluctuated slightly, overall it changed very little over 650,000 years, showing a median rate of change of only 0.0000057 haldanes (Gingerich 1993). Then, however, it evolved rapidly—at a rate of 0.00024 haldanes—during the 13,500-generation transition from *Hyracotherium grangeri* to a form called *H. aemulor* (Figure 4.25A). The history of this lineage exemplifies the pattern of punctuated equilibrium.

The "rapid" transition in *Hyracotherium*, like most such data from the fossil record, occurred at a far lower rate than those observed over the course of a few centuries (or less) in species that have been transported to new regions or otherwise affected by human-induced environmental change (see Chapter 13). North American house sparrows, introduced to the continent from Europe, manifested changes in wing and bill length of up to 0.024 haldanes, and the beak length of soapberry bugs increased at a rate of 0.010 to 0.035 haldanes as they adapted to feeding on introduced plants (Hendry and Kinnison 1999; see Figure 13.11). These rates are two orders of magnitude (a factor of 100) greater than tooth evolution was in the rapid transition between species of *Hyracotherium*. In general, rates of evolution are lower, the longer the time interval over which they are measured (Figure 4.25B), because as the time interval increases, it encompasses longer periods of stasis relative to the brief episodes of rapid adaptive evolution. If characteristics evolved in a single direction for thousands or millions of years, even at extremely low rates, organisms would be vastly more different from one another than they are, and mice would long since have become bigger than the largest dinosaurs. Evolutionary changes can be very rapid, but they are not sustained at high rates for very long.

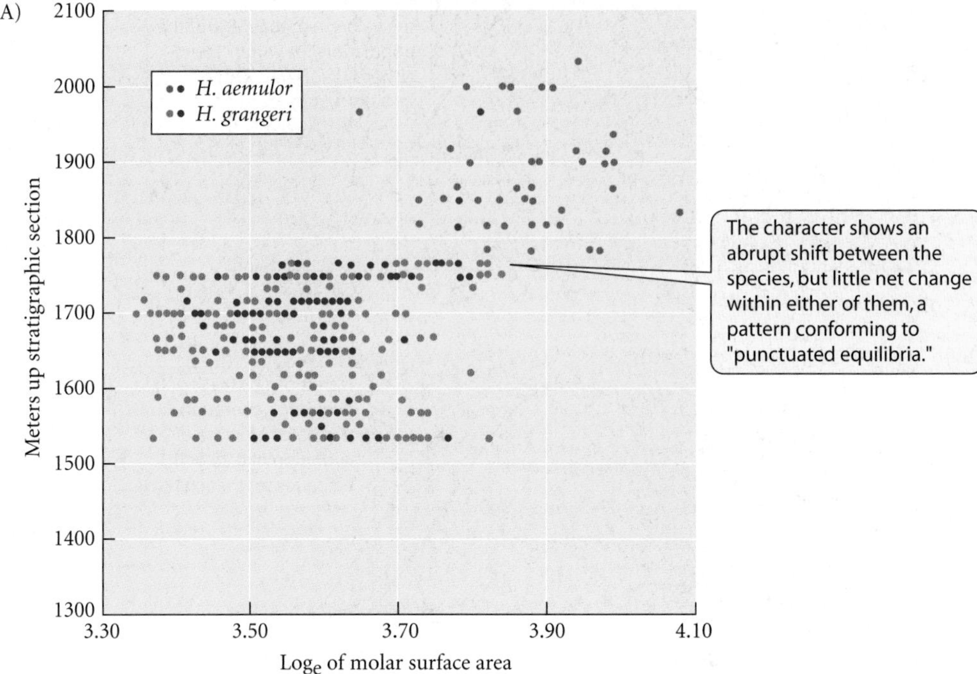

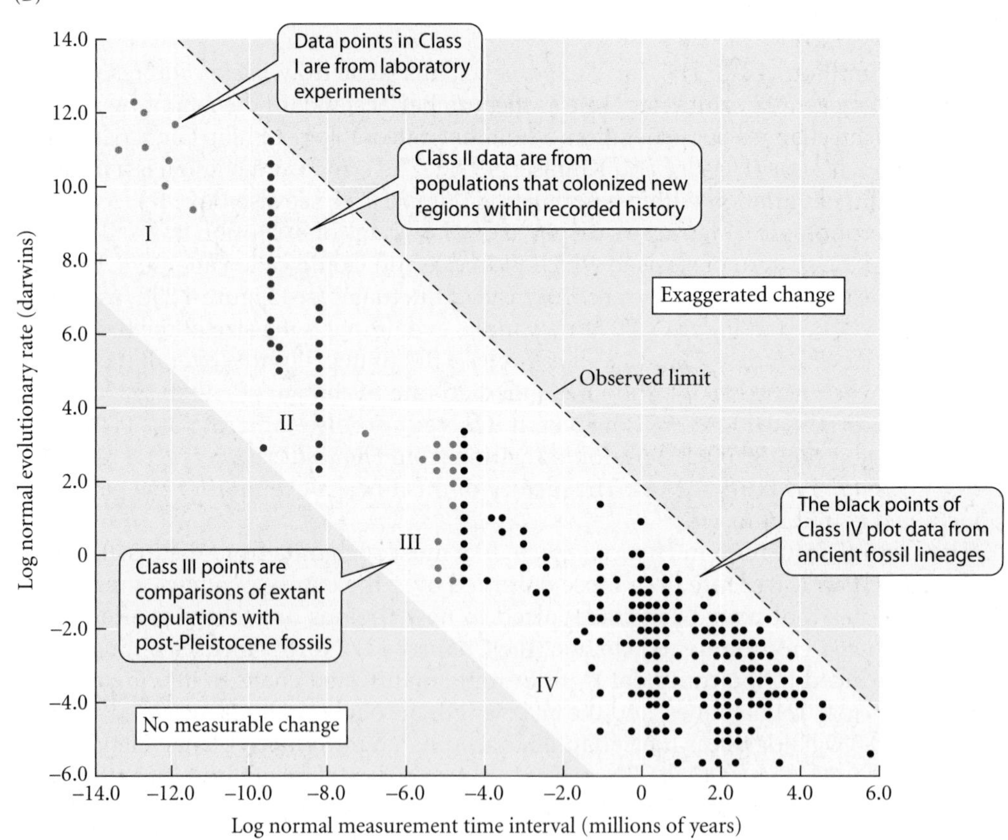

Figure 4.25 Measures of the rate of character evolution depend on time interval. (A) Evolution of the size (surface area) of the first molar of Eocene *Hyracotherium*, early members of the horse family. The data points show character values of individual specimens, plotted against depth in a stratigraphic section of rock. (Dark blue dots indicate multiple species with the same measurements.) *H. grangeri* extended for about 650,000 years (from meter 1520 to meter 1760) and evolved abruptly into a form different enough to be designated a different species, *H. aemulor*. (B) Each point shows an estimate of evolutionary rate of a character, plotted against the time interval over which evolution occurred. The data fall into four classes, as described on the figure. Lineages that changed so greatly that they would show "exaggerated change" are not plotted because their relationships would often not be recognized, and so they would not be compared. (A after Gingerich 1993; B after Gingerich 1983.)

Summary

1. Although many evolutionary histories are well known, the fossil record of most lineages of organisms is very incomplete.
2. Unusually detailed records of change within individual species show that although characters commonly remain relatively unchanged for long periods, when changes do occur they are rapid and may pass gradually through intermediate steps.
3. The origins of many higher taxa, such as tetrapods, birds, mammals, cetaceans, and the genus *Homo*, have been documented in the fossil record. These examples show mosaic evolution and gradual change in individual features. The decision of whether to classify intermediate fossils in one taxon or another is often arbitrary.
4. Changes in the form of characters are sometimes associated with major changes in their function.
5. The relative times of origin of taxa, as inferred from phylogenetic analysis, often correspond to their relative times of appearance in the fossil record.
6. Evolutionary trends, which can often be attributed to natural selection, are evident in the fossil record, but such trends may be reversed in related lineages.
7. Species may display very little change for long periods (stasis), and then shift rapidly to new phenotypes. The term "punctuated equilibria" refers both to this pattern and to the hypothesis, not widely accepted, that most changes in morphology occur in association with the evolution of new species (i.e., splitting of lineages).

GO TO THE
evolution
Companion Web Site at
www.sinauer.com/evolution
for quizzes, data analysis and simulation exercises, and other study aids.

Terms and Concepts

anagenesis
Cope's rule
epoch
era
evolutionary trend
geological time scale
grade
period
phyletic gradualism
plate tectonics
punctuated anagenesis
punctuated equilibria
punctuated gradualism
radiometric dating
stasis
strata

Suggestions for Further Reading

The second edition of S. M. Stanley's *Earth System History* (W. H. Freeman, New York, 2005) is a thorough introduction to geological processes, Earth history, and major events in the history of life, from a paleontologist's perspective. Other useful works on paleontology include R. L. Carroll's comprehensive and abundantly illustrated *Vertebrate Paleontology and Evolution* (W. H. Freeman, New York, 1988); and *Paleobiology: A Synthesis*, edited by D. E. G. Briggs and P. R. Crowther (Blackwell Publishing, Oxford, 1990), which contains a collection of brief, authoritative essays on numerous topics.

The history of human evolution is treated by R. Lewin in *Human Evolution: An Illustrated Introduction* (Blackwell Publishing, Oxford, 2005) and by C. J. Cela-Conde and F. J. Ayala in *Human Evolution: Trails from the Past* (Oxford University Press, New York, 2007).

Punctuated Equilibrium by Steven Jay Gould (Belknap Press of Harvard University Press, Cambridge, MA, 2007) is excerpted from his much longer book, *The Structure of Evolutionary Theory* (Belknap Press, 2002), published shortly before his death. For criticism of the hypothesis of punctuated equilibrium, see the second edition of J. S. Levinton's *Genetics, Paleontology, and Macroevolution* (Cambridge University Press, Cambridge, UK, 2001).

Excellent treatments of some of this chapter's topics, written for a general audience, include *Your Inner Fish: A Journey into the 3.5-Billion-Year History of the Human Body* by Neil Shubin (Allen Lane/Pantheon, New York, 2008), on anatomical homologies between humans and other vertebrates; *Evolution: What the Fossils Say and Why It Matters* by D. R. Prothero (Columbia University Press, New York, 2007), a useful sourcebook for counteracting creationism; *Gaining Ground: The Origin and Evolution of Tetrapods* by J. A. Clack (Indiana University Press, Bloomington, 2002); and *Glorified Dinosaurs: The Origin and Early Evolution of Birds* by L. M. Chiappe (John Wiley, New York, 2007).

Problems and Discussion Topics

1. "Time averaging" refers to the condensation of fossil samples from different time intervals into a single sample. Refer to Bell's study of sticklebacks (Figure 4.4). What would the data look like if he had collapsed samples spanning 20,000 years into single samples instead of analyzing separate samples at 5000-year intervals? What conclusions might Bell have drawn about the evolution of dorsal spine number if his only samples had been from 70 Kya and 30 Kya?
2. The rate of evolution of DNA sequences (and other features) is often calibrated by the age of fossil members of the taxa to which the living species belong (see Chapter 8). How do imperfections in the fossil record affect the estimates of evolutionary rates obtained in this way? Is there any way of setting limits to the range of possible rates?
3. An ideal fossil record would enable researchers to distinguish the patterns of phyletic gradualism, punctuated equilibria, and punctuated gradualism (see Figure 4.19). How would you do so? How do imperfections of the fossil record make it difficult to distinguish these patterns?
4. Creationists deny that the fossil record provides intermediate forms that demonstrate the origins of higher taxa. Of *Archaeopteryx*, some of them say that because it had feathers and flew, it was a bird, not an intermediate form. Evaluate this argument.
5. Consider the hypothesis that Eldredge and Gould advanced to explain the pattern they called punctuated equilibria. What would be the implications for evolution if the hypothesis were true versus false?
6. Changes in a phenotypic character in a population are considered evolutionary changes only if they have a genetic basis. Alterations of organisms' features directly by the environment they experience are not evolution. Because we cannot breed extinct organisms to determine whether differences have a genetic basis, how might we decide which phenotypic changes represent evolution and which do not? Consider (a) the difference in dorsal spine number of sticklebacks at 65 versus 60 Kya (Figure 4.4B), (b) the difference in the same character at 70 versus 25 Kya, and (c) the difference in the shape of the rear teeth in *Morganucodon* (Figure 4.10F) compared with *Procynosuchus* (Figure 4.10C). Can we be more confident that the difference is an evolved one in some cases than in others?
7. What are the possible causes of trends such as those illustrated by barnacles and horses in this chapter? How would you assess the validity of each cause you can think of?
8. What might be the possible causes of a history of stasis followed by rapid character change, as shown by tooth size in *Hyracotherium* (Figure 4.25A)?

CHAPTER 5

A History of Life on Earth

Mammals of the early Tertiary. This artist's rendering is a fanciful fauna of now-extinct mammals from several early Tertiary epochs. Included are some arboreal primates (upper right), a uintothere (center, with three pairs of stubby horns), and the massive *Arsinoitherium* with its pair of nasal horns (right foreground). In the far left background, *Diatryma*, a gigantic flightless bird, is devouring a captured *Hyracotherium*, an early member of the horse family. (Art © Publiphoto/Photo Researchers, Inc.)

If we could look at the Earth of 3,500,000,000 years ago, at the time of life's beginning, we would see only bacteria-like cells. Among these cells would be our most remote ancestors, unrecognizably different from ourselves. And if we then time-traveled through life's history toward the present, we would see played out before us a drama grander and more splendid than we can imagine: a planetary stage of many scenes, on which emerge and play—and usually die—millions and millions of species with features and roles more astonishing than any writer could conceive.

This chapter describes some high points in the grand history of life, especially the origin, diversification, and extinction of major groups of organisms. Most of this material treats geological and paleontological evidence, but phylogenetic studies of living organisms have also revealed critically important information. There is much more information in this chapter than you may wish to memorize. You may regard it largely as a source of information, or you may simply enjoy reading a sketch of one of the greatest stories of all time. A number of major events and important points that a well-trained biologist should know are highlighted in italics. As you read this chapter, moreover, notice examples of the following general patterns:

1. Climates and the distribution of oceans and land masses have changed over time, affecting the geographic distributions of organisms.
2. The taxonomic composition of the biota has changed continually as new forms originated and others became extinct.
3. At several times, extinction rates have been particularly high (so-called **mass extinctions**).
4. Especially after mass extinctions, the diversification of higher taxa has sometimes been relatively rapid.
5. The diversification of higher taxa has included increases both in the number of species and in the variety of their form and ecological habits.
6. Extinct taxa have sometimes been replaced by unrelated but ecologically similar taxa.
7. Of the variety of forms in a higher taxon that were present in the remote past, usually only a few have persisted in the long term.
8. The geographic distributions of many taxa have changed greatly.
9. Over time, the composition of the biota increasingly resembles that of the present.

Before Life Began

Most physicists agree that the current *universe came into existence about 14 billion (14,000,000,000) years ago* through an explosion (the "big bang") from an infinitely dense point. Elementary particles formed hydrogen shortly after the big bang, and hydrogen ultimately gave rise to the other elements. The collapse of a cloud of "dust" and gas formed our galaxy less than 10 billion years ago. Throughout the history of the universe, material has been expelled into interstellar space, especially during stellar explosions (supernovas), and has condensed into second- and third-generation stars, of which the Sun is one. *Our solar system was formed about 4.6 billion years ago*, according to radiometric dating of meteorites and moon rocks. The Earth is the same age, but because of geological processes such as the subduction of crust (see Figure 4.1), the oldest known rocks on Earth are younger, dating from about 4.06 billion years ago. The Earth was probably formed by the collision and aggregation of many smaller bodies, the impact of which contributed enormous heat.

The early Earth formed a solid crust as it cooled, releasing gases that *included water vapor but very little oxygen*. As Earth cooled, oceans of liquid water formed, probably by 4.5 billion years ago, and quickly achieved the salinity of modern oceans. By 4 billion years ago, there were probably many small protocontinents that gradually aggregated to form large land masses over the next billion years.

The Emergence of Life

The simplest things that might be described as "living" must have developed as complex aggregations of molecules. These aggregations, of course, would have left no fossil record, so it is only through chemical and mathematical theory, laboratory experimentation, and extrapolation from the simplest known living forms that we can hope to develop models of the emergence of life.

"Life" is a difficult concept to define. However, it is generally agreed that an assemblage of molecules is "alive" if it can capture energy from the environment, use that energy to replicate itself, and thus be capable of evolving. In the living things we know, these functions are performed by nucleic acids that carry information and by proteins, which replicate nucleic acids, transduce energy, and generate (and in part constitute) the phenotype. These components are held together in compartments—cells—formed by lipid membranes.

Although living or semi-living things might have originated more than once, *we can be quite sure that all organisms we know of stem from a single common ancestor,* because they all share certain features that are arbitrary as far as we can tell. For example, organisms synthesize and use only L optical isomers of amino acids as building blocks of proteins; L and D isomers are equally likely to be formed in abiotic synthesis, but a functional protein can

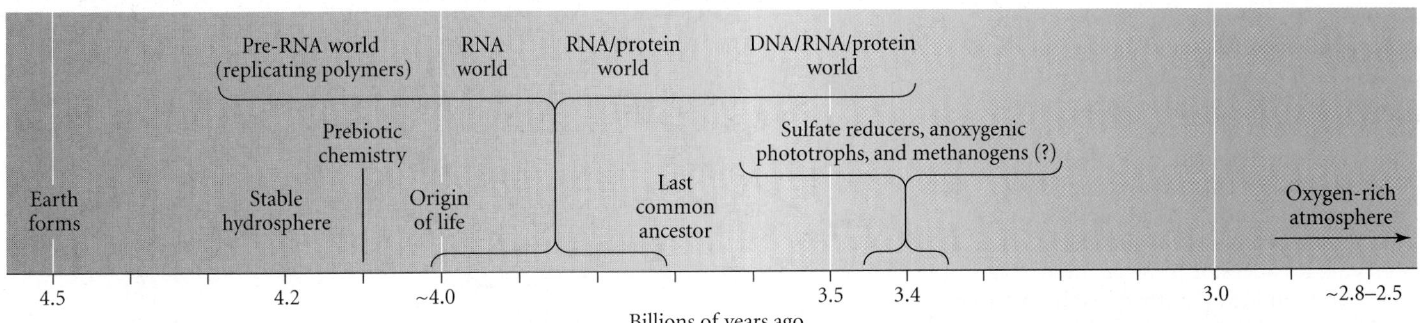

Figure 5.1 The approximate timing of some events in the early history of the Earth and life. (After Becerra et al. 2007.)

be made only of one type or the other. D isomers could have worked just as well. The genetic code, the machinery of replication and protein synthesis, and basic metabolic reactions are among the other features that are universal among organisms and thus imply that they all stem from a common ancestral form, or "last common ancestor" (LCA).

The most difficult problem in accounting for the origin of life is that in known living systems, only nucleic acids replicate, but their replication requires the action of proteins that are encoded by the nucleic acids. Despite this and other obstacles, progress has been made in understanding some of the likely steps in the origin of life (Figure 5.1; Maynard Smith and Szathmáry 1995; Rasmussen et al. 2004; Hazen 2005).

First, *simple organic molecules*, the building blocks of complex organic molecules, have been found in space, and *can be produced by abiotic chemical reactions.* In a famous experiment, Stanley Miller found that electrical discharges in an atmosphere of methane (CH_4), ammonia (NH_3), hydrogen gas (H_2), and water (H_2O) yield amino acids and compounds such as hydrogen cyanide (HCN) and formaldehyde (H_2CO), which undergo further reactions to yield sugars, amino acids, purines, and pyrimidines.

Some such simple molecules must have formed polymers that could replicate. *Once replication originated, prebiotic evolution by natural selection could occur*, because variants that replicated more prolifically and more faithfully would increase, relative to others. Polymerization may have been facilitated by adsorption to clay particles or by concentration caused by evaporation. The most likely early replicators were short RNA (or RNA-like) molecules. *RNA has catalytic properties, including self-replication* (Joyce 2002). Some RNA sequences (RIBOZYMES) can cut, splice, and elongate oligonucleotides, and short RNA template sequences can self-catalyze the formation of complementary sequences from free nucleotides. Recent experiments have shown that clay particles with RNA adsorbed onto their surface can catalyze the formation of a lipid envelope, which in turn catalyzes the polymerization of amino acids into short proteins. Thus aggregates that have some of the critical components of a protocell, including self-replicating RNA, can be formed by chemical processes alone. It is now thought that the first steps in the origin of life took place in an "RNA world," in which catalytic, replicating RNAs existed before proteins or DNA (see Figure 5.1). Within this RNA world, evolution occurred, since *natural selection and evolution can occur in nonliving systems of replicating molecules*. When Sol Spiegelman (1970) placed RNAs, RNA polymerase (isolated from a virus, phage Qβ), and nucleotide bases in a cell-free medium, different RNA sequences were replicated by the polymerase at different rates, so that their proportions changed. Because a replicating enzyme from a living system (i.e., the virus) was provided, these experiments do not bear on the very first steps in the origin of life, but they do show that populations of replicating macromolecules can evolve.

Long RNA sequences would not replicate effectively because the mutation rate would be too high for them to maintain any identity. A larger genome might evolve, however, if two or more coupled macromolecules each catalyzed the replication of the other. Replication probably was slow and inexact originally, and only much later acquired the fidelity that modern organisms display. Moreover, many different oligonucleotides undoubtedly could replicate themselves. Before proteins evolved, there were no phenotypes—only genotypes. So *the first "genes" need not have had any particular base pair sequence.*

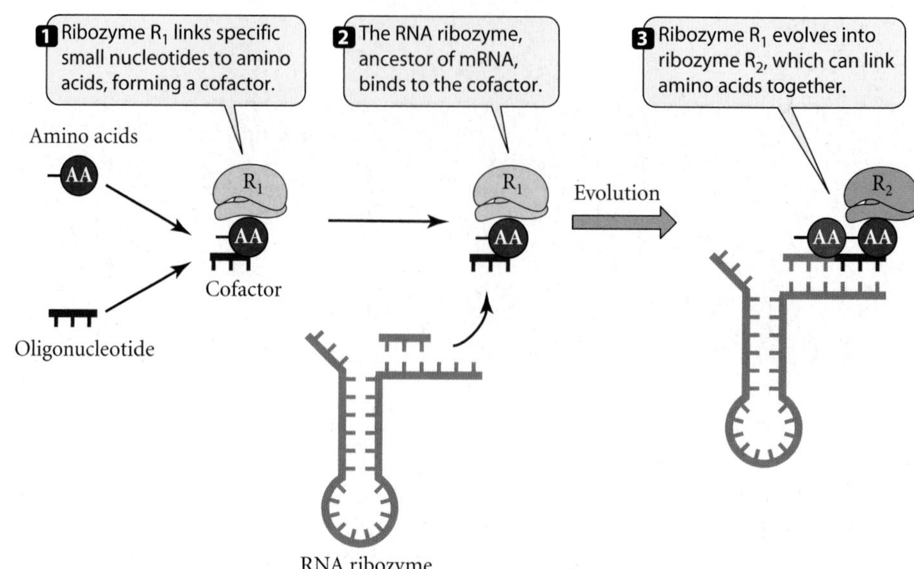

Figure 5.2 A hypothesis for the origin of polypeptide synthesis and the genetic code. An RNA ribozyme (green), an ancestor of mRNA, binds to a cofactor consisting of an amino acid (AA) and a short oligonucleotide, which have been joined by another ribozyme (R_1) that joins specific oligonucleotides to amino acids according to a primitive code. This system evolves to one in which ribozyme R_2, ancestor of the modern ribosome, links amino acids together. (After Maynard Smith and Szathmáry 1995.)

Thus there is no force to the argument, frequently made by skeptics, that the assembly of a particular nucleic acid sequence is extremely improbable (say, 1 chance in 4^{50} for a 50 base pair sequence).

How protein enzymes evolved is perhaps the greatest unsolved problem. Eörs Szathmáry (1993) has suggested that this process began when cofactors, consisting of an amino acid joined to a short oligonucleotide sequence, aided RNA ribozymes in self-replication (Figure 5.2). Many contemporary coenzymes have nucleotide components. The next step may have been the stringing together of several such amino acid–nucleotide cofactors. Ultimately, the ribozyme evolved into the ribosome, the oligonucleotide component of the cofactor into transfer RNA, and the strings of amino acids into catalytic proteins. Such ensembles of macromolecules, packaged within lipid membranes, may have been precursors of the first cells—although many other features evolved between that stage and the only cells we know.

Precambrian Life

The Archean, prior to 2.5 billion years ago, and the Proterozoic, from 2.5 billion to 542 Mya, are together referred to as **Precambrian time**. The oldest known rocks (3.8 billion years) contain carbon deposits that may indicate the existence of life. There is strong evidence of life by 3.0 billion years ago, and debated evidence as far back as 3.5 billion years ago, in the form of bacteria-like microfossils and layered mounds (stromatolites; Figure 5.3) with the same structure as those formed today along the edges of warm seas by cyanobacteria (blue-green bacteria).

The early atmosphere had little oxygen, so the earliest organisms were anaerobic. *When photosynthesis evolved* in cyanobacteria and other bacteria, *it introduced oxygen into the atmosphere*. The atmospheric concentration of oxygen increased greatly about 2.4 billion years ago, probably as a result of geological processes that buried large quantities of organic matter and prevented it from being oxidized (Knoll 2003). As oxygen built up in the atmosphere, many organisms evolved the capacity for aerobic respiration, as well as mechanisms to protect the cell against oxidation.

Living things today are classified into three "empires," or "domains": the Eucarya (all eukaryote organisms), and two groups of prokaryotes, the Archaea and Bacteria (see Figure 2.1). *For about 2 billion years—more than half the history of life—the two prokaryotic empires were the only life on Earth.* Today, many Archaea are anaerobic and inhabit extreme environments such as hot springs. [One such species is the source of the DNA polymerase enzyme (Taq polymerase) used for the PCR reaction that is the basis of much of modern molecular biology and biotechnology.] The Bacteria are extremely diverse in their metabolic capacities, and many are photosynthetic. Molecular phylogenetic studies of prokaryotes and of the nuclear genes of eukaryotes show that Archaea and Eucarya are more closely related to

Figure 5.3 Stromatolites formed by living cyanobacteria in Shark Bay, Australia. Similar structures are found in rocks of Archean age. (Photograph by the author.)

each other than to Bacteria (Figure 5.4). However, some DNA sequences provide conflicting evidence about relationships both among and within the empires, implying that there was extensive lateral transfer of genes among lineages during the early history of life, when well-defined, integrated genomes did not yet exist (Woese 2000; Boucher et al. 2003).

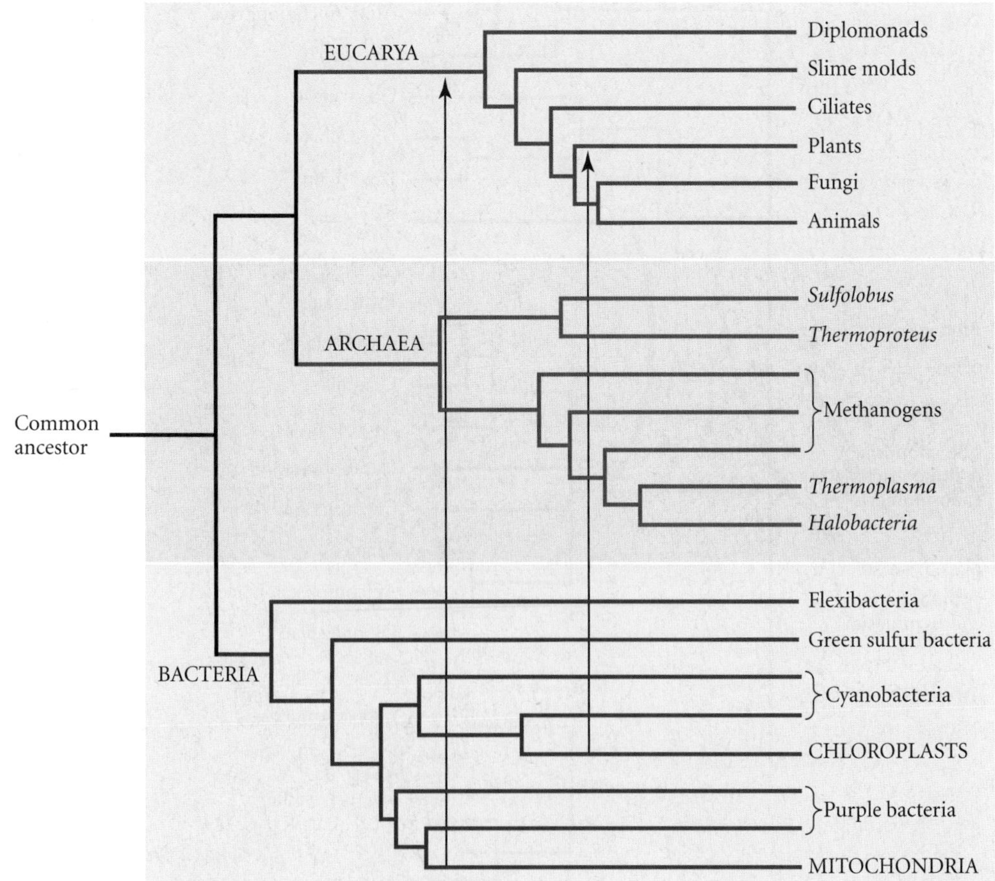

Figure 5.4 The basal branches of the tree of life: a current hypothesis of relationships among the three "empires," and among some lineages within each. Note the origin of mitochondria and chloroplasts from bacterial lineages. The vertical arrows indicate their endosymbiotic association with early eukaryotic lineages. (After Knoll 2003 and other sources.)

Thus, the early history of prokaryotes may have been more like a network than a phylogenetic tree, at least for many genes. The LCA of all modern organisms undoubtedly was the evolutionary outcome of a long history in which RNA gave way to DNA as the genetic material and genes with new functions arose (by gene duplication and mutation; see Chapters 8 and 20) and were mixed into new combinations (Becerra et al. 2007). The prokaryotes that descended from the LCA diversified greatly in their metabolic capacities (Cavalier-Smith 2006): photosynthetic, chemoautotrophic, sulfate-reducing, methanogenic, and other forms soon evolved, and continue today to be the prime movers of the biogeochemical cycles on which ecosystems depend.

A major event in the history of life was the origin of eukaryotes, which are distinguished by such features as a cytoskeleton, a nucleus with multiple chromosomes and a mitotic spindle, and a cell membrane rather than a rigid cell wall. Most eukaryotes undergo meiosis, the basis of highly organized recombination and sexual reproduction. Almost all eukaryotes have mitochondria, and many have chloroplasts.

Mitochondria and chloroplasts are descended from bacteria that were ingested, and later became **endosymbionts**, in protoeukaryotes (Margulis 1993; Maynard Smith and Szathmáry 1995; see Figure 5.4). Endosymbiosis has evolved many times, and has been

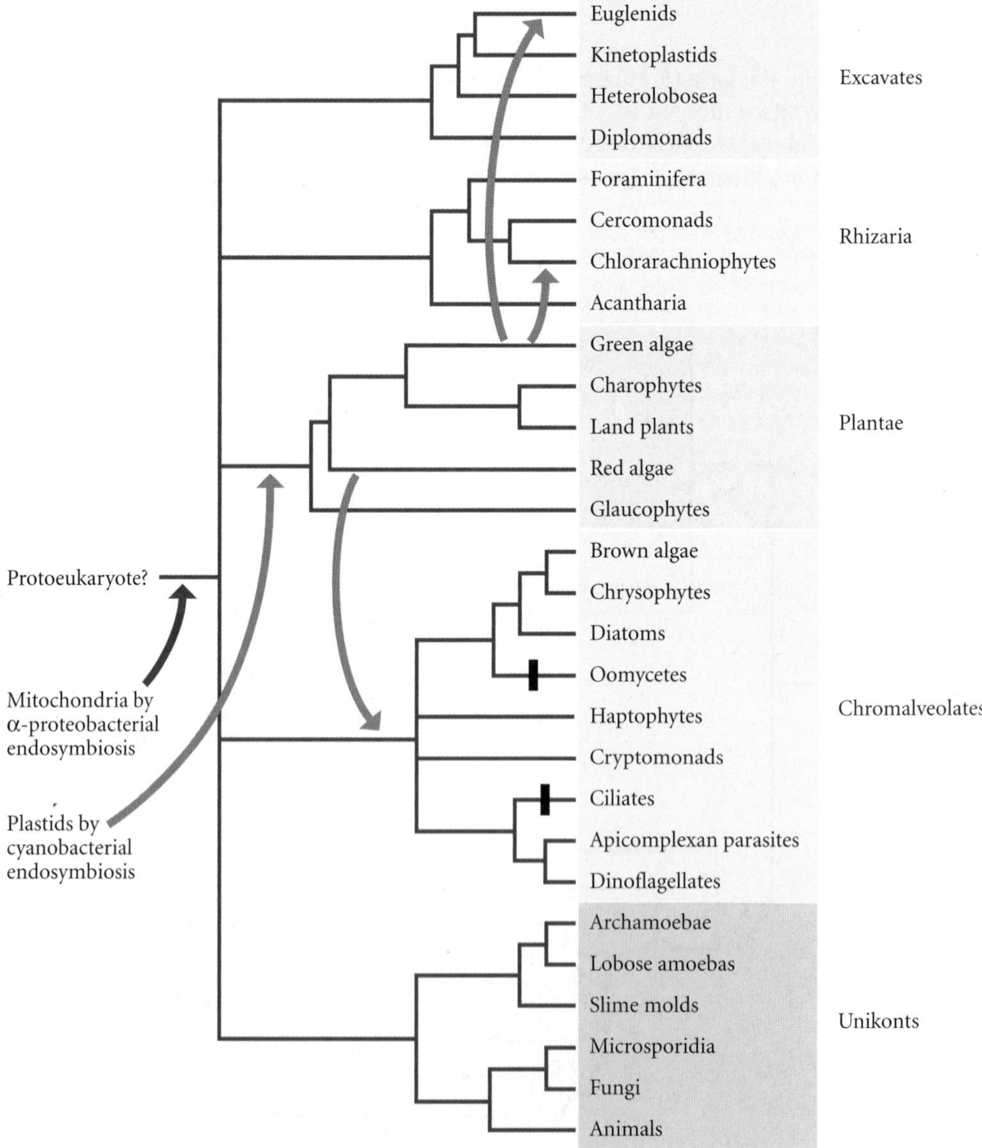

Figure 5.5 A recent estimate of the phylogeny of some of the major lineages of eukaryotes, inferred from DNA sequences. Only a few of these lineages were recognized as "kingdoms" in traditional classifications. Arrows show primary endosymbioses with bacteria that gave rise to mitochondria and chloroplasts, and secondary endosymbioses, in which unicellular eukaryotes (green algae and red algae) became the plastids of several other eukaryote groups. Black crossbars show loss of plastids in the oomycetes and ciliates. Plastids have lost photosynthetic function in apicomplexans (unicellular parasites that include the causal agent of malaria). Many aspects of this tree are still uncertain, including the relationships among the five "supergroups," shown here as an unresolved polytomy. (After Keeling 2007 and Palmer et al. 2004.)

an important source of functional complexity, in the history of life; its role in the origin of eukaryotes is surely the most important instance (Margulis and Fester 1991; Moran 2007). Mitochondria and chloroplasts are similar to bacteria both in their ultrastructure and in their DNA sequences. Mitochondria are derived from the purple bacteria (a group that includes *E. coli*), and chloroplasts from cyanobacteria. Unicellular eukaryotic green algae and red algae, in turn, have become endosymbionts of several other lineages, providing photosynthetic and other functions (Figure 5.5). The plastids of dinoflagellates (unicellular organisms that can be toxic to fish) and apicomplexans (parasites that include *Plasmodium*, the cause of malaria) may represent tertiary or quaternary endosymbioses, based on uptake of organisms that possessed algae-derived plastids (Keeling 2004). Almost nothing is known about the ancestral "host" of the bacteria that became mitochondria; it is unlikely that it had the features (e.g., cell wall) of a bacterium or archaean, but rather had protoeukaryotic features such as a cell membrane that could engulf smaller organisms, and an endoskeleton that prevented the cell from collapsing (Maynard Smith and Szathmáry 1995; Poole and Penny 2006).

Although chemical evidence suggests that eukaryotes may have evolved by about 2.7 billion years ago, the earliest eukaryote fossils are about 1.5 billion years old (Knoll 2003; Figure 5.6). Eukaryotes underwent rapid diversification, perhaps triggered by the increasing availability of nitrogen. They include far more lineages than the five kingdoms cited in most older textbooks: many lineages of "algae" and "protozoans" are more distantly related to one another than fungi and animals are (see Figure 5.5).

For nearly a billion years after their origin, almost all eukaryotes seem to have been unicellular, and most lineages remain so. A major event in the history of life was the *evolution*, independently in fungi, animals, and plants, *of multicellularity*. Multicellularity, which probably evolved because of the advantage of "divison of labor" between different cell types with different functions (Michod 1999, 2007), is a prerequisite for large size and for the development of elaborate organ systems. The evolution of tissues and organs required the evolution of cell signaling mechanisms, cell adhesion, and complex gene regulation: ways of controlling the expression of different genes in different cells. Of course, levels of transcription and translation are regulated in unicellular organisms, so the basic mechanisms already existed. In multicellular eukaryotes, however, genes may have multiple binding sites for different transcription factors, and contribute to diverse, complex developmental pathways (see Chapter 21).

The oldest fossils of multicellular animals are about 575 Myr old. Among the first animal fossils are the creatures known as the EDIACARAN FAUNA of the late Proterozoic (about 575–542 Mya) and Early Cambrian. Most of these animals were soft bodied, lacked skeletons, and appear to have been flat creatures that crept or stood on the sea floor (Figure 5.7). They are hard to classify with reference to later animals, but some may have been stem groups of the Cnidaria (corals and relatives) and the Bilateria (bilaterally symmetrical phyla with three embryonic germ layers). Ediacaran animals seem to have lacked features, such as mouthparts or locomotory appendages, that might be used in interacting with other animals, and there is little evidence that they were subject to predation.

(A)

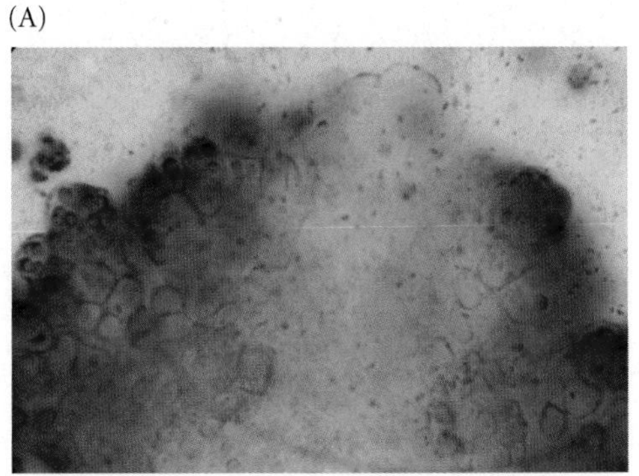

(B)

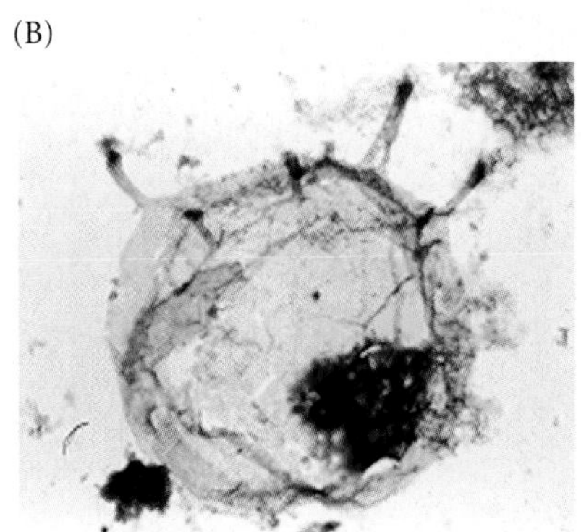

(C)

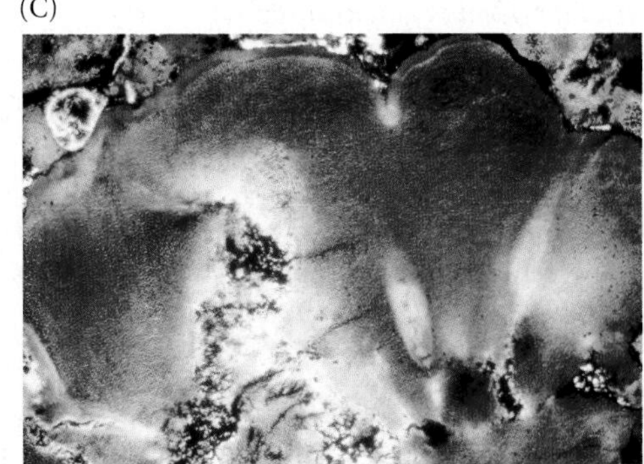

Figure 5.6 Some Proterozoic fossils. (A) A 1.5 billion-year-old colony of the cyanobacterium *Eoentophysalis*. (B) A unicellular eukaryote, the acritarch alga *Tappania*, from 1.5 billion-year-old strata in northern Australia. (C) A late Proterozoic (from about 590 Mya) multicellular red alga. (Photographs courtesy of A. H. Knoll.)

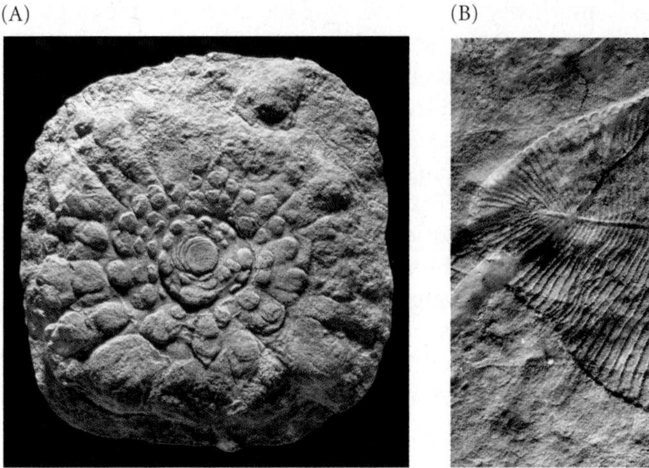

Figure 5.7 Members of the Ediacaran fauna. (A) *Mawsonites spriggi* may be a cnidarian relative of sea anemones, although this is not certain. (B) The actual relationship of the wormlike *Dickinsonia costata* to later animals is unknown. (A © The Natural History Museum, London; B courtesy of Martin Smith.)

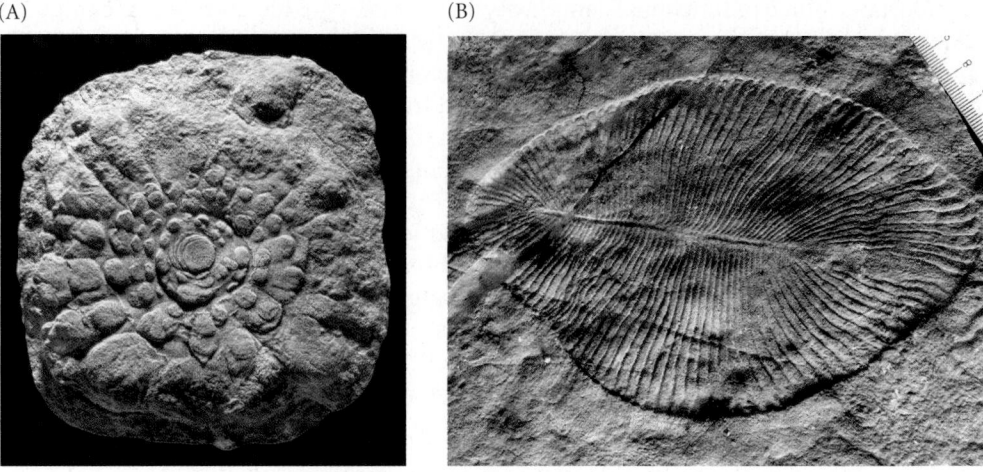

Paleozoic Life: The Cambrian Explosion

The Paleozoic era began with the Cambrian period, starting about 542 Mya. For the first 10 Myr or so, animal diversity was low; then, during a period of about 20 Myr, *almost all the modern phyla and classes of skeletonized marine animals*, as well as many extinct groups, *appeared in the fossil record*. This interval marks the first appearance of brachiopods, trilobites (Figure 5.8A), and other classes of arthropods, as well as echinoderms (Figure 5.8B), molluscs, and many others. The remarkably well preserved Cambrian fauna of the Burgess Shale of British Columbia (Figure 5.9) includes animals rather different from any that succeeded them. The Cambrian diversification included the earliest jawless (agnathan) vertebrates: the recently discovered Early Cambrian *Haikouichthys* had eyes, gill pouches, notochord, segmented musculature, and other features resembling those of larval lampreys (Figure 5.10A; Shu et al. 2003), and the Late Cambrian conodonts had teeth made of cellular bone (Figure 5.10B). Most of the fundamentally different body plans (often called **Baupläne**, German for "construction plans," or blueprints) known in animals apparently evolved during the Cambrian—perhaps the time of the most dramatic adaptive radiation in the history of life (Valentine 2004).

This diversification is generally called the **Cambrian explosion**, because it transpired over a short time relative to preceding and succeeding intervals. Some paleontologists, though, have pointed out that 20 Myr really is quite long, and so prefer a term such as "Cambrian fuse" (Prothero and Buell 2007). By whatever name, the Cambrian advent of animal phyla has long been a major puzzle and source of debate. First, how quickly did

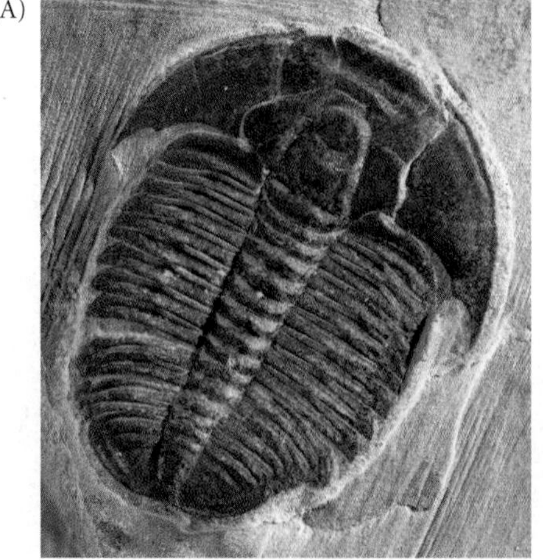

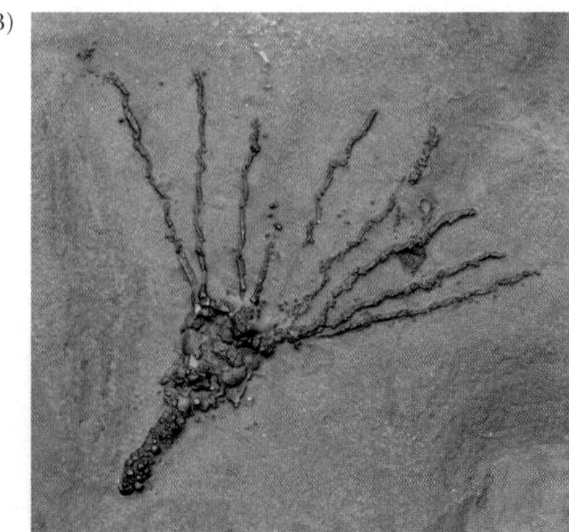

Figure 5.8 Two animal groups that first appeared during the Cambrian explosion. Both these fossils were uncovered in the sandy shales of southern Utah, an area that once was covered by shallow seas. (A) A trilobite (*Elrathii kingii*) from the middle Cambrian. Trilobites, an arthropod group, were very diverse throughout the Paleozoic but became extinct at the end of the Permian. (B) An echinoderm (*Gogia spiralis*) from the early Cambrian. Many echinoderm groups, which along with chordates comprise the deuterostome animals, flourish in the modern fauna. (A © Russell Shively/ShutterStock; B photo by David McIntyre.)

Figure 5.9 A scene on the Cambrian sea floor. This artist's rendering imagines the drama of life some 500 Mya, reconstructed from fossils found in the Burgess Shale. Species shown include the predatory clawed arthropod *Anomalocaris* and the burrowing chordate *Pikaia*. (Art © John Sibbick/NHMPL.)

evolution actually occur? Several investigators have applied a molecular clock to the DNA sequence divergence among the living animal phyla, and obtained widely variable estimates of how long ago these lineages originated, ranging from Ediacaran time (ca. 580 Mya) to as far back as 1000 Mya (Smith and Peterson 2002). Since most paleontologists think it unlikely that diverse molluscs, arthropods, and others have eluded detection in Precambrian deposits, they conclude either that the more extreme age estimates are unreliable, or that the Cambrian explosion represents the rapid evolution of shells and skeletons within clades that did not evolve fossilizable characteristics until long after they

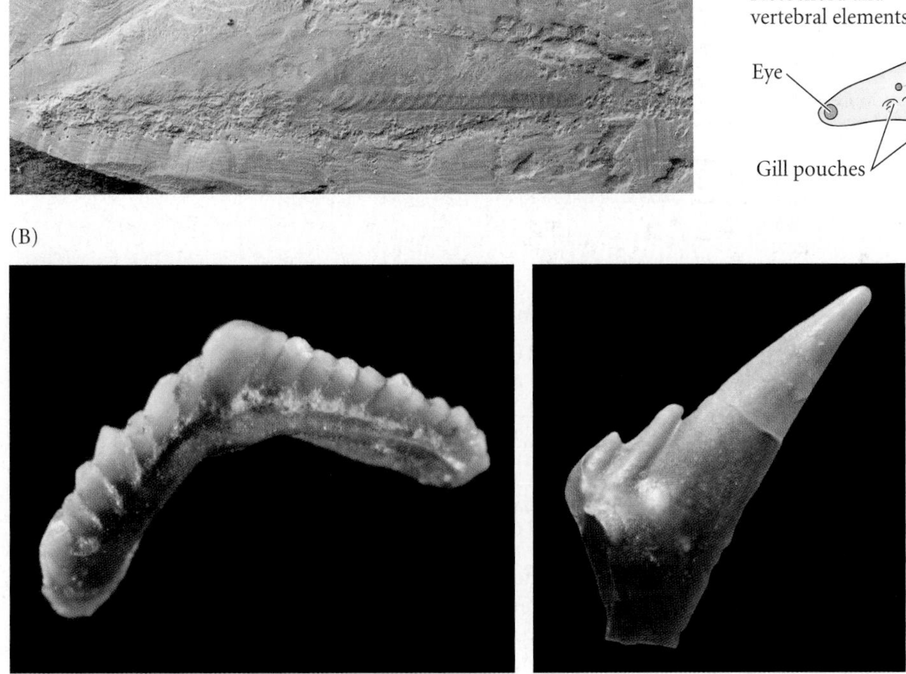

Figure 5.10 Cambrian vertebrates. (A) Photograph and drawing of one of the earliest known vertebrates, *Haikouichthys*, of the early Cambrian. The drawing calls attention to features interpreted as eye, notochord, vertebral elements, dorsal fin, esophagus, gill pouches, ventral fin, and anus, indicating a postanal tail region characteristic of vertebrates. (B) Bony, toothlike structures of Cambrian conodonts. Conodonts were slender, finless chordates believed to be related to agnathans (jawless vertebrates such as lampreys). (A courtesy of D.-G. Shu, from Shu et al. 2003; B © Custom Life Science Images.)

diverged from their common ancestors. Recent reanalyses of DNA sequences favor an Ediacaran origin of the bilaterian phyla, consonant with the possibly bilaterian nature of some Ediacaran animals (Peterson et al. 2008).

If, as seems almost indubitable, extraordinary morphological diversity evolved within about 20 Myr, how and why did so many great changes evolve at that time? A combination of genetic and ecological causes may account for this diversification (Knoll 2003; Marshall 2006). Regulatory genes that govern the differentiation of body parts (such as the Hox genes; see Chapter 21) may have undergone major evolutionary changes at this time, so that many new genetic combinations could arise. Some of the resulting morphological changes may have led to novel interactions among different organisms, such as predation, that further enhanced diversity by selecting for protective skeletons and new ways of overcoming such defenses. It is possible that environmental changes, such as an increase in oxygen, also played a role (Knoll 2003).

Molecular phylogenetic studies show that animals are most closely related to the unicellular choanoflagellates, which have cell-adhesion proteins and cell-signaling factors like those of animals (King et al. 2003; Abedin and King 2008). Sponges (phylum Porifera), which have many choanoflagellate-like cells, are the sister group of the other animals, known collectively as Metazoa (Figure 5.11). The radially symmetrical Cnidaria (jellyfishes, corals) and the Ctenophora (comb jellies) are basal branches relative to the Bilateria—

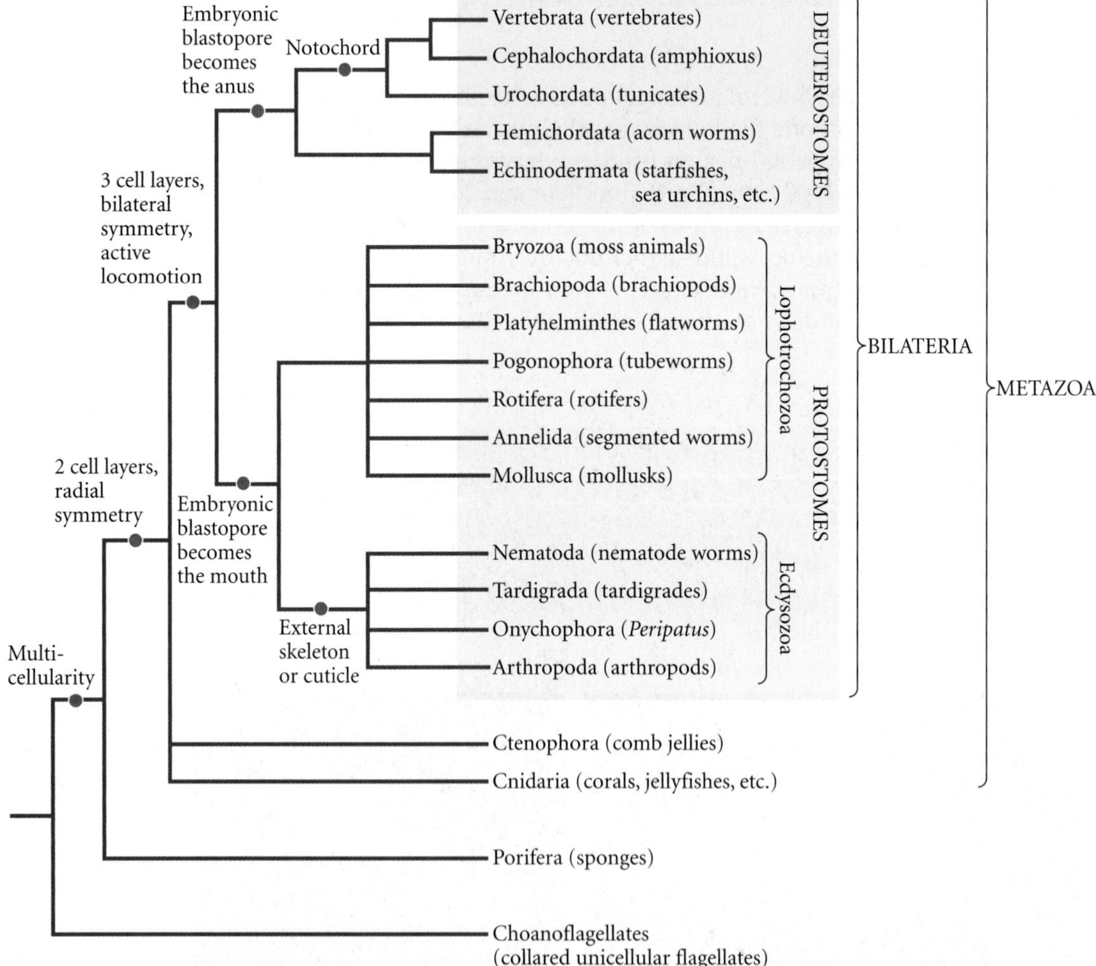

Figure 5.11 An estimate of relationships among animal phyla, based on the sequence of genes encoding ribosomal RNA. Some relationships among phyla are uncertain, and are shown as unresolved polytomies (i.e., multiple branches from a single stem, as opposed to a totally dichotomous branching pattern, seen here in the deuterostome lineage). (After Adoutte et al. 2000.)

bilaterally symmetrical animals with a head, often equipped with mouth appendages, sensory organs, and a brain. According to DNA-based phylogenetic studies, the Bilateria include three major branches: the deuterostomes, in which the blastopore formed during gastrulation becomes the anus, and two groups of protostomes, in which the blastopore becomes the mouth. The largest deuterostome phyla are the Echinodermata (sea urchins and relatives, in which a radially symmetrical adult form has evolved from a bilaterally symmetrical ancestor) and the Chordata (including the vertebrates, tunicates, and amphioxus). Protostomes form two major clades: the Ecdysozoa (arthropods, nematodes, and some smaller phyla) and the Lophotrochozoa (molluscs, annelid worms, brachiopods, and a variety of other groups).

The end of the Cambrian (488 Mya) was marked by a series of extinction events. The trilobites, of which there had been more than 90 Cambrian families, were greatly reduced, and several classes of echinoderms became extinct. As Stephen Jay Gould (1989) emphasized, if the early vertebrates had also succumbed, we would not be here today. The same may be said about every point in subsequent time: had our ancestral lineage been among the enormous number of lineages that became extinct, humans would not have evolved, and perhaps no other form of life like us would have, either.

Paleozoic Life: Ordovician to Devonian

Marine life

Many of the *animal phyla diversified greatly in the Ordovician* (488–444 Mya), giving rise to *many new classes and orders*, including as many as 21 classes of echinoderms (Figure 5.12). These classes differed greatly from one another and from the five classes of echinoderms (e.g., sea stars and sea urchins) that still exist. Many of the predominant Cambrian groups did not recover their earlier diversity, so the Ordovician fauna had a very different character from that of the Cambrian. Most Ordovician animals were EPIFAUNAL (i.e., living on the surface of the sea floor), although some bivalves (clams and relatives) evolved an INFAUNAL (burrowing) habit. The major large predators were sea stars and nautiloids (shelled cephalopods; that is, molluscs related to squids). Reefs were built by two groups of corals, with contributions from sponges, bryozoans, and cyanobacteria. The Ordovician ended with a mass extinction that in proportional terms may have been the second largest of all time. It may have been caused by a drop in temperature and a drop in sea level, for there were glaciers at this time in the polar regions of the continents.

Among the groups that survived the mass extinction were the agnathan (jawless), armored fishlike vertebrates which, except for one group, lacked paired fins (Figure 5.13A); the only living agnathans are hagfishes and lampreys. Agnathans were joined dur-

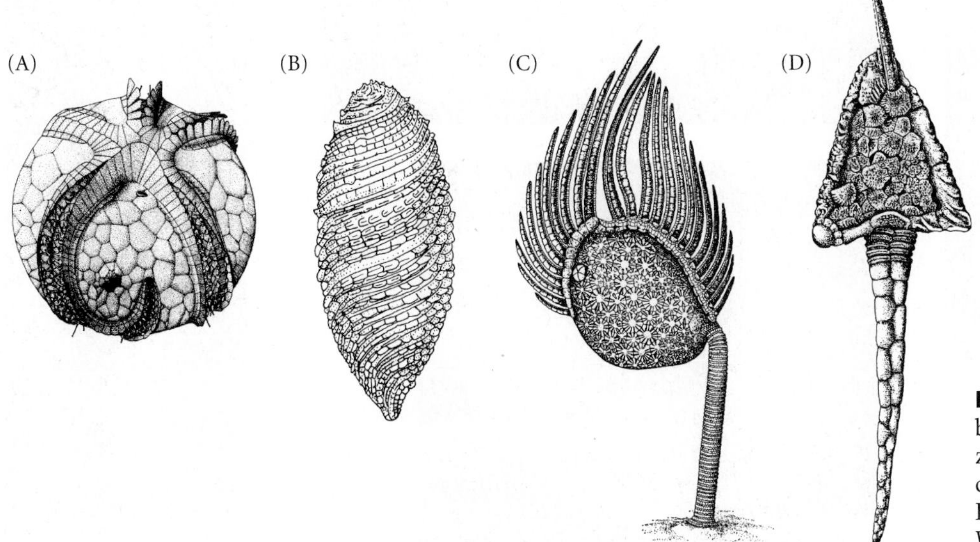

Figure 5.12 Classes of echinoderms that became extinct before the end of the Paleozoic include (A) Edrioasteroidea, (B) Helicoplacoidea, (C) Paracrinoidea, and (D) Homoiostelea. (A, C, D from Broadhead and Waters 1980; B after Moore 1966.)

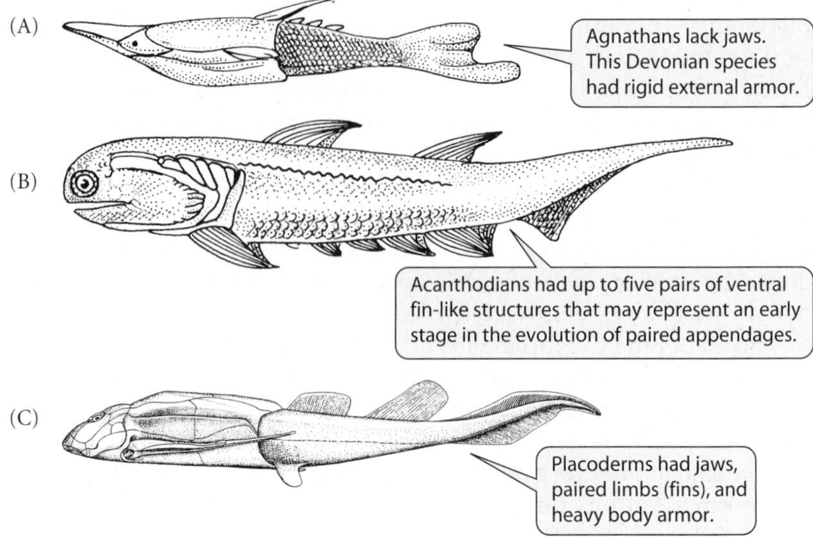

(A)
Agnathans lack jaws. This Devonian species had rigid external armor.
(B)
Acanthodians had up to five pairs of ventral fin-like structures that may represent an early stage in the evolution of paired appendages.
(C)
Placoderms had jaws, paired limbs (fins), and heavy body armor.

Figure 5.13 Extinct Paleozoic classes of vertebrates. (A) An agnathan (jawless vertebrate), class Heterostraci (*Pteraspis*, Devonian). (B) A gnathostome (jawed vertebrate), class Acanthodii (*Climatius*, Devonian). (C) A placoderm (*Bothriolepis*, Devonian). (A after Romer 1966; B after Romer and Parsons 1986; C after Carroll 1988.)

ing the Silurian (439–416 Mya) by the *first known gnathostomes*, marine vertebrates with jaws and two pairs of fins (Figure 5.13B,C). These groups continued to flourish as the Silurian was succeeded by a dramatic period in evolutionary history, the Devonian (416–354 Mya). In the same period, the nautiloids gave rise to the ammonoids, shell-bearing cephalopods that are among the most diverse groups of extinct animals (Figure 5.14).

The two subclasses of bony fishes (Osteichthyes) are both recorded first in the Late Devonian: the lobe-finned fishes (Sarcopterygii), which included diverse lungfishes and rhipidistians (see Chapter 4), and the ray-finned fishes (Actinopterygii), which would later diversify into the largest group of modern fishes, the teleosts.

Terrestrial life

Terrestrial plants, including mosses, liverworts, and tracheophytes (vascular plants), are a monophyletic group that evolved from green algae (Chlorophyta) (Figure 5.15; Judd et al. 2002). Living on land required the evolution of an external surface and spores imper-

Figure 5.14 Ammonoids and nautiloids. Shells housed the squid-like body of these cephalopod molluscs. (A–C) Three of the diverse forms of ammonoid shells. (A) *Aulacostephanus* (late Jurassic), a microconch. (B) *Nostroceras* (late Cretaceous), demonstrating an unusual coiling pattern. (C) *Kosmoceras* (Jurassic); this specimen was found in a region southeast of Moscow. (D) An orthoconic nautiloid, believed to belong to the same group as the modern *Nautilus*. These animals had noncoiled shells. (A,B,D © The Natural History Museum, London; C photo by David McIntyre.)

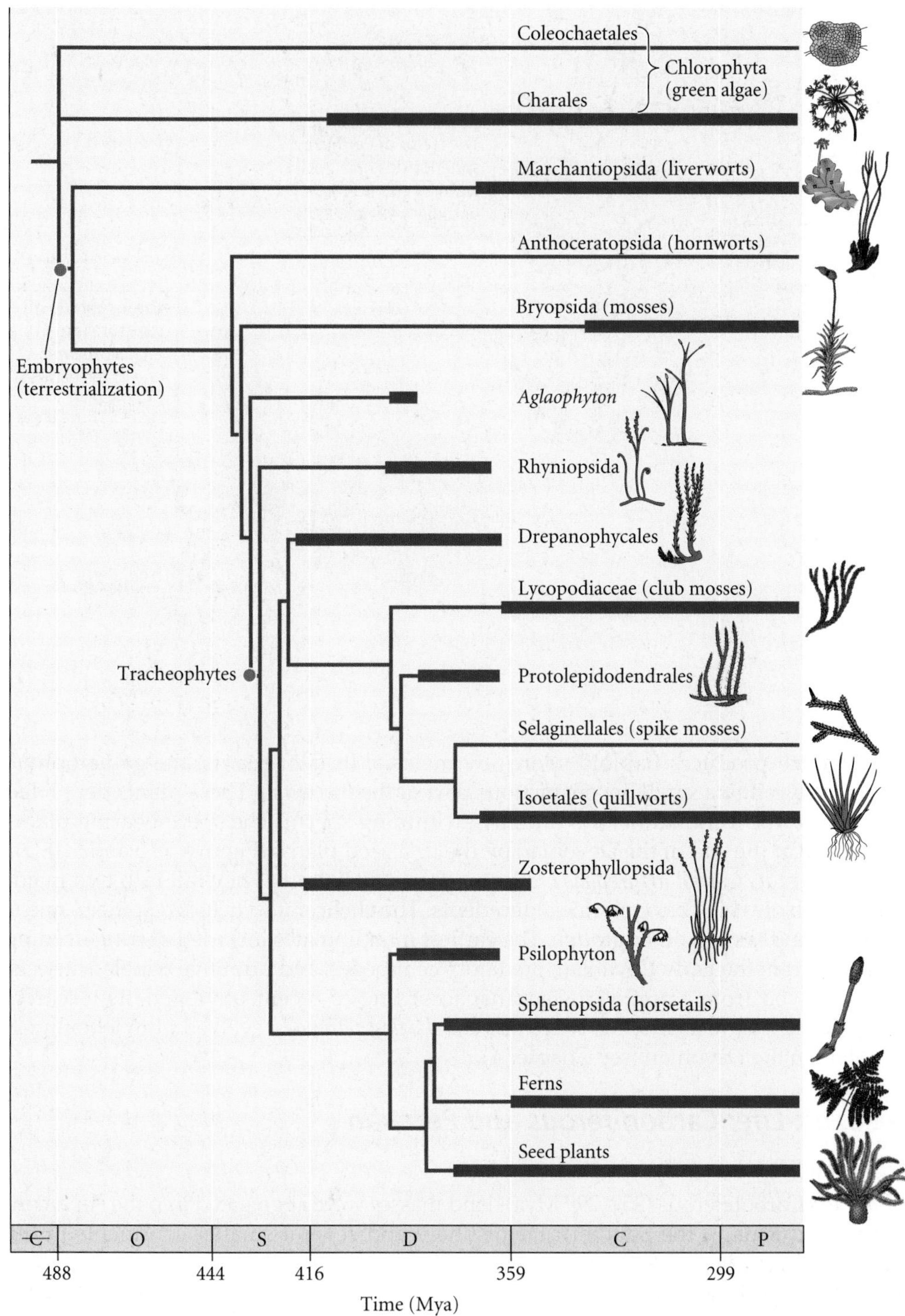

Figure 5.15 The phylogeny and Paleozoic fossil record of major groups of terrestrial plants and their closest relatives among the green algae (Chlorophyta). The broad bars show the known temporal distribution of each group in the fossil record. The Coleochaetales and Charales (two groups of green algae), liverworts, mosses, club mosses, Selaginellales, Isoetales, horsetails, ferns, and seed plants have living representatives. (After Kenrick and Crane 1997a.)

meable to water, structural support, vascular tissue to transport water within the plant body, and internalized sexual organs, protected from desiccation. The *first known terrestrial organisms are* mid-Ordovician spores and spore-bearing structures (sporangia) of *very small plants*, which were apparently related to today's liverworts (Wellman et al. 2003). By the mid-Silurian, there were small vascular plants, less than 10 centimeters high, that had sporangia at the ends of short, leafless, dichotomously branching stalks and lacked true roots (Figure 5.16A). By the end of the Devonian, land plants had greatly diversified: there were ferns, club mosses (Figure 5.16B), and horsetails (Figure 5.16C), some of which were large trees. In the earliest vascular plants, the haploid phase of the life cycle (gametophyte), which produces eggs and sperm, was as complex as the diploid (sporophyte)

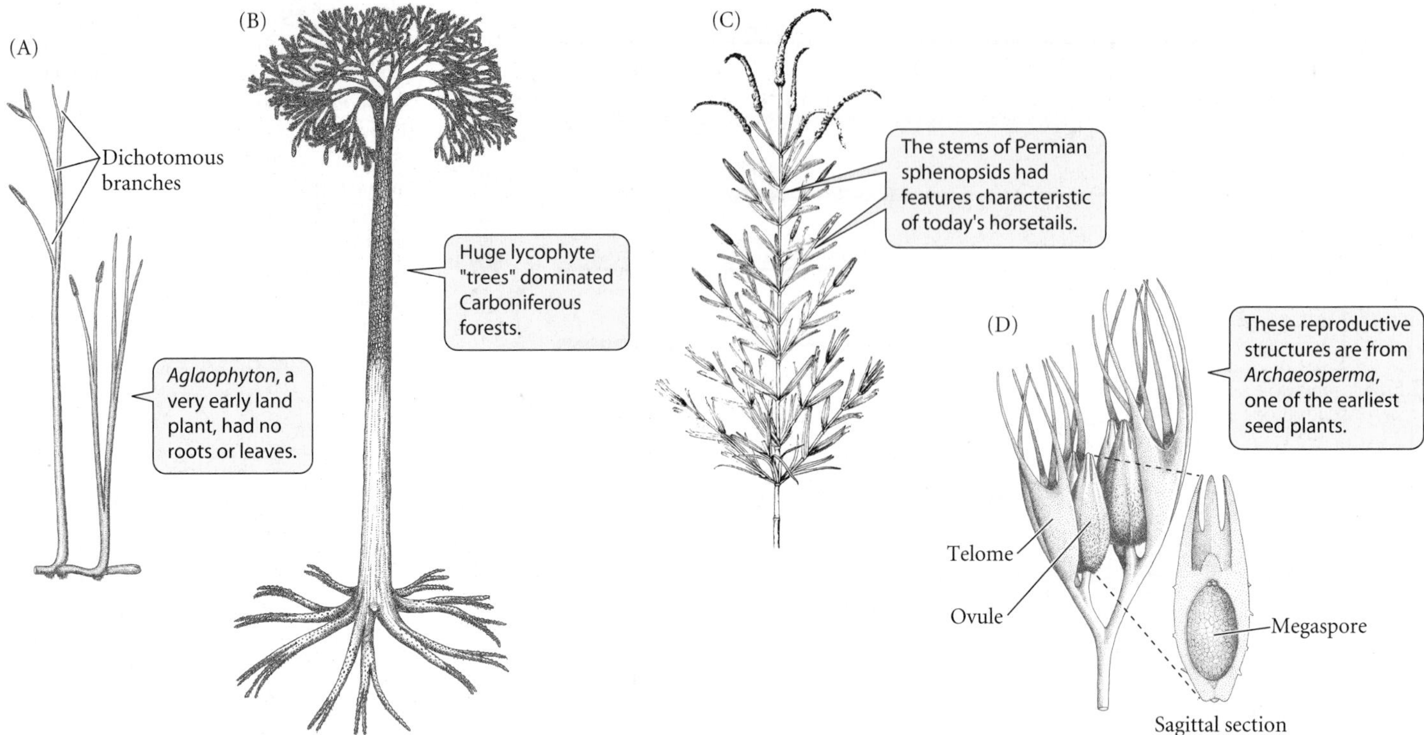

Figure 5.16 Paleozoic vascular plants, portrayed at different scales. (A) From the Devonian, *Aglaophyton* was less than 10 centimeters tall. (B) *Lepidodendron*, a large Carboniferous lycophyte tree. (C) Part of the stem of a Permian sphenopsid tree. (D) Reproductive structure of a Devonian seed plant, *Archaeosperma*. (A from Kidston and Lang 1921; B, D from Stewart 1983; C from Boureau 1964.)

phase, which produces haploid spores by meiosis. In later plants, the gametophyte became reduced to a small, inconspicuous part of the life cycle. These plants depended on water for the fertilization of ovules by swimming sperm. The spore-bearing plants were joined at the end of the Devonian by the first seed plants (Figures 5.15 and 5.16D).

The earliest terrestrial arthropods are known from the Silurian. They fall into two major groups, both of which have marine antecedents. The chelicerates included spiders, mites, scorpions, and several other groups. The earliest mandibulates included detritus-feeding millipedes from the Early Devonian, predatory centipedes, and primitive wingless insects, which evolved from aquatic crustaceans, according to recent phylogenetic research. Ichthyostegids, *the first terrestrial vertebrates and the first tetrapods*, evolved from lobe-finned fishes late in the Devonian (see Chapter 4).

Paleozoic Life: Carboniferous and Permian

Terrestrial life

During the Carboniferous (359–299 Mya), land masses were aggregated into the supercontinent Gondwana in the Southern Hemisphere and several smaller continents in the Northern Hemisphere. Widespread tropical climates favored the development of extensive swamp forests dominated by horsetails, club mosses, and ferns, which were preserved as the coal beds that we mine today. The *seed plants began to diversify* in the late Paleozoic. Some of them had wind-dispersed pollen, which freed them from depending on water for fertilization. The evolution of the seed provided the embryo with protection against desiccation, as well as a store of nutrients that enabled the young plant to grow rapidly and overcome adverse conditions. Bear in mind that none of these plants had flowers.

During the Carboniferous, the *first winged insects evolved*, and they rapidly diversified into many orders, including primitive dragonflies, orthopteroids (roaches, grasshoppers, and relatives), and hemipteroids (leafhoppers and their relatives). In the Permian, the first insect groups with complete metamorphosis (distinct larval and pupal stages) evolved, including beetles, primitive flies (Diptera), and the ancestors of the closely related Trichoptera (caddisflies) and Lepidoptera (moths and butterflies). Some orders of insects became extinct at the end of the Permian.

Tetrapod lineages ("amphibians") diversified in the Carboniferous, but most became extinct by the end of the Permian. Several late Carboniferous and early Permian lineages (anthracosaurs) have been variously classified as amphibians or reptiles. They gave rise to the *first known amniotes*, the captorhinomorphs. These primitive amniotes soon gave rise to the *synapsids*, which *included the ancestors of mammals and increasingly acquired mammal-like features* (see Figure 5.21). The first amniotes also gave rise to the diapsids, a major reptilian stock whose descendants dominated the Mesozoic landscape.

Aquatic life

During the Permian, the continents approached one another and formed *a single world continent*, **Pangaea** (Figure 5.17A), which started to break apart but then reformed during the Triassic. The sea level dropped to its lowest point in history, and climates were greatly altered by the arrangement of land and sea. The Permian ended 251 Mya with two episodes of extinction that together constitute one of the most significant events in the history of life, known as the *end-Permian mass extinction*. It is estimated that in this, *the most massive extinction event in the history of the Earth so far*, at least 52 percent of the families—and perhaps as many as 96 percent of all species—of skeleton-bearing marine invertebrates became extinct, probably within a few hundred thousand years. Groups such as ammonites, stalked echinoderms, brachiopods, and bryozoans declined greatly, and major taxa such as trilobites and several major groups of corals became extinct. On land, some orders of insects and many families of amphibians and mammal-like reptiles became extinct, and the composition of plant communities changed considerably (McElwain and Punyasena 2007). Paleontologists have not yet fully agreed on the cause of this event, but are tending to favor the hypothesis that it was triggered by vast volcanic eruptions in Siberia that covered 7 million square kilometers (2.7 million square miles) with layers of basalt as much as 6500 meters (4 miles) deep. These eruptions are thought to have released poisonous gases such as hydrogen sulfide and vast quantities of carbon dioxide, which in turn caused acid rain, increased acidity of ocean water (which interferes with the formation of calcium carbonate shells and skeletons), and global warming. The temperature change, in turn, may have caused turnover in the water column that reduced oxygen levels. Several lines of evidence are consistent with this combination of environmental changes (Erwin 2006; Knoll et al. 2007).

Mesozoic Life

The Mesozoic era, divided into the Triassic (251–200 Mya), Jurassic (200–145 Mya), and Cretaceous (145–65.5 Mya) periods, is often called the "age of reptiles." By its end, the Earth's flora and fauna were acquiring a rather modern cast, but this was preceded by the evolution of some of the most extraordinary creatures of all time. Pangaea began to break up, beginning with the Jurassic formation of the Tethyan Seaway between Asia and Africa, and then the full *separation of northern land masses, called* ***Laurasia****, from a southern continent known as* ***Gondwana*** (Figure 5.17B). Laurasia began to separate into several fragments during the Jurassic, but northeastern North America, Greenland, and western Europe remained connected until well into the Cretaceous. The southern continent, Gondwana, consisted of Africa, South America, India, Australia, New Zealand, and Antarctica. *These land masses slowly separated in the late Jurassic and the Cretaceous*, but even then the South Atlantic formed only a narrow seaway between Africa and South America (Figure 5.17C; see also Figure 6.11A). Throughout the Mesozoic, the sea level rose, and many continental regions were covered by shallow epicontinental seas. Although the polar regions were cool, *most of the Earth enjoyed warm climates*, with global temperatures reaching an all-time high in the mid-Cretaceous, after which substantial cooling occurred.

Marine life

During the Triassic, many of the marine groups that had been decimated during the end-Permian extinction again diversified; ammonoids (see Figure 5.14), for example, had increased from two genera to more than a hundred by the Middle Triassic. Planktonic

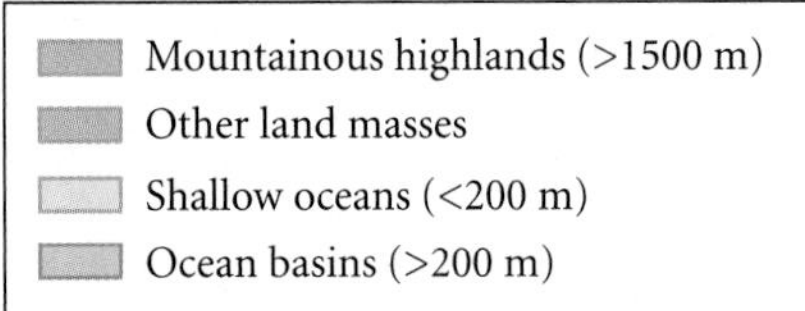

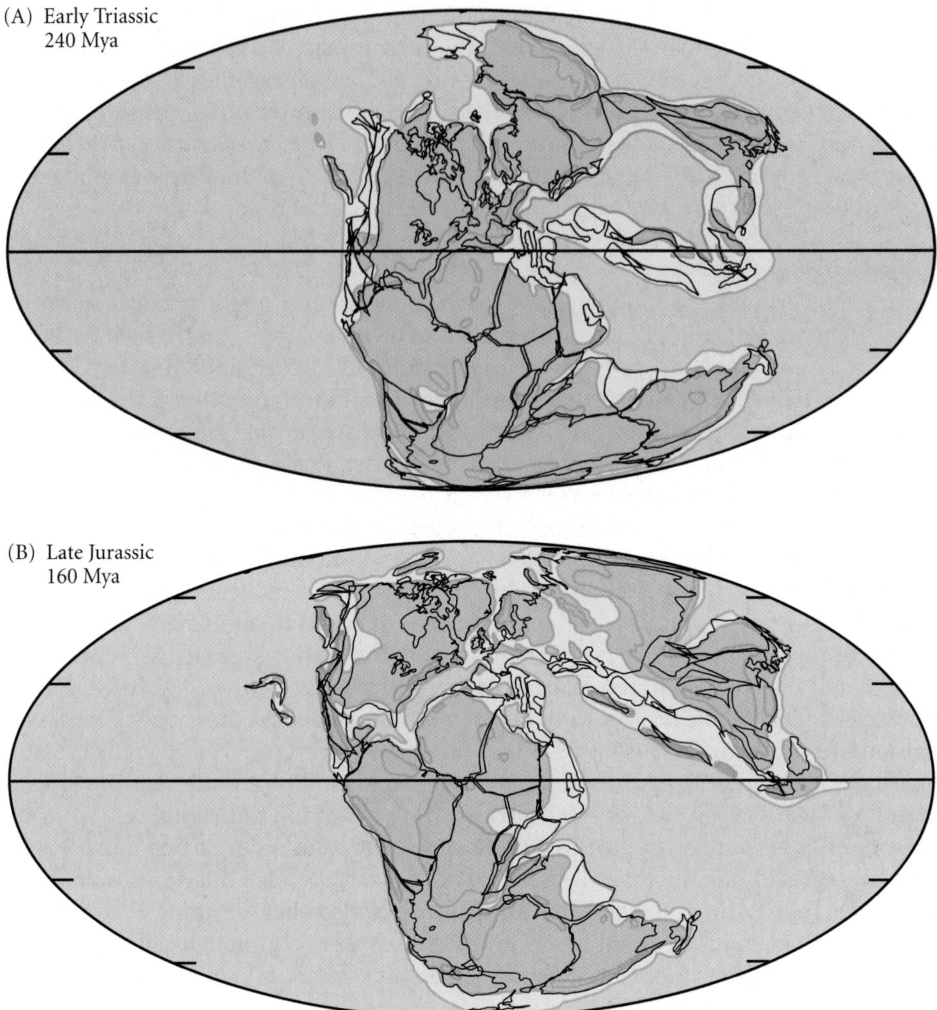

Figure 5.17 The distribution of land masses at several points in geological time. (A) In the earliest Triassic, most land was aggregated into a single mass (Pangaea). (B) Eurasia and North America were fairly separate by the late Jurassic. (C) Gondwana had become fragmented into most of the major southern land masses by the late Cretaceous. (D) By the late Oligocene, the land masses were close to their present configurations. The outlines of the modern-day continents are visible in all of the maps; other black lines delineate important tectonic plate boundaries. (Maps © 2004 by C. R. Scotese/ PALEOMAP Project.)

foraminiferans (shelled protists) and modern corals evolved, and bony fishes continued to radiate. Another *mass extinction occurred at the end of the Triassic,* again decimating groups such as ammonoids and bivalves, which then recovered and experienced yet another adaptive radiation. The teleosts, today's dominant group of bony fishes, evolved and began to diversify. During the Mesozoic and continuing into the early Cenozoic, predation seems to have escalated (Vermeij 1987; Huntley and Kowalewski 2007): during the so-called *Mesozoic marine revolution*, crabs and bony fishes evolved that could crush mollusc shells, and molluscs evolved with protective mechanisms such as thick shells and spines (Figure 5.18).

During the Jurassic and Cretaceous, modern groups of gastropods (snails and relatives), bivalves, and bryozoans rose to dominance, gigantic sessile bivalves (rudists) formed reefs, and the seas harbored several groups of marine "reptiles." The *end of the Cretaceous is marked by* what is surely the best-known *mass extinction* (often called the *K/T extinction*, using the abbreviations for Cretaceous and Tertiary), in which about 15 percent of marine animal families and up to 47 percent of genera became extinct (Jablonski 1995). Ammonoids, rudists, most marine "reptiles," and many families of invertebrates and planktonic protists became entirely extinct. The last of the nonavian dinosaurs became extinct at this time. Many paleontologists have concluded that this extinction was caused by the impact of an asteroid or some other extraterrestrial body, although others dispute this hypothesis (see Chapter 7).

Terrestrial plants and arthropods

For most of the Mesozoic, the flora was dominated by gymnosperms (i.e., seed plants that lack flowers). The major groups were the cycads (Figure 5.19A) and the conifers and their

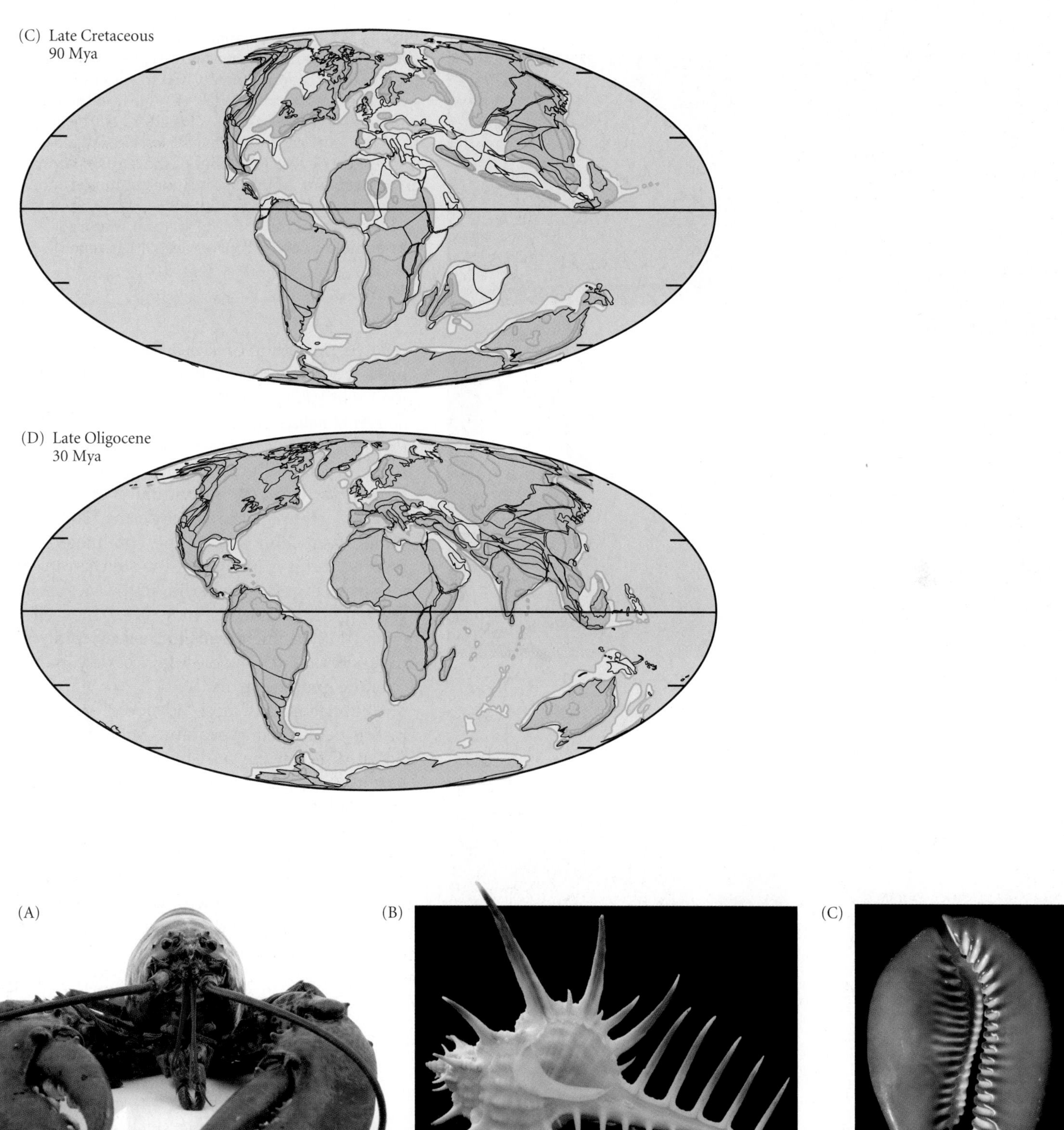

Figure 5.18 Features of marine predators and prey that escalated during and after the "Mesozoic marine revolution." (A) The huge claws of modern lobsters represent a trait found in several crustacean groups. The claws allow some lobsters and crabs to crush and rip mollusc shells. (B) Spines on both bivalves and gastropods (here *Murex*) prevent some fishes from swallowing prey and may reduce the effectiveness of crushing predators. (C) Thick shells and narrow apertures, as in *Cypraea mauritiana*, deter predators. (A © Rob Wilkinson/Alamy; B,C photos by David McIntyre.)

(A)

(B)

(C)

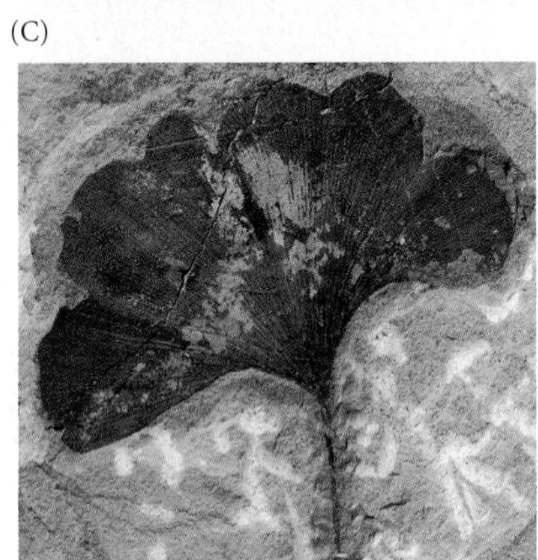

(D)

Figure 5.19 Seed plants. (A) A living cycad (*Cycas* sp.). Abundant and highly diverse during the Mesozoic, approximately 130 species of these gymnosperms survive today. (B) A leaf of the living ginkgo (*Ginkgo biloba*). (C) A fossilized *Ginkgo* leaf from the Paleocene. (D) *Protomimosoidea*, a Paleocene/Eocene fossil member of the legume family, an angiosperm group that includes mimosas and acacias. (A © Victoria Field/ShutterStock; B photo by David McIntyre; C © The Natural History Museum, London; D courtesy of W. L. Crepet.)

relatives, including *Ginkgo*, a Triassic genus that has left a single species that survives as a "living fossil" (Figure 5.19B,C). The *angiosperms, or flowering plants* (Figure 5.19D) *first appeared in the early Cretaceous*. Many of the anatomical features of angiosperms, including flowerlike structures, evolved individually in various Jurassic groups of gymnosperms, some of which were almost certainly pollinated by insects, and it is likely that the stem group of angiosperms evolved earlier in the Mesozoic. During the Cenozoic, long after they first evolved, the angiosperms increased rapidly in diversity and achieved ecological dominance over the gymnosperms.

The anatomically most "advanced" groups of insects made their appearance in the Mesozoic (Figure 5.20). By the late Cretaceous, most families of living insects, including ants and social bees, had evolved. Throughout the Cretaceous and thereafter, insects and angiosperms affected each other's evolution and may have augmented each other's diversity. As different groups of pollinating insects evolved, adaptive modification of flowers to suit different pollinators gave

(A)

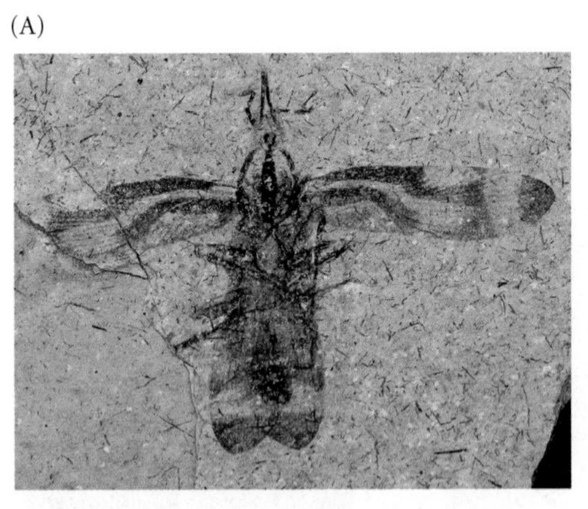

(B)

(C)

(D)

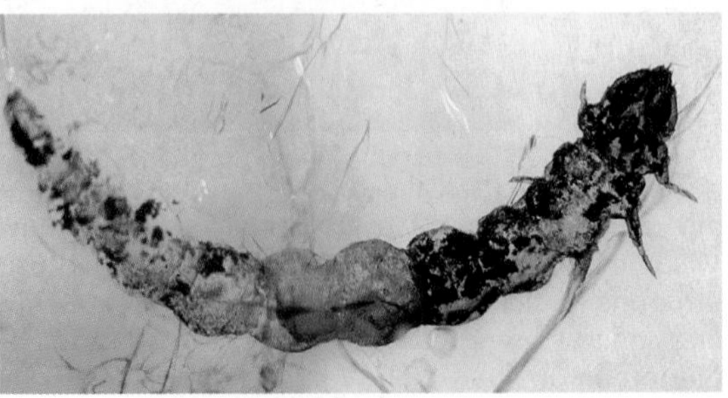

Figure 5.20 Some fossil insects. (A) A Jurassic relative of roaches (*Rhipidoblattina*) was a predator, unlike modern roaches. (B) An early Cretaceous beetle in one of the morphologically most primitive families (Cupedidae) that still has a few "living fossil" species. (C) One of the earliest known fossil bees (*Protobombus*, Eocene) was in the social bee family, Apidae. (D) Among living moths, the family Micropterigidae has the most ancestral features, such as biting mandibles. This Cretaceous micropterigid larva (100 Mya) is among the earliest lepidopteran fossils. (From Grimaldi and Engel 2005, courtesy of D. Grimaldi.)

rise to the great floral diversity of modern plants. *It is largely because of the spectacular increase of angiosperms and insects that terrestrial diversity is greater today than ever before.*

Vertebrates

The major groups of amniotes are distinguished by the number of openings in the temporal region of the skull (at least in the stem members of each lineage; Figure 5.21). One

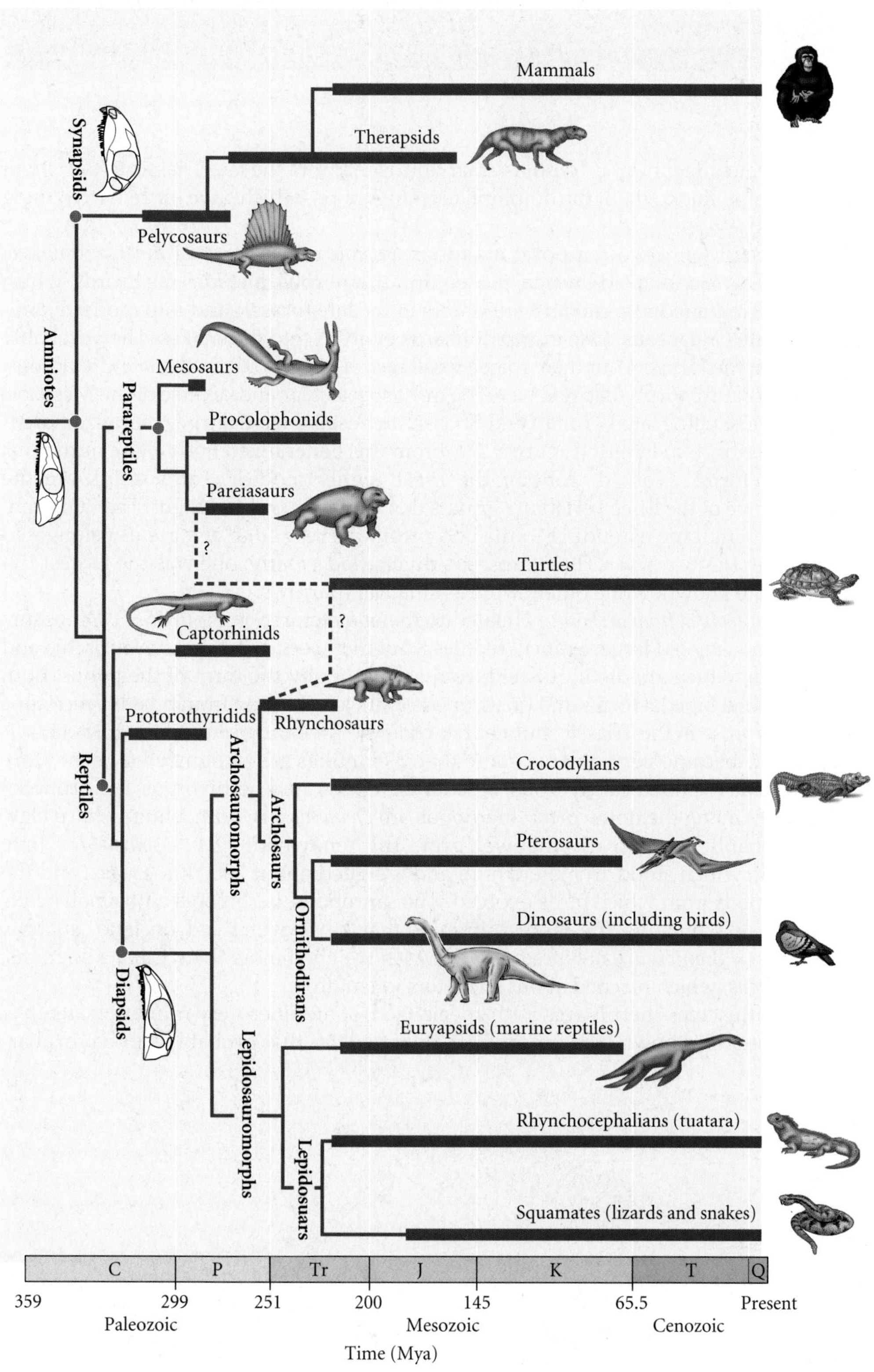

Figure 5.21 Phylogenetic relationships and temporal duration (thick bars) of major groups of amniote vertebrates. Some authors define "reptiles" as one of the two major lineages of amniotes, the other being the synapsids, which includes mammals. (After Lee et al. 2004.)

This fossil preserved the outline of the ichthyosaur's skin.

Figure 5.22 An ichthyosaur. The dorsal and tail fins of this extinct marine reptile are superficially similar to those of sharks and porpoises, although the tail fin of porpoises is horizontal. (Courtesy of The Field Museum, Chicago.)

such group included marine "reptiles" that flourished from the late Triassic to the end of the Cretaceous, among them the dolphinlike ichthyosaurs, which gave birth to live young (Figure 5.22).

The *diapsids*, with two temporal openings, *became one of the most diverse groups of amniotes*. One major diapsid lineage, the lepidosauromorphs, includes the lizards, which became differentiated into modern suborders in the late Jurassic and into modern families in the late Cretaceous. One group of lizards evolved into the snakes. They probably originated in the Jurassic, but their sparse fossil record begins only in the late Cretaceous.

The archosauromorph diapsids were the most spectacular and diverse of the Mesozoic amniotes. Most of the late Permian and Triassic archosaurs were fairly generalized predators a meter or so in length (Figure 5.23). From this generalized body plan, numerous specialized forms evolved. Among the most highly modified archosaurs are the pterosaurs, one of the three vertebrate groups that evolved powered flight. The wing consisted of a membrane extending to the body from the rear edge of a greatly elongated fourth finger (Figure 5.24). The pterosaurs diversified greatly: one was the largest flying vertebrate known, while others were as small as sparrows.

Dinosaurs evolved from archosaurs related to the one pictured in Figure 5.23. Dinosaurs are not simply any old large, extinct "reptiles," but members of the orders Saurischia and Ornithischia, which are distinguished from each other by the form of the pelvis. Both orders included bipedal forms and quadrupeds that were derived from bipedal ancestors. Both orders arose in the Triassic, but neither order became diverse until the Jurassic.

Dinosaurs became very diverse: more than 39 families are recognized (Figure 5.25). The Saurischia included carnivorous, bipedal theropods, and herbivorous, quadrupedal sauropods. Among the noteworthy theropods are *Deinonychus*, with a huge, sharp claw that it probably used to disembowel prey; the renowned *Tyrannosaurus rex* (late Cretaceous), which stood 15 meters high and weighed about 7000 kilograms; and the small theropods from which birds evolved. The sauropods, herbivores with small heads and long necks, include the largest animals that have ever lived on land, such as *Apatosaurus* (= *Brontosaurus*); *Brachiosaurus*, which weighed more than 80,000 kilograms; and *Diplodocus*, which reached about 30 meters in length.

The Ornithischia—herbivores with specialized, sometimes very numerous, teeth—included the well-known stegosaurs, with dorsal plates that probably served for ther-

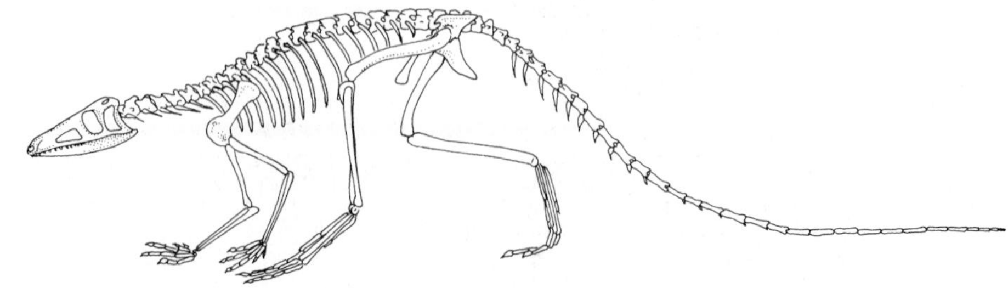

Figure 5.23 *Lagosuchus*, a Triassic thecodont archosaur, showing the generalized body form of the stem group from which dinosaurs evolved. (After Bonaparte 1978.)

(A)

(B)

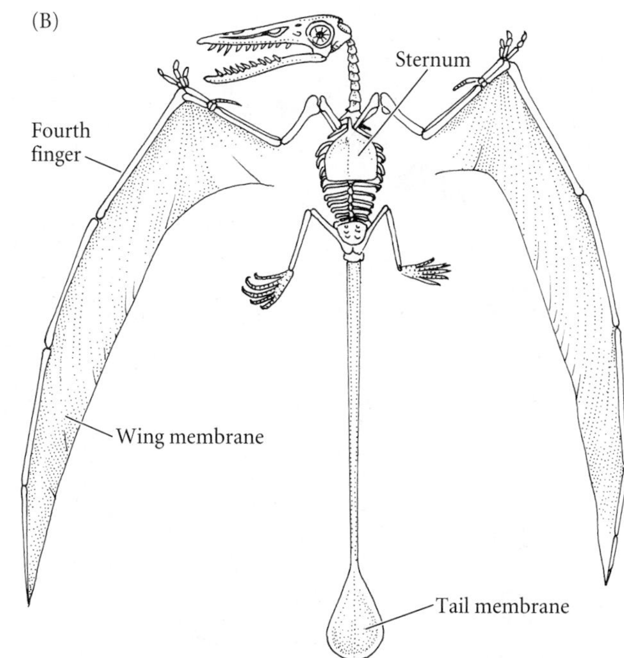

Figure 5.24 The pterosaurs, along with birds and bats, comprise the three groups of flying vertebrates. (A) The relatively small *Pterodactylus* was the first pterosaur genus discovered. Its name is Greek for "winged fingers." (B) A Jurassic pterosaur, *Rhamphorhynchus*, showing the wing membrane supported by the greatly elongated fourth finger, the large breastbone (sternum) to which flight muscles were attached, and the terminal tail membrane particular to this genus, possibly used for steering. (A © Andre Maslennikov/Photolibrary.com; B after Williston 1925.)

moregulation, and the ceratopsians (horned dinosaurs), of which *Triceratops* is the best known.

The extinction of the ceratopsians at the end of the Cretaceous left only one surviving lineage of dinosaurs, which radiated extensively in the late Cretaceous or early Tertiary and today includes about 10,000 species. Aside from these dinosaurs, more familiarly known as birds, the only living archosaurs are 22 species of crocodilians.

The late Paleozoic *synapsids* (see Figure 5.21), with a single temporal opening, *gave rise to the therapsids*, sometimes called "mammal-like reptiles," which flourished until the middle Jurassic. The first therapsid descendants that can be considered borderline mammals were the morganucodonts of the late Triassic and early Jurassic (see Figure 4.10F). Although most mammals in later Mesozoic deposits are known only from teeth and jaw fragments, it is clear that several lineages of small mammals proliferated. Most of them have left no living descendants, but one group gave rise to the *marsupials and the eutherian (placental) mammals in the early Cretaceous*. Many of the mammals of this time had the primitive "ground plan" of a generalized mammal, resembling living species such as opossums in overall form. Most of the modern orders of placental mammals are not recognizable in the fossil record until the early Cenozoic.

The Cenozoic Era

The Cenozoic era embraces six epochs, Paleocene through Pleistocene. Today we are still in the Pleistocene, but the last 10,000 years are often distinguished as a seventh epoch (the Holocene, or Recent). Traditionally, the first five epochs (from 65.5 to 1.8 Mya) are referred to as the Tertiary period, and the Pleistocene and Recent (1.8 Mya–present) as the Quaternary period. Paleontologists today generally divide the era into Paleogene (65.5–23 Mya) and Neogene (23 Mya–present) periods.

By the beginning of the Cenozoic, North America had moved westward, becoming separated from Europe in the east, but forming the broad *Bering Land Bridge* between Alaska and Siberia, which *remained above sea level* throughout most of the era (see Figure 5.17D). *Gondwana broke up* into the separate island continents of South America, Africa, India, and, far to the south, Antarctica plus Australia (which separated in the Eocene). About 18 to 14 Mya, during the Miocene, Africa made contact with southwestern Asia, India collided with Asia (forming the Himalayas), and Australia moved northward, approaching southeastern Asia. During the Pliocene, about 3.5 Mya, *the Isthmus of Panama arose, connecting North and South America for the first time.*

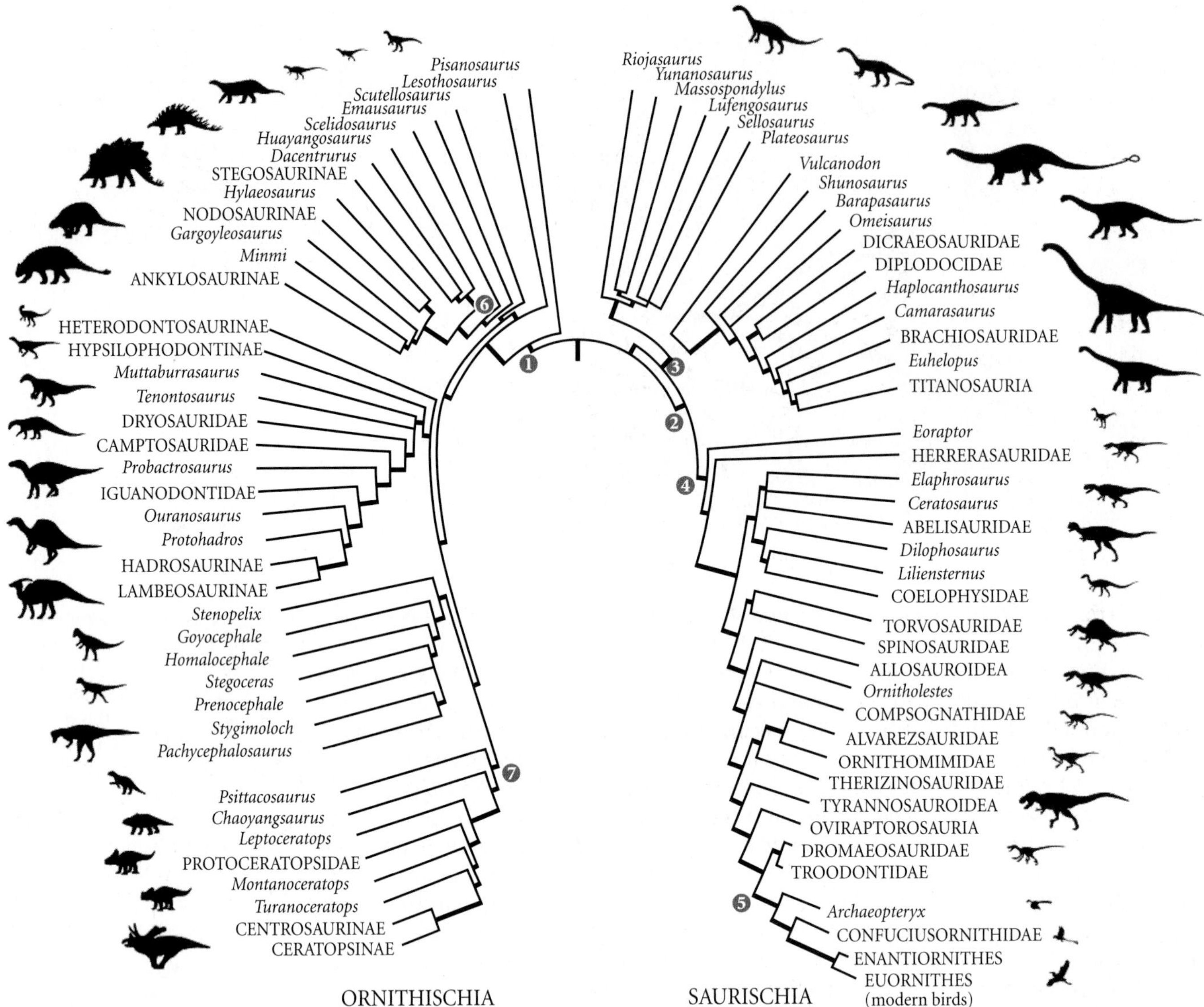

Figure 5.25 The great diversity of dinosaurs. The root of this proposed phylogeny is central, near the top. The two great clades of dinosaurs, Ornithischia (1; left) and Saurischia (2; right), are curved downward to fit the page. The Saurischia included the Sauropoda (3) and Therapoda (4), of which birds (Aves) are the only survivors (5). Ornithischia included stegosaurs (6) and ceratopsians (7). All ornithischian lineages are now extinct. (From Sereno 1999, *Science* 284: 2139. Copyright © AAAS.)

This reconfiguration of continents and oceans contributed to major climatic changes. In the late Eocene and Oligocene, there was *global cooling and drying*; extensive savannahs (sparsely forested grasslands) formed for the first time, and Antarctica acquired glaciers. Sea level fluctuated, dropping drastically in the late Oligocene (about 25 Mya). During the Pliocene, temperatures increased to some extent, but toward its end, *temperatures dropped, and a series of glaciations began* that have persisted throughout the Pleistocene. The most recent such "Ice Age" ended only about 8000 years ago.

Aquatic life

Most of the marine groups that survived the K/T mass extinction proliferated early in the Cenozoic, and some new higher taxa evolved, such as the burrowing sea urchins known as sand dollars. The taxonomic composition of Cenozoic marine communities was quite similar to that of modern ones. Teleost fishes continued to diversify throughout the Cenozoic, becoming by far the most diverse aquatic vertebrates.

During the Pleistocene glaciations, so much water was sequestered in glaciers that the sea level dropped as much as 100 meters below its present level. As many as 70 percent of the mollusc species along the Atlantic coast of North America became extinct in the early Pleistocene; these extinctions were especially pronounced in the tropics.

Terrestrial life

Most of the modern families of angiosperms and insects had become differentiated by the Eocene or earlier, and many fossil insects of the late Eocene and Oligocene belong to genera that still survive. The dominant plants of the savannahs that developed in the Oligocene were grasses (Poaceae), which underwent a major adaptive radiation at this time, and many groups of herbaceous plants, most of which evolved from woody ancestors. Among the most important of these is the family Asteraceae, which includes sunflowers, daisies, ragweeds, and many others. It is one of the two largest plant families today.

Many living orders and families of birds are recorded from the Eocene (55.8–33.9 Mya) and Oligocene (33.9–23 Mya). The largest order of birds, the perching birds (Passeriformes), first displayed its great diversity in the Miocene (23–5.3 Mya). Another great adaptive radiation was the snakes, which began an exponential increase in diversity in the Oligocene. Snakes today feed on a great variety of animal prey, from worms and termites to bird eggs and wild pigs, and they include marine, burrowing, and arboreal forms, some of which can glide.

The adaptive radiation of mammals

The Cenozoic is sometimes called "the age of mammals" (Prothero 2006). However, the time course of the diversification of placental mammals is highly controversial. Almost all mammalian fossils that can be assigned to modern orders occur after the K/T boundary (65.5 Mya), and most paleontologists maintain that most of the orders diverged shortly before or after this time (Wible et al. 2007). It has often been suggested, moreover, that the extinction of the large dinosaurs at the end of the Cretaceous relieved the subjugated mammals from competition and predation, and allowed them to undergo adaptive radiation. However, DNA sequence differences, analyzed by methods that use multiple fossils to calibrate the rate of sequence evolution, indicate that most of the orders diverged from one another in the Cretaceous, and that the major lineages within each order diverged from about 77 Mya (for the major living lineages of Primates) to 50 Mya (Figure 5.26; Springer et al. 2003). It is possible that although many lineages of mammals evolved during the Cretaceous, they remained small in size, were relatively uncommon, and perhaps did not evolve their distinctive ecological and morphological features until the Cenozoic. (As we have seen, a similar hypothesis has been suggested for the diversification of animal phyla.) One research group used a molecular clock to estimate the date of branch points throughout a phylogeny of 99 percent of the living species of mammals (see Chapter 7). They concluded that there was no acceleration in the rate of origin of new lineages immediately after the end-Cretaceous extinction, and challenged the hypothesis that the extinction of dinosaurs enabled the ancestors of today's mammals to diversify (Bininda-Emonds et al. 2007).

Marsupials are known as fossils from all the continents, including Antarctica. Today they are restricted to Australia and South America (with one exception, the North American opossum). The marsupial families that include kangaroos, wombats, and other living Australian marsupials evolved in the mid-Tertiary. In South America, marsupials experienced a great adaptive radiation; some resembled kangaroo rats, others saber-toothed cats. Most South American marsupials became extinct by the end of the Pliocene.

In addition to marsupials, many groups of placental mammals evolved in South America during its long isolation from other continents. These included an ancient placental group, the Xenarthra (or Edentata), which includes the giant ground sloths that sur-

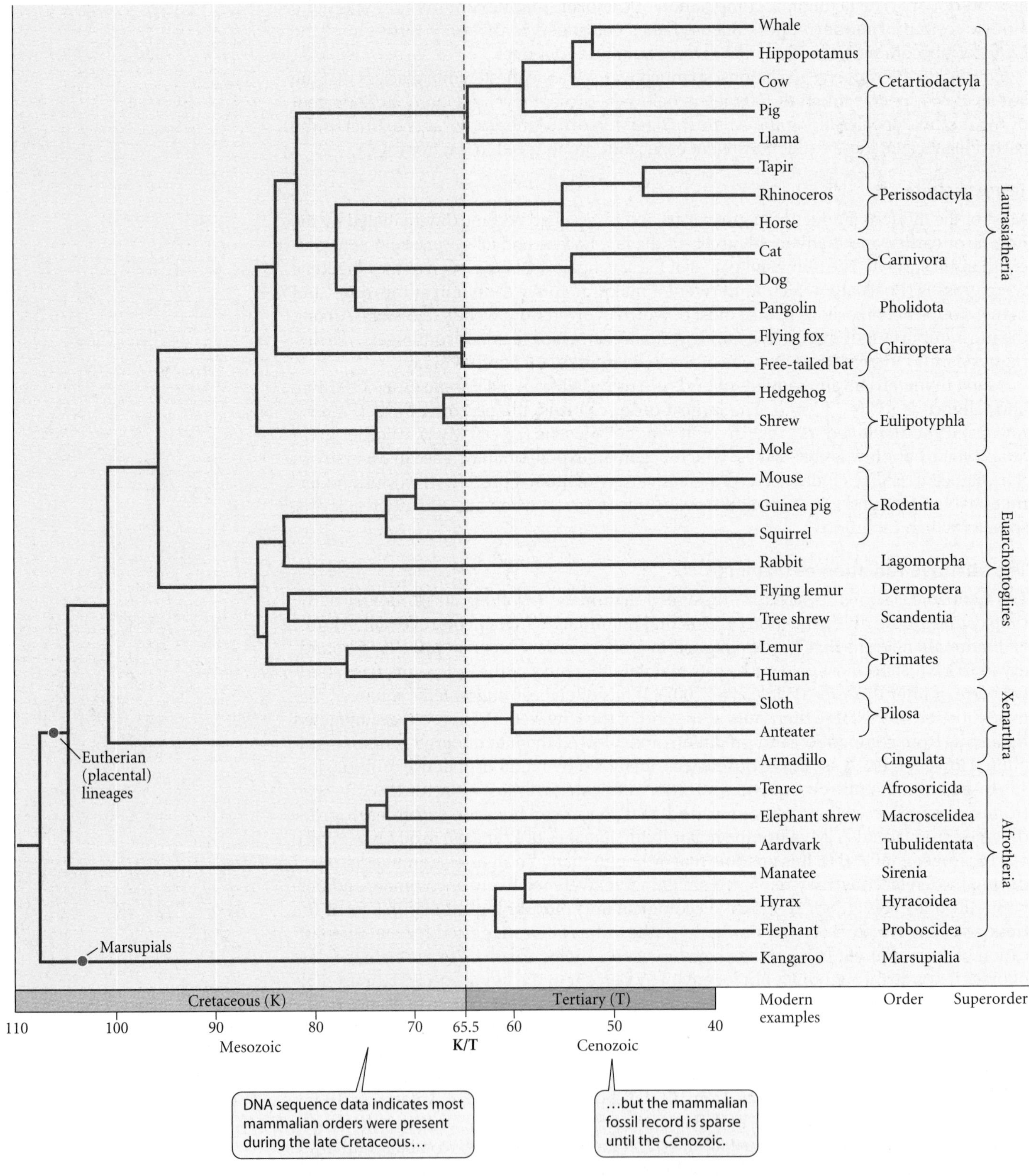

Figure 5.26 A phylogeny of living groups of marsupial and placental mammals, based on DNA sequence data. The timing of branch points is based on sequence divergence, calibrated by paleontological data. The data indicate that most orders diverged from one another during the Cretaceous. The dearth of mammalian fossils prior to the K/T boundary, however, suggests that early mammals in all these lineages may have been small and/or uncommon. Some relationships shown here are still tentative; note the different relationship of Xenarthra, as the sister group of Afrotheria, in Figure 2.22. (After Springer et al. 2003.)

Figure 5.27 The giant ground sloth *Megatherium* ("great beast") was a Pleistocene representative of the Xenarthra. (Photo © Nigel Reed QEDimages/Alamy.)

vived until the late Pleistocene (Figure 5.27) and the few armadillos, anteaters, and sloths that still persist. At least six orders of hoofed mammals that resembled sheep, rhinoceroses, camels, elephants, horses, and rodents evolved in South America, but declined and became extinct after South America became connected to North America in the late Pliocene. The extinction of many South American mammals has been attributed to the ecological impact of North American mammal groups, such as bears, raccoons, weasels, peccaries, and camels, that moved into South America at this time as part of what has been called the "Great American Interchange" of terrestrial organisms.

Among placental orders, one of the earliest and, in many ways, structurally most primitive is the Primates. The earliest fossils assigned to this order are so similar to basal eutherians that it is rather arbitrary whether they are called primates or not (Figure 5.28). The first monkeys are known from the Oligocene, and the first apes (superfamily Hominoidea) from the Miocene, about 22 Mya. The fossil record of hominin evolution, starting about 6 Mya, is described in Chapter 4.

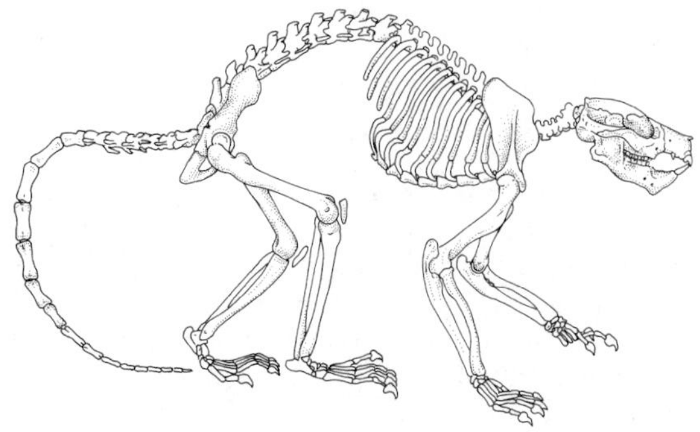

Figure 5.28 An early primate, the Paleocene *Plesiadapis*. (After Simons 1979.)

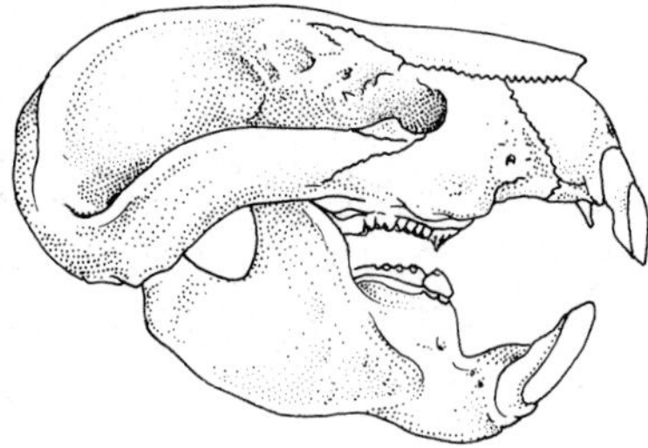

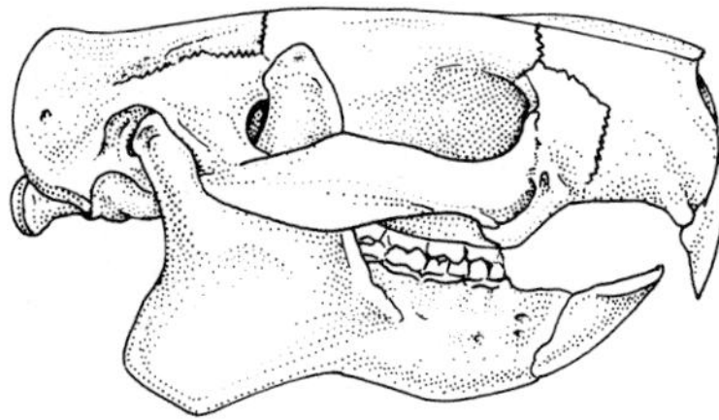

Figure 5.29 Convergent evolution in mammals: (A) A Paleocene multituberculate (*Taeniolabis*) and (B) an Eocene rodent (*Paramys*). Multituberculates, extending from the Cretaceous to the Oligocene, were nonplacental mammals that were ecologically similar to squirrels and other rodents (with similar morphological features such as large, ever-growing incisors). Competition with rodents may have driven them to extinction. (After Romer 1966.)

Rodents (Rodentia), related to Primates, are recorded first from the late Paleocene. Perhaps by direct competition, they replaced the Multituberculata, a nonplacental but ecologically similar group that had originated from stem mammals in the late Jurassic (Figure 5.29). The rodents became the most diverse order of mammals, in part because of an extraordinary proliferation of rats and mice (superfamily Muroidea) within the last 10 Myr.

A monophyletic group of orders ("Afrotheria"), recently proposed on the basis of DNA evidence, includes such very different-looking creatures as the shrewlike tenrecs, the aquatic manatees, and the elephants. This group is first recorded in the Eocene, but represents one of the earliest splits among the placental orders (see Figure 5.26). The elephants (Proboscidea) underwent the greatest diversification, differentiating into at least 40 genera. Woolly mammoths survived through the most recent glaciation, about 10,000 years ago, and two genera (the African and Indian elephants) persist today (Figure 5.30).

In the Eocene, a group of archaic hoofed mammals gave rise to a great radiation of carnivores and of hoofed mammals (ungulates) in the orders Perissodactyla and Artiodactyla, the early members of which differed only slightly. The Perissodactyla, or odd-toed ungulates, were very diverse from the Eocene to the Miocene, then dwindled to the few extant species of rhinoceroses, horses, and tapirs. The Artiodactyla (now called Cetartiodactyla by some authors) are first known in the Eocene as rabbit-sized animals that had the order's diagnostic ankle bones, but otherwise bore little similarity to the pigs, camels, and ruminants that appeared soon thereafter. In the Miocene, the ruminants began a sustained radiation, mostly in the Old World, that is correlated with the increasing prevalence of grasslands. Among the families that proliferated are the deer, the giraffes and relatives, and the Bovidae, the diverse family of antelopes, sheep, goats, and cattle. Recent research has shown that during the Eocene, an artiodactyl lineage related to today's hippopotamuses became aquatic and evolved into the cetaceans: the dolphins and whales (see Figure 4.11).

Pleistocene events

The last Cenozoic epoch, the *Pleistocene*, embraces a mere 1.8 million years, but because of its recency and drama, it *is critically important for understanding today's organisms*.

By the beginning of the Pleistocene, the continents were situated as they are now. North America was connected in the northwest to eastern Asia by the Bering Land Bridge, in

Figure 5.30 Proboscidea, the order of elephants, has only two living genera, but was once very diverse. A few of the extinct forms are (A) an early, generalized proboscidean, *Moeritherium* (late Eocene–early Oligocene); (B) *Phiomia* (early Oligocene); (C) *Gomphotherium* (Miocene); (D) *Deinotherium* (Miocene); and (E) *Mammuthus*, the woolly mammoth (Pleistocene). (After Romer 1966.)

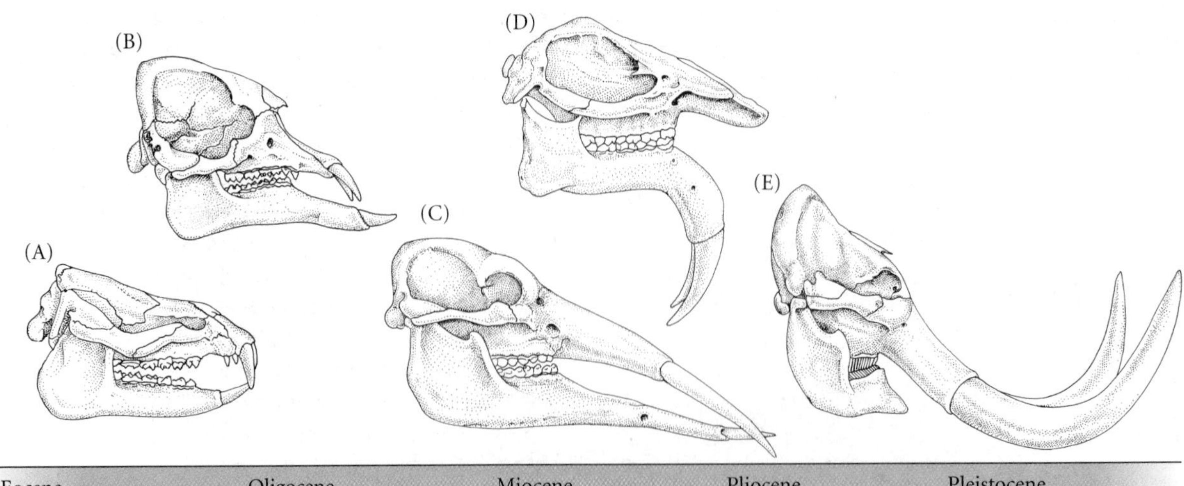

the region where Alaska and Siberia almost meet today (Figure 5.31A). North and South America were connected by the Isthmus of Panama. Except for those species that have become extinct, Pleistocene species are very similar to, or indistinguishable from, the living species that descended from them.

Global temperatures began to drop during the Pliocene, about 3 Mya, and then, in the Pleistocene itself, underwent violent fluctuations for a period of about 100,000 years. When temperatures cooled, continental glaciers as thick as 2 kilometers formed at high latitudes, receding during the warmer intervals. *At least four major glacial advances*, and many minor ones, occurred. The most recent glacial stage, termed the Wisconsin in North America and the Riss-Würm in Europe, *reached its maximum about 20,000 years ago* (see

(A)

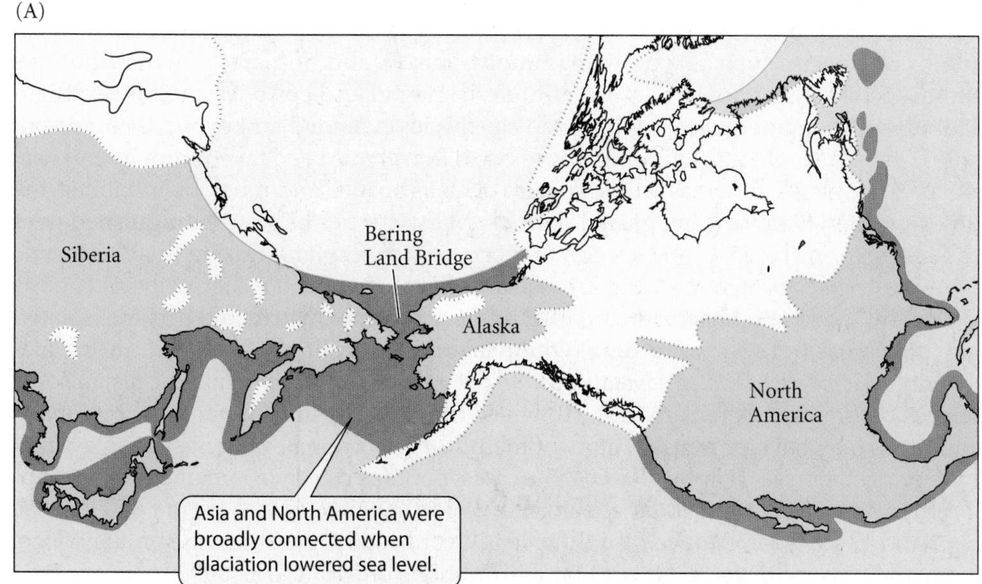

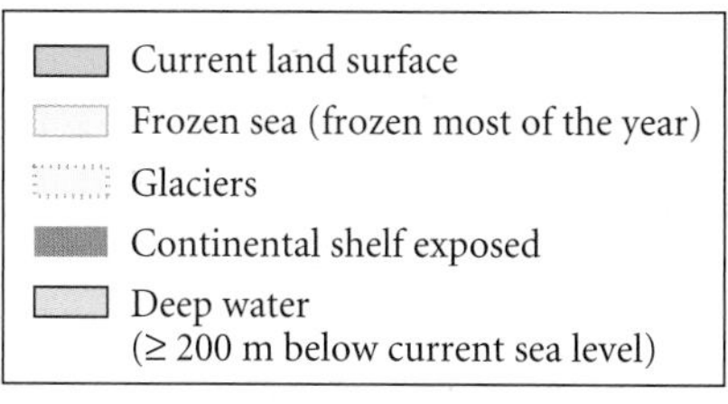

(B)

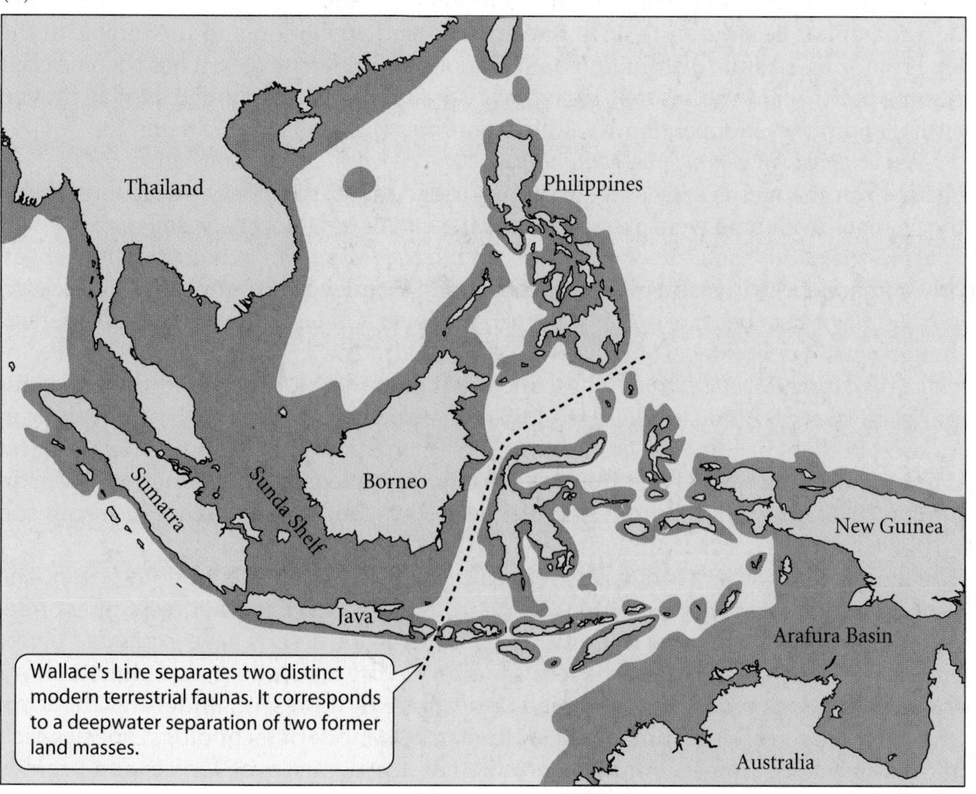

Figure 5.31 Pleistocene glaciers lowered sea levels by at least 100 meters, so that many terrestrial regions that are now separated by oceanic barriers were connected. (A) Eastern Asia and North America were joined by the Bering Land Bridge. Note the extent of the glacier in North America. (B) Indonesia and other islands were connected to either southeastern Asia or Australia. (After Brown and Lomolino 1998.)

Figure 5.31A), *and melted back 15,000 to 8000 years ago*. During glacial episodes, *sea level dropped* as much as 100 meters below its present level. This drop exposed parts of the continental shelves, extending many continental margins beyond their present boundaries and *connecting many islands to nearby land masses*. (Japan, for example, was a peninsula of Asia, New Guinea was connected to Australia, and the Malay Archipelago was an extension of Southeast Asia; Figure 5.31B.) Temperatures in equatorial regions were apparently about as high as they are today, so the latitudinal temperature gradient was much steeper than at present. *The global climate during glacial episodes was generally drier*. Thus *mesic and wet forests became restricted* to relatively small favorable areas, and grasslands expanded. During interglacial periods, the climate became warmer and generally wetter.

These events profoundly affected the distributions of organisms (see Chapter 6). When the sea level was lower, many terrestrial species moved between land masses that are now isolated; for example, the ice-free Bering Land Bridge was a conduit from Asia to North America for species such as woolly mammoths, bison, and humans. The distributions of many species shifted toward lower latitudes during glacial episodes and toward higher latitudes during interglacials, when tropical species extended far beyond their present limits. Fossils of elephants, hippopotamuses, and lions have been taken from interglacial deposits in England, whereas Arctic species such as spruce and musk ox inhabited the southern United States during glacial periods. Many species became extinguished over broad areas; for instance, beetle species that occurred in England during the Pleistocene are now restricted to such far-flung areas as northern Africa and eastern Siberia (Coope 1979). Many species that had been broadly, rather uniformly distributed became isolated in separate areas (refuges, or **refugia**) where favorable conditions persisted during glacial periods. *Some such isolated populations diverged* genetically and phenotypically, in some instances becoming different species. However, the frequent shifts in the distributions of populations may have prevented many of them from becoming different species by mixing them together (see Chapter 18). In some cases, populations have remained in their glacial refuge areas, isolated from the major range of their species (see Figure 6.8). On the other hand, many species have rapidly spread over broad areas from one or a few local refugia and have achieved their present distributions only in the last 8000 years or less. (See Pielou 1991 for more on postglacial history.)

The distributions of many species shifted rather gradually over the landscape as climate and habitat became suitable at one end of the distribution and unsuitable at the other. From studies of the distribution of fossil pollen of different ages, it has become clear that some of the plant species that are typically associated in communities today moved over the land rather independently, at different rates (Jackson and Williams 2004). Thus *the species composition of ecological communities changed kaleidoscopically* (Figure 5.32).

Aside from changes in species' geographic distributions, the most conspicuous effects of the changes in climate were extinctions. At the end of the Pliocene, *many shallow-water marine invertebrate species became extinct, especially tropical species*, which may have been poorly equipped to withstand even modest cooling (Stanley and Campbell 1981; Jackson 1995). No major taxa became entirely extinct, however. On land, the story was different. Although small vertebrates seem not to have suffered, *a very high proportion of large-bodied mammals and birds became extinct*. These included mammoths, saber-toothed cats, giant bison, giant beavers, giant wolves, ground sloths, and all the endemic South American ungulates. Both archaeological evidence and mathematical population models suggest that human hunting played an important role in this "megafaunal extinction" (Martin and Klein 1984; Alroy 2001; Burney and Flannery 2005), but climate change also seems to have played a role, especially in Europe (Koch and Barnosky 2006).

The most recent glaciers had hardly retreated when major new disruptions began. The advent of human agriculture about 11,000 years ago began yet another reshaping of the terrestrial environment. For the last several thousand years, deserts have expanded under the impact of overgrazing, forests have succumbed to fire and cutting, and climates have changed as vegetation has been modified or destroyed. At present, under the impact of an exponentially growing human population and its modern technology, species-rich tropical forests face almost complete annihilation, temperate-zone forests and prairies

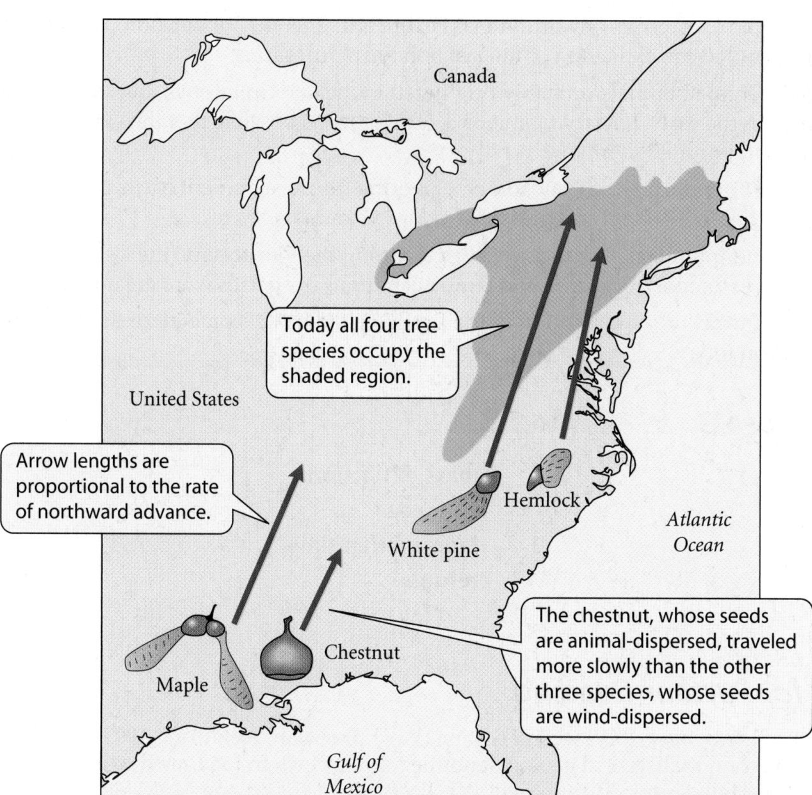
Canada
Today all four tree species occupy the shaded region.
United States
Arrow lengths are proportional to the rate of northward advance.
Hemlock
White pine
Atlantic Ocean
The chestnut, whose seeds are animal-dispersed, traveled more slowly than the other three species, whose seeds are wind-dispersed.
Chestnut
Maple
Gulf of Mexico

Figure 5.32 Different rates of northward spread of four North American tree species from refugia after the most recent glacial episode. After the glacial episode, maple and chestnut moved north from the Gulf region, and white pine and hemlock spread from the mid-Atlantic coastal plain. (After Pielou 1991.)

have been eliminated in much of the world, marine communities suffer pollution and appalling overexploitation, and global warming caused by combustion of fossil fuels threatens to change climates and habitats so rapidly that many species are unlikely to adapt (Wilson 1992; Kareiva et al. 1993). An analysis of sample regions covering 20 percent of the Earth's surface projected extinction of 18 to 35 percent of species in those regions, solely as a result of climate change, within just the next 50 years (Thomas et al. 2004). Even if these estimates were twice as pessimistic as they should be, it is clear that *one of the greatest ecological disasters, and one of the greatest mass extinctions of all time, is underway, and can be mitigated only if humankind acts decisively and quickly.*

Summary

GO TO THE
evolution
Companion Web Site at
www.sinauer.com/evolution
for quizzes, data analysis and simulation exercises, and other study aids.

1. Evidence from living organisms indicates that all living things are descended from a single common ancestor. Some progress has been made in understanding the origin of life, but a great deal remains unknown.
2. The first fossil evidence of life dates from about 3.5 billion years ago, about a billion years after the formation of the Earth. The earliest life forms of which we have evidence were prokaryotes.
3. Eukaryotes evolved about 1.5 billion years ago. Their mitochondria and chloroplasts evolved from endosymbiotic bacteria.
4. The fossil record displays an explosive diversification of the animal phyla near the beginning of the Cambrian period, about 542 Mya. The causes of this rapid diversification are debated, but may include a combination of genetic and ecological events.
5. Terrestrial plants evolved in the Ordovician, and terrestrial arthropods in the late Silurian; amphibians evolved in the late Devonian from lobe-finned fishes.
6. The most devastating mass extinction of all time occurred at the end of the Permian (about 252 Mya). It profoundly altered the taxonomic composition of the Earth's biota.
7. Seed plants and amniotes ("reptiles") became diverse and ecologically dominant during the Mesozoic era (251–65.5 Mya). Flowering plants and plant-associated insects diversified greatly

from the middle of the Cretaceous onward. A mass extinction (the "K/T extinction") at the end of the Mesozoic included the extinction of the last nonavian dinosaurs.

8. Most orders of placental mammals probably originated in the late Cretaceous, but underwent adaptive radiation in the early Tertiary (about 65.5–50 Mya). It is possible that the extinction of nonavian dinosaurs permitted them to diversify.
9. The climate became drier during the Cenozoic era, favoring the development of grasslands and the evolution of herbaceous plants and grassland-adapted animals.
10. A series of glacial and interglacial episodes occurred during the Pleistocene (the last 1.8 Myr), during which some extinctions occurred and the distributions of species were greatly altered.
11. With the passage of time, the composition of the Earth's biota has become increasingly similar to its composition today.

Terms and Concepts

Bauplan (pl., Baupläne)
Cambrian explosion
endosymbionts
Gondwana
Laurasia
mass extinction
Pangaea
Precambrian time
refugia

Suggestions for Further Reading

S. M. Stanley, *Earth and Life through Time*, second edition (W. H. Freeman, New York, 1993), is a comprehensive introduction to historical geology and the fossil record. In the fourth edition of *Life of the Past* (Prentice-Hall, Upper Saddle River, NJ, 1999), W. I. Ausich and N. G. Lane provide a well-illustrated introduction to the theme of this chapter. A good overview of the origin of life is R. M. Hazen's *Genesis: The Scientific Quest for Life's Origin* (Joseph Henry Press, Washington, D.C., 2005). J. Maynard Smith and E. Szathmáry, *The Major Transitions in Evolution* (W. H. Freeman, San Francisco, 1995), is an interpretation by leading evolutionary theoreticians of major events, ranging from the origin of life to the origins of societies and languages.

A. K. Behrensmeyer et al. (eds.), *Terrestrial Ecosystems through Time: Evolutionary Paleoecology of Terrestrial Plants and Animals* (University of Chicago Press, Chicago, 1992), presents detailed summaries of changes in terrestrial environments and communities in the past.

Useful books on the evolution of major taxonomic groups include J. W. Valentine, *On the Origin of Phyla* (University of Chicago Press, 2004); P. Kenrick and P. R. Crane, *The Origin and Early Diversification of Land Plants* (Smithsonian Institution Press, Washington, D.C., 1997); R. L. Carroll, *Vertebrate Paleontology and Evolution* (W. H. Freeman, New York, 1988); M. J. Benton (ed.), *The Phylogeny and Classification of the Tetrapods* (Clarendon, Oxford, 1988); D. B. Weishampel, P. Dodson, and H. Osmolska, *The Dinosauria* (University of California Press, Berkeley, 1990); D. R. Prothero, *After the Dinosaurs: The Age of Mammals* (Indiana University Press, Bloomington, 2006).

Problems and Discussion Topics

1. Why, in the evolution of ancestral eukaryotes, might it have been advantageous for separate organisms to become united into a single organism? Can you describe analogous, more recently evolved, examples of intimate symbioses that function as single integrated organisms?
2. Early in the origin of life, as it is presently conceived, there was no distinction between genotype and phenotype. What characterizes this distinction, and at what stage of organization may it be said to have come into being?
3. If we employ the biological species concept (see Chapter 17), when did species first exist? What were organisms before then, if not species? What might the consequences of the emergence of species be for processes of adaptation and diversification?
4. How would you determine whether the morphological diversity of animals has increased, decreased, or remained the same since the Cambrian?
5. Compare terrestrial communities in the Devonian and in the Cretaceous, and discuss what may account for the difference between them in the diversity of plants and animals.

6. Read papers on the different estimates of the timing of either the origin of bilaterian animal phyla or the orders of mammals, and discuss how the difference between paleontological and molecular phylogenetic estimates might be resolved.
7. Animals that are readily classified into extant phyla, such as Mollusca and Arthropoda, appeared in the Cambrian without transitional forms that show how their distinctive Baupläne evolved. What kinds of evidence would be useful for reconstructing the likely steps in the evolution of such body plans? Consult Valentine 2004 on a phylum of your choice.
8. Are the body cavities homologous among those phyla that have them? See Valentine 2004.
9. What evidence would be necessary to test the hypothesis that the megafaunal extinction in the Pleistocene was caused by humans?
10. Discuss the implications of the spread and change of species distributions after the retreat of the Pleistocene glaciers for evolutionary changes in species and for the species composition of ecological communities (see also Chapters 7, 17, and 19).

CHAPTER 6

The Geography of Evolution

New World and Old World monkeys. All New World monkeys (South and Central America), including the red howler *Alouatta seniculus* (above), belong to the taxon Platyrrhini. Douc langurs (*Pygathrix nemaeus*, right) are Old World monkeys (Africa and Asia) and belong to the entirely distinct taxon Catarrhini. (Howler © Morales/Photolibrary.com; langurs © Picture Press/Photolibrary.com.)

Where did humans originate, and by what paths did they spread throughout the world? Why are kangaroos found only in Australia, whereas rats are found worldwide? Why are there so many more species of trees, insects, and birds in tropical than in temperate-zone forests?

These questions illustrate the problems that **biogeography**, the study of the geographic distributions of organisms, attempts to solve. The evolutionary study of organisms' distributions is intimately related to geology, paleontology, systematics, and ecology. For example, geological study of the history of the distributions of land masses and climates often sheds light on the causes of organisms' distributions. Conversely, organisms' distributions have sometimes provided evidence for geological events. In fact, the geographic distributions of organisms were used by some scientists as evidence for continental drift long before geologists agreed that it really happens. In some instances, the geographic distribution of a taxon may best be explained by historical circumstances; in other instances, ecological factors operating at the present time may provide the best explanation. Hence the field of biogeography may be roughly divided into **historical biogeography** and **ecological biogeography**. Historical and ecological explanations of geographic distributions are complementary, and both are important (Myers and Giller 1988; Ricklefs and Schluter 1993; Brown and Lomolino 1998).

Biogeographic Evidence for Evolution

Darwin and Wallace initiated the field of biogeography. Wallace devoted much of his later career to the subject and described major patterns of animal distribution that are still valid today. The distributions of organisms provided both Darwin and Wallace with inspiration and with evidence that evolution had occurred. To us, today, the reasons for certain facts of biogeography seem so obvious that they hardly bear mentioning. If someone asks us why there are no elephants in the Hawaiian Islands, we will naturally answer that elephants couldn't get there. This answer assumes that elephants originated somewhere else: namely, on a continent. But in a pre-evolutionary world view, the view of special divine creation that Darwin and Wallace were combating, such an answer would not hold: the Creator could have placed each species anywhere, or in many places at the same time. In fact, it would have been reasonable to expect the Creator to place a species wherever its habitat, such as rain forest, occurred.

Darwin devoted two chapters of *The Origin of Species* to showing that many biogeographic facts that make little sense under the hypothesis of special creation make a great deal of sense if a species (1) has a definite site or region of origin, (2) achieves a broader distribution by dispersal, and (3) becomes modified and gives rise to descendant species in the various regions to which it migrates. (In Darwin's day, there was little inkling that continents might have moved. Today, the movement of land masses also explains certain patterns of distribution.) Darwin emphasized the following points:

First, he said, "*neither the similarity nor the dissimilarity of the inhabitants of various regions can be wholly accounted for by climatal and other physical conditions*." Similar climates and habitats, such as deserts and rain forests, occur in both the Old and the New World, yet the organisms inhabiting them are unrelated. For example, the cacti (family Cactaceae) are restricted to the New World, but the cactuslike plants in Old World deserts are members of other families (Figure 6.1). All the monkeys in the New World belong to one anatomically distinguishable group (Platyrrhini), and all Old World monkeys to another (Catarrhini), even if they have similar habitats and diets.

Darwin's second point is that "*barriers of any kind, or obstacles to free migration, are related in a close and important manner to the differences between the productions [organisms] of various regions*." Darwin noted, for instance, that marine species on the eastern and western coasts of South America are very different.

Darwin's "third great fact" is that *inhabitants of the same continent or the same sea are related, although the species themselves differ from place to place*. He cited as an example the aquatic rodents of South America (the coypu and capybara), which are structurally similar to, and related to, South American rodents of the mountains and grasslands, not to the aquatic rodents (beaver, muskrat) of the Northern Hemisphere.

(A)

(B)

(C)

Figure 6.1 Convergent growth form in desert plants. These plants, all leafless succulents with photosynthetic stems, belong to three distantly related families. (A) A North American cactus (family Cactaceae). This species, *Lophocereus schottii*, is native to Baja California. (B) A carrion flower of the genus *Stapelia* (Apocynaceae). These fly-pollinated succulents can be found from southern Africa to east India. (C) A species of *Euphorbia* (Euphorbiaceae) in the Namib Desert of Africa. (A–C © Photo Researchers, Inc. A by Richard Parker; B by Geoff Bryant; C by Fletcher and Baylis.)

"We see in these facts," said Darwin, "some deep organic bond, throughout space and time, over the same areas of land and water, independently of physical conditions.... The bond is simply inheritance [i.e., common ancestry], that cause which alone, as far as we positively know, produces organisms quite like each other."

For Darwin, it was important to show that a species had not been created in different places, but had a *single region of origin*, and had spread from there. He drew particularly compelling evidence from the inhabitants of islands. First, distant *oceanic islands generally have precisely those kinds of organisms that have a capacity for long-distance dispersal* and lack those that do not. For example, the only native mammals on many islands are bats. Second, *many continental species of plants and animals have flourished on oceanic islands to which humans have transported them*. Thus, said Darwin, "he who admits the doctrine of the creation of each separate species, will have to admit that a sufficient number of the best adapted plants and animals were not created for oceanic islands." Third, most of the species on islands are clearly *related to species on the nearest mainland*, implying that that was their source. This is the case, as Darwin said, for almost all the birds and plants of the Galápagos Islands. Fourth, the *proportion of endemic species on an island is particularly high when the opportunity for dispersal to the island is low*. Fifth, *island species often bear marks of their continental ancestry*. For example, Darwin noted, hooks on seeds are an adaptation for dispersal by mammals, yet on oceanic islands that lack mammals, many endemic plants nevertheless have hooked seeds.

It is a testimony to Darwin's knowledge and insight that all these points hold true today, after a century and a half of research. Our greater knowledge of the fossil record and of geological events such as continental movement and sea level changes has added to our understanding, but has not negated any of Darwin's major points.

Major Patterns of Distribution

The geographic distribution of almost every species is limited to some extent, and many higher taxa are likewise restricted (**endemic**) to a particular geographic region. For example, the salamander genus *Plethodon* is limited to North America, and *Plethodon caddoensis* occupies only the Caddo Mountains of western Arkansas. Some higher taxa, such as the pigeon family (Columbidae), are almost cosmopolitan (found worldwide), whereas others are narrowly endemic (e.g., the kiwi family, Apterygidae, which is restricted to New Zealand).

Wallace and other early biogeographers recognized that many higher taxa have roughly similar distributions, and that the taxonomic composition of the biota is more uniform within certain regions than between them. Based on these observations, Wallace designated several **biogeographic realms** for terrestrial and freshwater organisms that are still widely recognized today (Figure 6.2). These are the *Palearctic* (temperate and tropical Eurasia and northern Africa), the *Nearctic* (North America), the *Neotropical* (South and Central America), the *Ethiopian* (sub-Saharan Africa), the *Oriental* (India and Southeast Asia), and the *Australian* (Australia, New Guinea, New Zealand, and nearby islands). These realms are more the result of Earth's history than of current climate or land mass distribution. For example, WALLACE'S LINE separates islands that, despite their close proximity and similar climate, differ greatly in their fauna. These islands are on two lithospheric plates that approached each other only recently, and they are assigned to two different biogeographic realms: the Oriental and the Australian.

Each biogeographic realm is inhabited by many higher taxa that are much more diverse in that realm than elsewhere, or are even restricted to that realm. For example, the endemic taxa of the Neotropical realm include the Xenarthra (anteaters and allies), platyrrhine primates (such as spider monkeys and marmosets), many hummingbirds, a large assemblage of suboscine birds such as flycatchers and antbirds, many families of catfishes, and plant families such as the bromeliads (Figure 6.3). Within each realm, individual species may have more restricted distributions; regions that differ markedly in habitat, or that are separated by mountain ranges or other barriers, will have rather different sets of species.

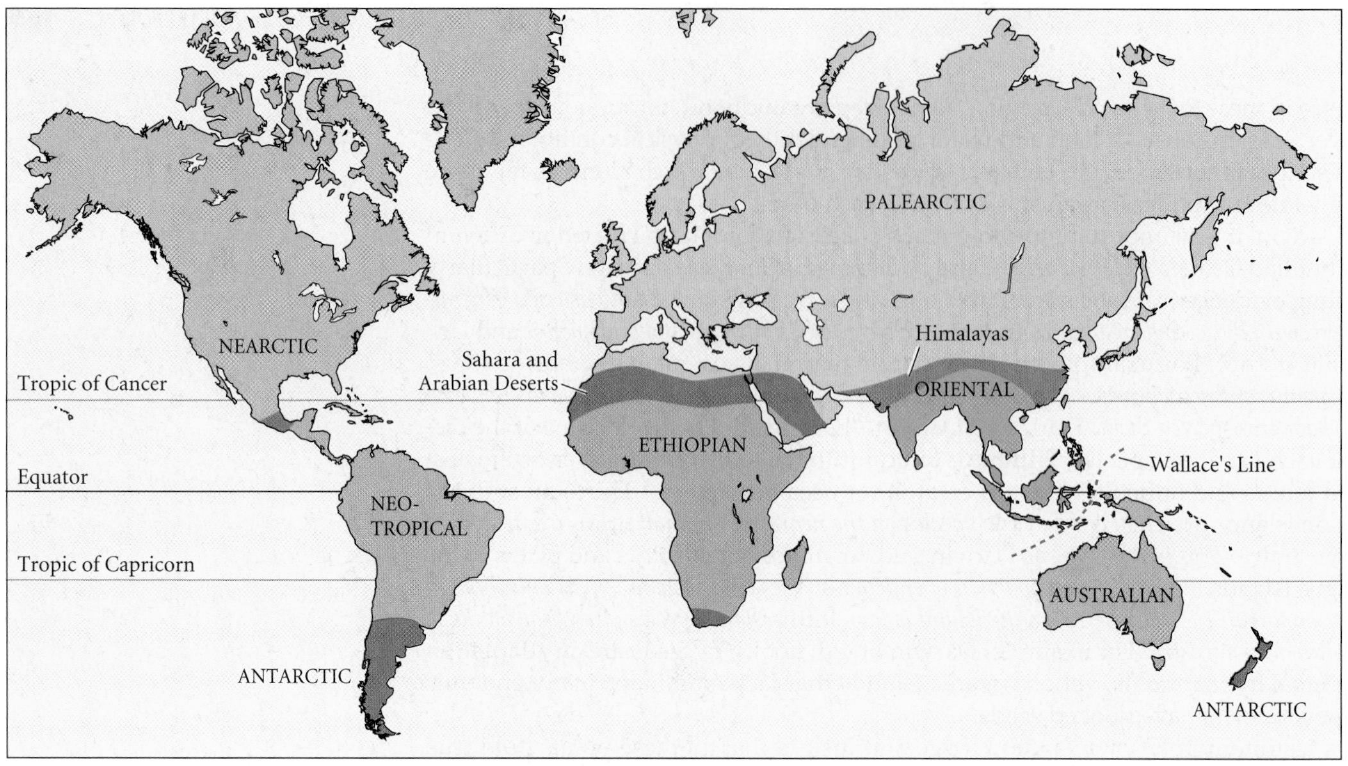

Figure 6.2 Biogeographic realms. The biogeographic realms recognized by A. R. Wallace are the Palearctic, Ethiopian, Oriental, Australian, Nearctic, and Neotropical. Some authors consider parts of southern South America, Africa, and New Zealand to be another realm, the Antarctic.

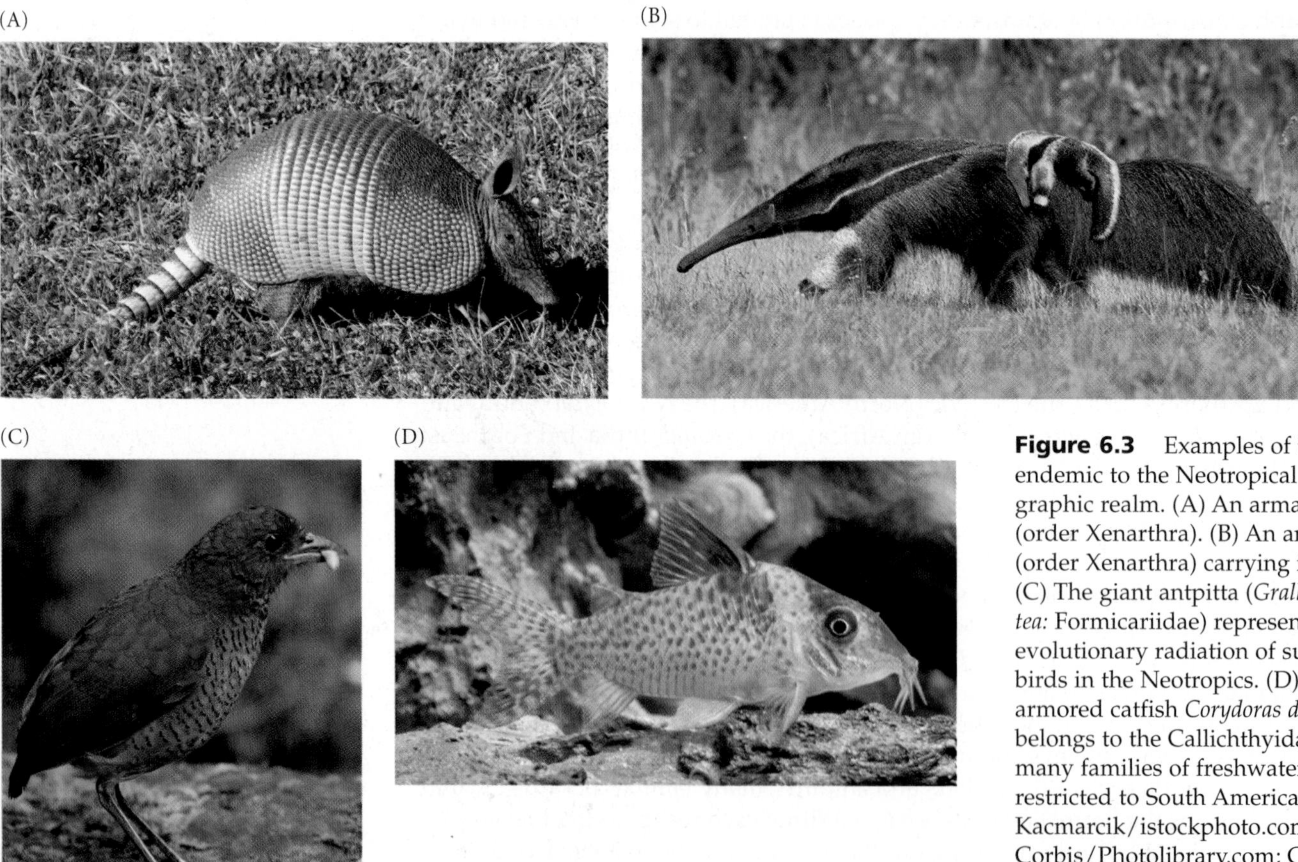

Figure 6.3 Examples of taxa endemic to the Neotropical biogeographic realm. (A) An armadillo (order Xenarthra). (B) An anteater (order Xenarthra) carrying its young. (C) The giant antpitta (*Grallaria gigantea:* Formicariidae) represents a huge evolutionary radiation of suboscine birds in the Neotropics. (D) The armored catfish *Corydoras delphax* belongs to the Callichthyidae, one of many families of freshwater catfishes restricted to South America. (A © Ron Kacmarcik/istockphoto.com; B © Corbis/Photolibrary.com; C © Danita Delimont/Alamy; D © Juniors Bildarchiv/Alamy.)

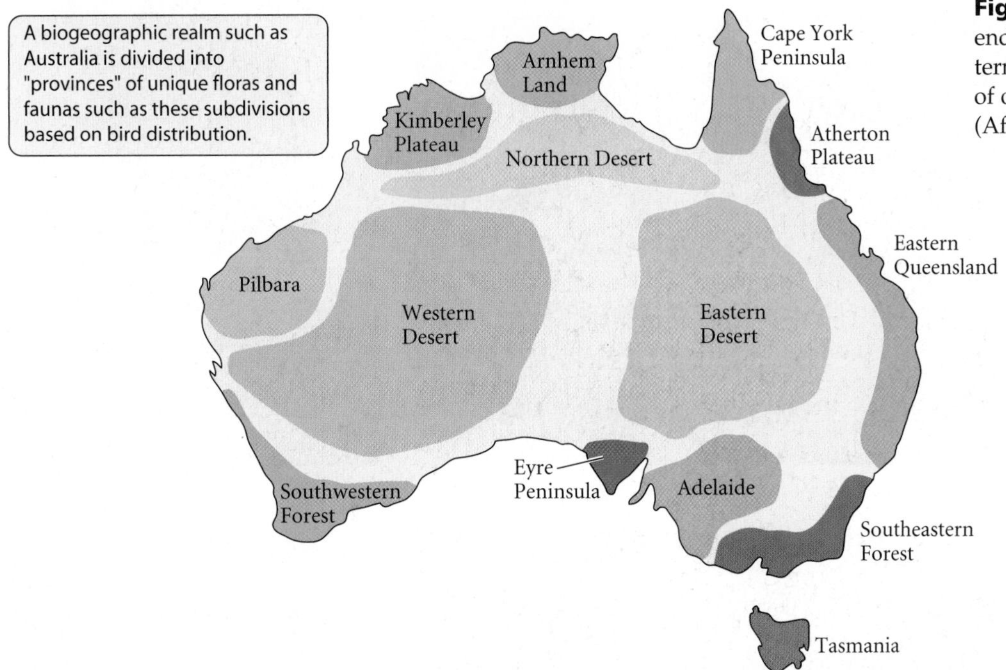

Figure 6.4 Provinces, or regions of endemism, in Australia, based on the pattern of distribution of birds. Distributions of other vertebrates form similar patterns. (After Cracraft 1991.)

Thus a biogeographic realm can often be divided into faunal and floral PROVINCES, or regions of endemism (Figure 6.4).

The borders between biogeographic realms (or provinces) cannot be sharply drawn because some taxa infiltrate neighboring realms to varying degrees. In the Nearctic realm (North America), for instance, some species, such as bison, trout, and birches, are related to Palearctic (Eurasian) taxa. But other Nearctic species are related to, and have been derived from, Neotropical stocks: examples include an armadillo, an opossum, and the Spanish moss (*Tillandsia usneoides*), a bromeliad that festoons southern trees.

Some taxa have **disjunct distributions**; that is, their distributions have gaps (Figure 6.5). Disjunctly distributed higher taxa typically have different representatives in each area they occupy. For example, many taxa are represented on two or more southern continents, including lungfishes, marsupials, cichlid fishes, and *Araucaria* pines (Goldblatt 1993). Another common disjunct pattern is illustrated by alligators (*Alligator*), skunk cabbages (*Symplocarpus*), and tulip trees (*Liriodendron*), which are among the many genera that are found both in eastern North America and in temperate eastern Asia, but nowhere in between (Wen 1999). Accounting for such distributions has long been a preoccupation of biogeographers.

Historical Factors Affecting Geographic Distributions

The geographic distribution of a taxon is affected by both contemporary and historical factors. The limits to the distribution of a species may be set by geological barriers that have not been crossed, or by ecological conditions to which the species is not adapted. In this section, we will focus on the historical processes that have led to the current distribution of a taxon: extinction, dispersal, and vicariance.

The distribution of a species may have been reduced by the extinction of some populations, and that of a higher taxon by the extinction of some constituent species. For example, the horse family, Equidae, originated and became diverse in North America, but it later became extinct there; only the African zebras and the Asian wild asses and horses have survived. (Horses were reintroduced into North America by European colonists.) Likewise, extinction is the cause of the disjunction between related taxa in eastern Asia and eastern North America. During the early Tertiary, many plants and animals spread throughout the northern regions of North America and Eurasia. Their spread was facili-

Figure 6.5 Examples of disjunct distributions. (A) A South American *Araucaria* pine growing in Chile. (B) *Araucaria bidwillii*, Queensland, Australia. (C) An American alligator (*Alligator mississipiensis*), distributed in the southeastern United States. (D) The endangered Chinese alligator (*Alligator sinensis*). (A, photo by the author; B © Dave and Sigrun Tollerton/Alamy; C © Juniors Bildarchiv/Photolibrary.com; D © Melvyn Longhurst/Alamy.)

tated by a warm, moist climate and by land connections from North America to both Europe and Siberia. Many of these taxa became extinct in western North America in the late Tertiary as a result of mountain uplift and a cooler, drier climate, and were extinguished in Europe by Pleistocene glaciations (Wen 1999; Sanmartín et al. 2001).

Species expand their ranges by **dispersal** (i.e., movement of individuals). Some authors distinguish two kinds of dispersal: RANGE EXPANSION, or movement across expanses of more or less continuous favorable habitat, and JUMP DISPERSAL, or movement across a barrier (Myers and Giller 1988). Some species of plants and animals can expand their range very rapidly. Within the last 200 years, many species of plants accidentally brought from Europe by humans have expanded across most of North America from New York and New England, and some birds, such as the European starling (*Sturnus vulgaris*) and the house sparrow (*Passer domesticus*), have done the same within a century (Figure 6.6). Other species have crossed major barriers on their own. The cattle egret (*Bubulcus ibis*) was found only in tropical and subtropical parts of the Old World until about 75 years ago, when it arrived in South America, apparently unassisted by humans (Figure 6.7). It has since spread throughout the warmer parts of the New World.

If a major barrier to dispersal breaks down, many species may expand their ranges more or less together, resulting in correlated patterns of dispersal (Lieberman 2003). For example, many plants and animals moved between South and North America when the Isthmus of Panama was formed in the Pliocene (see Chapter 5), and between Europe and North America over a trans-Atlantic land bridge in the early Tertiary (Sanmartín et al. 2001).

Vicariance refers to the separation of populations of a widespread species by barriers arising from changes in geology, climate, or habitat. The separated populations

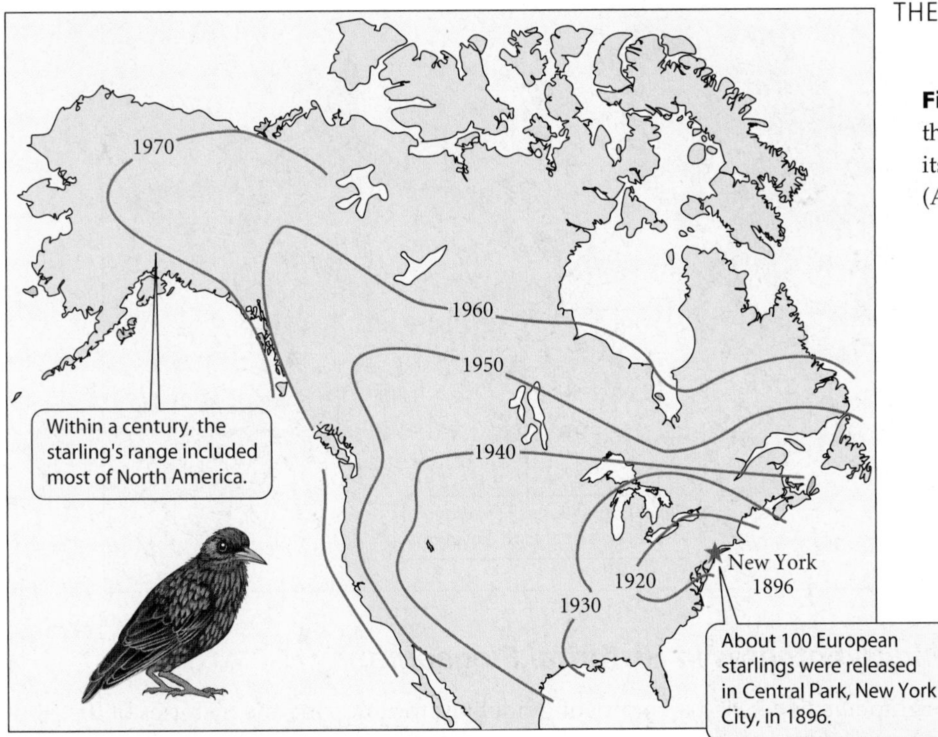
1970
1960
1950
1940
1930
1920
New York
1896
Within a century, the starling's range included most of North America.
About 100 European starlings were released in Central Park, New York City, in 1896.

Figure 6.6 The history of range expansion of the European starling (*Sturnus vulgaris*) following its introduction into New York City in 1896. (After Brown and Gibson 1983.)

diverge, and they often become different subspecies, species, or higher taxa. For example, in many fish, shrimp, and other marine animal groups, the closest relative of a species on the Pacific side of the Isthmus of Panama is a species on the Caribbean side of the isthmus. This pattern is attributed to the divergence of populations of a broadly distributed ancestral species that was sundered by the rise of the isthmus in the Pliocene (Lessios 1998). Vicariance sometimes accounts for the presence of related taxa in disjunct areas.

Dispersal and vicariance are both important processes, and neither can be assumed, a priori, to be the sole explanation of a taxon's distribution. In many cases, dispersal, vicariance, and extinction all have played a role. We have seen, for example, that during the Pleistocene glaciations, species shifted their ranges by dispersal into new regions (see Figure 5.32). Some northern, cold-adapted species became distributed far to the south. When the climate became warmer, southern populations became extinct, except for populations of some such species that survived on cold mountaintops (Figure 6.8). In this case, the vicariant disjunction of populations, caused by the formation of inhospitable intervening habitat, followed dispersal and went hand in hand with extinction.

Figure 6.7 A cattle egret (*Bubulcus ibis*) accompanying a cow in Alabama. This heron feeds on insects stirred up by grazing ungulates in both the Old World and the New World. (Photo © A. Morris/Visuals Unlimited.)

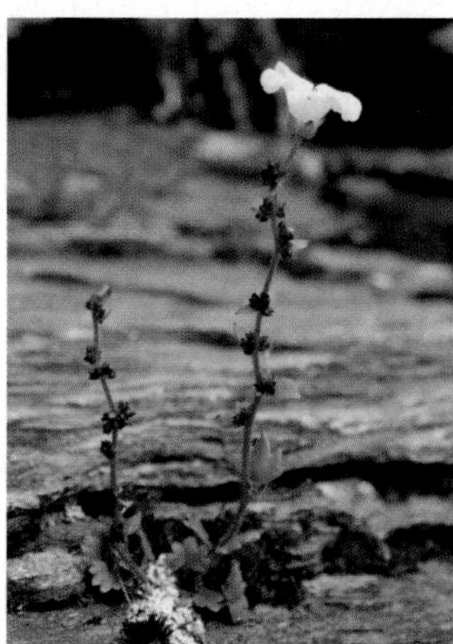

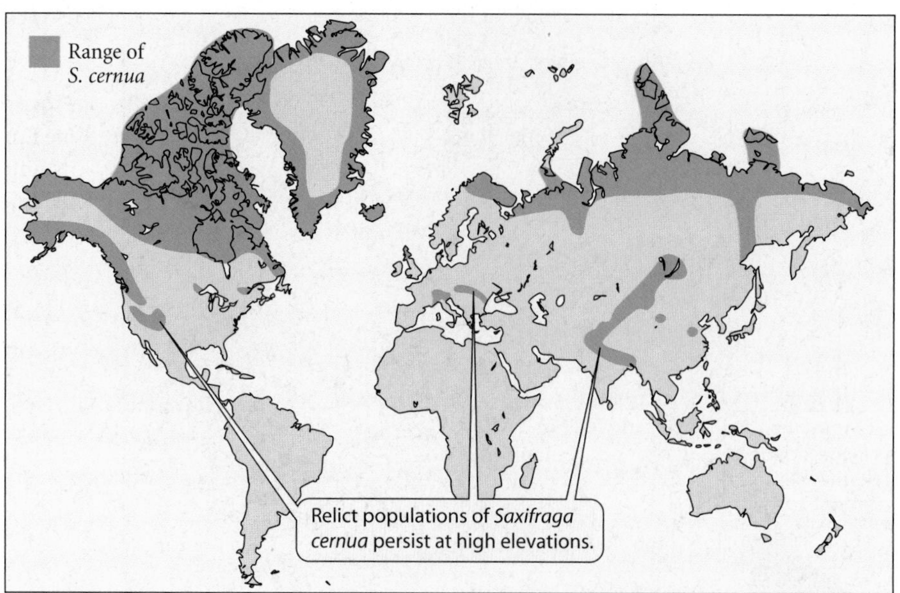

Figure 6.8 The disjunct distribution of a saxifrage (*Saxifraga cernua*) in northern and mountainous regions of the Northern Hemisphere. Relict populations persist at high elevations, following the species' retreat from the southern region that it occupied during glacial periods. (After Brown and Gibson 1983; photo courtesy of Egil Michaelsen and the Norwegian Botanical Association.)

Testing Hypotheses in Historical Biogeography

Biogeographers have used a variety of guidelines for inferring the histories of distributions. Some of these guidelines are well founded. For example, the distribution of a taxon cannot be explained by an event that occurred before the taxon originated: a genus that originated in the Miocene cannot have achieved its distribution by continental movements that occurred in the Cretaceous. Some other guidelines are more debatable. Some authors in the past assumed that a taxon originated in the region where it is presently most diverse. But this need not be so, as the horse family shows: although now native only to Africa and Asia, horses are descended from North American ancestors.

The major hypotheses accounting for a taxon's distribution are dispersal and vicariance. For example, one might ask whether the ratite birds (a group of flightless birds that includes African ostriches and the kiwi of New Zealand) dispersed from one continent to another, or whether they descended from ancestors on a single land mass that split into the several southern continents. Phylogenetic analysis plays a leading role in evaluating these hypotheses, but other sources of evidence can be useful as well. For example, an area is often suspected of having been colonized by dispersal if it has a highly "unbalanced" biota—that is, if it lacks a great many taxa that it would be expected to have if it had been joined to other areas. This assumption has been applied especially to oceanic islands that lack forms such as amphibians and nonflying mammals. The fossil record can also provide important evidence (Lieberman 2003)—for instance, it may show that a taxon proliferated in one area before appearing in another—and geological data may describe the appearance or disappearance of barriers. For example, fossil armadillos (see Figure 6.3A) are limited to South America throughout the Tertiary and are found in North American deposits only from the Pliocene and Pleistocene, after the Isthmus of Panama was formed. This pattern implies that armadillos dispersed into North America from South America. Paleontological data must be interpreted cautiously, however, because a taxon may be much older, and have inhabited a region longer, than a sparse fossil record shows.

Phylogenetic methods are the foundation of most modern studies of historical biogeography. Several such methods have been developed, especially by Daniel Brooks (1990), Roderick Page (1994), and Fredrik Ronquist (1997), to analyze geographic patterns. There are important differences among these methods, but they all use a parsimony approach to reconstruct the geographic distributions of ancestors from data on the distributions of living taxa. (Inferring ancestral distributions from a phylogeny resembles, to some extent, inferring ancestral character states; see Figure 3.3.) Ronquist's method, DISPERSAL-VICARIANCE ANALYSIS (DIVA), most fully accounts for the importance of dispersal and is therefore most biologically realistic. This method assumes that vicariance is the "null hypothesis" accounting for changes in distribution, in accord with the well-supported principle that new species are generally formed during geographic isolation (see Chapter 18). Whenever

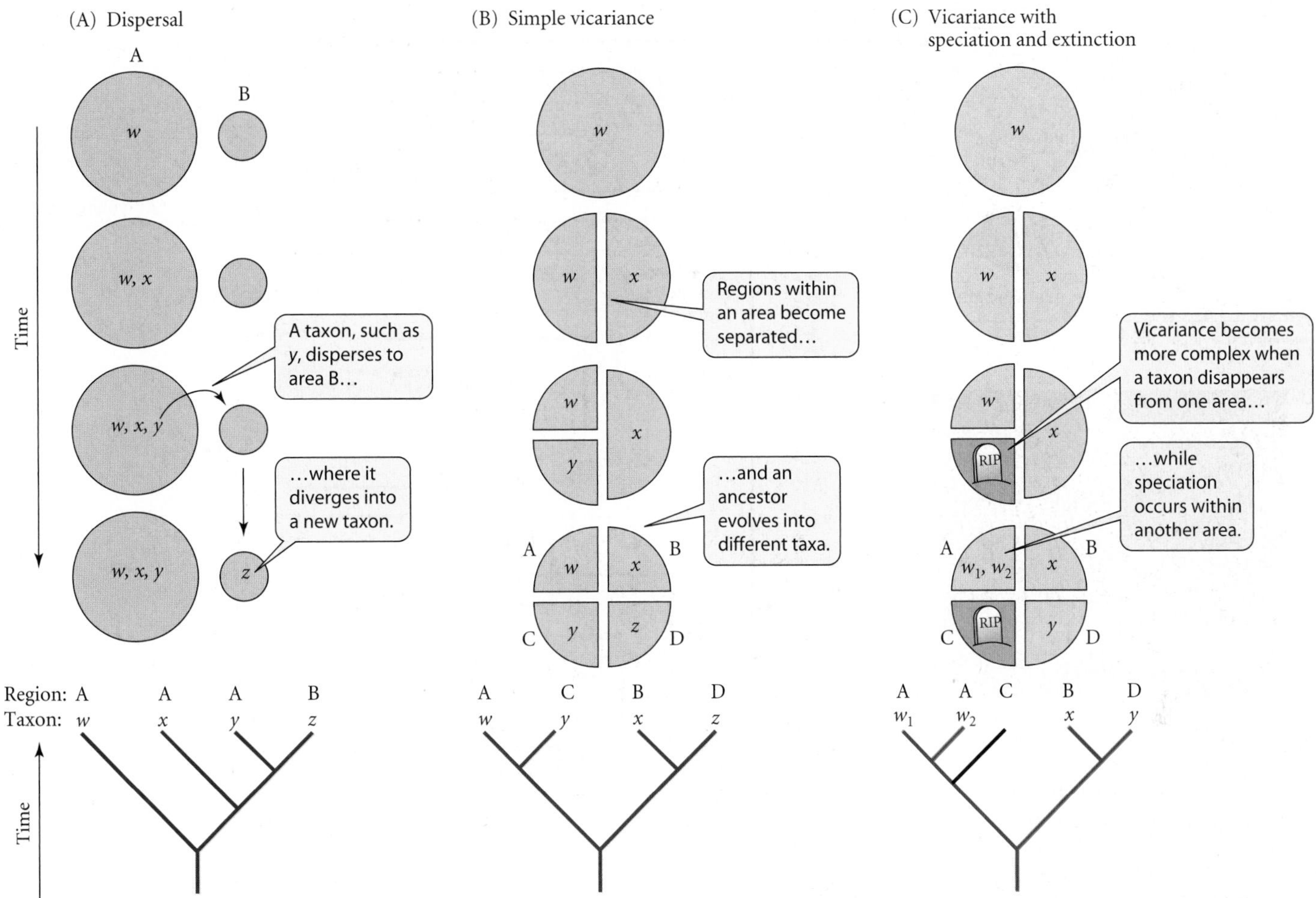

either dispersal or extinction must be invoked in order to explain a distribution, a "cost" is exacted. The historical hypothesis that accounts for the species' distributions with the lowest "cost" is considered the most parsimonious, or optimal, hypothesis.

Under the vicariance hypothesis, we expect monophyletic groups to occupy different areas, and we expect the sequence of geographic disjunctions implied by the phylogeny to match the sequence in which the areas themselves became separated (Figure 6.9). For example, in a clade distributed throughout Africa, Australia, and South America, species in Australia and South America should be more closely related to each other than to African species, because Africa was the first of these land masses to become separated from the rest of Gondwana (see Figure 6.11). In contrast, if species in area B are nested within a clade that otherwise is distributed in area A, dispersal from A to B may be likely (Figure 6.9A). Some biogeographers hold that vicariance should separate populations of many taxa simultaneously, so that the taxa should manifest common phylogenetic patterns of distribution. Dispersal can also engender common patterns among different taxa, especially when barriers to dispersal break down (Lieberman 2003).

Figure 6.9 Phylogenetic relationships as indicators of biogeographic history. (A) Dispersal from area A to area B, followed by divergence, is likely to yield a paraphyletic pattern of distribution of related species. (B) A vicariant history of successive separation of faunas is likely to yield a phylogeny of taxa that parallels the separation of the areas. (C) Complications in an otherwise vicariant history can arise for several reasons, including extinction (here, in area C) and speciation within an area (in area A).

Examples of historical biogeographic analyses

ORGANISMS IN THE HAWAIIAN ISLANDS. The Hawaiian Islands, in the middle of the Pacific Ocean, have been formed as a lithospheric plate has moved northwestward, like a conveyor belt, over a "hot spot," which has caused the sequential formation of volcanic cones. This process has been going on for tens of millions of years, and a string of submerged volcanoes that once projected above the ocean surface lies to the northwest of the present islands. Of the current islands, Kauai, at the northwestern end of the archipelago, is about 5.1 Myr old; the southeasternmost island, the "Big Island" of Hawaii, is the youngest, and is less than 500,000 years old (Figure 6.10A).

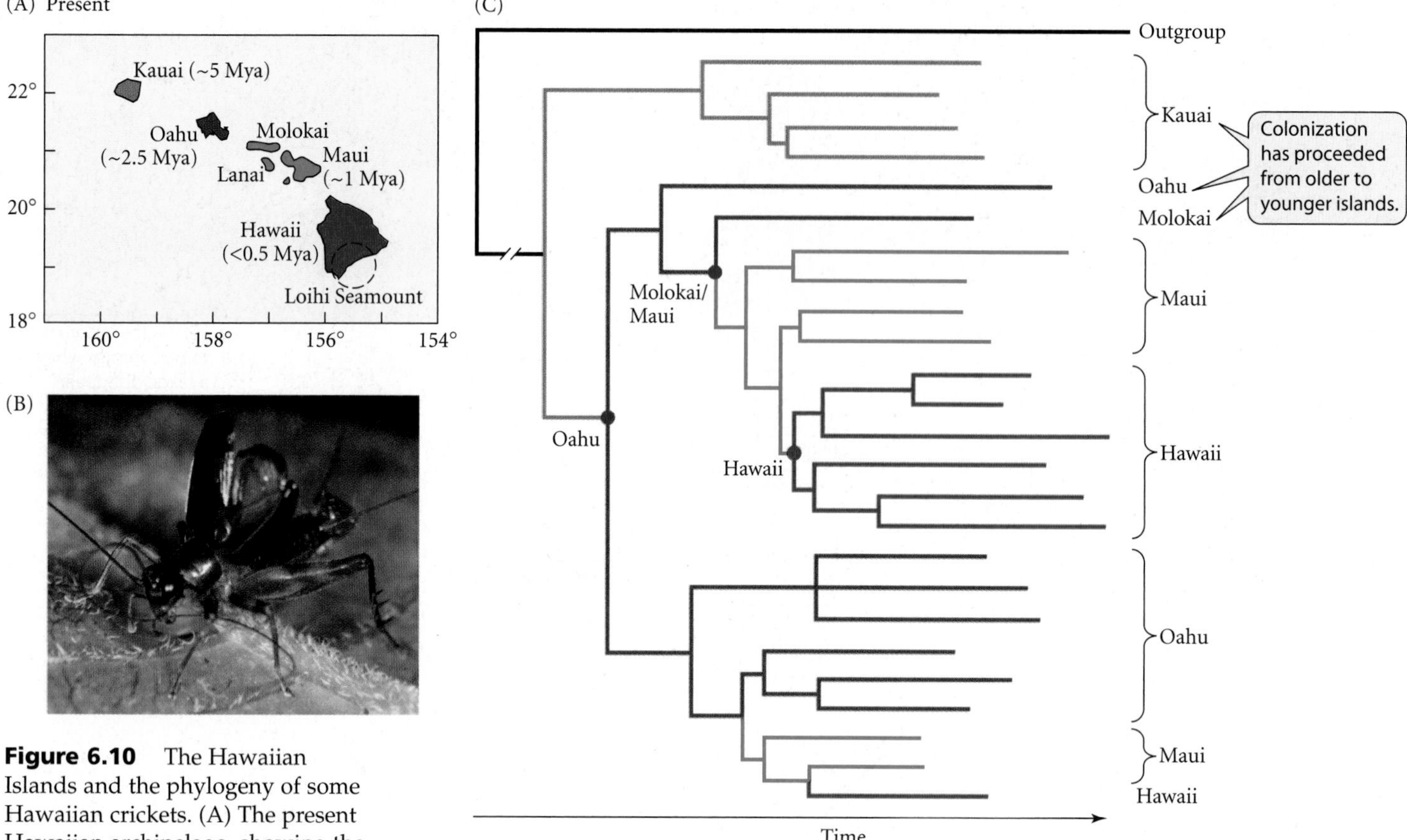

Figure 6.10 The Hawaiian Islands and the phylogeny of some Hawaiian crickets. (A) The present Hawaiian archipelago, showing the approximate dates of each island's formation. The Big Island, Hawaii, lies just to the northwest of the "hot spot" (broken circle), where islands have been successively formed over some 5 million years. (B) A Hawaiian cricket (genus *Laupala*). (C) Molecular phylogeny of *Laupala* species in the Hawaiian Islands; each branch represents a distinct species. Successively younger groups are found on younger islands. (After Mendelson and Shaw 2005; photo courtesy of Kerry Shaw.)

Given the geological history of the archipelago, the simplest phylogeny expected of a group of Hawaiian species would be a "comb," in which the most basal lineages occupy Kauai and the youngest lineages occupy Hawaii. This pattern would occur if species successively dispersed to new islands as they were formed, did not disperse from younger to older islands, and did not suffer extinction. Kerry Shaw's research group (Mendelson and Shaw 2005) found just this pattern when they performed a molecular phylogenetic analysis of a large genus of crickets (*Laupala*; Figure 6.10B,C). Colonization has proceeded from older to younger islands, and the rate of speciation within some of the islands has been higher than any other reported so far among arthropods.

ANIMALS IN MADAGASCAR. East of Africa lies the large island of Madagascar, whose highly endemic (and endangered) biota includes many groups, such as lemurs, that occur nowhere else. Together with India, Madagascar was the first land mass to split from Gondwana, having broken away from eastern Africa about 165 to 155 Mya (Figure 6.11). India became separated from Madagascar about 88 Mya, and collided with southern Asia about 50 Mya. There is some evidence that a land bridge may have connected Madagascar to the Antarctica-South America land mass until about 80 Mya (Noonan and Chippindale 2006).

For many years, biogeographers postulated that many of the endemic Madagascan taxa originated by vicariant separation from their relatives on other southern land masses. However, recent molecular phylogenetic studies indicate that most lineages of Madagascan plants and animals are too young to have been isolated by the division of Gondwana, and that dispersal has played the major role (Yoder et al. 2006). Madagascar's ants, frogs, snakes, several groups of birds, and all four groups of terrestrial mammals—including the lemurs, a primate group unique to Madagascar—appear to have evolved from African ancestors (Figure 6.12A). Chameleons (slow-moving lizards that catch insects

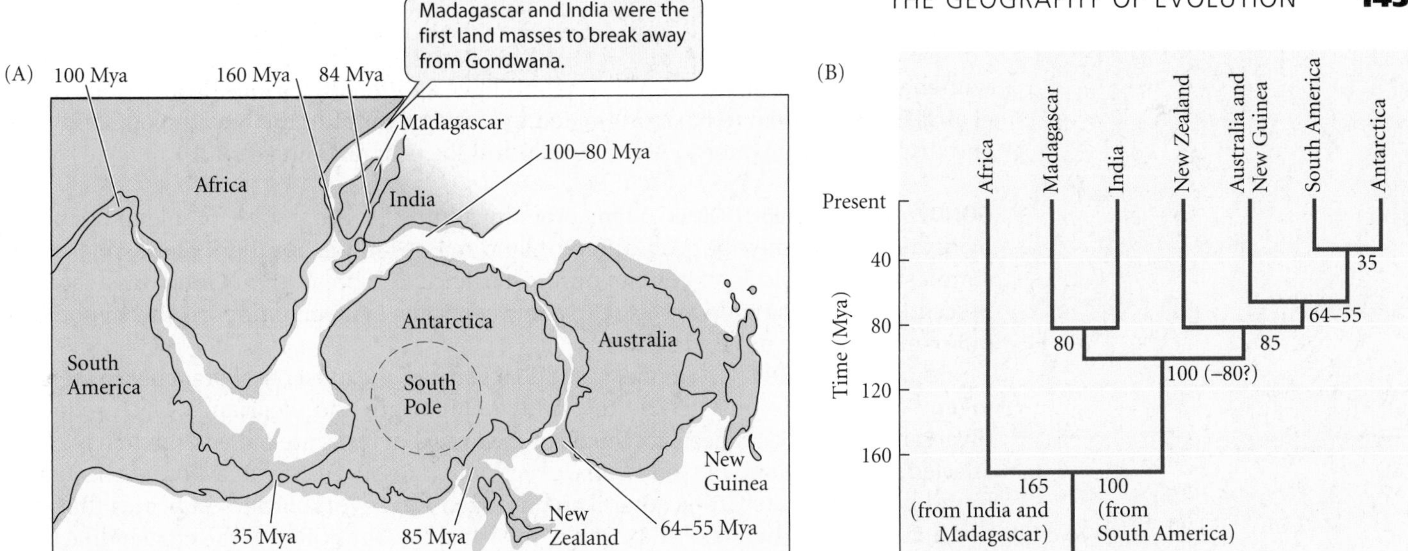

Figure 6.11 (A) A view of Gondwana in the early Cretaceous (120 Mya), centered on the present South Pole, indicating the approximate times at which connections among the southern land masses were severed. The current configurations of the continents are shown by the black lines; green areas beyond these lines were also exposed land during the early Cretaceous. (B) A branching diagram, sometimes called an area cladogram, that attempts to depict the history of the breakup of Gondwana. (A after Cracraft 2001.)

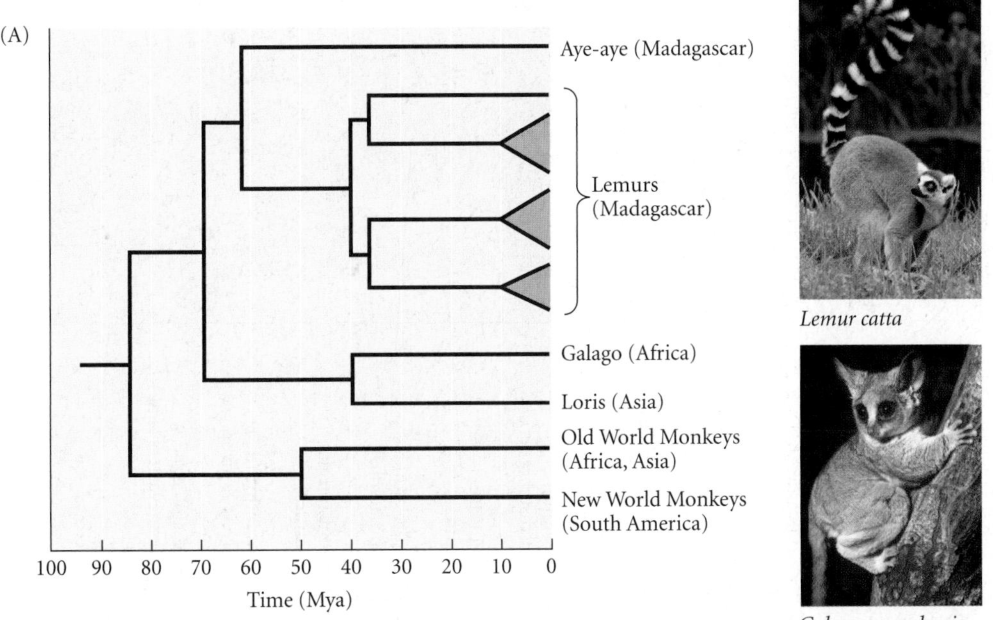

Lemur catta

Galago senegalensis

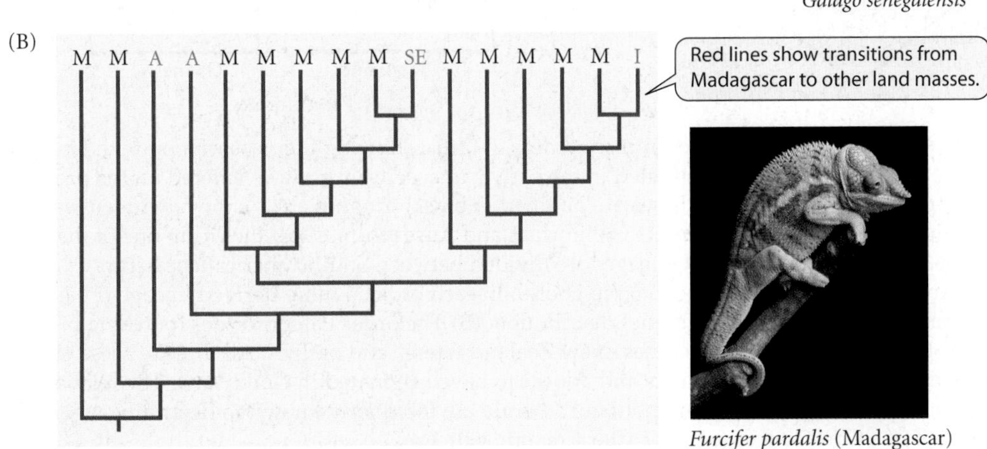

Furcifer pardalis (Madagascar)

Figure 6.12 Phylogenetic evidence of the history of two groups of animals in Madagascar. (A) The prosimians of Madagascar (lemurs, including the aye-aye) diverged from the African and Asian prosimians (galagos and lorises) well after Madagascar separated from the other Gondwanan land masses. (B) Chameleons are distributed in Africa (A), India (I), Madagascar (M), and the Seychelles Islands (SE) in the Indian Ocean. Because the phylogenetic distribution over these areas differs from the sequence by which the areas became separated (see Figure 6.11B), the distribution of chameleons is best explained by dispersal from Madagascar, rather than vicariance caused by the breakup of Gondwana. (A after Yoder and Yang 2004, photos © Vladimir Wrangel/istockphoto.com (lemur) and Rod Williams/Nature-pl.com (galago); B after Raxworthy et al. 2002, photo by David McIntyre.)

with their extraordinary projectile tongues) have a different, curious history (Raxworthy et al. 2002): they seem to have originated in Madagascar after the breakup of Gondwana and dispersed over water to Africa, India, and the islands (Figure 6.12B).

GONDWANAN DISTRIBUTIONS. Many other intriguing biogeographic problems are posed by taxa that have members on different land masses in the Southern Hemisphere. The simplest hypothesis is, of course, pure vicariance: the breakup of Gondwana isolated descendants of a common ancestor. In a few cases, the evidence supports this hypothesis, but many cases are not that simple.

Joel Cracraft (2001) has demonstrated that some of the most basal branches in the phylogeny of birds are consistent with Gondwanan origin and vicariance. DNA sequence divergence strongly suggests that most of the orders of birds are old enough to have been affected by the breakup of Gondwana, even though fossils of only a few orders have been found before the late Cretaceous. The phylogeny of several orders indicates that they originated in Gondwana. For example, the basal lineages of both the chickenlike birds (Galliformes) and the duck order (Anseriformes) are divided between South America and Australia (Figure 6.13A), and almost all of the basal lineages of the huge order of perching birds (Passeriformes) are likewise distributed among fragments of Gondwana (Figure 6.13B).

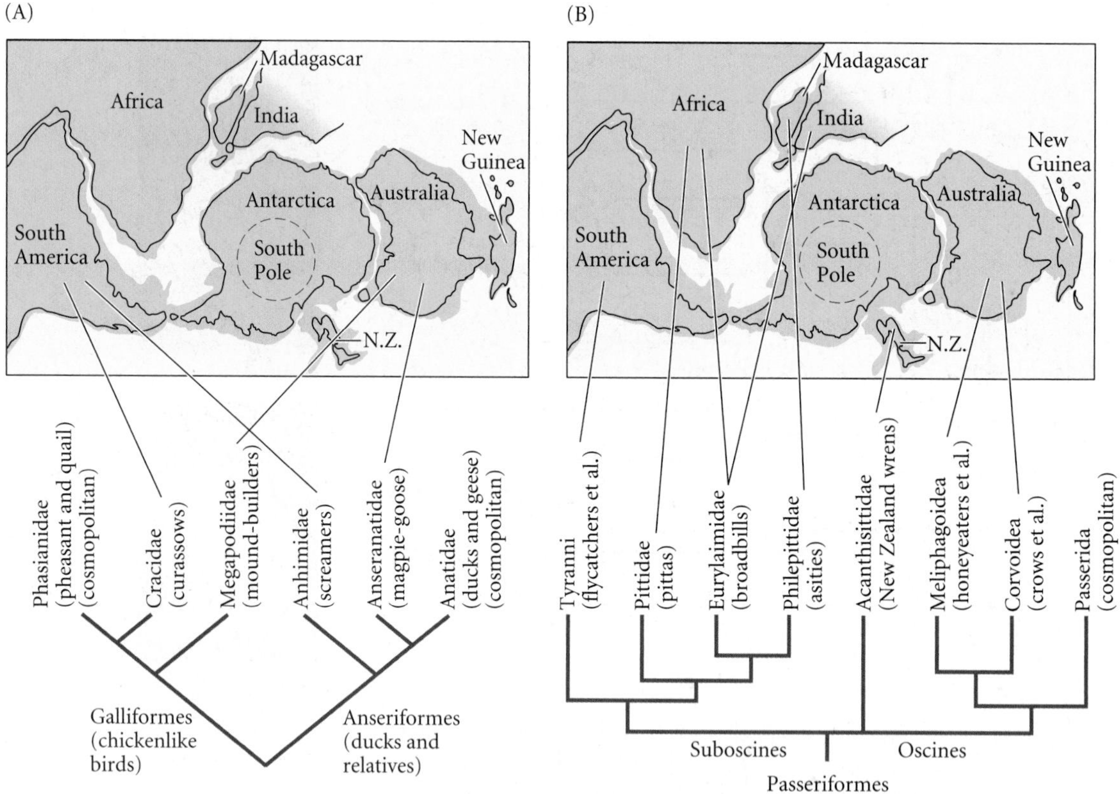

Figure 6.13 Phylogeny of major lineages in three orders of birds, showing their association with land masses, which are pictured as they were situated in the early Cretaceous, in a view centered on the present South Pole. The present continental boundaries are outlined in black; fringing areas shown in green were exposed during the Cretaceous. (A) The orders Galliformes and Anseriformes together form one of the oldest clades of birds. In each order, the basal lineages are divided between South America (curassows, screamers) and Australia (mound-builders, magpie-goose). In each order, a more derived lineage (Phasianidae, Anatidae) has a cosmopolitan (worldwide) distribution. (B) The order Passeriformes (perching birds, or songbirds) has three major clades: suboscines, New Zealand wrens, and oscines. All three of these clades have basal lineages in the southern continents and appear to have originated in Gondwana. The relationships among the many families of the cosmopolitan Passerida are too poorly known to determine whether they also originated in a Gondwanan region. (After Cracraft 2001.)

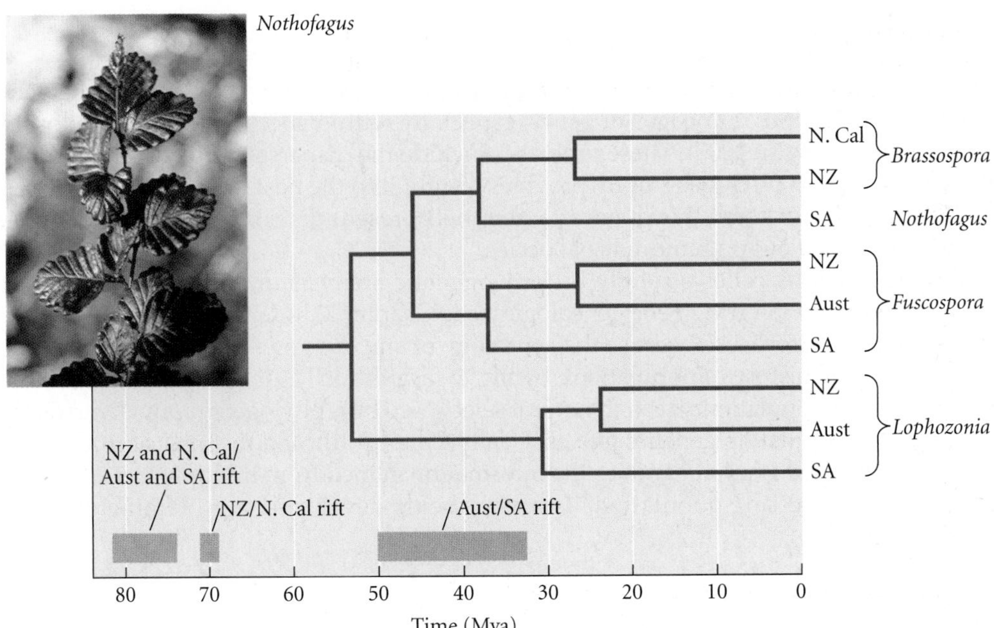

Figure 6.14 A simplified phylogeny of the four major lineages (subgenera) of southern beeches (*Nothofagus*), showing branching dates estimated by DNA sequence difference. In subgenera *Fuscospora* and *Lophozonia*, closely related species are in New Zealand and Australia, even though these land masses separated long before the rift between Australia and South America. The brown bars show estimated times at which Gondwanan plates separated, antedating the divergence of *Nothofagus* lineages. Consequently, vicariance by continental drift does not explain the disjunct distribution of these plants. (Based on Cook and Crisp 2005; photo by the author.)

The southern beeches (*Nothofagus*), distributed in southern South America, Australia, New Zealand, and the island of New Caledonia (near the Philippines), are a classic example of disjunct distribution that has been attributed to the rifting of Gondwana. Estimates of the age of divergence among different lineages, based on fossil-calibrated sequence divergence of several genes, were only partly consistent with this hypothesis (Cook and Crisp 2005). Two lineages (subgenera) gave rise to species in both New Zealand and Australian species, and in both cases the divergence occurred less than 50 Mya, more than 30 Myr after New Zealand had broken away (Figure 6.14). The closest relatives of the New Caledonian species are in New Zealand, but they diverged about 50 Mya, more than 70 Myr after these land masses separated. Whereas these examples suggest that the New Zealand and New Caledonian species evolved from ancestors that dispersed across ocean, the molecular dating of the disjunction between Australian and South American species in two different subgenera coincides well with the time of rifting between these continents, and may indeed represent a case of vicariance. On the whole, however, dispersal among the southern land masses, well after the rifting of Gondwana, has clearly played a major role in the distribution not only of the southern beeches but of many other plant and animal taxa as well (de Queiroz 2005), and dispersal seem likely to replace vicariance as the best explanation of the distribution of many or most groups of organisms.

The composition of regional biotas

These examples show that the geographic history of a clade is often complex and may include both vicariance and dispersal events. The taxonomic composition of the biota of any region is a consequence of diverse events, some ancient and some more recent. Certain taxa are **allochthonous**, meaning that they originated elsewhere. Others are **autochthonous**, meaning that they evolved within the region. For example, the biota of South America has (1) some autochthonous taxa that are remnants of the Gondwanan biota and are shared with other southern continents (e.g., lungfishes, *Araucaria* pines); (2) groups that diversified from allochthonous progenitors during the Tertiary, after South America became isolated by continental drift (e.g., New World monkeys, guinea pigs and related rodents); (3) some allochthonous species that entered from North America during the Pleistocene (e.g., the mountain lion, *Panthera concolor*, which also occurs in North America); and (4) species that have colonized South America within historical time, whether assisted by humans (e.g., many weeds) or not (e.g., the cattle egret, which apparently arrived from Africa in the 1930s; see Figure 6.7).

Phylogeography

Phylogeography is the description and analysis of the processes that govern the geographic distribution of lineages of genes, especially within species and among closely related species (Avise 2000). These processes include the dispersal of the organisms that carry the genes, so phylogeography provides insight into the past movements of species and the history by which they have attained their present distributions. Such analyses have been carried out on hundreds of species.

Phylogeography relies strongly on phylogenetic analysis of variant genes within species; that is, on inferring gene genealogies (see Chapter 2). Inferring population histories is not straightforward, because the genealogy of any one genetic locus may not accurately reflect the history of populations, owing to factors such as incomplete lineage sorting, changes in population size, and natural selection. Thus, phylogeography depends on the theory of population genetics (see especially Chapter 10), and the most accurate estimates are obtained from integrating the information gained from multiple loci. Methods for accurately inferring population history are being developed (e.g., Templeton 1998; Knowles 2004).

Pleistocene population shifts

Many authors have used the phylogeographic approach to infer the history of populations during the Pleistocene, when (as we know from paleontological evidence) many northern species occurred far to the south of their present distributions during glacial episodes, moving northward after the glaciers receded (see Chapter 5). Moreover, we know that different species occupied different glacial refuges and had different paths of movement. Many species have left no fossil traces of their paths, but phylogeographic analysis can help to reconstruct them (Taberlet et al. 1998; Hewitt 2000). For example, fossil pollen shows that refuges for deciduous vegetation in Europe during the most recent glacial period were located in Iberia (Spain and Portugal), Italy, and the Balkans (Figure 6.15A), and that the vegetation expanded most rapidly from the Balkans as the glacier retreated. The grasshopper *Chorthippus parallelus*, sampled from throughout Europe, has unique mitochondrial DNA haplotypes in Iberia and in Italy, whereas the haplotypes found in central and northern Europe are related to those in the Balkans (Figure 6.15B). Thus we can conclude that this herbivorous insect expanded its range chiefly from the Balkans, but did not cross the Pyrenees from Iberia, nor the Alps from Italy (Figure 6.15C).

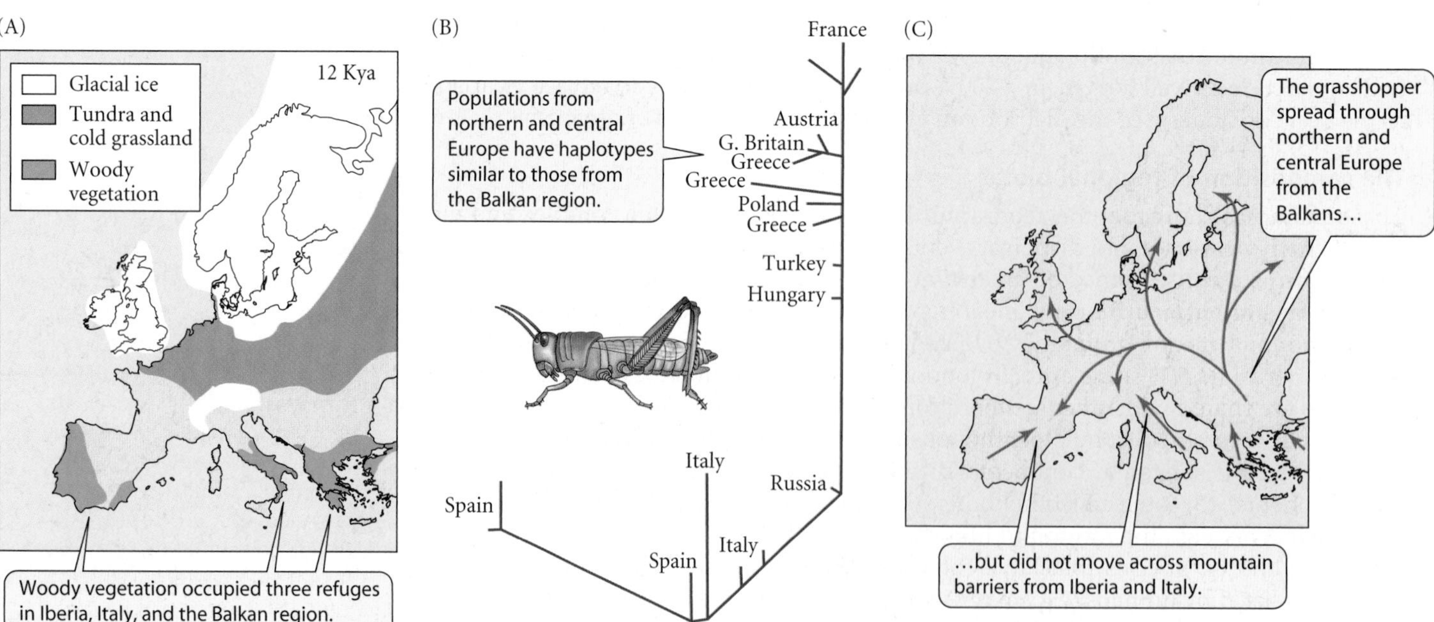
(A)
12 Kya
Glacial ice
Tundra and cold grassland
Woody vegetation
Woody vegetation occupied three refuges in Iberia, Italy, and the Balkan region.
(B)
Populations from northern and central Europe have haplotypes similar to those from the Balkan region.
France
Austria
G. Britain
Greece
Greece
Poland
Greece
Turkey
Hungary
Italy
Russia
Spain
Italy
Spain
(C)
The grasshopper spread through northern and central Europe from the Balkans...
...but did not move across mountain barriers from Iberia and Italy.

Figure 6.15 The recolonization of Europe from glacial refuges by the grasshopper *Chorthippus parallelus*, inferred from patterns of genetic variation. (A) Europe during the last glacial maximum (about 12 Kya). (B) Genetic relationships among contemporary grasshopper populations. The length of a line segment reflects both the number of mutational differences between haplotypes and the difference in proportions of those haplotypes among populations from the areas indicated. (C) Arrows show the inferred spread of *Chorthippus parallelus* after the glacier melted back. (A after Taberlet et al. 1998; B after Cooper et al. 1995; C after Hewitt 2000.)

A similar analysis of hedgehogs (*Erinaceus europaeus* and *E. concolor*) indicated, in contrast, that these insectivorous mammals colonized northern Europe from all three refugial areas.

Modern human origins

Phylogeography has also been applied to our own distribution. We saw in Chapter 4 that *Homo erectus* was broadly distributed throughout Africa and Asia by about 1 Mya and had evolved into "archaic *Homo sapiens*" by about 300,000 years ago (300 Kya). Fossil evidence indicates that these forms persisted until about 29 Kya. "Anatomically modern humans" date from 195 to 170 Kya in Africa, but are not found elsewhere for another 80,000 years. How these ancient populations are related to the different human populations of today is a controversial question that is being addressed with genetic evidence (Satta and Takahata 2002; Garrigan and Hammer 2006; Templeton 2007; Relethford 2008).

Based on the morphology of fossil specimens, advocates of the MULTIREGIONAL HYPOTHESIS hold that archaic *sapiens* populations in Africa and Eurasia (Europe and Asia) all evolved into modern *sapiens*, with gene flow spreading modern traits among the various populations (**Figure 6.16A**). According to this hypothesis, some of the genetic differences found among today's human populations in Europe and Asia should trace back to genetic differences that developed among populations of *erectus* and archaic *sapiens* nearly a million years ago. In this context, "archaic *sapiens*" includes the European Neanderthals, *Homo neanderthalensis*, which should have contributed to the modern gene pool according to the multiregional hypothesis.

In contrast, the REPLACEMENT HYPOTHESIS, or OUT-OF-AFRICA HYPOTHESIS, holds that after archaic *sapiens* spread from Africa to Asia and Europe, modern *sapiens* evolved from archaic *sapiens* in Africa, spread throughout the world in a second expansion, and replaced the populations of archaic *sapiens* without interbreeding with them to any substantial extent (**Figure 6.16B**). That is, the modern *sapiens* that evolved from archaic *sapiens* in Africa was reproductively isolated from the Neanderthals and other Eurasian populations of archaic *sapiens*—it was a distinct biological species. According to this hypothesis, the world's populations of archaic *sapiens* became almost entirely extinct as a result of competition or climate change (Finlayson 2005), and most genes in contemporary populations are descended from those carried by the population that spread from Africa.

Early genetic studies (Cann et al. 1987; Vigilant et al. 1991), which used mitochondrial DNA (mtDNA), supported (and indeed inspired) the replacement hypothesis. A later, more extensive study of mtDNA used the complete mitochondrial sequence of 53 humans of diverse geographic origin, using a chimpanzee as an outgroup (Ingman et al. 2000). The phylogenetic analysis showed several basal clades of African haplotypes and a derived

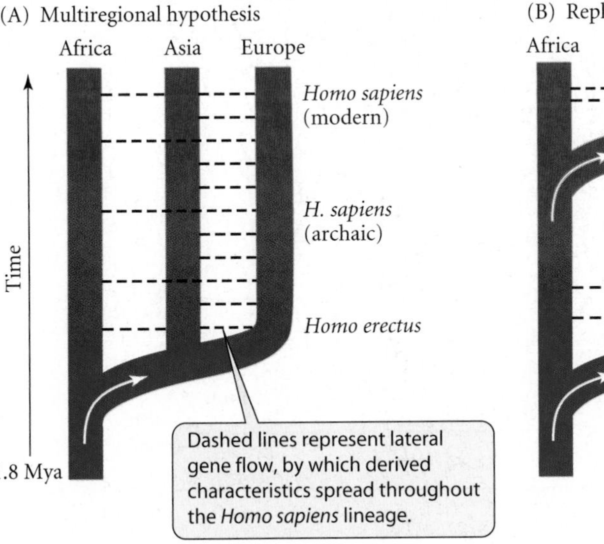

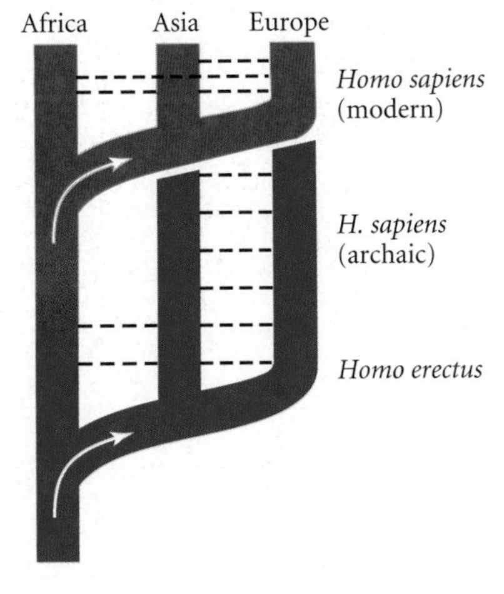

Figure 6.16 Two hypotheses on the origin of modern humans. (A) The multiregional hypothesis posits a single wave of expansion by *Homo erectus* from Africa to parts of Asia and Europe, and continuity of descent to the present day. Broken lines indicate gene flow among regional populations. (B) The replacement hypothesis proposes that populations of *H. erectus*, derived from African ancestors, gave rise to archaic *sapiens*, but that Asian and European populations of archaic *sapiens* became extinct when modern *sapiens* expanded out of Africa in a second wave of colonization.

clade that includes not only several African haplotypes but also all the non-African populations from throughout the world (Figure 6.17). Moreover, the non-African haplotypes vary less in nucleotide sequence than those found in Africa. These observations are consistent with the replacement hypothesis. If, as in the multiregional hypothesis, some contemporary Eurasian populations were descended from indigenous populations of archaic *Homo sapiens* (and from indigenous *H. erectus*), and thus had a separate ancestry extending back a million years, we would expect some of their genes to have accumulated far more mutational differences than are observed among living humans. We now know that both mitochondrial and nuclear DNA sequences from Neanderthal fossils are significantly different from modern human sequences and indicate that these lineages

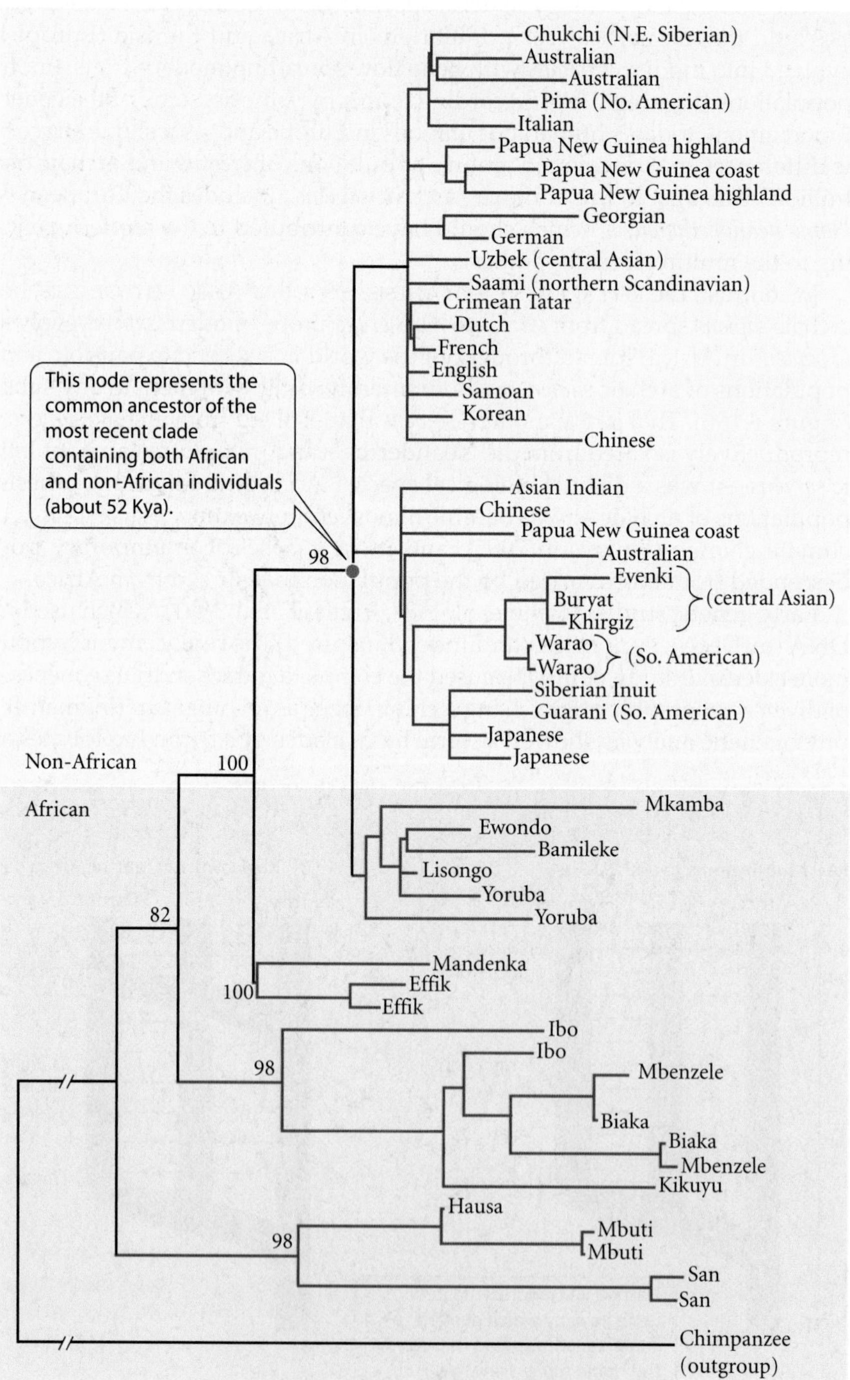

Figure 6.17 A gene tree based on complete sequences of mitochondrial genomes from human populations throughout the world. Haplotypes from individuals in Africa (green background) are phylogenetically basal, as expected given the African origin of the human species, and show high sequence diversity (represented by the lengths of the branches). Haplotypes taken from individuals in the rest of the world (yellow background) form a single clade of very similar haplotypes (denoted by short branches), as expected if these populations had been recently derived from a small ancestral population. Some populations (e.g., Australian) are represented by more than one individual. Numerals represent bootstrap values (see Box 2B). (After Ingman et al. 2000.)

(A)

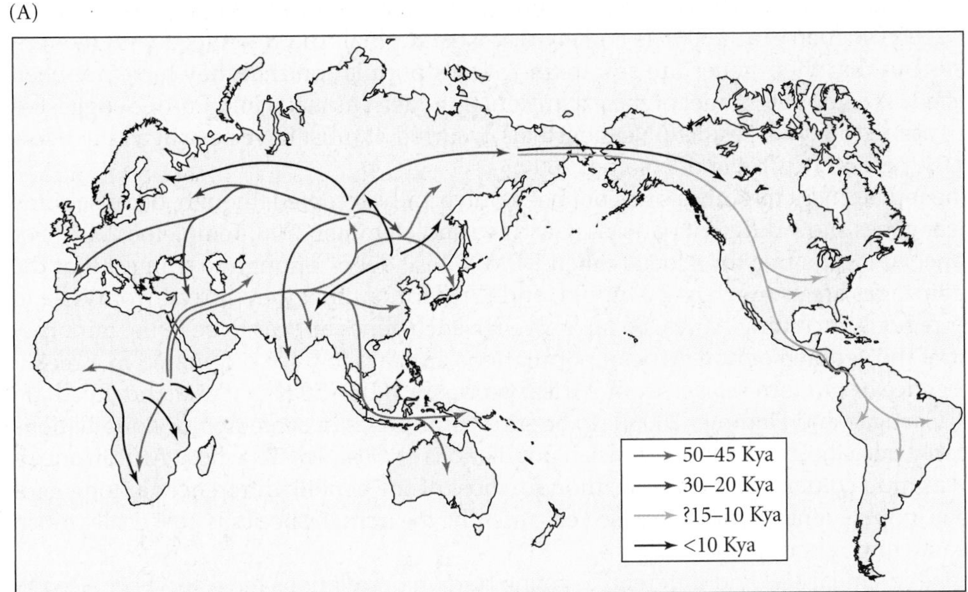

(B)

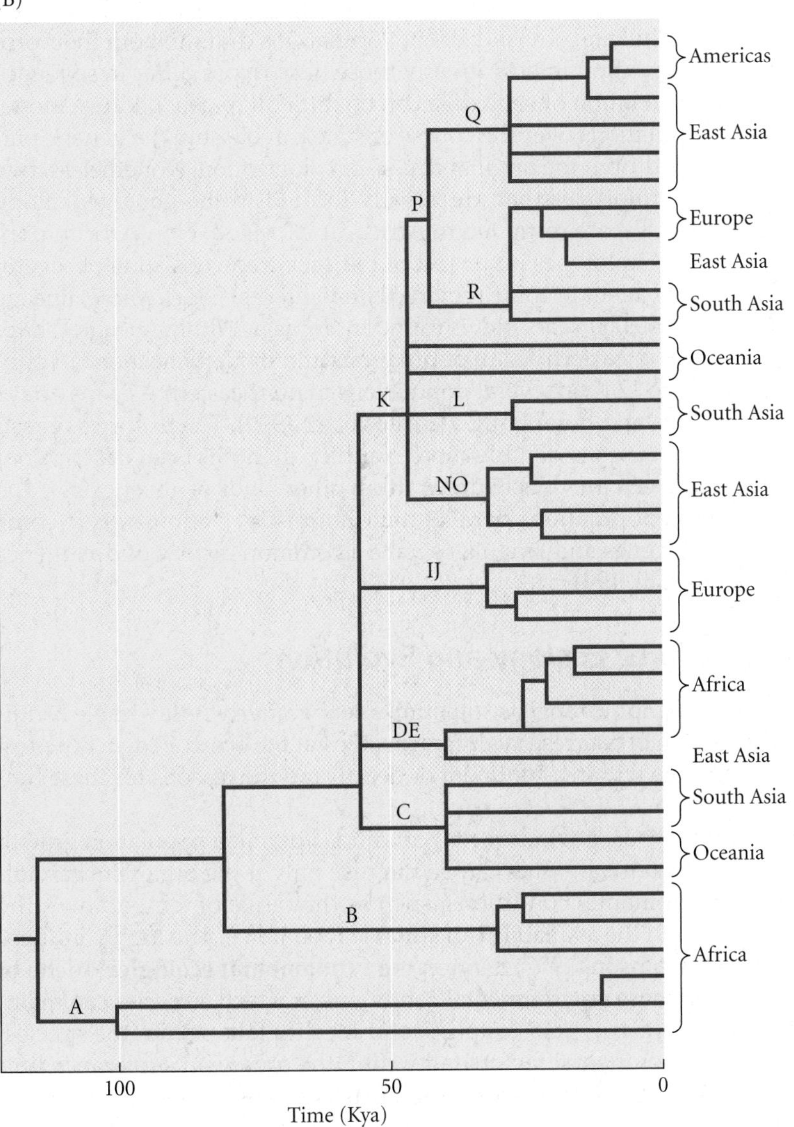

Figure 6.18 The movement of human populations from about 50 to 10 Kya. (A) Some of the routes postulated for human population dispersal at several times in the past, based on genetic data such as that from Y chromosomes. (B) A gene tree for Y chromosomes. Note the African distribution of phylogenetically basal chromosome lineages (groups A and B, bottom), and the derivation of chromosomes endemic to North and South America from lineages found in eastern Asia (group Q, top). Mitochondrial DNA also suggests a close relationship of native Americans (e.g., the South American Guarani in Figure 6.17) to some populations of eastern Asia. (A after Underhill et al. 2001; B after Underhill and Kivisild 2007.)

diverged about 500 Kya, before *sapiens* evolved into its anatomically modern form (Green et al. 2006; Noonan et al. 2006). Is it likely that Neanderthals did interbreed with modern *sapiens* but that their genes are so rare in today's population that they have not been detected? A computer model of the expansion of modern humans into Europe suggested that if admixture with resident Neanderthals occurred, it must have been at a rate of less than 0.1 percent (Currat and Excoffier 2004).

The replacement hypothesis has not been universally accepted, though, because some nuclear genes show different patterns (Garrigan and Hammer 2006; Templeton 2007). For instance, the gene tree for a locus called RRM2P4 is "deep" enough to suggest that the gene lineages are as much as 2 Myr old, and furthermore that they spread from Asia to other regions, including Africa. So far, however, such genes appear to be in the minority; most of the genome of non-African populations appears to have originated in a recent emergence of modern *sapiens* from Africa, probably 100 to 50 Kya (Satta and Takahata 2002; Garrigan and Hammer 2006)). To be sure, some genes in contemporary populations probably trace their ancestry to Eurasian populations of *Homo erectus* whose African ancestry was much older. On the whole, though, most of the genetic differences among geographic populations of humans arose very recently: the human species is genetically much the same throughout the world.

Genetic similarities and differences among human populations have also been used to trace later movements (Figure 6.18A, previous page). For example, sequence variation in a cluster of genes on the Y chromosome, which is carried only by men, has been studied on a worldwide basis (Underhill and Kivisild 2007). Populations differ in both their proportions of different haplotypes and in how greatly those haplotypes differ in sequence from one another. The interpretation of such data can be difficult, partly because movements of people among populations over the course of time can obscure the genetic patterns that may have developed from the original course of colonization. Nonetheless, two groups of Y chromosome haplotypes that are basally located in the gene genealogy (groups A and B in Figure 6.18B) are restricted to Africa, as expected. Non-African populations are characterized by haplotypes in the rest of the gene tree, consisting of several groups that are each more prevalent in some regions than in others. For example, lineage P, which appears to be about 50,000 years old, is found in Eurasia. Within lineage P, haplotypes in group Q are found in eastern Asian populations and in Native American populations. Starting perhaps 15 to 12 Kya, several populations in northeastern Asia may have dispersed into North America at different times (Santos et al. 1999). The history suggested by Y chromosomes (which is considerably more complex than this brief description) supports inferences that had been previously drawn from other kinds of genetic data. The genetic relationships among populations parallel their linguistic relationships to some extent, suggesting that both genes and languages have a common history of divergence in isolation (Cavalli-Sforza et al. 1994).

Geographic Range Limits: Ecology and Evolution

The border of a species' geographic range is sometimes set by utterly unfavorable conditions, as when the distribution of a terrestrial organism stops at the ocean's edge. But most species have a restricted range within a continent or ocean, and the reasons for these limits are less evident.

A species can persist only where environmental conditions permit a population growth rate greater than or equal to zero, and this can be the case only if the organism can tolerate each of several environmental conditions, such as the range of temperature, the amount of available water, and the availability of suitable food items. The highly influential ecologist G. Evelyn Hutchinson (1957) defined the **fundamental ecological niche** of a population as the set of all those environmental conditions in which a species can maintain a stable population size (Figure 6.19). A particular locality falls within the species' niche if all the relevant environmental factors fall within the organism's tolerance limits, but will fall outside the niche if any one variable, such as coldest winter temperature, falls outside these limits. Thus the *potential* niche space is dictated by the overlap of the fundamental niche and the realized (actual) environmental "space," or set of values. Even within the potential niche space, the range of environmental conditions that a species actually occupies—the **realized ecological niche**—may be further restricted if the species is excluded by competitors or predators.

The physiological tolerance of a species to some factors, such as temperature, can be determined by experiment, but it is nevertheless very difficult to identify all the environmental factors that might affect population growth. Furthermore, even if we find a correlation between the geographic range limit and a particular measured variable, such as maximal July temperature, this is likely to be correlated with other variables (such as temperature at other times of year, or the variability of temperature, or water deficit, or the availability of a required food item). Nevertheless, the distribution of many species is correlated with climatic variables (especially aspects of temperature and rainfall). Some species have shifted their geographic or altitudinal range in recent decades, apparently in response to human-caused climate change; for instance, both the northern (or higher) and southern (or lower) limits have shifted upward in several butterflies (Parmesan et al. 2005). In at least one species, the northern limit is set by the insect's intolerance to extreme low temperature; in another species, by the effect of temperature on the synchrony between the butterfly's life history and the seasonal development of its food plant. Some species borders are thought to be set by competition with other species. For example, three species of nectar-feeding honeyeaters occur in the mountains of New Guinea, but each mountain range has only two species, and those two have mutually exclusive altitudinal distributions. Which species is missing from a mountain range appears to be a matter of chance (Figure 6.20).

Related species often have similar ecological requirements, presumably derived from their common ancestor. This pattern, called **phylogenetic niche conservatism**, has been described with respect to both abiotic and biotic factors. A. Towne Peterson and colleagues (2002) predicted the geographic distribution of species of birds, mammals, and butterflies on each side of the Isthmus of Tehuantepec, in Mexico, from the climatic characteristics of sites occupied by their sister species (closest relatives) on the other side of the isthmus (Figure 6.21A). Almost all the predictions that were based on adequate samples were successful, implying that these animals' ecological requirements had not changed substantially since speciation. Many lineages of herbivorous insects have remained associated with the same genus or family of food plant (Figure 6.21B)—associations that, in some instances, have remained unchanged for more than 40 million years (Mitter and Farrell 1991; Winkler and Mitter 2008).

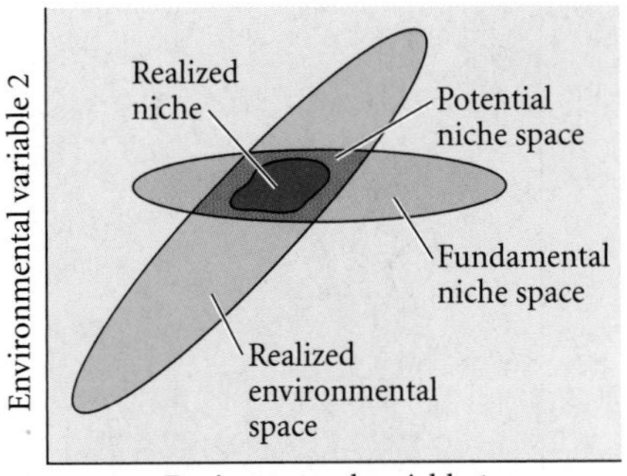

Figure 6.19 A conceptualization of ecological niches, showing the combinations of two environmental variables that would permit a particular species population to persist and grow. The fundamental niche is the set of possible environments in which persistence is possible. The set of environmental combinations that exists in a given region is the realized environmental space. The overlap of the realized environmental space and the fundamental niche is the potential niche space—the actual set of habitable combinations. However, the species might be excluded from some of these potentially habitable places (perhaps by competing species), so that it is limited to a smaller set of environmental conditions, the realized niche. In reality, many environmental variables (each of which would be represented by another dimension, which we cannot picture) determine population persistence. (After Jackson and Overpeck 2000.)

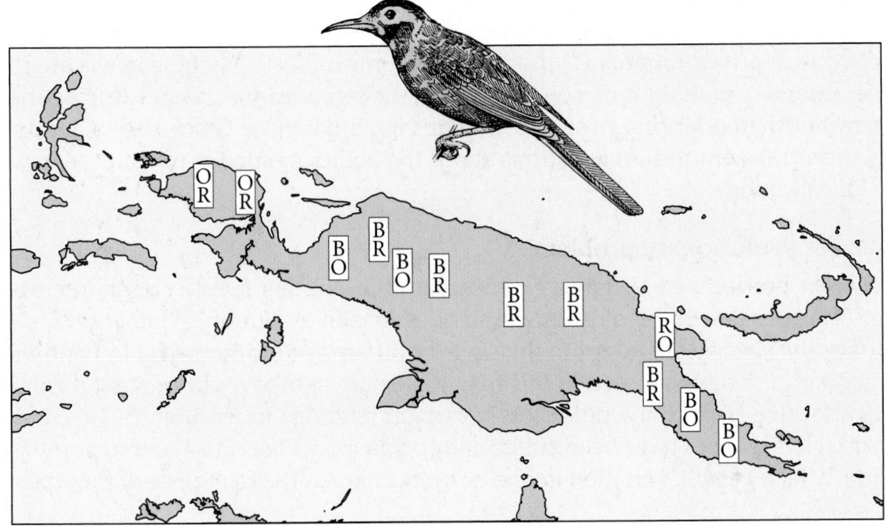

Figure 6.20 A "checkerboard" distribution in which species replace each other haphazardly. Among the various mountain ranges in New Guinea, three species of honeyeaters (*Melidectes*), denoted by letters O, R, and B, are distributed in pairs. Each pair has mutually exclusive altitudinal ranges, as shown by the stacked letters. The three species do not all coexist in any mountain range. This distributional pattern suggests that competition among these species limits their altitudinal ranges. (After Diamond 1975.)

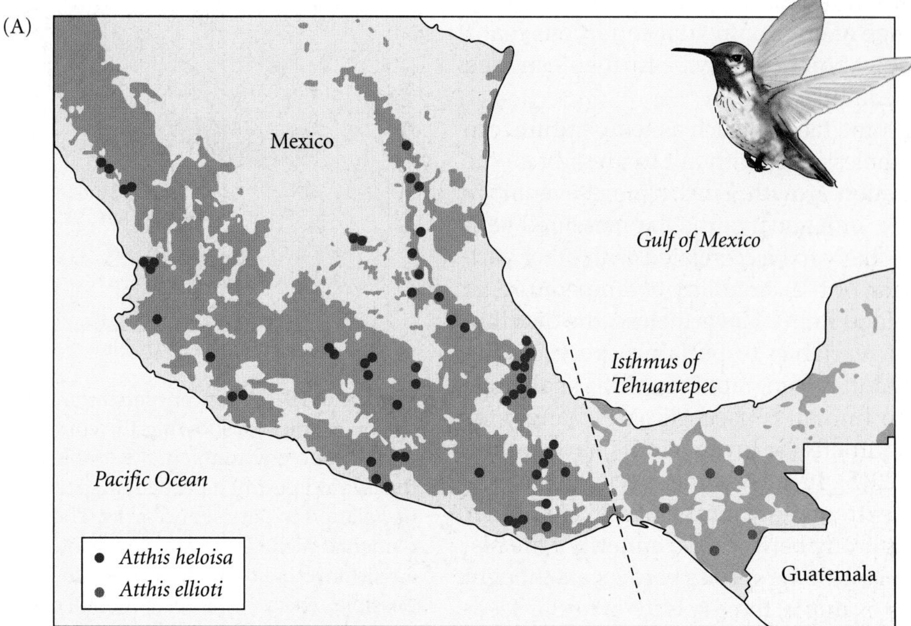

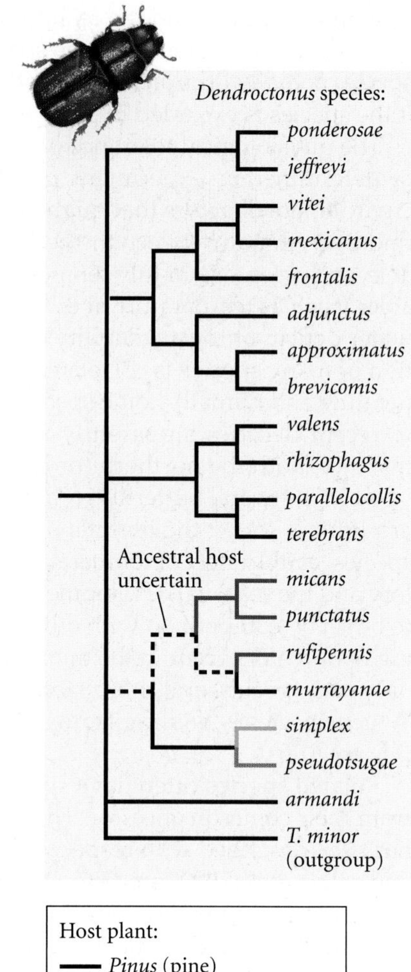

Figure 6.21 Two illustrations of phylogenetic niche conservatism. (A) The distribution of two sister species of hummingbirds, *Atthis heloisa* and *A. ellioti*, on either side of the Isthmus of Tehuantepec in Mexico. The climatic combinations at localities occupied by each species were used to predict the potential distribution of its sister species, on the assumption that their niches had not evolved since their common ancestor. Green areas west of the dashed line represet the geographic distribution of *A. heloisa*, predicted from the eastern species, and brown areas east of the line are the distribution of *A. ellioti*, predicted from the western species. The actual locations where each species was found, indicated by the colored circles, all fall within predicted range areas. Hence, these species have retained similar environmental tolerances or requirement. (B) Many species of herbivorous insects feed only on one or a few kinds of host plants. These associations are often phylogenetically conservative, as shown by bark beetles (*Dendroctonus*), in which lineages of related species tend to feed on the same genus of plant. The shadings of the branches of this phylogenetic tree indicate the host-plant genus of bark beetles today and the most parsimonious estimate of their ancestors' host associations. (A after Peterson et al. 1999; B after Kelley and Farrell 1998.)

Niche conservatism has important consequences (Wiens and Graham 2005). For instance, it contributes to our understanding of the geographic distributions of many clades: oaks (*Quercus*) and dogwoods (*Cornus*) are among the many plant taxa that occur in the temperate regions of eastern North America, Asia, and Europe but have not adapted to warm tropical environments (Donoghue and Smith 2004). Niche conservatism underlies the observation that many species shifted their geographic ranges during the Pleistocene, rather than adapting in situ to changes in climate (see Chapter 5). Clearly, these observations raise important questions about the ability of species to adapt to new environmental conditions.

Range limits: An evolutionary problem

Whether a species' border is caused by a climatic variable, another species, or an interaction of factors, it poses one of the most puzzling questions in evolutionary biology (Holt 2003): why does the species not adapt to the slightly different conditions just beyond its present range? And if it does succeed in this first small step, what would prevent it from further small adaptive steps that would slowly expand its range indefinitely?

Two principal hypotheses have been suggested (Bridle and Vines 2007). First, populations may simply lack genetic variation in one or more characteristics necessary for adaptation. For example, populations of two species of rain forest-dwelling *Drosophila* have no detectable genetic variation for tolerance to desiccation, which might prevent them from expanding into drier habitats (Kellermann et al. 2006). Second, incursion of genes from populations in favorable environments could prevent recipient populations from adapting to the unfavorable environment at the range margin, since this process of gene flow counteracts natural selection for local adaptation. Several mathematical models have shown that this could be a powerful limitation on adaptation (Holt and Gaines 1992; Kirkpatrick and Barton 1997; Lenormand 2002). Chapter 12 presents examples of the conflict between gene flow and natural selection, but this hypothesis on range limits needs much more research.

Evolution of Geographic Patterns of Diversity

The field of community ecology is concerned with explaining the species diversity, species composition, and trophic structure of assemblages of coexisting species (often called communities). These are partly biogeographical problems, since the geographic distributions of species determine whether or not they can overlap. Thus, historical biogeography bears on community ecology. Until recently, however, many ecologists have attempted to use only ecological theory, based on interactions among species, to predict some aspects of community structure. For example, competition should tend to prevent the coexistence of species that are too similar in their use of resources. This theory assumes that diversity and other features of communities have reached an equilibrium—which in turn assumes that evolution has provided a supply of potential community members with appropriate characteristics, or that such characteristics will rapidly evolve to suit the situation.

Community convergence

If community structure were predictable from ecological theories, we might expect that if two regions present a similar array of habitats and resources, species would evolve to use and partition them similarly. In other words, just as individual species often evolve convergent adaptations (see Figures 3.8 and 6.1), sets of species might evolve convergent community structure.

A striking example of community-level convergence has been described in the anoles (*Anolis*) of the West Indies (Williams 1972; Losos et al. 1998, 2006). Anoles are a species-rich group of insectivorous, mostly arboreal Neotropical lizards. Different species are known to compete for food, and this competition has influenced the structure of anole communities. Each of the small islands in the Lesser Antilles has either a single (solitary) species or two species. Solitary species are generally moderate in size, whereas larger islands have a small and a large species that can coexist because they take insect prey of different sizes and also differ in their microhabitats. The small species of the various islands are a monophyletic group, and so are the large species. Thus it appears that each island has a pair of species assembled from the small-sized and the large-sized clades.

The large islands of the Greater Antilles (Cuba, Hispaniola, Jamaica, Puerto Rico) harbor greater numbers of species. These anoles occupy certain microhabitats, such as tree crown, twig, and trunk, which are filled by different species on each island. The occupants of different microhabitats, called ECOMORPHS, have consistent, adaptive morphologies (**Figure 6.22**). These ecomorphs have evolved repeatedly, for the species on each of the islands form a monophyletic group that has radiated into species that ecologically and morphologically parallel those on the other islands. The most reasonable interpretation of this pattern is that as new species have arisen on each island, they have evolved in similar ways to avoid competition, by adapting to the same kinds of previously unused microhabitats.

Nonetheless, the *Anolis* assemblages on these islands are not entirely similar, because each island has multiple species of certain ecomorphs, and the numbers of species differ from one island to another (**Figure 6.23**; Losos et al. 2006). Moreover, the degree of con-

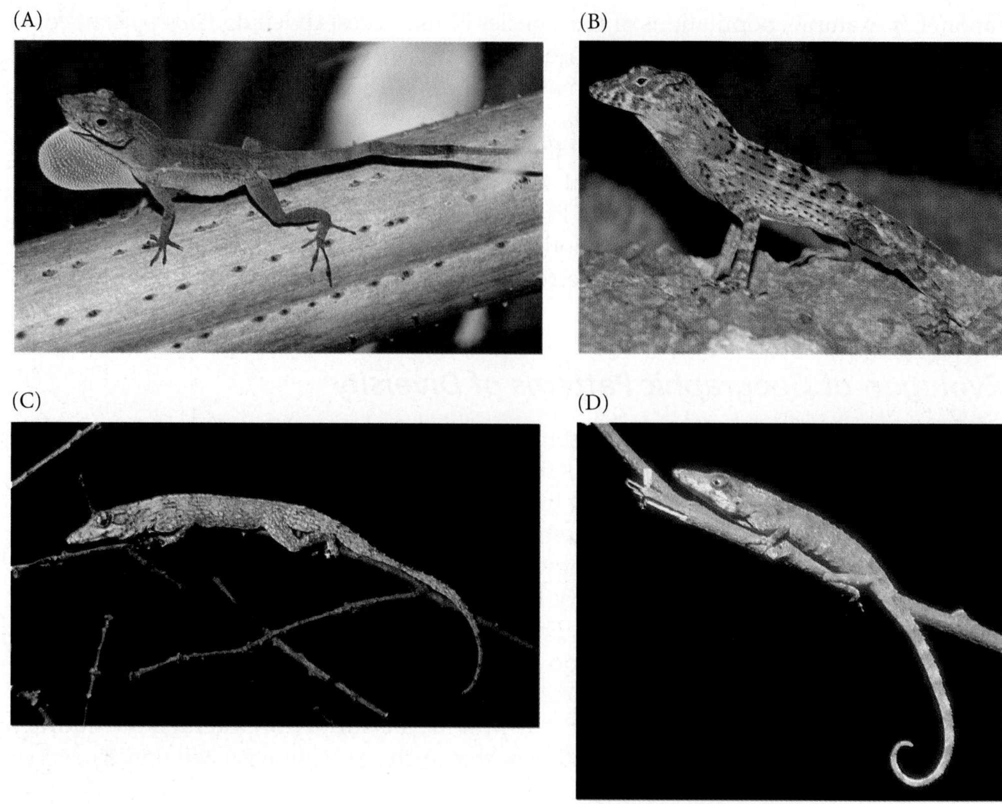

Figure 6.22 Convergent morphologies, or "ecomorphs," of *Anolis* lizards in the Greater Antilles, West Indies. (A) *Anolis lineatopus* from Jamaica. (B) *A. strahmi* from Hispaniola. Both species have independently evolved the stout head and body, long hind legs, and short tail associated with living on lower tree trunks and on the ground. (C) *Anolis valencienni* from Jamaica. (D) *A. insolitus* from Hispaniola. Both are twig-living anoles that have convergently evolved a more slender head and body, shorter legs, and long tail. (Photos by K. de Queiroz and R. Glor, courtesy of J. Losos.)

vergence shown by the anoles is unusual: ecological niches that are occupied in one region often seem unoccupied in other, climatically similar regions. Blood-feeding (vampire) bats occur in the New World tropics, but not in Africa despite abundant ungulate prey; sea snakes occur in the Indian and Pacific Oceans, but are absent from the Atlantic. The number of species of lizards is greater in Australia than in similar habitats in southern Africa, and in this respect is not convergent (**Figure 6.24A**). Differences between dry and wet habitats seem to have affected species number similarly on both continents, producing a convergent pattern. In contrast, the relationship between bird species diversity and the geometric complexity of vegetation shows no evidence of convergence between southern South America and North America or Australia (**Figure 6.24B**). The history of evolution of habitat use in the South American birds has resulted in a very different pattern.

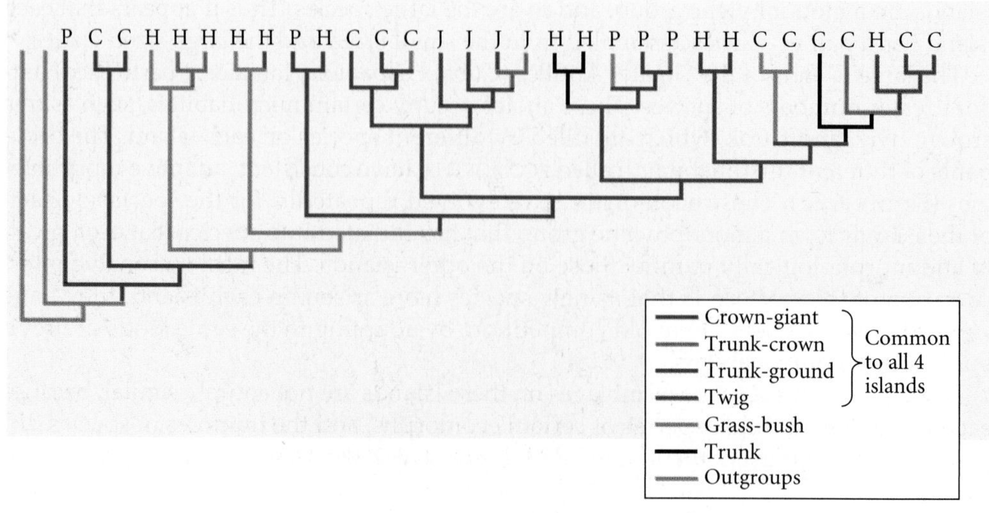

Figure 6.23 A molecular phylogeny of species of *Anolis* in the Greater Antilles. The letters at top indicate the island on which each species occurs (C, Cuba; H, Hispaniola; J, Jamaica; P, Puerto Rico). The branch colors indicate the ecomorph classes (see Figure 6.22) that occupy different microhabitats. Note that four ecomorphs have evolved independently on all four islands. The islands nevertheless differ in the number of species of anoles, and several ecomorphs are restricted to certain islands. (After Losos et al. 1998.)

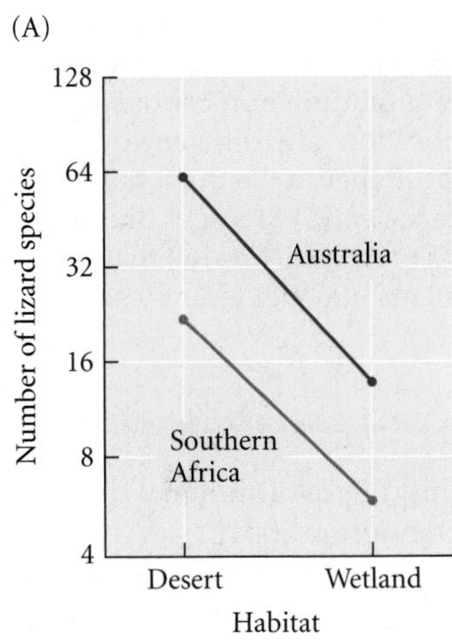

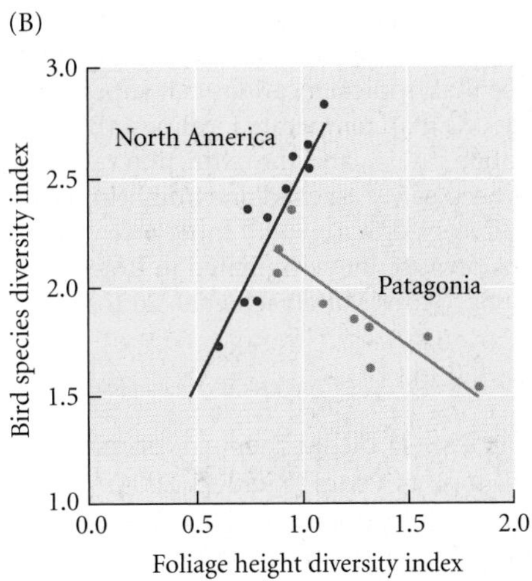

Figure 6.24 Convergence and divergence of community structure. (A) The number of species of lizards differs between Australia and southern Africa, reflecting different evolutionary histories. In both regions, however, deserts are richer in species than wetlands, and the similar slopes suggest that lizard evolution has been affected similarly by differences between these environments. (B) The number of coexisting species of birds in various locations, in relation to an index of the number of layers of foliage in the vegetation. The structure of the vegetation has a very different effect on bird diversity in southern South America (Patagonia) than in North America, evidence of divergent rather than convergent evolution of the bird communities. (After Schluter and Ricklefs 1993.)

Effects of History on Contemporary Diversity Patterns

There is enormous variation among geographic regions and among environments in the number of species of plants and animals. One of the most dramatic examples is the *latitudinal diversity gradient* of a decline in numbers of species (and of higher taxa such as genera and families) with increasing latitude, both on land and in the ocean (Figure 6.25). Most taxa of terrestrial animals and plants are far more diverse in tropical regions, especially in lowlands with abundant rainfall, than in extratropical regions. Ecological hypotheses proposed to account for this pattern focus on factors that might enable more tropical species to coexist in a stable community (Figure 6.26A), such as high productivity because of abundant solar energy, or fine partitioning of food resources among many species (Willig et al. 2003). However, recent studies increasingly suggest that the problem to be explained might be the low diversity at high latitudes, rather than why there are so

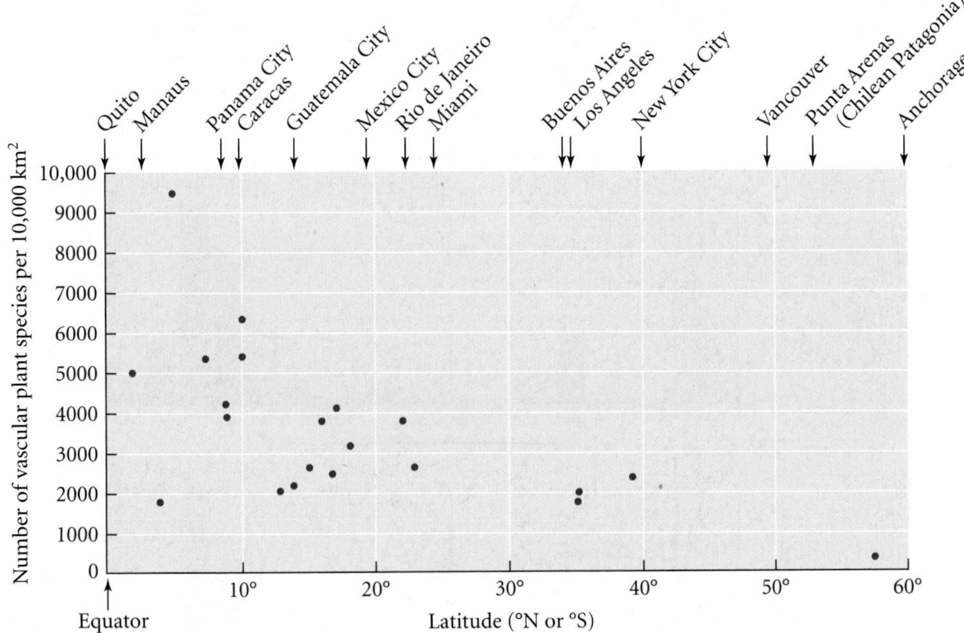

Figure 6.25 An example of the latitudinal gradient in species richness. The number of species of vascular plants in various regions of North and South America drops more than tenfold between equatorial and high-latitude regions. The latitudinal location of some cities is indicated, for easy reference. (After Huston 1994.)

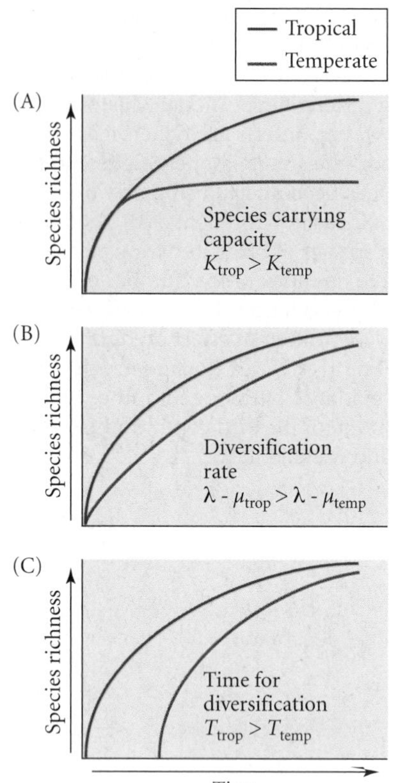

Figure 6.26 Three models of accumulation of species that have been proposed to account for the difference in species richness between tropical and temperate regions. (A) Some ecological hypotheses propose that tropical locations can support a higher equilibrium number of species ("carrying capacity," *K*) than temperate localities. (B) The diversification rate (difference between the speciation rate, λ, and the extinction rate, μ) might be higher in the tropics. Species numbers have not necessarily reached an equilibrium carrying capacity. (C) Lineages diversify at the same rate, but started to diversify more recently in the temperate zone (arrow) than in the tropics, perhaps because they originated in tropical environments but only recently adapted to the temperate zone. (After Mittelbach et al. 2007.)

many tropical species—and that the answer may best be found in evolutionary dynamics over many millions of years (Ricklefs 2004; Wiens and Donoghue 2004).

Two major historical hypotheses, which are not mutually exclusive, may account for the latitudinal diversity gradient (Mittelbach et al. 2007). The "diversification rate hypothesis" holds that the rate of increase in diversity has been greater in the tropics for a long time because of higher origination rate, lower extinction rate, or both (Figure 6.26B). There is considerable evidence for this hypothesis. For example, David Jablonski and colleagues (2006) determined that new genera of marine bivalves have arisen mostly in tropical areas throughout the last 11 Myr and have spread from there toward higher latitudes, while persisting in tropical regions as well (Figure 6.27). The "time and area hypothesis" holds that most lineages have originated in tropical environments throughout the Cenozoic era and even before, simply because for about the first 40 Myr of the Cenozoic, the Earth was

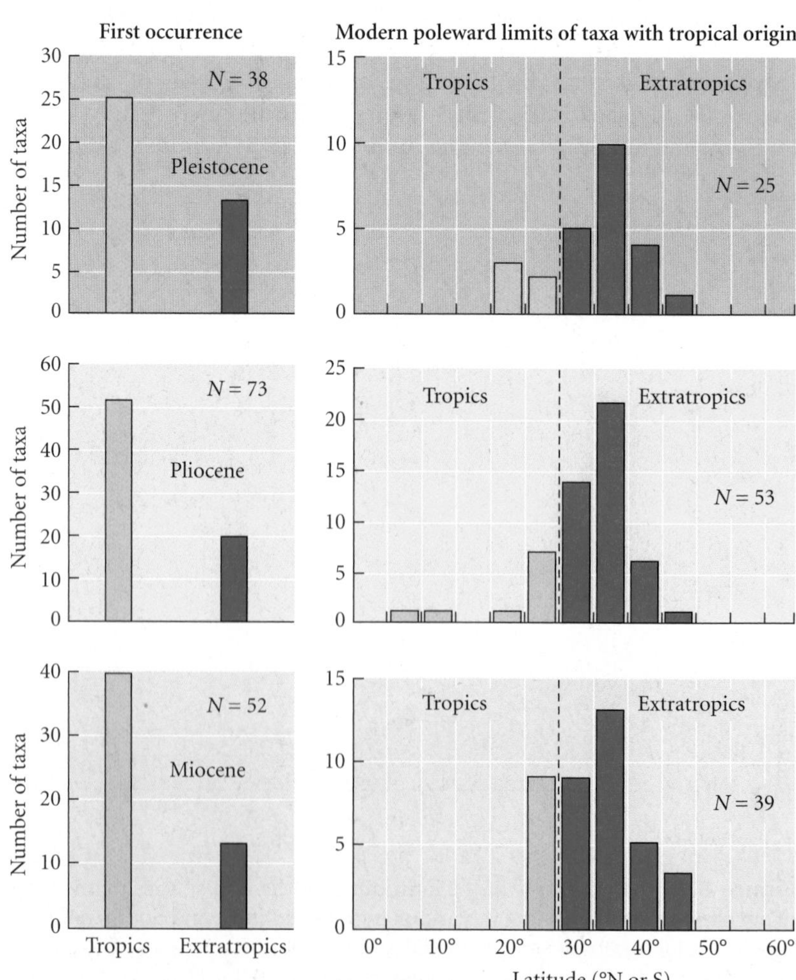

Figure 6.27 Genera of marine bivalves that first occurred in the (A) Pleistocene, (B) Pliocene, and (C) Miocene originated more often in tropical regions than outside the tropics (left panels). The present-day latitudinal range limits of species in those genera are shown in the panels at right: many have expanded beyond the tropics (right of the vertical line), while some species in those genera remain in the tropics. The tropics seem to be both a "cradle" of new lineages and a "museum" in which lineages persist. (After Jablonski et al. 2006.)

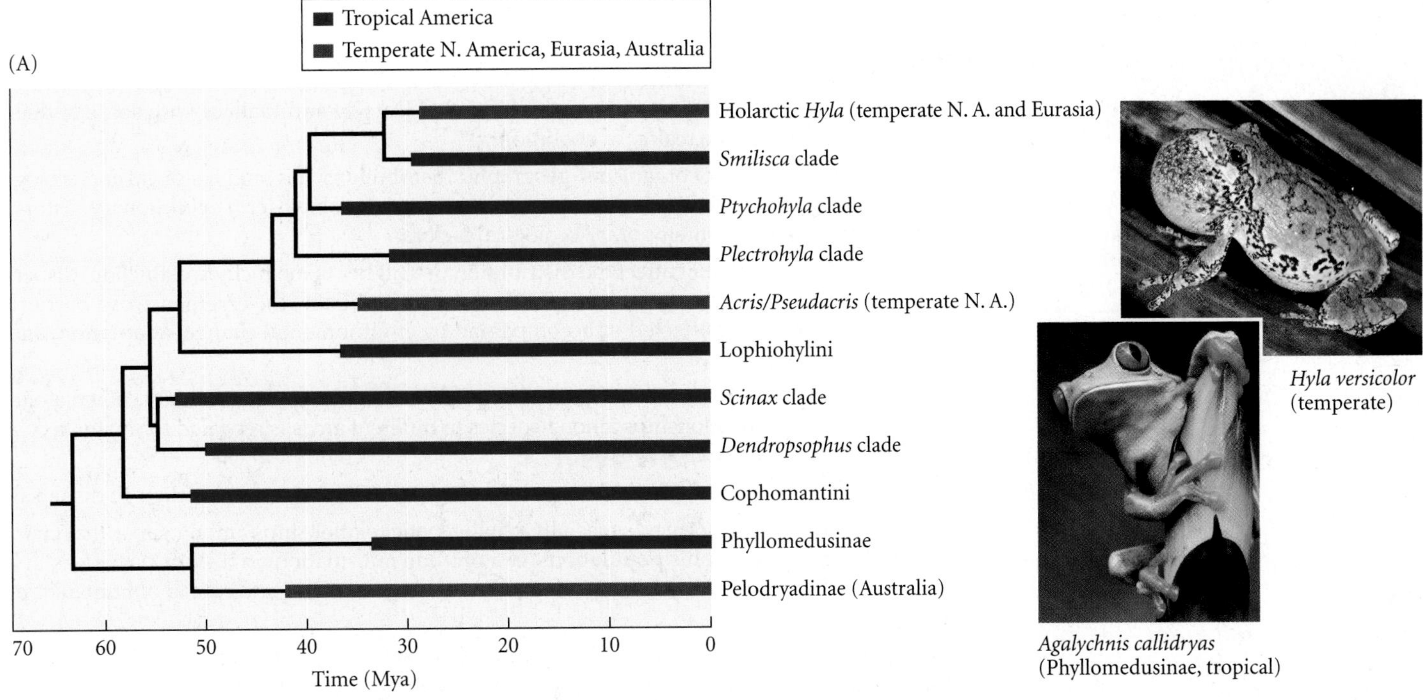

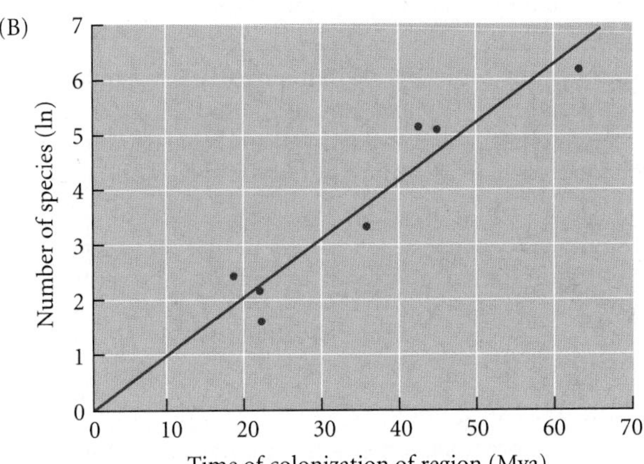

Figure 6.28 Treefrogs (Hylidae) are much less diverse in the temperate zone than in the tropics because only a few lineages have adapted to the temperate zone, and those only recently. (A) A molecular phylogeny of major lineages of Hylidae shows that this family has radiated in the American tropics, and rather recently gave rise to three lineages (in red) that invaded the temperate zone. (B) The number of treefrog species in a region (note log scale) is strongly correlated with the time since the lineage first started diversifying in that region, based on applying a molecular clock to the branch points in the phylogeny. (After Wiens et al. 2006; photos © Ryan M. Bolton and Mark Kostich/ShutterStock.)

warmer than it is today: more of the Earth had a tropical climate then than now. For that reason, most lineages are adapted to tropical climates, and the relatively few lineages that have evolved adaptations to the stressful temperatures and the seasonal fluctuations in food that are typical of the temperate zone are younger lineages that have not had time to become as diverse (Figure 6.26C). Thus this hypothesis is based on phylogenetic niche conservatism (Wiens and Donoghue 2004). In agreement with this hypothesis, Paul Fine and Richard Ree (2006) determined that the number of tree species in tropical, temperate, and boreal ecosystems on each continent is correlated with an index that integrates the area that each of these ecosystems has occupied since the Miocene, Oligocene, or even as far back as the Eocene. Thus, tropical environments and vegetation have occupied larger areas for a longer time than other environments, so this is where most genera and species arose. John Wiens and colleagues (2006) took an explicitly phylogenetic approach in a massive study of treefrogs (Hylidae). Their molecular phylogeny indicates that all the major lineages of treefrogs and their common ancestors were distributed in tropical America (Figure 6.28A). The temperate zone has been invaded by only two lineages in North America and Eurasia, and by an Australian clade that includes temperate-zone species. Moreover, the number of species in each major region is correlated with the time since treefrog clades have inhabited the region, based on the clade ages determined with a molecular clock (Figure 6.28B). It appears that the diversification rate has been much the same, and tropical regions have simply had more time to accumulate species.

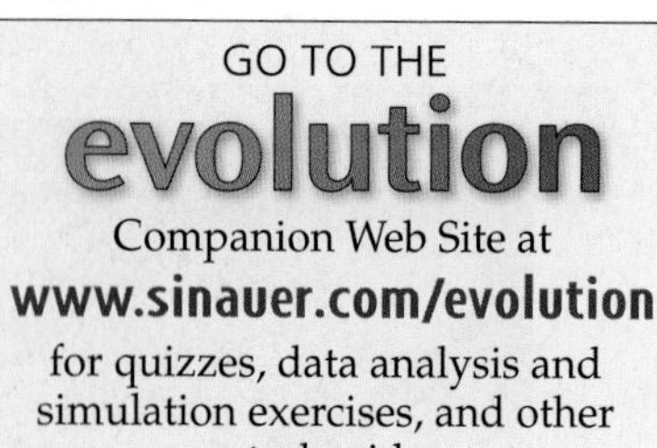

Summary

1. The geographic distributions of organisms provided Darwin and Wallace with some of their strongest evidence for the reality of evolution.
2. Biogeography, the study of organisms' geographic distributions, has both historical and ecological components. Certain distributions are the consequence of long-term evolutionary history; others are the result of contemporary ecological factors.
3. The historical processes that affect the distribution of a higher taxon include extinction, dispersal, and vicariance (fragmentation of a continuous distribution by the emergence of a barrier). These processes may be affected or accompanied by environmental change, adaptation, and speciation.
4. Histories of dispersal or vicariance can often be inferred from phylogenetic data. When a pattern of phylogenetic relationships among species in different areas is repeated for many taxa, a common history of vicariance is likely.
5. Disjunct distributions are attributable in some instances to vicariance and in others to dispersal.
6. Genetic patterns within species, especially phylogenetic relationships among genes that characterize different geographic populations, can provide information on historical changes in a species' distribution. Studies of this kind are illuminating the origin and spread of human populations.
7. The local distribution of species is affected by ecological factors, including both abiotic aspects of the environment and biotic features such as competitors and predators. Why species do not enlarge their range indefinitely, by incrementally adapting to conditions farther and farther away, is a major question in evolutionary ecology.
8. In some cases, sets of species have independently evolved to partition resources in similar ways, suggesting that competition may limit species diversity and may result in different communities with a similar structure. However, convergence of community structure is usually incomplete, suggesting that evolutionary history has had an important impact on ecological assemblages.
9. Geographic patterns in the number and diversity of species may stem partly from current ecological factors, but they probably cannot be understood without recourse to long-term evolutionary history.

Terms and Concepts

allochthonous
autochthonous
biogeographic realm
biogeography
disjunct distribution
dispersal
ecological biogeography
ecological niche (fundamental, potential, realized)
endemic
historical biogeography
phylogenetic niche conservatism
phylogeography
vicariance

Suggestions for Further Reading

M. V. Lomolino, B. R. Riddle, and J. H. Brown, *Biogeography* (second edition, Sinauer Associates, Sunderland, MA, 2006), is a comprehensive textbook of biogeography. A shorter textbook is C. B. Cox and P. D. Moore's *Biogeography: An Ecological and Evolutionary Approach* (Blackwell Scientific Publications, Oxford, 1993).

Phylogeography is treated in depth by J. C. Avise in *Phylogeography* (Harvard University Press, Cambridge, MA, 2000), and human phylogeography is included in J. Klein and N. Takahata, *Where Do We Come From? The Molecular Evidence for Human Descent* (Springer-Verlag, New York, 2002).

Problems and Discussion Topics

1. Until recently, the plant family Dipterocarpaceae was thought to be restricted to tropical Asia, where many species are ecologically dominant trees. Recently, a new species of tree in this family was discovered in the rain forest of Colombia, in northern South America. What hypotheses can account for its presence in South America, and how could you test those hypotheses?
2. The ratites are a very old clade of flightless birds that include the ostrich in Africa, rheas in South American, emu and cassowaries in Australia, and kiwis and recently extinguished moas in New Zealand. South American tinamous, which are capable of flight, are closely related. The "Gondwana distribution" of these birds has often been attributed to vicariance, but there is considerable uncertainty about their phylogeny and distributional history. Read several phylogenetic studies of the ratites and discuss how best to explain their distribution: A. Cooper et al., *Nature* 409:704–707, 2001; O. Haddrath and A. J. Baker, *Proc. Royal Soc. Lond. B* 268:939–945, 2001; and S. J. Hackett et al., *Science* 320:1763–1768, 2008.
3. Genetic variation among individuals of a species is the "raw material" of evolution (see Chapter 1). In Chapter 8, we will see that many or most characteristics of most species are genetically variable. Discuss whether or not this observation is inconsistent with the claim that many organisms display phylogenetic niche conservatism.
4. Some biogeographers, subscribing to the "cladistic vicariance" school of thought (Humphries and Parenti 1986), hold that vicariance should always be the preferred hypothesis, and dispersal should be invoked only when necessary, because the vicariance hypothesis can be falsified (if it is false), whereas dispersal can account for any pattern and therefore is not falsifiable. What are the pros and cons of this position? (See Endler 1983.)
5. In some cases, it can be shown that species are physiologically incapable of surviving temperatures that prevail beyond the borders of their range. Do such observations prove that cold regions have low species diversity because of their harsh physical conditions?
6. By far the most effective way of saving endangered species is to preserve large areas that include their habitat. For social, political, and economic reasons, the number and distribution of areas that can be allocated as preserves are highly limited. It might be easier to save more species if areas of endemism were correlated among different taxa, such as plants, birds, and mammals. Are they correlated? (See, for example, N. Myers et al., *Nature* 403:853–858, 2000; J. R. Prendergast et al., *Nature* 365:335–337, 1993; A. P. Dobson et al., *Science* 275:550–553, 1993.)
7. Would you expect large numbers of species in a region to have had similar histories of geographic distribution? Why or why not? How could you use phylogeographic analyses, such as illustrated in Figure 6.15, to address this question?
8. How might you test the hypothesis that adaptation to conditions beyond the edge of a species' range is prevented by migration and gene flow from populations well within the range?
9. Global climate change, owing largely to human production of greenhouse gases, is occurring faster than at almost any previous time in Earth's history. How do you think the geographic and altitudinal ranges of species will be affected? Will this differ among taxa or among geographic regions? Is it likely that many species will adapt to this climate change and maintain their current distribution? What will determine whether or not they will adapt?

CHAPTER 7

The Evolution of Biodiversity

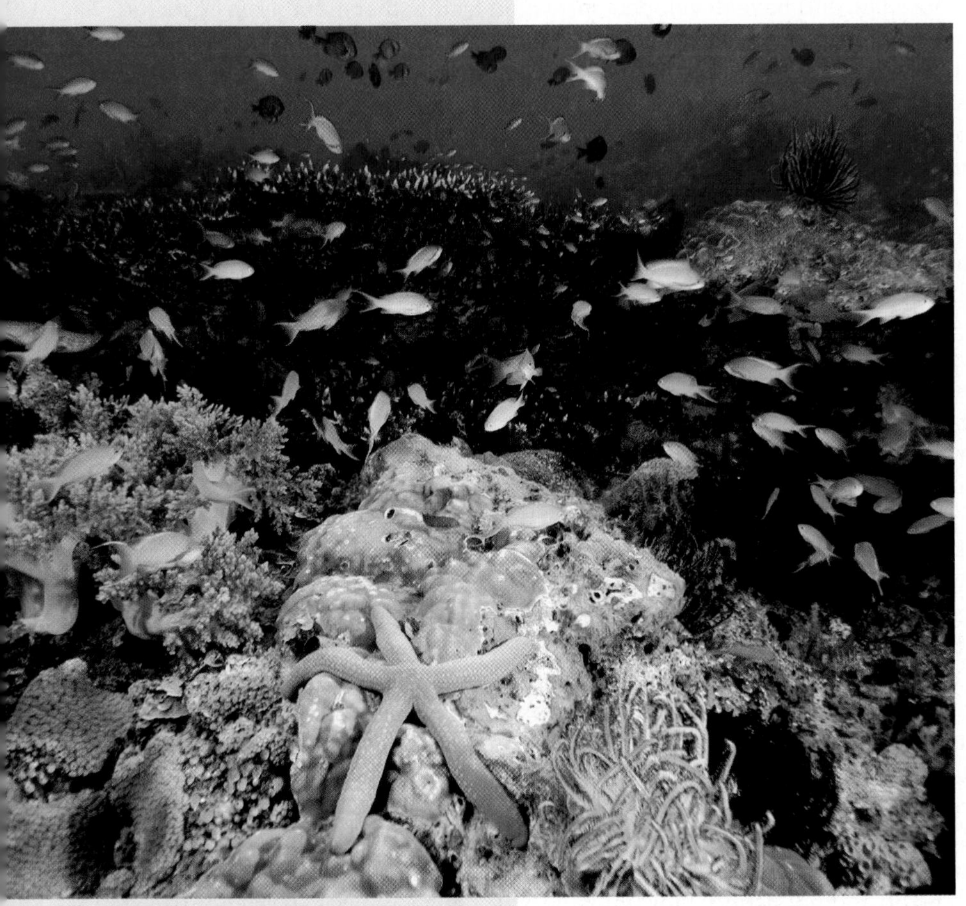

Modern biodiversity. Tropical coral reefs are the most species-rich marine environments, harboring hundreds of species. At Apo Island Marine Reserve in the Philippines, the Cogon reefs of soft and hard corals harbor a multitude of species including the sponges, tunicates, feather stars, sea stars, and fish seen here. (Photo © Andre Seale/Photolibrary.com.)

Biological diversity, or **biodiversity**, poses some of the most interesting questions in biology. Why are there more species of rodents than of primates, of flowering plants than of ferns? Why do some regions, such as the tropics, have more species than others, such as temperate areas? Why has the diversity of species changed over evolutionary time? Does diversity increase steadily, or has it reached a limit? Because so many factors can influence the diversity of species, these questions are both interesting and challenging.

Biodiversity can be studied from the complementary perspectives of ecology and evolutionary history. Ecologists focus primarily on factors that operate over short time scales to influence diversity within local habitats or regions. But factors that operate on longer time scales, such as climate change and evolution, also affect diversity. On a scale of millions of years, extinction, adaptation, speciation, climate change, and geological change create the potential for entirely different assemblages of species. In this chapter, we will examine long-term patterns of change in diversity, caused by originations and extinctions of taxa on a scale of millions of years. Ecological and evolutionary studies of contemporary species help to interpret these paleontological patterns. Conversely, understanding factors that have altered biodiversity in the past may help us predict how diversity will be affected by current and future environmental changes, such as the global climate change that is now underway as a result of human use of fossil fuels.

Estimating and Modeling Biological Diversity

Estimates of diversity

The simplest expression of taxonomic diversity is a simple count of species (SPECIES RICHNESS). Over long time scales or large areas, species diversity is often estimated by compiling records, such as the publications or museum specimens that have been accumulated by many investigators, into faunal or floral lists of species. Most paleontological studies of diversity employ counts of higher taxa, such as families and genera, because they generally provide a more complete fossil record than individual species do.

Diversity, like almost everything studied by scientists, is estimated from samples. Thus our picture is incomplete, and can often be systematically biased (misleading). Paleobiologists continue to study the effects of sampling on their estimates of diversity (e.g., Alroy et al. 2008), and have developed correction factors that often must be included in data analysis (Raup 1972; Signor 1985; Foote 2000a). For example, rare species are more likely to be included in large than small samples, so the estimated number of species increases with sample size. One way of correcting for this, called *rarefaction*, is to pick the same number of specimens at random from all the samples (perhaps the same number as are in the smallest sample). An important problem in paleontology is that records of fossil taxa are often described at the level of the stages (mostly about 5 to 6 Myr long) into which each geological period is divided. Thus the first and last occurrences of a taxon in the record are accurate only to about 5 Myr, so the estimated duration of the taxon is imprecise. Moreover, geological stages vary in duration, and more recent geological times are represented by greater volumes and areas of fossil-bearing rock. Therefore, it may be necessary to adjust the count of taxa by the amount of time and rock volume represented.

Because fossils are a small sample of the organisms that actually lived, a taxon is often recorded from several separated time horizons, but not from those in between. We therefore can deduce that in the same way, the actual time of origination of a taxon may have occurred before its earliest fossil record, and its extinction after its latest record. It follows that if many taxa actually became extinct in the same time interval, the last recorded occurrences of some are likely to be earlier, so their *apparent* times of extinction will be spread out over time. Conversely, if many taxa actually originated at the same time, some of them may appear to have originated at later times.

Since our count of living (Recent) species is much more complete than our count of past species, taxa that are still alive today have apparently longer durations and lower extinction rates than they would if they had been recorded only as fossils. That is, we can list a living taxon as present throughout the last 10 Myr, let us say, even if its only fossil occurrence was 10 Myr ago. Because the more recently a taxon arose, the more likely it is to still be extant, diversity will seem to increase as we approach the present, a bias called the **pull of the Recent**. This bias can be reduced by counting only fossil occurrences of each taxon and not listing it for time intervals between its last fossil occurrence and the Recent.

Because of unusually favorable preservation conditions or other chance events, a taxon may be recorded from only a single geological stage, even though it lived longer than that. Such "singletons" make up a higher proportion of taxa as the completeness of sampling decreases and therefore bias the sample; moreover, they can create a spurious correlation between rates of origination and extinction because they appear to originate and become extinct in the same time interval. Diversity may be more accurately estimated by ignoring such singletons and counting only those taxa that cross the border from one stage to another.

The rate of change in diversity depends on the rates at which taxa originate and become extinct. Several expressions for such rates have been used in the paleobiological literature. One is simply the number of taxa originating (or becoming extinct) per Myr (or other time unit). The most useful measure is the number of originations (or extinctions) per taxon ("per capita") per unit time (Foote 2000a).

Taxonomic Diversity through the Phanerozoic

The most complete fossil record has been left by marine animals with hard parts (shells or skeletons). Jack Sepkoski (1984, 1993) accomplished the heroic task of compiling data in the paleontological literature on the stratigraphic ranges of more than 4000 marine skeletonized families and 30,000 genera throughout the 542 Myr of the Phanerozoic. Using this database, he plotted the diversity of families throughout the Phanerozoic, in one of the most famous graphs in the literature of paleobiology (Figure 7.1A). The graph shows a rapid increase in the Cambrian and Ordovician, a plateau throughout the rest of the Paleozoic, and a steady, almost fourfold increase throughout the Mesozoic and Cenozoic. This pattern is interrupted by decreases in diversity caused by mass extinction events. The number of genera of skeletonized marine animals shows a similar pattern of change (Figure 7.1B), and diversity seems to have increased on land as well. The number of families of insects shows a steady increase since the Permian (Figure 7.2A), whereas the diversification of flowering plants and of birds and mammals accounts for dramatic increases in the diversity of vascular plants (Figure 7.2B) and of terrestrial vertebrates (Figure 7.2C) after the mid-Cretaceous.

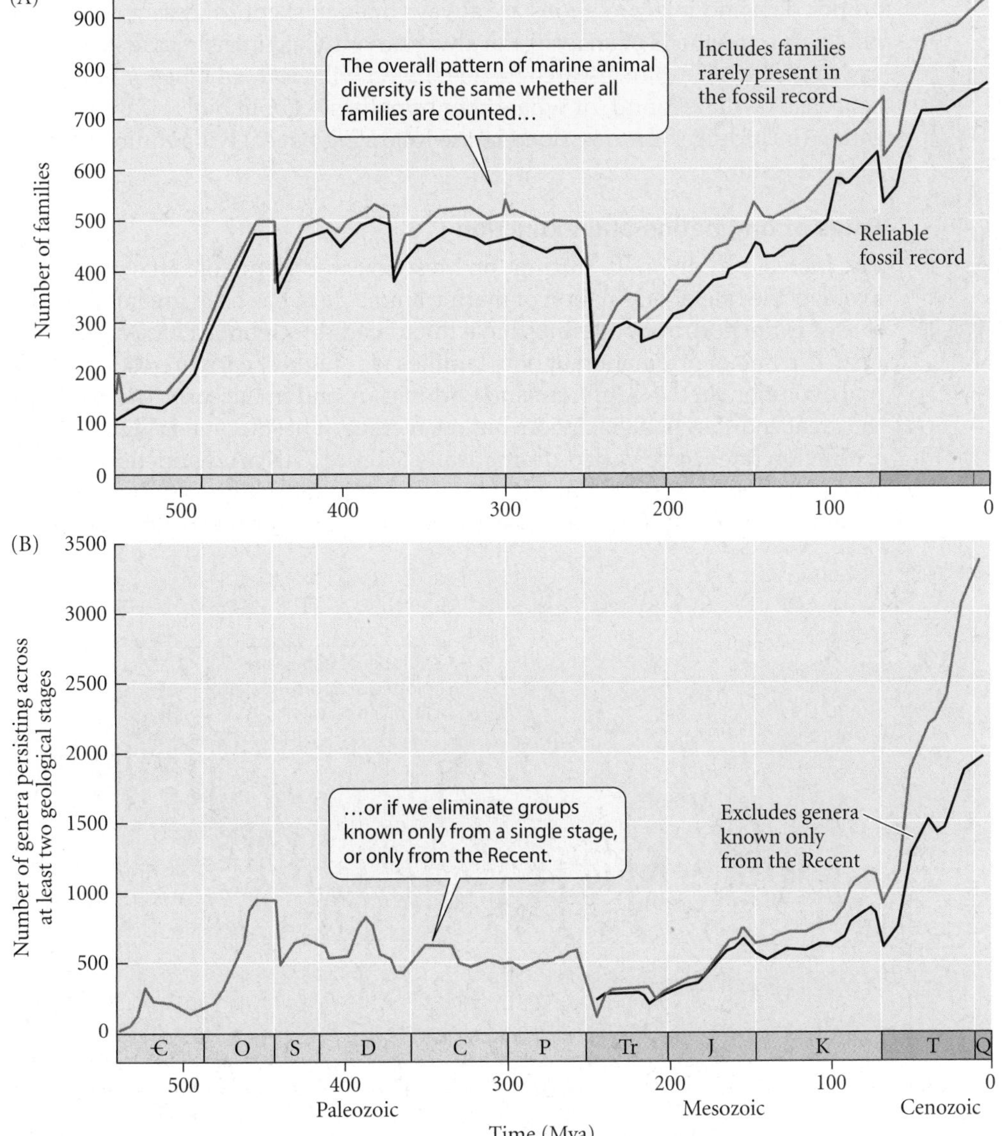

Figure 7.1 (A) Taxonomic diversity of skeletonized marine animal families during the Phanerozoic. The number of taxa entered for each geological stage (subdivisions of the geological periods; most stages represent 5–6 Myr) includes all those whose known temporal extent includes that stage. The blue curve includes families that are rarely preserved; the black curve represents only families that have a more reliable fossil record. There are approximately 1900 marine animal families alive today, including those rarely preserved as fossils. (B) Taxonomic diversity of 25,049 genera of skeletonized marine animals, counting only those that cross boundaries between two or more stages. The lower curve shows genera only as they are represented in the fossil record, excluding their known occurrences in the Recent. (A after Sepkoski 1984; B after Foote 2000a.)

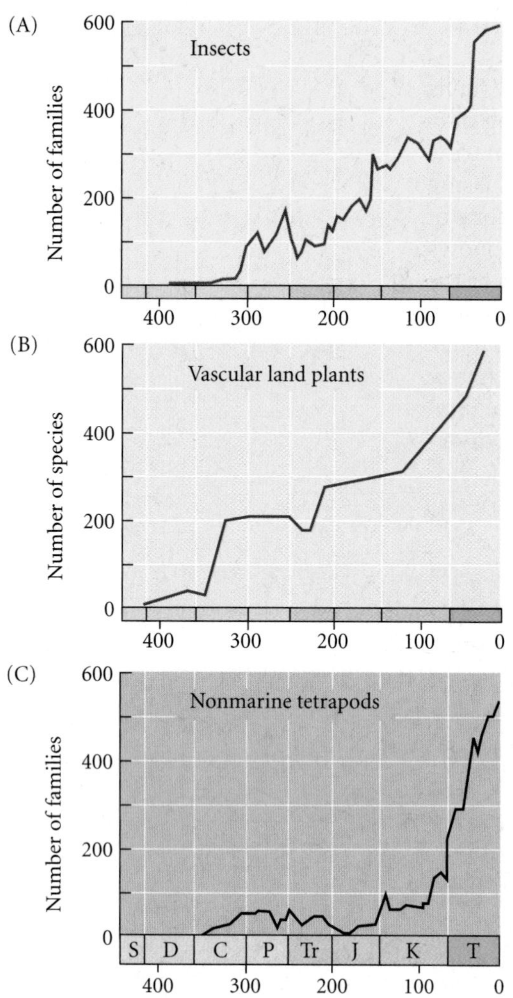

Figure 7.2 Changes in the number of known (A) families of insects, (B) species of vascular land plants, and (C) families of nonmarine tetrapod vertebrates throughout time. (A after Labandeira and Sepkoski 1993; B, C after Benton 1990.)

These plots do not include any of the corrections for sampling errors and biases that we have noted. A large team of researchers has reanalyzed the numbers of marine genera, using a rarefaction procedure on more than 44,000 samples from throughout the Phanerozoic record, as well as several other procedures to minimize systematic errors (Alroy et al. 2008). Their results differ from the Sepkoski plot in several ways (Figure 7.3): there is a decline in diversity in the Devonian instead of a Paleozoic plateau, a sharp increase in the Permian before the end-Permian extinction, and a much less steep post-Paleozoic increase, which seems to have leveled off after the mid-Cretaceous. This analysis has reopened an earlier debate about whether or not to take the pattern in the raw data, especially the dramatic post-Paleozoic increase, at face value. (See Miller 2000 for an intimate history of the debate.) The ongoing struggle to understand the history of biodiversity is a fine example of the way science works: data and interpretations are always open to skeptical questioning, and new data or methods of analysis can always modify old interpretations. It is too early to tell what future research on this subject will yield, whether or not a new consensus will be found, or what other hypotheses about biological diversification (including some described in the following pages) will require rethinking as a result.

Rates of origination and extinction

The increase in diversity during the Mesozoic and Cenozoic tells us that on average, the rate of origination of marine animal taxa has been greater than the rate of extinction, but both rates have fluctuated throughout Phanerozoic history. The rate of origination of new families was highest early in recorded animal evolution, in the Cambrian and Ordovician, and in the early Triassic, after the great end-Permian extinction, but on average, it has declined (Figure 7.4A). Extinction rates have varied dramatically (Figure 7.4B). A distinction is often made between episodes during which exceptionally high numbers of taxa became extinct, the so-called **mass extinctions**, and periods of so-called nor-

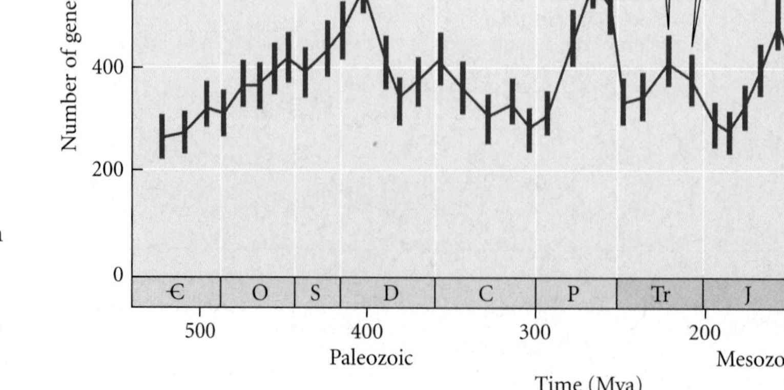

Figure 7.3 The number of marine genera through time, corrected for biases such as temporal differences in rock volume and the pull of the Recent. The vertical lines represent the range of values, within which the real number lies. Compare with Figure 7.1B. (After Alroy et al. 2008.)

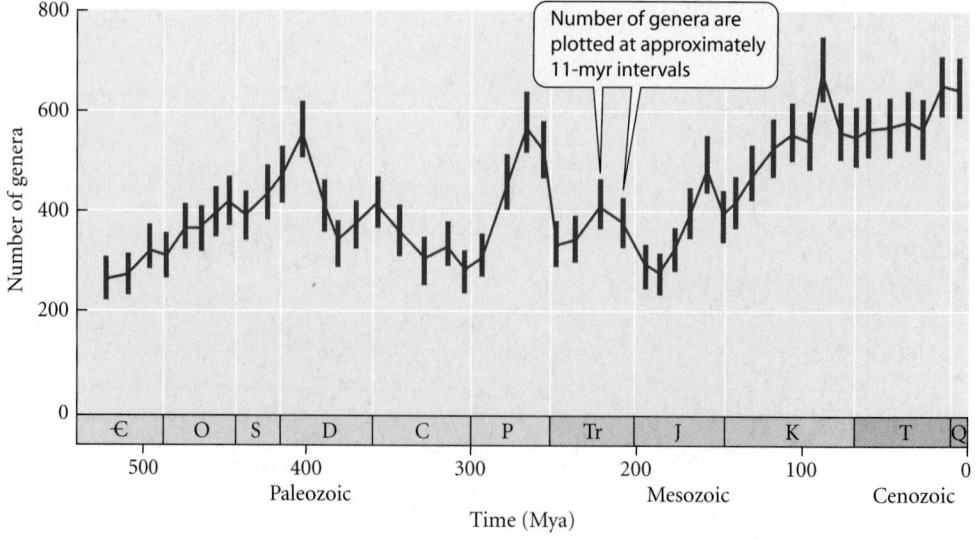

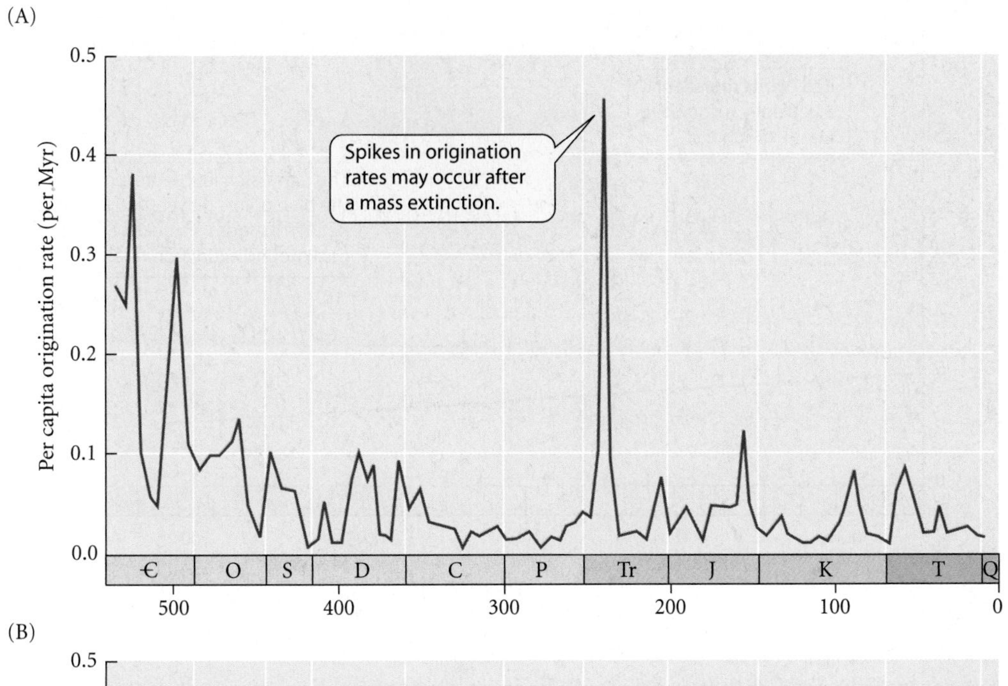

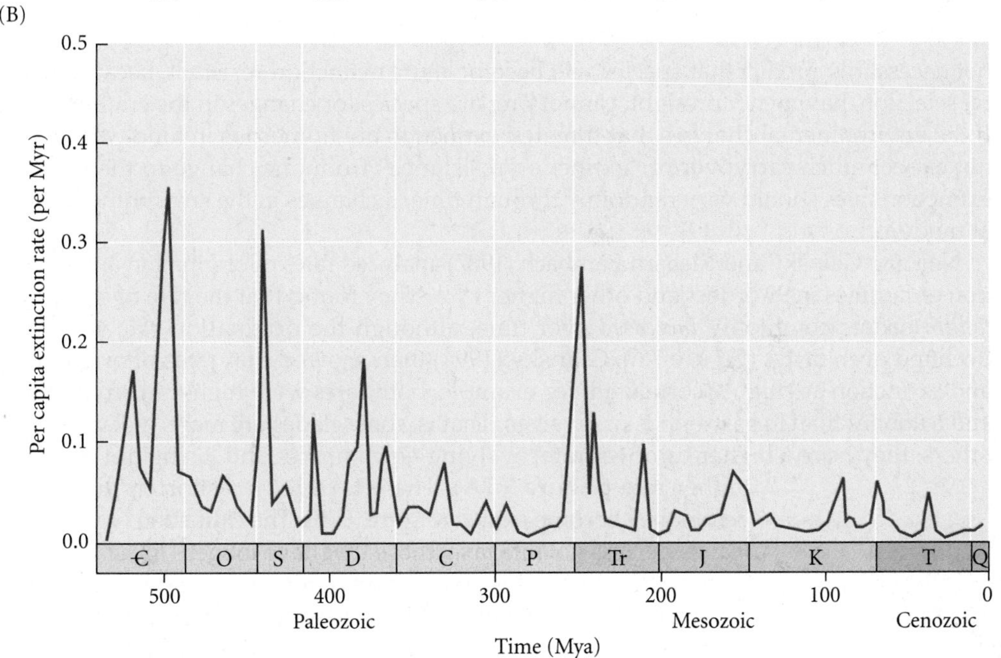

Figure 7.4 Rates of (A) origination and (B) extinction of marine animal genera in 107 stages of the Phanerozoic, expressed as the number of genera per capita per Myr. Only taxa that cross boundaries between stages are counted. Compare the extinction rates with those seen in Figure 7.5, which shows absolute numbers of extinct families, uncorrected for the number that were at risk of extinction at each time. (After Foote 2000a.)

mal or **background extinction** (Figure 7.5). Five mass extinctions are generally recognized: at the end of the Ordovician, in the late Devonian, at the Permian/Triassic (P/Tr) boundary (the end-Permian extinction), at the end of the Triassic, and at the Cretaceous/Tertiary (K/T) boundary (the K/T extinction). Several other episodes of heightened extinction occurred as well, however. We first discuss background extinction.

Extinction rates have declined over time

David Raup and Jack Sepkoski (1982) discovered that the background extinction rate, scored as absolute numbers of families becoming extinct per Myr, declined during the Phanerozoic (see Figure 7.5). This decline is also seen if extinction rate is measured per capita (see Figure 7.4B). What might account for the decline?

It would seem reasonable to expect lineages of organisms to become more resistant to extinction over the course of time as they become better adapted. However, extinction of species is caused by failure to adapt to environmental changes. Evolutionary theory does

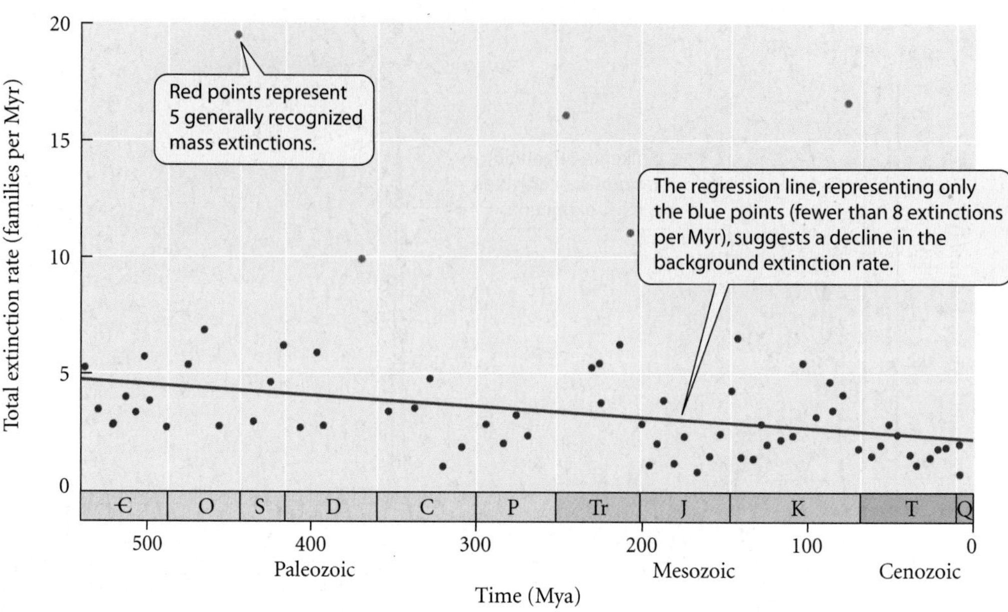

Figure 7.5 Extinction rates of marine animal families during the Phanerozoic, expressed as the number of families per Myr. The solid regression line fits the blue points, which represent fewer than 8 extinctions per Myr. The red points, which deviate significantly from the background cluster, mark the five major mass extinction events at the ends of the (1) Ordovician, (2) Devonian, (3) Permian (the end-Permian extinction), (4) Triassic, and (5) Cretaceous (the K/T extinction). The points in this graph represent absolute numbers of extinctions, not per capita rates, and thus do not control for differences in diversity at different times. (After Raup and Sepkoski 1982.)

not necessarily predict that species will become more extinction-resistant, because natural selection, having no foresight, cannot prepare species for changes in the environment. If the environmental changes that threaten extinction are numerous in kind, we should not expect much carryover of "extinction resistance" from one change to the next. So extinction rates should vary randomly through time, if changes in the environment occur at random.

Norman Gilinsky and Richard Bambach (1987) analyzed rates of origination and extinction of families in 99 orders and other higher taxa. They found that the rate of extinction *within* orders commonly *increased* over time, although the origination rate generally declined even faster (**Figure 7.6**). Gilinsky (1994) then showed that rates of origination and extinction are highly correlated; for example, both rates were higher in ammonoids and trilobites than in gastropods or bivalves. That is, some clades are more "volatile" than others: they have a higher turnover rate, evolving new families and losing old ones at a higher rate (**Figure 7.7A**). These taxa have a shorter "life span" before they become extinct (**Figure 7.7B**). The extinction of such taxa leaves the less volatile taxa, those that have longer "life spans" and lower extinction rates. This process, a form of natural selection among clades (see Chapter 11), results in a decline in the average extinction rate of clades over a long time, as long as highly volatile clades do not evolve anew—which has happened only rarely.

Why are extinction and origination rates correlated? Steven Stanley (1990) suggested that speciation rate may be correlated with extinction rate because both are influenced by certain features of organisms (see also Chapter 18). For these characteristics to deter-

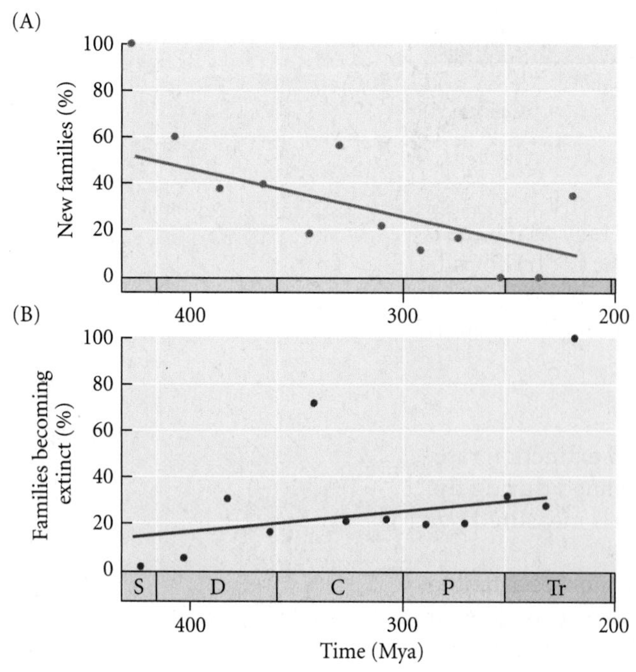

Figure 7.6 Changes in rates of (A) origination and (B) extinction in rugose corals since first appearance of the group. The vertical axis represents the number of families originating or becoming extinct as a percentage of the number of families present at the beginning of the interval. The origination rate declined over time and the extinction rate rose, although at a lower rate. Both effects suggest that as diversity increases, the rate of further increase in diversity (origination–extinction) decreases, so that diversity would tend toward an equilibrium. (After Gilinsky and Bambach 1987.)

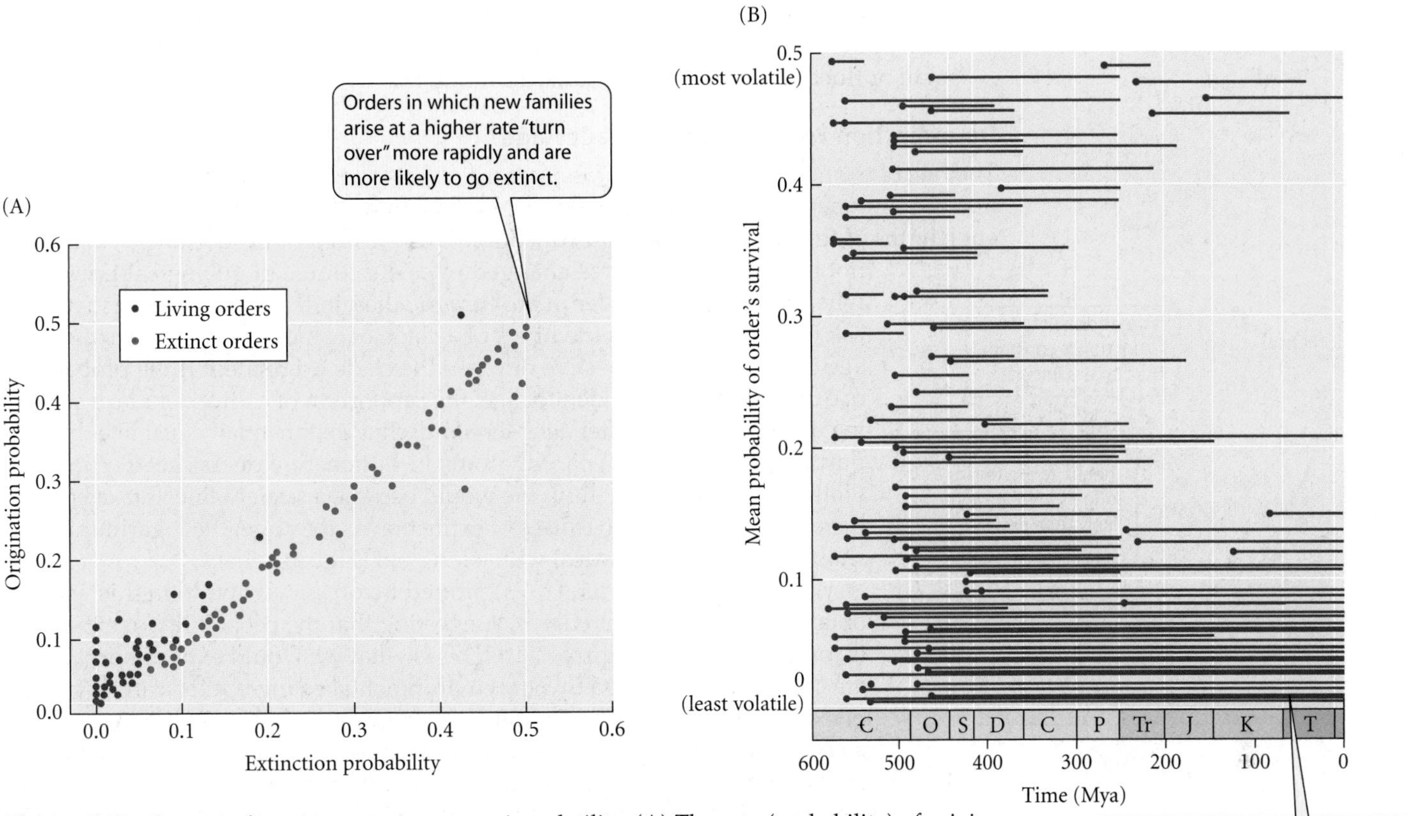

Figure 7.7 Groups of marine organisms vary in volatility. (A) The rate (probability) of origination of new families within an order, per unit time, is correlated with the rate (probability) of extinction. Orders with higher rates have higher turnover, and so are more likely to decrease greatly or become extinct. (B) The temporal duration of each order is plotted as a line extending through geological time (x-axis), in relation to the average of the origination and extinction rates of families within each order (y-axis). Note that the "high" (more volatile) lines are predominantly early in the Phanerozoic (i.e., at left of the diagram); such volatile taxa originated later in geological time. (After Gilinsky 1994.)

mine extinction rates of families, they must be fairly phylogenetically conservative—they should vary consistently among families. Some possible characteristics are:

1. *Degree of ecological specialization.* Ecologically specialized species are likely to be more vulnerable than generalized species to changes in their environment (Jackson 1974). They may also be more likely to speciate because of their patchier distribution, and newly formed species may be more likely to persist by specializing on different resources and thus avoiding competition with other species. Certain aspects of specialization are phylogenetically conservative in at least some taxa.
2. *Geographic range.* Species with broad geographic ranges tend to have a lower risk of extinction because they are not extinguished by local environmental changes (Gaston and Blackburn 2000). They also have lower rates of speciation (Figure 7.8; Jablonski and Roy 2003), probably because they have a high capacity for dispersal and perhaps broader environmental tolerances. Geographic range size is phylogenetically conserved in molluscs, birds, and mammals; that is, there are consistent differences among clades in average range size of their component species (Jablonski 1987; Brown 1995).

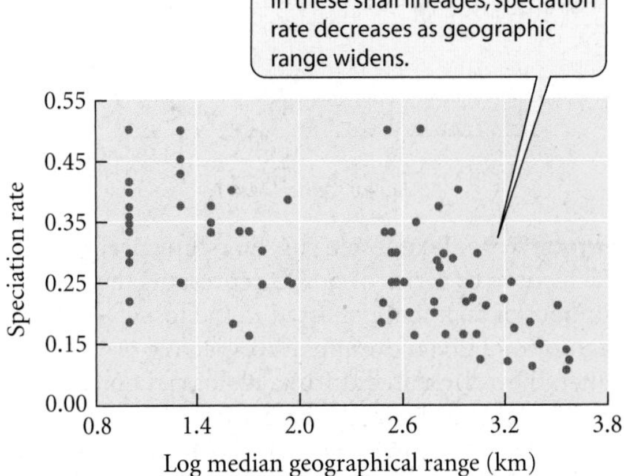

Figure 7.8 Speciation rate (expressed as the number of new species per lineage per Myr) is lower in lineages of gastropods (snails) in which the median geographical range is larger. Perhaps the average dispersal capability is greater in species of some genera than others, and this increases geographic range while lowering the rate at which geographically distant populations become different species (see Chapter 18). (After Jablonski and Roy 2003.)

3. *Population dynamics.* Species with low or fluctuating population sizes are especially susceptible to extinction. Some authors believe that speciation is also enhanced by small or fluctuating population sizes, although this hypothesis is controversial.

Do extinction rates change as clades age?

The rate of extinction of taxa in the fossil record can also be analyzed by plotting the fraction of component taxa (e.g. the fraction of genera within a family) that survive for different lengths of time (i.e., their age at extinction). This is different from asking whether or not extinction rates have changed over the course of geological time (e.g., whether they were lower in the Jurassic than in the Devonian); instead, we ask if the rate at which members of a clade become extinct changes over the age of the clade, irrespective of when the clade originated. If the probability of extinction is independent of age, the proportion of component taxa surviving to increasingly greater ages should decline exponentially (just like the proportion of "surviving" parent atoms in radioactive decay; see Figure 4.2). Plotted logarithmically, the curve would become a straight line. If taxa become increasingly resistant to causes of extinction as they age, the logarithmic plot should be concave upward, with a long tail (Figure 7.9A).

When Leigh Van Valen (1973) plotted taxon survivorship in this way, he obtained rather straight curves, suggesting that the probability of extinction is roughly constant (Figure 7.9B). This is what we would expect if organisms are continually assaulted by new environmental changes, each carrying a risk of extinction. One possibility, Van Valen suggested, is that the environment of a taxon is continually deteriorating because of the evolution of other taxa. He proposed the **Red Queen hypothesis**, which states that, like the Red Queen in Lewis Carroll's *Through the Looking-Glass*, each species has to run (i.e., evolve) as fast as possible just to stay in the same place (i.e., survive), because its competitors, predators, and parasites also continue to evolve. There is always a roughly constant chance that it will fail to do so.

The risks of extinction do not always remain constant with age, and of course they have changed throughout the Phanerozoic. In Figure 7.10, for example, we see survivorship curves of species in various genera of Cambrian and Ordovician trilobites (Foote 1988). The Cambrian curves (Figure 7.10A) are much steeper, showing that extinction rates were higher than in the Ordovician (Figure 7.10B). Especially in the Ordovician, the curves are concave rather than straight, showing that species within a genus became extinct at a lower rate as the family aged. Perhaps species evolved greater buffering against agents of extinction. However, another possibility is that if the number of species per genus increases over time, genera will have lower extinction rates because a genus persists until all its component species have become extinct.

(A) Hypothetical curves

If the probability of extinction is constant, the curve is linear.

If older taxa (i.e., those that have existed for a longer time) are less likely than newer ones to become extinct, the curve is concave.

Number of surviving taxa

1000

100

10

1

Age of taxon

(B) Data for Ammonoidea

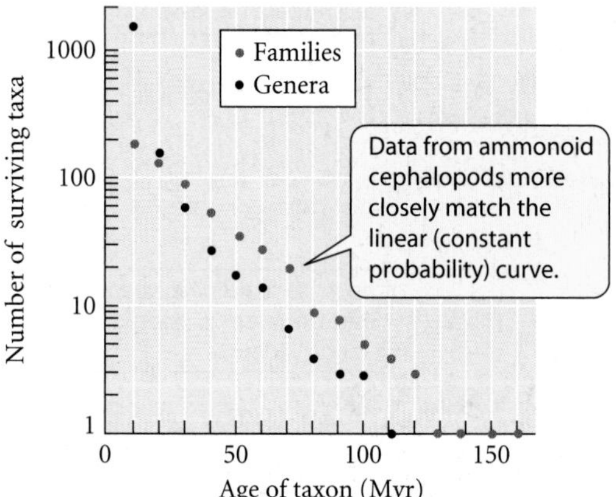

Figure 7.9 Taxonomic survivorship curves. Each curve or series of points represents the number of taxa that persisted in the fossil record for a given duration, irrespective of when they originated during geological time. (A) Hypothetical survivorship curves. In a semilogarithmic plot, the curve is linear if the probability of extinction is constant. It is concave if the probability of extinction declines as a taxon ages, as it might if adaptation lowered the long-term probability of extinction. (B) Taxonomic survivorship curves for families and genera of ammonoids. The plot for families suggests an extinction rate that is constant with age, whereas the plot for genera suggests that older survivors have a lower rate of extinction. (B after Van Valen 1973.)

Causes of extinction

Extinction has been the fate of almost all the species that have ever lived, but little is known of its specific causes. Biologists agree that extinction is caused by failure to adapt to changes in the environment. Ecological studies of contemporary populations and species point to habitat destruction as the most frequent cause of extinction by far, and some cases of extinction that are due to introduced predators, diseases, and competitors have been documented (Lawton and May 1995).

When a species' environment deteriorates, some populations may become extinct, and the geographic range of the species contracts, unless formerly unsuitable sites become suitable for colonists to establish new populations. Not all environmental changes cause populations to decline, but if they do, the survival of those populations—and perhaps of the entire species—depends on adaptive genetic change. Whether or not this suffices to prevent extinction depends on how rapidly the environment (and hence the optimal phenotype)

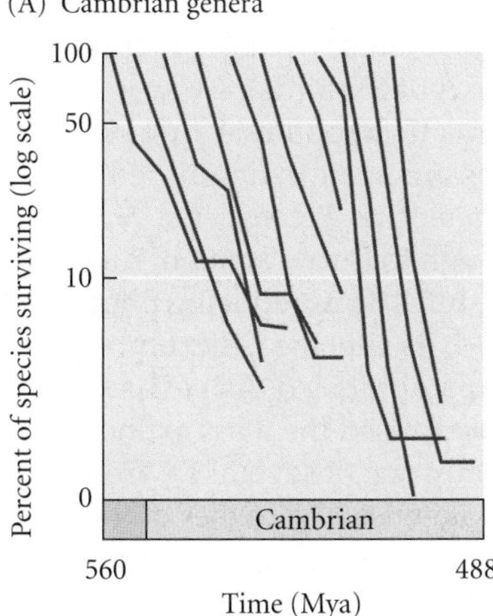

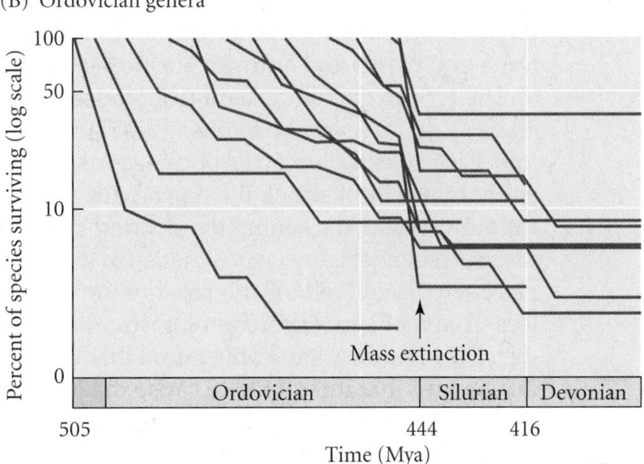

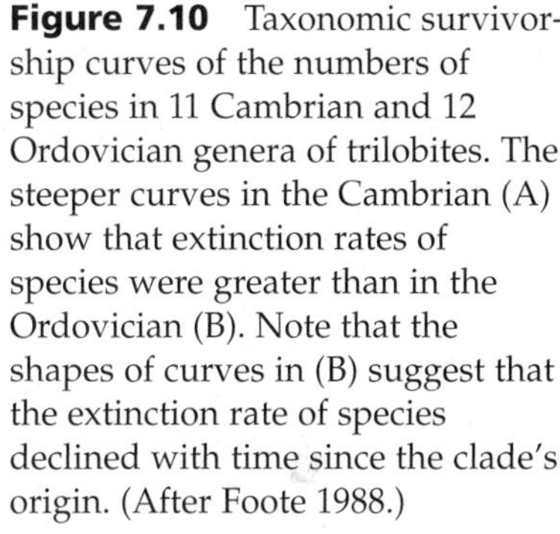

Figure 7.10 Taxonomic survivorship curves of the numbers of species in 11 Cambrian and 12 Ordovician genera of trilobites. The steeper curves in the Cambrian (A) show that extinction rates of species were greater than in the Ordovician (B). Note that the shapes of curves in (B) suggest that the extinction rate of species declined with time since the clade's origin. (After Foote 1988.)

changes relative to the rate at which a character evolves. The rate of evolution may depend on the rate at which mutation supplies genetic variation and on population size, because smaller populations experience fewer mutations. Thus an environmental change that reduces population size also reduces the chance of adapting to it (Lynch and Lande 1993). Because a change in one environmental factor, such as temperature, may bring about changes in other factors, such as the species composition of a community, the survival of a species may require evolutionary change in several or many features.

Both abiotic and biotic changes have doubtless caused extinction. For example, during the Pliocene, the rate of extinction of bivalves and many coral reef inhabitants increased, perhaps due to a decrease in temperature (Stanley 1986; Jackson and Johnson 2000). The role of competition in extinction is controversial, as we will see later in this chapter.

Mass extinctions

The history of extinction is dominated by mass extinctions at the end of the Ordovician, Devonian, Permian, Triassic, and Cretaceous periods (see Figure 7.5; Bambach 2006). The end-Permian extinction was the most drastic (Figure 7.11), eliminating about 54 percent of marine families, 84 percent of genera, and 80 to 90 percent of species (Erwin 2006). On land, major changes in plant assemblages occurred, several orders of insects became extinct, and the dominant amphibians and therapsids were replaced by new groups of therapsids (including the ancestors of mammals) and diapsids (including the ancestors of dinosaurs). The second most severe mass extinction, in terms of the proportion of taxa affected, occurred at the end of the Ordovician. Less severe, but much more famous, was

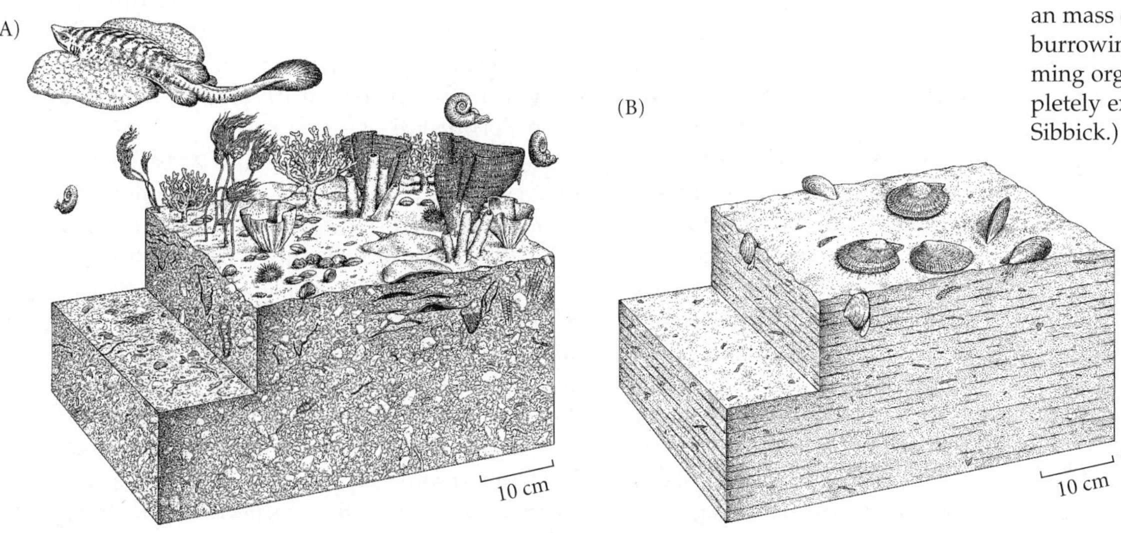

Figure 7.11 Reconstructions of an ancient seabed (A) immediately before and (B) after the end-Permian mass extinction. A rich fauna of burrowing, epifaunal, and swimming organisms was almost completely extinguished. (Artwork © J. Sibbick.)

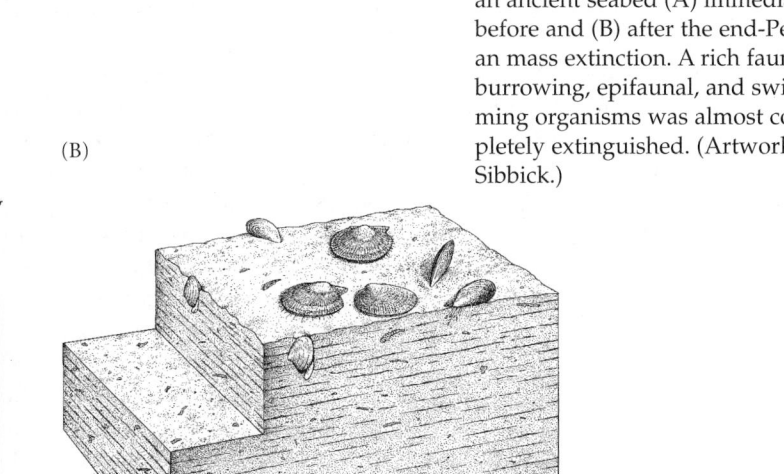

the K/T, or end-Cretaceous, extinction, which marked the demise of many marine and terrestrial plants and animals, including the dinosaurs (except for birds).

The K/T extinction is famous because of the truly dramatic hypothesis, first suggested by Walter Alvarez and colleagues (1980), that the dinosaurs were extinguished by the impact of an extraterrestrial body—an asteroid or large meteorite. Alvarez et al. postulated that this object struck the Earth with a force great enough to throw a pall of dust into the atmosphere, darkening the sky and lowering temperatures, thus reducing photosynthesis. Geologists now agree that such an impact occurred; its site, the Chicxulub crater, has been discovered off the coast of the Yucatán Peninsula of Mexico. As we have seen (see Chapter 5), most paleontologists agree that this impact caused the mass extinction at the K/T boundary, but some argue that the impact was only one of several environmental changes that interacted to cause the K/T extinction. Likewise, the causes of the end-Permian extinction are debated, although there is increasing consensus that gases released by massive volcanic eruptions in Siberia played a dominant role.

Mass extinctions were "selective" in that some taxa were more likely than others to survive. Survival of gastropods through the end-Permian extinction was greater for species with wide geographic and ecological distributions and for genera consisting of many species (Erwin 1993). Extinction appears to have been random with respect to other characteristics, such as mode of feeding. The pattern of selectivity was much the same as during periods of background extinction, when gastropods and other taxa with broad geographic distributions had lower rates of extinction than narrowly distributed taxa (Boucot 1975). Patterns of survival through the end-Cretaceous mass extinction, however, differed from those during "normal" times (Jablonski 1995). During times of background extinction, survivorship of late-Cretaceous bivalves and gastropods was greater for taxa with planktotrophic larvae (those that feed while being dispersed by currents) and for genera consisting of numerous species, especially if those genera had broad geographic ranges. In contrast, during the end-Cretaceous mass extinction, planktotrophic and nonplanktotrophic taxa had the same extinction rates, and the survival of genera, although enhanced by broad distribution, was not influenced by their species richness. Thus the characteristics that were correlated with survival seem to have differed from those during "normal" times.

During mass extinction events, taxa with otherwise superb adaptive qualities succumbed because they happened not to have some critical feature that might have saved them from extinction under those circumstances. Evolutionary trends initiated in "normal" times were cut off at an early stage. For example, the ability to drill through bivalve shells and feed on the animals inside evolved in a Triassic gastropod lineage, but it was lost when this lineage became extinct in the late-Triassic mass extinction (Fürsich and Jablonski 1984). The same feature evolved again 120 Myr later, in a different lineage that gave rise to diverse oyster drills. A new adaptation that might have led to a major adaptive radiation in the Triassic was strangled in its cradle, so to speak.

Both physical and biotic environmental conditions were probably very different after mass extinctions than before. Perhaps for this reason, many taxa continued to dwindle long after the main extinction events (Jablonski 2002), while others, often members of previously subdominant groups, diversified. Full recovery of diversity took millions of years—as much as 100 Myr after the end-Permian disaster.

The mass extinction events, especially the end-Permian and K/T extinctions, had an enormous effect on the subsequent history of life because, to a great extent, they wiped the slate clean. Stephen Jay Gould (1985) suggested that there are "tiers" of evolutionary change, each of which must be understood in order to comprehend the full history of evolution. The first tier is microevolutionary change *within populations and species*. The second tier is "species selection," the *differential proliferation and extinction of species* during "normal" geological times, which affects the relative diversity of lineages with different characteristics (see Chapter 11). The third tier is the *shaping of the biota by mass extinctions*, which can extinguish diverse taxa and reset the stage for new evolutionary radiations, initiating evolutionary histories that are largely decoupled from earlier ones.

Richard Bambach and colleagues (2002) found some support for Gould's idea when they classified Phanerozoic marine animal genera by three functional criteria: whether they were

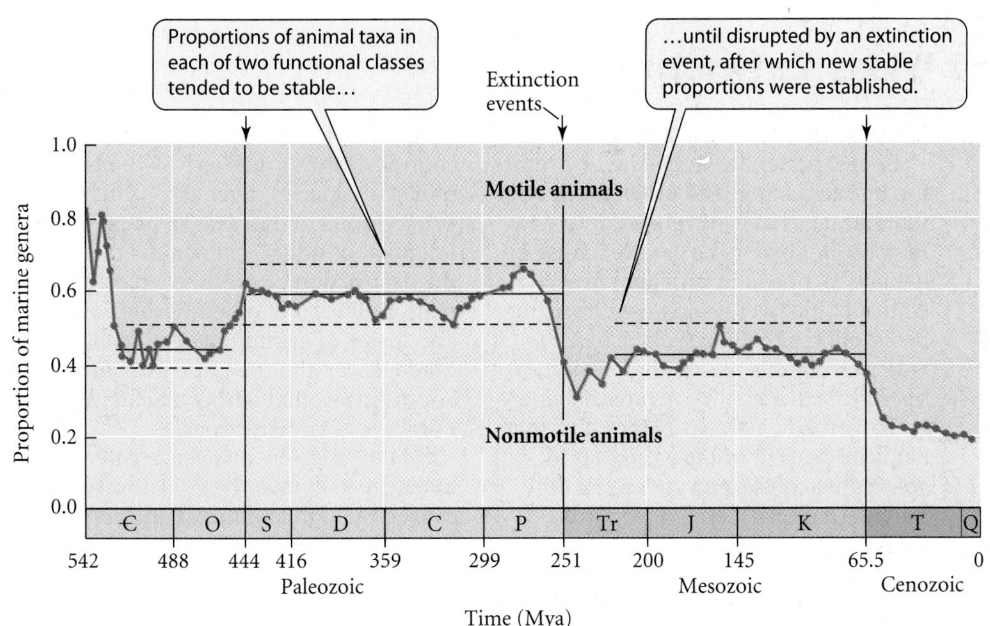

Figure 7.12 Changes in the proportions of genera of motile versus nonmotile marine animals during the Phanerozoic. The proportions were roughly stable (dashed lines) between mass extinctions, but changed abruptly to a new stable state after mass extinction events at the end of the Ordovician, Permian, and Cretaceous (black lines). Similar changes (not shown here) occurred in the proportions of predators versus nonpredators and in animals thought to be physiologically buffered versus unbuffered, based on anatomical criteria. (After Bambach et al. 2002.)

passive (nonmotile, such as barnacles) or active (motile), whether they were physiologically "buffered" (with well-developed gills and circulatory system, such as crustaceans) or not (such as echinoderms), and whether or not they were predatory. With respect to all three kinds of functional groupings, the proportions of taxa with alternative characteristics remained stable over intervals as long as 200 Myr, even though the total diversity and the taxonomic composition of the marine fauna changed greatly (Figure 7.12). However, shifts from one stable configuration to another occurred at the end of the Ordovician, Permian, and Cretaceous, suggesting that the extinction of long-prevalent (incumbent) taxa permitted the emergence of new community structures.

No truly massive extinction has occurred for 65 Myr; even the great climatic oscillations of the Pleistocene, though they altered geographic distributions and ecological assemblages, had a relatively small impact on the diversity of life. But it is depressingly safe to say that the next mass extinction has begun (Box 7A). The course of biodiversity has been altered for the foreseeable future by human domination of the Earth, and altered for the worse. Without massive, dedicated action, humanity will suffer profoundly, and much of the glorious variety of the living world will be extinguished as quickly as if another asteroid had smashed into the planet and again cast over it a pall of death.

Diversification

We turn now to the question of why increases in diversity have been greater in some lineages than in others and at some times than at others, and why diversity has tended to increase ever since the end-Permian extinction. We must ask, first, if ecosystems and the global biota tend to reach an equilibrium number of species, and second, what might change such an equilibrium.

Modeling rates of change in diversity

The number of taxa (N) changes over time by origination (as a result of branching of lineages) and extinction. These events are analogous to the births and deaths of individual organisms in a population, so models of population growth can be adapted to describe changes in taxonomic diversity. For each time interval Δt, suppose S new taxa originate per taxon present at the start of the interval, and suppose E is the number of taxa that become extinct, per original taxon, during the interval. Then ΔN, the change in N, equals

BOX 7A The Current Mass Extinction

For the first time in the history of life, a single species has precipitated a mass extinction. Within the next few centuries, the diversity of life will almost certainly plummet at a greater pace than ever before.

The human threat to Earth's biodiversity has accelerated steadily with the advent of ever more powerful technology and the exponential growth of the world's human population, which is approaching 7 billion as of mid-2009. The per capita rate of population growth is greatest in the developing countries, which are chiefly tropical and subtropical, but the per capita impact on the world's environment is greatest in the most highly industrialized countries. An average American, for example, has perhaps 140 times the environmental impact of an average Kenyan, because the United States is so profligate a consumer of resources harvested throughout the world, and of energy (with impacts ranging from strip mines and oil spills to insecticides and production of the "greenhouse gases" that cause global warming).

Some species are threatened by hunting or overfishing, and others by species that humans have introduced into new regions. But by far the greatest cause of extinction, now and probably over the course of the twenty-first century, is the destruction of habitat (Sala et al. 2000). It is largely for this reason that 29 percent of North American freshwater fishes are endangered or already extinct, and that about 10 percent of the world's bird species are considered endangered by the International Council for Bird Preservation.

The numbers of species likely to be lost are highest in tropical forests, which are being destroyed at a phenomenal and accelerating rate. As E. O. Wilson (1992) said, "in 1989 the surviving rain forests occupied an area about that of the contiguous forty-eight states of the United States, and they were being reduced by an amount equivalent to the size of Florida each year." Several authors have estimated that 10 to 25 percent of tropical rain forest species—accounting for as much as 5 to 10 percent of Earth's species diversity—will become extinct in the next 30 years. To this toll must be added extinctions caused by the destruction of species-rich coral reefs, pollution of other marine habitats, and losses of habitat in areas such as Madagascar and the Cape Province of South Africa, that harbor unusually high numbers of endemic species.

In the long run, an even greater threat to biodiversity may be global warming caused by high and increasing consumption of fossil fuels and production of carbon dioxide and other "greenhouse" gases. Earth's climate has warmed by a global average of 0.6°C during the last century, and the rate of warming is accelerating. Although climate change will vary in different regions (some will actually suffer a cooling trend), snow cover, glaciers, and polar ice caps are rapidly shrinking, and some tropical areas are becoming much drier. These changes are happening

the number of "births," SN, minus the number of "deaths," EN, and the rate of change in diversity (the **diversification rate**), $\Delta N/\Delta t$, is

$$\frac{\Delta N}{\Delta t} = SN - EN \quad \text{or} \quad \frac{\Delta N}{\Delta t} = RN$$

where $R = (S - E)$ and is the PER CAPITA RATE OF INCREASE. The growth in number of taxa is positive if $R > 1$.

Between the beginning (time t_0) and end (time t_1) of the time interval, the "population" grows by multiplying its original size by the per capita rate of increase: $N_1 = N_0R$. If the rates S and E remain constant, then after the next interval Δt, the population will be $N_2 = N_1R = N_0R^2$. In general, after t time intervals, the number of taxa will be

$$N_t = N_0R^t$$

as long as the per capita origination and extinction rates remain the same. This equation describes **exponential growth** of the number of taxa (or of a population of organisms) (Figure 7.13A). For continuous growth, rather than growth in discrete intervals, the equivalent expressions are

$$\frac{dN}{dt} = rN \quad \text{and} \quad N_t = N_0e^{rt}$$

where r is the INSTANTANEOUS PER CAPITA RATE OF INCREASE.

Let us pursue the analogy between the number of taxa and the number of organisms in a population (in which, as we know, birth and death rates are not constant). Factors such as severe weather may alter these rates by proportions that are unrelated to the density of the population. In contrast, **density-dependent factors**, which include competition

BOX 7A (Continued)

much faster than most of the climate changes that have occurred in the past. Some species may adapt by genetic change, but there is already evidence that many species will shift their ranges. Such shifts, however, are difficult or impossible for most mountaintop and Arctic species, and for many others because their habitats and the habitat "corridors" along which they might disperse have been destroyed. Computer simulations, based on various scenarios of warming rate and species' capacity for dispersal, suggest that within the next 50 years, between 18 and 35 percent of species will become "committed to extinction"—that is, they will have passed the point of no return (Thomas et al. 2004).

If mass extinctions have happened naturally in the past, why should we be so concerned? Different people have different answers, ranging from utilitarian to aesthetic to spiritual. Some point to the many thousands of species that are used by humans today, ranging from familiar foods to fiber, herbal medicines, and spices used by peoples throughout the world. Others cite the economic value of ecotourism and the enormous popularity of bird-watching in some countries. Biologists will argue that thousands of species may prove useful (as many already have) as pest-control agents, or as sources of medicinal compounds or industrially valuable materials. Except in a few well-known groups, such as vertebrates and vascular plants, most species have not even been described, much less been studied for their ecological and possible social value.

The rationale for conserving biodiversity is only partly utilitarian, however. Many people (including this author) cannot bear to think that future generations will be deprived of tigers, sea turtles, and macaws. They share with millions of others a deep renewal of spirit in the presence of unspoiled nature. Still others feel that it is in some sense cosmically unjust to extinguish, forever, the species with which we share the Earth.

Conservation is an exceedingly complicated topic; it requires not only a concern for other species, but compassion and understanding of the very real needs of people whose lives depend on clearing forests and making other uses of the environment. It requires that we understand not only biology, but also global and local economics, politics, and social issues ranging from the status of women to the reactions of the world's peoples and their governments to what may seem like elitist Western ideas. Anyone who undertakes work in conservation must deal with these complexities. But everyone can play a helpful role, however small. We can try to waste less; influence people about the need to reduce population growth (surely the most pressing problem of all); support conservation organizations; patronize environment-conscious businesses; stay aware of current environmental issues; and communicate our concerns to elected officials at every level of government. Few actions of an enlightened citizen of the world can be more important.

for food or space, cause the per capita birth rate to decline, and/or the death rate to increase, with population density, so that population growth slows down and the population density may reach a stable equilibrium (Figure 7.13B). Paleobiologists have suggested that changes in the number of species or higher taxa may similarly be affected by **diversity-dependent factors** that might reduce origination rates or increase extinction rates as the number of species increases. For example, competition among species for resources might limit the possible number of species to some maximal number, say K.

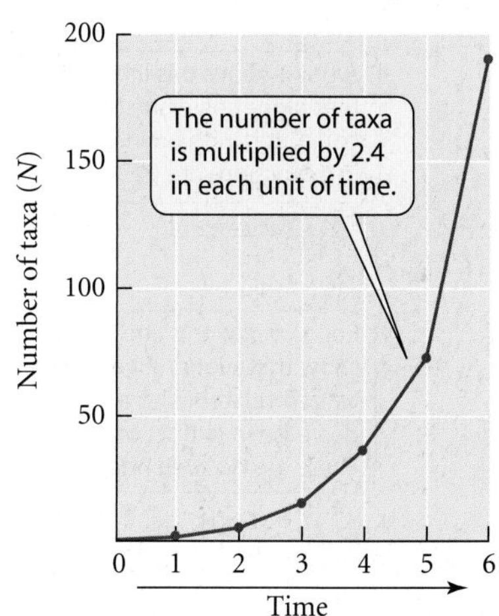

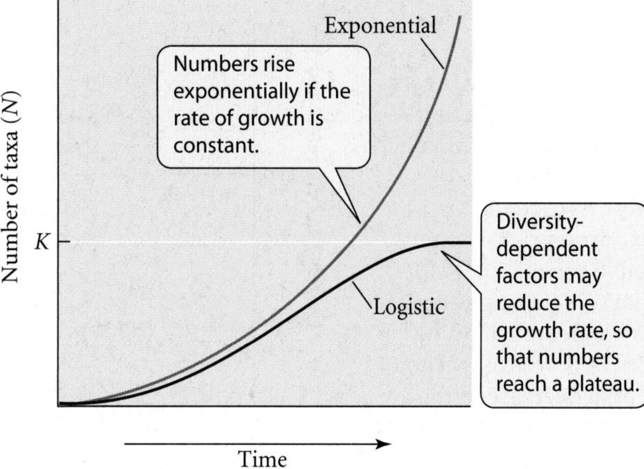

Figure 7.13 (A) Theoretical increase in the number of taxa (N), according to the equation $N_t = N_0R^t$, where R is the rate of increase in each discrete time interval (in this example, R = 2.4). (B) The increase in the number of taxa when change is continuous. The number of taxa grows exponentially if the rate of increase (r) is constant, but levels off if growth is logistic (i.e., if the rate of increase is negatively diversity-dependent).

Ecologists describe density-dependent population growth by the LOGISTIC EQUATION, which for growth in numbers of species may be expressed as

$$\frac{\Delta N}{\Delta t} = r_0 N\left(\frac{K-N}{K}\right)$$

where R_0 is the per capita rate of increase when the number of species is very low. Then the increase in number, $\Delta N/\Delta t$, declines as N increases and $(K - N)$ goes toward zero. At equilibrium, when $\Delta N/\Delta t = 0$, $N = K$.

Does species diversity reach equilibrium?

A huge ecological literature is concerned with whether or not the number of coexisting species (of some group such as plants or mammals) tends toward an equilibrium. This question is complex and not entirely resolved, but ecologists agree that some factors tend to limit species diversity. The space that plants compete for and the energy flux that organisms depend on are finite, so they can be divided among a limited number of species populations that are still large enough to persist. Moreover, phenomena such as competitive exclusion of species from each others' ranges (e.g., Figure 6.20) show that interactions among species can tend to limit local species diversity.

The fossil record supports the hypothesis that the per capita rate of increase in the number of species (or higher taxa) is lowered as the number grows. For example, Michael Foote (2000b), using Sepkoski's database, calculated changes (ΔS or ΔE) in the per capita rate of origination (S) or extinction (E) of genera from one stratigraphic stage to the next, and correlated these short-term changes with the number of genera present (D) at the beginning of the interval (Figure 7.14). For each of several phyla, correlations between ΔS and D were negative: the higher diversity is at the beginning of a time interval, the lower the rate of origination of new genera. Conversely, correlations between ΔE and D were positive: the higher diversity is, the higher the proportion of genera that become extinct. Whereas Foote found evidence of diversity-dependence in both origination and extinction rates, Alroy (1998) found that rates of origination of genera and species of mammals were diversity-dependent, but extinction rates were apparently not. Both these and other studies imply that the diversity of taxa should tend toward a stable equilibrium. Why, then, has diversity not remained constant?

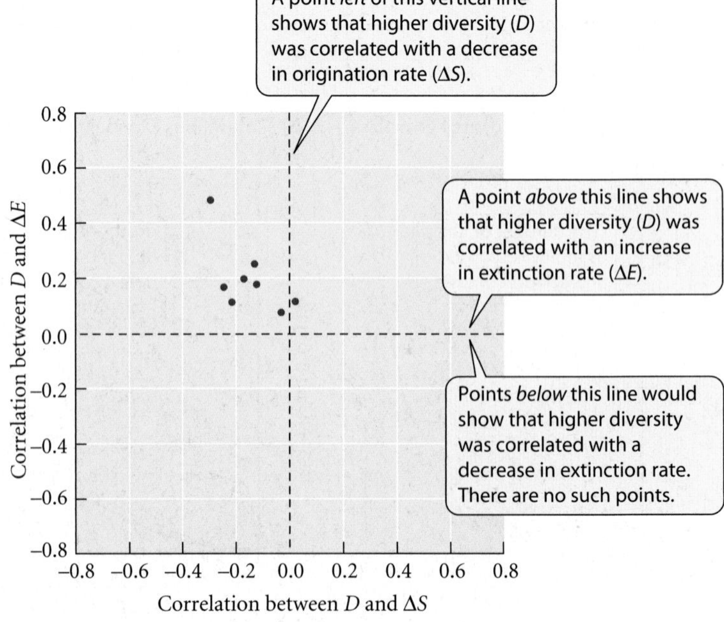

Figure 7.14 The correlation between diversity (D) in a geological stage and changes in origination (ΔS) and extinction (ΔE) rates from the mid-Jurassic to the Cenozoic. A point above the horizontal 0.0 line indicates a positive correlation (> 0) between diversity and the per capita rate of extinction, as expected if higher diversity increases extinction risk. A point left of the vertical 0.0 line shows a negative correlation (< 0) between diversity and per capita origination rate, suggesting that origination of new taxa is inhibited by greater standing diversity. The graph shows that high diversity was associated with an increase in the extinction rate and a decrease in the origination rate—evidence of diversity-dependent damping of growth in diversity. (After Foote 2000b.)

A system may shift from one equilibrium state to another when conditions change. At least three kinds of changes have altered conditions for organisms. First, changes have occurred in the physical environment, including changes in climate and in the configuration of land and sea. Second, the taxa that became dominant after mass extinctions were different from those that prevailed before, and would be expected to attain a different equilibrium level of diversity because of new patterns of competition and other interactions. There is no reason, for example, to expect the diversity of large herbivores and carnivores to be the same when they are mammals as when they are dinosaurs. Third, taxa have evolved to use new resources and habitats, changing the overall number of species that the planet's resources can support.

RELEASE FROM COMPETITION. Studies of both living and extinct organisms have shown that lineages often have diversified most rapidly when presented with ecological opportunity: what is often called "ecological space" or "vacant niches" not occupied by other species. In many isolated islands and bodies of water, descendants of just a few original colonizing species have diversified and filled ecological niches that are occupied in other places by unrelated organisms. Such adaptive radiations include the cichlid fishes in the Great Lakes of eastern Africa, the honeycreepers in the Hawaiian Islands, and Darwin's finches in the Galápagos Islands (see Figure 3.23). Islands and other habitats with taxonomically depauperate biotas typically harbor organisms that have evolved unusual new ways of life. For example, the larvae of almost all moths and butterflies are herbivorous, but in the Hawaiian Islands, the larvae of the moth genus *Eupithecia* are specialized for predation (Figure 7.15; Montgomery 1982). Probably such unusual forms are more prevalent where species diversity is reduced because they are not faced with as many predators or superior competitors in their early, relatively inefficient, stages of adaptation to new ways of life.

The fossil record provides many instances in which the reduction or extinction of one group of organisms has been followed or accompanied by the proliferation of an ecologically similar group. For example, conifers and other gymnosperms declined as angiosperms (flowering plants) diversified, and the orders of placental mammals appear in the fossil record after the late Cretaceous extinction of the nonavian dinosaurs.

Several hypotheses can account for these patterns (Benton 1996; Sepkoski 1996a). Two of these hypotheses involve competition between species in two clades. On one hand, the later group may have *caused* the extinction of the earlier group by competition, a process called **competitive displacement** (Figure 7.16A). On the other hand, an incumbent taxon may have *prevented* an ecologically similar taxon from diversifying, because it already "occupies" resources. Extinction of the incumbent taxon may then have vacated ecological "niche space," permitting the second taxon to radiate (Figure 7.16B). This process has been called **incumbent replacement** by Rosenzweig and McCord (1991), who argued that

Figure 7.15 A predatory moth caterpillar (*Eupithecia*) in the Hawaiian Islands, holding a *Drosophila* that it has captured with its unusually long legs. Predatory behavior is extremely unusual in the order Lepidoptera. (Photo by W. P. Mull, courtesy of W. P. Mull and S. L. Montgomery.)

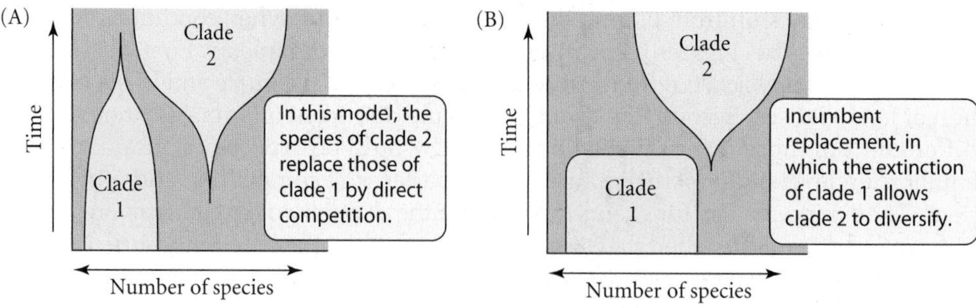

Figure 7.16 Models of competitive displacement and replacement. In each diagram, the width of a "spindle" represents the number of species. (A) Competitive displacement, in which the increasing diversity of clade 2 causes a decline in clade 1 by direct competitive exclusion. (B) Incumbent replacement, in which the extinction of clade 1 enables clade 2 to diversify. (After Sepkoski 1996a.)

the second taxon may have had superior adaptive features, but nevertheless could not have displaced the earlier taxon by competition.

There are good reasons to believe that competition among species has affected changes in diversity (Sepkoski 1996a). The rate of origination of new taxa has been much greater at times when diversity was unusually low—namely, during the Cambrian and after mass extinction events—than at other times (see Figure 7.4A). But how competition has affected changes in diversity is controversial.

Sepkoski et al. (2000) developed a mathematical model in which two clades increase in diversity following the logistic equation, but in which the increase in each clade is inhibited by both its own diversity and that of the other. They applied the model to data on the number of genera of two groups of bryozoans ("moss animals"), the cyclostomes and the cheilostomes. When these two groups meet, cheilostomes generally overgrow cyclostomes (Figure 7.17A). Especially after the end-Permian extinction, the diversity of cheilostomes has increased, whereas cyclostomes have not recovered (Figure 7.17B). When Sepkoski et

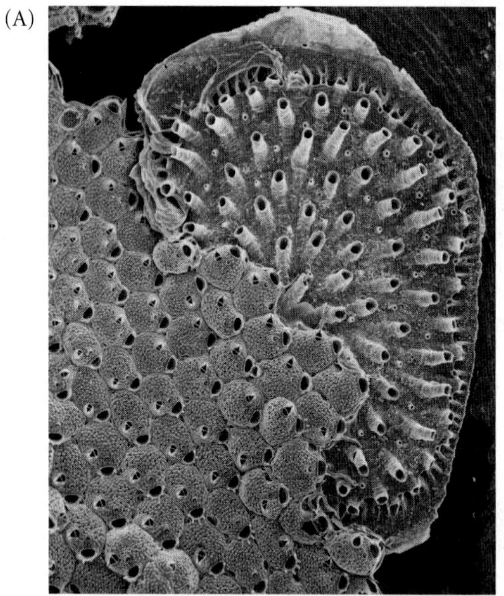

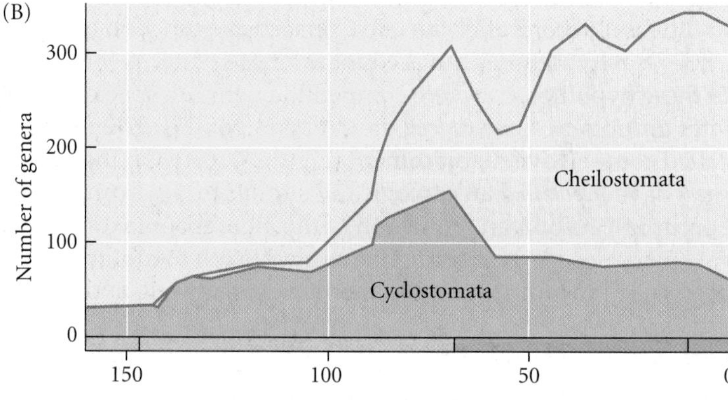

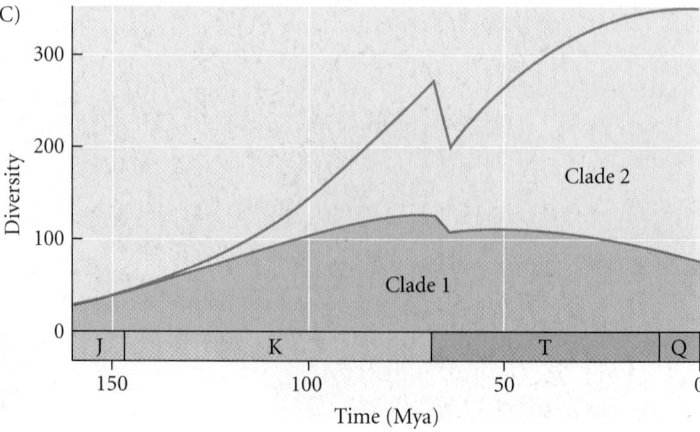

Figure 7.17 Bryozoans are sessile animals composed of component individuals that spread over hard surfaces. (A) Cheilostome bryozoans (left) can overgrow colonies of cyclostome bryozoans, causing their death. (B) Cheilostomes appeared in the late Jurassic and soon increased greatly in diversity, whereas the increase of cyclostomes was reversed in the Tertiary. (C) The changes in diversity in a model of competition among species in two clades, assuming that clade 2 species are competitively superior and that the diversity of both clades is reduced by an external perturbation at time 60. (A courtesy of Frank K. McKinney; B, C after Sepkoski et al. 2000.)

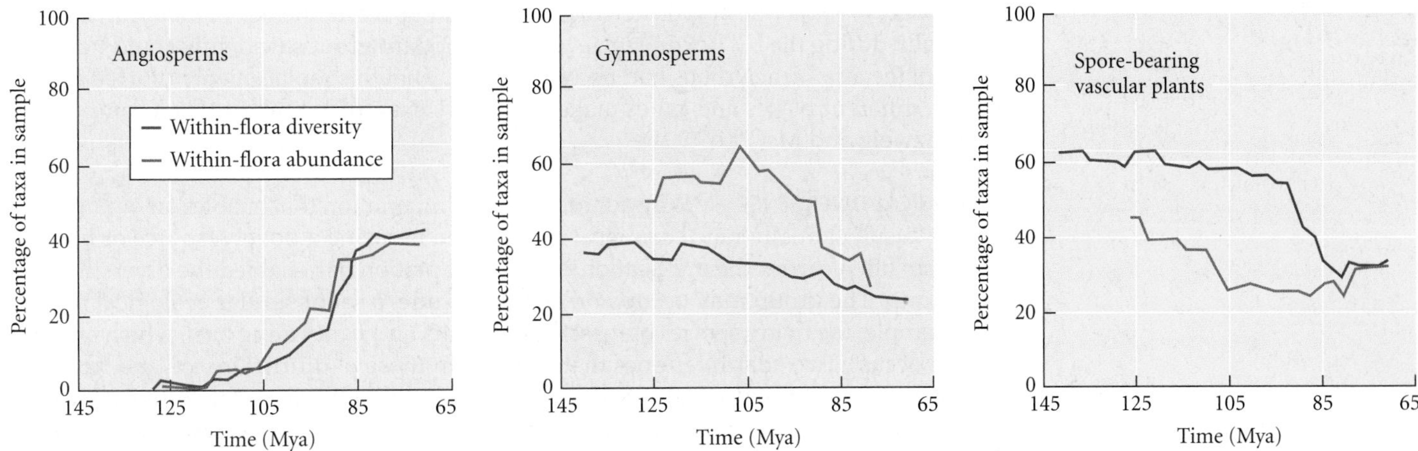

Figure 7.18 Changes in relative diversity (blue curves) and abundance (red curves) of major groups of vascular plants in fossil samples during the Cretaceous. The increase in both diversity and abundance of flowering plants was mirrored by the decline, in both respects, of the spore-bearing plants (e.g., ferns) and the decline in abundance of gymnosperms (e.g., conifers). This pattern is consistent with competitive displacement. (After Lupia et al. 1999.)

al. simulated the Permian drop in the diversity of both clades, their model rendered a profile of diversity change that closely matches the data (Figure 7.17C). This result does not prove that competition determined the history of bryozoan diversity, but it is consistent with that hypothesis.

A pattern of replacement is consistent with competitive displacement if the earlier and later taxa lived in the same place at the same time, if they used the same resources, if the earlier taxon was not decimated by a mass extinction event, and if the diversity and abundance of the later taxon increased as the earlier one declined (Lupia et al 1999). Vascular plants, which certainly compete for space and light, showed this pattern during the Cretaceous, when flowering plants increased in diversity and abundance at the expense of nonflowering plants, especially spore-bearing plants such as ferns (Figure 7.18).

Incumbent replacement has probably been more common than competitive displacement (Benton 1996). The great radiation of placental mammals in the early Cenozoic is often credited to the K/T extinction of the last nonavian dinosaurs and other large "reptiles," which may have suppressed mammals by both competition and predation. This is one of many examples of a long delay between the origin of a clade and its diversification and rise to ecological prominence, a delay that may often be caused by incumbent suppression of the new clade (Jablonski 2008). The best evidence of incumbency and release is supplied by repeated replacements. For instance, amphichelydians, the "stem group" of turtles, could not retract their heads and necks into their shells (Figure 7.19). Two

Figure 7.19 Pleurodiran and cryptodiran turtles replaced incumbent amphichelydian turtles, which became entirely extinct. (A) Amphichelydians, represented by the reconstructed skeleton of the earliest known turtle (*Proganochelys quenstedti*, upper Triassic), could not retract their heads for protection. (B) Snakeneck turtles such as *Chelodina longicollis* are pleurodiran turtles that flex the neck sideways beneath the edge of the carapace. (C) Cryptodiran turtles, represented here by an Eastern box turtle (*Terrapene carolina*), fully retract the head into the shell by flexing the neck vertically. (A courtesy of E. Gaffney, American Museum of Natural History; B © Juniors Bildarchiv/ Alamy; C © Constance McGuire/ istockphoto.com.)

(A)

(B)

(C)

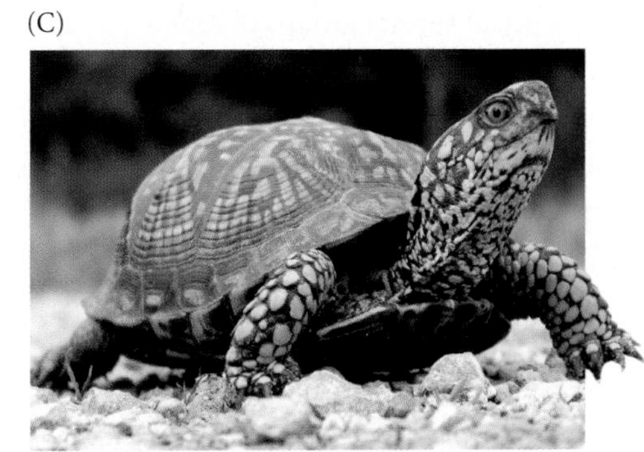

groups of modern turtles, which protect themselves by bending the neck under or within the shell, replaced the amphichelydians in different parts of the world four or five times, especially during the K/T extinction event. The modern groups evidently could not radiate until the amphichelydians had become extinct. That this replacement occurred in parallel in different places and times makes it a likely example of release from competition (Rosenzweig and McCord 1991).

ECOLOGICAL DIVERGENCE. A **key adaptation** is an adaptation that enables an organism to occupy a substantially new ecological niche, often by using a novel resource or habitat. The term often carries the implication that the adaptation has enabled the diversification of a group. The group may occupy an **adaptive zone**, a set of similar ecological niches. For example, the many species of insectivorous bats and fruit-eating bats, which are nocturnal, occupy two adaptive zones that differ from those of diurnal insect- and fruit-eating birds. The term **ecological space** is roughly equivalent to a set of adaptive zones.

The evolution of the ability to use new resources or habitats has certainly contributed importantly to the increase in diversity over time (Niklas et al. 1983; Bambach 1985). Among the sea urchins (Echinoidea), for example, three orders increased greatly in diversity beginning in the early Mesozoic (Figure 7.20). The order Echinacea evolved stronger jaws that enabled them to use a greater variety of foods, while the heart urchins (Atelostomata) and sand dollars (Gnathostomata) became specialized for burrowing in sand, where they feed on fine particles of organic sediment. The key adaptations allowing this major shift of habitat and diet include a flattened form and a variety of highly modified tube feet that can capture fine particles and transfer them to the mouth. Much of the history of increase in marine animal diversity throughout the Phanerozoic can be explained by increases in the occupancy of ecological space accomplished by evolutionary innovations such as those of the sand dollars (Bambach 1985; see Figure 7.12). Expansion into new habitats and feeding habits accounts for the diversification of most families of tetrapod vertebrates, such as various groups of frogs, snakes, and birds (Benton 1996).

Although, as in the sand dollars, the diversification of a clade can often be plausibly ascribed to a key adaptation, this is extremely difficult to demonstrate from a single case because the diversity might be due to other causes. Stronger evidence is provided if the rate of diversification is consistently associated with a particular kind of character that has evolved independently in a number of different clades. Such tests have been applied mostly to living organisms. The diversity of a number of clades with a novel character can be compared with the diversity of their sister groups that retain the ancestral character state. Since sister taxa are equally old, the difference between them in number of species must be due to a difference in rate of diversification, not to age. If the convergently evolved character is consistently associated with high diversity, we would have support for the hypothesis that it has caused a higher rate of diversification.

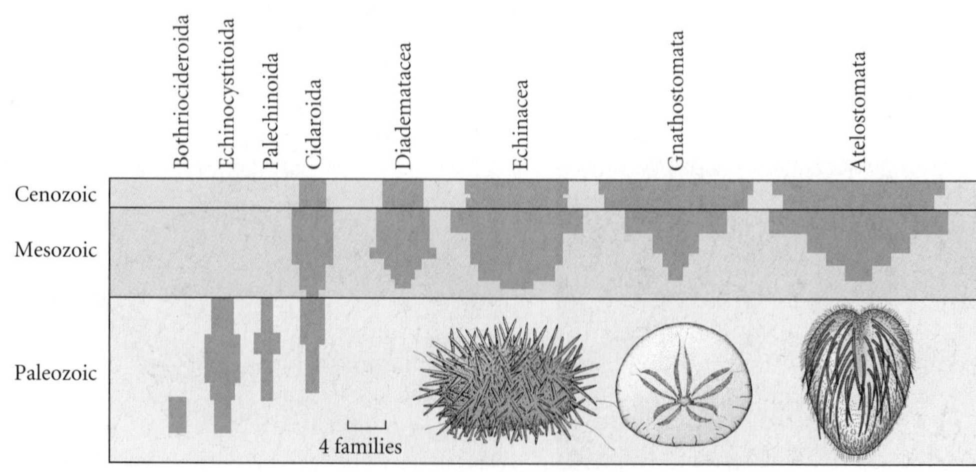

Figure 7.20 Changes in the diversity of several groups of echinoid echinoderms during the Mesozoic and Cenozoic. Diversification of sea urchins (order Echinacea), sand dollars (Gnathostomata), and heart urchins (Atelostomata) greatly increased, probably because of key adaptations described in the text. The width of the symmetrical profile of each group represents the number of families in that group at successive times. (After Bambach 1985.)

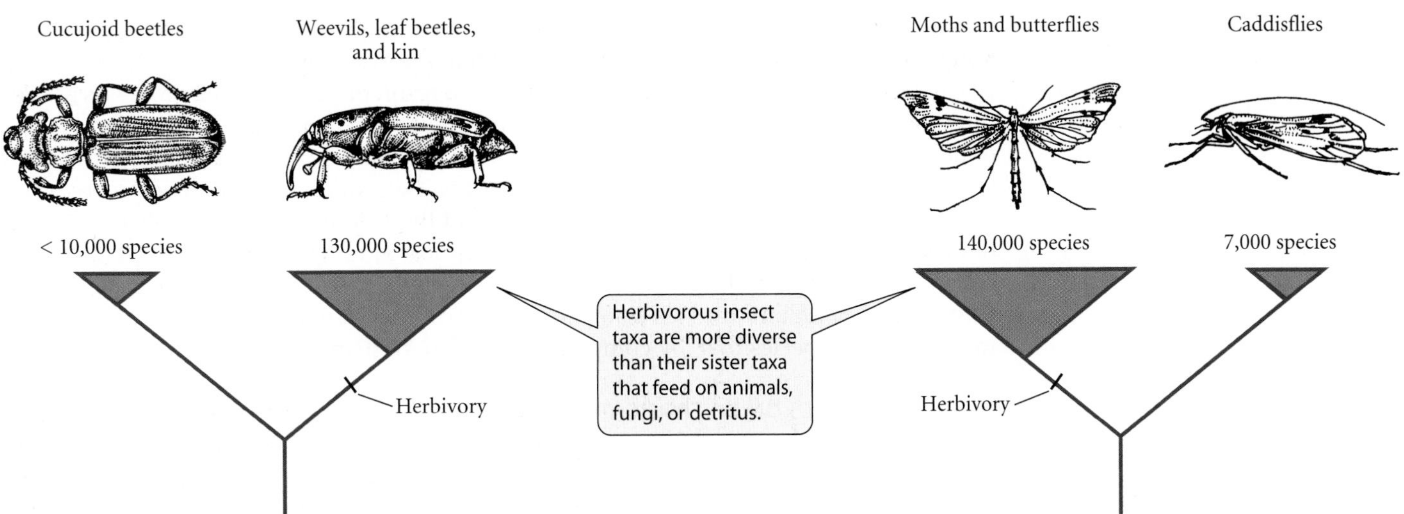

Figure 7.21 Two replicated sister-group comparisons of herbivorous clades of insects with their sister clades that feed on animals, fungi, or detritus. Herbivorous clades are consistently more diverse, demonstrating higher rates of diversification. (Data from Mitter et al. 1988.)

Charles Mitter and colleagues (Mitter et al. 1988; Farrell et al. 1991) applied this method, called REPLICATED SISTER-GROUP COMPARISON, to herbivorous insects and plants. The habit of feeding on the vegetative tissues of green plants has evolved at least 50 times in insects, usually from predatory or detritus-feeding ancestors. Phylogenetic studies have identified the nonherbivorous sister group of 13 herbivorous clades. In 11 of these cases, the herbivorous lineage has more species than its sister group (Figure 7.21). This significant correlation supports the hypothesis that entry into the herbivorous adaptive zone has promoted diversification. These researchers then examined the species diversity of 16 clades of plants that have evolved rubbery latex (as in milkweeds) or resin (as in pines), both of which deter attack by herbivorous insects. Thirteen of these clades have more species than their sister clades, which lack latex or resin. These defensive features may have fostered diversification.

From studies of modern organisms, we know that much diversity resides in the great numbers of related species that reduce competition with one another by subtle differences in resource use. Anoles living together on an island, for example, forage in different microhabitats (see Figure 6.23). Some paleontologists have suggested that subdivision of niches has increased over time because individual fossil deposits, which generally record local communities, contain more species in later than in earlier geological periods (Benton 1990). The increase in species number seems greater than the increase in the variety of major growth forms or adaptive zones, suggesting that the ever greater number of species coexisted by more finely partitioning similar resources.

PROVINCIALITY. The degree to which the world's biota is partitioned among geographic regions is called **provinciality**. A faunal or floral province is a region containing high numbers of distinctive, localized taxa (see Chapter 6). The fauna and flora of the contemporary world are divided into more provinces than ever before in the history of life. A trend from a cosmopolitan distribution of taxa to more localized distributions has persisted throughout much of the Mesozoic and Cenozoic and is thought by many paleontologists to be one of the most important causes of the increase in global diversity during this time (Valentine et al. 1978; Signor 1990).

Among marine animals, the number of faunal provinces was relatively low throughout much of the Paleozoic, and it dropped to an all-time low in the early Triassic, when most of the higher taxa that had survived the end-Permian mass extinction were so highly cosmopolitan that paleontologists recognize only a single, worldwide province during that time. During the Jurassic and Cretaceous, and especially the Tertiary, marine animals were distributed among an increasing number of latitudinally arranged provinces in both the Atlantic and Pacific regions. Similarly, among terrestrial vertebrates, a distinct fauna developed on each major land mass during the later Mesozoic and the Cenozoic, and the broad latitudinal distributions of many dinosaurs and other Mesozoic groups gave way to the much narrower latitudinal ranges of today's vertebrates.

Changes in the distribution of land masses as a result of plate tectonic processes are the fundamental cause of this trend. After the breakup of Pangaea in the Triassic, land masses ultimately became arrayed almost from pole to pole along a wider latitudinal span than ever before in Earth's history. This deployment of the continents created two increasingly disjunct ocean systems, the Indian-Pacific and the Atlantic, and established a pattern of ocean circulation that created a stronger latitudinal temperature gradient than ever before (Valentine et al. 1978). Not only did the variety of environments increase, but the fragmentation of land masses also allowed for divergent evolution and prevented the interchange of species that, by competition or predation, might lower diversity.

OTHER INFLUENCES ON DIVERSIFICATION. Interactions among species affect changes in diversity in many and complex ways that have not been fully analyzed. As we have seen, the diversification of one clade may suppress the diversification of competing organisms, or even reduce their diversity. Predators that have been introduced by humans into new regions have extinguished many species, but whether or not increases in the diversity of predators have affected the diversity of prey species on evolutionary time scales is uncertain. The Phanerozoic trends toward increased frequency of predatory marine animals and nonpredatory mobile and infaunal animals seem to be statistically independent; there is no evidence that one caused the other (Madin et al. 2006).

However, an increase in the number of species in a clade almost surely results, sooner or later, in an increase of affiliated species—especially parasites and mutualists—that use them as resources. The evidence that diversity thereby begets more diversity comes largely from studies of living organisms. For example, tropical forests contain far more species of hebivorous insects than temperate forests do, because tropical forests have more species of plants—resources on which the insects are specialized to some extent. The studies that have documented this pattern differ on whether or not the tropical insects are more specialized, more host-specific, than insects in temperate forests (Novotny et al. 2006, 2007; Dyer et al. 2007). Using a phylogeny of the largest family of butterflies, Niklas Janz and colleagues (2006) chose pairs of sister taxa that differ in the diversity of plants that their larvae eat. In 18 of 22 such pairs, the group with the higher diversity of host plants has the higher number of species, as expected if the diversification of plants has contributed to speciation and diversification of the insects (Figure 7.22). In many such cases, a molecular clock analysis has shown that the insects have diversified after the divergence among the plant lineages that they now use (Winkler and Mitter 2008).

Major changes in climate have been associated with increases in extinction and with changes in the distribution of habitats and vegetation types, which in turn have facilitat-

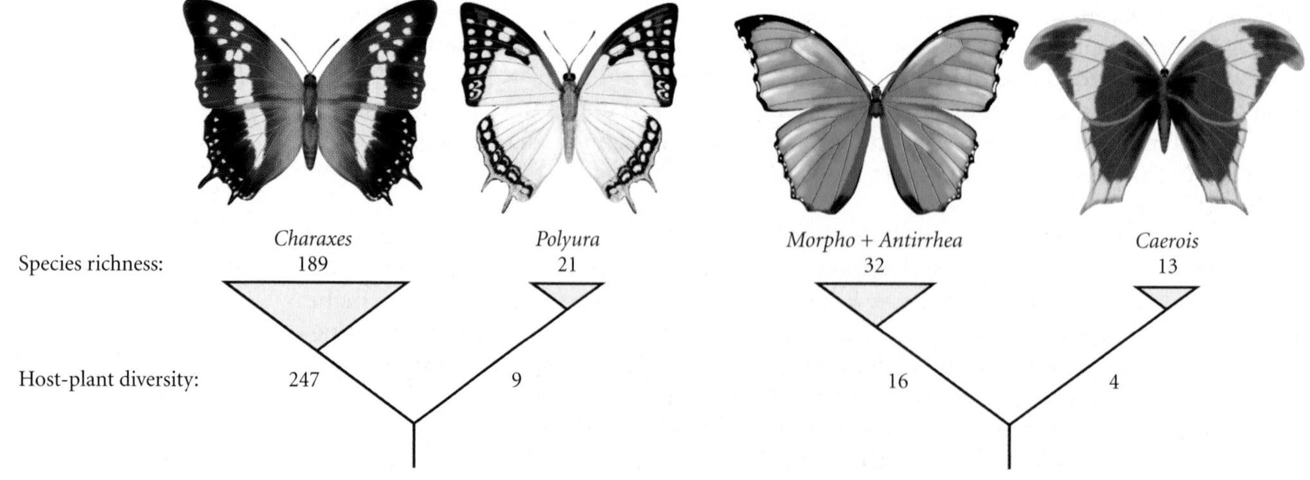

Figure 7.22 Examples of pairs of sister groups of nymphalid butterflies that differ in species richness (indicated at the branch tips). The more species-rich groups feed on a greater variety of larval host-plants, as indicated along the branches. Adaptation to a greater variety of food resources may have enhanced diversification. (Data from Janz et al. 2006.)

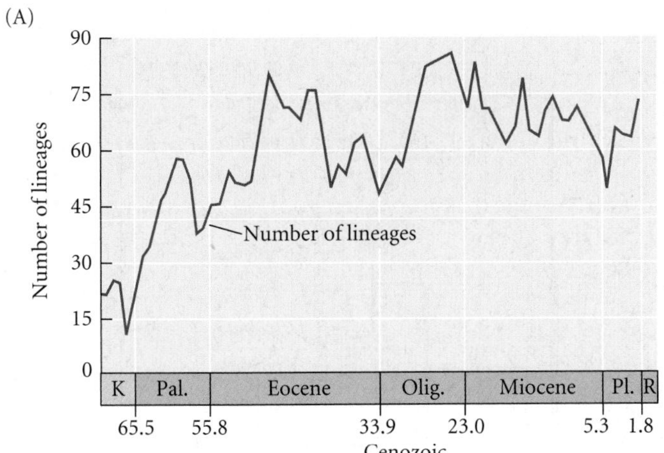

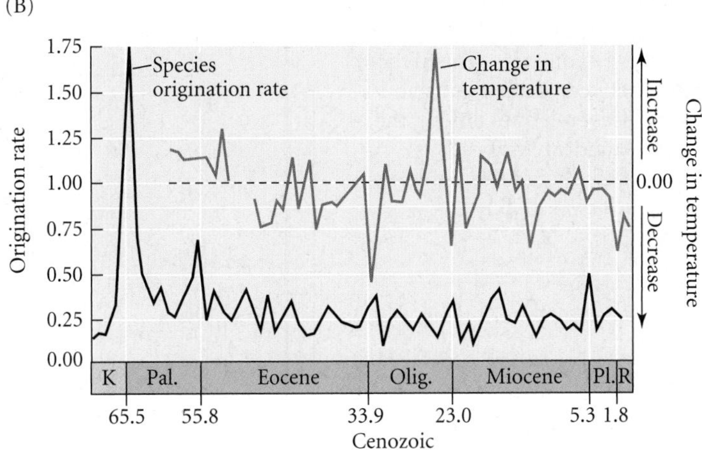

Figure 7.23 Changes in taxonomic diversity of Cenozoic mammals in North America. (A) The estimated history of diversity. (B) Fluctuations in the rate of origination of new lineages (black curve), compared with concurrent changes in temperature (red curve). The changes in temperature seem unrelated to origination rate, although some of them match declines in diversity that are evident in (A). (After Alroy et al. 2000.)

ed major changes in the distributions of taxa, often leading to diversification (Rothschild and Lister 2003). Thus, it is not known exactly how climate changes may have affected rates of origination and extinction; they may have had direct effects, but it is more likely that they have influenced diversification indirectly, through biotic changes. For example, in the mid-Eocene, about 50 to 40 Mya, the climate became cooler and drier, subtropical forests were widely replaced by savannahs in much of the temperate zone, and the diversity of primates and other arboreal mammals declined while that of large herbivores increased (Janis 1993). On the whole, however, changes in the rates of origination and extinction of mammals during the Tertiary are not closely correlated with changes in temperature (Figure 7.23). The importance of climate change in the evolution of diversity, relative to other factors such as key innovations and biotic interactions, is not yet well known (Alroy et al. 2000).

PHYLOGENETIC ANALYSES OF DIVERSITY TRENDS. It is possible to make inferences about the rate of increase in the number of species in a clade from a molecular phylogeny of living species, if the data are consistent with a molecular clock (Nee 2006). For example, Figure 7.24 is a phylogeny of the Hawaiian silverswords (see Figure 3.24), the diversification of which began at most 15 Mya (Baldwin and Sanderson 1998). The number of lineages increases from 1 (the common ancestor at 15 Mya) to 28 extant species. Now recall that if the rate of increase (r) in the number of lineages (N) has been constant, the number of lineages grows exponentially as $N_t = N_0e^{rt}$. Taking the natural logarithm (ln) of both sides of the equation, $\ln(N_t) = \ln(N_0) + rt$. That is, the log of the number of lineages increases linearly, with slope r. Slope r is the difference between the per capita rates of origination or speciation (S) and extinction (E), so the slope estimates the speciation rate if we assume there has been no extinction. If there have been species extinctions, however, the slope of the plot equals $S - E$ in the relatively distant past. But unless the extinction rate is high, species that arose recently have not yet had time to become extinct, so the slope must be closer to S in the very recent past (Figure 7.25A). Thus, it is possible to estimate both the speciation rate and the extinction rate from the slopes of a "lineage-through-time plot" of this kind. The data on Hawaiian silverswords fit the pure-speciation model better than the speciation-and-extinction model (Figure 7.25B), and the per capita rate of speciation estimated in this way was 0.56 species per Myr, a much higher rate than is typical of clades on continents. If a clade continued to increase indefinitely at a rate of 0.5 per year, it would grow from 1 to about 270,000 species in 25 Myr (Nee 2006).

Phylogenetic studies are also central in understanding why some taxa are more diverse than others—why, for example, there are more than 350,000 described species of Coleoptera (beetles) but only 20,000 Orthoptera (grasshoppers and relatives), or why there are fewer than 13,000 species of ferns but about 270,000 species of seed plants. As we have seen (see Figures 7.21, 7.22), sister taxa that differ in species richness must have differed

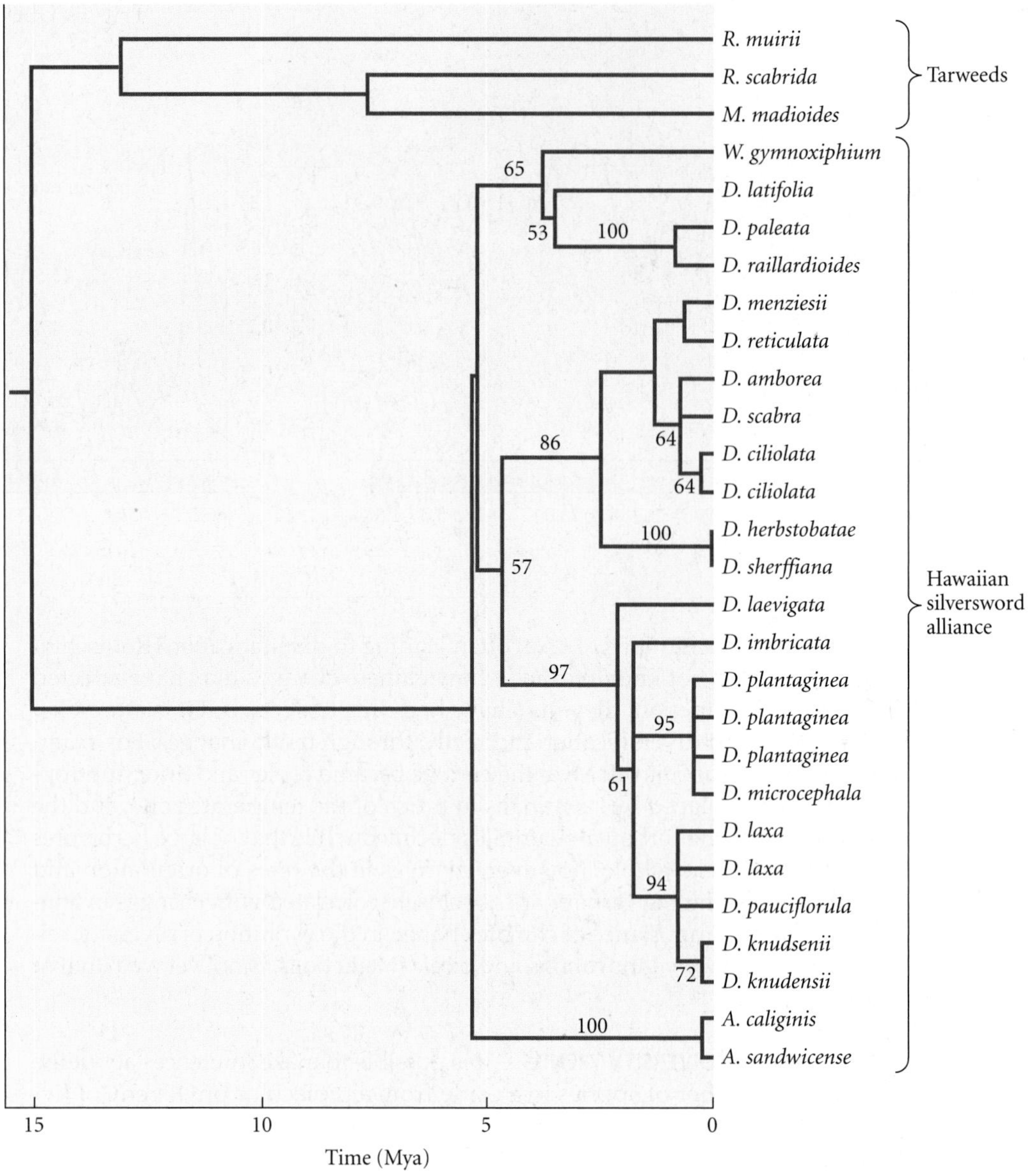

Figure 7.24 A phylogeny of Hawaiian silverswords, based on rDNA ITS sequences. The divergence between these plants and their closest relatives among the North American tarweeds is dated at 15 Myr. (After Baldwin and Sanderson 1998.)

in the rate of diversification since they are equal in age, and sometimes this rate difference is correlated with a key adaptation. However, Mark McPeek and Jonathan Brown (2007) compiled data from molecular phylogenies of 163 groups of animals in three phyla and found that, in general, species richness is correlated not with diversification rate but

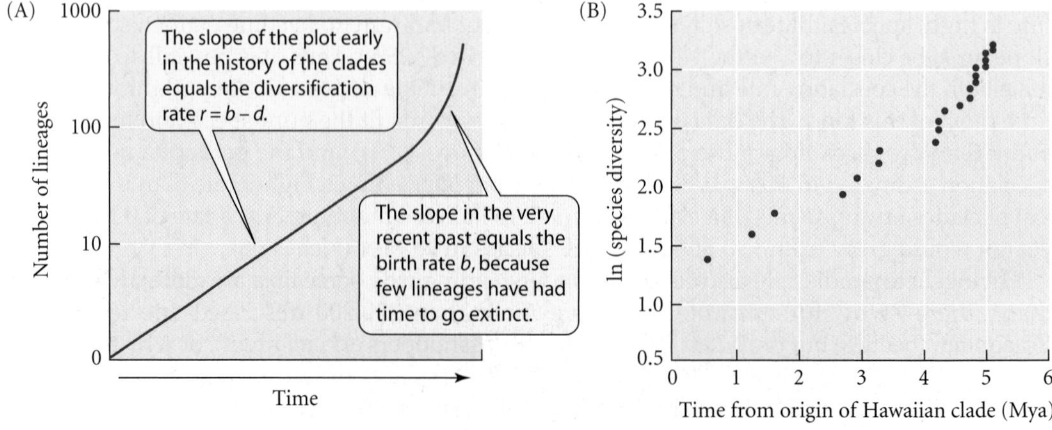

Figure 7.25 Estimating diversification rates from molecular phylogenies. (A) A theoretical lineage-through-time plot of the expected cumulative increase in the logarithm of the number of lineages if rates of "birth" (b) and "death" (d) of lineages are constant. (B) A lineage-through-time plot of the number of lineages of Hawaiian silverswords, based on the phylogeny in Figure 7.24, fits a model of approximately constant speciation rate without extinction. (A from Nee 2006; B after Baldwin and Sanderson 1998.)

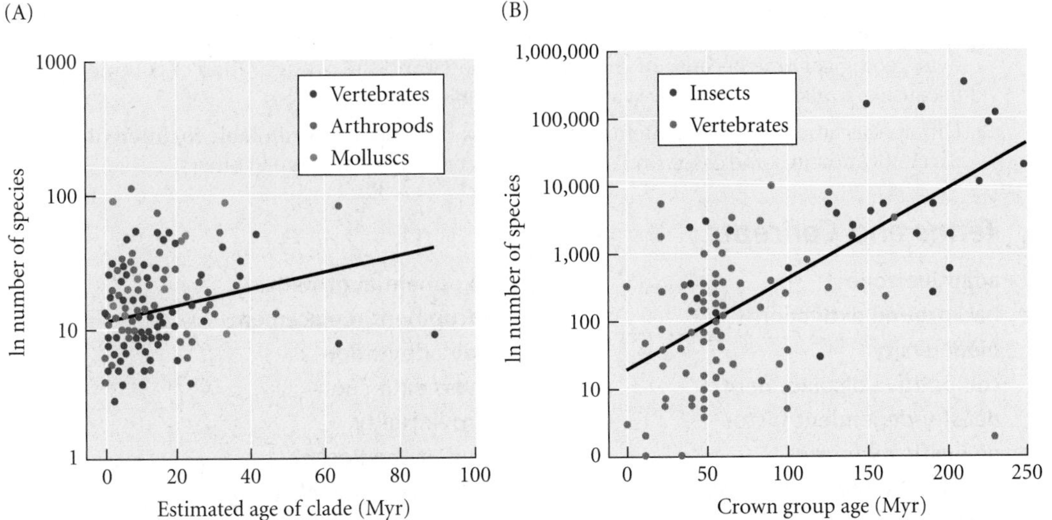

Figure 7.26 The species richness of clades in relation to their age. (A) Data from molecular phylogenies of living species in diverse clades of arthropods, molluscs, and vertebrates. (B) Number of extant species in orders of insects and vertebrates, in relation to age of the order (defined as first appearance of the crown group in the fossil record). These plots, especially of fossil-based data, show a correlation between species richness and clade age. (After McPeek and Brown 2007.)

with the age of the clade, estimated either by time-calibrated sequence differences (Figure 7.26A) or by earliest appearance in the fossil record (Figure 7.26B). Thus, many taxa have more species simply because they are older and have had more time to grow in number.

Summary

1. Analyses of diversity in the fossil record require procedures to correct for biases caused by the incompleteness of the record.
2. The diversity of skeletonized marine animals has increased during the Phanerozoic, but some aspects are uncertain. By direct count of numbers of taxa at each geological stage, diversity appears to have increased in the Cambrian to an approximate equilibrium that lasted for almost two-thirds of the Paleozoic; then, after a mass extinction at the end of the Permian, it appears to have increased (with interruptions) since the beginning of the Mesozoic, accelerating in the Cenozoic. Terrestrial plants and vertebrates show a similar pattern, except that their diversity was relatively stable for much of the Mesozoic. By the Pliocene, the diversity of families and lower taxa was apparently higher than ever before in the history of life. However, recent analyses that account for biases suggest considerably lower post-Permian increases.
3. The "background" rate of extinction (in between mass extinctions) declined during the Phanerozoic, perhaps because higher taxa that were particularly prone to extinction became extinct early.
4. Five major mass extinctions (at the ends of the Ordovician, Devonian, Permian, Triassic, and Cretaceous), as well as several less pronounced episodes of heightened extinction rates, are recognized. Although the cause of the incomparably devastating end-Permian extinction is unknown, it may have been the result of a rapid episode of major environmental changes initiated by the volcanic release of vast quantities of lava. The impact of a large extraterrestrial body at the end of the Cretaceous may have caused the extinction of many taxa, including the last of the nonavian dinosaurs.
5. Broad geographic and ecological distributions, rather than adaptation to "normal" conditions, enhanced the likelihood that taxa would survive mass extinctions. The diversification of many of the surviving lineages was probably released by extinction of other taxa that had occupied similar adaptive zones. Newly diversifying groups have sometimes displaced other taxa by direct competitive exclusion, but more often they have replaced incumbent taxa after these became extinct.
6. The increase in diversity over time appears to have been caused mostly by adaptation to vacant or underused ecological niches ("ecological space"), often as a consequence of the evolution of key adaptations, and by increasing provinciality (differentiation of the biota in different geographic regions) owing to the separation of land masses in the Mesozoic and Cenozoic and the consequent development of greater latitudinal variation in climate.

7. The rates of both extinction and origination of taxa have been diversity-dependent. Such observations imply that diversity tends toward an equilibrium. However, an equilibrium can change over geological time because of changes in climates and the configuration of continents, and because organisms evolve new ways of using habitats and resources.
8. Differences among taxa in contemporary species richness are attributable to different rates of diversification in some cases, and simply to differences in clade age in others.

Terms and Concepts

adaptive zone
background extinction
biodiversity
competitive displacement
density-dependent factor
diversification rate
diversity-dependent factor
ecological space
exponential growth
incumbent replacement
key adaptation
mass extinction
provinciality
pull of the Recent
Red Queen hypothesis

Suggestions for Further Reading

Most of the topics in this chapter are treated clearly in an excellent textbook on paleobiology, *Principles of Paleontology* (third edition) by M. Foote and A. I. Miller (W. H. Freeman, New York, 2007). See also D. Jablonski et al. (eds.), *Evolutionary Paleobiology* (University of Chicago Press, 1996). The end-Permian mass extinction is the subject of a popular book by D. H. Erwin, *Extinction: How Life on Earth Nearly Ended 250 Million Years Ago* (Princeton University Press, Princeton, NJ, 2006), and the consequences of mass extinctions are reviewed by R. K. Bambach, "Phanerozoic biodiversity mass extinctions," *Annual Review of Earth and Planetary Sciences* 34:127–155 (2006) and D. Jablonski, "Mass extinctions and macroevolution," *Paleobiology* 31 (Supplement):192–210 (2005).

A thoughtful and informed work about the future of biodiversity is found in E. O. Wilson's *The Diversity of Life* (W. W. Norton, New York, 1999). The same subject is discussed in a more academic manner in *Principles of Conservation Biology*, edited by M. J. Groom, G. K. Meffe, and C. R. Carroll (Sinauer Associates, Sunderland, MA, 2006).

Problems and Discussion Topics

1. Distinguish between the rate of speciation in a higher taxon and its rate of diversification. What are the possible relationships between the present number of species in a taxon, its rate of speciation, and its rate of diversification?
2. What factors might account for differences among taxa in their numbers of extant species? Suggest methods for determining which factor might actually account for an observed difference.
3. Ehrlich and Raven (1964) suggested that coevolution with plants was a major cause of the great diversity of herbivorous insects, and Mitter et al. (1988) presented evidence that the evolution of herbivory was associated with increased rates of insect diversification. However, the increase in the number of insect families in the fossil record was not accelerated by the explosive diversification of flowering plants (Labandeira and Sepkoski 1993). Suggest some hypotheses to account for this apparent conflict, and some ways to test them.
4. A factor that might contribute to increasing species numbers over time is the evolution of increased specialization in resource use, whereby more species coexist by more finely partitioning resources. Discuss ways in which, using either fossil or extant organisms, one might test the hypothesis that a clade is composed of increasingly specialized species over the course of evolutionary time.
5. In several phyla of marine invertebrates, lineages classified as new orders appear first in the fossil record in shallow-water environments and are recorded from deep-water environments only later in their history (Jablonski and Bottjer 1990). What might explain this observation? (Note: No one has offered a definitive explanation so far, so use your imagination.)

6. The analysis by McPeek and Brown (2007) suggests that clades with few living species may be very young. Is this necessarily the case? Are there alternative hypotheses? Can you find evidence for any of these hypotheses? What would constitute evidence?
7. The method of replicated sister-group comparison of species richness has been used to implicate certain adaptive characteristics as contributors to higher species richness. Is there any way, conversely, to test hypotheses on what factors may have contributed to the decline or extinction of groups?
8. Scientific debate continues about the history and interpretation of the diversity patterns of many taxa. Analyze such a debate, and decide whether either side has settled the issue. If not, what further research would be needed? An example is whether or not the enormous diversity of leaf beetles (Chrysomelidae) is due to a long history of co-diversification with their host plants. See Farrell 1998, Farrell and Sequeira 2004, and Gómez-Zurita et al. 2007.

CHAPTER 8

The Origin of Genetic Variation

Ladies of the court. In *Las Meninas*, the artist Velásquez portrayed two achondroplasic dwarves (at right) among the ladies attending the daughter of Philip IV of Spain. Achondroplasia is caused by dominant mutations that substitute arginine for glycine at a site in the receptor protein FGF-3, a signal transduction protein. The mutant receptor results in insufficient formation of bone from cartilage. Most mutations that contribute to evolution have less pronounced effects.

The fundamental process of evolution is a change in the inherited characteristics of a population or species. It is an alteration of the genetic composition of a population. To understand the processes of evolution, therefore, it is essential to know the fundamentals of genetics and to understand the several factors that can change organisms' characteristics at a genetic level.

Genetic changes in populations and species begin with changes in the genetic material carried by individual organisms: **mutations**. Every gene, every variation in DNA, every characteristic of a species, every species itself, owes its existence to processes of mutation. Mutation is not the cause of evolution, any more than fuel in a car's tank is the cause of the car's movement. But it is the sine qua non, the necessary ingredient of evolution, just as fuel is necessary—though not sufficient—for traveling down a highway. The fundamental role of mutation makes it a logical starting point for our analysis of the causes of evolution.

Technological advances in molecular biology have utterly transformed our knowledge of the nature and processes of mutation in the last several decades. Many kinds of alterations of DNA, and how they have affected individual genes, have been described, but the study of mutations has now entered a major new phase in which it is possible to describe how various kinds of mutations have shaped entire genomes. One of the landmarks in the history of science was the simultaneous publication in February 2001 of two "drafts" of the complete sequence of the human genome, one by the International Human Genome Sequencing Consortium (2001) and the other by a private company (Venter et al. 2001). Since then, complete genome sequences of other species are announced almost every week, so it is now possible to compare differences among a multitude of viruses, bacteria, and eukaryotes, ranging from unicellular parasites to rice, dog, and platypus. Comparisons among these sequences are providing unprecedented volumes of information about the processes and history of evolution (see Chapter 20). Still, as we will see, many questions about the role of mutation in evolution require more than molecular data, and are still very imperfectly answered.

Genes and Genomes

We will begin by describing mutation at the molecular level, but that requires a short review of the genetic material and its organization.

Except in certain viruses, in which the genetic material is RNA (ribonucleic acid), organisms' genomes consist of DNA (deoxyribonucleic acid), made up of a series of nucleotide **base pairs** (bp), each consisting of a purine (adenine, A, or guanine, G) and a pyrimidine (thymine, T, or cytosine, C). A haploid (gametic) genome of the fruit fly *Drosophila melanogaster* has about 1.5×10^8 bp, and that of a human about 3.2×10^9 bp (3.2 billion). However, DNA content varies greatly among organisms: it differs more than a hundredfold, for example, among species of salamanders, some of which have more than a hundred times as much DNA as humans. The genome of the single-celled protist *Amoeba dubia* is 200 times the size of a human's!

Each chromosome is a single long, often tightly coiled, DNA molecule, different portions of which constitute different genes, together with histone proteins. The term **gene** usually refers to a sequence of DNA that is transcribed into RNA, together with untranscribed regions that play roles in regulating its transcription. The term **locus** technically refers to the chromosome site occupied by a particular gene, but it is often used to refer to the gene itself. Thousands of genes in the human genome encode ribosomal and transfer RNAs (rRNAs and tRNAs) that are not translated into proteins. The number of protein-encoding genes is about 26,000 in the small flowering plant *Arabidopsis* (sometimes called mustard weed), 6000 in the fungus *Saccharomyces*, 16,000 in *Drosophila*, and 24,000 in mice and humans.

One strand of a protein-encoding gene is transcribed into RNA, a process regulated by **control regions**: untranscribed sequences (**enhancers** and **repressors**) to which regulatory proteins produced by other genes bind. A gene may have many different enhancer sequences. In eukaryotes, the transcribed sequence of a gene consists of coding regions (called **exons**) which in most genes are separated by noncoding regions (called **introns**; Figure 8.1). Control regions of a gene may be located at various distances from the coding regions, often "upstream" of them but sometimes within introns. The average human gene has 1340–1500 bp of coding sequence (encoding 445–500 amino acids) divided into 8.8 exons, which are separated by a total of 3365 bp of introns. After a gene is transcribed, the portions transcribed from exons are processed into a messenger RNA (mRNA) by splicing out the portions that were transcribed from introns. ALTERNATIVE SPLICING, whereby the mature mRNA corresponds to a variable number of the exons, can result in several proteins being encoded by a single gene. At least 35 percent of human genes appear to be subject to alternative splicing, so the number of possible proteins may greatly exceed the number of genes.

Through the action of ribosomes, enzymes, and tRNAs, mRNA is translated into a polypeptide or protein on the basis of the **genetic code**, whereby a triplet of bases (a **codon**) specifies a particular amino acid in the growing polypeptide chain. The RNA code (Figure 8.2), which is complementary to the DNA code, consists of $4^3 = 64$ codons, which, however, encode only the 20 amino acids of which proteins are composed. Most of the amino acids are encoded by two or more synonymous codons. The third position in a codon is the most "degenerate"; for example, all four CC– codons (CCU, CCC, CCA,

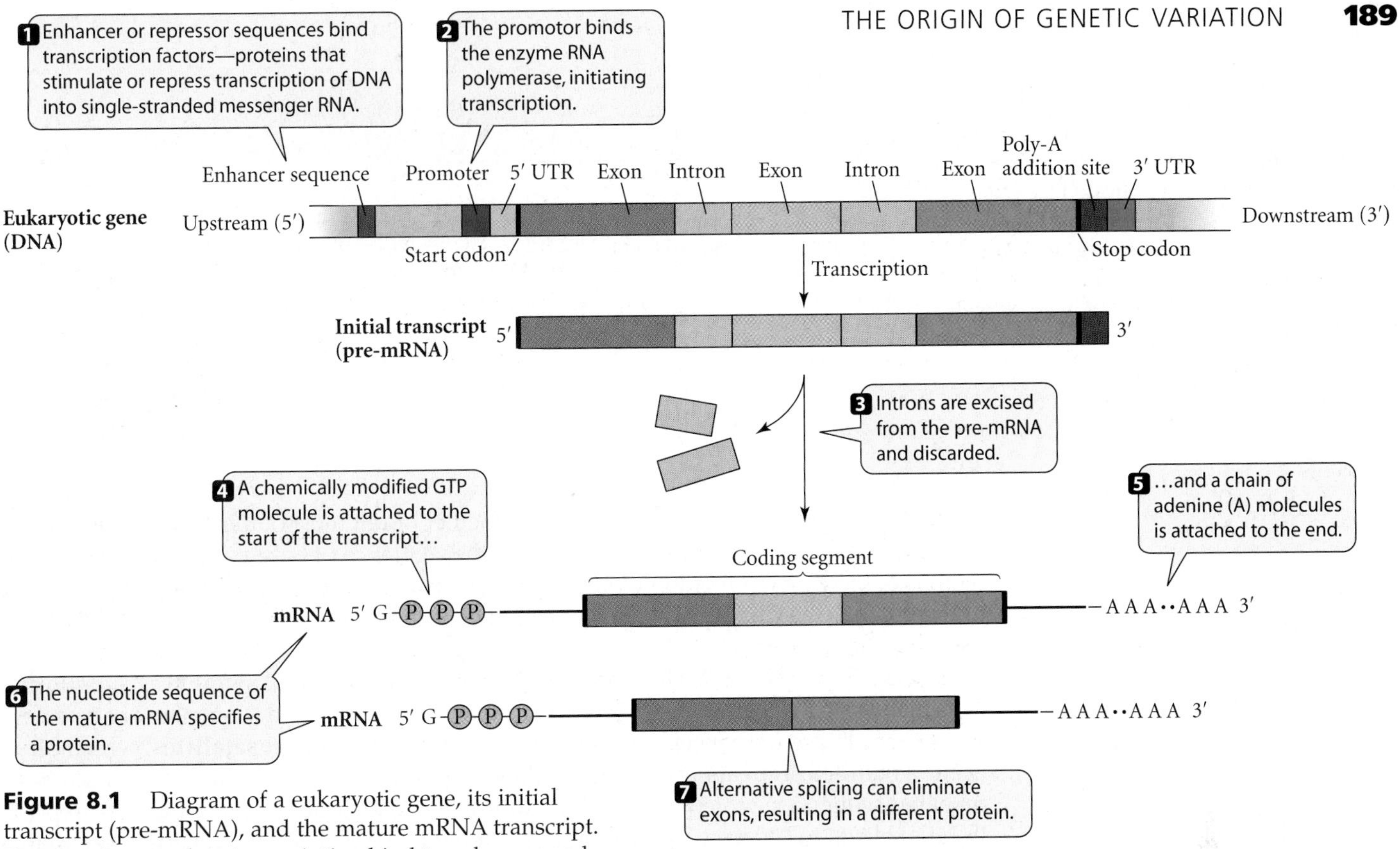

Figure 8.1 Diagram of a eukaryotic gene, its initial transcript (pre-mRNA), and the mature mRNA transcript. Proteins that regulate transcription bind to enhancer and repressor sequences, shown here upstream of the coding sequences. Transcription proceeds in the 5′ to 3′ direction. Introns are transcribed, but are spliced out of the pre-mRNA and discarded. The coding segment of the mature mRNA corresponds to the gene's exons. In some genes, alternative splicing may yield several different mRNAs. UTR refers to untranslated regions that sometimes include regulatory sequences.

CCG) specify proline. The second position is least degenerate—a substitution of one base for another in the second position usually results in an amino acid substitution in a protein. Three of the 64 codons are "stop" ("chain end") signals that terminate translation. Substitution in the third position of five other codons can therefore produce a "stop" codon, which may result in an incomplete, often nonfunctional protein.

It is a profoundly wonderful fact that the genetic code is nearly universal, from viruses and bacteria to pineapples and mammals. Moreover, the machinery of transcription and translation is remarkably uniform, so that sea urchin DNA or mRNA can be translated into a protein if injected into a frog. This uniformity is the basis of genetic engineering of crop plants with bacterial genes that encode natural insecticides, to give just one example.

Only about 2 percent of the human genome encodes proteins. There is reason to think that the portions of an intron near coding sequences may sometimes play a functional role. In eukaryotes, the vast majority of DNA has no known function, even though

First nucleotide	Second nucleotide: U	C	A	G	Third nucleotide
U	UUU Phe UUC Phe UUA Leu UUG Leu	UCU Ser UCC Ser UCA Ser UCG Ser	UAU Tyr UAC Tyr UAA Stop UAG Stop	UGU Cys UGC Cys UGA Stop UGG Trp	U C A G
C	CUU Leu CUC Leu CUA Leu CUG Leu	CCU Pro CCC Pro CCA Pro CCG Pro	CAU His CAC His CAA Gln CAG Gln	CGU Arg CGC Arg CGA Arg CGG Arg	U C A G
A	AUU Ile AUC Ile AUA Ile AUG Met	ACU Thr ACC Thr ACA Thr ACG Thr	AAU Asn AAC Asn AAA Lys AAG Lys	AGU Ser AGC Ser AGA Arg AGG Arg	U C A G
G	GUU Val GUC Val GUA Val GUG Val	GCU Ala GCC Ala GCA Ala GCG Ala	GAU Asp GAC Asp GAA Glu GAG Glu	GGU Gly GGC Gly GGA Gly GGG Gly	U C A G

Figure 8.2 The genetic code, as expressed in mRNA. Three of the 64 codons are "stop" ("chain end") signals; the other codons encode the 20 amino acids found in proteins. Note that many codons, especially those differing only in the third position, are synonymous. The three-letter abbreviations signify the amino acids.

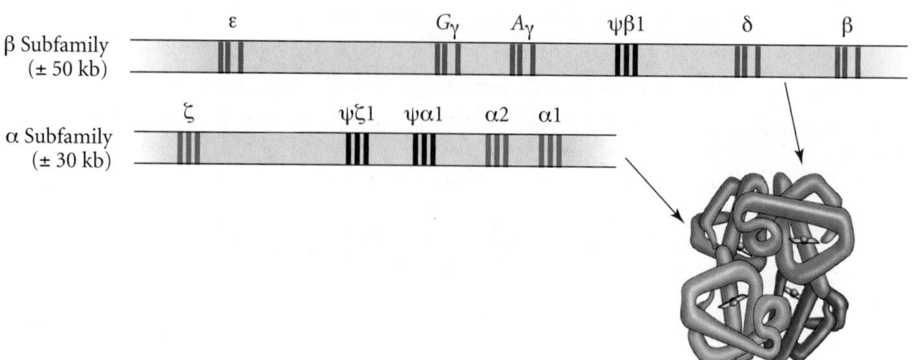

Figure 8.3 The human hemoglobin gene family has two subfamilies, α (green) and β (blue), located on different chromosomes. Each functional gene is indicated by three lines, representing its three exons. Pseudogenes are denoted by ψ. See Figure 3.29 for a history of how these genes have arisen by duplication and divergence. (After Hartwell et al. 2000.)

as much as 80 percent may be transcribed. Whether or not most of this DNA has any function is unresolved at this time. Mutations in the sequence of base pairs in nonfunctional regions of the genome would not affect the organism, and so would not be subject to natural selection (see Chapter 10). However, changes in the amount of noncoding DNA can and often do have effects.

At least 45 percent of the human genome consists of **repeated sequences**, amounting to as many as 4.3 million repetitive elements that include repeated sequences of a few base pairs each. These sequences are sometimes referred to as **microsatellites**, which often occur as **tandem repeats** that may number more than 2 billion in some species. Other repeated sequences, which may harbor coding DNA, include short interspersed repeats (SINEs) of 100 to 400 base pairs; long interspersed repeats (LINEs) that are more than 5 kilobases (kb) in length; and DNA transposons. These elements either are capable of undergoing or have been formed by the process of **transposition**: the production of copies that become inserted into new positions in the genome. DNA sequences that are capable of transposition are called **transposable elements (TEs)**.

Many protein-encoding genes (probably at least 40 percent of human genes) are members of **gene families**: groups of genes that are similar in sequence and often have related functions. For example, the human hemoglobin gene family includes two subfamilies, α and β, located on different chromosomes (Figure 8.3). Different α and β polypeptides are combined to form different types of hemoglobin before and after birth. Gene families are examples of repeated sequences that have diversified over time. Some gene families, such as that for mammalian olfactory receptors, have more than a thousand functional members. Both the α and β subfamilies also include hemoglobin **pseudogenes**: sequences that resemble the functional genes but that differ at several base pair sites and are not transcribed because they have internal "stop" codons. **Processed pseudogenes** are those that originated by REVERSE TRANSCRIPTION from an mRNA message into a DNA sequence that lacked introns, and later underwent silencing and further base pair changes. The DNA sequence of processed pseudogenes is recognizably similar to the coding portions of the gene from which they arose, but lacks passages that can be matched with the functional gene's introns.

Gene Mutations

The word "mutation" refers both to the process of alteration of a gene or chromosome and to its product, the altered state of a gene or chromosome. It is usually clear from the context which is meant.

Before the development of molecular genetics, a mutation was identified by its effect on a phenotypic character. That is, a mutation was a newly arisen change in morphology, survival, behavior, or some other property that was inherited and could be mapped (at least in principle) to a specific locus on a chromosome. In practice, many mutations are still discovered, characterized, and named by their phenotypic effects. Thus we will frequently use the term "mutation" to refer to an alteration of a gene from one form, or **allele**, to another, the alleles being distinguished by phenotypic effects. In a molecular context, however, a gene mutation is an alteration of a DNA sequence, independent of whether or not it has any phenotypic effect. A particular DNA sequence that differs by one or more mutations from homologous sequences is called a **haplotype**. We will often refer to **genetic markers**, which are detectable mutations that geneticists use to recognize specific regions of chromosomes or genes.

Mutations have evolutionary consequences only if they are transmitted to succeeding generations. Mutations that occur in somatic cells may be inherited in certain animals and plants in which the reproductive structures arise from somatic meristems, but in those animals in which the germ line is segregated from the soma early in development, a mutation is inherited only if it occurs in a germ line cell. Mutations are thought to occur mostly during DNA replication, which usually occurs during cell division. In humans, more than five times as many new mutations enter the population via sperm than via eggs because more cell divisions have transpired in the germ line before spermatogenesis than before oogenesis in individuals of equal age (Makova and Li 2002).

DNA is frequently damaged by chemical and physical events, and changes in base pair sequence can result. Many such changes are repaired by a variety of repair enzymes, but some are not. These alterations, or mutations, are considered by most evolutionary biologists to be *errors*. That is, *the process of mutation is generally thought to be not an adaptation,* but a consequence of unrepaired damage.

A particular mutation occurs in a single cell of a single individual organism. If that cell is in the germ line, it may give rise to a single gamete or, quite often, to a number of gametes that carry the mutation, and so may be inherited by several offspring. The occurrence of such "clusters" of mutations may have important evolutionary consequences (Woodruff et al. 1996). Initially, the mutation is carried by a very small percentage of individuals in the species population. If, because of natural selection or genetic drift, it ultimately becomes FIXED (i.e., is carried by nearly the entire population), the mutation may be referred to as a **substitution**. The distinction between a mutation, which is simply an altered form of a gene, compared with some standard sequence, and a substitution in an entire population or species is important. Most mutations do not become substitutions. Consequently, *mutation is not equivalent to evolution.*

Kinds of mutations

Mutational changes of DNA are many in kind; there is a full spectrum, from changes of a single base pair through alterations that affect longer DNA sequences that may encompass part or all of a gene, to alterations of entire chromosomes or even an entire genome. For convenience, we will arbitrarily distinguish gene mutations from chromosome changes.

GENE MUTATIONS. The simplest mutations are alterations of a single base pair. These include **base pair substitutions** (Figure 8.4). In classic genetics, a mutation that maps to a single gene locus is called a **point mutation**; in modern usage, this term is often restrict-

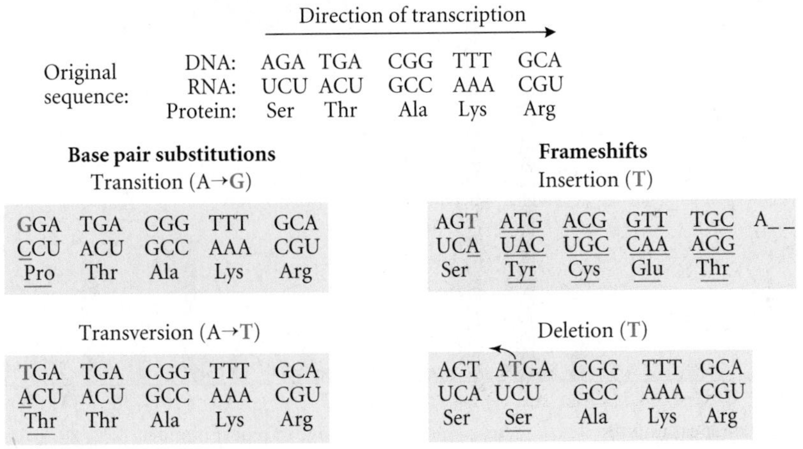

Direction of transcription →

Original sequence:

DNA:	AGA	TGA	CGG	TTT	GCA
RNA:	UCU	ACU	GCC	AAA	CGU
Protein:	Ser	Thr	Ala	Lys	Arg

Base pair substitutions

Transition (A→G)

GGA TGA CGG TTT GCA
CCU ACU GCC AAA CGU
Pro Thr Ala Lys Arg

Transversion (A→T)

TGA TGA CGG TTT GCA
ACU ACU GCC AAA CGU
Thr Thr Ala Lys Arg

Frameshifts

Insertion (T)

AGT ATG ACG GTT TGC A_ _
UCA UAC UGC CAA ACG
Ser Tyr Cys Glu Thr

Deletion (T)

AGT ATGA CGG TTT GCA
UCA UCU GCC AAA CGU
Ser Ser Ala Lys Arg

Figure 8.4 Examples of point mutations and their consequences for mRNA and amino acid sequences. (Only the transcribed, "sense" strand of the DNA is shown.) The boxes at the left show two kinds of base pair substitutions: a transition and a transversion at the first base position. The boxes at the right show two kinds of frameshift mutations. Note that a frameshift mutation affects the triplet series and translation "downstream," i.e., to the right.

ed to single base pair substitutions. A **transition** is a substitution of a purine for a purine (A ↔ G) or a pyrimidine for a pyrimidine (C ↔ T). **Transversions**, of eight possible kinds, are substitutions of purines for pyrimidines or vice versa (A or G ↔ C or T).

Mutations may have a phenotypic effect if they occur in genes that encode ribosomal and transfer RNA, nontranslated regulatory sequences such as enhancers, or protein-coding regions. Because of the redundancy of the genetic code, many mutations in coding regions are **synonymous mutations**, which have no effect on the amino acid sequence of the polypeptide or protein. **Nonsynonymous mutations**, in contrast, result in amino acid substitutions; they may have little or no effect on the functional properties of the polypeptide or protein, and thus no effect on the phenotype, or they may have substantial consequences. For example, a change from the (RNA) triplet GAA to GUA causes the amino acid valine to be incorporated instead of glutamic acid. This is the mutational event that in humans caused the abnormal β-chain in sickle-cell hemoglobin, which has many phenotypic consequences and is usually lethal in homozygotes.

INSERTIONS OR DELETIONS (together called "indels") of DNA into a DNA sequence are another common kind of mutation. Indels may be a single base pair or many. If a single base pair becomes inserted into or deleted from a coding sequence, the triplet reading frame is shifted by one nucleotide, so that downstream triplets are read as different codons and translated into different amino acids (see Figure 8.4). Thus, insertions or deletions often result in **frameshift mutations**. The greatly altered gene product is usually nonfunctional, although exceptions to this generalization are known. Base pair changes occur more frequently than insertions and deletions; for example, they account for more than ten times the number of differences between homologous processed pseudogenes of human and chimpanzee (Nachman and Crowell 2000).

REPLICATION SLIPPAGE is a process that alters the number of short repeats in microsatellites. The growing (3′) end of a DNA strand that is being formed during replication may become dissociated from the template strand and form a loop, so that the next repeat to be copied from the template is one that had already been copied. Thus extra repeats will be formed in the growing strand. Microsatellite "alleles" that differ in copy number arise by replication slippage at a high rate. For instance, the number of tetranucleotide (4 bp) repeats at a microsatellite locus was scored in 1615 offspring of 254 females and 288 males in a long-term study of superb fairy-wrens (*Malurus cyaneus*) in Canberra, Australia (Figure 8.5; Beck et al. 2003). Among the 3230 meioses represented, 45 mutational changes in repeat number were recorded, for a mutation rate of 1.4 percent. This is much higher than the rate of point mutation per base pair, as we will see.

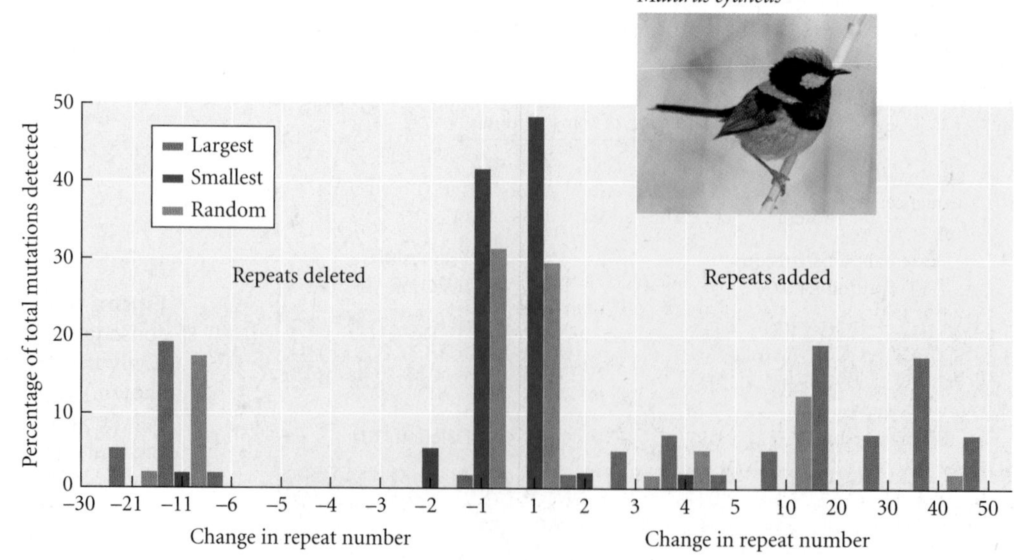

Figure 8.5 Mutational changes in the number of tetranucleotide repeats of a microsatellite in the superb fairy-wren (*Malurus cyaneus*), estimated by comparing repeats in offspring and their parents. Most frequently, only a single repeat was added or deleted. Results are based on alternative assumptions that the progenitor of an altered microsatellite was the largest among the possible parental alleles (red bars), the smallest (blue), or drawn at random (gold). The number of repeats varied from 32 to 93 in this population. (After Beck et al. 2003; photo © Greg Grace/Alamy.)

SEQUENCE CHANGES ARISING FROM RECOMBINATION. Recombination typically is based on precise alignment of the DNA sequences on homologous chromosomes in meiosis. When homologous DNA sequences differ at two or more base pairs, **intragenic recombination** between them can generate new DNA sequences, just as crossing over between genes generates new gene combinations. DNA sequencing has revealed many examples of variant sequences that apparently arose by intragenic recombination.

Recombination appears to be the cause of a peculiar mutational phenomenon called GENE CONVERSION, which has been studied most extensively in fungi. The gametes of a heterozygote should carry its two alleles (A_1, A_2) in a 1:1 ratio. Occasionally, though, they occur in different ratios, such as 1:3. In these cases, an A_1 allele has been replaced specifically by an A_2 allele, rather than by any of the many other alleles to which it might have mutated: it seems to have been converted into A_2. This seems to occur when a damaged DNA strand of one chromosome is repaired by enzymes that insert bases complementary to the sequence on the undamaged homologous chromosome.

Unequal crossing over (unequal exchange) can occur between two homologous sequences or chromosomes that are not perfectly aligned. Recombination then results in a TANDEM DUPLICATION on one recombination product and a deletion on the other (Figure 8.6). The length of the affected region may range from a single base pair to a large block of loci (a segmental duplication), depending on the amount of displacement of the two misaligned chromosomes. Most unequal crossing over occurs between sequences that already include tandem repeats (e.g., ABBC) because the duplicate regions can pair out of register:

$$\begin{pmatrix} \text{ABBC...} \\ \text{...ABBC} \end{pmatrix}$$

This pairing generates further duplications (ABBBC). Unequal crossing over is one of the processes that has generated the extremely high number of copies of nonfunctional sequences that constitute much of the DNA in most eukaryotes. It has given rise to many gene families, and it has been extremely important in the evolution of greater numbers of functional genes and of total DNA (see Chapter 20).

CHANGES CAUSED BY TRANSPOSABLE ELEMENTS. Most TEs produce copies that can move to any of many places in the genome, and sometimes they carry with them other genes near which they had been located. These DNA sequences include genes that encode enzymes that accomplish the transposition (movement). The several kinds of TEs include INSERTION SEQUENCES, which encode only enzymes that cause transposition; TRANSPOSONS, which encode other functional genes as well; and RETROELEMENTS, which carry a gene for the enzyme reverse transcriptase. Retroelements are first transcribed into RNA, which then is reverse-transcribed into a DNA copy (cDNA) that is inserted into the genome. Some retroelements are retroviruses (including the HIV virus that causes AIDS) whose RNA copies can cross cell boundaries.

TEs often become excised from a site into which they had previously been inserted, but leave behind sequence fragments that tell of their former presence. From such cases and from more direct evidence, TEs are known to have many effects on genomes (Bennetzen 2000; Kazazian 2004):

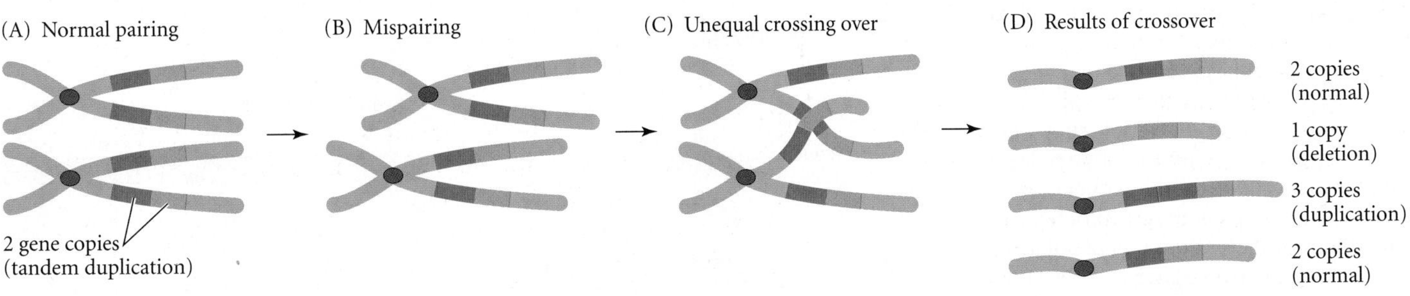

Figure 8.6 Unequal crossing over occurs most commonly when two repeated genes or sequences mispair with their homologues. Crossing over then yields a deletion of one of the sequences on one chromatid, and a duplication of the sequences (and hence three copies of that sequence) on the other chromatid. The duplicated sequence may encode part of a gene, or it may encode one or more complete genes. (After Hartl and Jones 2001.)

- When inserted into a coding region, they alter, and usually destroy, the function of the protein, often by causing a frameshift or by altering splicing patterns.
- When inserted into or near control regions, they can interfere with or alter gene expression (e.g., the timing or amount of transcription).
- They are known to increase mutation rates of host genes.
- They can cause rearrangements in the host genome, resulting from recombination between two copies of a TE located at different sites (Figure 8.7). Just as unequal crossing over between members of a gene family can generate duplications and deletions, so can recombination between copies of a TE located at nonhomologous sites. Recombination between two copies of a TE with the same sequence polarity can *delete* the region between them, whereas recombination between two copies with opposite polarity *inverts* the region between them.
- TEs that encode reverse transcriptase sometimes insert DNA copies (cDNA) not only of their own RNA but also of RNA transcripts of other genes, into the genome. These cDNA copies of RNA (**retrosequences**) resemble the exons of an ancestral gene located elsewhere in the genome, but they lack control regions and introns. Most retrosequences are processed pseudogenes, which do not produce functional gene products.
- By transposition and unequal crossing over, TEs can increase or decrease in number, and so change the size of the genome. About 10 percent of the human genome is composed of more than one million copies of a retrotransposable element called *Alu*, which are thought to have built up over the course of about 60 Myr (Petrov and Wendel 2006).

Most or all of these TE-induced effects have been observed in experimental populations. In laboratory stocks of *Drosophila melanogaster*, for example, most phenotypically visible mutations that geneticists have identified are caused by transposon insertions. The transposition rate of various retroelements ranges from about 10^{-5} to 10^{-3} per copy in inbred lines of *Drosophila*, resulting in an appreciable rate of mutation (Nuzhdin and Mackay 1994). All the kinds of changes engendered by TEs can be found by comparing genes and genomes of organisms of the same or different species. For instance, *L1* retrotransposon insertions are associated with many disease-causing mutations in both mice and humans (Kazazian 2004), and a difference in flower color between two species of *Petunia* has been caused by the insertion and subsequent incomplete excision of a transposon that disabled a gene that controls anthocyanin pigment production (Quattrocchio et al. 1999).

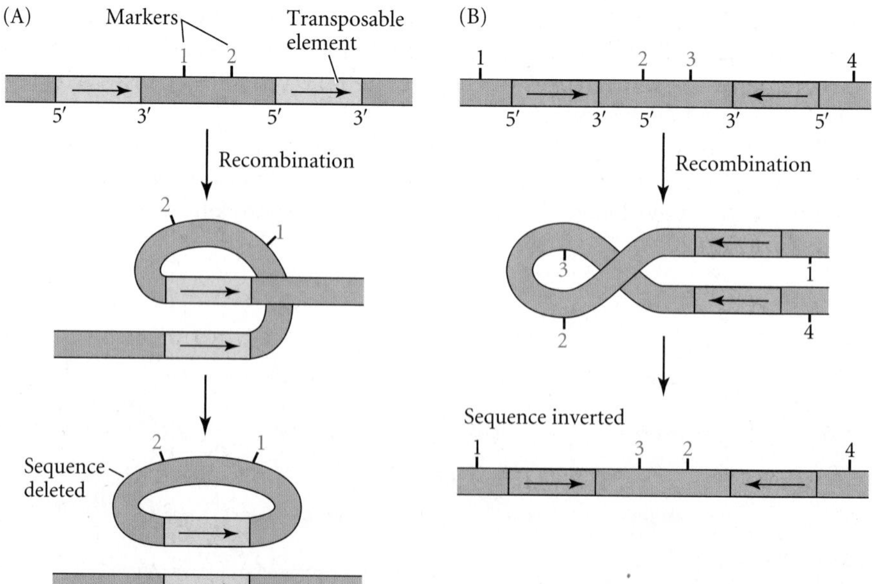

Figure 8.7 Recombination between copies of a transposable element can result in deletions and inversions. The boxes containing arrows represent TEs, with the polarity of base pair sequence indicated by the arrows. The numerals represent genetic markers. (A) Recombination between two direct repeats (i.e., with the same polarity) excises one repeat and deletes the sequence between the two copies. (B) Recombination between two inverted repeats (with opposite polarity) inverts the sequence between them. (After Lewin 1985.)

Examples of mutations

Geneticists have learned an enormous amount about the nature and causes of mutations by studying model organisms such as *Drosophila* and the bacterium *Escherichia coli*. Moreover, many human mutations have been characterized because of their effects on health. Human mutations are usually rather rare variants that can be compared with normal forms of the gene; in some instances, newly arisen mutations have been found that are lacking in both of a patient's parents.

Single base pair substitutions are responsible for conditions such as sickle-cell hemoglobin, described earlier, and for precocious puberty, in which a single amino acid change in the receptor for luteinizing hormone causes a boy to show signs of puberty when about 4 years old. Because many different alterations of a protein can diminish its function, the same phenotypic condition can be caused by many different mutations of a gene. For example, cystic fibrosis, a fatal condition afflicting 1 in 2500 live births in northern Europe, is caused by mutations in the gene encoding a sodium channel protein. The most common such mutation is a 3 bp deletion that deletes a single amino acid from the protein; another converts a codon for arginine into a "stop" codon; another alters splicing so that an exon is missing from the mRNA; and many of the more than 500 other base pair substitutions recorded in this gene are also thought to cause the disease (Zielenski and Tsui 1995). Mutations in any of the many different genes that contribute to the normal development of some characteristics can also result in similar phenotypes. For example, retinitis pigmentosa, a degeneration of the retina, can be caused by mutations in genes on 8 of the 23 chromosomes in the (haploid) human genome (Avise 1998).

Hemophilia can be caused by mutations in two different genes that encode blood-clotting proteins. In both genes, many different base pair substitutions, as well as small deletions and duplications that cause frameshifts, are known to cause the disease, and about 20 percent of cases of hemophilia-A are caused by an inversion of a long sequence within one of the genes (Green et al. 1995). Huntington's disease, a fatal neurological disorder that strikes in midlife, is caused by an excessive number of repeats of the sequence CAG: the normal gene has 10 to 30 repeats, the mutant gene more than 75. Unequal crossing over between the two tandemly arranged genes for α-hemoglobin (see Figure 8.3) has given rise to variants with three tandem copies (duplication) and with one (deletion). The deletion of one of the loci causes α-thalassemia, a severe anemia. Another case of deletion, which results in high cholesterol levels, is the lack of exon 5 in a low-density lipoprotein gene. This deletion has been attributed to unequal crossing over, facilitated by a short, highly repeated sequence called *Alu* that is located in the introns of this gene and in many other sites in the genome (Figure 8.8).

These examples might make it seem as if mutations are nothing but bad news. While this is close to the truth—far more mutations are harmful than helpful—these mutations represent a biased sample. Many advantageous mutations have become incorporated into

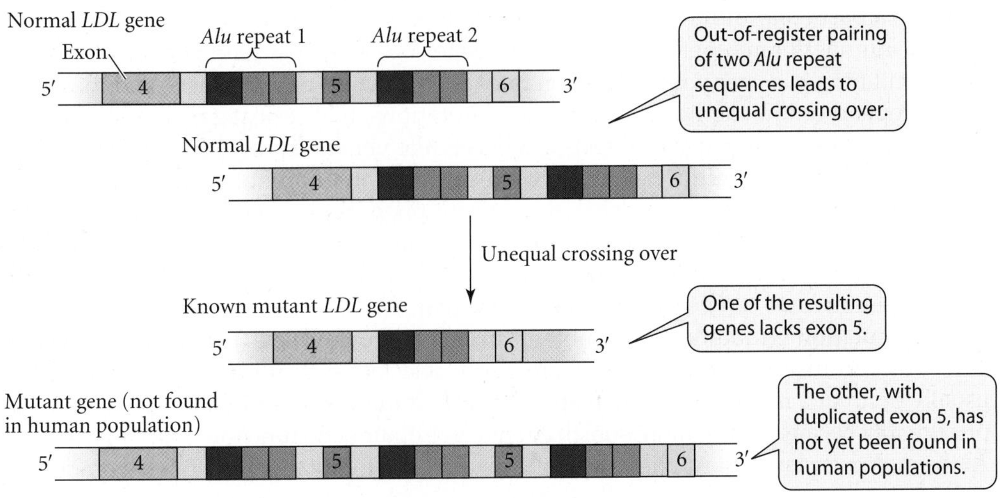

Figure 8.8 A mutated low-density lipoprotein (*LDL*) gene in humans lacks exon 5. It is believed to have arisen by unequal crossing over between two normal gene copies, as a result of out-of-register pairing between two of the repeated sequences (*Alu*, shown as blue boxes) in the introns. The numbered boxes are exons. (After Hobbs et al. 1986.)

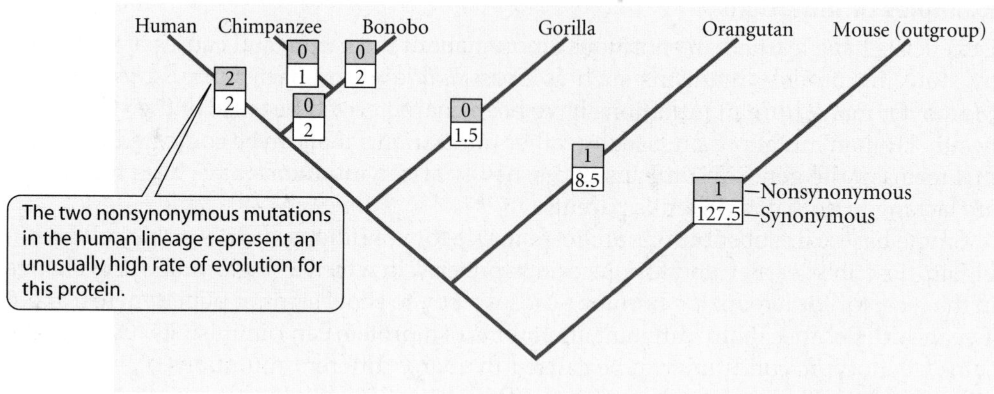

Figure 8.9 A phylogeny of the Hominoidea and its divergence from an outgroup, the mouse. Each box shows the number of nonsynonymous (yellow boxes) and synonymous (white boxes) substitutions in the *FOXP2* gene. The two nonsynonymous substitutions in the human lineage represent an unusually high rate of evolution of the FOXP2 protein, and may represent mutations that have been important in the evolution of language and speech. (After Zhang et al. 2002.)

species' genomes (fixed) and thus represent the current **wild-type**, or normal, genes. We will encounter many examples of beneficial mutations, and will only remark here that they may include not only point mutations but also some more complex genomic alterations. For example, most genes that have arisen by reverse transcription from mRNA are nonfunctional pseudogenes, but at least one has been found that is a fully functional member of the human genome. Phosphoglycerate kinase is encoded by two genes. One, on the X chromosome, has a normal structure of 11 exons and 10 introns. The other, on an autosome, lacks introns, and clearly arose from the X-linked gene by reverse transcription. It is expressed only in the testes, a novel pattern of tissue expression that seems to compensate for the lower level of transcription of the X-linked gene in this tissue (McCarrey et al. 1996).

When biologists seek those genes that have been involved in the evolution of a specific characteristic, they often use rare deleterious mutations of the kind described here as indicators of CANDIDATE GENES, those that may be among the genes they seek. For example, a rare mutation in the human *FOXP2* gene (*forkhead box 2*, which encodes a transcription factor) causes severe speech and language disorders. Two research groups, led by Jianzhi Zhang (Zhang et al. 2002) and Svante Pääbo (Enard et al. 2002), independently found that this gene has undergone two nonsynonymous (amino acid–changing) substitutions in the human lineage since the divergence of the human and chimpanzee lineages less than 7 Mya. This is a much higher rate of protein evolution than would be expected, considering that only one other such substitution has occurred between these species and the mouse, which diverged almost 90 Mya (Figure 8.9). Both research groups propose that these substitutions occurred in the human lineage less than 200 Kya and that they are among the important steps in the evolution of human language and speech.

Rates of mutation

Recurrent mutation refers to the repeated origin of a particular mutation, and the *rate* at which a particular mutation occurs is typically measured in terms of recurrent mutation: the number of independent origins per gene copy (e.g., per gamete) per generation or per unit time (e.g., per year). Mutation rates are estimates, not absolutes, and these estimates depend on the method used to detect mutations. In classical genetics, a mutation was detected by its phenotypic effects, such as white versus red eyes in *Drosophila*. Such a mutation, however, might be caused by the alteration of any of many sites within a locus; moreover, many base pair changes have no phenotypic effect. Thus phenotypically detected rates of mutation underestimate the rate at which all mutations occur at a locus. With modern molecular methods, mutated DNA sequences can be detected directly, so mutation rates can be expressed per base pair.

Back mutation is mutation of a "mutant" allele back to the allele (usually the wild type) from which it arose. Back mutations are ordinarily detected by their phenotypic effects. They usually occur at a much lower rate than "forward" mutations (from wild type to mutant), presumably because many more substitutions can impair gene function than can restore it. At the molecular level, most phenotypically detected back mutations are not restorations of the original sequence, but instead result from a second amino acid substitution, either in the same or a different protein, that restores the function that had been altered by the first mutation. Advantageous mutations arose and compensated for severely deleterious mutations within 200 generations in experimental populations of *E. coli* (Moore et al. 2000).

ESTIMATING MUTATION RATES. Rates of mutation are estimated in several ways that assess the rate averaged over different time scales (Drake et al. 1998; Houle and Kondrashov 2006). In the short term, mutations can be screened among offspring of genetically characterized parents or over several generations. This is usually done in the laboratory, but occasionally in field populations, as in the case of the study of fairy-wrens cited earlier. On a somewhat longer time scale are mutation-accumulation experiments, in which new mutations arising in a laboratory stock (which is usually initially homozygous) are scored after several generations. Because mutation rates per gene are usually very low, reasonably good estimates of the mutation rate require that very large numbers of progeny be scored. Thus many of these experiments use very fecund or very rapidly reproducing organisms, such as fruit flies or, especially, microorganisms.

There is a risk of underestimating the number of mutations if some cause death of their carriers and thus do not persist long enough to be counted. This is, of course, natural selection in action, so special efforts are needed to design experiments to minimize or control for the effects of natural selection.

An indirect method that estimates mutation rates averaged over many generations (Box 8A) is based on the number of base pair differences between homologous genes in different species, relative to the number of generations that have elapsed since they diverged from their common ancestor. This method depends on the neutral theory of molecular

BOX 8A Estimating Mutation Rates from Comparisons among Species

In Chapter 10, we will describe the *neutral theory of molecular evolution*. This theory describes the fate of purely neutral mutations—that is, those mutations that neither enhance nor lower fitness. One possible fate is that a mutation will become fixed—that is, attain a frequency of 1.0—entirely by chance. The probability that this event will occur equals u, the rate at which neutral mutations arise. In each generation, therefore, the probability is u that a mutation that occurred at some time in the past will become fixed. After the passage of t generations, the fraction of mutations that will have become fixed is therefore ut.

If two species diverged from a common ancestor t generations ago, the expected fraction of fixed mutations in both species is $D = 2ut$, since various mutations have become fixed in both lineages. If the mutations in question are base pair changes, a fraction $D = 2ut$ of the base pairs of a gene should differ between the species, assuming that all base pairs are equally likely to mutate. Thus the average mutation rate per base pair per generation is $u = D/2t$.

Thus we can estimate u if we can measure the fraction of base pairs in a gene that differ between two species (D), and if we can estimate the number of generations since the two species diverged from their common ancestor (t). This requires an estimate of the length of a generation, information on the absolute time at which the common ancestor existed (based on a fossil record or perhaps a geologically timed vicariance event, as described in Chapter 6), and an understanding of the phylogenetic relationships among the living and fossilized taxa.

In applying this method to DNA sequence data, it is necessary to assume that most base pair substitutions are neutral and to correct for the possibility that earlier substitutions at some sites in the gene have been replaced by later substitutions ("multiple hits"). Uncertainty about the time since divergence from the common ancestor is usually the greatest source of error in estimates obtained by this method.

The best estimates of mutation rates at the molecular level have been obtained from interspecific comparisons of pseudogenes, other nontranslated sequences, and fourfold-degenerate third-base positions (those in which all mutations are synonymous), since these are thought to be least subject to natural selection (although probably not entirely free of it). In comparisons among mammal species, the average rate of nucleotide substitution has been about 2.2 per nucleotide site per 10^9 years, for a mutation rate of 2.2×10^{-9} per site per year (Kumar and Subramanian 2002). If the average generation time were 2 years during the history of the lineages studied, the average rate of mutation per site would be about 1.1×10^{-9} per generation. As is described in the main text, comparison of human and chimpanzee sequences yielded an estimate of 2.5×10^{-8} per site per generation, and an average of 175 new mutations in each newborn person's genome.

Table 8.1 ***Spontaneous mutation rates of specific genes, detected by phenotypic effects***

Species and locus	*Mutations per 100,000 cells or gametes*
Escherichia coli	
Streptomycin resistance	0.00004
Resistance to T1 phage	0.003
Arginine independence	0.0004
Salmonella typhimurium	
Tryptophan independence	0.005
Neurospora crassa	
Adenine independence	0.0008–0.029
Drosophila melanogaster	
Yellow body	12
Brown eyes	3
Eyeless	6
Homo sapiens	
Retinoblastinoma	1.2–2.3
Achondroplasia	4.2–14.3
Huntington's disease	0.5

Source: After Dobzhansky 1970.

evolution, described in Chapter 10, which specifies that the per-generation rate of origin of selectively neutral mutations, per base pair, equals the proportion of base pairs that differ between two species, divided by twice the number of generations since their common ancestor. "Selectively neutral" mutations are those that do not alter their carriers' survival or reproductive success.

Mutation rates vary among genes and even among regions within genes, but on average, as measured by phenotypic effects, a locus mutates at a rate of about 10^{-6} to 10^{-5} mutations per gamete per generation (Table 8.1). The average mutation rate per base pair has been estimated at about 10^{-10} per replication in prokaryotes such as *E. coli* (Table 8.2) and about 10^{-9} per sexual generation in eukaryotes, based mostly on the indirect method of comparing DNA sequences of different species. For example, Michael Nachman and Susan Crowell (2000) sequenced 18 processed pseudogenes in two humans and one chimpanzee. Among 16,089 base pairs sequenced in each individual, they found 199 differences, of which 66 percent were transitions, 26 percent were transversions, and 8 percent were insertion-deletion variants. Averaging over several estimates of divergence time and length of a generation, Nachman and Crowell estimated the mutation rate to range between 1.3×10^{-8} and 3.4×10^{-8} per site per generation, and suggested a best estimate of 2.5×10^{-8}. A broader study, of changes at fourfold-degenerate third-base sites in 17,208 protein-coding DNA sequences representing 5669 different genes from 326 species of mammals, found considerable rate variation among genes, but the average mutation rate, 2.2×10^{-9} per base pair per year, was surprisingly similar among mammal lineages that differ greatly in generation time (Kumar and Subramanian 2002).

EVOLUTIONARY IMPLICATIONS OF MUTATION RATES. With such a low mutation rate per locus, it might seem that mutations occur so rarely that they cannot be important. However, summed over all genes, the input of variation by mutation is considerable. The diploid human genome has 7×10^9 bp, so if the mutation rate is 2.5×10^{-8} per bp per gen-

Table 8.2 ***Estimates of spontaneous mutation rates per base pair and per genome***

	Base pairs		*Mutation rate*			
Organism	*in haploid genome*	*in effective genome*[a]	*per base pair per replication*	*per replication per haploid genome*	*per replication per effective genomea*	*per sexual generation per effective genome*[b]
T2, T4 phage	1.7×10^5	—	2.4×10^{-8}	0.0040	—	—
Escherichia coli	4.6×10^6	—	5.4×10^{-10}	0.0025	—	—
Saccharomyces cerevisiae (yeast)	1.2×10^7	—	2.2×10^{-10}	0.0027	—	—
Neurospora crassa (bread mold)	4.2×10^7	—	7.2×10^{-11}	0.0030	—	—
Caenorhabditis elegans	8.0×10^7	1.8×10^7	2.3×10^{-10}	0.018	0.004	0.036
Drosophila melanogaster	1.7×10^8	1.6×10^7	3.4×10^{-10}	0.058	0.005	0.14
Mouse	2.7×10^9	8.0×10^7	1.8×10^{-10}	0.49	0.014	0.9
Human	3.2×10^9	8.0×10^7	5.0×10^{-11}	0.16	0.004	1.6

Source: After Drake et al. 1998.

[a] The effective genome is the number of base pairs in functional sequences that could potentially undergo mutations that reduce fitness.

[b] Calculated for multicellular organisms in which multiple DNA replication events occur in development between zygote and gametogenesis.

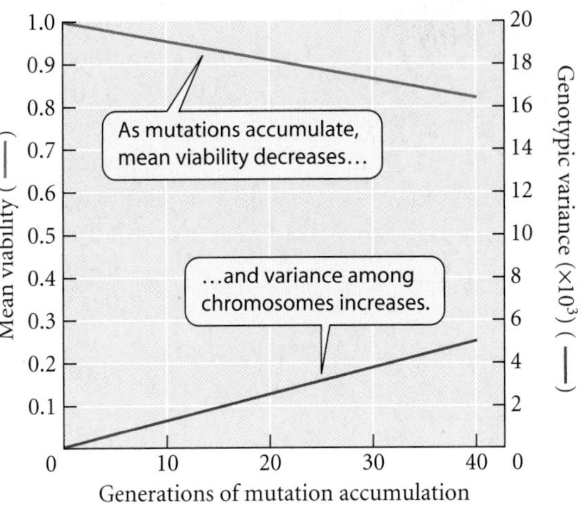

Figure 8.10 Effects of the accumulation of spontaneous mutations on the egg-to-adult survival of *Drosophila melanogaster*. The mean viability of flies made homozygous for chromosome 2 carrying new recessive mutations decreased, and the variation (variance) among those chromosomes increased. The rate of mutation was estimated from these data. (After Mukai et al. 1972.)

eration, as Nachman and Crowell estimate, then a zygote (or baby) carries about 175 new mutations on average. If only 2.5 percent of the genome consists of functional, transcribed sequences, about 4.3 of these mutations have the potential to alter proteins and affect phenotypic characters. So, in a population of 500,000 humans, more than 2 million new mutations with potential effects arise every generation. If even a tiny fraction of these mutations were advantageous, the amount of new "raw material" for adaptation would be substantial, especially over the course of thousands or millions of years. If the mutation rate is closer to 10^{-9} per base pair, the input of new mutations is still very high.

Mutation-accumulation experiments with *Drosophila* and the nematode worm *Caenorhabditis elegans* have confirmed that the total mutation rate per gamete is quite high. The first major experiment was performed by Terumi Mukai and colleagues (1972), who counted more than 1.7 million flies in order to estimate the rate at which the chromosome 2 accumulates mutations that affect egg-to-adult survival (VIABILITY). They used crosses (see Figure 9.10) in which copies of the wild-type chromosome 2 were carried in a heterozygous condition so that deleterious recessive mutations could persist without being eliminated by natural selection. Every 10 generations, they performed crosses that made large numbers of these chromosomes homozygous and measured the proportion of those chromosomes that reduced viability. The mean viability declined, and the variation (variance) among chromosomes increased steadily (Figure 8.10). From the changes in the mean and variance, Mukai et al. calculated a mutation rate of about 0.15 per chromosome 2 per gamete. This is the sum, over all loci on the chromosome, of mutations that affect viability. Because chromosome 2 carries about a third of the *Drosophila* genome, the total mutation rate is about 0.50 per gamete. Thus almost every zygote carries at least one new mutation that reduces viability. Subsequent studies have indicated that the mutation rate for *Drosophila* is at least this high, and that it reduces viability by 1 to 2 percent per generation (Lynch et al. 1999; Haag-Liautard et al. 2007). Indirect estimates indicate that humans likewise suffer about 1.6 new mutations per zygote that reduce survival or reproduction (Eyre-Walker and Keightley 1999).

Mutation rates vary among genes and chromosome regions, and they are also affected by environmental factors. MUTAGENS (mutation-causing agents) include ultraviolet light, X-rays, and a great array of chemicals, many of which are environmental pollutants. For example, mutation rates in birds and mice are elevated in industrial areas, and mice exposed to particulate air pollution in an urban-industrial site showed higher rates of mutation in repetitive elements than mice exposed only to filtered air at that site, or mice placed in a rural location (Figure 8.11). The worst recorded nuclear accident occurred in 1986, when the Chernobyl nuclear power plant in Ukraine (then part of the Soviet Union) exploded. Since then, studies have shown a much higher incidence of mutations in swallows from that region than elsewhere (Ellegren et al. 1997), as well as in the children of fathers who were exposed to the radiation (Dubrova et al. 1996).

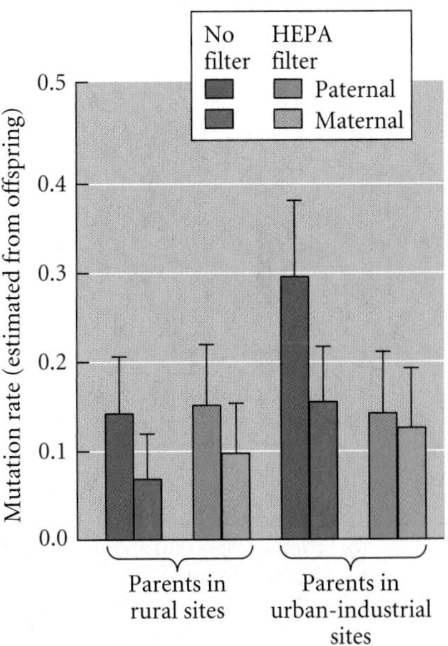

Figure 8.11 Mutation rates in mice, estimated from DNA sequences of two loci in their offspring. The mice were placed for 10 weeks in rural sites or in urban-industrial sites near steel mills and a major highway, where they were exposed either directly to the air or to air passed through a HEPA filter. Exposure to unfiltered urban-industrial air increased the mutation rate. (After Somers et al. 2004.)

As we will describe in Chapters 9 and 13, the variation in most phenotypic characters is **polygenic**: it is based on several or many different genetic loci. It is difficult to single out any of these loci to study the mutation rate per locus, but it is easy to estimate the **mutational variance** of the character—the increased variation in a population caused by new mutations in each generation. Studies of traits such as bristle number in *Drosophila* have shown that the mutational variance is high enough that an initially homozygous population would take only about 500 generations to achieve the level of genetic variation generally found in a natural population. The magnitude of the mutational variance varies somewhat among characters and species (Lynch 1988).

In summary, although any given mutation is a rare event, the rate of origination of new genetic variation in the genome as a whole, and for individual polygenic characters, is appreciable. However, mutation alone does not cause a character to evolve from one state to another, because the rate of mutation is too low. Suppose that alleles A_1 and A_2 determine alternative phenotypes (e.g., red versus purple) of a haploid species, that half the individuals carry A_1, and that the rate of recurrent mutation from A_1 to A_2 is 10^{-5} per gene per generation. In one generation, the proportion of A_2 genes will increase from 0.5 to $0.5 + [(0.5)(10^{-5})] = 0.50000495$. At this rate, it will take about 70,000 generations before A_2 constitutes 75 percent of the population, and another 70,000 generations before it reaches 87 percent. This rate is so slow that, as we will see, factors other than recurrent mutation usually have a much stronger influence on allele frequencies, and thus are responsible for whatever evolutionary change occurs.

Phenotypic effects of mutations

A mutation may alter one or more phenotypic characters, such as size, coloration, or the amount or activity of an enzyme. Alterations in such features may affect survival and/or reproduction, the major components of FITNESS (see Chapter 11). It is often convenient to distinguish between a mutation's effects on fitness and on other characters, even though they are connected.

The phenotypic effects of mutational changes in DNA sequence range from none to drastic. At one extreme, synonymous base pair changes are expected to have no evident phenotypic effect, and this is also apparently true of many amino acid substitutions, which seem not to affect protein function. The phenotypic effects of mutations that contribute to polygenic traits, such as bristle number in *Drosophila*, range from slight to substantial; in one study, mutations induced by insertion of transposable elements altered the number of abdominal bristles by about 0.9 bristle on average (Figure 8.12).

Even single base pair changes may have have novel, beneficial effects. A remarkable example is a mutation in the sheep blowfly (*Lucilia cuprina*) that confers resistance to the organophosphate insecticides that have been used to control this serious pest, and which has become prevalent in some blowfly populations. The resistance is attributable to a single mutation that substitutes aspartic acid for glycine at one position in the active site of a carboxylesterase, abolishing the enzyme's esterase activity and transforming it into an organophosphate hydrolase (Newcomb et al. 1997). Among the most fascinating mutations are those in the "master control genes" that regulate the expression of other genes in developmental pathways. (We will discuss these genes in detail in Chapter 21.) HOMEOTIC SELECTOR GENES, for example, determine the basic body plan of an organism, conferring a distinct identity on each segment of the developing body by producing DNA-binding proteins that regulate other genes that determine the features of each such segment. These genes derive their name from **homeotic mutations** in *Drosophila*, which redirect the development of one body segment into another. Mutations in the *Antennapedia*

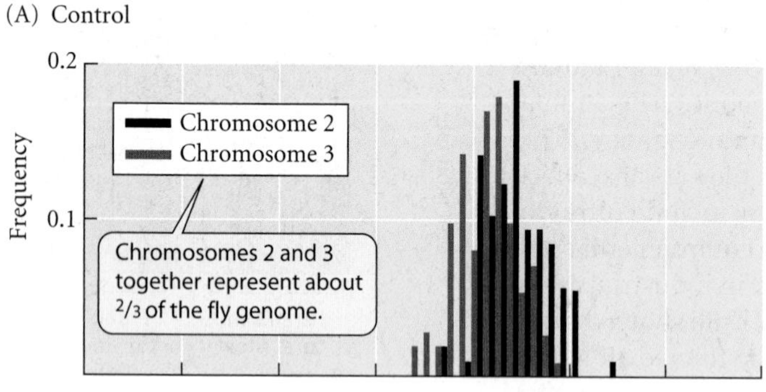

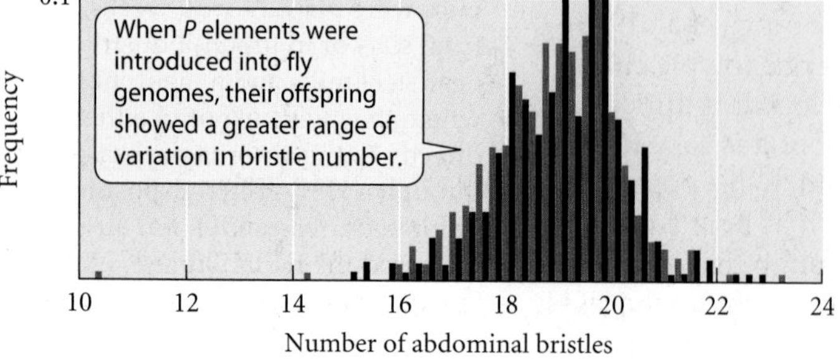

Figure 8.12 The frequency distribution of the number of abdominal bristles in (A) 392 homozygous control lines of *Drosophila melanogaster* and (B) 1094 homozygous experimental lines in which researchers used transposable elements (*P* elements) to induce mutations in chromosome 2 or chromosome 3. The mutations both increased and decreased bristle numbers compared with the control lines. (From Lyman et al. 1996.)

(A)

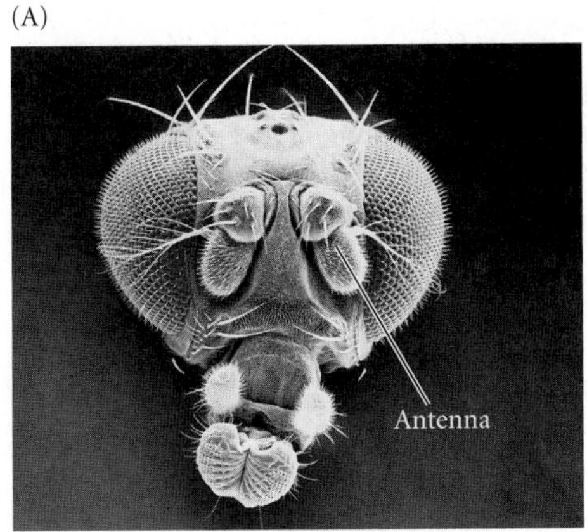

(B)

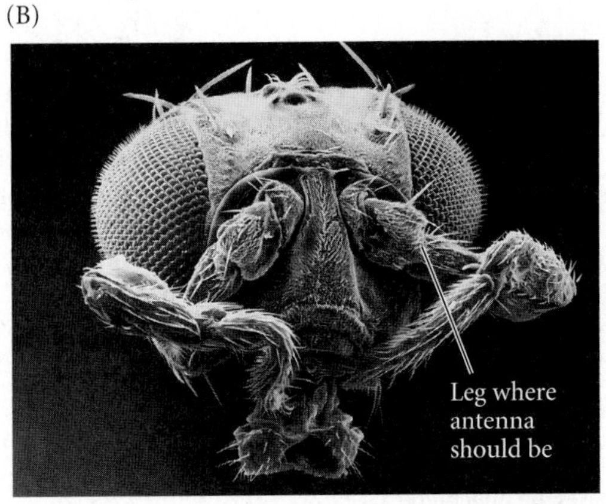

Figure 8.13 The drastic phenotypic effect of homeotic mutations that switch development from one pathway to another. (A) Frontal view of the head of a wild-type *Drosophila melanogaster*, showing normal antennae and mouthparts. (B) Head of a fly carrying the *Antennapedia* mutation, which converts antennae into legs. The *Antennapedia* gene is part of a large complex of Hox genes that confer identity on segments of the body (see Chapter 21). (Photos courtesy of F. R. Turner.)

gene complex, for example, cause legs to develop in place of antennae (Figure 8.13). Another master control gene, *Pax6*, switches on about 2500 other genes required for eye development in mammals, insects, and many other animals (Gehring and Ikeo 1999). Mutations in this gene cause malformation or loss of eyes.

Development of morphological features and many physiological functions depend on regulation of gene expression, usually by altering rates of transcription, and changes in gene regulation have been extremely important in evolution. Regulatory sequences such as enhancers undergo mutation and evolution, but these have so far been hard to study. However, differences in the abundance of RNA transcripts of a gene indicate differences in its regulation, and these can be identified on microarrays that assay the abundance of the transcripts of thousands of genes in a sample of tissue. This approach has been used to screen for changes in gene regulation in mutation-accumulation lines of *Drosophila*, *C. elegans*, and the yeast *Saccharomyces cerevisiae* (Landry et al. 2007). For example, Scott Rifkin and colleagues (2005) found mutational variance for level of gene expression in 4658 out of 11,798 genes assayed in 12 initially identical inbred lines of *Drosophila melanogaster* that had accumulated mutations for 200 generations (Figure 8.14). The study of mutation and variation in gene expression has just begun, and is an important new direction for evolutionary research.

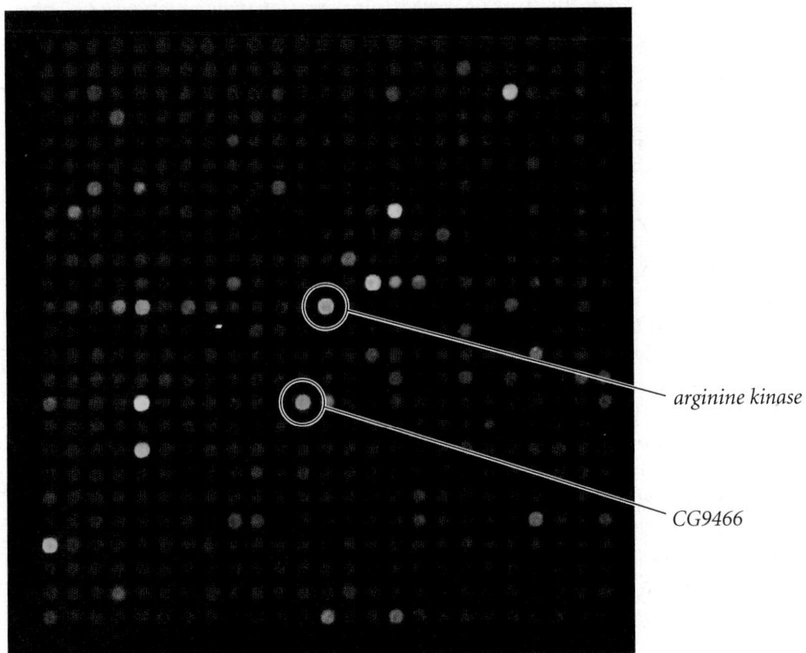

Figure 8.14 A microarray shows differences in gene expression that have arisen by spontaneous mutation in two inbred lines of *Drosophila melanogaster*. The figure shows a 25 × 25 block of a much larger array. Each position shows the difference in the level of expression, based on mRNA levels, of a single gene. Genes that are more highly expressed in line 71, due to mutation, are in red (such as the *arginine kinase* gene in row 12, column 13); those that are more highly expressed in line 51 are in green (such as the *CG9466* gene, which is involved in mannose metabolism). Genes that are expressed in both lines are in yellow. (Courtesy of Scott A. Rifkin.)

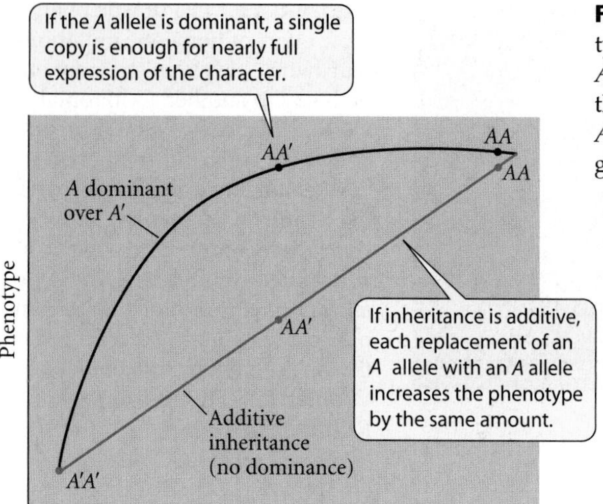

Figure 8.15 Two of the possible relationships between phenotype and genotype at a single locus with two alleles. If inheritance is additive, replacing each *A′* allele with an *A* allele steadily increases the amount of gene product, and the phenotype changes accordingly. If *A* is dominant over *A′*, the phenotype of *AA′* nearly equals that of *AA* because the single dose of *A* produces enough gene product for full expression of the character.

Dominance describes the effect of an allele on a phenotypic character when it is paired with another allele in the heterozygous condition. A fully dominant allele (say, A_1) produces nearly the same phenotype when heterozygous (A_1A_2) as when homozygous (A_1A_1), and its partner allele (A_2) in that instance is fully **recessive**. All degrees of INCOMPLETE DOMINANCE, measured by the degree to which the heterozygote resembles one or the other homozygote, may occur. Inheritance is said to be **additive** if the heterozygote's phenotype is precisely intermediate between those of the homozygotes. For example, A_1A_1, A_1A_2, and A_2A_2 may have phenotypes 3, 2, and 1, respectively; the effects of replacing each A_2 with an A_1 simply add up. LOSS-OF-FUNCTION mutations, in which the activity of a gene product is reduced, are often at least partly recessive, whereas dominant mutations often have enhanced gene product activity (Figure 8.15).

Many mutations are **pleiotropic**, meaning that they affect more than one character. For instance, new mutations that affected reproductive output early in life in mutation-accumulation lines of *C. elegans* also affected reproductive output later in life (Figure 8.16A). In a study of the genetic basis of differences between two geographic populations of monkeyflower (*Mimulus guttatus*), many genes ("quantitative trait loci," as described in Chapter 13) affected two or more traits (Figure 8.16B). In later chapters we will note some of the important evolutionary consequences of pleiotropy.

Effects of mutations on fitness

The effects of new mutations on survival and reproduction (i.e., fitness) may range from highly advantageous to highly disadvantageous. Undoubtedly, many mutations are neutral, or nearly so, having very slight effects on fitness (see Chapter 10). The *average*, or net, effect of those that do affect fitness is deleterious. Mutations that are identified by their visible phenotypic effects often have deleterious pleiotropic effects. For example, some mutations that affect *Drosophila* bristle number also disrupt the development of the nervous system

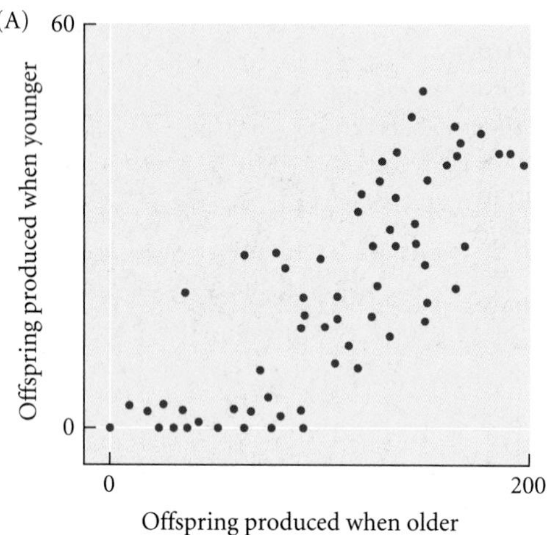

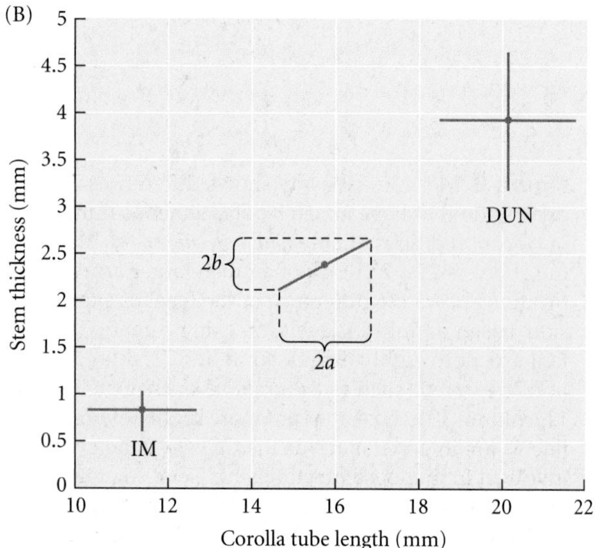

Figure 8.16 Two examples of pleiotropy. (A) In different cultures of the nematode *C. elegans*, each represented by a point, mutations accumulated that affected the number of offspring produced at young and at older ages similarly. (B) The mean stem thickness of monkeyflower plants from two geographic populations (DUN and IM) is plotted against the mean length of the corolla (petal) tube. The effects of one of the loci that contributes to these differences are represented by the central point. The position of this point (halfway between the parent means with respect to both features) shows that the locus affects both characters additively. The terms 2*a* and 2*b* measure the difference, with respect to both features, between homozygotes for the DUN population's allele and the IM population's allele. This locus may be a single gene or a combination of very closely linked genes. The crossbars on the parental means indicate standard deviations. (A after Estes et al. 2005; B after Hall et al. 2006.)

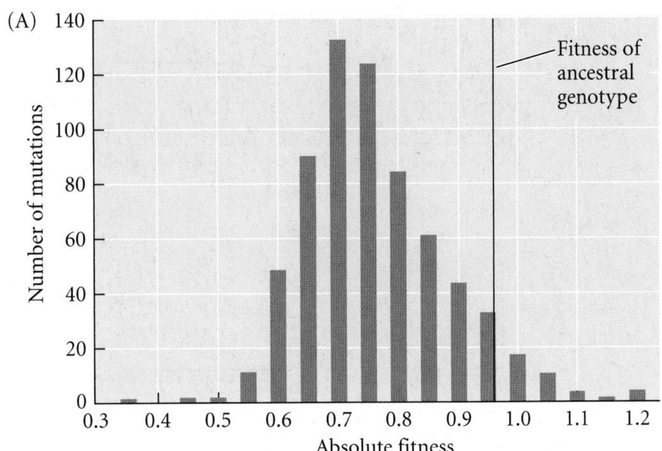

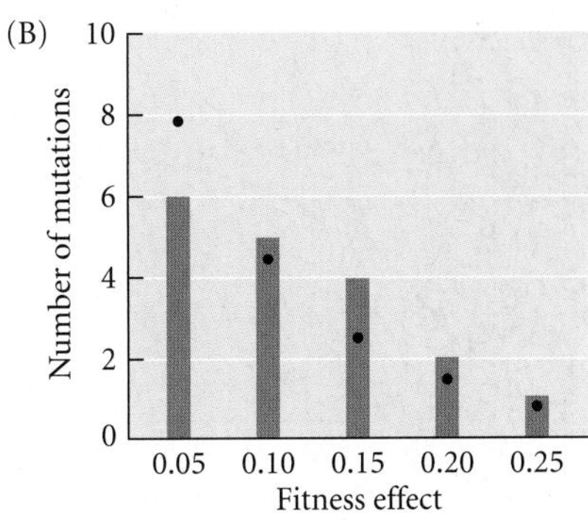

Figure 8.17 The distribution of fitness effects of 665 newly arisen mutations in laboratory cultures of the bacterium *Pseudomonas fluorescens*. (A) Fitness effects of all mutations. The solid vertical line marks the fitness of the ancestral genotype. (B) The fitness effects of the 28 beneficial mutations shown to the right of the vertical line in (A). The frequency distribution is approximately exponential, fitting expectations from theory (shown by dots) fairly well. (After Kassen and Bataillon 2006.)

and reduce the viability of larvae, which do not have bristles (Mackay et al. 1992). The net deleterious effect of mutation is evident in mutation-accumulation experiments such as Mukai's *Drosophila* experiment, described above, in which mean viability declined (see Figure 8.10). Among single mutations isolated in experimental cultures of the bacterium *Pseudomonas fluorescens*, a few slightly enhanced fitness, some greatly decreased it, and the majority had small deleterious effects (Figure 8.17). The beneficial mutations had an approximately exponential distribution of fitness effects.

The frequency distribution of mutational effects is not fixed, for the fitness consequences of many mutations depend on the population's environment and even on its existing genetic constitution. For example, the decline of fitness resulting from new mutations in some experimental *Drosophila* populations was more than 10 times greater if the flies were assayed under crowded, competitive conditions than under noncompetitive conditions (Shabalina et al. 1997).

Evolution would not occur unless some mutations were advantageous. Many experiments that demonstrate advantageous mutations have been done with microorganisms such as phage, bacteria, and yeast because of their short generation times and the ease with which huge populations can be cultured (Dykhuizen 1990; Elena and Lenski 2003; Elena and Sanjuán 2007).

Because bacteria can be frozen (during which time they undergo no genetic change) and later revived, samples taken at different times from an evolving population can be stored, and their fitness can later be directly compared. The fitness of a bacterial genotype is defined as its rate of increase in numbers relative to that of another genotype with which it competes in the same culture, but which bears a genetic marker and so can be distinguished from it. Suppose, for example, that a culture is begun with equal numbers of genotypes A and B, and that after 24 hours B is twice as abundant as A. If the bacteria have grown for x generations, each initial cell has produced 2^x descendants. Thus, if genotypes A and B have grown at the respective rates of 2^5 and 2^6 (i.e., B has produced one more generation per 24 hours), their relative numbers are 32:64, or 1:2. The relative fitnesses of the genotypes—that is, their relative growth rates—are measured by their rates of cell division per day: namely, 5:6, or 1.0:1.2. If the genotypes had the same growth rates—say, 2^5—both would increase in number, but their fitnesses would be equal.

Richard Lenski and colleagues used this method to trace the increase of fitness in populations of *E. coli* for an astonishing 20,000 generations. Each population was initiated with a single individual and was therefore genetically uniform at the start. Nevertheless, fitness increased substantially—rapidly at first, but at a decelerating rate later (Figure 8.18A). In a similar experiment (Bennett et al. 1992), *E. coli* populations adapted rapidly to several different temperatures (Figure 8.18B,C,D).

How frequently do beneficial mutations arise, and how advantageous are they? Because *E. coli* is mostly asexual, an advantageous mutation (call it x) that "sweeps" to fixation and replaces the previously common allele is accompanied by any other muta-

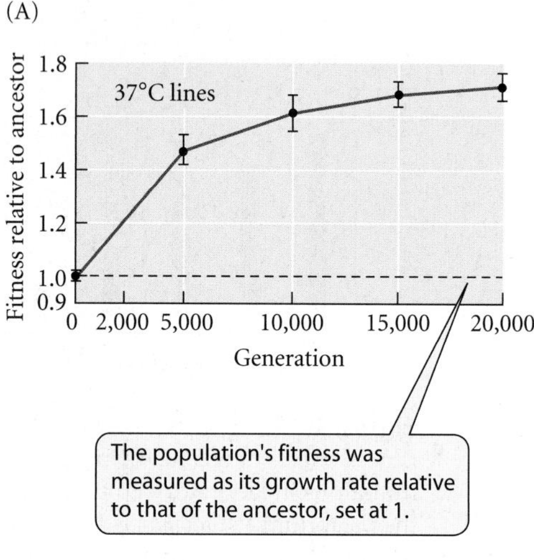

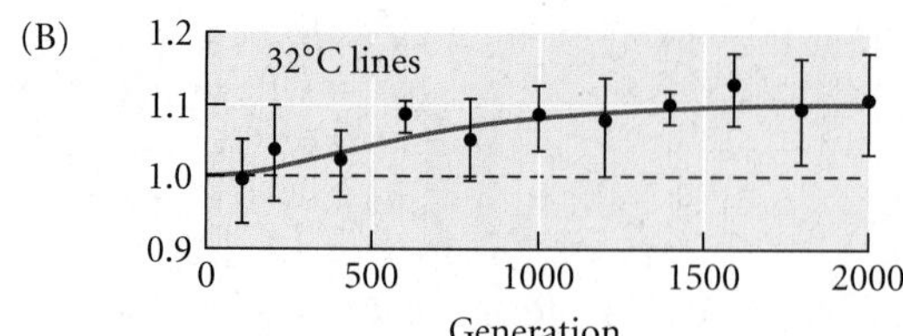

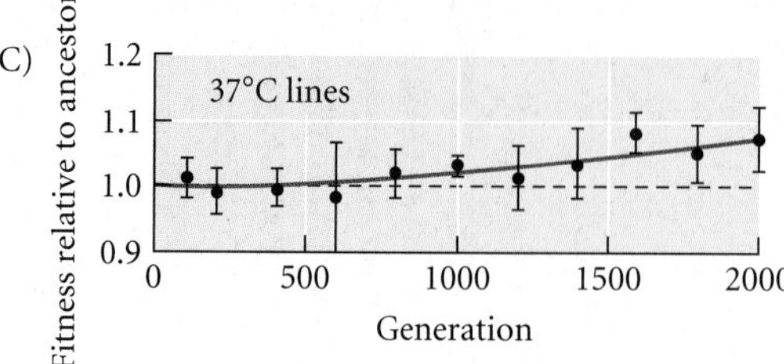

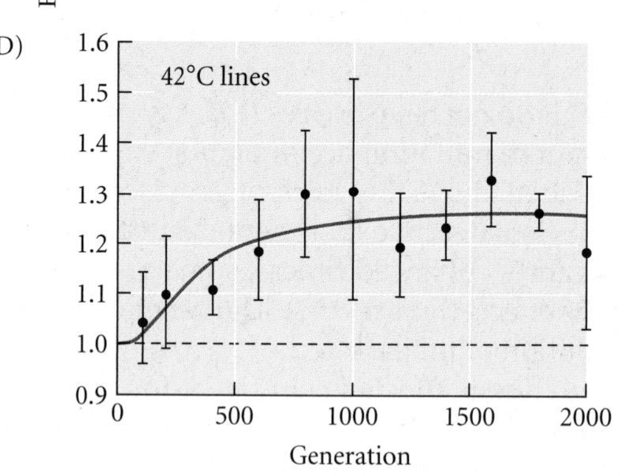

Figure 8.18 Adaptation in experimental populations of *Escherichia coli*. Vertical bars show a measure of variation (95 percent confidence interval) among replicate populations around the mean fitness. (A) Increase in fitness during 20,000 generations in populations kept at 37°C (the normal temperature of *E. coli*'s habitat). (B–D) Adaptation over a much shorter time (2000 generations) in populations kept at three different temperatures (32°C, 37°C, and 42°C). Because all of these populations initially lacked genetic variation, the increase in adaptation was due to natural selection acting on new advantageous mutations. (A after Cooper and Lenski 2000; B–D after Bennett et al. 1992.)

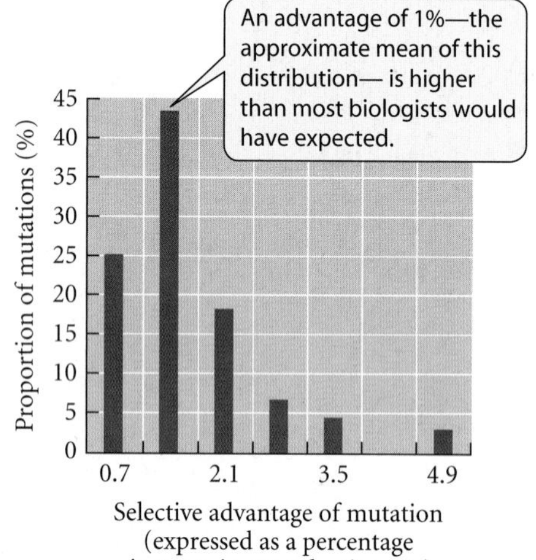

Figure 8.19 The frequency distribution of selective advantage conferred by new beneficial mutations in small ($N = 2 \times 10^4$) laboratory populations of *Escherichia coli*. These values estimate the proportional increase in reproductive rate of genotypes with the new mutation. (After Perfeito et al. 2007.)

tions that happen to be in the same individual genome in which the advantageous mutation occurred. If any such mutation is detectable (call it y), it can be used as a genetic marker of its genome, so even if we cannot determine which gene experienced the advantageous mutation, an increase in the frequency of y indicates that an advantageous mutation at some locus x has occurred. Lília Perfeito and colleagues (2007) used detectable microsatellite mutations, which arise at a high rate, as markers to track genetic changes resulting from new advantageous mutations that sweep to fixation. From the number of such events over the course of 1000 bacterial generations, Perfeito et al. estimated that 2×10^{-5} beneficial mutations arise per genome per generation, that these may account for 1 out of every 150 newly arising mutations, and that the average beneficial mutation increases a genotype's fitness by 1.3 percent (Figure 8.19). These estimates of the rate and magnitude of beneficial mutation are much higher than previously supposed, but remarkably high rates of beneficial mutation have also been reported in yeast (Dickinson 2008). These results suggest that the potential for adaptive evolution as a result of new mutations is quite high.

Bacteria can be screened for mutations that affect their biochemical capacities by placing them on a medium on which that bacterial strain cannot grow, such as a medium that lacks an essential amino acid or other nutrient. Whatever colonies do appear on the medium must have grown from the few cells in which mutations occurred that conferred a new biochemical ability. For example, Barry Hall (1982) studied a strain of *E. coli* that lacks the *lacZ* gene, which encodes β-galactosidase, the enzyme that enables *E. coli* to metabolize the sugar lactose as a source of carbon and energy. Hall screened populations for the ability to grow on lactose and recovered several mutations. A mutation in a different gene (*ebg*) altered an enzyme that normally performs another function so that it could break down lactose. Another mutation altered regulation of the *ebg* gene, and a third mutation altered the ebg enzyme so that it metabolized lactose into lactulose, which increased the cell's uptake of lactose from the environment. The three mutations together restored the metabolic capacities that had been lost by the deletion of the original *lacZ* gene. Thus mutation and selection in concert can give rise to complex adaptations.

Richard Lenski, who traced adaptation based on new mutations in populations of *E. coli* over the course of 20,000 generations (see Figure 8.18A), continued to maintain those populations in a medium in which availability of glucose limited population growth. The medium also contained citrate, a potential energy source that *E. coli* cannot use because it cannot transport citrate into the cell. Lenski's group observed that one population increased dramatically in density at generation 33,000 because of a mutation (Cit^+) that enabled it to use citrate (Blount et al. 2008). Long before the 33,000th generation, billions of mutations had transpired in these populations, yet the adaptation had not evolved earlier. Lenski's group showed that another mutation must have occurred in this culture that either facilitated the origin of the Cit^+ mutation or made it advantageous. That is, the evolution of this adaptation was historically contingent on previous genetic change. This example of the origin of a key adaptation—one that enabled use of a new ecological niche—suggests that evolution of novel characteristics might depend on extremely rare combinations of mutational events. That is, *the mutational process might limit evolution*.

The limits of mutation

The great majority of mutations that affect the phenotype alter one or more *pre-existing* traits. Mutations with phenotypic effects alter developmental processes, but they cannot alter developmental foundations that do not exist. We may conceive of winged horses and angels, but no mutant horses or humans will ever sprout wings from their shoulders, for the developmental foundations for such wings are lacking.

The direction of evolution may be constrained if some conceivable mutations are more likely to arise and contribute to evolution than others. In laboratory stocks of the green alga *Volvox carteri*, for example, new mutations affecting the relationship between the size and number of germ cells correspond to the typical state of these characters in other species of *Volvox* (Koufopanou and Bell 1991).

Mutation may not constrain the rate or direction of evolution very much if many different mutations can generate a particular phenotype, as may be the case when several or many loci affect the trait (polygeny). For example, when different copper-tolerant populations of the monkeyflower *Mimulus guttatus* are crossed, the variation in copper tolerance is greater in the F_2 generation than in either parental population, indicating that the populations differ in the loci that confer tolerance (Cohan 1984).

Nevertheless, certain advantageous phenotypes can apparently be produced by mutation at only a few loci, or perhaps only one. In such instances, the supply of rare mutations might limit the capacity of species for adaptation. The rarity of necessary mutations may help to explain why species have not become adapted to a broader range of environments, or why, in general, species are not more adaptable than they are (Bradshaw 1991). For instance, resistance to the insecticide dieldrin in different populations of *Drosophila melanogaster* is based on repeated occurrences of the same mutation, which, moreover, is thought to represent the same gene that confers dieldrin resistance in flies that belong to two other families (ffrench-Constant et al. 1990). This may mean that very few genes—perhaps only this one—undergo mutations that can confer dieldrin resistance, and that such a mutation is a very rare event. In Chapter 3, we saw that mutations of the same gene, *Mc1r*, have affected pigmentation in various species of reptiles, birds, and mammals.

Wichman et al. (2000) studied experimental populations of the closely related bacteriophage strains ϕX174 and S13 as they adapted to two species of host bacteria at high temperatures. Of the many amino acid substitutions that occurred in the populations, most occurred repeatedly, at a small number of sites, and many of those substitutions matched the variation found in natural populations and even the differences between the two kinds of phage (Figure 8.20). This result suggests that the natural evolution of these phage can take only a limited number of pathways, constrained by the possible kinds of advantageous mutations (Wichman et al. 2000).

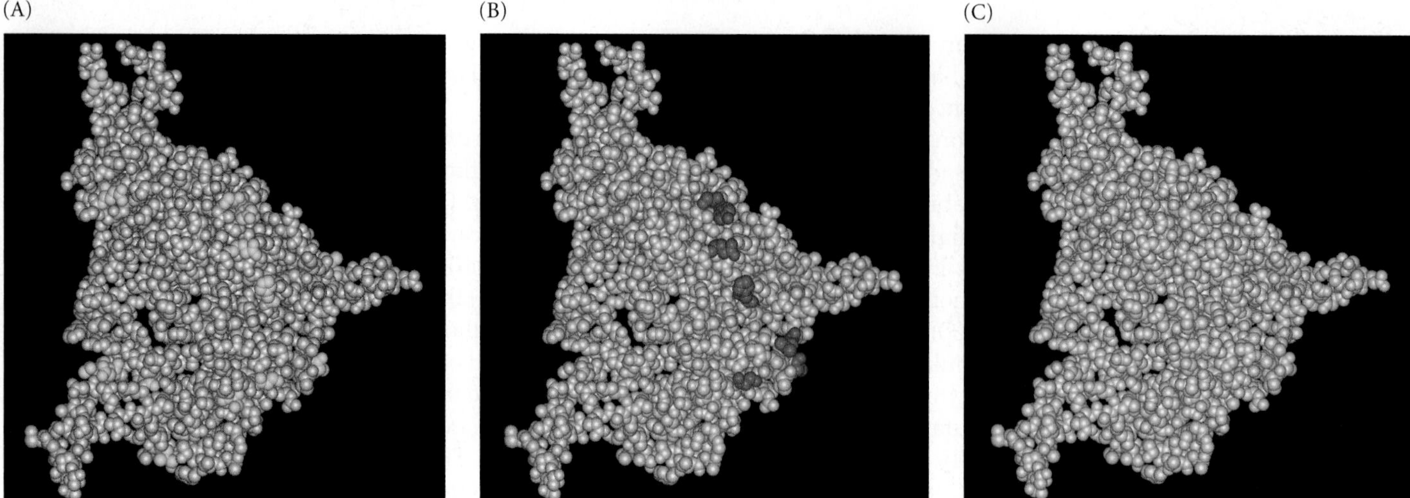

Figure 8.20 The surface of the major capsid protein (gpF) of phage strains ϕX174 and S13, showing parallel amino acid substitutions. (A) Amino acids that underwent substitution in the experimental lines are shown in yellow. (B) Amino acids shown in red are known to affect the fitness of wild phage in either of two bacterial host species. (C) Amino acids that represent differences between the two original strains, ϕX174 and S13, are shown in blue. Note that all the differences that have naturally evolved between these phage strains also occurred in the experimental lines. (After Wichman et al. 2000.)

Mutation as a Random Process

Mutations occur at random. It is extremely important to understand what this statement does and does not mean. It does not mean that all conceivable mutations are equally likely to occur, because, as we have noted, the developmental foundations for some imaginable transformations do not exist. It does not mean that all loci, or all regions within a locus, are equally mutable, for geneticists have described differences in mutation rates, at both the phenotypic and molecular levels, among and within loci. It does not mean that environmental factors cannot influence mutation rates: radiation and chemical mutagens do so.

Mutation *is* random in two senses. First, although we may be able to predict the *probability* that a certain mutation will occur, we cannot predict which of a large number of gene copies will undergo the mutation. The spontaneous process of mutation is stochastic rather than deterministic. Second, and more importantly, mutation is random in the sense that *the chance that a particular mutation will occur is not influenced by whether or not the organism is in an environment in which that mutation would be advantageous*. That is, the environment does not induce adaptive mutations. Indeed, it is hard to imagine a mechanism whereby most environmental factors could direct the mutation process by dictating that just the right base pair changes should occur.

The argument that adaptively directed mutation does not occur is one of the fundamental tenets of modern evolutionary theory. If it did occur, it would introduce a Lamarckian element into evolution, for organisms would then acquire adaptive hereditary characteristics in response to their environment. Such "neo-Lamarckian" ideas were expunged in the 1940s and 1950s by experiments with bacteria in which spontaneous, random mutation followed by natural selection, rather than mutation directed by the environment, explained adaptation.

One of these experiments was performed by Joshua and Esther Lederberg (1952), who showed that advantageous mutations occur without exposure to the environment in which they would be advantageous to the organism. The Lederbergs used the technique of REPLICA PLATING (Figure 8.21). Using a culture of *E. coli* derived from a single cell, the Lederbergs spread cells onto a "master" agar plate, without penicillin. Each cell gave rise to a distinct colony. They pressed a velvet cloth against the plate, and then touched the cloth to a new plate with medium containing the antibiotic penicillin, thereby transferring some cells from each colony to the replica plate, in the same spatial relationships as the colonies from which they had been taken. A few colonies appeared on the replica plate, having grown from penicillin-resistant mutant cells. When all the colonies on the master plate were tested for penicillin resistance, those colonies (and only those colonies) that had been the source of penicillin-resistant cells on the replica plate displayed resistance, showing that the mutations had occurred before the bacteria were exposed to penicillin.

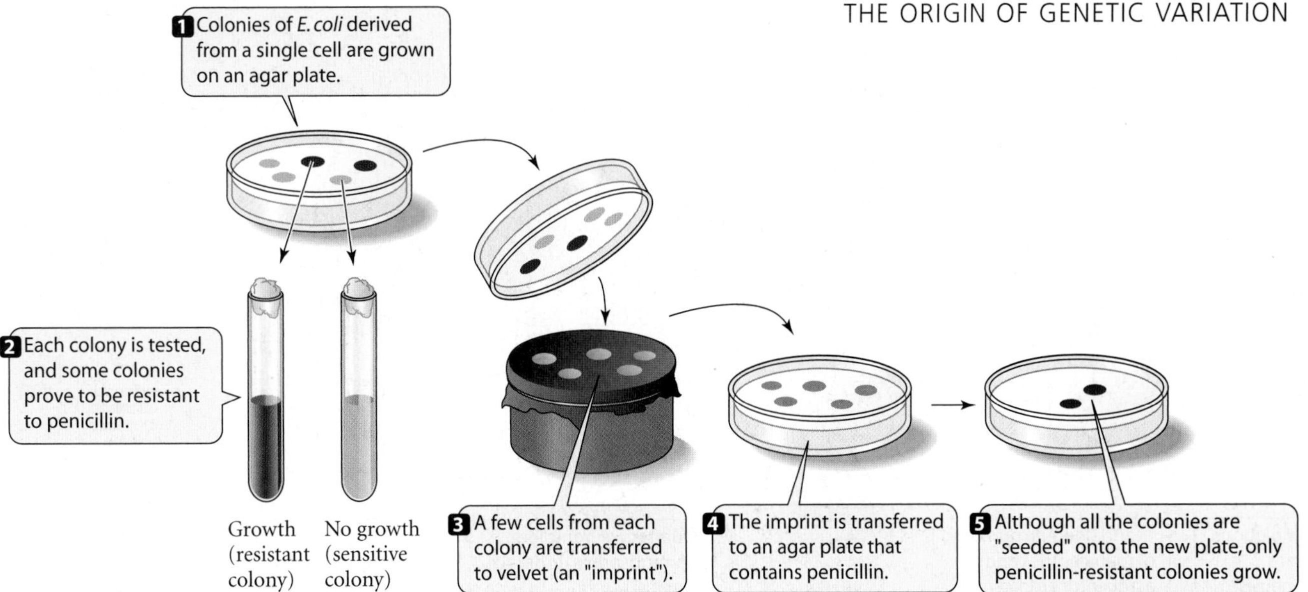

1 Colonies of E. coli derived from a single cell are grown on an agar plate.
2 Each colony is tested, and some colonies prove to be resistant to penicillin.
Growth (resistant colony)
No growth (sensitive colony)
3 A few cells from each colony are transferred to velvet (an "imprint").
4 The imprint is transferred to an agar plate that contains penicillin.
5 Although all the colonies are "seeded" onto the new plate, only penicillin-resistant colonies grow.

Figure 8.21 The method of replica plating, which the Lederbergs used to show that mutations for penicillin resistance arise spontaneously, before exposure to penicillin, rather than being induced by that exposure. They began (step 1) with an agar plate containing numerous colonies of *E. coli*, all derived from a single cell. They tested the colonies on this master plate for penicillin resistance (step 2). They pressed a velvet cloth against the plate (step 3), thereby transferring some cells from each colony to the velvet. They then touched the velvet to a new plate with medium containing the antibiotic penicillin (step 4), transferring bacteria to this replica plate in the same spatial relationships as the colonies on the master plate from which they had been taken. A few colonies appeared on the replica plate (step 5), having grown from penicillin-resistant mutant cells. Only those colonies that had been the source of penicillin-resistant cells on the master plate grew on the replica plate, showing that the mutations had occurred before the bacteria were exposed to penicillin.

Because of such experiments, biologists have generally accepted that mutation is adaptively random rather than directed. However, several investigators have reported results, again with *E. coli*, that at face value seem to suggest that some advantageous mutations might be directed by the environment. There is good evidence that mutation rates throughout the genome may be elevated by certain stressful environments, but no convincing evidence that specific mutations are more likely to occur in those environments in which they might be advantageous (Sniegowski and Lenski 1995; Foster 2000). We will return in Chapter 15 to the question of whether mutation rates might evolve to some optimal level.

Alterations of the Karyotype

An organism's **karyotype** is the description of its complement of chromosomes: their number, size, shape, and internal arrangement. In considering alterations of the karyotype, it is important to bear in mind that the loss of a whole chromosome, or a major part of a chromosome, usually reduces the viability of a gamete or an organism because of the loss of genes. Furthermore, a gamete or organism often is inviable or fails to develop properly if it has an **aneuploid**, or "unbalanced," chromosome complement—for example, if a normally diploid organism has three copies of one of its chromosomes. (For instance, humans with three copies of chromosome 21, a condition known as Down syndrome or trisomy-21, have brain and other defects.)

As we have seen, chromosome structure may be altered by duplications and deletions, which change the amount of genetic material (see Figure 8.6). Other alterations of the karyotype are changes in the number of whole sets of chromosomes (**polyploidy**) and rearrangements of one or more chromosomes.

Polyploidy

A diploid organism has two entire sets of homologous chromosomes ($2N$); a polyploid organism has more than two. (In discussing chromosomes, N refers to the number of different chromosomes in the gametic, or haploid, set, and the numeral refers to the number of representatives of each autosome.) Polyploids can be formed in several ways, especially when failure of the reduction division in meiosis produces diploid, or unreduced, gametes (Ramsey and Schemske 1998). The union of an unreduced gamete (with $2N$ chromosomes) and a reduced gamete (with N chromosomes) yields a TRIPLOID ($3N$) zygote. Triploids produce few offspring because most of their gametes have aneuploid chromosome complements. At segregation, each daughter cell may receive one copy of certain chromosomes and two copies of certain others (Figure 8.22A). However, tetraploid ($4N$) offspring can be formed if an unreduced ($3N$) gamete of a triploid unites with a normal

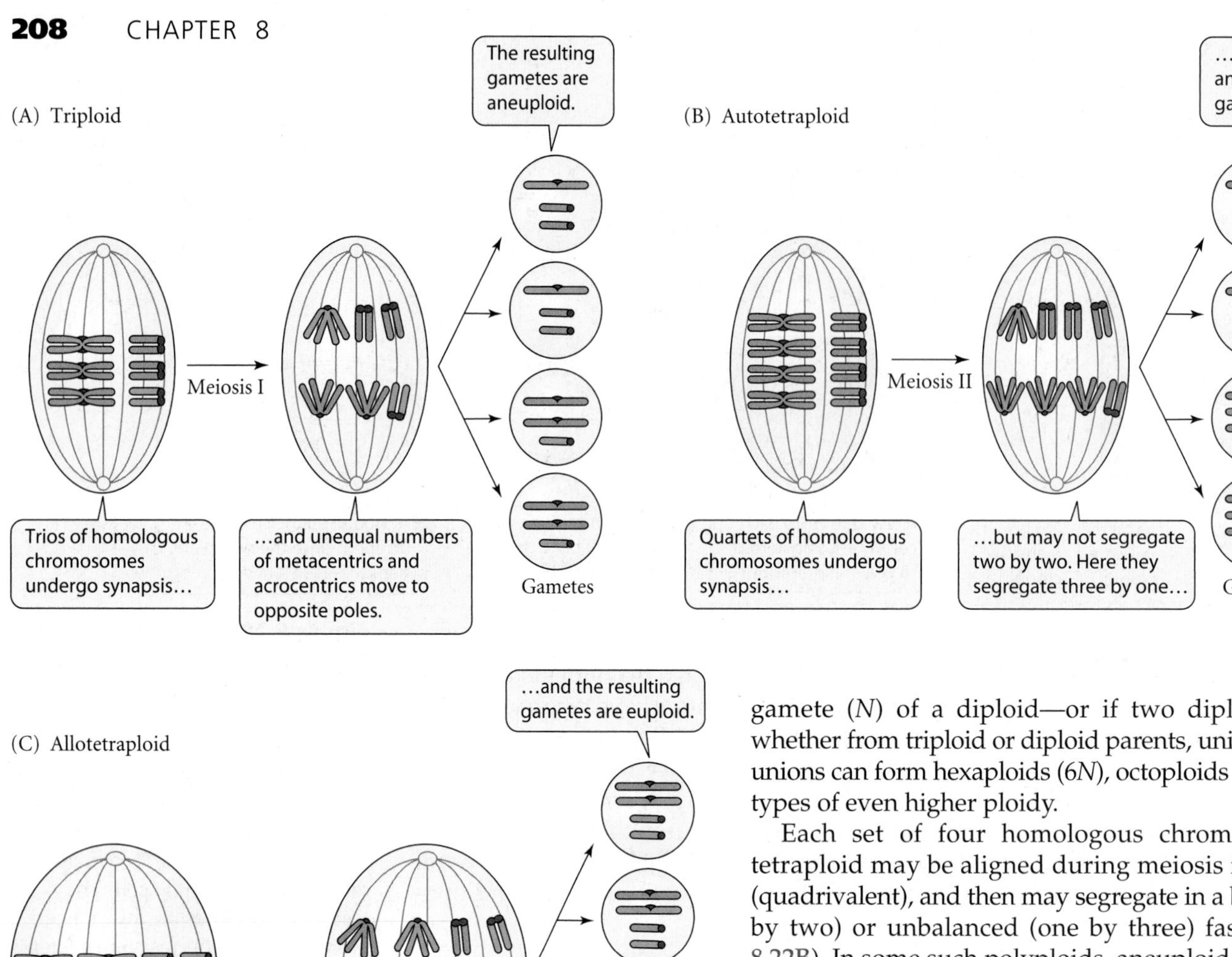

(A) Triploid
The resulting gametes are aneuploid.
Meiosis I
Trios of homologous chromosomes undergo synapsis…
…and unequal numbers of metacentrics and acrocentrics move to opposite poles.
Gametes
(B) Autotetraploid
…yielding aneuploid gametes.
Meiosis II
Quartets of homologous chromosomes undergo synapsis…
…but may not segregate two by two. Here they segregate three by one…
Gametes
(C) Allotetraploid
…and the resulting gametes are euploid.
Meiosis I
Each chromosome pairs with a single homologue from the same parental species.
Segregation is then normal…
Gametes

Figure 8.22 Segregation of chromosomes in meiosis in various polyploids. Each part shows synapsis and segregation on the spindle in meiosis I and the four haploid products of meiosis II. Two sets of homologous chromosomes, one metacentric and one acrocentric, are shown; each begins meiosis with two identical sister chromatids. All chromosomes with the same color (red or blue) are derived from one species.

gamete (*N*) of a diploid—or if two diploid gametes, whether from triploid or diploid parents, unite. Other such unions can form hexaploids (6*N*), octoploids (8*N*), or genotypes of even higher ploidy.

Each set of four homologous chromosomes of a tetraploid may be aligned during meiosis into a quartet (quadrivalent), and then may segregate in a balanced (two by two) or unbalanced (one by three) fashion (Figure 8.22B). In some such polyploids, aneuploid gametes may result and fertility may be greatly reduced. In other cases, the four chromosomes align not as a quartet but as two pairs that segregate normally, resulting in balanced (**euploid**), viable gametes, so that fertility is normal, or nearly so (Figure 8.22C). This requires that the chromosomes be differentiated so that each can recognize and pair with a single homologue rather than with three others.

Many species of plants and a few species of trout, tree frogs, and other animals have arisen by polyploidy, and there is evidence that extant vertebrates stem from a polyploid ancestor (see Figure 3.30 and Chapter 18). Estimates of the proportion of angiosperms that are polyploid range between 50 and 70 percent (Stace 1989). Some recently arisen polyploids have arisen by the union of unreduced gametes of the same species; these organisms are known as **autopolyploids**. But the majority are **allopolyploids**, which have arisen by hybridization between closely related species (Chapter 18). In allopolyploids, most of the parental species' chromosomes are different enough for the chromosomes from each parent to recognize and pair with each other, so that meiosis in an allotetraploid, for example, involves normal segregation of pairs rather than quartets of chromosomes (see Figure 8.22C).

Chromosome rearrangements

Changes in the structure of chromosomes constitute another class of karyotypic alterations. These changes are caused by breaks in chromosomes, followed by rejoining of the pieces in new configurations. Some such changes can affect the pattern of segregation in meiosis, and therefore affect the proportion of viable gametes. Although most chromosome rearrangements seem not to have direct effects on morphological or other phenotypic features, an alteration of gene sequence sometimes brings certain genes under the influence of the control regions of other genes, and so alters their expression. It is not certain that such "posi-

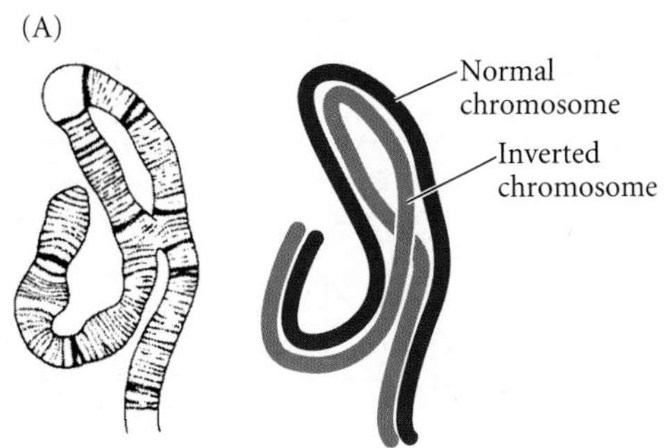

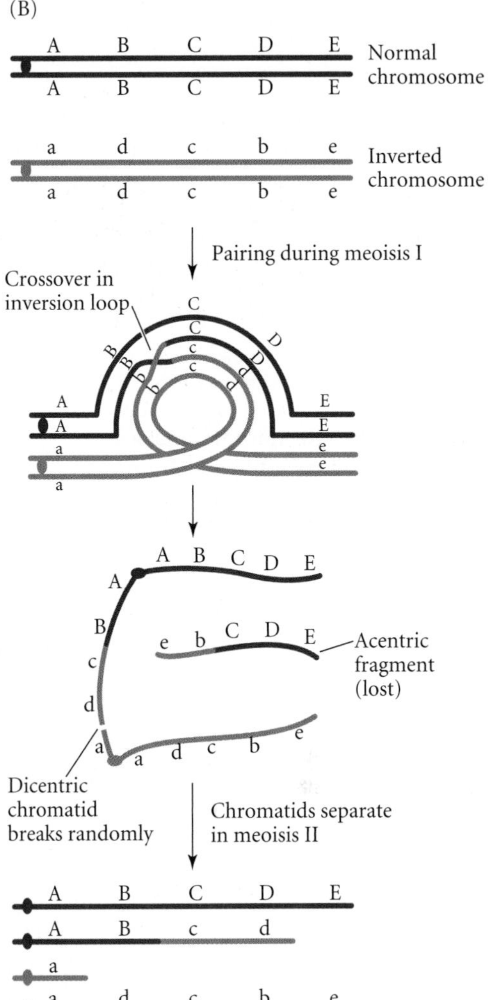

Figure 8.23 Chromosome inversions. (A) Synapsed chromosomes in a salivary gland cell of a larval *Drosophila pseudoobscura* heterozygous for *Standard* and *Arrowhead* arrangements. The two homologous chromosomes are so tightly synapsed that they look like a single chromosome. The "bridge" forms the loop shown in the diagram. Similar synapsis occurs in germ line cells undergoing meiosis. (B) Two homologous chromosomes differing by an inversion of the region B–D, and their configuration in synapsis. Crossing over between two chromatids (between B and C) yields products that lack a centromere or substantial blocks of genes. Because these products do not form viable gametes, crossing over appears to be suppressed. Only cells that receive the two chromatids that do not cross over become viable gametes. (After Strickberger 1968.)

tion effects" have contributed to evolutionary change. Individual organisms may be homozygous or heterozygous for a rearranged chromosome, and are sometimes referred to as **homokaryotypes** or **heterokaryotypes** respectively.

INVERSIONS. Consider a segment of a chromosome in which ABCDE denotes a sequence of markers such as genes. If a loop is formed, and breakage and reunion occur at the point of overlap, a new sequence, such as A<u>DCB</u>E, may be formed. (The inverted sequence is underlined.) Such an **inversion**, with a rearranged gene order, is PERICENTRIC if it includes the centromere and PARACENTRIC if it does not.

During meiotic synapsis in an inversion heterozygote, alignment of the genes on the normal and inverted chromosomes requires the formation of a loop, which can sometimes be observed under the microscope (Figure 8.23A). Now suppose that in a paracentric inversion, crossing over occurs between loci such as B and C (Figure 8.23B). Two of the four strands are affected. One strand lacks certain gene regions (A), and also lacks a centromere; it will not migrate to either pole, and is lost. The other affected strand not only lacks some genetic material but also has two centromeres, so the chromosome breaks when these centromeres are pulled to opposite poles. The resulting daughter cells lack certain gene regions and will not form viable gametes. Consequently, in inversion heterokaryotypes (but not in homokaryotypes), *fertility is reduced* because many gametes are inviable, and *recombination is effectively suppressed* because gametes carrying the recombinant chromosomes, which lack some genetic material, are inviable.

In *Drosophila* and some other flies (Diptera), however, the incomplete recombinant chromosomes enter the polar bodies during meiosis, so that female fecundity is not reduced. It is particularly easy to study inversions in *Drosophila* and some other flies because the larval salivary glands contain giant (polytene) chromosomes that remain in a state of permanent synapsis (so that inversion loops are easily seen), and because these chromosomes display bands, each of which apparently corresponds to a single gene. The banding patterns are as distinct as the bar codes on supermarket products, so an experienced investigator can identify different sequences. INVERSION POLYMORPHISMS are common in *Drosophila*—more than 20 different arrangements of chromosome 3 have been described for *Drosophila pseudoobscura*, for example—and they have been extensively studied from both population genetic and phylogenetic points of view.

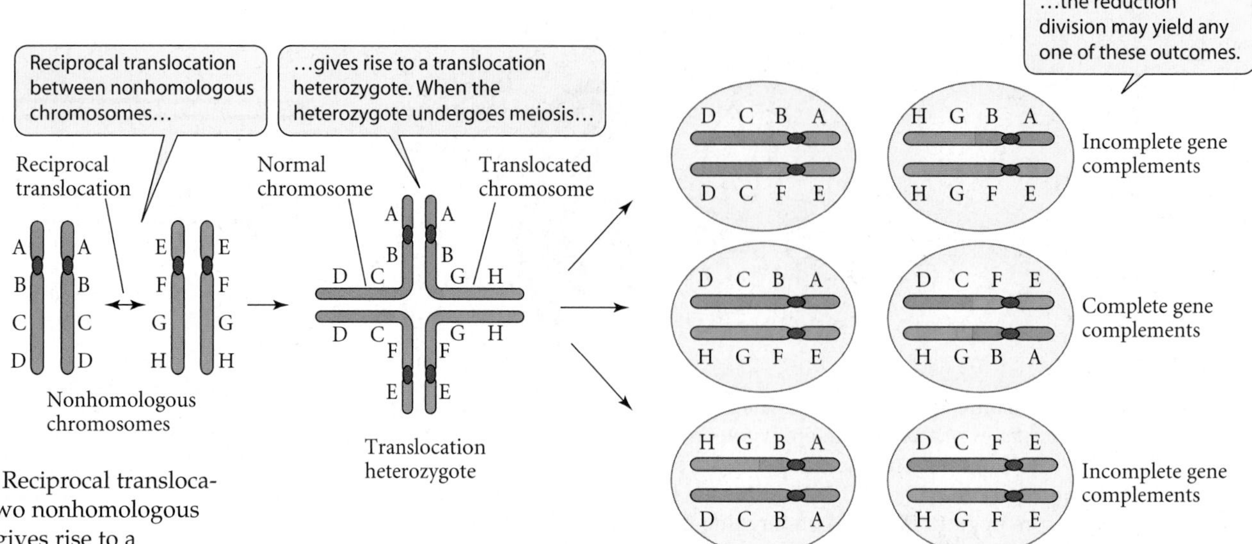

Figure 8.24 Reciprocal translocation between two nonhomologous chromosomes gives rise to a translocation heterozygote. When the heterozygote undergoes meiosis, many of the products will have incomplete gene complements.

TRANSLOCATIONS. By breakage and reunion, two nonhomologous chromosomes may exchange segments, resulting in a **reciprocal translocation** (Figure 8.24). Meiosis in a translocation heterokaryotype often results in a high proportion of aneuploid gametes, so the fertility of translocation heterokaryotypes is often reduced by 50 percent or more. Consequently, polymorphism for translocations is rare in natural populations. Nevertheless, related species sometimes differ by translocations, which have the effect of moving groups of genes from one chromosome to another. The Y chromosome of the male *Drosophila miranda*, for example, includes a segment that is homologous to part of one of the autosomes of closely related species.

FISSIONS AND FUSIONS. It is useful to distinguish ACROCENTRIC chromosomes, in which the centromere is near one end, from METACENTRIC chromosomes, in which the centromere is somewhere in the middle and separates the chromosome into two arms. In the simplest form of chromosome **fusion**, two nonhomologous acrocentric chromosomes undergo reciprocal translocation near the centromeres so that they are joined into a metacentric chromosome (Figure 8.25A). For example, two chromosomes found in chimpanzee and gorilla have fused to form human chromosome 2, resulting in one less pair of chro-

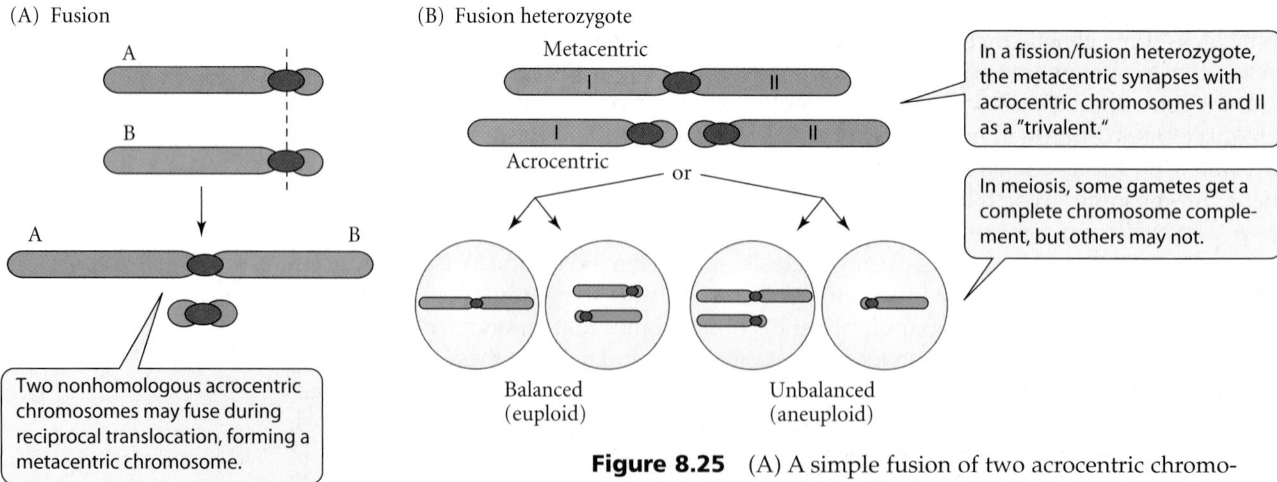

Figure 8.25 (A) A simple fusion of two acrocentric chromosomes to form arms A and B. (B) Segregation in meiosis of a fusion heterozygote can yield euploid (balanced) or aneuploid (unbalanced) complements of genetic material.

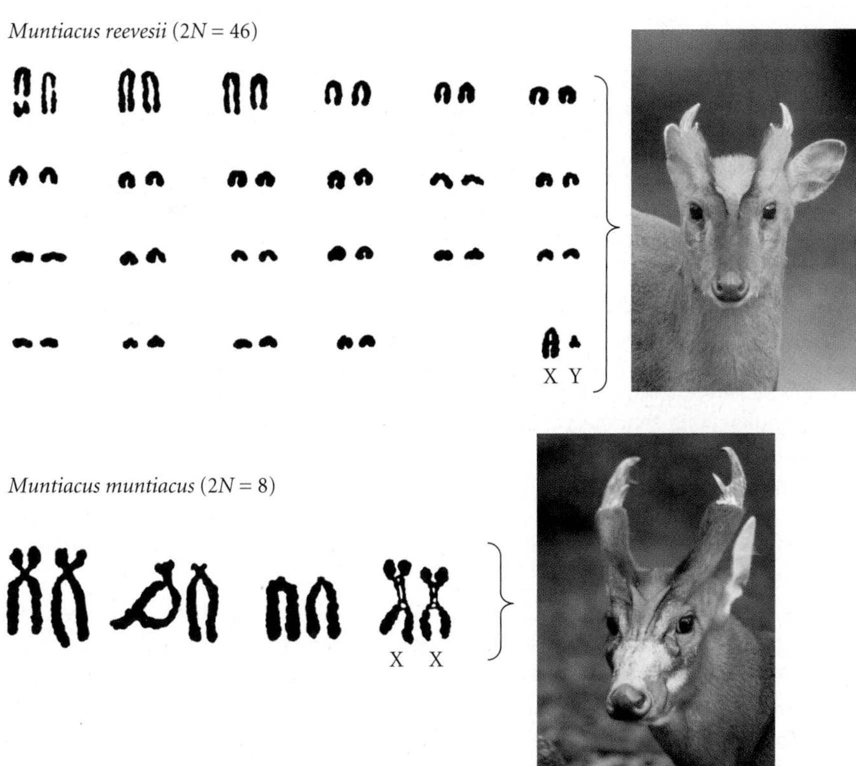

Figure 8.26 The diploid chromosome complements (taken from micrographs) of two closely related species of muntjacs (barking deer) represent one of the most extreme differences in karyotype known among closely related species. Despite the difference in karyotype, the species are phenotypically very similar. (*M. reevesii* photo © Mike Lane/Alamy Images; *M. muntiacus* © OSF/photolibrary.com.)

mosomes in the human genome (de Pontbriand et al. 2002). More rarely, a metacentric chromosome may undergo **fission**. A simple fusion heterokaryotype has a metacentric, which we may refer to as AB, with arms that are homologous to two acrocentrics, A and B. AB, A, and B together synapse as a "trivalent" (Figure 8.25B). Viable gametes and zygotes are often formed, but the frequency of aneuploid gametes can be quite high, especially for more complex patterns of fusion. Differences in chromosome number, whether caused by fusion or fission, often distinguish related species or geographic populations of the same species.

CHANGES IN CHROMOSOME NUMBER. Summarizing what we have covered so far, polyploidy (especially in plants), translocations, and fusions or fissions of chromosomes are the mutational foundations for the evolution of chromosome number. For example, the haploid chromosome number varies among mammals from 3 to 42 (Lande 1979), and among insects from 1 in an ant species to about 220 in some butterflies (the highest number known in animals). Related species sometimes differ strikingly in karyotype: in one of the most extreme examples, two very similar species of barking deer, *Muntiacus reevesii* and *M. muntiacus*, have haploid chromosome numbers of 23 and 3 or 4 (in different populations), respectively (Figure 8.26).

The spontaneous rate of origin of any given class of chromosome rearrangement (e.g., reciprocal translocation) is quite high: about 10^{-4} to 10^{-3} per gamete (Lande 1979). Only a small fraction of these become fixed in populations, though, and the rate at which species undergo karyotype divergence is highly variable. By DNA sequencing, the relative positions of many genes along chromosomes can be compared across species. On this basis, the number of breaks per megabase per million years has been estimated as 2.7×10^{-4} between human and mouse. *Drosophila* species appear to have unusually high rates of chromosome evolution, ranging from 6.5×10^{-2} to 10.1×10^{-2} for various species pairs (Ranz et al. 2001; Bartolomé and Charlesworth 2006). Human and chimpanzee genomes may differ by as many as 1500 inversions (Sharp et al. 2006).

Summary

1. Mutations of chromosomes or genes are alterations that are subsequently replicated. They ordinarily do not constitute new species, but instead are variant chromosomes or genes (alleles, haplotypes) within a species.
2. At the molecular level, mutations of genes include base pair substitutions, frameshifts, changes caused by insertion of various kinds of transposable elements, and duplications and deletions that can range from single base pairs to long segments of chromosome. New DNA sequences also arise by intragenic recombination.
3. Mutations that result in amino acid substitution in a protein (nonsynonymous mutations) and mutations in regulatory sequences may affect the phenotype and perhaps fitness. Synonymous mutations, those that do not alter the amino acid sequence, may not affect the phenotype or fitness, and so may be selectively neutral. The extent to which mutations in noncoding DNA, which dominates eukaryotic genomes, affect fitness is largely unknown.
4. The rate at which any particular mutation arises is quite low: on average, about 10^{-6} to 10^{-5} per gamete for mutations detected by their phenotypic effects, and about 10^{-9} per base pair. The mutation rate, by itself, is too low to cause substantial changes of allele frequencies. However, the total input of genetic variation by mutation, for the genome as a whole or for individual polygenic characters, is appreciable.
5. The magnitude of change in morphological or physiological features caused by a mutation can range from none to drastic. In part because most mutations have pleiotropic effects, the average effect of mutations on fitness is deleterious, but some mutations are advantageous.
6. Mutations alter pre-existing biochemical or developmental pathways, so not all conceivable mutational changes are possible. Some adaptive changes may not be possible without just the right mutation of just the right gene. For these reasons, the rate and direction of evolution may in some instances be affected by the availability of mutations.
7. Mutations appear to be random, in the sense that their probability of occurrence is not directed by the environment in favorable directions, and in the sense that specific mutations cannot be predicted. The likelihood that a mutation will occur does not depend on whether or not it would be advantageous.
8. Mutations of the karyotype (chromosome complement) include polyploidy and structural rearrangements (e.g., inversions, translocations, fissions, fusions). Many such rearrangements reduce fertility in the heterozygous condition.

Terms and Concepts

additive inheritance
allele
allopolyploid
aneuploid
autopolyploid
back mutation
base pairs (bp)
base pair substitution
codon
control regions (enhancers and repressors)
dominance
euploid
exon
fission (of chromosomes)
frameshift mutation
fusion (of chromosomes)
gene
gene family
genetic code
genetic marker
haplotype
heterokaryotype
homeotic mutation
homokaryotype
intragenic recombination
intron
inversion
karyotype
locus
microsatellite
mutation
mutational variance
nonsynonymous mutation
pleiotropic
point mutation
polygenic
polyploidy
processed pseudogene

pseudogene
recessive
reciprocal translocation
recurrent mutation
repeated sequence
retrosequence
substitution
synonymous mutation
tandem repeat
transition
transposable element (TE)
transposition
transversion
unequal crossing over
wild type

Suggestions for Further Reading

Evolutionary Genetics: Concepts and Case Studies (edited by C. W. Fox and J. B. Wolf, Oxford University Press, New York, 2006) is an excellent collection of essays by leading authorities on genetic aspects of evolution, written for students. The chapter on mutation, by D. Houle and A. Kondrashov (pp. 32–48), is perhaps the best overview of mutation as an evolutionary process. W.-H. Li, *Molecular Evolution* (Sinauer Associates, Sunderland, MA, 1997), treats evolutionary aspects of mutation at the molecular level. Reviews of mutation rates include J. W. Drake, B. Charlesworth, D. Charlesworth, and J. F. Crow, "Rates of spontaneous mutation" (1998, *Genetics* 148:1667–1686), and M. Lynch et al., "Perspective: Spontaneous deleterious mutation" (1999, *Evolution* 53: 645–663).

Problems and Discussion Topics

1. Consider two possible studies. (a) In one, you capture 3000 wild male *Drosophila melanogaster*, mate each with laboratory females heterozygous for the autosomal recessive allele *vg* (*vestigial*, which causes miniature wings when homozygous), and examine each male's offspring. You find that half the offspring of each of three males have miniature wings and have genotype *vgvg*. (b) In another study, you determine the nucleotide sequence of 1000 base pairs for 20 copies of the cytochrome *b* gene, taken from 20 wild mallard ducks. You find that at each of 30 nucleotide sites, one or another gene copy has a different base pair from all others. From these data, can you estimate the rate of mutation from the wild type to the *vg* allele (case a) or from one base pair to another (case b)? Why or why not?
2. From a laboratory stock of *Drosophila* that you believe to be homozygous wild type (++) at the *vestigial* locus, you obtain 10,000 offspring, mate each of them with homozygous *vg vg* flies, and examine a total of 1 million progeny. Two of these are *vgvg*. Estimate the rate of mutation from + to *vg* per gamete. What assumptions must you make?
3. The following DNA sequence represents the beginning of the coding region of the alcohol dehydrogenase (*Adh*) gene of *Drosophila simulans* (Bodmer and Ashburner 1984), arranged into codons:

   CCC ACG ACA GAA CAG TAT TTA AGG AGC TGC GAA GGT

   (a) Find the corresponding mRNA sequence, and use Figure 8.2 to find the amino acid sequence. (b) Again using Figure 8.2, determine for each site how many possible mutations (changes of individual nucleotides) would cause an amino acid change, and how many would not. For the entire sequence, what proportion of possible mutations are synonymous versus nonsynonymous? What proportion of the possible mutations at first, second, and third base positions within codons are synonymous? (c) What would be the effect on the amino acid sequence of inserting a single base, G, between sites 10 and 11 in the DNA sequence? (d) What would be the effect of deleting nucleotide 16? (e) For the first 15 (or more) sites, classify each possible mutation as a transition or transversion, and determine whether or not the mutation would change the amino acid. Does the proportion of synonymous mutations differ between transitions and transversions?
4. A genus of Antarctic fishes, *Channichthys*, lacks hemoglobin. In its relative *Trematomus*, hemoglobin serves its usual functions. Assuming that the gene encoding hemoglobin in *Channichthys* has no function, and is not transcribed, how might you expect the nucleotide sequence of this gene to differ between these two genera?
5. Ultraviolet light (UV) can induce mutations in organisms such as *Drosophila*. Because it damages DNA, and therefore essential physiological functions, it can also reduce survival. Suppose you expose a large number of *Drosophila* to UV, screen their adult offspring for new mutations,

and discover that a few offspring carry mutations that increase the amount of black pigment, which can protect internal organs from UV. The progeny of an equal number of control flies, not exposed to UV, show fewer or no mutations that increase pigmentation. Can you conclude that the process of mutation responds to organisms' need for adaptation to the environment?

6. Researchers have used artificial selection (see Chapter 9) to alter many traits in *Drosophila melanogaster*, such as phototactic behavior and wing length. No one has selected *Drosophila* (about 2 mm long) to be as large (ca. 30 mm) as bumblebees (although I'm not sure if anyone has tried). Do you suppose this could be done? How would you attempt it? If your attempts were unsuccessful, what hypotheses could explain your lack of success? What role might mutation play in your experiment?
7. The text above says that "two chromosomes found in chimpanzee and gorilla have fused to form human chromosome 2, resulting in one less pair of chromosomes in the human genome." What reason is there to think that the difference in chromosome number resulted from fusion rather than fission? In general, can we tell ancestral from mutant (derived) genes?
8. In a study described in this chapter, Nachman and Crowell (2000) estimated the neutral mutation rate from sequence differences between human and chimpanzee in several processed pseudogenes. One methodological detail not mentioned in this chapter was that they also sequenced several kilobases of flanking DNA on each side of the pseudogene. Any base pair differences found in these flanking sequences were not included in the calculation of mutation rates. (a) Why did they choose pseudogenes rather than some other kind of DNA sequence to estimate the neutral mutation rate? (b) Why did they sequence flanking DNA?
9. Most, if not all, attempts to estimate the rate of beneficial mutations are likely to produce underestimates. Why?

CHAPTER 9

Variation

Individuals of the same species often look different. Genetic variation within populations, as in the differing display feathers of the male ruff (*Philomachus pugnax*), is the basis of evolutionary change. (Photo © Arco Images/AGE Fotostock.)

Genetic variation is the foundation of evolution, for the great changes in organisms that have transpired over time and the differences that have developed among species as they diverged from their common ancestors all originated as genetic variants within species. Understanding the processes of evolution thus requires us to understand genetic variation and the ways in which it is transmuted into evolutionary change. Understanding genetic variation, moreover, provides insight into questions—ranging from the significance of "intelligence tests" to the meaning of "race"—that deeply affect human society. Of all the biological sciences, evolutionary biology is most dedicated to analyzing and understanding variation, a profoundly important feature of all biological systems.

Because genetic variation will be discussed throughout the rest of this book, we need a brief review of its vocabulary:

- *Phenotype* refers to a characteristic in an individual organism, or in a group of individuals that are alike in this respect. All features of an organism except DNA sequences may be called phenotypes, ranging from the level of transcription of a gene to aspects of the organism's morphology, behavior, and life history. Phenotypic variation is largely the result of genetic differences among individuals, but it can have several other sources, as described below, of which the most important are the direct effects of environmental variation on development.

- *Genotype* is the genetic constitution of an individual organism, or of a group of organisms that are alike in this respect, at one or more loci singled out for discussion.
- A *locus* (plural *loci*) is a site on a chromosome, or more usually, the gene that occupies a site.
- An *allele* is a particular form of a gene, usually distinguished from other alleles by its effects on the phenotype.
- A *haplotype* is one of the sequences of a gene or DNA segment that can be distinguished from homologous sequences by molecular methods such as DNA sequencing.
- *Gene copy* is the term we will use when counting the number of representatives of a gene. In a diploid population (such as humans), each individual carries two copies of each autosomal gene, so a sample of 100 individuals represents 200 gene copies. The term *gene copy* is used without distinguishing allelic or sequence differences among the copies. For that purpose, we will sometimes refer to *allele copies*. Thus a heterozygous individual, A_1A_2, has two copies of the A gene: one copy of allele A_1 and one copy of allele A_2.

The number of alleles or loci that contribute to genetic variation in a phenotypic trait differs from case to case. In the simplest cases, two alleles at a single locus account for most of the variation (e.g., the color forms of the snow goose; Figure 9.1A). In some cases, three or more alleles exist within a population. For instance, several forms of the African swallowtail butterfly (*Papilio dardanus*) are palatable mimics of different species of butterflies that birds find unpalatable (Figure 9.1B). In this case, the several different color patterns are inherited as if they were multiple alleles at one locus. Multiple loci contribute to the continuous variation typical of most phenotypic features, such as the familiar variation in human hair and skin color (Figure 9.1C).

Figure 9.1 Phenotypic variation caused by genetic differences within a species. In all cases, the forms interbreed freely. (A) "Blue" and white forms of the snow goose (*Chen caerulescens*) are caused by two alleles at a single locus. (B) Mimetic variation in the African swallowtail *Papilio dardanus*. Males are non-mimetic and all appear similar (top left individual). The other three individuals are females, each of which mimics a distantly related toxic species of butterfly. This female-limited variation in *P. dardanus* is inherited as if it were due to multiple alleles at one locus, but it is actually caused by several closely linked genes. (C) In *Homo sapiens*, alleles at several or many different loci contribute to variation in "quantitative characters" such as skin pigmentation, hair color and texture, and facial features. (A © Jim Zipp and John Bova/Photo Researchers, Inc.; B courtesy of Fred Nijhout; C © ShutterStock.)

Sources of Phenotypic Variation

Variation in a phenotypic character (including gene expression) can have several sources other than those encoded in DNA sequences. It is important to understand these sources, not only because they are interesting in themselves but because it is often necessary to distinguish them from genetic variation.

The *environment* directly affects the development or expression of many features. These may be permanent effects on an individual's phenotype, such as the determination of sex in many turtles by temperature at a critical stage of development, or the effect of nutrition, host availability, or crowding on wing development in insects (Figure 9.2). Some environmental factors act on an embryo before birth or hatching. Consumption of alcohol, tobacco, and many other drugs by pregnant women, for example, increases the risk of nongenetic birth defects in their babies. Thus, some CONGENITAL differences among individuals—those present at birth—do not have a genetic basis. Some environmental effects are temporary, as in physiological acclimation, enzyme induction, and the expression of many learned behaviors. Environmentally induced variation among individuals is often called **environmental variance**, which we will discuss at length later in this chapter).

Some variation persists even if variation in both genes and environment is experimentally reduced to the lowest feasible level. Some such variation seems to arise from random events at the molecular level in an organism and is called DEVELOPMENTAL NOISE. For instance, in fruit flies there are measurable differences between the left and right sides of the body, which have the same genes and surely the same environment. Whether wing length, for example, is greater on one side or the other seems random. This kind of asymmetry is called FLUCTUATING ASYMMETRY.

The term **maternal effect** refers to the effects of a mother on her offspring that are due not to the genes they inherit from her, but rather to nongenetic influences, such as the amount or composition of yolk in her eggs, the amount and kind of maternal care she provides, or her physiological condition while carrying eggs or embryos. Even the parent's environment can affect offspring phenotype; for example, young desert locusts (*Schistocerca gregaria*) develop dark coloration, long wings, and restless, gregarious behavior if either they or their mothers are crowded (see Figure 9.2B). Locusts that develop this gregarious syndrome form huge swarms that fly long distances and devastate vegetation wherever they land. Nongenetic "paternal effects" have also been described in a few organisms. Maternal effects are diverse and important in many contexts, and they can affect the dynamic of genetic adaptation (Rossiter 1996).

Phenotypic differences that are not based on DNA sequence differences are transmitted among generations of dividing cells in multicellular organisms (e.g., mitosis of liver cells yields liver cells) and also, sometimes, from parents to offspring. This phenomenon is called **epigenetic inheritance**, and its evolutionary consequences are the subject of considerable speculation (Jablonka and Lamb 2005, 2006). Cross-generational epigenetic inheritance is based on GENOMIC IMPRINTING, whereby the activity of a gene or chromosome—perhaps as many as 1 percent of all genes in mammals—depends on its parental origin. Among the several causes of genomic imprinting is DNA METHYLATION, in which a methyl group is joined to a cytosine (C) in a CG doublet. The methylation process is repeated during DNA

(A)

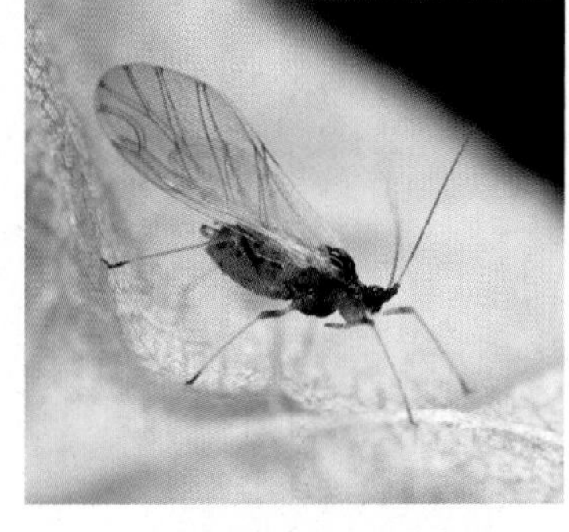

(B)

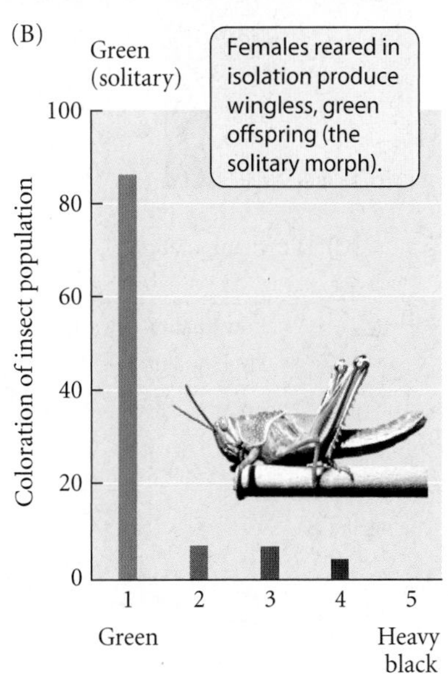

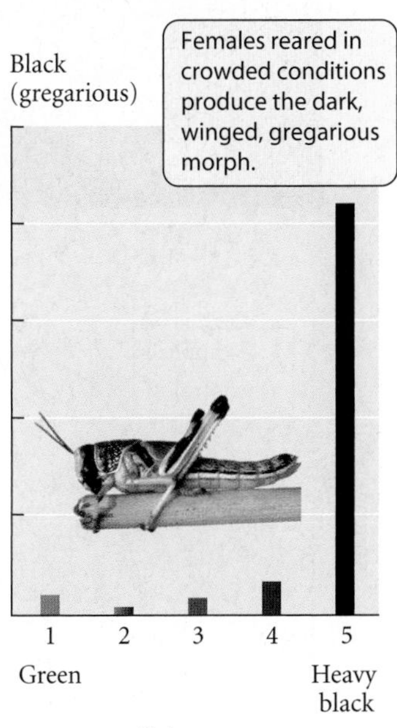

Figure 9.2 Examples of phenotypic variation caused by environmental conditions. (A) Many species of aphids develop wings only under certain environmental conditions. In early summer, parthenogenetically reproducing peach apids (*Myzus persicae*) develop as large, wingless adults (left). Their offpspring, however, are a winged morph (right) capable of dispersing to secondary host plants as summer wanes. (B) A maternal effect: the two morphs of *Schistocerca gregaria*. Offspring of mothers maintained in isolation during the mating and egg-laying stage develop the wingless, solitary form; when the mothers are reared in crowded conditions, the dark, winged morph results. (A left © Jeremy Burgess/Photo Researchers, Inc.; A right courtesy of Scott Bauer/USDA; B after Islam et al. 1994; photos courtesy of Compton Tucker/NASA.)

Standard *Linaria vulgaris*

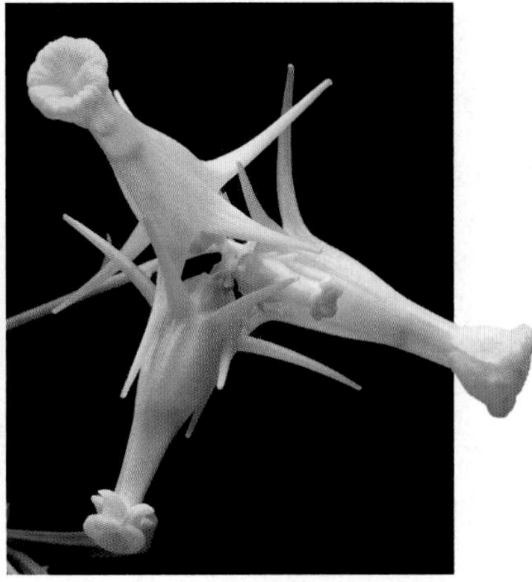

peloria form

Figure 9.3 Genomic imprinting produces the *peloria* mutant of toadflax (*Linaria vulgaris*), a heritable epigenetic mutation. This mutation occurs not in the nucleotide sequence of the gene, but in the methylation pattern that affects the expression of floral symmetry genes. (Courtesy of R. Grant-Downton.)

replication, and the methylated state, which often reduces or eliminates gene transcription, may be transmitted for few or many generations.

Genomic imprinting affects some species-typical characteristics (for instance, all the paternal chromosomes are eliminated in certain fungus gnats) and is responsible for some rare "mutant" phenotypes. For example, one of the five petals of the toadflax *Linaria vulgaris* normally has a long nectar-bearing spur, but in the *peloria* mutant, which was first discovered by Linnaeus, all five petals have this form, transforming the flower from bilaterally symmetrical to radially symmetrical (Figure 9.3). This recessive mutation is caused not by a sequence difference but by extensive methylation of a gene that controls flower symmetry (Cubas et al. 1999). There is no evidence yet that epigenetic variation contributes to evolutionary change, and considerable difference of opinion on whether or not it is likely to do so. (See articles pro and con in *Journal of Evolutionary Biology* 11:159–260, 1998.)

Many characteristics vary because of both genetic and environmental (and often maternal effect) sources of variation (Figure 9.4), and in many cases, both kinds of influences affect the same developmental pathway. When variation stems from both genetic and environmental sources, it is pointless to ask whether a characteristic is "genetic" *or* "environmental," as if it had to be exclusively one or the other. Moreover, the relative amounts of genetic and environmental variation can differ with different circumstances, even in the same population. This is especially true if genotypes differ in their **norm of reaction**, the variety of different phenotypic states that can be produced by a single genotype under different environmental conditions. For example, Gupta and Lewontin (1982) measured the mean number of abdominal bristles in each of ten genotypes of *Drosophila pseudoobscura* raised at three different temperatures. They found a **genotype × environment inter-**

(A)

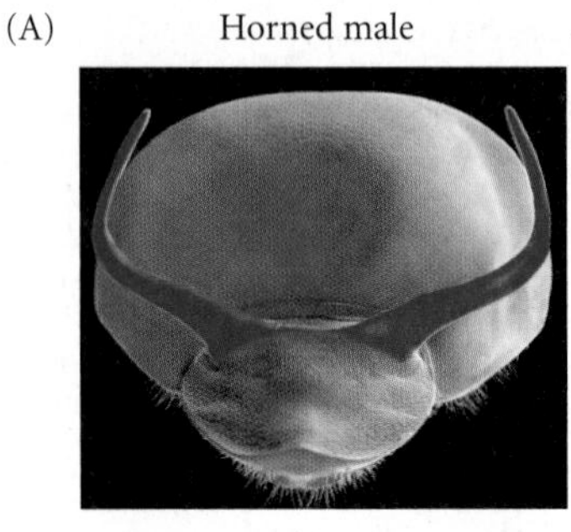

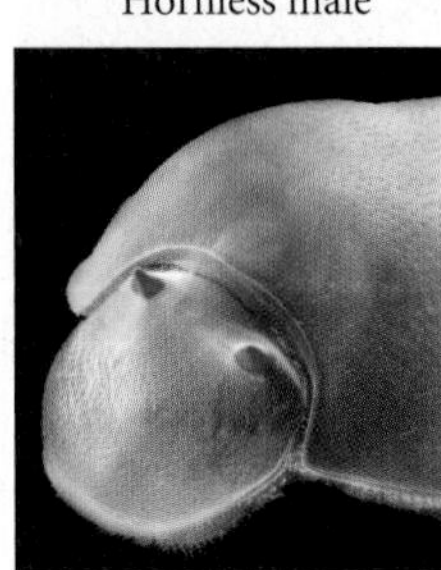

Figure 9.4 (A) The length of the horn in male *Onthophagus* beetles can be altered by both genetic and environmental factors. Long horns develop if body size during pupation exceeds a threshold. (B) Applying juvenile hormone (JH), which inhibits or reduces development of adult features in insects, reduced horn length, suggesting that this hormone plays an important role in development of this trait. (C) Longer horns were produced by rearing larvae on a higher-quality diet (an environmental change). (D) Artificial selection (breeding from shorter- or longer-horned males for several generations) produced a genetic change in horn length. (After Emlen 2000.)

(B)

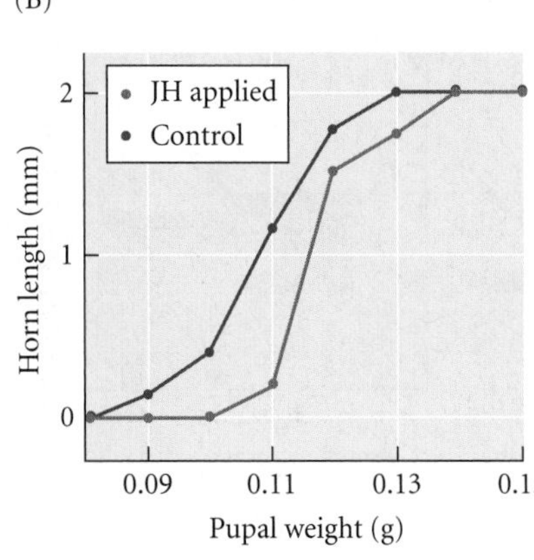

(C) Diet manipulation

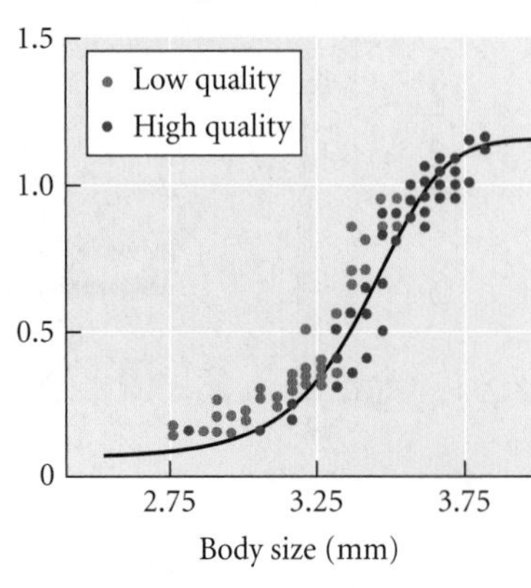

(D) Artificial selection

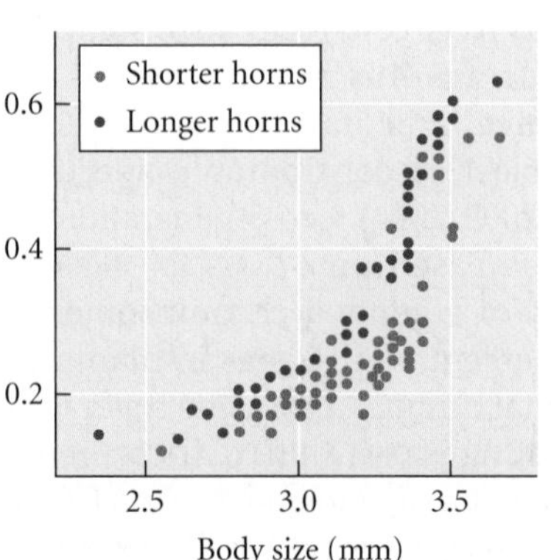

action, meaning that the effect of temperature on phenotype differed among genotypes (Figure 9.5). The amount of phenotypic variation in a population that is due to genetic differences among individuals thus depends on the particular range of temperatures at which the flies develop.

Because evolution depends on the genetic component of variation, it is often critically important to determine whether variation in a characteristic is genetic, environmental, or both. Several methods are frequently used:

1. Phenotypes can be experimentally crossed to produce F_1, F_2, and backcross progeny. Mendelian ratios among the phenotypes of the progeny (e.g., 3:1 or 1:2:1) are taken as evidence of simple genetic control.
2. Correlation between the average phenotype of offspring and that of their parents, or greater resemblance among siblings than among unrelated individuals, suggests that genetic variation contributes to phenotypic variation. However, we must take maternal effects into account, and we must be sure that siblings (or relatives in general) do

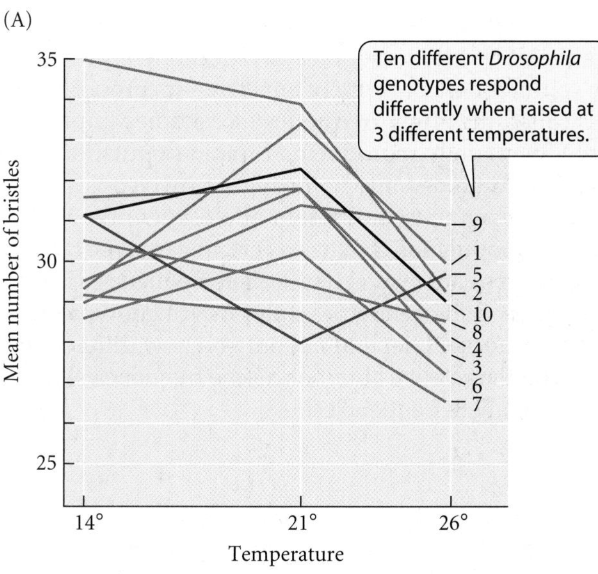

Figure 9.5 An example of genotype × environment interaction and how it can affect the relative contributions of genetic and environmental variation to phenotypic variation. (A) The number of bristles on the abdomen of male *Drosophila pseudoobscura* of ten genotypes, each reared at three different temperatures. Temperature affects the development of bristle number differently for each genotype, so there is an interaction between genotype and environment. (B) If a population of flies of genotypes 5 and 10 from part A developed at a normally distributed range of temperatures between about 17° and 20°C, the distribution of phenotypes would be bimodal, and most of the variation would be genetic, because at these temperatures the two genotypes differ greatly in phenotype. (Scan up along the vertical lines, which mark 17° and 20°C, find their intersection with a genotype's norm of reaction, and then read across to the *y*-axis to find the number of bristles expected of flies that developed in that temperature range.) Flies of each genotype are expected to vary in bristle number since they develop at a variety of temperatures. (C) If the same two genotypes developed at temperatures varying between about 23° and 25°C, most of the phenotypic variation would be environmental, because the two genotypes would respond similarly to these temperatures. (After Gupta and Lewontin 1982.)

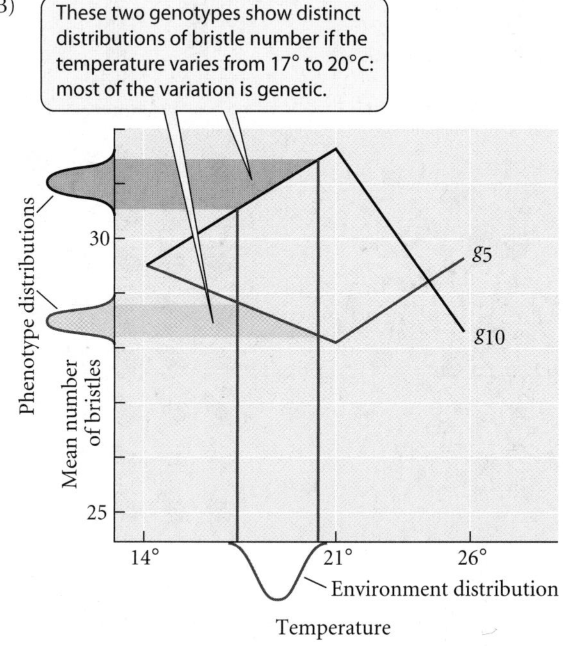

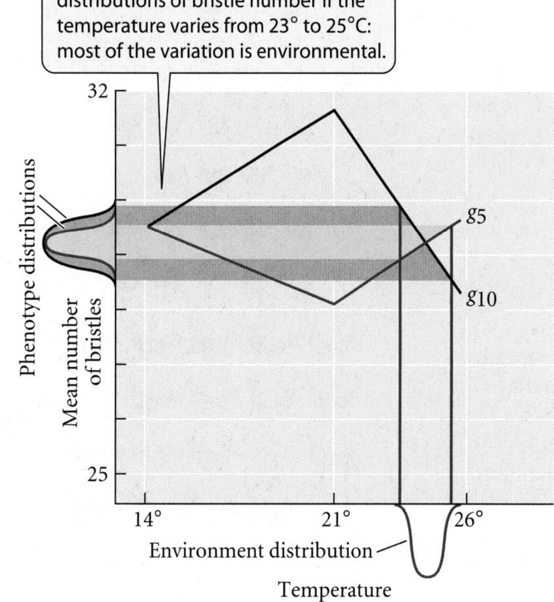

not share more similar environments than nonrelatives. For example, human geneticists rely strongly on studies of adopted children to determine whether behavioral or other similarities among siblings are due to shared genes rather than shared environments.

3. The offspring of phenotypically different parents can be reared together in a uniform environment, often referred to as a **common garden**. Differences among offspring from different parents that persist in such circumstances are likely to have a genetic basis. Propagation in the common garden for at least two generations is advisable to distinguish genetic from maternal effects.

Fundamental Principles of Genetic Variation in Populations

We embark now on our study of genetic variation and the factors that influence it—that is, the factors that cause evolution within species. *The definitions, concepts, and principles introduced here are absolutely essential for understanding evolutionary theory.* We begin with a short description of these ideas, followed by an explanation of a very important formal model.

At any given gene locus, a population may contain two or more alleles that have arisen over time by mutation. Sometimes one allele is by far the most common (and may be called the WILD TYPE), and the others are very rare; sometimes two or more of the alleles are each quite common. The relative commonness or rarity of an allele—its proportion of all gene copies in the population—is called the **allele frequency** (sometimes imprecisely referred to as the "gene frequency"). In sexually reproducing diploid populations, the alleles, carried in eggs and sperm, become combined into homozygous (two copies of the same allele) and heterozygous (one copy each of two different alleles) genotypes. The **genotype frequency** is the proportion of a population that has a certain genotype (Figure 9.6). The proportions of the different genotypes are related to the allele frequencies in simple but important ways, as we will soon see. If the genotypes differ in a phenotypic character, the amount of variation in that character will depend not only on how different the genotypes are from each other, but also on the relative abundance (the frequencies) of the genotypes—which in turn depends on the allele frequencies.

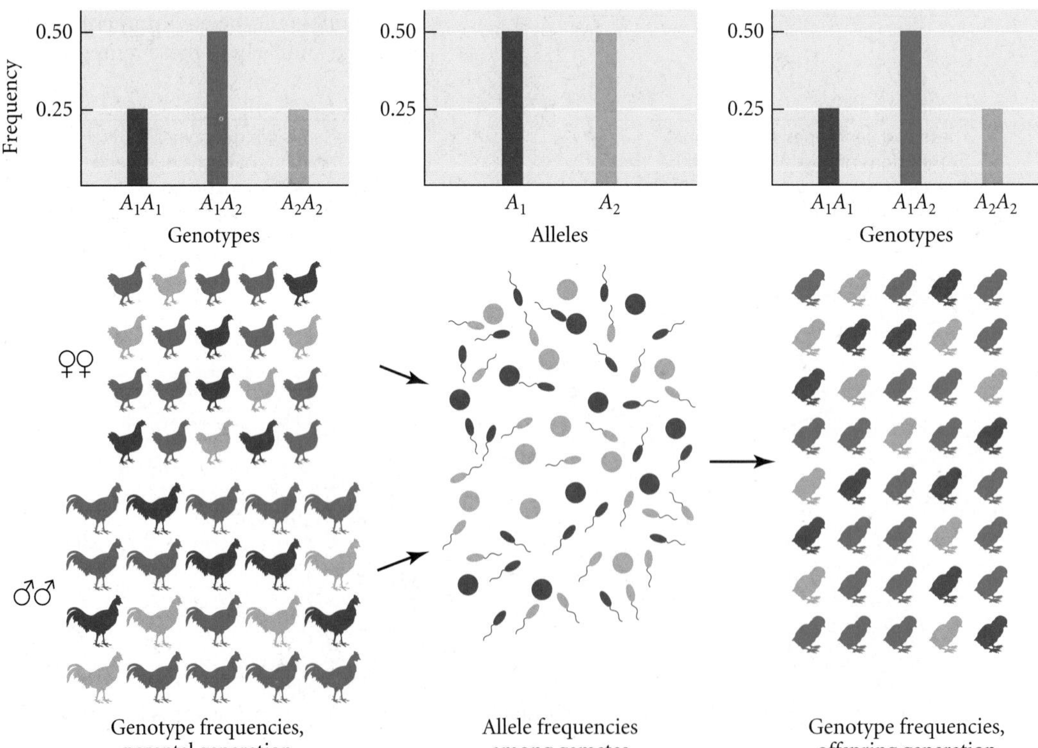

Figure 9.6 For a locus with two alleles (A_1, A_2), this diagram shows the frequency of three genotypes among females and males in one generation, the allele frequencies among their eggs and sperm, and the genotype frequencies among the resulting offspring.

An alteration of the genotype frequencies in one generation will usually alter the frequencies of the alleles carried by the population's gametes when reproduction occurs, so the genotype frequencies of the following generation will be altered in turn. *Such alteration, from generation to generation, is the central process of evolutionary change.* However, the genotype and allele frequencies do not change on their own; something has to change them. *The factors that can cause the frequencies to change are the causes of evolution.*

Frequencies of alleles and genotypes: The Hardy-Weinberg principle

Imagine that a diploid population has 1000 individuals, and that for a particular gene locus in that population, there exist only two alleles, A_1 and A_2. Thus there are three possible genotypes for this locus: A_1A_1, A_2A_2 (both homozygous), and A_1A_2 (heterozygous).

Let us say 400 individuals have the genotype A_1A_1, 400 are A_1A_2, and 200 are A_2A_2. Let the frequencies of genotypes A_1A_1, A_1A_2, and A_2A_2 be represented by D, H, and R, respectively, and let the frequencies of the alleles A_1 and A_2 be represented by p and q. Thus the genotype frequencies are $D = 0.4$, $H = 0.4$, and $R = 0.2$, and the allele frequencies are $p = 0.6$ and $q = 0.4$ (Figure 9.7A,B). The frequency of allele A_1 is the sum of all the alleles car-

(A) Parental genotype frequencies (not in equilibrium)

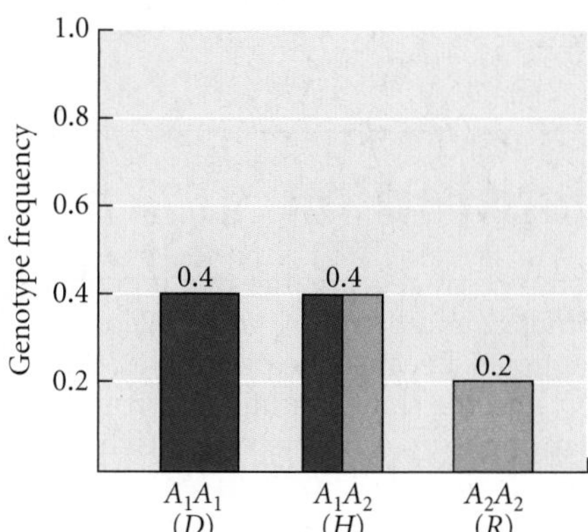

(B) Parental allele frequencies

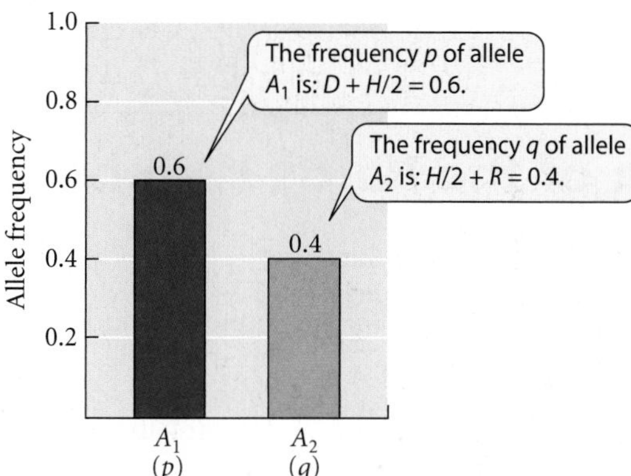

(C)

Offspring	Probability of a given mating producing the genotype		
A_1A_1	$\Pr[A_1 \text{ egg}] \times \Pr[A_1 \text{ sperm}] = p \times p = p^2$	0.6^2	$= 0.36$
A_1A_2	$\Pr[A_1 \text{ egg}] \times \Pr[A_2 \text{ sperm}] = p \times q = pq$	$0.6 \times 0.4 = 0.24$	$= 0.48$
	$\Pr[A_2 \text{ egg}] \times \Pr[A_1 \text{ sperm}] = q \times p = pq$	$0.4 \times 0.6 = 0.24$	
A_2A_2	$\Pr[A_2 \text{ egg}] \times \Pr[A_2 \text{ sperm}] = q \times q = q^2$	0.4^2	$= 0.16$

(D) Offspring genotype frequencies

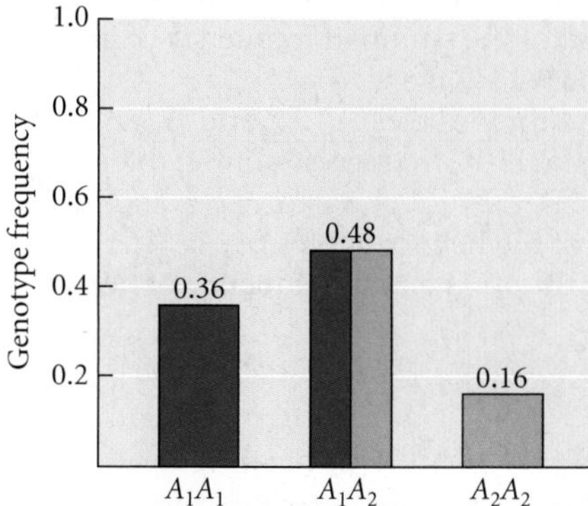

(E) Offspring allele frequencies

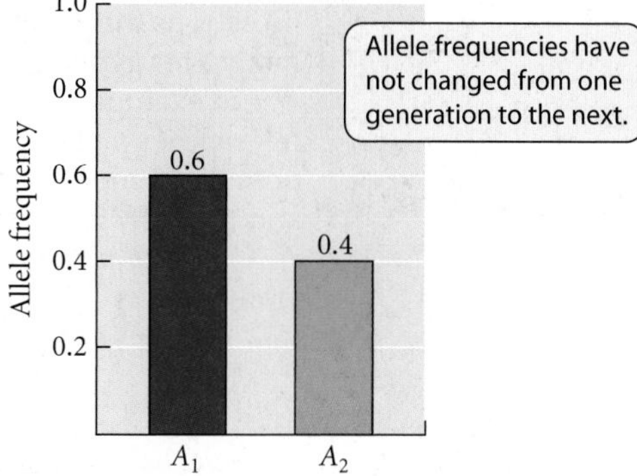

Figure 9.7 A hypothetical example illustrating attainment of Hardy-Weinberg genotype frequencies after a single generation of random mating. (A) Genotype frequencies in the parental population. (B) Allele frequencies in the parental population. (C) Calculation of genotype frequencies among the offspring. (D) Genotype frequencies among the offspring, if all assumptions of the Hardy-Weinberg principle hold. (E) Allele frequencies among the offspring.

ried by A_1A_1 homozygotes, plus one-half of those carried by heterozygotes. Hence $p = D + H/2 = 0.6$. Likewise, the frequency of allele A_2 is calculated as $q = R + H/2 = 0.4$.

Now let us assume that each genotype is equally represented in females and males, and that *individuals mate entirely at random*. (This is an entirely different situation from the crosses encountered in elementary genetics exercises, in which familiar Mendelian ratios are produced by nonrandomly crossing females of *one* genotype with males of *one* genotype.) Random mating is conceptually the same as mixing eggs and sperm at random, as might occur, for instance, in those aquatic animals that release gametes into the water. A proportion p of eggs carry allele A_1, q carry A_2, and likewise for sperm. The probabilities (Pr) for all possible allele combinations are shown in Figure 9.7C, and the resulting offspring genotype and allele frequencies are shown in Figure 9.7D and E.

Notice that in the new generation, the allele frequencies are still $p = D + H/2 = 0.6$ and $q = R + H/2 = 0.4$. *The allele frequencies have not changed from one generation to the next*, although the alleles have become distributed among the three genotypes in new proportions. These proportions, denoted as p^2, $2pq$, and q^2, constitute the HARDY-WEINBERG DISTRIBUTION of genotype frequencies.

The principle that genotypes will be formed in these frequencies as a result of random mating is named after G. H. Hardy and W. Weinberg (pronounced "vine-berg"), who independently calculated these results in 1908. The Hardy-Weinberg principle, broadly stated, holds that whatever the initial genotype frequencies for two alleles may be, *after*

BOX 9A Derivation of the Hardy-Weinberg Distribution

Let the frequencies of genotypes A_1A_1, A_1A_2, and A_2A_2 be D, H, and R, respectively, and let the frequencies of alleles A_1 and A_2 be p and q. The genotype frequencies sum to 1, as do the allele frequencies. The probability of a mating between a female of any one genotype and a male of any one genotype equals the product of the genotype frequencies. For reciprocal crosses between two different genotypes, we take the sum of the probabilities. The mating frequencies and offspring produced are:

Mating	*Probability of mating**	*Offspring genotype* A_1A_1	A_1A_2	A_2A_2
$A_1A_1 \times A_1A_1$	D^2	D^2		
$A_1A_1 \times A_1A_2$	$2DH$	DH	DH	
$A_1A_1 \times A_2A_2$	$2DR$		$2DR$	
$A_1A_2 \times A_1A_2$	H^2	$H^2/4$	$H^2/2$	$H^2/4$
$A_1A_2 \times A_2A_2$	$2HR$		HR	HR
$A_2A_2 \times A_2A_2$	R^2			R^2

*Note that there are a total of 9 mating possibilities: 3 female genotypes × 3 male genotypes. Reciprocal matings between two different genotypes count as 2.

Recalling that $p = D + H/2$ and $q = H/2 + R$, the frequency of each genotype among the offspring is:

A_1A_1: $D^2 + DH + H^2/4 = (D + H/2)^2 = p^2$

A_1A_2: $DH + 2DR + H^2/2 + HR = 2[(D + H/2)(H/2 + R)] = 2pq$

A_2A_2: $H^2/4 + HR + R^2 = (H/2 + R)^2 = q^2$

This result may also be obtained by recognizing that if genotypes mate at random, gametes, and therefore genes, also unite at random to form zygotes. Because the probability that an egg carries allele A_1 is p, and the same is true for sperm, the probability of an A_1A_1 offspring is p^2. A Punnett square shows the probability of each gametic union:

Eggs \ **Sperm**	A_1 (p)	A_2 (q)
A_1 (p)	A_1A_1 (p^2)	A_1A_2 (pq)
A_2 (q)	A_1A_2 (pq)	A_2A_2 (q^2)

The Hardy-Weinberg principle can be extended to more complicated patterns of inheritance, such as multiple alleles. For example, if there are k alleles (A_1, A_2, ... A_k), the H-W frequency of homozygotes for any allele A_i is p_i^2, and that of heterozygotes for any two alleles A_i and A_j is $2p_ip_j$, where p_i and p_j are the frequencies of the two alleles in question. The total frequency of all heterozygotes combined (H) is sometimes expressed as the complement of the summed frequency of all homozygous genotypes, or $H = 1 - \Sigma p_i^2$.

For example, if there are three alleles A_1, A_2, and A_3 with frequencies p_1, p_2, and p_3, the H-W frequencies of the three possible homozygotes (A_1A_1, A_2A_2, A_3A_3) are p_1^2, p_2^2, and p_3^2; and the H-W frequencies of the heterozygotes (A_1A_2, A_1A_3, A_2A_3) are $2p_1p_2$, $2p_1p_3$, and $2p_2p_3$. The total frequency of heterozygotes is

$$H = 1 - (p_1^2 + p_2^2 + p_3^2)$$

one generation of random mating, the genotype frequencies will be p^2*:*$2pq$*:*q^2. Moreover, both these genotype frequencies and the allele frequencies *will remain constant in succeeding generations* unless factors not yet considered should change them. When genotypes at a locus have the frequencies predicted by the Hardy-Weinberg principle, the locus is said to be in **Hardy-Weinberg (H-W) equilibrium**. Box 9A shows how these results are obtained by tabulating all matings and the proportions of genotypes among the progeny of each.

An example: The human *MN* locus

Among the many variations in proteins on the surface of human red blood cells are those resulting from variation at the *MN* locus. Two alleles (*M*, *N*) and three genotypes (*MM*, *MN*, *NN*) are distinguished by blood typing. A sample of 320 people in the Sicilian village of Desulo (R. Ceppellini, in Allison 1955) yielded the following numbers of people carrying each genotype:

MM	***MN***	***NN***
187	114	19

We can now estimate the *frequency* of each genotype as the proportion of the total sample (320) that carries that genotype. Thus,

$$\text{Frequency of } MM = D = 187/320 = 0.584$$

$$\text{Frequency of } MN = H = 114/320 = 0.356$$

$$\text{Frequency of } NN = R = 19/320 = 0.059$$

Note that these frequencies, or proportions, must sum to 1.

Now we can calculate the allele frequencies. Each person carries two gene copies, so the total sample represents $320 \times 2 = 640$ gene copies. Because *MM* homozygotes each have two *M* alleles and *MN* heterozygotes have one, the number of *M* alleles in the sample is $(187 \times 2) + (114 \times 1) = 488$. The number of *N* alleles is $(19 \times 2) + (114 \times 1) = 152$. Thus

$$\text{Frequency of } M = p = 488/640 = 0.763$$

$$\text{Frequency of } N = q = 152/640 = 0.237$$

As with the genotype frequencies, p and q must sum to 1.

These figures are *estimates* of the true genotype frequencies and allele frequencies in the population because they are based on a sample rather than on a complete census. The larger the sample, the more confident we can be that we are obtaining accurate estimates of the true values. (This assumes that the sample is a *random* one, i.e., that our likelihood of collecting a particular type is equal to the true frequency of that type in the population.)

Our hypothetical example (see Figure 9.7) showed that for a given set of allele frequencies, a population might or might not have H-W genotype frequencies. (The frequencies 0.36, 0.48, and 0.16 in the offspring generation were in H-W equilibrium, but the frequencies 0.40, 0.40, and 0.20 in the parental generation were not.) Our real example shows a close fit to the H-W frequency distribution. The allele frequencies of *M* and *N* were $p = 0.763$ and $q = 0.237$. If we calculate the *expected genotype frequencies* under the H-W principle and multiply them by the sample size (320), we obtain the *expected number* of individuals with each genotype. These values in fact closely fit the *observed* numbers:

	Genotype		
	MM	***MN***	***NN***
	p^2	$2pq$	q^2
Expected H-W frequency	0.582	0.362	0.056
Expected number (H-W frequency × sample size)	186	116	18
Observed number	187	114	19

The significance of the Hardy-Weinberg principle: Factors in evolution

The Hardy-Weinberg principle is the foundation on which almost all of the theory of population genetics of sexually reproducing organisms—which is to say, most of the genetic theory of evolution—rests. We will encounter it in the theory of natural selection and other causes of evolution. It has two important implications: First, genotype frequencies attain their H-W values after a single generation of random mating. If some factor in the past had caused genotype frequencies to deviate from H-W values, a single generation of random mating would erase the imprint of that history. Second, according to the H-W principle, not only genotype frequencies, but also allele frequencies, remain unchanged from generation to generation. A new mutation, for example, will remain at its initial very low allele frequency indefinitely.

Like any mathematical model, the Hardy-Weinberg principle holds only under certain assumptions. Since allele frequencies and genotype frequencies often do change (i.e., evolution occurs), the assumptions of the Hardy-Weinberg formulation must not always hold. Therefore *the study of genetic evolution consists of asking what happens when one or more of the assumptions are relaxed*. The most important assumptions of the Hardy-Weinberg principle are the following:

1. *Mating is random*. If a population is not **panmictic**—that is, if members of the population do not mate at random—the genotype frequencies may depart from the ratios $p^2{:}2pq{:}q^2$.
2. *The population is infinitely large* (or so large that it can be treated as if it were infinite). The calculations are made in terms of probabilities. If the number of events is finite, the actual outcome is likely to deviate, purely by chance, from the predicted outcome. If we toss an infinite number of unbiased coins, probability theory says that half will come up heads, but if we toss only 100 coins, we are likely not to obtain exactly 50 heads, purely by chance. Similarly, among a finite number of offspring, both the genotype frequencies and the allele frequencies may differ from those in the previous generation, *purely by chance*. Such random changes are called **random genetic drift**.
3. *Genes are not added from outside the population*. Immigrants from other populations may carry different frequencies of A_1 and A_2; if they interbreed with residents, this will alter allele frequencies and, consequently, genotype frequencies. Mating among individuals from different populations is termed **gene flow** or **migration**. We may restate this assumption as: There is no gene flow.
4. *Genes do not mutate from one allelic state to another*. Mutation, as we saw in Chapter 8, can change allele frequencies, although usually very slowly. The Hardy-Weinberg principle assumes no mutation.
5. *All individuals have equal probabilities of survival and of reproduction*. If these probabilities differ among genotypes (i.e., if there is a consistent difference in the genotypes' rates of survival or reproduction), then the frequencies of alleles and/or genotypes may be altered from one generation to the next. Thus the Hardy-Weinberg principle assumes that there is *no natural selection* affecting the locus.

Inasmuch as nonrandom mating, chance, gene flow, mutation, and selection can alter the frequencies of alleles and genotypes, these are the major factors that cause evolutionary change within populations.

Certain subsidiary assumptions of the Hardy-Weinberg principle can be important in some contexts. First, the principle, as presented, applies to autosomal loci; it can also be modified for sex-linked loci (which have two copies in one sex and one in the other). Second, the principle assumes that alleles segregate into a heterozygote's gametes in a 1:1 ratio. Deviations from this ratio are known and are called SEGREGATION DISTORTION or MEIOTIC DRIVE.

If the assumptions we have listed hold true for a particular locus, that locus will display Hardy-Weinberg genotype frequencies. But if we observe that a locus fits the Hardy-Weinberg frequency distribution, we cannot conclude that the assumptions hold true! For example, mutation or selection may be occurring, but at such a low rate that we cannot detect a deviation of the genotype frequencies from the expected values. Or, under some forms of natural selection, we might observe deviations from Hardy-Weinberg equilibrium if we measure genotype frequencies at one stage in the life history but not at other stages.

Frequencies of alleles, genotypes, and phenotypes

At Hardy-Weinberg equilibrium, the frequency of heterozygotes (**heterozygosity**) is greatest when the alleles have equal frequencies (Figure 9.8). When an allele is very rare, almost all its carriers are heterozygous. Because almost all copies of a rare recessive allele are carried by heterozygotes, the allele may not be detected readily. If, for example, the frequency of a recessive allele is $q = 0.01$, only $(0.01)^2 = 0.0001$ of the population displays this recessive allele in its phenotype. Thus populations can carry **concealed genetic variation**. However, a dominant allele may well be less common than a recessive allele. "Dominance" refers to an allele's phenotypic effect in the heterozygous condition, not to its numerical prevalence. In the British moth *Cleora repandata*, for example, black coloration is inherited as a dominant allele, and "normal" gray coloration is recessive. In a certain forest, 10 percent of the moths were found to be black (Ford 1971).

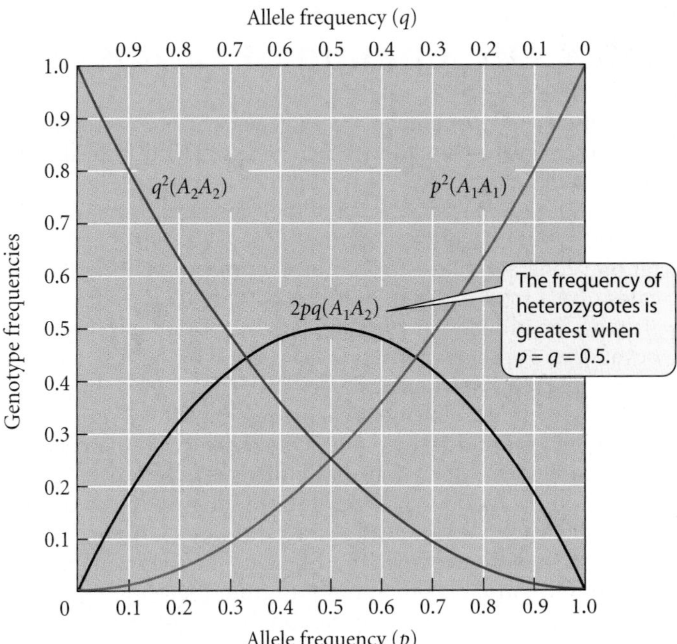

Figure 9.8 Hardy-Weinberg genotype frequencies as a function of allele frequencies at a locus with two alleles. Heterozygotes are the most common genotype in the population if the allele frequencies are between ⅓ and ⅔. (After Hartl and Clark 1989.)

Inbreeding

The Hardy-Weinberg principle assumes that a population is panmictic. What if mating does not occur at random? One form of nonrandom mating is **inbreeding**, which occurs when individuals are more likely to mate with relatives than with nonrelatives, or more generally, when the gene copies in uniting gametes are more likely to be identical by descent than if they joined at random. Gene copies are said to be **identical by descent** if they have descended, by replication, from a common ancestor, relative to other gene copies in the population. The probability that a random pair of gene copies are identical by descent is called F, the **inbreeding coefficient**. Box 9B shows that as inbreeding proceeds, the frequency of each homozygous genotype increases and the frequency of heterozygotes decreases by the same amount. The frequency of heterozygotes is $H = H_0(1 - F)$, where H_0 is the heterozygote frequency expected if the locus were in H-W equilibrium.

The ways in which genotype frequencies change are easily seen if we envision mating of individuals only with their closest relatives—namely, themselves—by self-fertilization. (You may have trouble imagining this if you think about people, but think about plants and you will see that it is not only plausible but quite common.) The homozygous genotype A_1A_1 can produce only A_1 eggs and A_1 sperm, and thus only A_1A_1 offspring; likewise, A_2A_2 individuals produce only A_2A_2 offspring. Heterozygotes produce A_1 and A_2 eggs in equal proportions, and likewise for sperm. When these eggs and sperm join at random, one-fourth of the offspring are A_1A_1, half are A_1A_2, and one-fourth are A_2A_2. (The two allele copies carried by the homozygous offspring are identical by descent.) Thus the frequency of heterozygotes is halved in each generation, and eventually reaches zero. Conversely, F increases as inbreeding continues; in fact, F can be estimated by the deficiency of heterozygotes relative to the H-W equilibrium value (see Box 9B). If CONSANGUINEOUS mating (mating among relatives) is a consistent feature of a population, F will increase over generations at a rate that depends on how closely related the average pair of mates are.

The most extreme form of inbreeding, **self-fertilization** or **selfing**, occurs in many species of plants and a few animals. The wild oat *Avena fatua*, for example, reproduces

BOX 9B Change of Genotype Frequencies by Inbreeding

Suppose we could label every gene copy at a locus and trace the descendants of each gene copy through subsequent generations. We would then be tracing alleles that are identical by descent. Some of the gene copies are allele A_1, and others are A_2. If we label one of the A_1 copies A_1*, then after one generation, a sister and brother may both inherit replicates of A_1* from a parent (with probability $0.5^2 = 0.25$). Among the progeny of a mating between sister and brother, both heterozygous for A_1*, one-fourth will be A_1*A_1*. These individuals carry two gene copies that are not only the same allele (A_1), but are also identical by descent (A_1*) (Figure 1). The A_1*A_1* individuals are said to be not only homozygous, but autozygous. Allozygous individuals, by contrast, may be either heterozygous or homozygous (if the two copies of the same allele are not identical by descent).

The inbreeding coefficient, denoted *F*, is the probability that an individual taken at random from the population will be autozygous. In a population that is not at all inbred, $F = 0$. In a fully inbred population, $F = 1$: all individuals are autozygous.

In a population that is inbred to some extent, *F* is the fraction of the population that is autozygous, and $1 - F$ is the fraction that is allozygous. If two alleles, A_1 and A_2, have frequencies *p* and *q*, the probability that an individual is allozygous and that it is A_1A_1 is $(1 - F) \times p^2$. Likewise, the fraction of the population that is allozygous and heterozygous is $(1 - F) \times 2pq$, and the fraction that is allozygous and A_2A_2 is $(1 - F) \times q^2$.

Turning our attention now to the fraction, *F*, of the population that is autozygous, we note that none of these individuals is heterozygous, because a heterozygote's alleles are not identical by descent. If an individual is autozygous, the probability that it is autozygous for A_1 is *p*, the frequency of A_1. Thus the fraction of the population that is autozygous and A_1A_1 is $F \times p$. Likewise, $F \times q$ is the fraction that is autozygous and A_2A_2.

Thus, taking into account the allozygous and autozygous fractions of the population, the genotype frequencies (Figure 2) are

	Allozygous		Autozygous		Genotype frequency
A_1A_1	$p^2(1 - F)$	+	pF	=	$p^2 + Fpq = D$
A_1A_2	$2pq(1 - F)$			=	$2pq\ (1 - F) = H$
A_2A_2	$q^2(1 - F)$	+	qF	=	$q^2 + Fpq = R$

Therefore, the consequence of inbreeding is that the frequency of homozygotes is higher, and the frequency of heterozygotes is lower, than in a population that is in Hardy-Weinberg equilibrium. Note that *H*, the frequency of heterozygotes in the inbred population, equals $(1 - F)$ multiplied by the frequency of heterozygotes we expect to find in a randomly mating population ($2pq$). Denoting $2pq$ as H_0, we have

$$H = H_0(1 - F) \quad \text{or} \quad F = (H_0 - H)/H_0$$

Thus we can estimate the inbreeding coefficient by two measurable quantities, the observed frequency of heterozygotes, *H*, and the "expected" frequency, $2pq$, which we can calculate from data on the allele frequencies *p* and *q*. In practice, then, the inbreeding coefficient *F* is a measure of the reduction in heterozygosity compared with a panmictic population with the same allele frequencies.

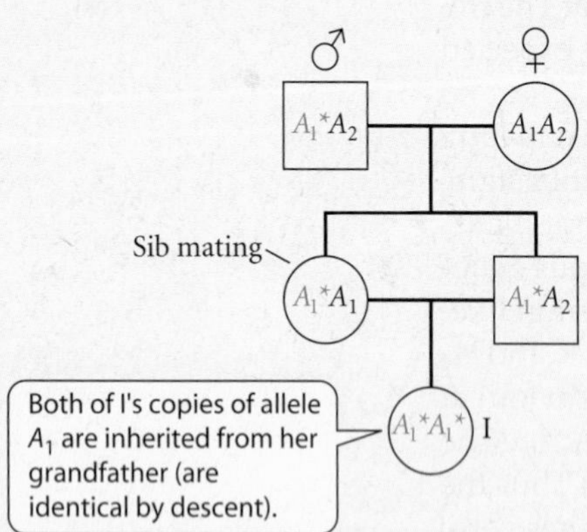

Figure 1 A pedigree showing inbreeding due to mating between siblings. Squares represent males, circles females. Copies of an A_1 allele, A_1* (red type), are traced through three generations. Individual I possesses two copies of A_1* that are identical by descent (she is autozygous). I's mother is homozygous for A_1, but the two copies are not identical by descent (she is allozygous).

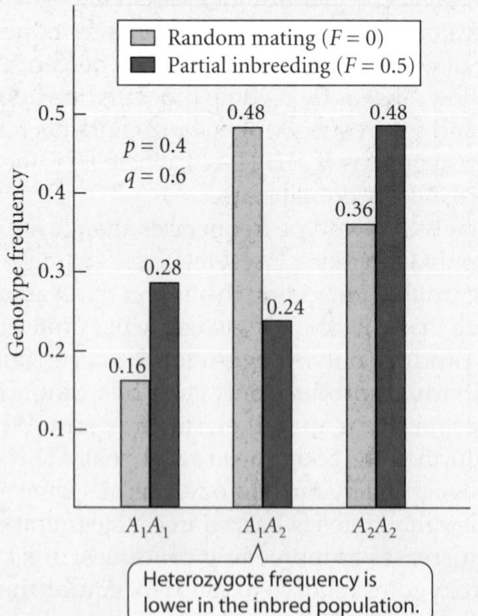

Figure 2 Genotype frequencies at a locus with allele frequencies $p = 0.4$ and $q = 0.6$ when mating is random ($F = 0$) and when the population is partially inbred ($F = 0.5$).

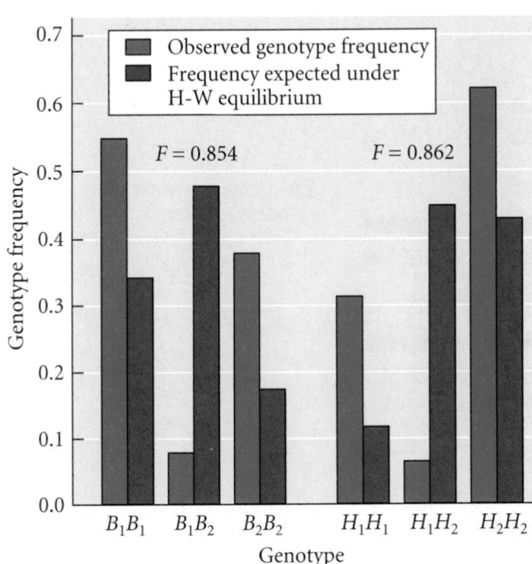

Figure 9.9 Genotype frequencies observed at two loci in a population of the self-fertilizing wild oat *Avena fatua* compared with those expected under Hardy-Weinberg equilibrium. Note that heterozygotes are deficient at both loci, and that calculated values of F are nearly the same for the two loci. (Data from Jain and Marshall 1967.)

mostly by selfing, and it has a low frequency of heterozygotes at the several loci studied (Figure 9.9). The inbreeding coefficients estimated from data on all the loci were nearly equal, which is as it should be, because inbreeding affects all loci in the same way.

Genetic Variation in Natural Populations: Individual Genes

Morphology and viability

Genetic **polymorphism** (*poly*, "many"; *morph*, "form") is the presence in a population of two or more variants (alleles or haplotypes). The term is usually qualified to mean that the rarer allele has a frequency greater than 0.01. A locus or character that is not polymorphic is **monomorphic**. Polymorphism originally referred to genetically determined phenotypes, but the term has been extended to include variation at the molecular level.

Until the 1960s, polymorphisms in natural populations could be detected and studied only by finding variable phenotypic features in species showing Mendelian segregation in experimental crosses—which limited study to few characteristics and only a few experimentally tractable species. Some dramatic variations were shown to be single-locus polymorphisms (e.g., Figure 9.1A), and some have been studied extensively (such as the butterfly polymorphism in Figure 9.1B).

In the 1920s, geneticists found that the same mutations of *Drosophila* that had been studied in laboratory populations—mutations affecting coloration, bristles, wing shape, and the like—also existed in wild populations. These recessive alleles could be revealed and studied by "extracting" chromosomes—that is, producing flies that are homozygous for a particular chromosome. This is accomplished by a series of crosses between a wild-type population and a laboratory stock that carries a dominant marker allele and an inversion that suppresses crossing over (Figure 9.10). The inversion causes the entire chromosome to be inherited as a unit, and the genetic marker enables the experimenter to trace its inheritance and to bring two copies of the chromosome together in homozygous form. A chromosome that has been made homozygous in this way may carry not only recessive alleles that affect morphological features, but also alleles that affect traits such as viability (survival from egg to adult). The crosses produce an F_3 generation of flies, of which one-fourth are expected to be homozygous for the wild chromosome. If no wild-type adult offspring (i.e., offspring without any of the dominant markers) appear, then the wild chromosome must carry at least one recessive **lethal allele**—that is, one that causes death before the flies reach the adult stage. Performing such crosses with many different wild flies makes it easy to determine what fraction of wild chromosomes cause complete lethality or reduce the likelihood of survival.

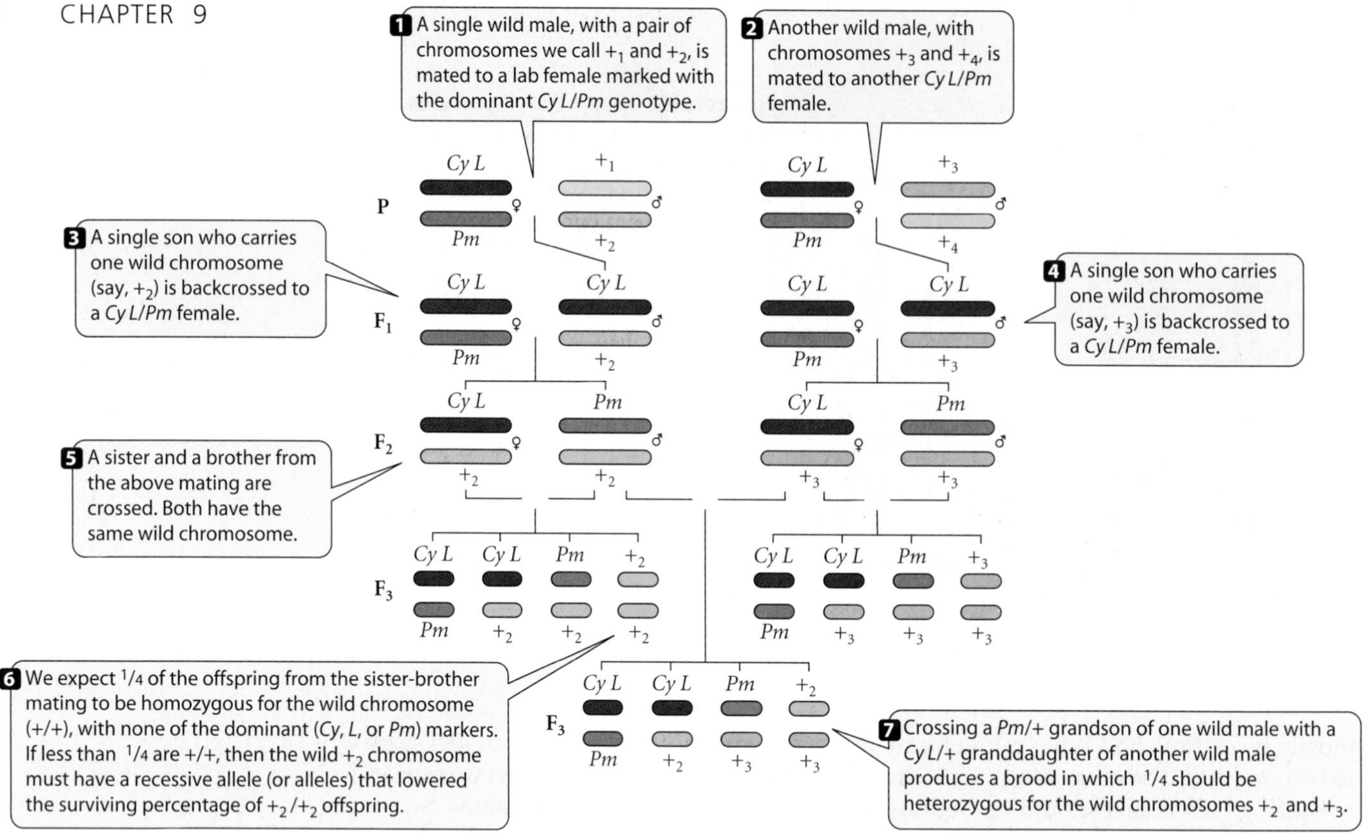

1 A single wild male, with a pair of chromosomes we call $+_1$ and $+_2$, is mated to a lab female marked with the dominant *Cy L/Pm* genotype.
2 Another wild male, with chromosomes $+_3$ and $+_4$, is mated to another *Cy L/Pm* female.
3 A single son who carries one wild chromosome (say, $+_2$) is backcrossed to a *Cy L/Pm* female.
4 A single son who carries one wild chromosome (say, $+_3$) is backcrossed to a *Cy L/Pm* female.
5 A sister and a brother from the above mating are crossed. Both have the same wild chromosome.
6 We expect 1/4 of the offspring from the sister-brother mating to be homozygous for the wild chromosome (+/+), with none of the dominant (*Cy*, *L*, or *Pm*) markers. If less than 1/4 are +/+, then the wild $+_2$ chromosome must have a recessive allele (or alleles) that lowered the surviving percentage of $+_2/+_2$ offspring.
7 Crossing a *Pm*/+ grandson of one wild male with a *Cy L*/+ granddaughter of another wild male produces a brood in which 1/4 should be heterozygous for the wild chromosomes $+_2$ and $+_3$.
P
F_1
F_2
F_3

Figure 9.10 Crossing technique for "extracting" a chromosome from a wild-type male *Drosophila melanogaster* and making it homozygous in order to detect recessive alleles. In this figure, plus signs denote wild-type chromosomes, and the subscript number identifies a particular chromosome. The process is shown for two wild-type males (with chromosome pairs $+_1,+_2$ and $+_3,+_4$), each of which is crossed to a laboratory stock that has dominant mutant markers on the homologous chromosomes: *Cy* (curly wing) and *L* (lobed wing) on one and *Pm* (plum eye color) on the other. Each of these chromosomes also has inversions (see Chapter 8) that prevent crossing over. Consequently, one of each male's wild-type chromosomes (say, $+_2$) is transmitted intact to the F_3 generation. When crosses are performed as shown, the F_3 generation consists, in principle, of equal numbers of four genotypes, including a +/+ homozygote and the *Cy L*/+ and *Pm*/+ heterozygotes. The viability of these genotypes is measured by their proportion in the F_3 generation, relative to the expected 1:1:1:1 ratio. The lowermost family illustrates how flies heterozygous for two wild-type chromosomes ($+_2$ and $+_3$) can be produced. Their viability is also measured by deviation from the 1:1:1:1 ratio expected in this cross. (After Dobzhansky 1970.)

Theodosius Dobzhansky and his collaborators examined hundreds of copies of chromosome 2* sampled from wild *Drosophila pseudoobscura*. They discovered that about 10 percent of those copies were lethal when made homozygous (Figure 9.11). About half of the remaining chromosomes reduced survival to at least some extent. The other chromosomes in the genome yielded similar results, leading to the conclusion that almost every wild fly carries at least one chromosome that, if homozygous, substantially reduces the likelihood of survival. Moreover, many chromosomes cause sterility as well. Using a very different analytical method based on the mortality of children from marriages between relatives, Morton, Crow, and Muller (1956) concluded that humans are just like flies: "the average person carries heterozygously the equivalent of 3–5 recessive lethals [lethal alleles] acting between late fetal and early adult stages."

Such data, which have been confirmed by later work, point to a staggering incidence of life-threatening genetic defects. They imply, moreover, that *natural populations carry an enormous amount of concealed genetic variation* that is manifested only when individuals are homozygous. However, when flies carrying two different lethal chromosomes (i.e., derived from two wild flies) are crossed, the heterozygous progeny generally have nearly normal viability (see Figure 9.11). Hence the two chromosomes must carry recessive lethal alleles at different loci: one lethal homozygote, for example, is *aaBB*, the other is *AAbb*, and the heterozygote, *AaBb*, has normal viability because each recessive lethal is masked by a dominant "normal" allele. From such data, it can be determined that the lethal allele at any one locus is very rare ($q < 0.01$ or so), and that the high proportion of lethal chromosomes is caused by the summation of rare lethal alleles at many loci.

***Drosophila melanogaster*, *D. pseudoobscura*, and many other species of *Drosophila* have four pairs of chromosomes, denoted X/Y, 2, 3, and 4. Chromosome 4 is very small and has few genes. The X chromosome and the two major autosomes, chromosomes 2 and 3, carry roughly similar numbers of genes.

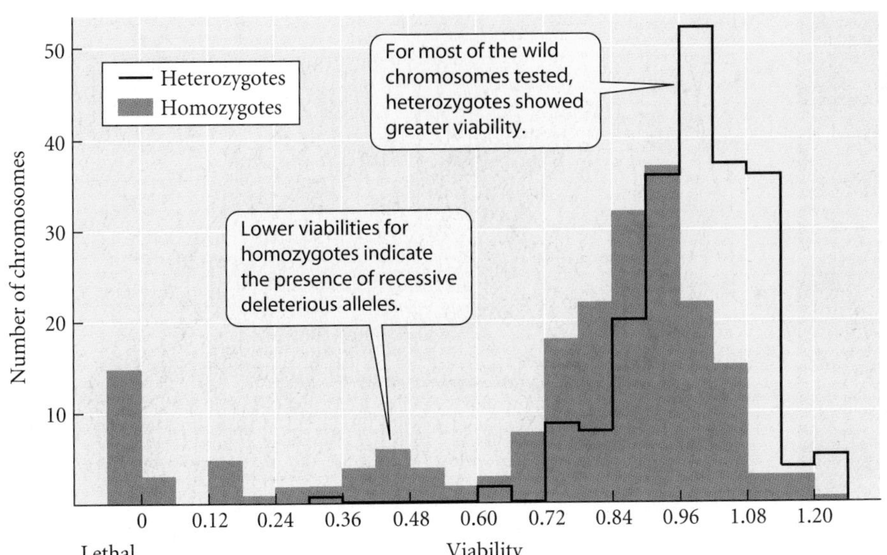

Figure 9.11 The frequency distribution of relative viabilities of chromosomes extracted from a wild population of *Drosophila pseudoobscura* by the method illustrated in Figure 9.10. A viability value of 1.00 indicates conformity to the expected ratios of laboratory and wild-type genotypes in the F_3 generation, as explained in Figure 9.10. The colored distribution shows the viability (relative survival from egg to adult) of homozygotes for 195 wild chromosomes. The great majority of chromosomes lower viability when made homozygous, indicating that they carry deleterious recessive alleles. Heterozygotes for various wild chromosomes, shown by the black line, have higher average viability. (After Lewontin 1974.)

Inbreeding depression

Because populations of humans and other diploid species harbor recessive alleles that have deleterious effects, and because inbreeding increases the proportion of homozygotes, populations in which many matings are consanguineous often manifest a decline in components of fitness, such as survival and fecundity. Such a decline is called **inbreeding depression**. This effect has long been known in human populations (Figure 9.12). For example, a very high proportion (27 to 53 percent) of individuals with Tay-Sachs disease, a lethal neurodegenerative disease caused by a recessive allele that is especially prevalent in Ashkenazic Jewish populations, are children of marriages between first cousins (Stern 1973).

Inbreeding depression is a well-known problem in the small, captive populations of endangered species that are propagated in zoos for reestablishing wild populations. Special breeding designs are required to minimize inbreeding in these populations (Frankham et al. 2002; Figure 9.13). Inbreeding also may increase the risk of extinction of small populations in nature. Thomas Madsen and colleagues (1995, 1999) studied an isolated Swedish population of a small poisonous snake, the adder *Vipera berus*, that consisted of fewer than 40 individuals. The snakes were found to be highly homozygous (we will see shortly how this can be determined), the females had small litter sizes (compared with outbred adders in other populations), and many of the offspring were deformed or stillborn. The authors introduced 20 adult male adders from other populations, left them there for four mating seasons, and then removed them. Soon thereafter, the population increased dramatically (Figure 9.14), owing to the improved survival of the outbred offspring.

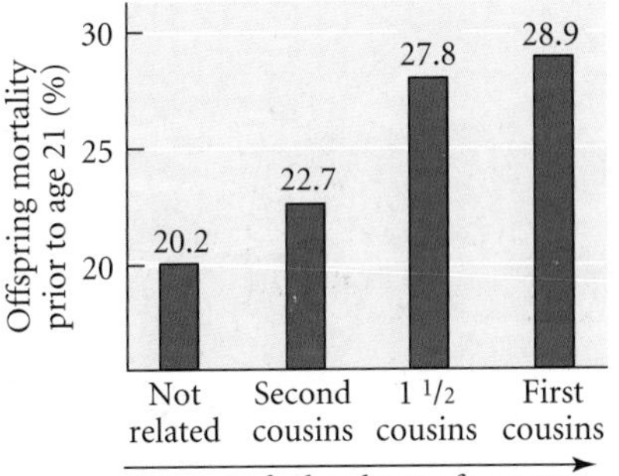

Figure 9.12 Inbreeding depression in humans: the more closely related the parents, the higher the mortality rate among their offspring. The data are for offspring up to 21 years of age from marriages registered in 1903–1907 in Italian populations. (After Stern 1973.)

Genetic variation at the molecular level

Evolution would be very slow if populations were genetically uniform, and if only occasional mutations arose and replaced pre-existing genotypes. In order to know what the potential is for rapid evolutionary change, it would be useful to know how much genetic variation natural populations contain.

Answering this question requires that we know what fraction of the loci in a population are polymorphic, how many alleles are present at each locus, and what their frequencies are. To know this, we need to count both the monomorphic (invariant) and polymorphic loci in a random sample of loci. Ordinary phenotypic characters cannot provide this information because we cannot count how many genes contribute to phenotypically uniform traits.

In 1966, Richard Lewontin and John Hubby, working with *Drosophila pseudoobscura*, addressed this question in a landmark paper. They reasoned that because most loci code for proteins (including enzymes), an invariant enzyme should signal a monomorphic

(A)

(B)

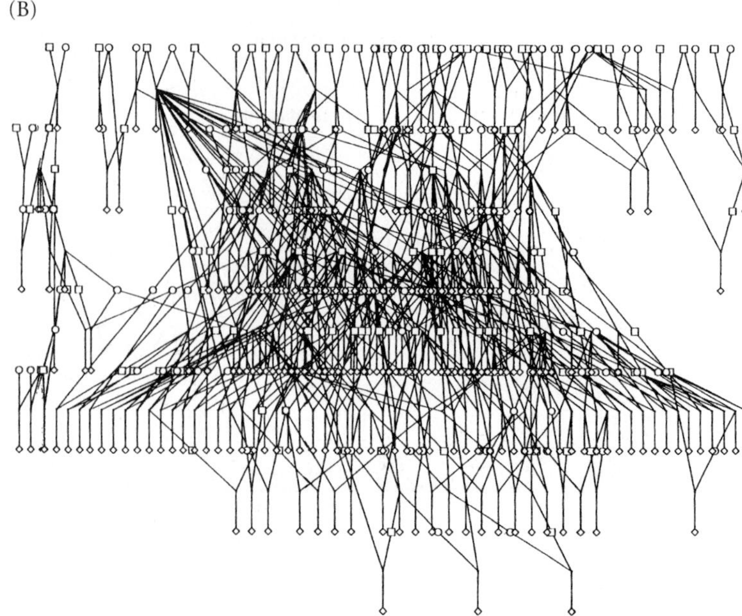

Figure 9.13 (A) The golden lion tamarin (*Leontopithecus rosalia*) is a small, highly endangered Brazilian monkey. Offspring from a captive breeding program in 140 zoos, which collectively maintain about 500 individuals, are now being reintroduced into natural preserves. (B) Inbreeding is reduced by an elaborate outcrossing scheme, illustrated by this pedigree. Lines show offspring of individual females (circles) and males (squares). (A © Tom & Pat Leeson/Photo Researchers, Inc.; B from Frankham et al. 2002.)

locus, and a variable enzyme should signal a polymorphic locus. At that time, DNA sequencing had not yet been developed, but biochemists could visualize certain proteins with **electrophoresis**. In this technique, proteins in a tissue extract move slowly through a starch gel or some other medium to which an electrical current is applied. Certain amino acid substitutions alter the net charge of a protein, so some variants that are encoded by different alleles differ in mobility; their positions can be visualized by reacting the enzyme with a substrate, then making the product visible as a colored blot. The various homozygotes and heterozygotes can then be distinguished (Figure 9.15). Electrophoretically distinguishable forms of an enzyme that are encoded by different alleles are called **allozymes**. The amount of genetic variation is often underestimated by this technique because not all amino acid substitutions alter electrophoretic mobility.

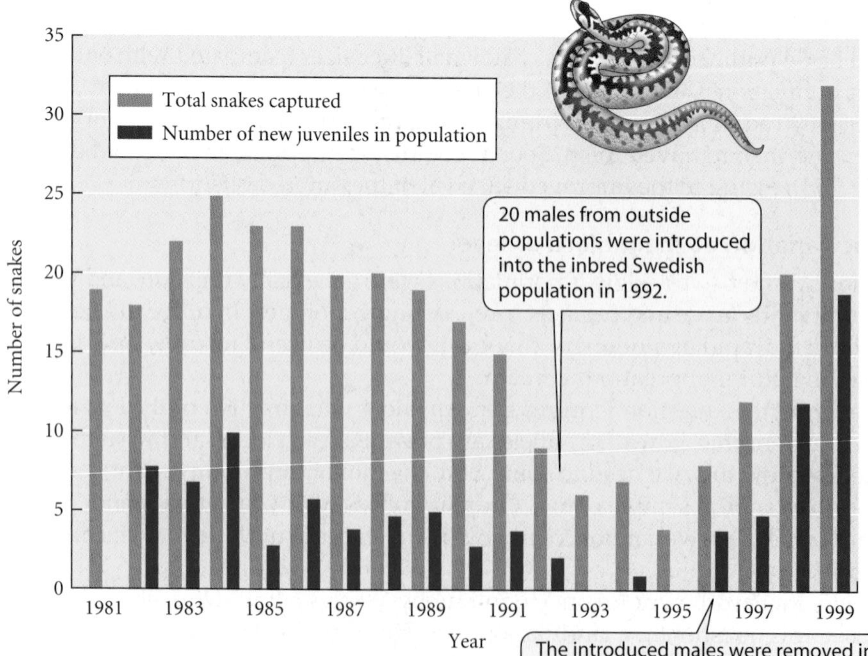

Figure 9.14 Population decline and increase in an inbred population of adders in Sweden. The gold bars represent the total number of males found in the population each year; the blue bars show the number of juveniles recruited into the population. (After Madsen et al. 1999.)

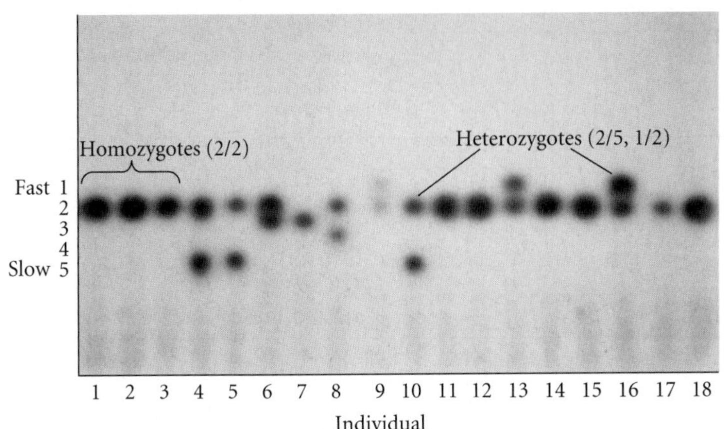

Figure 9.15 An electrophoretic gel, showing genetic variation in the enzyme phosphoglucomutase among 18 individual killifishes (*Fundulus zebrinus*). Five allozymes (alleles) can be distinguished by differences in mobility. The fastest, at top, is allele 1; the slowest, at bottom, is allele 5. Homozygotes display a single band, and heterozygotes two bands. (Courtesy of J. B. Mitton.)

Lewontin and Hubby examined 18 loci in populations of *Drosophila pseudoobscura*. In every population, about a third of the loci were polymorphic, represented by two to six different alleles, and these alleles segregated at high frequencies. The proportion of heterozygotes is a good measure of how nearly equal in frequency the alleles are (see Figure 9.8 and Box 9A). Assuming Hardy-Weinberg equilibrium, the frequency of heterozygotes (H) at a locus is $1 - \Sigma p_i^2$, where p_i is the frequency of the ith allele and p_i^2 is the frequency of homozygote A_iA_i. Averaged over all 18 loci (including the monomorphic ones), the frequency of heterozygotes was about 0.12 in each of Lewontin and Hubby's populations. This is equivalent to saying that an average individual is heterozygous at 12 percent of its loci. (This calculation is called the AVERAGE HETEROZYGOSITY, $\overline{H}$.) In a similar study of a human population, Harris and Hopkinson (1972) found that 28 percent of 71 loci were polymorphic, and that the average heterozygosity ($\overline{H}$) was 0.07. Since these pioneering studies, other investigators have examined hundreds of other species, most of which have proved to have similarly high levels of genetic variation.

Lewontin and Hubby's paper (see also Lewontin 1974) had a great impact on evolutionary biology. Their data, and Harris's, confirmed that *almost every individual in a sexually reproducing species is genetically unique.* (Even with only two alleles each, 3000 polymorphic loci—the estimate for humans—could generate $3^{3000} = 10^{1431}$ genotypes, an unimaginably large number.) Populations are far more genetically diverse than almost anyone had previously imagined. Lewontin and Hubby asked what factors could be responsible for so much variation, and their answers set in motion a research agenda that kept population geneticists busy for three decades. Their central question was, "Do forces of natural selection maintain this variation, or is it neutral, subject only to the operation of random genetic drift?"

This approach resulted in an enormous advance in the study of genetic variation, because it revealed, in almost every species, abundant polymorphisms with a clear genetic basis. These polymorphisms could be studied in their own right (e.g., to study natural selection), or they could be used simply as genetic markers to determine, for example, which individuals mate with each other, or how genetically different related species or populations are from each other. However, this approach has been largely superseded by DNA sequencing—which at first was an extraordinarily laborious procedure. Because of technical advances, studies today routinely include much larger samples.

The first study of genetic variation using complete DNA sequencing was carried out by Martin Kreitman (1983), who studied the locus that encodes alcohol dehydrogenase (ADH) in *Drosophila melanogaster*. Throughout the world, populations of this species are polymorphic for two common electrophoretic alleles, "fast" (Adh^F) and "slow" (Adh^S).

Kreitman sequenced 11 copies of a 2721–base pair region that includes four exons and three introns, as well as noncoding flanking regions on either side (Figure 9.16). Among only 11 gene copies, he found 43 variable base pair sites, as well as six insertion/deletion polymorphisms (presence or absence of short runs of base pairs). Fewer sites were vari-

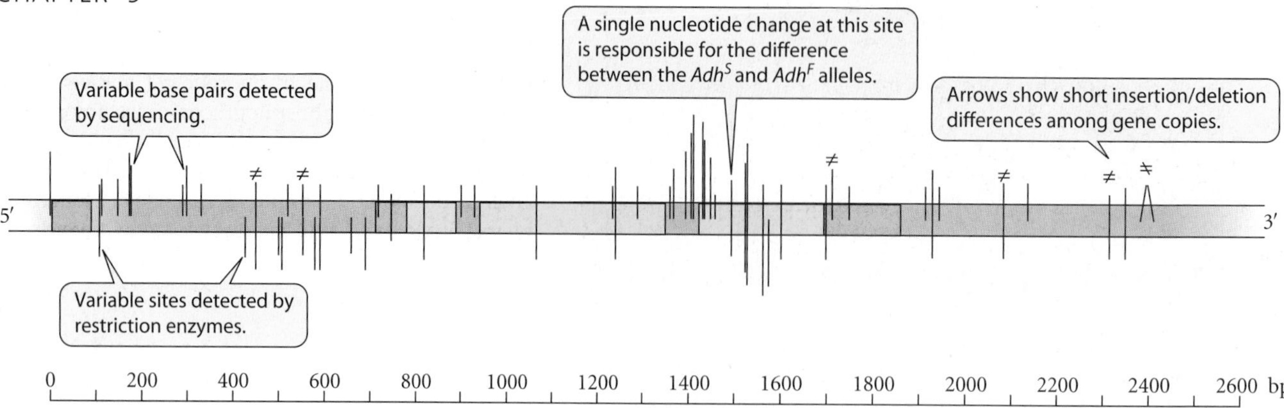

Figure 9.16 Nucleotide variation at the *Adh* locus in *Drosophila melanogaster*. Four exons (colored blocks) are separated by introns (gray blocks). The yellow blocks represent the coding regions of the exons. The lines intersecting the diagram from above show the positions of 43 variable base pairs and 6 insertions or deletions (indicated by arrows) detected by sequencing 11 copies of this gene. The lines intersecting the diagram from below show the positions of 27 variable sites detected by another method (restriction enzymes) among 87 gene copies. The scale below shows base pair (bp) position along the sequence. (After Kreitman 1983.)

able in the exons (1.8 percent) than in the introns (2.4 percent). The most striking discovery was that all but 1 of the 14 variations in the coding region were synonymous substitutions, the exception being a single nucleotide change responsible for the single amino acid difference between the Adh^S and Adh^F alleles.

Considerable sequence variation, especially synonymous variation, has been found in most of the genes and organisms that have been examined since Kreitman's pioneering study. A common measure of variation, analogous to the average heterozygosity ($\overline{H}$) used to quantify allozyme variation, is the average NUCLEOTIDE DIVERSITY PER SITE (π), the proportion of nucleotide sites at which two gene copies ("sequences") randomly taken from a population differ. For Kreitman's *Adh* sample, $\pi = 0.0065$, although over the genome as a whole, *D. melanogaster* is much more variable ($\pi = 0.05$; Li 1997). In human populations, the average nucleotide diversity per site across loci is $\pi = 0.0015$. DNA sequence variation in natural populations includes all the kinds of mutations described in Chapter 8, and it provides abundant genetic markers for studying mating patterns, population structure (Chapter 10), phylogeography (Chapter 6), natural selection (Chapter 12), and many other topics.

Genetic Variation in Natural Populations: Multiple Loci

All genetic variation owes its origin ultimately to mutation, but in the short term, a great deal of the genetic variation within populations arises through recombination. In sexually reproducing eukaryotes, genetic variation arises from two processes: the union of genetically different gametes, and the formation of gametes with different combinations of alleles, owing to independent segregation of nonhomologous chromosomes and to crossing over between homologous chromosomes.

The potential genetic variation that can be released by recombination is enormous. To cite a modest example: if an individual is heterozygous for only one locus on each of five pairs of chromosomes, independent segregation alone generates $2^5 = 32$ allele combinations among its gametes, and mating between two such individuals could give rise to $3^5 = 243$ genotypes among their progeny. If each locus affects a different feature, this represents a great variety of character combinations. However, recombination can both increase and decrease genetic variation. In sexually reproducing populations, genes are transmitted to the next generation, but genotypes are not: they end with organisms' deaths, and are reassembled anew in each generation. Thus an unusual, favorable gene combination may occasionally arise through recombination, but if the individuals bearing it mate with other members of the population, it will be lost immediately by the same process. Consequently, recombination can either enhance or retard adaptation.

If alleles at one locus (say A_1 and A_2) were generally associated with specific alleles at another locus (say B_1, B_2), any factor (such as natural selection) that caused a change in the frequencies of alleles at one locus would result in a correlated change at the other locus. For instance, if eggs and sperm generally carried either the haploid combination A_1B_1 or A_2B_2, an increase in the frequency of A_1 would automatically cause B_1 to increase (a phenomenon called **genetic hitchhiking**). If the loci affected different characteristics, an adaptive change in one feature would then result in a nonadaptive change in another feature. So it is important to know if such associations between alleles are likely to be common.

Suppose two populations, one consisting only of genotype $A_1A_1B_1B_1$ and the other only of $A_2A_2B_2B_2$, are mixed together and mate at random. In the next generation, there will be three genotypes in the population, A_1B_1/A_1B_1, A_1B_1/A_2B_2, and A_2B_2/A_2B_2 (the slash separates the allele combinations an individual receives from its two parents). In this generation, there is a perfect association, or correlation, between specific alleles at the two loci: A_1 with B_1 and A_2 with B_2. This association is referred to as **linkage disequilibrium (LD)**. If there is no such association, the loci are in **linkage equilibrium**. "Linkage" means the physical association of genes on the same chromosome. (As we will see, alleles at loci on different chromosomes may be nonrandomly associated, i.e., in a state of LD, so this term is slightly misleading.)

Recombination during meiosis reduces the association, the level of LD, between specific alleles at the two loci, and brings the loci toward linkage equilibrium. If there were no recombination in our example, gametes would carry only the allele combinations A_1B_1 and A_2B_2, which, when united, could generate only the same three genotypes as before. However, recombination in the double heterozygote (A_1B_1/A_2B_2) gives rise to some A_1B_2 and A_2B_1 gametes, which, by uniting with A_1B_1 or A_2B_2 gametes, produce genotypes such as A_1B_1/A_1B_2. Let us denote the frequencies of the gamete types as g_{ij}, where i represents the A allele and j the B allele (for example, g_{12} is the frequency of the A_1B_2 combination). A COEFFICIENT OF LINKAGE DISEQUILIBRIUM may be defined as $D = (g_{11} \times g_{22}) - (g_{12} \times g_{21})$. If $D > 0$, then gametes A_1B_1 and A_2B_2, as well as genotypes formed by their union (such as $A_1A_1B_1B_1$), will be more frequent than expected. As recombination occurs, then, D declines because the "deficient" allele combinations (A_1B_2, A_2B_1) and genotypes increase in frequency slowly from generation to generation, until the alleles at each locus are randomized with respect to alleles at the other locus and there is no correlation between them (Figure 9.17). The higher the recombination rate between the loci is, the faster the approach to linkage equilibrium proceeds: it is most rapid for unlinked genes (Figure 9.18).

When the loci are in linkage equilibrium, knowing which A allele an egg or sperm carried would not help us predict its B allele. In this case, the probability (frequency) of a gamete carrying any one of the four possible allele combinations is the product of the probabilities of the alleles considered separately. These probabilities are the allele frequencies. If we denote the frequencies of alleles A_1 and A_2 as p_A and q_A, ($p_A + q_A = 1$), and of B_1 and B_2 as p_B and q_B, ($p_B + q_B = 1$), the frequency of the A_1B_1 combination among the eggs (or sperm) produced by the population as a whole is then p_Ap_B, and the frequency (proportion) of genotype $A_1A_1B_1B_1$ (A_1B_1/A_1B_1) in the next generation is $(p_Ap_B)^2$, or $p_A^2p_B^2$. If the observed frequencies of all the genotypes conform to the expected frequencies thus calculated, we can conclude that the loci are in a state of linkage equilibrium. Whether loci are in linkage equilibrium or linkage disequilibrium, the genotype frequencies at each locus, viewed individually, conform to Hardy-Weinberg frequencies. At linkage equilibrium, the loci could be closely linked, but the probability that a copy of the A_1 allele would have a B_1 as its neighbor on the same chromosome would be p_B, its frequency in the population at large.

Why would we ever find loci in linkage disequilibrium, if recombination breaks it down? There are several possible reasons.

1. Nonrandom mating can maintain LD. In an extreme case, the sample of organisms that display LD may actually include two reproductively isolated species.
2. When a new mutation arises, the single copy is necessarily associated with specific alleles at other loci on the chromosome, and therefore is in LD with those alleles. The copies of this mutation in subsequent generations will retain this association until it is broken down by recombination.
3. The population may have been formed recently by the union of two populations with different allele frequencies, and LD has not yet decayed.
4. Recombination may be very low or nonexistent. Chromosome inversions and parthenogenesis (asexual reproduction) have this effect.

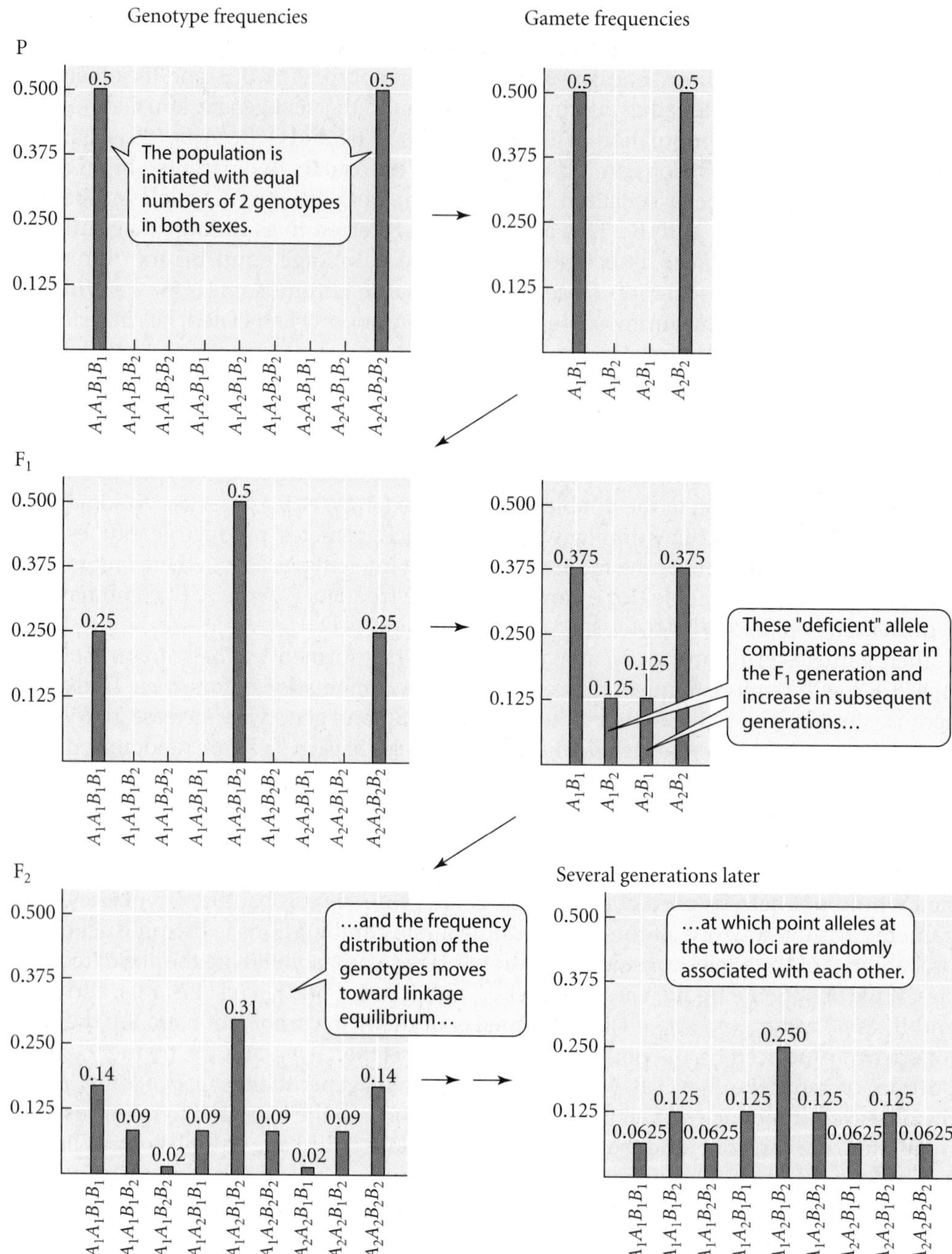

Figure 9.17 The decay of linkage disequilibrium between two unlinked loci over three generations in a population initiated with equal numbers of two genotypes ($A_1A_1B_1B_1$ and $A_2A_2B_2B_2$) in both sexes. The genotype frequencies among population members are shown at left; the frequencies of allele combinations among their gametes, which will join to produce the next generation, are shown at right. Beginning in the first offspring generation (F_1), the frequency of deficient allele combinations in the gametes (A_1B_2, A_2B_1) increases, and that of excessive combinations (A_1B_1, A_2B_2) decreases, because of recombination in double heterozygotes (A_1B_1/A_2B_2). Therefore, the frequency distribution of genotypes among the offspring changes. This process continues until linkage equilibrium (final graph), when the alleles at the two loci are randomly associated with each other.

5. LD may be caused by genetic drift (see Chapter 10). If the recombination rate is very low, the four gamete types in the example above may be thought of as if they were four alleles at one locus. One of these "alleles" may drift to high frequency by chance, creating an excess of that combination relative to others.
6. Natural selection may cause LD if two or more gene combinations are much fitter than recombinant genotypes.

In panmictic sexually reproducing populations, pairs of polymorphic loci often are found to be in linkage equilibrium, or nearly so. There are some interesting exceptions, however. For example, the European primrose *Primula vulgaris* is HETEROSTYLOUS, meaning that plants within a population differ in the lengths of stamens and style (pistil). Almost all plants have either the "pin" phenotype, with long style and short stamens, or the "thrum" phenotype, with short style and long stamens (Figure 9.19). Thus the anthers

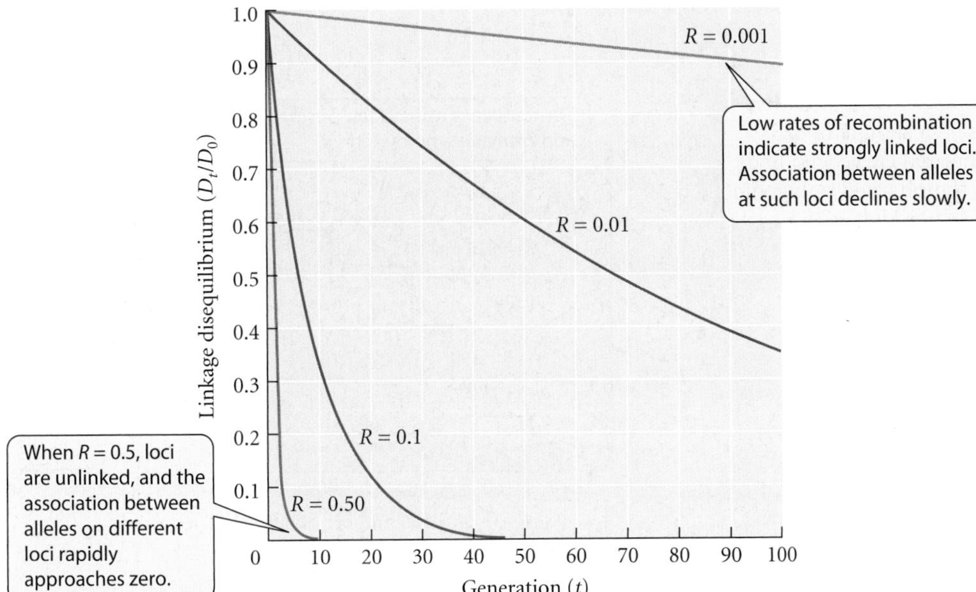
R = 0.001
Low rates of recombination indicate strongly linked loci. Association between alleles at such loci declines slowly.
R = 0.01
R = 0.1
R = 0.50
When R = 0.5, loci are unlinked, and the association between alleles on different loci rapidly approaches zero.
Linkage disequilibrium (D_t/D_0)
Generation (t)
1.0
0.9
0.8
0.7
0.6
0.5
0.4
0.3
0.2
0.1
0
10
20
30
40
50
60
70
80
90
100

Figure 9.18 The decrease in linkage disequilibrium (D) over time, relative to its initial value (D_0), for pairs of loci with different recombination rates (R). $R = 0.50$ if loci are unlinked. (After Hartl and Clark 1989.)

of each morph (at the ends of the stamens) are at the same level as the other morph's stigma. In most experimental crosses, this difference is inherited as if it were due to a single pair of alleles, with *thrum* dominant over *pin*. Rarely, however, "homostylous" progeny are produced, in which the female and male structures are equal in length (either short or long). Thus style and stamen length are actually determined by separate, closely linked loci: alleles G and g determine short and long style, respectively, and alleles A and a determine long and short stamens, respectively. Thrum plants have genotype GA/ga, and pin plants ga/ga. The gamete combinations Ga and gA, which give rise to homostylous plants, are very rare—partly because thrum and pin phenotypes are most successful in cross-pollination, since pollen from one is placed on an insect in a position that corresponds to the stigmatic surface of the other.

Linkage disequilibrium is common in asexual populations (see Chapter 15) because they undergo little recombination. It is also found among very closely situated molecular markers, such as sites within genes. LD provides important ways of detecting processes such as natural selection from DNA sequence data (see Chapter 12), and researchers in human genetics use molecular markers, in a procedure called LINKAGE DIS-

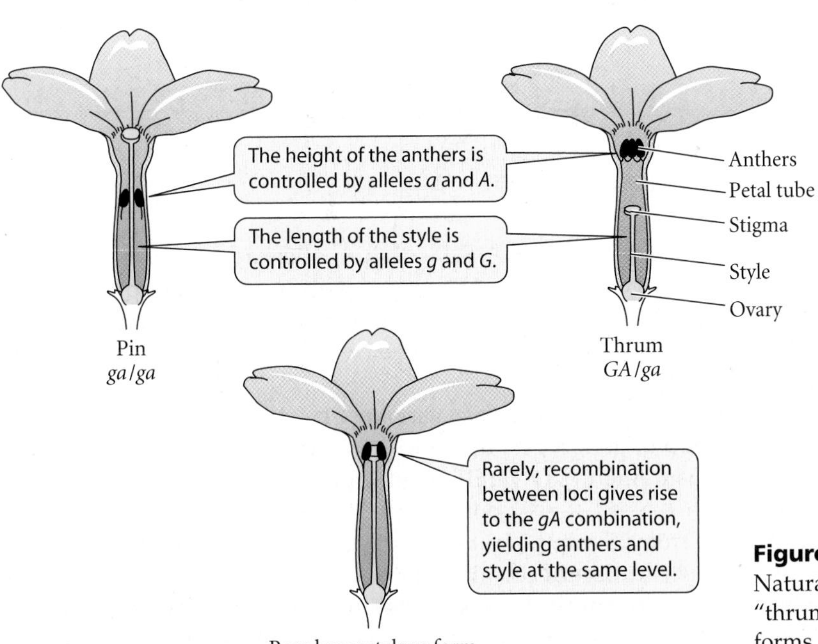
The height of the anthers is controlled by alleles a and A.
The length of the style is controlled by alleles g and G.
Anthers
Petal tube
Stigma
Style
Ovary
Pin
ga/ga
Thrum
GA/ga
Rarely, recombination between loci gives rise to the gA combination, yielding anthers and style at the same level.
Rare homostylous form
gA/ga

Figure 9.19 Heterostyly in the primrose *Primula vulgaris*. Natural populations consist almost entirely of "pin" and "thrum" plants. Rarely, crossing over produces homostylous forms, in which the stigma and anthers are situated at the same level. (After Ford 1971.)

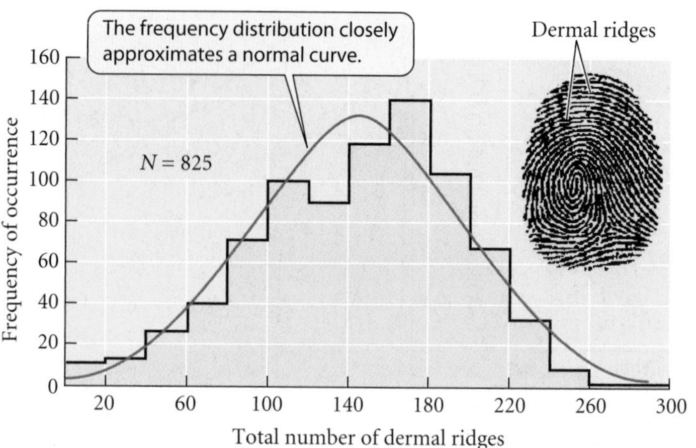

Figure 9.20 The frequency distribution of the number of dermal ridges, summed over all ten fingertips, in a sample of 825 British men. The distribution nearly fits a normal curve (red line). The number of dermal ridges is an additively inherited polygenic character, with a heritability of about 0.95. (After Holt 1955.)

EQUILIBRIUM MAPPING, to find nearby mutations that cause genetic diseases. A great deal of such research, based on sequence variation in the human genome, is under way, and evolutionary geneticists are playing a major role in developing methods for analyzing such data.

Variation in quantitative traits

Discrete genetic polymorphisms in phenotypic traits, such as pin or thrum flowers, are much less common than slight differences among individuals, such as variation in the number of bristles on the abdomen of *Drosophila* or in weight or nose shape among humans. Such variation, called **quantitative variation**—or CONTINUOUS or METRIC variation—often approximately fits a **normal distribution** (Figure 9.20). The genetic component of such variation is often **polygenic**: that is, it is due to variation at several or many loci, each of which contributes to the variation in phenotype. Figure 9.21 shows a typical pattern of variation and inheritance when homozygous strains that differ in a polygenic character are crossed. Note that in this case, the average character of offspring is about midway between the parents' means, and that variation increases in the F_2 generation, because of recombination among many loci that generates a great variety of genotypes.

A simple model of the relation between genotype and phenotype in a case of quantitative variation, in which we envision only two variable loci, might be:

	A_1A_1	A_1A_2	A_2A_2
B_1B_1	5	6	7
B_1B_2	7	8	9
B_2B_2	9	10	11

In this example, relative to the genotype $A_1A_1B_1B_1$, each A_2 allele adds one unit, on average, and each B_2 allele adds two units to the phenotype. This is a model of purely **additive allele effects**. In Figure 9.21, for example, the "small" and "large" parental strains might be $A_1A_1B_1B_1$ and $A_2A_2B_2B_2$, respectively, the F_1 would be $A_1A_2B_1B_2$, and the F_2 would include all nine genotypes.

Recombination can either increase or decrease variation in a quantitative trait, depending on the initial distribution of genotypes. Imagine that at five loci, + and – alleles add or subtract one unit of measurement, say 1 millimeter (mm). If we begin with quintuply heterozygous parents (both +–+–+/–+–+–), both of size 20 mm, recombination can result in offspring ranging in size from 15 mm (–/– at all five loci) to 25 mm (+/+ at all five loci). (Compare the F_1 and F_2 distribution of corolla lengths in Figure 9.21.) However, if we begin with parents that are quintuply homozygous for just + or just – alleles (genotypes +++++/+++++ and – – – – –/– – – – –), with respective sizes 25 and 15 mm, the F_2 generation will show lower variance than the parental generation, because most offspring inherit various mixtures of + and – alleles. (Compare the P and F_2 generations in Figure 9.21.)

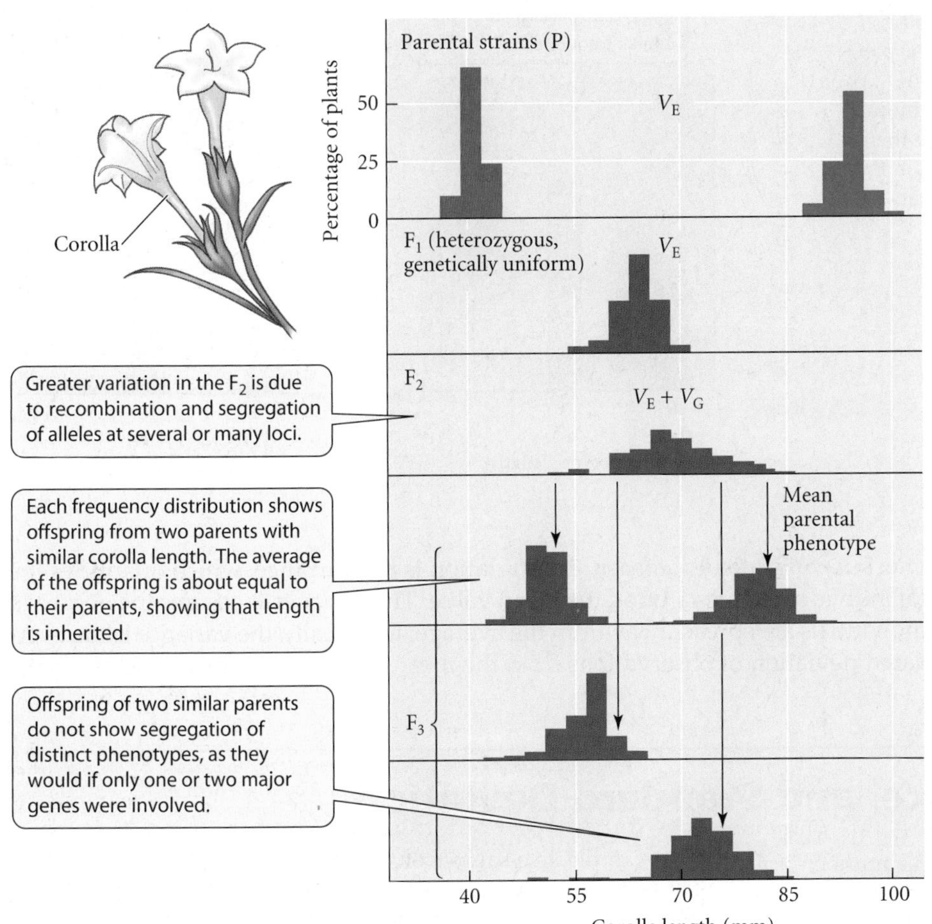
Corolla
Percentage of plants
50
25
0
Parental strains (P)
V_E
F_1 (heterozygous, genetically uniform)
V_E
F_2
$V_E + V_G$
Mean parental phenotype
F_3
40
55
70
85
100
Corolla length (mm)
Greater variation in the F_2 is due to recombination and segregation of alleles at several or many loci.
Each frequency distribution shows offspring from two parents with similar corolla length. The average of the offspring is about equal to their parents, showing that length is inherited.
Offspring of two similar parents do not show segregation of distinct phenotypes, as they would if only one or two major genes were involved.

Figure 9.21 Inheritance of a continuously varying trait, length of the corolla (petals), in the tobacco plant *Nicotiana longiflora*. V_G and V_E represent variation that is due to genes and environment, respectively. The crosses show that the genetic variation is due to multiple genes (polygenic variation) rather than one or two loci. The two parental strains (P) are homozygous genotypes; the F_1 is heterozygous but genetically uniform. The F_2 shows expanded, continuous variation resulting from recombination among the loci that affect the trait. If only one or two loci differed between parental strains, the F_2 would show discrete length categories. Four F_3 families are shown, from parents whose means are indicated by arrows. The mean of the offspring in each family is close to that of their parents, indicating that differences among F_2 phenotypes are inherited. (After Mather 1949.)

In order to judge how much variation is released by recombination, a team led by the great population geneticist Theodosius Dobzhansky studied the effects of chromosomes they had "extracted" (see Figure 9.10) from a wild population of *Drosophila pseudoobscura* (Spassky et al. 1958). Homozygous chromosomes from natural populations of this species show tremendous variation in their effects on survival from egg to adult (see Figure 9.11). However, the Dobzhansky team chose ten homologous chromosomes that conferred almost the same, nearly normal viability when homozygous and made all possible crosses between flies bearing those chromosomes. From the F_1 female offspring, in which crossing over had occurred, they then extracted recombinant chromosomes and measured their effect on viability when homozygous. Even though the original ten chromosomes had differed little in their effect on viability, the variance in viability among the recombinant chromosomes was more than 40 percent of the variance found among homozygotes for much larger samples of chromosomes from natural populations. Thus a single episode of recombination among just ten chromosomes generates a large fraction of a wild population's variability. Some of the recombinant chromosomes were "synthetic lethals," meaning that recombination between two chromosomes that yield normal viability produced chromosomes that were lethal when they were made homozygous. This finding implies that each of the original chromosomes carried an allele that did not lower viability on its own, but did cause death when combined with another allele, at another locus, on the other chromosome.

HERITABILITY. The description and analysis of quantitative variation are based on statistical measures because until recently, the loci that contribute to quantitative variation could not be singled out for study. The amount of genetic variation in a character depends on the number of variable loci, the genotype frequencies at each locus (Figure 9.22), and the phenotypic difference among genotypes.

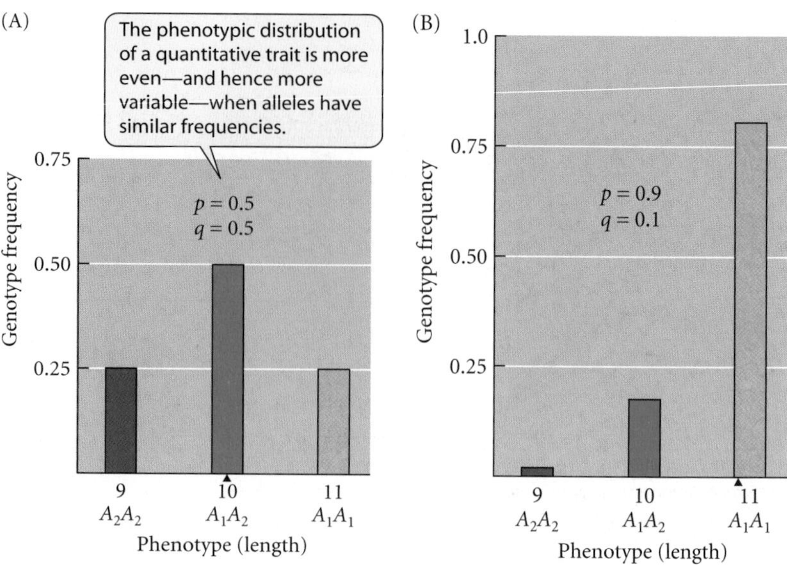

Figure 9.22 Variation in a quantitative trait, such as body length, due to alleles with frequencies (A) $p = 0.5$, $q = 0.5$ or (B) $p = 0.9$, $q = 0.1$. The black triangle denotes the mean. The distribution of lengths is more even, hence more variable, when the alleles have the same frequencies. (In the more variable population B, it would be harder to guess the phenotype of a randomly chosen individual.) The genetic variance, V_G, equals 0.500 in A and 0.472 in B. Hardy-Weinberg equilibrium is assumed in both cases.

The most useful statistical measure of variation is the **variance**, which quantifies the spread of individual values around the mean value. The variance measures the degree to which individuals are spread away from the average; technically, the variance is the average squared deviation of observations from the mean (see Box 9C). In simple cases, the

BOX 9C Mean, Variance, and Standard Deviation

Suppose we have measured a character in a number of specimens. The character may vary continuously, such as body length, or discontinuously, such as the number of fin rays in a certain fin of a fish. Let X_i be the value of the variable in the ith specimen (e.g., $X_3 = 10$ cm in fish number 3). If we have measured n specimens, the sum of the values is $X_1 + X_2 + \ldots + X_n$, or

$$\sum_{i=1}^{n} X_i$$

(or simply ΣX_i). The **arithmetic mean** (or average) is

$$\bar{x} = \frac{\sum X_i}{n}$$

If the variable is discontinuous (e.g., fin rays), we may have n_1 individuals with value X_1, n_2 with value X_2, and so on for k different values. The arithmetic mean is then

$$\bar{x} = \frac{n_1X_1 + n_2X_2 + \cdots + n_k}{n_1 + n_2 + \cdots + n_k}$$

The sum of n_i equals n, so we may write this as

$$\bar{x} = \frac{n_1X_1}{n} + \frac{n_2X_2}{n} + \ldots + \frac{n_kX}{n}$$

If we denote $n_i/n = f_i$, the frequency of individuals with value X_i, this becomes

$$\bar{x} = \sum_{i=1} (f_iX_i)$$

For example, in a sample of $n = 100$ fish, we may have $n_1 = 16$ fish with 9 fin rays ($X_1 = 9$), $n_2 = 48$ fish with 10 rays ($X_2 = 10$), and $n_3 = 36$ fish with 11 rays ($X_3 = 11$). There are thus three phenotypic classes ($k = 3$). The mean is

$$\bar{x} = \sum_{i=1} (f_iX_i) = (0.16)(9) + (0.48)(10) + (0.36)(11) = 10.2$$

Because we have only a sample from the fish population, this sample mean is an *estimate* of the true (parametric) mean of the population, which we can know only by measuring every fish in the population.

How shall we measure the amount of variation? We might measure the *range* (the difference between the two most extreme values), but this measure is very sensitive to sample size. A larger sample might reveal, for example, rare individual fish with 5 or 15 fin rays. These rare individuals do not contribute to our impression of the degree of variation. For this and other reasons, the most commonly used measures of variation are the **variance** and its close relative, the standard deviation. The true (parametric) variance is estimated by the mean value of the square of an observation's variation from the arithmetic mean:

$$V = \frac{(X_1 - \bar{x})^2 + (X_2 - \bar{x})^2 + \cdots + (X_n - \bar{x})^2}{n_1 - 1}$$
$$= \frac{1}{n-1} \sum_{i=1} n_i(Xi - \bar{x})^2$$

For statistical reasons, the denominator of a sample variance is $n - 1$ rather than n. In our hypothetical data on fin ray counts, variance in a phenotypic character (V_P) is the sum of **genetic variance** (V_G) and the **environmental variance** (V_E) caused by direct effects of environmental differences among individual organisms. That is, $V_P = V_G + V_E$. Oversimplifying, we can imagine that each genotype in a population has an average phenotypic value (of, say, body length), but that individuals with that genotype vary in their phenotypes because of environmental effects or developmental noise. The genetic variance, V_G, measures the amount of variation among the averages of the different genotypes, and the environmental variance, V_E, measures the average amount of variation among individuals with the same genotype (at the relevant loci). The proportion of the phenotypic variance that is genetic variance is the **heritability** of a trait, denoted h^2. Thus,

$$h^2 = V_G/(V_G + V_E)$$

One way of detecting a genetic component of variation, and of estimating V_G and h^2, is to measure correlations* between parents and offspring, or between other relatives. For example, suppose that in a population, the mean value of a character in the members of each brood of offspring were exactly equal to the value of that character averaged between their two parents (the MIDPARENT MEAN; Figure 9.23A). So perfect a correlation clearly would imply a strong genetic basis for the trait. In fact, in this instance, V_G/V_P (i.e., the

*More properly, the *regression* of offspring mean on the mean of the two parents. The regression coefficient measures the slope of the relationship, and is conceptually related to the correlation.

BOX 9C (Continued)

$$V = \frac{16(9-10.2)^2 + 48(10-10.2)^2 + 36(11-10.2)^2}{99} = 0.485$$

The variance is a very useful statistical measure, but it is hard to visualize because it is expressed in squared units. It is easier to visualize its square root, the **standard deviation** (see also Figure 4.22):

$$s = \sqrt{V}$$

For our hypothetical data, $s = \sqrt{0.485} = 0.696$. The meaning of this number can perhaps be best understood by contrast with a sample of 1 fish with 9 rays, 18 with 10 rays, and 81 with 11 rays—an intuitively less variable sample. Then $\bar{x} = 10.8$, $V = 0.149$, and $s = 0.387$. V and s are smaller in this sample than in the previous sample because more of the individuals are closer to the mean.

A continuous variable, such as body length, often has a bell-shaped, or normal, frequency distribution (Figure 1). In the mathematically idealized form of this distribution (which many real samples approximate quite well), about 68 percent of the observations fall within one standard deviation on either side of the mean, 96 percent within two standard deviations, and 99.7 percent within three. If, for example, body lengths in a sample of fish are normally distributed, then if $\bar{x} = 100$ cm and $s = 5$ cm (hence $V = 25$ cm^2), 68 percent of the fish are expected to range between 95 and 105 cm, and 96 percent to range between 90 and 110 cm. If the standard deviation is greater—say, $s = 10$ cm ($V = 100$ cm^2)—then, for the same mean, the range limits embracing 68 percent of the sample are broader: 90 cm and 100 cm.

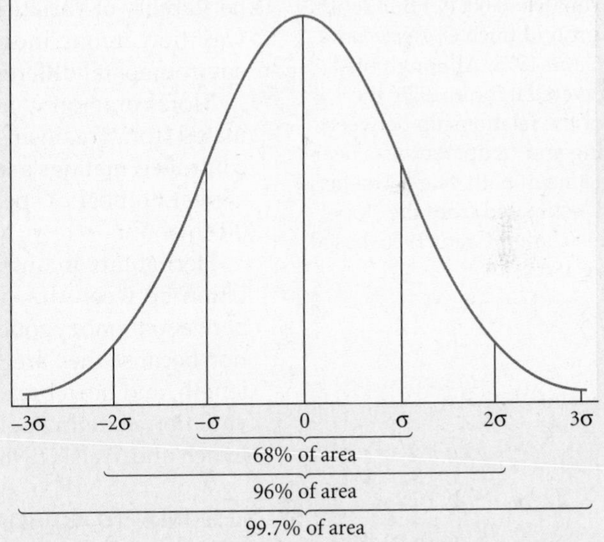

Figure 1 The normal distribution curve, with the mean taken as a zero-reference point, showing how the variable represented on the *x*-axis can be measured in standard deviations (σ). The bracketed areas show the fraction of the area under the curve (that is, the proportion of observations) embraced by one, two, and three standard deviations on either side of the mean. The true (parametric) value of the standard deviation is denoted by σ; the estimate of σ based on a sample is denoted *s* in this book.

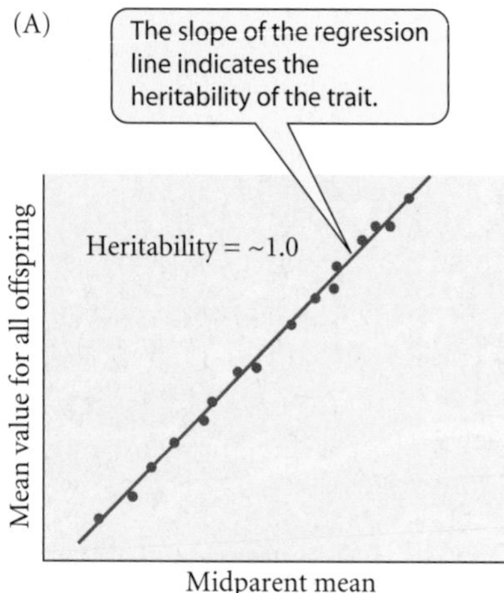

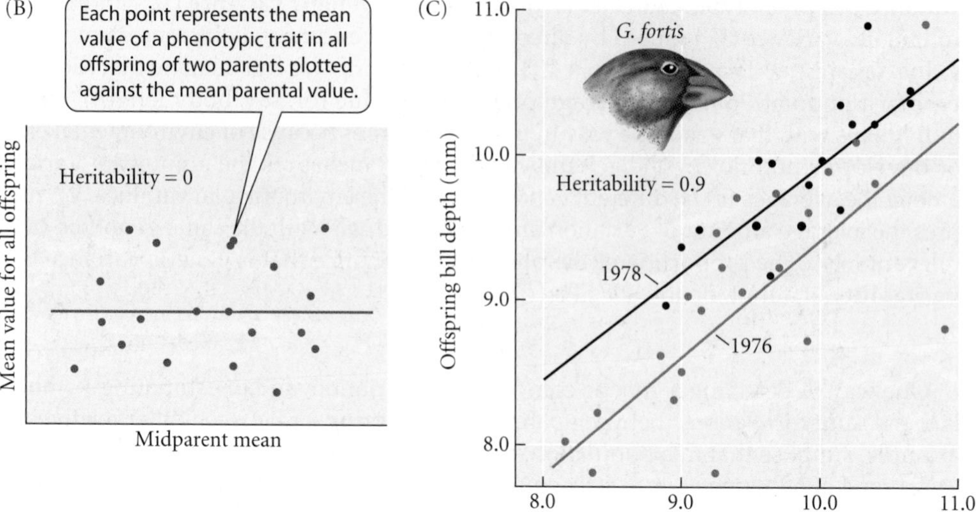

Figure 9.23 The relationship between the phenotypes of offspring and parents. Each point represents the mean of a brood of offspring, plotted against the mean of their two parents (midparent mean). (A) A hypothetical case in which offspring means are nearly identical to midparent means. The heritability is nearly 1.00. (B) A hypothetical case in which offspring and midparent means are not correlated. The slope of the relationship and the heritability are approximately 0.00. (C) Bill depth in the ground finch *Geospiza fortis* in 1976 and 1978. Although offspring were larger in 1978, the slope of the relationship between offspring and midparent was nearly the same in both years. The heritability, estimated from the slope, was 0.9. (C after Grant 1986, based on Boag 1983.)

heritability, h^2) would equal 1.0: all the phenotypic variation would be accounted for by genetic variation. If the correlation were lower, with some of the phenotypic variation perhaps due to environmental variation, the heritability would be lower.

In a real-life example, Peter Boag (1983) studied heritability in the medium ground finch (*Geospiza fortis*) of the Galápagos Islands. He kept track of mated pairs of *G. fortis* and their offspring by banding them so that they could be individually recognized. He measured the phenotypic variance of bill depth and several other features, and correlated the phenotypes of parents and their offspring (Figure 9.23C). Boag estimated that the heritability of variation in bill depth was 0.9; that is, about 90 percent of the phenotypic variation among individuals was attributable to genetic differences, and 10 percent to environmental differences.

More commonly, genetic variance and the heritability of various traits have been estimated from organisms reared in a greenhouse or laboratory, in which it is easier to set up controlled matings and to keep track of progeny. Most of the characteristics reported, for a great number of species, are genetically variable, with h^2 usually in the range of about 0.1 to about 0.9 (e.g., Mousseau and Roff 1987).

Heritability in human populations has often been estimated by comparing the correlation between dizygotic ("fraternal," or "nonidentical") twins with the correlation between monozygotic ("identical") twins, which are expected to have a higher correlation because they are genetically identical. Physical characteristics such as height, finger length, and head breadth are highly heritable (ca. 0.84–0.94) within populations, and the variation in dermatoglyphic traits (fingerprints) is nearly entirely genetically based (0.96; Lynch and Walsh 1998).

RESPONSES TO ARTIFICIAL SELECTION. Because a character can be altered by selection only if it is genetically variable, **artificial selection** can be used to detect genetic variation in a character. To do this, an investigator (or a plant or animal breeder) breeds only those individuals that possess a particular trait (or combination of traits) of interest. Artificial selection may grade into natural selection, but the conceptual difference is that under artificial selection, the reproductive success of individuals is determined largely by a single characteristic chosen by the investigator, rather than by their overall capacity (based on all characteristics) for survival and reproduction.

Hundreds of such experiments have been performed. For example, Theodosius Dobzhansky and Boris Spassky (1969) bred the offspring of 20 wild female *Drosophila pseudoobscura* flies to form a base population, and from this population drew flies to establish several selected populations, each maintained in a large cage with food. Two of these populations were selected for positive phototaxis (a tendency to move toward light) and two for negative phototaxis (a tendency to move away from light). To select for phototaxis, the investigators put males and virgin females in a maze in which the flies had to make 15 successive choices between light and dark pathways, ending up in one of 16 tubes (Figure 9.24A). Flies that made 15 turns toward light arrived at tube 1, those that made 15 turns toward dark arrived at tube 16, and those that made equal numbers of turns toward light and dark ended up in tubes 8 and 9. The mean and variance of phototactic scores were estimated from the number of flies in each of the 16 tubes. In each generation, 300 flies of each sex, from each population, were released into the maze, and the 25 flies of each sex that had the most extreme high score (in the positively selected population) or low score (in the negatively selected population) were saved to initiate the next generation. This procedure was repeated for 20 generations.

Initially, the flies were neutral, on average: the mean scores were 8 to 9 (Figure 9.24B). Very quickly, however, both the positively and negatively selected populations diverged, in opposite directions, from the initial mean. We may therefore infer that variation among flies in their response to light is partly hereditary. From the rapidity of the change, Dobzhansky and Spassky calculated that the heritability of phototaxis is about 0.09.

Experiments of this kind have shown that *Drosophila* species are genetically variable for almost every trait, including features of behavior (e.g., mating speed), morphology, life history (e.g., longevity), physiology (e.g., resistance to insecticides), and even features of the genetic system (e.g., rate of crossing over). Artificial selection has been the major tool of breeders who have produced agricultural varieties of corn, tomatoes, pigs, chickens, and every other domesticated species, which often differ extremely in numerous characteristics. Such evidence has led many evolutionary geneticists to conclude that *species contain genetic variation that could serve as the foundation for the evolution of many of their characteristics*, and that *many or most characters should be able to evolve quite rapidly*—far more quickly than Darwin ever imagined (Barton and Partridge 2000).

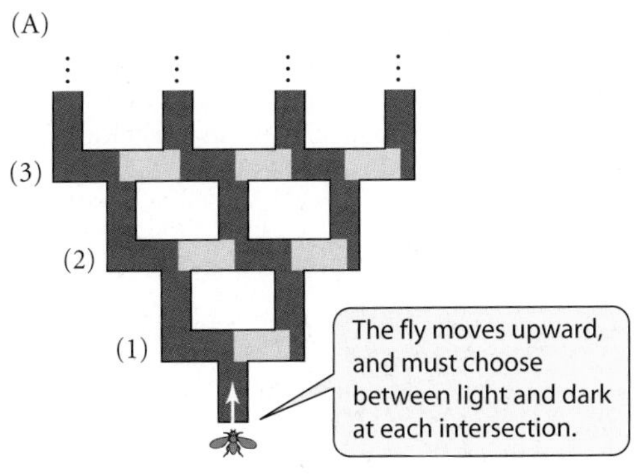

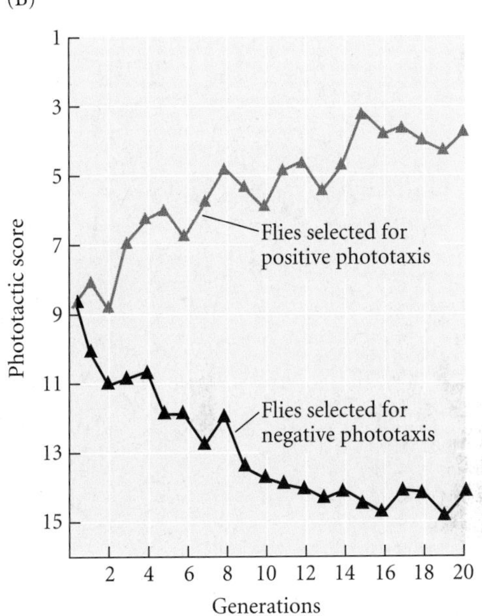

Figure 9.24 Selection for movement in response to light (phototaxis) in *Drosophila pseudoobscura*. (A) Diagram of part of a maze. The maze is oriented vertically below a light source, and flies introduced at the bottom move upward, choosing light or dark at each intersection. Lateral movement across intersections is prevented by barriers. The diagram shows the first 3 of the 15 choices made by the flies in the selection experiment. (B) The mean phototactic scores of flies in the first 20 generations of selection in two populations selected for positive and negative phototaxis. (After Dobzhansky and Spassky 1969.)

Variation among Populations

Almost all species are subdivided into several or many separate populations, with most mating taking place between members of the same population. Such populations of a single species often differ in genetic composition. Studies of differences among populations in different geographic areas, or **geographic variation**, have provided many insights into the mechanisms of evolution.

Patterns of geographic variation

If distinct forms or populations have overlapping geographic distributions, such that they occupy the same area and can frequently encounter each other, they are said to be **sympatric** (from the Greek *syn*, "together," and *patra*, "fatherland"). Populations with adjacent but nonoverlapping geographic ranges that come into contact are **parapatric** (Greek *para*, "beside"). Populations with separated distributions are **allopatric** (Greek *allos*, "other").

A **subspecies**, or GEOGRAPHIC RACE, in zoological taxonomy means a recognizably distinct population, or group of populations, that occupies a different geographic area from other populations of the same species. (In botanical taxonomy, subspecies names are sometimes given to sympatric, interbreeding forms.) In some instances, subspecies differ in a number of features with concordant patterns of geographic variation. For example, in the northern flicker (*Colaptes auratus*), the subspecies *auratus* and *cafer*, distributed in eastern and western North America, respectively, differ in the color of the underwing, in the crown and mustache marks, in the presence or absence of several other plumage marks, and in size (Short 1965; Moore and Price 1993). Despite these differences, the two

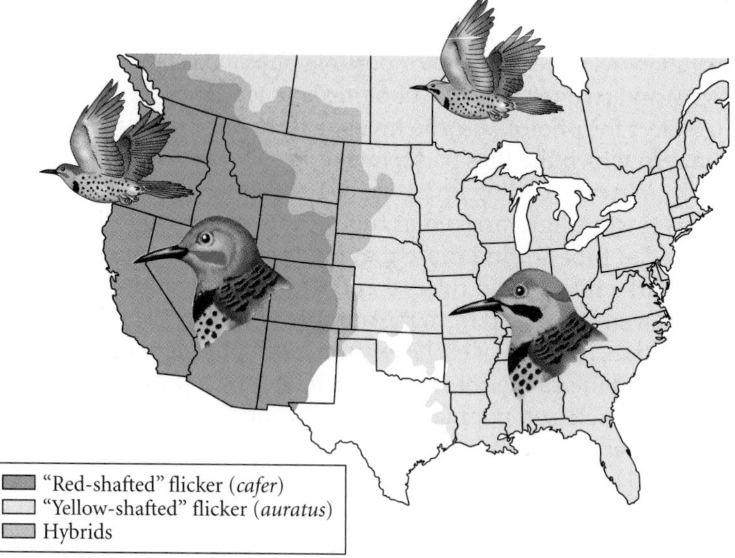

Figure 9.25 Two subspecies of a common North American woodpecker, the northern flicker (*Colaptes auratus*). The eastern ("yellow-shafted") subspecies (*C. a. auratus*) and the western ("red-shafted") subspecies (*C. a. cafer*) form a broad hybrid zone. (After Moore and Price 1993.)

populations interbreed in the Great Plains (Figure 9.25), forming a wide **hybrid zone** (a region in which genetically distinct parapatric forms interbreed).

Different characters often have discordant patterns of geographic variation. For example, in the rat snake *Elaphe obsoleta* (Figure 9.26), some subspecies are distinguished by color and others by the pattern of stripes or blotches. These features, in turn, are not well corre-

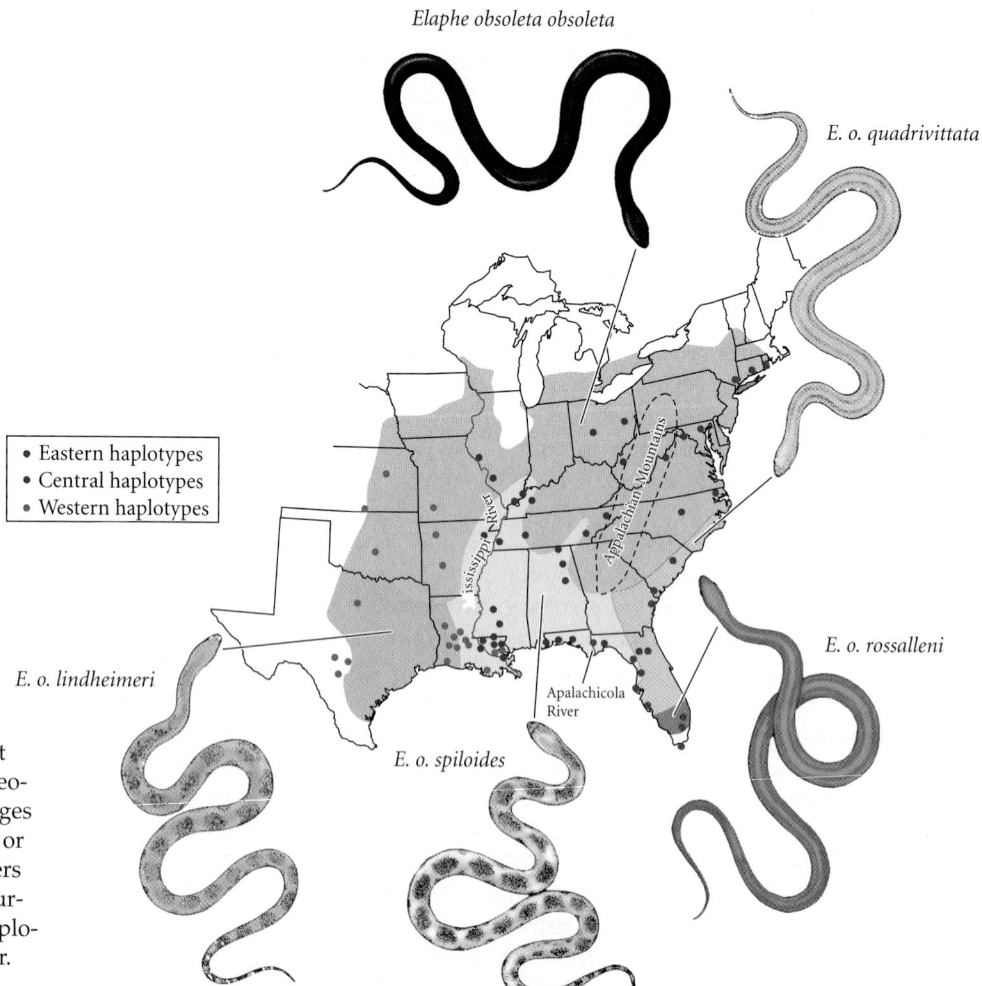

Figure 9.26 Five subspecies of the rat snake *Elaphe obsoleta*. These parapatric geographic races interbreed where their ranges meet. The races differ in pattern (stripes or blotches) and in color, but these characters have discordant distribution patterns. Furthermore, the distribution of mtDNA haplotypes differs from that of either character. (After Conant 1958.)

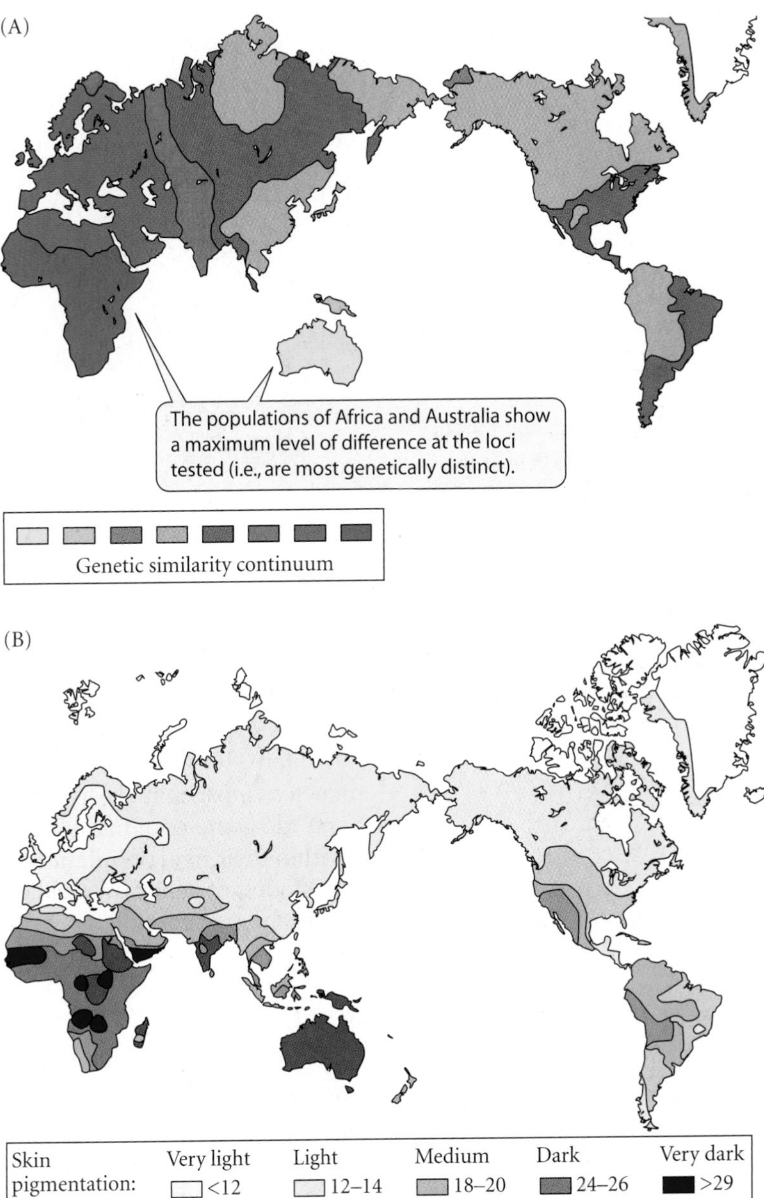

Figure 9.27 (A) A division of the world's human populations into eight classes of genetic similarity, based on numerous enzyme and blood group loci. The eight classes represented in the key are arrayed in order of increasing difference. (B) The geographic distribution of skin color, classified in eight grades of pigmentation intensity. Some considerable differences between the two maps are evident. (After Cavalli-Sforza et al. 1994.)

lated with the geographic distribution of mitochondrial DNA haplotypes, which distinguish western, central, and eastern groups of populations (Burbrink et al. 2000). Among human populations, likewise, some characteristics have geographic patterns that differ from the average pattern of genetic variation among populations (Figure 9.27). Many systematists opine that populations that differ by such discordant patterns should not be named as subspecies.

A gradual change in a character or in allele frequencies over geographic distance is called a **cline**. For instance, body size in the white-tailed deer (*Odocoileus virginianus*) increases with increasing latitude over much of North America, a pattern that is so common in mammals and birds that it has been named BERGMANN'S RULE. This pattern, which is too consistent to be attributed to chance, provides important evidence of *adaptive geographic variation* resulting from natural selection. Larger body size is thought to be advantageous for homeotherms in colder climates because it reduces the surface area, relative to body mass, over which body heat is lost. Adaptation is also strongly suggested by the consistency of clines in allele frequencies at the alcohol dehydrogenase locus in *Drosophila melanogaster* (Figure 9.28): on several continents, the Adh^F allele gradually increases from low to high frequency as one moves toward higher latitudes (Oakeshott et al. 1982).

Ecotypes are phenotypes that are associated with a particular habitat, often in a patchy, mosaic pattern. In an important early study, Jens Clausen, David Keck, and William Hiesey (1940) set out to determine whether the differences among ecotypes are genetically based or are directly caused by differences in the environment. They worked with several plants, including *Potentilla glandulosa*, the sticky cinquefoil, which is distributed in western North America from sea level to above timberline and displays variation among populations that is correlated with altitude. Clausen et al. divided, or cloned, each of a number of plants from populations of several different ecotypes from sites at different altitudes and grew them in common gardens at three different altitudes in California. The differences among ecotypes in some features, such as flower color, remained unchanged, irrespective of altitude. Clausen et al. concluded that these features differ genetically and are not substantially affected by the environment. Other features, such as plant height, varied among the three gardens, showing that they were affected by the environment, but the ecotypes nevertheless differed one from another in each garden, implying that these features were also influenced by genetic differences (Figure 9.29). In further studies, Clausen et al. showed that the genetic differences among populations in these features are polygenic.

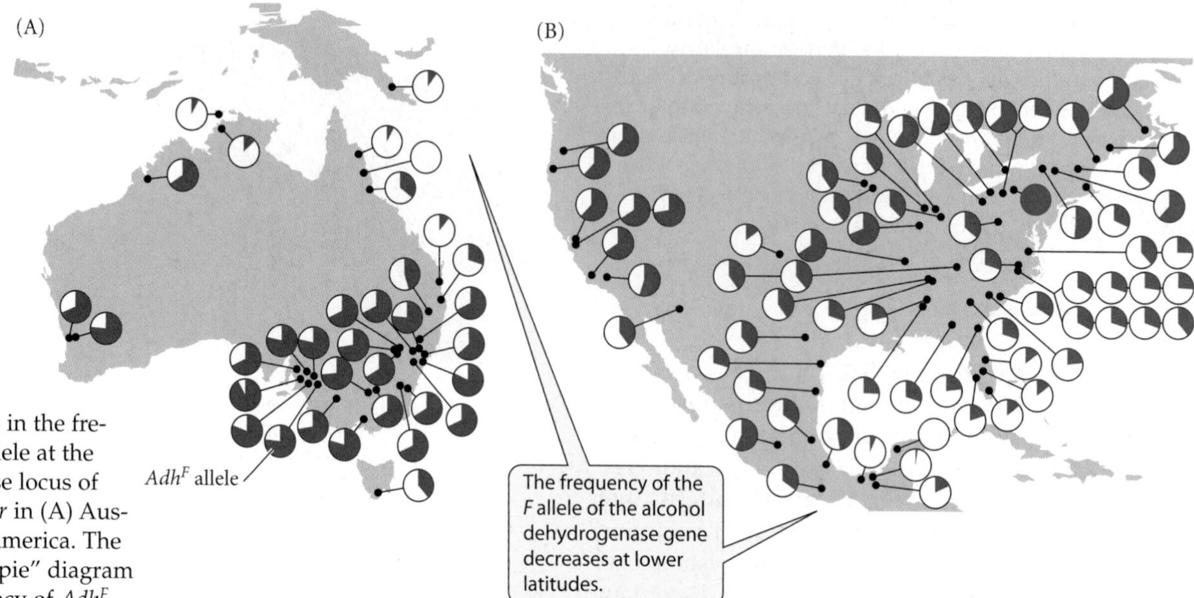

Figure 9.28 Clines in the frequency of the Adh^F allele at the alcohol dehydrogenase locus of *Drosophila melanogaster* in (A) Australia and (B) North America. The colored area of each "pie" diagram represents the frequency of Adh^F, which increases at higher latitudes on both continents. (After Oakeshott et al. 1982.)

Gene flow

Populations of a species typically exchange genes with one another to a greater or lesser extent. This process is called gene flow. Genes can be carried by moving individuals (such as most animals, as well as seeds and spores) or by moving gametes (such as pollen and the gametes of many marine animals). Migrants that do not succeed in reproducing within their new population do not contribute to gene flow.

Models of gene flow treat organisms as if they formed either discrete populations (e.g., on islands) or continuously distributed populations (**isolation by distance** models). In an

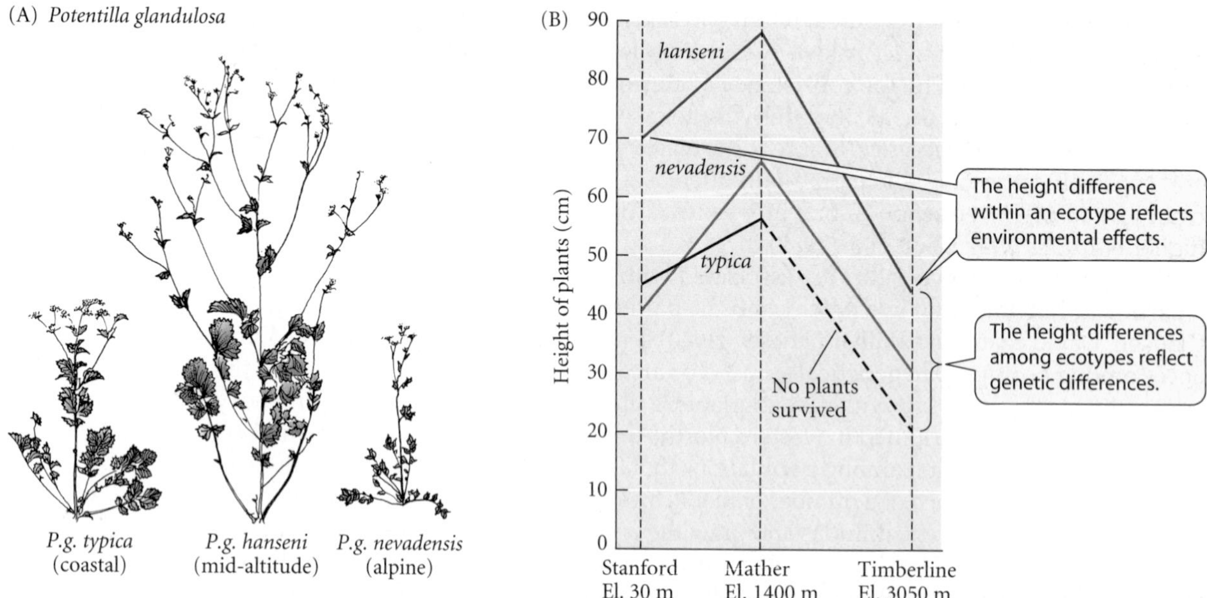

Figure 9.29 A common-garden study of ecotypic variation in the sticky cinquefoil (*Potentilla glandulosa*). (A) Representative specimens of coastal, mid-altitude, and alpine ecotypes, grown together at a site 1400 m above sea level. (B) The mean heights of the three ecotypes when grown in common gardens at three elevations. The differences among ecotypes at the same elevation reflect genetic differences, whereas the differences within each ecotype at different elevations reflect environmental effects. (After Clausen et al. 1940.)

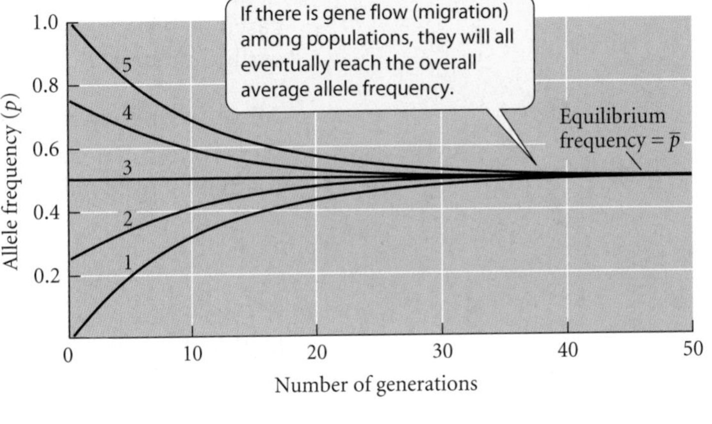

Figure 9.30 Gene flow causes populations to converge in allele frequencies. This model shows the changes over time in the frequency of one allele in five populations that exchange genes equally at the rate $m = 0.1$ per generation. (After Hartl and Clark 1989.)

isolation by distance model, each individual is the center of a NEIGHBORHOOD, within which the probability of mating declines with distance from the center. The population as a whole consists of overlapping neighborhoods.

Gene flow, if unopposed by other factors, homogenizes the populations of a species—that is, it brings them all to the same allele frequencies unless it is sufficiently counterbalanced by the divergent forces of genetic drift or natural selection (see Chapters 10 and 12). For example, if migrants are equally likely to disperse to any of a group of discrete populations, all of equal size, each population will ultimately reach the average allele frequency among the group of populations (Figure 9.30). The rate at which this process occurs is proportional to the RATE OF GENE FLOW (m), the proportion of gene copies of the breeding individuals in each generation that have been carried into that population by immigrants from other populations.

In this model, there is a fairly constant rate of migration among established populations. In some cases, another kind of gene flow may well be more important (McCauley 1993). If local populations at some sites become extinct, and those sites are then colonized by individuals from several other populations, the allele frequencies in the new colonies are a mixture of those among the source populations. Similarly, different populations will be genetically similar if they have all been recently founded by colonists from the same source population.

The characteristics of a species greatly affect its capacity for dispersal and gene flow. For example, animals such as land snails, salamanders, and wingless insects generally move little, and they are divided into relatively small, genetically distinct populations. Gene flow is greater among more mobile organisms, such as far-flying monarch butterflies (*Danaus plexippus*) and the many mussels and other marine invertebrates with planktonic larvae that are carried long distances by currents. However, even seemingly mobile species often display remarkably restricted dispersal. Despite their capacity for long-distance movement, many migratory species of salmon and birds breed near their birthplace, forming genetically distinct populations.

Rates of gene flow among natural populations can be estimated directly by following the dispersal of marked individuals or their gametes. A. J. Bateman (1947) studied gene flow by pollen in maize (corn, *Zea mays*) by counting the number of heterozygous progeny of homozygous recessive plants ("seed parents") situated at various distances from a stand of homozygous dominant plants ("pollen parents"). At only 30 to 50 feet away, less than 1 percent of the progeny were fathered by the dominant plants (Figure 9.31). Similar studies have shown that gene flow by both pollen and seed dispersal is very restricted in many species of plants (Levin 1984). However, estimates of gene flow

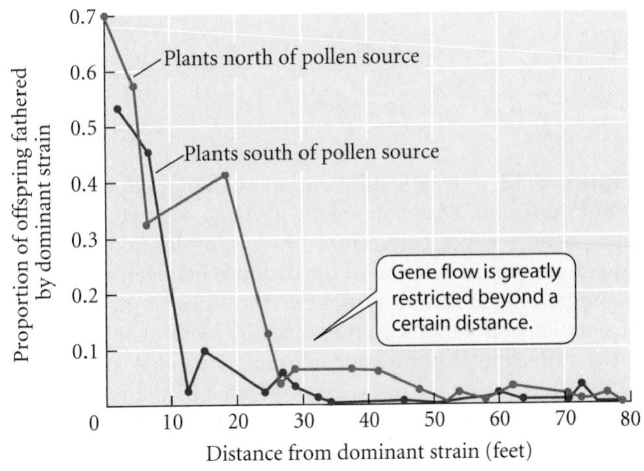

Figure 9.31 Gene flow in maize (corn), a wind-pollinated plant. The y-axis gives the proportion of offspring of recessive plants, grown at different distances from a strain of plants with a dominant allele, that were fathered by that dominant strain. The curves for plants situated north and south of the pollen source differ because of the effect of the prevailing wind on the dispersal of pollen. Most pollen is dispersed only a short distance. (After Bateman 1947.)

based on genetic differences among populations, as we will describe in Chapter 10, generally are more reliable than direct observations of this kind.

Allele frequency differences among populations

Variation in allele frequencies among populations can be quantified in several ways. A commonly used measure, for a locus with two alleles, is

$$F_{ST} = \frac{V_q}{(\bar{q})(1-\bar{q})}$$

where $\bar{q}$ is the mean frequency of one of the alleles and V_q is the variance among populations in its frequency. A comparable measure, G_{ST}, may be calculated for a locus with more than two alleles. Both F_{ST} and G_{ST} range from 0 (no variation among populations) to 1 (populations are fixed for different alleles).

Armbruster et al. (1998) estimated G_{ST}, averaged over five allozyme loci, in samples of the mosquito *Wyeomyia smithii*, the larvae of which develop only in the water-holding leaves of pitcher plants. Samples were taken from southern sites in North America (New Jersey, North Carolina, and the Gulf Coast) and from northern localities (Manitoba, Ontario, Maine, and Pennsylvania) that had been covered by the most recent Pleistocene glacier. In both the southern and northern regions, more distant populations differed more in allele frequency (i.e., had a greater G_{ST}), an illustration of isolation by distance (Figure 9.32A). However, northern populations were genetically more similar than southern populations that were separated by comparable distances. The most likely interpretation is

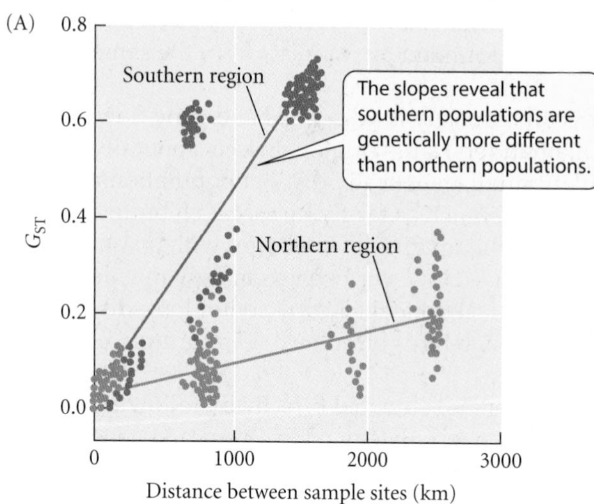

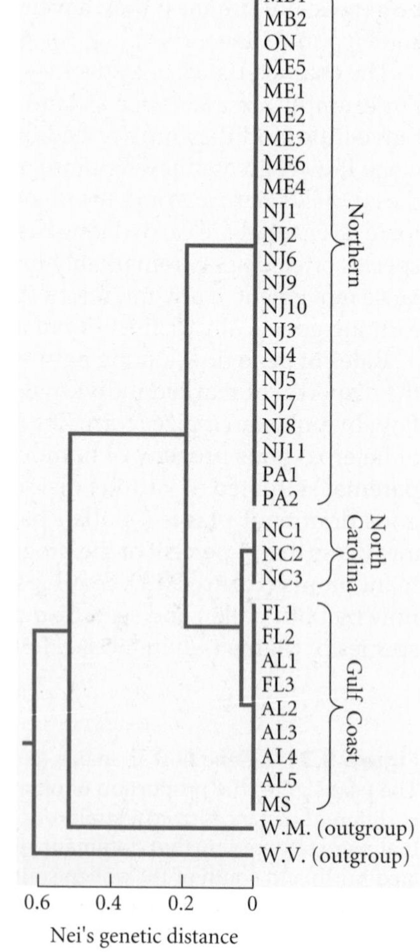

Figure 9.32 Genetic differentiation among populations of the North American pitcher-plant mosquito, *Wyeomyia smithii*. (A) G_{ST}, a measure of allele frequency differences, averaged over five loci, between southern samples (red points) and between northern samples (green points), in relation to the distance between sample localities. Each point represents a comparison of samples from two localities. (B) A phenogram, in which genetically more dissimilar populations, as measured by Nei's genetic distance, are joined at deeper levels than more similar populations. Populations of *Wyeomyia* from Canada (MB, ON), Maine (ME), New Jersey (NJ), and Pennsylvania (PA) are all more similar to one another than to those from North Carolina (NC) and the Gulf Coast (FL, AL, MS). W.M. and W.V. are closely related species of *Wyeomyia*. (After Armbruster et al. 1998.)

that there has not been enough time for northern populations to become substantially differentiated since the species colonized the formerly glaciated region.

Another measure of the genetic difference between two populations is Nei's index of **genetic distance** (Nei 1987), which measures how likely it is that gene copies taken from two populations will be different alleles, given data on allele frequencies.* This index is often used in concert with a "clustering algorithm" to construct diagrams (**phenograms**) that portray the relative difference (or similarity) among populations (or species). A phenogram, for example, can be used to illustrate how similar the northern populations of pitcher-plant mosquitoes are compared with the difference between North Carolina and Gulf Coast populations (Figure 9.32B). Phenograms look like phylogenetic trees, but they represent a phylogeny only if the rate of divergence among the populations or species has been constant.

DNA sequences provide information not only on allele frequency differences among populations, but also on the genealogical (phylogenetic) relationships among alleles, which may cast light on the history of the populations (see Chapters 2 and 6). For example, populations of MacGillivray's warbler (*Oporornis tolmiei*) from temperate western North America all share one common haplotype (*A*) at the mitochondrial cytochrome *b* locus (Figure 9.33A,B). Rare variants, differing from haplotype *A* by single mutations, have been found in some populations. In contrast, samples from northern Mexico carry a variety of haplotypes, differing from one another by as many as five mutations (Figure 9.33B). This pattern indicates that the Mexican population has been stable, accumulating genetic diversity for a long time. The temperate North American populations, however, probably stem from recent (postglacial) colonization by a few original founders that carried haplotype *A*. The rare haplotypes have only recently arisen from haplotype *A* by new mutations (Milá et al. 2000). (See Chapter 6 for other examples of such phylogeographic studies.)

Almost all MacGillivray's warblers in Mexico have different haplotypes from those in temperate North America, so that almost 90 percent of the total genetic variation is *between* the two regions, and only 10 percent is *within* populations. The situation is almost the opposite if we compare human populations.

*Nei's distance is calculated as $D = -\log\left(\frac{\Sigma p_{i1}p_{i2}}{\sqrt{\Sigma p_{i1}p_{i2}}}\right)$, where p_{i1} and p_{i2} are the frequencies of allele i in populations 1 and 2.

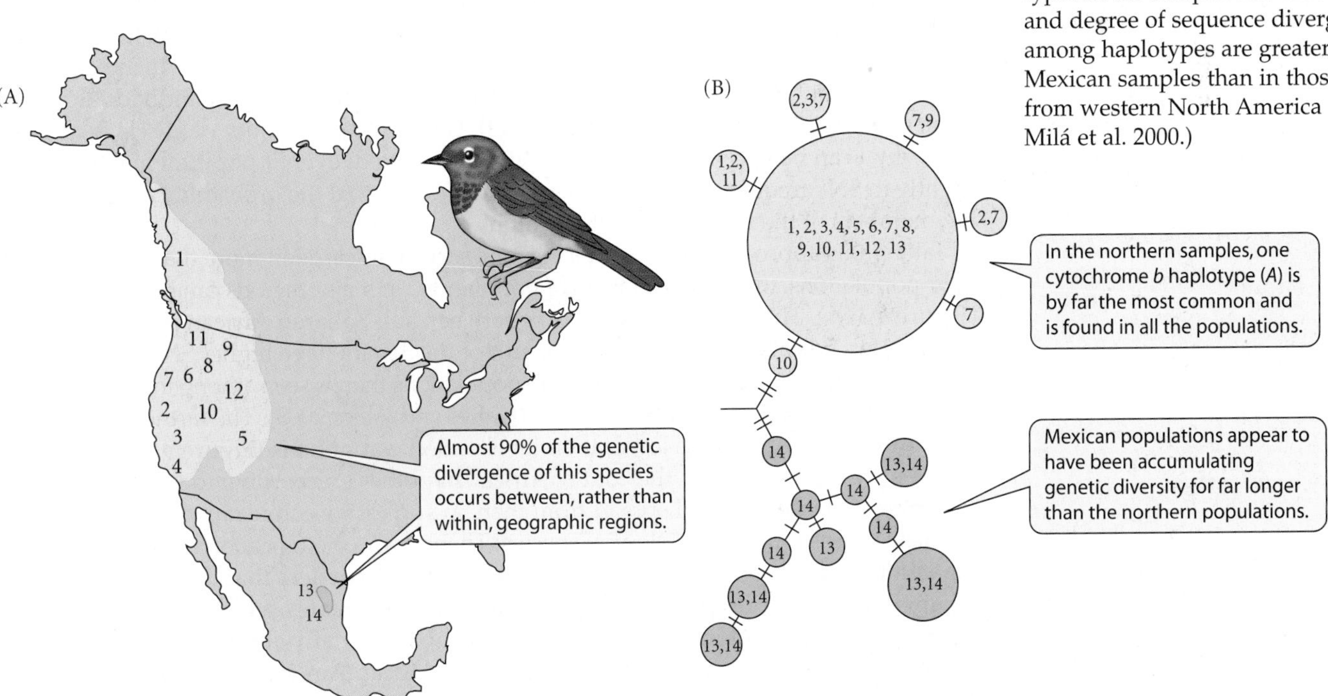

Figure 9.33 Geographic variation in mtDNA in MacGillivray's warbler, showing regional differences in genetic diversity. (A) Sample localities in the western United States and Canada (1–12) and in Mexico (13, 14). (B) An evolutionary tree of the 17 cytochrome b haplotypes found. Each haplotype is represented by a circle, labeled with the localities in which it was found. The size of each circle is proportional to the total frequency of the haplotype it represents. Haplotypes are connected to one another by sequence similarity, with bars across branches indicating single nucleotide changes. Some haplotypes are placed at branch points, implying that they are the ancestors of other haplotypes in the sample. The diversity and degree of sequence divergence among haplotypes are greater in Mexican samples than in those from western North America (After Milá et al. 2000.)

Human genetic variation

Since the nearly complete DNA sequence of the human genome was first reported in 2001, massive efforts have been made to characterize sequence variation in populations throughout the world. The HapMap Project, for example, has found **single nucleotide polymorphisms (SNPs)**, or single-site nucleotide variations, at more than 3 million sites in the genome. There are now more data on genetic variation in *Homo sapiens* than any other species, and both the databases and analyses of them grow steadily.

More than 60 percent of SNPs are rare, with the frequency of the less common nucleotide less than 0.05. Nonetheless, the average heterozygosity (π) across the genome is 0.001 to 0.002: an average individual is heterozygous at 1 out of every 1000 to 2000 sites (Crawford et al. 2005). The frequencies of these variants vary among human populations, so we may decompose the genetic variation in the entire human species into variation *within populations*, *among populations* within regions (such as continents), and *among regions*. Many such analyses have been performed, and they consistently show that the great majority of variation is within populations (Kittles and Weiss 2003). For example, J. Z. Li and collaborators (2008) analyzed 642,690 autosomal SNPs in 938 individuals from 51 populations throughout the world, and calculated that 89.9 percent of the variation is among individuals in an average population, 2.1 percent is among populations, and 9.0 percent is among major geographic regions. To update a conclusion reached by Lewontin et al. (1984), who performed a similar analysis using allozyme frequencies: "If everyone on earth became extinct except for the Kikuyu of East Africa, about 90 percent of all human variability would still be present in the reconstituted species." On the whole, then, human populations across the globe are genetically very similar.

At no known loci are "races" or other regional populations fixed for different alleles (Cavalli-Sforza et al. 1994). Some variations, however, do differ considerably in frequency among human populations, including genetic diseases. Mutations that cause cystic fibrosis are most prevalent in northern Europe, for example, and the mutation that causes Tay-Sachs disease has highest frequency in the Ashkenazic Jewish population, originally located in eastern Europe. SNPs also vary in frequency among populations. Some rare SNP alleles are restricted to specific local or regional populations; in other cases, alleles are shared among populations, but differ slightly in frequency. If two populations differ slightly in allele frequency at enough sites, however, they can be distinguished. Suppose, for example, that at each of n SNP sites, a rare variant has frequency q in population A, but is absent from population B. Perhaps $q = 10^{-3}$. The probability that a randomly chosen member of population A lacks the variant at any specific site is $1 - q$, and the probability that he or she lacks the rare variant at all n sites is $(1 - q)^n$, which becomes a very small number if n is very large. But this is the probability that the individual has the same genotype as a member of population B. Hence the probability of misclassifying an individual can be very low if enough SNPs are assayed, even if populations differ only slightly in SNP frequencies—which means, conversely, that the probability of distinguishing members of the two populations is high.

Taking this approach, Li and collaborators, who characterized more than 642,000 SNPs in 51 populations, found that clusters of populations in major regions can be distinguished (Figure 9.34A). The greatest difference is between sub-Saharan African populations and non-African populations. Populations south of the Sahara have higher SNP heterozygosity (π) and greater differences (F_{ST}) among populations than is seen among indigenous populations in Europe, Asia, or the Americas (Kittles and Weiss 2003). The farther populations are from eastern Africa, the lower their level of SNP variation is (Figure 9.34B). All these features are consistent with the "out of Africa" hypothesis, according to which *Homo sapiens* originated in Africa and spread from there in successive colonizations.

Homo sapiens is a single biological species. There are no biological barriers to interbreeding among human populations, and even the cultural barriers that do exist often break down. For instance, despite racist barriers in the United States, a blood group allele, Fy^a, that is fairly abundant in European populations but virtually absent in African populations was found to have a frequency of 0.11 in the African-American population of Detroit, Michigan, from which it was calculated that 26 percent of the population's genes have

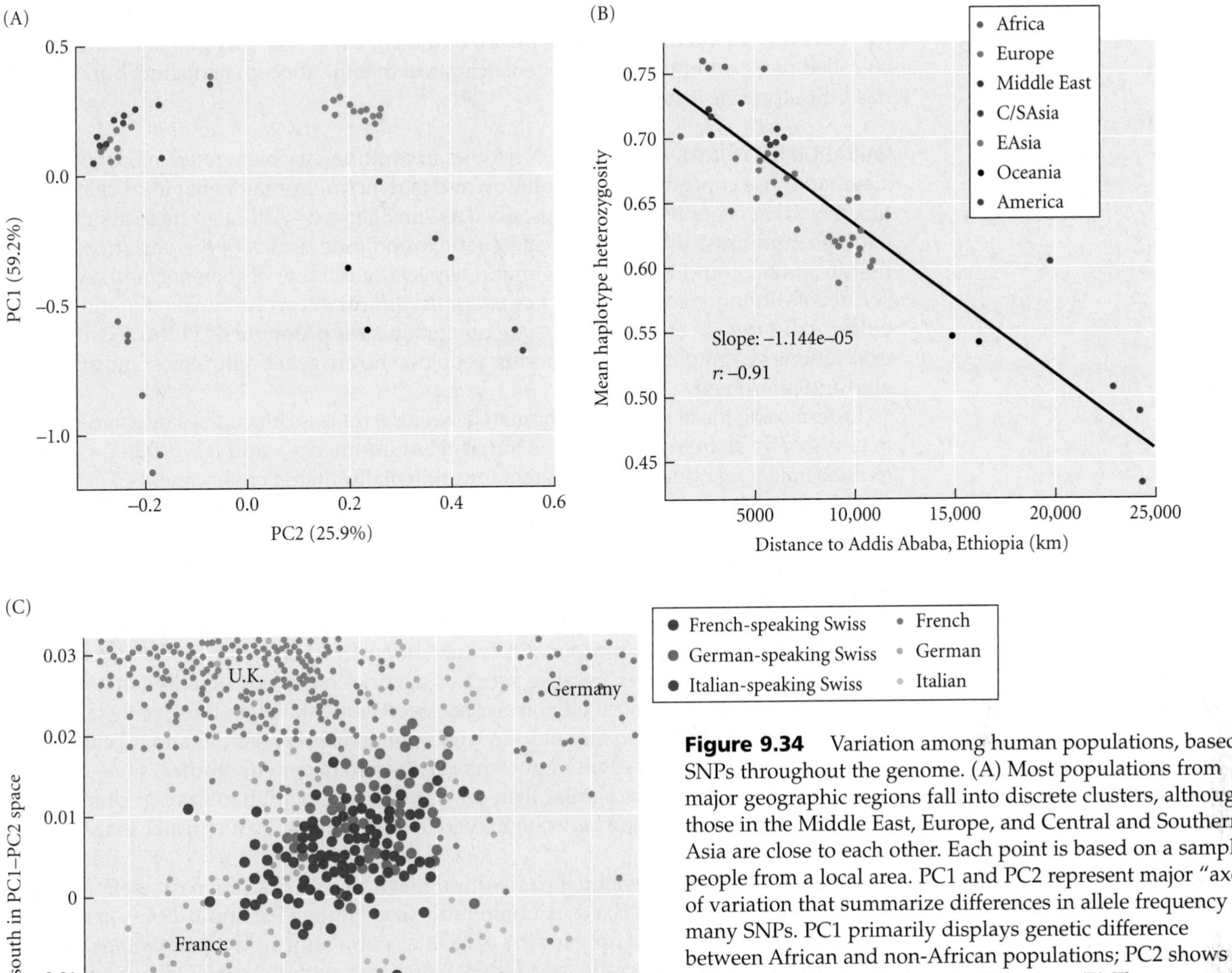

Figure 9.34 Variation among human populations, based on SNPs throughout the genome. (A) Most populations from major geographic regions fall into discrete clusters, although those in the Middle East, Europe, and Central and Southern Asia are close to each other. Each point is based on a sample of people from a local area. PC1 and PC2 represent major "axes" of variation that summarize differences in allele frequency for many SNPs. PC1 primarily displays genetic difference between African and non-African populations; PC2 shows mostly a west/east gradient in difference. (B) The average heterozygosity for SNPs, a measure of genetic diversity within a population, decreases as populations become more distant from eastern Africa, represented by Addis Ababa, Ethiopia. (C) Based on many SNPs, individuals can be assigned to even close local regions within Europe. Each symbol represents an individual. Even French-, German-, and Italian-speaking Swiss can be separated to some extent. (A, B after Li et al. 2008; C after Novembre et al. 2008.)

been derived from the white population, at a rate of gene flow (*m*) of about 1 percent per generation (Cavalli-Sforza and Bodmer 1971). Because of interbreeding, and because the expansion from Africa was quite recent, the genetic differences among populations are gradual: there are no sharp geographic breaks in genetic composition. Consequently, the number of human races that have been defined by a few features such as skin color and hair texture is arbitrary, and indeed various authors have recognized anywhere from 3 to more than 60 "races." Each supposed racial group can be subdivided into an indefinite number of distinct populations. Among Africans, for example, Congo pygmies are the shortest of humans, and Masai are among the tallest. Using SNPs, Novembre et al. (2008) found they could distinguish populations in Europe that are less than 500 miles apart (Figure 9.34C). Thus there is a continuum of increasing genetic difference with increasing distance. And, as we have seen, the geographic variation at many loci does not conform to the average pattern of genetic differentiation (see Figure 9.27). For all these reasons, there is little reason to presume, a priori, that most traits will vary substantially among so-called races, and many human geneticists feel that the concept of distinct human races has little scientific justification.

VARIATION IN COGNITIVE ABILITIES. No topics in evolutionary biology are more controversial than those concerning the evolution and genetics of human behavioral characteristics, including cognitive abilities described as "intelligence." Although these abilities have evolved, and must therefore have a genetic foundation, much of the variation in their manifestation could well be nongenetic, especially in view of the enormous effects of social conditioning and learning. For example, almost everyone agrees that most of the cultural differences among groups are not genetic, for pronounced cultural differences exist among geographically neighboring peoples who are genetically almost indistinguishable and interbreed.

Determining the heritability of human behavioral traits is difficult because family members typically share not only genes but also environments—and it would be unethical to deliberately rear humans in different experimentally altered environments. For this reason, studies of people adopted as children are critically important. The genetic component of variation is estimated by correlations between twins or other siblings reared apart, or by adoptees' correlation with their biological parents. In a variant of this method, the correlation of adopted children with their adoptive parents is contrasted with that of the biological (nonadopted) children of the same parents, which is expected to be higher if the variation has a genetic component. There is a risk that heritability will be overestimated in such studies, however, because adoption agencies often place children in homes that are similar (in factors such as religion and socioeconomic status) to those of their siblings. Some modern studies try to measure such environmental correlations and take them into account. Twins have played an important role in human genetic studies, since monozygotic twins should be more similar than dizygotic twins if variation has a genetic component. The genetic correlation between dizygotic twins should be no greater than between non-twin siblings.

It is exceedingly important to bear in mind that a genetic basis for a characteristic does not mean that the trait is fixed or unalterable (see Figures 9.5 and 9.29). A heritability value, therefore, may hold only for the particular population and the particular environment in which it was measured, and cannot be reliably extrapolated to other populations. By the same token, high heritability of variation *within* populations does not mean that differences *among* populations have a genetic basis. A character may display high heritability within a population, yet be altered dramatically by changing the environment. Twin studies have suggested that the heritability of human height is 0.8 or more, yet in many industrial nations, mean height has increased considerably within one or two generations as a result of nutritional and other improvements.

To a greater extent than in almost any other context, the "nature versus nurture" debate has centered on variations in the cognitive abilities collectively called "intelligence"—or more properly, on IQ ("intelligence quotient") scores. IQ tests are supposed to be "culture-free," but they have been strongly criticized as favoring white, middle-class individuals. IQ testing has a long, sordid history of social abuse (Gould 1981). Even in the recent past, some individuals have wrongly argued that IQ cannot be improved by compensatory education because it is highly heritable (Jensen 1973), and others have suggested that African-Americans, who on average score lower than European-Americans on IQ tests, have genetically lower intelligence (Herrnstein and Murray 1994). (For counterarguments, see Fraser 1995; Fischer et al. 1996.)

Recent studies of the heritability of cognitive abilities, based on twins reared apart and on adopted compared with nonadopted children, have corrected many of the flaws of earlier studies; for example, they have assessed the similarity of the family environments in which separated twins were reared (Bouchard et al. 1990; McClearn et al. 1997; Plomin et al. 1997). "General cognitive ability" (IQ) appears to be substantially heritable, and, surprisingly, the heritability increases with age, from about 0.40 in childhood to 0.50 in adolescence to about 0.60 in adulthood and old age. The genetic component of specific cognitive abilities, such as verbal comprehension, spatial visualization, perceptual speed, and accuracy, is somewhat lower, and these features are only partly correlated with one another.

Despite the substantial heritability of IQ, there is abundant evidence that education and an enriched environment can substantially increase IQ scores (Fraser 1995). For example, one study found that children who remained in their parents' homes had an average IQ score of 107, those adopted into different homes an average of 116, and those who were returned to their biological mothers after a period of adoption, only 101 (Tizard 1973).

Probably the most incendiary question about IQ is whether genetic differences account for differences in average IQ scores among "racial" or ethnic groups, such as the 15-point difference (one standard deviation) between the average scores of European-Americans and African-Americans. Virtually all the evidence indicates that the average difference between blacks and whites is due to their very different social, economic, and educational environments (Nisbett 1995). In studies of black children adopted into white homes or reared in the same residential institution with white children, the two "racial" groups had similar scores. One study correlated individuals' IQ with their degree of admixture of alleles for several blood groups that differ in frequency between European and African populations. The correlation between IQ and degree of European ancestry was nearly zero. Finally, the average IQ of German children fathered by white American soldiers during World War II was nearly identical to the IQ of those with black American fathers.

Summary

1. Evolution occurs by the replacement of some genotypes by others. Hence evolution requires genetic variation.
2. The all-important concepts of allele frequency and genotype frequency are central to the Hardy-Weinberg principle, which states that in the absence of perturbing factors, allele and genotype frequencies remain constant over generations. For two alleles with frequencies p and q at an autosomal locus, the Hardy-Weinberg genotype frequencies are in the ratio $p^2{:}2pq{:}q^2$.
3. The potential causes of allele frequency changes at a single locus are those factors that can cause deviations from the Hardy-Weinberg equilibrium. These factors are (a) nonrandom mating; (b) finite population size, resulting in random changes in allele frequencies (genetic drift); (c) incursion of genes from other populations (gene flow); (d) mutation; and (e) consistent differences among genes or genotypes in reproductive success (natural selection).
4. Inbreeding occurs when related individuals mate and have offspring. Inbreeding increases the frequency of homozygous genotypes and decreases the frequency of heterozygotes. Most diploid populations contain rare recessive deleterious alleles at many loci, so inbreeding causes a decline in components of fitness (inbreeding depression).
5. Populations of most species contain a great deal of genetic variation. This variation includes rare alleles at many loci, which usually appear to be deleterious. But it also includes many common alleles, so that many loci—perhaps up to a third of them—are polymorphic, as revealed by enzyme electrophoresis. Most genes are variable when analyzed at the level of DNA sequence.
6. Many phenotypic traits, including morphological, physiological, and behavioral features, exhibit polygenic variation.
7. Alleles at different loci, affecting the same or different traits, sometimes are nonrandomly associated within a population, a condition called linkage disequilibrium, or LD.
8. Variation in most phenotypic traits includes both a genetic component and a nongenetic ("environmental") component. Variation in some traits may also include components caused by nongenetic maternal effects and epigenetic inheritance. The proportion of the phenotypic variance that is due to genetic variation (genetic variance) is the heritability of the trait. Genetic variance and heritability can be estimated by breeding experiments and by artificial selection. Many characters appear to be so genetically variable that we should expect them to be able to evolve quite rapidly.
9. Genetic differences among different populations of a species take the form of differences in the frequencies of alleles that may also be polymorphic within populations. Unless countered by

natural selection or genetic drift, gene flow among populations will cause them to become homogeneous.

10. Patterns of allele frequency differences and the phylogeny of alleles or haplotypes can shed light on the history that gave rise to geographic variation.

Terms and Concepts

additive allele effects
allele frequency
allopatric
allozyme
arithmetic mean
artificial selection
cline
common garden
concealed genetic variation
ecotype
electrophoresis
environmental variance
epigenetic inheritance
gene flow
genetic distance
genetic hitchhiking
genetic variance
genotype x environment interaction
genotype frequency
geographic variation
Hardy-Weinberg (H-W) equilibrium
heritability
heterozygosity
hybrid zone
identical by descent
inbreeding
inbreeding coefficient
inbreeding depression
isolation by distance
lethal allele
linkage disequilibrium (LD)
linkage equilibrium
maternal effect
migration
monomorphism
norm of reaction
normal distribution
panmictic
parapatric
phenogram
polygenic variation
polymorphism
quantitative variation
random genetic drift
self-fertilization (= selfing)
single nucleotide polymorphism (SNP)
standard deviation
subspecies
sympatric
variance

Suggestions for Further Reading

The study of variation builds on a rich history, much of which is summarized in three classic works by major contributors to evolutionary biology. Theodosius Dobzhansky's *Genetics of the Evolutionary Process* (Columbia University Press, New York, 1970) and its predecessors, the several editions of *Genetics and the Origin of Species*, are among the most influential books on evolution from the viewpoint of population genetics. Ernst Mayr's *Animal Species and Evolution* (Harvard University Press, Cambridge, MA, 1963) is a classic work that contains a wealth of information on geographic variation and the nature of species in animals, with important interpretations of speciation and other evolutionary phenomena. Richard C. Lewontin's *The Genetic Basis of Evolutionary Change* (Columbia University Press, New York, 1974) is an insightful analysis of the study of genetic variation by classic methods and by electrophoresis.

Excellent contemporary treatments of variation include *Principles of Population Genetics* by D. L. Hartl and A. G. Clark (fourth edition, Sinauer Associates, Sunderland, MA, 2007), which includes extensive treatment of much of the material in this and the next four chapters, and *Molecular Markers, Natural History, and Evolution* by J. C. Avise (second edition, Sinauer Associates, Sunderland, MA, 2004), an outstanding description of the ways in which molecular variation is used to study topics such as mating patterns, kinship and genealogy within species, speciation, hybridization, phylogeny, and conservation.

Problems and Discussion Topics

1. In an electrophoretic study of enzyme variation in a species of grasshopper, you find 62 A_1A_1, 49 A_1A_2, and 9 A_2A_2 individuals in a sample of 120. Show that $p = 0.72$ and $q = 0.28$ (where p and q are the frequencies of alleles A_1 and A_2), and that the genotype frequencies of A_1A_1, A_1A_2, and A_2A_2 are approximately 0.51, 0.41, and 0.08, respectively. Demonstrate that the genotype frequencies are in Hardy-Weinberg equilibrium.
2. In a sample from a different population of this grasshopper species, you find four alleles at this locus. The frequencies of A_1, A_2, A_3, and A_4 are $p_1 = 0.50$, $p_2 = 0.30$, $p_3 = 0.15$, $p_4 = 0.05$. Assuming Hardy-Weinberg equilibrium, calculate the expected proportion of each of the ten possible genotypes (e.g., that of A_2A_3 should be 0.09). Show that heterozygotes, of all kinds, should constitute 63.9 percent of the population. In a sample of 100 specimens, how many would you expect to be heterozygous for allele A_4? How many would you expect to be homozygous A_4A_4?
3. In the peppered moth (*Biston betularia*), black individuals may be either homozygous (A_1A_1) or heterozygous (A_1A_2), whereas pale gray moths are only homozygous (A_2A_2). Suppose that in a sample of 250 moths from one locality, 108 are black and 142 are gray. (a) Which allele is dominant? (b) Assuming that the locus is in Hardy-Weinberg equilibrium, what are the allele frequencies? (c) Under this assumption, what *proportion* of the sample is heterozygous? What is the *number* of heterozygotes? (d) Under the same assumption, what proportion of black moths is heterozygous? (Answer: approximately 0.85.) (e) Why is it necessary to assume Hardy-Weinberg genotype frequencies in order to answer parts b–d? (f) For a sample from another area consisting of 287 black and 13 gray moths, answer all the preceding questions.
4. In an experimental population of *Drosophila*, a sample of males and virgin females includes 66 A_1A_1, 86 A_1A_2, and 28 A_2A_2 flies. Each genotype is represented equally in both sexes, and each can be distinguished by eye color. Determine the allele and genotype frequencies, and whether or not the locus is in Hardy-Weinberg equilibrium. Now suppose you discard half the A_1A_1 flies and breed from the remainder of the sample. Assuming the flies mate at random, what will be the genotype frequencies among their offspring? (Hint: The proportion of A_2A_2 should be approximately 0.23.) Now suppose you discarded half of the A_1A_2 flies instead of A_1A_1. What will be the allele and genotype frequencies in the next generation? Why is the outcome so different in this case?
5. In an electrophoretic study of a species of pine, you can distinguish heterozygotes and both homozygotes for each of two genetically variable enzymes, each with two alleles (A_1, A_2 and B_1, B_2). A sample from a natural population yields the following numbers of each genotype: 8 $A_1A_1B_1B_1$, 19 $A_1A_2B_1B_1$, 10 $A_2A_2B_1B_1$, 42 $A_1A_1B_1B_2$, 83 $A_1A_2B_1B_2$, 44 $A_2A_2B_1B_2$, 48 $A_1A_1B_2B_2$, 97 $A_1A_2B_2B_2$, 49 $A_2A_2B_2B_2$. (a) Determine the frequencies of alleles A_1 and A_2 (p_A, q_A) and B_1 and B_2 (p_B, q_B). (b) Determine whether locus A is in Hardy-Weinberg equilibrium. Do the same for locus B. (c) *Assuming* linkage equilibrium, calculate the expected *frequency* of each of the nine genotypes. (Hint: The expected frequency of $A_1A_1B_1B_2$ is 0.106.) (d) From the results of part c, calculate the expected *number* of each genotype in the sample, and determine whether or not the loci actually are in linkage equilibrium. (e) From these calculations, can you say whether or not the loci are linked?
6. Until a few decades ago, most population geneticists believed that populations are genetically uniform, except for rare deleterious mutations. We now know that most populations are genetically very variable. Contrast the implications of these different views for evolutionary processes.
7. Different characters may vary more or less independently among geographic populations of a species (as in the rat snake example in Figure 9.26), or may vary concordantly (as in the flicker example in Figure 9.25). Suggest processes that could produce each pattern.
8. Suppose two alleles additively affect variation in a trait such as finger length. Contrast two populations with the same allele frequencies: mating is random in one ($F = 0$), whereas the other is undergoing inbreeding (perhaps $F = 0.25$). Explain why the variance in finger length is *greater* in the inbreeding population. As a challenge, algebraically determine the ratio of the variance in the inbreeding population to that in the panmictic population.
9. Why would estimates of gene flow from genetic differences among populations be better than those based on directly following the movement of organisms or their gametes?

CHAPTER 10

Genetic Drift: Evolution at Random

Cheetahs have experienced a population bottleneck. African cheetahs (*Acinonyx jubatus*) experienced a population bottleneck in the recent past, which led to a severe decrease in the amount of genetic variation within the species. (Photo © Gerry Ellis/DigitalVision.)

One of the first and most important lessons a student of science learns is that many words have very different meanings in a scientific context than in everyday speech. The word "chance" is a good example. Many nonscientists think that evolution occurs "by chance." What they mean is that evolution occurs without purpose or goal. But by this token, everything in the natural world—chemical reactions, weather, planetary movements, earthquakes—happens by chance, for none of these phenomena has a purpose. In fact, scientists consider purposes or goals to be unique to human thought and do not view any natural phenomena as purposeful. But scientists don't view chemical reactions or planetary movements as chance events, either—because in science, "chance" has a very different meaning.

Although the meaning of "chance" is a complex philosophical issue, scientists use chance, or **randomness**, to mean that when physical causes can result in any of several outcomes, we cannot predict what the outcome will be in any particular case. Nonetheless, we may be able to specify the *probability*, and thus the *frequency*, of one or another outcome. Although we cannot predict the sex of someone's next child, we can say with considerable certainty that there is a probability of 0.5 that it will be a daughter.

Almost all phenomena are affected simultaneously by both chance (unpredictable) and nonrandom, or DETERMINISTIC (predictable), factors. Any of us may experience a car accident resulting from the unpredictable behavior of another driver, but we are predictably more likely to do so if we drive after drinking. So it is with evolution. As we will see in the next chapter, natural selection is a deterministic, nonrandom process. But at the same time, there are important random processes in evolution, including mutation (as discussed in Chapter 8) and random fluctuations in the frequencies of alleles or haplotypes: the process of **random genetic drift**.

Genetic drift and natural selection are the two most important causes of allele substitution—that is, of evolutionary change—in populations. Genetic drift occurs in all natural populations because, unlike ideal populations at Hardy-Weinberg equilibrium, natural populations are finite in size. Random fluctuations in allele frequencies can result in the replacement of old alleles by new ones, resulting in **nonadaptive evolution**. That is, while natural selection results in adaptation, genetic drift does not—so this process is not responsible for those anatomical, physiological, and behavioral features of organisms that equip them for reproduction and survival. Genetic drift nevertheless has many important consequences, especially at the molecular genetic level: it appears to account for much of the difference in DNA sequences among species.

Because all populations are finite, alleles at all loci are potentially subject to random genetic drift—but all are not necessarily subject to natural selection. For this reason, and because the expected effects of genetic drift can be mathematically described with some precision, some evolutionary geneticists hold the opinion that genetic drift should be the "null hypothesis" used to explain an evolutionary observation unless there is positive evidence of natural selection or some other factor. This perspective is analogous to the "null hypothesis" in statistics: the hypothesis that the data do not depart from those expected on the basis of chance alone.* According to this view, we should not assume that a characteristic, or a difference between populations or species, is adaptive or has evolved by natural selection unless there is evidence for this conclusion.

The theory of genetic drift, much of which was developed by the American geneticist Sewall Wright starting in the 1930s and by the Japanese geneticist Motoo Kimura starting in the 1950s, includes some of the most highly refined mathematical models in biology. (But fear not! We shall skirt around almost all the math.) We first explore the theory and then see how it explains data from real organisms. In our discussion of the theory of genetic drift, we will describe random fluctuations in the frequencies (proportions) of two or more kinds of self-reproducing entities that do not differ *on average* (or differ very little) in reproductive success (fitness). For the purposes of this chapter, those entities are alleles. But the theory applies to any other self-replicating entities, such as chromosomes, asexually reproducing genotypes, or even species.

The Theory of Genetic Drift

Genetic drift as sampling error

That chance should affect allele frequencies is readily understandable. Imagine, for example, that a single mutation, A_2, appears in a large population that is otherwise A_1. If the population size is stable, each mating pair leaves an average of two progeny that survive to reproductive age. From the single mating $A_1A_1 \times A_1A_2$ (for there is only one copy of A_2), the probability that one surviving offspring will be A_1A_1 is ½; therefore, the probability that two surviving progeny will both be A_1A_1 is $½ \times ½ = ¼$—which is the probability that the A_2 allele will be immediately lost from the population. We may assume that mating pairs vary at random, around the mean, in the number of surviving offspring they leave (0, 1, 2, 3, ...). In that case, as the pioneering population geneticist Ronald Fisher calcu-

*For example, if we measure height in several samples of people, the null hypothesis is that the observed means differ from one another only because of random sampling, and that the parametric means of the populations from which the samples were drawn do not differ. A statistical test, such as a *t*-test or analysis of variance, is designed to show whether or not the null hypothesis can be rejected. It will be rejected if the sample means differ more than would be expected if samples had been randomly drawn from a single population.

lated, the probability that A_2 will be lost, averaged over the population, is 0.368. He went on to calculate that after the passage of 127 generations, the cumulative probability that the allele will be lost is 0.985. This probability, he found, is not greatly different if the new mutation confers a slight advantage: as long as it is rare, it is likely to be lost, just by chance.

In this example, the frequency of an allele can change (in this instance, to zero from a frequency very near zero) because the one or few copies of the A_2 allele may happen not to be included in those gametes that unite into zygotes, or may happen not to be carried by the offspring that survive to reproductive age. The genes included in any generation, whether in newly formed zygotes or in offspring that survive to reproduce, are a *sample* of the genes carried by the previous generation. Any sample is subject to random variation, or **sampling error**. In other words, the proportions of different kinds of items (in this case, A_1 and A_2 alleles) in a sample are likely to differ, by chance, from the proportions in the set of items from which the sample is drawn.

Imagine, for example, a population of land snails (*Cepaea nemoralis*) in which (for the sake of argument) offspring inherit exactly the brown or yellow color of their mothers. Suppose 50 snails of each color inhabit a cow pasture. (The proportion of yellow snails is $p = 0.50$.) If two yellow and four brown snails are stepped on by cows, p will change to 0.511. Since it is unlikely that a snail's color affects the chance of its being squashed by cows, the change might just as well have been the reverse, and indeed, it may well be the reverse in another pasture, or in this pasture in the next generation. In this random process, the chances of increase or decrease in the proportion of yellow snails are equal in each generation, so the proportion will fluctuate. But an increase of, say, 1 percent in one generation need not be compensated by an equal decrease in a later generation—in fact, since this process is random, it is very unlikely that it will be. Therefore the proportion of yellow snails will wander over time, eventually ending up near, and finally at, one of the two possible limits: 0 and 1.0. It seems reasonable, too, that if the population should start out with, say, 80 percent brown and 20 percent yellow snails, it is more likely that the proportion of yellow will wander to zero than to 100 percent. In fact, the probability of yellow being lost from the population is exactly 0.80, which is the same as the probability that brown will reach 100 percent—that is, that it will reach **fixation**.

Coalescence

The concept of random genetic drift is so important that we will take two tacks in developing the idea. Figure 10.1 shows a hypothetical, but realistic, history of gene lineages. First, imagine the figure as depicting lineages of individual asexual organisms, such as bacteria, rather than genes. We know from our own experience that not all members of our parents' or grandparents' generations had equal numbers of descendants; some had none. Figure 10.1 diagrams this familiar fact. We note that the individuals in generation t (at the right of the figure) are the progeny of only some of those that existed in the previous generation ($t - 1$): purely by chance, some individuals in generation $t - 1$ failed to leave descendants. Likewise, the population at generation $t - 1$ stems from only some of those individuals that existed in generation $t - 2$, and similarly back to the original population at time 0.

Now think of the objects in Figure 10.1 as copies of genes at a locus, in either a sexual or an asexual population. Figure 10.1 shows that as time goes on, more and more of the original gene lineages become extinct, so that the population consists of descendants of fewer and fewer of the original gene copies. In fact, if we look backward rather than forward in time, *all the gene copies* in the population ultimately *are descended from a single ancestral gene copy*, because given long enough, all other original gene lineages become extinct. The genealogy of the genes in the present population is said to **coalesce** back to a single common ancestor. Because that ancestor represents one of the several original alleles, the population's genes, descended entirely from that ancestral gene copy, must eventually become monomorphic: one or the other of the original alleles becomes fixed (reaches a frequency of 1.00). The smaller the population, the more rapidly all gene copies in the current population coalesce back to a single ancestral copy, since it takes longer for many than for few gene lineages to become extinct by chance.

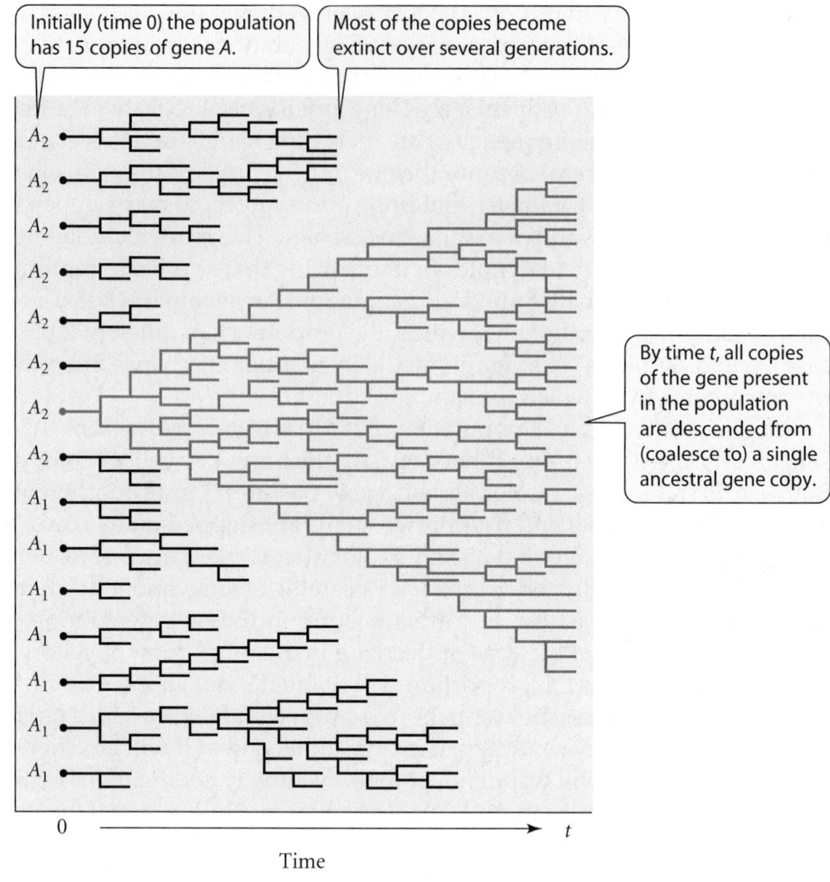
Initially (time 0) the population has 15 copies of gene A.
Most of the copies become extinct over several generations.
By time t, all copies of the gene present in the population are descended from (coalesce to) a single ancestral gene copy.
A2
A2
A2
A2
A2
A2
A2
A2
A1
A1
A1
A1
A1
A1
A1
0
t
Time

Figure 10.1 A possible history of descent of gene copies in a population that begins (at time 0, at left) with 15 copies, representing two alleles. Each gene copy has 0, 1, or 2 descendants in the next generation. The gene copies present at time t (far right) are all descended from (coalesce to) a single ancestral copy, which happens to be an A_2 allele (the lineage shown in red). Gene lineages descended from all other gene copies have become extinct. If the failure of gene copies to leave descendants is random, then the gene copies at time t could equally likely have descended from any of the original gene copies present at time 0. (After Hartl and Clark 1989.)

In our example, all gene copies have descended from a copy of an A_2 allele, but because this is a random process, A_1 might well have been the "lucky" allele if the sequence of random events had been different. If, in the generation that included the single common ancestor of all of today's gene copies, A_1 and A_2 had been equally frequent ($p = q = 0.5$), then it is equally likely that the ancestral gene copy would have been A_1 or A_2; but if A_1 had had a frequency of 0.9 in that generation, then the probability is 0.9 that the ancestral gene would have been an A_1 allele. *Our analysis therefore shows that by chance, a population will eventually become monomorphic for one allele or the other, and that the probability that allele A_1 will be fixed, rather than another allele, equals the initial frequency of A_1.*

According to this analysis, for example, all the mitochondria of the entire human population are descended from the mitochondria carried by a single woman, who has been called "mitochondrial Eve," at some time in the past. (Mitochondria are transmitted only through eggs.) This does *not* mean, however, that the population had only one woman at that time: "mitochondrial Eve" happened to be the one among many women to whom all mitochondria trace their ancestry (in a pattern like that seen in Figure 10.1). Various nuclear genes likewise are descended from single gene copies in the past that were carried by many different members of the ancestral human population.

If this process occurs in a large number of independent, non-interbreeding populations, each with the same initial number of copies of each of two alleles at, say, locus A, then we would expect a fraction p of the populations to become fixed for A_1 and a fraction $1 - p$ to become fixed for A_2. Thus the genetic composition of the populations would diverge by chance. If the original populations had each contained three (or more) different alleles, rather than two, each of those alleles would become fixed in some of the populations, with a probability equal to its initial frequency (say, p_i).

As allele frequencies in a population change by genetic drift, so do the genotype frequencies, which conform to Hardy-Weinberg equilibrium among the new zygotes in each generation. If, for example, the frequencies p and q (that is, p and $1 - p$) of alleles A_1 and A_2 change from 0.5: 0.5 to 0.45: 0.55, then the frequencies of genotypes A_1A_1, A_1A_2, and A_2A_2 change from 0.25:0.50:0.25 to 0.2025:0.4950:0.3025. As was described in Chapter 9, the frequency of heterozygotes, H, declines as one of the allele frequencies shifts closer to 1 (and the other moves toward 0):

$$H = 2p(1 - p)$$

Bear in mind that this model, as developed so far, includes only the effects of random genetic drift. It assumes that other evolutionary processes—namely, mutation, gene flow, and natural selection—do not operate. Thus the model does not describe the evolution of adaptive traits—those that evolve by natural selection. We will incorporate natural selection in the following chapters.

Random fluctuations in allele frequencies

Let us take another, more traditional, approach to the concept of random genetic drift. Assume that the frequencies of alleles A_1 and A_2 are p and q in each of many independent populations, each with N breeding individuals (representing $2N$ gene copies in a diploid species). Small independent populations are sometimes called **demes**, and an ensemble of such populations may be termed a **metapopulation**. As before, we assume that the genotypes do not differ, *on average*, in survival or reproductive success—that is, the alleles are **neutral** with respect to fitness.

In each generation, the large number of newborn zygotes is reduced to N individuals of reproductive age by mortality that is random with respect to genotype. By sampling error, the proportion of A_1 (p) among the survivors may change. The new p (call it p') could take on any possible value from 0 to 1.0, just as the proportion of heads among N tossed coins could, in principle, range from no heads to all heads. The probability of each possible value—whether it be the proportion of heads or the proportion of A_1 allele copies—can be calculated from the binomial theorem, generating a PROBABILITY DISTRIBUTION. Among a large number of demes, the new allele frequency (p') will vary, by chance, around a mean—namely, the original frequency, p.

Now if we trace one of the demes, in which p has changed from 0.5 to, say, 0.47, we see that in the following generation, it will change again from 0.47 to some other value, *either higher or lower with equal probability*. This process of random fluctuation continues over time. Since no stabilizing force returns the allele frequency toward 0.5, p will eventually wander (drift) either to 0 or to 1: *the allele is either lost or fixed*. (Once the frequency of an allele has reached either 0 or 1, it cannot change unless another allele is introduced into the population, either by mutation or by gene flow from another population.) The allele frequency describes a **random walk**, analogous to a New Year's Eve reveler staggering along a very long train platform with a railroad track on either side: if he is so drunk that he doesn't compensate steps toward one side with steps toward the other, he will eventually fall off the edge of the platform onto one of the two tracks (Figure 10.2).

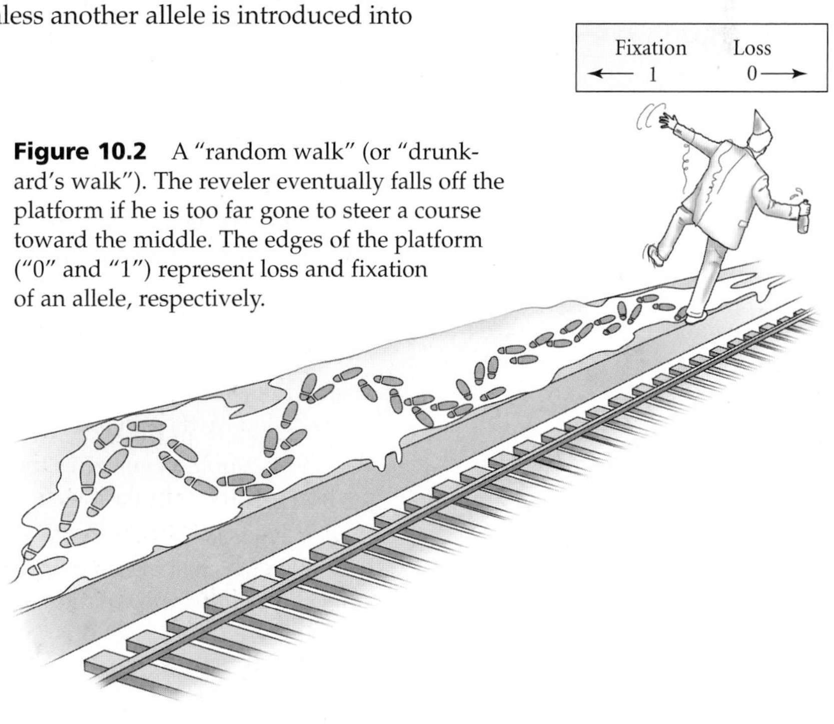

Fixation ← 1 | Loss 0 →

Figure 10.2 A "random walk" (or "drunkard's walk"). The reveler eventually falls off the platform if he is too far gone to steer a course toward the middle. The edges of the platform ("0" and "1") represent loss and fixation of an allele, respectively.

Just as an allele's frequency may increase by chance in some demes from one generation to the next, it may decrease in other demes. As a result, allele frequencies may vary among the demes. The *variance* in allele frequency among the demes continues to increase from generation to generation (Figure 10.3). Some demes reach $p = 0$ or $p = 1$ and can no longer change. Among those in which fixation of one or the other allele has not yet occurred, the allele frequencies continue to

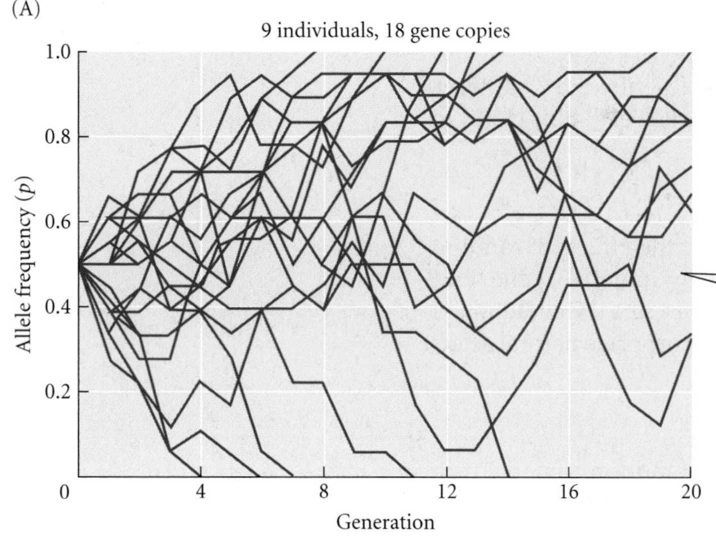

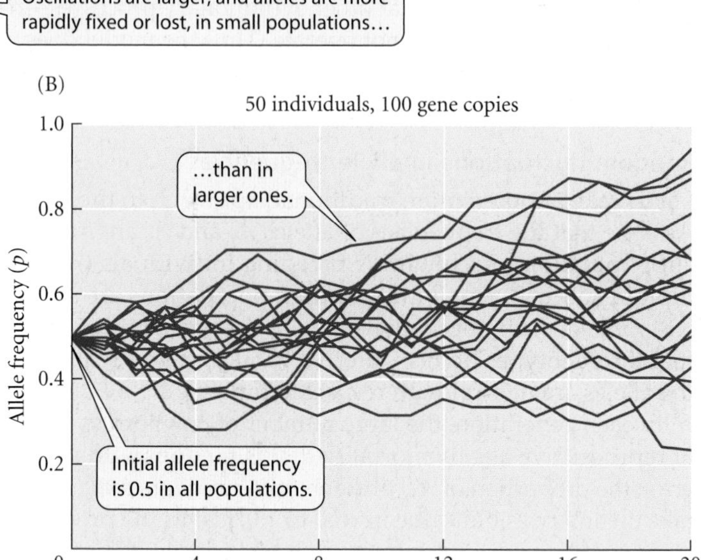

Figure 10.3 Computer simulations of random genetic drift in populations of (A) 9 diploid individuals ($2N = 18$ gene copies) and (B) 50 diploid individuals ($2N = 100$ gene copies). Each line traces the frequency (p) of one allele for 20 generations. Each panel shows allele frequency changes in 20 replicate populations, all of which begin at $p = 0.5$ (i.e., half the gene copies are A_1 and half A_2). (After Hartl and Clark 1989.)

spread out, with all frequencies between 0 and 1 eventually becoming equally likely (Figure 10.4). The number of populations fixed for one or another allele ($p = 0$ or $p = 1$) continues to increase until all demes in the metapopulation have become fixed. Thus, *demes that initially are genetically identical evolve by chance to have different genetic constitutions*. (Remember, though, that we are assuming that the alleles have identical effects on fitness—that is, that they are neutral.)

Evolution by Genetic Drift

The following points, which follow from the previous discussion, are some of the most important aspects of evolution by genetic drift:

1. Allele (or haplotype) frequencies fluctuate at random within a population, and eventually one or another allele becomes fixed.
2. Therefore, the genetic variation at a locus declines and is eventually lost. As the frequency of one allele approaches 1.0, the frequency of heterozygotes, $H = 2p(1 - p)$, declines. The rate of decline in heterozygosity is often used as a measure of the rate of genetic drift within a population.
3. At any time, an allele's probability of fixation equals its frequency, and is not affected or predicted by its previous history of change in frequency.
4. Therefore, populations with the same initial allele frequency (p) diverge, and a proportion p of the populations is expected to become fixed for that allele. A proportion $1 - p$ of the populations becomes fixed for alternative alleles.
5. If an allele has just arisen by mutation, and is represented by only one among the $2N$ gene copies in the population, its frequency is

$$p_t = \frac{1}{2N}$$

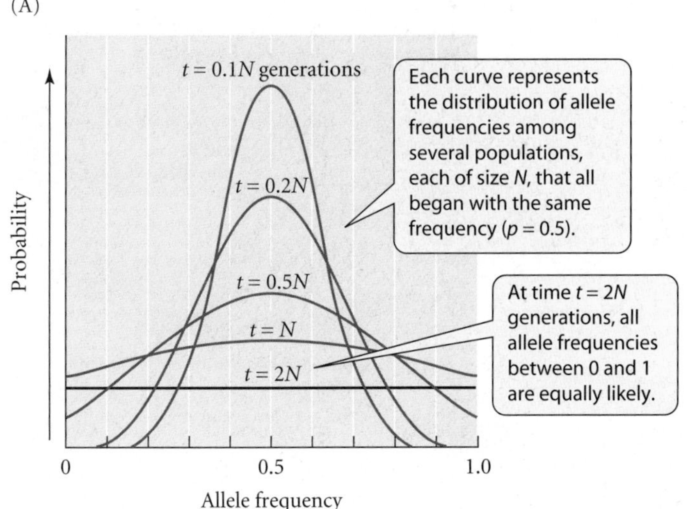

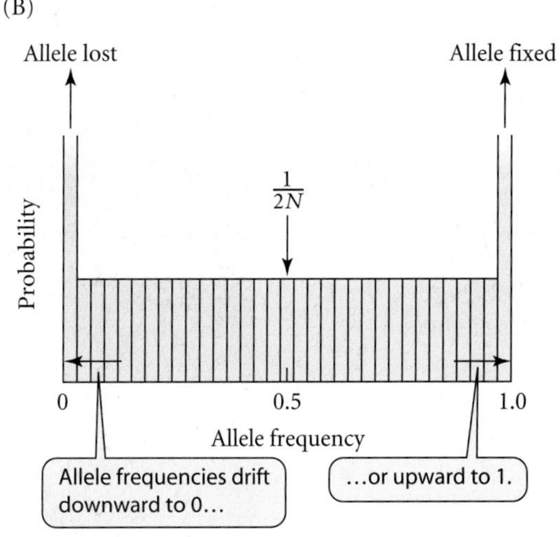

Figure 10.4 Changes in the probability that an allele will have various possible frequencies as genetic drift proceeds over time. (A) Each curve shows the probability distribution of allele frequencies between 0 and 1 at different times. The number of generations that elapse (t) is measured in units of the initial population size (N). For example, if the population begins with $N = 50$ individuals, $t = 2N$ represents the frequency distribution after 100 generations. The probability distribution after $t = 0.1N$ generations is shown by the uppermost curve. This curve may be thought of as the distribution of allele frequencies among several populations, each of size N, that began with the same allele frequency. With the passage of generations, the curve becomes lower and broader as the allele frequencies in all populations drift toward either 0 or 1. At $t = 2N$ generations, all allele frequencies between 0 and 1 are equally likely. (This panel does not show the proportion of populations in which the allele has been fixed or lost.) (B) The proportion of populations with different allele frequencies after $t = 2N$ generations have elapsed, including populations in which the allele has been fixed ($p = 1$) or lost ($p = 0$). The proportion of populations in which the allele is lost or fixed increases at the rate of $1/(4N)$ per generation, and each allele frequency class between 0 and 1 decreases at the rate of $1/(2N)$ per generation. (A after Kimura 1955; B after Wright 1931.)

and this is its likelihood of reaching $p = 1$. Clearly, it is more likely to be fixed in a small than in a large population. Moreover, if the same mutation arises in each of many demes, each of size N, the mutation should eventually be fixed in a proportion $1/(2N)$ of the demes. Similarly, of all the new mutations (at all loci) that arise in a population, a proportion $1/(2N)$ should eventually be fixed.

6. Evolution by genetic drift proceeds faster in small than in large populations. In a diploid population, the average time to fixation of a newly arisen neutral allele that does become fixed is $4N$ generations, on average. That is a long time if the population size (N) is large.
7. Among a number of initially identical demes in a metapopulation, the average allele frequency ($\bar{p}$) does not change, but since the allele frequency in each deme does change, eventually becoming 0 or 1, the frequency of heterozygotes (H) declines to 0 in each deme and in the metapopulation as a whole.
8. Just as a low-frequency advantageous allele may be lost from a population by genetic drift, a mildly disadvantageous allele may increase in frequency by genetic drift, and may even be fixed. Consequently, fitness may decline in small populations. This may be viewed as a form of inbreeding depression, because matings in small populations are frequently between relatives, whose genes are identical by descent.
9. If we consider two alleles at two linked loci and ignore recombination, the four gamete types (A_1B_1, A_1B_2, A_2B_1, A_2B_2) may be considered as if they were four alleles. Their frequencies will fluctuate by random genetic drift, and one gamete type (A_iB_j) may increase to a frequency greater than that expected (p_ip_j) at linkage equilibrium. Therefore, linkage disequilibrium (nonrandom association of alleles) may develop by genetic drift. Recombination breaks down linkage disequilibrium even as genetic drift recreates it, so the equilibrial level of linkage disequilibrium depends on the recombination rate and population size (Hill 1981).

Effective population size

The theory presented so far assumes highly idealized populations of N breeding adults. If we measure the actual number (N) of adults in real populations, however, the num-

Figure 10.5 The effective population size among northern elephant seals (*Mirounga angustirostris*) is much lower than the census size because only a few of the large males compete successfully for the smaller females. The winner of the contest here will father the offspring of an entire "harem" of females. (Photo © Richard Hansen/Photo Researchers.)

ber we count (the CENSUS SIZE) may be greater than the number that actually contribute genes to the next generation. Among elephant seals, for example, a few dominant males mate with all the females in a population, so the alleles those males happen to carry contribute disproportionately to following generations; from a genetic point of view, the unsuccessful subdominant males might as well not exist (Figure 10.5). Thus the rate of genetic drift of allele frequencies, and of loss of heterozygosity, will be greater than expected from the population's census size, corresponding to what we expect of a smaller population. In other words, the population is *effectively* smaller than it seems. The **effective population size** (denoted N_e) of an actual population is the number of individuals in an ideal population (in which every adult reproduces) in which the rate of genetic drift (measured by the rate of decline in heterozygosity) would be the same as it is in the actual population. For instance, if we count 10,000 adults in a population, but only 1000 of them successfully breed, genetic drift proceeds at the same rate as if the population size were 1000, and that is the effective population size, N_e.

The effective population size can be smaller than the census size for several reasons:

1. *Variation in the number of progeny* produced by females, males, or both reduces N_e. The elephant seal represents an extreme example.
2. Similarly, a *sex ratio* different from 1:1 lowers the effective population size.
3. *Natural selection* can lower N_e by increasing variation in progeny number; for instance, if larger individuals have more offspring than smaller ones, the rate of genetic drift may be increased at all neutral loci because small individuals contribute fewer gene copies to subsequent generations.
4. If *generations overlap*, offspring may mate with their parents, and since these pairs carry identical copies of the same genes, the effective number of genes propagated is reduced.
5. Perhaps most importantly, *fluctuations in population size* reduce N_e, which is more strongly affected by the smaller than by the larger sizes. For example, if the number of breeding adults in five successive generations is 100, 150, 25, 150, and 125, N_e is approximately 70 (the harmonic mean*) rather than the arithmetic mean, 110.

N_e can be estimated by direct demographic studies of the factors listed, but it is more often estimated by genetic data. The current N_e is often estimated by changes in allele frequencies among generations, since these are greater in smaller populations (Waples 1989; Wang and Whitlock 2003). We will soon see that the average N_e over the long term can be esti-

*The HARMONIC MEAN is the reciprocal of the average of a set of reciprocals. If the number of breeding individuals in a series of t generations is $N_0, N_1, \ldots N_t$, N_e is calculated from $1/N_e = (1/t)(1/N_0 + 1/N_1 + \ldots + 1/N_t)$.

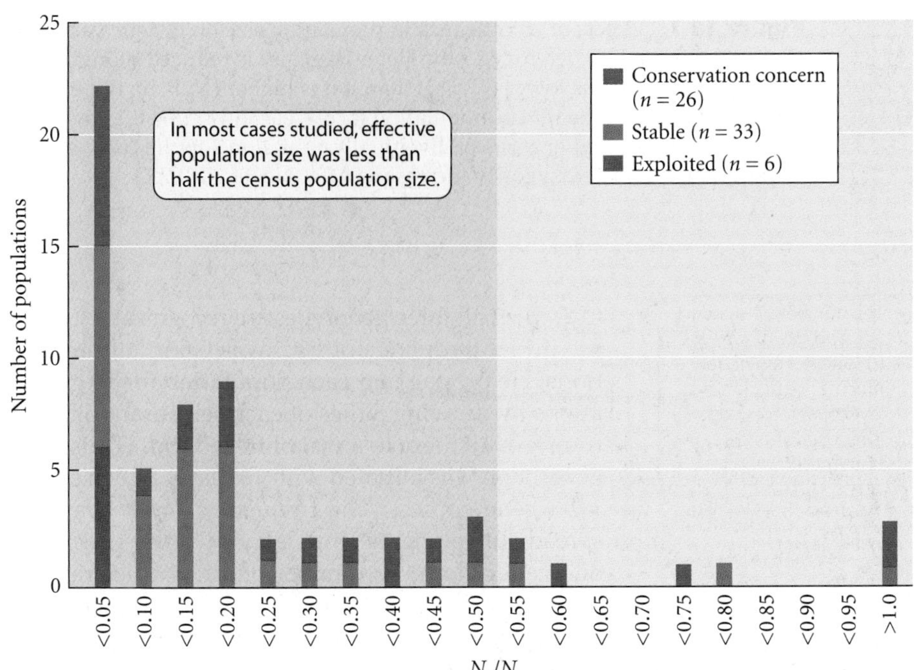

Figure 10.6 The ratio of effective population size to census population size (N_e/N) in 65 animal populations. Purple segments represent species of special conservation concern; the red segment represents six exploited fish populations. (After Palstra and Ruzzante 2008.)

mated from heterozygosity, given an estimate of the neutral mutation rate, because the equilibrial level of genetic variation is set by the balance between the gain of variation attributable to mutation and the loss attributable to genetic drift. Current N_e values have been estimated for many species of animals, especially managed wildlife populations and species of conservation concern. For example, the greater prairie-chicken (*Tympanuchus cupido*) has been largely extinguished from the former prairies of central North America and persists only as fragmented populations. The number of microsatellite alleles and mitochondrial DNA (mtDNA) haplotypes has declined over the last 50 years, and the effective sizes of local populations (15–32 birds in each area) are ten times lower than census sizes (Johnson et al. 2004). The ratio of the short-term effective population size to the census size (N_e/N) of an important game fish in the Gulf of Mexico, red drum (*Sciaenops ocellatus*), estimated from changes in allele frequencies, was only 0.001, probably because some of the estuaries in which the larvae develop are much less productive than others. The comparable ratio for long-term N_e/N, based on heterozygosity, was an order of magnitude lower (Turner et al. 2002). In general, the ratio N_e/N averages about 0.10 to 0.14 (Frankham 1995; Palstra and Ruzzante 2008; Figure 10.6).

Founder effects

Restrictions in size through which populations may pass are called **bottlenecks**. A particularly interesting bottleneck occurs when a new population is established by a small number of colonists, or founders—sometimes as few as a single mating pair (or a single inseminated female, as in insects in which females store sperm). The random genetic drift that ensues is often called a **founder effect**. If the new population rapidly grows to a large size, allele frequencies (and therefore heterozygosity) will probably not be greatly altered from those in the source population, although some rare alleles will not have been carried by the founders. If the colony remains small, however, genetic drift will alter allele frequencies and erode genetic variation. If the colony persists and grows, new mutations eventually restore heterozygosity to higher levels (Figure 10.7). As is true of genetic drift in general, bottlenecks increase levels of linkage disequilibrium (Charlesworth et al. 2003).

Genetic drift in real populations

LABORATORY POPULATIONS. Peter Buri (1956) described genetic drift in an experiment with *Drosophila melanogaster*. He initiated 107 experimental populations of flies, each with 8 males and 8 females, all heterozygous for two alleles (*bw* and bw^{75}) that affect eye color

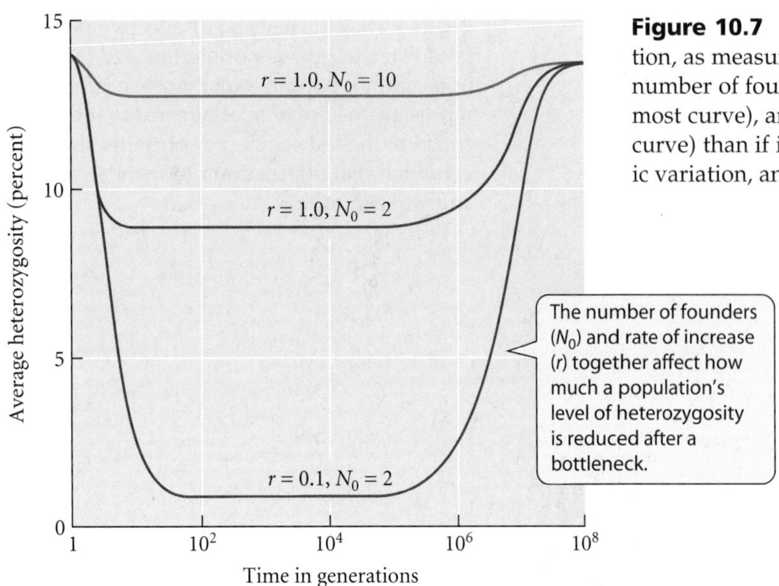

Average heterozygosity (percent)
15
10
5
0
$r = 1.0, N_0 = 10$
$r = 1.0, N_0 = 2$
$r = 0.1, N_0 = 2$
The number of founders (N_0) and rate of increase (r) together affect how much a population's level of heterozygosity is reduced after a bottleneck.
1
10^2
10^4
10^6
10^8
Time in generations

Figure 10.7 Effects of a bottleneck in population size on genetic variation, as measured by heterozygosity. Heterozygosity is reduced more if the number of founders is lower ($N_0 = 2$) than if it is higher ($N_0 = 10$; uppermost curve), and if the rate of population increase is lower ($r = 0.1$; lowest curve) than if it is higher ($r = 1.0$). Eventually, mutation supplies new genetic variation, and heterozygosity increases. (After Nei et al. 1975.)

(by which all three genotypes are recognizable). Thus the initial frequency of bw^{75} was 0.5 in all populations. He propagated each population for 19 generations by drawing 8 flies of each sex at random and transferring them to a vial of fresh food. (Thus each generation was initiated with 16 flies × 2 gene copies = 32 gene copies.) The frequency of bw^{75} rapidly spread out among the populations (Figure 10.8); after one generation, the number of bw^{75} copies ranged from 7 ($q = 7/32 = 0.22$) to 22 ($q = 0.69$). By generation 19, 30 populations had lost the bw^{75} allele and 28 had become fixed for it; among the unfixed populations, intermediate allele frequencies were quite evenly distributed. The results nicely matched those expected from genetic drift theory (see Figure 10.4).

More recently, McCommas and Bryant (1990) established four replicate laboratory populations, using houseflies (*Musca domestica*) taken from a natural population, at each of three bottleneck sizes: 1, 4, and 16 pairs. Each population rapidly grew to an equilibrium size of about 1000 flies, after which the populations were again reduced to the same bottleneck sizes. This procedure was repeated as many as five times. After each recovery from

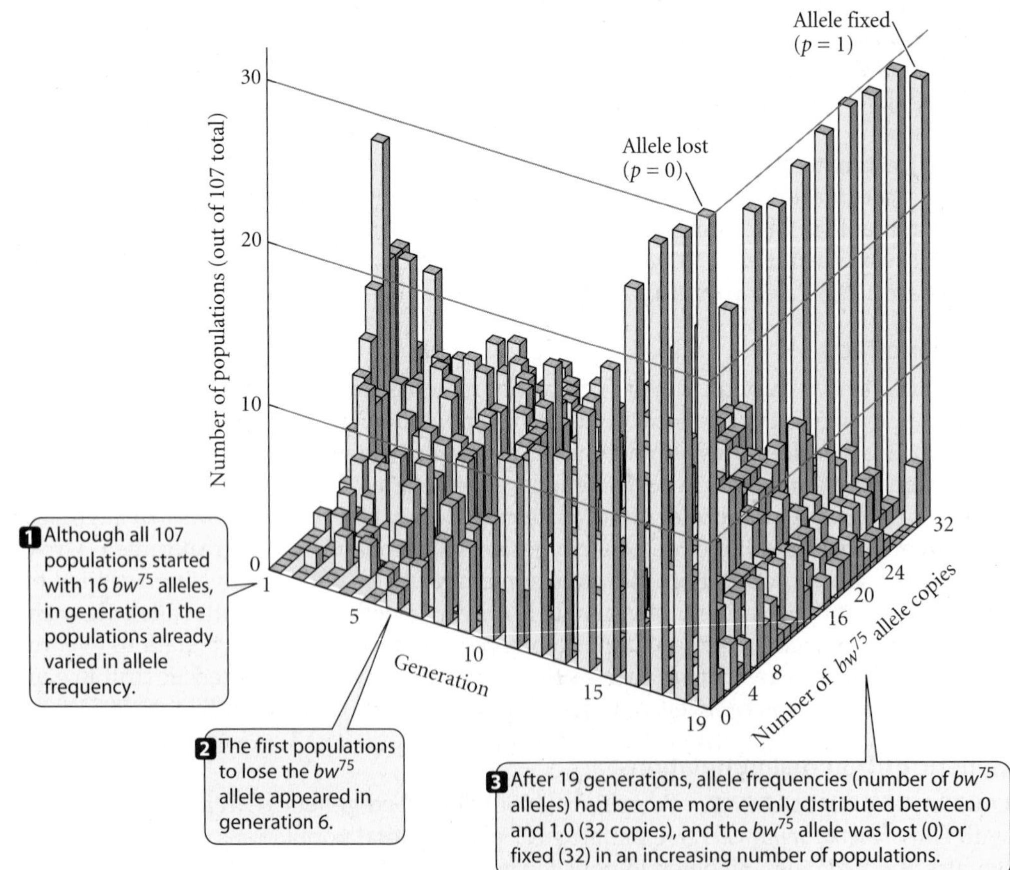

Number of populations (out of 107 total)
30
20
10
0
Allele fixed ($p = 1$)
Allele lost ($p = 0$)
1
5
10
15
19
Generation
0
4
8
16
20
24
32
Number of bw^{75} allele copies
1 Although all 107 populations started with 16 bw^{75} alleles, in generation 1 the populations already varied in allele frequency.
2 The first populations to lose the bw^{75} allele appeared in generation 6.
3 After 19 generations, allele frequencies (number of bw^{75} alleles) had become more evenly distributed between 0 and 1.0 (32 copies), and the bw^{75} allele was lost (0) or fixed (32) in an increasing number of populations.

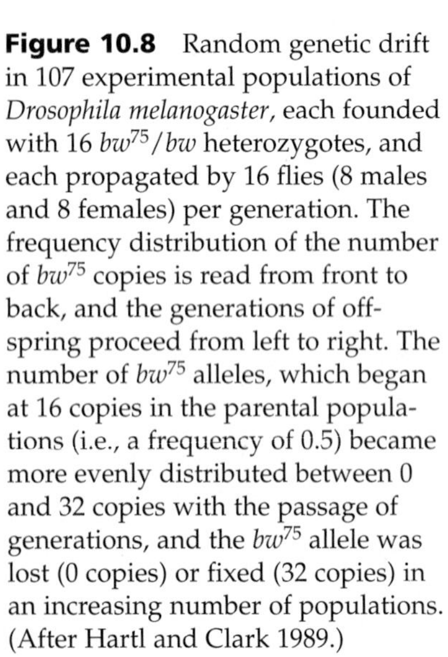

Figure 10.8 Random genetic drift in 107 experimental populations of *Drosophila melanogaster*, each founded with 16 bw^{75}/bw heterozygotes, and each propagated by 16 flies (8 males and 8 females) per generation. The frequency distribution of the number of bw^{75} copies is read from front to back, and the generations of offspring proceed from left to right. The number of bw^{75} alleles, which began at 16 copies in the parental populations (i.e., a frequency of 0.5) became more evenly distributed between 0 and 32 copies with the passage of generations, and the bw^{75} allele was lost (0 copies) or fixed (32 copies) in an increasing number of populations. (After Hartl and Clark 1989.)

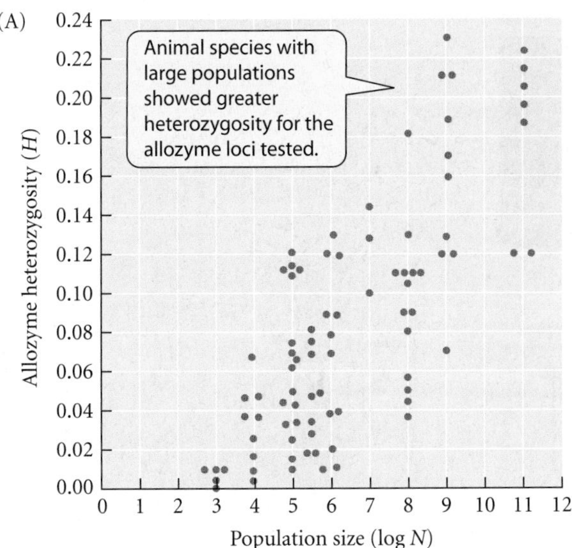

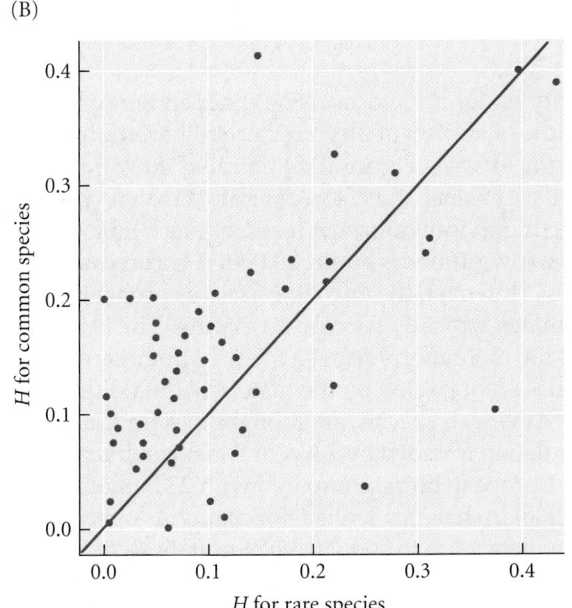

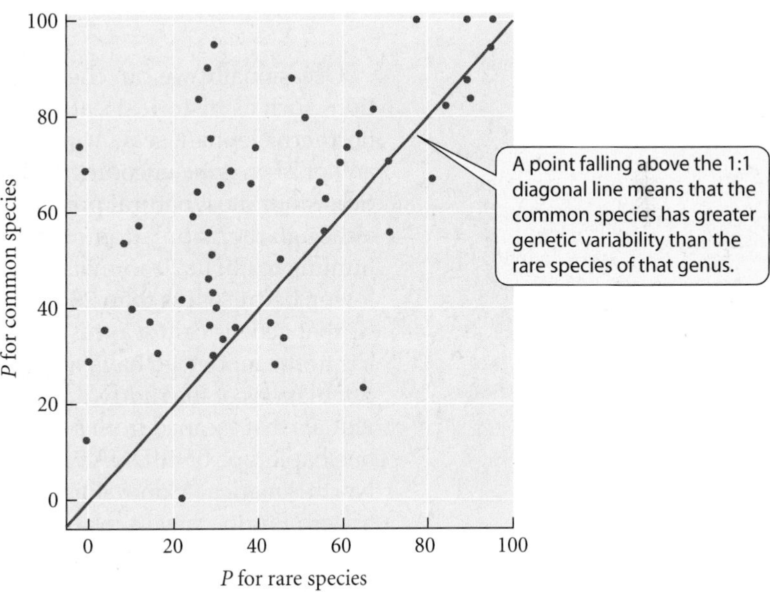

Figure 10.9 Genetic variation is reduced in smaller populations. (A) The relation between heterozygosity (H) for allozymes and estimated population sizes of some species of animals. (B) Each circle plots either heterozygosity (H; left diagram) or the proportion of polymorphic loci (P; right diagram) for a common and a rare species in the same genus of plant (i.e., each circle represents two species—one common and one rare—in the sample plant genus). (A after Soulé 1976; B after Cole 2003.)

a bottleneck, the investigators estimated the allele frequencies at four polymorphic enzyme loci for each population, using electrophoresis (see Chapter 9). They found that average heterozygosity ($\overline{H}$) declined steadily after each bottleneck episode, and that the smaller the bottlenecks were, the more rapidly it declined. On the whole, $\overline{H}$ closely matched the values predicted by the mathematical theory of genetic drift.

NATURAL POPULATIONS. When we describe the genetic features of natural populations, the data usually are not based on experimental manipulations, nor do we usually have detailed information on the populations' histories. We therefore attempt to *infer causes* of evolution (such as genetic drift or natural selection) by *interpreting patterns. Such inferences are possible only on the basis of theories* that tell us what pattern to expect if one or another cause has been most important.

Patterns of molecular genetic variation in natural populations often correspond to what we would expect if the loci were affected by genetic drift. For example, many studies of both animals and plants have shown that allozyme variation is generally lower in rare than in common species (Figure 10.9; Frankham et al. 2002).

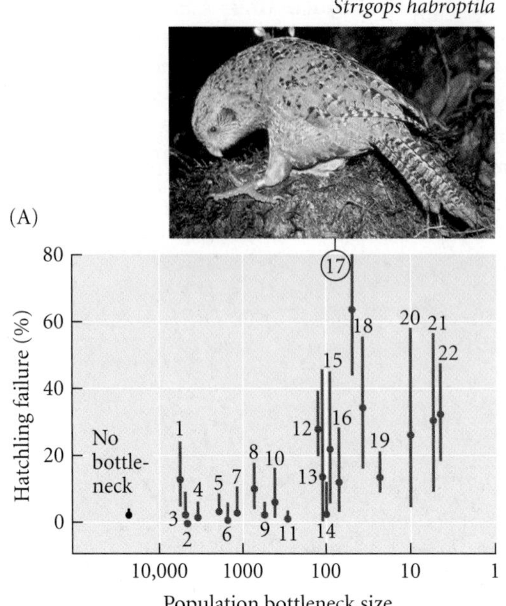

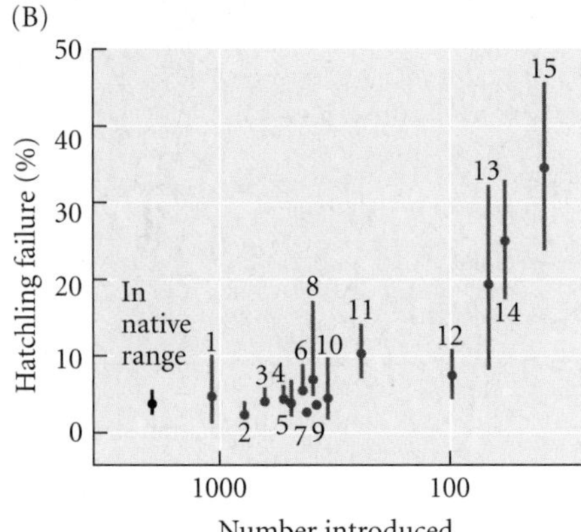

Figure 10.10 The failure rate for egg hatch is greater in New Zealand bird species that have suffered a bottleneck in population size. In both graphs, bottleneck size decreases logarithmically to the right. (A) Data for 22 species of native birds, many of which approached extinction after European settlement. Some are still highly endangered, such as #17, the kakapo (a large flightless parrot, pictured). (B) Data for 15 species of introduced birds, showing hatching failure in relation to the number of individuals released in New Zealand. Most of these are European species. (After Briskie and Mackintosh 2004; photo © Robin Bush/OSF/Photolibrary.com.)

Occasionally, we can check the validity of our inferences using independent information, such as historical data. For example, a survey of electrophoretic variation in the northern elephant seal (*Mirounga angustirostris;* see Figure 10.5) revealed no variation at any of 24 enzyme-encoding loci (Bonnell and Selander 1974)—a highly unusual observation, since most natural populations are highly polymorphic (see Chapter 9). However, although the population of this species now numbers about 30,000, it was reduced by hunting to about 20 animals in the 1890s. Moreover, the effective size was probably even lower, because less than 20 percent of males typically succeed in mating. The hypothesis that genetic drift was responsible for the monomorphism—a likely hypothesis according to the model we have just described—is supported by the historical data. European populations of the ruddy duck, a North American species, are thought to stem from individuals that escaped from a captive population founded by 7 North American birds. Only one haplotype of mtDNA was found in European birds, compared with 23 haplotypes in North America (Muñoz-Fuentes et al. 2006). Ashkenazi Jewish populations, especially in eastern Europe, have a relatively high frequency (up to about 0.03) of mutations that cause Tay-Sachs and other diseases. Patterns of linkage disequilibrium among closely linked genetic markers, as well as other aspects of sequence variation in these genes, correspond to known bottlenecks in the history of the Ashkenazim, such as the diaspora in 70 CE (AD) and the Black Death in 1348 (Risch et al. 2003; Slatkin 2004). Thus bottlenecks can increase the frequency of deleterious mutations, as theory predicts and has been documented in many natural populations of animals (Figure 10.10).

The Neutral Theory of Molecular Evolution

Whether or not random genetic drift has played an important role in the evolution of many of the morphological and other phenotypic features of organisms is a subject of considerable debate. There is no question, however, that at the levels of DNA and protein sequences, genetic drift is a major factor in evolution.

From the evolutionary synthesis of the late 1930s until the mid-1960s, most evolutionary biologists believed that almost all alleles differ in their effects on organisms' fitness, so that their frequencies are affected chiefly by natural selection. This belief was based on numerous studies of genes with morphological or physiological effects. But in the 1960s, the theory of evolution by random genetic drift of selectively neutral alleles became important as two kinds of molecular data became available. In 1966, Lewontin and Hubby showed that a high proportion of enzyme loci are polymorphic (see Chapter 9). They argued that natural selection could not actively maintain so much genetic variation, and suggested that much of it might be selectively neutral. At about the same time, Motoo Kimura (1968) calculated the rates of evolution of amino acid sequences of several proteins, using the phylogenetic approach described in Chapter 2 (see Figure 2.15). He concluded that a given protein evolved at a similar rate in different lineages. He argued that such constancy would not be expected to result from natural selection, but would be expected if most evolutionary changes at the molecular level are caused by mutation and genetic drift. These authors and others (King and Jukes 1969) initiated a controversy, known as the "neutralist-selectionist debate," that is still not entirely resolved. Although everyone now agrees that some molecular variation and evolution are neutral (i.e., a result of genetic drift), "selectionists" think a larger fraction of molecular evolutionary changes are due to natural selection than "neutralists" do.

The **neutral theory of molecular evolution** holds that although a small minority of mutations in DNA or protein sequences are advantageous and are fixed by natural selection, and although many mutations are disadvantageous and are eliminated by natural selection, the great majority of those mutations that are fixed are effectively neutral with respect to fitness and are fixed by genetic drift. According to this theory, most genetic variation at the molecular level—whether revealed by DNA sequencing or by enzyme electrophoresis—is selectively neutral and lacks adaptive significance. This theory, moreover, holds that evolutionary substitutions at the molecular level proceed at a roughly constant rate, so that the degree of sequence difference between species can serve as a MOLECULAR CLOCK, enabling us to determine the divergence time of species (see Chapter 2).

It is important to recognize that the neutral theory does *not* hold that the morphological, physiological, and behavioral features of organisms evolve by random genetic drift. Many—perhaps most—such features may evolve chiefly by natural selection, and they are based on base pair substitutions that (according to the neutralists) constitute a very small fraction of DNA sequence changes. Furthermore, the neutral theory acknowledges that many mutations are deleterious and are eliminated by natural selection, so that they contribute little to the variation we observe. Thus the neutral theory does not deny the operation of natural selection on *some* base pair or amino acid differences. It holds, though, that *most* of the variation we observe at the molecular level, both within and among species, has little effect on fitness, either because the differences in base pair sequence are not translated into differences at the protein level, or because most variations in the amino acid sequence of a protein have little effect on the organism's physiology.

Principles of the neutral theory

Let us assume that mutations occur in a gene at a constant rate of u_T per gamete per generation, and that because of the great number of mutable sites, every mutation constitutes a new DNA sequence (or allele, or haplotype). Of all such mutations, some fraction (f_0) are effectively neutral, so the **neutral mutation rate**, $u_0 = f_0 u_T$, is less than the total mutation rate, u_T. By *effectively neutral*, we mean that the mutant allele is so similar to other alleles in its effect on survival and reproduction (i.e., fitness) that changes in its frequency are governed by genetic drift alone, not by natural selection. (It is, of course, possible that the mutation does affect fitness slightly. Then, as we will see in Chapter 12, natural selection and genetic drift operate simultaneously, but because genetic drift is stronger in small than in large populations, the changes in the mutant allele's frequency will be governed almost entirely by genetic drift if the population is small enough. Therefore a particular allele may be effectively neutral, relative to another allele, when the population is small, but not when the population is large.)

The rate of origin of effectively neutral alleles by mutation, u_0, depends on the gene's function. If many of the amino acids in the protein it encodes cannot be altered without seriously affecting an important function—perhaps because they affect the shape of a protein that binds to DNA or to other proteins—then a majority of mutations in the gene will be deleterious rather than neutral, and u_0 will be much lower than the total mutation rate, u_T. Such a locus is said to have many **functional constraints**. On the other hand, if the protein can function well despite any of many amino acid changes (i.e., it is less constrained), u_0 will

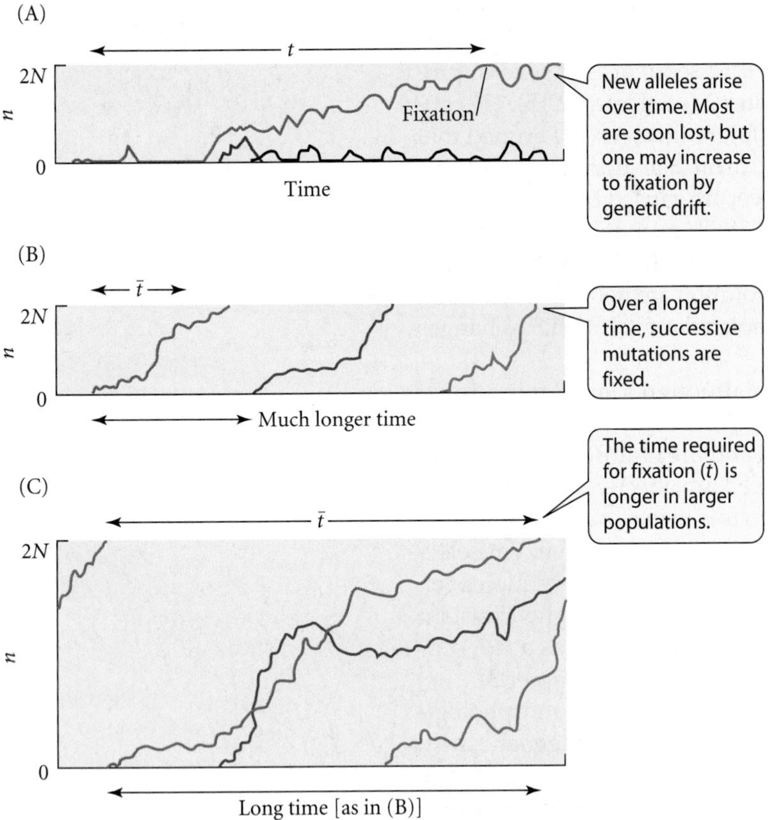
(A)
t
2N
n
0
Fixation
Time
New alleles arise over time. Most are soon lost, but one may increase to fixation by genetic drift.
(B)
t̄
2N
n
0
Much longer time
Over a longer time, successive mutations are fixed.
The time required for fixation (t̄) is longer in larger populations.
(C)
t̄
2N
n
0
Long time [as in (B)]

Figure 10.11 Evolution by genetic drift. Each graph plots the number of copies (n) of a mutation in a diploid population of N individuals ($2N$ gene copies) against time. (A) Most new mutations are lost soon after they arise, but occasionally an allele increases toward fixation by genetic drift. The average time required for such fixation is $\bar{t}$. (B) Over a longer time, successive mutations become fixed at this locus. (C) The time required for fixation is longer in larger populations. Hence at any time there will be more neutral alleles in a larger population. (After Crow and Kimura 1970.)

be higher. Within regions of DNA that encode proteins, we would expect the neutral mutation rate to be highest at third base-pair positions in codons and lowest at second base-pair positions, because those positions have the highest and lowest redundancy, respectively (see Figure 8.2). We would expect constraints to be least, or even nonexistent, and the fraction of neutral mutations (or substitutions) to be greatest, for DNA sequences that are not transcribed and have no known function, such as many introns and pseudogenes. (By applying the principles of the neutral theory, geneticists have found unexpected evidence that some portions of certain introns and pseudogenes may actually be functional; see Chapter 20.)

Now consider a population of effective size N_e in which the rate of neutral mutation at a locus is u_0 per gamete per generation (Figure 10.11). The *number* of new mutations is, on average, $u_0 \times 2N_e$, since there are $2N_e$ gene copies that could mutate. From genetic drift theory, we have learned that the probability that a mutation will be fixed by genetic drift is its frequency, p, which equals $1/(2N_e)$ for a newly arisen mutation. Therefore the number of neutral mutations that arise in any generation *and will someday be fixed* is

$$2N_e u_0 \times 1/(2N_e) = u_0$$

Since, on average, it will take $4N_e$ generations for such mutations to reach fixation, about the same number of neutral mutations should be fixed in every generation: *the rate of fixation of mutations is theoretically constant, and equals the neutral mutation rate.* This is the theoretical basis of the molecular clock. Notice that, surprisingly, the rate of substitution does not depend on the population size: each mutation drifts toward fixation more slowly if the population is large, but this is compensated for by the greater number of mutations that arise.

If two species diverged from their common ancestor t generations ago, and if each species has experienced u_0 substitutions per generation (relative to the allele in the common ancestor), then the number of base pair differences (D) between the two species should be $D = 2u_0t$, because each of the two lineages has accumulated u_0t substitutions. Therefore, if we have an estimate of the number of generations that have passed (see Box 8A), the neutral mutation rate can be estimated as

$$u_0 = D/2t$$

This formula requires qualification, however. Over a sufficiently long time, some sites experience repeated base substitutions: a particular site may undergo substitution from, say, A to C and then from C to T or even back to A. Thus the observed number of differences between species will be less than the number of substitutions that have transpired. As the time since divergence becomes greater, the number of differences begins to plateau, as is evident from the number of differences per base pair between the mtDNA of different mammalian taxa (Figure 10.12; see also Figure 2.15).

In Figure 10.12, each point represents a pair of taxa for which the age of the common ancestor has been estimated from the fossil record. The number of base pair differences increases linearly for about 5 to 10 Myr, then begins to level off; after about 40 Myr, little further divergence is evident because of "multiple hits"—that is, successive substitutions. From the linear part of the curve, the mutation rate can be readily calculated, assuming that all the base pair differences represent neutral substitutions (in Figure 10.12, it is about 0.01 mutations per base pair per lineage per Myr, or about 10^{-8} per year). As the curve begins to level off, however, the rate can be estimated only by making corrections for multiple substitutions (Li 1997). Data from taxa on the plateau cannot be used to estimate the substitution rate.

Note that if u_0 has been accurately estimated in some lineages, it might be used to estimate the time since related organisms diverged from their common ancestors, from the relationship $t = D/2u_0$ (see Chapter 2). It can also be used to estimate the time since gene lineages on a gene tree diverged (see Figures 2.20 and 10.1).

Within a population, there is turnover, or flux, of alleles or haplotypes (see Figure 10.11). As one or another allele approaches fixation (about every $4N$ generations, on average), other alleles are lost. But new neutral alleles arise continually by mutation, and although many are immediately lost by genetic drift, others drift to higher frequency and persist for some time in a polymorphic state before they are lost or fixed. Although the identity of the several or many alleles present in the population changes over time, the level of variation reaches an equilibrium when the rate at which alleles arise by mutation is balanced by the rate at which they are lost by genetic drift. This equilibrium level of variation, represented by the frequency of heterozygotes, H, is higher in a large population than in a small one. It can be shown mathematically that at equilibrium,

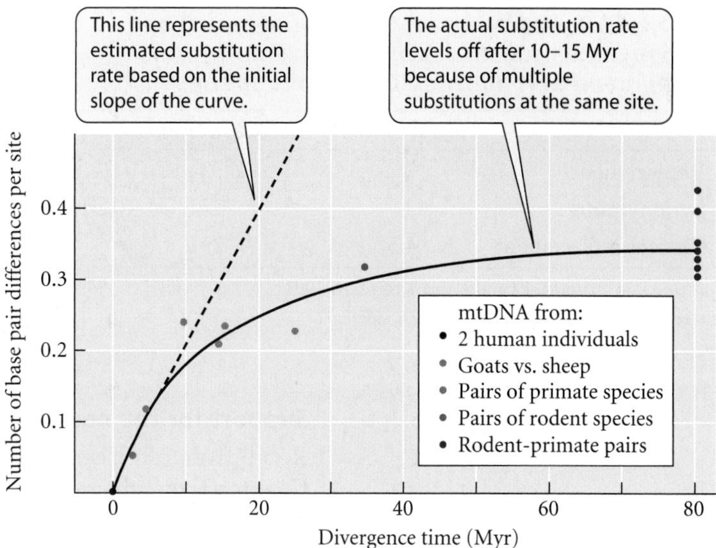

Figure 10.12 The number of base pair differences per site between the mtDNA of pairs of mammalian taxa, plotted against the estimated time since their most recent common ancestor. (After Brown et al. 1979.)

$$H = \frac{4N_e u_0}{4N_e u_0 + 1}$$

(Figure 10.13). For example, given the observed mutation rate for allozymes, 10^{-6} per gamete (Voelker et al. 1980), the equilibrium frequency of heterozygotes would be 0.004 if the effective population size N_e were 1000, but would be 0.50 if N_e were 250,000.

Variation within and among species

According to the neutral theory, the rate of allele substitution over time and the equilibrium level of heterozygosity are both proportional to the neutral mutation rate, u_0. If, because of differences in constraint or other factors, various kinds of DNA sequences or base pair sites differ in their rate of neutral mutation, those sequences or sites that differ more *among* related species should also display greater levels of variation *within* species. That is, *there should be a positive correlation between the heterozygosity at a locus and its rate of evolution.*

John McDonald and Martin Kreitman (1991) applied this principle in their analysis of DNA sequences of 6 to 12 copies of the coding region of the *Adh* (alcohol dehydrogenase) gene (see Figure 9.16) in each of three closely related species of *Drosophila*. Polymorphic sites (differences within species) and substitutions (differences among species) were classified as either synonymous or amino acid–replacing (nonsynonymous). If the neutral mutation rate is u_R for replacement changes and u_S for synonymous changes, then according to the neutral theory, the ratio of replacement to synonymous differences should be the same—u_R:u_S—for both polymorphisms and substitutions, if indeed the replacement changes are subject only to genetic drift. The data (Table 10.1)

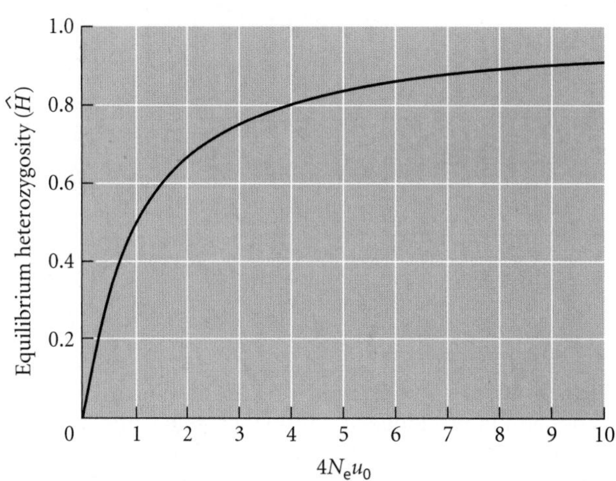

Figure 10.13 The equilibrium level of heterozygosity at a locus increases as a function of the product of the effective population size (N_e) and the neutral mutation rate (u_0). (After Hartl and Clark 1989.)

Table 10.1 ***Replacement (nonsynonymous) and synonymous substitutions and polymorphisms within and among three*** **Drosophila** ***species***[a]

	Polymorphisms	*Substitutions*
Replacement	2	7
Synonymous	42	17
Percent replacement	4.5	29.2

Source: Data from McDonald and Kreitman 1991.
[a]*D. melanogaster*, *D. simulans*, and *D. yakuba*.

showed, however, that only 5 percent of the polymorphisms, but fully 29 percent of the substitutions, that distinguish species are replacement changes. McDonald and Kreitman considered this result to be evidence that the evolution of amino acid–replacing substitutions is an adaptive process governed by natural selection. If most replacement substitutions are advantageous rather than neutral, they will increase in frequency and be fixed more rapidly than by genetic drift alone. They will therefore spend less time in a polymorphic state than selectively neutral synonymous changes do, and will thus contribute less to polymorphic variation within species. This is called the **McDonald-Kreitman test for selection** in molecular evolution.

Support for the neutral theory

Recall from our discussion of the uses of molecular data in phylogenetic inference (see Chapter 2) that the rate of nucleotide or amino acid substitution can be estimated from the number of sequence differences among species in two ways. First, an *absolute* rate can be estimated by a calibration based on fossil evidence of the time since two or more taxa diverged from their common ancestor (see Figure 10.12). Second, the *relative* rates of evolution among different lineages can be estimated simply from the number of differences that have accumulated in each member of a monophyletic group, relative to an outgroup (see Figure 2.15).

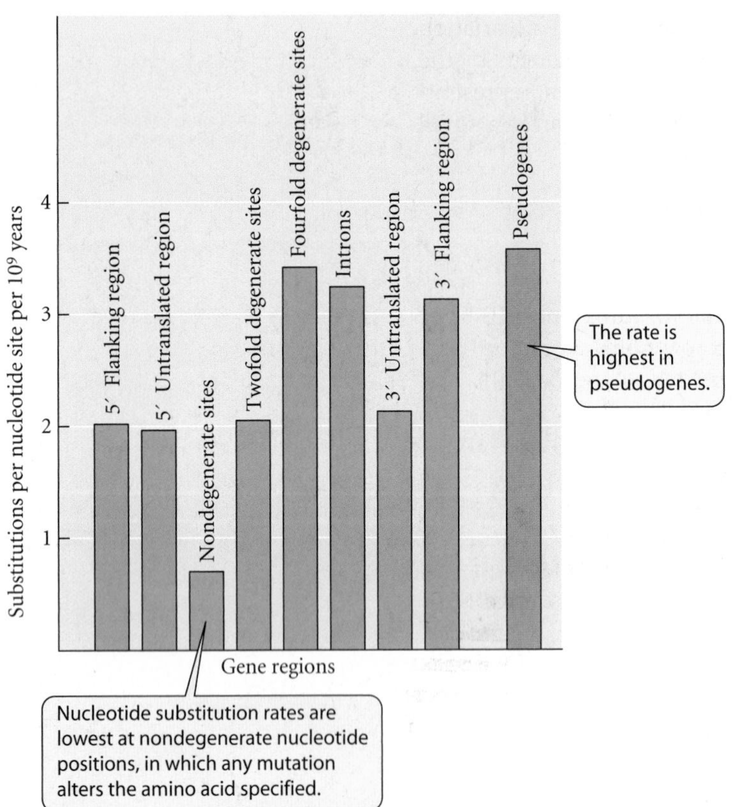

Figure 10.14 Average rates of substitution, in different parts of genes and in pseudogenes, estimated from comparisons between humans and rodents. Differences in the rate of molecular evolution among these classes are consistent with the neutral theory. Nucleotide positions are nondegenerate if no mutation would be synonymous; twofold degenerate if one other nucleotide would be synonymous; and fourfold degenerate if mutation to any of the other three nucleotides is synonymous (see Figure 8.2) (After Li 1997.)

Sequencing of DNA has provided a wealth of data on rates of molecular evolution. These data have provided evidence that most—although not all—DNA sequence evolution has been neutral (Figure 10.14). First, the rate of synonymous substitutions is generally greater than the rate of replacement substitutions. That is, substitutions occur most frequently at third-base positions in codons and least frequently in second-base positions. Second, rates of substitution are higher in introns than in coding regions of the same gene, and even higher in pseudogenes, the nonfunctional genes related in sequence to functional genes. Third, some genes, such as histone genes, evolve much more slowly than others. The genes that evolve most slowly are those thought to be most strongly constrained by their precise function. A striking example is that of the peptide hormone insulin, which is formed by the splicing of two segments of a proinsulin chain, the third segment of which (the C peptide) is removed, apparently playing no role other than in the formation of the mature insulin chain. Among mammals, the average rate of amino acid substitution has been six times greater in the C peptide locus than in the loci coding for the other portions of proinsulin (Kimura 1983).

Some proteins physically interact with each other. Protein interaction networks have been determined for yeast and *Drosophila*. Fraser et al. (2002) compiled an interaction network for 2245 yeast proteins, and then chose a subset of 164 proteins for which they could reliably estimate the number of amino acid substitutions between homologous genes of yeast and a very distantly related organism, the nematode *Caenorhabditis elegans*. As the neutral theory predicts, they found a significant negative correlation of the rate of a protein's amino acid substitution and the number of other proteins with which it interacts (Figure 10.15). Moreover, the effect of a given protein on the organism's fitness, as measured by the reduction in yeast growth rate when the gene was experimentally deleted, is positively correlated with the number of proteins with which it interacts. Fraser et al. also found that proteins in the same interaction cluster evolve at more similar rates than expected by chance, suggesting that interacting proteins coevolve.

These classes of evidence all indicate that the *rate of evolution is greatest for DNA alterations that are least likely to affect function*, and therefore least likely to alter the organism's fitness. This conclusion provides strong support for the neutral theory.

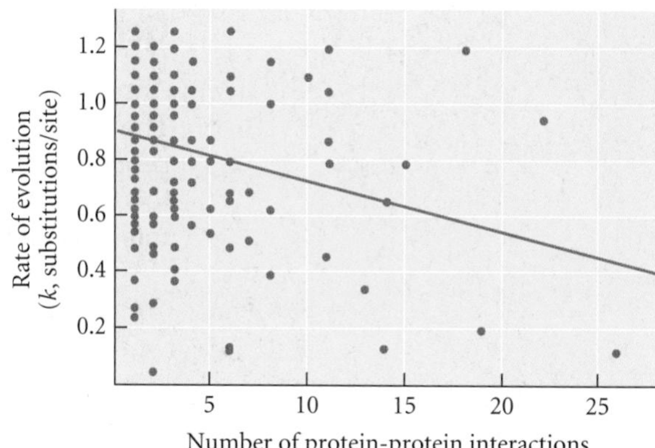

Figure 10.15 Inverse correlation between the rate of protein evolution and number of protein-protein interactions in a network. This relationship implies that such interactions impose functional constraints that reduce the rate of a protein's evolution. (After Fraser et al. 2002.)

Support for the neutral theory's prediction that rates of sequence evolution should be constant among phyletic lineages is more equivocal: some rates have been constant, and others have not. For example, rates of synonymous substitution have been about the same in the lineages leading to rodents, primates, and artiodactyls (even-toed hoofed mammals such as sheep and pigs), based on the relative rate test (Figure 10.16A). However, nonsynonymous substitutions have occurred at a significantly lower rate in primates than in the other two orders, and rodents have shown the highest rate (Figure 10.16B). The constancy of synonymous substitutions, which are presumably neutral, implies that the total rate of mutation (u_T) has been the same, per unit time, in these lineages. If so, then the higher rate of nonsynonymous substitutions in rodents could be due to either positive selection of advantageous mutations or lower effective population sizes, which would render some slightly deleterious mutations effectively neutral. This is only one of many examples of differences in the rate of sequence evolution among higher taxa. For instance, mtDNA sequences have evolved more slowly in turtles than in other vertebrates (Avise et al. 1992). The causes of variation in rates of sequence evolution are still not fully understood.

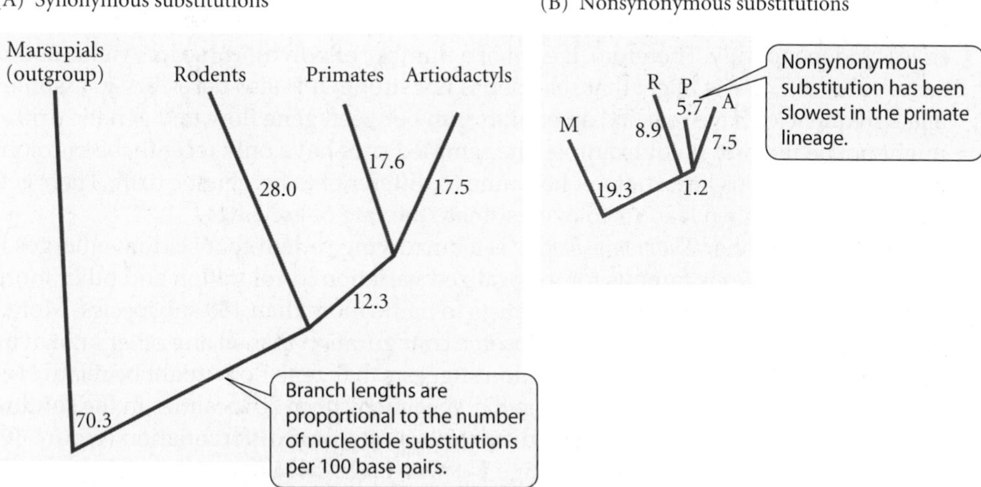

Figure 10.16 Constancy and inconstancy in the rate of sequence evolution. The number of nucleotide substitutions in 14 nuclear genes among marsupials (M), rodents (R), primates (P), and artiodactyls (A) is represented by the branch lengths, which are proportional to the number of substitutions per 100 base pairs. (A) Synonymous substitutions have occurred no faster in the rodent lineage (28.0) than in the lineages leading to either primates (12.3 + 17.6 = 29.9) or artiodactyls (12.3 + 17.5 = 29.8). (B) Nonsynonymous substitutions have occurred at a higher rate in rodents (8.9) than in the lineage from the rodent-primate ancestor to modern primates (1.2 + 5.7 = 6.9). The rate has been lower in primates (5.7) than in artiodactyls (7.5). In both diagrams, marsupials are an outgroup, and the substitutions along this branch cannot be partitioned into those between the lineage leading to marsupials and the lineage leading to placental mammals. (After Easteal and Collett 1994.)

Gene Flow and Genetic Drift

A measure of the variation in allele frequency among populations is F_{ST} (see Chapter 9). The rate at which populations drift toward fixation of one allele or another is inversely proportional to the effective population size, N_e (or N, for simplicity). However, the drift toward fixation is counteracted by gene flow from other populations, at rate m. These factors strike a balance, or equilibrium, at which the fixation index (F_{ST}) is approximately

$$F_{ST} = \frac{1}{4Nm + 1}$$

The quantity Nm is the *number* of immigrants per generation. If $m = 1/(N)$ (i.e., only 1 breeding individual per population is an immigrant, per generation), then $Nm = 1$, and $F_{ST} = 0.20$. That is, even a little gene flow keeps all the demes fairly similar in allele frequency, and heterozygosity remains high. Recasting this equation as

$$Nm = \frac{(1/F_{ST}) - 1}{4}$$

enables us to indirectly estimate average rates of gene flow among natural populations, since F_{ST} can be estimated from the variation of allele frequencies. In fact, such indirect estimates may be better than direct estimates of gene flow such as those described in Chapter 9, because direct observations may be made when m is unusually high or low, and are usually insufficient to detect long-distance migration, rare episodes of massive gene flow, and the perhaps rare (but nevertheless important) processes of population extinction and recolonization (Slatkin 1985). We must assume that the alleles for which we calculate F_{ST} are selectively neutral. (F_{ST} would underestimate gene flow if natural selection favored different alleles in different areas, and it would overestimate gene flow if selection favored the same allele everywhere.) This assumption can be evaluated by the degree of consistency among different loci for which F_{ST} is estimated. Genetic drift and gene flow affect all loci the same way, whereas natural selection affects different loci more or less independently. Therefore, if each of a number of polymorphic loci yields about the same value of F_{ST}, it is likely that selection is not strong. It is also necessary to assume that allele frequencies have reached an equilibrium between gene flow and genetic drift. This might not be the case if, for example, the sampled sites have only recently been colonized and the populations have not yet had time to differentiate by genetic drift. Their genetic similarity would then lead us to overestimate the rate of gene flow.

The pocket gopher *Thomomys bottae* is a burrowing rodent that seldom emerges from the soil. This species is famous for its localized variation in coloration and other morphological features, which has led taxonomists to name more than 150 subspecies. Moreover, local populations differ more in chromosome configuration than in any other known mammalian species. Such geographic variation suggests that gene flow might be relatively low. Indeed, 21 polymorphic enzyme loci in 825 specimens from 50 localities in the southwestern United States and Mexico showed extreme geographic differentiation (Figure 10.17). Among all 50 populations, the average F_{ST} was 0.412 (which might imply $Nm = 0.36$); among localities in Arizona, it was 0.198 (implying $Nm = 1.01$). The genetically most different populations were most geographically distant or segregated by expanses of unsuitable habitat—both factors that would reduce gene flow (Patton and Yang 1977).

Gene trees and population history

Earlier in this chapter, we introduced the principle of genetic drift by showing that because gene lineages within a population become extinct by chance over the course of time, all gene copies in a population today are descended from one gene copy that existed at some time in the past. The genealogical history of genes in populations is the basis of COALESCENT THEORY. This theory, applied to DNA sequence data, provides inferences about the structure and the effective size (N_e) of species populations (Hudson 1990). Recall

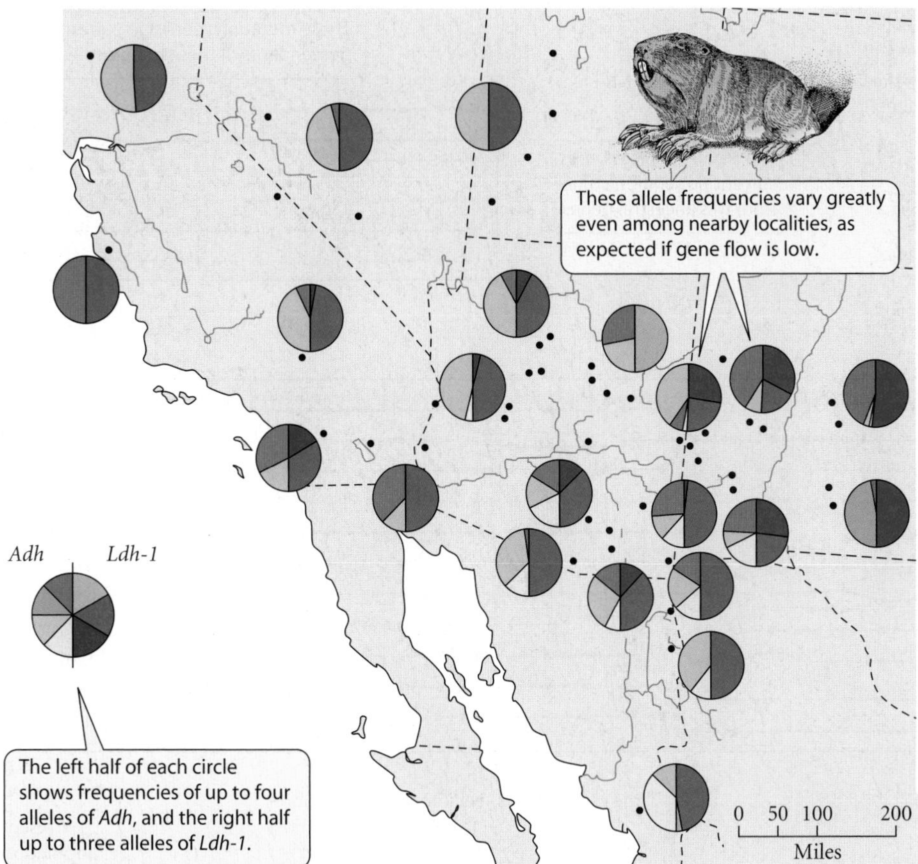
These allele frequencies vary greatly even among nearby localities, as expected if gene flow is low.
Adh
Ldh-1
The left half of each circle shows frequencies of up to four alleles of Adh, and the right half up to three alleles of Ldh-1.
0 50 100 200
Miles

Figure 10.17 Geographic variation in allele frequencies at two electrophoretic loci in the pocket gopher *Thomomys bottae*. The left half of each circle shows frequencies of up to four alleles of the gene for the enzyme alcohol dehydrogenase (*Adh*); the right half, those of up to three alleles of the lactate dehydrogenase-1 gene *Ldh1*. The allele frequencies vary greatly even among localities located close to one another, as expected if gene flow is low. (After Patton and Yang 1977.)

that if the number of breeding individuals changes over time, N_e is approximately equal to the harmonic mean, which is much closer to the smallest number the population has experienced than to the arithmetic mean. If, for example, the population has rapidly grown to its present large size from a historically much smaller size, N_e is close to the latter size, and this value can be estimated from coalescent theory.

Because the smaller the effective size (N_e) of a population, the more rapidly genetic drift transpires, the existing gene copies in a small population must stem from a more recent common ancestor than the gene copies in a large population (compare parts A and B in Figure 10.18). That is, if we look back in time from the present, it takes longer for the present genes in the larger population to coalesce to their common ancestor. Mathematical models show that in a haploid population of N_e individuals, the average time back to the common ancestor of all the gene copies (t_{CA}) is $2N_e$ generations, and in a diploid population, $t_{CA} = 4N_e$ generations. In a diploid population, the common ancestor of a random pair of gene copies occurred $2N_e$ generations ago. (For mitochondrial genes, carried in an effectively haploid state and transmitted only by females, $t_{CA} = N_e$ generations.)

If two randomly sampled gene copies had a common ancestor t generations ago, and each lineage experiences on average u mutations per generation, then each will have accumulated $u \times t$ mutations since the common ancestor. Therefore, the expected number of base pair differences between them (π) will be $2ut$, because there are two gene lineages. Since $t = 2N_e$, $\pi = 4N_e u$. We therefore *expect the average number of base pair differences between gene copies to be greater in large than in small populations.* (This difference is illustrated by the tick marks, representing mutations, on the gene trees in Figure 10.18.) In fact, if we have an estimate of the mutation rate per base pair (u), and if we measure the average proportion of sites that differ between random pairs of gene copies (π), *we can estimate the effective population size* as

$$N_e = \pi/4u$$

(For mitochondrial genes, the estimate of N_e is $N_e = \pi/u$.)

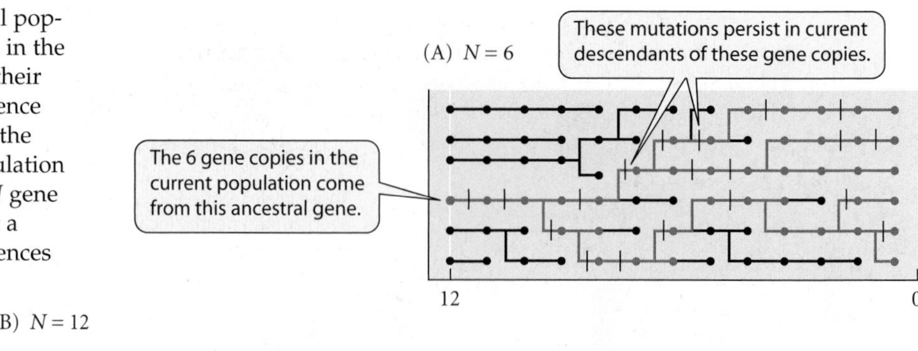

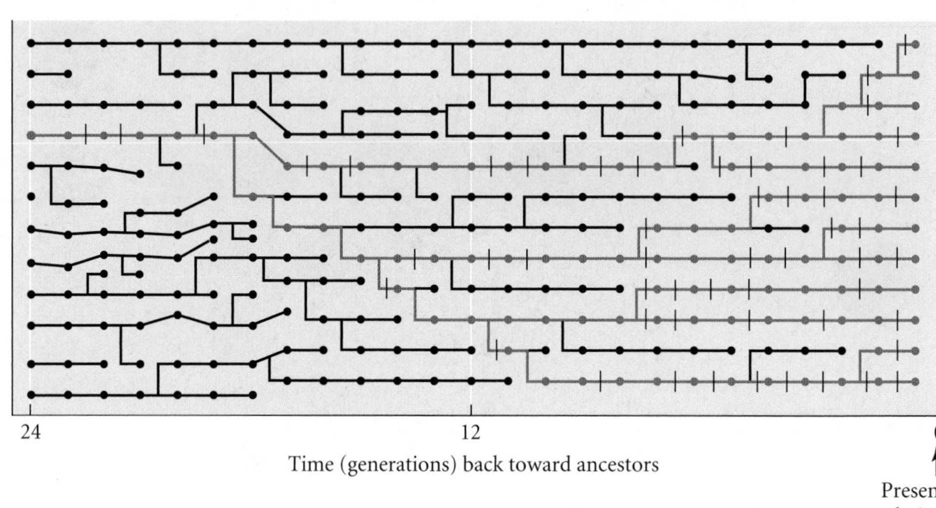

Figure 10.18 Coalescence time in (A) a small population and (B) a large population. Gene copies in the present-day population are shown in red, and their genealogy is shown back as far as their coalescence (common ancestry). The expected time back to the common ancestor of all gene copies in the population is $2N$ generations for a haploid population of N gene copies. Tick marks represent mutations, each at a unique site in the gene. More mutational differences among gene copies are expected if the coalescence time has been longer.

The origin of modern *Homo sapiens* revisited

The gene tree for human mtDNA, sampled from populations throughout the world, is rooted between most African mtDNA lineages and a clade of lineages that includes some African haplotypes and all non-African haplotypes (see Figure 6.17). Moreover, there are far fewer nucleotide differences among non-African haplotypes, on average, than among African haplotypes. We noted in Chapter 6 that these and other such data tend to support the "replacement," or "out-of-Africa," hypothesis, according to which the world's human population outside Africa is descended from a relatively small population that spread from Africa rather recently, replacing Eurasian populations of archaic *Homo sapiens* without interbreeding with them to any significant extent.

The theory described above has been applied to human data of this kind. Several studies of mtDNA have concluded that all human mitochondrial genes, both African and non-African, are descended from a common ancestral gene that existed at t_{CA} = 156 to 250 Kya (Vigilant et al. 1991; Horai et al. 1995; Ingman et al. 2000). Similar conclusions were reached in analyses of Y chromosome sequences (carried by males) and autosomal microsatellite loci (Goldstein et al. 1995; Hammer 1995). Of course, this does not mean that the human population at that time consisted of one woman and one man; it means only that all other mitochondria and Y chromosomes in the population at that time failed to leave descendants in today's population.

The most recent common ancestor of all mitochondrial or Y-linked genes existed before the various modern populations diverged from each other. The number of mutations distinguishing non-African from African sequences provides an estimate of when those populations diverged. Initial estimates, mostly based on mtDNA, suggested that this divergence occurred 40 to 143 Kya. A recent study of mtDNA and genes on the X and Y chromosomes concluded that the ancestral Eurasian population was founded only about 40 Kya (Garrigan et al. 2007). These estimates are much more recent than they would be if the major regional populations of humans were descended from the archaic *Homo* populations that inhabited those regions.

From data on variation in sequences, together with estimates of mutation rates, several studies have estimated the historical effective size of the human population. All such studies have concluded that the world's human population is descended from an effective breeding population of only 4600 to 11,200 people (see Hammer 1995; Rogers 1995)! For example, a recent study of sequence variation in nuclear genes estimated the long-term N_e of humans as 10,400, lower than that of bonobo (12,300), chimpanzee (21,300), or gorilla (25,200; Yu et al. 2004). These ancestors must have formed a fairly localized population in Africa, for if they had been distributed across Europe and Asia as well, the population density would have been so low that the demes could not have formed a cohesive species. The lower level of variation among non-African than among African sequences, moreover, indicates that the non-African population was founded by perhaps as few as 1500 individuals (Rogers and Harpending 1992; Garrigan et al. 2007). As humans expanded across Eurasia and into North America (see Figure 6.18), successive founder events lowered both DNA sequence variation (see Figure 9.34B) and genetic variation for features such as skull measurements (Manica et al. 2007). Moreover, the average level of linkage disequilibrium between SNPs (single nucleotide polymorphisms; see Chapter 9) is greater, the farther the population is from East Africa (**Figure 10.19**). A higher proportion of SNPs is thought to cause potentially damaging alterations of proteins in European-American than in African-American individuals, suggesting that historical bottlenecks may have elevated the frequencies of deleterious mutations in Eurasian populations (Lohmueller et al. 2008). All these aspects of human genetic variation are consistent with the theory of genetic drift and with the hypothesis that the human population outside Africa has descended recently from a rather small number of East African ancestors.

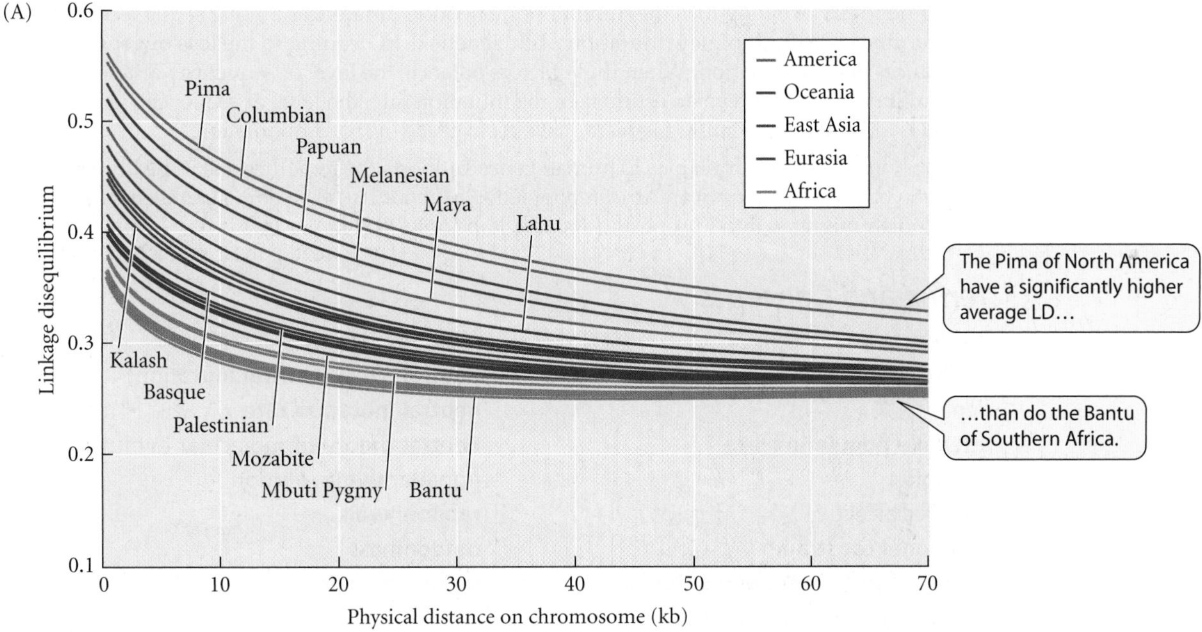

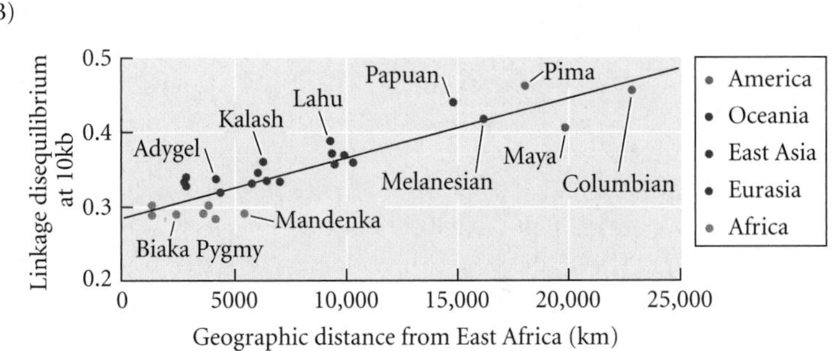

Figure 10.19 Linkage disequilibrium (LD) between genetic markers (SNPs) in human populations. (A) The average LD between many pairs of markers declines, the farther apart (in kilobases, kb) the markers are on a chromosome, because of increasing recombination rate. At any distance, some human populations have higher average LD than others. (B) The average LD is greater the farther these populations are from East Africa, as expected because of successive founder events during colonization of farther regions. (After Jakobsson et al. 2008.)

Summary

1. The frequencies of alleles that differ little or not at all in their effect on organisms' fitness (neutral alleles) fluctuate at random. This process, called random genetic drift, reduces genetic variation and leads eventually to the random fixation of one allele and the loss of others, unless it is countered by other processes, such as gene flow or mutation.
2. Different alleles are fixed by chance in different populations.
3. The probability, at any time, that a particular allele will be fixed in the future equals the frequency of the allele at that time. For example, if a newly arisen mutation is represented by one copy in a diploid population of N individuals, the probability that it will increase to fixation is $1/(2N)$.
4. The smaller the effective size of a population, the more rapidly random genetic drift operates. The effective size is often much smaller than the apparent population size.
5. Patterns of allele frequencies at some loci in both experimental and natural populations conform to predictions from the theory of genetic drift.
6. The theory of genetic drift has been applied especially to variation at the molecular level. The neutral theory of molecular evolution holds that, although many mutations are deleterious, and a few are advantageous, most molecular variation within and among species is selectively neutral. The fraction of mutations that are neutral varies: it is high for proteins that lack strong functional constraints and for sequences that are not transcribed. Likewise, it is higher for synonymous than for nonsynonymous (amino acid–replacing) nucleotide substitutions.
7. As the neutral theory predicts, synonymous mutations and mutations in less constrained genes are fixed more rapidly than those that are more likely to affect function. The neutral theory also predicts that over long spans of time, substitutions will occur at an approximately constant rate for a given gene (providing a basis for the "molecular clock"). The rate of molecular evolution, as measured by differences among species, appears to be more nearly constant for synonymous than for nonsynonymous substitutions.
8. For neutrally evolving loci, the number of nucleotide differences among sequences increases over time as a result of new mutations, but genetic drift, leading to the loss of gene lineages, reduces genetic variation. When these factors balance, the level of sequence variation reaches equilibrium. Thus, given an estimate of the mutation rate, the level of sequence variation provides a basis for estimating the historical effective size (N_e) of a population.
9. Applying the above principles to human genes supports the hypothesis that the human population has descended from an African population of about 10,000 breeding members, from which colonists migrated into Europe and Asia perhaps only 40 Kya.

Terms and Concepts

bottleneck
coalescence
deme
effective population size
fixation
founder effect
functional constraint
genetic drift (= random genetic drift)
McDonald-Kreitman test for selection
metapopulation
neutral allele (neutral mutation)
neutral mutation rate
neutral theory of molecular evolution
nonadaptive evolution
random walk
randomness
sampling error

Suggestions for Further Reading

Fundamentals of Molecular Evolution by D. Graur and W.-H. Li (Sinauer Associates, Sunderland, MA, 2000) is a leading textbook on the topics discussed in this chapter, and covers many other topics as well. *The Neutral Theory of Molecular Evolution* by M. Kimura (Cambridge University Press, Cambridge, UK, 1983) is a comprehensive (although somewhat dated) discussion of the neutral theory by its foremost architect.

Problems and Discussion Topics

1. For a diploid species, assume one set of 100 demes, each with a constant size of 50 individuals, and another set of 100 demes, each with 100 individuals. (a) If in each deme the frequencies of neutral alleles A_1 and A_2 are 0.4 and 0.6 respectively, what fraction of demes in each set is likely to become fixed for allele A_1 versus A_2? (b) Assume that a neutral mutation arises in each deme. Calculate the probability that it will become fixed in a population of each size. In what *number* of demes (approximately) do you expect it to become fixed? (c) If a fixation occurs, how many generations do you expect it to take?
2. Assuming that the average rate of neutral mutation is 10^{-9} per base pair per gamete, how many generations would it take, on average, for 20 base pair substitutions to be fixed in a gene with 2000 base pairs? Suppose that the number of base pair differences in this gene between species A and B is 92, between A and C is 49, and between B and C is 91. Assuming that no repeated replacements have occurred at any site in any lineage, draw the phylogenetic tree, estimate the number of fixations that have occurred along each branch, and estimate the number of generations since each of the two speciation events.
3. Some evolutionary biologists have argued that the neutral theory should be taken as the null hypothesis to explain genetic variation within species or populations and genetic differences among them. In this view, adaptation and natural selection should be the preferred explanation only if genetic drift cannot explain the data. Others might argue that since there is so much evidence that natural selection has shaped species' characteristics, selection should be the explanation of choice and that the burden of proof should fall on advocates of the neutral theory. Why might one of these points of view be more convincing than the other?
4. Some critics of Darwin's theory of evolution by natural selection have claimed that the concept of natural selection is tautologous (i.e., it is a "vicious circle"). They say, "Natural selection is the principle of the survival of the fittest. But the fittest are defined as those that survive, so there is no way to prove or disprove the theory." Argue against this statement, based on the contents of this chapter.
5. How can gene trees be used to estimate rates of gene flow among geographic populations of a species? What assumptions would have to be made? (See Slatkin and Maddison 1989.)
6. Suppose that several investigators want to use genetic markers such as allozymes or SNPs to estimate gene flow (the average number of migrants per generation) among populations of several species. One wants to study movement of howler monkeys among forest patches left by land clearing in Brazil; another plans to study movement among populations of mink frogs in lakes in Ontario, north of Lake Superior; a third intends to study the warbler finch on the 17 major islands of the Galápagos archipelago, near the Equator. For which of these species is this approach most likely, or least likely, to yield a valid estimate of gene flow? Why?
7. Could the principles of this chapter be used to estimate when an infectious disease organism first entered the human population? What conditions or characteristics of the organism would help to make this feasible? See, for example, "Timing the ancestor of the HIV-1 pandemic strains" by B. Korber et al. (*Science* 288:1789–1796, 2000).

CHAPTER 11

Natural Selection and Adaptation

Adapting an adaptation. Nudibranchs such as *Flabellina iodinea* are marine gastropod molluscs that lack shells. Many nudibranchs are unpalatable or dangerous because of stinging nematocysts they acquire by feeding on coral tissue and storing the noxious structures in their own bodies as a defense against predators. Bright "warning coloration" like this individual's is adaptive in toxic animal species, a signal to would-be predators that consuming this particular prey is not a good idea. (Photo © Ralph A. Clevenger/Photolibrary.com.)

The theory of natural selection is the centerpiece of *The Origin of Species* and of evolutionary theory. It is this theory that accounts for the adaptations of organisms, those innumerable features that so wonderfully equip them for survival and reproduction; it is this theory that accounts for the divergence of species from common ancestors and thus for the endless diversity of life. Natural selection is a simple concept, but it is perhaps the most important idea in biology. It is also one of the most important ideas in the history of human thought—"Darwin's dangerous idea," as the philosopher Daniel Dennett (1995) has called it—for it explains the apparent design of the living world without recourse to a supernatural, omnipotent designer.

An **adaptation** is a characteristic that enhances the survival or reproduction of organisms that bear it, relative to alternative character states (especially the ancestral condition in the population in which the adaptation evolved). Natural selection is the only mechanism known to cause the evolution of adaptations, so many biologists would simply define an adaptation as a characteristic that has evolved by natural selection. The word "adaptation" also refers to the process whereby the members of a population become better suited to some feature of their environment through change in a characteristic that affects their survival or reproduction. These definitions, however, do not fully incorporate the complex issue of just how adaptations (or the process of adaptation) should be defined or measured. We will touch on some of these complexities later in this chapter.

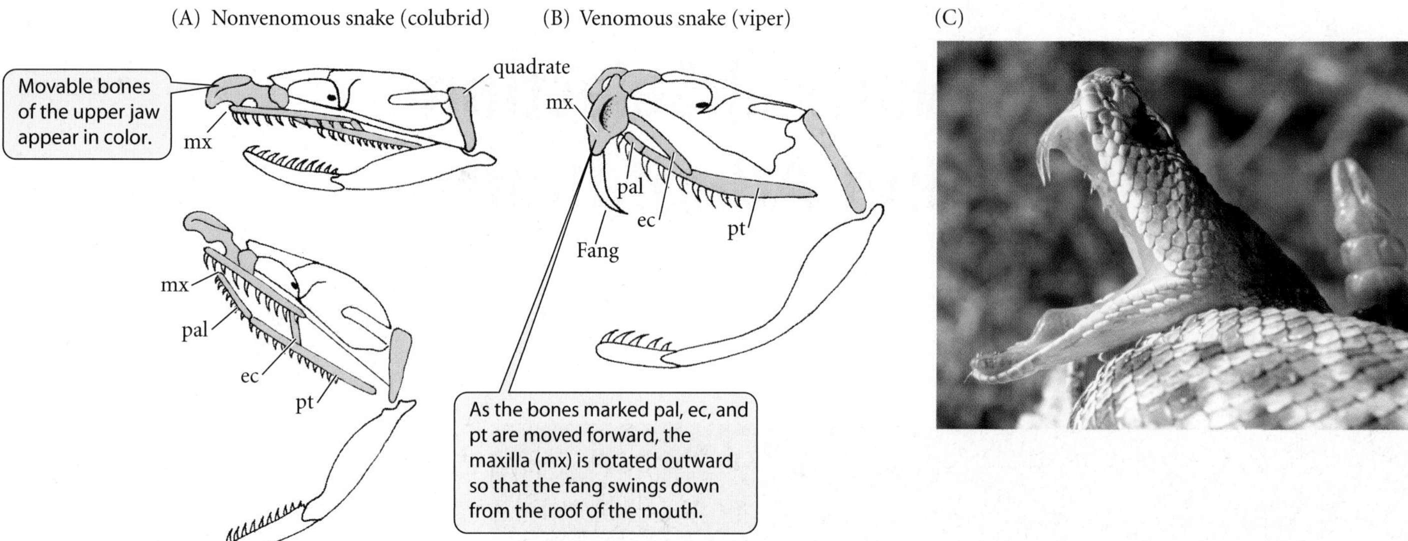
(A) Nonvenomous snake (colubrid)
(B) Venomous snake (viper)
(C)
Movable bones of the upper jaw appear in color.
mx
quadrate
mx
mx
pal
ec
pt
pal
ec
pt
Fang
As the bones marked pal, ec, and pt are moved forward, the maxilla (mx) is rotated outward so that the fang swings down from the roof of the mouth.

Figure 11.1 The kinetic skull of snakes. The movable bones of the upper jaw are shown in gold. (A) The skull of a nonvenomous snake with jaws closed (top) and open (bottom). (B) A viper's skull. (C) The head of a red diamondback rattlesnake (*Crotalus ruber*) in strike mode. (A, B after Porter 1972; C © Tom McHugh/Photo Researchers, Inc.)

Adaptations in Action: Some Examples

We can establish a few important points about adaptations by looking at some striking examples.

- In most terrestrial vertebrates, the skull bones are rather rigidly attached to one another, but in snakes they are loosely joined. Most snakes can swallow prey much larger than their heads, manipulating them with astonishing versatility. The lower jawbones (mandibles) articulate to a long, movable quadrate bone that can be rotated downward so that the mandibles drop away from the skull; the front ends of the two mandibles are not fused (as they are in almost all other vertebrates), but are joined by a stretchable ligament. Thus the mouth opening is greatly increased (Figure 11.1A). Both the mandibles and the tooth-bearing maxillary bones, which are suspended from the skull, independently move forward and backward to pull the prey into the throat. In rattlesnakes and other vipers, the maxilla is short and bears only a long, hollow fang, to which a duct leads from the massive poison gland (a modified salivary gland). The fang lies against the roof of the mouth when the mouth is closed. When the snake opens its mouth, the same lever system that moves the maxilla in nonvenomous snakes rotates the maxilla 90 degrees (Figure 11.1B), so that the fang is fully erected. Snakes' skulls, then, are complex mechanisms, "designed" in ways that an engineer can readily analyze. Their features have been achieved by modifications of the same bones that are found in other reptiles.
- Among the 18,000 to 25,000 species of orchids, many have extraordinary modifications of flower structure and astonishing mechanisms of pollination. In pseudocopulatory pollination, for example (Figure 11.2), part of the flower is modified to look somewhat like a female insect, and the flower emits a scent that mimics the attractive sex pheromone (scent) of a female bee, fly, or thynnine wasp, depending on the orchid species. As a male insect "mates" with the flower, pollen is deposited precisely on that part of the insect's body that will contact the stigma of the next flower visited. Several points are of interest. First, adaptations are found among plants as well as animals. For Darwin, this was an important point, because Lamarck's theory, according to which animals inherit characteristics altered by their parents' behavior, could not explain the adaptations of plants. Second, the floral form and scent are adaptations to promote reproduction rather than survival. Third, the plant achieves reproduction by deceiving, or exploiting, another organism; the insect gains nothing from its interaction with the flower. In fact, it would surely be advantageous to resist the flower's deceptive allure, since copulating with a flower probably reduces the insect's opportunity to find proper mates. So organisms are not necesssarily as well adapted as they could be.

Figure 11.2 Pseudocopulatory pollination. (A) *Ophrys apifera*, one of the "bee orchids," uses pheromones to attract male bees and is shaped such that, in attempting to copulate with the flower, pollen adheres to the insect's body. (B) A long-horned bee (*Eucera longicornis*) attempts to mate with an *Ophrys scolopax* flower. A yellow pollen mass adheres to the bee's head. (A © E. A. Janes/Photolibrary.com; B © Perennou Nuridsany/Photo Researchers, Inc.)

- After copulation, male redback spiders (*Latrodectus hasselti*; relatives of the "black widow" spider), often somersault into the female's mouthparts and are eaten (Figure 11.3A). This suicidal behavior might be adaptive, because males seldom have the opportunity to mate more than once, and it is possible that a cannibalized male fathers more offspring. Maydianne Andrade (1996) tested this hypothesis by presenting females with two males in succession, recording the duration of copulation, and using genetic markers to determine the paternity of the females' offspring. She found that females that ate the first male with whom they copulated were less likely to mate a second time, so these cannibalized males fertilized all the eggs. Furthermore, among females that did mate with both males, the percentage of offspring that were fathered by the second male was greater if he was eaten than if he survived. (Figure 11.3B). Both outcomes support the

Figure 11.3 (A) The small male redback spider somersaults into the large female's mouthparts after copulation. (B) The proportion of eggs fertilized by the second male that copulated with a female was correlated with the duration of his copulation. On average, copulation by cannibalized males lasted longer than that by noncannibalized males. (A after Forster 1992; B after Andrade 1996.)

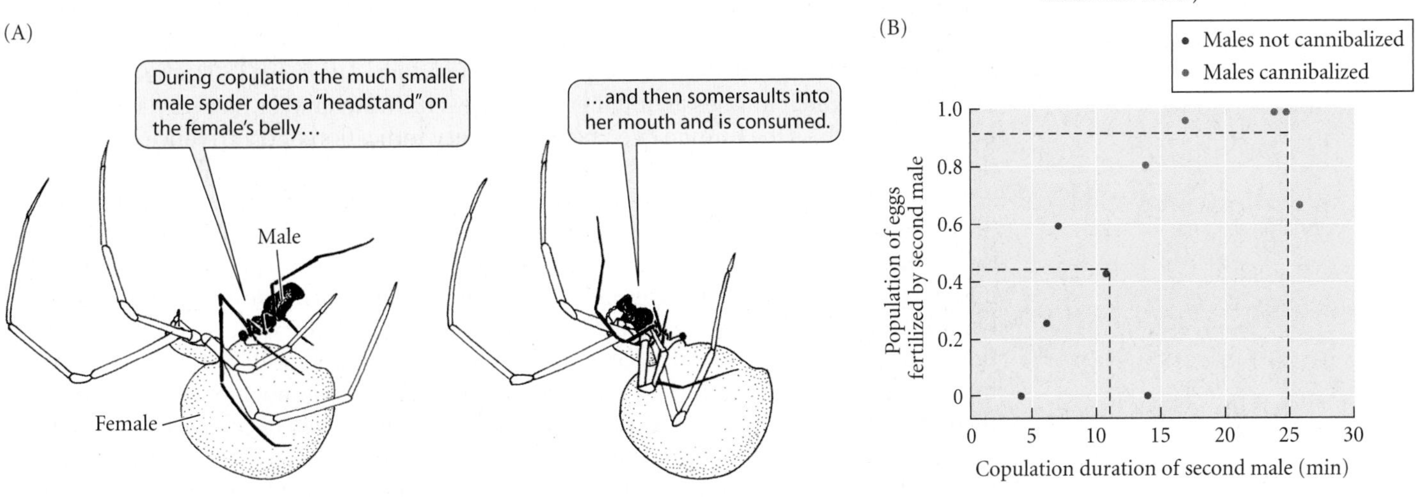

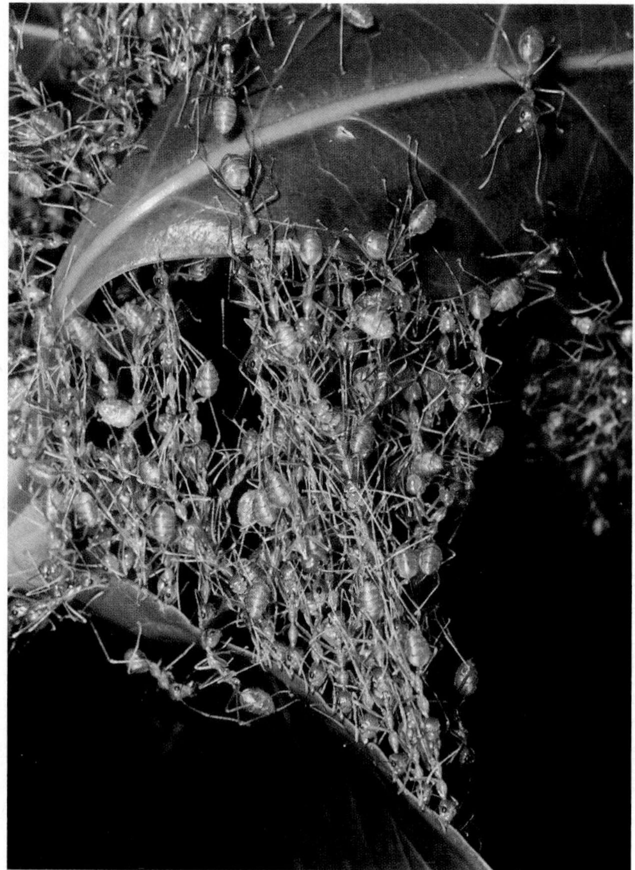

Figure 11.4 Weaver ants (*Oecophylla*) constructing a nest. Chains of workers, each seizing another's waist with her mandibles, pull leaves together. (Photo from Hölldobler and Wilson 1983, courtesy of Bert Hölldobler.)

hypothesis that sexual suicide enhances reproductive success. This example suggests that prolonged survival is not necessarily advantageous, and illustrates how hypotheses of adaptation may be formulated and tested.

- Many species of animals engage in cooperative behavior, but it reaches extremes in some social insects. An ant colony, for example, includes one or more inseminated queens and a number of sterile females, the workers. Australian arboreal weaver ants (genus *Oecophylla*) construct nests of living leaves by the intricately coordinated action of numerous workers, groups of which draw together the edges of leaves by grasping one leaf in their mandibles while clinging to another (Figure 11.4). Sometimes several ants form a chain to collectively draw together distant leaf edges. The leaves are attached to one another by the action of other workers carrying larvae that emit silk from their labial glands. (The adult ants cannot produce silk.) The workers move the larvae back and forth between the leaf edges, forming silk strands that hold the leaves together. In contrast to the larvae of other ants, which spin a silk cocoon in which to pupate, *Oecophylla* larvae produce silk only when used by the workers in this fashion. These genetically determined behaviors are adaptations that enhance the reproductive success not of the worker ants that perform them, since the workers do not reproduce, but rather of their mother, the queen, whose offspring include both workers and reproductive daughters and sons. In some species, then, individuals have features that benefit other members of the same species. How such features evolve is a topic of special interest.

The Nature of Natural Selection

Design and mechanism

Most adaptations, such as a snake's skull, are *complex*, and most have the appearance of *design*—that is, they are constructed or arranged so as to accomplish some *function*, such as growth, feeding, or pollination, that appears likely to promote survival or reproduction. In inanimate nature, we see nothing comparable—we would not be inclined to think of erosion, for example, as a process designed to shape mountains.

The complexity and evident function of organisms' adaptations cannot conceivably arise from the random action of physical forces. For hundreds of years, it seemed that adaptive design could be explained only by an intelligent designer; in fact, this "argument from design" was considered one of the strongest proofs of the existence of God. For example, the Reverend William Paley wrote in *Natural Theology* (1802) that, just as the intricacy of a watch implies an intelligent, purposeful watchmaker, so every aspect of living nature, such as the human eye, displays "every indication of contrivance, every manifestation of design, which exists in the watch," and must, likewise, have had a Designer.

Supernatural processes cannot be the subject of science, so when Darwin offered a purely natural, materialistic alternative to the argument from design, he not only shook the foundations of theology and philosophy, but brought every aspect of the study of life into the realm of science. His alternative to intelligent design was design by the completely mindless process of natural selection, according to which organisms possessing variations that enhance survival or reproduction replace those less suitably endowed, which therefore survive or reproduce in lesser degree. This process cannot have a goal, any more than erosion has the goal of forming canyons, for *the future cannot cause material events in the present*. Thus the concepts of goals or purposes have no place in biology (or in any other of the natural sciences), except in studies of human behavior. According to Darwin and contemporary evolutionary theory, the weaver ants' behavior has the appearance of design because among many random genetic variations (mutations) affecting the behavior of an ancestral ant species, those displayed by *Oecophylla* enhanced survival and reproduction under its particular ecological circumstances.

Adaptive biological processes *appear* to have goals: weaver ants act as if they have the goal of constructing a nest; an orchid's flower develops toward a suitable shape and stops developing when that shape is attained. We may loosely describe such features by TELEOLOGICAL statements, which express goals (e.g., "She studied *in order* to pass the exam"). But no conscious anticipation of the future resides in the cell divisions that shape a flower or, as far as we can tell, in the behavior of weaver ants. Rather, the apparent goal-directedness is caused by the operation of a program—coded or prearranged information, residing in DNA sequences—that controls a process (Mayr 1988). A program likewise resides in a computer chip, but whereas that program has been shaped by an intelligent designer, the information in DNA has been shaped by a historical process of natural selection. Modern biology views the development, physiology, and behavior of organisms as the results of purely mechanical processes, resulting from interactions between programmed instructions and environmental conditions or triggers.

Definitions of natural selection

It is important to recognize that *"natural selection" is not synonymous with "evolution."* Evolution can occur by processes other than natural selection, especially genetic drift. And natural selection can occur without any evolutionary change, as when natural selection maintains the status quo by eliminating deviants from the optimal phenotype.

Many definitions of natural selection have been proposed (Endler 1986). For our purposes, we will define natural selection as *any consistent difference in fitness among phenotypically different classes of biological entities*. Let us explore this definition in more detail.

The **fitness**—often called the **reproductive success**—of a biological entity is its average per capita rate of increase in numbers. When we speak of natural selection among genotypes or organisms, the components of fitness generally consist of (1) the probability of survival to the various reproductive ages, (2) the average number of offspring (e.g., eggs, seeds) produced via female function, and (3) the average number of offspring produced via male function. "Reproductive success" has the same components, since survival is a prerequisite for reproduction.

Variation in the number of offspring produced as a consequence of competition for mates is often referred to as **sexual selection**, which some authors distinguish from natural selection. We will follow the more common practice of regarding sexual selection as a kind of natural selection.

Because the *probability* of survival and the *average* number of offspring enter into the definition of fitness, and because these concepts apply only to *groups* of events or objects, fitness is defined for a *set* of like entities, such as all the individuals with a particular genotype. That is, natural selection exists if there is an average (i.e., statistically consistent) difference in reproductive success. It is not meaningful to refer to the fitness of a single individual, since its history of reproduction and survival may have been affected by chance to an unknown degree, as we will see shortly.

Differences in survival and reproduction obviously exist among individual organisms, but they also exist below the organismal level, among genes, and above the organismal level, among populations and species. In other words, different kinds of biological entities may vary in fitness, resulting in different **levels of selection**. The most commonly discussed levels of selection are genes, individual organisms that differ in genotype or phenotype, populations within species, and species. Of these, selection among individual organisms (**individual selection**) and among genes (**genic selection**) are by far the most important.

Natural selection can exist only if different classes of entities differ in one or more features, or traits, that affect fitness. Evolutionary biologists differ on whether or not the definition of natural selection should require that these differences be inherited. We will adopt the position taken by those (e.g., Lande and Arnold 1983) who define selection among individual organisms as a consistent difference in fitness among phenotypes. Whether or not this variation in fitness alters the frequencies of genotypes in subsequent generations depends on whether and how the phenotypes are inherited—but that determines the *response to selection*, not the process of selection itself. Although we adopt the phenotypic perspective, we will almost always discuss natural selection among heritable phenotypes because selection seldom has a lasting evolutionary effect unless there is inheritance. Most of our discussion will assume that inheritance of a trait is based on genes. However, many of the principles of evolution by natural selection also apply if inheritance is epigenetic (based on, for example, differences in DNA methylation; see Chapter 9) or is based on cultural transmission, especially from parents to offspring. CULTURE has been defined as "information capable of affecting individuals' behavior that they acquire from other members of their species through teaching, imitation, and other forms of social transmission" (Richerson and Boyd 2005, p. 5).

Notice that according to our definition, natural selection exists whenever there is variation in fitness. Natural selection is not an external force or agent, and certainly not a purposeful one. It is a name for statistical differences in reproductive success among genes, organisms, or populations, and nothing more.

Natural selection and chance

If one neutral allele replaces another in a population by random genetic drift (see Chapter 10), then the bearers of the first allele had a greater rate of increase than the bearers of the other. However, natural selection has not occurred, because the genotypes do not differ *consistently* in fitness: the alternative allele could just as well have been the one to increase. There is no *average* difference between the alleles, no *bias* toward the increase of one relative to the other. Fitness differences, in contrast, are *average* differences, *biases*, differences in the *probability* of reproductive success. This does not mean that every individual of a fitter genotype (or phenotype) survives and reproduces prolifically while every individual of an inferior genotype perishes; some variation in survival and reproduction occurs independent of—that is, at random with respect to—phenotypic differences. But natural selection resides in the difference in rates of increase among biological entities that is *not* due to chance. *Natural selection is the antithesis of chance.*

If fitness and natural selection are defined by consistent, or average, differences, then we cannot tell whether a difference in reproductive success between two *individuals* is due to chance or to a difference in fitness. We cannot say that one identical twin had lower fitness than the other because she was struck by lightning (Sober 1984), or that the genotype of the Russian composer Tchaikovsky, who had no children, was less fit than the genotype of Johann Sebastian Bach, who had many. We can ascribe genetic changes to natural selection rather than random genetic drift only if we observe consistent, nonrandom changes in replicate populations, or measure numerous individuals of each phenotype and find an average difference in reproductive success.

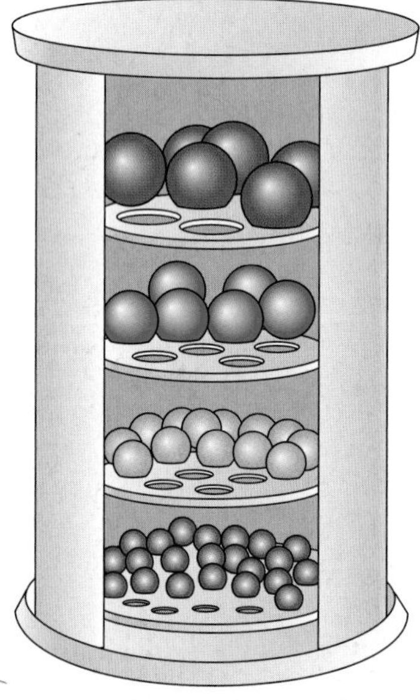

Figure 11.5 A child's toy that selects small balls, which drop through smaller and smaller holes from top to bottom. In this case there is selection *of* red balls, which happen to be the smallest, but selection is *for* small size. (After Sober 1984.)

Selection of and selection for

In the child's "selection toy" pictured in Figure 11.5, balls of several sizes, when placed in the top compartment, fall through holes in partitions, the holes in each partition being smaller than in the one above. If the smallest balls in the toy are all red, and the larger ones are all other colors, the toy will select the small, red balls. Thus we must distinguish *selection of objects* from *selection for properties* (Sober 1984). Balls are selected *for* the property of small size—that is, *because of* their small size. They are not selected for their color, or because of their color; nonetheless, here there is selection *of* red balls. Natural selection may similarly be considered a sieve that selects *for* a certain body size, mating behavior, or other feature. There may be incidental selection *of* other features that are correlated with those features.

The importance of this semantic point is that when we speak of the **function** of a feature, we imply that there has been natural selection *of* organisms with that feature and *of* genes that program it, but selection *for* the feature itself. We suppose that the feature *caused* its bearers to have higher fitness. The feature may have other **effects**, or conse-

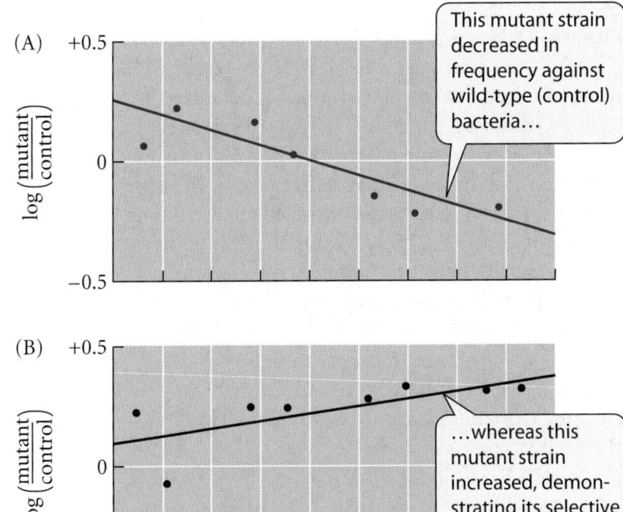

Figure 11.6 Natural selection on mutations in the β-galactosidase gene of *Escherichia coli* in laboratory populations maintained on lactose. In each case, a strain bearing a mutation competed with a control strain bearing the wild-type allele. Populations were initiated with equal numbers of cells of each genotype; i.e., with log (ratio of mutant/control) initially equal to zero. Without selection, no change in the log ratio would be expected. (A) One mutant strain decreased in frequency, showing a selective disadvantage. (B) Another mutant strain increased in frequency, demonstrating its selective (adaptive) advantage. (After Dean et al. 1986.)

quences, that were not its function, and *for* which there was no selection. For instance, there was selection for an opposable thumb and digital dexterity in early hominins, with the incidental effect, millions of years later, that we can play the piano. Similarly, a fish species may be selected for coloration that makes it less conspicuous to predators. The *function* of the coloration, then, is predator avoidance. An *effect* of this evolutionary change might well be a lower likelihood that the population will become extinct, but *avoidance of extinction is not a cause of evolution* of the coloration.

Examples of Natural Selection

We can illustrate the foregoing rather abstract points by several examples of natural selection, some of which show how natural selection can be studied.

Bacterial populations

Bacteria and other microbes are useful for experimental evolutionary studies because of their very rapid population growth. Anthony Dean and colleagues (1986) studied competition between a wild-type strain of *Escherichia coli* and each of several strains that differed from the wild type only by mutations of the gene that codes for β-galactosidase, the enzyme that breaks down lactose. Pairs of genotypes, each consisting of the wild type and a mutant, were cultured together in vessels with lactose as their sole source of energy. The populations were so large that changes in genotype frequencies attributable to genetic drift alone would be almost undetectably slow. Indeed, in certain populations the ratio of mutant to wild type did not change for many generations, indicating that these mutations were selectively neutral. One mutant strain, however, decreased in frequency, and so had lower fitness than the wild type, apparently because of its lower enzyme activity. Another mutant strain, with higher enzyme activity, increased in frequency, displaying a greater rate of increase than the wild type (Figure 11.6).

This experiment conveys the essence of natural selection: it is a completely mindless process without forethought or goal. Adaptation—evolution of a bacterial population with a higher average ability to metabolize lactose—resulted from a difference in the rates of reproduction of different genotypes caused by a phenotypic difference (enzyme activity).

Another experiment with bacteria illustrates *the distinction between "selection of" and "selection for."* In *E. coli*, the wild-type allele *his*$^+$ codes for an enzyme that synthesizes histidine, an essential amino acid, whereas *his*$^-$ alleles are nonfunctional. The *his*$^-$ alleles are selectively neutral if histidine is supplied so that cells with the mutant allele can grow. Atwood and colleagues (1951) observed, to their surprise, that every few hundred generations, the allele frequencies changed rapidly and drastically in experimental cultures that were supplied with histidine (Figure 11.7). The experimenters showed that the *his* alleles were **hitchhiking** with advantageous mutations at other loci—a phenomenon readily observed in bacteria because their rate of recombination is extremely low. Occasionally, a genotype (say, *his*$^-$) would increase rapidly in frequency because of linkage to an advantageous mutation that had occurred at another locus. Subsequently, the alternative allele (*his*$^+$) might increase because of linkage to a new advantageous mutation at another locus

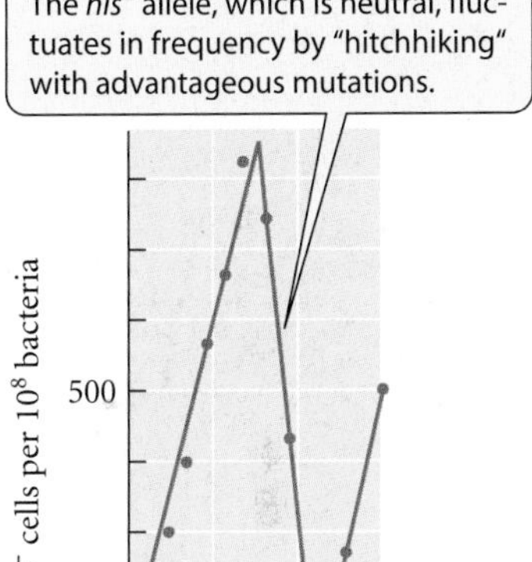

Figure 11.7 Allele frequency fluctuates because of hitchhiking in a laboratory population of *Escherichia coli*. The *y*-axis represents the frequency of the selectively neutral *his*$^-$ allele compared with that of the wild-type *his*$^+$ allele. The frequency of the *his*$^-$ allele increases if a cell bearing it experiences an advantageous mutation at another locus, then decreases if a different, more advantageous mutation occurs in a wild-type cell. (After Nestmann and Hill 1973.)

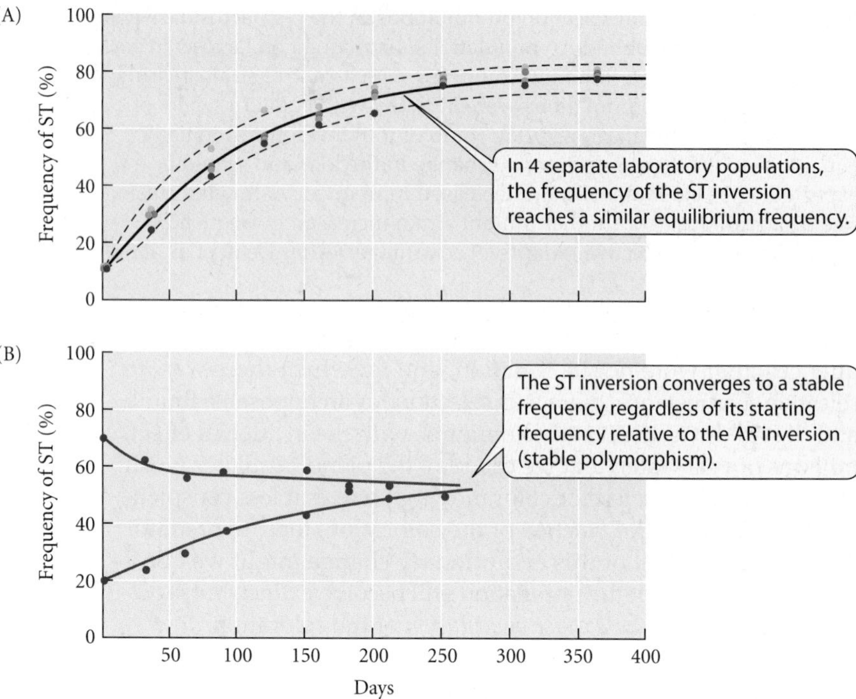
(A)
100
80
60
40
20
0
Frequency of ST (%)
In 4 separate laboratory populations, the frequency of the ST inversion reaches a similar equilibrium frequency.
(B)
100
80
60
40
20
0
Frequency of ST (%)
The ST inversion converges to a stable frequency regardless of its starting frequency relative to the AR inversion (stable polymorphism).
50
100
150
200
250
300
350
400
Days

Figure 11.8 Changes in the frequencies of chromosome inversions in laboratory populations of *Drosophila pseudoobscura*. (A) The frequency of the ST inversion increased in much the same way in four laboratory populations, leveling off at an equilibrium frequency. (B) The frequency of ST arrived at about the same equilibrium level irrespective of starting frequencies of ST and AR. The convergence of the populations toward the same frequency shows that the ST and AR inversions affect fitness, and that natural selection maintains both in a population in a stable, or balanced, polymorphism. (A after Dobzhansky and Pavlovsky 1953; B after Dobzhansky 1948.)

altogether. Thus there was selection *for* new advantageous mutations in these bacterial populations, and selection *of* neutral alleles at the linked *his* locus.

Inversion polymorphism in *Drosophila*

Natural populations of *Drosophila pseudoobscura* are highly polymorphic for inversions—that is, differences in the arrangement of genes along a chromosome resulting from 180 degree reversals in the orientation of chromosome segments (see Chapter 8). Theodosius Dobzhansky suspected that these inversions affected fitness when he observed that the frequencies of several such arrangements in natural populations displayed a regular seasonal cycle that suggested changes in their relative fitnesses as a consequence of environmental changes, perhaps in temperature. He followed these observations with several experiments using POPULATION CAGES: boxes in which populations of several thousand flies are maintained by periodically providing cups of food in which larvae develop. In one such experiment, Dobzhansky and Pavlovsky (1953) used flies with two inversions, called Standard (ST) and Chiricahua (CH), that can be distinguished under the microscope by their banding patterns. They set up four populations, each with 20 percent ST and 80 percent CH chromosome copies (Figure 11.8A). In all the populations, the ST chromosome increased in frequency and leveled off at about 80 per cent. In another experiment (Figure 11.8B), Dobzhansky (1948) initiated one cage with 1119 ST and 485 AR (Arrowhead inversion) chromosome copies (i.e., frequencies of 0.70 and 0.30, respectively). A second cage was initiated with ST and AR frequencies of 0.19 and 0.81, respectively. Within about 15 generations, the frequency of ST had dropped to and leveled off at about 0.54 in the first cage, and it had risen to almost the same frequency (0.50) in the second cage.

The important feature of these experiments is that the chromosome frequencies changed consistently in different populations, approaching a *stable equilibrium*. Replicate populations followed the same trajectory in the first experiment, and different populations approached the same equilibrium in the second experiment, despite different initial frequencies. These results can only be due to natural selection, for random genetic drift would not show such consistency. Moreover, natural selection must be acting in such a way as to *maintain variation* (polymorphism); it does not necessarily cause fixation of a single best genotype. When the genotype frequencies reach equilibrium, natural selection continues to occur, but evolutionary change does not.

(A)

(B)

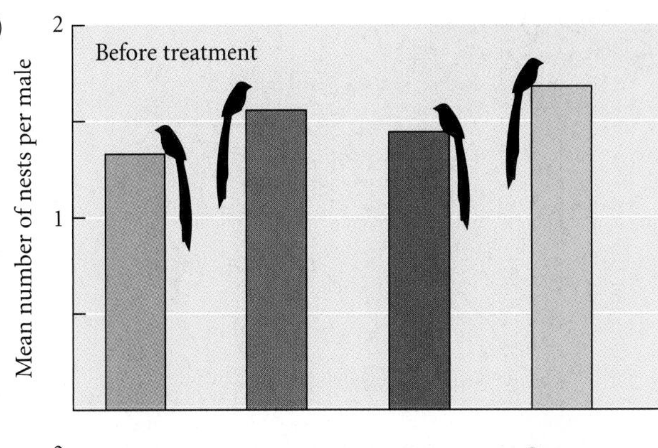

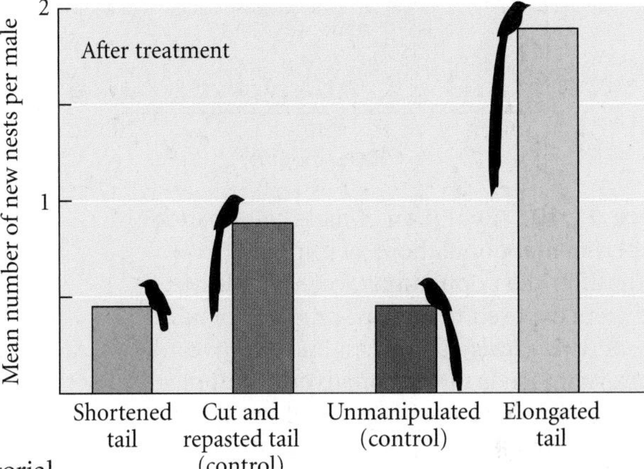

Figure 11.9 (A) A male long-tailed widowbird (*Euplectes progne*) in territorial flight. (B) Effects of experimental alterations of tail length on males' mating success, measured by the number of nests in each male's territory. Nine birds were chosen for each of four treatments: shortening or elongating the tail feathers, or controls of two types: one in which the tail feathers were cut and repasted, and one in which the tail was not manipulated. (A © Jason Gallier/Alamy; B after Andersson 1982.)

Male reproductive success

The courting males of many species of animals have elaborate morphological features and engage in conspicuous displays; roosters provide a familiar example. Some such features appear to have evolved through female choice of males with conspicuous features, which therefore enjoy higher reproductive success than less elaborate males (see Chapter 15). For example, male long-tailed widowbirds (*Euplectes progne*) have extremely long tail feathers. Malte Andersson (1982) shortened the tail feathers of some wild males and attached the clippings to the tail feathers of others, thus elongating them well beyond the natural length. He then observed that males with shortened tails mated with fewer females than did normal males, and that males with elongated tails mated with more females (Figure 11.9).

Male guppies (*Poecilia reticulata*) have a highly variable pattern of colorful spots. In Trinidad, males have smaller, less contrasting spots in streams inhabited by their major predator, the fish *Crenicichla*, than in streams without this predator. John Endler (1980) moved 200 guppies from a *Crenicichla*-inhabited stream to a site that lacked the predator. About two years (15 generations) later, he found that the newly established population had larger spots and a greater diversity of color patterns, so that the population now resembled those that naturally inhabit *Crenicichla*-free streams. Endler also set up populations in large artificial ponds in a greenhouse. After six months of population growth, he introduced *Crenicichla* into four ponds, released a less dangerous predatory fish (*Rivulus*) into four others, and maintained two control populations free of predators. In censuses after four and ten generations, the number and brightness of spots per fish had increased in the ponds without *Crenicichla* and had declined in those with it (Figure 11.10).

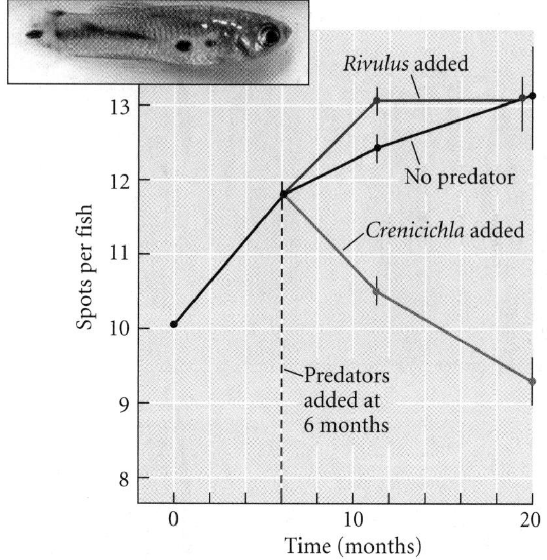

Figure 11.10 Evolution of male color pattern in experimental populations of guppies. Six months after the populations were established, some were exposed to a major predator of adult guppies (*Crenicichla*), some to a less effective predator that feeds mainly on juvenile guppies (*Rivulus*), and some were left free of predators (controls). Numbers of spots were counted after 4 and 10 generations. The vertical bars measure the variation among males. (After Endler 1980; photo courtesy of Anne Houde.)

Males with more and brighter spots have greater mating success, but they are also more susceptible to being seen and captured by *Crenicichla*.

These experiments show that natural selection may consist of differences in reproductive rate, not survival. Differences in mating success, which Darwin called *sexual selection*, result in adaptations for obtaining mates, rather than adaptations for survival. *The guppy experiments also show that a feature may be subjected to conflicting selection pressures* (such as sexual selection and predation), and that the direction of evolution may then depend on which is stronger. Many advantageous characters, in fact, carry corresponding disadvantages, often called COSTS or TRADE-OFFS: the evolution of male coloration in guppies is governed by a trade-off between mating success and avoidance of predation.

Population size in flour beetles

The small beetle *Tribolium castaneum* breeds in stored grains and can be reared in containers of flour. Larvae and adults feed on flour but also eat (cannibalize) eggs and pupae. Michael Wade (1977, 1979) set up 48 experimental populations under each of three treatments. Each population was propagated from 16 adult beetles each generation. The control populations (treatment C) were propagated simply by moving beetles to a new vial of flour: each population in one generation gave rise to one population in the next. In treatment A, Wade deliberately selected for high population size by initiating each generation's 48 populations with sets of 16 beetles taken only from those few populations (out of the 48) in which the greatest number of beetles had developed. In treatment B, low population size was selected in the same way, by propagating beetles only from the smallest populations (Figure 11.11A).

Over the course of nine generations, the average population size declined in all three treatments, most markedly in treatment B and least in

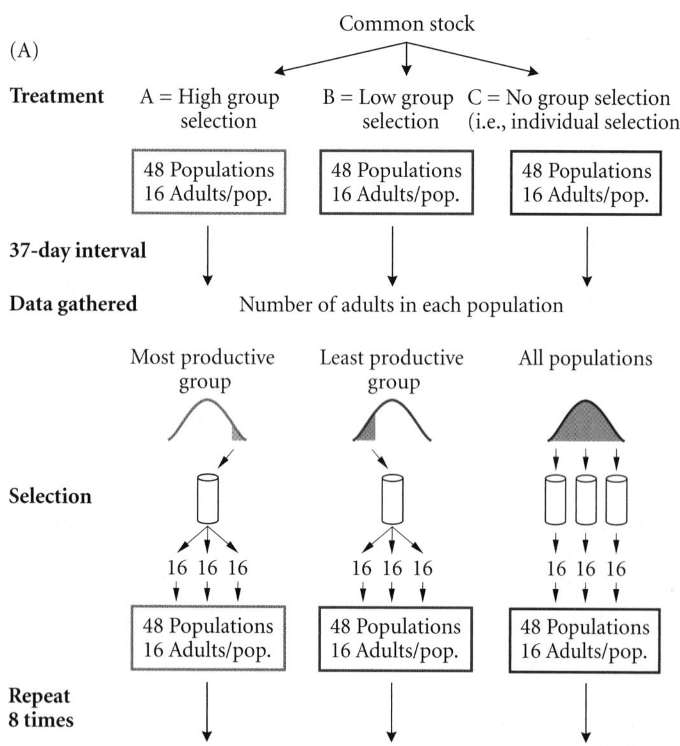

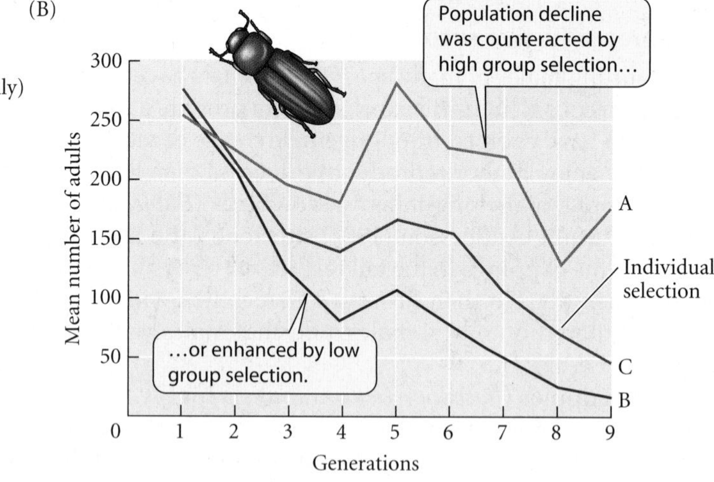

Figure 11.11 Effects of individual selection and group selection on population size in the flour beetle *Tribolium castaneum*. (A) The experimental design. In each generation, group selection was imposed by propagating new populations only from the most productive (treatment A) or least productive (treatment B) populations. In treatment C, all populations were propagated, so there was no group selection: any changes were due to natural selection among individuals within populations. (B) Changes in the mean number of adult beetles in the three treatments. (After Wade 1977.)

treatment A (Figure 11.11B). The net reproductive rate also declined. In treatment C, these declines must have been due to evolution *within* each population, with natural selection acting on the genotypes of individual beetles within a population. This process is *individual selection*, of the same kind we have assumed to operate on, say, the color patterns of guppies. But in treatments A and B, Wade imposed another level of selection by allowing some populations, or groups, but not others, to persist based on a phenotypic characteristic of each *group*—namely, its size. This process, called **group selection**, or **interdemic selection**, operates *in addition to* individual selection among genotypes within populations. We must distinguish selection *within* populations from selection *among* populations.

The decline of population size in the control (C) populations seems like the very antithesis of adaptation. Wade discovered, however, that compared with the foundation stock from which the experimental populations had been derived, adults in the C populations had become more likely to cannibalize pupae, and females were prone to lay fewer eggs when confined with other beetles. For an individual beetle, cannibalism is an advantageous way of obtaining protein, and it may be advantageous for a female not to lay eggs if she perceives the presence of other beetles that may eat them. But although these features are advantageous to the individual, they are disadvantageous to the population, whose growth rate declines.

By selecting groups for low population size (treatment B), Wade *reinforced* these same tendencies. In treatment A, however, selection at the group level for large population size *opposed* the consequences of individual selection within populations. Compared with the C populations, beetles from treatment A had higher fecundity in the presence of other beetles, and they were less likely to cannibalize eggs and pupae. Thus selection among groups had affected the course of evolution.

This experiment shows that the size or growth rate of a population may decline as a result of natural selection even as individual organisms become fitter. *It also illustrates that selection might operate at two levels*: among individuals and among populations. In this case, selection at the group level was imposed by the investigator, so the experiment shows that it is possible, but whether or not it occurs in nature is an open question.

Kin discrimination in cannibalistic salamanders

To continue the cannibal theme: aquatic larvae of the tiger salamander (*Ambystoma tigrinum*) often develop into a distinctive phenotype that eats smaller larvae. Most, although not all, cannibals tend to avoid eating close relatives, such as siblings. One hypothesis for the evolution of such kin discrimination is that an allele that influences its bearer to spare its siblings' lives will increase in frequency, because the siblings are likely to carry other copies of that same allele, which are identical by descent. This is the concept of *kin selection*, to which we will soon return. There are alternative hypotheses, however, of which the most likely is that cannibals are at risk of contracting infectious diseases, especially if they share with their relatives a genetic susceptibility to certain pathogens—in which case it would be advantageous not to eat kin.

David Pfennig and colleagues (1999) tested these and other hypotheses. They reasoned that if the disease hypothesis were correct, cannibals ought to avoid eating diseased larvae, and that non-kin prey would be more likely to transmit disease to a cannibal than would kin prey. But when they offfered cannibals a choice between diseased and healthy prey, or exposed them to diseased kin versus diseased non-kin, neither of the predictions was confirmed. However, the kin selection hypothesis holds (see Chapter 16) that a behavior is advantageous if its genetic benefit/cost ratio is greater than the coefficient of relationship between the actor (cannibal) and the subject of the action (e.g., a sibling whose coefficient of relationship is 0.5). Pfennig et al. compared discriminating and nondiscriminating cannibals, each of which was presented with several related and unrelated prey. They found that discriminators suffered no reduction in growth rate (i.e., no evident cost), whereas sparing their siblings' lives results in a high genetic benefit, i.e., survival of copies of the cannibal's own genes (Figure 11.12). They

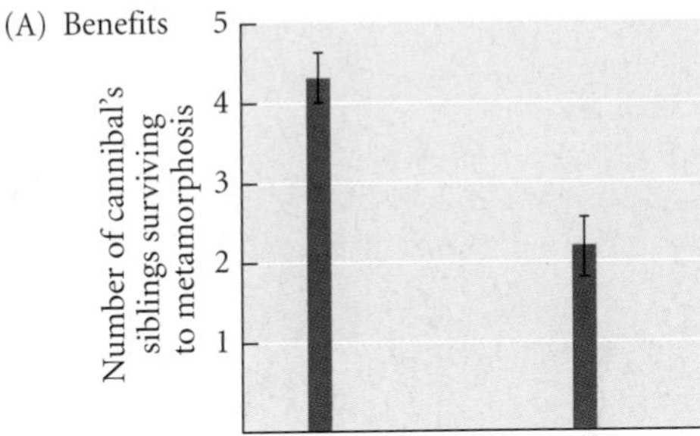

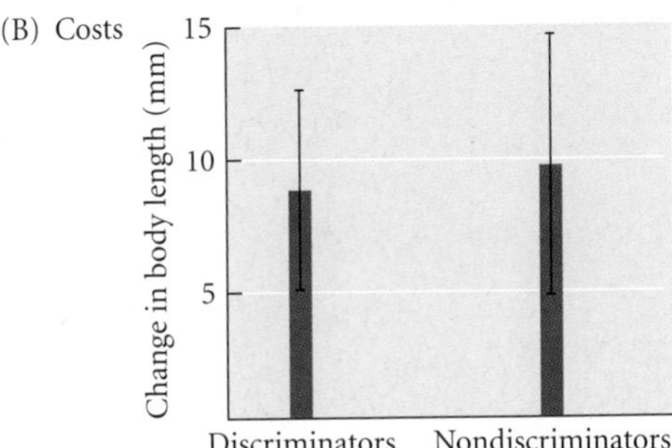

Figure 11.12 Genetic benefit and cost of kin discrimination in cannibalistic tiger salamanders (*Ambystoma tigrinum*). (A) Discriminators benefit by sparing their siblings' lives, since siblings share half of their genes. (B) At least as indicated by the rate of growth in length, discriminators suffer no reduction in fitness, or cost. (After Pfennig et al. 1999.)

concluded that kin selection explains the discriminatory behavior. The concept of natural selection invoked here is that *a gene may change in frequency because of its effect on the survival of copies of itself*, even if these copies are carried by other individuals of the species.

Selfish genetic elements

In many species of animals and plants, there exist **"selfish" genetic elements**, which are transmitted at a higher rate than the rest of an individual's genome and are detrimental (or at least not advantageous) to the organism (Hurst and Werren 2001; Burt and Trivers 2006). Many of these elements exhibit **segregation distortion**, a form of **gametic selection**. For example, the *t* locus of the house mouse (*Mus musculus*) has several alleles that, in a male heterozygous for one of these alleles and for the normal allele *T*, are carried by more than 90 percent of the sperm. In the homozygous condition, certain of the *t* alleles are lethal, and others cause males to be sterile. Despite these disadvantages to the individual, segregation distortion is so great that disadvantageous *t* alleles can reach a high frequency in the population. Another selfish element is a small chromosome called *psr* (which stands for "paternal sex ratio") in the parasitic wasp *Nasonia vitripennis*. It is transmitted mostly through sperm rather than eggs. When an egg is fertilized by a sperm containing this genetic element, it causes the destruction of all the other paternal chromosomes, leaving only the maternal set. In *Nasonia*, as in all Hymenoptera, diploid eggs become females and haploid eggs become males. The *psr* element thus converts female eggs into male eggs, thereby ensuring its own future propagation through sperm, even though this could possibly so skew the sex ratio of a population as to threaten its survival. Some cases of segregation distortion are caused by **meiotic drive**, in which an allele is carried by more than half of a heterozygote's gametes.

Selfish genetic elements forcefully illustrate the nature of natural selection: it is nothing more than differential reproductive success (of genes in this case), which need not result in adaptation or improvement in any sense. These elements also exemplify different levels of selection: in these cases, genic selection acts in opposition to individual selection. Selection among genes may not only be harmful to individual organisms, but might also cause the extinction of populations or species.

Levels of Selection

The last three examples introduced the idea of different levels of selection, corresponding to levels of biological organization. The philosopher of science Samir Okasha (2006) refers to higher-level units as collectives, and the included units as particles. Thus collective/particle pairs might be clade/species, species/population, population/individual organism, organismal genotype/gene. If the particles (e.g., genotypes) in a collective (e.g., population) vary in some character, then collectives will differ in their mean (average) character. Sometimes this is a straightforward relationship: the distribution of body sizes among individuals determines the mean body size in a population, which is therefore an "aggregate" property of its members. But there may also be features of a population (e.g., its abundance or geographic distribution) that cannot be measured on an individual, even if they are the consequences of individual organisms' properties. Such features, sometimes called "emergent" or "relational" characteristics (Damuth and Heisler 1988), may affect the rates at which populations become extinct or give rise to new populations. Note, then, that we might distinguish two measures of fitness of a collective: the mean reproductive success of its constituent members, or the rate at which collectives produce "offspring collectives" (Figure 11.13).

Conceptualizing the level at which selection acts is a philosophically complex issue, on which a great deal has been written. If, for instance, alleles *A* and *a* determine dark versus pale color in an insect, and populations of dark insects have a lower extinction rate than populations of pale insects, does this represent selection at the level of populations, of individual insects, or of genes? Okasha suggests that we can take a "gene's-eye view," in which allele *A* increases as a consequence of both the extinction of populations and the reproductive success of individuals. This "view," though, is not the same as genic selec-

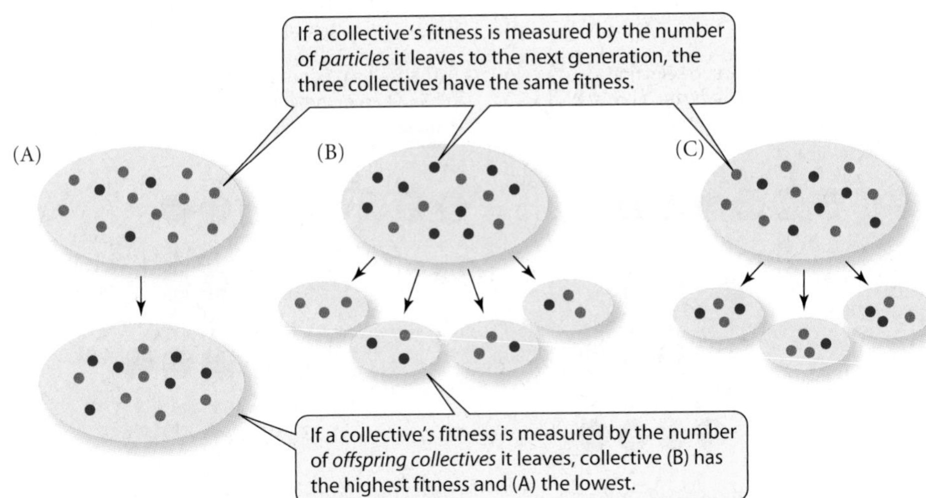

If a collective's fitness is measured by the number of *particles* it leaves to the next generation, the three collectives have the same fitness.
(A)
(B)
(C)
If a collective's fitness is measured by the number of *offspring collectives* it leaves, collective (B) has the highest fitness and (A) the lowest.

Figure 11.13 Concepts of the fitness of "particles" and "collectives." The small circles represent individual particles, such as two different asexual genotypes. Each of the three ovals represents a collective of individuals. In collective (A), the fitness (per capita rate of increase) of blue particles is greater than the fitness of red particles. In collective (B), red particles have higher fitness. In collective (C), the fitness of blue and red particles is equal. Collectives (B) and (C) both produce "offspring collectives" by colonization, but (B) has more offspring collectives than (C). (After Okasha 2006, in part.)

tion, which requires that the cause of gene frequency change operate at the level of the gene, not at the level of individual genotypes (such as greater susceptibility of pale insects to predators) or of the population. In contrast, an "outlaw gene," such as the *t* allele in mice, increases because of the activity of the gene itself, not of the collective (the genotype of the mouse) of which the gene is a part, and so it exhibits genic selection.

Evolutionary biologists have extensively discussed selection at the level of gene, genotype of the individual organism, population, and species (Sober 1984; Okasha 2006). Selection can occur at other levels, such as among cell lineages within a multicellular organism, which is the basis of cancer. Evolutionary biologists have begun to study this topic as well (Michor et al. 2003).

Selection of organisms and groups

By "natural selection," both Darwin and contemporary evolutionary biologists usually mean consistent differences in fitness among genetically different organisms within populations. However, it is common to read statements to the effect that oysters have a high reproductive rate "to ensure the survival of the species," or that antelopes with sharp horns refrain from physical combat because combat would lead to the species' extinction. These statements betray a misunderstanding of natural selection as it is usually conceived. If traits evolve by individual selection—by the replacement of less fit by more fit individuals, generation by generation—then the possibility of future extinction cannot possibly affect the course of evolution. Moreover, an **altruistic trait**—a feature that reduces the fitness of an individual that bears it for the benefit of the population or species—cannot evolve by individual selection. An altruistic genotype amid other genotypes that were not altruistic would necessarily decline in frequency, simply because it would leave fewer offspring per capita than the others. Conversely, if a population were to consist of altruistic genotypes, a selfish mutant—a "cheater"—would increase to fixation, even if a population of such selfish organisms had a higher risk of extinction (Figure 11.14).

Figure 11.14 The mythical self-sacrificial behavior of lemmings, which (according to popular belief) rush en masse into the sea to prevent overpopulation. Cartoonist Gary Larson, in *The Far Side*, illustrates the "cheater" principle, and why such altruistic behavior would not be expected to evolve. (Reprinted with permission of Chronicle Features, San Francisco.)

There is a way, however, in which traits that benefit the population at a cost to the individual might evolve, which is by group selection: differential production or survival of groups that differ in genetic composition. For instance, populations made up of selfish genotypes, such as those with high reproductive rates that exhaust their food supply, might have a higher extinction rate than populations made up of altruistic

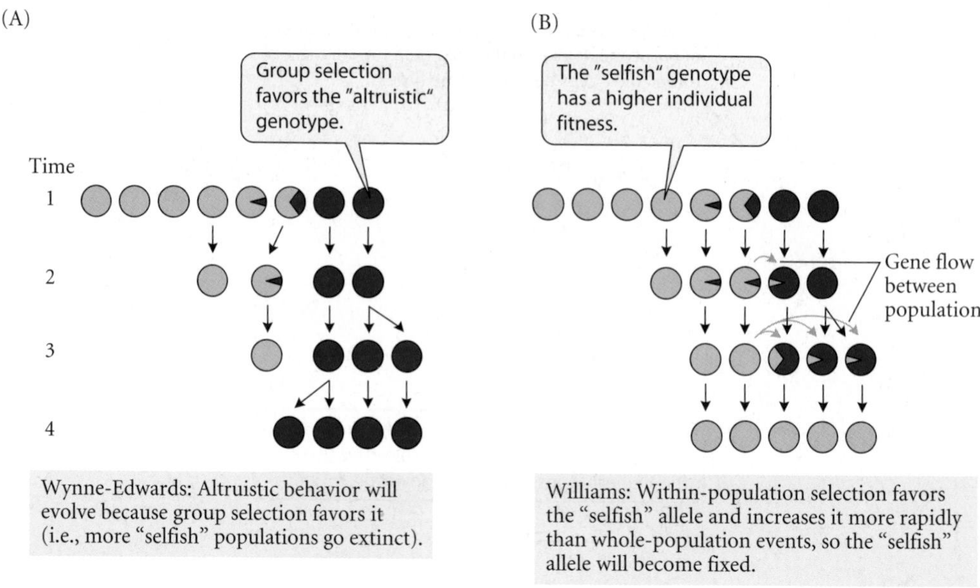
(A)
(B)
Group selection favors the "altruistic" genotype.
The "selfish" genotype has a higher individual fitness.
Time
1
2
3
4
Gene flow between populations
Wynne-Edwards: Altruistic behavior will evolve because group selection favors it (i.e., more "selfish" populations go extinct).
Williams: Within-population selection favors the "selfish" allele and increases it more rapidly than whole-population events, so the "selfish" allele will become fixed.

Figure 11.15 Conflict between group and individual selection. Each circle represents a population of a species, traced through four time periods. Some new populations are founded by colonists from established populations, and some populations become extinct. The proportions of blue and pink areas in each circle represent the proportions of an "altruistic" and a "selfish" genotype in the population, the selfish genotype having a higher reproductive rate (individual fitness). Lateral arrows indicate gene flow between populations. (A) An altruistic trait may evolve by group selection if the rate of extinction of populations of the selfish genotype is very high. (B) Williams's argument: because individual selection operates so much more rapidly than group selection, the selfish genotype increases rapidly within populations and may spread by gene flow into populations of altruists. Thus the selfish genotype becomes fixed, even if it increases the rate of population extinction.

genotypes that have lower reproductive rates. If so, then the species as a whole might evolve altruism through the greater survival of groups of altruistic individuals, even though individual selection within each group would act in the opposite direction (Figure 11.15A).

The hypothesis of group selection was criticized by George Williams (1966) in an influential book, *Adaptation and Natural Selection*. Williams argued that supposed adaptations that benefit the population or species, rather than the individual, do not exist: either the feature in question is not an adaptation at all, or it can be plausibly explained by benefit to the individual or the individual's genes. For example, females of many species lay fewer eggs when population densities are high, but not to ensure a sufficient food supply for the good of the species. At high densities, when food is scarce, a female simply cannot form as many eggs, so her reduced fecundity may be a physiological necessity, not an adaptation. Moreover, an individual female may indeed be more fit if she forms fewer eggs in these circumstances and allocates the energy to surviving until food becomes more abundant, when she may reproduce more prolifically.

Williams based his opposition to group selection on a simple argument. Individual organisms are much more numerous than the populations into which they are aggregated, and they turn over—are born and die—much more rapidly than populations, which are born (formed by colonization) and die (become extinct) at relatively low rates. Selection at either level requires differences—among individuals or among populations—in rates of birth or death. Thus the rate of replacement of less fit by more fit individuals is potentially much greater than the rate of replacement of less fit by more fit populations, so individual selection will generally prevail over group selection (Figure 11.15B). Among evolutionary biologists, the majority view is that *few characteristics have evolved because they benefit the population or species.* We will consider an alternative position, that group selection is important in evolution, in Chapter 16.

If adaptations that benefit the population are so rare, how do we explain worker ants that labor for the colony and do not reproduce, or birds that emit a warning cry when they see a predator approaching the flock? Among the possible explanations is one posited by William Hamilton (1964): such seemingly altruistic behaviors have evolved by **kin selection**, which is best understood from the "viewpoint" of a gene (see Chapter 16). An allele for altruistic behavior can increase in frequency in a population if the beneficiaries of the behavior are usually related to the individual performing it. Since the altruist's relatives are more likely to carry copies of the altruistic allele than are members of the population at random, when the altruist enhances the fitness of its relatives, even at some cost to its

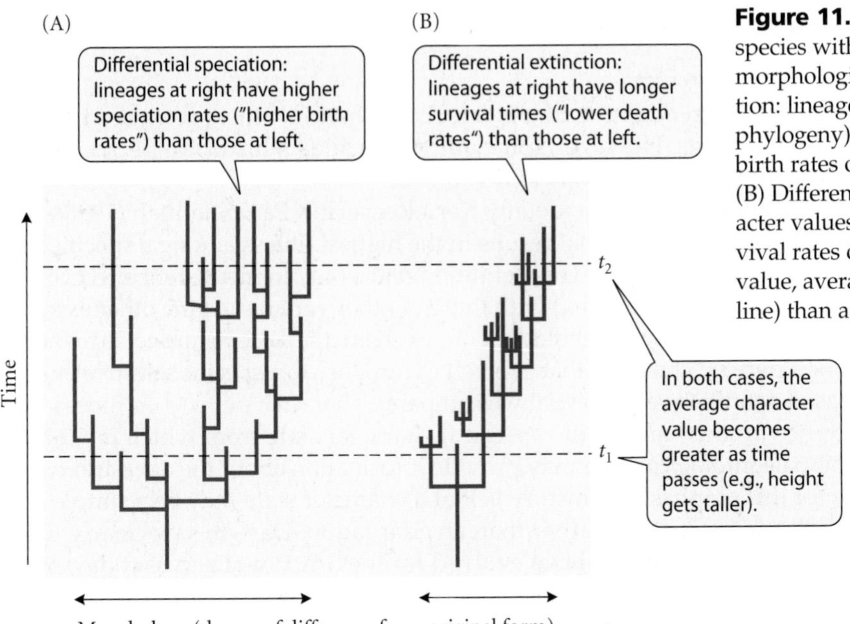

Figure 11.16 Species selection (differential proliferation of species with different character states). The x-axis represents a morphological character, such as body size. (A) Differential speciation: lineages with higher character values (toward the right of the phylogeny) have higher rates of speciation—analogous to higher birth rates of individual organisms—than those with lower values. (B) Differential extinction and survival: lineages with higher character values have longer survival times—analogous to higher survival rates of individual organisms. In both cases, the character value, averaged across species, is greater at time t_2 (upper dashed line) than at time t_1 (lower dashed line). (After Gould 1982.)

own fitness, it can increase the frequency of the allele. We may therefore define kin selection as a form of selection in which alleles differ in fitness by influencing the effect of their bearers on the reproductive success of individuals (kin) who carry the same allele by common descent.

Species selection

Selection among groups of organisms is called **species selection** when the groups involved are species and there is a correlation between some characteristic and the rate of speciation or extinction. Species selection became a major topic of interest when Niles Eldredge and Stephen Jay Gould (1972), in their theory of punctuated equilibria (Chapter 4), proposed that most characteristics are static within species lineages, and evolve mostly or entirely in "daughter" species that arise as small, localized populations (see Figure 4.19C). A consequence of this view, they said, is that long-term trends in a feature, such as body size in horses (see Figure 4.17B), may be attributed to differences in rates of speciation (Figure 11.16A) or extinction (Figure 11.16B), so that the mean character among the species that make up a clade changes over time (see also Stanley 1979). Some authors use "species selection" to refer to both "aggregate" and "emergent" collective features, whereas others restrict it to the few "emergent" characteristics of a species. The size of a species' geographic range, which might be considered an emergent property, is correlated with the species' geological duration in late Cretaceous molluscs (Figure 11.17A). Moreover, related species have a similar range size, which therefore is "heritable" at the species level (Figure 11.17B; Jablonski 1987; Jablonski and Hunt 2006). The combination of species selection and species-level heritability might have resulted in a long-term increase in the average range size of species, but the K-T mass extinction event cut short any possible trend, and range size did not affect the chance of species' survival at that time.

Another likely example of the effects of species selection is the prevalence of sexual species compared with closely related asexual forms. Many groups of plants and animals have given rise to asexually reproducing lineages, but almost all such lineages are very young, as indicated by their very close genetic similarity to sexual forms. This observation implies that asexual forms have a higher rate of extinction than sexual populations, since asexuals that arose long ago have not persisted (Normark et al. 2003; see Figure 15.2).

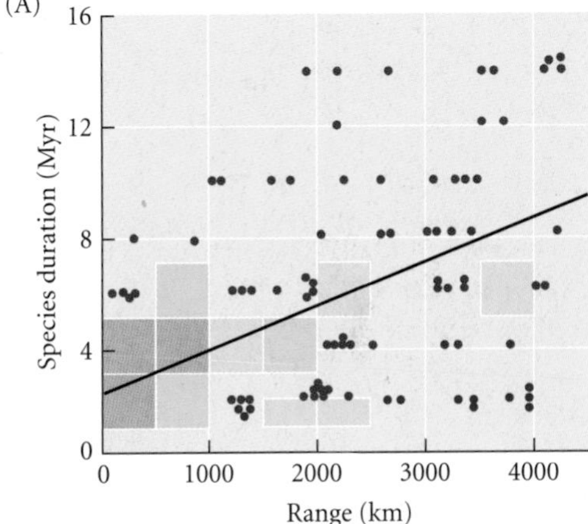

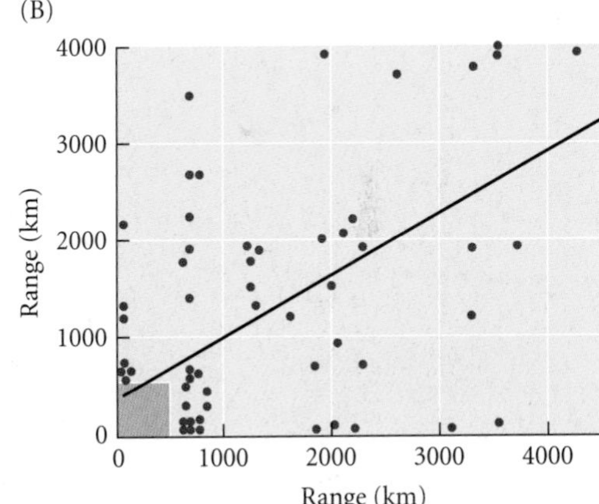

Figure 11.17 An example of species selection. (A) The geological duration of species of late Cretaceous gastropods is correlated with the size of their geographic range (kilometers of coastline). Points represent single species; green, blue, and beige squares represent groups of 6–10, 11–20, and >20 species, respectively. (B) Geographic range size is correlated between pairs of closely related species, so range size is a highly "heritable" species-level trait. Each point represents a pair of species. (After Jablonski 1987.)

The Nature of Adaptations

Definitions of adaptation

All biologists agree that an adaptive trait is one that enhances fitness compared with at least some alternative traits. However, some authors include a historical perspective in their definition of adaptations, and others do not.

An ahistorical definition was provided by Kern Reeve and Paul Sherman (1993): "An adaptation is a phenotypic variant that results in the highest fitness among a specified set of variants in a given environment." This definition refers only to the current effects of the trait on reproductive success, compared with those of other variants. At the other extreme, Paul Harvey and Mark Pagel (1991) hold that "for a character to be regarded as an adaptation, it must be a derived character that evolved in response to a specific selective agent." This history-based definition requires that we compare a character's effects on fitness with those of a specific variant; namely, the ancestral character state from which it evolved. Phylogenetic or paleontological data may provide information about the ancestral state.

One reason for this emphasis on history is that a character state may be a simple consequence of phylogenetic history, rather than an adaptation. Darwin saw clearly that a feature might be beneficial, yet not have evolved for the function it serves today, or for any function at all: "The sutures in the skulls of young mammals have been advanced as a beautiful adaptation for aiding parturition [birth], and no doubt they facilitate, or may be indispensable for this act; but as sutures occur in the skulls of young birds and reptiles, which have only to escape from a broken egg, we may infer that this structure has arisen from the laws of growth, and has been taken advantage of in the parturition of the higher animals" (*The Origin of Species*, chapter 6). Whether or not we should postulate that a trait is an adaptation depends on such insights. For example, we know that pigmentation varies in many species of birds (see Figure 9.1A), so it makes sense to ask whether there is an adaptive reason for color differences among closely related species. But it is not sensible to ask whether it is adaptive for a goose to have four toes rather than five, because the ancestor of birds lost the fifth toe and it has never been regained in any bird since. Five toes are probably not an option for birds because of genetic developmental constraints. Thus, if we ask why a species has one feature rather than another, the answer may be adaptation, or it may be phylogenetic history.

A **preadaptation** is a feature that fortuitously serves a new function. For instance, parrots have strong, sharp beaks, used for feeding on fruits and seeds. When domesticated sheep were introduced into New Zealand, some were attacked by an indigenous parrot, the kea (*Nestor notabilis*), which pierced the skin and fed on the sheep's fat. The kea's beak was fortuitously suitable for a new function and may be viewed as a preadaptation for slicing skin.

Preadaptations that have been co-opted to serve a new function have been termed **exaptations** (Gould and Vrba 1982). For example, the wings of alcids (birds in the auk family) may be considered exaptations for swimming: these birds "fly" underwater as well as in air (Figure 11.18A). An exaptation may be further modified by selection so that the modifications are adaptations for the feature's new function: the wings of penguins have been modified into flippers and cannot support flight in air (Figure 11.18B). Some proteins have been "exapted" to serve new functions, and some play a dual role (Piatigorsky 2007). For instance, the diverse crystallin proteins of animal eye lenses have been co-opted from several phylogenetically widespread proteins, such as stress proteins that stabilize cellular function, lactate dehydrogenase, and other enzymes (Figure 11.19). In some cases, exactly the same protein serves both its ancestral and new roles, such as the τ-crystallin of reptiles and birds which doubles as α-enolase; in other cases, the ancestral gene was duplicated, and the crystallin encoded by one of the duplicates has undergone some amino acid substitutions.

Recognizing adaptations

Not all the traits of organisms are adaptations. There are several other possible explanations of organisms' characteristics. First, a trait may be a necessary consequence of physics or chemistry. Hemoglobin gives blood a red color, but there is no reason to think that red-

Figure 11.18 Exaptation and adaptation. (A) The wing might be called an exaptation for underwater "flight" in members of the auk family, such as this common murre (*Uria aalge*). (B) The modifications of the wing for efficient underwater locomotion in penguins (these are Humboldt penguins, *Spheniscus humboldti*) may be considered adaptations. (A © Chris Gomersall/ Alamy; B © Christian Musat/ ShutterStock.)

ness is an adaptation; it is a by-product of the structure of hemoglobin. Second, the trait may have evolved by random genetic drift rather than by natural selection. Third, the feature may have evolved not because it conferred an adaptive advantage, but because it was correlated with another feature that did. (Genetic hitchhiking, as exemplified in the bacterial experiment by Atwood et al. described on page 285, is one cause of such correlation; pleiotropy—the phenotypic effect of a gene on multiple characters—is another.) Fourth, as we saw in the previous section, a character state may be a consequence of phylogenet-

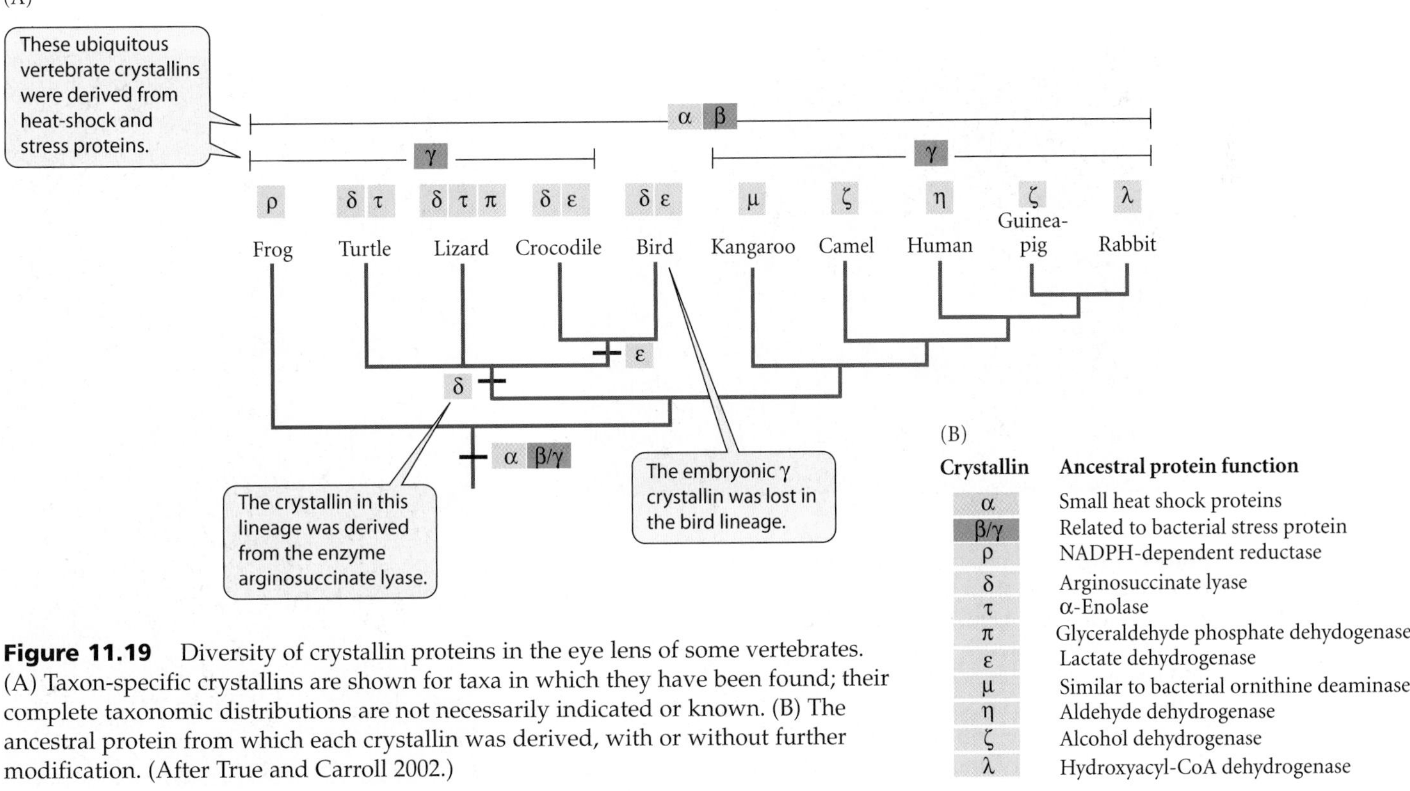

Figure 11.19 Diversity of crystallin proteins in the eye lens of some vertebrates. (A) Taxon-specific crystallins are shown for taxa in which they have been found; their complete taxonomic distributions are not necessarily indicated or known. (B) The ancestral protein from which each crystallin was derived, with or without further modification. (After True and Carroll 2002.)

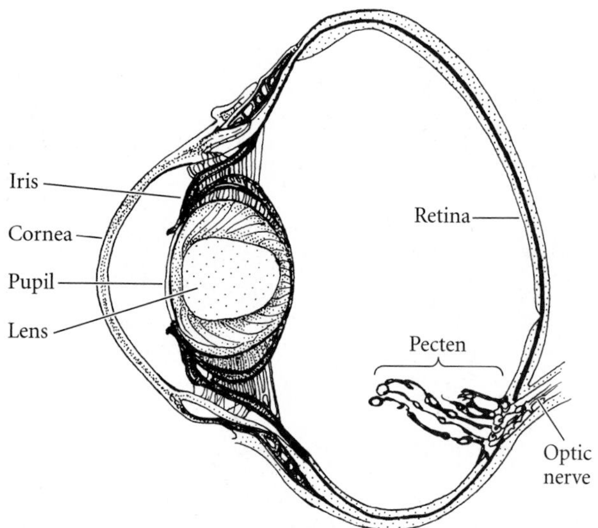

Figure 11.20 Sagittal section of a bird's eye, showing the pecten. Among the 30 or so hypotheses that have been proposed for the function of the pecten, the most likely is that it supplies oxygen to the retina. (After Gill 1995.)

ic history. For instance, it may be an ancestral character state, as Darwin recognized in his analysis of skull sutures.

Because there are so many alternative hypotheses, many authors believe that we should not assume that a feature is an adaptation unless the evidence favors this interpretation (Williams 1966). This is not to deny that a great many of an organism's features, perhaps the majority, are adaptations. Several methods are used to infer that a feature is an adaptation for some particular function. We shall note these methods only briefly and incompletely at this point, exemplifying them more extensively in later chapters. The approaches described here apply to phenotypic characters; in the next chapter, we will describe how selection can be inferred from DNA sequence data.

COMPLEXITY. Even if we cannot immediately guess the function of a feature, *we often suspect it has an adaptive function if it is complex*, for complexity cannot evolve except by natural selection. For example, a peculiar, highly vascularized structure called a pecten projects in front of the retina in the eyes of birds (Figure 11.20). Only recently has evidence been developed to show that the pecten supplies oxygen to the retina, but it has always been assumed to play some important functional role because of its complexity and because it is ubiquitous among bird species.

DESIGN. The function of a character is often inferred from its *correspondence with the design* an engineer might use to accomplish some task, or with the *predictions of a model* about its function. For instance, many plants that grow in hot environments have leaves that are finely divided into leaflets, or which tear along fracture lines (Figure 11.21). These features conform to a model in which the thin, hot "boundary layer" of air at the surface of a leaf is more readily dissipated by wind passing over a small rather than a large surface, so that the leaf's temperature is more effectively reduced. The fields of functional morphology and ecological physiology are concerned with analyses of this kind.

(A)

(B)

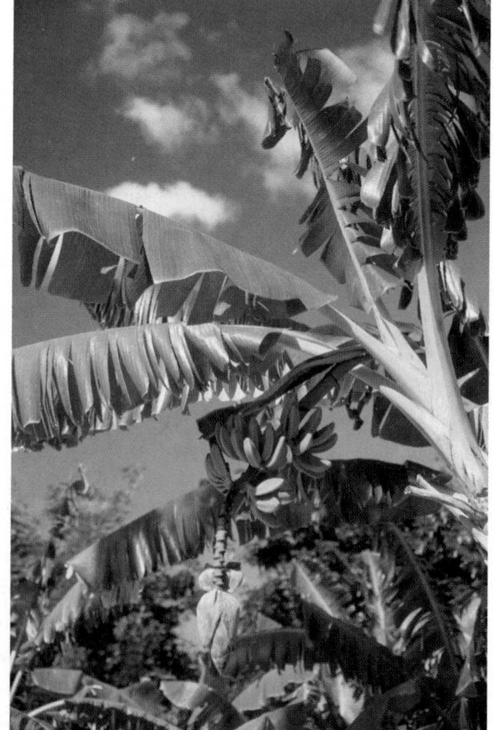

Figure 11.21 Functional morphological analyses have shown that small surfaces shed the hot "boundary layer" of air that forms around them more readily than large surfaces. Many tropical and desert-dwelling plants have large leaves, but they are broken up into leaflets, as in *Acacia dealbata* (A), or split into small sections, as in the banana (B). The form of these leaves is therefore believed to be an adaptation for reducing leaf temperature. (A © Bartomeu Borrell/Photolibrary.com; B © Mireille Vautier/Alamy Images.)

EXPERIMENTS. Experiments may show that a feature enhances survival or reproduction, or enhances performance (e.g., locomotion or defense) in a way that is likely to increase fitness, relative to individuals in which the feature is modified or absent. Andersson's (1982) alteration of the tail length of male widowbirds (see Figure 11.9) illustrates how artificially created variation may be used to demonstrate a feature's adaptive function—in this case, its role in mating success.

THE COMPARATIVE METHOD. A powerful means of inferring the adaptive significance of a feature is the **comparative method**, which consists of *comparing sets of species to pose or test hypotheses on adaptation and other evolutionary phenomena*. This method takes advantage of "natural evolutionary experiments" provided by convergent evolution. If a feature evolves independently in many lineages because of a similar selection pressure, we can often infer the function of that feature by determining the ecological or other selective factor with which it is correlated.

For instance, a long, slender beak has evolved in at least six lineages of birds that feed on nectar (see Figure 3.10). Human digestion of milk provides another example. Most adult humans are sickened by lactose, the principal carbohydrate in milk, and cannot digest it because the digestive enzyme lactase-phloridzin hydrolase (LPH) is regulated at low levels. However, people in several populations around the world have persistently high LPH levels in adulthood, especially in northern Europe and in certain populations in Africa. Milk and milk products have traditionally been an important part of the diet of all these populations. Adult lactose digestion has evolved at least three times in different populations (Holden and Mace 1997; Tishkoff et al. 2007), based on three different mutations, marked by different SNPs, in a DNA sequence upstream of the gene for LPH. DNA sequence evidence suggests that these evolutionary changes occurred after dairying was adopted. That a similar feature arose independently in a similar ecological context—a diet of dairy products—strongly suggests that it is an adaptation to that context.

Biologists often predict such correlations by postulating, perhaps on the basis of a model, the adaptive features we would expect to evolve repeatedly in response to a given selective factor. For example, in species in which a female mates with multiple males, the several males' sperm compete to fertilize eggs. Males that produce more abundant sperm should therefore have a reproductive advantage. In primates, the quantity of sperm produced is correlated with the size of the testes, so large testes should be expected to provide a greater reproductive advantage in polygamous than in monogamous species. Paul Harvey and collaborators compiled data from prior publications on the mating behavior and testes size of various primates and confirmed that, as predicted, the weight of the testes, relative to body weight, is significantly higher among polygamous than monogamous taxa (Figure 11.22).

This example raises several important points. First, although all the data needed to test this hypothesis already existed, the relationship between the two variables was not known until Harvey and collaborators compiled the data, because no one had had any reason to do so until an adaptive hypothesis had been formulated. Hypotheses about adaptation can be fruitful because they suggest investigations that would not otherwise occur to us.

Second, because the consistent relationship between testes size and mating system was not known a priori, the hypothesis generated a prediction. The predictions made by evolutionary theory, as in many other scientific disciplines, are usually predictions of what we will find when we collect data. Prediction in evolutionary theory does *not* usually mean that we predict the future course of evolution of a species. Predictions of what we will find, deduced from hypotheses, constitute the **hypothetico-deductive method**, of which Darwin was one of the first effective exponents (Ghiselin 1969; Ruse 1979).

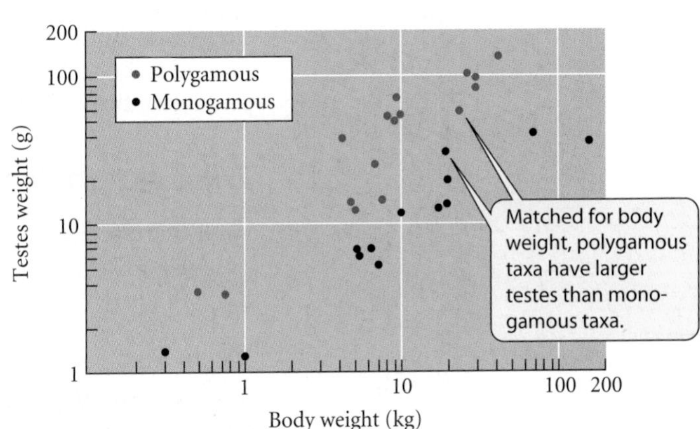

Figure 11.22 The relationship between weight of the testes and body weight among polygamous and monogamous primate taxa. (After Harvey and Pagel 1991.)

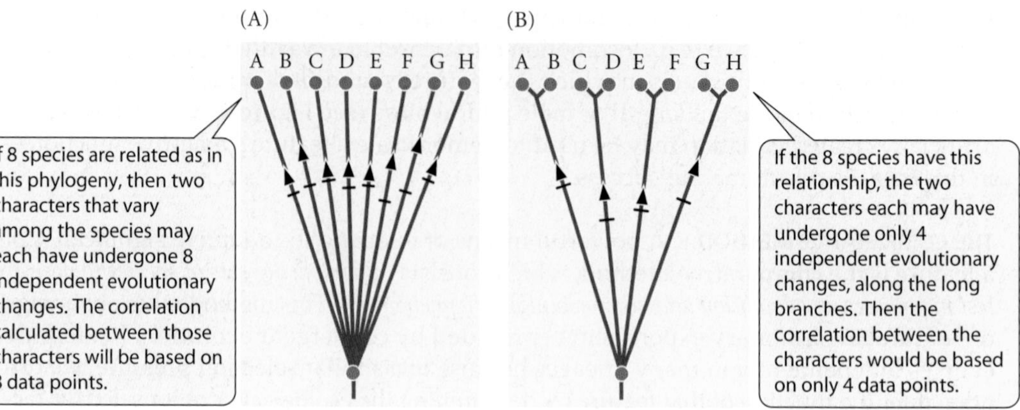

(A)
(B)
A B C D E F G H
A B C D E F G H
If 8 species are related as in this phylogeny, then two characters that vary among the species may each have undergone 8 independent evolutionary changes. The correlation calculated between those characters will be based on 8 data points.
If the 8 species have this relationship, the two characters each may have undergone only 4 independent evolutionary changes, along the long branches. Then the correlation between the characters would be based on only 4 data points.

Figure 11.23 The problem of phylogenetic correlation in employing the comparative method. Suppose we test a hypothesis about adaptation by calculating the correlation between two characters, such as testes size (arrowheads) and mating system (ticks), in eight species (A–H). (A) If the species are related as shown in this phylogenetic tree, the character states in each species have evolved independently, as shown by ticks and arrowheads, and we have a sample of eight. (B) If the species are related as shown in this phylogenetic tree, the states of both characters may be similar in each pair of closely related species because of their common ancestry, rather than independent adaptive evolution. Some authors maintain that the two species in each pair are not independent tests of the hypothesis; we would have four samples in this case. (After Felsenstein 1985.)

Third, the hypothesis was supported by demonstrating that the average testes sizes of polygamous and monogamous taxa show a STATISTICALLY SIGNIFICANT difference. To do this, it is necessary to have a sufficient number of data points—that is, a large enough sample size. For a statistical test to be valid, each data point must be INDEPENDENT of all others. Harvey et al. could have had a larger sample size if they had included, say, 30 species of marmosets and tamarins (Callithricidae) as separate data points, rather than using only one. All the members of this family are monogamous. That suggests that monogamy evolved only once, and has been retained by all callithricids for unknown reasons: perhaps monogamy is advantageous for all the species, or perhaps an internal constraint of some kind prevents the evolution of polygamy even if it would be adaptive. Because our hypothesis is that testes size *evolves* in response to the mating system, we must suspect that the different species of callithricids represent only one evolutionary change, and so provide only one data point (Figure 11.23). Therefore, if we use convergent evolution (i.e., the comparative method) to test hypotheses of adaptation, we should count the *number of independent convergent evolutionary events* by which a character state evolved in the presence of one selective factor versus another (Ridley 1983; Felsenstein 1985; Harvey and Pagel 1991). Consequently, methods that employ PHYLOGENETICALLY INDEPENDENT CONTRASTS (Felsenstein 1985) are usually used in comparative studies of adaptation. Often, this approach confirms a conclusion reached by a simple correlation among all species, and some biologists argue that counting only phylogenetically independent character changes may not be necessary if the characters are genetically variable, because the characters should be unconstrained and able to evolve to a different adaptive optimum quite rapidly (Westoby et al. 1997). However, the more "conservative" method of phylogenetically independent contrasts sometimes calls into question conclusions from the simpler approach. For example, the temperature preferred by 12 species of Australian skinks is correlated with the temperature at which these lizards run fastest, but this correlation proved to be statistically nonsignificant when the data were analyzed by the method of phylogenetically independent contrasts (Garland et al. 1991).

What Not to Expect of Natural Selection and Adaptation

We conclude this discussion of the general properties of natural selection and adaptation by considering a few common misconceptions of, and misguided inferences from, the theory of adaptive evolution.

The necessity of adaptation

It is naïve to think that if a species' environment changes, the species must adapt or else become extinct. Not all environmental changes reduce population size. Nonetheless, an environmental change that does not threaten extinction may set up selection for change in some characteristics. Thus white fur in polar bears may be advantageous, but it is not necessary for survival (Williams 1966). Just as a changed environment need not set in motion selection for new adaptations, new adaptations may evolve in an unchanging environment if new mutations arise that are superior to any pre-existing genetic variations. We have already stressed that the probability of future extinction of a population or species does not in itself constitute selection on individual organisms, and so cannot cause the evolution of adaptations.

Perfection

Darwin noted that "natural selection will not produce absolute perfection, nor do we always meet, as far as we can judge, with this high standard in nature" (*The Origin of Species*, Chapter 6). Selection may fix only those genetic variants with a higher fitness than other genetic variants in that population at that time. It cannot fix the best of all conceivable variants if they do not arise, or have not yet arisen, and the best possible variants often fall short of perfection because of various constraints. For example, with a fixed amount of available energy or nutrients, a plant might evolve higher seed numbers, but only by reducing the size of its seeds or some other part of its structure (see Chapter 14).

Progress

Whether or not evolution is "progressive" is a complicated question (Nitecki 1988; Ruse 1996). The word "progress" has the connotation of a goal, and as we have seen, evolution does not have goals. But even if we strip away this connotation and hold only that progress means "betterment," the possible criteria for "better" depend on the kind of organism. Better learning ability or greater brain complexity has no more evident adaptive advantage for most animals—for example, rattlesnakes—than an effective poison delivery system would have for humans. Measurements of "improvement" or "efficiency" must be relevant to each species' special niche or task. There are, of course, many examples of adaptive trends, each of which might be viewed as progressive within its special context. We will consider this topic in depth in Chapter 22.

Harmony and the balance of nature

As we have seen, selection at the level of genes and individual organisms is inherently "selfish": the gene or genotype with the highest rate of increase increases at the expense of other individuals. The variety of selfish behaviors that organisms inflict on conspecific individuals, ranging from territory defense to parasitism and infanticide, is truly stunning. Indeed, cooperation among organisms requires special explanations. For example, a parent that forages for food for her offspring, at the risk of exposing herself to predators, is cooperative, but for an obvious reason: her own genes, including those coding for this parental behavior, are carried by her offspring, and the genes of individuals that do not forage for their offspring are less likely to survive than the genes of individuals that do. This is an example of kin selection, an important basis for the evolution of cooperation within species (see Chapter 16).

Because the principle of kin selection cannot operate across species, "natural selection cannot possibly produce any modification in a species exclusively for the good of another species" (*The Origin of Species*, Chapter 6). If a species exhibits behavior that benefits another species, either the behavior is profitable to the individuals performing it (as in bees that obtain food from the flowers they pollinate), or they have been duped or manipulated by the species that profits (as are insects that copulate with orchids). Most mutualistic interactions between species consist of reciprocal exploitation (see Chapter 19).

The equilibrium we may observe in ecological communities—the so-called balance of nature—likewise does not reflect any striving for harmony. We observe coexistence of predators and prey not because predators restrain themselves, but because prey species are well enough defended to persist, or because the abundance of predators is limited by some factor other than food supply. Nitrogen and mineral nutrients are rapidly and "efficiently" recycled within tropical wet forests not because ecosystems are selected for or strive for efficiency, but because under competition for sparse nutrients, microorganisms have evolved to decompose litter rapidly, while plants have evolved to capture the nutrients released by decomposition. Selection of individual organisms for their ability to capture nutrients has the *effect*, in aggregate, of a dynamic that we measure as ecosystem "efficiency." There is no scientific foundation for the notion that ecosystems evolve toward harmony and balance (Williams 1992a).

Morality and ethics

Natural selection is just a name for differences among organisms or genes in reproductive success. Therefore, it cannot be described as moral or immoral, just or unjust, kind or cruel, any more than wind, erosion, or entropy can. Hence it cannot be used as a justification or model for human morality or ethics. Nevertheless, evolutionary theory has often been misused in just this way. Darwin expressed distress over an article "showing that I have proved 'might is right,' and therefore that Napoleon is right, and every cheating tradesman is also right." In the late nineteenth and early twentieth century, Social Darwinism, promulgated by the philosopher Herbert Spencer, considered natural selection to be a beneficent law of nature that would produce social progress as a result of untrammeled struggle among individuals, races, and nations. Evolutionary theory has likewise been used to justify eugenics and racism, most perniciously by the Nazis. But neither evolutionary theory nor any other field of science can speak of or find evidence of morality or immorality. These precepts do not exist in nonhuman nature, and science describes only what *is*, not what *ought to be*. The **naturalistic fallacy**, the supposition that what is "natural" is necessarily "good," has no legitimate philosophical foundation.

Summary

1. A feature is an adaptation for a particular function if it has evolved by natural selection for that function by enhancing the relative rate of increase—the fitness—of biological entities with that feature.
2. Natural selection is a consistent difference in fitness among phenotypically different biological entities, and is the antithesis of chance. Natural selection may occur at different levels, such as genes, individual organisms, populations, and species.
3. Selection at the level of genes or organisms is likely to be the most important because the numbers and turnover rates of these entities are greater than those of populations or species. Therefore, most features are unlikely to have evolved by group selection, the one form of selection that could in theory promote the evolution of features that benefit the species even though they are disadvantageous to the individual organism.
4. Not all features are adaptations. Methods for identifying and elucidating adaptations include studies of function and design, experimental studies of the correspondence between fitness and variation within species, and correlations between the traits of species and environmental or other features (the comparative method). Phylogenetic information may be necessary for proper use of the comparative method.
5. Natural selection does not necessarily produce anything that we can justly call evolutionary progress. It need not promote harmony or balance in nature, and, utterly lacking any moral content, it provides no foundation for morality or ethics in human behavior.

Terms and Concepts

adaptation
altruistic trait
comparative method
exaptation
fitness
function (vs. effect)
genic selection
group selection (= interdemic selection)
hitchhiking
hypothetico-deductive method
individual selection
kin selection
levels of selection
meiotic drive (= segregation distortion)
naturalistic fallacy
preadaptation
reproductive success
selfish genetic elements
sexual selection
species selection

Suggestions for Further Reading

Adaptation and Natural Selection by G. C. Williams (Princeton University Press, Princeton, NJ, 1966) is a classic: a clear, insightful, and influential essay on the nature of individual and group selection. See also the same author's *Natural Selection: Domains, Levels, and Challenges* (Oxford University Press, New York, 1992).

Two books by R. Dawkins, *The Selfish Gene* (Oxford University Press, Oxford, 1989) and *The Blind Watchmaker* (W. W. Norton, New York, 1986), explore the nature of natural selection in depth, as well as treating many other topics in a vivid style for general audiences. Levels of selection and related issues are treated in *The Nature of Selection: Evolutionary Theory in Philosophical Focus* by E. Sober (MIT Press, Cambridge, MA, 1984), *Evolution and the Levels of Selection* by S. Okasha (Oxford University Press, Oxford, 2006), and *Levels of Selection in Evolution*, edited by L. Keller (Princeton University Press, Princeton, NJ, 1999). *The Comparative Method in Evolutionary Biology* by P. H. Harvey and M. D. Pagel (Oxford University Press, Oxford, 1991) treats the use and phylogenetic foundations of the comparative method.

Problems and Discussion Topics

1. Discuss criteria or measurements by which you might conclude that a population is better adapted after a certain evolutionary change than before.
2. Consider the first copy of an allele for insecticide resistance that arises by mutation in a population of insects exposed to an insecticide. Is this mutation an adaptation? If, after some generations, we find that most of the population is resistant, is the resistance an adaptation? If we discover genetic variation for insecticide resistance in a population that has had no experience of insecticides, is the variation an adaptation? If an insect population is polymorphic for two alleles, each of which confers resistance against one of two pesticides that are alternately applied, is the variation an adaptation? Or is each of the two resistance traits an adaptation?
3. Adaptations are features that have evolved because they enhance the fitness of their carriers. It has sometimes been claimed that fitness is a tautological, and hence meaningless, concept. According to this argument, adaptation arises from the "survival of the fittest," and the fittest are recognized as those that survive; consequently there is no independent measure of fitness or adaptiveness. Evaluate this claim. (See Sober 1984.)
4. It is often proposed that a feature that is advantageous to individual organisms is the reason for the great number of species in certain clades. For example, wings have been postulated to be a cause of the great diversity of winged insects compared with the few species of primitively wingless insects. How could an individually advantageous feature cause greater species diversity? How can one test a hypothesis that a certain feature has caused the great diversity of certain groups of organisms?
5. Provide an adaptive and a nonadaptive hypothesis for the evolutionary loss of useless organs, such as eyes in many cave-dwelling animals. How might these hypotheses be tested?
6. Could natural selection, at any level of organization, ever cause the extinction of a population or species?
7. If natural selection has no foresight, how can it explain features that seem to prepare organisms for future events? For example, deciduous trees at high latitudes drop their leaves before winter arrives, male birds establish territories before females arrive in the spring, and animals such as squirrels and jays store food as winter approaches.
8. List the possible criteria by which evolution by natural selection might be supposed to result in "progress," and search the biological literature for evidence bearing on one or more of these criteria.

CHAPTER 12

The Genetical Theory of Natural Selection

Abundant reproduction. Of the millions of reproductive spores released by this puffball mushroom (*Lycoperdon perlatum*), only a few will survive to reproduce. Such reproductive excess provides opportunity for evolution by natural selection. (Photo © Manfred Delpho/Photolibrary.com.)

Natural selection is the most important concept in the theory of evolutionary processes. It is surely the explanation for most of the characteristics of organisms that we find most interesting, ranging from the origin of DNA to the complexities of the human brain. Natural selection—differential reproductive success—is fundamentally a very simple concept. But its explanatory power is much greater if we take into account the many ways in which it can act and the ways in which its outcome is affected by genetic factors such as recombination and the relationship between phenotype and genotype. When we take into account these complexities, we can begin to address a great variety of questions: Why are some characteristics, but not others, variable within species? How great a difference can we expect to see among different populations of a species, such as our own? Do populations of a species always evolve the same adaptation to a particular environmental challenge? How do cooperative and selfish behaviors evolve? Why do some species reproduce sexually and others asexually? How can we explain the extraordinary display feathers of the peacock, the immense fecundity of mushrooms and oysters, the brevity of a mayfly's life, the pregnancy of the male sea horse, the abundance of transposable elements in our own genome?

Darwin fully realized that a truly complete theory of evolutionary change would require understanding the mechanism of inheritance. That understanding began to develop only in 1900, when Mendel's publication was discovered. Modern evolutionary theory started to develop as the growing understanding of Mendelian genetics was synthesized with Darwin's theory of selection. The "genetical theory of natural selection" (as the pioneering population geneticist R. A. Fisher titled his seminal 1930 book) is the keystone of contemporary evolutionary theory, on which our understanding of adaptive evolution depends.

As we delve into the genetical theory of natural selection, we should keep in mind the following important points about natural selection:

- *Natural selection is not the same as evolution.* Evolution is a two-step process: the origin of genetic variation by mutation or recombination, followed by changes in the frequencies of alleles and genotypes, caused chiefly by genetic drift or natural selection. Neither natural selection nor genetic drift accounts for the origin of variation.
- *Natural selection is different from evolution by natural selection.* In some instances, selection occurs—that is, in each generation, genotypes differ in survival or fecundity—yet the proportions of genotypes and alleles stay the same from one generation to another.
- Although natural selection may be said to exist whenever different phenotypes vary in average reproductive success, *natural selection can have no evolutionary effect unless phenotypes differ in genotype.* For instance, selection among genetically identical members of a clone, even though they differ in phenotype, can have no evolutionary consequences. Therefore, it is useful to describe the reproductive success, or fitness, of genotypes, even though genotypes differ in fitness only because of differences in phenotype.
- A feature cannot evolve by natural selection unless that feature affects reproduction or survival. The long-haired tail of a horse, used as a fly-switch, could not have evolved merely because it increases horses' comfort; it must have resulted in increased reproductive success, perhaps by lowering mortality caused by fly-borne diseases.

Unlike genetic drift, inbreeding, and gene flow, which act at the same rate on all loci in a genome, the allele frequency changes caused by natural selection in a sexually reproducing species proceed largely independently at different loci. Moreover, different characteristics of a species evolve at different rates (mosaic evolution), as we would expect if natural selection brings about changes in certain features while holding others constant (see Chapter 3). Thus we are justified in beginning our analysis of natural selection with a single variable locus that alters a phenotypic character.

Fitness

Unless otherwise specified, the following discussion of natural selection concerns selection at the level of individual organisms in populations. The consequences of natural selection depend on (1) the relationship between phenotype and fitness, and (2) the relationship between phenotype and genotype. These relationships, then, yield (3) a relationship between fitness and genotype, which determines (4) whether or not evolutionary change occurs.

Modes of selection

Let us denote the fitness of a genotype as w. Supposing there are two alleles (A_1, A_2) at a locus (Figure 12.1A), the fitnesses of the three genotypes A_1A_1, A_1A_2, and A_2A_2 are then w_{11}, w_{12}, and w_{22}. Selection is said to be **directional** if $w_{11} \geq w_{12} > w_{22}$. If the heterozygote has greatest fitness ($w_{11} < w_{12} > w_{22}$), it is said to be **overdominant**, and if it has lowest fitness ($w_{11} > w_{12} < w_{22}$), it is **underdominant**. In the case of directional selection, the disadvantageous allele (A_2 in the example given) is lowered in frequency and perhaps entirely eliminated, and selection on this allele is sometimes said to be **purifying**. Some literature uses POSITIVE and NEGATIVE selection to refer to whether a new mutation is selected for or against.

For a quantitative (continuously varying) trait, such as size, selection is directional if one extreme phenotype is fittest, **stabilizing** (NORMALIZING) if an intermediate phenotype is fittest, or **diversifying** (**disruptive**) if two or more phenotypes are fitter than the intermediates between them (Figure 12.1B). Which *genotype* has the highest fitness under a given selection regime depends on the relationship between phenotype and genotype.

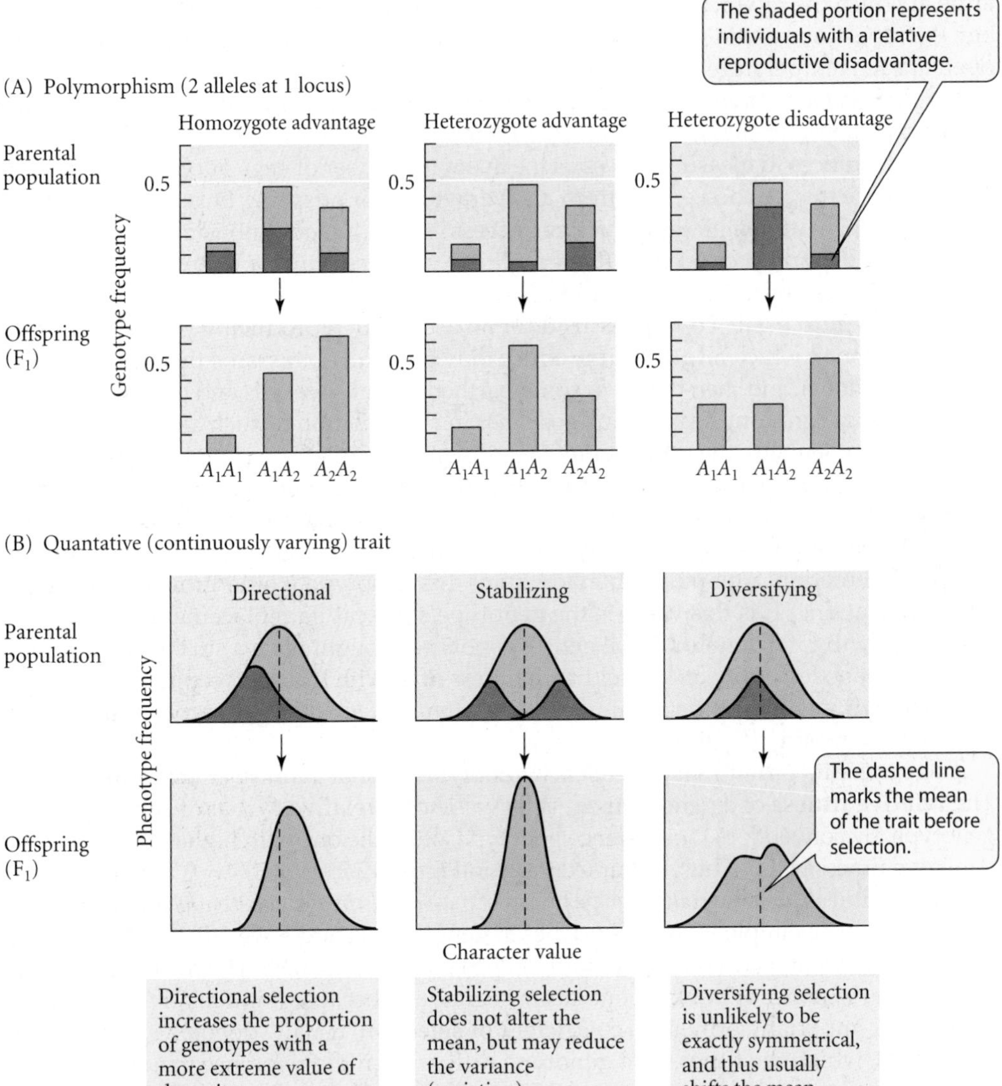

Figure 12.1 Modes of selection on (A) a polymorphism consisting of two alleles at one locus, and (B) a heritable quantitative (continuously varying) character. The upper graphs in both (A) and (B) show the distribution in the parental generation, before selection occurs. The shaded portions represent individuals with a relative disadvantage (lower fitness). The dashed line in (B) represents the mean character value in the parental generation. The lower graphs in both (A) and (B) show the frequency distribution in the F_1 generation, after selection has occurred. (After Endler 1986.)

For example, under directional selection for large size, genotype A_1A_1 would be most fit if it were largest, but A_1A_2 would be favored if it were larger than either homozygote. As we will soon see, this difference would have important evolutionary results: the population would become fixed for the largest phenotype if the homozygote were largest, but not if the heterozygote were largest.

The relationship between phenotype and fitness can depend on the environment, since different environmental conditions can favor different phenotypes. It also depends on how the mean and variation in a character are distributed relative to the fitness/phenotype relationship. Thus, if the mean body size is below the optimum, it will be directionally selected until it corresponds to the optimum (at least approximately); after that, it is subject to stabilizing selection.

Defining fitness

Because we are concerned with only those effects of selection that depend on inheritance, we will use models in which an average fitness value is assigned to each genotype. A genotype is likely to have different phenotypic expressions as a result of environmental influences on development, so the fitness of a genotype is the mean of the fitnesses of its several phenotypes, weighted by their frequencies. For example, a particular genotype of *Drosophila pseudoobscura* has a variable number of bristles, depending on the temperature at which the fly develops (see Figure 9.5). Thus, if fitness depended on bristle number, the fitness of a given genotype would depend on the proportions of flies that developed at each temperature.

The fitness of a genotype is the average lifetime contribution of individuals of that genotype to the population after one or more generations, measured at the same stage in the life history. Often it suffices to measure fitness as the average number of eggs or offspring one generation hence that are descended from an average egg or offspring born. A general term for this average number is **reproductive success**, which includes not simply the average number of offspring produced by the reproductive process, but the number that survive, since survival is prerequisite to reproduction.

Fitness is most easily conceptualized for an asexually reproducing (PARTHENOGENETIC) population in which all adults reproduce only once, all at the same time (nonoverlapping generations), and then die, as in some parthenogenetic weevils and other insects that live for a single growing season. Suppose that in a population of such an organism, consisting only of females, the proportion of eggs of genotype A that survive to reproductive age is 0.05, and that each reproductive adult lays an average of 60 eggs (her FECUNDITY). Then the fitness of A is (proportion surviving) × (average fecundity) = $0.05 \times 60 = 3$. This is the number of offspring an average newborn individual of genotype A contributes to the next generation. A population made up of this genotype would grow by a factor of 3 per generation. Thus this value is the genotype's per capita replacement rate, or population growth rate, denoted R. Likewise, genotype B might have a survival rate of 0.10 and an average fecundity of 40, yielding a fitness of 4. With these per capita growth rates, both genotypes will increase in number, but the *proportion* of the genotype with higher R will increase rapidly (Figure 12.2).

The per capita growth rate, R_i, of each genotype i is that genotype's **absolute fitness**. The **relative fitness** of a genotype, w_i, is its value of R relative to that of some reference genotype. By convention, the reference genotype, often the one with highest R, is assigned a relative fitness of 1.0. Thus, in our example in Figure 12.2, $w_A = 3/4 = 0.75$ and $w_B = 1.0$. The **mean fitness**, $\bar{w}$, is then *the average fitness of individuals in a population relative to the fittest genotype*. In our example, if the frequencies of genotypes A and B were 0.2 and 0.8, respectively, the mean fitness would be $\bar{w} = (0.2)(0.75) + (0.8)(1.0) = 0.95$. The mean fitness does not indicate whether or not the population is growing, because it is a relative measure.

Another important term is the **coefficient of selection**, usually denoted s, which is the amount by which the fitness of a genotype differs from the reference genotype. In our example, $w_A = 0.75$, so $s = 0.25$. The coefficient of selection measures the **selective advantage** of the fitter genotype, or the intensity of selection against the less fit genotype.

It is easy to show mathematically that *the rate of genetic change under selection depends on the relative*, not the absolute, *fitnesses of genotypes*. The rate at which a population would become dominated by genotype B in our hypothetical example would be the same whether genotypes A and B had R values of 0.6 versus 0.8, or 15 versus 20, or 300 versus 400.

Components of fitness

Survival and female fecundity are only two of the possible **components of fitness**. Fitness may be more complex if a species reproduces sexually and if it reproduces repeatedly during the individual's lifetime. Often, differences among males in reproductive success contribute to differences in fitness. When generations overlap, as in humans and many other species that reproduce repeated-

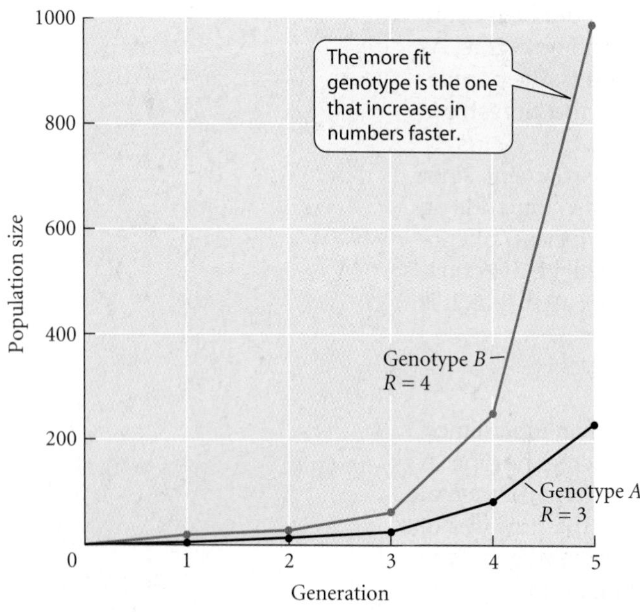

Figure 12.2 The growth of two genotypes with different per capita growth rates in an asexually reproducing population with nonoverlapping generations.

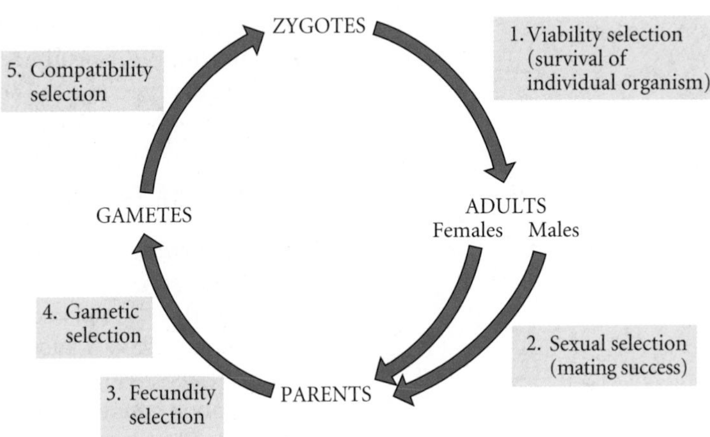

Figure 12.3 Components of natural selection that may affect the fitness of a sexually reproducing organism over the life cycle. Beginning with newly formed zygotes, (1) genotypes may differ in survival to adulthood; (2) they may differ in the numbers of mates they obtain, especially males; (3) those that become parents may differ in fecundity (number of gametes produced, especially eggs); (4) selection may occur among the haploid genotypes of gametes, as in differential gamete viability or meiotic drive; and (5) union of some combinations of gametic genotypes may be more compatible than others. (After Christiansen 1984.)

ly, the absolute fitness of a genotype may be measured in large part by its per capita rate of population increase per unit time, *r* (see Chapter 14). This rate of increase depends on the proportion of individuals surviving to each age class and on the fecundity of each age class. Moreover, *r* is strongly affected by the age at which females have offspring, not just by their number; that is, genotypes may differ in the length of a generation. If females of genotypes *A* and *B* have the same number of offspring when they are 6 months and 12 months old, respectively, the rate of increase (the fitness) of *A* is about twice that of *B*, because *A* will have twice as many generations of descendants per unit time.

In sexually reproducing species, genotypes do not simply make copies of themselves; instead, they transmit haploid gametes. Therefore genotype frequencies depend on the allele frequencies among uniting gametes. These allele frequencies are affected by several components of selection at the "zygotic" (organismal) stage, and sometimes by selection at the gametic stage as well (Figure 12.3; Christiansen 1984). Table 12.1 summarizes the components of selection in a sexual species.

Evolution by natural selection depends on the way in which changes in allele frequencies are determined by the components of fitness of each zygotic and each gametic genotype. These components of fitness are combined (usually by multiplying them) into the overall fitness of each genotype. For instance, the fitness of the genotypes in the simple

Table 12.1 *Components of selection in sexually reproducing organisms*

I. Zygotic selection

A. Viability. The probability of survival of the genotype through each of the ages at which reproduction can occur. After the age of last reproduction, the length or probability of survival does not usually affect the genotype's contribution to subsequent generations, and so does not usually affect fitness.

B. Mating success. The number of mates obtained by an individual. Mating success is a component of fitness if the number of mates affects the individual's number of progeny, as is often the case for males, but less often for females, all of whose eggs may be fertilized by a single male. Variation in mating success is the basis of sexual selection.

C. Fecundity. The average number of viable offspring per female. In species with repeated reproduction, the contribution of each offspring to fitness depends on the age at which it is produced (see Chapter 14). The fertility of a mating may depend only on the maternal genotype (e.g., number of eggs or ovules), or it may depend on the genotypes of both mates (e.g., if they display some reproductive incompatibility).

II. Gametic selection

D. Segregation advantage (meiotic drive or segregation distortion). An allele has an advantage if it segregates into more than half the gametes of a heterozygote.

E. Gamete viability. Dependence of a gamete's viability on the allele it carries.

F. Fertilization success. An allele may affect the gamete's ability to fertilize an ovum (e.g., if there is variation in the rate at which a pollen tube grows down a style).

example above was found by multiplying the survival and fecundity of each genotype. In that example, one genotype had superior fecundity and the other had superior survival: a genotype may be superior to another in certain components of fitness and inferior in others, but its *overall fitness determines the outcome of natural selection*.

Models of Selection

In the following discussion, we make the simplifying assumptions that the population is very large, so genetic drift may be ignored; that mating occurs at random; that mutation and gene flow do not occur; and that selection at other loci does not affect the locus we are considering. We will later consider the consequences of changing these unrealistic assumptions. We also assume, for the sake of simplicity, that selection acts through differential survival among genotypes in a species with discrete generations. The principles are much the same for other components of selection and for species with overlapping generations.

If a locus has two alleles (A_1, A_2) with frequencies p and q, respectively, the change in frequency of A_1 from one generation to the next is expressed by Δp, which is positive if the allele is increasing in frequency, negative if it is decreasing, and zero if it is at equilibrium (Δq would refer to a change in frequency of A_2). In any model of selection, the change in allele frequencies depends on the relative fitnesses of the different genotypes and on the allele frequencies themselves. Box 12A provides a mathematical framework for several models of selection.

Directional selection

THEORY. The replacement of relatively disadvantageous alleles by more advantageous alleles is the fundamental basis of adaptive evolution. This replacement occurs when the homozygote for an advantageous allele has a fitness equal to or greater than that of the heterozygote or of any other genotype in the population (Box 12A, Cases 1,2).

An advantageous allele may initially be fairly common if under previous environmental circumstances it was selectively neutral or was maintained by one of several forms of BALANCING SELECTION (see p. 315). However, an advantageous allele is likely to be initially very rare if it is a newly arisen mutation or if it was disadvantageous before an environmental change made it advantageous.

An advantageous allele that increases from a very low frequency is often said to INVADE a population. *Unless an allele can increase in frequency when it is very rare, it is unlikely to become fixed in the population*. According to this principle, some conceivable adaptations are unlikely to evolve because they could not increase if they were initially very rare. For instance, venomous coral snakes (*Micrurus*) are brilliantly patterned in red, yellow, and black (Figure 12.4). This pattern is presumed to be APOSEMATIC (warning) coloration, which is beneficial because predators associate the colors with danger and avoid attacking such snakes. How this coloration initially evolved has long been a puzzle, however, since the first few mutant snakes with brilliant colors would presumably have been easily seen and killed by naive predators. Given that all coral snakes are aposematically colored, it is understandable that predators might evolve an aversion to them (and, indeed, some predatory birds seem to have an innate aversion to coral snakes)—but how the snake's adaptation "got off the ground" is uncertain. (One possibility is that predators generalized from other brilliantly colored unpalatable or dangerous organisms, such as wasps, and avoided aposematic snakes from the beginning.)

Figure 12.4 Warning (aposematic) coloration in a western coral snake (*Micrurus euryxanthus*). If a population of dangerous or unpalatable organisms has a high frequency of such a color pattern, predators may rapidly learn to avoid the aposematically colored organisms, or may evolve to avoid them. It is less obvious how a new, rare mutation for such coloration, if it makes the organisms conspicuous to naive predators, can increase in frequency. (Photo © John Cancalosi/naturepl.com.)

BOX 12A Selection Models with Constant Fitnesses

We first present a general model of allele frequency change under natural selection (Hartl and Clark 1997) and then modify it for specific cases. Suppose three genotypes at a locus with two alleles differ in relative fitness due to differences in survival:

	A_1A_1	A_1A_2	A_2A_2
Frequency at birth	p^2	$2pq$	q^2
Relative fitness	w_{11}	w_{12}	w_{22}

The ratio of A_1A_1: A_1A_2: A_2A_2 among surviving adults is

$$p^2w_{11}:2pqw_{12}:q^2w_{22}$$

and the ratio of the alleles (A_1: A_2) among their gametes is

$$[p^2w_{11} + ½\,(2pqw_{12})]:[½\,(2pqw_{12}) + q^2w_{22}]$$

which simplifies to

$$p(pw_{11} + qw_{12}):q(pw_{12} + qw_{22})$$

The gamete frequencies, which are the allele frequencies among the next generation of offspring, are found by dividing each term by the sum of the gametes, which is

$$p(pw_{11} + qw_{12}) + q(pw_{12} + qw_{22})$$
$$= p_2w_{11} + 2pqw_{12} + q_2w_{22}$$
$$= \overline{w}$$

Thus the allele frequencies after selection (p', q') are the gamete frequencies, or

$$p' = \frac{p(pw_{11} + qw_{12})}{\overline{w}}$$

$$q' = \frac{q(pw_{12} + qw_{22})}{\overline{w}}$$

The change in allele frequency between generations is $\Delta p = p' - p$, or

$$\Delta p = \frac{p(pw_{11} + qw_{12}) - p\overline{w}}{\overline{w}}$$

Substituting for $\overline{w}$ and doing the algebra yields

$$\Delta p = \frac{pq[p(w_{11} - w_{12}) + q(w_{12} - w_{22})]}{\overline{w}} \quad \text{(A1)}$$

We can analyze various cases of selection by entering explicit fitness values for the w's. A few important cases are the following:

1. Advantageous dominant allele, disadvantageous recessive allele ($w_{11} = w_{12} > w_{22}$).

For w_{11}, w_{12}, and w_{22}, substitute 1, 1, and $1 - s$ respectively in Equation A1. The mean fitness is $p^2(1) + 2pq(1) + q^2(1 - s) = 1 - sq^2$ (bearing in mind that $p^2 + 2pq + q^2 = 1$). The equation for allele frequency change is

$$\Delta p = \frac{spq^2}{1 - sq^2}$$

or, equivalently,

$$\Delta q = \frac{-spq^2}{1 - sq^2} \quad \text{(A2)}$$

2. Advantageous allele partially dominant, disadvantageous allele partially recessive ($w_{11} > w_{12} > w_{22}$).

Let h, lying between 0 and 1, measure the degree of dominance for fitness, and substitute 1, $1 - hs$, and $1 - s$ for w_{11}, w_{12}, and w_{22}. (If $h = 0$, allele A_2 is fully recessive.) After sufficient algebra, we find that

$$\Delta p = \frac{spq[h(1 - 2q) + sq]}{1 - 2pqhs - sq^2} \quad \text{(A3)}$$

which is positive for all $q > 0$, so allele A_1 increases to fixation. If $h = ½$, Equation A3 reduces to $\Delta p = spq/[2(1 - sq)]$.

$$\Delta p = \frac{spq}{[2(1 - sq)]}$$

3. Fitness of heterozygote is greater than that of either homozygote ($w_{11} < w_{12} > w_{22}$).

Using s and t as selection coefficients, let the fitnesses of A_1A_1, A_1A_2, and A_2A_2 be $1 - s$, 1, and $1 - t$ respectively. Substituting these in Equation A1, we obtain

$$\Delta p = \frac{pq(-sp + tq)}{1 - sp^2 - tq^2} \quad \text{(A4)}$$

There is a stable "internal equilibrium" that can be found by setting $\Delta p = 0$; then $sp = tq$. Substituting $1 - p$ for q, the equilibrium frequency p is $t/(s + t)$. Thus the frequency of A_1 is proportional to the relative strength of selection against A_2A_2.

4. Fitness of heterozygote is less than that of either homozygote (that is, $w_{11} > w_{12} < w_{22}$).

As this is the reverse of the preceding case, let $1 + s$, 1, and $1 + t$ be the fitnesses of A_1A_1, A_1A_2, and A_2A_2. The equation for allele frequency change is

$$\Delta p = \frac{pq(sp - tq)}{1 + sp^2 + tq^2} \quad \text{(A5)}$$

Δp is positive if $sp > tq$, and negative if $sp < tq$. Setting $\Delta p = 0$ and solving for p, we find an internal equilibrium at $p = t/(s + t)$, but this is an unstable equilibrium. For example, if $s = t$, the unstable equilibrium is $p = 0.5$, but then Δp is positive if $p > q$ (i.e., if $p > 0.5$), and negative if $p < q$. The allele frequency therefore arrives at either of two stable equilibria, $p = 1$, or $p = 0$.

A simple example of directional selection occurs if the fitness of the heterozygote is precisely intermediate between that of the two homozygotes (i.e., neither allele is dominant with respect to fitness). The frequencies and fitnesses of the three genotypes may be denoted as follows:

Genotype	A_1A_1	A_1A_2	A_2A_2
Frequency	p^2	$2pq$	q^2
Fitness	1	$1 - (s/2)$	$1 - s$

Figure 12.5 A cryptic katydid (family Tettigoniidae) from Ecuadoran Amazonia. The wing veins, modified to resemble those of a leaf, contribute to the insect's extraordinarily cryptic appearance. (Photo © Morley Read/naturepl.com.)

These fitnesses may be entered into Equation A1 in Box 12A, which, when solved, shows that the advantageous allele A_1 increases in frequency, per generation, by the amount

$$\Delta p = \frac{\frac{1}{2}spq}{1-sq} \qquad (12.1)$$

where $(1 - sq)$ equals the mean fitness, $\overline{w}$.

Equation 12.1 tells us that Δp is positive whenever p and q are greater than zero. Therefore allele A_1 increases to fixation ($p = 1$), and $p = 1$ is a *stable equilibrium*. The rate of increase (the magnitude of Δp) is proportional to both the coefficient of selection s and the allele frequencies p and q, which appear in the numerator. Therefore the rate of evolutionary change increases as the variation at the locus increases. (It is approximately proportional to $2pq$, the frequency of heterozygotes, when selection is weak.)

Another important aspect of Equation 12.1 is that Δp is positive as long as s is greater than zero, even if it is very small. Therefore, as long as no other evolutionary factors intervene, *a character state with even a minuscule advantage will be fixed by natural selection*. Hence even very slight differences among species, in seemingly trivial characters such as the distribution of hairs on a fly or veins on a leaf, could conceivably have evolved as adaptations. This principle explains the extraordinary apparent "perfection" of some features. Some katydids, for example, resemble dead leaves to an astonishing degree, with transparent "windows" in the wings that resemble holes and blotches that resemble spots of fungi or algae (Figure 12.5). One might suppose that a less detailed resemblance would provide sufficient protection against predators, and some species are indeed less elaborately cryptic; but if an extra blotch increases the likelihood of survival by even the slightest amount, it may be fixed by selection (providing, we repeat, that no other factors intervene).

The same equations that describe the increase of an advantageous allele describe the fate of a disadvantageous allele: If A_1 and A_2 are advantageous and disadvantageous alleles, respectively, with frequencies p and q, and if $p + q = 1$, then $\Delta p = -\Delta q$. That is, purifying selection reduces the frequency of a deleterious mutation or eliminates it altogether.

The number of generations required for an advantageous allele to replace one that is disadvantageous depends on the initial allele frequencies, the selection coefficient, and the degree of dominance (Figure 12.6). An advantageous allele can increase from low frequency more

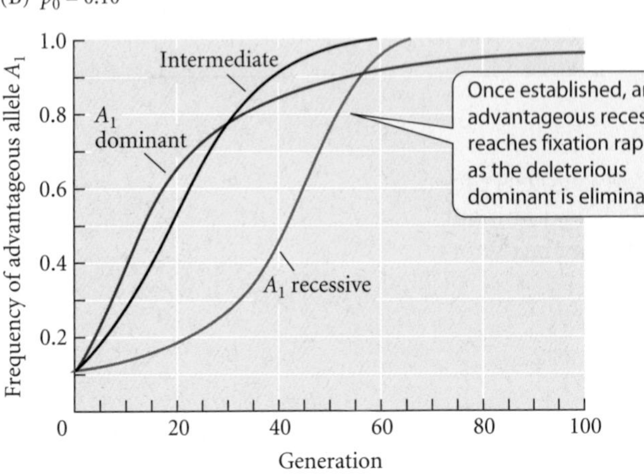

Figure 12.6 Increase of an advantageous allele (A_1) from initial frequencies of (A) $p_0 = 0.01$ and (B) $p_0 = 0.10$. The three curves in each graph show the increase of a fully dominant allele (green), an allele with intermediate dominance (black), and a recessive allele (red). For the advantageous dominant A_1, the fitnesses of genotypes A_1A_1, A_1A_2, and A_2A_2 are 1.0, 1.0, and 0.8 respectively; for the "intermediate" case they are 1.0, 0.9, and 0.8; and for the advantageous recessive A_1 they are 1.0, 0.8, and 0.8.

rapidly if it is dominant than if it is recessive because it is expressed in the heterozygous state, and until it reaches a fairly high frequency, it is carried almost entirely by heterozygotes. After a dominant advantageous allele attains high frequency, the deleterious recessive allele is eliminated very slowly, because a rare recessive allele occurs mostly in heterozygous form, and is thus shielded from selection.

The denominator in Equation 12.1 is the mean fitness of individuals in the population, $\bar{w}$, which increases as the frequency (q) of the deleterious allele decreases. The mean fitness therefore increases as natural selection proceeds. In a graphical representation of this relationship (Figure 12.7), we may think of the population as climbing up a "hillside" of increasing mean fitness until it arrives at the summit. (The arrowheads on the curves in Figure 12.7 show the direction of change in p for four different cases.)

Equation 12.1, finally, can be used to draw an interesting inference from data. If we have data on the frequencies of genotypes at a locus (and therefore also have estimates of

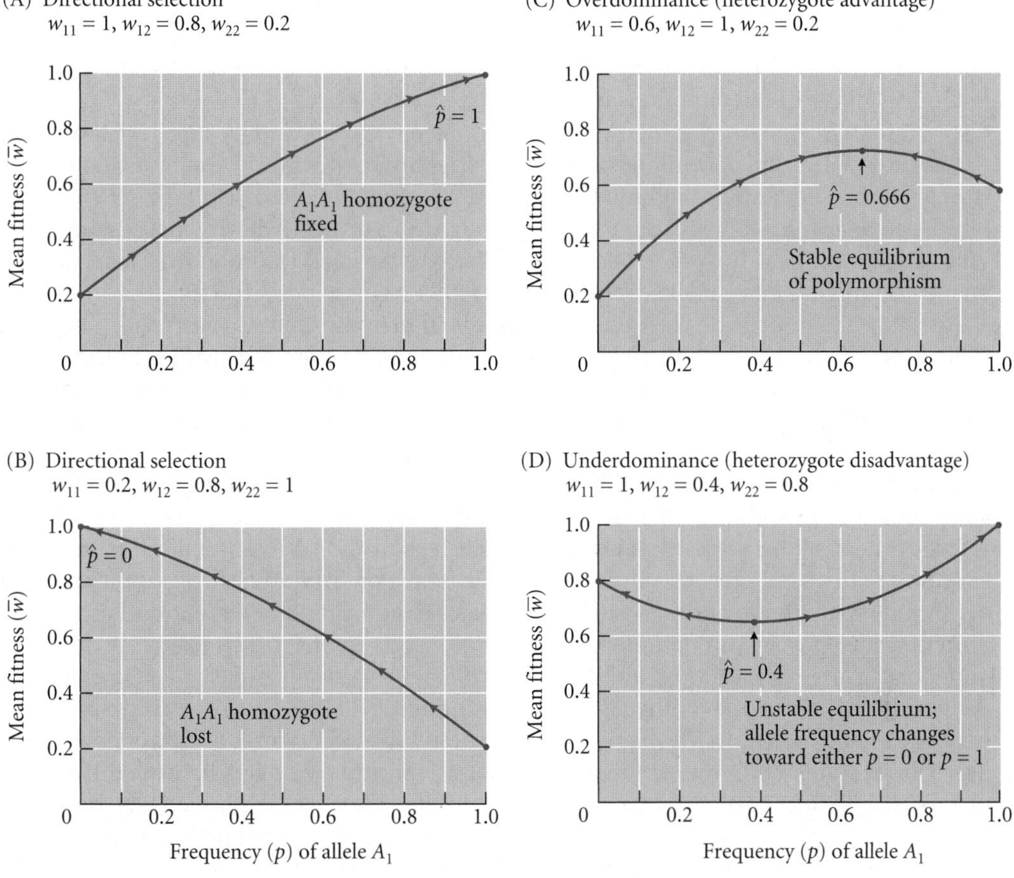

Figure 12.7 Plots of mean fitness ($\bar{w}$) against allele frequency (p) for one locus with two alleles when genotypes differ in survival. Each of these plots represents an "adaptive landscape," and may be thought of as a surface, or hillside, over which the population moves. From any given frequency of allele A_1 (p), the allele frequency moves in a direction that increases mean fitness ($\bar{w}$). The arrowheads show the direction of allele frequency change. (A) Directional selection. Here, A_1A_1 is the favored genotype. The equilibrium ($\hat{p} = 1$) is stable: the allele frequency returns to $p = 1$ if displaced. (B) Directional selection, in which the relative fitnesses are reversed compared with (A), perhaps because of changed environmental conditions. A_2A_2 is now the favored genotype. (C) Overdominance (heterozygote advantage). From any starting point, the population arrives at a stable polymorphic equilibrium ($\hat{p}$). (D) Underdominance (heterozygote disadvantage). The interior equilibrium ($\hat{p} = 0.4$ in this example) is unstable because even a slight displacement initiates a change in allele frequency toward one of two stable equilibria: $\hat{p} = 0$ (loss of A_1) or $\hat{p} = 1$ (fixation of A_1). Therefore this curve represents an adaptive landscape with two peaks. (After Hartl and Clark 1989.)

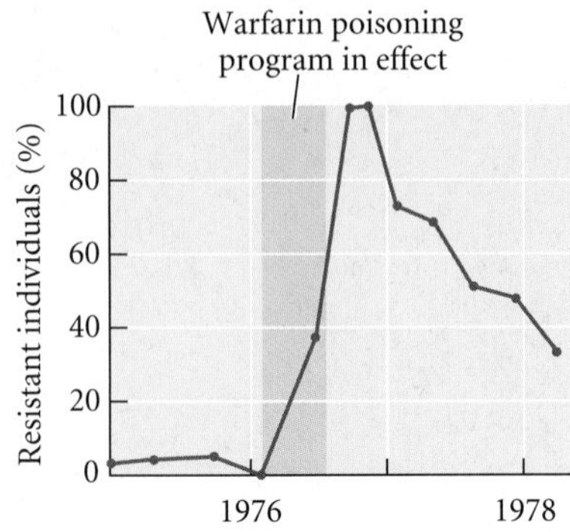

Figure 12.8 Proportion of warfarin-resistant individuals in a population of rats in Wales. The proportion increased when warfarin poisoning was practiced in 1976, but decreased after the poisoning program ended. (After Bishop 1981.)

the frequencies of alleles, p and q), and if we also have data on allele frequencies in successive generations (i.e., an estimate of Δp), we can solve for s in the equation. This is one way of estimating the strength of natural selection. Several other methods are used to estimate selection coefficients, such as estimating the survival rates (or other components of fitness) of different genotypes in natural populations (Endler 1986).

EXAMPLES OF DIRECTIONAL SELECTION. If a locus has experienced consistent directional selection for a long time, the advantageous allele should be near equilibrium—that is, near fixation. Thus the dynamics of directional selection are best studied in recently altered environments, such as those altered by human activities. Many examples of rapid evolution under such circumstances have been observed. Many are changes in polygenic traits, as described in the next chapter.

An example of rapid evolution of a single-locus trait is the case of warfarin resistance in brown rats (*Rattus norvegicus*; Bishop 1981). Warfarin is an anticoagulant: it inhibits an enzyme responsible for the regeneration of vitamin K, a necessary cofactor in the production of blood-clotting factors. Susceptible rats poisoned with warfarin often bleed to death from slight wounds. A mutation confers resistance by altering the enzyme to a form that is less sensitive to warfarin, but also less efficient in regenerating vitamin K, so that a higher dietary intake of the vitamin is necessary.

Warfarin has been used as a rat poison in Britain since 1953, and by 1958 resistance was reported in certain rat populations. Under exposure to warfarin, resistant rats have a strong survival advantage, and the frequency of the mutation has been known to increase rapidly to nearly 1.0 (Figure 12.8). Resistant rats suffer a strong disadvantage compared with susceptible rats, however, because of their greater need for vitamin K, and the frequency of the resistance allele drops rapidly if the poison is not administered.

Resistance to insecticides has evolved in many insects and mites (Roush and Tabashnik 1990; Metcalf and Luckmann 1994). Resistance appeared in many species in the 1940s, when synthetic pesticides came into wide use. By 1990, populations of more than 500 species were known to be resistant to one or more insecticides (Figure 12.9). Some species, such as the Colorado potato beetle (*Leptinotarsa decemlineata*), have evolved resistance to all the major classes of pesticides (Roush and McKenzie 1987). The evolution of resistance adds immensely to the cost of agriculture and is a major obstacle in the fight against insect-borne diseases such as malaria. For these reasons, as well as the devastating toxic effects of many pesticides on natural ecosystems and on human health, supplementary or alternative methods of pest control are a major topic of research in entomology.

Insecticide resistance in natural populations of insects is often based primarily on single mutations of large effect (Roush and McKenzie 1987). The resistance allele (R) is usually partially or fully dominant over the allele for susceptibility. R alleles generally have very low frequencies in populations that have not been exposed to insecticides, but they often increase nearly to fixation within two or three years after a pesticide is applied, because the mortality of susceptible genotypes is extremely high. In the absence of the insecticide, however, resistant genotypes are about 5 to 10 percent less fit than susceptible genotypes, and they decline in frequency. Like warfarin resistance, resistance to insecticides illustrates a **cost of adaptation**, or **trade-off**: advantageous traits often have "side effects" that are disadvantageous, at least in some environments.

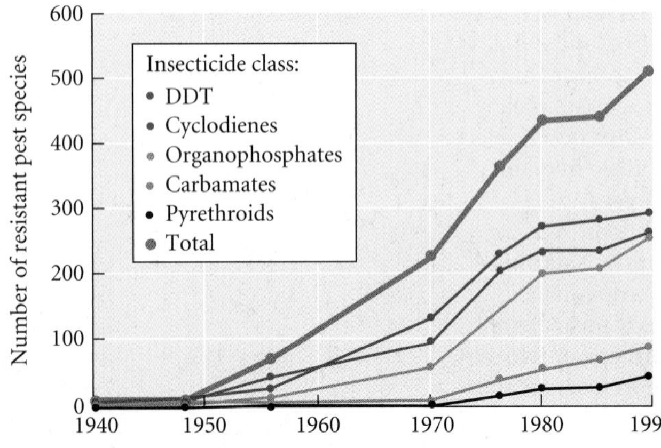

Figure 12.9 Cumulative numbers of arthropod pest species known to have evolved resistance to five classes of insecticides. The upper curve shows the total number of insecticide-resistant species. (After Metcalf and Luckmann 1994.)

Deleterious alleles in natural populations

Although the most advantageous allele at a locus should in theory be fixed by directional selection, deleterious alleles often persist because they are repeatedly reintroduced, either by recurrent mutation or by gene flow from other populations in which they are favored by a different environment. In either case, the frequency of the deleterious allele moves toward a *stable equilibrium* that is a *balance between the rate at which it is eliminated by selection and the rate at which it is introduced by mutation or gene flow.*

SELECTION AND MUTATION. Suppose that a deleterious recessive allele A_2 with frequency q arises at a mutation rate u from other alleles that have a collective frequency of $p = 1 - q$. The increase in frequency of A_2 that is due to mutation in each generation is up, whereas the decrease in its frequency that is due to selection (from Equation A2 in Box 12A) is $-spq^2/\overline{w}$. At equilibrium, the rate of increase equals the rate of decrease:

$$up = \frac{spq^2}{\overline{w}}$$

where $\overline{w} = 1 - sq^2$.

Let us assume that A_2 is rare, so that $\overline{w}$ is approximately equal to 1, and solve for q. The result is the equilibrium frequency, denoted $\hat{q}$. We find that $\hat{q}^2 = u/s$, and

$$\hat{q} = \sqrt{\frac{u}{s}}$$

The equilibrium frequency of a deleterious recessive allele, therefore, is directly proportional to the mutation rate and inversely proportional to the strength of selection. Thus, if s is much greater than u, the allele will be very rare. For example, if $s = 1$ (i.e., A_2 is a recessive lethal allele) and the mutation rate is 10^{-6}, the equilibrium frequency will equal 0.001, but almost all the zygotes with this allele will be heterozygous. If the deleterious allele is partly or entirely dominant, its equilibrium frequency will be even lower because of selection against both homozygous and heterozygous carriers. Such MUTATION-SELECTION BALANCE explains why many chromosomes in *Drosophila* populations (as well as in humans and many other species) carry rare mutations that slightly reduce fitness in heterozygous condition and are strongly deleterious or even lethal when homozygous (Crow 1993; see Figure 9.10).

SELECTION AND GENE FLOW. Very often, different environmental conditions favor different alleles in different populations of a species. Thus, in the absence of gene flow, the frequency q of an allele A_2 will be 1 in certain populations and 0 in others. Gene flow among populations can introduce each allele into populations in which it is deleterious, and the allele frequency in each population thus arrives at an equilibrium ($\hat{q}$) set by the balance between the incursion of alleles by gene flow and their elimination by selection. Thus gene flow can contribute to genetic variation within populations. If gene flow is much greater than selection, a population inhabiting a small patch of a distinctive environment will not become genetically differentiated from surrounding populations (Lenormand 2002).

If local populations are distributed along an environmental gradient over which the fitness of different genotypes changes (Figure 12.10A), then even if the environment changes gradually, we should expect an abrupt shift in allele frequencies (a STEP CLINE) in the absence of gene flow among populations along the gradient (Figure 12.10B). This is true as long as one homozygote or the other has highest fitness in each population. However, a *smooth cline in allele frequencies may be established if there is gene flow* among populations along the gradient. The width of the cline (the distance over which q changes from, say, 0.2 to 0.8) is proportional to V/s, where V measures the distance that genes disperse, and s is the strength of selection against them (the coefficient of selection). Thus, if selection is strong relative to gene flow, steep clines in allele frequencies will result, so that populations are strongly differentiated.

Alleles at the aminopeptidase I locus in the blue mussel (*Mytilus edulis*) display a cline in Long Island Sound, a body of water about 160 kilometers long in which salinity increases from west to east, where the Sound meets the ocean. The frequency of one of several allo-zymes, ap^{94}, increases from about 0.12 in the west to 0.55 at the oceanic sites (Figure 12.11). The aminopeptidase I enzyme cleaves terminal amino acids from proteins, increas-

(A)

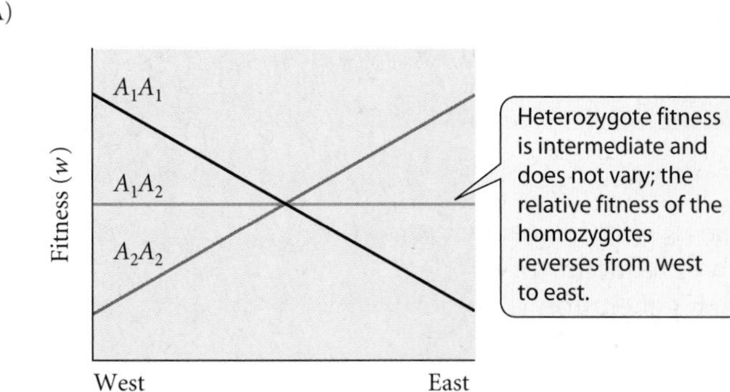

Figure 12.10 Geographic variation arising from selection and gene flow when genotype fitnesses differ gradually along a geographic gradient in an environmental factor. (A) Relative fitnesses W_{11}, W_{12}, and W_{22} of genotypes A_1A_1, A_1A_2, and A_2A_2 are plotted for populations along an east–west gradient. A_1A_1 has higher fitness than A_2A_2 in the "west," and vice versa in the "east." (B) Frequency of the A_2 allele in an array of populations along the gradient. Each curve represents a different level of gene flow (g), ranging from 0 to 100 percent (g = 1). The lower the level of gene flow, the steeper the cline in allele frequency. (After Endler 1973.)

(B)

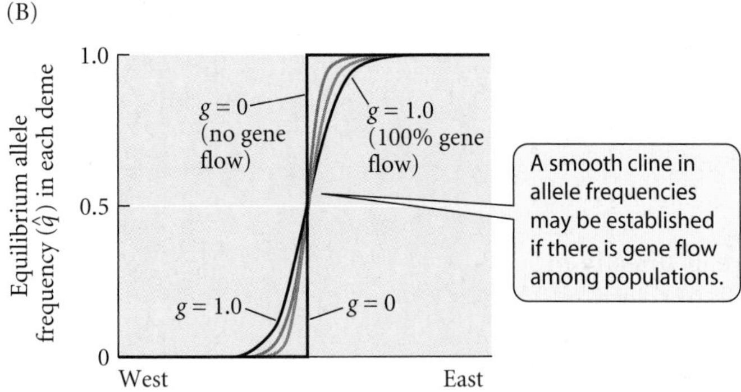

ing the intracellular concentration of free amino acids and thereby helping to maintain osmotic balance in salt water. Richard Koehn and Jerry Hilbish (1987) and their colleagues found that aminopeptidase I activity is higher in mussels carrying the ap^{94} allele than in other genotypes. Because ap^{94} genotypes have higher intracellular amino acid levels, the ap^{94} allele is favored at oceanic salinity. In less saline waters, however, a lower concentration of amino acids suffices for osmotic balance, and high aminopeptidase I activity is disadvantageous because the breakdown of protein is costly in terms of both energy and nitrogen, and must be compensated by feeding. The investigators found that in Long Island Sound, ap^{94} has a high frequency among recently settled young mussels, but that its frequency decreases as they age, indicating that ap^{94} mussels have a higher mortality rate than others. The persistence of ap^{94} in the Sound therefore seems to be due to incursions of larvae from the ocean in each generation.

Gene flow can reduce the level of adaptation of populations to their local environment. For example, aquatic larvae of the salamander *Ambystoma barbouri* that occupy streams

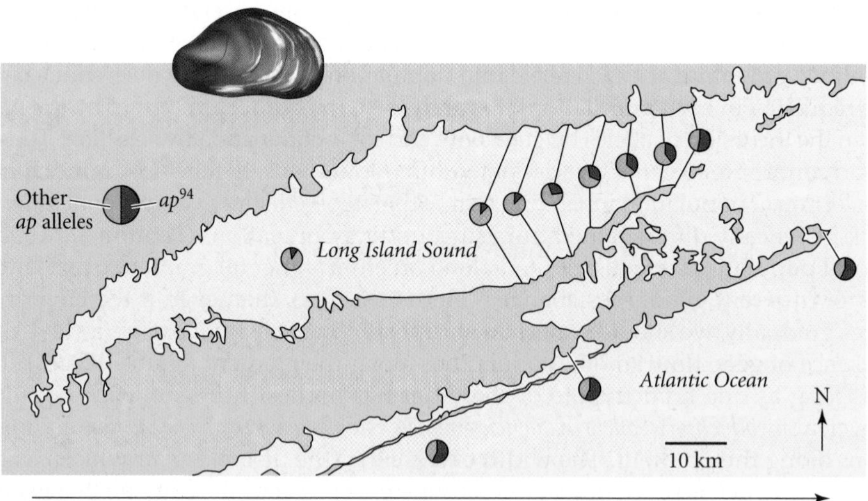

Figure 12.11 Frequency of the ap^{94} allele, indicated by the black portion of each circle, in samples of the mussel *Mytilus edulis* in Long Island Sound and nearby sites. The frequency of the allele increases rapidly over a 30-kilometer distance in the Sound, where salinity increases from west to east. The cline is maintained by selection against ap^{94} in the Sound, countered by high gene flow from oceanic populations where the allele is advantageous. (After Koehn and Hilbish 1987.)

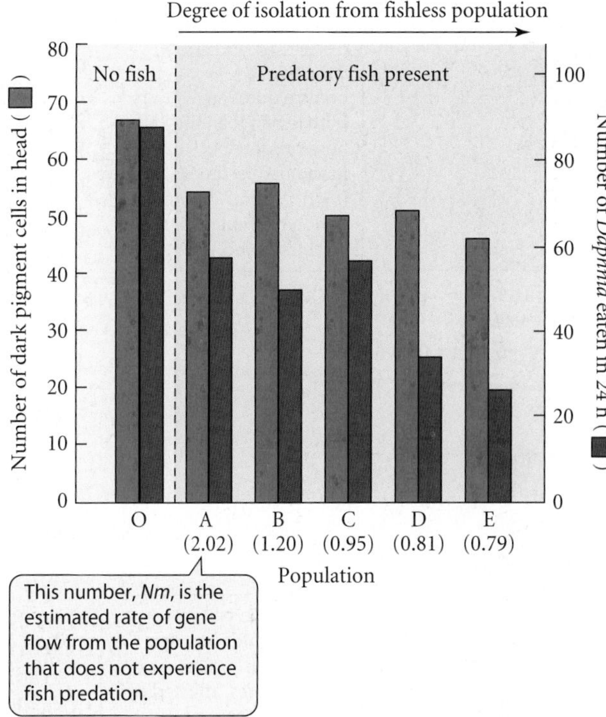

Figure 12.12 Gene flow reduces adaptation to fish predation among larvae of the salamander *Ambystoma barbouri*. Both sets of bars show mean character values in a "fishless" population (O) and in five populations that coexist with fish and are increasingly isolated from the fishless population. *Nm* estimates gene flow into populations A through E from fishless populations, based on F_{ST} values calculated from allozyme allele frequencies (see Chapter 10). Darkness of coloration (blue bars) is estimated from the density of dark pigment cells on the head; paler coloration is favored in streams with fish. The red bars represent feeding rates when salamander larvae in experimental containers were provided with *Daphnia* crustaceans as food, but were then exposed to chemical cues (odorants) from fish. A low feeding rate indicates larval populations that were more immobile in the presence of fish odor, and thus more likely to escape the notice of predatory fish. (After Storfer and Sih 1998; Storfer et al. 1999.)

with predatory fish have paler coloration than populations in sites without fish; the pale coloration matches the background and thus reduces predation. Salamander populations from fish-inhabited streams also have a heightened response to the odor of fish: they seek shelter and become inactive, even though this reduces their feeding activity. However, neither of these adaptations is as well developed in populations that have high levels of gene flow from "fishless" populations (Figure 12.12; Storfer and Sih 1998; Storfer et al. 1999).

Polymorphism Maintained by Balancing Selection

Until the 1940s, the prevalent, or classic, view had been that at each locus, a best allele (the "wild type") should be nearly fixed by natural selection, so that the only variation should consist of rare deleterious alleles, recently arisen by mutation and fated to be eliminated by purifying selection. As we saw in Chapter 9, studies of natural populations revealed instead a wealth of variation. The factors that might be responsible for this variation are: (1) recurrent mutation producing deleterious alleles, subject to only weak selection; (2) gene flow of locally deleterious alleles from other populations in which they are favored by selection; (3) selective neutrality (i.e., genetic drift); and (4) maintenance of polymorphism by natural selection. The last of these hypotheses was championed by British ecological geneticists led by E. B. Ford and American population geneticists influenced by Theodosius Dobzhansky. They represented the BALANCE SCHOOL, holding that a great deal of genetic variation is maintained by **balancing selection** (which is simply selection that maintains polymorphism).

There is still uncertainty about how much of the extensive genetic variation in natural populations fits the models represented by these contrasting points of view. Several models of natural selection can account for persistent, stable polymorphism, but we do not know the extent to which each accounts for the observed genetic variation within populations.

Heterozygote advantage

If the heterozygote has higher fitness than either homozygote, both alleles will be propagated in successive generations, in which union of gametes will yield all three genotypes among the zygotes. Such **heterozygote advantage** is also termed overdominance or SINGLE-LOCUS HETEROSIS. It results in a stable equilibrium at which the allele frequencies depend on the balance between the fitness values (hence, the selection coefficients) of the two homozygotes (Figure 12.13; see also Box 12A and **Figure 12.7C**).

Genotypes that are heterozygous at several or many loci often appear to be fitter than more homozygous genotypes. For example, inbreeding depression is commonly

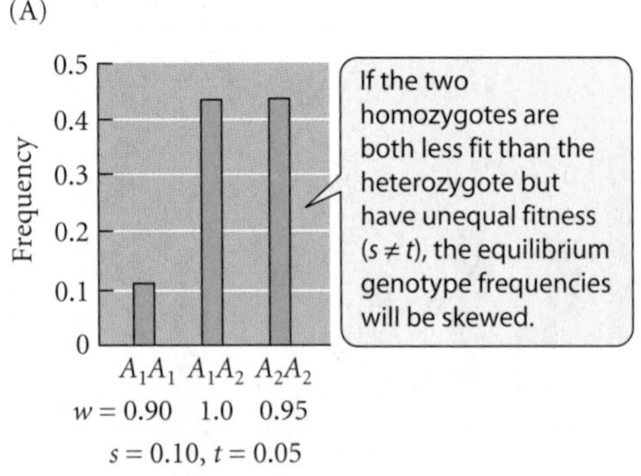
(A)
Frequency
0.5
0.4
0.3
0.2
0.1
0
If the two homozygotes are both less fit than the heterozygote but have unequal fitness ($s \neq t$), the equilibrium genotype frequencies will be skewed.
A_1A_1 A_1A_2 A_2A_2
w = 0.90 1.0 0.95
s = 0.10, t = 0.05

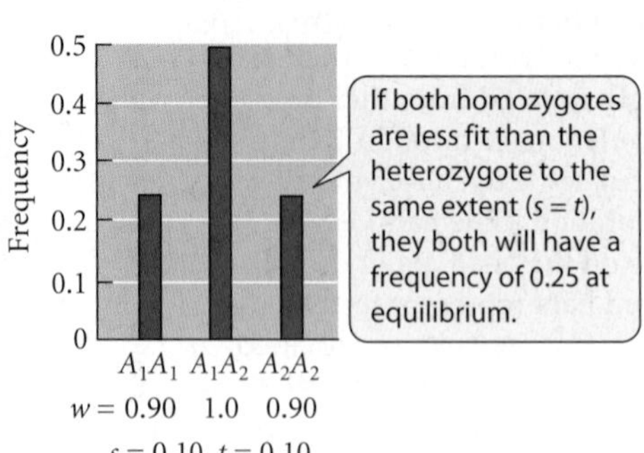
Frequency
0.5
0.4
0.3
0.2
0.1
0
If both homozygotes are less fit than the heterozygote to the same extent ($s = t$), they both will have a frequency of 0.25 at equilibrium.
A_1A_1 A_1A_2 A_2A_2
w = 0.90 1.0 0.90
s = 0.10, t = 0.10

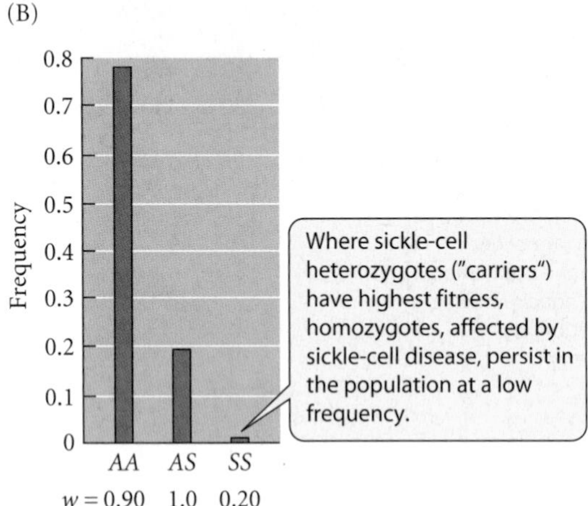
(B)
Frequency
0.8
0.7
0.6
0.5
0.4
0.3
0.2
0.1
0
Where sickle-cell heterozygotes ("carriers") have highest fitness, homozygotes, affected by sickle-cell disease, persist in the population at a low frequency.
AA *AS* *SS*
w = 0.90 1.0 0.20
s = 0.10, t = 0.80

Figure 12.13 (A) Genotype frequencies among newborn zygotes at a locus with heterozygote advantage (overdominance) for several relative fitnesses of the two homozygotes. The fitnesses of A_1A_1, A_1A_2, and A_2A_2 are $1 - s$, 1, and $1 - t$, respectively; these values, and those of s and t (the coefficients of selection), are shown below each graph. (B) Expected frequencies for the sickle-cell hemoglobin polymorphism, using fitness values estimated for an African population exposed to malaria. *AA* is the normal homozygote, *AS* carries the sickle-cell trait, and *SS* expresses sickle-cell anemia.

observed when organisms become more homozygous under inbreeding (see Chapter 9). However, this phenomenon can be explained by dominance rather than overdominance: inbred lines become homozygous for deleterious recessive alleles. Likewise, it can be very difficult to show that the heterozygote (*Aa*) at a particular locus has higher fitness than both homozygotes (*AA*, *aa*). Suppose most of the chromosomes in a population are *Ab* and *aB*, where *A* and *B* are favorable dominant alleles and *a* and *b* are unfavorable recessive alleles at two closely linked loci. The homozygotes *Ab*/*Ab* and *aB*/*aB*, each expressing an unfavorable recessive, would have lower fitness than the heterozygote *Ab*/*aB*. If we were aware of the phenotypic effect of only one of the loci, it would appear to display overdominance for fitness. Such apparent but spurious superiority of the heterozygote at the observed locus is called **associative overdominance**.

Few cases of overdominance for fitness have been well documented. The best understood case of heterozygote advantage is the β-hemoglobin locus in some African and Mediterranean human populations (Cavalli-Sforza and Bodmer 1971). One allele at this locus encodes sickle-cell hemoglobin (*S*), which is distinguished by a single amino acid substitution from normal hemoglobin (encoded by the *A* allele). At low oxygen concentrations, *S* hemoglobin forms elongate crystals, which carry oxygen less effectively, causing the red blood cells to adopt a sickle shape and to be broken down more rapidly. Heterozygotes (*AS*) suffer slight anemia; homozygotes (*SS*) suffer severe anemia (sickle-cell disease) and usually die before reproducing. "Normal" homozygotes (*AA*) suffer higher mortality from malaria than heterozygotes (*AS*), however, because the protist that causes the most serious form of malaria (*Plasmodium falciparum*) develops in red blood cells. Since the red blood cells of heterozygotes are broken down more rapidly, the growth of the protist is curtailed. Thus heterozygotes survive at a higher rate than either homozygote, and the frequency of the *S* allele is quite high (about $q = 0.13$), in some parts of Africa with a high incidence of malaria. (The relative fitnesses have been estimated as $w_{AA} = 0.89$, $w_{AS} = 1.0$, $w_{SS} = 0.20$.) The heterozygote advantage therefore arises from a balance of OPPOSING SELECTIVE FACTORS: anemia and malaria. In the absence of malaria, balancing selection gives way to directional selection, because then the *AA* genotype has the highest fitness. In the African-American population, which is not subject to malaria, the frequency of *S* is about 0.05, and is presumably declining as a result of mortality.

Sickle-cell hemoglobin is one of several polymorphisms associated with resistance to malaria, a disease that causes 2 million deaths annually. Thalassemia, characterized by partial or complete suppression of synthesis of the α- or β-hemoglobin chain, is caused by many different point mutations and large deletions that reach high frequencies in many tropical and subtropical human populations. Thalassemia is associated with resistance to malaria, but it is not yet certain that heterozygotes have higher fitness than homozygotes (Clegg and Weatherall 1999).

Antagonistic and varying selection

The opposing forces acting on the sickle-cell polymorphism are an example of **antagonistic selection**, which in this instance maintains polymorphism because the heterozygote happens to have the highest fitness. Unless this is the case, antagonistic selection usually does not maintain polymorphism (Curtsinger et al. 1994). Suppose, for example, that the survival rates of an insect in the larval stage are 0.5 for genotypes A_1A_1 and A_1A_2 and 0.4 for A_2A_2, whereas the proportions of surviving larvae that then survive through the pupal stage are 0.6 and 0.9, respectively. In other words, the A_1 allele improves larval survival, but reduces pupal survival, compared with the A_2 allele. For genotypes A_1A_1 and A_1A_2, the proportion surviving to reproductive adulthood is (0.5)(0.6) = 0.30, and for A_2A_2 it is (0.4)(0.9) = 0.36. Thus, A_2A_2 has a net selective advantage, and allele A_2 will be fixed.

Within a single breeding population, a fluctuating environment may favor different genotypes in different generations (TEMPORAL FLUCTUATION), or different genotypes may be best adapted to different microhabitats or resources (SPATIAL VARIATION). Like antagonistic selection, variation in the environment *does not necessarily maintain genetic variation*, although under some circumstances it may do so (Felsenstein 1976; Hedrick 1986).

Temporal fluctuation in the environment may slow down the rate at which one or another allele approaches fixation, but usually it does not preserve multiple alleles indefinitely. Spatial variation in the environment is most likely to maintain polymorphism if different homozygotes in a single population are best adapted to different microhabitats or resources—that is, if they have different "niches." (This phenomenon is sometimes called **multiple-niche polymorphism**.) A stable multiple-niche polymorphism is more likely if each individual organism usually experiences only one of the environments. It also is more likely if selection is "soft" rather than "hard." SOFT SELECTION occurs when the number of survivors in a patch of a particular microenvironment is determined by competition for a limiting factor, such as space or food, and the *relatively* superior genotype has a higher probability of survival. Then selection determines the genotype frequencies among the surviving adults, but not their total number. HARD SELECTION occurs when the likelihood of survival of an individual in a microenvironment depends on its *absolute* fitness, not on the density of competitors. Selection then determines not only the genotype frequencies but also the total number of survivors. (For example, selection for insecticide resistance may be hard, since the survival of an individual insect may depend only on whether it has a resistant or susceptible genotype. In contrast, selection for long mouthparts that enable a bee to obtain nectar from flowers more rapidly than shorter-tongued bees would be soft selection if the shorter-tongued bees could obtain nectar, and survive and reproduce prolifically, in the absence of superior longer-tongued competitors.)

The black-bellied seedcracker (*Pyrenestes ostrinus*) provides a nice example of multiple-niche polymorphism (Smith 1993; Figure 12.14). Populations of this African finch have a bimodal distribution of bill width. The difference between wide and narrow bills seems to be due to a single allele difference. The two morphs differ in the efficiency with which they process seeds of different species of sedges, their major food: wide-billed birds process hard seeds, and narrow-billed birds soft seeds, more efficiently. Smith banded more than 2700 juvenile birds and found that survival to

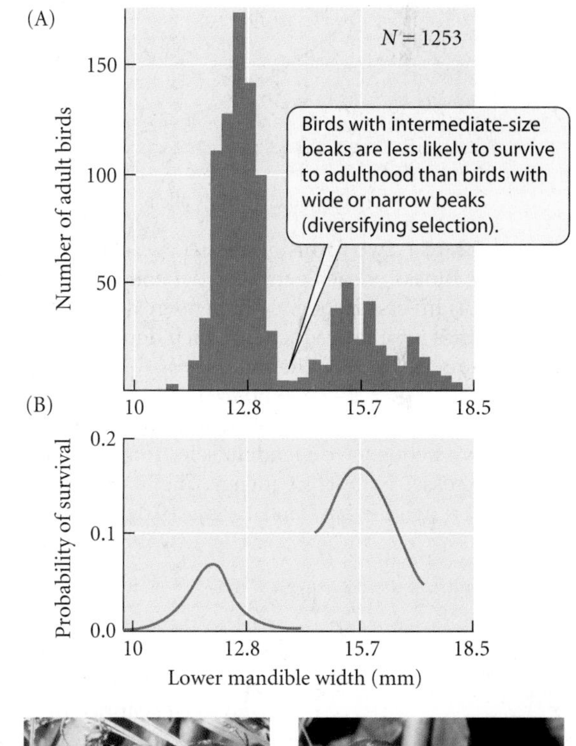

Figure 12.14 Multiple-niche polymorphism in the black-bellied seedcracker. (A) Probability of survival to adulthood of banded juvenile birds in relation to their lower mandible width, a measure of bill size. (B) The distribution of lower mandible width among adults is bimodal. The small- and large-billed morphs are shown at left and right, respectively. (After Smith 1993; photos courtesy of Thomas B. Smith.)

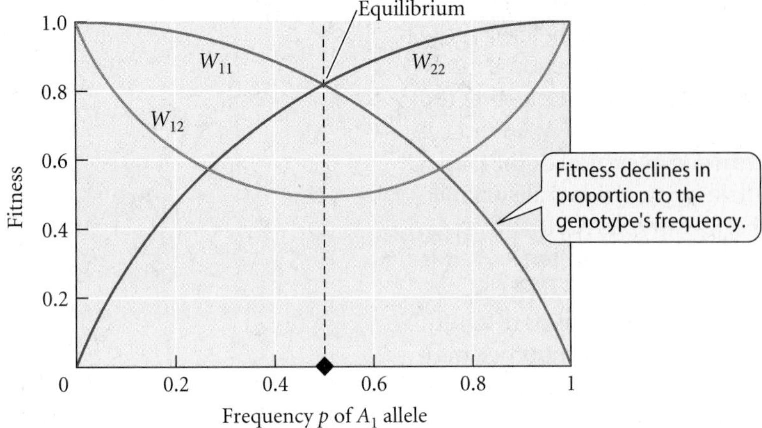

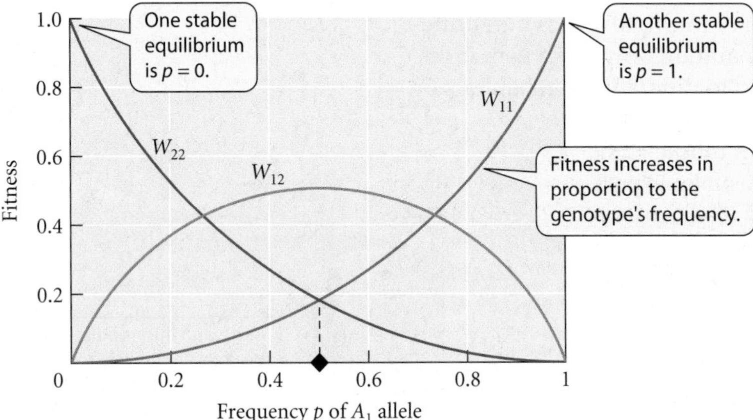

Figure 12.15 Two forms of frequency-dependent selection, in which the fitness of each genotype depends on its frequency in a population. (A) Inverse frequency-dependent selection. A genotype's fitness declines with its frequency, which depends on allele frequencies (p, the frequency of A_1). The fitnesses of A_1A_1, A_1A_2, and A_2A_2 are $W_{11} = 1 - sp^2$, $W_{12} = 1 - spq$, and $W_{22} = 1 - sq^2$ in this model; the curves are calculated for $s = 1$. A stable equilibrium exists at $p = q = 0.5$. (B) Positive frequency-dependent selection, in which a genotype's fitness increases with its frequency. The fitnesses of A_1A_1, A_1A_2, and A_2A_2 are $1 + sp^2$, $1 + 2spq$, and $1 + sq^2$. (After Hartl and Clark 1989.)

adulthood was lower for birds with intermediate bills than for either wide- or narrow-billed birds. Thus diversifying selection, arising from the superior fitness of different genotypes on different resources, appears to maintain the polymorphism.

Frequency-dependent selection

In the models we have considered so far, the fitness of each genotype is assumed to be constant within a given environment. Very often, however, *the fitness of a genotype depends on the genotype frequencies in the population*. The population then undergoes **frequency-dependent selection**. This kind of balancing selection appears to maintain many polymorphisms, and it has many other important consequences, especially for the evolution of animal behavior, as we will see in Chapter 16.

INVERSE FREQUENCY-DEPENDENT SELECTION. In **inverse frequency-dependent selection**, the rarer a phenotype is in the population, the greater its fitness (Figure 12.15A). For example, the per capita rate of survival and reproduction of a dominant phenotype (with genotype A_1A_1 or A_1A_2) may be greatest when it is very rare and may decrease as it becomes more common; the same may be true of the recessive phenotype (with genotype A_2A_2). Thus when A_2 is at high frequency, it declines because A_2A_2 has lower fitness than A_1A_1 and A_1A_2, and likewise for A_1. Whatever the initial allele frequencies may be, they shift toward a stable equilibrium value, which in this case occurs when the frequencies of the two phenotypes are equal (i.e., when $q^2 = 0.5$). At this point, the mean fitnesses of both phenotypes are the same: neither has an advantage over the other.

Many biological phenomena can give rise to inverse frequency-dependent selection. The self-incompatibility alleles of many plant species are a striking example. These alleles enforce outcrossing because pollen carrying any one allele cannot effectively grow down the stigma of the plant that produced it or of any other plant that carries the same allele. Thus, if there are three alleles, S_1, S_2, and S_3, in a plant population, pollen of type S_1 can grow on stigmas, and fertilize ovules, of genotype S_2S_3, but not on S_1S_2 or S_1S_3; S_2 pollen can fertilize only S_1S_3 plants, and S_3 pollen can fertilize only S_1S_2 plants. (Notice that plants cannot be homozygous at this locus.) A new S allele that arises by mutation can fertilize almost all plants since it is rare at first, and so can increase in frequency until it reaches a frequency of about $1/k$, where k is the number of S alleles in the population. We should expect to find a large number of self-incompatibility alleles, all about equal in frequency, at equilibrium. In some plant species, indeed, hundreds of such alleles exist.

Michio Hori (1993) analyzed a fascinating example of frequency-dependent selection in the cichlid fish *Perissodus microlepis* in Lake Tanganyika, which feeds by approaching other cichlids from behind and snatching a mouthful of scales from the prey's flank. This cichlid's mouth is twisted to the right (dextral) or left (sinistral; Figure 12.16A). Dextral individuals always approach the left side of their prey, and sinistral individuals the right. The two morphs, which appear to be based on two alleles at a single locus, fluctuate in

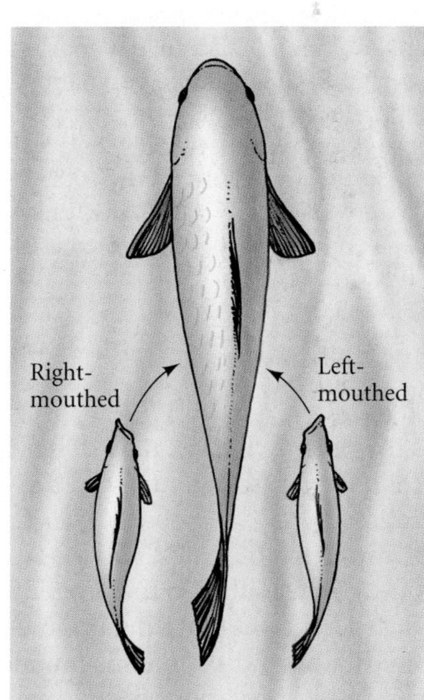

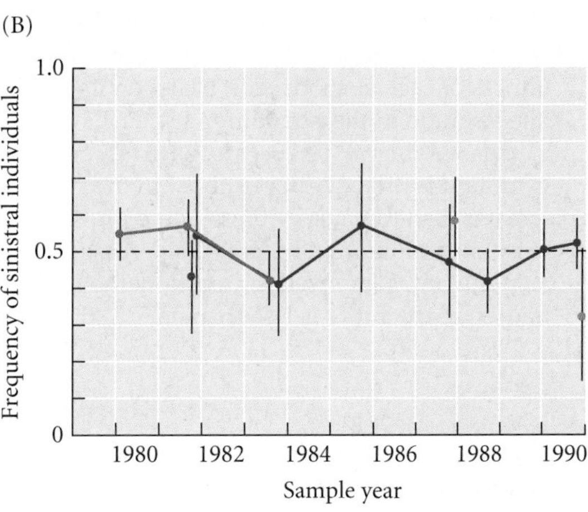

Figure 12.16 An inverse frequency-dependent polymorphism in the scale-eating cichlid *Perissodus microlepis*. (A) Dextral (right-mouthed) and sinistral (left-mouthed) *P. microlepis* attack prey from opposite sides. (B) Fluctuations in the frequency of the sinistral form at two nearby locations along the shore of Lake Tanganyika. (B after Hori 1993.)

frequency around 0.5 (Figure 12.16B). The frequencies are stabilized by the prey's escape behavior: when the dextral form is more abundant, the prey are more wary of approaches on their left side, and consequently are attacked more frequently on the right, by the rarer sinistral form. Conversely, the dextral form has greater feeding success when the sinistral form is more common.

Selection is often frequency-dependent when genotypes compete for limiting resources, as in the model of soft selection described above. Suppose that two genetically determined phenotypes, P_1 and P_2, can survive on either resource 1 or resource 2, but that P_1 is a superior competitor for resource 1 and P_2 is a superior competitor for resource 2. If P_1 is rare, each P_1 individual will compete for resource 1 chiefly with inferior P_2 competitors, and so will have a higher per capita rate of increase than P_2 individuals. As P_1 increases in frequency, however, each P_1 individual competes with more individuals of the same genotype, so its per capita advantage, relative to P_2, declines. The same pattern applies to phenotype P_2.

This effect is illustrated in an experiment on the grass *Anthoxanthum odoratum*, which can be propagated asexually by vegetative cuttings. In a natural environment, Norman Ellstrand and Janis Antonovics (1984) planted small "focal" cuttings and surrounded each with competitors. The surrounding plants were either the same genotype as the focal individual, taken as vegetative shoots from the same parent plant, or different genotypes, obtained from sexually produced seed. After a season's growth, focal individuals on average were larger and produced more seed (i.e., had higher fitness) if surrounded by different genotypes than if surrounded by the same genotype. This pattern is just what we would expect if different genotypes use somewhat different resources (perhaps certain mineral nutrients are partitioned differently by adjacent plants that differ in genotype). The more intense competition among individuals of the same genotype would then impose inverse frequency-dependent selection. In the same vein, agricultural researchers have sometimes found that fields planted to a mixture of crop varieties yield more than fields of a single variety.

THE EVOLUTION OF THE SEX RATIO BY FREQUENCY-DEPENDENT SELECTION. Why is the sex ratio about even (1:1) in many species of animals? This is quite a puzzle, because from a group-selectionist perspective, we might expect that a female-biased sex ratio (i.e., production of more females than males) would be advantageous because such a population could grow more rapidly. If sex ratio evolves by individual selection, however, and if all females have the same number of progeny, why should a genotype producing an even sex ratio have an advantage over any other?

The solution to the puzzle was provided by the great population geneticist R. A. Fisher (1930), who realized that because every individual has both a mother and a father, females and males must contribute equally to the ancestry of subsequent generations, and must there-

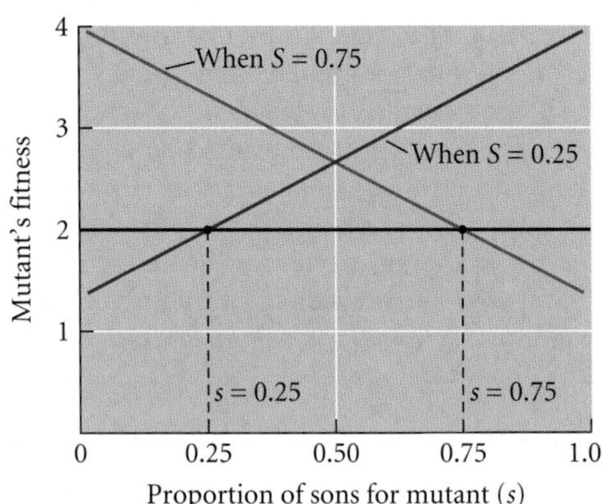

Figure 12.17 Frequency-dependent selection on sex ratio. Mutants may arise that vary in individual sex ratio (s, the proportion of sons among a female's progeny). The fitness of each such mutant, based on its average number of grandchildren, depends on the sex ratio in the population (S). The average fitness of individuals in the population equals 2. When $S = 0.25$—i.e., when 25 percent of the population is male—the fitness of a mutant is directly proportional to the proportion of sons among its offspring, and it is greater than the mean fitness of the prevalent genotype if the mutant's s exceeds 0.25. Any such mutant will therefore increase in frequency. Conversely, if $S = 0.75$, the fitness of a mutant is inversely proportional to its individual sex ratio, and its fitness increases in frequency if its s is less than 0.75. (After Charnov 1982.)

fore have the same average fitness. Therefore individuals that vary in the sex ratio of their progeny can differ in the number of their grandchildren (and later descendants), and thus differ in fitness if this is measured over two or more generations.

To see why this is so, let us define the sex ratio as the proportion of males, and distinguish the sex ratio in a population (the POPULATION SEX RATIO, S) from that in the progeny of an individual female (an INDIVIDUAL SEX RATIO, s). In a large, randomly mating population, the fitness of a genotype that produces a given individual sex ratio depends on the population sex ratio, which in turn depends on the frequencies of the various individual-sex-ratio genotypes. Because the average per capita reproductive success of the minority sex is greater than that of the majority sex, selection favors genotypes whose individual sex ratios are biased toward that sex that is in the minority in the population as a whole.

For example, suppose the population sex ratio is 0.25 (1 male to 3 females) because it consists of a genotype with this individual sex ratio. Suppose each female has 4 offspring. The average number of progeny per female is 4, but since every offspring has a father, the average number of progeny sired by a male is 12 (since each male mates with 3 females on average). Thus a female has 4 grandchildren through each of her daughters and 12 through each son, for a total of $(3 \times 4) + (1 \times 12) = 24$ grandchildren. Now suppose a rare genotype with an individual sex ratio of 0.50 (2 daughters and 2 sons) enters the population. Each such individual has $2 \times 4 = 8$ grandchildren through her daughters and $2 \times 12 = 24$ grandchildren through her sons, for a total of 32. Since this is a greater number of descendants than the mean number per individual of the prevalent female-biased genotype, any allele that causes a more male-biased progeny in this female-biased population will increase in frequency. Likewise, if the population sex ratio were male-biased, an allele for female-biased individual sex ratios would spread. By this reasoning, a genotype that produces an even sex ratio (0.5) has highest fitness and cannot be replaced by any other genotype (Figure 12.17). (A genotype that produces a sex ratio of 0.5 represents an EVOLUTIONARILY STABLE STRATEGY, or ESS, as we will see in Chapter 14.)

Alexandra Basolo (1994) tested this theory using the platyfish *Xiphophorus maculatus*, which has three kinds of sex chromosomes, W, X, and Y. Females are XX, WX, or WY, and males are XY or YY. Of the six possible crosses, four yield a sex ratio of 0.5, but XX mated with YY yields all sons, and WX mated with XY yields 0.25 sons, so there is variation among genotypes in individual sex ratios. Basolo set up experimental populations with different frequencies of these chromosomes, which carried different color-pattern alleles so that chromosome frequencies could be followed. Two populations were initiated with sex ratios (proportion of males) of 0.25 percent and 0.78 percent. Within only two generations, the sex ratio in both populations evolved nearly to 0.5, as Fisher's theory predicts (Figure 12.18).

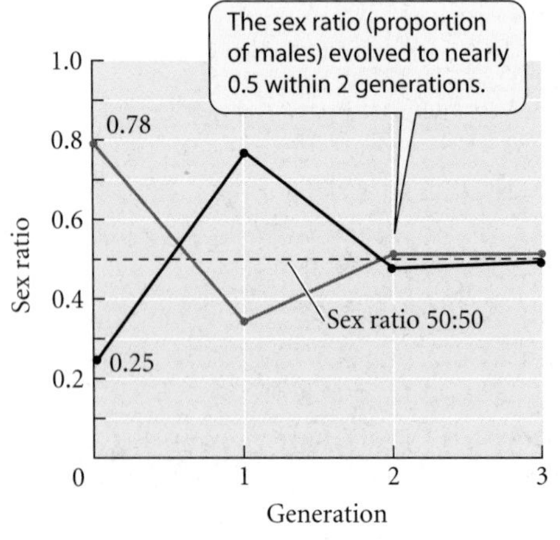

Figure 12.18 Changes in sex ratio (proportion of males) in two experimental populations of platyfish, initiated with sex ratios (proportions of males) of 0.25 and 0.78. In both populations, the sex ratio evolved nearly to 0.5 within only two generations. (After Basolo 1994.)

Multiple Outcomes of Evolutionary Change

One of the most important principles in evolution is that initial genetic conditions often determine which of several paths, or trajectories, of genetic change a population will follow. Thus *the evolution of a population often depends on its previous evolutionary history.* Positive frequency-dependent selection and heterozygote disadvantage are two important factors that can give rise to multiple outcomes—**multiple stable equilibria**.

Positive frequency-dependent selection

In **positive frequency-dependent selection**, the fitness of a genotype is greater the more frequent it is in a population. As a result, *whichever allele is initially more frequent will be fixed* (see Figure 12.15B).

For example, the unpalatable tropical butterfly *Heliconius erato* has many different geographic races that differ markedly in color pattern. Each race is monomorphic. Adjacent geographic races interbreed at zones only a few kilometers wide. The geographic variation in color pattern in this species is closely paralleled by that in another unpalatable species of *Heliconius*—a spectacular example of Müllerian mimicry (Figure 12.19). Predators that have had an unpleasant experience with one or a few individuals of either species tend thereafter to avoid similar butterflies, of either species.

James Mallet and Nicholas Barton (1989) showed that within *Heliconius erato*, gene flow from one geographic race to another is countered by positive frequency-dependent selection: immigrant butterflies that deviate from the locally prevalent color pattern are selected against because predators have not learned to avoid attacking butterflies with unusual color patterns. On either side of a contact zone between two geographic races in Peru, Mallet released *H. erato* of the other race and, as a control, butterflies of the same race from a different locality. The butterflies, marked so they could be recognized, were repeatedly recaptured for some time thereafter. Compared with control butterflies, with the same color pattern as the population into which they were released, far fewer of those with the allopatric color pattern were captured. From bill marks left on the wings of butterflies that had escaped from birds, the authors concluded that the missing butterflies were lost to bird predation, and calculated an average selection coefficient of 0.52 against the "wrong" color pattern in either population. This amounts to a selection coefficient of about $s = 0.17$ at each of the three major loci that control the differences in color pattern between the races—very strong selection indeed.

Heterozygote disadvantage

The case in which the heterozygote has lower fitness than either homozygote is called **heterozygote disadvantage** or underdominance. If a population is initially monomorphic for A_1A_1, and A_2 then enters at low frequency by mutation or gene flow, almost all A_2 alleles are carried by heterozygotes (A_1A_2); since their fitness is lower than that of A_1A_1, selection reduces the frequency (q) of A_2 to zero. Likewise, A_1 is eliminated if it enters a monomorphic population of A_2A_2, which also has a greater fitness than A_1A_2. Monomorphism for either A_1A_1 or A_2A_2 is therefore a stable equilibrium, and the initially more frequent allele is fixed by selection. This model applies to some chromosome rearrangements, such as

Figure 12.19 An extraordinary case of geographic variation and Müllerian mimicry. The butterflies on the right represent geographic races of a single species, *Heliconius erato*, from regions in Central and South America. The butterflies on the left are races of a rather distantly related species, *Heliconius melpomene*, from the corresponding regions. Both species are unpalatable to predators. Races of *H. erato* interbreed with one another to some extent where they meet, but the hybrid zone between them is very narrow; the same is true of *H. melpomene*. (From Cornell University Insect Collection, courtesy of Andrew Brower.)

inversions and translocations, for which heterozygotes may have reduced fertility as a result of improper segregation in meiosis (see Figures 8.23, 8.24, and the discussion below).

If the homozygotes' fitnesses are different, but both greater than that of the heterozygote, the mean fitness in a population that has become fixed for the less fit homozygote is less than if it were fixed for the other homozygote, but *selection cannot move the population from the less fit to the more fit condition*. Thus *a population is not necessarily driven by natural selection to the most adaptive possible genetic constitution*.

Adaptive landscapes

Recall that we can calculate the mean fitness ($\overline{w}$) of individuals in a population with any conceivable allele frequency (p) and plot a curve showing $\overline{w}$ as a function of p (see Figure 12.7). When fitnesses are constant, natural selection changes allele frequencies in such a way that mean fitness ($\overline{w}$) increases, so that the population moves up the slope of this curve. The current location of the population on this slope is then a simple guide to how allele frequencies will change under selection: simply see which direction of allele frequency change will increase $\overline{w}$. For an underdominant locus, the curve dips in the middle and slopes upward to $p = 0$ and $p = 1$ (see Figure 12.7D). Thus natural selection decreases or increases p depending on whether a population begins to the left or the right of the minimum of the $\overline{w}$ curve.

Curves such as those in Figure 12.7 are often referred to as **adaptive landscapes**, or ADAPTIVE TOPOGRAPHIES. The curve in Figure 12.7D represents two **adaptive peaks** separated by an **adaptive valley**. This metaphor, introduced by Sewall Wright, is widely used in evolutionary biology, and we will refer to it in Chapter 18. Each point on the curve (the adaptive landscape) represents the *average* fitness of individuals in a hypothetical population made up of (in this case) three genotypes with frequencies p^2, $2pq$, and q^2. Each possible value of p—each possible hypothetical population—yields a different value on the x-axis, and therefore a different point on the landscape. When, as in Figure 12.7D, the relationship of $\overline{w}$ to p has two or more maxima, two genetically different populations (e.g., with $p = 0$ or $p = 1$ in this case) can (but need not) have the same mean fitness ($\overline{w}$) *under the same environmental conditions*. Different environments, which might alter the fitnesses of the genotypes relative to one another, are represented not by different points on any one landscape, but rather by *different landscapes*—different relationships between $\overline{w}$ and p.

Interaction of selection and genetic drift

In developing the theory of selection so far, we have assumed an effectively infinite population size. However, *in a finite population, allele frequencies are simultaneously affected by both selection and chance*. As the movement of an airborne dust particle is affected both by the deterministic force of gravity and by random collisions with gas molecules (Brownian movement), so the effective size (N_e) of a population and the strength of selection (s) both affect changes in allele frequencies. The effect of random genetic drift is negligible if selection on a locus is strong relative to the population size—that is, if s is much greater than $1/(4N_e)$, i.e., if $4N_e s > 1$. Conversely, if s is much less than $1/(4N_e)$ (that is, if $4N_e s < 1$), selection is so weak that the allele frequencies change mostly by genetic drift: the alleles are *nearly neutral*. The critical value is $4N_e s$: genetic drift predominates if selection is weak or the effective population size is small.

This principle might be important if heterozygotes are inferior in fitness, so that the adaptive landscape has two peaks (see Figure 12.7D). Selection alone cannot move a population down the slope of one peak and across a valley to the slope of another peak, even if the second peak is higher: a population does not first become poorly adapted so that it can then become better adapted (Figure 12.20A). But during episodes of very low population size, allele frequencies may fluctuate so far by genetic drift that they cross the adaptive valley—after which selection can move the population "uphill" to the other peak (Figure 12.20B). The probability that such a **peak shift** will occur (Barton and Charlesworth 1984) depends on the population size and on the difference in height (mean fitness) between the valley and the initially occupied peak.

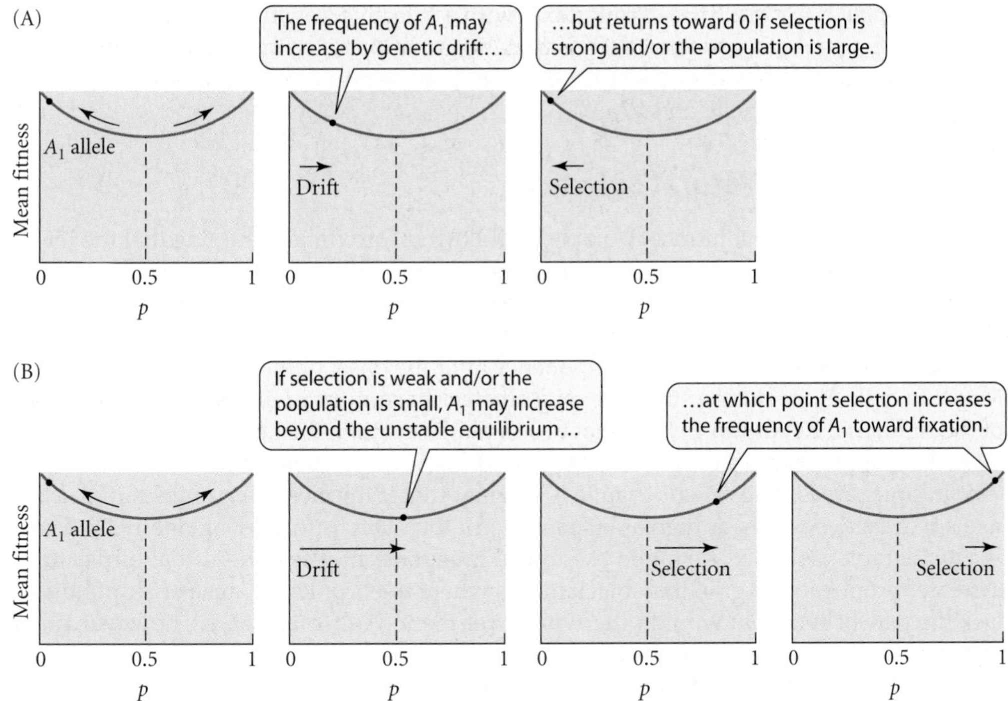

Figure 12.20 A peak shift resulting from the joint action of genetic drift and natural selection. For two alleles or chromosome rearrangements that lower the fitness of the heterozygote, the fitnesses of A_1A_1, A_1A_2, and A_2A_2 are 1, $1 - s$, and 1, respectively. The curve shows the mean fitness ($\bar{w}$) for each possible value of p, the frequency of A_1, and may be considered an adaptive landscape with two peaks (at $p = 0$ and $p = 1$). An unstable equilibrium exists at $p = 0.5$, indicated by the dashed line. The value of p at a particular time is shown by a colored dot. (A) The frequency of A_1, initially near 0, may increase by genetic drift, but returns toward 0 if the population is large. (B) If the population becomes small, however, p may increase beyond the unstable equilibrium frequency (0.5) as a result of genetic drift. If this occurs, selection increases the frequency of A_1 toward fixation.

Thus, when there are multiple stable equilibria, *genetic drift and selection may act in concert to accomplish what selection alone cannot*, moving a population from one adaptive peak to another. This theory explains how populations can come to differ in underdominant chromosome rearrangements such as translocations and pericentric inversions. Some chromosome rearrangements are thought to conform to the model of underdominance because heterozygotes have lower fertility than either homozygote (see Chapter 8). For example, local populations of the Australian grasshopper *Vandiemenella viatica* are monomorphic for different chromosome fusions and pericentric inversions. Heterozygotes for these chromosomes have many aneuploid gametes. Any such chromosome, introduced by gene flow into a population monomorphic for a different arrangement, is reduced in frequency by natural selection, so no two "chromosome races" are sympatric; instead, they meet in "tension zones" only 200 to 300 meters wide (White 1978). Because these grasshoppers are flightless and quite sedentary, local populations are small, providing the opportunity for genetic drift to occasionally initiate a peak shift whereby a new chromosome arrangement is fixed.

The effect of population size on the efficacy of selection has several important consequences. First, a population may not attain exactly the equilibrium allele frequency predicted from its genotypes' fitnesses; instead, it is likely to wander by genetic drift in the vicinity of the equilibrium frequency. Second, *deleterious mutations can become fixed by genetic drift*, especially if selection is weak and the effective population size is small, as it may be if population bottlenecks occur. Most importantly, *a slightly advantageous mutation is less likely to be fixed by selection if the population is small than if it is large, because it is more likely*

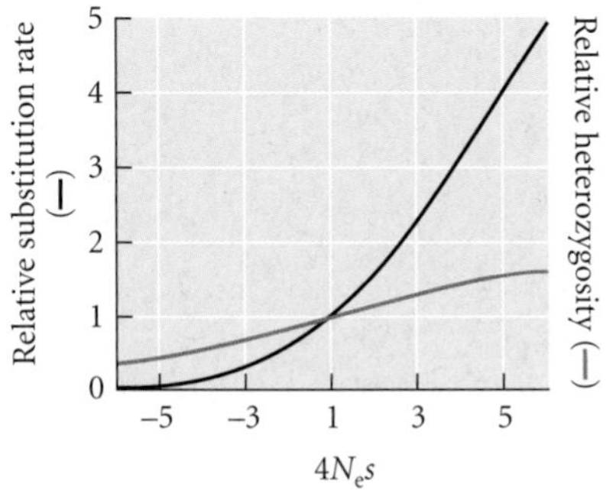

Figure 12.21 The relative substitution rate of selected to neutral mutations (black line) and the average frequency of heterozygotes (red line) both increase as a function of the product of effective population size and the selection coefficient, $4N_es$. (After Fay and Wu 2003.)

to be lost simply by chance. Thus if mutations with a selective advantage *s* arise repeatedly, both the rate of fixation of advantageous mutations compared with neutral mutations and the relative levels of heterozygosity for the two types of mutations increase with increasing $4N_es$ (Figure 12.21; Fay and Wu 2003).

The Strength of Natural Selection

Until the 1930s, most evolutionary biologists followed Darwin in assuming that the intensity of natural selection is usually very slight. By the 1930s, however, examples of very strong selection came to light. One of the first examples was INDUSTRIAL MELANISM in the peppered moth (*Biston betularia*). In some parts of England, a black form of the moth carrying a dominant allele increased in frequency after the onset of the Industrial Revolution. Museum collections dating since the mid-nineteenth century in England, after the onset of the Industrial Revolution, show that in less than a century, the "typical" pale gray form declined and the black (melanic) form increased from about 1 percent to more than 90 percent in some areas. The rate of change is so great that it implies a very substantial selective advantage, possibly as high as 50 percent, for the melanic form (Haldane 1932). There is considerable evidence, obtained by several independent researchers, that birds attack a greater proportion of gray than black moths where tree trunks, because of air pollution, lack the pale lichens that would otherwise cover them (Figure 12.22A,C); however, other factors also appear to affect the allele frequencies (Majerus 1998).

As air pollution has become regulated, conditions have reverted to favor the typical gray phenotype, and the frequency of the melanic form has declined rapidly in Great Britain, Europe, and the United States (Figure 12.22B; Grant and Wiseman 2002; Cook 2003). By applying the appropriate equation for allele frequency change (see Box 12A) to the data, Lawrence Cook has estimated selection coefficients (*s*) against the melanic form of 0.05 to more than 0.20 in various British sites.

In the last few decades, components of fitness have been quantified for polymorphic traits in many species. Although cases in which *s* is close to zero are probably underrepresented because investigators may not report studies that yielded no evidence of selection, the selection coefficients estimated in natural populations range from low to very

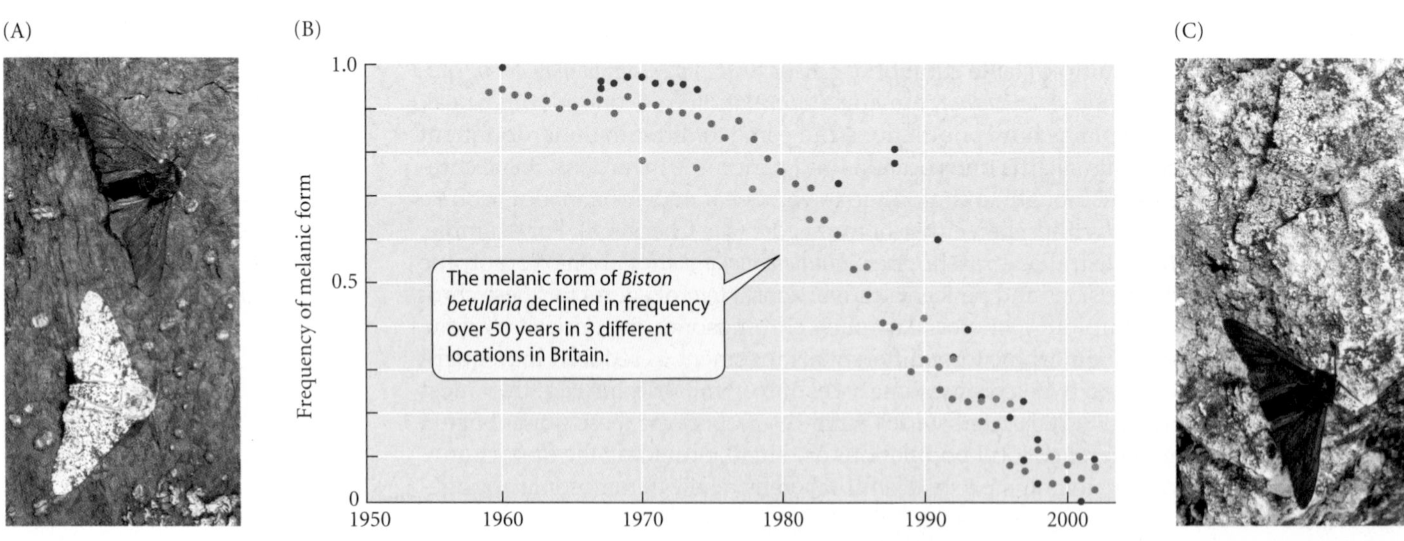

Figure 12.22 (A) The pale "typical" form and the dark melanic form of the peppered moth (*Biston betularia*) on a dark tree trunk. The British biologist H. B. D. Kettlewell pinned these and other freshly killed specimens to both dark and pale trunks, and determined that birds took a higher proportion of the more conspicuous phenotype in each situation. (B) The decline in the frequency of the melanic form in three British localities, indicated by different symbols. (C) The "typical" and melanic forms of the moth on a pale tree trunk covered with lichens, which grow in areas with clean air. (A, C © David Fox/photolibrary.com; B after Cook 2003.)

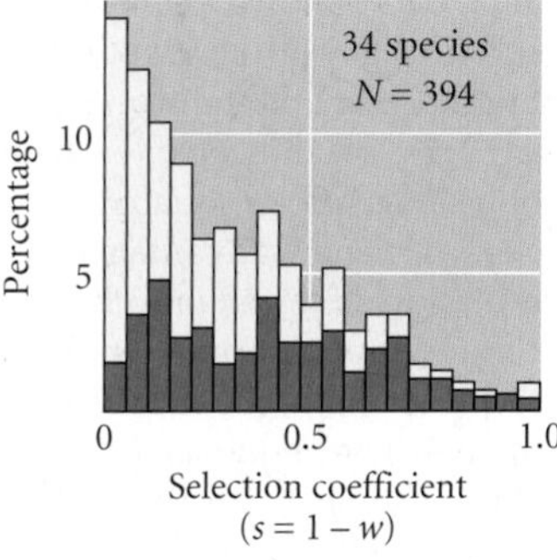

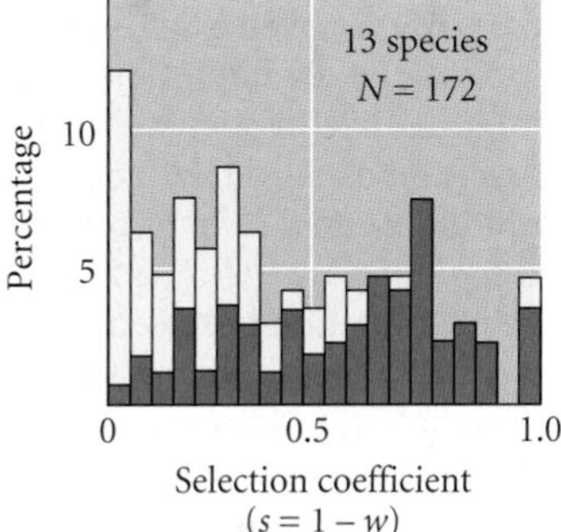

Figure 12.23 A compilation of selection coefficients (s) reported in the literature for discrete, genetically polymorphic traits in natural populations of various species. The total height of each bar represents the percentage of all reported values in each interval, and the red portion represents the percentage shown by statistical analysis to be significantly different from zero. N is the number of values reported, based on one or more traits from each of several species. (A) Selection based on differences in survival. (B) Selection based on differences in reproduction (fecundity, fertility, and sexual selection). (After Endler 1986.)

high (Figure 12.23). Moreover, selection acting through both survival and reproduction can be very strong. Thus natural selection is a powerful factor in evolution and is often far stronger than Darwin ever would have imagined.

Molecular Signatures of Natural Selection

Until recently, studies of natural selection were pursued mostly on field or laboratory populations, and focused on phenotypic characteristics. Today, such studies have been joined by a very different approach: inferring selection from variation in DNA sequences. The advantages of this approach are many: in almost any species, a great many genes can be studied in a very short time without lengthy and often difficult observations; and the genes themselves, rather than inherited phenotypes, are directly studied. Without further biological or ecological information, this approach does not reveal the causes of selection or, often, even the function of the gene. Furthermore, the methods depend on assumptions that may not hold in particular cases, and it can be difficult to distinguish the effects of selection from other factors such as population structure and growth. Nevertheless, these are powerful approaches that are providing astonishing insights into selection in humans and other species.

Theoretical expectations

Selection on DNA sequences may be inferred if the pattern of variation differs from patterns expected under the neutral theory of molecular evolution. We saw in Chapter 10, for example, that at equilibrium between mutation and genetic drift, the expected amount of sequence variation in a diploid population, as expressed by the frequency of heterozygotes per nucleotide site, is

$$\frac{4N_e u_0}{4N_e u_0 + 1}$$

where N_e is the effective population size and u_0 is the rate of neutral mutation. Furthermore, the nucleotides at different variable sites in a DNA sequence should not be correlated with one another at equilibrium because recombination, even between closely linked sites, should eventually lead to linkage equilibrium (see Chapter 9).

Now suppose that an advantageous mutation occurs at a particular base pair site in a gene, and consider the *effects of this selection on neutral variation at sites that are closely linked to the selected site*. As the frequency of the new mutation increases, all of the gene copies with the mutation have descended from the single copy in which the mutation first occurred. As John Maynard Smith and J. Haigh (1974) pointed out long before DNA sequencing was feasible, the neutral variant nucleotides in that ancestral gene copy, linked to the mutation, will also increase, by *hitchhiking*. Thus POSITIVE DIRECTIONAL SELECTION (selection for an advantageous mutation) reduces variation at closely linked sites (reviewed by Nielsen 2005). If the advantageous mutation is fixed by selection, *all neutral variation in the gene is eliminated* by a **selective sweep** (Figure 12.24), and variation is only slowly reconstituted as new neutral mutations occur among the copies of the advantageous gene. If, furthermore, we observe the population when the advantageous mutation has reached intermediate frequency, but before it is fixed, we will observe *linkage disequilibrium (LD) among various neutral polymorphic sites,* because the advantageous mutation has remained associated with particular nucleotides at linked sites, which therefore are correlated with one another.

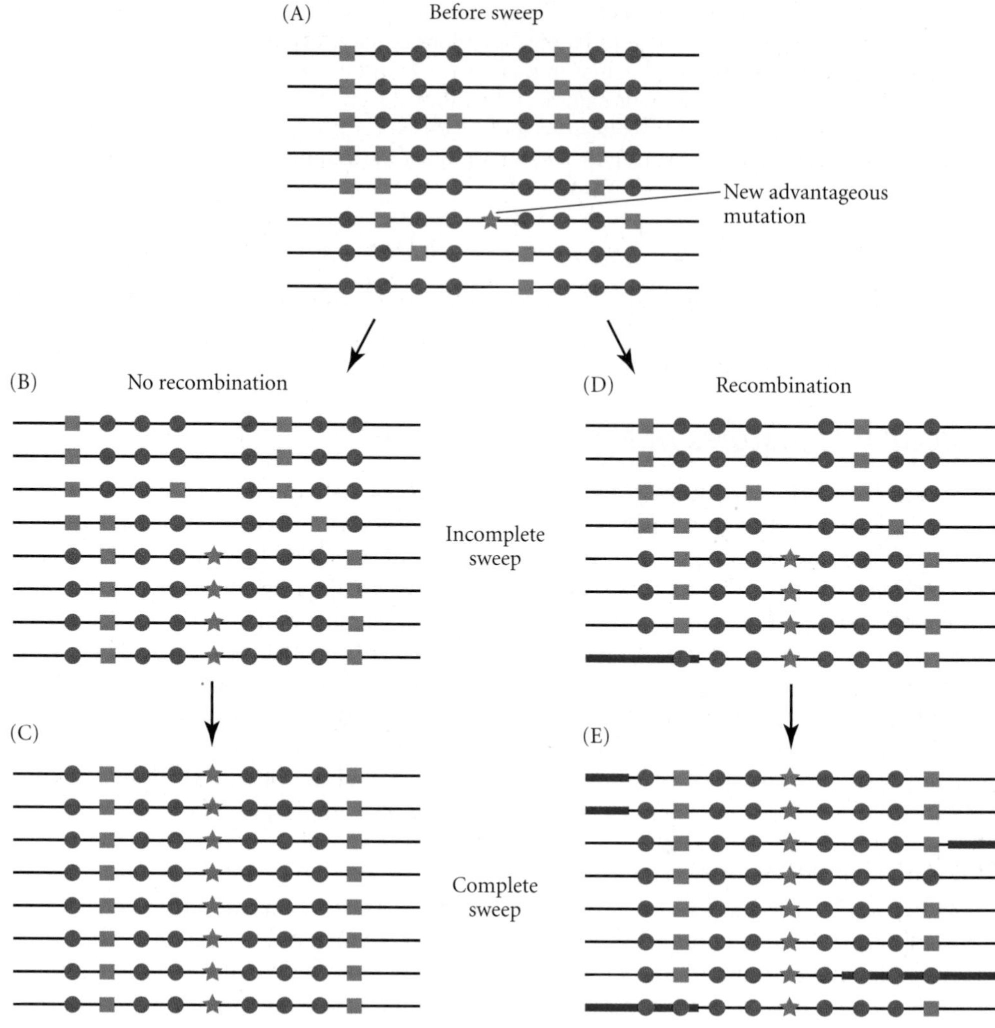

Figure 12.24 Selective sweeps. Each panel shows a sample of eight chromosomes from a population. Eight neutral polymorphisms, such as SNPs, are indicated, each with two states (green square or red circle). (A) A new advantageous mutation arises on one chromosome. (B) In a population that lacks recombination, the mutation increases in frequency (lower four chromosomes), causing a partial sweep that reduces the amount of linked variation. Note that the SNP variants linked to the mutation are perfectly correlated with each other, forming a long haplotype that has intermediate frequency. The SNPs on chromosomes that lack the mutation show little or no linkage disequilibrium. (C) The advantageous mutation has been fixed, and has completely swept away linked variation. (D) If there is recombination, the increase in frequency of the advantageous mutation has much the same effect as in (B), but note that recombination has partly broken down linkage disequilibrium of SNPs with the mutation. The colored segment of the lowermost chromosome has been acquired from another chromosome by crossing over. (E) Fixation of the advantageous allele results in considerable linkage diseqeuilibrium and high frequency of long haplotypes, but to a lesser extent than if there were no recombination. (After Nielsen et al. 2007.)

How far do these effects extend along the DNA sequence from the selected site? The recombination rate (r) is greater between the selected site and more distant sites. As r increases, the initial hitchhiking effect is weaker, and the faster the LD between a given site and the advantageous mutation breaks down (Figure 12.25). However, the higher the selective advantage (s) of the mutation, the more rapidly the selected *haplotype* (the mutation plus its linked neutral variants) increases in frequency, so that closer sites have less opportunity to recombine with other haplotypes in the population. Thus, the haplotype is longer, the higher s and the lower r is. Since r is defined on a per-generation basis, one might use the length of LD to estimate when an advantageous mutation arose, given an estimate of r from genetic data, and an assumed or estimated value of s.

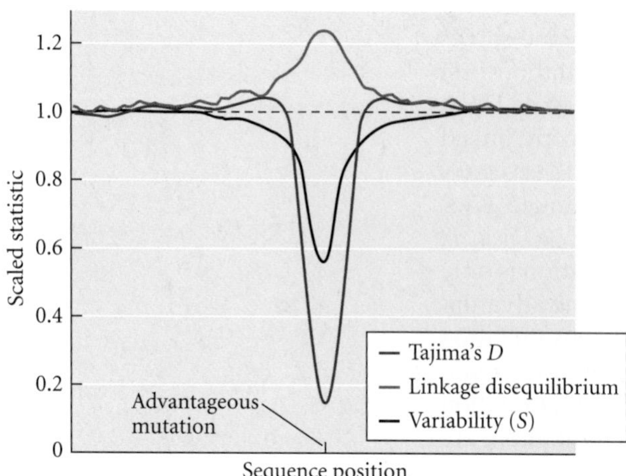

Figure 12.25 A selective sweep causes an excess of the most common allele and a deficiency of intermediate-frequency alleles. Variability (S), as measured by the number of variable nucleotide sites in a sample of sequences, is lowered near the site of an advantageous mutation, as is Tajima's D, an index of the departure of the frequency distribution of different sequences (alleles) from that expected under the neutral theory. Linkage disequilibrium is greatest near the selected site. (After Nielsen 2005.)

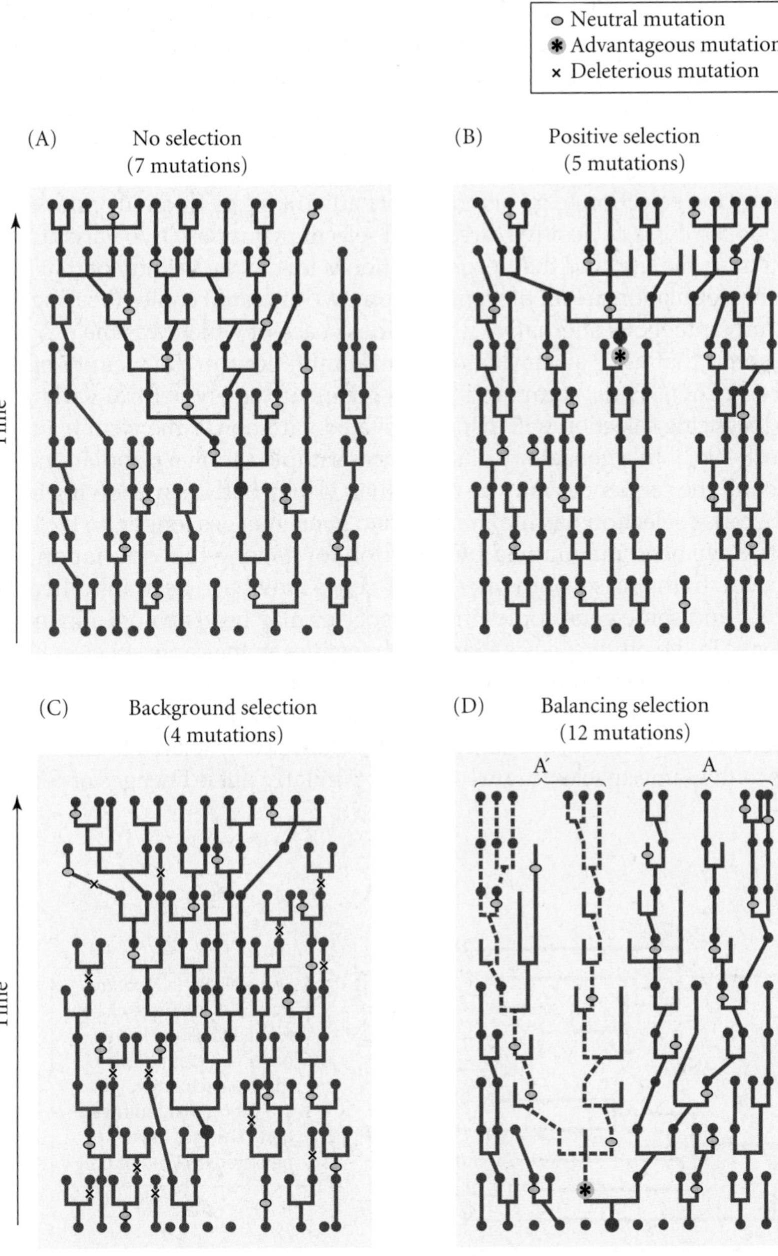

Figure 12.26 Schematic genealogies of gene copies in a population, showing the effects of three modes of selection on nucleotide diversity, compared with the neutral model (compare with Figure 10.17). The contemporary population (represented by 12 gene copies) is at the top of each diagram, and the ancestry of contemporary gene copies is marked by the red gene trees. In each diagram, some gene lineages (blue lines) become extinct by chance. These diagrams assume that no recombination occurs in the gene. Ovals (yellow) represent selectively neutral mutations, each at a different site in the gene. (A) Neutral mutations only. Contemporary gene copies vary by 7 mutations. (B) Positive selection of an advantageous mutation, marked by an asterisk. In a selective sweep, this gene lineage replaces all others, so contemporary copies differ only in the 5 mutations that have occurred since the advantageous mutation. (C) Background selection. Deleterious mutations, marked by ×, eliminate some gene copies, hence reducing the number of surviving neutral mutations. (D) Balancing selection of alleles *A* and *A*′, the latter lineage (dashed lines) arising from the mutation marked with an asterisk. The two gene lineages persist for a long time, and so accumulate more mutations than in the neutral case.

The effects of various forms of selection are evident in gene genealogies. Consider two unlinked loci, one that has been evolving solely by genetic drift (Figure 12.26A) and one that has experienced a selective sweep (Figure 12.26B). Compared with the neutrally evolving gene, the copies of the gene that was fixed by selection are descended from a more recent common ancestor (the one in which the favorable mutation occurred); they have had less time to accumulate different neutral mutations, and so are more similar in sequence. *A selective sweep resembles a bottleneck in population size in that it reduces variation and increases the genealogical relatedness among gene copies, but a bottleneck will affect the entire genome, not just the portion surrounding an advantageous mutation.*

Like directional selection, purifying selection against deleterious mutations reduces neutral polymorphism at closely linked sites. Brian Charlesworth and colleagues (Charlesworth et al. 1993; Charlesworth 1994a), who have termed this effect **background selection**, pointed out that when a copy of a deleterious mutation is eliminated from a population, selectively neutral mutations linked to it are eliminated as well (Figure 12.26C). Thus, for this region of DNA, the effective population size is reduced to the proportion of gametes that are free of deleterious mutations. The reduction in the level of heterozygosity for neutral mutations is greatest if the rate of deleterious mutation is high, if the mutations are strongly deleterious, and if the recombination rate is very low. Charles Aquadro and colleagues (1994) surveyed sequence variation in 19 genes of *Drosophila melanogaster* and found that sequence variation (π; see Chapter 9) was lowest in parts of the genome with the lowest recombination rate.

The effects of balancing selection (e.g., heterozygote advantage or frequency-dependent selection) are opposite those of positive directional selection. Suppose two variants are maintained at a polymorphic site, and that recombination is low in the vicinity of that site. The gene copies in the population are all descended from two ancestral copies (bearing the original, selectively advantageous alternative nucleotides), each of which was the progenitor of a lineage of genes that have accumulated neutral mutations in the vicinity of the selected site (Figure 12.26D). Thus, compared with a gene with solely neutral variation, a gene subjected to balancing selection will display elevated variation in the vicinity of the selected site (Strobeck 1983). In a genealogy of sequences sampled from a population, the common ancestor of all the sequences may be older than if they had been evolving solely by genetic drift because selection has maintained two gene lineages longer. In fact, the polymorphism may have been maintained by selection for so long that speciation has occurred in the interim. In that case, both lineages of genes may have been inherited by two (or more) species, and some gene copies in each species may be genealogically more closely related to genes in the other species than to other genes in the same species.

Exactly this pattern has been found at the self-incompatibility locus in the family Solanaceae, in which many of the alleles in different genera of plants that diverged more than 30 Mya (e.g., petunia and tobacco) are genealogically more closely related to each other than to other alleles in the same species (Figure 12.27A). Similarly, allele lineages of certain

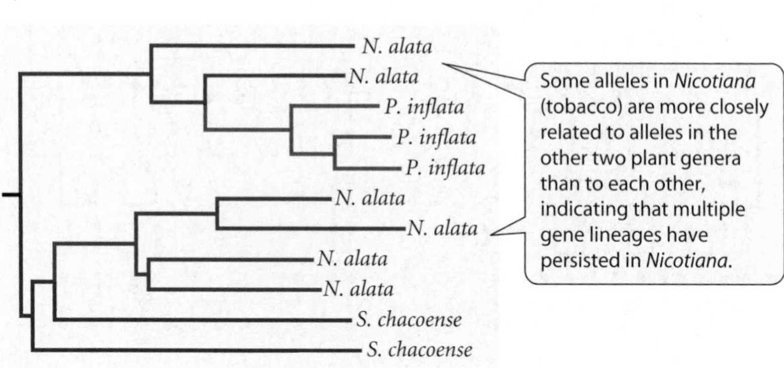

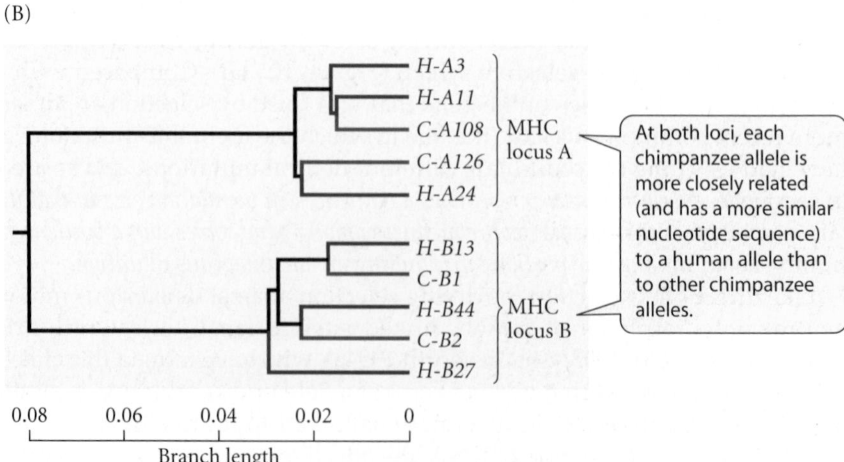

Figure 12.27 Because of long-lasting balancing selection, a polymorphism may be inherited by two or more species from their common ancestor, and certain haplotypes in each species may be most closely related to haplotypes in the other species. These trees show relationships among gene sequences (haplotypes) from two or more species; each species is indicated by a different color. (A) Alleles at the self-incompatibility locus in the potato family, Solanaceae. The common ancestor of the six alleles sequenced in the tobacco *Nicotiana alata* is older than the common ancestor of the three genera *Nicotiana*, *Petunia*, and *Solanum*. (B) The phylogenetic relationships among six alleles in humans and four alleles in chimpanzees at the major histocompatibility (MHC) loci A and B. Both species have loci A and B, which form monophyletic clusters, indicating that the two loci arose by gene duplication before speciation gave rise to the human and chimpanzee lineages. At both loci, each chimpanzee allele is more closely related (and has more similar nucleotide sequence) to a human allele than to other chimpanzee alleles. Thus polymorphism at each locus in the common ancestor has been carried over into both descendant species. These polymorphisms are at least 5 Mya. (A after Ioerger et al. 1990; B after Nei and Hughes 1991.)

of the major histocompatibility (MHC) genes are older than the divergence between humans and chimpanzees (Figure 12.27B). The proteins encoded by these genes bind foreign peptides (antigens), a key step in the immune response. Because variant MHC proteins may differ in specificity for different antigens, heterozygotes may have a broader spectrum of responses and may therefore have higher fitness than homozygotes (Nei and Hughes 1991).

Signatures of selection

Selection, as we have seen, can leave several imprints, or SIGNATURES, in genomes. These are the basis for many tests for selection, as departures from the patterns expected under the neutral theory. These tests apply to selection on different evolutionary time scales. They have been used with data from many organisms, especially *Homo sapiens* (Nielsen 2005; Sabeti et al. 2006; Nielsen et al. 2007). The stastistical methods used in practice are more complicated than the following descriptions, which convey only the reasoning behind them.

PROPORTION OF FUNCTIONAL CHANGES. Nonsynonymous (amino acid-changing) mutations in protein-encoding genes are thought to be more likely to reduce than to enhance function, and so to be deleterious on average (see Chapter 10). For a homologous gene from different species, the ratio of nonsynonymous differences (d_n) to synonymous differences (d_s), each scaled by the number of such differences that are possible between the sequences, is expected to equal 1 if all the nonsynonymous differences are selectively neutral. A ratio $d_n/d_s < 1$ implies that many possible mutations are deleterious and have been purged by purifying selection, whereas $d_n/d_s > 1$ implies a preponderance of fixed advantageous mutations. This is a very conservative test, because it reveals positive selection only if enough advantageous mutations have occurred to stand out against the neutral mutations. It is suitable for selection over long time scales. This test has consistently shown that reproductive proteins, such as gamete surface proteins and seminal fluid proteins, evolve in amino acid sequence much faster than the neutral expectation in a wide variety of animals (Swanson and Vacquier 2002), including humans and other apes (Wyckoff et al. 2000). At the other extreme, more than 480 sequences longer than 200 base pairs (bp) have been found in the human genome that are completely identical to sequences in rat and mouse, and more than 95 percent identical to chicken sequences (Bejerano et al. 2004). These are probably not protein-coding sequences, but they clearly must have critically important functions. They reside near genes that are involved in RNA processing or developmental regulation, and so may play roles in these processes.

A closely related approach is the McDonald-Kreitman test, in which selection is implied by different nonsynonymous/synonymous ratios for intraspecific polymorphisms than for between-species differences (see Chapter 10). This is one of the strongest tests for selection (Nielsen 2005).

REDUCED SEQUENCE DIVERSITY. A selective sweep, by increasing the frequency of one haplotype at the expense of others, reduces π, the average heterozygosity, at linked sites, so the frequencies of linked variants become more uneven. The number, S, of variable sites in a sample of DNA sequences is less affected by selection than π is, since π depends on population frequencies of variant nucleotides but S does not. The difference $D = \pi - S$, known as TAJIMA'S D (Tajima 1989), is thus one of several indices of departure from the frequency distribution of variation expected under the neutral theory. However, these indices also can be affected by population history or population subdivision. Thus a negative value of D ($D < 0$) can be caused by a selective sweep or by a bottleneck in population size (which has similar effects on genetic variation, as noted above). Demographic factors such as bottlenecks will affect all loci, so indices of genetic diversity such as D may provide evidence of selection at loci that stand out from others.

In a study of variation at 132 loci in both European-Americans and African-Americans, Joshua Akey and colleagues (2004) found evidence for selective sweeps, over and above effects of population growth, at 22 loci in European-Americans. The most pronounced effect was at a 115-kilobase-long region that included four genes, among them one named *TRPV6* (Figure 12.28). In this region, especially near three amino acid–changing substitu-

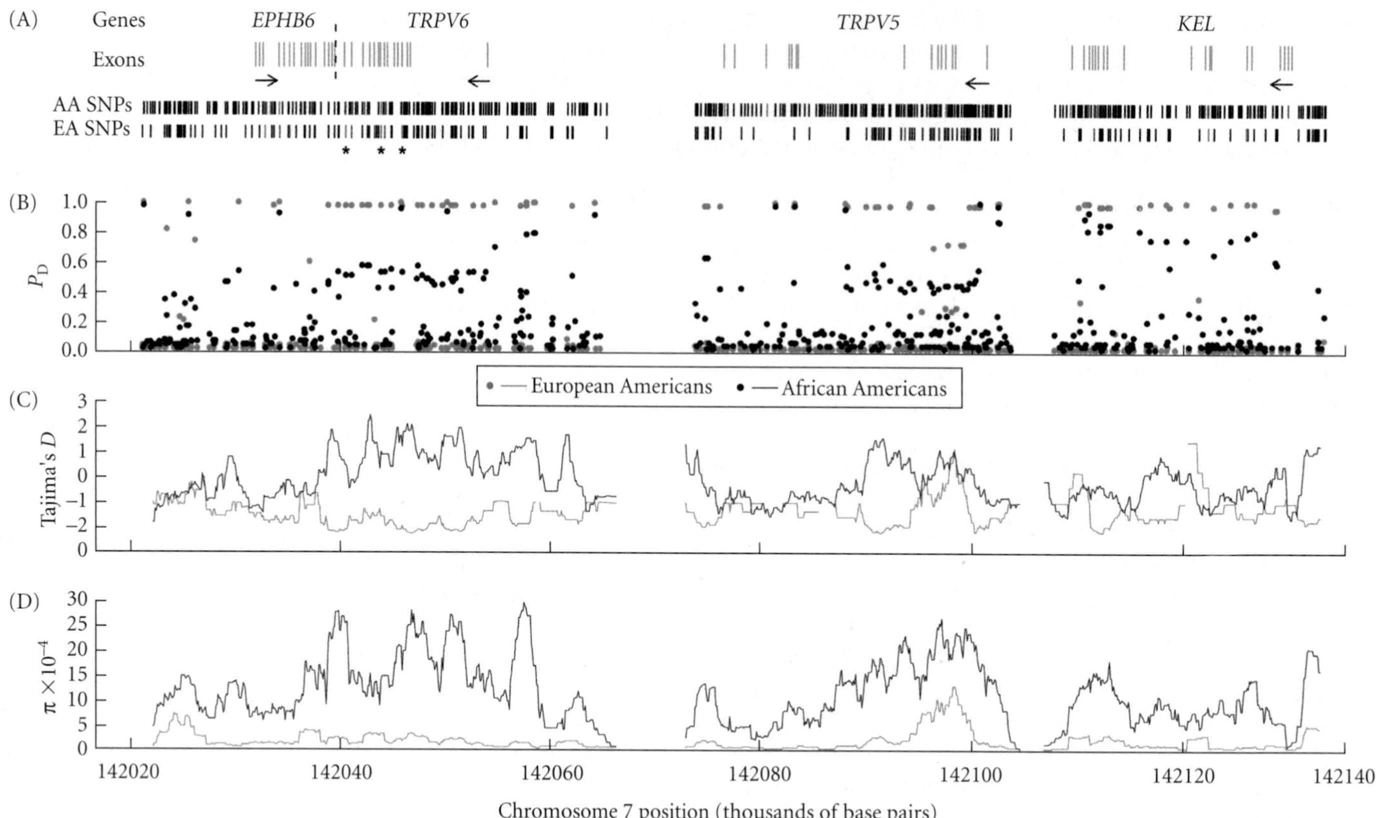

Figure 12.28 Evidence of a recent selective sweep on human chromosome 7. (A) The vertical gray lines represent exons in four genes in a 115-kilobase region. The vertical lines below these show SNPs in European-American (EA) and African-American (AA) populations. Three asterisks show the position of nonsynonymous SNPs in the *TRPV6* gene. (B) The frequencies of derived SNP alleles, p_D, shown by dots, are greatest in the *TRPV6* gene. (C, D) Low values of Tajima's *D* and of nucleotide diversity (π) also provide evidence of directional selection. (From Akey et al. 2004.)

tions in *TRPV6*, Tajima's *D* is strongly negative and nucleotide diversity (π) is reduced. The *TRPV6* gene affects calcium absorption by several organs. The high frequency of the nonsynonymous substitutions in the European population may reflect recent adaptation to a diet that includes milk—as does evolution at the *LCT* gene, which affects digestion of the milk sugar lactose, as described below (and see p. 297 in Chapter 11).

LONG HAPLOTYPES AND LINKAGE DISEQUILIBRIUM. One indication of a selective sweep is an allele (which might be indicated by a single genetic marker) that has high frequency and is strongly associated, by LD, with distant genetic markers, thus forming a long haplotype that has high frequency. Although high frequency is typical of an old allele, long-distance LD in a chromosome region with normal recombination is likely to indicate a young allele, because recombination dissociates increasingly close alleles as time passes. The strength of selection can be inferred from the frequency and length of the longest haplotypes, and the age of a "core" haplotype (a short region of interest) can be gauged by the pattern of decay of LD with alleles at various distances (Sabeti et al. 2002, 2006).

For example, persistence of the ability to digest lactose in milk is based on different mutations of the gene for the enzyme lactase-phloridzin hydrolase (confusingly, the gene is denoted *LCT*) in Eurasia and Africa. Todd Bersaglieri and colleagues (2004) found that in Eurasia, the lactase persistence mutation, denoted *T-13910*, resides in a haplotype that has high frequency (0.77) and extends with little breakdown for about 1000 kilobases (1 megabase, Mb). In contrast, most chromosomes with the nonpersistence allele (*C-13910*)

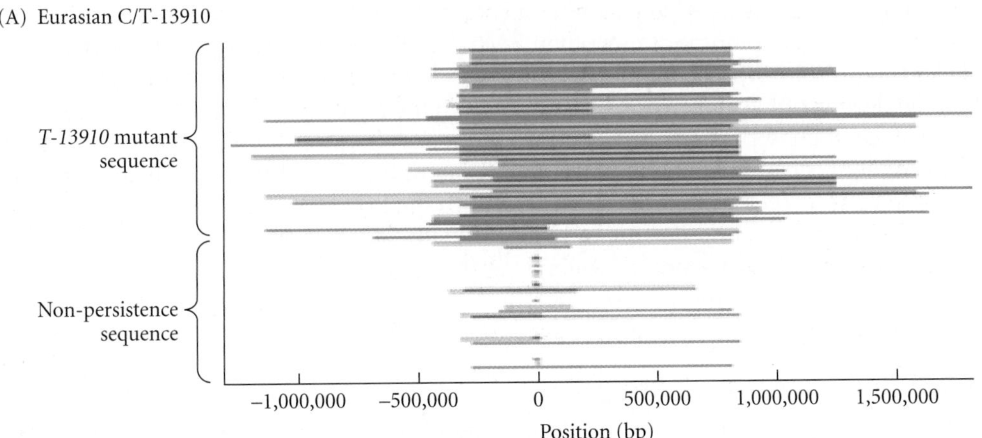

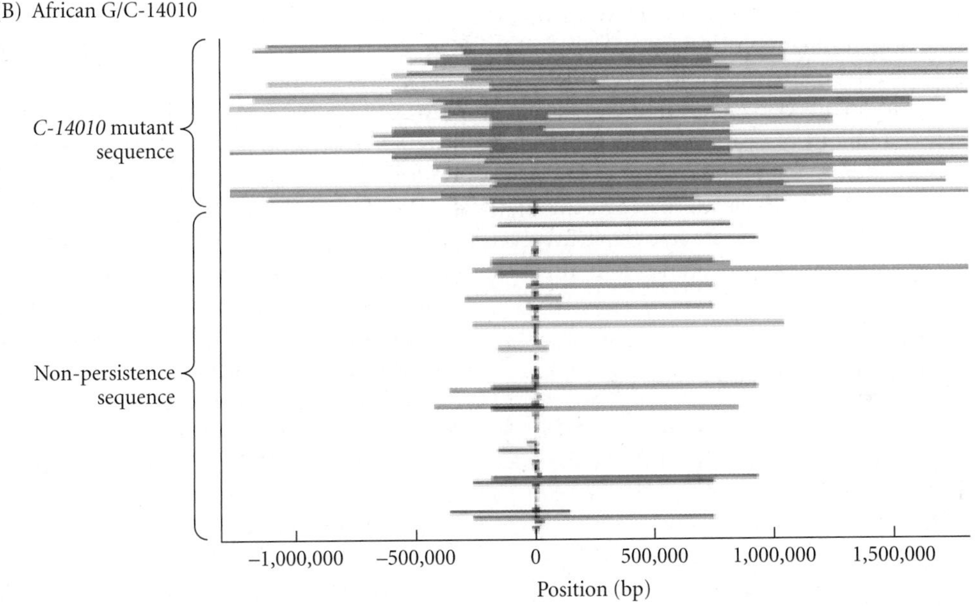

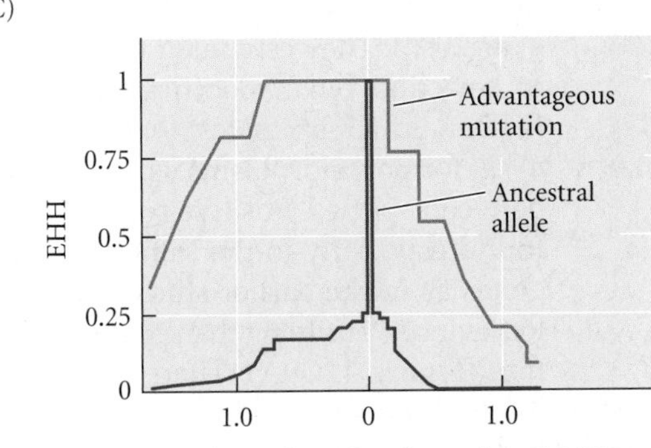

Figure 12.29 Linkage disequilibrium marks selection for lactase persistence in human populations. (A) Haplotypes in European and Asian populations are longer in sequences with the lactase-persistence mutation *T-13910* (green; at position 0) than in nonpersistence sequences (orange), as expected if the lactase-persistence allele has recently increased in a partial selective sweep. (B) The same pattern is seen in a comparison of sequences with the lactase-persistence mutation *C-14010* (red) with nonpersistence sequences (blue) in Kenya and Tanzania. (C) EHH is a measure of the relative frequency with which pairs of randomly chosen sequences in a population are homozygous. This is plotted for sequences of different length, measured by distance of SNP markers from the position of the advantageous lactase-persistence mutation in a Kenyan population. The farther a site is from 0, the lower the EHH, indicating that longer haplotypes are less frequent than shorter ones, because recombination has broken down LD. The drop-off is much more gradual for sequences with the derived, advantageous mutation than for sequences with the ancestral allele (nonpersistence). There is little LD among sites in ancestral sequences. (After Tishkoff et al. 2007; A based on data of Bersaglieri et al. 2004.)

have much shorter distances of LD (Figure 12.29A). The investigators estimated the strength of selection to be $s = 0.014–0.150$ and proposed, on the basis of the decay of LD at increasing distance from the core region, that the selective sweep occurred between 2200 and 20,600 years ago. This is consistent with the origin of dairy farming in northern Europe, about 9000 years ago. In pastoral populations in eastern Africa, Sarah Tishkoff and colleagues (2007) found that lactase persistence is based on two independent mutations. The *C-14010* allele, which has a frequency as high as 0.46 in some tribal populations, is in an extended haplotype of more than 2 Mb, far greater than that of any nonpersistence alleles (Figure 12.29B). The frequency distribution of haplotypes of various lengths shows that the decline of LD at increasing distances from the selected site is much steeper in nonpersistence- than in persistence-associated sequences (Figure 12.29C), as theory predicts. Tishkoff et al. postulate *s* to be between 0.035 and 0.097, and that the allele started to increase 6000 to 7000 years ago, consistent with the origin of pastoralism in Egypt less than 9000 years ago, and its spread to eastern sub-Saharan Africa within the last 4500 years. Because milk is a source of water as well as carbohydrates, lipids, and protein, and because milk consumption causes diarrhea in lactose-intolerant people, the ability to digest milk could have had a large selective advantage.

Adaptive evolution across the genome

It is now possible to apply tests for selection to thousands of human genes, using increasingly large databases on human sequence variation, as well as complete genome sequences for chimpanzee, house mouse, and other species (including other primates) that enable us to distinguish derived from ancestral alleles and to describe large-scale patterns of genome evolution (see Chapter 20). Several genome-level scans for signatures of selection have already yielded fascinating results.

Comparisons between human (H) and chimpanzee (C) genes test for selection on a time scale of millions of years. Andrew Clark and colleagues (2003) compared 8079 homologous H and C genes, using the mouse as outgroup, and found d_n/d_s significantly greater than 1 (indicating positive selection) in 35. Among the classes of genes in which evidence for selection was most prominent were those involved in immune defense, olfaction, and spermatogenesis, as well as some genes that affect the cell cycle and which may be related to cancer. Genes that show their maximal expression (transcription) in testes are prone to positive selection, but (perhaps surprisingly) genes that are expressed in the brain are among the most conservative genes of all. Equal numbers of positively selected genes were found in the human and chimpanzee lineages. The same research group followed with a survey of 11,624 genes sequenced in 39 humans and 1 chimpanzee (Bustamante et al. 2005). They estimated the strength of selection by a modification of the McDonald-Kreitman test that compares d_n/d_s for polymorphisms vs. divergence between species. Positive selection was found especially in the same classes of genes as in the earlier study, but more loci evinced purifying than positive selection (Figure 12.30A). There is a strong

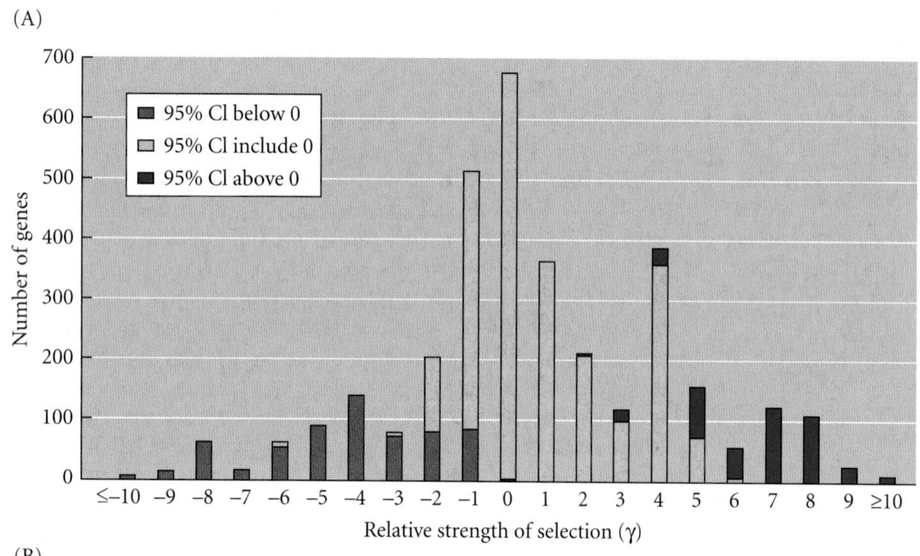

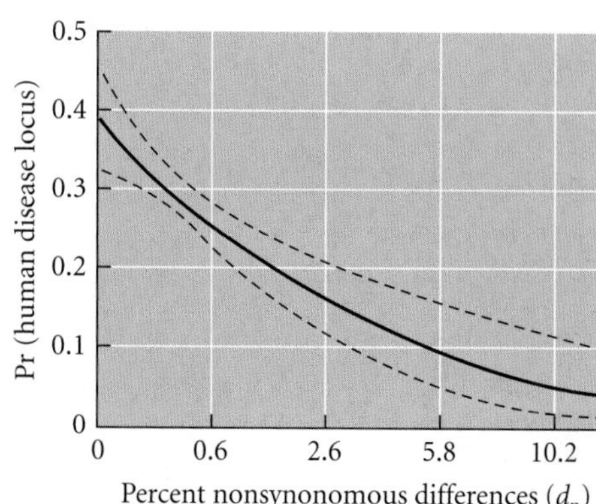

Figure 12.30 Comparison of homologous human and chimpanzee genes indicates positive selection throughout the human genome. (A) The numbers of genes showing negative and positive selection of different strengths γ, based on differences in d_n/d_s between human polymorphisms versus fixed differences. Red and blue regions of the histogram represent selection significantly different from zero (neutrality). (B) A low d_n (fixed nonsynonymous differences between human and chimpanzee) indicates that most mutations in a gene are deleterious and have been eliminated by natural selection. Such genes were predicted to cause human disease if mutated, and indeed have a greater probability of having been recorded as a "disease locus." (After Bustamante et al. 2005.)

correlation between the strength of purifying selection on a gene, as judged from the H-C comparison, and the likelihood that the gene is known to display disease-causing mutations in humans (Figure 12.30B).

As we have seen, a powerful test for recent selection is the detection of long haplotypes, defined by linkage disequilibrium, that have intermediate frequency. Benjamin Voight and colleagues (2006) detected many such instances in a scan for more than 800,000 genome-wide SNPs in samples from sub-Saharan Africa (the Yoruba of Nigeria), East Asia, and Europe. Evidence of partial selective sweeps was strongest in most of the same gene categories as in other studies, as well as some others, such as genes involved in metabolism of carbohydrates and lipids. Although some selective sweeps are shared among two or three of the sample populations, most are restricted to one (Figure 12.31A). For example, four genes involved in skin pigmentation have undergone positive selection in the European population. On the whole, selective sweeps have been more recent in non-African populations than in the Yoruba. This approach has been extended by John Hawks and collaborators (2007), who used a 3.9-million-SNP dataset for the same populations, and used linkage disequilibrium decay to distinguish haplotypes that signify positive selection, rather than expansion from a population bottleneck. They detected more than 3000 selection events in the Yoruba sample, and more than 2000 in one European and two Asian samples. The decay of LD with distance between SNPs indicates that the age distribution of selective sweeps peaks at about 5000 (in Europe) to 8000 (in Africa) years ago (Figure 12.31B), and that *the rate at which genes have undergone evolution by natural selection has greatly increased in the last 40,000 years*. Hawks et al. attribute this acceleration to the great increase of human population size, which has greatly increased the supply of new mutations, and perhaps to the massive changes that humans have created in their own ecological and cultural environment.

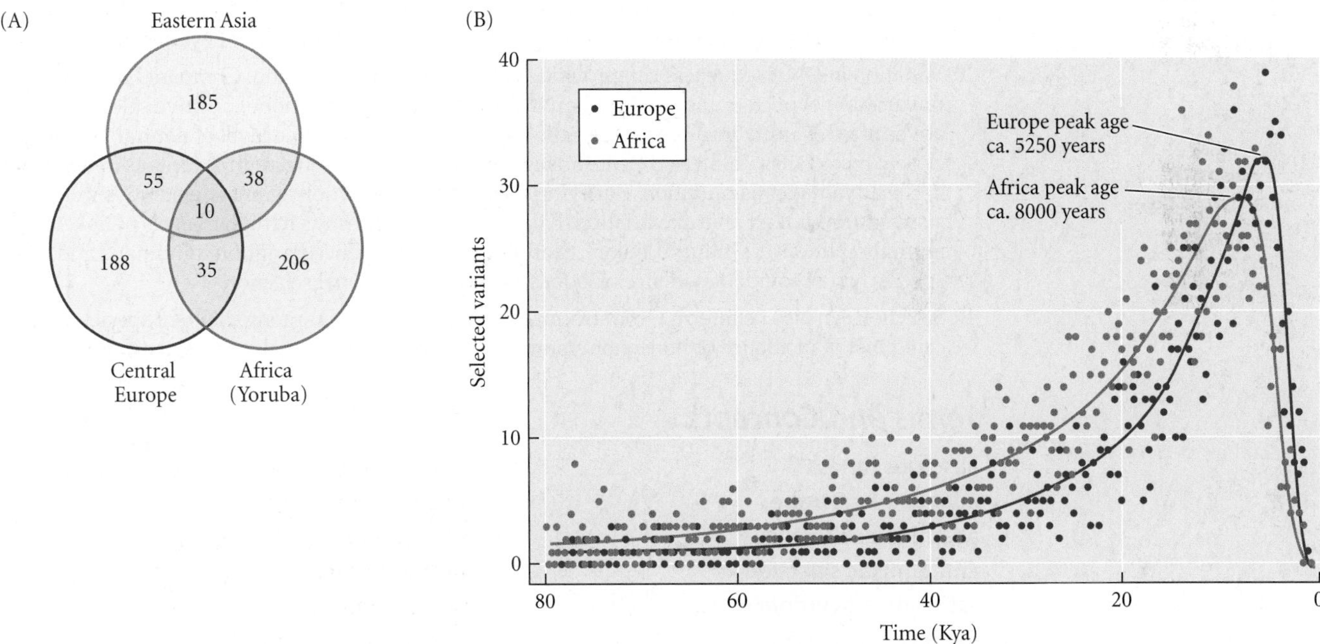

Figure 12.31 Recent selective sweeps in the human genome, based on linkage disequilibrium among SNPs. (A) Numbers of genomic regions showing long haplotypes and strong selection in African, European, and East Asian populations. The numbers in the overlapping sectors are shared among populations; these are fewer than expected by chance, showing that most sweeps have been specific to a region. (B) The length of haplotypes showing linkage disequilibrium was used to estimate the age of selective sweeps. The lines are fitted curves calculated on the assumption of constant population size and an average positive selection coefficient of 0.022 in Africa and 0.034 in Europe. The curves indicate an acceleration in the number of sweeps, peaking at the times indicated. (A after Voight et al. 2006; B after Hawks et al. 2007.)

Summary

1. Even at a single locus, the diverse genetic effects of natural selection cannot be summarized by the slogan "survival of the fittest." Selection may indeed fix the fittest genotype, or it may maintain a population in a state of stable polymorphism, in which inferior genotypes may persist.
2. The absolute fitness of a genotype is measured by its rate of increase, the major components of which are survival, female and male mating success, and fecundity. In sexual species, differences among gametic (haploid) genotypes may also contribute to selection among alleles.
3. Rates of change in the frequencies of alleles and genotypes are determined by differences in their relative fitness, and are also affected by genotype frequencies and the degree of dominance at a locus.
4. Much of adaptive evolution by natural selection consists of replacement of previously prevalent genotypes by a superior homozygote (directional selection). However, genetic variation at a locus often persists in a stable equilibrium condition, owing to a balance between selection and recurrent mutation, between selection and gene flow, or because of any of several forms of balancing selection.
5. The kinds of balancing selection that maintain polymorphism include heterozygote advantage, inverse frequency-dependent selection, and variable selection arising from variation in the environment.
6. Often the final equilibrium state to which selection brings a population depends on its initial genetic constitution: there may be multiple possible outcomes, even under the same environmental conditions. This is especially likely if the genotypes' fitnesses depend on their frequencies, or if two homozygotes both have higher fitness than the heterozygote.
7. When genotypes differ in fitness, selection determines the outcome of evolution if the population is large; in a sufficiently small population, however, genetic drift is more powerful than selection. When the heterozygote is less fit than either homozygote, genetic drift is necessary to initiate a shift from one homozygous equilibrium state to the other.
8. Studies of variable loci in natural populations show that the strength of natural selection varies greatly, but that selection is often strong, and is thus a powerful force of evolution.
9. Variation in DNA sequences can provide evidence of natural selection. Compared with the level of variation expected under neutral mutation and genetic drift alone, positive selection (of an advantageous mutation) causes "selective sweeps" that reduce the level of neutral variation at closely linked sites. It can also create linkage disequilibrium among neutral variants in the region of the advantageous mutation. Purifying (background) selection against deleterious mutations also reduced linked neutral variation. Balancing selection results in higher levels of linked variation than under the neutral theory. Studies of DNA sequence variation in humans and other species have provided evidence of extensive recent directional selection.
10. Selection can also be inferred from DNA sequence comparisons among different species, as indicated by the incidence of nonsynonymous versus synonymous nucleotide substitutions.

Terms and Concepts

absolute fitness
adaptive landscape
adaptive peak/valley
antagonistic selection
associative overdominance
background selection
balancing selection
coefficient of selection
components of fitness
cost of adaptation (= trade-off)
directional selection
diversifying (= disruptive) selection
frequency-dependent selection (inverse or positive)
heterozygote advantage/disadvantage
mean (average) fitness
multiple-niche polymorphism
multiple stable equilibria
overdominance
peak shift
purifying selection
relative fitness
reproductive success
selective advantage
selective sweep
stabilizing selection
underdominance

Suggestions for Further Reading

Natural Selection in the Wild by J. A. Endler (Princeton University Press, Princeton, NJ, 1986) analyzes methods of detecting and measuring natural selection and reviews studies of selection in natural populations. Textbooks of population genetics, such as D. L. Hartl and A. G. Clark's *Principles of Population Genetics* (third edition, Sinauer Associates, Sunderland, MA, 1997) and P. W. Hedrick's *Genetics of Populations* (Jones and Bartlett, Sudbury, MA, 2000), present the mathematical theory of selection in depth. R. Nielsen, "Molecular signatures of natural selection" (*Annual Review of Genetics* 39:197–218, 2005) reviewed methods of studying natural selection at the DNA sequence level, and applications of these approaches to human genetic variation are reviewed by P. C. Sabeti et al., "Positive natural selection in the human lineage" (*Science* 312:1614–1620, 2006).

Problems and Discussion Topics

1. If a recessive lethal allele has a frequency of 0.050 in newly formed zygotes in one generation, and the locus is in Hardy-Weinberg equilibrium, what will be the allele frequency and the genotype frequencies at this locus at the beginning of the next generation? (Answer: $q = 0.048$; $p^2 = 0.9071$, $2pq = 0.0907$, $q^2 = 0.0023$.) Calculate these values for the succeeding generation. If the lethal allele arises by mutation at a rate of 10^{-6} per gamete, what will be its frequency at equilibrium?
2. Suppose the egg-to-adult survival of A_1A_1 is 80 percent as great as that of A_1A_2, and the survival of A_2A_2 is 95 percent as great. What is the frequency (p) of A_1 at equilibrium? What are the genotype frequencies among zygotes at equilibrium? Now suppose the population has reached this equilibrium, but that the environment then changes so that the relative survival rates of A_1A_1, A_1A_2, and A_2A_2 become 1.0, 0.95, and 0.90. What will the frequency of A_1 be after one generation in the new environment? (Answer: 0.208.)
3. If the egg-to-adult survival rates of genotypes A_1A_1, A_1A_2, and A_2A_2 are 90, 85, and 75 percent, respectively, and their fecundity values are 50, 55, and 70 eggs per female, what are the approximate absolute fitnesses (R) and relative fitnesses of these genotypes? What are the allele frequencies at equilibrium? Suppose the species has two generations per year, that the genotypes do not differ in survival, and that the fecundity values are 50, 55, and 70 in the spring generation and 70, 65, and 55 in the fall generation. Will polymorphism persist, or will one allele become fixed? What if the fecundity values are 55, 65, 75 in the spring and 75, 65, 55 in the fall?
4. In pines, mussels, and other organisms, investigators have often found that components of fitness such as growth rate and survival are positively correlated with the number of allozyme loci at which individuals are heterozygous rather than homozygous (Mitton and Grant 1984; Zouros 1987). The interpretation of such data has been controversial (see references in Avise 2004). Provide two hypotheses to explain the data, and discuss how the hypotheses might be distinguished.
5. Considering the principles of mutation, genetic drift, and natural selection discussed in this and previous chapters, do you expect *adaptive* evolution to occur more rapidly in small or in large populations? Why? Answer the same question with respect to nonadaptive (*neutral*) evolution.
6. Discuss whether or not natural selection would be expected to (a) increase the abundance (population size) of populations or species; (b) increase the rate at which new species evolve from ancestral species, thus increasing the number of species.
7. Both creationists (e.g., Wells 2000) and a science writer (Hooper 2002) have charged that H. B. D. Kettlewell's famous evidence that predatory birds exert natural selection on the color forms of the peppered moth was deeply flawed and possibly deceitful. Grant (2002), Cook (2003), and others have rebutted this claim, and provide an entrée into the extensive research literature on this topic. Discuss whether or not this charge, if true, would weaken the case for evolution by natural selection. Read some of the relevant literature and write an essay on whether or not there is evidence for Kettlewell's claim.
8. What would be the difference in the rapidity of a population's adaptation to an environmental change if an advantageous allele were already present at low frequency (say, 10^{-4} to 10^{-3}), compared with adaptation based on a newly arisen mutation at that locus? Would both of these events be accompanied by a selective sweep? Would they both be detectable by studying DNA sequence variation?

9. Suppose that in an asexually reproducing population of bacteria, a single advantageous mutation occurs that enhances the ability to metabolize a carbohydrate and obtain energy, and that at about the same time, another single advantageous mutation at another locus reduces thermal stress. Both result in an increased rate of cell division (asexual reproduction). Will evolution at either locus be affected by the mutation at the other locus? (Hint: these two mutations occur in different members of the population, since the chance that both occur in the same cell are low.) Now suppose the same two mutations occur in a sexually reproducing species, at different genes in a chromosome region that has a low recombination rate. How will these mutations affect each other's increase, if at all?
10. Describe the conditions under which evolution does not occur even though natural selection on a genetically variable character is occurring.

CHAPTER 13

Phenotypic Evolution

When we seek to understand how a morphological, physiological, or other phenotypic feature of a species evolves by natural selection, the one-locus models described in the previous chapter provide an indispensable framework. However, there are many characteristics they do not adequately describe, and many questions they do not adequately address. The variation in most features is based on allelic variation at several or many loci, as well as direct effects of environment. Many characteristics are complex, their function depending on the coaction of several component features. Some characteristics are correlated with each other. From these considerations, a host of questions arises: How does evolution of a feature proceed if it is based on many genes? Do the genes that contribute to phenotypic evolution have small or large effects? What determines the speed of evolution of a trait? Are there limits to how much it can change? How do correlated characters affect each others' evolution? Does nongenetic variation play any evolutionary role? Can nongenetic variation be modified by genetic change? Considering that the vast majority of species that have ever lived are extinct, what limits the ability of species to adapt to changing environments?

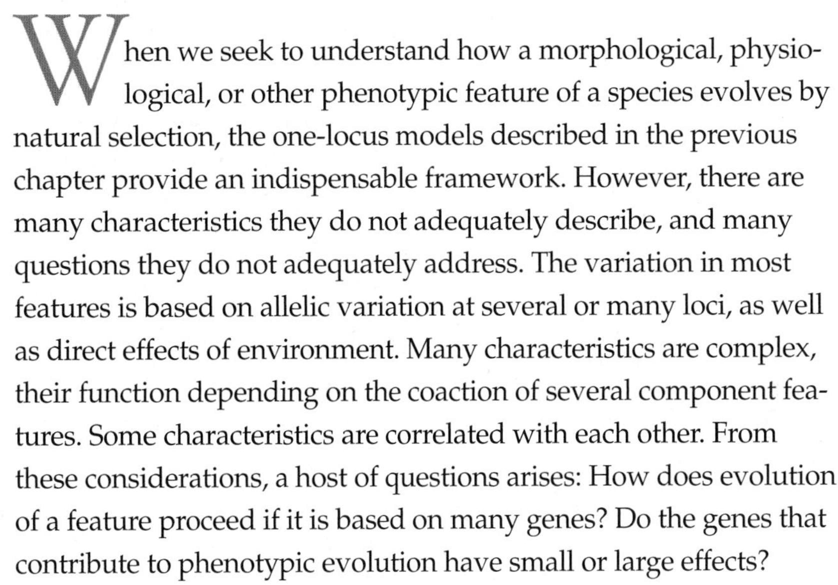

Some results of artificial selection. Domesticated breeds of pigeons all were developed from the wild rock pigeon (*Columba livia*; top) by selective breeding. Each breed has features quite unlike its wild ancestor, such as the greatly inflated esophagus, erect stance, elongated body and legs, and feathered toes of the English pouter pigeon, and the fantail with its 32 (or more) tail feathers. Charles Darwin bred pigeons himself, and studied information from pigeons and other domesticated species in developing his argument for evolution. (Rock pigeon © Luis César Tejo/Shutterstock; pouter © Duncan Usher/Alamy; fantail © Arco Images GmbH/Alamy.)

Genetic Architecture of Phenotypic Traits

The term "genetic architecture" refers to the genetic basis of a trait and its relationship to other traits (Hansen 2006). The components of genetic architecture are the *number of loci* that contribute to its development, its variation within species, and sometimes the differences between species or populations; the *magnitude of the effect* on the trait of different alleles at each locus (*allelic effects*); the pattern of *additive effects* and *dominance* at the various loci, and the phenotypic effect of *synergistic interactions* among loci (**epistasis**); and the *pleiotropic effects* of the loci—the extent to which individual genes affect more than one characteristic. The variation in some traits is simple, based on just one allelic difference, but more often, as we have noted, variation is continuous (quantitative), because of effects of several genes and the environment. The field of **quantitative genetics** was developed to analyze quantitative characters, and its methods are used by biologists who study the evolution of morphology, life history characteristics, behavior, and other phenotypic traits.

The number of loci that contribute to variation in a character may be less than the number that actually contribute to its development. However, only variable loci can be detected—and detecting those that have small phenotypic effects is not easy. The principal method of detection relies on the association of phenotypic differences with genetic markers. Suppose, as a simple hypothetical example, that we have two inbred strains of a species of fly that has two pairs of chromosomes, that the strains differ in wing length, and that they also carry different alleles at one marker locus on each chromosome (Figure 13.1). Strain 1 is homozygous for alleles X_1 and Y_1 on chromosomes 1 and 2, respectively, while strain 2 is homozygous for alleles X_2 and Y_2. A backcross between the F_1 and strain 2 ($X_1X_2Y_1Y_2 \times X_2X_2Y_2Y_2$) yields progeny with four combinations of markers. Some prog-

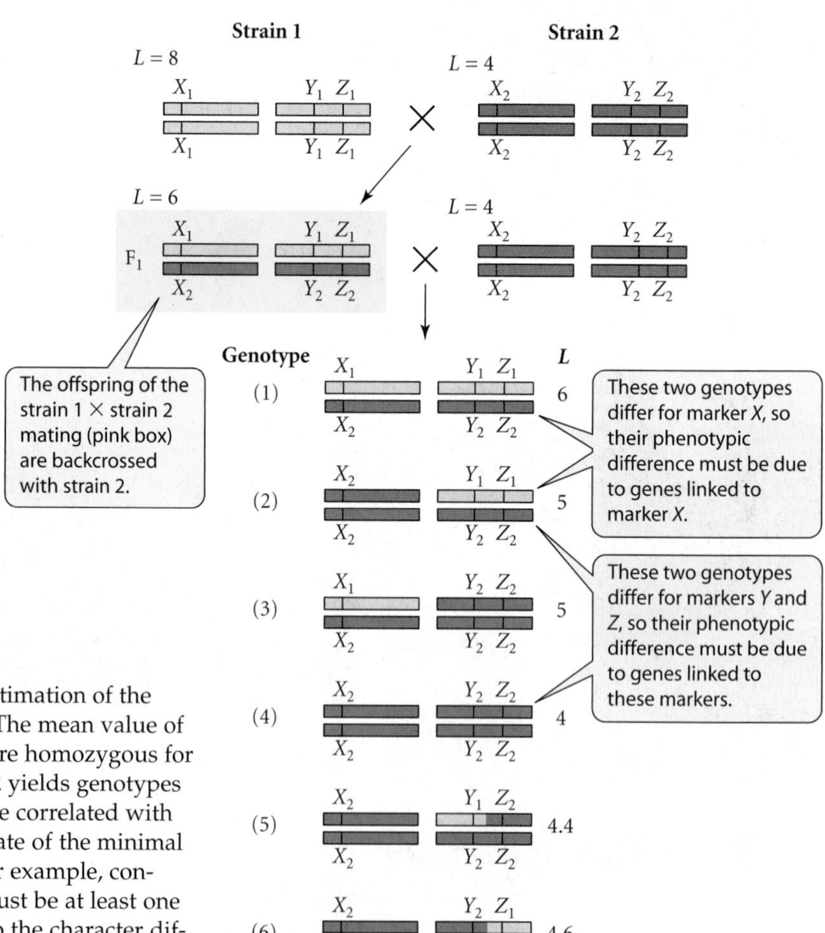

Figure 13.1 Illustration of the principle underlying estimation of the number of loci affecting variation in a quantitative trait. The mean value of the trait (L) is shown for each genotype. Strains 1 and 2 are homozygous for three marker loci (X, Y, Z). Backcrossing the F_1 to strain 2 yields genotypes 1–6 (among others). Differences in L among genotypes are correlated with differences in the markers they carry, providing an estimate of the minimal number of genes contributing to variation in the trait. For example, contrasting genotype 1 with either 2 or 3 shows that there must be at least one locus on each of the two chromosomes that contributes to the character difference between strains 1 and 2.

eny, such as $X_1X_2Y_2Y_2$ and $X_2X_2Y_2Y_2$, differ only with respect to chromosome 1. If they also differ in wing length, then chromosome 1 must carry *at least one* locus, linked to locus *X*, that contributes to variation. (Compare genotypes 3 and 4, or 1 and 2, in Figure 13.1.) Likewise, if genotypes distinguished by the marker *Y*, such as $X_2X_2Y_1Y_2$ and $X_2X_2Y_2Y_2$, differ in wing length, then chromosome 2 must also carry at least one gene that affects this character. (Compare genotypes 2 and 4, or 1 and 3, in Figure 13.1.)

Suppose, further, that two marker loci (*Y* and *Z*) on chromosome 2 also distinguish the inbred strains. Then some progeny differ at locus *Y* only, or at locus *Z* only, as a result of crossing over in the F_1 parent. If wing length differs between flies that differ only at marker *Y*, and also differs between flies that differ only at marker *Z*, then each of the chromosome regions marked by these genes must have at least one gene that affects wing length. Such results would imply that at least three loci affect the character. (Compare genotypes 4 and 5, and also genotypes 4 and 6, in Figure 13.1; note also that these two contrasts indicate that these chromosome regions differ in the magnitude of their effects on the phenotype.)

The more markers there are, distributed along all the organism's chromosomes, the greater the number of chromosome regions that can be compared in this way, and the greater the potential number of trait-affecting loci that may be detected. (Of course, some markers may not be associated with any difference in the trait studied, suggesting that no closely linked genes affect the character, or at least none with effects large enough to detect.) Traditionally, geneticists used morphological mutations as markers for studies of this kind, but they now increasingly use molecular markers (such as transposable elements or single-nucleotide polymorphisms, known as SNPs), because many more such markers can be found in well-studied organisms. This method is called **QTL mapping.** ("QTL" stands for "quantitative trait locus" or "loci.") A QTL is not necessarily a single gene, but rather a chromosome interval that may include many genes, of which one or more affect the character under study. In order to detect and locate many potential loci by the use of many molecular markers, very large numbers of progeny must be compared, and the data are subjected to special statistical analyses.

Trudy Mackay and her colleagues have performed especially intensive studies of the genetics of a representative quantitative trait, bristle number in *Drosophila melanogaster* (e.g., Nuzhdin et al. 1999; Dilda and Mackay 2002). The bristles they studied, on the abdomen and the sternopleuron (part of the side of the thorax), have a sensory function and are part of the peripheral nervous system. The investigators selected for high and low bristle numbers in laboratory populations founded by large numbers of wild flies (Long et al. 1995). They used transposable elements on the X chromosome and chromosome 3 as markers to analyze the genetic differences between the high- and low-selected strains. They detected 53 QTL, of which 33 affected sternopleural bristle number, 31 affected abdominal bristle number, and 11 affected both characters (Figure 13.2A). Some QTL alleles had rather large effects, adding or subtracting several bristles, but many had smaller effects (Figure 13.2B,C). Many QTL displayed both additive effects on the phenotype (i.e., alleles were neither dominant nor recessive; see below). Some had strong EPISTATIC INTERACTIONS, meaning that the joint effect of some pairs of loci differed from the sum of their individual effects. The phenotypic effects of many of the QTL differed strongly between the sexes, and the effects of some QTL depended on the temperature at which the flies were raised (that is, these QTL displayed genotype × environment interactions, as described in Chapter 9; see Figure 9.5). The 53 QTL detected are certainly only a moderate fraction of the loci that can affect these features, since the two chromosomes scanned represent only 60 percent of the genome. Moreover, when mutations were induced by inserting transposable elements at random throughout the genome, 262 mutated loci were estimated to have affected bristle number (Mackay and Lyman 2005). Evidently, then, not all the loci that affect development of bristles varied in the population used for QTL mapping.

Most of the QTL mapped to chromosome sites where CANDIDATE LOCI were located: genes known from past studies to affect bristle morphology or development of the peripheral nervous system. DNA sequence variation at some of these loci has been shown to cor-

(A) QTL for sternopleural bristles: Chromosome 3

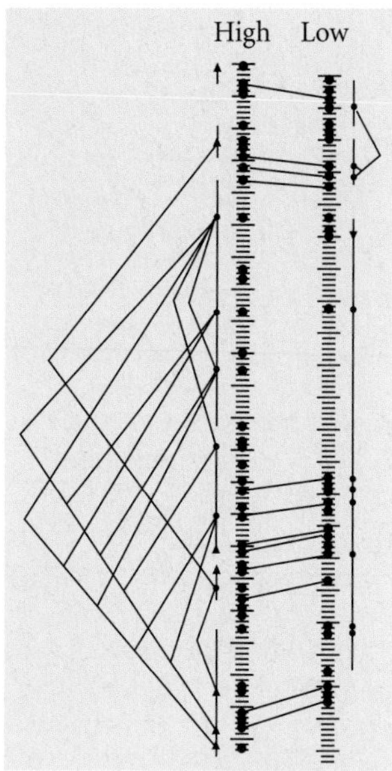

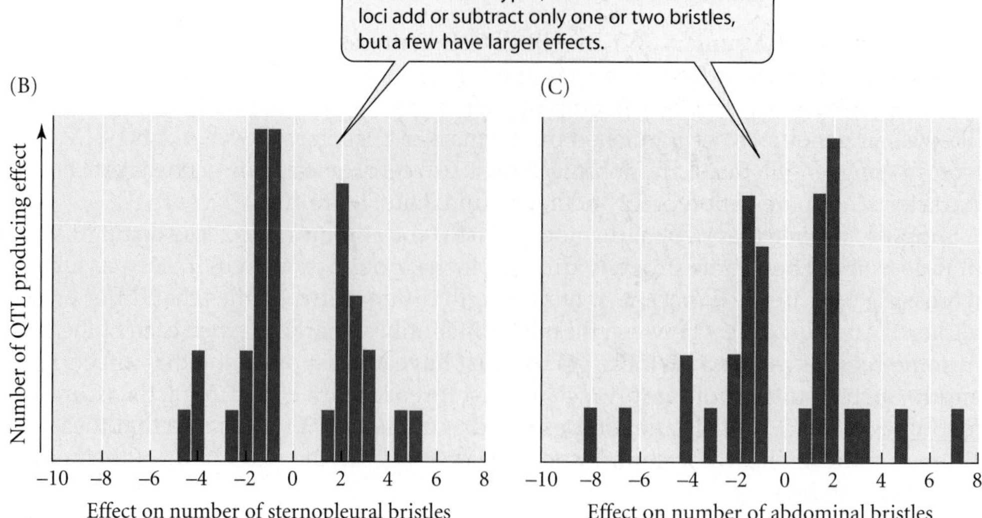

Figure 13.2 (A) The locations of QTL on chromosome 3 that contribute to differences in the number of sternopleural bristles between populations of *Drosophila melanogaster* selected for more ("High") or fewer ("Low") bristles. Lines between chromosomes join markers found in both the "High" and "Low" populations. Angled lines join QTL that interact to affect bristle number. (B) Frequency distribution of allelic effects on sternopleural bristles. (C) The same, for abdominal bristles. (A from Dilda and Mackay 2002; B, C after Mackay 2001b.)

relate with bristle variation, and alleles of these loci distinguished by DNA sequence segregate at high frequencies in natural populations of *Drosophila* (Lai et al. 1994). The high frequency of these alleles suggests that they are either selectively neutral or are maintained by balancing selection (Chapter 12). However, many of the QTL that affect bristle number have been shown to reduce fitness (Mackay et al. 2001a).

These and other such studies have important implications for the study of quantitative traits (Mackay 2001a; Mackay and Lyman 2005). Models of quantitative traits generally assume that variation in a trait is due to many loci, each with alleles that have small effects. However, some of the variation may in fact be due to loci with large effects. The statistical models have often assumed that variation stems from the sum of additively acting alleles at functionally interchangeable loci. Variation in bristle number, however, is due to loci that are not interchangeable: they have different developmental roles, and they interact with one another in producing the phenotype. However, models that take into account such complex gene interactions—although important for researchers in this field—are beyond the scope of this book.

Components of Phenotypic Variation

In order to discuss the evolution of quantitative traits, we must recall and expand on some concepts introduced in Chapter 9. An important measure of variation is the *variance* (V), defined as the average of the squared deviations of observations from the *arithmetic mean* of a sample. The square root of the variance is the *standard deviation* ($s = \sqrt{V}$), measured in the same units as the observations. If a variable has a normal (bell-shaped) frequency distribution, about 68 percent of the observations lie within one standard deviation on either side of the mean, 96 percent within two standard deviations, and 99.7 percent within three (see Box 9C).

The **phenotypic variance** (V_P) in a phenotypic trait is the sum of the variance that is due to differences among genotypes (the **genetic variance**, V_G) and the variance that is due to direct effects of the environment and developmental noise (the **environmental variance**, V_E). Thus $V_P = V_G + V_E$. Considering for the moment the effect of only one locus on the phenotype, we may take the midpoint between the two homozygotes' means as a point of reference (Figure 13.3). Then the mean phenotype of A_1A_1 individuals deviates from that midpoint by $+a$, and that of A_2A_2 by $-a$. The quantity a, the ADDITIVE EFFECT of an allele, measures how greatly the phenotype is affected by the genotype at this locus. If the inheritance of the phenotype is entirely additive, the heterozygote's phenotype is exactly in between that of the two homozygotes. That is, neither allele is dominant or recessive. Such additive effects are responsible for a critically important component of genetic variance, the **additive genetic variance**, denoted V_A. (The genetic variance V_G may also include nonadditive components, resulting from dominance and epistatic gene interactions, that we will not concern ourselves with here.)

The additive genetic variance depends both on the magnitude of the additive effects of alleles on the phenotype and on the genotype frequencies. If one genotype is by far the most common, most individuals are close to the average phenotype, so the variance is lower than if the several genotypes have more equitable frequencies, as they will if the allele frequencies are more nearly equal. When, as in Figure 13.3, two alleles (with frequencies p and q) have purely additive effects, V_A at a single locus is

$$V_A = 2pqa^2$$

When several loci contribute additively to the phenotype, the average phenotype of any particular genotype is the sum of the phenotypic values of each of the loci. Likewise, V_A is the sum of the additive genetic variance contributed by each of the loci: $V_A = 2 \Sigma p_i q_i a_i^2$. The values of p, q, and a may differ for each locus, i. The expression given here for V_A would be modified if there were dominance at any of the loci.

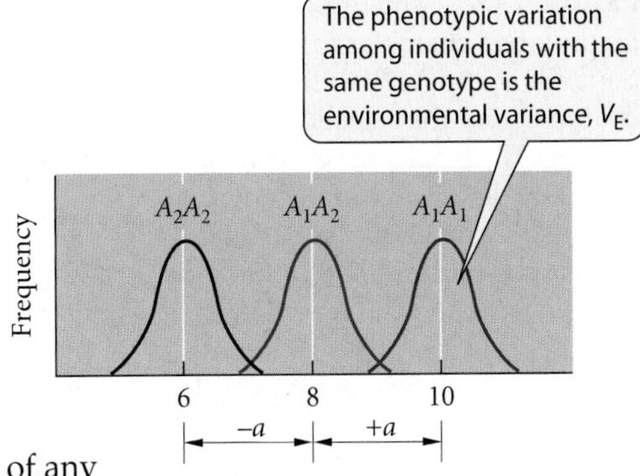

Figure 13.3 Additive effects of two alleles at one locus on the value of a character. When inheritance is purely additive, the heterozygote lies at the midpoint between the homozygotes. The additive effect of substituting an A_1 or an A_2 allele in the genotype is a (in this case, $a = 2$). The magnitude of a affects that of the additive genetic variance, V_A. The phenotypic variation within each genotype is measured by the environmental variance, V_E.

The additive genetic variance plays a key role in evolutionary theory because *the additive effects of alleles are responsible for the degree of similarity between parents and offspring* and therefore *are the basis for response to selection within populations*. When alleles have additive effects, the expected average phenotype of a brood of offspring equals the average of their parents' phenotypes.* Evolution by natural selection requires that selection among phenotypically different parents be reflected in the mean phenotype of the next generation. Therefore, V_A enables a **response to selection**—a change in the mean character state of one generation as a result of selection in the previous generation.

The proportion of phenotypic variance that is due to additive genetic differences among individuals is referred to as the character's **heritability in the narrow sense**, h^2_N. The heritability is determined by the additive genetic variance (V_A), which depends on allele frequencies; and by the environmental variance (V_E), which depends in part on how variable the environmental factors are that affect the development or expression of the character. That is,

$$h^2_N = \frac{V_A}{V_P}$$

where $V_P = V_G + V_E$ and V_G is the sum of V_A and nonadditive genetic components. Because allele frequencies and environmental conditions may vary among populations, an estimate of heritability is strictly valid only for the population in which it was measured, and only in the environment in which it was measured. Moreover, it is wrong to think that if a character has a heritability of 0.75, the feature is ¾ "genetic" and ¼ "environmental," as if a character is formed by mixing genes and environment the way one would mix paints to achieve a desired color. It is the *variation* in the character that is statistically partitioned, and even this partitioning is a property of the particular population, not a fixed property of the feature. For evolutionary studies, moreover, the additive genetic variance is often more informative than the heritability (Houle 1992).

The additive genetic variance is estimated from resemblances among relatives. In order to achieve accurate estimates, it is important that relatives not develop in more similar environments than nonrelatives; otherwise, it may not be possible to distinguish similarity that is due to shared genes from similarity that is due to shared environments.

The narrow-sense heritability (h^2_N, or simply h^2) of a trait equals the slope (b) of the *regression* of offspring phenotypes (y) on the average of the two parents of each brood of offspring (x) (Figure 13.4; see also Figure 9.23). The slope of this relationship is calculated so as to minimize the sum of the squared deviations of all the values of y from the line. As the correspondence between offspring and parents is reduced, the slope of the regression declines (compare Figure 13.4A and B). Therefore, anything that reduces the similarity between parents and their offspring (such as environmental effects on phenotype, or dominance) reduces heritability. There are also ways of estimating V_A and other

*The correlation between parent and offspring phenotypes is lower if there is dominance at a locus.

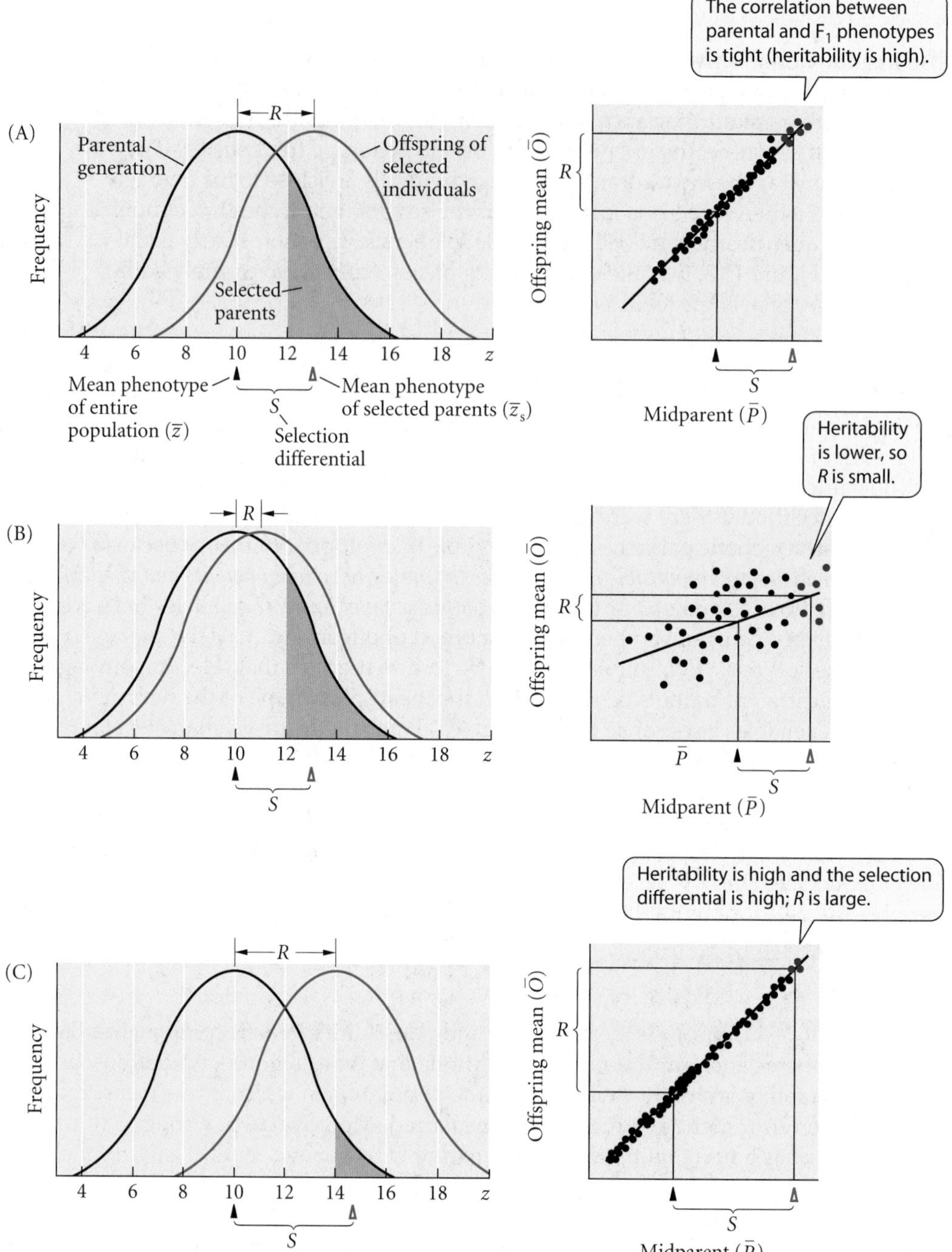

Figure 13.4 The response of a quantitative character to selection depends on the heritability of the character and the selection differential. (A) The black curve shows a normally distributed trait, z, with an initial mean of 10. Truncation selection is imposed, such that individuals with $z > 12$ reproduce. The graph at the right shows a strong correspondence between the mean phenotype of pairs of selected parents ($\bar{P}$) and that of their broods of offspring ($\bar{O}$), i.e., high heritability. Blue circles represent the selected parents and offspring, black circles the rest of the population (were it to have bred). Because heritability is high, the selection differential S results in a large response to selection (R). The blue curve is the distribution of z in the next generation, whose mean lies R units to the right of the mean of the parental generation. (B) The circumstances are the same, but the relationship between phenotypes of parents and offspring is more variable and has a lower slope in the graph at the right, i.e., lower heritability. Consequently, the selection differential S translates into a smaller response (R). The frequency distribution of offspring (blue curve) shifts only slightly to the right. (C) Here heritability is very high, as in (A), but the selected parents have $z > 14$. Thus the selection differential S is larger, resulting in a greater response to selection (R).

components of genetic variance from similarities among other relatives, such as siblings ("full sibs") or half sibs (individuals that have only one parent in common).

The additive genetic variance can be affected by linkage disequilibrium (LD), nonrandom associations among loci (Chapter 9). Suppose that at two loci that additively affect the character, the frequencies of alleles A_1 and A_2 are p_A and q_A, and those of alleles B_1 and B_2 are p_B and q_B, and denote the frequencies of the four possible gamete types (A_1B_1, A_1B_2, A_2B_1, A_2B_2) as g_{11}, g_{12}, g_{21}, and g_{22}, respectively. If the alleles at the two loci A and B are randomly associated, then the frequency of any gamete type will be the simple product of the allele frequencies. (For instance, the expected frequency of A_1B_2 gametes is $g_{12} = p_Aq_B$.) If the loci are in linkage disequilibrium, the gamete frequencies depart from their expected values; they cannot be predicted from the allele frequencies, but must be measured instead. A COEFFICIENT OF LINKAGE DISEQUILIBRIUM may be defined as $D = (g_{11} \times g_{22}) - (g_{12} \times g_{21})$. If $D > 0$, then gametes A_1B_1 and A_2B_2, as well as genotypes formed by their union (such as $A_1A_1B_1B_1$ and $A_2A_2B_2B_2$), will be more frequent than expected. If these are the most extreme phenotypes (e.g., largest and smallest size), then *positive linkage disequilibrium would increase the phenotypic (and genetic) variance*. Conversely, if $D < 0$, these genotypes are less frequent than expected, and the phenotypic variance is correspondingly reduced. Recombination, which tends to bring the frequencies toward linkage equilibrium, would therefore either increase or decrease variation, depending on whether D is negative or positive. If the two loci affect different characters, linkage disequilibrium can create a correlation between the two features, as we shall see below.

Genetic variance in natural populations

Heritable variation has been reported for the great majority of traits in which it has been sought, in diverse species (Lynch and Walsh 1998). Commonly, h^2 is in the range of about 0.4 to 0.6, but not all characters are equally variable. For example, characters strongly correlated with fitness (such as fecundity) tend to have lower heritability than characters that seem unlikely to affect fitness as strongly (Mousseau and Roff 1987). However, the low h^2 of fitness components arises from the greater magnitude of other variance components, especially V_E, in the denominator of the expression $h^2 = V_A/V_P$. The additive genetic variance (V_A) of components of fitness is actually higher than that of morphological and other traits, probably because many physiological and morphological characteristics affect fitness (Houle et al. 1996).

In some cases, though, traits do not appear to be genetically variable at all. The paucity of genetic variation would then be a genetic constraint that could affect the direction of evolution (for example, an insect might adapt to some species of plants rather than to others) or prevent adaptation altogether. For instance, in sites near mines where the soil concentration of copper or zinc is high, a few species of grasses have evolved tolerance for these toxic metals, but most species of plants have not. Populations of various species of grasses growing on normal soils were screened for copper tolerance by sowing large numbers of seeds in copper-impregnated soil. Small numbers of tolerant seedlings were found in those species that had evolved tolerance in other locations, but no tolerant seedlings were found in most of the species that had not formed tolerant populations (Macnair 1981; Bradshaw 1991). Genetic variation that might enable the evolution of copper tolerance seems to be rare or absent in those species. Similarly, genetic variation for resistance to desiccation seems lacking in a rainforest-inhabiting species of *Drosophila* (Hoffmann et al. 2003), and several species of host-plant-specific leaf beetles would not feed on certain plants related to their normal hosts, and revealed no genetic variation in their feeding response (Futuyma et al. 1996).

Genetic Drift or Natural Selection?

Quantitative characters will evolve by genetic drift if alleles at the underlying loci are selectively neutral. Selection might be inferred if data do not fit a model of neutral evolution. Bear in mind that variation in the trait itself might be selectively neutral, yet variation at the underlying loci might nevertheless affect fitness via pleiotropic effects on other traits, so that selection would indirectly affect the trait.

First, consider changes in a trait over a relatively short time, during which new mutations have little effect. Changes in the frequency of alleles at variable loci will cause fluctuations in the mean trait value, which will eventually drift away from its original value. In this "short-term" model, the change in the mean will be faster, the smaller the population, since allele frequency changes by genetic drift transpire more rapidly in smaller populations.

Alternatively (Turelli et al. 1988), consider a very long time, during which variation and evolution of the trait are affected only by new mutations (which increase variation) and genetic drift (which erodes it). The variance that arises per generation by mutation, V_m, is proportional to the number of mutating loci, the mutation rate per locus, and the average phenotypic effect of a mutation. As we have seen, V_m is often about 0.001 times V_E (Lynch 1988). Over the long term, mutation will be balanced by genetic drift, so the genetic variance should, theoretically, reach a stable value ($2N_eV_m$). As mutations that affect the character arise and are fixed by genetic drift, the mean will fluctuate at random. If a number of isolated populations are derived from an initially uniform ancestral population, mutation and genetic drift can cause genetic divergence among them in a polygenic character, just as they do at a single locus (see Chapter 10). In this "long-term" model, the rate of divergence among the population means equals V_m per generation: it depends only on the mutation rate, not on the population size. (Recall that this is also true of the rate at which populations diverge by substitution of neutral mutations at a single locus; see Chapter 10.) If a trait has diverged among populations or species much faster than the expected neutral rate, it is likely to have evolved by directional selection. If it has changed much more slowly than the neutral rate, one may suspect that stabilizing selection has operated—either on the trait itself or on other characteristics that are pleiotropically affected by the same genes.

Sonya Clegg and colleagues (2002) applied these models to morphological features, such as bill and leg length, of a small Australian songbird, *Zosterops lateralis* (the silver-eye), and its close relative *Z. tephropleurus*, which inhabits distant Lord Howe Island. DNA sequence differences between them suggest that they have been separated for hundreds of thousands of years. Clegg et al. also analyzed populations of *Z. lateralis* formed by colonization of New Zealand and its coastal islands within the last 170 years. Applying the short-term model to differences between the recently formed populations and Australian birds, Clegg et al. concluded that the rate of evolution was so high that evolution by genetic drift would have required effective population sizes much smaller than historical records indicate. They argued, then, that directional selection has contributed to evolution. In contrast, Clegg et al. used the long-term model to analyze the differences between the Australian and Lord Howe species, assuming that V_m is about 0.001. They concluded that the rate of change has been slow—perhaps even slower than predicted by the mutation rate. Stabilizing selection may have maintained the similarity between these species. In the same vein, Michael Lynch (1990) studied the rates of evolution of skeletal features of several groups of mammals, based on fossils (e.g., horse and hominid lineages), as well as on differences between living species (such as lion and cheetah) for which there are good estimates of the time since the common ancestor. Lynch also assumed that the mutation rate (V_m) for such features is about 0.001, as has been estimated for skeletal features of mice and other living species of mammals. He found that almost all the features had evolved at much *lower* rates than expected under mutation and genetic drift (Figure 13.5). Only the cranial capacity of *Homo sapiens* has evolved at rates that may be higher than expected from the neutral model. The very low rate at which most characters seem to have evolved suggests that stabilizing selection has maintained them at roughly constant values for long periods. Furthermore, fossils show that many features fluctuate rather rapidly, but show little net change (see Figure 4.22), so that the rate of evolution over the long term is much less than it is over shorter intervals of time.

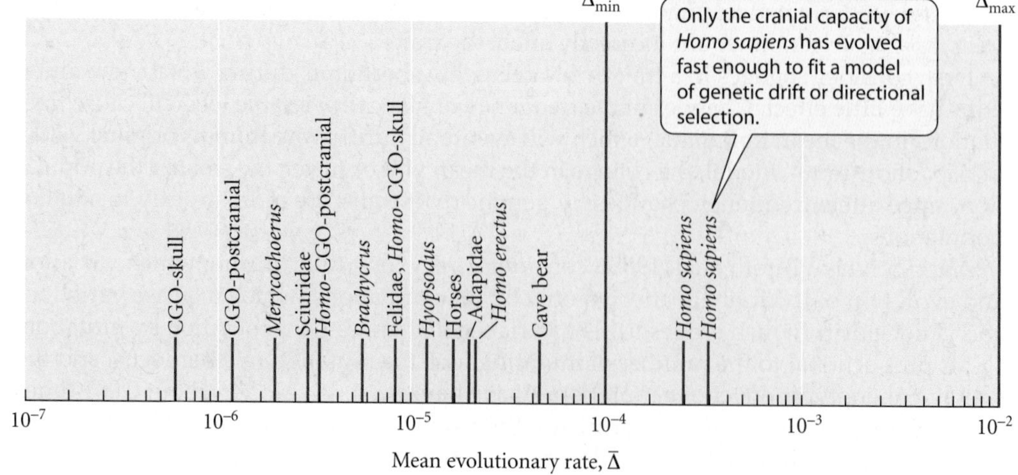

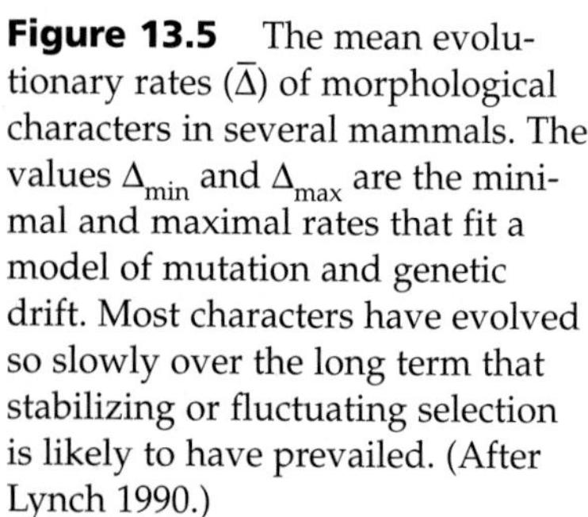
Figure 13.5 The mean evolutionary rates ($\bar{\Delta}$) of morphological characters in several mammals. The values Δ_{min} and Δ_{max} are the minimal and maximal rates that fit a model of mutation and genetic drift. Most characters have evolved so slowly over the long term that stabilizing or fluctuating selection is likely to have prevailed. (After Lynch 1990.)

Natural Selection on Quantitative Traits

Response to directional selection

Suppose an experimenter imposes selection for greater tail length by breeding only those rats in a captive population with tails longer than a certain value. This form of selection is called TRUNCATION SELECTION, and it can be used as an approximation of the way directional natural selection acts. The mean tail length of the selected parents differs from that of the population from which they were taken ($\bar{z}$) by an amount S, the **selection differential** (see Figure 13.4A). The average tail length among the offspring of the selected parents ($\bar{z}'$) differs from that of the parental generation as a whole ($\bar{z}$) by an amount R, the response to selection (Figure 13.4A, right-hand graph). The magnitude of R is proportional to the heritability of the trait (compare Figure 13.4A and B) and to the selection differential S (compare Figure 13.4A and C). In fact, the change in mean phenotype in offspring, R, that is due to selection of parents, S, can be read directly from the regression line as

$$R = h^2_{\mathrm{N}}S$$

Since this equation* can be rearranged as $h^2_{\mathrm{N}} = R/S$, heritability can be estimated from a selection experiment in which S (which is under the experimenter's control) and R are measured. Such an estimate of h^2_{N} is called the **realized heritability**. This is how (as you may recall from Chapter 9) Dobzhansky and Spassky estimated that the heritability of phototaxis in *Drosophila pseudoobscura* was about 0.09 (see Figure 9.24).

As selection proceeds, it increases the frequencies of those alleles that produce phenotypes closer to the optimal value. As those frequencies increase, multilocus genotypes (combinations of alleles at different loci) that had been extremely rare become more common, so phenotypes arise that had been effectively absent before. *Thus the mean of a polygenic character shifts beyond the original range of variation as directional selection proceeds*, even if no further mutations occur.

If alleles at different loci differ in the magnitude of their effects on the phenotype, those with the largest favorable effects are likely to be fixed first (Orr 1998). In the absence of complicating factors, prolonged directional selection should ultimately fix all favored alleles, eliminating genetic variation. Further response to selection would then require new variation, arising from mutation. As we have seen, for many features the mutational variance, V_{m}, is on the order of $10^{-3} \times V_{\mathrm{E}}$ per generation, or about 0.001 that of environmental variance. Thus a fully homozygous population could, by mutation, attain $V_{\mathrm{A}}/(V_{\mathrm{A}} + V_{\mathrm{E}}) = h^2 = 0.5$, for an entirely additively inherited trait, in about 1000 generations. If, however, many of these mutations have harmful pleiotropic effects and have a net selective disadvantage, the "usable" mutational variance may be much lower (Hansen and Houle 2004). Indeed, the equation $R = h^2_{\mathrm{N}}S$ assumes that none of the alleles that contribute genetic variance are deleterious.

Responses to artificial selection

Animal and plant breeders have used artificial selection to alter domesticated species in extraordinary ways (see chapter-opening illustration). Darwin opened *The Origin of Species* with an analysis of such changes, and evolutionary biologists have drawn useful inferences about evolution from artificial selection ever since then. Artificial selection differs from natural selection because the human experimenter focuses on one trait rather than on the organism's overall fitness. Nevertheless, natural selection often operates much like artificial selection.

Responses to artificial selection over just a few generations generally are rather close to those predicted from estimates of heritability based on correlations among relatives, such

---

*The simple equation presented here is the "breeders' equation," used for predicting responses to artificial selection in domesticated species. A conceptually related equation is used more often by evolutionary biologists: $\Delta\bar{z} = V_{\mathrm{A}}/\bar{w} \times d\bar{w}/d\bar{z}$, where z is a character state, V_{A} is the additive genetic variance, and w is fitness. Thus the rate of evolution of the character mean ($\Delta\bar{z}$) is proportional to V_{A} and to the derivative of fitness with respect to the character value (Lande 1976a). Because of its direct role in determining the rate of evolution, V_{A} is better than the heritability, h^2_{N}, as a measure of "evolvability" (Houle 1992).

as parents and offspring. These heritability estimates seldom predict accurately the change in a trait over many generations of artificial selection, however, because of changes in linkage disequilibrium and genetic variance, input of new genetic variation by mutation, and the action of natural selection, which often opposes artificial selection (Hill and Caballero 1992). Such effects were found, for example, in an experiment by B. H. Yoo (1980), who selected for increased numbers of abdominal bristles in lines of *Drosophila melanogaster* taken from a single laboratory population. For 86 generations, Yoo scored bristle numbers on 250 flies, and bred the next generation from the top 50 flies of each sex. In the base population from which the selection lines were drawn, the mean bristle number was 9.35 in females, and more than 99 percent of females had fewer than 14 bristles (i.e., three standard deviations above the mean). After 86 generations, mean bristle numbers in the experimental populations had increased to 35 to 45 (Figure 13.6). This represents an average increase of 316 percent, or 12 to 19 phenotypic standard deviations—far beyond the original range of variation. In a very short time, selection had accomplished an enormous evolutionary change, at a rate far higher than is usually observed in the fossil record.

This progress was not constant, however. Some of the experimental populations showed periods of little change, followed by episodes of rapid increase. Such irregularities in response are partly due to the origin and fixation of new mutations with rather large effects (Mackay et al. 1994). Several populations eventually stopped responding: they reached a **selection plateau**. This cessation of response to selection was not caused by loss of genetic variation, because when Yoo terminated ("relaxed") selection after 86 generations, mean bristle number declined, proving that genetic variation was still present.

A selection plateau and a decline when selection is relaxed are commonly observed in selection experiments. These patterns are caused by natural selection, which opposes artificial selection: genotypes with extreme values of the selected trait have low fitness. The changes in fitness are due to both hitchhiking of deleterious alleles (linkage disequilibrium) and to pleiotropy. Yoo found, for example, that lethal alleles had increased in frequency in the selected populations because they were closely linked to alleles that increased bristle number. Other investigators have shown that some alleles affecting bristle number have pleiotropic effects that reduce viability (Kearsey and Barnes 1970; Mackay et al. 1992).

Several investigators have found that artificially selected traits change faster in large than in small experimental populations (e.g., Weber and Diggins 1990; López and López-Fanjul 1993). This is because more genetic variation is introduced by mutation in large

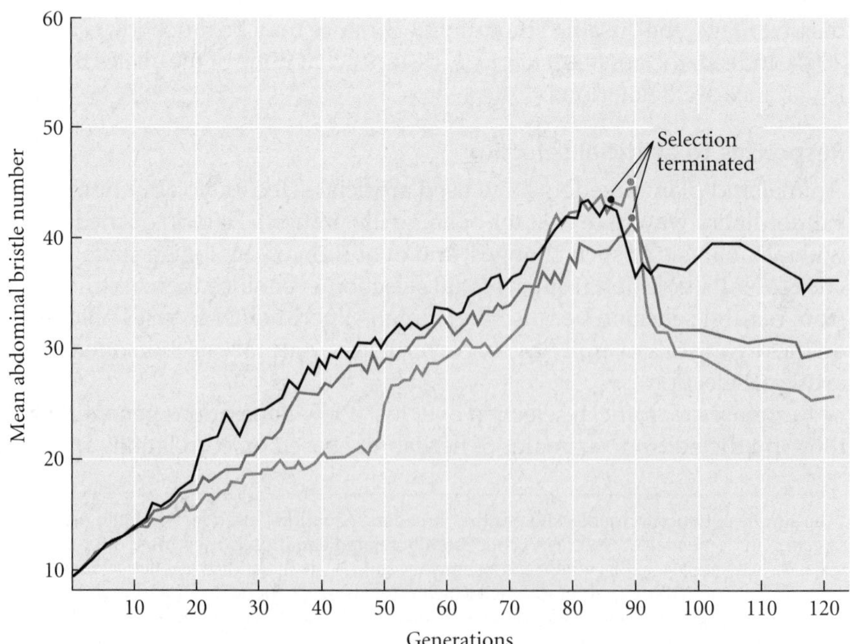

Figure 13.6 Responses to artificial selection for increased numbers of abdominal bristles in three laboratory populations of *Drosophila melanogaster*. After about 86 generations, the means had increased greatly. Selection was terminated at the points indicated by colored circles, and bristle number declined thereafter, indicating that genotypes with fewer bristles had higher fitness. (After Yoo 1980.)

than in small populations, large populations lose variation by genetic drift more slowly, and selection is more efficient in large populations. (Recall from Chapter 12 that whether allele frequency change is affected more by selection or by genetic drift depends on the relationship between the coefficient of selection and the population size.) In some of these experiments, the progress of artificial selection has not been impeded by counteracting natural selection. For example, after more than 100 generations of selection for faster flight, populations of *Drosophila melanogaster* showed no evidence of diminished fitness (Weber 1996), suggesting that the underlying alleles did not have deleterious pleiotropic effects.

Directional selection in natural populations

In studies of natural populations, several measures of the strength of natural selection on quantitative traits have been used. The simplest indices of selection can be used if the mean ($\bar{z}$) and variance (V) of a trait are measured in a single generation before ($\bar{z}_b$, V_b) and then again after ($\bar{z}_a$, V_a) selection has occurred. (For instance, these measurements may be made on juveniles and then on those individuals that successfully survive to adulthood and reproduce.) Then, if selection is directional, an index of the **intensity of selection** is

$$i = \frac{\bar{z}_a - \bar{z}_b}{\sqrt{V_P}}$$

where V_P is the phenotypic variance.

Another measure of selection is the **selection gradient**: the slope, b, of the relation between phenotype values (z) and the fitnesses (w) of those phenotypes.* This measure is especially useful when several characters are correlated with one another to some degree, such as, say, beak length (z_1) and body size (z_2). Selection on each such feature (say z_1) can be estimated while (in a statistical sense) holding the other (say z_2) constant by using the equation

$$w = a + b_1 z_1 + b_2 z_2$$

(Lande and Arnold 1983). The slopes (partial regression coefficients) b_1 and b_2 enable one to estimate, for instance, how greatly variations in beak length affect fitness among individuals with the same body size. In most studies, certain components of fitness (such as juvenile-to-adult survival) are estimated, rather than fitness in its entirety.

Peter and Rosemary Grant (1986, 1989) and their colleagues have carried out long-term studies on some of the species of Darwin's finches on certain of the Galápagos Islands. They have shown that birds with larger (especially deeper) bills feed more efficiently on large, hard seeds, whereas there is some evidence that small, soft seeds are more efficiently used by birds with smaller bills. When the islands suffered a severe drought in 1977, seeds, especially small ones, became sparse, medium ground finches (*Geospiza fortis*) did not reproduce, and their population size declined greatly as a result of mortality. Compared with the pre-drought population, the survivors were larger and had larger bills (Figure 13.7). From the differences in morphology between the survivors ($\bar{z}_a$) and the pre-drought population ($\bar{z}_b$), the intensity of selection i and the selection gradient b were calculated for three characters:

Character	i	b
Weight	0.28	0.23
Bill length	0.21	−0.17
Bill depth	0.30	0.43

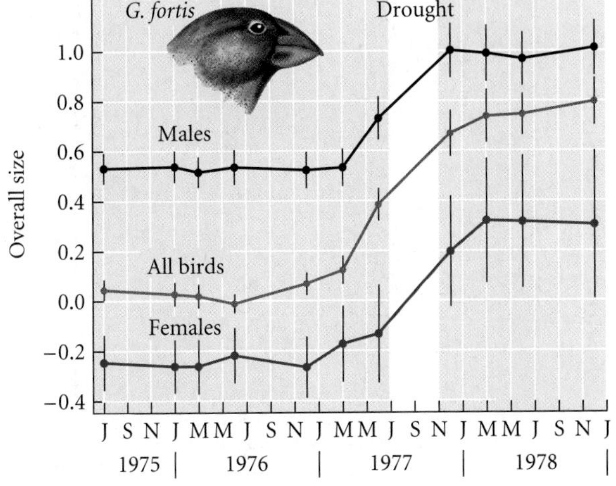

Figure 13.7 Changes in the mean size of the ground finch *Geospiza fortis* on Daphne Island in the Galápagos archipelago resulting from mortality during a drought in 1977. Changes occurring in 1977 and 1978 were all the result of mortality; no reproduction occurred during this period. "Overall size" is a composite of measurements of several characters. (After Grant 1986.)

*The slope is the derivative of fitness with respect to character state and is closely related to $d\bar{w}/d\bar{z}$, one of the determinants of the rate of character evolution in Lande's equation (see the previous footnote, p. 345).

The values of i show that each character increased by about 0.2 to 0.3 standard deviations, a very considerable change to have occurred in one generation. The values of b show the strength of the relationship between survival and each character while holding the other characters constant. Selection strongly favored birds that were larger and had deeper bills because they could more effectively feed on large, hard seeds, virtually the only available food. The negative b value shows that selection favored shorter bills. Nevertheless, bill length increased, in opposition to the direction of selection, because bill length is correlated with bill depth. Thus a feature can evolve in a direction opposite to the direction of selection if it is strongly correlated with another trait that is more strongly selected. (We will soon return to this theme.)

Why don't these finches evolve ever larger bills? The Grants found that during normal years, birds with smaller bills survive better in their first year of life, probably because they feed more efficiently on abundant small seeds; in addition, small females tend to breed earlier in life than large ones. Thus conflicting selection pressures create stabilizing selection that, on average, favors an intermediate bill size.

Stabilizing and disruptive selection

If selection is stabilizing or diversifying, the change within a generation in the phenotypic variance provides a measure of the intensity of SELECTION ON VARIANCE:

$$j = \frac{V_a - V_b}{V_b}$$

(Endler 1986). This index is negative if selection is stabilizing, positive if it is diversifying.

Many traits are subject to stabilizing selection, so the mean changes little, if at all. For example, human infants have lower rates of mortality if they are near the population mean for birth weight than if they are lighter or heavier (Figure 13.8; Karn and Penrose 1951). Stabilizing selection often occurs because of **trade-offs**, antagonistic agents of selection (Travis 1989). Arthur Weis and colleagues (1992) found that different natural enemies impose conflicting selection on the size of galls induced by the goldenrod gall fly (*Eurosta solidaginis*). The larva of this fly induces a globular growth (gall) on the stem of its host plant, goldenrod. Much of the variation in gall size is due to genetic variation in the fly, which feeds and pupates within the gall, emerging in spring. During the summer, parasitoid wasps, *Eurytoma gigantea* and *E. obtusiventris*, lay eggs on the fly larvae by inserting their ovipositors through the gall wall. During winter, woodpeckers and chickadees open galls and feed on the fly pupae. Mortality caused by parasitoids and birds can be determined by examining galls in the spring.

The researchers found that the parasitoid *E. gigantea* consistently selected for wide galls (i.e., mortality was greatest for fly larvae in narrow galls) because the wasp's ovipositor cannot penetrate thick gall tissue. *E. obtusiventris* generally selected for intermediate-sized galls, whereas birds most frequently attacked wide galls, selecting for narrower gall diameter. Taken together, these enemies imposed rather strong stabilizing selection ($j = -0.30$), but because selection by parasitoids was weaker than selection by birds, a directional component ($i = 0.34$) was detected as well (Figure 13.9).

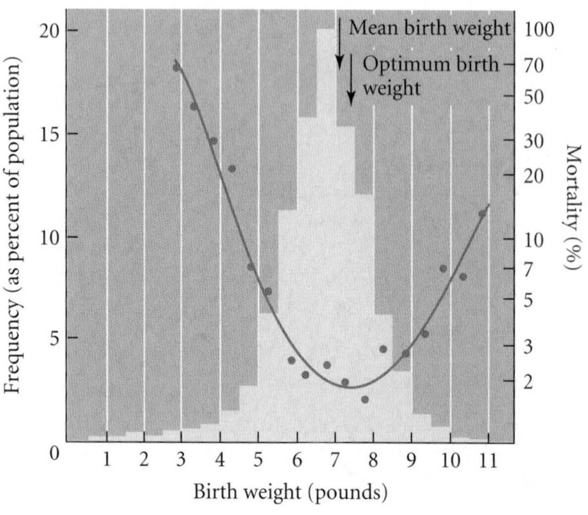

Figure 13.8 Stabilizing selection for birth weight in humans. The rate of infant mortality is shown by the points and the line fitting them. The histogram shows the distribution of birth weights in the population. (After Cavalli-Sforza and Bodmer 1971.)

Evolution observed

Adaptive evolution, at rates far greater than the average evolutionary rates documented in the fossil record, has been documented for many morphological, physiological, and behavioral traits, especially in species that have been introduced into new regions or subjected to human alterations of their environment (Hendry and Kinnison 1999; Palumbi 2001; Smith and Bernatchez 2008). For instance, the codling moth (*Cydia pomonella*), a major pest of apples, is a European species that was first recorded in New England in 1750, and has since expanded its range over 12 degrees of latitude. As in many insects, the cue for entering diapause, a state of low metabolic activity that is neces-

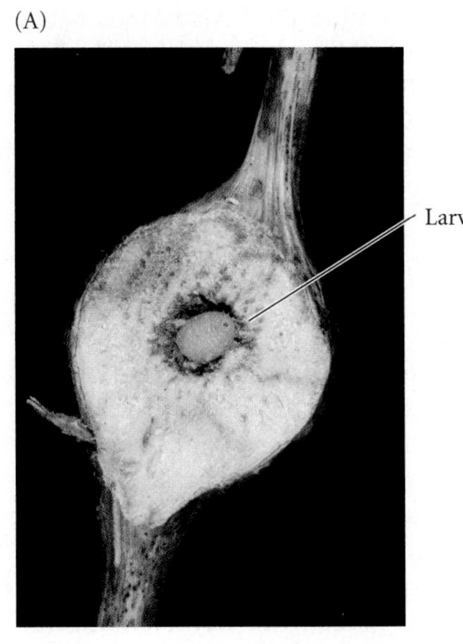

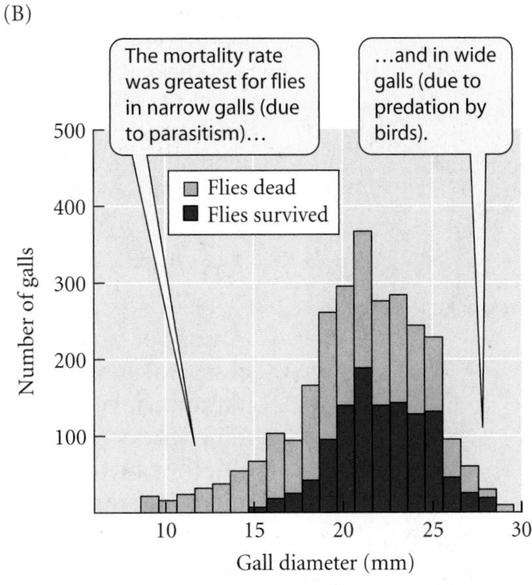

Figure 13.9 (A) Larva of the gall fly (*Eurosta solidaginis*), shown inside the dissected stem gall of a goldenrod plant. (B) Stabilizing selection on the size of galls made by *E. solidaginis*. The height of each bar shows the proportion of each plant gall size in the population, and the blue portion shows the proportion of fly larvae that survived. Flies in intermediate-sized galls had the highest survivorship. (A, photo by David McIntyre; B after Weis et al. 1992.)

sary for surviving the winter, is a critical photoperiod (day length). Northern populations of native species are typically genetically programmed to enter diapause at longer day lengths than southern populations because winter arrives at northern latitudes sooner, when days are still relatively long. Since their spread in North America, codling moth populations have diverged genetically so that they display the same adaptive cline in critical photoperiod as native moth species that have occupied the same latitudinal span for a much longer time (Figure 13.10).

Until the 1950s, German populations of the blackcap (*Sylvia atricapilla*), a European songbird, migrated only to the western Mediterranean region for the winter. Since that time, more and more German blackcaps have overwintered in Britain, migrating northwest rather than southwest to the Mediterranean. When feeling the urge to migrate, a caged bird flutters in the right direction for its migration if it can see the night sky. (Most small birds migrate at night and use star patterns to determine direction.) Peter Berthold and colleagues (1992) used this behavior to show that offspring of blackcaps from populations that overwinter in Britain (who have never made the migratory journey themselves) orient toward the northwest, whereas offspring from populations that overwinter in the Mediterranean orient toward the southwest. The difference between the populations is genetically based and has evolved in a few decades. The authors suggest that the selective advantage of this change in behavior lies in improved winter weather and other conditions in Britain and in earlier spring return from Britain than from the Mediterranean.

Figure 13.10 Geographic variation in the critical photoperiod for entering diapause in two species of moths at various latitudes. *Chilo suppressalis* is native to Japan. *Cydia pomonella* is an introduced pest in North America, where it has spread within the last 250 years. The two species now occupy the same latitudinal range. (After Tauber et al. 1986.)

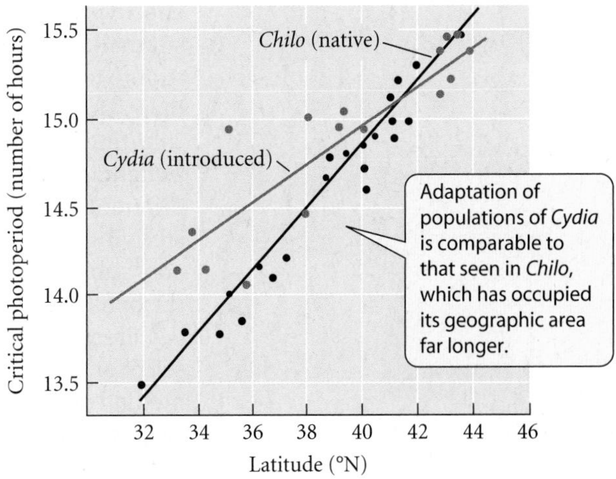

Several species of insects, such as the soapberry bug (*Jadera haematoloma*; Carroll and Boyd 1992; Carroll et al. 1997), have adapted rapidly to new food plants. This insect feeds on seeds of plants in the soapberry family (Sapindaceae) by piercing the enveloping seed pod with its slender beak (Figure 13.11). In Texas, its natural host plant—the soapberry tree—has a small pod. Its major contemporary host plant, however, is the Asian golden rain tree, which has become common only within the last 20 to 50 years and has a larger pod. Beak length of local populations of the bug has evolved to be 8 percent longer. In Florida, conversely, the beak has become 25 percent shorter in soapberry bug populations that feed on the introduced flat-podded rain tree (*Koelreuteria elegans*), also from Asia, which has a flatter, smaller seed pod than the native host. Adaptation to an altered environment—to new food plants—has resulted in large, rapid changes in morphology.

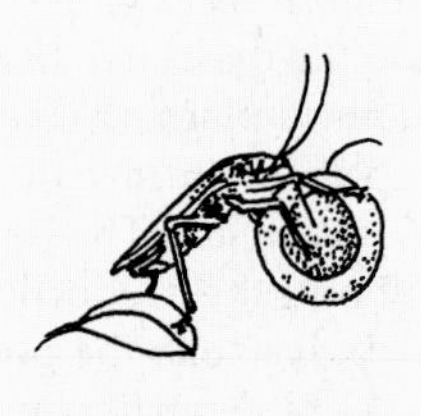

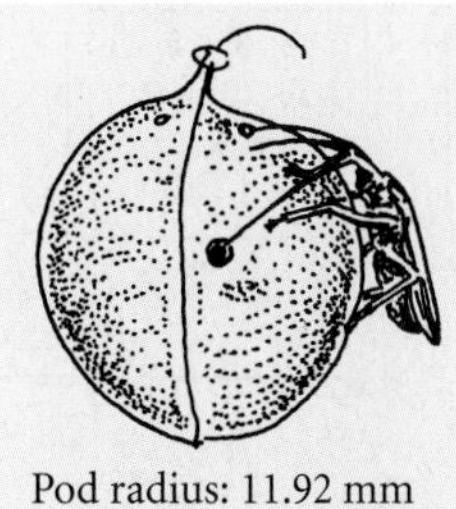

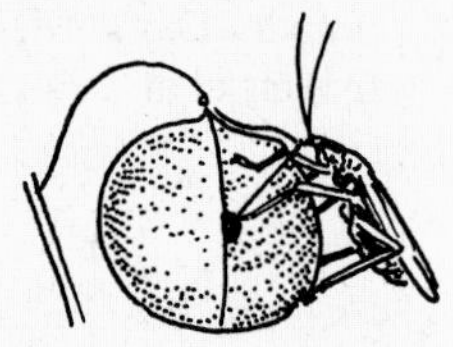

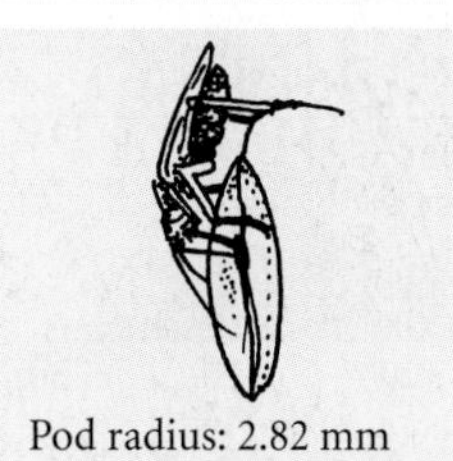

Figure 13.11 Soapberry bugs (*Jadera haematoloma*) and their native and introduced host plants in Texas and Florida, drawn to scale. The bug's beak is the needlelike organ projecting from the head at right angles to the body. The average pod radius of each host species and the average (with standard deviation) beak length of associated *Jadera* populations are given. Beak length has evolved rapidly as an adaptation to the new host plant. (After Carroll and Boyd 1992.)

Copper, zinc, and other heavy metals are toxic to plants, but in several species of grasses and other plants, metal-tolerant populations have evolved where soils have been contaminated by mine works that range from more than 700 to less than 100 years old. In some cases, tolerance has evolved within decades on a microgeographic scale, such as in the vicinity of a zinc fence. Tolerance is based on a variable number of genes, depending on the species and population. When tolerant and nontolerant genotypes of a species are grown in competition with other plant species in the absence of the metal, the relative fitness of the tolerant genotypes is often much lower than that of the nontolerant genotypes, implying a cost of adaptation (Antonovics et al. 1971; Macnair 1981).

Commercial overexploitation has severely depleted populations of many species of fish, and has resulted in evolutionary change as well (Kuparinen and Merilä 2007; Hutchings and Fraser 2008). In many species, there has been a trend toward earlier sexual maturation at a smaller size, as predicted if larger age classes are more subject to predation (see Chapter 14). In some species, such as Atlantic cod (*Gadus morhua*), these changes clearly have a genetic basis (**Figure 13.12A,B**). Similarly, trophy-hunting for bighorn sheep (*Ovis canadensis*) with the largest horns has resulted in the evolution of smaller horns (**Figure 13.12C**). In both instances, the very quality that adds value to the resource has been diminished by the response to selection. Evolution has been rapid in both of these cases: the rate of decline of cod body length at maturation was –1.2 to –1.5 haldanes, and of horn length of bighorn sheep –0.15 haldanes. By contrast, the highest rate of evolution of tooth size in the fossil record of *Hyracotherium* was 0.00024 haldanes (see Chapter 4, pp. 92–93).

What Maintains Genetic Variation in Quantitative Characters?

Accounting for the high levels of genetic variation in quantitative traits is a difficult problem (Barton and Turelli 1989). Although the neutral theory predicts high genetic variance in large populations, estimates of h^2 for characters of species that are thought to have small effective population sizes (e.g., Darwin's finches, salamanders, mice) are often just as substantial as those for more populous species (Houle 1989). Furthermore, alleles that contribute to quantitative variation are probably seldom selectively neutral. As we have seen, many quantitative characters are subject to fairly intense selection. Moreover, genes contributing to quantitative traits have pleiotropic effects on survival and other fitness components, as we know from studies of *Drosophila* bristles. Indeed, these genes may have different primary functions, and only incidentally affect the quantitative trait we may be studying. Thus even characters that might in themselves be selectively "trivial" are probably subject to indirect selection because of the pleiotropic effects of the underlying genes (Dobzhansky 1956).

Directional selection diminishes genetic variation. So, less obviously, does stabilizing selection, because among the many polygenic genotypes that may yield the optimal intermediate genotype, some are homozygous mixtures of + and – alleles, and one such mixture will eventually be fixed. Of the several hypotheses that have been advanced to account for quantitative genetic variation, the most likely may be VARIABLE SELECTION and

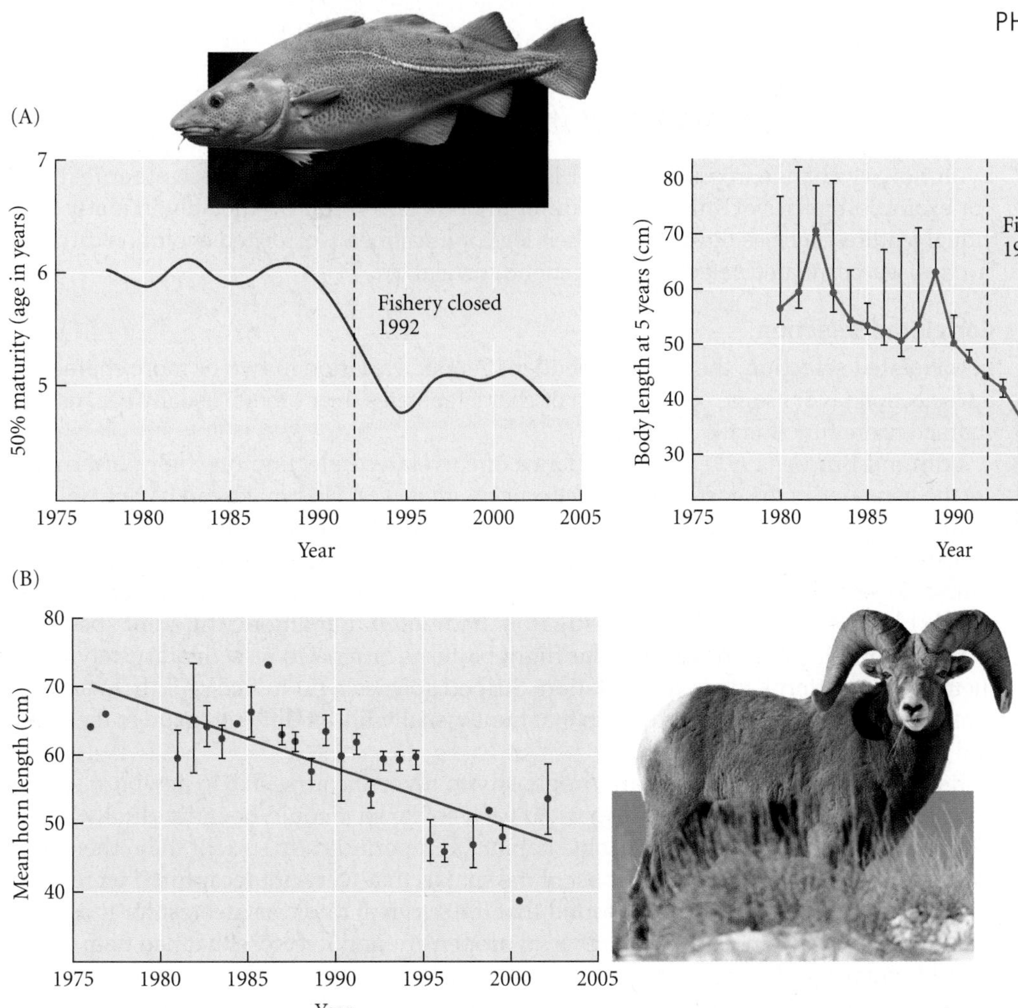

Figure 13.12 Evolutionary changes caused by human harvesting. (A) The age at which 50 percent of Atlantic cod reached maturity (left graph) declined until 1992, when the fishery was closed because of overfishing. Body length in the Atlantic cod at this age (right graph) also declined until 1992. (B) Mean horn length of 4-year-old bighorn sheep rams declined because of selection imposed by hunting. (A after Olsen et al. 2004, photo © Blickwinkel/Alamy; B after Coltman et al. 2003; photo © B. Wilson/ShutterStock.)

MUTATION-SELECTION BALANCE. Fluctuation in the optimal phenotype from one generation to another can delay the loss of genetic variation, although it does not prevent it indefinitely. Gene flow between populations with different optimal phenotypes can maintain variation in each population. Moreover, populations in which stabilizing selection favors the same phenotype can diverge in genetic composition as mutation and genetic drift create turnover in alleles at the contributing loci. Gene flow among such populations can help to maintain genetic variation (Goldstein and Holsinger 1992). However, laboratory populations, maintained under rather uniform conditions and isolated from gene flow, do not differ substantially in heritable variation from natural populations, casting doubt on the importance of variable selection and gene flow (Bürger et al. 1989).

A currently favored hypothesis is that levels of polygenic variation reflect a balance between the erosion of variation by stabilizing selection and the input of new variation by mutation (V_m; Lande 1976b; Houle et al. 1996). There is some doubt that V_m is high enough to counter the strong stabilizing selection that acts on many traits, a point against this hypothesis (Turelli 1984). Moreover, some of the alleles that contribute substantially to the variance in traits such as bristle number have higher frequencies than predicted from a balance between mutation and purifying selection (Lai et al. 1994). On the other hand, far more loci contribute to a fitness-related trait than to a single morphological trait, so V_m should be higher for fitness-related traits, and should maintain higher genetic variance (V_A). In fact, V_A is considerably greater for fitness-related traits than for morphological traits, as predicted by the mutation–selection balance hypothesis (Houle et al. 1996). Many data are consistent with a model of mutation/selection balance in which stabilizing selection on a character acts on mutations of large effect at a few loci, while much of the variation consists of small pleiotropic effects of slightly deleterious mutations at many other loci that have a variety of other primary functions (Zhang and Hill 2005).

Correlated Evolution of Quantitative Traits

Evolutionary change in one character is often correlated with change in other features. For example, species of animals that differ in body size differ predictably in many individual features, such as the length of their legs or intestines. Correlated evolution can have two causes: correlated selection and genetic correlation.

Correlated selection

In **correlated selection**, there is independent genetic variation in two or more characters, but selection favors some combination of character states over others, usually because the characters are functionally related.

Edmund Brodie (1992) found evidence of correlated selection on color pattern and escape behavior in the garter snake *Thamnophis ordinoides*. This snake can have a uniform color, spots, or lengthwise stripes. Brodie found that when chased down an experimental runway, some snakes fled in a straight line, while others repeatedly reversed their course. Resemblance among siblings indicated that both coloration and escape behavior are highly heritable. Other investigators had noted that among different species of snakes, spotted species have irregular flight patterns or tend to be sedentary, relying on their cryptic patterns to avoid predation. Striped species tend to flee rapidly in a single direction, probably because predators that hunt visually find it difficult to judge the speed and position of a moving stripe.

Brodie scored color pattern and propensity to reverse course in 646 newborn snakes born to 126 pregnant females. He marked each snake so it could be individually recognized, released the snakes in a suitable habitat, and periodically sought them thereafter. Brodie had reason to believe that most of the snakes that were not recaptured were eaten by crows and other predators. He found that the survival rate was greatest for those that had both strong striping and a low reversal propensity, and for those that had both a nonstriped pattern and a high reversal propensity (Figure 13.13). Other phenotypes, such as striped snakes that reversed course when chased, had lower survival rates. Thus there was correlated selection on color pattern and escape behavior in the direction that had been predicted from comparisons among species of snakes and from the theory of visual perception.

Figure 13.13 A fitness surface for combinations of two traits in the garter snake *Thamnophis ordinoides*, based on survival in the field. The height of a point on the surface represents the fitness (relative survival) of individuals with particular values of stripedness and reversal (tendency to reverse course when fleeing). (After Brodie 1992.)

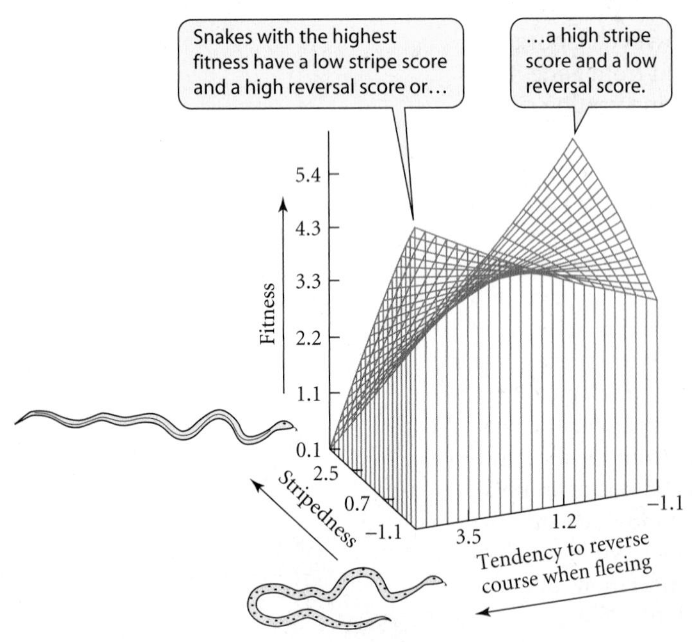
Snakes with the highest fitness have a low stripe score and a high reversal score or...
...a high stripe score and a low reversal score.
5.4
4.3
3.3
2.2
1.1
0.1
Fitness
2.5
0.7
−1.1
Stripedness
3.5
1.2
−1.1
Tendency to reverse course when fleeing

Genetic correlation

We often observe that characters are correlated within a species: tall people, for example, tend to have long arms. The magnitude of a correlation between two characters—the degree to which they vary in concert—is expressed by the CORRELATION COEFFICIENT (r), which ranges from +1.0 (for a perfect correlation in which both features increase or decrease together) to –1.0 (when one feature decreases exactly in proportion to the other's increase). For uncorrelated characters, $r = 0$.

The **phenotypic correlation**, r_P, between, say, body size and fecundity is simply what we measure in a random sample from a population. Just as the phenotypic variance may have both genetic and environmental components, so too may the phenotypic correlation. Two features of individuals with the same genotype may vary together because both are affected by environmental factors, such as nutrition. Such features display an **environmental correlation**, r_E. In genetically variable populations, the correlated variation may also be caused by genetic differences that affect both characters, causing a **genetic correlation**, r_G.

Genetic correlations can have two causes. One cause is linkage disequilibrium among the genes that independently affect each character. The other cause is pleiotropy—the influence of the same genes on different characters. A genetic correlation that is due to pleiotropy will be perfect (r_G = 1.0 or –1.0) if the alleles that increase one character all increase (or all decrease) the other character. If some genes affect only one of the characters, or if some alter both characters in the same direction (+,+ or –,–) while others have opposite effects on the two characters (+,– or –,+), the genetic correlation will be imperfect.

Genetic correlations need not be constant, and they may evolve (Turelli 1988). A genetic correlation caused by linkage disequilibrium, such as the correlation between style (pistil) length and stamen height in the primrose *Primula vulgaris* (see Figure 9.19), will decline as a result of recombination unless selection for the adaptive gene combinations maintains it. Correlations caused by pleiotropy may also change, although more slowly than those caused by linkage disequilibrium. Some alleles may become fixed, so that the loci contribute no variation and therefore no correlation, whereas other loci, perhaps affecting only one character, remain variable—and changes in allele frequency at these loci will also affect r_G.

Another cause of change in genetic correlation is natural selection, which may favor MODIFIER ALLELES that alter the pleiotropic effects of other loci. For example, natural populations of Australian blowflies (*Lucilia cuprina*) exposed to the insecticide diazinon rapidly evolved resistance as a result of an increase in the frequency of a resistance allele, *R*. At first, resistant flies had reduced viability (when tested in the absence of diazinon) and a high incidence of bilateral asymmetry (which is thought to reflect disruptions of development). These features were pleiotropic effects of the *R* allele. Over the course of several years, however, viability increased and the incidence of asymmetry decreased. These changes were not due to change at the *R* locus, because the *R* allele still evinced pleiotropic effects when it was backcrossed from the resistant field population into nonresistant laboratory stocks. Rather, natural selection had increased the frequency of modifier alleles at other loci, which altered and mitigated the deleterious effects of the *R* allele (McKenzie and Clarke 1988).

Genetic correlation between traits can be estimated from correlations among relatives in the same way that genetic variance can. For example, Juan Fornoni and collaborators (2005) examined the relationship between resistance to herbivorous insects and tolerance of insect damage in jimsonweed (*Datura stramonium*) in central Mexico. Resistance, the capacity to repel herbivores by chemical or physical defenses, was measured as the proportion of leaf area that was not eaten by insects in plants that were grown in a common garden in a natural environment. Tolerance means a plant's ability to sustain high fitness even if damaged; the higher the number of seeds a plant produced, relative to the amount of leaf tissue that had been consumed by insects, the higher the plant's tolerance. From each of 25 plants (male, or "pollen parents"), Fornoni et al. placed pollen on flowers of two other plants (females, or "seed parents"), and planted several offspring from each seed parent. The variation among family means from different pollen parents represents additive genetic variance in each trait, and the correlation between the means for two characters estimates the genetic correlation. (Calculations based on pollen-parent means are free of whatever nongenetic maternal effects might exist.) The results in one such common garden (**Figure 13.14A**) revealed a negative genetic correlation: more resistant genotypes proved to be less tolerant of insect damage. This result is consistent with the hypothesis that the plant allocates energy or resources to each of these traits at the expense of the other. The same plant families, though, planted in another common garden in a different locality, displayed a positive genetic correlation (**Figure 13.14B**), perhaps because the environment differed in such a way as to relieve the trade-off in allocation to these two functions.

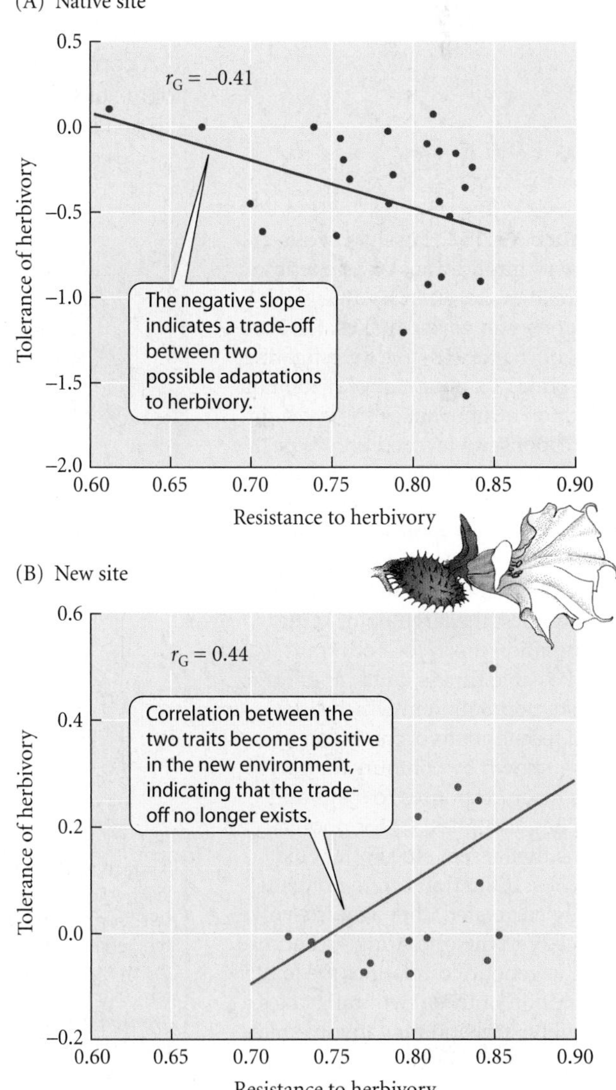

Figure 13.14 Genetic correlation between tolerance of and resistance to insect herbivory in jimsonweed (*Datura stramonium*). Each point represents the joint mean of the two characters in offspring of a single pollen parent. A negative correlation implies a trade-off between these two possible adaptations to herbivory. The same set of families was grown (A) in its native site of Ticumán, Mexico, and (B) in another site, Santo Domingo, where the environment altered the sign of the genetic correlation (r_G) from negative to positive. (After Fornoni et al. 2003.)

How genetic correlation affects evolution

Genetic correlations among characters can cause them to evolve in concert. They can also either enhance or retard the rate of adaptive evolution, depending on the circumstances. In extreme cases, they may severely constrain adaptation.

If two characters, z_1 and z_2, are genetically correlated, the rate and direction of evolution of z_1 depend on both direct selection (on z_1 itself) and selection on z_2. If selection on z_2 is much stronger than on z_1, z_1 may change mostly as a result of its correlation with z_2, and the change may not be in an adaptive direction. For example, Stevan Arnold (1981) estimated a genetic correlation of 0.89 between the feeding reaction of newborn garter snakes to slugs (z_1) and to leeches (z_2; Figure 13.15). If both slugs and leeches are present in the environment, there may be direct selection in favor of slug-feeding (increased z_1), but stronger selection against leech-feeding (decreased z_2), since a leech can kill a snake by biting its digestive tract after being swallowed. If adaptive aversion to leeches evolves (decreased z_2), maladaptive aversion to slugs (decreased z_1) may also evolve, at least temporarily, as a correlated effect (Figure 13.15C). (We assume that other kinds of food are available, so that the population can persist even if it evolves aversion to both leeches and slugs.) Conversely, if leeches are rare and slugs are the most abundant food, selection for feeding on slugs may be stronger, the population may evolve a slug-feeding habit, and it may also evolve the maladaptive habit of occasionally eating leeches (Figure 13.15D). After the more strongly selected trait approaches its optimum (e.g., z_2, strong aversion to leeches, in Figure 13.15C), the weakly selected trait can evolve to its optimum (z_1, positive feeding response to slugs). This change is based mostly on those genes that affect only z_1.

A conflict may therefore exist between the genetic correlation of characters and directional selection on those characters. When such a conflict exists, the two characters may evolve to their optimal states only slowly, and may even evolve temporarily in a maladap-

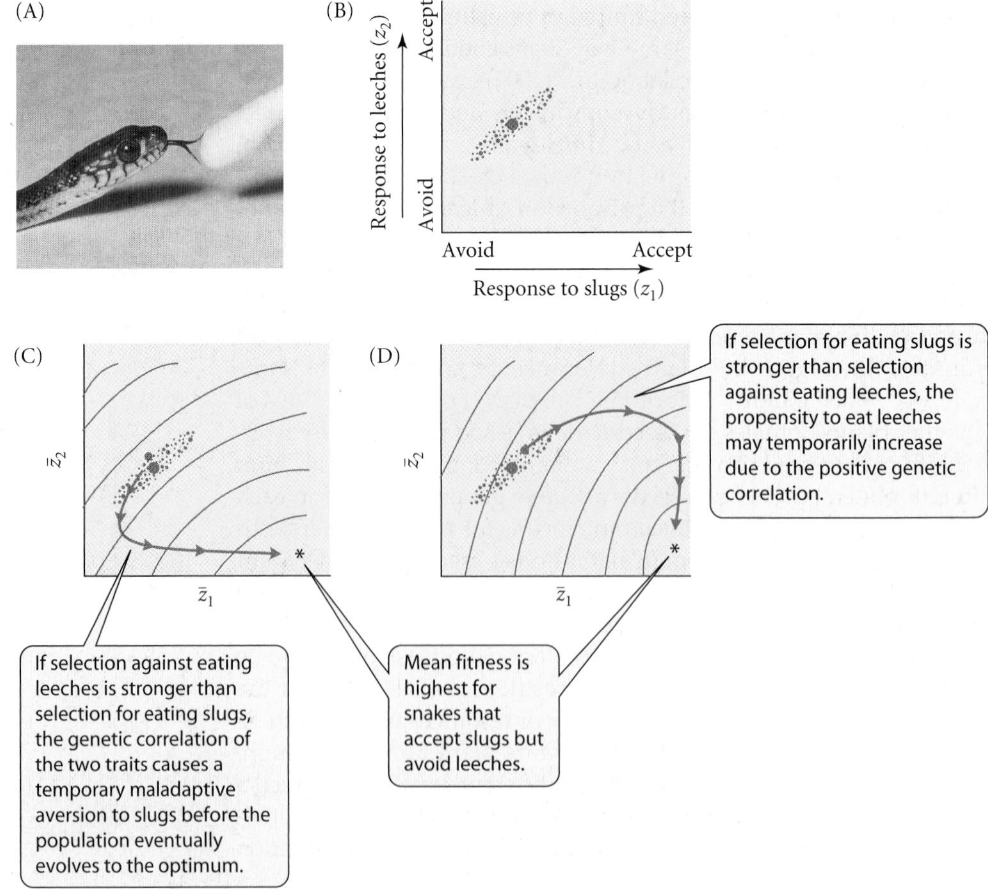
(A)
(B)
Response to leeches (z_2)
Accept
Avoid
Avoid
Accept
Response to slugs (z_1)
(C)
(D)
$\bar{z}_2$
$\bar{z}_1$
*
If selection for eating slugs is stronger than selection against eating leeches, the propensity to eat leeches may temporarily increase due to the positive genetic correlation.
If selection against eating leeches is stronger than selection for eating slugs, the genetic correlation of the two traits causes a temporary maladaptive aversion to slugs before the population eventually evolves to the optimum.
Mean fitness is highest for snakes that accept slugs but avoid leeches.

Figure 13.15 Possible evolutionary implications of a genetic correlation between feeding responses in garter snakes. (A) A newborn garter snake investigating the odor of a potential prey item on a cotton swab. Snakes show a higher response to favored prey types by flicking the tongue more often. (B) The positive genetic correlation (r_G) between response to slugs (z_1) and to leeches (z_2). The "cloud" of small points represents individual genotypes; the large point is the population mean for both traits. (C, D) Possible fitness landscapes for a snake population in which slugs and leeches both occur. Mean fitness, shown by contours for various mean responses to slugs and leeches, is highest (*) for populations that accept slugs but avoid leeches. If the traits were not genetically correlated, they would evolve directly to the optimum (*). Because of the genetic correlation, the joint evolution of these two traits takes a curved path, and may involve maladaptation in one trait. (Photo courtesy of S. J. Arnold.)

tive direction. (We have already seen that selection for a deeper bill in the Galápagos finch *Geospiza fortis* caused average bill length to increase, even though selection favored a shorter bill.) In some cases, a genetic correlation may be so strong that one or both traits cannot reach their optimal states. For example, there is a necessary trade-off between the number and the size of eggs (or seeds) that an organism can produce because the resources that it can allocate to reproduction are limited. This trade-off creates a negative genetic correlation, with some genotypes producing more but smaller eggs and others producing fewer but larger ones (see Chapter 14). Although selection might favor both more and larger eggs, increasing both is possible to only a very limited extent. Thus genetic correlations, owing in some cases to trade-offs of this kind, can sometimes act as genetic constraints on evolution. Whether or not a genetic correlation acts as a long-term constraint depends on several factors, such as how readily the genetic correlation changes.

Genetic correlations may enhance adaptive evolution, rather than constrain it, in some cases (Wagner 1988). Multiple characters may evolve as an integrated ensemble more rapidly if they are genetically correlated, as when they are subject to the same developmental controls. This is especially true if the characters are functionally related. For instance, the size of each organ (e.g., lungs, gut, bones) must match the overall body size if an animal is to function properly. Body size would evolve much more slowly in response to selection if every organ had to undergo independent genetic change than if there were coordinated increases or decreases in the sizes of the various organs. During development, in fact, the various organs grow in concert, and alleles that change body size have correlated effects on most body parts. (See the discussion of allometric growth in Chapter 3).

Can Genetics Predict Long-Term Evolution?

If the response to selection in natural populations were never limited by the availability of genetic variation in single characters or combinations of characters, the rate and direction of adaptive evolution would depend only on the strength and direction of natural selection. There is reason, however, to think that in some instances, evolution may proceed along "genetic lines of least resistance" (Stebbins 1974; Schluter 1996). Some characters may have very little genetic variation, and therefore may constrain, or at least bias, the direction of evolution. For example, when species of host-specific herbivorous beetles were screened for their propensity to feed on novel plants that they do not normally eat, genetic variation was found in the feeding responses to only certain plants, especially those most closely related to the insects' normal host plants. This pattern suggested that these beetles might more readily adapt to feed on closely related plants than on distantly related plants—which is exactly what has occurred in the evolution of these beetles and many other groups of herbivorous insects (Futuyma et al. 1995).

If two characters are genetically correlated, the greatest genetic variation lies along the long axis of the ellipse formed by a plot of genotypes' joint character values (Figure 13.16A). Dolph Schluter (2000) referred to this axis as the "maximal genetic variation" (g_{max}), and predicted that, at least in the short term, adaptive evolution should be greatest along this axis because of the constraining effect of genetic correlation between characters under directional selection. Over time, however, g_{max} should have less of an effect, because characters that are not perfectly genetically correlated can eventually evolve to their optimal values (as described earlier). (The same principle holds for multiple characters. A table of values of V_A for each character and the genetic covariance, closely related to the correlation, between each pair of characters is called the genetic variance-covariance matrix, usually referred to simply as the **G matrix**, or **G**.)

For several adaptive morphological characters in stickleback fishes, sparrows, and other vertebrates, Schluter determined the direction of divergence of a species from a closely related species and compared it to g_{max}, the direction of evolution predicted by the genetic correlation between characters (Figure 13.16B). As predicted, the deviation between these two directions was least for species that had diverged only recently, and it increased with time since common ancestry. The rather close initial correspondence between the actual and genetically predicted directions suggests that patterns of genetic variation and

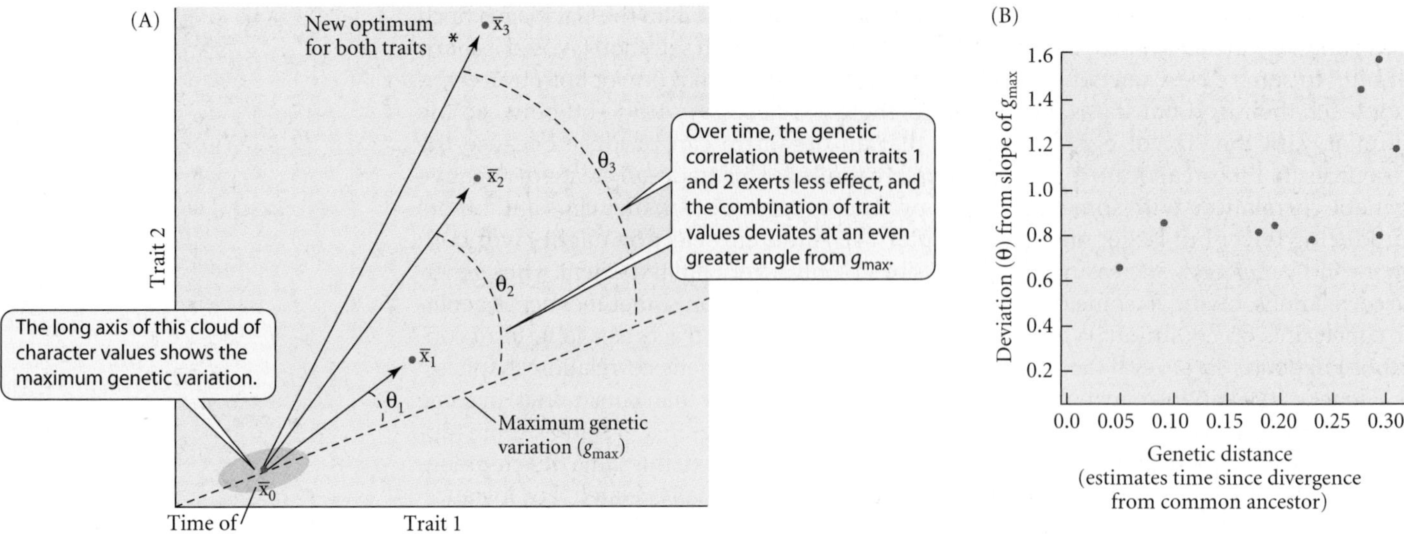

Figure 13.16 Evolution along genetic lines of least resistance. (A) The ellipse shows the distribution of values of the genetically correlated characters 1 and 2. The maximal genetic variation (g_{max}) is for combinations of trait values along the long axis of the ellipse. If the ancestral population mean is at point $\bar{x}_0$ and it experiences directional selection toward a new optimum (*), the genetic correlation will cause evolution toward $\bar{x}_1$ at first. The direction of evolution (arrow) deviates by angle θ_1 from g_{max}. Over time, the genetic correlation will exert a lesser effect, so that the line between $\bar{x}_0$ and $\bar{x}_2$ deviates more from g_{max} (angle θ_2). The deviation is still greater (θ_3) at a later time, when the population approaches the optimum for both traits. (B) Genetic correlations among several adaptive traits (e.g., bill dimensions) were estimated in song sparrows (*Melospiza melodia*) and used to determine g_{max}. Points show the deviation (angle θ in panel A) between g_{max} and the difference in these traits between the song sparrow and nine other species of sparrows, in relation to the molecular genetic distance between these species and the song sparrow. The low deviation of closely related species (those at low genetic distance) from g_{max} shows that g_{max} in the song sparrow predicts the initial direction of evolution fairly well. The theory of genetic correlation predicts that the characters of more distantly related species should display a larger deviation from g_{max}, as is observed in these data. (After Schluter 1996.)

correlation can persist for long periods of time and can influence the direction of evolution. Schluter estimated that this influence may last for up to four million years.

The pattern Schluter describes suggests that genetic correlations among characters may remain consistent for a long time. Whether or not this is generally true is one of the most important, most poorly understood aspects of phenotypic evolution. Some investigators have found that the genetic correlations among certain characters are fairly similar among geographic populations of a species or among closely related species, but other studies have found lower similarities, suggesting that the strength of genetic correlations can evolve rather rapidly (Steppan et al. 2002). Mattieu Bégin and Derek Roff (2004) found that additive genetic variances and genetic covariances (a statistic related to correlation) of several morphological measurements are quite similar among species of field crickets and, furthermore, that the pattern of correlation of characters among species closely matched the average pattern of genetic correlation within species (Figure 13.17). They suggested that this correspondence indicates genetic constraint. However, it is not certain that the genetic constraint was strong enough to resist natural selection, since there is no evidence that the divergence among the species was caused by natural selection. Katrina McGuigan and collaborators (2005) estimated **G** for various measurements in two species of Australian rainbow fish (*Melanotaenia*), each of which has both stream- and pond-dwelling populations that have evolved parallel, adaptive differences in shape. The **G** matrix was similar in both species, and it predicted the difference between the ecologically equivalent forms of the two species quite well. However, the shape difference between the stream and pond forms did not correspond at all to the "genetic lines of least resistance," indicating that evolution by natural selection had not been genetically constrained.

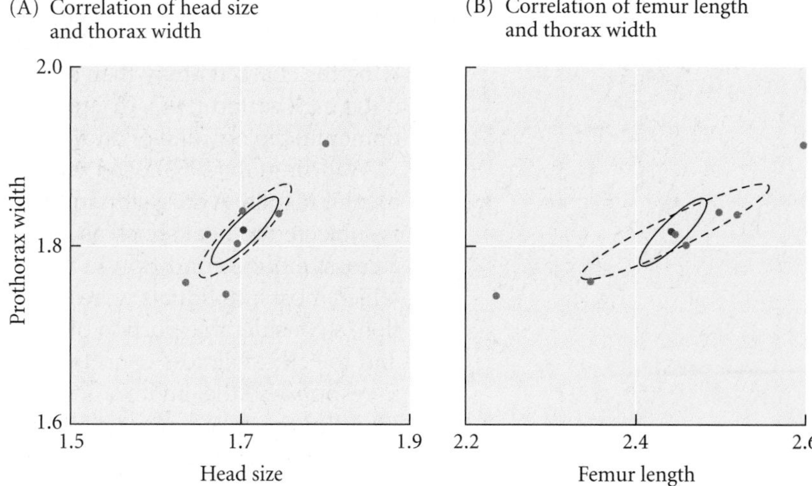
(A) Correlation of head size and thorax width
(B) Correlation of femur length and thorax width
Prothorax width
2.0
1.8
1.6
1.5
1.7
1.9
Head size
2.2
2.4
2.6
Femur length

Figure 13.17 Orientation of the genetic correlation between pairs of morphological measurements within species of crickets (solid ellipses), compared with the correlation among species for the same pairs of characters (broken ellipses). The joint mean values for each of the seven species are indicated by red circles. (A) The axis of the genetic correlation between head size and prothorax width is nearly identical to the axis of variation among species, suggesting that divergence among them has followed the genetic line of least resistance. (B) Variation among species in the relation between femur length and prothorax width does not match the within-species genetic correlation as closely. (After Bégin and Roff 2004.)

Norms of Reaction

The correspondence between genotypic differences and phenotypic differences depends on developmental processes. In some cases, these processes may reduce the phenotypic expression of genetic differences; in other instances, a single genotype may produce radically different phenotypes in response to environmental stimuli, a phenomenon called **phenotypic plasticity**. The **norm of reaction** of a genotype is the set of phenotypes that genotype is capable of expressing under different environmental conditions (Figure 13.18; also see Figure 9.5). The norm of reaction can be visualized by plotting the genotype's phenotypic value in each of two or more environments.

When the effect of environmental differences on the phenotype differs from one genotype to another in a population, the reaction norms of the genotypes are not parallel, and the phenotypic variance includes a variance component (referred to as $V_{G \times E}$) that is due to **genotype × environment (G × E) interaction** (Figure 13.18B,C). If all the genotypes have parallel reaction norms (Figure 13.18A), there is no G × E interaction ($V_{G \times E} = 0$). With genotype × environment interaction, the reaction norm may be capable of evolving.

Canalization

For many characteristics, the most adaptive norm of reaction may be a constant phenotype, buffered against alteration by the environment (Figure 13.18C). It may be advantageous, for example, for

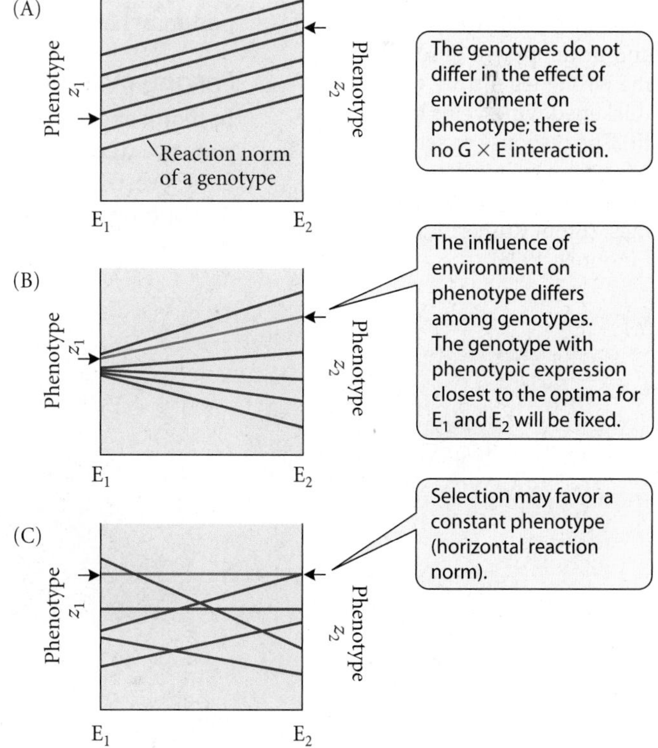
(A)
Phenotype z_1
Phenotype z_2
Reaction norm of a genotype
E_1
E_2
The genotypes do not differ in the effect of environment on phenotype; there is no G × E interaction.
(B)
The influence of environment on phenotype differs among genotypes. The genotype with phenotypic expression closest to the optima for E_1 and E_2 will be fixed.
(C)
Selection may favor a constant phenotype (horizontal reaction norm).
Environment

Figure 13.18 Genotype × environment interaction and the evolution of reaction norms. Each line represents the reaction norm of a genotype—its expression of a phenotypic character in environments E_1 and E_2. The character states expressed are labeled z_1 and z_2. The arrows indicate the adaptively optimal phenotype in each environment. (A) The genotypes do not differ in the effect of environment on phenotype; there is no G × E interaction. The optimal norm of reaction cannot evolve in this case, because no genotype matches the arrows. (B) The effect of environment on phenotype differs among genotypes; G × E interaction exists. The genotype with the norm of reaction closest to the optima in E_1 and E_2 (red line) will be fixed. New mutations that bring the phenotype closer to the optimum for each environment may be fixed thereafter. (C) Selection may favor a constant phenotype, irrespective of environment. A genotype with a horizontal reaction norm (red line) may be optimal.

an animal to attain a fixed body size at maturity or metamorphosis, despite variations in nutrition or temperature that affect the rate of growth. The developmental system underlying the character may then evolve so that it resists environmental influences on the phenotype (Scharloo 1991). Conrad Waddington, one of the first biologists to integrate developmental biology and evolutionary biology, referred to this phenomenon as **canalization**.

Waddington (1953) used the concept of canalization to interpret some curious experimental results. A crossvein in the wing of *Drosophila* sometimes fails to develop if the fly is subjected to heat shock as a pupa. By selecting and propagating flies that developed a crossveinless condition in response to heat shock, Waddington bred a population in which most individuals were crossveinless when treated with heat. But after further selection, a considerable portion of the population was crossveinless even without heat shock, and the crossveinless condition was heritable. A character state that initially developed in response to the environment had become genetically determined, a phenomenon that Waddington called **genetic assimilation**.

Although this result is reminiscent of the discredited theory of inheritance of acquired characteristics, it has a simple genetic interpretation. Genotypes of flies differ in their susceptibility to the influence of the environment (heat shock)—that is, they differ in the degree of canalization, so that some are less easily deflected into an aberrant developmental pattern. Selection for this pattern favors alleles that canalize development into the newly favored pathway. As such alleles accumulate, less environmental stimulus is required to produce the new phenotype. The finding that genetic assimilation does not occur in inbred populations that lack genetic variation supports this interpretation (Scharloo 1991). More recently, genes that confer canalization have been identified. One is *heat shock protein 90* (*Hsp90*), the protein product of which stabilizes a wide variety of signal-transducing proteins in eukaryotes. When *Hsp90* function is impaired by mutation or chemical treatment, morphological abnormalities of all kinds appear in *Drosophila* and the wild mustard *Arabidopsis* (Rutherford and Lindquist 1998; Queitsch et al. 2002; Figure 13.19). The variation in any such abnormal feature has a polygenic basis, as shown by the response to selection for increased or decreased expression in experimental populations of *Drosophila* (Figure 13.19B). Suzanne Rutherford and Susan Lindquist (1998), who described these effects, suggest that the buffering effect of *Hsp90* enables genetic variation to accumulate because it is not expressed and is sheltered from selection, but that release of such variation may have promoted evolutionary change in otherwise canalized characteristics.

Phenotypic plasticity

In many species, *adaptive phenotypic plasticity* has evolved: a genotype has the capacity to produce different phenotypes, suitable for different environmental conditions (Pigliucci 2001; West-Eberhard 2003). Phenotypic plasticity includes rapidly reversible changes in morphology, physiology, and behavior, as well as "developmental switches" that cannot

(A)

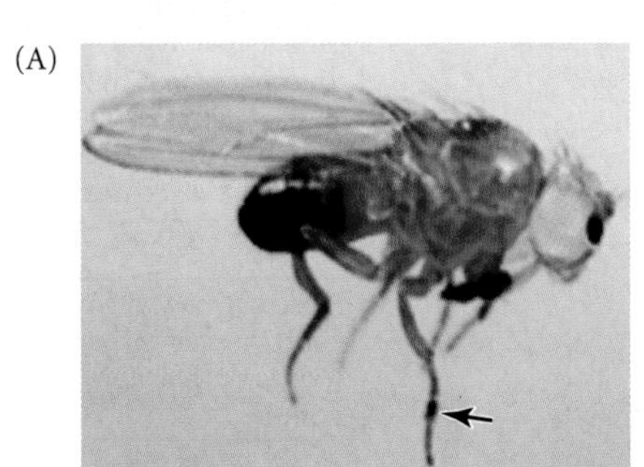

(B)

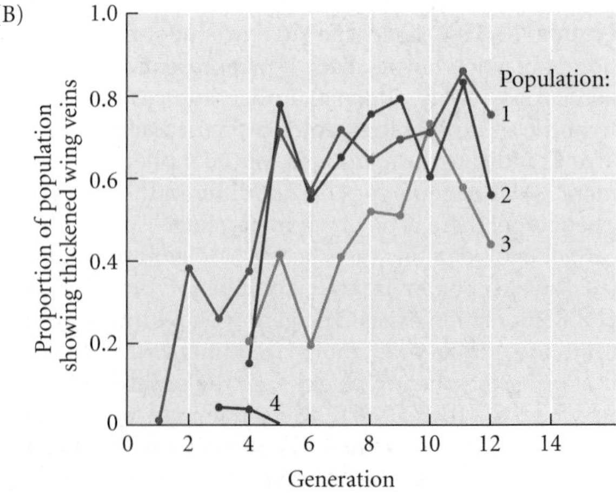

Figure 13.19 Revelation of cryptic genetic variation when *Hsp90* gene activity is deficient. (A) Abnormalities include deformed foreleg and development of sex comb on the wrong leg (arrow, left) and thickened wing veins (right). (B) The response to selection for the wing-vein trait in four different populations of *Hsp90*-compromised flies. (From Rutherford and Lindquist 1998).

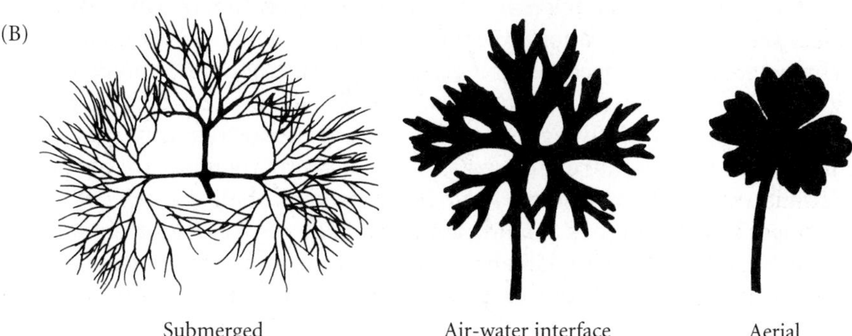

Figure 13.20 Examples of phenotypic plasticity. (A) Larvae of the geometrid moth *Nemoria arizonaria* that hatch in the spring (left) resemble the oak flowers (catkins) on which they feed. Those that hatch in the summer (right) feed on oak leaves and resemble twigs. (B) The form of a leaf of the water-crowfoot *Ranunculus aquatilis* depends on whether it is submerged, aerial, or situated at the air-water interface during development. (A, photos courtesy of Erick Greene; B from Cook 1968.)

be reversed during the organism's lifetime (Figure 13.20A). In some semiaquatic plants, for instance, the form of a leaf depends on whether it develops below, above, or on the surface of water (Figure 13.20B). Adaptive phenotypic plasticity of this kind has evolved by natural selection for those genotypes with norms of reaction that most nearly yield the optimal phenotype for the various environments the organism commonly encounters (Schlichting and Pigliucci 1998).

If an adaptive reaction norm includes the optimal phenotypes for a wide range of environments, phenotypic plasticity may shield the genotype from selection for further change, and thus reduce the rate of evolution. But might phenotypic plasticity also lead the way to new adaptations? Several authors, including Mary Jane West-Eberhard (2003), have proposed that phenotypic changes induced directly by a new environment may sometimes prove adaptive, and may later be genetically assimilated (or genetically accommodated, in her terminology). Thus adaptation would be initiated not by selection on genetic variation but by phenotypic plasticity. West-Eberhard cites many cases in which a phenotype that is part of a plastic reaction norm of one species is a fixed, constant phenotype in a related species. For example, the larva of the sphinx moth *Manduca quinquemaculata* is black if it develops at low temperature and green at high temperature, whereas the larva of its close relative *M. sexta* is green at all temperatures (Suzuki and Nijhout 2006). Species of *Anolis* lizards that typically perch on narrow twigs have much shorter hindlimbs than species that perch on broad surfaces (see Figure 6.22), and a similar difference, although of much smaller magnitude, was found between samples of a single species (*Anolis sagrei*) that were reared in containers with narrow versus broad sticks to perch on (Losos et al. 2000). Phenotypic plasticity is likely to aid adaptation to a new

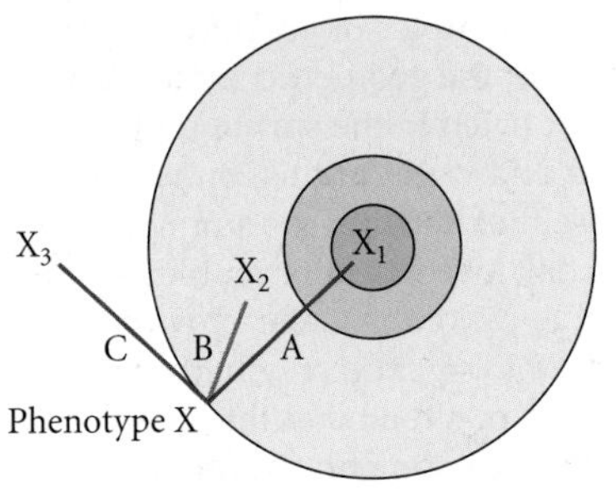

Figure 13.21 Consequences of phenotypic plasticity in a population with phenotype X when placed in a novel environment, represented by the concentric circles; the optimal phenotype for the new environment is within the center circle. The phenotype is implicitly understood to be a composite of several traits. Vector A represents a genotype with perfect phenotypic plasticity. It develops the optimal phenotype, and no evolutionary change ensues. Vector B is a genotype whose plasticity does not encompass the optimum. Selection may favor enhanced and somewhat reoriented plasticity. Vector C is a genotype with nonadaptive plasticity that moves the phenotype away from the optimum, and which does not facilitate adaptation to the novel environment. (After Ghalambor et al. 2007.)

environment if the ancestral norm of reaction points toward the new optimum (e.g., vector B in Figure 13.21). But a reaction norm that points toward the new optimum, such as the plasticity of leg length in *Anolis sagrei*, may be an ancestral adaptation that evolved by the traditional mechanism of selection on genetic variation. If so, the pioneering role of phenotypic plasticity may be limited. Few, if any, cases of adaptation initiated by phenotypic plasticity have been well documented so far (de Jong 2005; Ghalambor et al. 2007).

Evolution of variability

Whereas *variation* refers to the differences actually present in a sample or a species, the word **variability**, in the strict sense, refers to the ability, or potential, to vary (Wagner et al. 1997). For example, in insects, the number of compound eyes (two, or in a few species, none) seems able to vary much less than the number of units (ommatidia) that compose each compound eye. In mammals, because of the developmental correlation between the size of their bodies and the size of their brains and intestines, some conceivable variations—large bodies with tiny brains, for instance—are seldom or never seen. Developmental processes therefore affect variability, the extent to which genetic variation can be expressed as phenotypic variation. Does variability depend solely on immutable "laws" of development, or does it evolve by natural selection? This question applies to both variability in individual characters and correlations among characters.

The variability of individual characters is affected by the evolution of canalization. A character that is insensitive to alteration by environmental factors is ENVIRONMENTALLY CANALIZED. A character may also become GENETICALLY CANALIZED; that is, it may acquire low sensitivity to the effects of mutations. In such instances, the phenotype may remain unchanged even if the genes underlying its development vary considerably.

Threshold traits, for example, are expressed as discrete alternatives but are controlled by polygenic variation rather than by single loci. The polygenic variation is not expressed phenotypically unless development is perturbed substantially (beyond a threshold) by a large enough genetic or environmental change. For example, in natural populations of *Drosophila melanogaster* and related species, there is almost no variation in the number of bristles (four) on the scutellum (part of the thorax). In homozygotes for the *scute* (*sc*) mutation, however, bristle number is variable, because of the expression of polygenic variation at other loci (Figure 13.22A). Thus this mutation breaks down canalization, and conversely, the normal allele at the *scute* locus may be considered to exert genetic canalization. It has an *epistatic* effect on the quantitative loci, overseeing their phenotypic expression. Using an experimental population that was homozygous for *scute* and was therefore variable in bristle number, James Rendel and colleagues (1966) imposed selection against variation by breeding from the least variable families (i.e., those that most consistently had two bristles). Within about 30 generations, the phenotypic variation became greatly reduced (Figure 13.22B). The investigators apparently had selected for genes that canalized development into a new pathway.

Can natural selection produce the same result? Wagner et al. (1997) and Kawecki (2000) have explored the evolution of canalization mathematically. According to their studies, alleles for environmental canalization should increase if there is prolonged stabilizing selection against deviations from an optimal phenotype. The evolution of genetic canalization, however, would be expected only under rather restricted conditions. If directional selection fluctuates rapidly in direction, canalization of the phenotype may be advantageous because it prevents a response to selection in one generation that is maladaptive a few generations later. Under consistent, long-term stabilizing selection, canalization evolves only if selection is neither too weak nor too strong. If stabilizing selection is weak, alleles that prevent the phenotypic expression of mutations have too slight a selective advantage to increase. If stabilizing selection is strong, it eliminates new mutations so fast that few individuals deviate from the optimum, so there is little selection for alleles that prevent the phenotypic expression of the mutations.

The theory of canalization may explain why some characters—such as some synapomorphies of higher taxa—have remained unchanged for vast periods of time. For example, the earliest known Devonian tetrapods had a variable number of about eight or nine

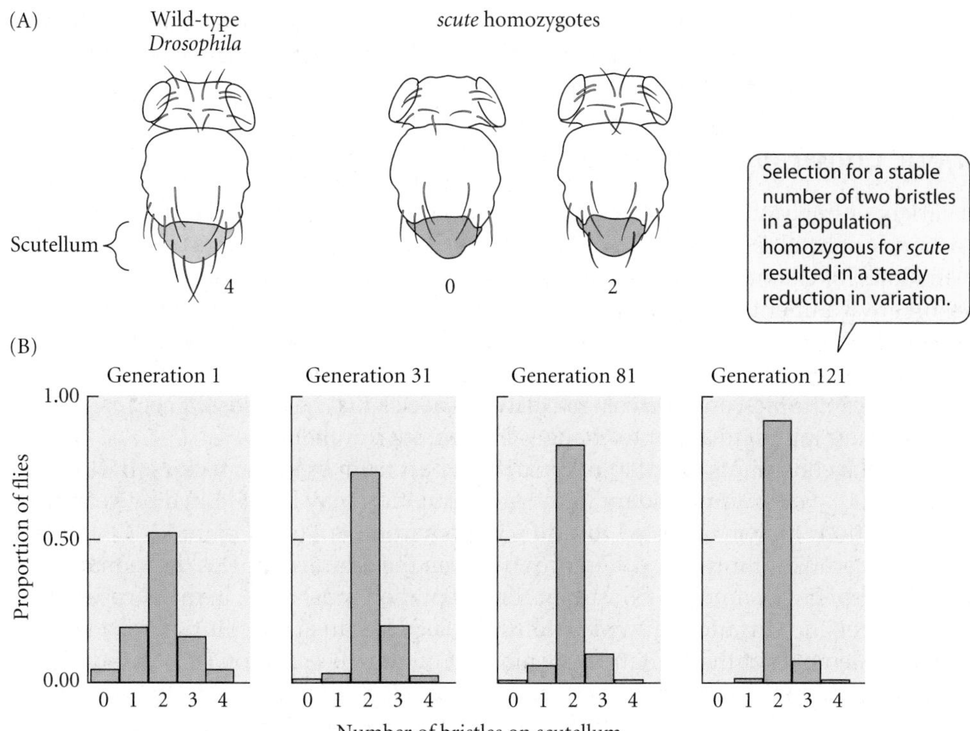

Figure 13.22 Canalization by artificial selection. (A) The top of the head and thorax of *Drosophila melanogaster*. The posterior part of the thorax, the scutellum, bears four bristles in wild-type flies, but a variable number (e.g., zero or two) in *scute* homozygotes. (B) Selection for a stable number of two bristles in a population homozygous for *scute* resulted in a steady reduction of variation, shown here at 1, 31, 81, and 121 generations of selection. (Based on data from Rendel et al. 1966.)

toes (see Chapter 4). Soon afterward, however, tetrapods "settled on" five-digit (pentadactyl) limbs, and almost no tetrapod vertebrates since then have had more than five digits. Therefore, there was no inescapable rule that feet could have no more than five toes, but the developmental processes evolved so that the maximum digit number came to be constrained. We return to the problem of phylogenetically "conservative" features in Chapter 22.

The hypothesis of morphological integration (Olson and Miller 1958), or more generally, **phenotypic integration** (Pigliucci and Preston 2004), holds that functionally related characteristics should be genetically correlated with one another; thus the characters should remain coordinated even as they vary within species and as they evolve. Günter Wagner and Lee Altenberg (1996) have shown theoretically that prolonged directional selection favors modifier alleles that enhance a pleiotropic correlation between functionally related traits along an axis pointing toward the optimum for the characters (marked with an asterisk in Figure 13.16A). For example, if it were functionally important for the upper and lower mandibles of a bird's bill to be the same length, then selection for a longer bill would include selection for alleles that coordinate the development of the two mandibles, creating a pleiotropic correlation between them.

Much of the evidence bearing on this hypothesis comes from studies of correlations among the various parts of flowers, since the match between flower structure and the size and position of a pollinating animal is thought to affect plants' reproductive success (Armbruster et al. 2000, 2004). There is more information from phenotypic than genetic correlations, but flower structure usually shows a strong correspondence between them. In some cases, the hypothesis of phenotypic integration appears to be upheld. For example, the sexual structures of the amazing Australian triggerplants (*Stylidium*) form a column that snaps forward to hit an insect on the back when the insect is properly positioned on the landing platform formed by the lower petals and touches a sensitive point on the column (Figure 13.23). The phenotypic correlation between the length of the column and of the landing platform is very high (0.81 in one population studied). In general, floral structures are more highly correlated with one another than with vegetative structures. On the whole, however, the evidence that genetic correlations reflect selection for proper function, rather than developmental

Figure 13.23 The flower of the triggerplant *Stylidium bicolor*. The column consists of fused stamens and pistil. When an insect lands on the platform formed by the lower petals and touches the base of the column, the column snaps forward. Thus the precision of pollen placement (and pickup of pollen from other flowers) depends on the dimensions of the column and the landing platform. The phenotypic correlation between these characters is quite high (0.81). (From Armbruster et al. 2004.)

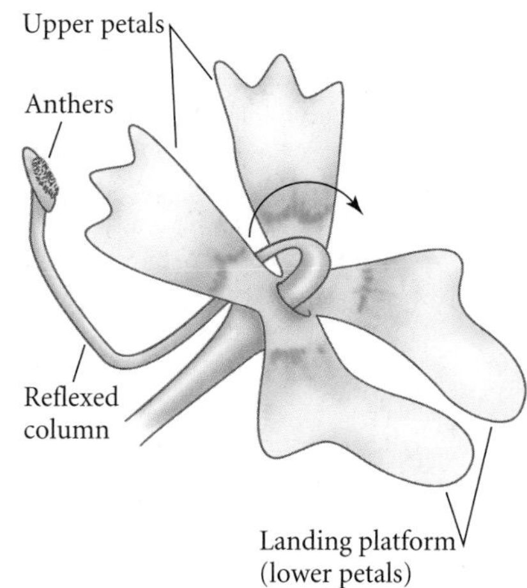

pathways that may or may not be adapted to the species' special ecological situation, is equivocal (Armbruster et al. 1999; Herrera et al. 2002). Just how prevalent adaptive phenotypic integration is remains to be seen.

Genetic Constraints on Evolution

The variety of finely attuned, often astonishing adaptations of organisms has evolved by the action of natural selection on genetic variation. But failures of adaptation also require explanation. For example, relatively few species of plants have adapted to saline or other stressful environments, even if they grow nearby (Bradshaw 1991); there appear to be "empty niches" for kinds of organisms that have evolved less often than ecological opportunity might suggest (e.g., blood-feeding bats have evolved only once and are limited to the American tropics, and the vast majority of species that have existed became extinct because of their failure to adapt to changes in their environment.

Some of the constraints on adaptation arise in part from extrinsic factors (Barton and Partridge 2000). For example, adaptation of a population may be inhibited by gene flow from populations that are adapted to a different environment (see Figure 12.12). But this occurs only because of intrinsic differences between the effects of alleles that enhance fitness in different environments. Constraints on adaptation must arise, in many cases, from the kind of genetic variation on which natural selection can act. Evolution may be constrained by elements of the **G** matrix: the amount of genetic variation (V_A) in individual characters and genetic correlations (r_G) among characters (Blows and Hoffmann 2005).

Most characters display genetic variance, but there are exceptions; for example, no V_A was detected in studies of the feeding response of some beetle species (*Ophraella*) to certain plants (Futuyma et al. 2005) or in heavy metal tolerance in species of grasses that have failed to evolve tolerance (Macnair 1997). Some artificial selection experiments have achieved very little change, or none at all: for example, no increase in resistance to desiccation occurred in laboratory populations of a rainforest-inhabiting species of *Drosophila* (Hoffmann et al. 2003). Adaptation in some cases appears to have been based on rare mutations rather than on the store of genetic variation that most populations harbor. For example, resistance to organophosphate insecticide in the mosquito *Culex pipiens* is based on a single mutation that spread worldwide by gene flow, instead of evolving independently in different populations exposed to similar selection (Raymond et al. 2001). Such examples raise the question of whether or not existing genetic variation is a sufficient basis for adaptation to novel selection pressures. Because many of the alleles that contribute to V_A for quantitative characters have deleterious pleiotropic effects, they may not contribute to adaptation, so the "usable" V_A may be far less than meets the eye (Houle et al. 1996). Adaptation may not occur if it requires specific novel mutations that fail to arise.

Several lines of evidence can indicate whether recent adaptive evolution has been based on standing genetic variation or new mutations (Barrett and Schluter 2008). For example, reduced body armor has evolved independently in many freshwater populations of three-spined sticklebacks (*Gasterosteus aculeatus*), but is based on the same allele at the *ectodysplasin* (*Eda*) locus (**Figure 13.24**). This allele arose more than 2 Mya, based on its sequence divergence from the ancestral allele that characterized fully armored populations, and has been found at low frequency in these populations (Colosimo et al. 2005). In contrast, evidence that a mutation was beneficial as soon as it occurred is provided by reduced synonymous variation in the vicinity of a beneficial allele (a selective sweep; see Chapter 12), because the copies of an advantageous allele that persisted in the population at low frequency for a long time before it became advantageous would have been separated by recombination from specific linked SNP variants, and would show little evidence of a selective sweep (Przeworski et al. 2005). By this criterion, some recent adaptations appear to have been shaped from standing genetic variation, but new mutations have also contributed. Recall, for example, that many genes in the human genome show evidence of selective sweeps.

Because of pleiotropy, variation in any one character is seldom independent of other characters. For instance, in a study of 70 skeletal measurements that vary among inbred strains of mice, Günter Wagner and colleagues (2008) detected 102 QTL loci; the aver-

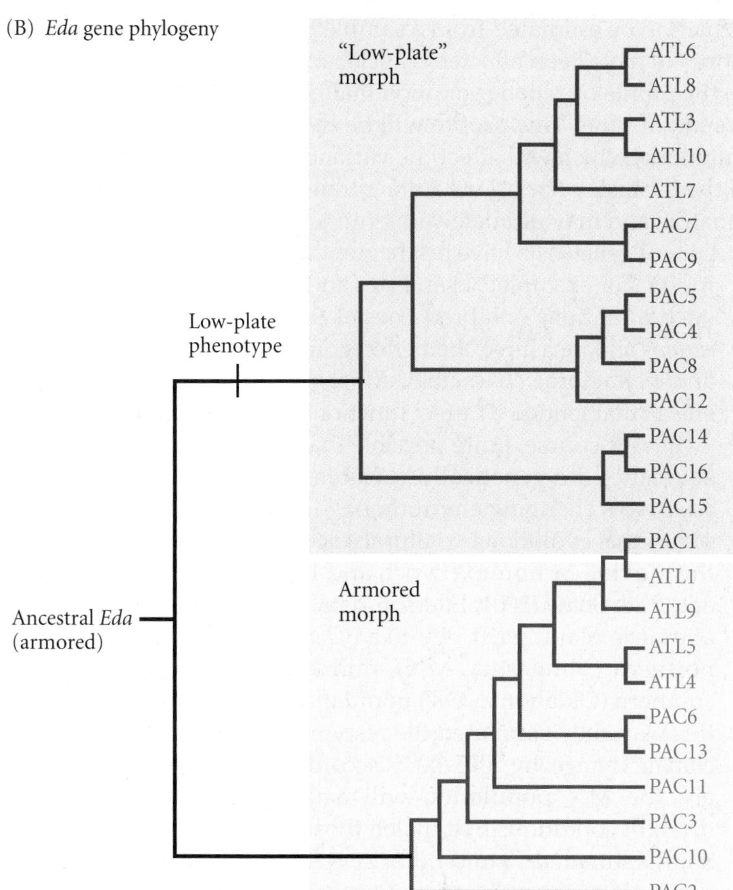

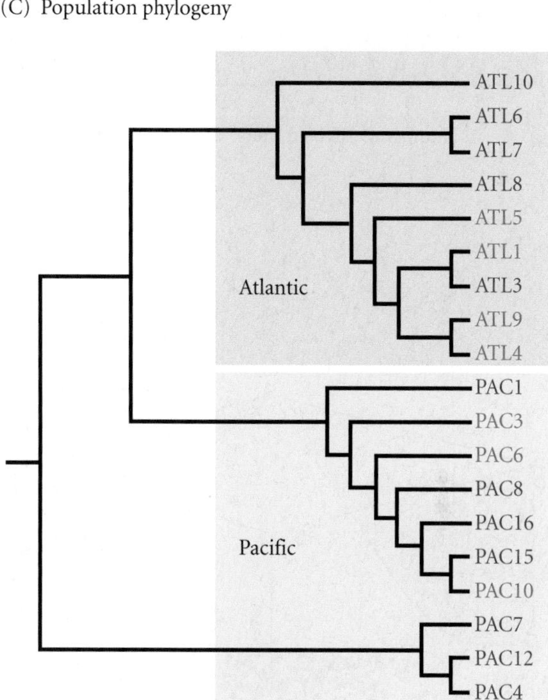

Figure 13.24 Evolution based on standing variation in the stickleback *Gasterosteus aculeatus*. (A) Cleared and stained specimens of (left) the ancestral, completely armored type, found in marine and some freshwater populations, and (right) the "low-plate" morph, found in many populations in northern Eurasia and North America. The low-plate phenotype is caused by an allele of the *Ectodysplasin* (*Eda*) gene, which encodes a signaling protein that is required for differentiation of ectodermal features. (B) A phylogeny of *Eda* genes from Pacific and Atlantic regions of the species' distribution shows that alleles from almost all the low-plate populations (blue numbers) have a single origin. Almost all low-plate populations occupy fresh water. (C) A phylogeny of the populations based on SNPs at 25 random loci shows that low-plated populations (blue) have evolved independently. Therefore the *Eda* allele has been maintained in the species, and has increased in frequency independently in different populations. (After Colosimo et al. 2005.)

age locus affected 7.8 traits, and some affected as many as 30 (Figure 13.25). Pleiotropy creates genetic correlations among characters, and the result is that there will exist more genetic variation for some combinations of trait values than for others (see Figures 13.15B and 13.16). As we have seen, divergence among populations and species has been found (in some studies, although not all) to have occurred along the directions of greatest genetic variation. Lower genetic variation in other directions (i.e., other character combinations) is not an absolute constraint on selection, but evolution toward the optimum would in that case be slow and indirect (see Figure 13.15C,D). It is, however, quite possible that each of several characteristics may display V_A individually, yet for there to exist no genetic variation at all for certain character combinations. Jason Mezey and David Houle (2005) detected no such constraint in a study of 20 aspects of wing shape in *Drosophila melanogaster*, suggesting that there exist no constraints on the evolution of wing shape. In contrast, Mark Blows and colleagues (2004) have described a case that does imply constraint. Female *Drosophila serrata* select males on the basis of their relative concentration of nine cuticular hydrocarbons (CHCs) in the male's body surface. Blows et al. found that all nine components are genetically variable, but the axis of greatest

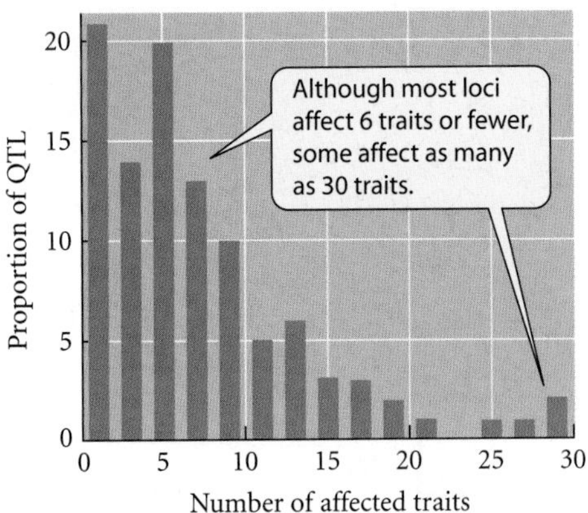

Figure 13.25 A frequency distribution of the number of skeletal traits affected by pleiotropic QTL in the mouse. (After Wagner et al. 2008.)

genetic variation in CHC combinations ran nearly at a right angle to the direction of sexual selection (Figure 13.26). Similar disparity could well exist between the genetic variation and the direction of selection on combinations of traits that are necessary for adaptation to a change in ecological circumstances.

The **G** matrix that can be estimated from a sample taken from a natural or laboratory population will have been affected by selection and genetic drift, and so might differ from the profile of genetic variation that may be available for adaptation over a long span of time. This profile will be determined by the rates and effects of new mutations, which will affect the variance (V_m, referred to earlier in this chapter) and the intrinsic correlations among traits. The MUTATION MATRIX, **M**, will affect the **G** matrix, and may ultimately determine whether there are absolute constraints. Evolutionary geneticists have just begun the arduous work of estimating **M**. For example, Susan Estes and coworkers (2005) accumulated mutations in inbred lines of the nematode *Caenorhabditis elegans* and measured their effects on life history, morphological, and behavioral characters. Most mutations had pleiotropic effects that tended to affect functionally related traits.

It is, of course, quite possible that even if the relevant characteristics are genetically variable, a population may fail to adapt to a changing environment, if the environment changes so fast that evolution by natural selection cannot keep pace with the moving optimum (Lynch and Lande 1993). Julie Etterson and Ruth Shaw (2001; Etterson 2004) grew seedlings of an annual prairie plant, partridge-pea (*Chamaecrista fasciculata*), from northern (Minnesota, MN), mid-latitude (Kansas, KS), and southern (Oklahoma, OK) populations at all three latitudes. In this way, they simulated the response of populations to global climate change: by 2025–2035, according to climate change models, the MN population will experience temperature and drought conditions that match those of KS today. Etterson and Shaw estimated V_A and r_G for several traits in each population grown in each environment, and estimated the direction and intensity of selection on each trait by its correlation with seed production. Most features displayed heritable variation, but the rate of evolution of each trait toward the expected future optimum was significantly reduced by genetic correlations among traits that were antagonistic to the direction of selection (Figure 13.27A). The investigators concluded that in the MN popula-

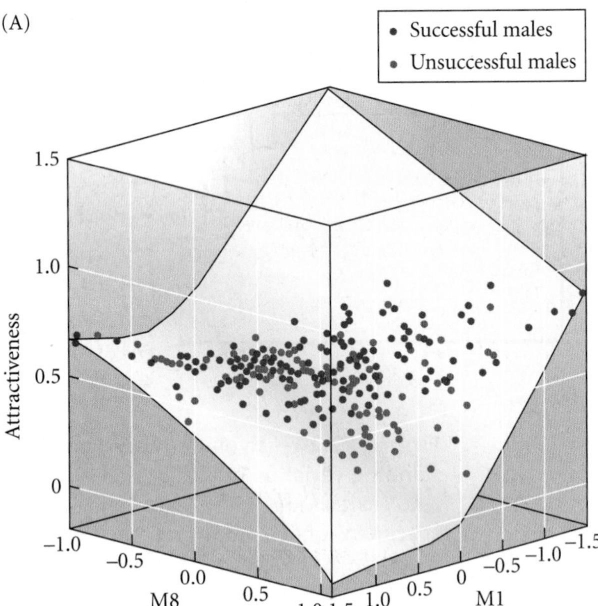

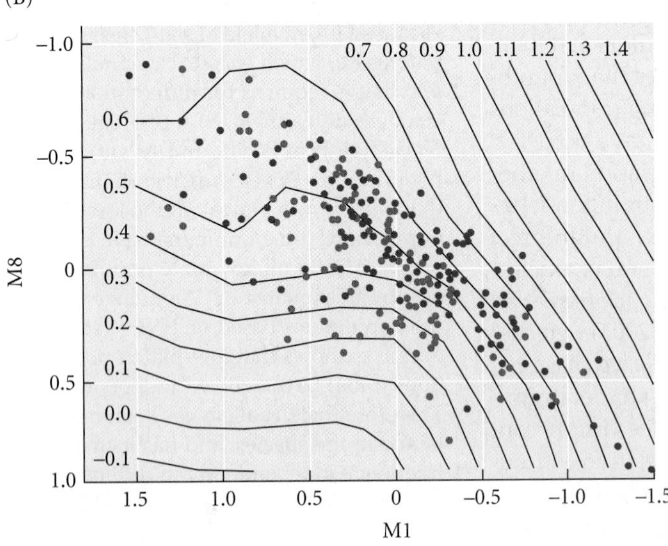

Figure 13.26 A possible genetic constraint on the response to selection. The attractiveness of male *Drosophila serrata* to females is plotted in relation to genetic variation in nine cuticular hydrocarbons, expressed as two axes of variation, M1 and M8. Fitness, defined as attractiveness, is displayed as (A) a three-dimensional surface and (B) as a topography. Fitness increases from bottom to top in A, and from the lower left to the upper right in B. The points show the genetic value of M1 and M8 of successful males (blue circles) and unsuccessful males (red circles). The axis of maximal genetic variation is almost perpendicular to the slope of the fitness surface. Therefore, these traits are not likely to respond to the directional selection imposed by female preference. (After Blows et al. 2004.)

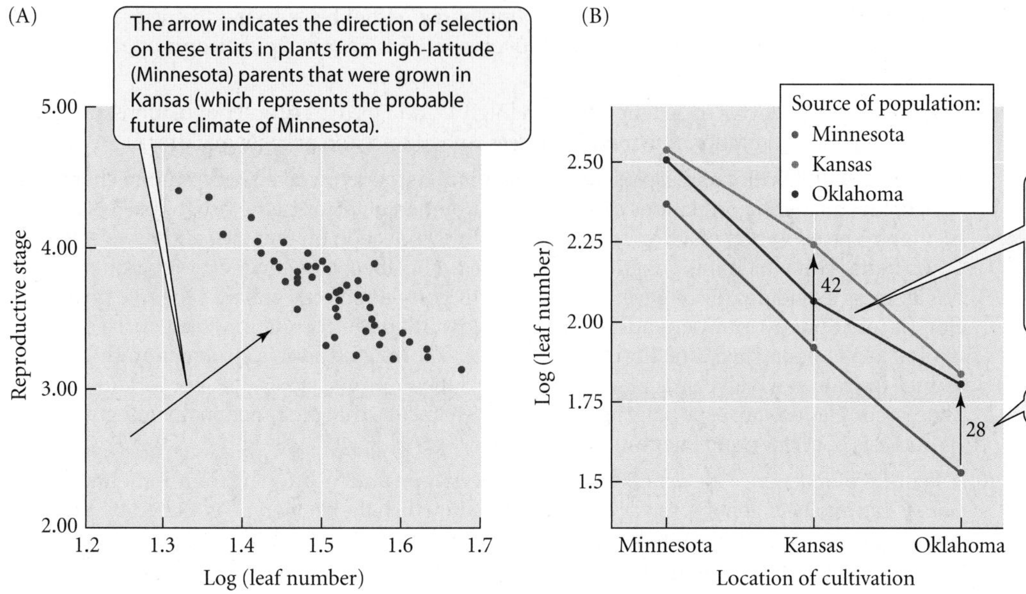

Figure 13.27 Why the partridge-pea might not adapt to climate change. (A) The genetic correlation between two traits (leaf number and reproductive stage, representing rate of development) is shown by the points. There is little genetic variation along the axis of selection. (B) Average leaf number of plants derived from populations in Minnesota, Oklahoma, and Kansas, grown in each of those locations. (After Etterson and Shaw 2001.)

tion, some key traits would not evolve to match the mean of today's KS population by the time the MN population suffers a Kansas-like climate (Figure 13.27B). They suspect that adaptation would take even longer than calculated from the genetic parameters, because seed production of the MN population in the hot, dry KS environment is so greatly diminished that the population will probably dwindle and lose genetic variation by genetic drift. Because of CO_2 produced by human combustion of fossil fuels, climates are changing at rates more than 100 times greater than during the Pleistocene glacial/interglacial oscillations, and it seems likely that thousands of species, like the partridge-pea, will fail to adapt fast enough to avert extinction.

Summary

1. Quantitative trait loci (QTL) can be mapped using molecular or other markers. The variation in many traits is caused by variation at several or many loci, some with large and others with small effects. For certain characters, some of the genes have been identified and their function is known.
2. Variation (variance) in a phenotypic trait (V_P) may include genetic variance (V_G) and variance due to the environment (V_E). Genetic variance may include both additive genetic variance (V_A) due to the additive effects of alleles and nonadditive genetic variance due to dominance and epistasis. Only the additive variance creates a correlation between parents and offspring (and it can be measured by this correlation). Thus only V_A enables response to selection.
3. If alleles that contribute to variation in a polygenic trait are selectively neutral and change in frequency by genetic drift, the short-term rate of evolution depends on the effective population size, but the long-term rate is proportional to the rate of polygenic neutral mutation. Evolutionary rates are often lower than the neutral model predicts, implying that stabilizing selection or purifying selection has acted.
4. The ratio V_A/V_P is the heritability (h^2_N, or simply h^2) of a trait. Heritability is not fixed, but depends on allele frequencies and on the amount of phenotypic variation induced by environmental variation. The short-term effect of selection ("response" to selection) on a character can be predicted from the heritability and the strength of selection.
5. Most, although not all, characters show substantial genetic variance in natural populations and may therefore evolve rapidly if selection pressures change. Many examples of rapid evolution, within a century or less, have been described.
6. Artificial selection experiments show that traits can often be made to evolve far beyond the initial range of variation. The response to selection is based on both genetic variation in the original population and new mutations that occur during the experiments.

7. Stabilizing selection is common in natural populations, either because the character is nearly at its optimum or because conflicting selection pressures or negative pleiotropic effects prevent further change.
8. The causes of high levels of genetic variation (high V_A and h^2_N) in natural populations are uncertain, but input by mutation may balance losses due to selection and genetic drift.
9. Linkage disequilibrium and especially pleiotropy cause genetic correlations among characters, which, together with correlations caused by environmental factors, give rise to phenotypic correlations. The evolution of a trait is governed both by selection on that trait and by selection on other traits with which it is genetically correlated. The effect of a genetic correlation depends on its strength and degree of permanence. Genetic correlations can enhance the rate of adaptation (if functionally interdependent features show adaptive correlation), can cause a trait to evolve in a maladaptive direction (if selection on a correlated trait is strong enough), or may reduce the rate at which characters evolve toward their optimal states. It is not certain whether genetic correlations are especially strong among characters that are functionally integrated (the hypothesis of phenotypic integration).
10. The norm of reaction—the expression of the phenotype under different environmental conditions—can evolve if genotypes vary in the degree to which the phenotype is altered by the environment in which an individual develops. Some characters exhibit adaptive phenotypic plasticity, whereas selection in other cases favors constancy of phenotype despite differences in environment.
11. Canalization is the buffering of development against alteration by environmental or genetic variation. Canalized characters include threshold characters, in which underlying polygenic variation is not phenotypically expressed unless a drastic mutation or environmental perturbation breaks down canalization. Canalization can evolve under some circumstances. The evolution of canalization may explain the constancy of some characters over vast periods of evolutionary time.
12. Although many characters are genetically variable and can evolve rapidly, evolution appears often to be constrained, partly because of limitations on genetic variation, some of which may be due to the origin of variation by mutation. Genetic variation appears to be rare or lacking in some traits, but more often, pleiotropic genetic correlations may be antagonistic to the direction of selection on multiple characters. Understanding genetic constraints is the key to understanding phenomena that range from phylogenetically conservative characters to extinction, both in the past and in the near future.

Terms and Concepts

additive genetic variance
canalization
correlated selection
environmental correlation
environmental variance
epistasis
G matrix
genetic assimilation
genetic correlation
genetic variance
genotype × environment (G × E) interaction
heritability
intensity of selection
norm of reaction
phenotypic correlation
phenotypic integration
phenotypic plasticity
phenotypic variance
QTL mapping
quantitative genetics
realized heritability
response to selection
selection differential
selection gradient
selection plateau
threshold trait
trade-off
variability

Suggestions for Further Reading

Introduction to Quantitative Genetics by D. S. Falconer and T. F. C. Mackay (fourth edition, Longman Group Ltd., Harlow, UK, 1996) is a widely read, clear introduction to the subject. D. A. Roff, in *Evolutionary Quantitative Genetics* (Chapman and Hall, New York, 1997), provides a comprehensive treatment of the evolution of quantitative traits. An advanced treatment is *Genetics and Analysis of Quantitative Traits* by M. Lynch and J. B. Walsh (Sinauer Associates, Sunderland, MA, 1998).

An important review of some recent research is T. F. C. Mackay's "The genetic architecture of quantitative traits" (*Annual Review of Genetics* 35: 303–339, 2001). Phenotypic plasticity, canalization, and related topics are the subject of C. D. Schlichting and M. Pigliucci's *Phenotypic Evolution: A Reaction Norm Perspective* (Sinauer Associates, Sunderland, MA, 1998).

A useful overview of genetic constraints on adaptation is "A reassessment of genetic limits to evolutionary change" by M. W. Blows and A. A. Hoffmann (*Ecology* 86: 1371–1384, 2005.)

Problems and Discussion Topics

1. Under artificial selection for increased body weight, what will be the response to selection (R), after one generation, for the following values of phenotypic variance (V_P), additive genetic variance (V_A), environmental variance (V_E), and selection differential (S)? (a) $V_P = 2.0$ grams^2, $V_A = 1.25$ g^2, $V_E = 0.75$ g^2, $S = 1.33$ g; (b) $V_P = 2.0$ g^2, $V_A = 0.95$ g^2, $V_E = 1.05$ g^2, $S = 1.33$ g; (c) $V_P = 2.0$ g^2, $V_A = 1.25$ g^2, $V_E = 0.75$ g^2, $S = 2.67$ g. (Answer for part a: mean weight will increase by about 0.83 g.) If the parameters remain the same for successive generations of selection, and the initial mean weight is 10 g, what is the expected mean after two generations of selection in each case?
2. If most quantitative genetic variation within populations were maintained by a balance between the origin of new mutations and selection against them, then most mutations might be eliminated before environments could change and favor them. If the "residence times" of most mutations were short enough, the alleles that distinguish different populations or species would generally not be those found segregating within populations (Houle et al. 1996). How might one determine whether the alleles that contribute to among-population differences in the means of quantitative traits are also polymorphic within populations?
3. It has been suggested that genetic correlation between the expressions of a trait in the two sexes is responsible for some apparently nonadaptive traits, such as nipples in men and the muted presence in many female birds of the bright colors used by males in their displays (Lande 1980). Suggest ways of testing this hypothesis. What other traits might have evolved because they are genetically correlated with adaptive traits, rather than being adaptive themselves?
4. Debate the proposition that paucity of genetic variation and genetic correlations do not generally constrain the rate or direction of evolution.
5. Consider a character that is typical of the species in a higher taxon and may indeed be an important synapomorphic character for the clade. (For example, the number of petals is such a character in many genera and families of plants.) How might you decide whether this consistency is a result of intrinsic, unchangeable developmental "rules" or of a history of selection for canalization?
6. Traditional quantitative genetics is based on a theory of multiple anonymous loci, the functional roles of which are unknown. Using the "candidate loci" approach, the sequence and function of some of these loci is now being discovered. In what ways is this important for understanding the evolution of phenotypic traits?
7. How can adaptive phenotypic plasticity be distinguished from nonadaptive phenotypic plasticity?
8. Would you expect the pleiotropic effects of mutations to be most prevalent among functionally related traits, as Estes et al. described in *Caenorhabditis*? Why or why not?
9. Consider the number of traits a gene pleiotropically affects, and the magnitude of the effect of different alleles on each of these characteristics. How would these variables affect the rate of adaptation to a novel environment? (See Orr 2000 and Wagner et al. 2008.)

CHAPTER 14

The Evolution of Life Histories

An unusual life history. In small, carnivorous Australian marsupials of the genus *Antechinus*, all males die in their first year of life, after a mating event that is highly synchronized throughout the population. Male mortality is the result of diverse physiological effects of extremely high levels of testosterone and cortisols. The hormonal effects leading to male death, and the very short life span, are unusual among mammals, and have evidently evolved by sexual selection. (Photo © Gordon Garradd/Photolibrary.com.)

Much of the richness of biology lies in the diverse, often astonishing, and sometimes bizarre adaptations of organisms—features that have evolved because they increased fitness relative to ancestral characteristics. "Fitness" appears as an abstract variable in much of the theory of natural selection as we have considered it thus far, but as we have seen, fitness has several major components. These include survival, female fecundity, and male mating success, which in turn are the results of real anatomical, physiological, and cellular features. When we turn our attention to the components of fitness, we encounter variations and even paradoxes that cry out for explanation.

Consider, for example, some differences among species in various aspects of their life cycles. Sea anemones and corals can live for close to a century, bristlecone pine trees (*Pinus aristata*) have survived for 4600 years, and vegetatively propagating clones of quaking aspen (*Populus tremuloides*) can live for more than 10,000 years. In contrast, annual plants die less than a year after germinating, and many small animals, such as some rotifers, live for at most a few weeks. Fecundity likewise varies. Many bivalve molluscs and other marine invertebrates release thousands or millions of tiny eggs in each spawning, whereas a blue whale (*Balaenoptera musculus*) gives birth to a single offspring that weighs as much as an adult elephant, and a kiwi (*Apteryx*)

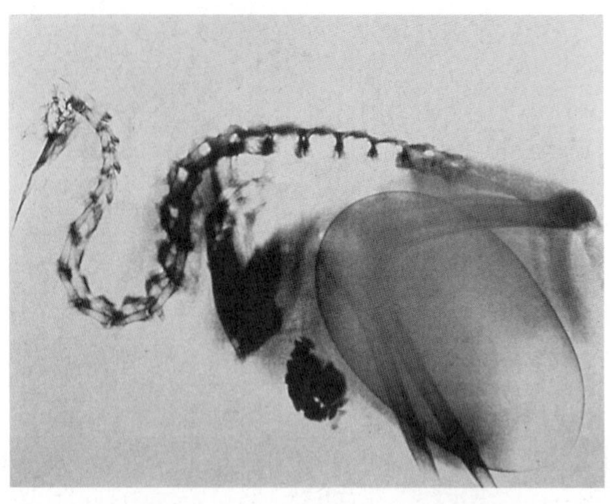

Figure 14.1 This X-ray of a kiwi (*Apteryx*) shows the bird's enormous egg, which weighs 25 percent of the female's body weight. (Photo courtesy of Otorohanga Zoological Society.)

lays a single egg that weighs 25 percent as much as its mother (Figure 14.1). Some birds, such as cranes and parrots, form lifelong pair-bonds, while in other birds, such as many grouse and hummingbirds, a male may mate with many females and form a pair-bond with none (Figure 14.2). In some such species, males have baroque ornaments that may be necessary for mating success. Many species, such as humans, reproduce repeatedly, whereas others, such as century plants (*Agave*) and some species of salmon (*Oncorhynchus*), reproduce only once and then die. Reproductive age may be reached rapidly or slowly. A newly laid egg of *Drosophila melanogaster* may be a reproducing adult 10 days later, and a parthenogenetic aphid may carry an embryo even before she herself is born. In contrast, periodical cicadas feed underground for 13 or 17 years before they emerge, reproduce, and then die within a month (Figure 14.3). Many animals and plants have separate sexes, but earthworms and many plants are SIMULTANEOUS HERMAPHRODITES that have both sexual functions at the same time, whereas certain sea bass, squash, and other SEQUENTIAL HERMAPHRODITES develop first as one sex and later switch to the other sex. The process of reproduction often involves sex, accompanied by genetic recombination, but many organisms reproduce by parthenogenesis—development from an unfertilized egg.

What accounts for such extraordinary variation in species' survival and reproduction—the very features that we would expect to be most intimately related to their fitness? There are apparently many ways to achieve high fitness, or reproductive success. To make sense

Figure 14.2 (A) Inflatable sacs, white breast feathers, and spiky tail feathers are displayed by a male greater sage grouse (*Centrocercus urophasianus*). In this species, males compete for females and do not form a pair-bond or care for the offspring. (B) Atlantic puffins (*Fratercula arctica*) are sexually monomorphic seabirds that form pair-bonds. Both parents care for the offspring. (A © Carol Walker/Naturepl.com; B © INTERFOTO Pressebildagentur/Alamy.)

(A)

(B)

(A)

(B)

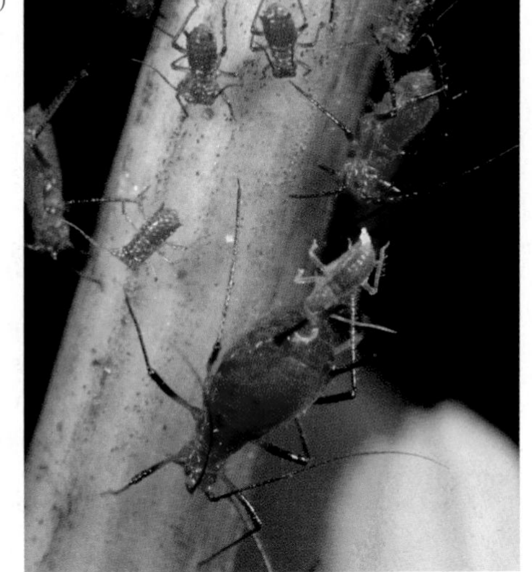

Figure 14.3 Two insects that differ in generation time and rate of increase due to a great difference in the respective ages at which reproduction begins. (A) Adult and nymph of a periodical cicada (*Magicicada septendecim*). Emerging after 17 years spent underground as a nymph, feeding on sap in plant roots, the adult cicada is now ready to reproduce. Periodical cicadas have the longest generation time known among insects. (B) Aphids (*Uroleucon nigrotuberculatum*) belonging to several generations cluster on a goldenrod stalk. These insects descend by parthenogenetic reproduction from a single female. Aphids give birth to live young that, in some species, may start to develop offspring of their own even before they are born. (A © Jim Zipp/Photo Researchers, Inc.; B photo by David McIntyre.)

of all this diversity, evolutionary ecologists and evolutionary geneticists have developed theories of the evolution of **life histories**: the age-specific probabilities of survival and reproductive success characteristic of a species.

Individual Selection and Group Selection

Why do codfish produce hundreds of thousands of eggs? Is it to compensate for the high mortality of both eggs and juveniles, thus helping to ensure the survival of the species? Why do people die of "old age"? Is it to make room for the vigorous new generation that will propagate the species? Why do so many species reproduce sexually? Do sex and recombination enhance genetic variation, and therefore enable the species to adapt to environmental changes?

Even some professional biologists have been known to answer "yes" to these questions. But either they have assumed that these characteristics did not evolve by Darwinian natural selection (selection among individuals), or they have not realized that the good of the species does not affect the course of selection among individuals. That is, they have not fully understood the meaning of natural selection.

Because they are components of fitness, differences in fecundity and life span must have evolved at least partly by natural selection. Selection among populations—the only possible cause of evolution of a trait that is harmful to the individual but beneficial to the population or species—is generally a weaker force than selection among individuals, as we saw in Chapter 11. This must be especially true for **life history traits**, which are components of individuals' fitness.

The possibility of a species' future extinction as a result of excessive population growth or inadequate reproduction is irrelevant to, and cannot affect, the course of natural selection among individuals. A mutation that increased the fecundity of humans (or any other species) would increase individual fitness and (if it had no other effects) would therefore become fixed—even if overpopulation and mass starvation were to ensue. Instead of supposing that a species' fecundity evolves to balance mortality, *we should consider the level of mortality to be the ecological consequence of the level of fecundity*, since most populations are regulated by density-dependent factors such as limited food and other resources (Williams 1966).

At first surmise, then, we should expect any species to evolve ever greater fecundity and an ever longer life span. The problem, therefore, is to understand what advantage low fecundity or a short life span—or the genes that underlie them—might provide to individual organisms, rather than to entire populations or species. By the same token, we must beware of supposing that sexual reproduction has evolved because it benefits species by enabling them to adapt to changes in the environment.

Modeling Optimal Phenotypes

Optimality theory, also called optimization theory, is an important approach to understanding particular classes of life history trait adaptations. It consists of specifying, often on the basis of mathematical models, which state of some character, among a specified set of plausible states (often called "strategies"), would maximize individual fitness, subject to specified constraints (Parker and Maynard Smith 1990). (In some cases, INCLUSIVE FITNESS, as described in Chapter 16, is used instead of individual fitness.) Often the criterion of optimization in a model is a measurable variable, such as rate of food acquisition, that is assumed to be correlated with fitness. The "character" to be optimized may be a reaction norm rather than a single state. This is the case for many aspects of behavior, in which the optimum may be modulated to environmental variables. For instance, the optimal time an animal spends foraging in a patch of habitat may depend on the travel time between suitable patches. Optimality theory is used extensively in the study of animal behavior (the field of BEHAVIORAL ECOLOGY), and often is applied to the ideal expression of a flexible behavior that is modulated by experience and environmental conditions, including the behavior of conspecific individuals. For many features, however, the optimum may be envisioned as a genetically determined state.

Optimality theory rests on the assumption that sufficient genetic variation has been available for natural selection to shape a feature, and so it ignores the history of genetic change and examines only the expected outcome of selection. Optimality theory has been criticized on the grounds that constraints such as limited genetic variation or genetic correlations may prevent attainment of an optimum (e.g., Gould and Lewontin 1979). Defenders of optimality theory (e.g., Parker and Maynard Smith 1990) reply that the approach does not actually assume that organisms are perfectly adapted, but instead aims to understand specific examples of adaptation. "General" optimal models provide qualitative insights into the kinds of broad "solutions" to adaptive problems that might be expected to evolve, and often predict the direction of difference between populations that face different problems. "Specific" models may use data on particular species to make quantitative predictions. For instance, the relation of surface area (A) to volume (V), $A \propto V^{2/3}$, dictates that gases, water, and heat are exchanged at higher rates between small objects and the environment than between large objects and the environment. Consequently, we might expect, and do observe, that birds and mammals in cold environments frequently have greater body size and shorter appendages than closely related forms in hot environments, and that many plant species in dry environments have thick leaves or none at all (Figure 14.4). The fields of functional morphology and physiology often use this and other principles of physics to calculate expected values of morphological features.

Simple optimal models are frequency-independent. A special class of optimal models, however, treats situations in which the optimal feature of an individual depends on other individuals with which it interacts; fitness is then frequency-dependent (see Chapter 12). These models, based on the mathematical theory of games, use the concept of an **evolutionarily stable strategy** (**ESS**), which John Maynard Smith (1982) defined as "*a strategy such that, if all the members of a population adopt it, then no mutant strategy could invade under the influence of natural selection.*" That is, an ESS is a phenotype that cannot be replaced by any other phenotype under the prevailing conditions. A strategy may be PURE, meaning that an individual always has the same phenotype, or MIXED, meaning that an individual's phenotype varies over time, as is often the case with behavior. ESS models frequently

(A)

(B)

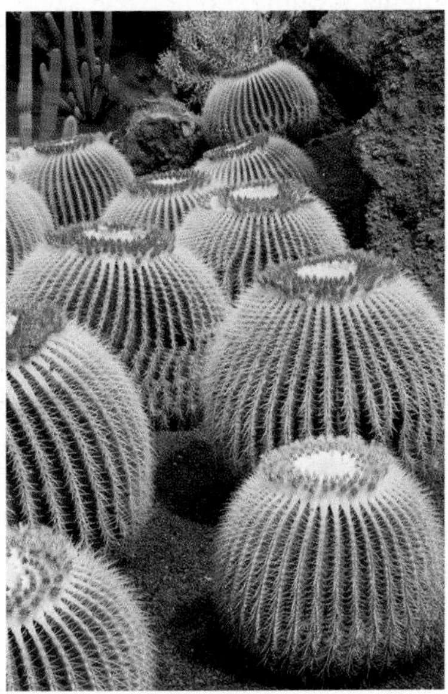

Figure 14.4 Adaptations based on surface-to-volume relationships. (A) Appendages such as ears and legs are relatively shorter, reducing the surface area over which heat may be lost, in the Arctic hare (*Lepus arcticus*, left) than in the antelope jack rabbit (*L. alleni*, right), which occupies hot deserts. (B) Water loss is reduced by plants such as the "living stone" (*Aloinopsis schooneesii*, left) in which masses of small, thick, water-storing succulent leaves grow close to the ground. In members of the large cactus family such as the golden barrel (*Echinocactus grusonii*, right), globular stems and the lack of leaves minimize the surface/volume ratio and thus minimize water loss. (A © Thorsten Milse/Photolibrary.com and Robert Shantz/Alamy; B © David McIntyre and Lagui/ShutterStock.)

describe interactions between two individuals, each of which has one of two or more phenotypes. For each possible pairwise combination of strategies, an individual has a different payoff—the increment or decrement of fitness that it receives. The payoff depends not only on the individual's own phenotype but also on that of the individual with which it interacts. We have already encountered an example of an ESS model when we considered the optimal sex ratio among a female's offspring, which depends on the sex ratio in the population at large (see Figure 12.17). Box 14A provides examples of both simple and ESS approaches. The ESS example shows, incidentally, that evolution under frequency-dependent selection may not result in maximal population fitness.

Life History Traits as Components of Fitness

Among the most important components of fitness are survival to and through the reproductive ages, the number of offspring (fecundity) of females, and the number of offspring a male fathers. Here we will focus on the evolution of life history traits, especially the potential life span, the ages at which reproduction begins and ends, and female fecundity at each age (Stearns 1992; Charlesworth 1994b; Roff 2002). These traits affect the growth rates of populations.

BOX 14A Optimal Models: Examples

A simple model: Optimal diet choice

If an individual predator encounters two kinds of prey (1 and 2) that differ in energy content (E_1 and E_2), would the predator do best to attack one or both, assuming that fitness would be increased by maximizing the rate of energy gain? The classic optimal foraging model (Krebs and McCleery 1984) assumes that prey types 1 and 2 are encountered at rates L_1 and L_2, and that if captured they require handling times H_1 and H_2. The model assumes that handling time, which may depend on various aspects of the predator's behavior and morphology, is a fixed constraint (i.e., it does not evolve); that handling and searching cannot be done simultaneously; that prey are recognized immediately and without error; and that prey are encountered at random.

A nonselective predator (a GENERALIST) foraging for time T_s will obtain $E = L_1T_s E_1 + L_2T_s E_2$ calories if it does not reject any prey, and the total time expended will be $T = (T_s) + (L_1T_s H_1 + L_2T_s H_2)$, where the terms in parentheses represent total search time and total handling time, respectively. Thus the average rate of intake will be

$$\frac{E}{T} = \frac{L_1E_1 + L_2E_2}{1 + L_1H_1 + L_2H_2}$$

Now if prey type 1 yields more calories per unit handling time ($E_1/H_1 > E_2/H_2$), the optimal diet should be only type 1 if

$$\frac{L_1E_1}{1 + L_1H_1} > \frac{L_1E_1 + L_2E_2}{1 + L_1H_1 + L_2H_2}$$

that is, if the caloric intake is greater for a specialist on type 1 than for a generalist, which expends time handling less profitable prey. With a bit of algebra, this reduces to

$$\frac{L_1E_1}{1 + L_1H_1} > \frac{E_2}{H_2}$$

So specialization is favored if the specialist's average rate of intake is greater than the calories gained per unit time spent handling a less profitable prey item. The abundance of the more profitable prey (L_1) affects whether or not a predator should specialize on it, whereas the abundance of a less profitable prey (L_2) has dropped out of the inequality, and so has no bearing on the optimal diet.

An ESS model: Tree height

Hanna Kokko (2007) provides a simple example of an ESS model, which we present here in barest outline. Suppose that the vegetative biomass of a plant can be apportioned between stem (h, ranging from 0 to 1) and leaves (f, which also ranges from 0 to 1), that leaves have photosynthetic rate g, and that fitness is proportional to fg, the amount of carbon fixed (which may then be used for reproduction). If $h = 1$, fitness is 0, because the plant has no photosynthesis. Because taller stems must be thicker to support the added weight without buckling, f must decline with h more rapidly at larger values of h. Now suppose that plant A has a neighbor, plant B, which reduces A's rate of photosynthesis, g_A, if it is taller and casts shade, and suppose further that g_A declines as the difference in height ($h_B - h_A$) increases. Two plants of the same height are assumed to reduce each other's fg to some extent. Is there a strategy—a height h—such that a population of plants with that strategy would not be replaced by a genotype with a different ("mutant") strategy?

The answer can be found mathematically, but Kokko provides a numerical example that illustrates the reasoning. She envisions a "payoff matrix" (below) in which fg values, based on the above considerations, are specified for two competing plants A and B when they have heights h_A and h_B, respectively, and there are four possible heights. In each cell, the left and right numbers are the payoffs, fg, to plants A and B, respectively.

Height of A (h_A)	Height of B (h_B): 0	1/3	2/3	1
0	0.625; 0.625	0.369; 0.848	0.276; 0.686	0.255; 0
1/3	0. 848; 0.369	0.602; 0.602	0.356; 0.620	0.266; 0
2/3	0.685; 0.276	0.620; 0.356	0.440; 0.440	0.256; 0
1	0; 0.255	0; 0.266	0; 0.260	0; 0

Female fecundity, semelparity, and iteroparity

In Chapter 12, we defined the fitness of a genotype for the simple case in which females reproduce once and then die (a **semelparous** life history). In this case, reproductive success—the number of descendants of an average female after one generation—is R, the product of the probability of a female's survival to reproductive age (L) and the average number of offspring per survivor (M):

$$R = LM$$

For **iteroparous** species—those in which females reproduce more than once—the calculation of fitness is more complex. The average number of offspring per female is the sum of the offspring an average female produces at each age, weighted by the probability that a female survives to that age. We use x to denote age, l_x to denote the probability of sur-

BOX 14A (Continued)

Let us move through this matrix to an equilibrium, if one exists. If both plants have strategy $h = 0$ (upper left cell of the table), they have equal payoffs (0.625), less than maximal because they shade each other equally. If B has strategy $h = 1/3$ (one cell to the right), it gains ($fg = 0.848$) and A's payoff is reduced to 0.369, so $h = 1/3$ is a superior strategy. But if the entire population (i.e., both plants) have $h = 1/3$ (go down one cell), both have a payoff of 0.602. If B "mutates" to $h = 2/3$, its fg rises to 0.620 and A's payoff drops to 0.356. So $h = 2/3$ is a superior strategy. If both A and B (the entire population) have this strategy, they both drop to a payoff of 0.440 as a result of mutual shading. As seen in Figure 1A, this is the ESS because the only possible further change, to $h = 1$, reduces fitness to 0. The same principle illustrated here by two individuals holds if a population of plants is modeled, and if h is a continuous instead of discrete variable (Figure 1B). The ESS is found at the intersection of the "best response" curves.

Notice that in this model, the total amount of carbon a plant fixes (fg) at equilibrium, a measure of average fitness, is lower (0.440) than the maximum possible (0.625, when $h = 0$). Evolution in response to competition has led to lower average fitness. This is a common result in models of frequency-dependent selection.

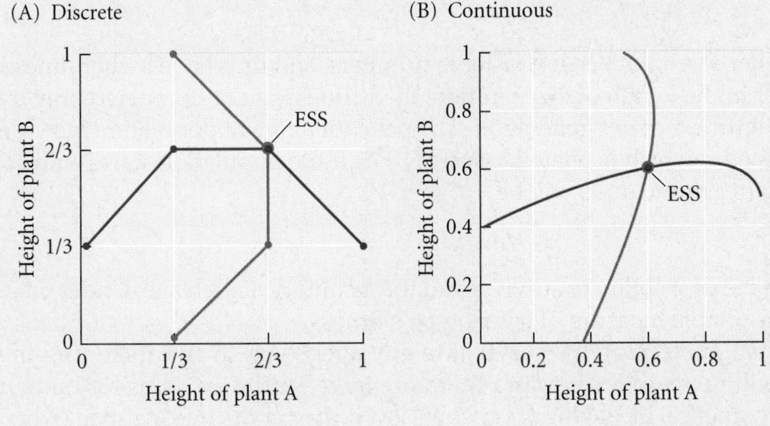

Figure 1 A graphical depiction of the "best responses" of competing plants to each other's height, according to the game-theoretical approach. (A) Here the plants display one or the other of four possible heights, ranging from 0 to 1. (B) The corresponding curves when height is a continuous variable.

vival to age x (i.e., the proportion of eggs or newborns that survive to age x), and m_x to denote the average fecundity (number of eggs or newborns) at age x. (Figure 14.5 illustrates l_x and m_x for several populations of a species of *Drosophila*.) Suppose that at ages 1, 2, 3, and 4 years, females lay 0, 4, 8, and 0 eggs, respectively, and that their chances of surviving to those ages are 0.75, 0.50, 0.25, and 0.10, respectively. Then we can write a simple life table:

x	l_x	m_x	$l_x m_x$
0	1.00	0	0
1	0.75	0	0
2	0.50	4	2
3	0.25	8	2
4	0.10	0	0
5	0.00	0	0
$\Sigma = R$			4

R is calculated as

$$R = \Sigma\, l_x m_x$$

Thus each female is replaced, on average, by $R = 4$ offspring. (By convention, ecologists generally count just daughters in such analyses, and assume that sons are produced in equal numbers.) This sum is also the growth rate, per generation, of the genotype: if the population starts with N_0 individuals, its size after g generations will be

$$N = N_0 R^g$$

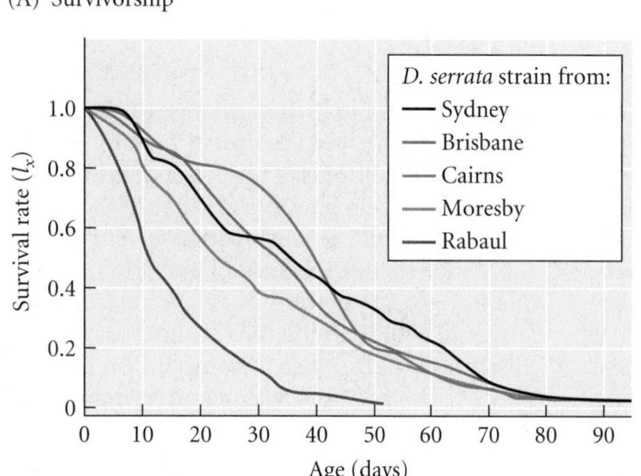

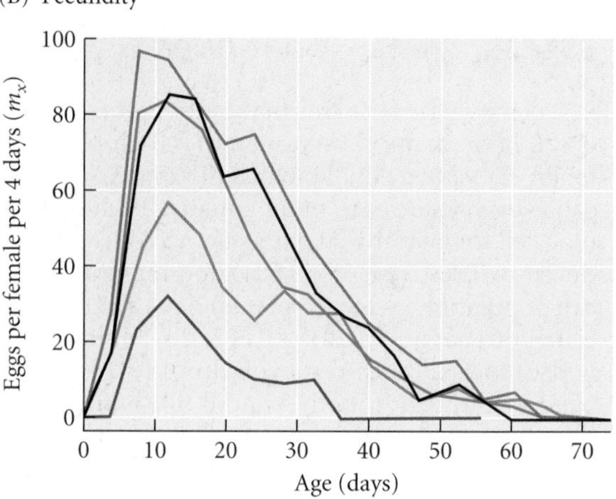

Figure 14.5 Genetic variation in life history characteristics. The graphs show (A) age-specific probability of survival, l_x, and (B) fecundity, m_x, in strains of the Australian fruit fly *Drosophila serrata* from five localities when raised in the laboratory. The survival curves show the fraction of newborns that survive to each age, and the fecundity curves show the average egg production per female at each age. Egg production peaks a few days after the flies transform from the pupal to the adult stage. These curves show only the adult (post-pupa) stage of the life history. (After Birch et al. 1963.)

A genotype with higher R would increase faster in numbers, and thus have higher fitness.

If genotypes differ in the length of a generation, their fitness can be compared only by their increase per unit time, not per generation. The per capita rate of population increase per unit time is denoted r, which is related to R by $R = e^r$. If the population starts with N_0 individuals, its size after t time units will be

$$N = N_0 e^{rt}$$

Like R, r depends on the probability of survival and the fecundity at each age. Under most circumstances, r is a suitable measure of a genotype's fitness.

All else being equal, increasing l_x—survival to any age x—up to and including the reproductive ages will increase R (or r) and therefore increase fitness. If, as for human females (or in the hypothetical life table above), there is a postreproductive life span (when $m_x = 0$), changing the probability of survival to advanced postreproductive ages does not alter R. If, however, the reproductive period is extended into older ages, then increasing survival to those ages does increase R. Similarly, increasing m_x (fecundity at any age x) increases fitness, all else being equal.

Age structure and reproductive success

Offspring produced at an early age increase fitness more because they contribute more to population growth (r) than the same number of offspring produced at a later age. That is, they have greater "value" in terms of fitness. For instance, suppose females reproduce either at age 2 or age 3, but have the same fecundity. Then 2-year-old females contribute more to the future population size than do 3-year-olds. Because fewer individuals survive to age 3 than to age 2, 2-year-olds collectively will leave more offspring. Moreover, population growth is like compound interest. Just as your bank account grows faster if you make a deposit now than if you wait, the offspring of 2-year-olds will themselves contribute offspring (i.e., "gain interest") before the offspring of 3-year-olds do so. Thus a genotype that reproduces earlier in life has a shorter generation time, and higher fitness (as measured by r), than a genotype that delays reproduction until a later age.

Because reproduction early in life contributes more to the rate of population growth than an equal production of offspring later in life, the SENSITIVITY of fitness to small changes in life history traits—the effect of a given magnitude of change in fecundity or survival on fitness (r)—depends on the age at which the change is expressed (Charlesworth 1994). For the hypothetical data in Table 14.1, for example, an increase in survival to the first age class (l_1) would obviously increase fitness. On the other hand, increasing survival from age 5 to age 6 (l_5) would not alter fitness at all, because this species does not reproduce beyond age 5 ($m_6 = 0$). Therefore, *natural selection usually does not favor postreproductive survival.* (Postreproductive survival may be advantageous, however, if postreproductive parents care for their offspring, as in humans.)

Table 14.1 *A hypothetical example of a life table and of the sensitivity of fitness (r) to changes in age-specific survival and fecundity*[a]

Age class	Number of survivors	Fraction of survivors	Average fecundity	Survival \ Fecundity			Sensitivity of r to m_x	Sensitivity of r to l_x
x		l_x	m_x	$l_x m_x$	e^{-rx}	$e^{rx} l_x m_x$	$S_m(x)$	$S_s(x)$
0	1000	1.000	0.00	0.00	1.000	0.000	0.335	0.334
1	750	0.750	0.00	0.000	0.796	0.000	0.200	0.334
2	600	0.600	1.20	0.720	0.634	0.456	0.128	0.182
3	480	0.480	1.40	0.672	0.505	0.339	0.081	0.068
4	360	0.360	1.03	0.396	0.402	0.159	0.049	0.018
5	180	0.180	0.96	0.144	0.320	0.046	0.019	0.018
6	100	0.100	0.00	0.000	0.255	0.000	0.011	—
Sums:					1.932 = R			1.000

Source: After Stearns 1992.

[a] The instantaneous rate of increase, r, is calculated by "trial and error" from the equation $1=\sum_{x=a}^{x=z} e^{-rx} l_m m_x$ and is found to be 0.228.

The sensitivity coefficients $S_m(x)$ and $S_s(x)$ indicate the effect on r of a small change in fecundity (m_x) or survival (l_x), respectively, at age x.

They are calculated, respectively, as $S_m(x)=\frac{e^{-rx} l_m}{T}$ and $S_s(x)=\sum_{y=x}^{y=z}\frac{e^{-ry} l_y m_y}{T}$.

Furthermore, the selective advantage of a slight increase in survival or fecundity at an advanced age increases fitness (r) less than an equal increase at an early age, because the contribution of a cohort to population growth declines with age. (Note the decline in the "sensitivity values" $S_t(x)$ and $S_m(x)$ with age in Table 14.1.) A simple reason is that older females are less likely to be alive to reproduce. Another reason, as we have seen, is that offspring born to older females contribute less to the rate of population growth than do those born to young females.

Trade-Offs

Because traits evolve so as to maximize fitness, we might naively expect organisms to evolve ever greater fecundity, ever longer life, and ever earlier maturation. That all organisms are in fact limited in these respects may be attributed to various constraints.

PHYLOGENETIC CONSTRAINTS arise from the history of evolution, which has bequeathed to each lineage certain features that constrain the evolution of life history traits and other characters. For instance, although many insects feed as adults, and so obtain energy and protein that enable them to form successive clutches of eggs, adult silkworm moths and some other insect groups lack functional mouthparts, so their fecundity is limited by the resources they stored when they fed as larvae. Most such insects lay only one batch of eggs and then die. In most groups of birds, the number of eggs per clutch varies within and among species, but all species in the order Procellariiformes (albatrosses, petrels, and relatives) lay only a single egg.

Other constraints, termed PHYSIOLOGICAL or GENETIC CONSTRAINTS, are less well understood, but may be detected by comparisons among different genotypes or phenotypes. Some such constraints constitute **trade-offs**, whereby the advantage of a change in a character is correlated with a disadvantage in other respects. For example, the reproductive activities of animals often increase their risk of predation, so there is a trade-off between reproduction and survival.

Trade-offs can influence the evolution of virtually all characteristics. They are a major component of most optimal models and of many genetic models as well. For instance, selection for different alleles in different environments occurs if there are trade-offs in fitness among genotypes across environments.

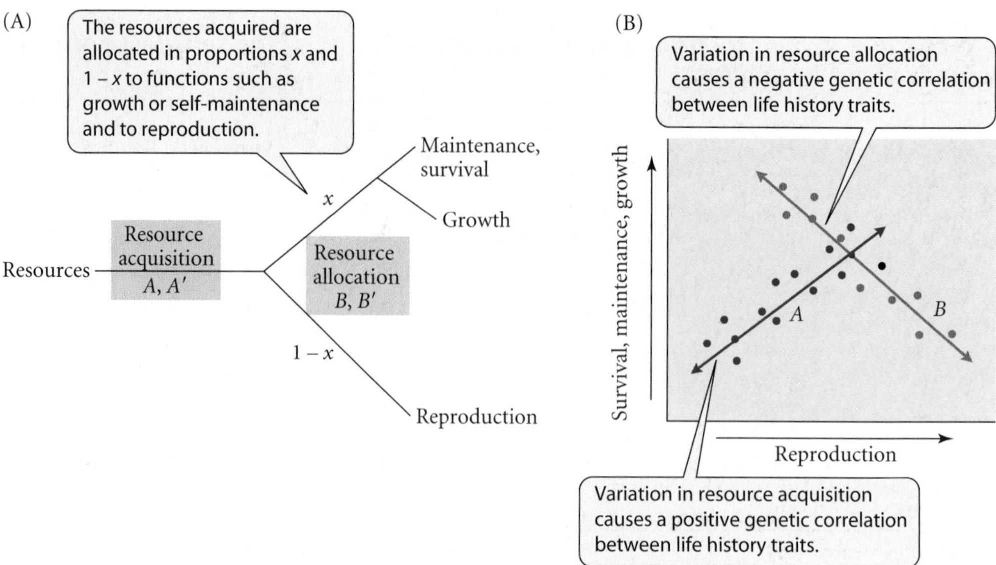

Figure 14.6 Factors giving rise to positive or negative genetic correlations between life history traits such as survival (or growth) and reproduction. (A) Variation at locus *A* affects the amount of energy or other resources that an individual acquires from the environment. Variation at locus *B* affects allocation of resources to functions such as growth or self-maintenance and to reproduction, in proportions x and $1 - x$. (B) Genotypes that differ at locus *A* are represented by green circles, those that differ at locus *B* by red circles. The overall genetic correlation between survival and reproduction depends on the relative magnitudes of variation in resource acquisition versus resource allocation.

Some physiological trade-offs may result in **antagonistic pleiotropy**, wherein genotypes manifest an inverse relationship between different components of fitness. For instance, genotypes that allocate more energy or nutrients to reproduction (often called **reproductive effort**) and less to their own maintenance or growth may display decreased subsequent survival or growth, an example of a **cost of reproduction**. This ALLOCATION TRADE-OFF would be manifested as a *negative genetic correlation* between reproduction and survival. If there were also genetic variation in the amount of resources individuals acquired from the environment, however, this could give rise to a *positive correlation* between reproduction and survival (Figure 14.6; Bell and Koufopanou 1986; van Noordwijk and deJong 1986). The allocation trade-off might still constrain evolution, but the trade-off might be difficult to detect in this case.

There are several ways to detect trade-offs (Reznick 1985):

1. Correlations between the means of two or more traits in different populations or species can strongly suggest a trade-off, although such correlations might result from other, unknown differences among the populations. We would expect, and can often document, an allocation trade-off between many small versus fewer large offspring if parents must allocate limited resources (Figure 14.7).
2. Phenotypic or, better, genetic correlations between traits within populations can be useful indicators of the extent to which enhancement of one component of fitness would be immediately accompanied by reduction of another (see Chapter 13). For instance, Law and colleagues (1979) grew individuals of each of many families of meadow grass (*Poa annua*) in a randomized array. They found that families that, on average, produced more inflorescences in their first season produced fewer in the second, and also achieved less vegetative growth. This experiment demonstrated a genetic basis for a cost of reproduction.
3. Correlated responses to artificial or natural selection provide some of the most consistent evidence of trade-offs (Reznick 1985; Stearns 1992). Linda Partridge and colleagues (1999) set up ten selection lines of *Drosophila melanogaster* from the same base population. They selected five "young" populations by rearing

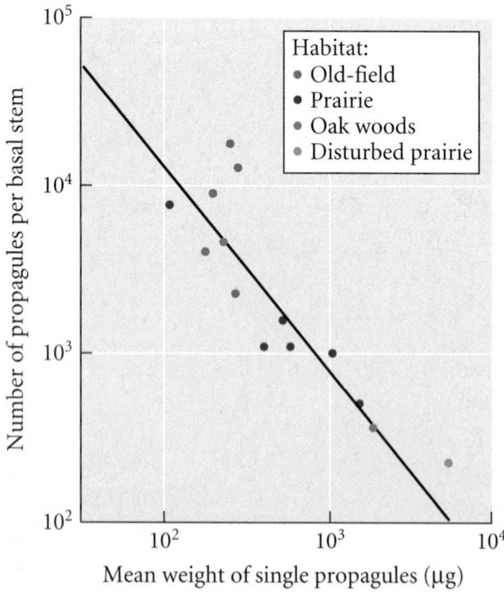

Figure 14.7 The relationship between number and weight of propagules (seeds) among species of goldenrods (*Solidago*) suggests an allocation trade-off. The colonizing species, growing in old-field habitats, tend to produce smaller seeds than species that grow in more stable prairies, where competition may be intense and favor larger offspring. (After Werner and Platt 1976.)

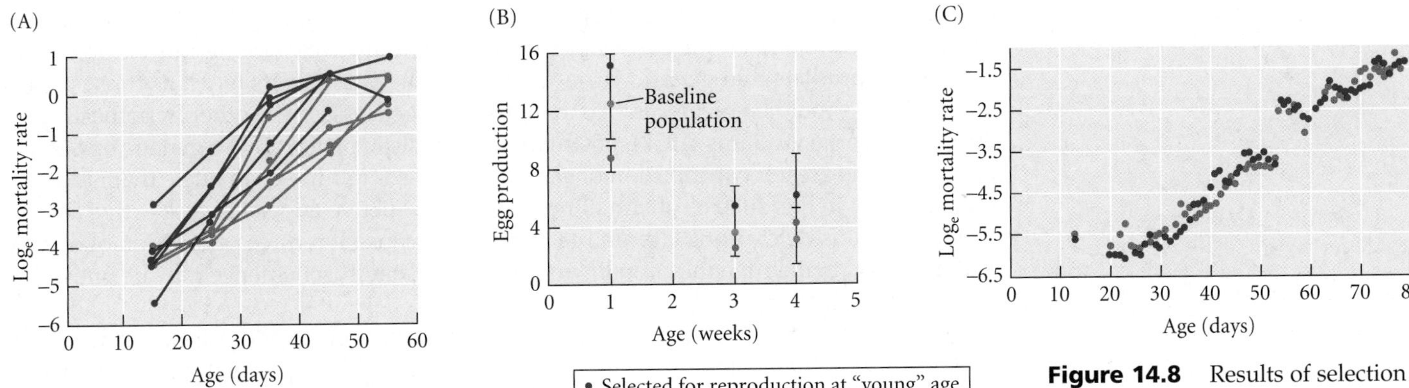

Figure 14.8 Results of selection of laboratory populations of *Drosophila* for age at reproduction. (A) The mortality rate, per 10-day interval, is lower at older ages for populations that were selected for reproduction at "old" age than for populations that were selected for reproduction at "young" age. (B) Relative to the "young" populations and the baseline population, the "old" populations had lower egg production when young. (C) The difference in mortality rate between "old" and "young" populations disappeared when a gene that prevents female reproduction was crossed into the populations, suggesting that the costs of reproduction increase mortality. (After Partridge et al. 1999 and Sgrò and Partridge 1999.)

offspring from eggs laid by females less than 1 week old, and five "old" populations by propagating from eggs laid when females were 3 to 4 weeks old. After 19 generations, the mean life span of the "young" populations did not differ from that of the base population, but the longevity of the "old" populations had increased—as we would expect, since only flies that lived at least 3 weeks could contribute genes to subsequent generations (Figure 14.8A). However, the fecundity of 1-week-old females in the "old" populations decreased compared with that of the base or "young" populations (Figure 14.8B). Thus survival to greater age seems to have been achieved at the expense of reproduction early in life—an important result, as we will see shortly.

4. Experimental manipulation of one trait and observation of the effect on other traits often reveal trade-offs. For instance, Sgrò and Partridge (1999) followed the selection experiment on *Drosophila* longevity by experimentally sterilizing females from both "young" and "old" populations, either by gamma radiation or by inheritance of a dominant allele that causes female sterility. In both experimental treatments, the difference in longevity between the "young" and "old" populations disappeared, proving that longevity is affected by a physiological cost of reproduction (Figure 14.8C). These results are consistent with other evidence that in *Drosophila* and many other insects, mating activity and egg production reduce the longevity of both sexes, and that virgins live longer than nonvirgins (Bell and Koufopanou 1986; Fowler and Partridge 1989).

The Evolution of Life History Traits

Life span and senescence

Most organisms in which germ cells are distinct from somatic tissues undergo physiological degeneration with age, a process called aging or **senescence**. Why does this occur? Many physiological processes may contribute to senescence; for example, it has been suggested that a contributing factor may be the shortening of TELOMERES, the DNA caps at the ends of chromosomes, with age. Telomeres shorten more slowly in long-lived than in short-lived species, and even lengthen in at least one long-lived species of bird (Haussmann et al. 2003). Senescence and life span, and their immediate physiological causes, differ among species, and clearly evolve. Thus, telomere loss or any other single physiological cause of senescence does not explain why species have the life span they have, since they could have evolved a different life span. This is a clear example of the difference between a *proximate*, mechanistic, explanation and an *ultimate*, evolutionary, explanation of a biological phenomenon.

There are two related hypotheses on the evolution of senescence and limited life span (Rose 1991). Both rest on the principle that the selective advantage of an enhanced probability of survival declines with age. Peter Medawar (1952) proposed that deleterious mutations that affect later age classes accumulate in populations at a higher frequency than those that affect earlier age classes because selection against them is weak. If there are many such loci, then the causes of senescence should vary among individuals. This hypothesis predicts that genetic variation in fitness-related traits (such as those affecting survival) should be greater in late than in early age classes. The other hypothesis, proposed by George Williams (1957), postulates antagonistic pleiotropy, or genetic trade-offs. Because of the greater contribution of earlier age classes to fitness, an allele that is advantageous early in life, such as one that increases reproductive effort, has a selective advantage even if it is deleterious later in life (perhaps because it reduces allocation of energy and materials to maintenance, repair, and defense). That is, senescence may be one of the costs of reproduction.

The evidence for Medawar's mutation accumulation hypothesis is limited, but Williams's hypothesis of antagonistic pleiotropy is strongly supported by selection experiments that provide evidence of a negative relationship between early reproduction and both longevity and later reproduction (Partridge 2001; see Figure 14.8). These experiments are among the most striking confirmations of evolutionary hypotheses that had been posed long before the experiments took place.

Age schedules of reproduction

If survival contributes to fitness only as long as reproduction continues, why don't organisms reproduce indefinitely? The answer is that all else being equal, there is always an advantage to reproducing earlier in life. Since early reproduction is correlated with lowered subsequent reproduction, we should expect organisms to be semelparous, allocating all their resources to a single early burst of reproduction rather than to maintaining themselves. We must therefore ask why so many species are nevertheless iteroparous.

Reproducing at an early age may increase the risk of death, decrease growth, or decrease subsequent fecundity so as to lower r compared with what it would be if reproduction were deferred. Fecundity, for example, is often correlated with body mass in species that grow throughout life, such as many plants and fishes. In such species, allocating resources to growth, self-maintenance, and self-defense rather than to immediate reproduction is an investment in the much greater fecundity that may be attained later in life. Accordingly, mathematical models show that *repeated reproduction is more likely to evolve if adults have high survival rates from one age class to the next, and if the rate of population increase is low.*

Figure 14.9 Reproductive effort—an index of the proportion of biomass allocated to inflorescences—in annual (semelparous) and perennial (iteroparous) species of British grasses. Allocation to reproduction is greater in the semelparous annual species. (After Wilson and Thompson 1989.)

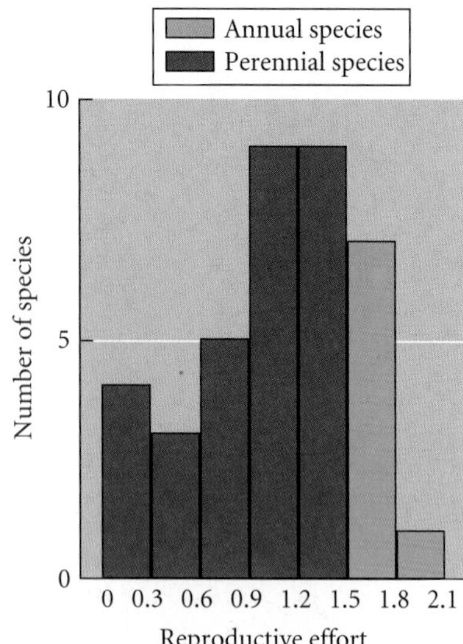

The same factors also favor later, rather than earlier, reproductive maturity in an iteroparous species, and can favor the evolution of a genetically longer life span. Since some resources are used for growth, maintenance, and defense, the effort devoted by an iteroparous species to reproduction will be less at each reproductive age than what a semelparous species devotes to reproduction in its single "big bang" reproductive episode. As individuals age, however, the benefit of withholding energy from reproduction declines because the prospects of subsequent reproduction decline. Therefore, we would expect that at some point in life, *the proportion of energy or other resources devoted to reproduction by iteroparous species should increase with age* (Williams 1966; Charlesworth 1994b).

These theoretical predictions have been supported by many studies. Comparisons among species within several taxa support the prediction that reproductive effort, in each reproductive episode, should be lower in iteroparous than in semelparous organisms (Roff 2002). For example, inflorescences make up a lower proportion of plant weight in perennial than in annual species of grasses (Figure 14.9; Wilson and Thompson 1989). Likewise, the prediction that high adult survival rates favor delayed maturation and high reproductive effort later in life is upheld by studies of mammals, fishes, lizards and snakes, and other groups: species that have long life spans in nature also mature at a later age (Figure 14.10; Promislow and Harvey 1991; Shine and Charnov 1992).

David Reznick and colleagues have studied guppies (*Poecilia reticulata*) in Trinidad (e.g., Reznick et al. 1990; Reznick and Travis 2002). In some streams, the cichlid fish *Crenicichla altaw* preys heavily on large (mature) guppies (see Chapter 11). In other streams, or above waterfalls, *Crenicichla* is absent, and there is much less predation. Predation by *Crenicichla* should favor the evolution of maturity and reproduction early in life, and guppies from

Crenicichla-dominated streams indeed mature faster and at smaller sizes, reproduce more frequently, have higher reproductive effort (measured as weight of embryos relative to weight of mother), and have more and smaller offspring than guppies from low-predation streams. In two streams, Reznick and colleagues moved guppies from below a waterfall, where they were preyed on by *Crenicichla*, to sites above the waterfall, where guppies and *Crenicichla* were absent. After several generations, the researchers took guppies from the sites of origin and introduction and reared their offspring in a common laboratory environment. As predicted by life history theory, the populations relieved of predation on large adults had evolved delayed maturation and larger adult size, and they tended toward fewer, larger offspring and lower reproductive effort (Figure 14.11). Recall that the opposite effects—earlier maturation at smaller size—have been seen in fish populations that have experienced increased human predation (overfishing) on adults (see Figure 13.12A,B).

Figure 14.10 Among species of snakes and lizards, the lower the annual mortality rate of adults, the later reproduction begins. This pattern conforms to the prediction that delayed onset of reproduction is most likely to evolve in species with high rates of adult survival. (After Shine and Charnov 1992.)

Number and size of offspring

All else being equal, a genotype with higher fecundity has higher fitness than one with lower fecundity. Why, then, do some species, such as humans, albatrosses, and kiwis, have so few offspring?

OPTIMAL CLUTCH SIZE. The British ecologist David Lack (1954) proposed that the *optimal clutch size* for a bird is the number of eggs that yields the greatest number of surviving offspring. The number of surviving offspring from larger broods may be less than the number from more modest clutches because parents are unable to feed larger broods adequately. This decrease in offspring survival has proved to be one of several costs of large clutch size in birds; excessively large clutches may also reduce the parents' subsequent clutch size and survival (Stearns 1992).

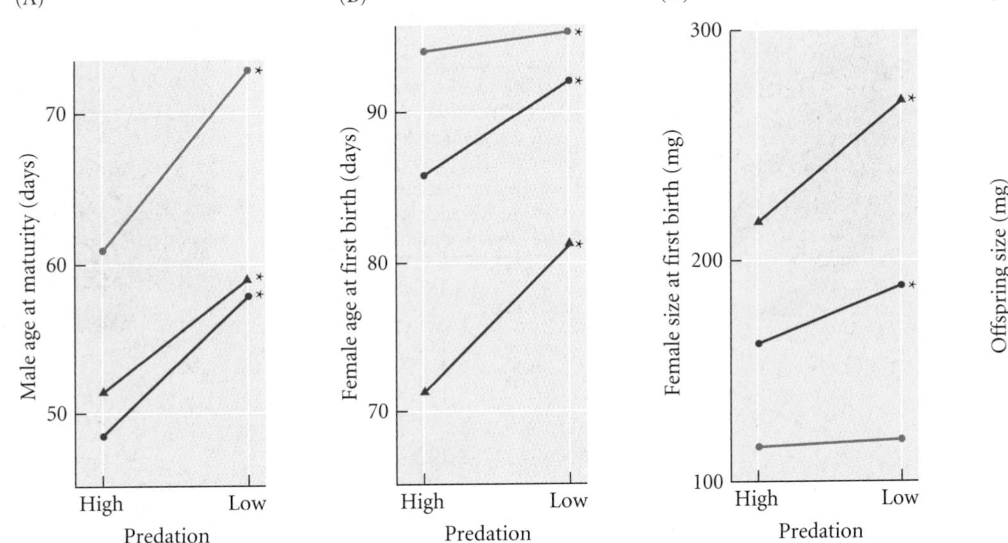

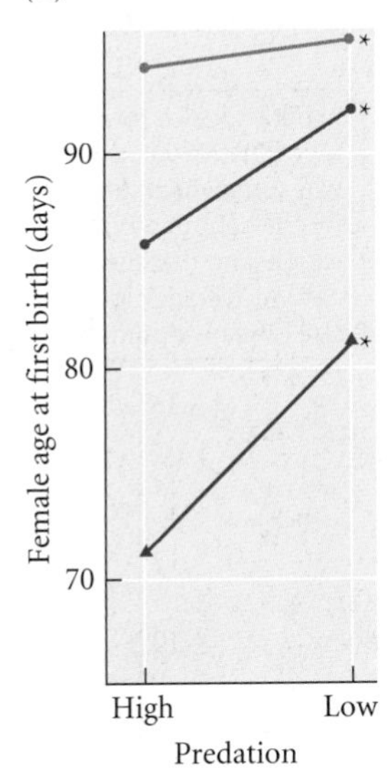

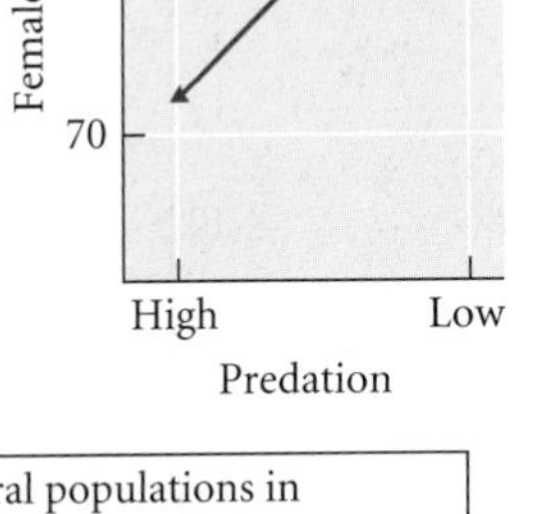

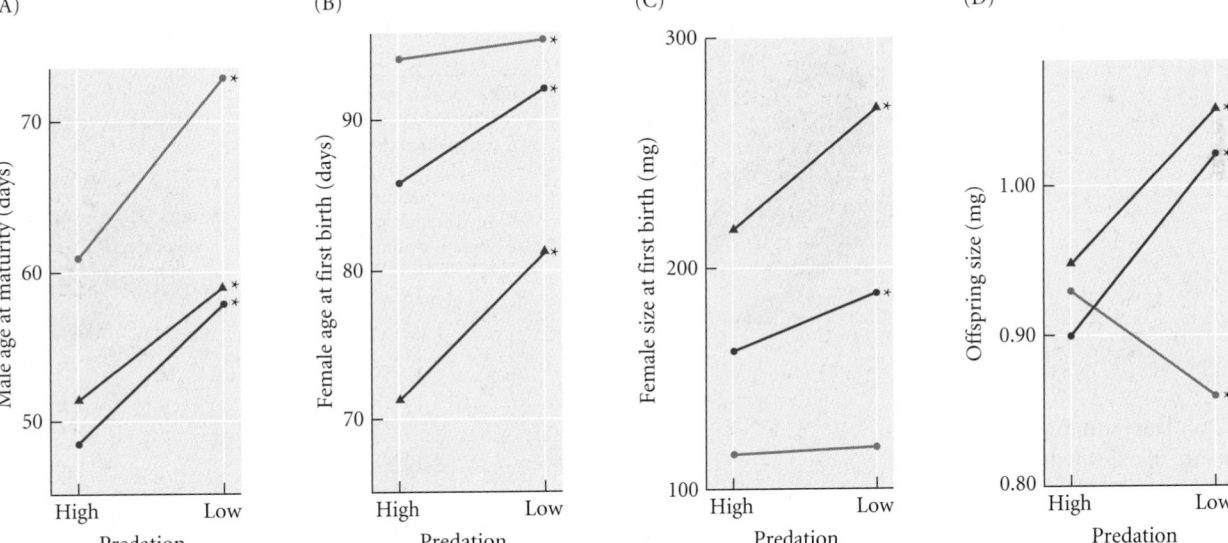

▲ Contrasts the means of two natural populations in high-predation streams and two in low-predation streams.

• Contrasts an experimental population, isolated from predators for 7 generations, with a downstream control population that experiences high predation.

• Contrasts an experimental population isolated for about 18 generations, with a high-predation downstream control.

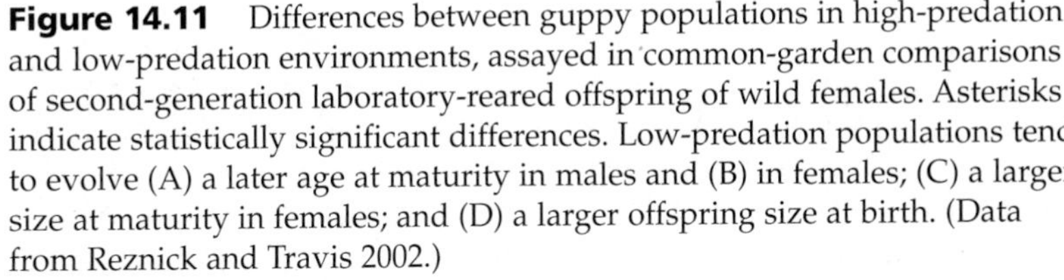

Figure 14.11 Differences between guppy populations in high-predation and low-predation environments, assayed in common-garden comparisons of second-generation laboratory-reared offspring of wild females. Asterisks indicate statistically significant differences. Low-predation populations tend to evolve (A) a later age at maturity in males and (B) in females; (C) a larger size at maturity in females; and (D) a larger offspring size at birth. (Data from Reznick and Travis 2002.)

Joost Tinbergen and Serge Daan (1990) analyzed data from a long-term study, performed in Holland, of survival and reproduction in great tits (*Parus major*), including observations of unmanipulated nests and a 5-year program of decreasing and increasing brood size by moving hatchlings among nests. They estimated the effects of these treatments, relative to unmanipulated controls, on the REPRODUCTIVE VALUE of both the parents and of the brood of young. The reproductive value (V) of individuals of a certain age is their expected future contribution to population growth, taking into account their expected future age-specific survival and reproduction (which are assumed to be the l_x and m_x values estimated from the long-term study of the population). Tinbergen and Daan found that artificially increasing brood size decreased V because it lowered survival in the nest, survival from fledging to the next breeding season, and the probability that the parents would lay a second clutch of eggs in the same year.

Decreasing brood size also reduced V, simply because the nests produced fewer fledglings (Figure 14.12A). Tinbergen and Daan calculated that the optimal clutch size was 8.9 eggs, close to the natural mean, 9.2. Curiously, the birds that naturally had highest V were those few that produced clutches of 15 eggs. The authors suggested that individual pairs may adjust their clutch size to what would be best for them individually, based on such factors as their physiological condition and the quality of their territory. The same research group (Daan et al. 1990) performed a similar study of a small falcon, the Eurasian kestrel (*Falco tinnunculus*). Artificially increased clutch size reduced the subsequent survival (and V) of parents, but did not diminish that of offspring. Total reproductive value was greatest for unmanipulated clutches, again suggesting that clutch size is close to the optimum (Figure 14.12B).

PARENTAL INVESTMENT. When parents can provide only a limited amount of yolk, endosperm, nourishment, or other forms of parental care (collectively referred to as **parental investment**), there may be a negative correlation between number and size of offspring, and greater initial size usually enhances survival and growth rate. (For example, human twins

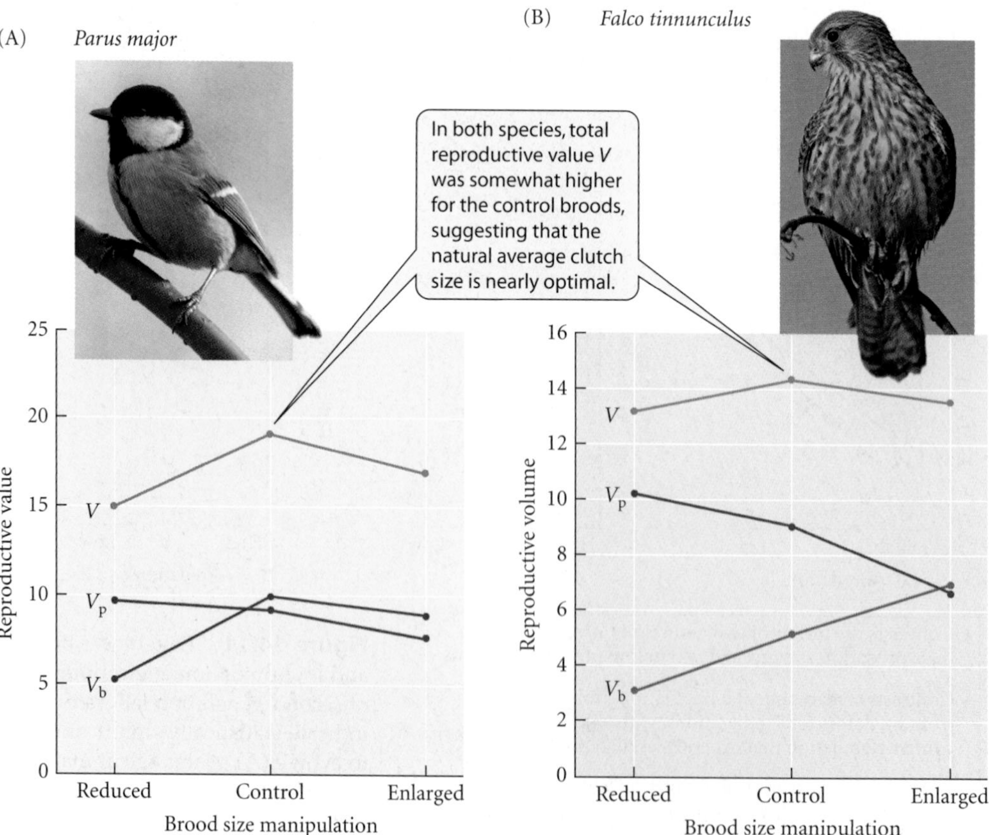

Figure 14.12 (A) The reproductive value, a measure of expected future reproductive success, of broods (V_b), their parents (V_p), and the sum (V) for (A) great tits and (B) Eurasian kestrels in which egg clutch sizes were experimentally reduced, enlarged, or left unchanged (control). (A after Tinbergen and Daan 1990, photo © Maslov Dmitry/ShutterStock; B after Daan et al. 1990, photo © Imago/Zuma Press.)

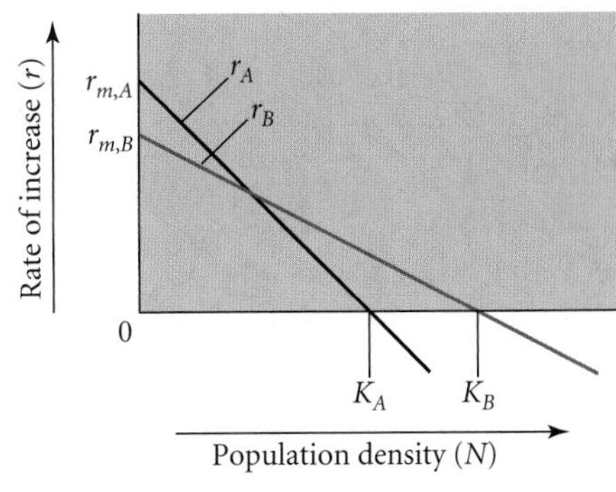

Figure 14.13 A model of density-dependent selection of rates of increase. The instantaneous per capita rate of increase, r, declines for genotypes A and B as population density (N) increases. The intrinsic rate of increase (r_m), the population growth rate at very low density, is lower for genotype B, but this genotype has a selective advantage at high density. (After Roughgarden 1971.)

have a lower birth weight, on average, than single-born infants, and infant survival of human quadruplets and quintuplets is notoriously precarious.) Barry Sinervo (1990) experimentally manipulated hatchling size in the lizard *Sceloporus occidentalis* by removing yolk from eggs with a hypodermic needle. Smaller hatchlings ran more slowly, which probably would reduce their survival in the wild. In such cases, females' reproductive success—the number of *surviving* offspring they leave to the next generation—should be maximized by producing a modest number of offspring that are larger and better equipped for survival. This is the most plausible explanation of the very low fecundity of species such as humans, whales, and elephants.

The Evolution of the Rate of Increase

Because the per capita rate of increase (r) of a genotype is the measure of its fitness, we might suppose that species would always evolve higher rates of increase. We have seen, however, that a shorter life span, lower fecundity, and delayed maturation, all of which lower r, can each be advantageous. Thus the potential rate of population growth can evolve—and certainly has evolved—to lower levels in many species. One simple reason is that most evolution is likely to occur when the actual rate of increase of a population, r, is lower than the **intrinsic** (potential) **rate of increase** (r_m) because density-dependent factors such as resource limitation or predation reduce birth rates or increase death rates. Different genotypes are likely to have higher r under crowded conditions than when the population density is low (when r_m is realized), as illustrated in Figure 14.13.

As the population density approaches equilibrium (K), a more competitive genotype may sustain positive population growth while inferior competitors decline in density (have negative r). Precisely this difference has been found between laboratory populations of *Drosophila* that were maintained at high versus low density for 25 and 200 generations, respectively, and then assayed for population growth rate (r) and productivity at each of several densities (Figure 14.14A). These populations evolved so that the high-density-selected populations had higher productivity under crowded conditions, but lower growth rate under uncrowded conditions, than the low-density-selected populations. Notice, further, that Figure 14.13 implies that genotypes that can attain higher equilibri-

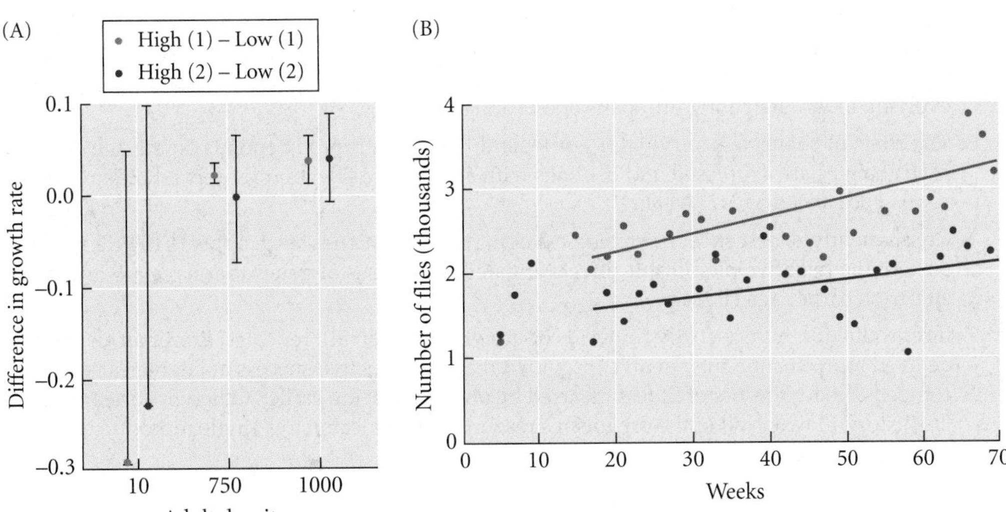

Figure 14.14 (A) Each point shows the difference in per capita population growth rate between two sets of *Drosophila* populations [Low (1) and Low (2)] that were maintained at low density for 200 generations; and two sets of populations [High (1) and High (2)] derived from the first sets but maintained at high density for 25 generations. The per capita growth rates were measured at several adult population densities. A negative difference in growth rate shows that Low strains had higher growth rate than High strains when tested at low density. Positive differences in growth rate show that High strains had higher growth rate when tested at high density. Thus, selection at high density for 25 generations resulted in evolution similar to the difference between the hypothetical genotypes A and B in Figure 14.13. (B) The density of two experimental populations (red and blue points) of *Drosophila serrata* increased over many generations, implying adaptation to high density and improved conversion of food (supplied at a constant rate) into flies. The potential rate of increase of this species is so great that the population would have reached carrying capacity in less than 10 weeks. (A after Mueller et al. 1991; B after Ayala 1968.)

um population density (K) at high density have higher fitness, so we might expect selection to have the effect of increasing abundance. This prediction has been verified in experimental *Drosophila* populations that displayed genetically based trends toward higher population density (Figure 14.14B). Selection does not always maximize abundance, however (Prout 1980).

In populations that are regulated by density-dependent factors (see Chapter 7) and occupy relatively stable environments, predation or competition for resources often causes heavier mortality among juveniles than adults. The mortality of seedling trees in a mature forest, for example, is exceedingly high, but if a tree does survive beyond the sapling stage—perhaps because a treefall has opened a light gap—it is likely to have a long life. As we have seen, these conditions favor the evolution of iteroparous reproduction late in life, and thus favor the evolution of long life spans. Moreover, under competitive conditions, juvenile survival can be enhanced by large size, so producing large eggs or offspring—and fewer of them—may be advantageous. Thus many authors have concluded that traits associated with a low intrinsic rate of increase—delayed maturation, production of few, large offspring, a long life span—are likely to evolve in species that occupy stable, competitive, or resource-poor environments. For example, species of beetles, fish, and other animals that inhabit caves generally develop very slowly and produce large eggs at an extraordinarily low rate (Culver 1982).

The evolution of the life history traits discussed in this chapter is an important topic in population ecology, for these traits affect the growth and abundance of populations. But they are only some of the components of fitness, the basis of natural selection. Although variation in male reproductive success seldom affects population growth, it is as important a component of fitness as female reproductive success in sexually reproducing species. The subject of sex deserves a discussion of its own, which we provide in the following chapter.

GO TO THE

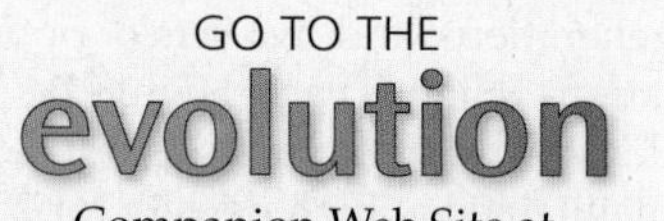

Companion Web Site at
www.sinauer.com/evolution
for quizzes, data analysis and simulation exercises, and other study aids.

Summary

1. Life history features such as reproductive rates and longevity do not evolve to perpetuate the species. They can best be understood from the perspective of individual selection. Life history traits are components of the fitness of individual genotypes, the basis for natural selection.
2. Models of the evolution of adaptive characteristics include both population genetic models and optimal models, which attempt to determine which character states might be expected to evolve under specified conditions, and under specified constraints. Optimal models include those that find the ESS (evolutionarily stable strategy) when fitness depends on the frequencies of different phenotypes among interacting individuals.
3. The major components of fitness (the per capita rate of increase of a genotype, r) are the age-specific values of survival, female fecundity, and male mating success. Natural selection on morphological and other phenotypic characters results chiefly from the effects of the characters on these life history traits.
4. Constraints, especially trade-offs between reproduction and survival and among the several components of reproduction such as the number and size of offspring, prevent organisms from evolving indefinitely long life spans and infinite fecundity.
5. The effect of changes in survival (l_x) or fecundity (m_x) on fitness depends on the age at which such changes are expressed and declines with age. Hence selection for reproduction and survival at advanced ages is weak.
6. Consequently, senescence (physiological aging) evolves. Senescence appears to be a result, in part, of the negative pleiotropic effects on later age classes of genes that have advantageous effects on earlier age classes.
7. If reproduction is very costly (in terms of growth or survival), repeated (iteroparous) and/or delayed reproduction may evolve, provided that reproductive success at later ages more than compensates for the loss of fitness incurred by not reproducing earlier. Otherwise a semelparous life history, in which all of the organism's resources are allocated to a single reproductive effort,

is optimal. Iteroparity is especially likely to evolve if juvenile mortality is high relative to adult mortality and if population density is stable.

8. Because lower fecundity and delayed reproduction can evolve, the intrinsic rate of population increase, the maximal rate of increase that is expressed at low population density, may evolve to be lower. In such populations, however, the rate of population growth is often close to zero because of density-dependent limitations.

Terms and Concepts

antagonistic pleiotropy
cost of reproduction
evolutionarily stable strategy (ESS)
intrinsic rate of increase
iteroparous
life history
life history trait
optimality theory
parental investment
reproductive effort
semelparous
senescence
trade-off

Suggestions for Further Reading

The Evolution of Life Histories by S. C. Stearns (Oxford University Press, Oxford, 1992) and *Life History Evolution* by D. A. Roff (Sinauer Associates, Sunderland, MA, 2002) are comprehensive treatments of the topics discussed in this chapter.

Problems and Discussion Topics

1. Female parasitoid wasps search for insect hosts in which to lay eggs, and they can often discriminate among individual hosts that are more or less suitable for their offspring. Behavioral ecologists have asked whether or not the wasps' willingness to lay eggs in less suitable hosts varies with age. On the basis of life history theory, what pattern of change would you predict? Does life history theory make any other predictions about animal behavior?
2. Suppose that a mutation in a species of annual plant increases allocation to chemical defenses against herbivores, but decreases production of flowers and seeds (i.e., there is an allocation trade-off). What would you have to measure in a field study in order to predict whether or not the frequency of the mutation will increase?
3. In many species of birds and mammals, clutch size is larger in populations at high latitudes than in populations at low latitudes. Species of lizards and snakes at high latitudes often have smaller clutches, and are more frequently viviparous (bear live young rather than laying eggs), than low-latitude species (see references in Stearns 1992). What selective factors might be responsible for these patterns?
4. An important life history characteristic, not discussed in this chapter, is dispersal between hatching and reproductive age. The extent of dispersal varies considerably among different organisms. What are the advantages versus disadvantages of dispersing? How might the evolution of dispersal be affected by group selection versus individual selection? (See Olivieri et al. 1995 and references therein.)
5. Another consequence of dispersal might be selection among genotypes that differ in characteristics that affect dispersal. What might be some effects of dispersal on life history or other characters? (See the July 2008 supplement in *The American Naturalist*, Vol. 172.)

CHAPTER 15 Sex and Reproductive Success

Mating ritual. Male satin bowerbirds (*Ptilonorhynchus violaceus*) in eastern Australia build elaborate bowers on the rain forest floor and decorate them with blue objects (such as blue feathers). The duller, green-plumaged females "inspect" the bowers and their builders while males engage in elaborate courtship displays to induce a female to enter the bower and mate. Once mated, the female builds a nest and raises young without the male's help. (Photo © Lloyd Nielsen/OSF/Photolibrary.com.)

In the previous chapter, we treated those components of fitness, age-specific schedules of survival, and female reproduction that affect the increase or decrease in numbers of individual organisms with different genotypes, and indeed of populations. Changes in the relative numbers of variant alleles, however, stem not only from those fitness components but from many others. The most obvious of these is variation in male mating success, or sexual selection. Why are there males, anyway? There are plenty of all-female populations. For that matter, many species do not have separate sexes, but consist of hermaphroditic, or cosexual, members. Why should some species have separate sexes?

Organisms vary greatly in what is sometimes called their GENETIC SYSTEM: whether they reproduce sexually or asexually, self-fertilize or outcross, are hermaphroditic or have separate sexes. Discovering why and how each of these characters evolved poses some of the most challenging problems in evolutionary biology and is the subject of some of the most creative contemporary research on evolution.

The genetic system affects genetic variation, which of course is necessary for the long-term survival of species. This fact has been cited for more than a century as the reason for the existence of recombination and sexual reproduction. But, as we have seen, arguments that invoke benefits to the species are suspect because they rely on group selection, which is ordinarily a weak agent of evolution. The question, then, is whether or not natural selection *within* populations can account for features of the genetic system.

The Evolution of Mutation Rates

If group selection and the long-term survival of a species do not account for genetic systems, how does natural selection do so? We can first address this question by thinking about the evolution of mutation rates. Two hypotheses have been proposed: either the mutation rate has evolved to some optimal level, or it has evolved to the minimal possible level. Variation in factors such as the efficacy of DNA repair provides potential genetic variation in the genomic mutation rate. According to the hypothesis of optimal mutation rate, selection has favored somewhat inefficient repair enzymes. According to the hypothesis of minimal mutation rate, mutation exists only because the repair system is as efficient as it can be, or because selection is not strong enough to favor investment of energy in a more efficient repair system. According to this hypothesis, the process of mutation is not an adaptation.

An optimal (greater than zero) mutation rate could be favored because populations, or lineages within populations, that experience beneficial mutations—and thus adapt faster to changing environments—might persist longer than populations or lineages that do not experience mutation. We do not know how fast this process of long-term selection would occur, because the faster the environment changes, the higher the mutation rate must be to avert extinction (Lynch and Lande 1993).

Alternatively, we can ask how evolution within populations affects the mutation rate, by postulating a "mutator locus" that affects the mutation rate of other genes. Let us assume that mutator alleles that increase the mutation rate affect the fitness of their bearers only indirectly, via the mutations they cause. It turns out that the fate of such a mutator allele depends on the level of recombination. In an asexual population, the mutator allele is likely to decline in frequency because copies of the allele are permanently associated with the mutations they cause, and far more mutations reduce than increase fitness. But occasionally, a mutator allele causes a beneficial mutation. It will then increase in frequency by hitchhiking with the mutation it has caused.

In a sexual population, however, recombination will soon separate the mutator allele from a beneficial mutation it has caused, so it will not hitchhike to high frequency. Because deleterious mutations occur at so many loci, the mutator will usually be associated with one or another of them at first, and will therefore decline in frequency. Therefore, *natural selection in sexual populations will tend to eliminate any allele that increases mutation rates, and mutation rates should evolve toward the minimal achievable level*—even though mutation is necessary for the long-term survival of a species. Consequently, most evolutionary biologists believe that the existence of mutation is not an adaptation in most organisms, but rather is a by-product of imperfect DNA replication that either cannot be improved or can be improved only at too great a cost in fitness (Leigh 1973; Sniegowski et al. 2000). Greater fidelity of replication does seem to be costly, since genetic changes in an RNA virus that reduced the mutation rate also reduced the rate of virus proliferation (Elena and Sanjuán 2007).

Mutator alleles actually occur at considerable frequencies in some natural populations of the bacterium *Escherichia coli*. Because *E. coli* reproduces mostly asexually, it is understandable that mutator alleles sometimes increase to fixation in experimental populations (Figure 15.1). It has been shown that these increases are caused by hitchhiking with new beneficial mutations (Shaver et al. 2002).

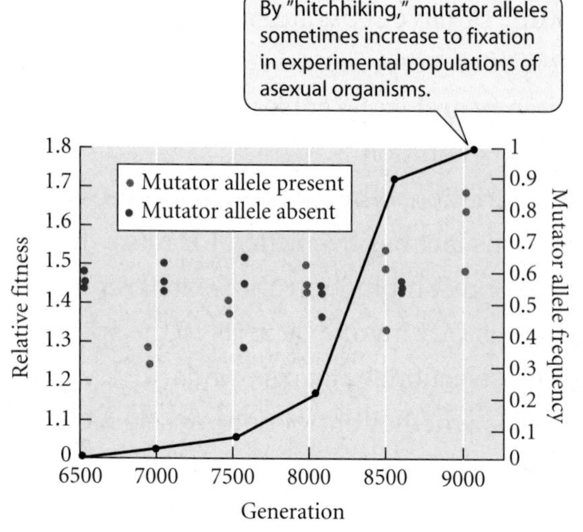

Figure 15.1 A mutator allele that increases the mutation rate throughout the genome increased in frequency (solid line) and eventually became fixed in an experimental population of *Escherichia coli* by hitchhiking with one or more advantageous mutations. Samples of clones carrying the mutator allele (red circles) showed an increase in fitness over time, whereas the fitness of clones that lacked the allele (green circles) did not change. A mutator allele would not be expected to increase in frequency in a sexual population. (After Shaver et al. 2002.)

Sexual and Asexual Reproduction

Sex usually refers to the union (SYNGAMY) of two genomes, usually carried by gametes, followed at some later time by REDUCTION, ordinarily by the process of meiosis and gametogenesis. Sex often, but not always, involves outcrossing between two individuals, but it can occur by self-fertilization in some organisms. Sex almost always includes SEGREGATION of alleles and RECOMBINATION among loci, although the extent of recombination varies greatly. Most sexually reproducing species have distinct female and male sexes, which are defined by a difference in the size of their gametes (ANISOGAMY). In ISOGAMOUS organisms, such as *Chlamydomonas* and many other algae, the uniting cells are the same size; such species have MATING TYPES but not distinct sexes. Species in which individuals are either female or male, such as willow trees and mammals, are termed **dioecious** or GONOCHORISTIC; species such as roses and earthworms, in which an individual can produce both kinds of gametes, are **hermaphroditic** or COSEXUAL.

Asexual reproduction may be carried out by **vegetative propagation**, in which an offspring arises from a group of cells, as in plants that spread by runners or stolons, or by **parthenogenesis**, in which the offspring develops from a single cell. The most common kind of parthenogenesis is **apomixis**, whereby meiosis is suppressed and an offspring develops from an unfertilized egg. The offspring is genetically identical to its mother, except for whatever new mutations may have arisen in the cell lineage from which the egg arose. A lineage of asexually produced, and thus genetically nearly identical, individuals may be called a CLONE.

In some taxa, recombination and the mode of reproduction can evolve rather rapidly. Using artificial selection in laboratory populations of *Drosophila*, investigators have altered the rate of crossing over between particular pairs of loci, and have even developed parthenogenetic strains from sexual ancestors (Carson 1967; Brooks 1988). Asexual populations are known in many otherwise sexual species of plants and animals, such as crustaceans and insects. However, many of the features required for sexual reproduction seem to degenerate rapidly in populations that have evolved asexual reproduction, so reversal from asexual to sexual reproduction becomes very unlikely (Normark et al. 2003).

The Paradox of Sex

Parthenogenesis versus the cost of sex

The traditional explanation of the existence of recombination and sex is that they increase the rate of adaptive evolution of a species, either in a constant or a changing environment, and thereby reduce the risk of extinction. That there is indeed a long-term advantage of sex is indicated by phylogenetic evidence that most asexual lineages of eukaryotes have arisen quite recently from sexual ancestors, and by the observation that such lineages often retain structures that once had a sexual function. A typical example is the dandelion *Taraxacum officinale*, which is completely apomictic but is very similar to sexual species of *Taraxacum*, retaining nonfunctional stamens and the brightly colored petal-like structures that in its sexual relatives serve to attract cross-pollinating insects (Figure 15.2A). If a parthenogenetic lineage were able to persist for many millions of years, it should have

(A)

(B)

Figure 15.2 (A) The dandelion *Taraxacum officinale* reproduces entirely asexually, by apomixis. The brightly colored flower, which evolved for attracting pollinating insects, suggests that asexual reproduction in this species has evolved recently. In fact, some species of this genus reproduce sexually. (B) Scanning electron micrograph of a bdelloid rotifer, *Rotaria tardigrada*. This group of rotifers is unusual among metazoans because it has apparently been parthenogenetic for a very long time. (A photo © Tiax/ShutterStock; B from Diego et al. 2007.)

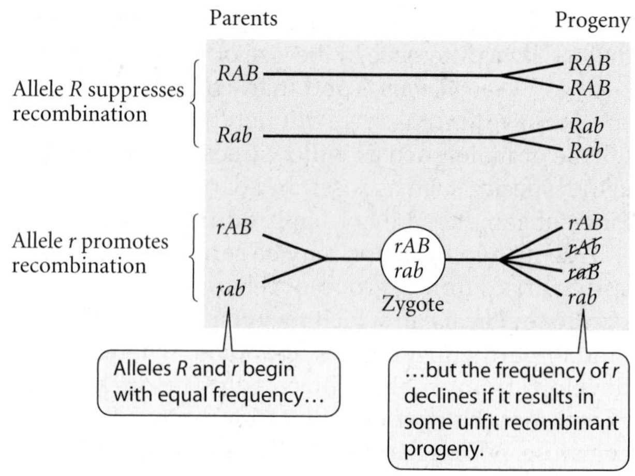

Figure 15.3 Selection against an allele (r) that promotes recombination, if allele combinations Ab and aB have lower fitnesses than AB and ab. The parental generation has two copies of R and two copies of r. Allele R, which suppresses sex and recombination, increases in frequency because of its association with the favored genotypes AB and ab. The organism may be one that, like some algae, is haploid in the dominant part of the life cycle. Asexual genotypes (R) compete with sexual genotypes (r) in which a diploid zygote undergoes meiosis. The same principle holds for organisms in which the diploid phase of the life cycle is dominant.

diverged greatly from its sexual relatives and given rise to a morphologically and ecologically diverse clade. Such diversity, betokening persistence since an ancient origin of asexuality, exists in only a few eukaryote groups, such as the bdelloid rotifers (Figure 15.2B; Normark et al. 2003). Most asexual lineages that arose a very long time ago must have become extinct, since there are so few ancient asexual forms.

The recency of most parthenogenetic lineages suggests that sex reduces the risk of extinction. If this were the reason for its prevalence, sex might be one of the few characteristics of organisms that has evolved by group selection. But recombination and sex also have serious disadvantages. One is that *recombination destroys adaptive combinations of genes*. For example, recall that in the primrose *Primula vulgaris*, plants with a short style and long stamens carry a chromosome with the gene combination *GA*, whereas those with a long style and short stamens are homozygous for the combination *ga* (see Figure 9.19). These are adaptive combinations because each can successfully cross with the other; but recombination produces combinations such as *Ga*, which has both short stamens and a short style, and is less successful in pollination. In general, asexual reproduction preserves adaptive combinations of genes, whereas sexual reproduction breaks them down and reduces linkage disequilibrium between them. An allele that promotes recombination, perhaps by increasing the rate of crossing over, may decline in frequency if it is associated with gene combinations that reduce fitness (Figure 15.3).

Sexual reproduction has a second disadvantage that is great enough to make its existence one of the most difficult puzzles in biology. This disadvantage is the **cost of sex**. Imagine two genotypes of females with equal fecundity, one sexual and one asexual. In many sexual species, only half of all offspring are female. However, all the offspring of an asexual female are female (because they inherit their mother's sex-determining genes). If sexual and asexual females have the same fecundity, then a sexual female will have only half as many grandchildren as an asexual female (Figure 15.4). Therefore, the rate of increase of an asexual genotype is approximately twice as great as that of a sexual genotype (all else being equal), so an asexual mutant allele would very rapidly be fixed if it occurred in a sexual population. In the long term, of course, the evolution of asexuality

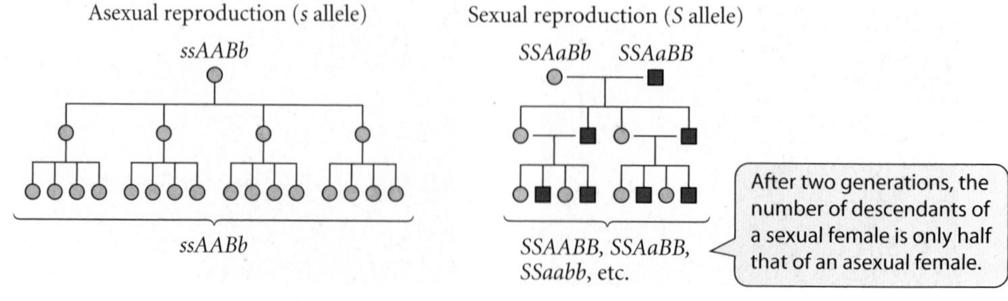

Figure 15.4 The cost of sex. The disadvantage of an allele S, which codes for sexual reproduction, compared with an allele s, coding for asexual reproduction. Circles represent females, squares males. Each of two females in the same population produces four equally fit offspring, but the frequency of the S allele drops rapidly from two-thirds in the first generation to one-third by the third generation because of the production of males. However, the S allele would increase if the environment were to change so that a recombinant genotype such as *aabb* had a much higher survival rate than other genotypes.

Figure 15.5 How linkage disequilibrium affects the response to selection. On the left, there is negative LD between favored alleles (green bars), and most of the genetic variation is aligned at right angles to the direction of selection, so adaptive evolution is slow. On the right, positive LD maximizes the genetic variation along the direction of selection. A possible outcome is rapid fixation of the advantageous two-locus genotype. (After Poon and Chao 2004.)

Advantageous allele
Disadvantageous allele
Hexagon size is proportional to the frequency of the gene combination in a population.
Direction of selection
Negative linkage disequilibrium of favored alleles. Adaptive evolution is slow.
Positive linkage disequilibrium of favored alleles. Variation along the direction of selection is maximized.

might doom a population to extinction, but given this twofold advantage of asexuality, it is doubtful that extinction occurs frequently enough to prevent the replacement of sexual with apomictic genotypes within populations. The problem, therefore, is to discover whether there are any *short-term advantages* of sex that can overcome its short-term disadvantages (Maynard Smith 1978; Charlesworth 1989; Kondrashov 1993). One can envision, for example, that sexual lineages within a population might evolve higher fitness than the asexual lineages in the same population, and replace them if the difference in fitness offset the cost of sex.

Hypotheses for the advantage of sex and recombination

Of the many explanations for the prevalence of recombination and sex that have been proposed (Kondrashov 1993), several are generally considered most likely. A common theme is that the variance in fitness is small if deleterious mutations are in different asexual genomes: they are in negative linkage disequilibrium. Recombination breaks down linkage disequilibrium, so that combinations of deleterious alleles on the one hand, and of advantageous alleles on the other, arise and thus increase the variance in fitness, so that selection can increase fitness more effectively (Figure 15.5).

PREVENTING MUTATIONAL DETERIORATION. Herman Muller (1964), who won a Nobel Prize for his role in discovering the mutagenic effect of radiation, proposed a hypothesis that has been named MULLER'S RATCHET (Figure 15.6). In an asexual population, deleterious mutations at various loci create a spectrum of genotypes carrying 0, 1, 2, ... m mutations. Individuals can carry more mutations than their ancestors did (because of new mutations), but not fewer. Thus the zero-mutation class declines over time because its members experience new deleterious mutations. Moreover, because of genetic drift in a finite population, the zero class may be lost by chance, despite its superior fitness. (The smaller the population, the more likely this is to happen.) Thus all remaining genotypes have at least one deleterious mutation. Sooner or later, by the same process of drift, the one-mutation class is lost, and all remaining individuals carry at least two mutations. The accidental loss of

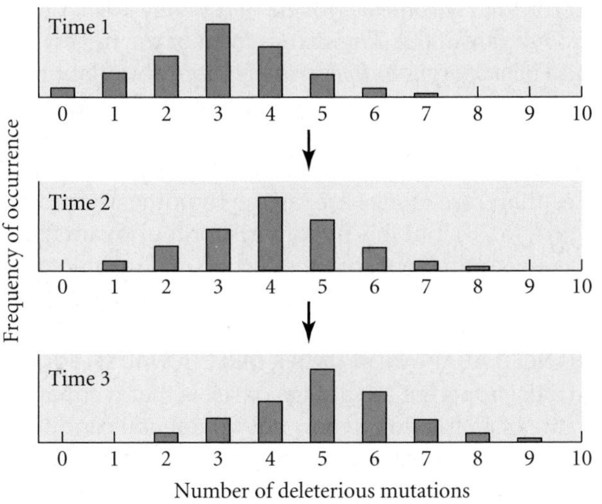

Figure 15.6 Muller's ratchet. The frequency of individuals with different numbers of deleterious mutations (0–10) is shown for an asexual population at three successive times. The class with the lowest mutation load (0 in top graph, 1 in middle graph) is lost over time, both by genetic drift and by its acquisition of new mutations. In a sexual population, class 0 can be reconstituted, since recombination between genomes in class 1 that bear different mutations can generate progeny with none. (After Maynard Smith 1988.)

superior genotypes continues, and is an irreversible process—a ratchet. This reduction of fitness is likely to lower population size, and this, in turn, increases the rate at which the least mutation-laden genotypes are lost by genetic drift. Thus there may be an accelerated decline of fitness—a "mutational meltdown"—leading to extinction (Lynch et al. 1993). In contrast, *recombination in a sexual population reconstitutes the least mutation-laden classes of genotypes* by generating progeny with new combinations of favorable alleles.

Nancy Moran (1996) has suggested that Muller's ratchet explains several features of the genome of the endosymbiotic bacteria (*Buchnera*) of aphids, such as accelerated rates of nonsynonymous mutations, but not synonymous mutations, compared with free-living relatives. Whether or not the advantage of sex under this hypothesis can counterbalance the twofold cost of sex is uncertain, but the reality of mutational degeneration is undeniable.

A hypothesis related to Muller's ratchet was proposed by Alexey Kondrashov (1988), who pointed out that deleterious mutations at multiple loci will be eliminated from a sexual population more rapidly than from an asexual population because recombination in the sexual population brings them together, and so eliminates them simultaneously. Recent theoretical work has shown that the resulting increase in fitness can more than balance the cost of sex (Keightley and Otto 2006).

ADAPTATION TO FLUCTUATING ENVIRONMENTS. Suppose a polygenic character is subject to stabilizing selection, but the optimal character state fluctuates because of a fluctuating environment (Maynard Smith 1980). Let us assume that alleles *A*, *B*, *C*, *D* ... additively increase a trait such as body size, and alleles *a*, *b*, *c*, *d* ... decrease it. Stabilizing selection for intermediate size reduces the variance and creates negative linkage disequilibrium, so that combinations such as *AbCd* and *aBcD* are present in excess (see Chapter 13). If selection changes so that larger size is favored, combinations such as *ABCD* may not exist in an asexual population, but they can arise rapidly in a sexual population. This capacity can provide not only a long-term advantage for sex (a higher rate of adaptation of the population) but a short-term advantage as well, because sexual parents are likely to leave more surviving offspring than asexual parents. For this hypothesis to work, the selection regime must fluctuate rather frequently, and some factor must maintain genetic variation, because a long-term regime of stabilizing selection for a constant optimal phenotype would fix a homozygous genotype (such as *AABBccdd*; see Chapter 13).

One popular possibility is that genetic variation is maintained, and sex is favored, by parasites. As a resistant host genotype (e.g., *ABCD*) increases in frequency, a parasite might evolve to attack it. Parasite genotypes that can attack less common host genotypes (such as *abCD*) would then become rare, so the uncommon host genotypes would acquire higher fitness and increase in frequency. Continuing cycles of coevolution between host and parasite might select for sex, which would continually regenerate rare combinations of alleles.

Curtis Lively and collaborators have provided support for this hypothesis (Jokela and Lively 1995; Lively and Dybdahl 2000). The sexual form of the freshwater snail *Potamopyrgus antipodarum* is more abundant than the asexual form in sites where a trematode parasite is abundant (Figure 15.7A), and common asexual genotypes are more heavily infected by sympatric trematodes than rare clones are, as the hypothesis predicts (Figure 15.7B). But this hypothesis for the advantage of sex may require very strong selection and may not provide a general explanation (Otto and Nuismer 2004).

Figure 15.7 Evidence that selection by a parasitic trematode may favor sexual reproduction in a freshwater snail species that has both sexual and asexual genotypes. (A) The proportion of sexual genotypes was greater in local populations that were exposed to a high incidence of the parasite, as shown by the decreased proportion of females in such populations. (B) When snails of different asexual genotypes were exposed to trematodes, individuals with a rare genotype were less likely to become infected than were those with four common genotypes. (A after Jokela and Lively 1995; B after Lively and Dybdahl 2000.)

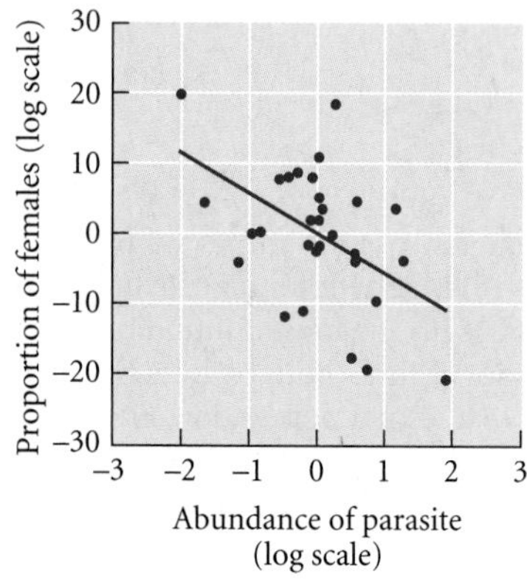

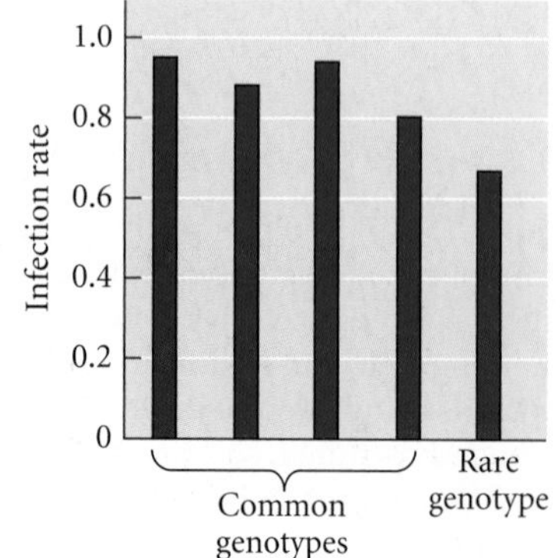

ENHANCED ADAPTATION UNDER DIRECTIONAL SELECTION. The most apparent advantage of sex is that it enhances the rate of adaptation to new environmental conditions by combining new mutations or rare alleles (Figure 15.8). In an asexual population, beneficial mutations *A*

Population 1: Large, asexual

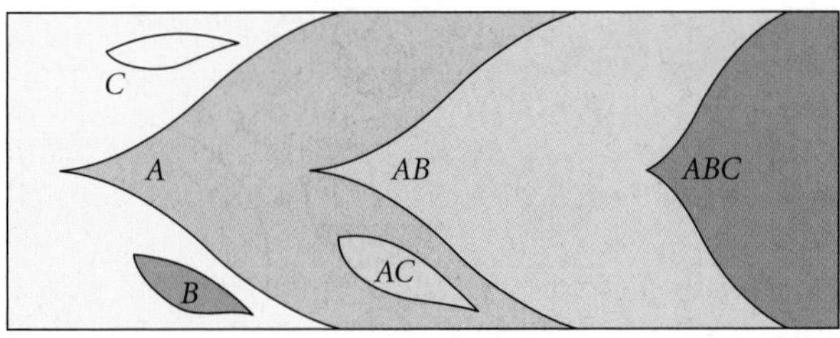

Population 2: Large, sexual

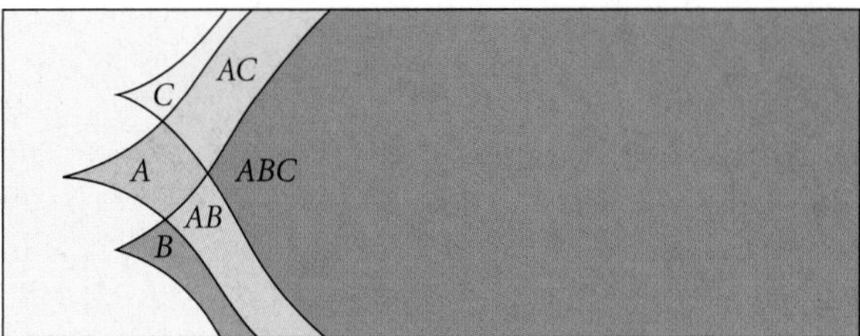

Population 3: Small, asexual

Population 4: Small, sexual

Figure 15.8 Effects of recombination on the rate of evolution. *A*, *B*, and *C* are new mutations that are advantageous in concert. In asexual populations (1 and 3), combination *AB* (or *ABC*) is not formed until a second mutation (such as *B*) occurs in a lineage that already bears the first mutation (*A*). In a large sexual population (2), independent mutations can be brought together in a lineage more rapidly by recombination, so adaptation is more rapidly achieved. In a small sexual population (4), however, the interval between the occurrence of favorable mutations is so long that the population does not adapt more rapidly than an asexual population. (After Crow and Kimura 1965.)

and *B* would be combined only when a second mutation (*B*) occurs in a growing lineage that has already experienced mutation *A* (or vice versa). Furthermore, a clone with one advantageous mutation might begin to increase but then dwindle when a slightly more advantageous mutation arises in a different clone, as illustrated by the genotypes AC and AB in Population 1 in Figure 15.8. This process of CLONAL INTERFERENCE (Gerrish and Lenski 1998) has been described in experimental populations of viruses and bacteria (e.g., Miralles et al. 1999). In a sexual population, in contrast, mutations that have occurred in different lineages can be combined more rapidly (see Figure 15.8). However, this difference in evolutionary rate depends on population size. In small populations, mutations are so few that the first (*A*) is likely to be fixed by selection before the second (*B*) arises, whether the population is asexual or sexual.

This hypothesis has been experimentally confirmed. For example, sexual populations of the unicellular green alga *Chlamydomonas reinhardtii* evolved higher fitness in the laboratory more rapidly than asexual populations, and as predicted, this effect was seen in large, but not in small, populations (Colegrave 2002). Slower adaptation by asexual populations is likely to be a major reason for their high rate of extinction, as documented by the recent origin of most asexual eukaryotes. But it is unlikely that directional selection is frequent enough to counter the cost of sex.

It has been difficult to demonstrate that any single hypothesis accounts for the prevalence of sex, and it might well be that a combination of factors is usually at play (West et al. 1999 and associated commentaries). Why sex is so widespread remains an open question.

Sex Ratios and Sex Allocation

We turn now to the question of why some species are hermaphroditic and others are dioecious, and what accounts for variation in sex ratio among dioecious species. The theory of sex allocation has been developed to explain such variation (Charnov 1982; Frank 1990).

The **sex ratio** is defined as the proportion of males. As in Chapter 12, we distinguish the sex ratio in a population (the POPULATION SEX RATIO) from that in the progeny of an individual female (an INDIVIDUAL SEX RATIO). In Chapter 12, we saw that in a large, randomly mating population, a genotype with a male-biased individual sex ratio has an advantage if the population sex ratio is female-biased, and vice versa, because each individual of the minority sex has a higher number of offspring than each individual of the majority sex. A genotype with an individual sex ratio of 0.5 is an ESS (evolutionarily stable strategy) because it has, per capita, the greatest number of grandchildren (see Figure 12.17).

In many species, however, mating occurs not randomly among members of a large population, but within small local groups descended from one or a few founders. After one or a few generations, progeny emerge into the population at large, then colonize patches of habitat and repeat the cycle. In many species of parasitoid wasps, for example, the progeny of one or a few females emerge from a single host and almost immediately

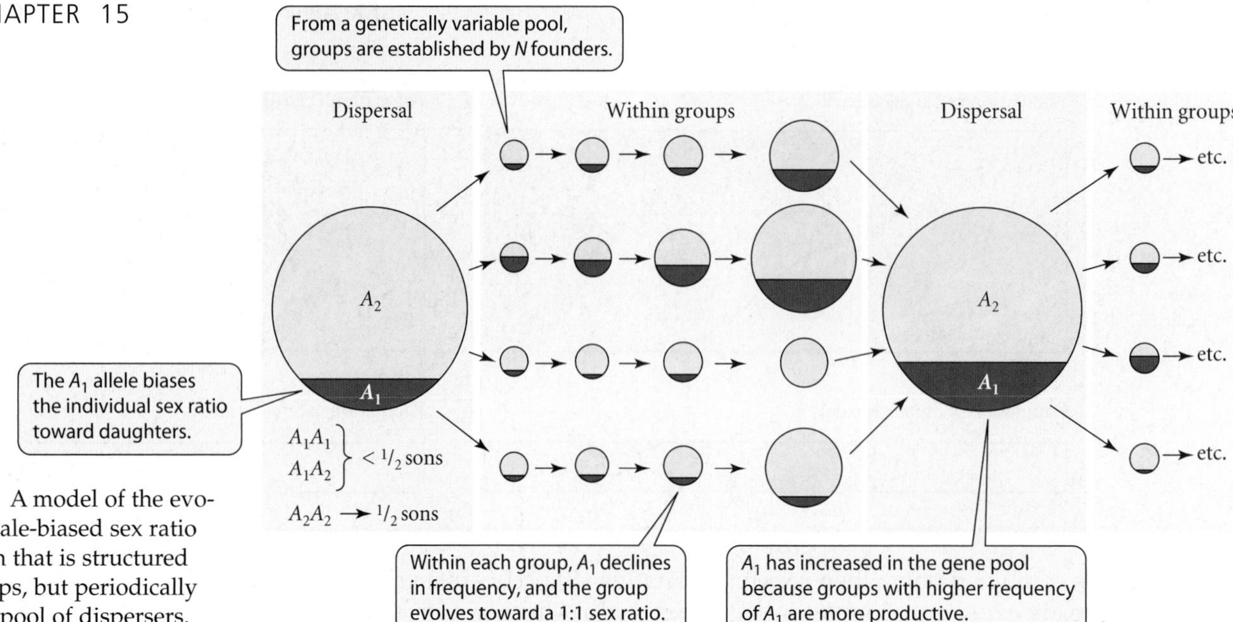
From a genetically variable pool, groups are established by N founders.
Dispersal
Within groups
Dispersal
Within groups
etc.
etc.
etc.
etc.
A_2
A_1
A_2
A_1
The A_1 allele biases the individual sex ratio toward daughters.
A_1A_1
A_1A_2 $< 1/2$ sons
$A_2A_2 \rightarrow 1/2$ sons
Within each group, A_1 declines in frequency, and the group evolves toward a 1:1 sex ratio.
A_1 has increased in the gene pool because groups with higher frequency of A_1 are more productive.

Figure 15.9 A model of the evolution of a female-biased sex ratio in a population that is structured into local groups, but periodically forms a single pool of dispersers. The frequency of A_1, an allele that biases the individual sex ratio toward daughters, is indicated by the dark portion of each circle. The size of each circle represents the size of a group or population. From a genetically variable pool, groups are established by one or a few founders. The frequency of A_1 varies among groups by chance. Although A_1 declines in frequency within each group over the course of several generations, the growth in group size is greater the higher the frequency of A_1 (because of the greater production of daughters). When individuals emerge from the groups, they form a pool of dispersers, in which A_1 has increased in frequency since the previous dispersal episode. (After Wilson and Colwell 1981.)

mate with each other; the daughters then disperse in search of new hosts. Such species often have female-biased sex ratios.

William Hamilton (1967) explained such "extraordinary sex ratios" by what he termed LOCAL MATE COMPETITION (Antolin 1993). Whereas in a large population a female's sons compete for mates with many other females' sons, they compete only with one another in a local group founded by their mother. Thus the founding female's genes can be propagated most prolifically by producing mostly daughters, with only enough sons to inseminate all of them. Additional sons would be redundant, since they all carry their mother's genes. Another way of viewing this situation is to recognize that groups founded by genotypes whose individual sex ratio is biased in favor of females contribute more individuals (and genes) to the population as a whole than groups founded by unbiased genotypes. The difference among local groups in their production of females increases the frequency, in the population as a whole, of female-biasing alleles (Figure 15.9; Wilson and Colwell 1981). The greater the number of founders of a group, the more nearly even the optimal sex ratio will be.

Some of the best evidence for this theory is provided by the adaptive plasticity of the sex ratio in parasitoid wasps such as *Nasonia vitripennis*. The progeny of one or more females develop in a fly pupa and mate with each other immediately after emerging. Based on the theory described above, we would expect females to produce more daughters than sons, but we would expect the sex ratio to increase with the number of families developing in a host. Moreover, if the second wasp that lays eggs in a host can detect previous parasitization, we would expect her to adjust her individual sex ratio to a higher value than that of the first female. John Werren (1980) calculated the theoretically optimal individual sex ratio of a second female, then measured the individual sex ratios of second females by exposing fly pupae successively to strains of *Nasonia* that were distinguishable by an eye-color mutation. On the whole, his data fit the theoretical prediction very well (Figure 15.10).

The theory of sex allocation, which explains why some species are cosexual and others dioecious, builds on the same principle as the explanation of even (0.5) sex ratio, namely that fitness through female and male functions must be equal. Assume that there is a trade-off in how a potentially hermaphroditic individual allocates energy and resources to female functions (e.g., eggs, seeds) and to male functions (e.g., sperm, pollen, mate-seeking). In dioecious species, individuals allocate all of their reproductive energy to one sexual function or the other. The reproductive success that individuals achieve might be a linear function of their proportional allocation to, say, male function (curve 2 in Figure 15.11A); it might be an accelerating function (curve 3); or it might decelerate, showing

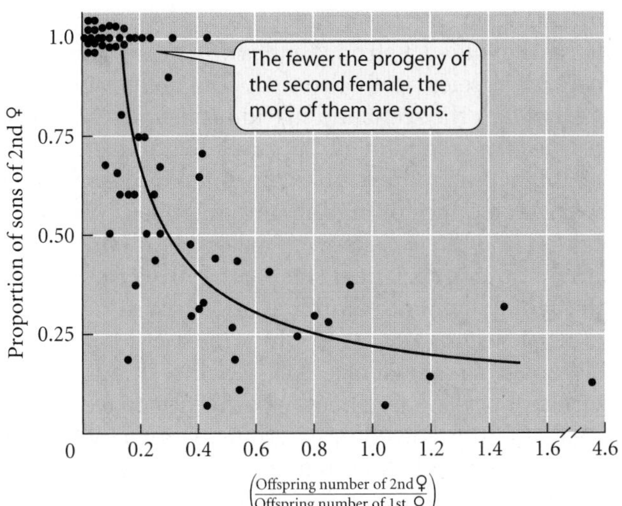

Figure 15.10 Adaptive adjustment of individual sex ratio by a parasitoid wasp in which females usually mate with males that emerge from the same host. The points show the relationship between the proportion of sons among the offspring of a "second" female (i.e., one that lays eggs in a fly pupa in which another female has already laid eggs) and the proportion of her offspring in the host. The curved line is the theoretically predicted individual sex ratio in second females' broods, as a function of the sex ratio in the first female's brood and the relative number of the two females' offspring. If the second female's offspring make up only a small fraction of the total, that female's optimal "strategy" should be to produce mostly sons, which could potentially inseminate many female offspring of the first female. As predicted, the fewer the progeny of the second female, the more of them are sons. (After Werren 1980.)

diminishing returns (curve 1). It can be shown that in the last case, the theoretically optimal strategy is a hermaphroditic one (Figure 15.11B), whereas it is optimal to be one sex or the other if reproductive success accelerates with increasing allocation. One factor that may affect the shape of the trade-off curve in Figure 15.11B is the cost of developing the structures required for the two sexual functions. If most structures associated with reproduction enhance both male and female function (e.g., petals), hermaphroditism may be favored, because with relatively little extra expenditure the individual can produce offspring via both sexual functions. If male and female functions require different structures, however, a pure male or female incurs the cost of developing only one set of structures, whereas a hermaphrodite incurs both costs and thus is likely to have lower fitness.

Inbreeding and Outcrossing

Self-fertilization (selfing), which results in inbreeding, occurs in some hermaphroditic animals (such as certain snails) and in a great many plants. Many plants can both self-fertilize and export pollen to the stigmas of other plants (**outcrossing**). Characteristics that promote outcrossing include dioecy (separate sexes), asynchronous male and female function (maturation and dispersal of pollen either before or after the stigma of the same flower is receptive), and several kinds of self-incompatibility (Matton et al. 1994). DNA sequences of self-incompatibility alleles, which prevent pollen from fertilizing plants that carry the pollen's allele, show that these polymorphisms are stable and very old,

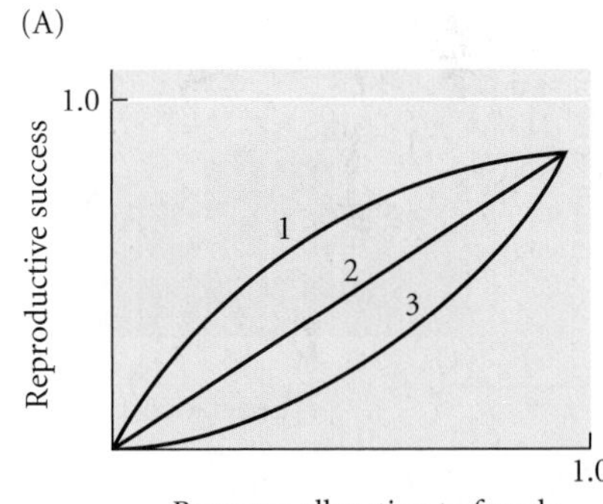

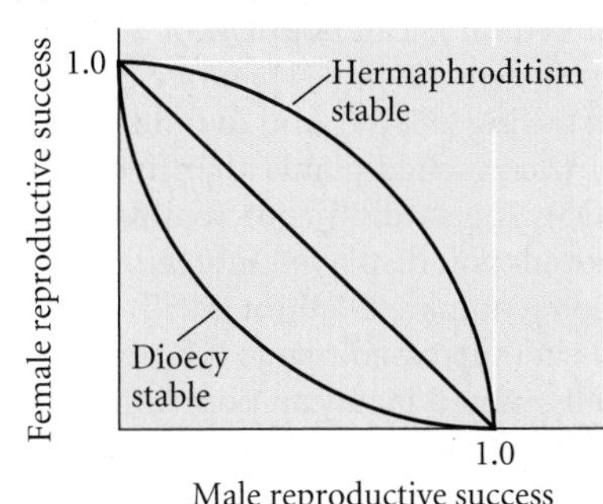

Figure 15.11 The theory of sex allocation. (A) The reproductive success of an individual as a function of the fraction of resources allocated to one sexual function (say, female) rather than the other. Increasing allocation to that sexual function may yield decelerating (1), linear (2), or accelerating (3) gains in reproductive success. (B) Reproductive success gained through female function plotted against that gained through male function. Because resources are allocated between these functions, there is a trade-off between the reproductive success an individual achieves through either sexual function. This trade-off is linear, and the sum of female reproductive success and male reproductive success equals 1.0 at each point on the trade-off curve, if reproductive success is linearly related to allocation (curve 2 in A). If the gain in reproductive success is a decelerating function of allocation to one sex or the other (curve 1 in A), then the fitness of a hermaphrodite exceeds that of a unisexual individual (with reproductive success = 1.0 for one sexual function and 0 for the other). If reproductive success is an accelerating function of allocation (curve 3 in A), then the trade-off curve is concave, and dioecy (separate sexes) is stable—that is, a hermaphrodite's fitness is lower than that of either dioecious type. (After Thomson and Brunet 1990.)

Outcrossing
Selfing

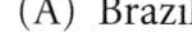

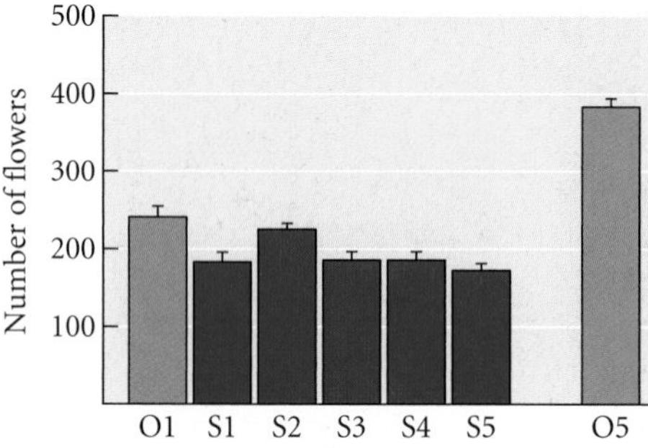

(B) Jamaica

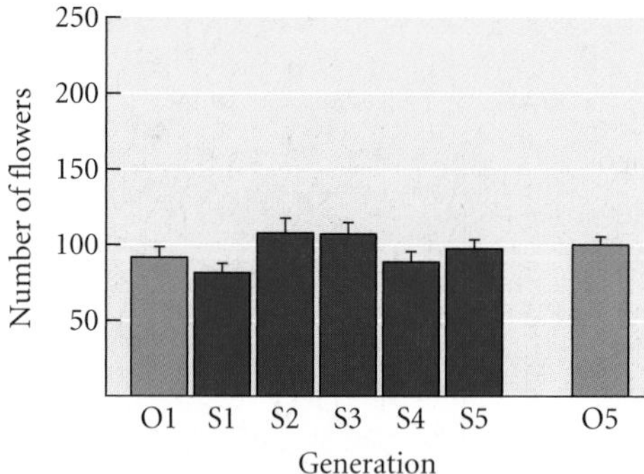

Figure 15.12 The mean number of flowers produced by *Eichhornia paniculata* plants from (A) a naturally outcrossing Brazilian population and (B) a naturally inbred Jamaican population in the first outcross generation (O1), in five generations of selfing (S1–S5), and in outcrosses between selfed plants after five generations (O5). The naturally outcrossing Brazilian population displayed inbreeding depression (compare O1 through S5) and heterosis in outcrossed plants (O5), but the naturally inbred Jamaican population did not. (After Barrett and Charlesworth 1991; photo courtesy of S. C. H Barrett.)

and have been maintained by balancing selection (see Figure 12.27A). In contrast, many plants, such as wheat, have evolved a strong tendency toward self-fertilization within flowers. Many such species produce only a little pollen and have small, inconspicuous flowers that lack the scent and markings to which pollinators are attracted. In some cases, flowers remain budlike and do not open at all.

In animals, the major factors that may reduce inbreeding are prereproductive dispersal and INCEST AVOIDANCE: avoidance of mating with relatives (Thornhill 1993), which inevitably brings to mind the well-known "incest taboo," often codified in religious and civil law, in human societies. This is a highly controversial topic with respect to both the actual incidence of inbreeding ("incest") and the interpretation of the social taboo. Societies vary as to which kin matings are prohibited (Ralls et al. 1986). Moreover, the incidence of closely incestuous sexual activity is evidently far higher than society has generally been ready to recognize, especially in the form of sexual attention forced upon young women by fathers and uncles. These observations raise doubt about whether a strong, genetically based aversion to incest has evolved in our species (at least in males). Some anthropologists hold that outbreeding is a social device (not an evolved genetic trait) to establish coalitions between families or larger groups in order to reap economic and other benefits of cooperation.

Inbreeding increases homozygosity, and it is often accompanied by inbreeding depression (see Chapter 9). Inbreeding depression usually seems to be caused by homozygosity for deleterious recessive (or nearly recessive) alleles, but it may occasionally be caused by homozygosity at overdominant loci, at which heterozygotes have highest fitness (Uyenoyama et al. 1993). If inbreeding depression is caused by homozygous recessive alleles, selection may be expected to "purge" the population of those alleles as inbreeding continues, so genetic variation should decline and mean fitness should increase. The mean fitness of a long-inbred population might therefore equal or even exceed that of the initial outbreeding population (Lande and Schemske 1985).

Spencer Barrett and Deborah Charlesworth (1991) provided some evidence for this hypothesis, using a highly outcrossing Brazilian population and an almost exclusively self-fertilizing Jamaican population of the aquatic plant *Eichhornia paniculata*. For five generations, Barrett and Charlesworth self-pollinated plants in each population, then cross-pollinated the inbred lines. In the naturally outcrossing Brazilian lines, flower number declined as inbreeding proceeded (Figure 15.12A), but it increased dramatically in crosses between the inbred lines. In contrast, the naturally selfing Jamaican lines conformed to the purging theory: they showed neither inbreeding depression nor heterosis when crossed with each other (Figure 15.12B).

The most important advantage of characteristics such as self-incompatibility is thought to be avoidance of inbreeding depression (Charlesworth and Charlesworth 1978; Lloyd 1992). The evolution of obligate or predominant self-fertilization would have to overcome two obstacles: inbreeding depression and the loss of reproductive success through outcross pollen. Several possible advantages of selfing might outweigh these disadvantages. First, exclusive selfers may save energy and resources because they usually have small flowers and produce little pollen. However, a slight initial reduction of flower size seems unlikely to offset the severe disadvantages of selfing (Jarne and Charlesworth 1993).

Second, even if selfing is disadvantageous on average, it may occasionally produce a highly fit homozygous genotype that sweeps to fixation, carrying with it the alleles that increase the selfing rate (Holsinger 1991). A related possibility is that selfing may "protect" locally adapted genotypes from OUTBREEDING DEPRESSION as a result of gene flow and recombination. For example, Nikolas Waser and Mary Price (1989) crossed scarlet gilia (*Ipomopsis aggregata*) plants separated by various geographic distances, planted the seeds under uniform conditions, and measured fitness-related characteristics in the progeny. Fitness was greatest in the progeny of moderately distant plants; it was lower in the progeny of plants situated close together (presumably an example of inbreeding depression) and in the progeny of distant plants (outbreeding depression). Waser and Price postulated that distant populations have different coadapted gene combinations, perhaps adapted to different environmental conditions, and that recombination between such gene combinations would be prevented by selfing.

Third, and perhaps most important, is REPRODUCTIVE ASSURANCE: a plant is almost certain to produce some seeds by selfing, even if scarcity of pollinators, low population density, or other adverse environmental conditions prevent cross-pollination. There is abundant support for this hypothesis (Wyatt 1988; Jarne and Charlesworth 1993). For example, adaptations for self-fertilization are especially common in plants that grow in harsh environments, where insect visitation is low or unpredictable, and on islands, where populations are sparse for some time after colonization.

The Concept of Sexual Selection

Mating success is a major component of fitness in sexually reproducing species (see Figure 12.3). Darwin introduced the concept of **sexual selection** to describe differences among individuals of a sex in the number or reproductive capacity of mates they obtain, although it may be extended to mean competition among individuals for access to the gametes of the opposite sex (Andersson 1994; Kokko et al. 2006). Variation in reproductive success among males may depend on the number of their mates, the fecundity of their mates, and the proportion of the females' eggs that they fertilize when they mate. Sexual selection was Darwin's solution to the problem of why conspicuous traits such as the bright colors, horns, and displays of males of many species have evolved. He proposed two forms of sexual selection: contests between males for access to females and female choice (or "preference") of some male phenotypes over others. Several other bases for sexual selection, including sperm competition, have been recognized (Table 15.1).

Sexual selection exists because females produce relatively few, large gametes (eggs) and males produce many small gametes (sperm). *This difference creates an automatic conflict* between the reproductive strategies of the sexes: a male can mate with many females, and often suffers little reduction in fitness if he mates with an inappropriate female,

Table 15.1 ***Mechanisms of competition for mates and characters likely to be favored***

Mechanism	*Characters favored*
Same-sex contests	Traits improving success in confrontation (e.g., large size, strength, weapons, threat signals); avoidance of contests with superior rivals
Mate preference by opposite sex	Attractive and stimulatory features; offering of food, territory, or other resources that improve mate's reproductive success
Scrambles	Early search and rapid location of mates; well-developed sensory and locomotory organs
Endurance rivalry	Ability to remain reproductively active during much of season
Sperm competition	Ability to displace rival sperm; production of abundant sperm; mate guarding or other ways of preventing rivals from copulating with mate
Coercion	Adaptations for forced copulation and other coercive behavior
Infanticide	Similar traits as for same-sex contests
Antagonistic coevolution	Ability to counteract the other sex's resistance to mating (by, e.g., hyperstimulation); egg's resistance to sperm entry

Source: After Andersson 1994; Andersson and Iwasa 1996.

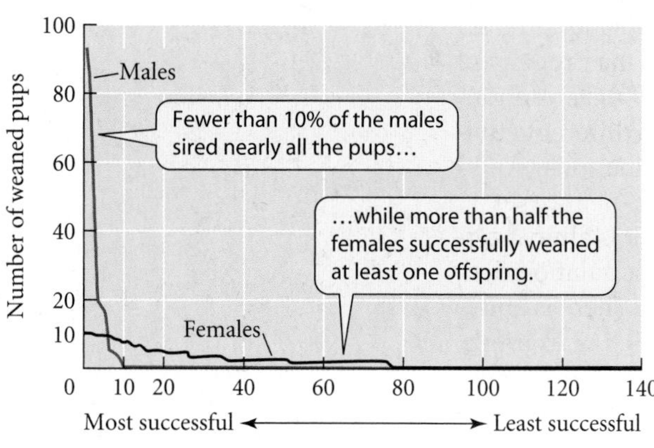

Figure 15.13 Among elephant seals, a few dominant males defend harems of females against other males (see Figure 10.5). A plot of the number of offspring produced by each of 140 elephant seals over the course of several breeding seasons shows much greater variation in reproductive success among males than among females. (After Gould and Gould 1989, based on data of B. J. LeBoeuf and J. Reiter.)

whereas all of a female's eggs can potentially be fertilized by a single male, and her fitness can be significantly lowered by inappropriate matings. Thus encounters between males and females often entail conflicts of reproductive interest (Trivers 1972; Parker 1979). Furthermore, the OPERATIONAL SEX RATIO, the relative numbers of males and females in the mating pool at any time, is often male-biased because males mate more frequently and because the cost of breeding is greater for females, which may have little opportunity to mate and reproduce repeatedly during a breeding season because of energetic and time constraints (Kokko et al. 2006). Commonly, then, *females are a limiting resource for males*, which compete for mates, but males are not a limiting resource for females. Thus *variation in mating success is generally greater among males than among females* (Figure 15.13), and indeed, is a measure of the intensity of sexual selection (Wade and Arnold 1980). Likewise, eggs are a limiting resource for sperm, engendering competition among sperm of different males, or among pollen from different plants. In some species, however, the tables are turned. In cases of so-called SEX ROLE REVERSAL, as in phalaropes and sea horses (Figure 15.14), males care for the young and can tend fewer offspring than a female can produce. In such instances, females may compete for males.

Contests between Males and between Sperm

Male animals often compete for mating opportunities through visual displays of bright colors or other ornaments, many of which make a male look larger. The males of some species fight outright and possess weapons, such as horns or tusks, that can inflict injury (Figure 15.15; Emlen 2008). Plumage patterns and songs are used to establish dominance by many birds, such as the red-winged blackbird (*Agelaius phoeniceus*), males of which have a bright red shoulder patch (Andersson 1994). Male blackbirds that were experimentally silenced were likely to lose their territories to intruders; however, intruders were deterred by tape recordings of male song. When territorial males were removed and replaced with stuffed and mounted specimens whose shoulder patches had been paint-

(A)

(B)

Figure 15.14 Sex role reversal. (A) Two female red phalaropes (*Phalaropus fulicarius*) fight over the smaller, duller plumaged male on their breeding grounds in Alaskan tundra. In contrast to most birds, female phalaropes court males, which care for the eggs and young. (B) A male Australian sea horse (*Hippocampus breviceps*) giving birth. Males of this species carry and nurture developing young in their pouch. (A © Roy Mangersnes/Naturepl.com; B © Paul A. Zahl/Photo Researchers, Inc.)

Figure 15.15 Hornlike structures have evolved independently in the males of diverse animals, which use them to compete for access to females. Such structures are illustrated here for 13 species, including several that are extinct (marked with an asterisk). 1, narwhal (*Monodon monoceros*); 2, chameleon (*Chamaeleo* [*Trioceros*] *montium*); 3, trilobite (*Morocconites malladoides**); 4, unicorn fish (*Naso annulatus*); 5, ceratopsid dinosaur (*Styracosaurus albertensis**); 6, horned pig (*Kubanochoerus gigas**); 7, protoceratid ungulate (*Synthetoceras* sp.*); 8, dung beetle (*Onthophagus raffrayi*); 9, brontothere (*Brontops robustus**); 10, rhinoceros beetle (*Allomyrina* [*Trypoxylus*] *dichotomus*); 11, isopod (*Ceratocephalus grayanus*); 12, horned rodent (*Epigaulus* sp.*); 13, giant rhinoceros (*Elasmotherium sibiricum**). (Illustration courtesy of Douglas Emlen.)

ed over to varying extents, the specimens with the largest red patches were avoided by potential trespassers.

In sexual selection by male contest, directional selection for greater size, weaponry, or display features can cause an "arms race" that results in evolution of ever more extreme traits. Such "escalation" becomes limited by opposing ecological selection (i.e., selection imposed by ecological factors) if the cost of larger size or weaponry becomes sufficiently great (West-Eberhard 1983). The equilibrium value of the trait is likely to be greater than it would be if only ecological selection were operating. As Darwin noted, the duller coloration and lack of exaggerated display features in females and nonbreeding males of many species imply that these features of breeding males are ecologically disadvantageous. Among antelopes and their relatives (Bovidae), the size of males' horns is greater in species that experience stronger sexual selection as indicated by the number of females in the "harem" of a socially dominant male, and thus the more male-biased the operational sex ratio is (Bro-Jorgensen 2007).

Closely related to contests among males for mating opportunities are the numerous ways in which males reduce the likelihood that other males' sperm will fertilize a female's eggs (Thornhill and Alcock 1983; Birkhead and Møller 1992; Simmons 2001). Males of many species of birds defend territories, keeping other males away from their mates (although studies of DNA markers show that females of many such species nevertheless engage in high rates of extra-pair copulation). The males of many frogs, crustaceans, and insects clasp the female, guarding her against other males for as long as she produces fertilizable eggs. In some species of *Drosophila*, snakes, and other animals with internal fertilization, the seminal fluid of a mating male reduces the sexual attractiveness of the female to other males, reduces her receptivity to further mating, or forms a copulatory plug in the vagina (Partridge and Hurst 1998; see the discussion below on chase-away sexual selection).

Sexual competition between males may continue during and after copulation. In many damselflies, the male's genitalia are adapted to remove the sperm of previous mates from the female's reproductive tract (Figure 15.16). In many animals, **sperm competition** occurs when the sperm of two or more males have the opportunity to fertilize a female's eggs (Parker 1970; Birkhead 2000). In some such cases, a male can achieve greater reproductive success than other males simply by producing more sperm. This explains why polygamous species of primates tend to have larger testes than monogamous species (see Figure 11.22). Sperm precedence, whereby most of a female's eggs are fertilized by the sperm of only one of the males with which she has mated, occurs in many insects and other species. In

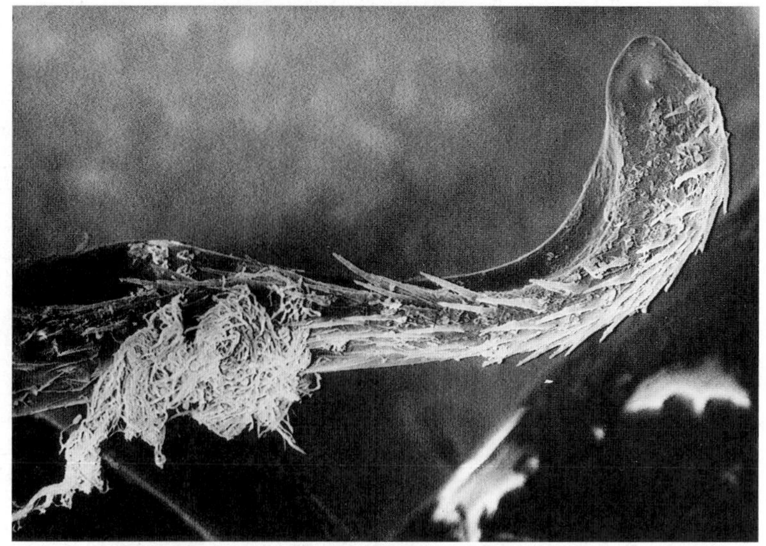

Figure 15.16 An elaborate mechanism for improving a male's likelihood of paternity: a structure near the end of the penis of the black-winged damselfly *Calopteryx*, showing spinelike hairs and a clump of a rival's sperm. (Courtesy of J. Waage.)

Drosophila melanogaster, the degree of sperm precedence is affected by genetic variation among females and by genetic variation among males in the ability of their sperm to displace other males' sperm and to resist displacement (Clark et al. 1994; Clark and Begun 1998). Competition among sperm may explain variation among species in sperm morphology, such as giant sperm in some insects, such as *Drosophila hydei*, in which the 23-millimeter-long sperm are ten times the length of the body (Pitnick and Markow 1994).

Sexual Selection by Mate Choice

In many species of animals, individuals of one sex (usually the male) compete to be chosen by the other. The evolution of sexually selected traits by mate choice presents some of the most intriguing problems in evolutionary biology and is the subject of intense research (Andersson 1994; Johnstone 1995; Andersson and Simmons 2006).

Females of many species of animals mate preferentially with males that have larger, more intense, or more exaggerated characters such as color patterns, ornaments, vocalizations, or display behaviors. For example, female peafowl, widowbirds, barn swallows, and other birds prefer males with longer back or tail feathers (see Figure 11.9). The preferred male characters are often ecologically disadvantageous. For example, males of the cricket *Teleogryllus oceanicus* strike their wings together to produce a calling song that attracts females, followed by a courtship song that is essential for inducing the female to mate. On the Hawaiian island Kauai, an introduced fly (*Ormia ochracea*) uses the cricket's calling song to find its prey. Marlene Zuk and colleagues (2006) have found that over the course of fewer than 20 generations, the cricket population has become almost entirely silent: most males have modified wings, probably based on a single mutation, that do not produce sound (Figure 15.17). These males gather near the few remaining calling males and mate with approaching females, which have evolved to accept males without hearing the courtship song. Perhaps it is not surprising that phylogenetic studies have revealed many examples in which sexually selected ornaments have been lost in evolution (Wiens 2001).

Subject to limits imposed by ecological selection, male traits will obviously evolve to exaggerated states if they enhance mating success. But why should females have a preference for these traits, especially for features that seem so arbitrary—and even dangerous for the males that bear them? The several hypotheses that have been proposed include (1) *direct* and (2) *indirect benefits* to choosy females, (3) *sensory bias*, and closely related to sensory bias, (4) *antagonistic coevolution*. The category of indirect benefits includes two principal hypotheses.

Direct benefits of mate choice

The least controversial hypothesis applies to species in which the male provides a direct benefit to the female or her offspring, such as nutrition, a superior territory with resources for rearing offspring, or parental care. Under these circumstances there is selection pres-

(A)

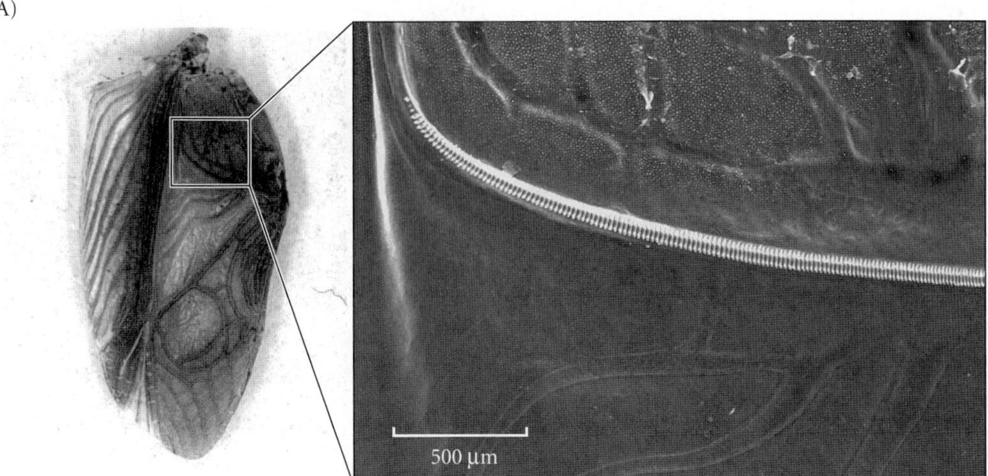

(B)

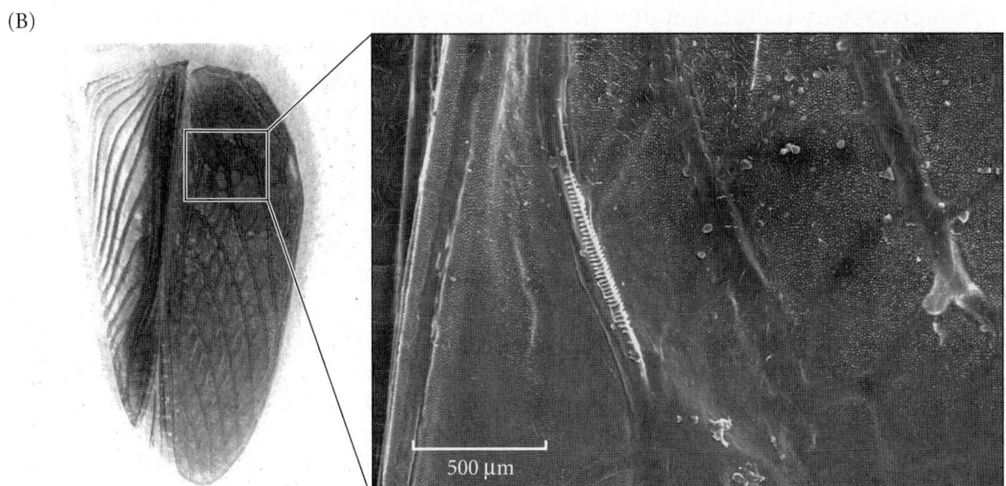

Figure 15.17 The forewing of (A) normal and (B) silent male crickets, *Teleogryllus oceanicus*. The normal wing has a file, a modification of a wing vein with many teeth, that is rubbed against a scraper on the opposite wing to produce sound. The file is greatly reduced and relocated in the silent phenotype, and cannot be used to produce sound. The frequency of the silent phenotype has increased in the presence of a parasitic fly. (From Zuk et al., 2006.)

sure on females to recognize males that are superior providers by some feature that is correlated with their ability to provide. Once this capacity has evolved in females, their preference selects for males with the distinctive, correlated character. For example, the males of many insects provide "nuptial gifts" to females, often in the form of nutrients and diverse chemical compounds that they transfer to females, together with sperm (Gwynne 2008). In the moth *Utetheisa ornatrix*, these compounds include alkaloids that the insect sequestered from its food plant as a larva, and that are toxic and repellant to predators (but not to the moth). Females provide their eggs both with alkaloids that they themselves sequestered and that they acquired by mating. Females prefer to mate with males that release more of a sex pheromone that the male produces from metabolized plant alkaloids, and larger males release more of this pheromone. Vikram Iyengar and Thomas Eisner (1999) demonstrated experimentally that females preferred larger males when given a choice, and that fewer of the eggs from matings with preferred males were eaten by ladybird beetles than eggs from "forced" matings with nonpreferred males, presumably because they contained more alkaloids (Figure 15.18A).

Male investment in nuptial gifts can sometimes become so great that males become a limiting resource for females, and sexual selection becomes reversed. The male Mormon cricket (*Anabrus simplex*), like other male katydids (family Tettigoniidae), produces a spermatophore, a mass of proteins and other compounds that may account for a fifth of his weight, that he transfers to his mate. The female eats the spermatophore (Figure 15.18B), resulting in an increase in the number and size of her eggs. Darryl Gwynne (1984, 1993)

Figure 15.18 Males that provide direct benefits to females. (A) The male tiger moth *Utetheisa ornatrix* transfers pyrrholizidine alkaloids to the female during mating. (B) This female Mormon cricket (*Anabrus simplex*) will eat the large white spermatophore her mate has placed in her genital opening as a "nuptial gift." (A © Thomas R. Fletcher/ Alamy; B courtesy of John Alcock.)

showed that where food is abundant and all males can make spermatophores, males compete for females, which often reject males, as is common in many species of animals. But in a high-density population in which the crickets compete for food, many males lack spermatophores because they produce them at a low rate, so the operational sex ratio is female-biased, females compete aggressively for males and vary in the number of matings (and nutritious spermatophores) they obtain, and males mate preferentially with larger females.

Indirect benefits of mate choice

The most difficult problem in accounting for the evolution of female preferences is presented by species in which the male provides no direct benefit to either the female or her offspring, but contributes only his genes. In this case, alleles affecting female mate choice increase or decrease in frequency depending on the fitness of the females' offspring. Thus females may benefit indirectly from their choice of mates (Kirkpatrick and Barton 1997; Kokko et al. 2002).

The two prevalent models of such indirect benefits are RUNAWAY SEXUAL SELECTION (sometimes called the "sexy son" hypothesis), in which the sons of females that choose a male trait have improved mating success because they inherit the trait that made their fathers appealing to their mothers, and GOOD GENES MODELS, or INDICATOR MODELS, in which the preferred male trait indicates high viability, which is inherited by the offspring of females who choose such males. Good genes models are sometimes called "handicap models" because the male trait indicates high viability despite the ecological handicap that it poses to survival (Zahavi 1975). The runaway and good genes models are closely related, since both depend on an association between alleles that affect female preference and alleles that affect the fitness—whether viability or male mating success—of offspring.

RUNAWAY SEXUAL SELECTION. In runaway sexual selection, as proposed by R. A. Fisher (1930), the evolution of a male trait and a female preference, once initiated, becomes a self-reinforcing, snowballing, or "runaway" process (Lande 1981; Kirkpatrick 1982; Pomiankowski and Iwasa 1998). This process is often referred to as the "Fisherian model" of sexual selection (even though Fisher discussed both kinds of indirect benefits). In the simplest form of the model, haploid males of genotypes T_1 and T_2 have frequencies of t_1 and t_2, respectively. T_2 has a more exaggerated trait, such as a longer tail, that carries an ecological disadvantage, such as increasing the risk of predation. Females of genotype P_2 (with frequency p_2) prefer males of type T_2, whereas P_1 females exhibit little preference (or prefer T_1). It is assumed that alleles P_1 and P_2 do not affect survival or fecundity, and thus are selectively neutral.

Although the expression of genes P and T is sex-limited, both sexes carry both genes and transmit them to offspring. Because P_2 females and T_2 males tend to mate with each other, linkage disequilibrium will develop: their offspring of both sexes tend to inherit both the P_2 and T_2 alleles. The male trait and the female preference thus become geneti-

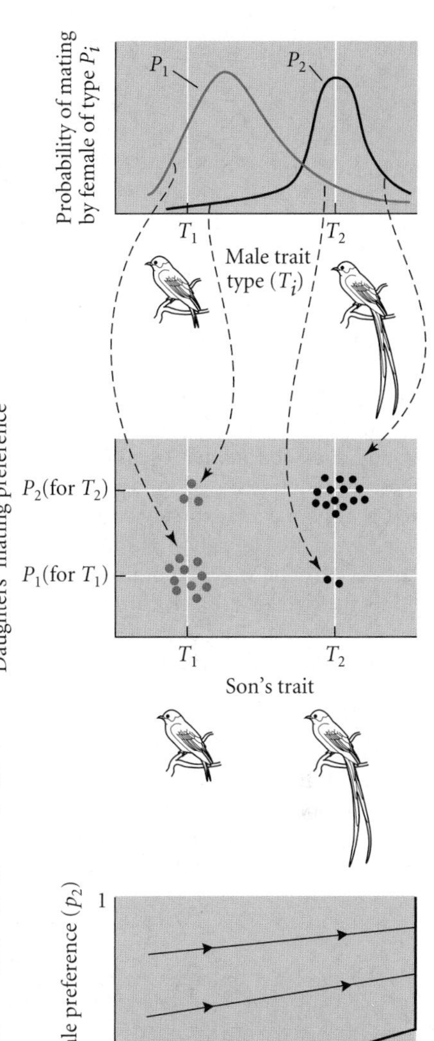

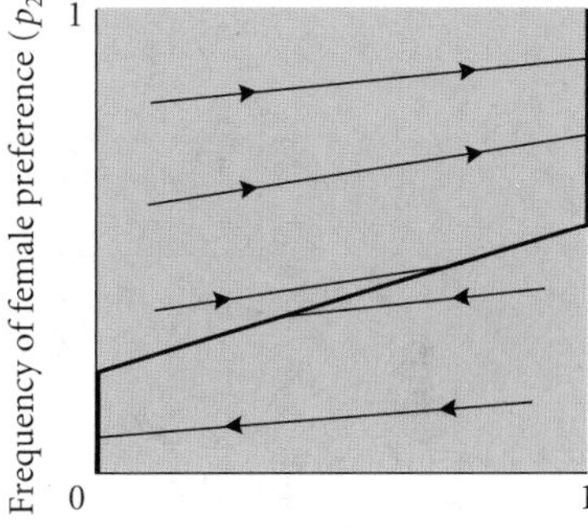

Figure 15.19 A model of runaway sexual selection by female choice. T_1 and T_2 represent male genotypes that differ in some trait, such as tail length. P_1 and P_2 females have different preferences for T_1 versus T_2 males, as shown in the top graph. The resulting pattern of mating creates a correlation in the next generation between the tail length of sons and the mate preference of daughters (middle graph). Thus a genetic correlation is established in the population, in which alleles P_2 and T_2 are associated to some extent. Any change in the frequency (t_2) of the allele T_2 thus causes a corresponding change in p_2. In the bottom graph, each point in the space represents a possible population with some pair of frequencies p_2 and t_2, and the vectors show the direction of evolution. When p_2 is low, t_2 declines as a result of ecological selection, so p_2 declines through hitchhiking. When p_2 is high, sexual selection for T_2 males outweighs ecological selection, so t_2 increases, and p_2 also increases through hitchhiking. Along the solid line, allele frequencies are not changed by selection but may change by genetic drift. (Bottom graph after Pomiankowski 1988.)

cally correlated (see Chapter 13), so that any increase in the frequency of the male trait is accompanied by an increase in the frequency of the female preference through hitchhiking (Figure 15.19).

Perhaps T_2 males have a slight mating advantage over T_1 because they are both acceptable to P_1 females and preferred by the still rare P_2 females. Whether for this or another reason, suppose t_2 increases slightly. Because of the genetic correlation between the loci, an increase in t_2 is accompanied by an increase in p_2. That is, T_2 males have more progeny, and their daughters tend to inherit the P_2 allele, so P_2 also increases in frequency. As P_2 increases, T_2 males have a still greater mating advantage because they are preferred by more females; thus the association of the P_2 allele with the increasing T_2 allele can increase P_2 still further.

If both the male trait and the female preference are polygenically inherited, the same principle holds, but new genetic variation resulting from mutation can enable indefinite evolution of both characters, and if the genetic correlation between them is strong enough, the evolutionary process "runs away" toward preference for an ever more exaggerated male trait.

As we have seen, many exaggerated sexually selected traits carry ecological costs for the males that bear them. Female choice may also carry an ecological cost; for instance, rejecting a male in search of a more acceptable one may delay reproduction or entail other risks. Such costs may prevent the runaway process from occurring, unless recurrent mutations tend to reduce the expression of the male trait. In that case, the cost of female choice may be offset by having attractive, reproductively successful sons (Pomiankowski et al. 1991). Still, indirect selection may be weak, and capable of sustaining female choice only if it carries a low cost.

If females have genetically variable responses to each of several or many male traits, different traits or combinations of traits may evolve, depending on initial genetic conditions (Pomiankowski and Iwasa 1998). Thus runaway sexual selection can follow different paths in different populations, so that *populations may diverge in mate choice and become reproductively isolated*. Sexual selection is therefore a powerful potential cause of speciation (see Chapter 18). Runaway sexual selection of this kind could explain the extraordinary variety of male ornaments among different species of hummingbirds and many other kinds of animals.

INDICATORS OF GENETIC QUALITY. Because females risk substantial losses of fitness if their offspring do not survive or reproduce, an appealing hypothesis is that females should evolve to choose males with high genetic quality, so that their offspring will inherit "good genes" and so have a superior prospect of survival and reproduction. Any male trait that is correlated with genetic quality—any INDICATOR of "good genes"—could be used by females as a guide to advantageous matings, so selection would favor a genetic propensity in females to choose mates on this basis. Female preference for male indicator traits should be most likely to evolve if the trait is a **condition-dependent indicator** of fitness—

as many male display traits indeed are. In this case, males with allele T can develop the indicator trait to a fuller extent if they are in good physiological condition because they also carry "good genes"; say, allele B at another locus. Because females with a "preference allele" (P) mate with males that carry both T and B, their offspring inherit all three alleles. Thus linkage disequilibrium develops among the three loci, and both P and T increase in frequency because of their association with the advantageous allele B. (Single-gene control of the three traits is assumed here for simplicity, but any of the traits could be polygenic.)

In this model, as for the runaway process, costly female preference is unlikely to evolve unless indirect selection, due to the association with good genes, is strong. The strength of indirect selection on female preference is proportional to the genetic variance in fitness in the population. However, natural selection should reduce variance in fitness. Several authors have addressed the question of how genetic variance could be maintained and therefore provide a foundation for female preference. For instance, it has been proposed that continual change in the genetic composition of parasite populations may maintain genetic variation in resistance traits in populations of their hosts, and that sexually selected male traits may be indicators of resistance to parasites (Hamilton and Zuk 1982).

A simpler hypothesis is that genetic variation for fitness components is replenished by recurrent mutation. The mutational variance for fitness components is quite high, as we saw in Chapter 13, and indeed, these traits are variable in natural populations. Locke Rowe and David Houle (1996) developed a model of "capture" of genetic variance by condition-dependent sexually selected traits. In this model, a male's condition, and thus the size of the display ornament he can develop, is inversely proportional to the number of deleterious mutations he carries. The high mutational variance in condition will sustain genetic variance in the display trait and enable long-sustained evolution of both more extreme ornaments and more extreme female preferences. The assumptions of this model are realistic. A high courtship rate in male dung beetles, for example, is a condition-dependent trait that is preferred by females; both courtship rate and condition are genetically variable, as the model predicts (Kotiaho et al. 2001).

Figure 15.20 Benefits and costs in sexual selection. (A) Evidence for the "good genes" hypothesis of female choice. The percentage of young sticklebacks that became infected when exposed to tapeworm larvae was inversely correlated with the coloration of their fathers. (B) The survival of unmated female crickets (black curve) was higher than that of mated females, and the survival of females mated to attractive males (red curve) was lower than that of females mated to unattractive males (blue curve). A male's attractiveness was scored by the responses of other females. (A after Barber et al. 2001; B after Head et al. 2005.)

(A)

Offspring infected by tapeworm (%)

Pond 1

Pond 2

The offspring of more brightly colored males displayed higher resistance to infection.

"Redness" of father

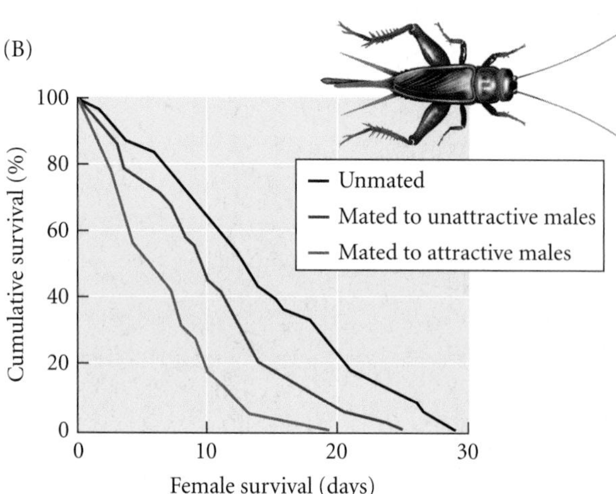

EVIDENCE FOR THE ROLE OF INDIRECT EFFECTS. There is considerable evidence for the good genes hypothesis (Møller and Alatalo 1999; Head et al. 2005). For instance, Barber and colleagues (2001) studied a character preferred by female sticklebacks (*Gasterosteus aculeatus*): the intensity of red coloration of the male's belly. They found that young sticklebacks with bright red fathers were more resistant to infection by tapeworms than their half-sibs who had dull red fathers (Figure 15.20A). The red coloration is based on carotenoid pigments, which are obtained from food and appear to enhance development of an effective immune system. In one of the most comprehensive studies, Megan Head and coworkers (2005) confined female house crickets (*Acheta domestica*) with males that had been scored as attractive or unattractive, based on the responses of other females. The investigators scored the lifetime survival of these females, the lifetime survival and egg production of their daughters, and the survival and mating success of their sons, and used these data to estimate the intrinsic rate of increase of each female and her descendants, a measure of overall fitness. They found that mating with attractive males imposed a substantial cost to female survival (Figure 15.20B), but that this was outweighed by the higher rate of increase of descendants of attractive males, whose daughters were more fecund and whose sons were more successful in attracting females. This experiment provided evidence both for direct costs of mating with attractive males, and for substantial indirect benefits.

There is evidence, too, for the Fisherian runaway process. For example, male sandflies (*Lutzomyia longipalpis*) aggregate in LEKS (areas where males gather and compete for females that visit in order to mate). Females typically reject several males before choosing one with

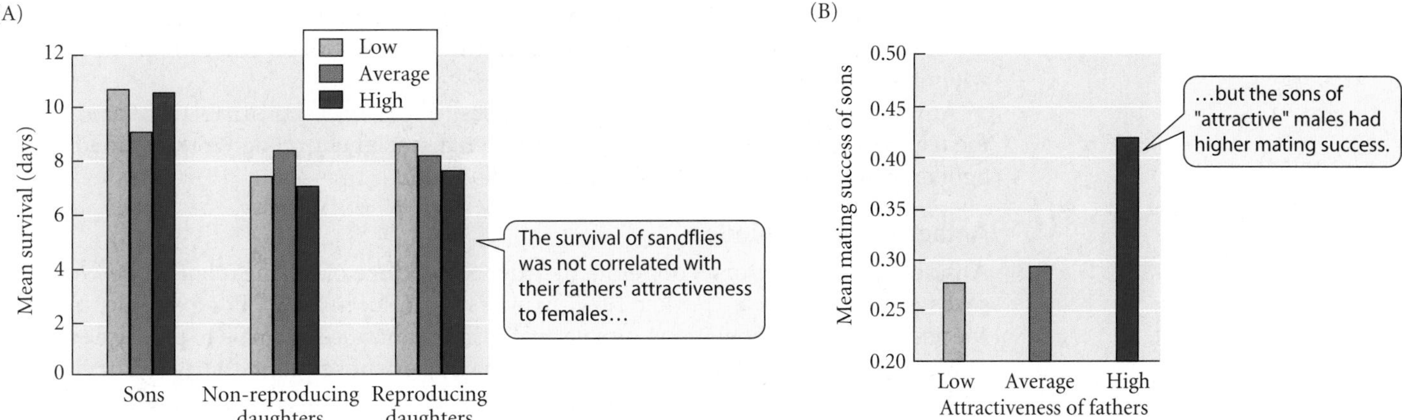

Figure 15.21 Evidence for Fisher's "sexy son" hypothesis. (A) Survival is shown for the sons and for both the reproducing and nonreproducing daughters of three classes of sandfly fathers that differed in attractiveness to females. There is no indication that attractive fathers contribute "good genes" to their offspring. (B) The *sons* of highly attractive male sandflies, however, show strikingly high mating success when females choose between them and the sons of less attractive fathers. Thus the indirect benefit to females of mating with an attractive male is that her sons will enjoy greater reproductive success. (After Jones et al. 1998.)

whom to mate. T. M. Jones and coworkers (1998), using experimental groups of males, determined which were most often, and which least often, chosen by females. Females who mated with the most "attractive" versus the least "attractive" males did not differ in survival or in the number of eggs they laid; that is, there was no evidence of a direct benefit of mating with attractive males. Nor was the father's attractiveness correlated with his offsprings' survival or with his daughters' fecundity; that is, there was no evidence that attractive fathers transmitted good genes for survival or egg production (Figure 15.21A). However, when the investigators provided virgin females with trios of sons—one from a highly attractive father, one from an "average" father, and one from an unattractive father—the sons of attractive fathers were by far the most successful in mating, as predicted by the Fisherian "sexy son" hypothesis (Figure 15.21B).

Sensory bias

Phylogenetic studies have shown that female preferences sometimes evolve before the preferred male trait (Figure 15.22A; Ryan 1998). Certain traits may be intrinsically stimulating and evoke a greater response simply because of the organization of the sensory system. The occurrence of such **sensory bias** is well supported by studies of sensory physiology and animal behavior. For instance, in some species of the fish genus *Xiphophorus* (swordtails), part of the male's tail is elongated into a "sword" (Figure 15.22B). Alexandra Basolo (1995, 1998) found that females preferred males fitted with plastic swords not only in sword-bearing species of *Xiphophorus*, but also in a species of *Xiphophorus* that lacks the sword and in the swordless genus *Priapella*, the sister group of *Xiphophorus*. In fact, female *Priapella* had a stronger preference for swords than females

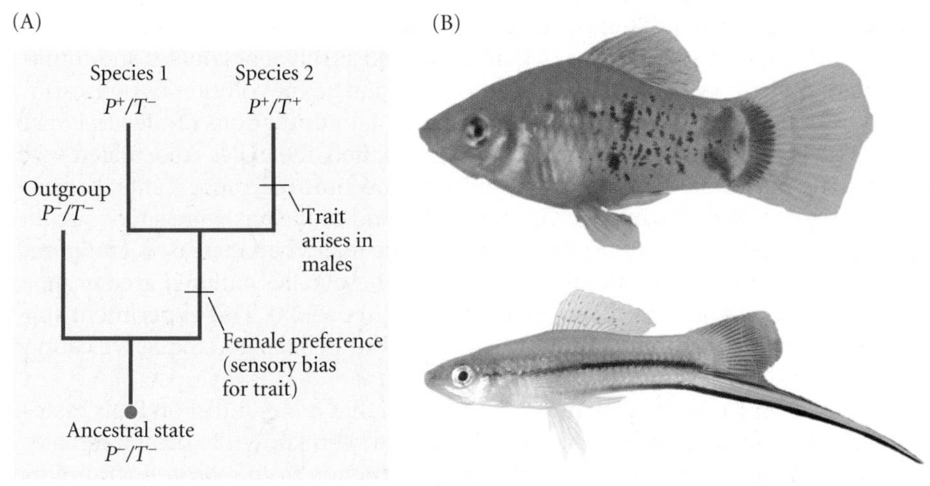

Figure 15.22 Evidence supporting sexual selection as a result of sensory bias in female mate choice. (A) A hypothetical phylogeny of three species that differ in presence (+) or absence (–) of a male trait (T) and in female preference (P) for that trait. Both T and P are absent in the outgroup species, evidence that their presence is derived. That P is present in both species 1 and 2, whereas T is present in 2 only, is evidence that a female preference, or sensory bias, for the trait T evolved before the trait did. (B) Such evidence has been found in poeciliid fishes that lack (top) or possess (bottom) a swordlike tail ornament in the male. (A after Ryan 1998; B photos courtesy of Alexandra Basolo.)

of a sword-bearing species of *Xiphophorus*. The female preference evidently evolved in the common ancestor of these genera, and thus provided selection for a male sword when the mutations for this feature arose.

Animals frequently show greater responses to SUPERNORMAL STIMULI that are outside the usual range of stimulus intensity, and even artificial neural networks trained to recognize patterns show biases for exaggerated stimuli.

Antagonistic coevolution

Although the sexes must cooperate in order to produce offspring, conflict between the sexes is also pervasive (Parker 1979; Arnqvist and Rowe 2005). For example, unless a species is strictly monogamous, a male profits if he can cause a female to produce as many of his offspring as possible, even if this reduces her subsequent ability to survive and reproduce. Thus genes that govern male versus female characters may conflict, resulting in **antagonistic coevolution** (Rice and Holland 1997; Holland and Rice 1998). Such evolution may consist of a protracted "arms race" in which change in a male character is parried or neutralized by evolution of a female character, which in turn selects for change in the male character, and so on in a chain reaction.

Consider interactions between gametes in externally fertilizing species such as abalones (large gastropods, *Haliotis*). Sperm compete to fertilize eggs, so selection on sperm always favors a greater ability to penetrate eggs rapidly. But selection on eggs should favor features that slow sperm entry, or else POLYSPERMY (entry by multiple sperm) may result. Polyspermy disrupts development, and eggs have elaborate mechanisms to prevent it. The conflict between the interests of egg and sperm may result in continual antagonistic coevolution of greater penetration ability by sperm and countermeasures by eggs. Such coevolution might be the cause of the extraordinarily rapid evolution of the amino acid sequence of the sperm lysin protein of abalones, a protein released by sperm that bores a hole through the vitelline envelope surrounding the egg. Nonsynonymous differences between the lysin genes of different species of abalones have evolved much faster than synonymous differences, a sure sign of natural selection (Vacquier 1998; see also Palumbi 1998 and Chapter 20).

Genetic conflict between the sexes has been studied in greatest detail in *Drosophila melanogaster*, in which mating reduces female survival, both because females are harassed by courting males and because the male seminal fluid contains toxic proteins (Chapman et al. 1995). A study of genetic variation showed that males whose sperm best resist displacement by a second male's sperm cause the greatest reduction of their mates' life span: a clear case of sexual conflict (Civetta and Clark 2000). In a series of clever experiments, William Rice and Brett Holland have shown that male and female *Drosophila* are locked in a coevolutionary "arms race" (Rice 1992, 1996; Holland and Rice 1999). In one experiment, Rice crossed flies for 29 generations so that genetically marked segments of two autosomes were inherited only through females. Thus alleles in these chromosome regions that enhanced female fitness could increase in frequency, irrespective of their possible effect on males. Rice observed that the proportion of females among adult flies increased in the course of the experiment, because of both improved survival of females and diminished survival of males—just as predicted by the antagonistic coevolution hypothesis.

In a second experiment, Rice constructed experimental populations of flies in which males could evolve but females could not. (In each generation, the males were mated with females from a separate, non-evolving stock with chromosome rearrangements that prevented recombination with the males' chromosomes, and only sons were saved for the next generation.) After 30 generations, the fitness of these males had increased, compared with control populations, but females that mated with these males suffered greater mortality, perhaps because of enhanced semen toxicity (Figure 15.23). This experiment suggests that males and females are continually evolving, but in balance, so that we cannot see the change unless evolution in one sex is prevented.

Semen toxicity is only one of many examples of harm that males inflict on their mates. Forced copulation by groups of male mallard ducks has been known to drown females, and female bedbugs suffer reduced survival and reproduction from repeated "traumatic

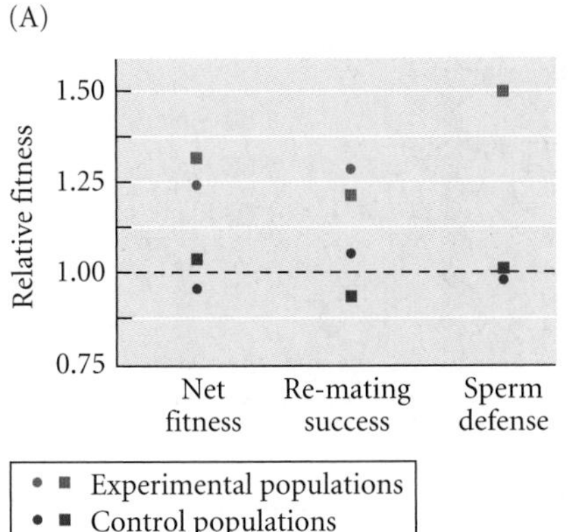

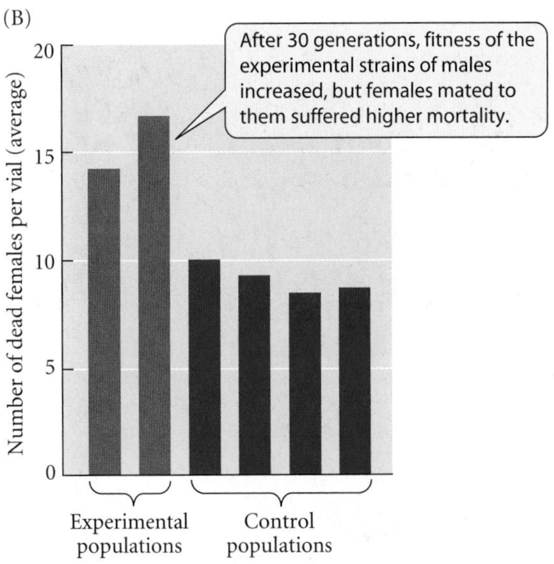

Figure 15.23 Experimental evidence of genetic conflict between the sexes. (A) Measures of fitness of males from two experimental populations in which only males could evolve, relative to that of males from control populations. The measures are net fitness (number of sons per male), the rate at which the males re-mated with previously mated females, and sperm defense (the ability of a male's sperm to fertilize eggs of a female that was given the opportunity to mate with a second male). (B) When males from these experimental populations were mated with females from another stock, they caused higher female mortality than did males from four control populations. (After Rice 1996.)

insemination." (The male bedbug pierces his mate's abdominal wall with his genital structure, rather than inserting it into her genital opening; see Stutt and Siva-Jothy 2001.) Thus females evolve resistance to males' inducements to mate, and their resistance selects for features that enable males to overcome the females' reluctance—a dynamic that has been termed CHASE-AWAY SEXUAL SELECTION (Holland and Rice 1998). Among species of water striders, for example, in which males force females to copulate, the evolution of features that enable males to grip females has been accompanied by changes that enable females to break the males' grip (Figure 15.24).

Under chase-away selection, males may evolve increasingly strong stimuli, such as brighter colors or more elaborate song, to induce reluctant females to mate. In such cases, more elaborate male traits evolve not because females evolve to prefer them, but because females are selected to resist less elaborate traits. If this hypothesis is true, then males may accumulate characteristics that were once sufficient to induce mating, but are sufficient no longer. Recall, for example, the surprising observation that the swordtail's sword stimulates conspecific females less than female *Priapella*, the swordless relative. This is just

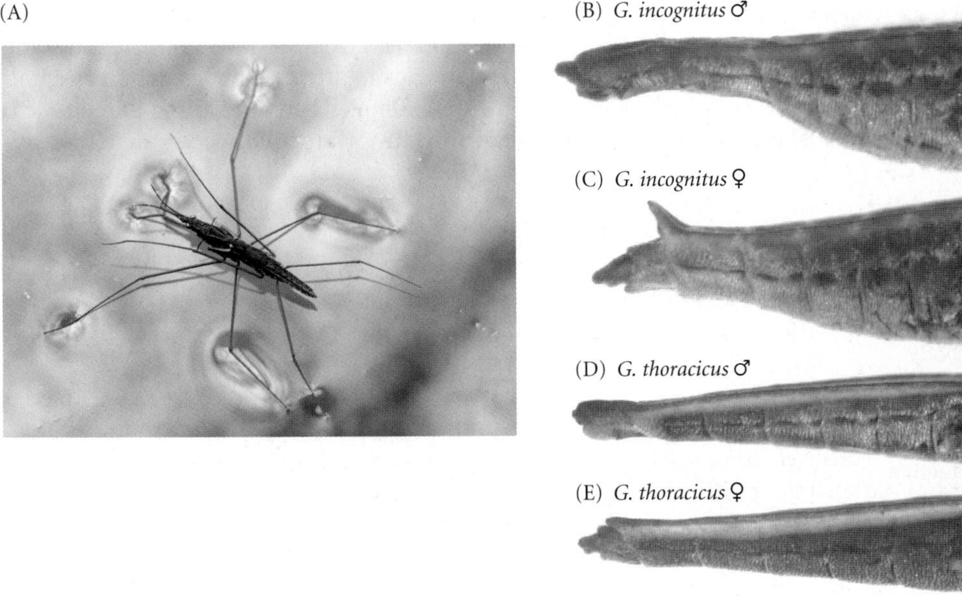

Figure 15.24 Among water striders (genus *Gerris*), which live on the surface of water, males forcibly copulate with females. (A) A mated pair of *G. lacustris.* (B,C) In *G. incognitus*, the abdomen of the male has exaggerated grasping adaptations (extended, flattened genital segments), and that of the female has features (prolonged, erect abdominal spines; curved abdomen tip) that obstruct the male's grip during premating struggles. (D,E) The closely related species *G. thoracicus* shows the ancestral condition, in which these adaptations have not evolved. (A © Dave Bevan/Alamy; B–E from Arnqvist and Rowe 2002.)

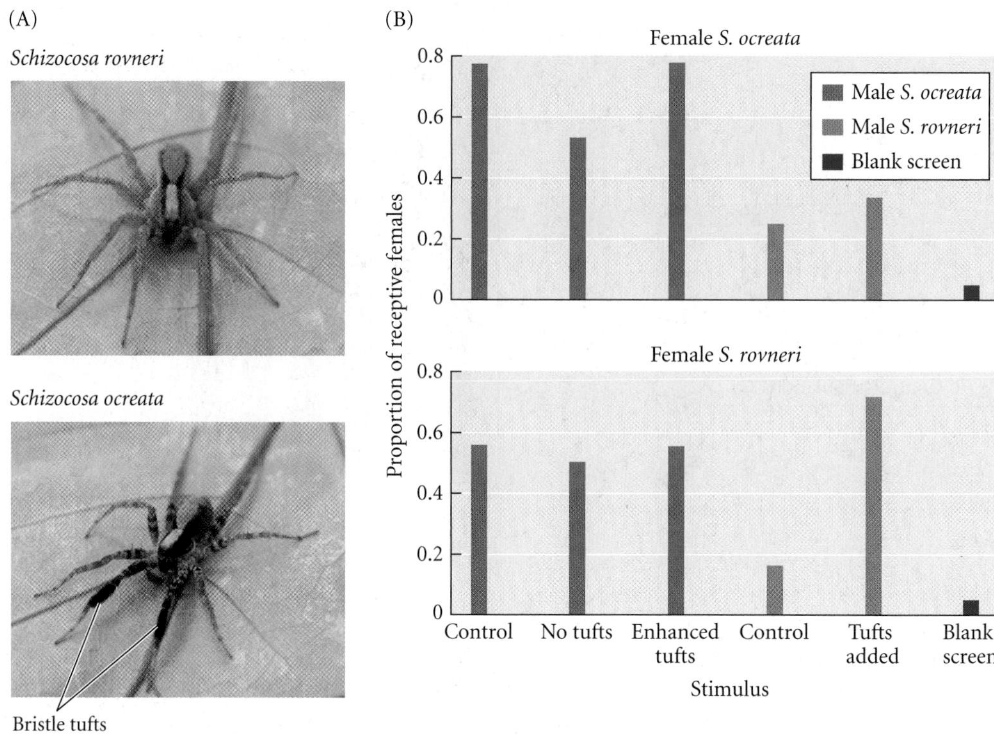

Figure 15.25 A possible example of chase-away sexual selection. (A) The male wolf spider *Schizocosa rovneri* (top) lacks tufts of bristles on the forelegs, whereas male *S. ocreata* (bottom) possesses this derived character. (B) Responses of females of both species to video clips of displaying males of both species, including images of apparently tuft-bearing *rovneri* males. Female *S. rovneri* responded more to this image (bottom graph) than did female *S. ocreata* (top graph), suggesting that *S. ocreata* may have evolved female resistance to this display trait. (Photos courtesy of George Uetz; B after McClintock and Uetz 1996.)

what we would expect if the ancestral attraction to swords, still evident in *Priapella*, had been replaced by female resistance after the sword evolved. Similarly, tufts of leg bristles in the wolf spider *Schizocosa ocreata* are a derived male character, relative to their absence in a closely related species, *S. rovneri* (Figure 15.25A). Female spiders were presented with video images manipulated to show courting males with added, enlarged, or removed tufts (McClintock and Uetz 1996). These manipulations did not affect the mating receptiveness of female *S. ocreata*, but adding tufts greatly enhanced the receptiveness of females of the tuftless species (Figure 15.25B). These results are consistent with the hypothesis that male tufts evolved in *S. ocreata* to stimulate females, but female resistance evolved thereafter.

Alternative Mating Strategies

Not only females, but also males are subject to costs of reproduction, which underlie some interesting variations in male life histories such as ALTERNATIVE MATING STRATEGIES. In some species, large males display and/or defend territories to attract females, whereas small males do not, but rather "sneak" about, intercepting females and attempting to mate with them. In some instances, "sneaker" males have lower reproductive success than display-

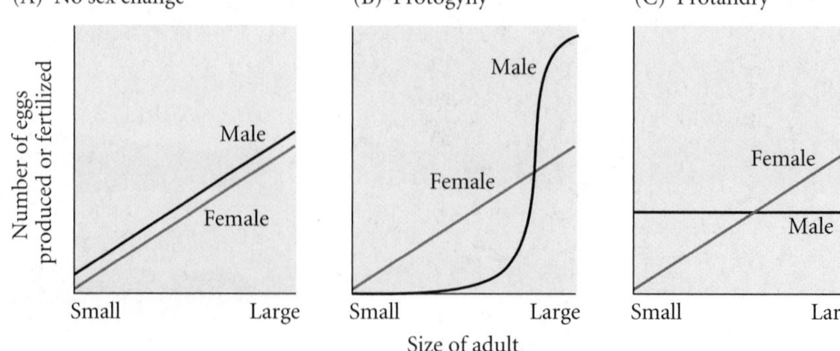

Figure 15.26 A model for the evolution of sex change in sequential hermaphrodites. (A) When reproductive success increases equally with body size in both sexes, there is no selection for sex change. (B) A switch from female to male (protogyny) is optimal if male reproductive success increases more steeply with size than female reproductive success. (C) The opposite relationship favors the evolution of protandry, in which males become females when they grow to a large size. (After Warner 1984.)

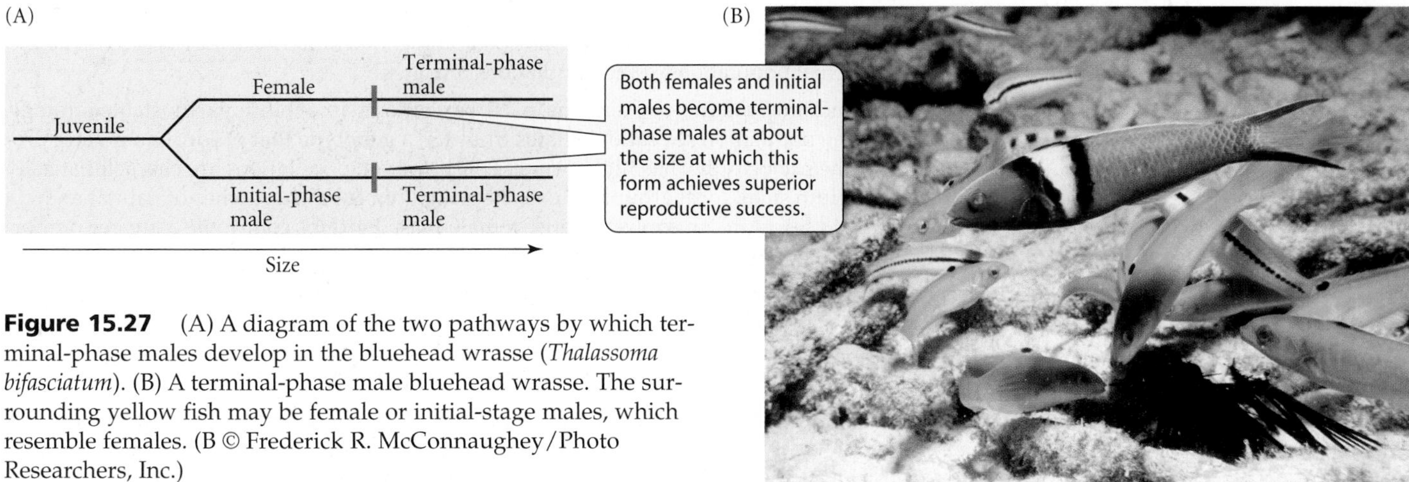

Figure 15.27 (A) A diagram of the two pathways by which terminal-phase males develop in the bluehead wrasse (*Thalassoma bifasciatum*). (B) A terminal-phase male bluehead wrasse. The surrounding yellow fish may be female or initial-stage males, which resemble females. (B © Frederick R. McConnaughey/Photo Researchers, Inc.)

ing males, so their behavior is probably not an adaptation; they are probably making the best of a bad situation, as they are unable to compete successfully. In other instances, however, the sneaker strategy appears to be an alternative adaptation, yielding the same fitness as the display strategy. For example, in the Pacific coho salmon (*Oncorhynchus kisutch*), large, red "hooknose" males develop hooked jaws and enlarged teeth, and fight over females, whereas "jack" males are smaller, resemble females, do not fight, and breed when they are only about a third as old as the hooknose males. Based on data on survival to breeding age and frequency of mating, the fitnesses of these two types of males appear to be nearly equal (Gross 1984).

Some plants, annelid worms, fishes, and other organisms change sex over the course of their life spans (SEQUENTIAL HERMAPHRODITISM). In species that grow in size throughout reproductive life, a sex change can be advantageous if reproductive success increases with size to a greater extent in one sex than in the other (Figure 15.26). In the bluehead wrasse (*Thalassoma bifasciatum*), for instance, some individuals start life as females and later become brightly colored "terminal-phase males" that defend territories. Other individuals begin as "initial-phase males," which resemble females and spawn in groups (Figure 15.27). A female usually does not produce as many eggs as a large terminal-phase male typically fertilizes. Both females and initial-phase males become terminal-phase males at about the size at which this form achieves superior reproductive success (Warner 1984).

Summary

1. Alleles that increase mutation rates are generally selected against because they are associated with the deleterious mutations they cause. Therefore, we would expect mutation rates to evolve to the minimal achievable level, even if this should reduce genetic variation and increase the possibility of a species' extinction.
2. Asexual populations have a high extinction rate, so sex has a group-level advantage in the long term. But this is unlikely to offset the short-term advantage of asexual reproduction.
3. In a constant environment, alleles that decrease the recombination rate are advantageous because they lower the proportion of offspring with unfit recombinant genotypes. In addition, asexual reproduction has approximately a twofold advantage over sexual reproduction because only half of the offspring of sexuals (i.e., the females) contribute to population growth, whereas all of the (all-female) offspring of asexuals do so. Therefore, the prevalence of recombination and sex requires explanation.
4. Among the several hypotheses for the short-term advantage of sex are: (a) in asexual populations, fitness declines because genotypes with few deleterious mutations, if lost by genetic drift, cannot be reconstituted, as they are in populations with recombination (Muller's ratchet); (b) deleterious mutations can be more effectively purged by natural selection in sexual than in asexual populations, thus maintaining higher mean fitness; (c) recombination enables the mean of

a polygenic character to evolve to new, changing optima in a fluctuating environment; (d) the rate of adaptation, by fixing combinations of advantageous mutations, may be higher in sexual than asexual populations, if the populations are large.

5. In large, randomly mating populations, a 1:1 sex ratio is an evolutionarily stable strategy, because if the population sex ratio deviates from 1:1, a genotype that produces a greater proportion of the minority sex has higher fitness. If, however, populations are characteristically subdivided into small local groups whose offspring then colonize patches of habitat anew, a female-biased sex ratio can evolve because female-biased groups contribute a greater proportion of offspring to the population as a whole.
6. The evolution of hermaphroditism versus dioecy (separate sexes) depends on how reproductive success via female or male function is related to the allocation of an individual's energy or resources. Dioecy is advantageous if the reproductive "payoff" from one or the other sexual function increases disproportionately with allocation to that function.
7. Outcrossing can be advantageous because it prevents inbreeding depression in an individual's progeny. Conversely, self-fertilization may evolve if fewer resources need to be expended on reproduction, if an allele for selfing becomes associated with advantageous homozygous genotypes, or if selfing ensures reproduction despite low population density or scarcity of pollinators.
8. Differences between the sexes in the size and number of gametes give rise to conflicts of reproductive interest and to sexual selection, in which individuals of one sex compete for mates (or for opportunities to fertilize eggs). The several forms of sexual selection include direct competition between males, or between their sperm, and female choice among male phenotypes.
9. Females may prefer certain male phenotypes because of sensory bias, direct contributions of the male to the fitness of the female or her offspring, or indirect contributions to female or offspring fitness. Indirect benefits may include fathering offspring that are genetically superior with respect to mating success ("runaway sexual selection") or with respect to components of viability ("good genes models"). Sexually selected male features may also evolve by antagonistic coevolution: selection on females to resist mating, and on males to overcome female resistance with irresistible stimuli.
10. The evolution of reproductive effort by males is governed by similar principles as in females. Delayed maturation may evolve if larger males are more successful in attracting or competing for mates. Similar principles explain phenomena such as sequential hermaphroditism (sex change with age) and alternative mating strategies.
11. The evolution of features of genetic systems, such as rates of mutation and recombination, sexual versus asexual reproduction, and rates of inbreeding, can usually be understood best as consequences of selection at the level of genes and individual organisms, rather than group selection.

Terms and Concepts

antagonistic coevolution
apomixis
condition-dependent indicator
cost of sex
dioecious
hermaphroditic
outcrossing
parthenogenesis
sensory bias
sex
sex ratio
sexual selection
sperm competition
vegetative propagation

Suggestions for Further Reading

The Evolution of Sex, edited by R. E. Michod and B. R. Levin (Sinauer Associates, Sunderland, MA, 1988), and articles by P. Jarne and D. Charlesworth (1993), A. S. Kondrashov (1993), M. K. Uyenoyama et al. (1993), and N. H. Barton and B. Charlesworth (1998) treat the evolution of genetic systems. An informative and provocative book for a general audience by a prominent evolutionary biologist is J. Roughgarden's *Evolution's Rainbow: Diversity, Gender, and Sexuality in Nature and People* (University of California Press, Berkeley, 2004). The most comprehensive review of sexual selection, although somewhat dated, is *Sexual Selection* by M. Andersson (Princeton University Press, Princeton, NJ, 1994). *Sexual Conflict* by G. Arnqvist and L. Rowe (Princeton University Press, Princeton, NJ, 2005) is a comprehensive analysis of that subject.

Problems and Discussion Topics

1. Populations of some species of fish, insects, and crustaceans consist of both sexually and obligately asexually reproducing individuals. Would you expect such populations to become entirely asexual or sexual? What factors might maintain both reproductive modes? How might studies shed light on the factors that maintain sexual reproduction?
2. Some mites, scale insects, and gall midges display "paternal genome loss" (or *pseudoarrhenotoky*, a fine word for parlor games). Males develop from diploid (fertilized) eggs, but the chromosomes inherited from the father become heterochromatinized and nonfunctional early in development, so that males are functionally haploid. How might this peculiar genetic system have evolved?
3. Many parthenogenetic "species" of plants and animals are known to be genetically highly variable. Determine from the literature whether this genetic variation is due to mutation within asexual lineages or to multiple origins of asexual genotypes from a sexually reproducing ancestor.
4. The number of chromosome pairs in the genome ranges from one (in an ant species) to several hundred (in some butterflies and ferns). In a single genus of butterflies (*Lysandra*), the haploid number varies from 24 to about 220 (White 1978). Is it likely that chromosome number evolves by natural selection? Is there evidence for or against this hypothesis?
5. Would you expect sexual selection to increase or decrease adaptation of a population to its environment? Might this depend on the form of sexual selection (e.g., male conflict, female mate choice)? Do the several hypotheses for the evolution of female choice (e.g., runaway selection versus good genes hypotheses) differ in their implications for adaptation to the environment?
6. In many socially monogamous species of parrots and other birds, both sexes are brilliantly colored or highly ornamented. Is sexual selection likely to be responsible for the coloration and ornaments of both sexes? How can there be sexual selection in pair-bonding species with a 1:1 sex ratio, since every individual presumably obtains a mate? Which of the models of sexual selection described in this chapter might account for these species' characteristics? (See Jones and Hunter 1999 for an experimental study of this topic.)
7. There is an enormous and very controversial literature on whether behavioral features in humans have a genetic basis and have evolved for reasons postulated in evolutionary models, or whether they are culturally formed. Analyze reasons for and against the proposition that men are more aggressive than women because of sexual selection for competitiveness. (See, for example, Daly and Wilson 1983 versus Kitcher 1985.)

CHAPTER 16

Conflict and Cooperation

Social animals. The meerkats (*Suricata suricatta*) of South Africa's Kalahari Desert live in colonies of 20 to 30 animals, most of whom are siblings. Colony members work together to forage, care for the young, guard their underground network of burrows, and defend their territory from incursions by other meerkat families. Altruistic behavior includes "sentry" animals that warn the group of approaching predators. (Photo courtesy of A. Sinauer.)

Darwin first conceived of natural selection when he read the economist Thomas Malthus's theory that population growth must inevitably cause competition for food and other resources. Malthus's *Essay on Population* inspired Darwin's realization that in all species, only a fraction of those born survive to reproduce, and that the survivors must often be those best equipped to compete for limiting resources. Thus conflict has been inherent in the idea of natural selection from the start. Darwin soon realized, however, that not all of natural selection stems from overt struggle among members of a species. We also observe cooperation in nature: among the cells and organs of an individual organism, between mates in the act of sexual reproduction, among individuals in many social species of animals, and even between some mutualistic pairs of species. Yet conflict is so ubiquitous, even inevitable, that the question of how evolutionary theory can account for cooperation has occupied evolutionary biologists ever since Darwin.

Members of a species may struggle for more than food or space, as we shall see, and social interactions among individuals, even between mates or between parents and their offspring, may entail surprising elements of conflict. More surprisingly, evolutionary battles may rage even within individual organisms, between different genes. Almost all instances of cooperation include at least the possibility of conflict as well. Conflict and cooperation are fundamental for understanding the evolution of a vast variety of biological phenomena, ranging from mating displays to sterile social insects, from cannibalism to some most peculiar features of human pregnancy.

Characteristics that enhance the survival or reproduction of populations or species, but not of individual organisms or genes, can evolve only by group selection, as we saw in Chapter 11. Group selection as traditionally conceived (as in Figure 11.15) is generally a weak agent of evolutionary change, so most evolutionary biologists seek explanations of the evolution of cooperation and conflict based on selection at the level of the individual organism or the gene. This approach has been very fruitful, and has deepened our understanding of many otherwise puzzling phenomena.

Conflict

Many of the world's fisheries, such as the cod fishery in the north Atlantic, have been devastated by overfishing. It profits each boat to harvest as many fish as possible, and the outcome of such individually rational behavior is the destruction of the resource on which all participants depend. This outcome has been called the "tragedy of the commons," in reference to communal grazing lands that are similarly destroyed by overgrazing. Similar problems arise in the natural world, and pose the question of how restraint can evolve, especially when a "cheater" genotype that maximizes its personal gain has higher fitness, at least in the short run. Explaining how cooperation can evolve, when selfish genotypes would be expected to have higher fitness, is an extensive field of both theoretical and empirical inquiry.

Conflict and selfishness are ubiquitous. For example, more than 150 species of birds are known to practice "conspecific brood parasitism," wherein a female lays eggs in other

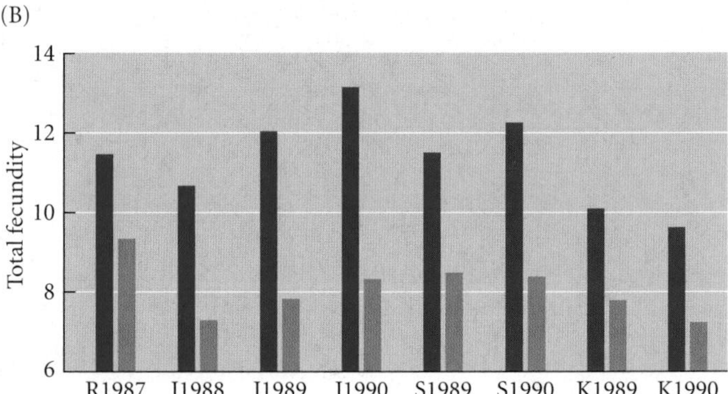

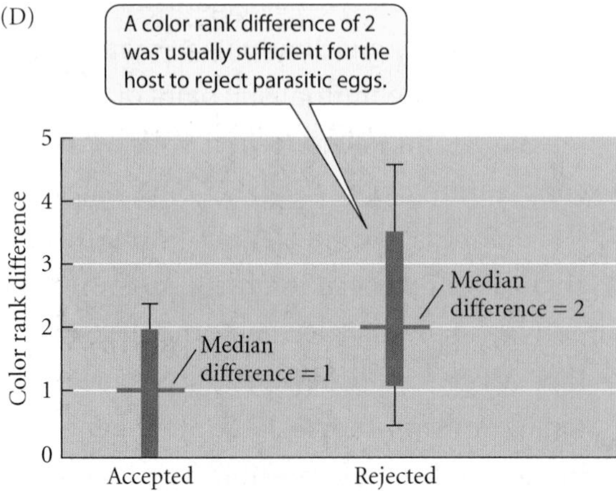

Figure 16.1 Intraspecific conflict: brood parasitism in the American coot. (A) The average number of eggs laid in their own nests by parasitic and nonparasitic females is not significantly different, so there is no evidence that parasitic females pay a cost. (B) The total fecundity of parasitic females (eggs in own nest plus those laid in other females' nests) is much greater than that of nonparasitic females. The notations along the *x*-axis indicate data from four wetlands (R, J, S, and K) and different years. (C) Eggs of host and parasitic females from a single nest show variation in color that can be ranked from 1 (lightest) through 7 (darkest); subtracting the lower number from the higher gives the "color rank diffence." (D) Parasitic eggs rejected by host females differed more from the host's own eggs than those that were accepted by the host. (A and B after Lyon 1993, photo © Laure Neish/istockphoto.com; C © Bruce Lyon; D after Lyon 2003.)

females' nests, in addition to those she herself incubates (Lyon and Eadie 2008). In the American coot (*Fulica americana*), this behavior greatly increases the reproductive output of parasitic females compared with that of nonparasitic females (Figure 16.1A,B). The host female's fitness is reduced because her chicks compete with the parasitic chicks for limited food, and so, as might be expected, coots have evolved the capacity to detect and evict foreign eggs from their nests, based on variation in egg color (Figure 16.1C).

A conflict between individual and group benefit has been described in the slime mold *Dictyostelium discoideum* (an amoeba-like eukaryote) and in myxobacteria such as *Myxococcus xanthus* (Dao et al. 2000). Although distantly related, these organisms have similar, very odd life histories. Individual cells in the soil aggregate when starved for food, and form a single moving organism (a "slug") that settles and forms an erect stalk with a cap within which spores develop. The spore-forming cells reproduce, but the stalk-forming cells die. In both species, "cheater" mutations have been described that preferentially make spores. The mutant *Dictyostelium* cells cause wild-type cells to move to the front of the slug and form the stalk (Figure 16.2A). The frequency of this "selfish" allele increased over the course of several life cycles in laboratory culture (Figure 16.2B).

The fitness of contending genotypes in social interactions such as competition is often frequency-dependent, and is often modeled by finding the evolutionarily stable strategy, or ESS: a phenotype that cannot be replaced by any other phenotype under the specified conditions (Chapter 14). In all the theory discussed in this chapter, costs and benefits figure importantly, and are always measured in terms of fitness. ESS theory has been applied to conflict among animals, such as males competing for a territory or a mate. In many species, aggressive encounters consist largely of displays that seldom escalate into physical conflicts that could cause injury or death. The question is whether it is individually advantageous not to escalate conflicts. John Maynard Smith (1982), one of the first to apply game theory to evolution, postulated two strategies: a "Hawk" escalates the conflict either until it is injured or until its opponent retreats, whereas a "Dove" retreats as soon as its opponent escalates the conflict. Is it better to be a Hawk, more likely to win the prize, or a Dove, more likely to escape injury and live to compete another day? The Dove strategy is never an ESS, because a Hawk genotype can always increase in a population of Doves. The Hawk strategy is an ESS if the fitness gained in a successful contest (V), even with another Hawk, is greater than the cost of injury (C). If, however, $C > V$, ESS analysis shows that a pure Hawk phenotype can be replaced by a "mixed strategy," such that an individual will adopt Hawk behavior with probability $p = V/C$. Thus this model predicts that the optimal behavior is variable and is contingent on conditions such as the value of the resource.

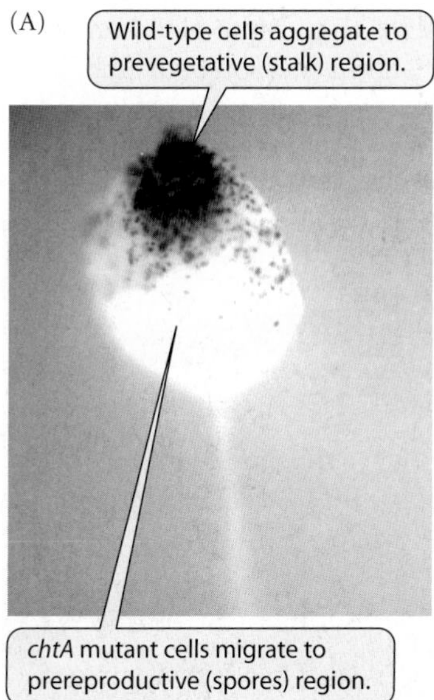

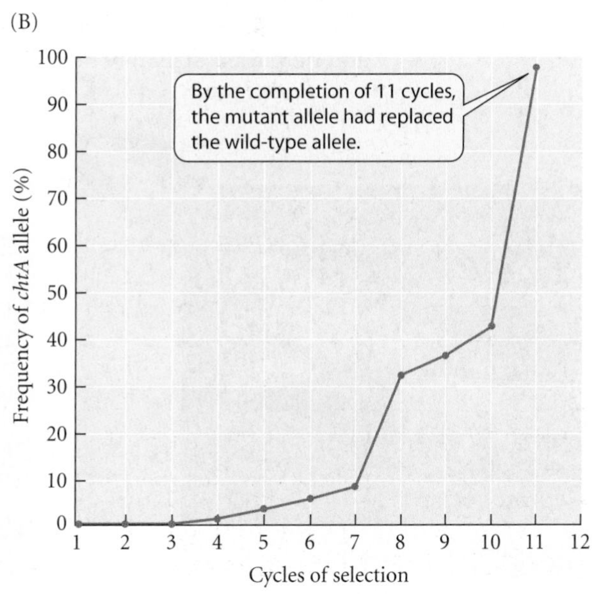

Figure 16.2 Effects of the selfish mutation of the gene *chtA* in the slime mold *Dictyostelium*. (A) In a mixture of mutant and wild-type cells, which are shown as darker because of a genetic label, wild-type cells moved to the upper cup of the aggregate, a region that becomes the stalk of the fruiting body rather than the reproductive spores. (B) Over the course of 12 growth and development cycles, the frequency of the mutant *chtA* allele increased in laboratory culture. (A courtesy of H. L. Ennis; B after Dao et al. 2000.)

(A)

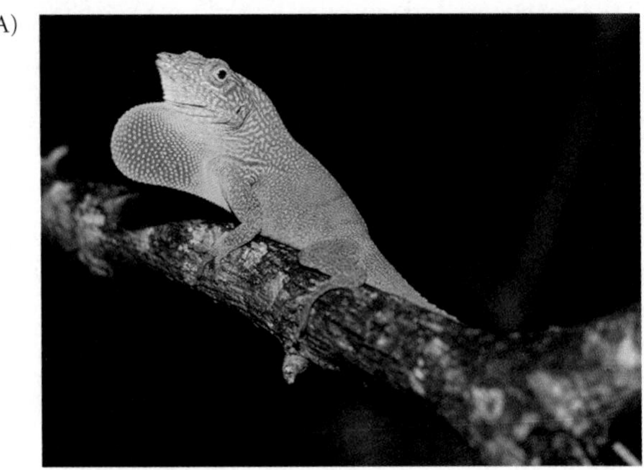

(B)

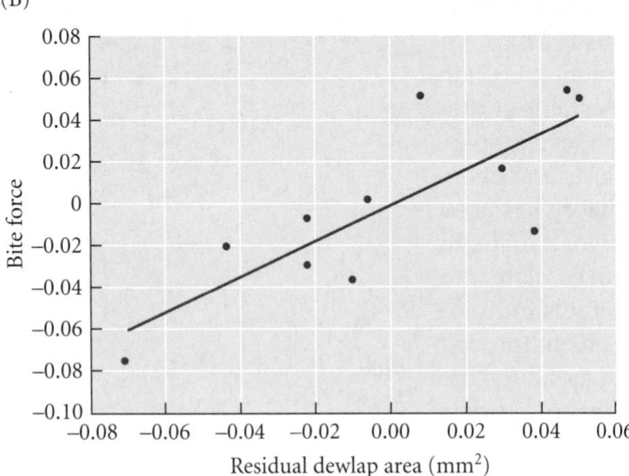

Figure 16.3 An honest signal used in threat displays of male *Anolis* lizards. (A) A male *Anolis grahami* displays a brightly colored dewlap. (B) Dewlap size is correlated with the force of the male's bite. Because both dewlap size and bite force are correlated with body size, the points in this graph have been statistically adjusted for variation in body size. Some values appear negative because of this adjustment. (A courtesy of J. Losos; B after Vanhooydonck et al. 2005.)

We can also postulate an Assessor strategy, in which the individual escalates the conflict if it judges the opponent to be smaller or weaker, but retreats if it judges the opponent to be larger or stronger. In most theoretical cases, the Assessor strategy is an ESS. In accord with these models, many animals react aggressively or not depending on their opponent's size or correlated features. In some species of *Anolis* lizards, for example, the size of a territorial male's dewlap, a brighly colored throat ornament that is displayed in aggressive encounters, is correlated with the force of the male's bite (Figure 16.3; Vanhooydonck et al. 2005).

Characters such as the lizard's dewlap are HONEST SIGNALS of the individual's fighting ability or resource-holding potential. Theoretically, deceptive signals, indicating greater fighting ability than the individual actually has, should be unstable in evolutionary time because selection would favor genotypes that ignored the signals, which, having then lost their utility, would be lost in subsequent evolution. Thus, many existing signals of resource-holding potential are probably honest assessment signals (Johnstone and Norris 1993). However, dishonest signals are not uncommon (Bradbury and Vehrencamp 1998). The greatly enlarged claw of a male fiddler crab, for example, is used for display and fighting with other males. In *Uca annulipes*, males that lose their large claw regenerate one that is almost the same size, but it has less muscle and is much weaker. Although such males lose physical fights with intact males, they effectively bluff and deter potential opponents, and they are apparently just as successful in attracting females (Figure 16.4; Backwell et al. 2000). Male Australian crayfish (*Cherax dispar*) also use enlarged claws for display and fighting. The size of the claws, but not their strength, is correlated with the ability of males to achieve dominance, although males have smaller, less powerful claw muscles than same-sized females, and here their size may be a dishonest signal (Wilson et al. 2007).

(A)

(B)

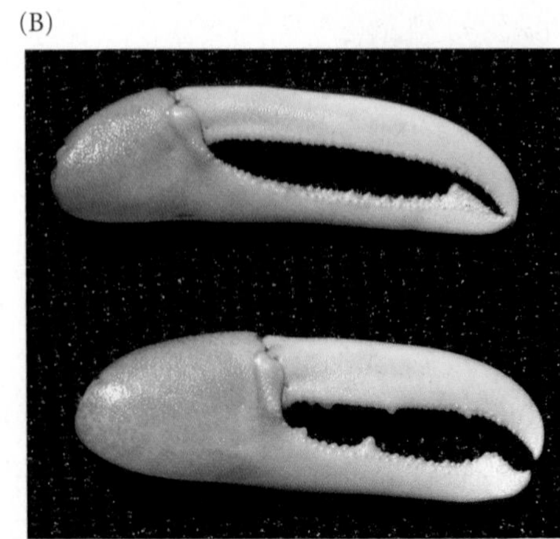

Figure 16.4 (A) A male fiddler crab (*Uca* sp.) signaling with its greatly enlarged claw. (B) A slender regenerated claw (top) and a robust original claw of *Uca annulipes*. The regenerated claw is an effective bluff, even though it does not have the strength needed to win a fight with an intact male. (B from Backwell et al. 2000; photos courtesy of Patricia Backwell.)

Social Interactions and Cooperation

HELPING activities are of two kinds, cooperation and altruism (Lehmann and Keller 2006). **Cooperation** is activity that provides a benefit to other individuals and to the actor as well; **altruism** is activity that enhances the fitness of other individuals but lowers the fitness of the actor. DEFECTORS or CHEATERS receive such benefit but do not provide any. In this chapter, we will often be concerned with entities at one level that interact to form higher-level entities. For instance, individual cells may form a multicellular organism; female and male organisms may form mated pairs, at least briefly; individual organisms may be organized into flocks or colonies. Both cooperation and conflict may be, and usually are, inherent in such interactions. The reduction of conflict sometimes results in the emergence of higher-level entities with distinct identity, such as multicellular organisms (Maynard Smith and Szathmáry 1995; Michod 1999).

Many models of the evolution of cooperation and conflict have been developed and elaborated. Most of them fit into several major categories, as described by Sachs and colleagues (2004), Nowak (2006), and Lehmann and Keller (2006).

Cooperation based on direct benefits

Associations or groups of organisms may be formed and maintained entirely out of self-interest. An individual that joins a flock or herd, for instance, lowers the risk of predation simply by finding safety in numbers. It is often advantageous to each individual to be as close as possible to the center of the group, thus using other group members as shields against approaching predators. The effect of this behavior is to increase the compactness and cohesion of the group (Hamilton 1971).

Cooperative behavior often has a delayed benefit. For example, subordinate male long-tailed manakins (*Chiroxiphia linearis*; Figure 16.5) form a long-lasting association with an unrelated dominant (alpha) male. The team performs a coordinated "leapfrogging" display. Females strongly prefer male teams with highly coordinated displays, but the dominant alpha male obtains almost all the copulations (McDonald and Potts 1994). Subordinate males rarely obtain copulations, but they succeed the alpha male when he dies—although they may have to wait as long as 13 years. Among unrelated female paper wasps (*Polistes dominulus*) that build a nest together, one aggressively achieves dominance over the others and prevents them from reproducing; but dominant females may and do get killed, in which case a subordinate female takes over the nest and achieves reproductive success (Queller et al. 2000). In the fire ant *Solenopsis invicta* and some other species of ants, several unrelated queens found a nest together (Bernasconi and Strassmann 1999). By jointly producing more workers, they greatly increase the likelihood that the nest will survive in competition with other colonies. After the first brood of workers has matured, the queens fight to the death, leaving a single survivor. Joining other queens in founding a nest is risky, but it can evolve if the chance of surviving is greater than for queens that initiate a nest alone.

Figure 16.5 Two, or occasionally more, male long-tailed manakins perform a joint leapfrogging display to females. The dominant male obtains all, or almost all, copulations. (From Alcock 2005.)

Reciprocity: Cooperation based on repeated interactions

An individual X may profit from benefiting another individual Y if there is a high likelihood that Y will provide reciprocal aid in the future (Trivers 1971). The mathematical condition for helping to evolve is then

$$\omega m \beta B - C > 0$$

where ω is the probability that an individual will interact with the same partner again, m is the ability to remember the partner's previous cooperation, β is the individual's response to the partner's previous level of cooperation, and B is the total fitness benefit from helping (Lehmann and Keller 2006). The critical points are that there must be repeated interactions, individuals must recognize and remember each other, and the benefit must be great enough to outweigh the cost of helping.

This is a complex topic, because of the obvious possibility that Y will defect and X will have paid the cost without recompense. The conditions under which cooperation can evolve have been analyzed by game theory, especially analyses of the "prisoner's dilemma" (Axelrod and Hamilton 1981). This is named for the situation in which two imprisoned suspects could cooperate by agreeing not to give evidence against each other (cooperation that would benefit both), but each is rewarded by the police if he or she selfishly acts as informant. The simplest strategy is "tit for tat," in which each "player" starts with cooperation and then does whatever the other has done in the previous round. However, this strategy does not allow for accidental mistakes. Among various alternatives, a perhaps more stable strategy that maintains cooperation is one in which each individual repeats its previous move whenever it is doing well, and changes otherwise (Nowak 2006). Bear in mind that the term "strategy" does not imply conscious reasoning or planning; a strategy is simply a phenotypic character, such as a particular behavior, of a certain genotype.

Reciprocity of this kind, based on "partner choice," appears to be uncommon, perhaps because it requires rather complex information processing to distinguish different individuals and remember their past behavior (Hammerstein 2003). Reciprocity is obviously a major feature of human social systems, and has been shown for some other species of mammals as well. For example, vampire bats (*Desmodus rotundus*), which feed on mammalian blood, form roosting groups in which members that have fed successfully sometimes feed regurgitated blood to other group members that have been unsuccessful in foraging. The recipients reciprocate at other times (Wilkinson 1988). Partnerships among individuals are a major feature of many primate societies, and are reinforced by such activities as reciprocal grooming or sexual contact (Figure 16.6).

The theory of reciprocity by partner choice is part of a more general TRANSACTIONAL MODEL OF REPRODUCTIVE SKEW, first developed by Sandra Vehrencamp (1983). The basic idea is that dominant individuals gain from the assistance of subordinate helpers, and "pay" those helpers by allowing them to reproduce just a little more than they could if they left the group and reproduced on their own. Considerable evidence supports this idea (Keller and Reeve 1994, 1999). For example, a dominant dwarf mongoose (a weasel-like African carnivore) suppresses reproduction by older subordinates less than younger subordinates. As subordinates grow older, they are better able to disperse and breed in another group, so the model predicts that dominants should offer them a greater share of the group's reproductive output as an incentive to stay (Creel and Waser 1991). From this perspective, many interactions may be viewed as BIOLOGICAL MARKETS, in which two classes of "traders" exchange "commodities." Market models can be applied to intraspecific cooperation, mutualism between species, and sexual selection (Noë and Hammerstein 1995).

In many situations, cooperation is enhanced if selfish noncooperators are punished: punishment alters the ratio of benefit to cost (Frank 2003). For example, Stuart West and collaborators (2002) have modeled the interaction between legumes (such as beans) and the bacteria—called rhizobia—that invade legume roots, induce formation of an enveloping nodule, and fix nitrogen (N_2) that benefits the plant. The rhizobia obtain photosynthate (carbon) from the plant. This is a potentially unstable interaction, because a rhizobial genotype that cheats, and receives carbon without expending energy on N_2 fixation,

(A)

(B)

(C)

Figure 16.6 Mammal species that display reciprocity between individuals. (A) Vampire bats (*Desmodus rotundus*) form roosting groups in which members that have fed successfully sometimes feed regurgitated blood to other members of the group. (B,C) Among primates, social alliances between individuals are reinforced by activities such as (B) grooming in chacma baboons (*Papio hamadryas ursinus*) and (C) genital rubbing in female bonobos (*Pan paniscus*). (A © Nick Upton/Naturepl.com; B photo by Anne Engh, courtesy of Dorothy Cheney; C © Frans Lanting/Minden Pictures.)

would have an advantage. West et al. showed that the advantage of N_2 fixation depends partly on kin selection (see below), based on a high degree of relationship (r) among bacteria. But the level of N_2 fixation that evolves is greater if the plant preferentially supplies carbon to nodules that fix more N_2, or punishes nodules that fix less (Figure 16.7). Experiments have shown that legumes do impose such "sanctions" againt poorly performing nodules by limiting the supply of oxygen to those nodules (Kiers et al. 2003).

Many interactions are based on "partner choice." Reciprocity can also be favored by "partner fidelity feedback" (Sachs et al. 2004), in which the association between individuals is so long-lasting that the benefits each partner provides to the other feed back to the individual's own benefit. In *The Defiant Ones* (1958), Tony Curtis and Sidney Poitier portray two escaped convicts who, because they are chained to each other, must cooperate even though they dislike each other. When the fitness of each member of a group depends on the fitness of the other members—of the group as a whole—cooperation is clearly in every individual's interest. This principle is important in many biological contexts; perhaps the simplest examples are cooperation among the genes in a cell and among the cells in an organism. If the cell dies, so do all the included genes; if the organism dies, so do its cells. Selection at the higher level—cell or organism—thus eliminates outlaw genes or renegade cells that selfishly diminish the survival of the group. It also favors mechanisms that suppress or destroy such outlaws when they arise. We will see clear examples of this principle later in this chapter.

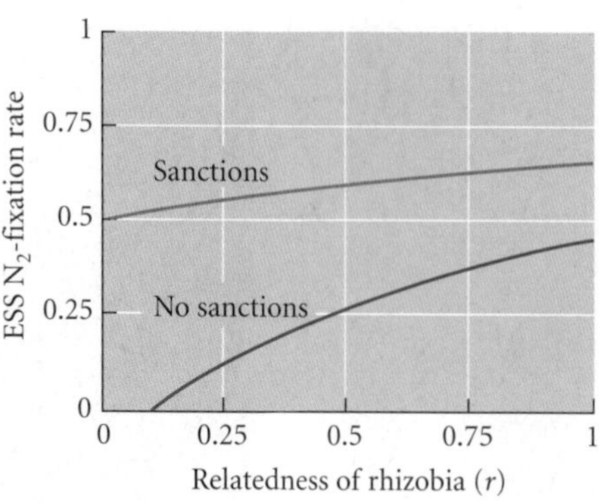

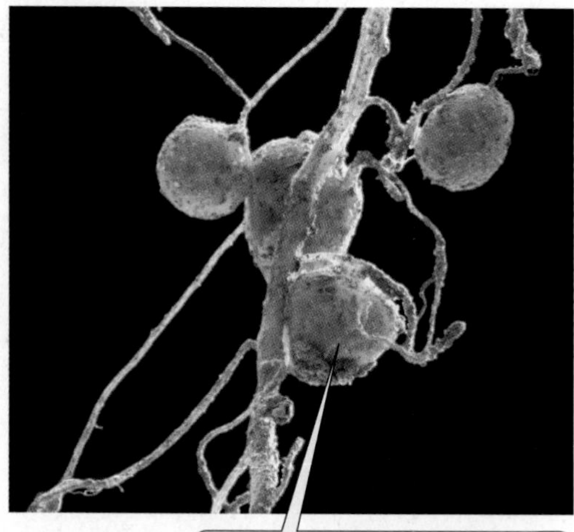

Root nodules form in the symbiotic relationship between legumes and bacteria of the genus *Rhizobium*.

Figure 16.7 A model of the evolution of nitrogen (N_2) fixation in legume-associated rhizobia. The blue trace shows an evolutionarily stable strategy (ESS), the relative level of N_2 fixation that would evolve if the average coefficient of relationship among rhizobia coinfecting a plant were r. The red trace shows the same when plant sanctions are modeled (i.e., the plant supplies carbon to a nodule in proportion to the rate of N_2 fixation by the rhizobia in that nodule). Plant sanctions increase the ESS level of N_2 fixation. (After West et al. 2002; photo by David McIntyre.)

The evolution of altruism by shared genes

If we think of selection at the level of the gene, we will recognize that an allele replaces another allele in a population if it leaves more copies of itself in successive generations by whatever effect it may have. Commonly, the effect is on the individual organism that carries the allele, but an allele may also leave more copies of itself by enhancing the fitness of other individual organisms that carry copies of that same allele. For instance, an allele for maternal care of offspring increases because it increases the chance of survival of copies of the mother's allele that are carried by her offspring. In a pathbreaking paper, William D. Hamilton (1964) introduced the concept of the **inclusive fitness** of an allele, its effect on both the fitness of the individual bearing it (DIRECT FITNESS) and the fitness of other individuals that carry copies of the same allele (INDIRECT FITNESS). An individual organism, likewise, has inclusive fitness, with both direct and indirect components. At this level, selection based on inclusive fitness is called **kin selection** because these other individuals are the bearer's relatives, or kin.

Kin selection is one of the most important explanations for cooperation (Hamilton 1964; Michod 1982; Grafen 2006; Lehmann and Keller 2006). Let us suppose that an individual performs an act that benefits another individual, but incurs a cost to itself: a reduction in its own (direct) fitness. The fundamental principle of kin selection is that an allele for such an altruistic trait can increase in frequency only if the number of extra copies of the allele passed on by the altruist's beneficiary (or beneficiaries) to the next generation as a result of the altruistic interaction is greater, on average, than the number of allele copies lost by the altruist. This principle is formalized in **Hamilton's rule**, which states that *an altruistic trait can increase in frequency if the benefit* (*b*) *received by the donor's relatives, weighted by their relationship* (*r*) *to the donor, exceeds the cost* (*c*) *of the trait to the donor's fitness*. That is, altruism spreads if $rb > c$.

The **coefficient of relationship**, r, is the fraction of the donor's genes that are identical by descent to any of the recipient's genes (Grafen 1991). ("Identical by descent" means derived from common ancestral genes, usually within a few previous generations; see Chapter 9.) For example, at an autosomal locus in a diploid species, an offspring inherits one of its mother's two gene copies, so $r = 0.5$ for mother and offspring (Figure 16.8). For two full siblings, $r = 0.5$ also, because the probability is 0.25 that both siblings inherit copies of the same gene from their mother, and likewise from their father.

The simplest example of a trait that has evolved by kin selection is parental care. If females with allele A enhance the survival of their offspring by caring for them, whereas females lacking this allele do not, then if parental care results in more than two extra surviving offspring, A will increase in frequency, even if parental care should cost the mother her life. If $c = 1$ (death of mother) and $b = 1$ (survival of an extra offspring), then since a mother's A allele has a probability of $r = 0.5$ of being carried by her offspring, Hamilton's rule is satisfied by the survival of more than two extra offspring, relative to noncaring mothers.

Interactions among other ("collateral") relatives also follow Hamilton's rule. For instance, the relationship between an individual and her niece or nephew is $r = 0.25$, so

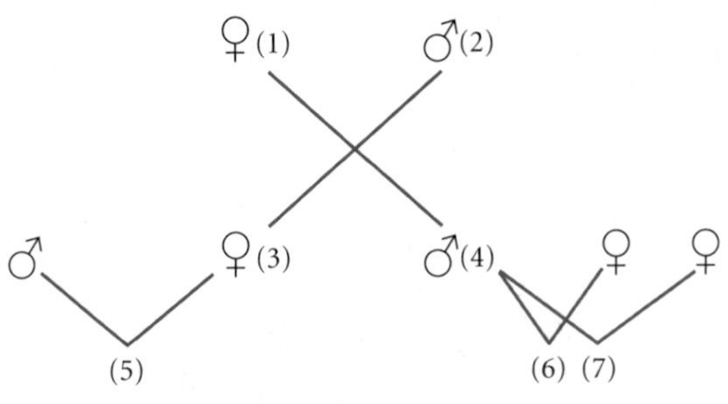

Actor, Recipient	r
Mother (1), offspring (3 or 4)	0.5
Father (2), offspring (3 or 4)	0.5
Full siblings (3, 4)	0.5
Half-siblings (6, 7)	0.25
Aunt (3), niece (6)	0.25
Full cousins (5, 6)	0.125

The coefficient of relationship (r) between actor and the recipient of the action is the average proportion of the actor's genes that are present and identical by descent in the recipient.

Figure 16.8 Coefficients of relationship (r) among some relatives in a diploid species. Individuals 3 and 4 are full siblings, the offspring of female 1 and male 2. Individuals 3 and 4 mate with unrelated individuals, producing offspring 5, 6, and 7. The table gives the values of r between an actor who provides a benefit and the recipient of that benefit.

alleles that cause aunts to care for nieces and nephews will spread only if they increase the fitness benefit by more than fourfold the cost of care. The more distantly related the beneficiaries are to the donor, the greater the benefit to them must be for the allele for an altruistic trait to spread.

Parental care illustrates why indiscriminate altruism cannot evolve by individual selection. If allele *A* caused a female to dispense care to young individuals in the population at random, it could not increase in frequency because, on average, the fitness of all genotypes in the population, whether they carried *A* or not, would be equally enhanced. Thus the only difference in fitness among genotypes would be the reduction in fitness associated with dispensing care.

Interacting individuals often form groups that last for a generation or less, and that may differ in the frequencies of alleles that determine social behaviors (such as cooperation or defection). David Sloan Wilson (1975) termed these TRAIT GROUPS, and proposed that groups with many cooperators may produce more offspring than groups with fewer. When the groups dissolve and random mating takes place before the establishment of new groups, the productive groups of cooperators contribute more genes to the total gene pool, so the allele for cooperation increases in frequency. This process was diagrammed in Figure 15.9, which shows how this process may affect the evolution of sex ratio. Wilson (1997; Wilson and Wilson 2007) has argued that this process vindicates a role for group selection in evolution. Some other authors, however, have pointed out that kin selection can operate only when most interactions are among relatives—in other words, relatives form groups. They argue that trait groups really are made up of individuals that, with respect to genes for cooperation, are more closely related to each other than random. Thus, the view is widely held that kin selection and trait-group selection are two equivalent ways of describing the same process (Frank 1998; Foster et al. 2006; Lehmann and Keller 2006). We return to this point on p. 422.

FACTORS THAT AFFECT THE EFFICACY OF KIN SELECTION. Kin selection can operate only if individuals are more likely to help kin than nonkin. This can be achieved in two ways. First, individuals may be able to distinguish related from unrelated individuals, perhaps based on assessing their similarity with respect to one or more characteristics that are highly variable in the population (Sherman et al. 1997). This variation might be genetically based, or it might be caused by a shared environmental imprint; for example, individual colonies of many ants and other social insects have a distinctive "colony odor" that is apparently derived from food or other environmental factors (Wilson 1971). An example of a kin-selected characteristic, based on kin recognition, is provided by tadpoles of the spadefoot toads *Spea bombifrons* and *S. multiplicata*, which develop into detritus- and plant-feeding omnivores if they eat these materials early in life, or into cannibalistic carnivores, with large horny beaks and large mouth cavities, if they eat animal prey (Pfennig and Frankino 1997). Tadpoles are less likely to develop the cannibalistic phenotype if they develop in the company of full siblings than if they develop alone or in a mixture of related and nonrelated individuals. Moreover, omnivores associate more with their siblings

Figure 16.9 Competition among kin can counteract the kin-selected advantage of altruism, which may reduce the fitness of some kin more than it increases the benefit to others. (Illustration © John Norton.)

than with nonrelatives, whereas carnivores do the opposite, and carnivores eat siblings much less frequently than unrelated individuals.

Second, kin selection may operate if individuals are usually associated with kin, at least during the period in their life history, or under environmental circumstances, when helping behavior is expressed. Such a population structure requires that individuals have not become randomly mixed before the time of dispersal. For example, local colonies and troops of many primates, prairie dogs, and other mammals are composed largely of relatives (Manno et al. 2007). In such cases, kin recognition would not be necessary. However, aggregation also can increase competition among kin for food and other resources, counteracting kin selection for cooperation because the benefits accrued by recipients of altruistic help are balanced by the reduced fitness of the relatives with which the recipients compete (Figure 16.9; Queller 1994; Frank 1998). Ashleigh Griffin and colleagues (2004) showed experimentally that local competition of this kind can counteract kin selection for cooperation. The bacterium *Pseudomonas aeruginosa* releases compounds called siderophores into the environment to scavenge iron and make it available for bacterial metabolism. This is an altruistic trait because siderophore production is costly, but other individuals can take up siderophore-bound iron and benefit from it. Griffin et al. initiated cultures with mixtures of a siderophore-producing genotype and a cheater genotype, distinguished by color, that does not produce siderophore (Figure 16.10A). Each population consisted of four subpopulations that were combined and mixed at intervals. Relatedness (r) was varied by initiating a culture with either one (high r) or both (low r) of the genotypes. Competition was "global" if the subpopulations were mixed in their entirety, and new subpopulations were then sampled from the mixture; this procedure enabled the productivity of a subpopulation to affect the genetic composition of the entire population. Altruism (siderophore production) might increase a subpopulation's productivity, and so evolve by kin selection. Competition was "local" if the subpopulations contributed equal numbers to the population when mixed; this procedure removes the advantage to a genotype of being in a more productive subpopulation. Over the course of 16 days, the proportion of cooperators (siderophore producers) changed as predicted (Figure 16.10B). In accordance with Hamilton's rule, cooperators increased or stayed at high frequency in the high r treatments but decreased when r was low. But local competition counteracted kin selection, causing the frequency of cooperators to remain unchanged or to decline.

KIN SELECTION OR GROUP SELECTION? In the experiment by Griffin et al., the "global competition" treatment was one in which temporary subpopulations grew to different sizes, based on differences in their genetic composition, and so contributed differentially to the allele frequencies in the pooled population when they were mixed. This conforms precisely to D. S. Wilson's (1975) model of TRAIT-GROUP SELECTION, and suggests that altruism can evolve by group selection, or in general, that **multilevel selection** (selection at mul-

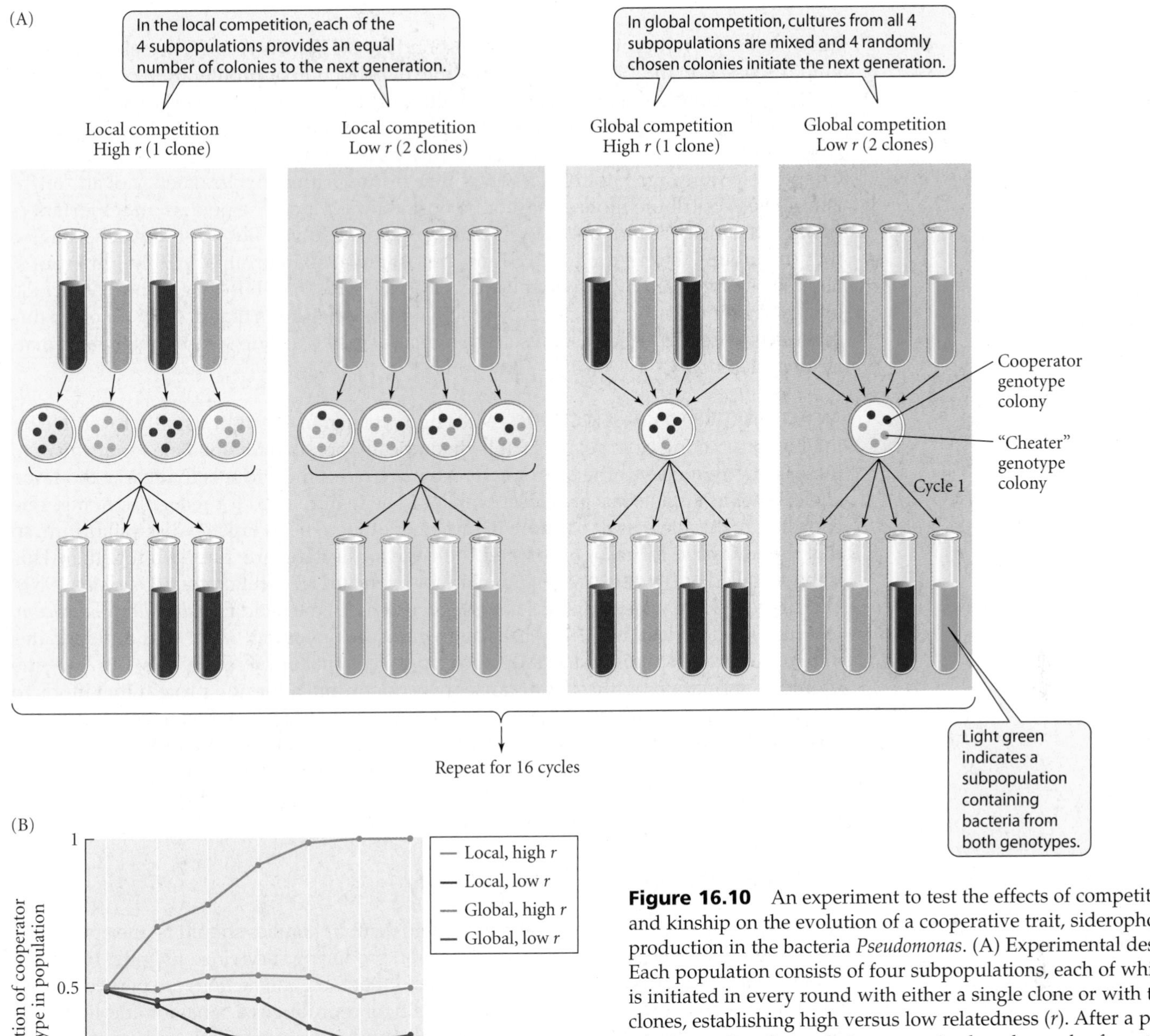

Figure 16.10 An experiment to test the effects of competition and kinship on the evolution of a cooperative trait, siderophore production in the bacteria *Pseudomonas*. (A) Experimental design. Each population consists of four subpopulations, each of which is initiated in every round with either a single clone or with two clones, establishing high versus low relatedness (r). After a period of growth, subpopulations are mixed, and samples from the mixture initiate a new set of subpopulations. (B) Changes in the proportion of the cooperator genotype after 16 cycles, showing that cooperation increases when r is high and competition is global, and decreases when r is low and competition is local. (After Griffin et al. 2004.)

tiple levels, i.e., gene, organism, population) may influence trait evolution. Trait-group selection, however, differs importantly from the kind of group selection that was rejected in the 1960s as too weak to counteract selection at the level of genes or organisms.

It may be useful to recognize two kinds of multilevel selection (MLS), denoted MLS1 and MLS2, which correspond to the two concepts of fitness that were introduced in Chapter 11. The "collectives" and "particles" described in Chapter 11 (see Figure 11.13) are groups and genes, respectively. MLS1 focuses on the frequency of an allele or trait in the entire population, and the groups, or subpopulations, differ in fitness by virtue of how many genes or individuals they contribute to the population (fitness$_1$). MLS2 focuses on the numbers of groups that are distinguished by the composition of their genes, and the fitness of groups of a certain type is measured by the number of like groups they produce (fitness$_2$). Trait-group selection corresponds to MLS1, whereas group selection of the kind discussed in the 1960s, in which the rate of extinction or proliferation of entire populations determines the prevalence of an altruistic trait, corresponds to MLS2.

Most students of the subject agree that group selection in the shape of MLS2 is unlikely to explain the origin or persistence of an altruistic trait, for the reasons described in Chapter 11. Many agree that MLS1, in which groups with a high frequency of an "altruistic allele" contribute more genes to the population-wide gene pool, is an important—perhaps the most important—way in which altruism evolves. But most researchers agree that the groups that figure in MLS1 are almost always kin groups. Within kin groups, altruism is disfavored by individual selection (see Figure 16.9), but kin groups that include altruistic genotypes may be more productive than kin groups that do not. Whether this process is better described as kin selection or as a form of group selection may be a matter of preference.

AN ALTERNATIVE TO KIN SELECTION. Although altruism is thought usually to evolve by interactions among close relatives, it is theoretically possible for an allele for altruism to increase in frequency if the bearer can recognize other individuals that carry the same allele, whether or not they are closely related. The feature used for recognition might be encoded by the allele itself or, more likely, by another gene in linkage disequilibrium, so that the trait used for recognition and the gene for helping are inherited together. This is the so-called GREEN BEARD MODEL, named for a fanciful recognition trait (Dawkins 1989). One of the few known examples is the *csA* gene of the slime mold *Dictyostelium discoideum*, in which (as described on p. 415) cells aggregate into a moving "slug" that differentiates into reproductive spores and a stalk made up of cells that altruistically die. The *csA* gene promotes formation of the slug because it encodes a cell adhesion protein that binds to the protein in the membrane of other cells. When David Queller and collaborators (2003) mixed wild-type cells and cells in which *csA* was knocked out, the knockout cells were excluded from the slug, and so few of them became spores. Wild-type *csA* genes promote their own reproduction by forming colonies of cells in which some *csA*-bearers sacrifice themselves for the reproductive benefit of others.

A Genetic Battleground: The Family

At first surmise, it would seem that relationships within families should be the epitome of cooperation, since the parents' fitness depends on producing surviving offspring. However, evolutionary biologists have come to understand that these interactions are pervaded with potential conflict, and that much of the diversity of reproductive behavior and life histories among organisms stems from the balance between conflict and cooperation. (Incidentally, the ways in which some animal species behave toward family members starkly show that natural selection utterly lacks morality, as pointed out in Chapter 11.)

Mating systems and parental care

Whether or not one or both parents care for offspring varies greatly among animal species and partly determines the MATING SYSTEM, the pattern of how many mates individuals have and whether or not they form pair-bonds (Clutton-Brock 1991; Davies 1991). Providing care (such as guarding eggs against predators, or feeding offspring) increases offspring survival, which enhances the fitness of both parents (and of the offspring). But parental care is also likely to have a cost. It entails risk, and it requires the expenditure of time and energy that the parent might instead allocate to further reproduction: a female might lay more eggs, and a male might find more mates.

In most animals, neither parent provides care to the eggs or offspring after eggs are laid, and one or both sexes, especially in long-lived species, may mate with multiple partners (PROMISCUOUS MATING). In many birds and mammals, females provide care but males do not, and males may mate with multiple females (POLYGYNY). In some fishes and frogs and a few species of birds, only males guard eggs or care for offspring, and in some such species, the female may mate and lay eggs with different males (POLYANDRY). (These terms stem from

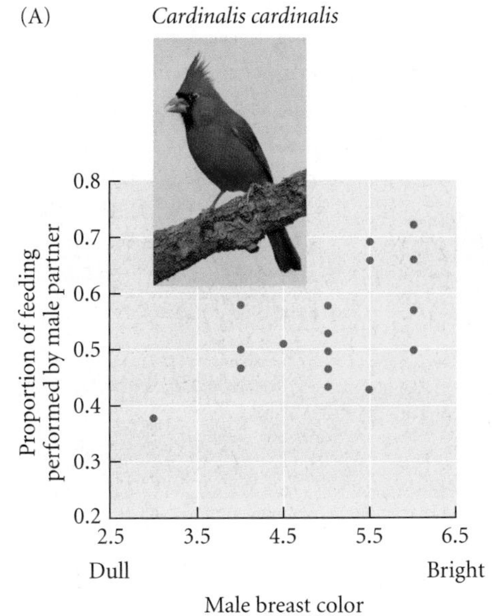

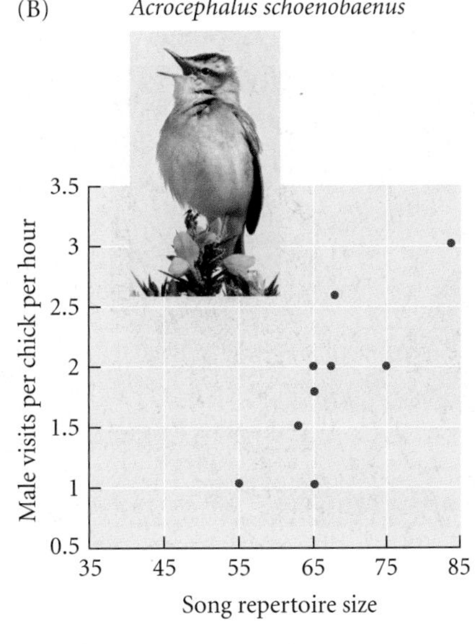

Figure 16.11 Examples of variation in paternal care correlated with signal traits in songbirds. (A) Paternal care is correlated with the brightness of the male's breast in the northern cardinal. (B) Among sedge warblers, males that sing a larger repertoire of songs also bring food to offspring more frequently. (A after Linville et al. 1998, photo © Purestock/Alamy; B after Buchanan and Catchpole 2000, photo © Mark Hicken/Alamy.)

the Greek *polys*, "many"; *gyne*, "woman"; *andros*, "man.") In many species of birds, some mammals, and a few insects such as dung beetles, a female and male form a "socially monogamous" pair-bond and contribute biparental care of the offspring. However, many such birds engage in frequent "extra-pair copulation" and are not sexually monogamous. As we have seen, females may also increase their reproductive success by laying eggs in the nests of unwitting foster parents. Females in some pair-bonding species of birds appear to favor males with characteristics that serve as honest signals of paternal caregiving (Figure 16.11). This is as some theory predicts (Kokko 1998), but it has also been argued that highly ornamented males might devote less time to offspring care and more to extra-pair copulation, in which they may be especially likely to succeed (Houston et al. 2005).

Each parent in a mated pair will maximize her or his fitness by some combination of investing care in current offspring and attempting to produce still more offspring. In a species with biparental care, each parent would profit by leaving as much care as possible to the other partner, as long as any resulting loss of her (or his) fitness owing to the death of her (or his) current offspring were more than compensated by the offspring she (or he) would have from additional matings. If offspring survival were almost as great with uniparental care as with biparental care, selection would favor females that defected, abandoning the brood to the care of the male—or vice versa (Figure 16.12). Thus a conflict between mates arises as to which will evolve a promiscuous habit and which will

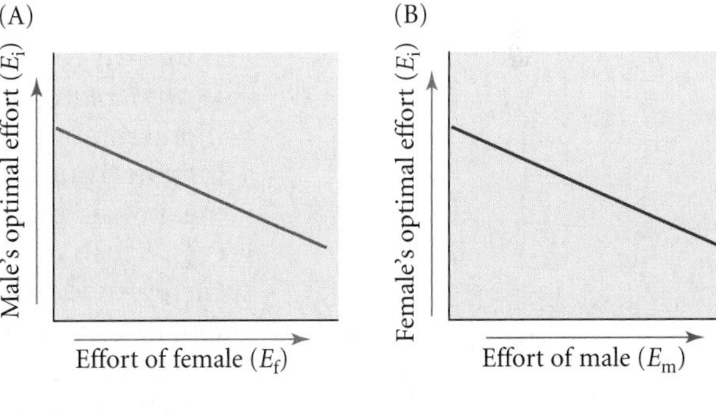

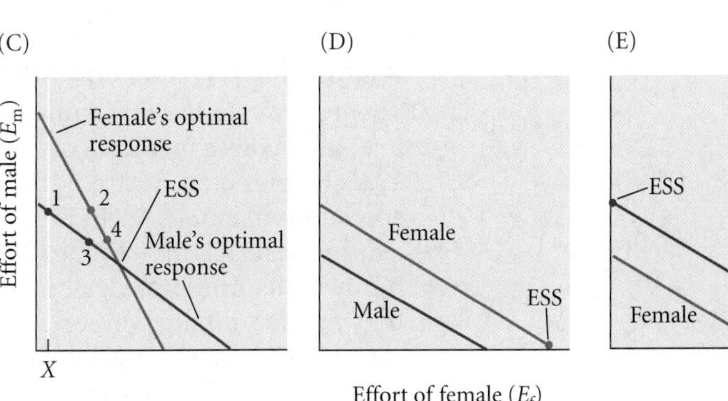

Figure 16.12 An ESS model of parental care. (A,B) The optimal parental effort expended by each sex declines, the more effort its partner expends. (C) Curves for males and females plotted together. Their intersection marks the ESS, the evolutionarily stable strategy. If, for example, the population starts with female effort (E_f) equal to X, male effort (E_m) evolves to point 1; but then the optimal E_f is at point 2 on the female's optimality line. When E_f evolves to point 2, E_m evolves to point 3; but then E_f evolves to point 4. Eventually, E_m and E_f evolve to the intersection (the ESS), no matter what the initial conditions are. (D,E) Conditions can be envisioned in which the optimal curves for the sexes do not intersect and the ESS is care by only the female (D) or the male (E). (After Clutton-Brock and Godfray 1991.)

Figure 16.13 Parental care. (A) A male stickleback (*Gasterosteus aculeatus*) builds and cares for a nest containing egg clutches fathered by him. Such nest-brooding activity can attract additional females to mate with the male, and the resulting eggs are added to his nest. (B) Red-necked grebes (*Podiceps grisegena*) exemplify the many bird species in which both parents care for the young. (A © blickwinkel/Alamy; B © Tom Ulrich/OSF/Photolibrary.com.)

care for the eggs or young. Selection favors defection more strongly in the sex for which parental care is more costly (in terms of lost opportunities for further reproduction). However, if the survival of offspring depends strongly on care, both parents may maximize their fitness by forming a socially monogamous pair-bond and cooperating in offspring care.

This theory may explain why in birds and mammals, parental care is generally provided by females or by both mates, whereas in fishes and frogs it is usually provided by males (Clutton-Brock 1991). Fishes and frogs guard the eggs or young but do not feed them. Males can often mate with multiple females and guard all their eggs in a single nest (Figure 16.13A); thus they pay a smaller cost than females, whose subsequent reproduction depends on replenishing the massive resources they expend in egg production. In birds and mammals that must feed their young, parental care is more costly for males than for females, since males could potentially obtain many matings in the time they must spend rearing one brood exclusively (Figure 16.13B).

The strength of natural selection for parental care is greater if the opportunity for reproduction outside the pair-bond is restricted, if offspring survival greatly depends on maximal care, and if individuals are likely to be actually caring for their own offspring rather than someone else's. Male parental care, for example, is expected to be proportional to a male's "confidence of paternity"—which may be quite low in species with frequent extra-pair copulation (Whittingham et al. 1992; Westneat and Stewart 2003). The dunnock (*Prunella modularis*) is a songbird with a variable mating system: some individuals form monogamous pairs, some polygynous trios (two females and one male), and some polyandrous trios (two males and one female). Although male reproductive success is maximized by polygyny, females in polyandrous trios have the greatest reproductive success, which may account for the persistent variation in mating system. Polyandrous males each provide parental care only in proportion to the amount of time they spent with the female during her fertile period, presumably using this as an estimate of their paternity (Davies 1992). In some other species, males simply evict or kill offspring that are unlikely to be their own. In other words, they practice INFANTICIDE.

Infanticide, abortion, and siblicide

There are a number of circumstances in which an individual's fitness may be enhanced by killing young members of the same species (Hausfater and Hrdy 1984; Clutton-Brock 1991; Borries et al. 1999). For example, infanticide may be sexually selected. In many species of mammals, including lions and some primates and rodents, males that replace a mated male kill the existing offspring of his mates. The infanticidal male can then father his own offspring faster, because females become fertile and sexually receptive sooner if they are not nursing young. DNA analyses of wild langur monkeys (*Presbytis entellus*) showed that males that kill infants are the fathers of the infants that are subsequently born to the same mothers (Borries et al. 1999).

Parents sometimes kill offspring as a way of adaptively regulating brood size (Mock 2004). A parent's fitness is proportional (all else being equal) to the number of surviving offspring, which equals (size of egg clutch or brood) × (per capita probability of survival). In species with parental care, the probability of survival may decrease as brood or clutch size increases, because of competition among offspring for food, and the parental care expended on an excessively large brood can reduce the parent's survival and subsequent reproduction, reducing its lifetime fitness (see Chapter 14). Thus it can be adaptive for a parent to reduce the number of offspring to an optimal number (possibly even zero). For example, females of some species of mice kill some or all of their young if food is scarce or the number born is too high. Likewise, shortage of food or disturbance by predators can induce many birds to desert their eggs or young, especially early in the breeding season (when the parents have the opportunity to renest elsewhere). In the same vein, plants typically abort the development of many of their offspring (seeds), allocating limited resources to fewer but larger seeds with a greater chance of survival.

When the offspring in a brood (sibs) share an environment, such as young birds in a nest or larval parasitic wasps within a caterpillar, they may actively fight for resources, and larger individuals may kill smaller siblings (SIBLICIDE). Siblicide is the norm in some species of eagles and boobies, in which the female lays two eggs but one of the nestlings always kills the other (Figure 16.14). (The second egg may be the female's "insurance" in case the first is inviable.) In other species, such as owls and herons, the number of survivors of sibling competition depends on the food supply. The young of some shark species eat their siblings while they share their mother's uterus.

Figure 16.14 Siblicide in the brown booby (*Sula leucogaster*). The parent is sheltering a large chick that has forced its sibling out of the nest. The parent ignores the dying chick (foreground). (Photo by John Alcock.)

Parent-offspring conflict

Parents and offspring typically differ with respect to the optimal level of parental care (Trivers 1974; Godfray 1999). A parent's investment of energy in caring for one offspring may reduce her production of other (e.g., future) offspring. Production of other offspring increases the inclusive fitness of both the parent and each current offspring. However, the parent is equally related to all her offspring ($r = 0.5$), whereas each offspring values parental investment in itself twice as much as investment in a full sib ($r = 0.5$) and four times as much as investment in a half sib ($r = 0.25$). Therefore offspring are expected to try to obtain more resources from a parent than it is optimal for the parent to give, resulting in **parent-offspring conflict**. Animals may evolve features that enable offspring to extract extra resources from their parents; for example, baby American coots (*Fulica americana*) have unusual reddish plumes that stimulate their parents to feed them (Lyon et al. 1994).

From this perspective, siblicide is more advantageous to a sib-killing offspring than to its parents, and so it is surprising that parent birds usually stand by without interfering while one of the offspring slaughters the others (see Mock 2004). The theory of parent-offspring conflict, however, does explain why young birds and mammals often beg for more food than the parents are "willing" to give, and why adults must aggressively wean their offspring from further care. We will return to parent-offspring conflict when we discuss genetic conflict later in the chapter.

Cooperative breeding

In many species of birds and in several mammals and fishes, young are reared not only by their parents but also by other individuals that are physiologically able to reproduce but do not. In most such species of birds, the helpers are young males that are prevented by competition from acquiring mates or territories. Several factors may explain cooperative breeding, including various direct benefits as well as increased inclusive fitness by help-

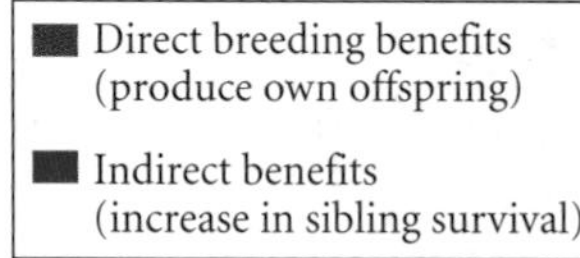

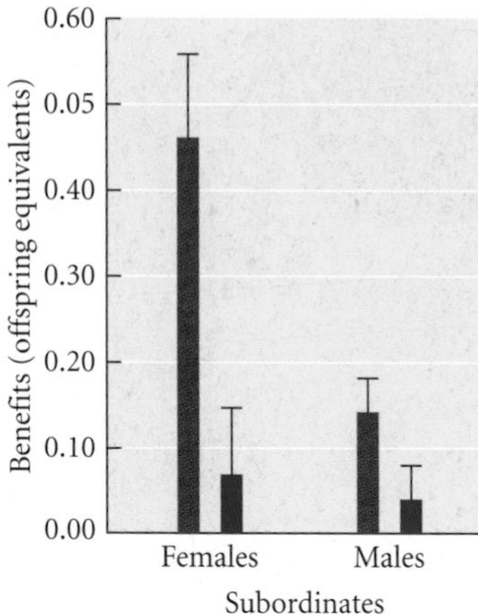

Figure 16.15 Fitness benefits of cooperative breeding in the Seychelles warbler. The brown columns show that subordinate females (i.e., those who helped their parents raise their siblings) eventually produced, on average, about 0.45 offspring of their own (the direct genetic benefit). Subordinate males produced a very few offspring of their own. The indirect genetic benefit subordinates received is represented by the green columns, which measure the average increase in survival of their parents' offspring. Subordinate females increased the survival of their parents' offspring more than subordinate males did. (After Richardson et al. 2002.)

ing kin (Emlen 1991; Cockburn 1998; Clutton-Brock 2002). In many species, nonbreeding individuals aid their parents in rearing their siblings. In the Seychelles warbler (*Acrocephalus sechellensis*), for example, many individuals, especially females, do not breed in their first year, but instead stay as subordinates in their natal territory and help their putative parents rear offspring. For almost 15 years, all the birds on one island have been fitted with colored rings that make them individually recognizable, and the parentage of every individual was determined by matches between highly variable microsatellite DNA markers during a 3-year period (Richardson et al. 2002). The investigators discovered that subordinate females derive a direct benefit from staying in their parents' territory: 44 percent of them produced offspring! (Moreover, 40 percent of all offspring resulted from extra-pair paternity.) Nevertheless, both male and female subordinates also gained indirect fitness by increasing the survival of nestlings to which they were related (Figure 16.15).

Social insects

The most extreme altruism is found in EUSOCIAL animals: those in which nearly or completely sterile individuals (workers) rear the offspring of reproductive individuals, usually their parent (or parents). Eusociality is known in one species of mammal (the naked mole-rat, *Heterocephalus glaber*; Figure 16.16A), in all species of termites (Isoptera; Figure 16.16B), in many Hymenoptera (Figure 16.16C,D), and in a few other kinds of insects (Wilson 1971; Keller 1993; Bourke and Franks 1995; Crozier and Pamilo 1996).

(A)

(B)

(C)

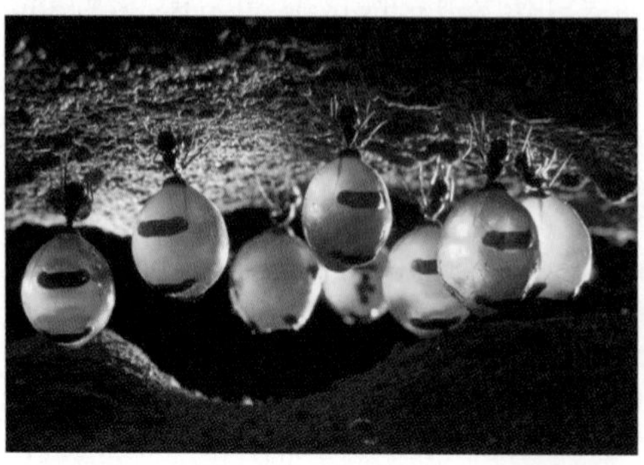

(D)

Figure 16.16 Eusociality is found primarily in insects. (A) Naked mole-rats (*Heterocephalus glaber*) are the only mammalian species known to be eusocial. (B) This queen termite is attended by small, sterile workers and large-headed, sterile soldiers. (C) Australian honeypot ants (*Camponotus inflatus*), engorged with nectar, hang from roof of their nest's larder. These "repletes" regurgitate nectar on demand to their worker nestmates. (D) Common wasps (*Vespula vulgaris*) at their nest. (A © Gregory Dimijian/Photo Researchers, Inc.; B © Colin Carson/AGE Fotostock; © Reg Morrison/Auscape/Minden Pictures; D © Juniors Bildarchiv/Alamy.)

Eusociality has evolved independently many times among the wasps, bees, and ants in the order Hymenoptera. All of these lineages are phylogenetically nested within, and have arisen from, solitary, stinging, wasplike insects in which females provide parental care to their young, usually by constructing a burrow or nest and provisioning it with food (e.g., paralyzed insects) for their larvae. In all eusocial Hymenoptera, the workers are females that do not mate. The workers and one or more egg-laying females (queens) constitute a colony that may persist for a single season or for many years, depending on the species. Most of the queen's eggs develop into workers, but at certain times, eggs are produced that develop into reproductive females and males. In most species, whether a female develops into a queen or worker depends on diet, which is often controlled by the workers. These insects are HAPLODIPLOID: females develop from fertilized eggs and are diploid, but males develop from unfertilized eggs and are haploid. Consequently, coefficients of relationship (r) among relatives differ from those in diploid species (Figure 16.17). Whereas r in diploid species is 0.5 both between parent and offspring and between full siblings, a female hymenopteran is more closely related to her sister (r = 0.75) than to her daughter (r = 0.5), and is less closely related to her brother (r = 0.25).

In a colony with a single, singly mated queen, the inclusive fitness of a female (a worker) may be greater if she devotes her energy to rearing reproductive sisters (future queens) than to rearing daughters. For this reason, Hamilton (1964), who introduced the theory of kin selection, suggested that the haplodiploid condition in Hymenoptera is one of the factors that has predisposed these insects to evolve eusociality. This hypothesis is now considered unlikely, because a female hymenopteran that produces equal numbers of daughters (r = 0.75) and sons (r = 0.25) has the same average relationship to her offspring as in a diploid species. Moreover, in many species, a colony has multiple queens, or a single queen that mated with and carries the sperm of multiple males, and so the relationships among the workers and between them and future queens are lower than 0.75. The single queen in a honeybee (*Apis mellifera*) colony, for example, mates with ten males, on average, before she takes up queenship of a hive, never to mate again. It appears more likely that the ancestors of social Hymenoptera were preadapted (see Chapter 11) for social life because they practiced parental care of defenseless, almost immobile larvae, and had stings that could be "exapted" for nest defense (Strassmann and Queller 2007). Sociality may have first evolved because daughters that stayed at the natal nest would increase the chance of their younger siblings' survival if their mother perished while foraging. Lone female wasps have a high risk of dying before their offspring are fully grown, and helpers in a facultatively eusocial species of wasp rear the older offspring of their fallen nestmates to maturity (Field et al. 2000).

Kin selection is a powerful explanation of many aspects of cooperation and conflict in eusocial insects (Foster et al. 2006; Ratnieks et al. 2006; Strassmann and Queller 2007). Colonies of all such species are family-based, and normally exclude members of other colonies. Much of the evidence for the role of kin selection is based on the consequences of conflict within colonies. In most eusocial Hymenoptera, workers can lay haploid eggs

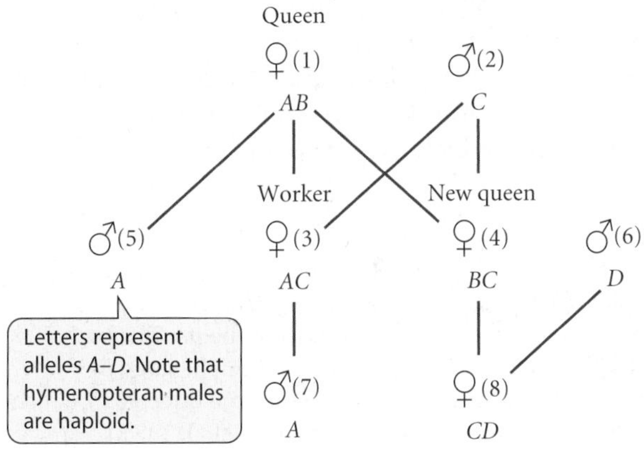

Actor, Recipient	r
Female (1), Daughter (3, 4)	0.5
Female (1 or 3), Son (5 or 7)	0.5
Male (2), Daughter (3, 4)	1.0
Male (2), Mate's son (5)	0.0
Sister (3), Sister (4)	0.75
Sister (3), Brother (5)	0.25
Female (3), Niece (8)	0.375
Female (4), Nephew (7)	0.375

Figure 16.17 Some coefficients of relationship among relatives in hymenopteran species, assuming a single, singly mated queen per colony. Individuals in the pedigree are numbered as in the table, and one of the possible genotypes of each is labeled with letters representing alleles *A–D*. The coefficient of relationship (r) is calculated by finding the proportion (1.0 or 0.5) of an actor's genes that are inherited from one of its parents, multiplying by the probability that a copy of the same gene has also been inherited by the recipient, and summing these products over the actor's one or two parents. Notice that in these haplodiploid species, sisters are more closely related than mother and daughter or son. Compare with the analogous situation for diploid species shown in Figure 16.8. (After Bourke and Franks 1995.)

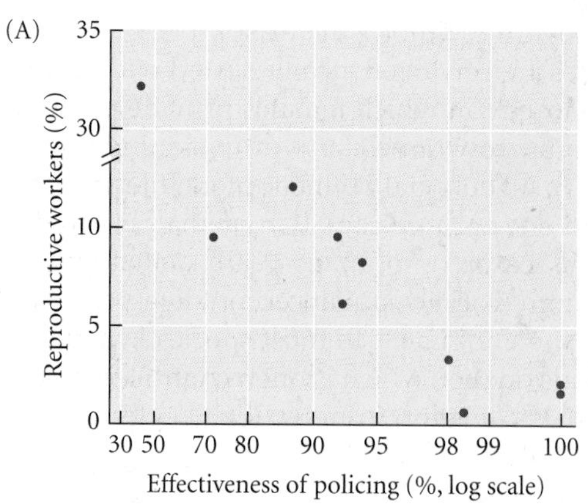

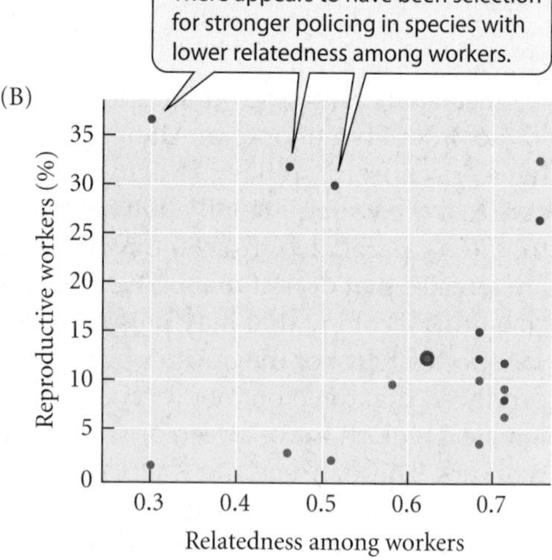

Figure 16.18 Policing in eusocial insects, illustrated by comparisons of nine wasp species and the honeybee. (A) Fewer workers attempt to reproduce if the eggs they lay are more frequently killed by nestmates. This implies that policing is effective. (B) For the same ten species, the red points show that in queenless colonies, in which policing does not occur, fewer workers lay eggs in species in which they are more closely related to each other. That is, they are more altruistic, as predicted by Hamilton's rule. However, when queens are present and policing occurs (blue points), the relationship is reversed, probably because there has been selection for stronger policing in species with lower relatedness among workers. (After Ratnieks and Wenseleers 2007.)

that develop into males. A worker is more closely related to her own sons ($r = 0.5$) than to her mother's (the queen's) sons ($r = 0.25$). Therefore, workers gain direct fitness by laying eggs, and are altruistic if they do not. Because the queen is more closely related to her own sons ($r = 0.5$) than to her daughters' son's ($r = 0.25$), and because male offspring do not contribute to colony growth and welfare, queens practice "policing" in many species, by destroying workers' eggs—and in some species, workers destroy the eggs of other workers (Ratnieks et al. 2006; Wenseleers and Ratnieks 2006). The level of reproduction by workers is lower in species in which a high degree of policing has evolved. Policing usually does not occur if the queen dies or is removed, and under these conditions, fewer workers lay eggs—more workers retain their altruistic behavior—in those species in which the level of relatedness among workers is higher (Figure 16.18).

In most social insects, workers can control the sex ratio among the larvae destined to be queens or males, by withholding care from male larvae and by altering the proportion of female larvae that develop into queens versus workers. In a colony with a single queen, a worker's inclusive fitness is maximized by rearing queens and males in a 3:1 ratio, because they are related as $r = 0.75$ to their queen sisters, but only as $r = 0.25$ to their brothers. In colonies with multiple queens, however, many of the female larvae workers would have a coefficient of relationship lower than 0.75 to many of the offspring they rear (since they are not full sisters), so the sex ratio should be closer to 0.5 (1:1). A resident queen's inclusive fitness is maximized by a 1:1 sex ratio, since she is equally related to her daughters and sons.

Many data support these predictions (Bourke and Franks 1995; Crozier and Pamilo 1996). The average sex ratio is closer to 3:1 in single-queen than in multiple-queen species of ants, and it is almost exactly 1:1 in slave-making species of ants, in which the workers that rear the slave-maker's brood are members of other species, captured by the slave-maker's own workers. The slaves have no genetic interest whatever in the captor's reproductive success, and so are not expected to alter the sex ratio by preferential treatment.

Moreover, there is direct evidence in the wood ant (*Formica exsecta*) that workers manipulate the sex ratio as kin selection theory predicts (Sundström et al. 1996). Although all colonies have about the same sex ratio among eggs, the sex ratio among pupae becomes shifted toward females in single-queen colonies, but toward males in multiple-queen colonies, which would be advantageous for the queens (Figure 16.19). This is some of the most convincing evidence of the role of kin selection in social animals.

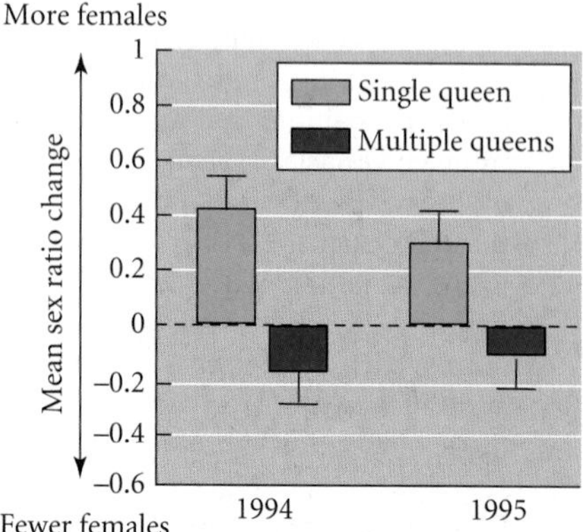

Figure 16.19 Evidence that workers in hymenopteran colonies manipulate the sex ratio to increase their inclusive fitness, as predicted by kin selection theory. The bars show the change in the proportion of females between the egg and pupal stage in colonies of wood ants with single queens and in colonies with multiple queens, in which each worker is unrelated to many of the offspring. (After Sundström et al. 1996.)

Genetic Conflict

Thyme (*Thymus vulgaris*), a familiar food seasoning, is a member of the mint family. Like most mints, some thyme plants have hermaphroditic flowers, with both male (stamens) and female (pistil) parts. Many thyme plants, however, lack anthers; they are "male-sterile," or female. Male sterility is caused by a mitochondrial gene. The interesting point is that all thyme plants carry this "cytoplasmic male sterility" (CMS) factor. Hermaphrodites have male function only because they carry a "restorer" gene, inherited on a nuclear chromosome, that counteracts CMS. Remarkably, there exist hermaphroditic plant species in which all individuals carry both a CMS gene and a nuclear restorer. Why should a gene exist, only to be counteracted by another gene?

The answer appears to be that there exist conflicts between different genes as an outcome of selection at the gene level. Such conflicts can arise whenever a gene has a transmission advantage over other genes, perhaps by a segregation advantage during meiosis or perhaps by not following the rules of meiosis at all. In Chapter 11, we encountered "selfish genes" that increase in frequency not because they provide an advantage to the organism that carries them, but simply because they have an advantage in transmission from one generation to the next. One example is the *t* allele in mice, which causes sterility or death in homozygotes, but increases in frequency due to meiotic drive. This particular allele is transmitted to more than 90 percent of a heterozygous male's sperm. Such "outlaw genes," which promote their own spread at a faster rate than other parts of the genome, can create a context in which there is selection for genes at other loci to suppress their effects. When this is the case, a **genetic conflict** is said to exist (Hurst et al. 1996; Burt and Trivers 2006).

Cytoplasmic male sterility in plants illustrates how such conflicts arise (Figure 16.20). Since mitochondria are maternally inherited, any mitochondrial allele that can increase the number of female relative to male offspring will increase in frequency relative to an allele that does not alter the sex ratio from 1:1. Male-sterile plants produce more seeds than hermaphroditic plants because protein and other resources are diverted from pollen to seed production. Thus mitochondrially inherited cms^+ alleles increase in frequency. In contrast, recall (from Chapter 12) that because every individual has a mother and a father, nuclear alleles that produce an even sex ratio are advantageous. Thus production of excess females is disadvantageous to nuclear genes, since they are not transmitted by pollen from male-sterile plants. A mutation of a nuclear gene that nullifies the effect of a cms^+ allele may therefore increase in frequency. The only function of such a gene, then, is to counteract the effect of a selfish gene at another locus.

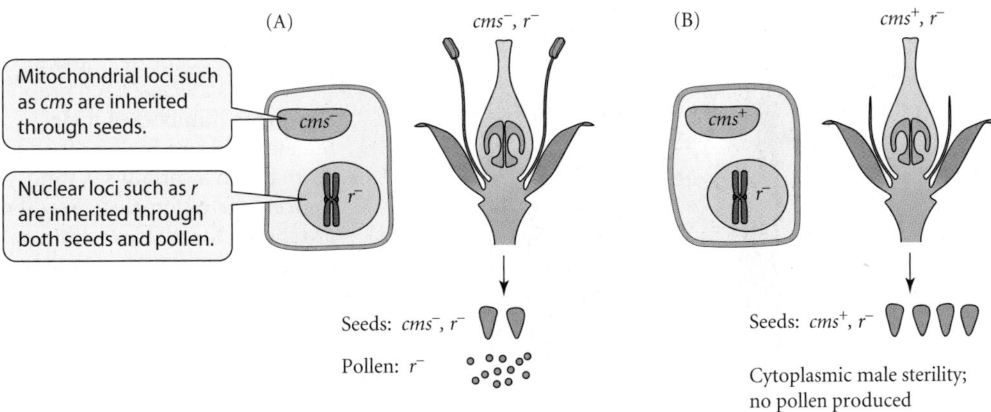

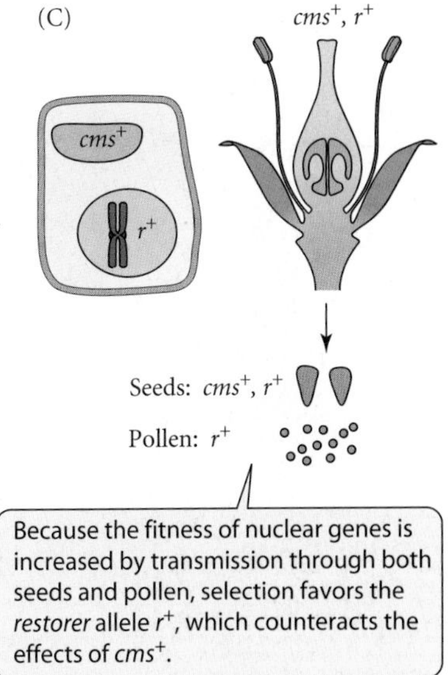

Figure 16.20 Evolution of cytoplasmic male sterility (CMS) and restoration of male fertility: a genetic conflict. The diagrams portray a flower and a cell of each genotype. (The nucleus is depicted as haploid for simplicity.) (A) Mitochondrial loci such as *cms* are inherited only through eggs, and nuclear loci such as *r* are inherited through both eggs and pollen. (B) A cms^+ mutation eliminates pollen production, and may increase seed number due to the plant's reallocation of energy and resources from pollen to eggs. Thus cms^+ increases in frequency. (C) Because the fitness of nuclear genes is increased by transmission through both eggs and pollen, selection favors the restorer allele r^+, which counteracts the effects of cms^+. The population may become fixed for antagonistic alleles cms^+ and r^+, but may be phenotypically indistinguishable from the ancestral population, cms^-, r^-.

Mitochondrial and other cytoplasmically inherited genes commonly conflict with nuclear genes, as cytoplasmic male sterility illustrates. For example, they may exaggerate sexual conflict, since they benefit from enhancing female, but not male, fitness (Zeh 2004). Mitochondrial genes that reduce sperm competitive ability, for instance, could be fixed if they increase female fitness.

Because of their effects on sex ratio, genes on the sex chromosomes of animals may similarly initiate genetic conflicts. In many species, the sex ratio among the progeny of a single mating depends on the proportion of eggs fertilized by X-bearing versus Y-bearing sperm, yielding daughters and sons, respectively. X-linked genes that can increase the proportion of successful X-bearing sperm (by meiotic drive), distorting the sex ratio of the progeny, have a transmission advantage and may increase in frequency. If they do, there is selection for autosomal genes that achieve highest fitness by restoring the 1:1 sex ratio.

In matings of *Drosophila simulans* from the Seychelles Islands, the progeny sex ratio is about 1:1, but the F_1 males from a cross between Seychelles females and males from elsewhere in the world produce a great excess of female offspring (Figure 16.21). This skewed sex ratio is the result of a *segregation-distorter* gene on the X chromosomes of Seychelles flies that causes degeneration of the male's Y-bearing sperm. The effect of this chromosome is not ordinarily seen because the Seychelles population also has recessive autosomal genes that suppress the distortion. Other populations of *D. simulans* lack both the X-linked *distorter* and the suppressor alleles, so the effects of *distorter* become evident when it is crossed into their genetic background. In the Seychelles and other areas where the *distorter* gene has increased in frequency, there has been selection for autosomal suppression of its effects (Atlan et al. 1997).

Our final example of genetic conflicts may be a surprise: it concerns pregnancy in humans and other mammals (Haig 1993, 2002). Here, the cooperation one might expect between mother, father, and offspring is compromised by parent-offspring conflict and a genetic conflict between the parents that is played out in the offspring.

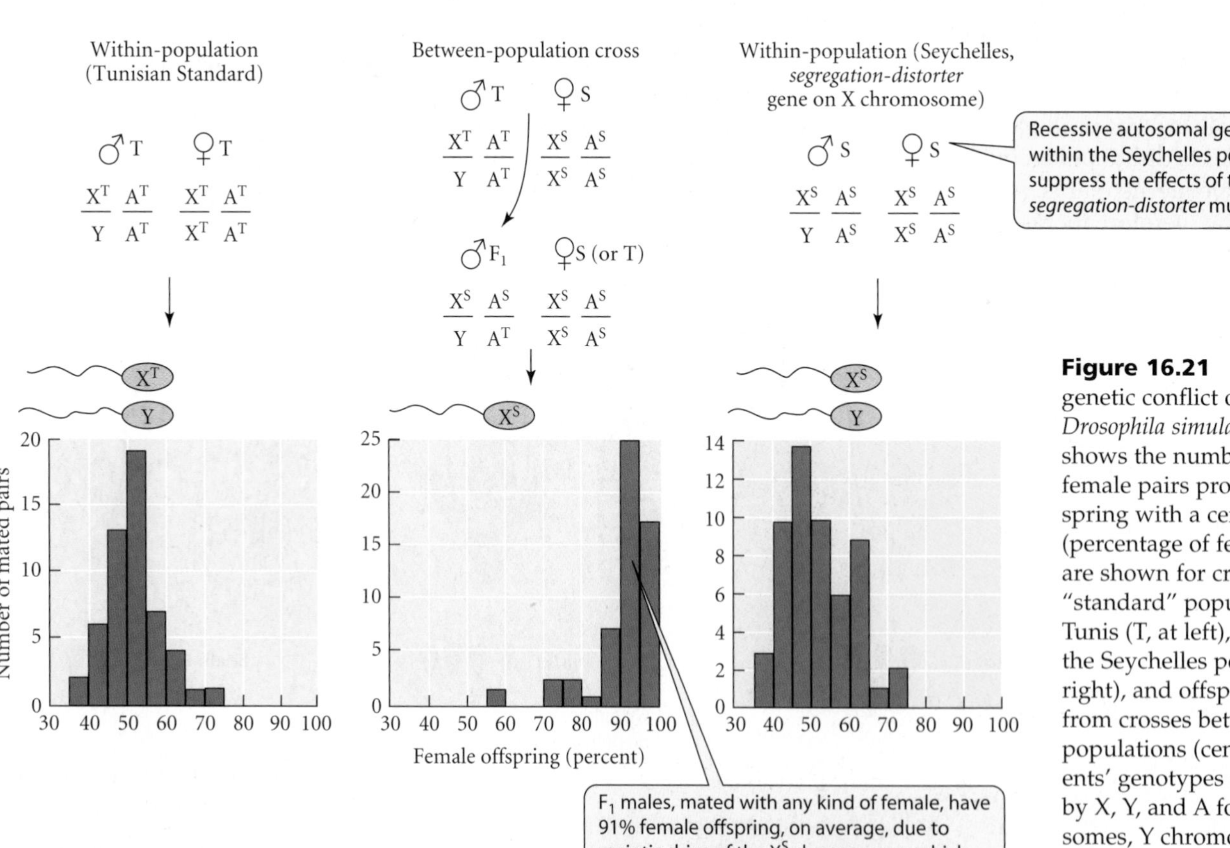

Figure 16.21 Effects of genetic conflict on sex ratio in *Drosophila simulans*. Each graph shows the numbers of male-female pairs producing offspring with a certain sex ratio (percentage of females). Results are shown for crosses within a "standard" population from Tunis (T, at left), crosses within the Seychelles population (S, at right), and offspring of F_1 males from crosses between these two populations (center). The parents' genotypes are symbolized by X, Y, and A for X chromosomes, Y chromosomes, and autosomes, respectively. (After Atlan et al. 1997.)

We have seen that an offspring is expected to try to get more resources from its mother than the mother should willingly give, since she profits more than the offspring does from reserving some resources for subsequent offspring. Some of the interactions between the human fetus and the mother follow this prediction. For example, mothers increase their production of insulin (which causes cells to remove glucose from the blood) during pregnancy—the very time you might expect them to decrease insulin levels in order to provide sugar to the fetus. However, the fetus produces extremely high levels of a hormone (hPL) that counteracts insulin, so that the net result is no alteration of blood glucose concentration. This escalation of opposing hormones seems to serve no purpose, but is just what one might expect of a parent-offspring conflict.

Even more remarkable is the expression of certain genes that are IMPRINTED so that whether or not they are transcribed in the embryo depends on whether they were inherited from the mother or the father. (For example, the maternal copy of a gene may be methylated, in which case it is not expressed in the offspring.) In terms of selection at the level of genes, the fitness of paternally inherited genes will be greatest if the embryo survives, since copies of those genes will not be carried by the mother's subsequent offspring if she mates with a different male. In contrast, maternally transmitted genes will also be carried by the mother's subsequent offspring, so their inclusive fitness is greatest if they prevent the mother from nurturing the current embryo exclusively, to the detriment of later offspring. Therefore, paternally inherited genes should enhance the embryo's ability to obtain nourishment from the mother, whereas maternally inherited genes should temper the embryo's ability to do so. In mice, a pair of conflicting genes shows just this pattern. Only the paternal copy of the gene for IGF-2 (insulin-like growth factor 2), which promotes the early embryo's ability to obtain nutrition from the uterus, is expressed in the embryo. IGF-2 is degraded by another protein (IGF-2R, insulin-like growth factor 2 receptor); only the maternal allele of the gene for this protein is expressed.

Parasitism, Mutualism, and Levels of Organization

The concept of genetic conflict bears on many topics in evolutionary biology (Hurst et al. 1996; Burt and Trivers 2006). Consider, for example, a genetic element—a gene or set of genes—that can replicate faster than the rest of the genome with which it is associated. It might be, for example, a bacterium that lives within a host organism's cells. (Such an organism is called an ENDOSYMBIONT.) If the population of endosymbionts within a single host is genetically variable, selection within that population favors (by definition) the genotype that increases in numbers faster than others. Because growth and reproduction of the endosymbionts depend on resources obtained from the host, excessive growth in the number of symbionts may reduce the host's fitness.

The fitness of an endosymbiont genotype is measured by the number of new hosts it infects per generation. If the endosymbionts exhibit **horizontal transmission**—that is, they are transmitted laterally among members of the host population (Figure 16.22A), the number of new hosts infected may be proportional to the number of symbiont progeny released from each old host. If symbionts escape to new hosts before the old host dies, their fitness does not depend strongly on the reproductive success of the individual host in which the parent symbionts reside. Therefore, selection favors symbiont genotypes with a high reproductive rate, even if they kill the host. In other words, selection favors evolution of a PARASITE that may be highly virulent.

Suppose, in contrast, that symbionts exhibit **vertical transmission**, from a mother host to daughter hosts (Figure 16.22B). Symbiont and host are now chained to each other, and the reproductive success of the symbionts now depends entirely on the fitness of the host. Selection for high proliferation *within* the symbiont populations occupying each host is opposed by selection *among* the populations of symbionts that occupy different hosts. On balance, selection at the group level favors genotypes with restrained reproduction—those that do not extract so many resources from the host as to cause its death before it can transmit the endosymbionts to its progeny. Selection may even favor alleles in the symbiont that enhance the host's fitness, since that also enhances the fitness of the symbionts

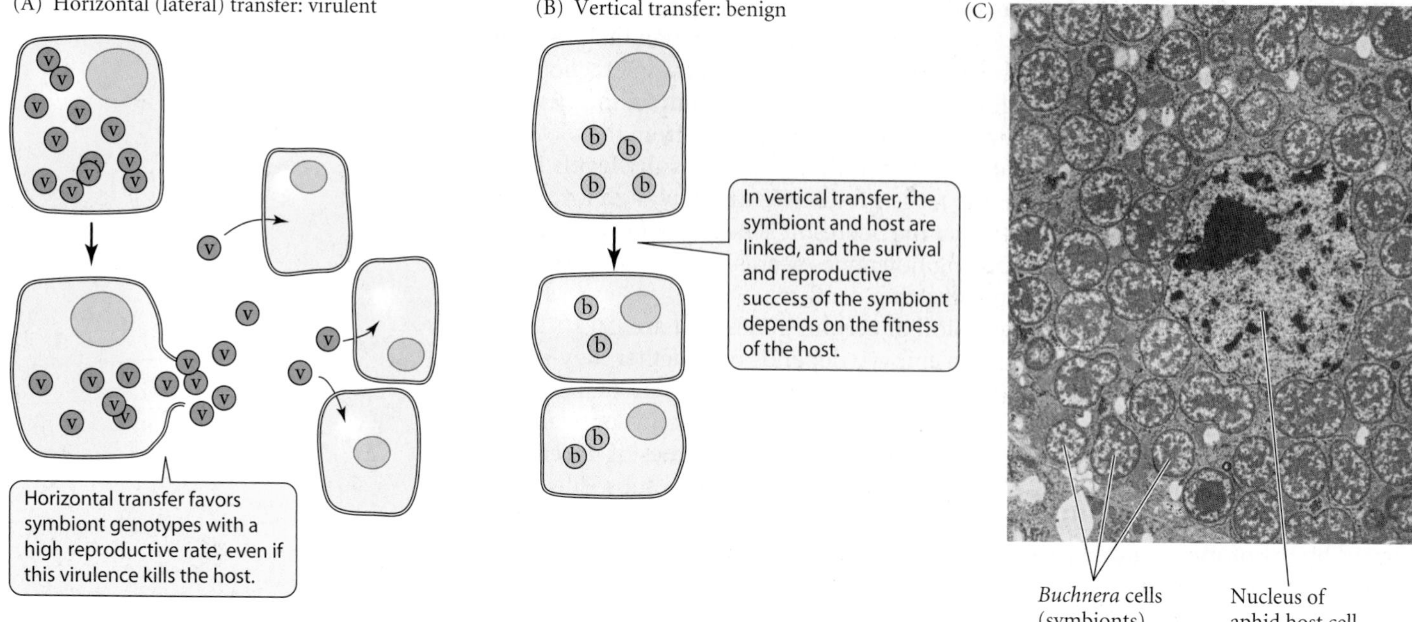
(A) Horizontal (lateral) transfer: virulent
(B) Vertical transfer: benign
(C)
In vertical transfer, the symbiont and host are linked, and the survival and reproductive success of the symbiont depends on the fitness of the host.
Horizontal transfer favors symbiont genotypes with a high reproductive rate, even if this virulence kills the host.
Buchnera cells (symbionts)
Nucleus of aphid host cell

Figure 16.22 (A) Horizontal transmission of endosymbiotic elements (v) from one host to unrelated hosts selects for a high level of virulence. (B) Vertical transmission of endosymbiotic elements from a host to its descendants favors relatively benign endosymbionts (b) with a lower reproductive rate. (C) In the extreme case, a vertically transmitted symbiont may become an integral part of the host. These intracellular bacteria (*Buchnera*) in specialized cells (bacteriocytes) of an aphid supply essential amino acids to the host. (C, photo courtesy of N. Moran and J. White.)

carried by that host. Furthermore, selection favors host alleles that control or inhibit the symbiont (an instance of genetic conflict). Evolution in both the symbiont and its host may therefore result in **mutualism:** an interaction in which two genetic entities enhance each other's fitness. (We return to the evolution of parasitism and mutualism in Chapter 19.)

In the extreme case, the symbiont may become an integral, essential part of the host. Many eukaryotes harbor vertically transmitted intracellular bacteria that play indispensable biochemical roles, such as those that reside in special cells in aphids and synthesize essential amino acids (Figure 16.22C). Mitochondria and chloroplasts are considered organelles in eukaryotic cells, but they were originally endosymbiotic bacteria. In these cases, the host and the symbiont have highly correlated reproductive interests: any advantage to one party provides an advantage to the other. The common interest of associated endosymbiont and host genomes resulted in the evolution of a new kind of collective entity, a higher LEVEL OF ORGANIZATION.

The strong correlation between the reproductive interests of associated elements—such as the organelle and nuclear genomes of eukaryotes—leads us to understand why the most familiar level of organization, the individual organism, exists at all (Buss 1989; Maynard Smith and Szathmáry 1995; Frank 1998; Michod 1999). An organism is more than a group of cells; for example, dividing bacteria that remain loosely attached, but physiologically independent of one another, do not constitute an organism. Rather, the cells of a multicellular organism cooperate and play different roles—including the distinction between cells that give rise to gametes (germ cells) and those that do not (the soma). Why should unicellular ancestors, in which each cell had a prospect of reproduction, have given rise to multicellular descendants in which some cells sacrifice this prospect?

The fundamental answer is kin selection: if the cell lineages in a multicellular organism arise by mitosis from a single-celled egg or zygote, the genes of cooperative cells that sacrifice reproduction for the good of the cell "colony" are propagated by closely related reproductive cells. However, the coefficient of relationship is reduced to the extent that mutational differences arise among cells. A mutation that increases the rate of cell division has a selective advantage *within* the colony, and may have a further advantage if it increases the chance that those cells that carry it will be included in the reproductive tissues and give rise to gametes. But unregulated cell division—as in cancer—usually harms the organism. Selection at the level of whole colonies of cells—organisms—therefore opposes selection among cells within colonies. It has favored mechanisms of "policing" that regulate cell division and prevent renegade cell genotypes from disrupting the integrated function of the organism. In animals, it has resulted in the evolution of a germ line that is segregated early in development from the soma, thereby excluding most disruptive mutations from the gametes and reducing their possible fitness advantage. Selection for organismal integration may be responsible for the familiar but remarkable fact that almost all organisms begin life as a single cell, rather than as a group of cells. This feature increases the kinship among all the cells of the developing organism, reducing genetic variation and competition within the organism and increasing the heritability of fitness. The result, then, has been the emergence of the "individual," and with it, the level of organization at which much of natural selection and evolution take place.

Human Behavior and Human Societies

Few subjects are more controversial than evolutionary approaches to human behavior, on which thousands of articles and books have been written. These topics are controversial partly because the questions are intrinsically difficult to answer, concerning as they do a species with extraordinarily complex behavior that cannot be studied in the same experimental ways as other species; partly because it can be difficult to sort out the well-informed hypotheses and studies from a great mass of poorly informed speculation and poorly analyzed data; and partly because many people (including many social scientists) are skeptical and often hostile to biological interpretations of human behavior. This reaction is easy to understand, because there is a long, ugly history of citing biology to justify human prejudice and oppression of non-Western peoples, "racial" groups, women, homosexuals, and others. Male dominance over women, prejudice against gay people, and imperialistic domination of other cultures did not start with Darwinism, but after 1859, in an abuse of evolutionary theory, they were justified by reference to so-called human nature. Racism has long been justified by arguments that its targets were genetically inferior, and the horrendous extremes to which this ideology was taken by the Nazis caused a revulsion against the idea of genetic determinism that has had a lasting effect.

Nevertheless, it is possible and useful to study human biology, including its implications for understanding human behavior, if it is done rigorously and with careful explication of the inferences that can and cannot be legitimately drawn from data. All the behaviors of any species, including humans, are biological phenomena at some level, since they originate in a physical, biological structure, the brain. Evolutionary biology, applied judiciously, is among the disciplines that can lead us to greater understanding. Finding a biological basis for a behavior or any other trait does not justify ill treatment. We do not abandon our ideals or lessen our respect for people by diagnosing the biological basis of a physical or mental disability; likewise, we need not abandon progressive ideals by finding a biological component in a widespread characteristic.

Much of the literature that takes an evolutionary approach to human behavior addresses three major topics. One is the degree to which *variation* among people, within and among populations, has a genetic versus nongenetic basis, and how the variation is best explained. Another is how to explain features that are said to be *universal characteristics* of the human species: did they arise by biological evolution (and therefore have a genetic basis), and if so, how and why? Third, how did the human capacity for culture arise, and how do cultural traits, in all their astonishing variety, change? In all of these contexts, it is critical to bear in mind that most human behaviors, like those of most other animals, are likely to be phenotypically plastic expressions, or norms of reaction to environmental conditions, of a genotype (see Chapters 9 and 13).

Variation in sexual orientation

An example of a variable behavioral trait that has prompted considerable evolutionary speculation is the continuum from exclusive heterosexual through degrees of bisexual to exclusive homosexual orientation. Much of the literature on this topic was strongly influenced by the extreme prejudice against gay men and lesbians that has been widespread in most Western cultures, and it is only relatively recently that sexual orientation has even become viewed as a legitimate topic for research. Many environmentalist psychological hypotheses have been advanced to account for homosexuality (although seldom for heterosexuality), such as the supposed influence of a distant father or a doting mother, but many researchers have concluded that none of these have convincing evidence (Wilson and Rahman 2005). There is considerable evidence that sexual orientation is partly inherited, although the heritability is clearly closer to 0.2 than to 1.0, so experiential factors, largely unknown, also affect it (Camperio-Ciani et al. 2004). What is clear is that sexual orientation begins to develop early in life; no contemporary researchers think that sexual orientation (as a self-identity) is chosen (Wilson and Rahman 2005). At least a small percentage of the male population in almost all cultures studied, throughout the world and throughout history, have expressed homosexuality (although not necessarily exclusively so), and in some cultures this has been the norm (Duberman et al. 1989). It seems possible, then, that sexual orientation has been variable since the spread of *Homo sapiens* throughout the world, if not earlier. Homosexual behavior has been recorded in more than 100 species of mammals (Roughgarden 2004), and is important in maintaining social alliances in bonobos (especially between females) and gorillas (especially between males; Vasey 1995).

To the extent that homosexual orientation has a genetic basis, it superficially seems to present an evolutionary enigma like sterile worker insects, since homosexuals are generally supposed to reproduce less than heterosexuals. This is actually a weak assumption, since it might well be the case that throughout most of human history, as in some societies both today and in the past (e.g., ancient Greece), homosexuals were accepted or tolerated as long as they played a socially expected procreative role, and thus may have been as reproductively prolific as heterosexuals. The frequency of alleles affecting sexual orientation today would be largely a consequence of a long history, when they might have been nearly selectively neutral. (This discussion concerns mostly males, since there is much less information, especially cross-cultural information, on female homosexuality.) Assuming, however, that male homosexuals have a lower average reproductive rate, several population genetic models could account for a stable polymorphism in sexual orientation (Gavrilets and Rice 2006; Camperio-Ciani et al. 2008). The models that qualitatively match the data best ascribe homosexual inclination to at least two loci, including at least one on the X chromosome, with the reproductive disadvantage of male homosexuality balanced by increased fecundity of females with the same X-linked allele. In a study in northern Italy, gay male subjects reported a significantly higher proportion of homosexual maternal relatives (5 percent) than paternal relatives (2 percent), consistent with inheritance on the X chromosome. Moreover, both the mothers and maternal aunts of gay men had significantly more children than those of heterosexual men (Figure 16.23; Camperio-Ciani et al. 2004).

These and other such data do not definitively demonstrate a genetic component of sexual orientation, nor the advantage any such alleles may have, but they are fully consistent with such a model. Some gay people will welcome such a conclusion, and others will not. In 1993, when a study purporting to demonstrate a "gay gene" on the X chromosome was published (data that subsequently were not replicated), many gay-rights advocates welcomed the news, since they considered it useful evidence against the widespread belief that homosexual orientation is freely chosen and can be altered. Others feared that identifying such a gene would enable genetic engineers to alter a person's future sexual orientation in utero, or enable parents to abort prospective gay offspring. However, environmentalist hypotheses (doting mother, etc.) have long been the basis of psychiatric interventions aimed at "curing" homosexuals (which the author and other LGBT* advocates strenuously oppose). We may infer that whatever the cause(s) of homosexual orientation may be, whether genetic, environmental, or both, the knowledge may well be used to the detriment of gay people if existing social prejudice remains and legal protections are lacking. In itself, scientific knowledge is neither good nor bad; whether it is put to good or bad ends depends on the ethical standards and compassion of those empowered to use it.

*LGBT: lesbian, gay, bisexual, transsexual

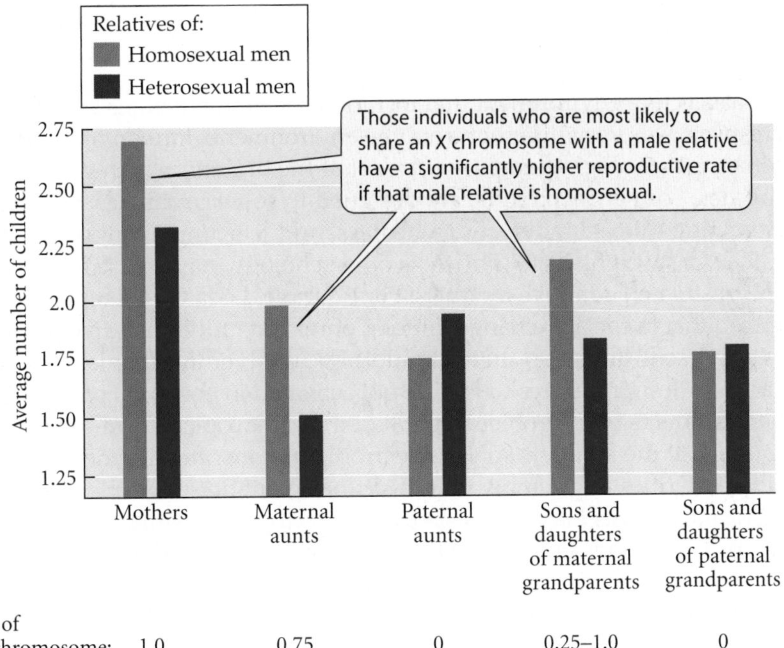

Figure 16.23 Average number of children of relatives of homosexual and heterosexual men in northern Italy. (Based on data in Camperio-Ciani et al. 2004.)

The question of human nature

Some human behavioral and cognitive traits display heritable variation within populations, as we saw in the case of IQ scores (Chapter 9) and probably sexual orientation. In contrast, culture, defined by Richerson and Boyd (2005) as "information capable of affecting individuals' behavior that they acquire from other members of their species through teaching, imitation, and other forms of social transmission," is responsible for the enormous variation in social practices among human populations. As we have seen, genetic differences among human populations are slight compared with variation within populations (Chapter 9), and there is abundant evidence that behavioral differences among populations are due to cultural divergence. The most difficult question is how to account for behavioral traits that are said to be universal among human populations, not those that vary. These include body adornment, cooperation, crying, death rituals, division of labor, various facial expressions, gossip, language, males more politically dominating and more aggressive, marriage, music, reciprocity, and many others (Pinker 2002). Especially since World War II, many or most social scientists have strongly favored the view that most universal social behaviors are culturally determined. The contrasting view is that many such behaviors have an innate, genetically determined foundation, the expression of which is often modulated, depending on cultural or other environmental influences: in other words, that behaviors are genotypic reaction norms with greater or lesser phenotypic plasticity, depending on the trait. For example, the capacity for language clearly has a genetic foundation that has evolved since the human lineage separated from our common ancestor with chimpanzees. Although captive chimpanzees, bonobos, and gorillas can learn and use sign language and other representational modes of communication, with an apparent command of syntax and representational meaning, they do not begin to match the abilities of even very young human children. So human language ability has evolved, but its expression—the actual language an individual speaks—is completely a matter of learning.

The evolutionary approach to studying human behavior was termed SOCIOBIOLOGY by E. O. Wilson (1975), an authority on social insects who favored applying evolutionary theories of behavior to humans as to other species. This field is more often referred to today as "human behavioral ecology," which aims to develop and test hypotheses about the adaptive function of behaviors that the proponents of this approach assume have evolved by the action of natural selection on genetic variation. "Evolutionary psychology" is closely related, but aims to understand how selection has shaped the psychological characteristics that underlie behavior. Both approaches emphasize that the genetic foundations of behavior were established early in the history of *Homo sapiens,* and perhaps its earlier ancestors, in a selective environment that included social interactions but which in most ecological respects was very different from the environments humans experience today.

Evolutionary psychologists employ a model of the brain in which genetically determined "modules," or neural circuits, are designed to solve certain classes of problems, such as interacting with kin, detecting cheaters, and learning a language (Tooby and Cosmides 1992; Cartwright 2000). The tools of psychology may be used to test hypotheses about the postulated adaptive modules. For example, Leda Cosmides and John Tooby (1992) proposed that in social exchanges among unrelated individuals (reciprocity), cheating is an ever-present threat, so there should have evolved mechanisms for detecting cheaters that have design features which are not activated in nonsocial contexts. They presented college students with problems that had the same logical form but different content, and found that the students solved the problem more often if it described cheating than if it did not. Critics might reply that this experiment does not rule out the possibility that American students *learn* to be especially sensitive to cheating, rather than having a genetically programmed sensitivity. Still, hypotheses gain support in science by making predictions, not otherwise expected, that are then matched by new data.

Adaptive hypotheses in human behavioral ecology are sometimes tested by looking for correlations among culturally different populations, a variant of the comparative method of testing for adaptation (see Chapter 11). This approach is particularly important in analyzing supposedly universal characteristics, since results from any one culture—such as the one to which the investigator belongs—may provide culture-specific results. An example in a particularly controversial area is a cross-cultural study by David Buss of mate-choice preferences among 37 diverse cultures in different parts of the world (Buss 1989). Buss aimed to test several hypotheses based on expectations from sexual selection: that females, more than males, should prefer mates with indicators (such as ambition and earning potential) of high potential for guarding and care of both mate and offspring; that men should place higher value than females on physically attractive potential mates (a supposed indicator of age-specific reproductive potential); and that males should value chastity in their mates more than women do (since extra-pair copulation places males at greater risk than females of unknowingly caring for another's offspring). Buss reported that a large majority of cultures conformed to every prediction. For example, women are said to value earning potential more than men do in 97 percent of the cultures, men prefer women younger than themselves in 100 percent, and men value chastity in their mates more than women do in 62 percent. This and many other such studies are consistent with adaptive evolutionary hypotheses about sex differences in behavior, but they do not rule out nonevolutionary interpretations. Another possible interpretation of Buss's results, for example, is that they reflect male domination of females, which might be universal simply because men are generally larger and stronger. Barbara Smuts (1995), though, notes that humans exhibit greater male dominance and male control of female sexuality than most other primates, and proposes several evolutionary hypotheses for this difference.

Adaptive evolutionary hypotheses have been proposed for many human behavioral and social features, from murder and jealousy to music-making and religious belief and ritual (see, e.g., Pinker 2002; Wilson 2002). It remains difficult, in many cases, to determine whether these characteristics are best explained by evolutionary (genetic) adaptation or by cultural adaptation, which might be expected to produce much the same phenotype, but by different means.

Cultural evolution and gene-culture coevolution

Differences in language, technologies, social customs, and way of life among the world's many different cultural groups clearly reflect a history of divergent change (Figure 16.24). Of the many scientists who have endeavored to understand how cultures change, Peter Richerson and Robert Boyd, of the University of California, have perhaps been the most inspired by models of genetic evolution to develop a judiciously analogous model of CULTURAL EVOLUTION, based on processes within and among populations (Richerson and Boyd 2005). Their theory, furthermore, includes the notion of coevolutionary effects of genetics and culture on each other. The hallmark of human culture, they say, is that it enables the

(A)

(B)

(C)

Figure 16.24 Wardrobe, customs, and technology are some of the cultural differences among contemporary human populations such as (A) the Huli of montane Papua New Guinea, (B) the Inuit of northern Canada, and (C) suburban professionals in the United States and western Europe. (A, photo by the author; B © Sue Flood/Photolibrary.com; C © O. Ortakcioglu/istockphoto.com.)

accumulation of adaptive behaviors over the course of many generations, so that societies acquire knowledge and capabilities far beyond what any one individual could invent or learn de novo. Only the human species has evolved this ability.

Just as individual genes mutate, Richerson and Boyd say, *nongenetic variations* of cultural traits are produced by individual people in a population. Some such changes are random, such as a mispronunciation of a word or misremembering a cultural item. Others are nonrandom, perhaps transformed during learning, or intentionally altered by an individual, possibly in an effort to improve a method. The frequencies of cultural variants may change by random fluctuation, "*cultural drift*" (e.g., the few practitioners of a special craft in a small population may die before having trained apprentices), or by *nonrandom transmission* or *selection*. A cultural variant increases in frequency if it is copied or learned by other individuals, i.e., is transmitted to them. Individuals may adopt or imitate a variant because of its content (it is perceived to be advantageous, or is simply easy to remember), because of features of the individuals who already exhibit the trait (e.g., copying prestigious or successful persons), or because the trait is already common and people perceive a reason to conform to the norm. Often, copying others' behavior is advantageous in the absence of any other information; or the society may actively enforce group norms by coercion.

Cultural traits are also subject to *natural selection*, if they affect people's behavior in ways that increase the probability of transmitting the trait to other people. (Recall that natural selection was defined in Chapter 11 as a consistent difference in reproductive success of phenotypically different classes of biological entities. Here, the entities are cultural traits, biological in the sense that they are produced by brains.) Selection may occur within populations if, for example, people acquire traits from their parents, and the traits affect the parents' survival, or reproduction. Hygienic or food-gathering traits can affect survival, for instance; religious beliefs and doctrines affect reproductive traits, even in modern societies. Traits that are acquired by imitating celebrities, teachers, and other nonparental figures may be selected if they affect the likelihood of becoming such a figure. Richerson and Boyd emphasize that cultural differences among groups, such as tribes, can evolve by group selection as a result of competition among groups. Even if groups are very large, they suggest, group selection can be more effective in cultural than in genetic evolution because cultural traits are often very homogeneous within groups, because group norms may be forcibly maintained. A homogeneous culture of Christianity, for example, has been maintained in many populations not only by inheritance of parents' beliefs but also by policing that ranges from the Inquisitions of the past to social exclusion of nonconformists in some societies today. The ways in which traits may be group-selected include differential population growth based on agricultural practice and superiority in warfare. Darwin envisioned such group selection in *The Descent of Man* (1874), writing that "although a high standard of morality gives but a slight or no advantage to each individual man and his children over other men of the same tribe,...an increase in the number of well-endowed men and an advancement in the standard of morality will certainly give an immense advantage to one tribe over another."

Many traits, however, will spread by biased transmission rather than selection, and although they may well be advantageous in a particular ecological or social environment, they need not be. Maladaptive traits can spread if they have a conformist bias (e.g., belief in witchcraft) or a prestige bias (e.g., smoking among teenagers), and most people at most times in history have not had any information that would enable them to tell whether or not a belief (e.g., in witches) or cultural practice (e.g., sacrificing to the gods) was true or beneficial.

Many cultural traits concern marriage rules, cooperation, conflict resolution, and other social interactions that are the subject of adaptive hypotheses in human behavioral ecology and evolutionary psychology. Richerson and Boyd suggest that genes and culture may have coevolved. They note that large brains and cumulative cultural adaptation both seem to have developed about 500,000 years ago, so cultural environments could have imposed selection on genes for about 20,000 generations. For example, mutations that enable adults to digest lactose were selected in dairying cultures (see Chapter 12).

Language may epitomize gene-culture coevolution. Humans have not only an innate capacity for language that is far greater than in other apes, but also physical adaptations such as the position of the larynx, which is lower in the throat than in apes, enabling a greater variety of sounds. These features must have evolved because there was selection for greater linguistic versatility, probably based on advantages in social interactions. This selection could have occurred only in a population that already was using rudimentary language of some kind. The subsequent changes in larynx position and in language centers of the brain in turn enabled languages to develop greater complexity by cultural change.

Similarly, there may have been selection for genotypes with a greater predisposition to adopt certain kinds of cultural traits or for characteristics that influence the processes of cultural change. If cultural traits that promote cooperation were advantageous to survival or reproduction (of individuals or of groups), genotypes more inclined to cooperate would increase in frequency. If conforming to group beliefs generally enhanced survival, a genetic tendency toward conformity could evolve. To some extent, then, some cognitive features such as those postulated by evolutionary psychologists might have evolved, but in cultural contexts that themselves could change. Richerson and Boyd suggest that we might have two sets of innate predispositions that evolved genetically. The older set, shared with many of our primate relatives, evolved by kin selection and reciprocity, and concerns interactions among family members, mate preferences, and relatively simple social interactions. The younger set of such predispositions, evolved during the last few hundred thousand years or less, may involve features such as true empathy and altruism that are advantageous in the context of large tribes and societies, in which simple reciprocity is ineffective. But these are predispositions only, and allow for an immense range of cultural expressions and individual potentialities. No one should fear that such evolved biological foundations of our psychology and social constructions dictate what we can, much less what we should, do with our lives.

Summary

GO TO THE **evolution** Companion Web Site at **www.sinauer.com/evolution** for quizzes, data analysis and simulation exercises, and other study aids.

1. Many biological phenomena result from conflict or cooperation among organisms or among genes. The evolution of most interactions can be explained best by selection at the level of individual organisms or genes.
2. Characteristics that contribute to conflict and cooperation often evolve by frequency-dependent selection. Such features can sometimes be modeled by calculating the evolutionarily stable strategy (ESS): the phenotype that, once established, cannot be replaced by mutant phenotypes. In animals, interactions are often mediated by signals that may or may not honestly indicate the individual's strength, potential parental caregiving, or other relevant variable.
3. Altruism benefits other individuals and reduces the fitness of the cooperative individual, whereas cooperative behavior need not reduce the actor's fitness. Cooperation can evolve because it is directly beneficial to the actor, although the benefit may be delayed, or by reciprocity, based on repeated interactions between individuals or on long-lasting associations in which the fitness interests of the associates are aligned. Altruism generally evolves by kin selection, based on differences among alleles or genotypes in inclusive fitness: the combination of an allele's direct effects (on its carrier's fitness) and its indirect effects (on other copies of the allele, borne by the carrier's kin). Hamilton's rule describes the condition for increase of an allele for an altruistic trait in terms of the coefficient of relationship, the benefit to the beneficiary, and the cost to the donor. Under multilevel selection, short-lived groups containing altruistic genotypes may contribute disproportionately to the entire population, but these groups generally consist of kin. Thus this form of group selection is generally the same as kin selection.
4. Cooperative interactions, both within and between species, are often maintained in part by "policing," or punishment of cheaters.
5. Conflict and kin selection together affect the evolution of many interactions among family members. The genetic benefit of caring for offspring is an increase in the number of current offspring that survive. The genetic cost is the number of additional offspring that the parent could expect to have if she/he abandoned the offspring and reproduced again. Parental care is expected to evolve only if its genetic benefit exceeds its genetic cost. Whether or not one or both parents evolve to provide care can depend on the male's confidence of paternity and on the relative cost/benefit ratio for each parent.
6. Conflicts between parents and offspring may arise because a parent's fitness may be increased by allocating some resources to its own survival and future reproduction, thus providing fewer resources to current offspring than would be optimal from the offspring's point of view. This principle may be one of several reasons why in many species, parents may reduce their brood size by aborting some embryos, killing some offspring, or allowing siblicide among their offspring.
7. Conflicts may exist among different genes in a species' genome. For example, a gene may spread at a faster rate than other parts of the genome, engendering selection for other genes to prevent it from doing so. At loci that are transmitted through only one sex, gene-level selection favors alleles that alter the sex ratio in favor of that sex. Such alteration creates selection at other loci for suppressors that restore the 1:1 sex ratio.
8. Other phenomena explained by genetic conflicts include genomic imprinting, which affects the expression of maternally and paternally derived alleles in mammalian embryos, and the evolution of integration among cells—the very essence of multicellular organisms.
9. The most extreme examples of cooperation and altruism are in eusocial insects and humans. In eusocial insects, in which unmated workers rear other workers and reproductive individuals (queens and males), many interactions are governed by kin selection and by policing.
10. The extent to which human behaviors, including social behaviors, have an evolved genetic foundation is highly controversial. It is most readily addressed in studies of traits that vary within populations. Variation in some behavioral traits, such as sexual orientation, appears to have a heritable component, but unknown environmental factors also contribute to variation. Behavioral differences among human populations appear to be attributable to nongenetic factors, including culture.
11. Human traits that are claimed to be universal, although usually variable in expression (e.g., language, body adornment), are likely to have some genetic foundation that has evolved since our common ancestor with other apes. How trait-specific the genetic bases may be is unknown.

Evolutionary psychologists propose that the brain includes many rather problem-specific psychological "modules." Another view is that we have broad genetic predispositions, some of which are held in common with other apes (e.g., altruism toward close kin) and others of which evolved more recently by natural selection stemming from cultural environments (e.g., capacity for empathy with strangers in large societies). Cultures evolve nongenetically by processes that are partly analogous to genetic evolution, and cultural and genetic changes may influence each other. Such genetic predispositions as may constitute so-called human nature, however, do not evidently constrain or limit any groups of people more than others.

Terms and Concepts

altruism
coefficient of relationship
cooperation
genetic conflict
Hamilton's rule
horizontal transmission
inclusive fitness
kin selection
multilevel selection
mutualism
parent-offspring conflict
reciprocity
vertical transmission

Suggestions for Further Reading

An outstanding, easily readable introduction to the study of behavior, emphasizing evolutionary explanations, is *Animal Behavior: An Evolutionary Approach*, by J. Alcock (ninth edition, Sinauer Associates, Sunderland, MA, 2009). An excellent set of essays on many aspects of cooperation and conflict is *Levels of Selection*, edited by L. Keller (Princeton University Press, Princeton, NJ, 1999); see also J. L. Sachs et al., "The evolution of cooperation" (*Quarterly Review of Biology* 79:135–160, 2004) and L. Lehmann and L. Keller, "The evolution of cooperation and altruism—a general framework and a classification of models" (*Journal of Evolutionary Biology* 19:1365–1376, 2006). Genetic conflict is reviewed in A. R. Burt and R. Trivers, *Genes in Conflict: The Biology of Selfish Genetic Elements* (Harvard University Press, Cambridge, MA, 2006). J. Maynard Smith and E. Szathmáry provide an outstanding analysis of the origin of levels of organization, from cells to societies, in *The Major Transitions in Evolution* (Oxford University Press, New York, 1995). A few of the many books on evolution and human behavior are J. Cartwright, *Evolution and Human Behavior* (MIT Press, Cambridge, MA, 2000), J. H. Barkow, L. Cosmides, and J. Tooby (eds.), *The Adapted Mind: Evolutionary Psychology and the Generation of Culture* (Oxford University Press, New York, 1992), S. Pinker, *The Blank Slate: The Modern Denial of Human Nature* (Penguin, New York, 2002), P. J. Richerson and R. Boyd, *Not by Genes Alone: How Culture Transformed Human Evolution* (University of Chicago Press, 2005), and J. Roughgarden, *Evolution's Rainbow: Diversity, Gender, and Sexuality in Nature and People* (University of California Press, Berkeley, 2004).

Problems and Discussion Topics

1. Describe the contexts in which animals might evolve signals used in interactions among members of the same species or in interactions with different species. Would you expect each kind of signal to be honest or dishonest? Why?
2. Many species of albatrosses and other seabirds nest in colonies on islands. The adults search for food at sea, sometimes over great distances. Some ecologists who study seabirds turn to their favorite topic of conversation in a bar one evening, and one says, "I can imagine an albatross genotype that destroys the eggs or nestlings of its colonymates while the parents are away foraging, and leaves them dead without eating them. Do you think such a behavior would evolve?" One of his companions says, "Yes, because it would make more food available for the albatross and its offspring." Another says, "No, because it would threaten the survival of the population." The fourth says, "You're both wrong. I have a different explanation for why albatrosses don't kill others' chicks." What is the fourth ecologist's explanation, and why does she think her companions are wrong?
3. Speculate about why some genetic elements (e.g., mitochondria and chloroplasts) are usually inherited through only one parent rather than biparentally, as are most nuclear genes. (See Hurst et al. 1996.)

4. Kin selection explains why organisms may provide benefits to relatives. Is there a conflict between the principle of kin selection and the evolution of siblicide and abortion?
5. Analyze reasons for and against the proposition that men are more aggressive than women because of sexual selection for competitiveness. (See, for example, Daly and Wilson 1983 versus Kitcher 1985.)
6. Standardized for body weight, testes are larger in polygamous than in monogamous species of primates and several other groups of animals (see Figure 11.22). This approach has been used to infer what the predominant mating system of *Homo sapiens* has been for most of its evolutionary history (see, e.g., Harcourt et al. 1981; Harvey and May 1989). What has been the conclusion, and how might we evaluate its validity?
7. Why is it more difficult to test the hypothesis that a species-typical trait (such as the supposedly universal human behavioral traits discussed in this chapter) has an evolved genetic foundation than a variable trait? Is it sufficient to conclude that a trait has a genetic foundation if it differs consistently between two closely related species, such as human and chimpanzee?
8. Many cultural traits that do not have a genetic basis are adaptive, such as various methods of obtaining and preparing food and shelter. Is the propensity to believe in powerful immaterial entities, such as spirits and gods, a cultural adaptation? How might it have (culturally or biologically) evolved, and how might the hypotheses be tested? See Wilson 2002 and Dennett 2006.
9. What are the similarities and differences between the likely processes of biological evolution and cultural evolution?
10. Traditional group selection (referred to as "multilevel selection 2" on p. 423) explained cooperative behaviors as the consequence of the enhanced survival of populations of cooperative or altruistic genotypes, compared with populations of selfish genotypes. Describe at least three kinds of observations that are more consistent with kin selection, reciprocity, or direct benefit than with traditional group selection.

CHAPTER 17

Species

Two species that can hybridize. Although the indigo bunting (*Passerina cyanea*) and the lazuli bunting (*Passerina amoena*) differ in plumage, they hybridize where their ranges overlap. (*P. cyanea* © William Leaman/Alamy; *P. amoena* © Jim Zipp/Photo Researchers, Inc.)

Speciation forms the bridge between the evolution of populations and the evolution of taxonomic diversity. The diversity of organisms is the consequence of CLADOGENESIS, the branching or multiplication of lineages, each of which then evolves (by ANAGENESIS, or evolution within species) along its own path. Each branching point in the great phylogenetic tree of life marks a speciation event: the origin of two species from one. In speciation lies the origin of diversity, and the study of speciation bridges microevolution and macroevolution. The most important consequence of speciation is that different species undergo largely independent divergence, maintaining separate identities, evolutionary tendencies, and fates (Wiley 1978). It is also possible that speciation facilitates the evolution of new morphological and other phenotypic characters.

Many events in the history of evolution are revealed to us only by virtue of speciation. If a single lineage evolves great changes but does not branch, the record of all steps toward its present form is erased, unless they can be found in the fossil record. But if the lineage branches frequently, and if intermediate stages of a character are retained in some branches that survive to the present, then the history of evolution of the feature may be represented, at least in part, among living species. This fact is used routinely to infer phylogenetic relationships among living taxa and to trace the evolution of characteristics on phylogenetic trees (see Figure 3.3).

Table 17.1 *Some species concepts*

Biological species concept Species are groups of actually or potentially interbreeding natural populations that are reproductively isolated from other such groups (Mayr 1942).

Evolutionary species concept A species is a single lineage (an ancestor–descendant sequence) of populations or organisms that maintains an identity separate from other such lineages and which has its own evolutionary tendencies and historical fate (Wiley 1978).

Phylogenetic species concepts (1) A phylogenetic species is an irreducible (basal) cluster of organisms that is diagnosably distinct from other such clusters, and within which there is a parental pattern of ancestry and descent (Cracraft 1989). (2) A species is the smallest monophyletic group of common ancestry (de Queiroz and Donoghue 1990).

Genealogical species concept Species are "exclusive" groups of organisms, where an exclusive group is one whose members are all more closely related to one another than to any organism outside the group (Baum and Shaw 1995).

Recognition species concept A species is the most inclusive population of individual biparental organisms that share a common fertilization system (Paterson 1985).

Cohesion species concept A species is the most inclusive population of individuals having the potential for phenotypic cohesion through intrinsic cohesion mechanisms (Templeton 1989).

Some steps toward speciation may occur fast enough for us to study directly, but the full history of the process is usually too prolonged for one generation, or even a few generations, of scientists to observe. Conversely, speciation is often too fast to be fully documented in the fossil record, and even an ideal fossil record could not document some of the genetic processes in speciation that are still inadequately understood. Thus the study of speciation is based largely on inferences from living species.

What Are Species?

Many definitions of "species"—which is Latin for "kind"—have been proposed (Table 17.1). It is important to bear in mind that a definition is not true or false, because the definition of a word is a convention. Still, if a conventional definition of a word has been established, one can apply it in error. Although a rose by any other name would smell as sweet, we would be wrong to call a rose a skunk cabbage. A definition can be more or less useful, and it can be more or less successful in accurately characterizing a concept or an object of discussion. Probably no definition of "species" suffices for all the contexts in which a species-like concept is used. Jerry Coyne and Allen Orr (2004), authors of a comprehensive book on speciation, note that species can be defined in a way that (1) enables us to classify organisms systematically, (2) corresponds to discrete groups of similar organisms, (3) helps us understand how discrete clusters of organisms arise in nature, (4) represents products of evolutionary history, and/or (5) applies to the largest possible variety of organisms. It turns out that no two of these possible goals always coincide, and as Coyne and Orr point out, it is unlikely that any one species concept will serve most of these purposes.

Linnaeus and other early taxonomists held what Ernst Mayr (1942, 1963) has called a TYPOLOGICAL or ESSENTIALIST notion of species. Individuals were members of a given species if they sufficiently conformed to that "type," or ideal, in certain morphological characters that were "essential" fixed properties—a concept descended from Plato's "ideas" (see Chapter 1). Thus a bird specimen was a member of the species *Corvus corone*, the carrion crow, if it looked like a carrion crow, and was a common raven (*Corvus corax*) if it had certain different features (Figure 17.1A,B). But typological notions of species were challenged by variation. Carrion crows, which are entirely black, are readily distinguished from hooded crows (*Corvus cornix*), which are black and gray (Figure 17.1C)—except in a narrow region in central Europe, where crows have various amounts of gray (Figure 17.1D). Are there two species or one?

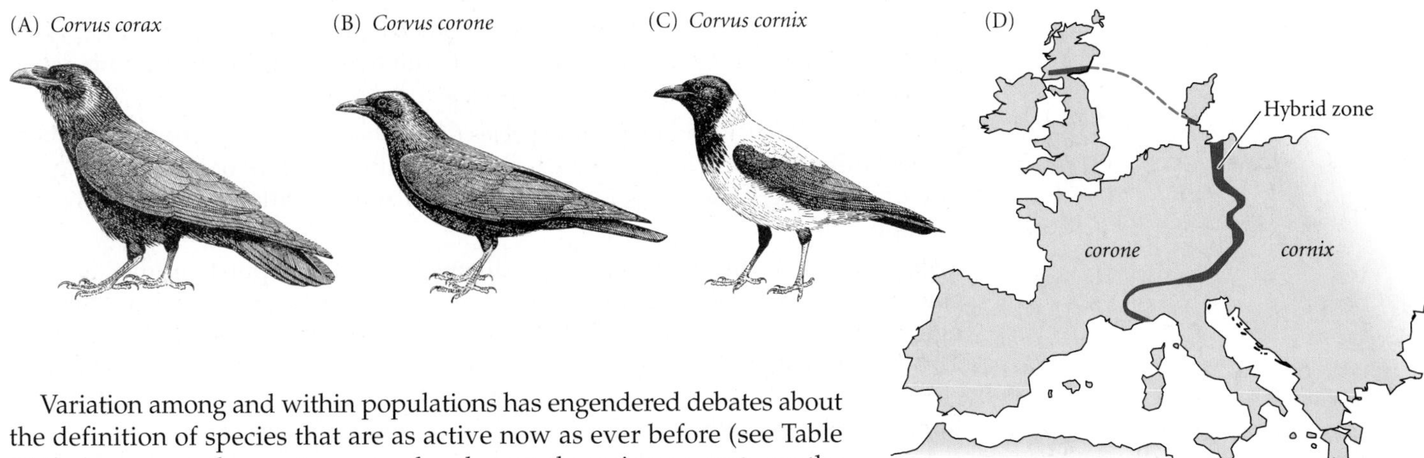

Figure 17.1 Three closely related birds. (A) Common ravens (*Corvus corax*) are distinguished from crows by their larger size, heavier bill, shaggy throat feathers, more pointed tail (when viewed from above or below), and voice. Even though (B) carrion crows (*Corvus corone*) and (C) hooded crows (*C. cornix*) are superficially more different from each other than from the common raven, hybrids showing various mixtures of their plumage patterns are found in central Europe (D). *C. corone* and *C. cornix* have been classified by some taxonomists as subspecies of a single species, but because they appear to exchange genes only to a very limited extent, they might better be considered species. (A–C from Goodwin 1986; D after Mayr 1963.)

Variation among and within populations has engendered debates about the definition of species that are as active now as ever before (see Table 17.1). At present, the most commonly advocated species concepts are the **biological species concept (BSC)** and several variations on the **phylogenetic species concept (PSC)**. These differ chiefly in that phylogenetic species concepts emphasize species as the outcome of evolution—the *products* of a history of evolutionary divergence—whereas the biological species concept emphasizes the *process* by which species arise and takes a prospective view of the future status of populations (Harrison 1998). The biological species concept and closely related definitions have been and continue to be most frequently used by evolutionary biologists who are concerned with processes of evolution. No matter which species concept is adopted, however, some populations of organisms will not be unambiguously assigned to one species or another. There are always borderline cases, because the species properties advanced in every definition evolve gradually.

Phylogenetic species concepts

Phylogenetic species concepts, which are gaining acceptance, especially among systematists, emphasize the phylogenetic history of organisms. One of several definitions of a phylogenetic species is "an irreducible (basal) cluster of organisms diagnosably different from other such clusters, and within which there is a parental pattern of ancestry and descent" (Cracraft 1989). This definition would presumably apply to both sexual and asexual organisms. According to this definition, speciation would occur whenever a population undergoes fixation of a genetic difference—even a single DNA base pair—that distinguishes it from related populations. The study of speciation, then, would be simply the study of divergence between populations.

The biological species concept

The biological species concept used throughout this book was defined by Ernst Mayr (1942): "*Species are groups of actually or potentially interbreeding populations, which are reproductively isolated from other such groups.*" "Reproductive isolation" means that any of several biological differences between the populations greatly reduce gene exchange between them, even when they are not geographically separated. These differences might or might not include mortality or sterility of hybrids between the two forms. Mayr (1942) and other advocates of the BSC have not insisted that populations must be 100 percent reproductively isolated in order to qualify as species; they recognize that, as in the case of the crows in Figure 17.1, there can be a little genetic "leakage" between some species.

The roots of the biological species concept are very old, for it has always been recognized that morphologically very different organisms (e.g., different sexes) might be born of the same parents and thus be **conspecific** (members of the same species). Nevertheless, the BSC arose from studies of variation that showed that morphological similarity and difference do not suffice to define species. Several critical observations contributed to its development:

1. *Variation within populations.* Characteristics vary among the members of a single population of interbreeding individuals. The white and blue forms of the snow goose (see

Figure 9.1A), known to be born to the same mother, represent a genetic polymorphism, not different species. A mutation that causes a fruit fly to have four wings rather than two is just that: a mutation, not a new species.

2. *Geographic variation.* Populations of a species differ; there exists a spectrum from slight to great difference; and intermediate forms, providing evidence of interbreeding, are often found where such populations meet. Human populations are a conspicuous example.
3. *Sibling species.* **Sibling species** are reproductively isolated populations that are difficult or impossible to distinguish by morphological features, but which are often recognized by differences in ecology, behavior, chromosomes, or other such characters. The discovery that the European mosquito *Anopheles "maculipennis"* was actually a cluster of six sibling species had great practical importance because some transmit human malaria and others do not. (Box 17A provides definitions of sibling species and some other terms related to the biological species concept; later in the chapter, a second box will describe the diagnosis of a sibling species.)

Domain and application of the biological species concept

All concepts have limitations. A concept may have a limited *domain of application* (for example, "matter" is ambiguous on a subatomic scale). A concept may inadequately describe BORDERLINE CASES. (The concept of an "individual organism," for instance, is ambiguous in a grove of aspen trees that have grown by vegetative propagation from a

BOX 17A Some Terms Encountered in the Literature on Species

Some of the following terms are used frequently in the literature on species, others infrequently. These definitions conform to usage by adherents to the biological species concept (e.g., Mayr 1963).

Geographic isolation Reduction or prevention of gene flow between populations by an extrinsic barrier to movement, such as topographic features or unfavorable habitat.

Reproductive isolation Reduction or prevention of gene flow between populations by genetically determined differences between them.

Allopatric populations Populations occupying separated geographic areas.

Parapatric populations Populations occupying adjacent geographic areas, meeting at the border.

Sympatric populations Populations occupying the same geographic area and capable of encountering one another.

Hybrid zone A region where genetically distinct populations meet and interbreed to some extent, resulting in some individuals of mixed ancestry ("hybrids").

Introgression The movement, or incorporation, of genes from one genetically distinct population (usually considered a species or semispecies) into another.

Sibling species Reproductively isolated species that are difficult to distinguish by morphological characteristics.

Sister species Species that are thought, on the basis of phylogenetic analysis, to be each other's closest relatives, derived from an immediate common ancestor. (Cf. sister groups in phylogenetic systematics.)

Chronospecies Phenotypically distinguishable forms in an ancestor-descendant series in the fossil record that are given different names.

Subspecies Populations of a species that are distinguishable by one or more characteristics and are given subspecific names (see the *Elaphe* subspecies, Figure 9.26). In zoology, subspecies have different (allopatric or parapatric) geographic distributions and are equivalent to "geographic races." In botany, they may be sympatric forms.

Race A vague term, sometimes equivalent to subspecies and sometimes to polymorphic genetic forms within a population.

Semispecies Usually, one of two or more parapatric, genetically differentiated groups of populations that are thought to be partially, but not fully, reproductively isolated; nearly, but not quite, different species.

Superspecies Usually, the aggregate of a group of semispecies. Sometimes designates a group of closely related allopatric or nearly allopatric forms that are designated different taxonomic species.

Ecotype Used mostly in botany to designate a phenotypic variant of a species that is associated with a particular type of habitat; may be designated a subspecies.

Polytypic species A geographically variable species, often divided into subspecies. (Most species are polytypic, whether or not subspecies have been named.)

single seed.) There may also be *practical limitations* in applying the concept. (The "human population of the world" is a presumably valid concept, but technological and economic limitations prevent us from counting the population accurately.)

DOMAIN. The domain of the BSC is restricted to sexual, outcrossing organisms. It is also limited to short intervals of time, since it is meaningless to ask whether an ancestral population could have interbred with its descendants a million years later. To be sure, asexually reproducing organisms are given names, such as *Escherichia coli*, and putative ancestral and descendant populations in the fossil record may be distinguished by names such as the mid-Pleistocene *Homo erectus* and the later *Homo sapiens*. These cases illustrate that the word "species" has two overlapping but distinct meanings in biology. One meaning is embodied in the BSC. The other meaning is *a taxonomic category*, just like "genus" and "family." Some organisms that bear binomial names (such as *Escherichia coli*) are taxa in the species category, but are not biological species.

BORDERLINE CASES. Interbreeding versus reproductive isolation is not an either/or, all-or-none distinction. There exist graded levels of gene exchange among adjacent (parapatric) populations and sometimes between sympatric, more or less distinct populations. Several such situations are encountered frequently.

Narrow **hybrid zones** exist where genetically distinct populations meet and interbreed to a limited extent, but in which there exist partial barriers to gene exchange (see Figure 17.1D). The hybridizing entities are often recognized as species but may be called **semispecies**. A collection of semispecies is a **superspecies** (Mayr 1963).

Partial, but not completely free, gene exchange sometimes occurs between populations that are rather broadly sympatric (sympatric hybridization). Sympatric hybridization seems more common in plants than in animals, and it is one reason why many botanists have been reluctant to adopt the biological species concept. For example, different species of oaks (*Quercus*) often hybridize to some extent (Figure 17.2). However, advocates of the BSC recognize that a low level of gene exchange may occur between closely related species. Frequently, species that hybridize do so in some localities but not others, often depending on environmental factors.

PRACTICAL DIFFICULTIES. The greatest practical limitation of the BSC lies in determining whether or not geographically segregated (allopatric) populations belong to the same species, for applying the BSC requires us to assess whether or not they would *potentially* interbreed if they should someday encounter each other. "Potential interbreeding" is

Figure 17.2 An example of sympatric hybridization. The gray oak (*Quercus grisea*) and Gambel's oak (*Q. gambelii*) have broadly overlapping ranges in the southwestern United States, including much of Texas, New Mexico, Arizona, and Colorado. Hybrids showing variation in leaf shape and other features are found in many of these places. (Photos courtesy of M. Cain.)

an integral part of the BSC because the heart of this species concept lies in the idea that populations with intrinsic barriers to gene exchange can undergo independent evolutionary change, even if they should become sympatric. Moreover, it clearly would be absurd to distinguish as a species every population that is geographically, but perhaps very temporarily, isolated.

In many instances, range extension or colonization could well bring presently separate populations into contact, so the evolutionary future of the populations depends on whether or not they have evolved reproductive isolation. Humans have inadvertently or purposely introduced many species into new areas, some of which have hybridized with native populations (Abbott 1992), and many native species have extended their ranges considerably over the course of decades. After the Pleistocene glaciations, disjunct populations of a great many species expanded their ranges, met one another, and in many instances now interbreed. Future changes in climate will undoubtedly affect some currently disjunct populations in similar ways.

In principle, one could experimentally test for reproductive isolation between allopatric populations, and such tests have been performed with *Drosophila* and many other organisms by bringing them together in a laboratory or garden. For many organisms, however, such tests are not feasible (although the mere impracticality of such studies does not invalidate the concept of reproductive isolation). In practice, therefore, the classification (i.e., naming) of allopatric populations is often somewhat arbitrary. Commonly, allopatric populations have been classified as species if their differences in phenotype or in DNA sequence are as great as those usually displayed by sympatric species in the same group.

Taxonomic versus biological species

Loren Rieseberg and collaborators (2006) synthesized data on phenotypic variation and crossability of named taxonomic species in more than 400 genera of plants and animals. They found that in more than 80 percent of genera, specimens fall into discrete phenotypic clusters, as expected if gene exchange between clusters is strongly reduced. In most major groups of both animals and plants, the correspondence between phenotypic clusters and named species was rather poor, but this is not due to hybridization or gene exchange. Instead, it is attributable to asexual reproduction (in which case the BSC does not apply), to polyploidy (forms that differ in the number of sets of chromosomes are reproductively isolated, but usually have not been named as distinct species), or to "oversplitting" by taxonomists who have named species on the basis of slight morphological differences that do not constitute discretely different populations. Moreover, more than 70 percent of phenotypically discrete forms of both animals and plants show postzygotic reproductive isolation when experimentally tested. Rieseberg et al. conclude that biological species are equally prevalent in both plants and animals. In fact, contrary to widespread belief, taxonomic species of plants are even more likely than those of animals to consist of reproductively isolated lineages (Figure 17.3). Although natural hybridization between plant species is not uncommon, it is concentrated in a small minority of plant families and genera (Ellstrand et al. 1996).

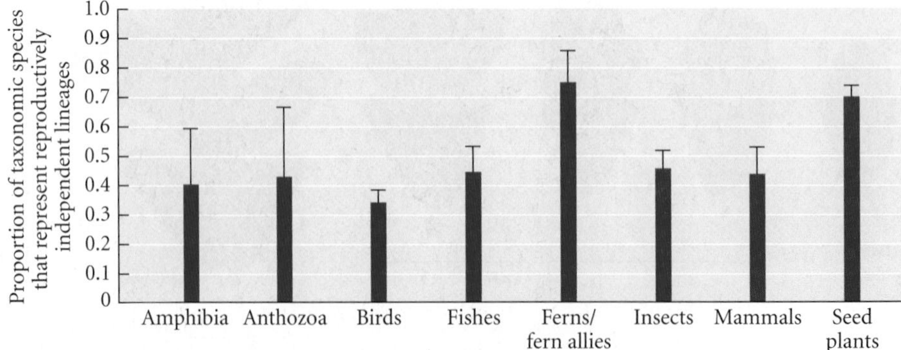

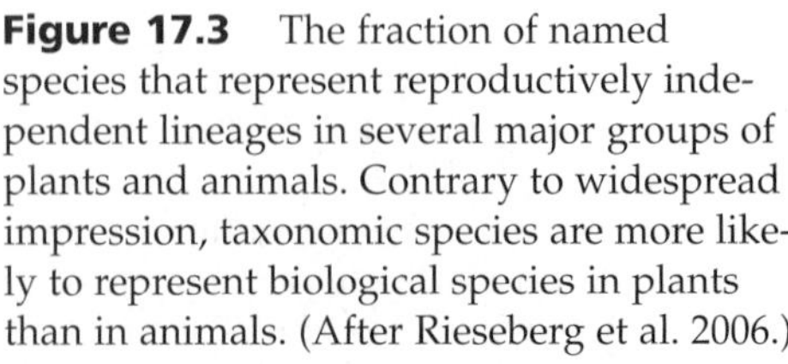
Figure 17.3 The fraction of named species that represent reproductively independent lineages in several major groups of plants and animals. Contrary to widespread impression, taxonomic species are more likely to represent biological species in plants than in animals. (After Rieseberg et al. 2006.)

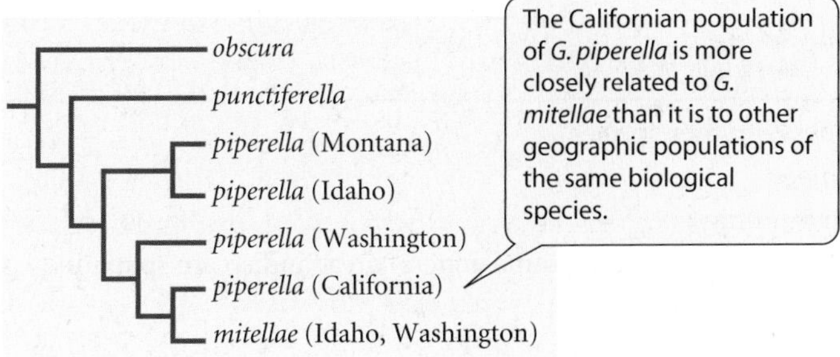
obscura
punctiferella
piperella (Montana)
piperella (Idaho)
piperella (Washington)
piperella (California)
mitellae (Idaho, Washington)
The Californian population of G. piperella is more closely related to G. mitellae than it is to other geographic populations of the same biological species.

Figure 17.4 The phylogeny of some species and populations in the moth genus *Greya*, based on mitochondrial DNA sequence data. This pattern would be expected if a local population of ancestral *G. piperella* evolved reproductive isolation and became the distinct biological species *G. mitellae*. According to the biological species concept, *G. mitellae* and *G. piperella* would be recognized as two species, one of which (*G. piperella*) is paraphyletic. According to the phylogenetic species concept, if *G. mitellae* is recognized as a species, the various populations of *G. piperella* should also be called species. (After Harrison 1998.)

When species concepts conflict

Advocates of the BSC and the PSC are most likely to classify populations differently in two circumstances. First, allopatric populations that can be distinguished by fixed characters are species according to the PSC, but if the diagnostic differences are slight, advocates of the BSC may recognize the populations as geographic variants of a single species.

Second, in some cases a local population of a widespread species evolves reproductive isolation from other populations, which remain reproductively compatible with one another. Phylogenetic study may then show that the "new" species is more closely related to some populations of the "old" species than some of the "old" populations are to one another. Under the BSC, two species would be recognized, one of which is paraphyletic (Figure 17.4). (Recall from Chapter 3 that a paraphyletic taxon lacks one or more of the descendants of the common ancestor of the members of that taxon, and that paraphyletic taxa are not acceptable under cladistic rules of classification.) Under the PSC, the various distinguishable populations of the paraphyletic group might be named as distinct species.

Barriers to Gene Flow

Gene flow between biological species is largely or entirely prevented by biological differences that have often been called ISOLATING MECHANISMS, but which we will term **isolating barriers**, or BARRIERS TO GENE FLOW. Under the BSC, therefore, *speciation—the origin of two species from a common ancestral species—consists of the evolution of biological barriers to gene flow*. As we noted above, mere physical isolation does not define populations as different species, although isolation by topographic or other barriers is considered to be instrumental in the formation of species.

It is an error to think that sterility of hybrids is the criterion of species as conceived in the BSC. There are many kinds of isolating barriers (Table 17.2). The most important distinction is between **prezygotic** and **postzygotic barriers**. Most prezygotic barriers are premating barriers, although some species are isolated by postmating, prezygotic barriers. Any of these barriers may be incomplete; for example, interspecific mating may occur at a low rate, or hybrid offspring may have reduced fertility (be "partially sterile").

Premating barriers

Premating barriers prevent (or reduce the likelihood of) transfer of gametes to members of other species.

ECOLOGICAL ISOLATION. Many species breed at different times of year (seasonal isolation). Two closely related field crickets (*Gryllus pennsylvanicus* and *G. veletis*), for example, reach reproductive age in the fall and spring, respectively, in the northeastern United States (Harrison 1979). Some species are isolated by habitat, so potential mates seldom meet. For instance, two Japanese species of herbivorous ladybird beetles (*Henosepilachna nipponica* and *H. yasutomii*) feed on thistles (*Cirsium*) and blue cohosh (*Caulophyllum*), respectively. Each species mates exclusively on its own host plant, and this ecological segregation appears to be the only barrier to gene exchange (Katakura and Hosogai 1994). Disturbance of habitat sometimes results in a breakdown of ecological isolation; for example, the wild

Table 17.2 *A classification of isolating barriers*

I. *Premating barriers:* Features that impede transfer of gametes to members of other species
- A. Ecological isolation: Potential mates (although sympatric) do not meet
  1. Temporal isolation (populations breed at different seasons or times of day)
  2. Habitat isolation (populations have propensities to breed in different habitats in the same general area, and so are spatially segregated)
- B. Potential mates meet but do not mate
  1. Behavioral (sexual or ethological) isolation (in animals, differences prevent populations from mating)
  2. Pollinator isolation (in plants, populations transfer pollen by different animal species or on different body parts of a single pollinator; may also be classified as ecological isolation)

II. *Postmating, prezygotic barriers:* Mating or gamete transfer occurs, but zygotes are not formed
- A. Mechanical isolation (copulation occurs, but no transfer of male gametes takes place because of failure of mechanical fit of reproductive structures)
- B. Copulatory behavioral isolation (failure of fertilization because of behavior during copulation or because genitalia fail to stimulate properly)
- C. Gametic isolation [failure of proper transfer of gametes or of fertilization, either due to intrinsic incompatibility or to competition between conspecific and heterospecific gametes (conspecific sperm precedence or pollen tube precedence)]

III. *Postzygotic barriers:* Hybrid zygotes are formed but have reduced fitness
- A. Extrinsic (hybrid fitness depends on context)
  1. Ecological inviability (hybrids do not have an ecological niche in which they are competitively equal to parent species)
  2. Behavioral sterility (hybrids are less successful than parent species in obtaining mates)
- B. Intrinsic (hybrid fitness is low because of problems that are relatively independent of environmental context)
  1. Hybrid inviability (developmental problems cause reduced survival)
  2. Hybrid sterility (usually due to reduced ability to produce viable gametes; also "behavioral sterility," neurological incapacity to perform normal courtship)

Source: After Coyne and Orr 2004.

irises *Iris fulva* and *I. hexagona* hybridize in Louisiana where their habitats (bayous and marshes, respectively) have been disturbed (Nason et al. 1992).

BEHAVIORAL ISOLATION. In animals, **behavioral isolation** (also called SEXUAL ISOLATION or ETHOLOGICAL ISOLATION) is an important barrier to gene flow among sympatric species that may frequently encounter each other but simply do not mate. The **specific mate recognition system** of a species consists of signals and responses between potential mates (Paterson 1985), and one sex (often the female) will not respond to inappropriate signals. In three morphologically indistinguishable species of green lacewings (*Chrysoperla*), for example, a male and a female engage in a duet, initiated by the male, of low-frequency songs produced by vibrating the abdomen (Martínez Wells and Henry 1992b). Mating does not occur unless the female sings back to the male. The species produce very different songs (Figure 17.5), and females respond much more often to tape recordings of their own species than to those of other species. Moreover, they discriminate against the intermediate songs produced by hybrids.

Sexual isolation in many animals (e.g., many mammals and insects) is based on differences in chemical mating signals (sex pheromones). Many other groups (e.g., many birds, fishes, jumping spiders) use visual signals, sometimes accompanied by acoustic or chemical signals, in their courtship displays (Figure 17.6). Differences in such signals often underlie sexual isolation. In many organisms, the courtship signals have not been identified, but it is nevertheless possible to measure sexual isolation by comparing the frequency of conspecific and heterospecific matings in experimental settings.

In plants, the nearest equivalent of behavioral isolation is pollination of different species by different pollinating animals that respond to differences in the color, form, or scent of flowers (Grant 1981). For example, the monkeyflower *Mimulus lewisii*, like most mem-

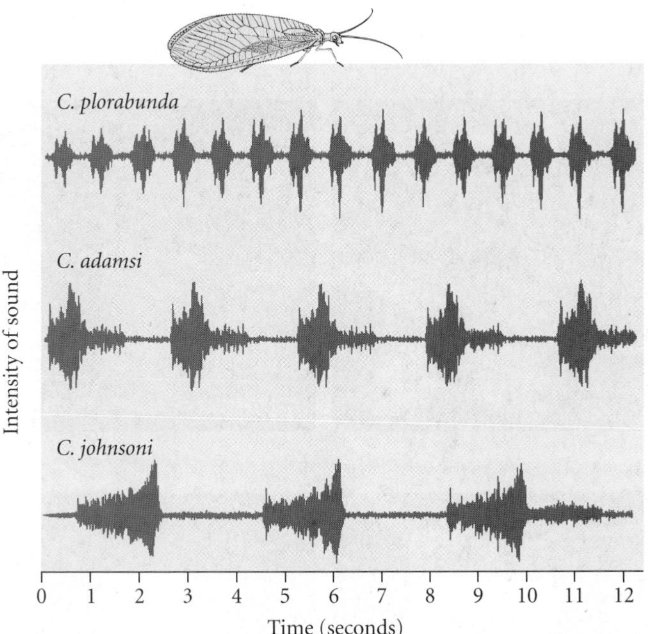

Figure 17.5 Oscillograms of the songs of three morphologically indistinguishable species of green lacewings (*Chrysoperla*). Each oscillogram displays a plot of amplitude against time. Two of the species, *C. adamsi* and *C. johnsoni*, were distinguished and named only after studies of songs and DNA differences made it clear that they are distinct species. (After Martínez Wells and Henry 1992a.)

bers of the genus, is pollinated by bees, and has pink flowers with a wide corolla. Its close relative *M. cardinalis* has a narrow, red, tubular corolla and is pollinated by hummingbirds (Schemske and Bradshaw 1999; Figure 17.7).

Postmating, prezygotic barriers

On the border between premating and postzygotic barriers are features that prevent successful formation of hybrid zygotes even if mating takes place. In many groups of insects and some other taxa, the genitalia of related species differ in morphology, and it was suggested long ago that each species' male genitalia are a special "key" that can open only a conspecific female's "lock." Only a few studies support this hypothesis, but there is good

Figure 17.6 Secondary sexual characteristics, such as bright color patterns and elaborate crests and tail feathers, vary greatly among male hummingbirds. Featured prominently in courtship displays, they undoubtedly contribute to reproductive isolation. Females of different species are much more similar to one another. All of the males shown here belong to species found in equatorial South and Central America. (A) Violet-tailed sylph (*Aglaiocercus coelestis*). (B) Rufous-crested coquette (*Lophornis delattrei*). (C) Booted racket-tail (*Ocreatus underwoodii*). (D) Crimson topaz (*Topaza pella*). (A © Danita Delimont/Alamy; B © David Tipling/Alamy; C © Pete Oxford/Minden Pictures; D © Morales/Photolibrary.com.)

Figure 17.7 (A) *Mimulus lewisii* has the broadly splayed petals characteristic of many bee-pollinated flowers. (B) An F_1 hybrid between *M. lewisii* and *M. cardinalis*. (C) *M. cardinalis* has the red coloration and narrow, tubular form that have evolved independently in many bird-pollinated flowers. (D–F) Some F_2 hybrids, showing the variation that Schemske and Bradshaw used to analyze the genetic basis of differences between these two species. (From Schemske and Bradshaw 1999.)

reason to believe that females terminate mating, and prevent transfer of sperm, if a male's genitalia do not provide suitable tactile stimulation (Eberhard 1996; Figure 17.8).

Transfer of sperm is no guarantee that the sperm will fertilize a female's eggs. In some insects, such as ground crickets (*Nemobius*), fertilization by a heterospecific male's sperm will occur if the female mates only with that male, but if she also mates with a conspecific male, only the conspecific sperm will be successful in fertilizing her eggs. This phenomenon is known as CONSPECIFIC SPERM PRECEDENCE (Howard 1999). A similar effect is seen in some plants, in which heterospecific pollen cannot compete well against conspecific pollen in growing down through the style to reach the ovules.

GAMETIC ISOLATION occurs when gametes of different species fail to unite. This barrier is important in many externally fertilizing species of marine invertebrates that release eggs and sperm into the water. Because cell surface proteins determine whether or not sperm can adhere to and penetrate an egg, divergence in such proteins can result in gametic isolation (Palumbi 1998). Among species of abalones (large gastropods), the sperm protein lysin dissolves the vitelline membrane of only conspecific eggs, enabling

Figure 17.8 The posterior lobe of the genital arch in males of three closely related species of *Drosophila*: (A) *D. simulans*, (B) *D. sechellia*, and (C) *D. mauritiana*. This is almost the only morphological feature by which these species differ. Differences in genitalia can contribute to reproductive isolation between species if copulation between them occurs. (Photos courtesy of J. R. True.)

(A)

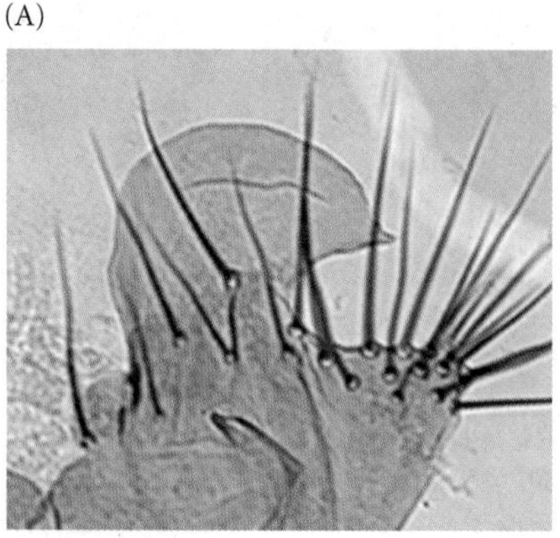

(B)

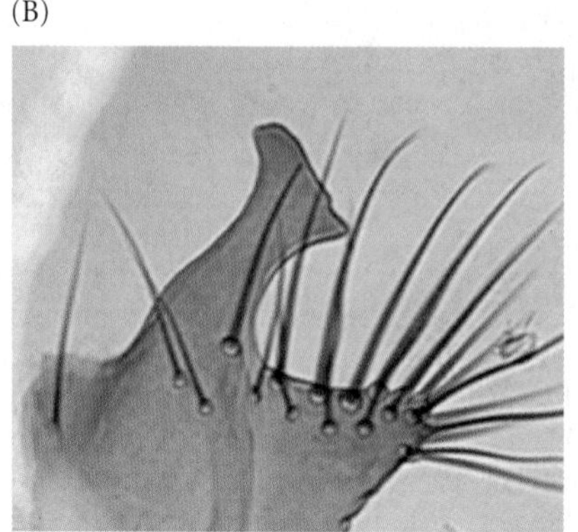

(C)

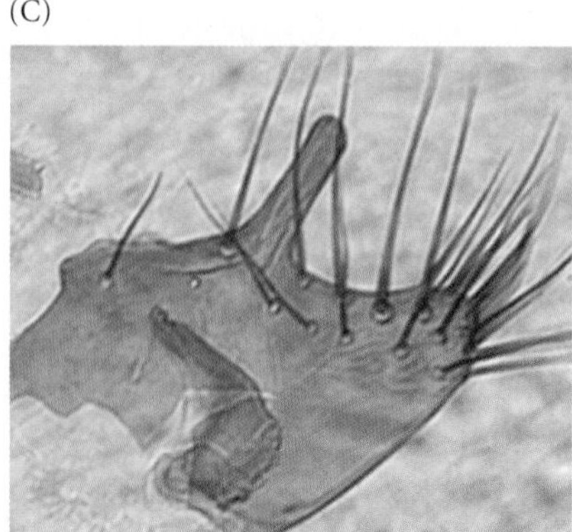

the sperm to enter. The failure of heterospecific eggs and sperm to unite is related to the high rate of divergence among abalone species of the amino acid sequence of both lysin and the vitelline membrane protein with which it interacts (Galindo et al. 2003; see Chapter 20).

Postzygotic barriers

Postzygotic barriers consist of reduced survival or reproductive rates of hybrid zygotes that would otherwise backcross to the parent populations, introducing genes from each to the other. These barriers are sometimes classified as either "extrinsic" or "intrinsic," depending on whether or not their effect depends on the environment.

HYBRID INVIABILITY. Hybrids between species often (though by no means always) have lower survival rates than nonhybrids. Quite often mortality is intrinsic, occurring in the embryonic stages because of failure of proper development, irrespective of the environment. Little is known about the malfunctions in development that cause mortality in hybrids. Especially in plants, hybrid inviability may be extrinsic, in that hybrids may have higher survival in intermediate or disturbed habitats than in those occupied by the parent species (Anderson 1949; Cruzan and Arnold 1993).

HYBRID STERILITY. The reduced fertility of many hybrids is an intrinsic barrier that can be caused by *structural differences between the chromosomes* that cause segregation of some ANEUPLOID gametes during meiosis (i.e., gametes with an unbalanced complement of chromosomes) or by *differences between the genes* from the two parents, which interact disharmoniously. These two causes are difficult to distinguish, and it is not clear how often differences in chromosome structure reduce fertility (King 1993; Rieseberg 2001).

As is sometimes true of hybrid inviability, hybrid sterility is often limited to the HETEROGAMETIC sex. (The heterogametic sex is the one with two different sex chromosomes, or with only one sex chromosome; the HOMOGAMETIC sex has two sex chromosomes of the same type. The male is heterogametic in mammals and most insects; the female is heterogametic in birds and butterflies.) This generalization is called **Haldane's rule**, and it appears to be one of the most consistent generalizations that can be made about speciation (Coyne and Orr 1989b).

Hybrid sterility and inviability may be manifested not only in F_1 hybrids but also in F_2 and backcross offspring. This phenomenon has been observed in crosses both between species and among different geographic populations of the same species, in which case it is referred to as **F_2 breakdown** in fitness. For example, survival of F_2 larvae in a cross between *Drosophila pseudoobscura* from California and from Utah was lower than that in either "pure" population. This observation was interpreted to mean that recombination in the F_1 generated various combinations of alleles that were "disharmonious." In contrast, alleles at different loci within the same population have presumably been selected to form harmonious combinations. They are said to be COADAPTED, and each population is said to have a **coadapted gene pool** (Dobzhansky 1955).

MULTIPLE ISOLATING BARRIERS. Commonly, related species differ by several kinds of isolating barriers. These are likely to come into play in a temporal sequence, so that a potentially strong barrier may be inconsequential in nature. For example, Justin Ramsey and colleagues (2003) used distributional data, field observations of pollinator transitions between plants, and experimental data from greenhouse-grown plants to estimate the strength of several barriers to gene exchange between the two species of monkeyflowers in Figure 17.7. These species are about 58 percent separated by geography or habitat. In sympatric populations, 97.6 percent of pollinator transitions are between conspecific plants. As a result of conspecific pollen precedence, 70 to 95 percent of ovules are fertilized by conspecific pollen, even if cross-pollination occurs. F_1 hybrids have somewhat reduced seed germination, and the viability of their pollen is reduced by more than 60 percent. Despite these multiple barriers to gene exchange, pollinator fidelity accounts for more than 97 percent of the total reproductive isolation between sympatric populations of the two species.

How Species Are Diagnosed

Biological species are *defined* as reproductively isolated populations, but *diagnosing* species—distinguishing them in practice—is seldom done by directly testing their propensity to interbreed or their ability to produce fertile offspring. Morphological and other phenotypic characters are the usual evidence used for diagnosing species (Figure 17.9), even though species are not defined by their degree of phenotypic difference. Morphological and other phenotypic characters, judiciously interpreted, serve as *markers* for reproductive isolation or community among sympatric populations. If a sample of sympatric organisms falls into two discrete clusters that differ in two or more characters, it is likely to represent two species.

How can phenotypic differences of this kind indicate a barrier to gene exchange? From elementary population genetics (see Chapter 9), we know that a locus in a single population with random mating should conform fairly closely to Hardy-Weinberg genotype frequencies. Furthermore, two or more loci should be nearly at linkage equilibrium, unless very strong selection or suppression of recombination exists. If these loci affect a more or less additively inherited character, its variation will have a single-peaked, more or less normal distribution. If they affect different characters, variation in those characters is likely not to be strongly correlated. Conversely, if a sample includes two (or more) reproductively isolated populations with different allele frequencies, then a single locus should show a deficiency of heterozygotes compared with the Hardy-Weinberg expectation; variation in a polygenic character may have a bimodal distribution; and variation in different, genetically independent characters may be strongly correlated. (Box 17B describes an application of these principles.) On the basis of these principles, molecular markers at two or more loci can provide an even clearer indication of reproductive isolation than can phenotypic traits.

Differences among Species

Some of the differences among species include those that are responsible for reproductive isolation. Others are adaptive differences related to ecological factors, such as temperature tolerance and habitat use. Still others are presumably neutral differences that have arisen by mutation and genetic drift. Any such character difference may have evolved

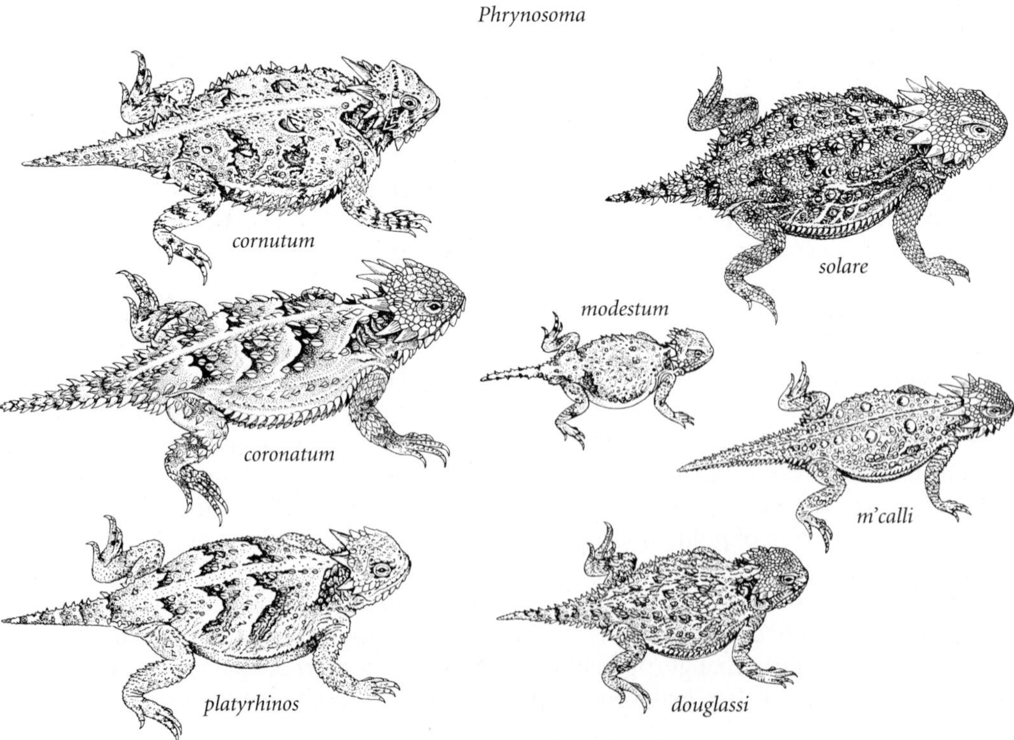

Figure 17.9 An example of species distinguished by morphological characters. These seven species of horned lizards (*Phrynosoma*) from western North America can be distinguished by differences in the number, size, and arrangement of horns and scales as well as body size and proportions, color pattern, and habitat. (From Stebbins 1954.)

BOX 17B Diagnosis of a New Species

Each species in the leaf beetle genus *Ophraella* feeds on one species or a few related species of plants. *O. notulata*, for example, has been recorded only from two species of *Iva* along the eastern coast of the United States. This species is most readily distinguished from other species of *Ophraella* by the number and pattern of dark stripes on each wing cover.

Some leaf beetles found in Florida closely resembled *O. notulata*, but were collected on ragweed, *Ambrosia artemisiifolia*. This host association suggested the possibility that these beetles were a different species. In a broader study of the genus, I collected samples of beetles from both *Ambrosia* and *Iva* throughout Florida and examined them by enzyme electrophoresis (Futuyma 1991). I found consistent differences in allele frequencies between samples from *Iva* and *Ambrosia* at three loci, even in samples from both plants in the same locality. In the most extreme case, one allele had an overall frequency of 0.968 in *Ambrosia*-derived specimens, but was absent in *Iva*-derived specimens, in which a different allele had a frequency of 0.989. No specimens had heterozygous allele profiles that would suggest hybridization. Thus these genetic markers were evidence of two reproductively isolated gene pools.

A careful examination then revealed average differences between *Ambrosia*- and *Iva*-associated beetles in a few morphological characters, such as the shape of one of the mouthparts and the relative length of the legs. None of these morphological differences distinguished all individuals of one species from all individuals of the other. Later studies showed that adults and newly hatched larvae strongly prefer their natural host plant (*Ambrosia* or *Iva*) when given a choice, and that the beetles mate preferentially with their own species. In laboratory crosses, viable eggs were obtained by mating female *Ambrosia* beetles with males from *Iva*, but not the reverse. Few of the hybrid larvae survived to adulthood, and none laid viable eggs. Based on all of this evidence, I concluded that the *Ambrosia*-associated form, which I named *Ophraella slobodkini* in honor of the ecologist Lawrence Slobodkin, is a sibling species of *O. notulata*.

Ophraella slobodkini

partly in geographically segregated populations *before* they became different species, partly *during* the process of speciation, and partly *after* the reproductive barriers evolved.

The genetic distance (D) between populations, measured by the degree of difference in their allozyme allele frequencies, can be used as an approximate "molecular clock" to estimate the relative times of divergence of various pairs of populations or species. Coyne and Orr (1989a, 1997) used such information to plot the temporal pattern by which reproductive isolation evolves. They compiled data on reproductive isolation from reports, published over the previous 60 years, of experiments on numerous combinations of populations or species of *Drosophila*. They arrived at several important conclusions:

1. The strength of both prezygotic and postzygotic isolation increases gradually with the time since the separation of the populations (Figure 17.10). That is, speciation is a gradual process.

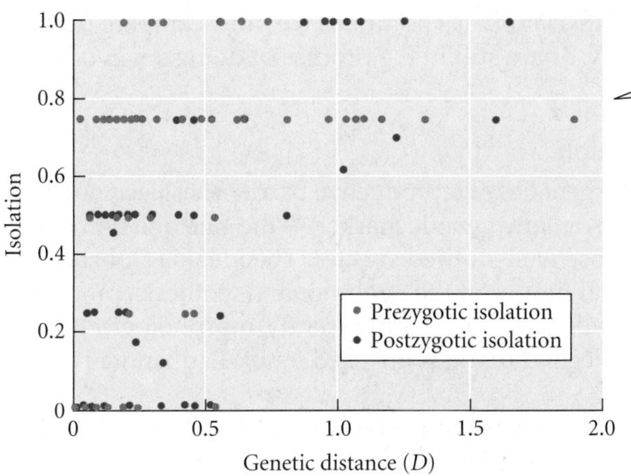

Figure 17.10 The level of prezygotic or postzygotic reproductive isolation between pairs of populations and species of *Drosophila*, plotted against genetic distance (D). Time since divergence can be inferred from genetic distance. Prezygotic isolation was measured by observing mating versus failure to mate between flies from different populations when confined together in the laboratory. Postzygotic isolation was based on survival and fertility of hybrid individuals, measured in the laboratory. (After Coyne and Orr 1997.)

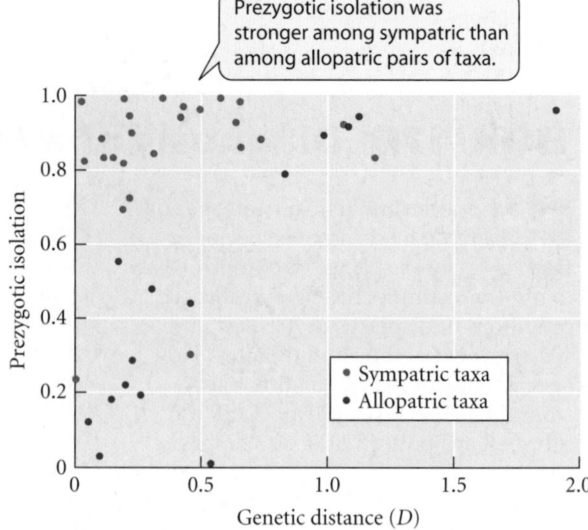

Figure 17.11 The level of prezygotic isolation among allopatric and sympatric pairs of *Drosophila* populations, plotted against genetic distance (*D*). At low genetic distances—interpreted as recent divergence—prezygotic isolation is greater among sympatric than among allopatric populations. (After Coyne and Orr 1997.)

2. The time required for full reproductive isolation to evolve is highly variable, but on average, it is achieved when *D* is about 0.30–0.53, which (based on a molecular clock calibrated from the few fossils of *Drosophila*) corresponds to about 1.5–3.5 Myr. However, a considerable number of fully distinct species appear to have evolved in less than 1 Myr.
3. Among recently diverged populations or species, premating isolation is, overall, a stronger barrier to gene exchange than postzygotic isolation (hybrid sterility or inviability). However, in Coyne and Orr's results, this effect was entirely due to sympatric taxa. Prezygotic isolation was stronger among sympatric than among allopatric pairs of taxa (Figure 17.11). This finding bears on a controversy about whether or not sexual isolation evolves to prevent hybridization, as we will see in the following chapter.
4. In the early stages of speciation, hybrid sterility or inviability is almost always seen in males only; female sterility or inviability appears only when taxa are older. Thus postzygotic isolation evolves more rapidly in males than in females.

Because genetic differences continue to accumulate long after two species achieve complete reproductive isolation, some of the genes, and even some of the traits, that now confer reproductive isolation may not have been instrumental in forming the species in the first place. Thus it can be very difficult to tell which of several demonstrable isolating barriers, or which of the many gene differences that may confer hybrid sterility, was the cause of speciation. Such information can be obtained by studying populations that have achieved reproductive isolation only very recently (see below).

The Genetic Basis of Reproductive Barriers

In analyzing barriers to gene exchange, we wish to know whether the genetic differences required for speciation consist of few or many genes and how those genes act. Because some genetic differences accrue after speciation has occurred, we must compare populations that have speciated very recently, or are still in the process of doing so, in order to answer these questions.

Genes affecting reproductive isolation

The most extensive information on the genetics of reproductive barriers has been obtained for certain *Drosophila* species because so many genetic markers—the sine qua non of any genetic analysis—are available for those well-studied species. Theodosius Dobzhansky (1936, 1937), one of the most influential figures of the evolutionary synthesis, pioneered the use of genetic markers to study hybrid sterility. Until recently, morphological mutations were used as markers; today, molecular markers are used for QTL (quantitative trait locus) mapping (see Chapter 13).

Jerry Coyne (1984) analyzed hybrid sterility in crosses between *D. simulans* and its close relative *D. mauritiana*. F_1 hybrid males are sterile, but hybrid females are fertile. Coyne

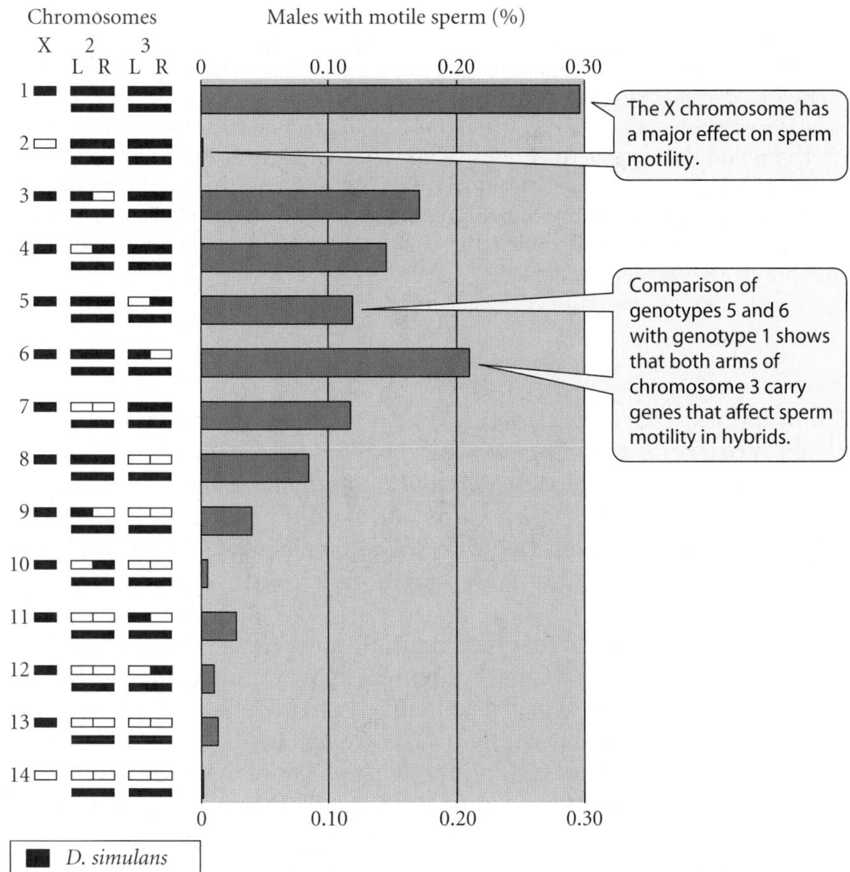

Figure 17.12 The proportion of males with motile sperm in nonhybrid *Drosophila simulans* and in backcross hybrids with various combinations of chromosome arms from *D. simulans* and *D. mauritiana*. All genotypes have a *D. simulans* Y chromosome (not shown). Note that every chromosome arm derived from *D. mauritiana* reduced sperm motility compared with that of the standard *D. simulans* genotype (1). (After Coyne 1984.)

used a *D. simulans* stock with recessive visible mutations on the X chromosome and on each arm of both of the autosomes. Female *D. simulans* were crossed to *D. mauritiana* males, and the fertile F_1 hybrid females were backcrossed to *D. simulans* males. Among the resulting male progeny, a recessive mutant phenotype showed which *D. simulans* chromosome arms were carried in homozygous condition. Coyne scored males for motile versus immotile sperm (immotility is correlated with male sterility). Sperm motility differed for each pair of genotype classes distinguished by one or more recessive markers (Figure 17.12). Therefore, each chromosome arm carries at least one gene difference between *D. simulans* and *D. mauritiana* that contributes to male hybrid sterility.

Coyne used only five genetic markers in this experiment, so he could detect no more than five linked genetic factors contributing to hybrid sterility. Chung-I Wu and his coworkers have used multiple molecular markers to mark small segments of the X chromosomes of *Drosophila mauritiana* and *D. sechellia* and to backcross them, individually or in combination, into a genome otherwise derived from *D. simulans*. Males had lowered fertility whenever they carried two or more such segments. By extrapolation from these short segments of chromosomes, Wu and his coworkers have suggested that as many as 40 gene differences on the X chromosome and 120 in the genome as a whole might cause hybrid male sterility among these closely related species (Wu and Hollocher 1998). A similar study of *Drosophila simulans* and *D. melanogaster* suggested that about 200 genes can contribute to hybrid inviability (Presgraves 2003). However, far fewer gene differences suffice to confer postzygotic isolation. Male sterility of F_1 hybrids between *Drosophila pseudoobscura* and *D. bogotana*, which are thought to have diverged less than 200,000 years ago, appears to be based on only about five gene regions, of which four are required for any sterility at all (Orr and Irving 2001). It therefore appears that the early acquisition of hybrid sterility requires few gene differences.

Genes that contribute to hybrid sterility or inviability do not have these effects in nonhybrid individuals; these effects must therefore stem from interactions between genes

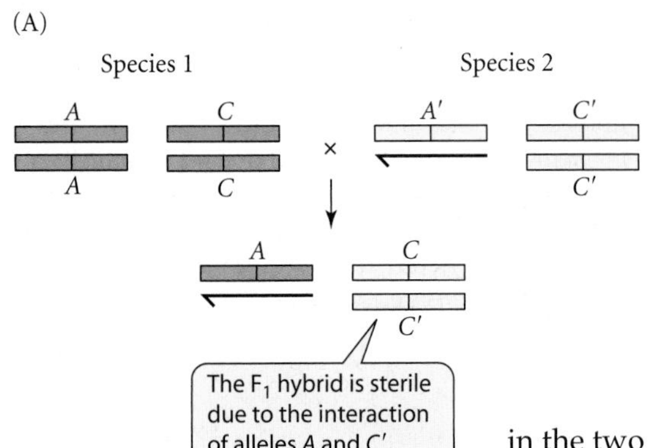

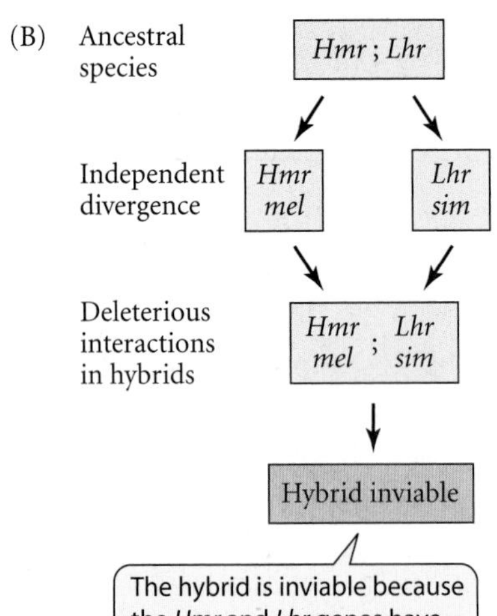

Figure 17.13 Gene interactions that cause sterility or inviability in hybrids between species. (A) The X chromosome and one autosome of males of two species are shown above the F_1 genotype. The interaction between complementary loci *A* and *C* causes sterility or inviability. Such an interaction between loci is called a Dobzhansky-Muller incompatibility. (B) The genes *Hmr* and *Lhr* have undergone changes, because of natural selection, in *Drosophila melanogaster* and *D. simulans*, respectively. Together, these changes cause inviability of the male hybrid. (B after Brideau et al. 2006.)

in the two different species. That is, *epistatic interactions* contribute to postzygotic isolation (Figure 17.13A). A pair of such genes causes lethality of F_1 hybrids between *Drosophila melanogaster* and *D. simulans* (Figure 17.13B). Relative to the common ancestor, mutations have been fixed in the *Hmr* gene of *D. melanogaster* and the *Lhr* gene of *D. simulans* that together cause the death of hybrid males. The *Hmr* ("hybrid male rescue") and *Lhr* ("lethal hybrid rescue") genes were first found because mutations in either gene enable hybrid males to survive.

Studies of this kind generally show that the X chromosome has a greater effect than any of the autosomes. This result sheds some light on the more rapid evolution of sterility in male than in female *Drosophila* hybrids, an instance of Haldane's rule. It has been proposed that X-linked genes diverge faster than autosomal genes because favorable X-linked recessive alleles are most exposed to natural selection (since males carry only one X). In addition, autosomal genes affecting male sterility have diverged faster than those affecting female sterility, possibly because of sexual selection.

Like postzygotic isolation, premating isolation is frequently based on polygenic traits, although in some cases only a few genes are involved (Ritchie and Phillips 1998). Differences in the shape of the male genitalia of *Drosophila simulans* and *D. mauritiana*, which may cause interspecific matings to terminate prematurely (see Figure 17.8), are affected by at least 19 loci (Zeng et al. 2000). Sexual isolation based on pheromones, in contrast, may be based on a few genes with major effects. For example, the European corn borer moth (*Ostrinia nubilalis*) includes two "pheromone races" (or sibling species) that differ in the ratio of two components (E- and Z-11 tetradecenyl acetate) of the sex pheromone emitted by females. A single autosomal gene difference accounts for most of the difference in the E:Z ratio. Both in the field and in wind tunnels, males of each race are almost exclusively attracted to the E:Z ratio produced by females of their own race. A single autosomal gene, not linked to the gene for the female's pheromone ratio, controls the difference in the physiological responses of the males' antennae to the two pheromone components, but the behavioral response of males to different E:Z ratios is sex-linked (Roelofs et al. 1987). This and similar studies show that the male and female components of communication that result in sexual isolation are usually genetically independent.

In a massive QTL study of F_2 progeny from crosses between the bee-pollinated monkeyflower *Mimulus lewisii* and its hummingbird-pollinated relative *M. cardinalis* (see Figure 17.7), Bradshaw and colleagues (1998) found that each of 12 characters that distinguish the species' flowers differed by between one and six loci. For most of these features, one locus accounted for at least 25 percent of the difference between the species. Schemske and Bradshaw (1999) then placed a full array of F_2 hybrids in an area where the two species are sympatric and observed pollination. At least 4 of the 12 floral traits affected the relative frequency of visitation by bees versus hummingbirds. Two QTL underlying these characters could be shown to have a significant effect on pollinator isolation. Alleles at one locus that strongly affect flower color were backcrossed from each species into the other, resulting in flowers that were normal in every repect except for color (Bradshaw and Schemske 2003). The normally hummingbird-pollinated *M. cardinalis* then received 74 times more bee visits than the wild type, and the normally bee-pollinated *M. lewisii* received 68 times as many hummingbird visits as the wild type. This suggests that a single mutation can greatly alter pollination and contribute substantially to reproductive isolation.

Functions of genes that cause reproductive isolation

What are the nature and function of the genes that cause reproductive isolation? For most prezygotic isolating barriers, answering this question is a matter of understanding the genetic mechanisms of development of ordinary phenotypic characters, such as the features used in courtship or the behaviors involved in mate choice or habitat preference. The genes that underlie the breakdown of fertility or viability in hybrids are much more mysterious, and only in the last few years has there been progress toward understanding what these genes actually do.

In one of the best-understood cases, Daven Presgraves and collaborators (2003, 2007) showed that one of the *Drosophila simulans* genes that causes hybrid inviability when crossed into a *D. melanogaster* genetic background encodes a nucleoporin protein (Nup96), one of about 30 such proteins composing the nuclear pore complexes that regulate the passage of proteins and RNA between the cell nucleus and the cytoplasm. This gene interacts with several *D. melanogaster* genes that encode other proteins in the nuclear pore complex. In both the *Nup96* gene and the genes with which it interacts, amino acid–replacing nucleotide substitutions have occurred at a high rate relative to synonymous substitutions, a clear indication that natural selection, rather than genetic drift, has driven divergence. Why selection favored changes in this protein is not yet known.

Chromosome differences and postzygotic isolation

Chromosome differences among species include alterations of chromosome structure (see Chapter 8) and differences in the number of chromosome sets (polyploidy, treated in Chapter 18). The role of structural alterations in postzygotic isolation and speciation is controversial (King 1993; Rieseberg 2001; Coyne and Orr 2004). An important question is whether heterozygosity for chromosome rearrangements causes reduced fertility (postzygotic isolation) in hybrids as a result of segregation of aneuploid gametes in meiosis.

A RECIPROCAL TRANSLOCATION is an exchange between two homologous chromosomes. Suppose, for example, that 1.2 and 3.4 represent two metacentric chromosomes in one population, with 1 and 2 representing the arms of one and 3 and 4 the arms of the other. A second population that is fixed for a translocation might have chromosomes 1.4 and 3.2. The F_1 hybrid would have all four chromosome types (1.2, 3.4, 1.4, 3.2). Only if the two parental combinations segregated (1.2 and 3.4 to one pole, 1.4 and 3.2 to the other) would balanced (euploid) gametes be formed (see Figure 8.24). Other patterns of segregation (e.g., 1.2 and 3.2 to one pole, 3.4 and 1.4 to the other) would yield unbalanced (aneuploid) gametes that lack considerable genetic material.

Some species differ by multiple translocations. For example (Dobzhansky 1951), the jimsonweeds *Datura stramonium* and *D. discolor* both have 12 pairs of chromosomes. In the F_1 hybrid, seven pairs form normal synapsed pairs in meiosis. The other five pairs have undergone multiple translocations. If these chromosomes in *D. stramonium* are designated 1.2, 3.4, 5.6, 7.8, and 9.10, those of *D. discolor* represent 1.3, 2.7, 4.10, 5.9, and 6.8. In synapsis, a ring of ten chromosomes is formed, with the two arms of each *stramonium* chromosome aligned with an arm of each of two *discolor* chromosomes (Figure 17.14). The opportunities for aneuploid segregation are numerous.

Perhaps for this reason, such chromosome rearrangements are seldom found as polymorphisms within populations. More often, they are nearly or entirely monomorphic in different populations, except where those populations meet in narrow hybrid zones. For example, parapatric races of the burrowing mole-rat *Spalax ehrenbergi* in Israel differ in the number of chromosome pairs due to chromosome fusions. Hybrids between these races are found in zones that range from 2.8 km to only 0.3 km wide (Figure 17.15).

(A)

Datura stramonium	*Datura discolor*
1 2	1 3
3 4	2 7
5 6	4 10
7 8	5 9
9 10	6 8

(B) F_1 hybrid

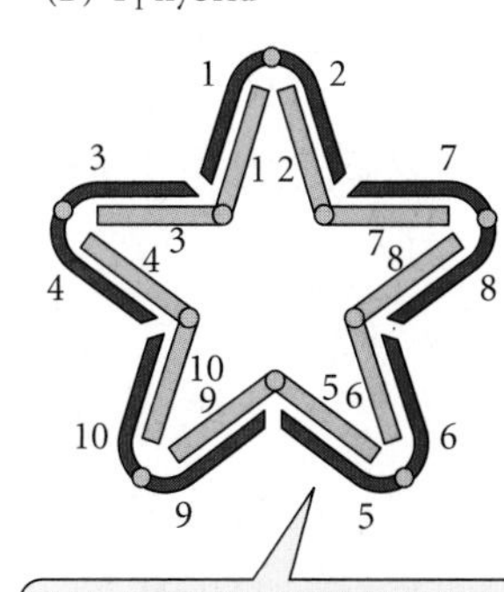

In synapsis, the 10 chromosomes form a ring, with the two arms of each *stramonium* chromosome aligned with the corresponding arm from each of two *discolor* chromosomes.

Figure 17.14 (A) Five chromosomes of the jimsonweeds *Datura stramonium* and *D. discolor*, which differ by five reciprocal translocations. Homologous chromosome arms are correspondingly numbered. Only one member of each chromosome pair is shown. (B) A diagram of the possible arrangement of these chromosomes in synapsis in an F_1 hybrid.

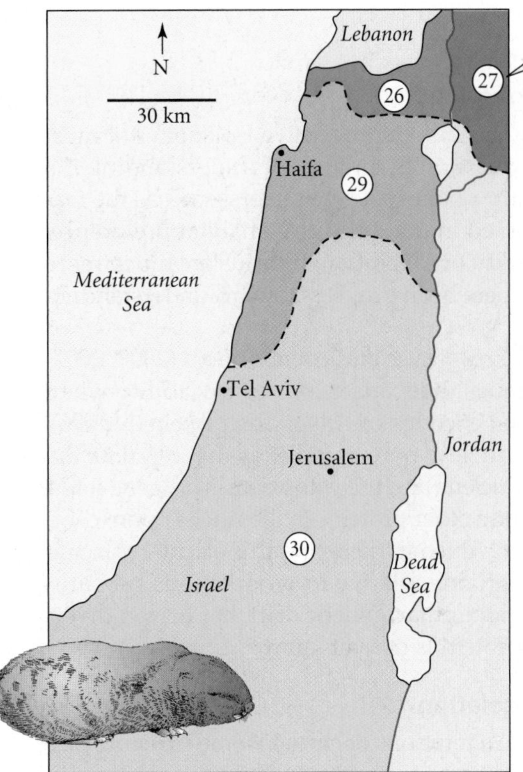

Figure 17.15 The distribution of four "races" of the mole-rat *Spalax ehrenbergi* with different chromosome numbers. Pairs of races meet at very narrow hybrid zones, indicated by the dashed lines. (After Nevo 1991.)

This pattern is expected if chromosomal heterozygotes have lower fitness than homozygotes (are UNDERDOMINANT), perhaps owing to reduced fertility caused by aneuploidy. If so, a chromosome introduced by gene flow from one population into another would seldom increase in frequency, because its initial frequency would be low, it would occur mostly in heterozygous condition, and it would probably be eliminated by selection (see Chapter 11). However, it is very difficult to tell whether the reduced fitness of hybrids is caused by the structural differences between the chromosomes or by differences between the genes of the parent populations. Certainly, chromosome rearrangements reduce gene exchange between populations. For example, the sunflowers *Helianthus annuus* and *H. petiolaris* differ by inversions and translocations that affect some chromosomes but not others (Rieseberg et al. 1999). The rearranged chromosomes show a more abrupt transition in a hybrid zone between these species than do the chromosomes lacking rearrangements. But as we will see in our discussion of hybrid zones below, this pattern may be due to genetic incompatibility, not to production of aneuploid gametes. According to Coyne and Orr (2004), evidence that chromosome heterozygotes have reduced fertility because of meiotic irregularities is convincing for some plants and mammals but not for other organisms, at least so far.

The significance of genetic studies of reproductive isolation

Both prezygotic and postzygotic isolation are generally due to differences between populations at several gene loci. A mismatch between genes affecting sexual communication, such as a male courtship signal and a female response, creates reproductive isolation. Similarly, a functional mismatch between genes gives rise to hybrid sterility or inviability. In both cases, reproductive isolation requires that populations diverge by at least two allele substitutions. Thus an $A_1A_1B_1B_1$ ancestral population may give rise to populations with genotypes $A_1A_1B_2B_2$ and $A_2A_2B_1B_1$, the incompatibility between A_2 and B_2 being the cause of reproductive isolation. This kind of epistatic incompatibility between loci is called a **Dobzhansky-Muller incompatibility**, because Theodosius Dobzhansky, in 1934, and Hermann Muller, in 1940, postulated that such interactions underlie postzygotic isolation. The interaction between the *Hmr* and *Lhr* genes that causes lethality of male hybrids between *Drosophila melanogaster* and *D. simulans* (see Figure 17.13B) is exactly what Dobzhansky and Muller envisioned.

As we have seen, the number of gene differences that suffice for postzygotic isolation may be rather small, but more such differences accrue over time. Thus reproductive isolation eventually becomes irreversible, and the evolutionary lineages undergo independent genetic change thereafter.

Finally, genetic studies have shown that differences among species, including characters that confer reproductive isolation, have the same kinds of genetic foundations as variation within species. Thus there is no foundation for the opinion, held by some earlier biologists such as the geneticist Richard Goldschmidt (1940), that species and higher taxa arise through qualitatively new kinds of genetic and developmental repatterning. Moreover, reproductive isolation, like the divergence of any other character, usually evolves by the gradual substitution of alleles in populations.

Molecular Divergence among Species

Many of the differences between species in allozymes and DNA sequences are presumably selectively neutral, or nearly so. Such differences increase over time, but no specific level of allozyme or DNA divergence enables us to declare that populations have become different species. Some reproductively isolated populations display little or no divergence in molecular markers, presumably because reproductive isolation, perhaps based on genetic change in one or a few characters, has evolved very recently. This is especially true in some groups of cichlids and other fishes (McCune and Lovejoy 1998).

The pattern of molecular variation within and between **sister species**—two species with an immediate common ancestor—can shed light on their history of divergence, especially when data from DNA sequences or from restriction site differences among gene copies are used to estimate the phylogeny of DNA sequences (the "gene tree"). Two populations (or species) that become isolated from each other at first share many of the same gene lineages, inherited from their polymorphic common ancestor (Figure 17.16). Each population is at first polyphyletic with respect to gene lineages (i.e., its genes are derived from several ancestral genes). Thus, with respect to some loci, individuals in each population are genealogically less closely related to one another than they are to some individuals in the other population (Figure 17.16, time t_1).

According to the COALESCENT THEORY described in Chapter 10, genetic drift in each species eventually results in the loss of all the ancestral lineages of DNA sequence variants except one; that is, coalescence to a common ancestral gene copy occurs in each species. (This process can also be caused by directional selection for a favorable mutation.) Gene lineages are lost by genetic drift at a rate inversely proportional to the effective population size. At some point, one population (population 1 in Figure 17.16) will become monophyletic for a single gene lineage, while the other population, if it is larger, retains both this and other gene lineages (population 2 in Figure 17.16, at time t_2). At this time, the more genetically diverse population will be paraphyletic with respect to this gene, and some gene copies sampled from population 2 will be more closely related to population 1's gene

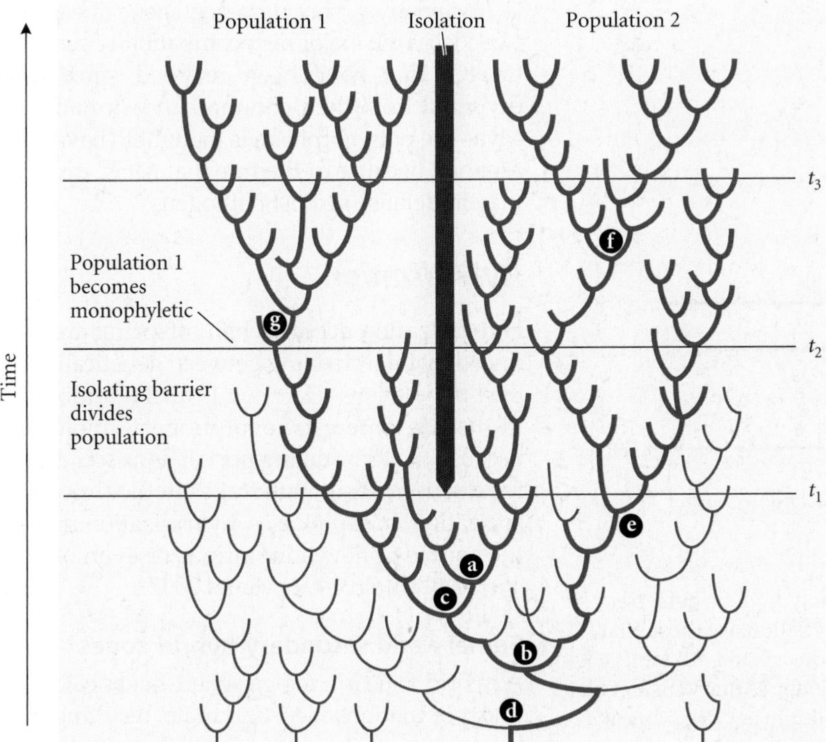

Figure 17.16 The transition from genetic polyphyly to paraphyly to monophyly in speciation. Colored branches show lineages of haplotypes at a single locus in a population that becomes divided into two populations at time t_1. As a result of genetic drift and/or natural selection, all ancestral lineages except one are eventually lost, so that all the gene copies in the contemporary populations (at the top) coalesce to ancestral gene copy b. Population 1 loses gene lineages more rapidly than population 2, perhaps because it is smaller. Between times t_1 and t_2, both populations are polyphyletic for haplotype lineages, since some gene copies in each population, such as those derived from gene copy a, are more closely related to some gene copies in the other population than to some other gene copies (e.g., those derived from b) in their own population. Population 1 becomes monophyletic (coalescing to gene copy g) sooner—at time t_2—than population 2, perhaps because of its smaller size. Between times t_2 and t_3, population 2 is genetically paraphyletic, since its gene copies derived from ancestor c are more closely related to population 1's gene copies than they are to gene copies in population 2 that have been derived from ancestor e. At time t_3, population 2 also becomes monophyletic, with all gene copies derived from f. (After Avise and Ball 1990.)

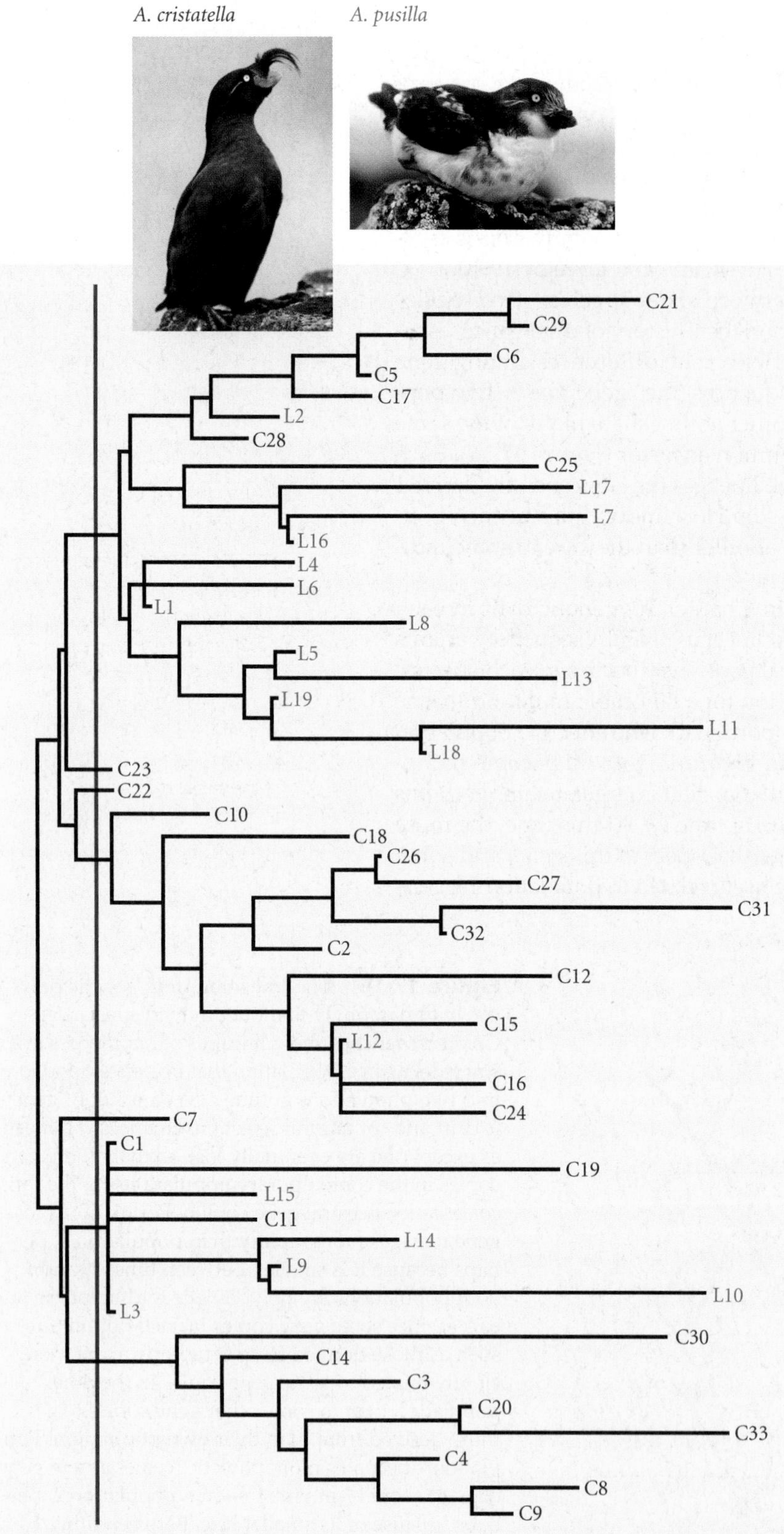

Figure 17.17 Incomplete lineage sorting results in a polyphyletic gene tree for the α-enolase locus. Allelic lineages present before speciation are those that cross the red vertical line. C and L mark alleles found in the crested auklet (*Aethia cristatella*) and the least auklet (*A. pusilla*), respectively. (After Walsh et al. 2005; *A. cristatella* courtesy of Art Sowls, U.S. Fish and Wildlife Service; *A.pusilla* © Steven J. Kazlowski/Alamy.)

copies than they are to other copies from population 2. Thus the phylogenetic relationships among genes from organisms in both populations will not correspond to the relationships among the individual organisms or the populations. Eventually, however, both populations will become monophyletic for gene lineages (Figure 17.16, time t_3), and the relationships among genes will reflect the relationships among populations. This process of sorting of gene lineages into species, or **lineage sorting**, is faster if effective population sizes (N_e) are small (Neigel and Avise 1986; Pamilo and Nei 1988).

Closely related species often share ancestral polymorphisms; in other words, lineage sorting has not gone to completion (Funk and Omland 2003). For example, severeal haplotype lineages of the nuclear α-enolase gene are shared between the crested auklet (*Aethia cristatella*) and the least auklet (*A. pusilla*), two small seabirds that split from a common ancestor at least 2.8 Mya (Figure 17.17; Walsh et al. 2005). Likewise, several species of grasshoppers show paraphyletic relationships at several loci (see Figure 2.21). This pattern is expected if species have arisen very recently and/or have had large population sizes, so that genetic drift, resulting in coalescence, has proceeded slowly (see Chapter 10). Moreover, shared polymorphisms can persist for a very long time if natural selection maintains the variation in both species (see Figure 12.27). For example, humans and chimpanzees are each other's closest relatives, sharing several gene lineages at two loci in the major histocompatibility complex (MHC) that have been retained since they diverged from their common ancestor about 5 Mya. The polymorphism is thought to have been retained because of the role that MHC proteins play in defense against pathogens.

Hybridization

Hybridization occurs when offspring are produced by interbreeding between genetically distinct populations (Harrison 1990). Hybridization in nature interests evolutionary biologists because the hybridizing populations sometimes represent intermediate stages in the process of speciation. In some cases, hybridization may be the source of new adaptations or even of new species (Arnold 1997; Mallet 2007).

Primary and secondary hybrid zones

A hybrid zone is a region where genetically distinct populations meet and mate, resulting in at least some offspring of mixed ancestry (Harri-

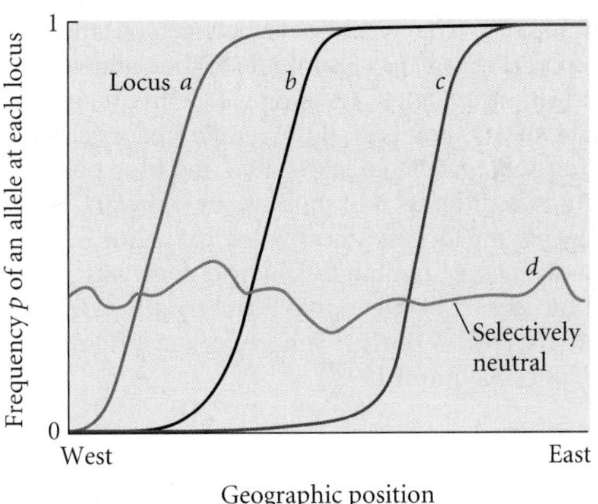

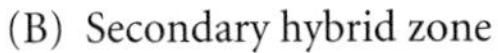

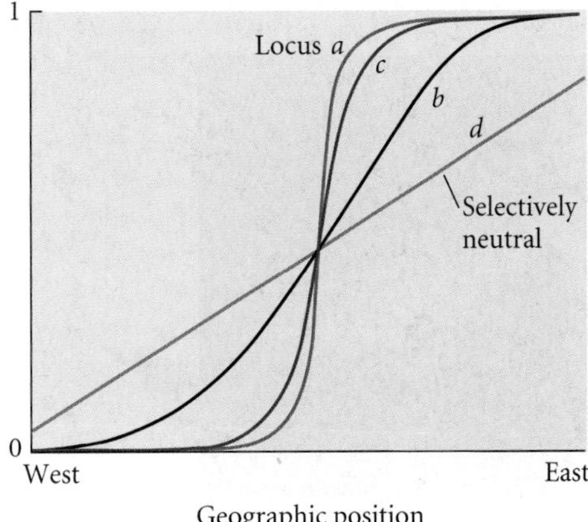

Figure 17.18 Expected patterns of variation in the frequency (p) of alleles or characters across a hybrid zone originating by (A) divergent selection along an environmental gradient (a primary zone) and (B) secondary contact. "Stepped clines" are shown for four loci. In (A), the loci are assumed to be affected differently by the environmental gradient, so the clines have different geographic positions. Allelic variation at locus d (orange line) is assumed to be nearly selectively neutral. (B) In a secondary hybrid zone, clines for all loci, including that for the nearly neutral alleles at locus d, are expected to have about the same location. The width of the cline depends on the strength of selection, relative to gene flow, at the locus or at closely linked loci. After a long enough time, the cline at locus d will come to resemble that in (A) because of gene flow. In some circumstances, the two causes of hybrid zones can result in indistinguishable patterns.

son 1990, 1993). A character or locus that changes across a hybrid zone exhibits a CLINE that may be quite steep; for example, the alleles for the gray hood in hybridizing hooded and carrion crows display a steep cline in frequency across the zone of hybridization between these forms (see Figure 17.1D).

Hybrid zones are thought to be caused by two processes. PRIMARY HYBRID ZONES originate in situ as natural selection alters allele frequencies in a series of more or less continuously distributed populations. Thus the position of the zone is likely to correspond to a sharp change in one or more environmental factors. SECONDARY HYBRID ZONES are formed when two formerly allopatric populations that have become genetically differentiated expand so that they meet and interbreed (**secondary contact**). It is not always easy to determine whether a hybrid zone is primary or secondary (Endler 1977). However, we might expect that in a primary hybrid zone, natural selection on different loci or characters would result in clines with different geographic positions, and that selectively neutral variation would not display a clinal pattern (Figure 17.18A). In contrast, the hybrids between populations that meet at secondary hybrid zones often have low intrinsic fitness, because of heterozygote disadvantage or breakdown of coadapted gene complexes. (Such hybrid zones are often referred to as **tension zones**.) The clines in characters that differentiate the populations, therefore, need not match changes in the environment and are expected to be COINCIDENT (i.e., located in the same place; Figure 17.18B). Clines in selectively neutral markers should be coincident with the others, although they may become broader over time because of gene flow.

The eastern European fire-bellied toad (*Bombina bombina*) and the western European yellow-bellied toad (*B. variegata*) meet in a long hybrid zone that is only about 6 km wide. The two species (or semispecies) differ in allozymes and in several morphological features (Figure 17.19), and hybrids have lower rates of survival than nonhybrids because of epistatic incompatibility. These taxa are thought to have arisen during the Pliocene, and to have formed a secondary hybrid zone after spreading from different refuges in southeastern and southwestern Europe that they occupied during the Pleistocene glacial periods (Szymura 1993).

Genetic dynamics in a hybrid zone

Dispersal, selection, and linkage all affect the distribution of alleles and phenotypic characters in hybrid zones. Let us consider how these factors affect clines in tension zones, in which hybrids have low fitness because of epistatic incompatibility or heterozygote disadvantage at certain loci (Barton and Gale 1993). Because the hybrids have low fitness irrespective of variation in environmental conditions, the geographic position of a tension zone is not determined by ecological factors. Its position may move if, for example, more individuals disperse across the hybrid zone in one direction than the other.

Suppose populations (semispecies) 1 and 2 have come into contact, that they are fixed for alleles A_1 and A_2, respectively, and that the fitness of A_1A_2 is reduced. Dispersal of individuals of each semispecies into the range of the other, followed by random mating, con-

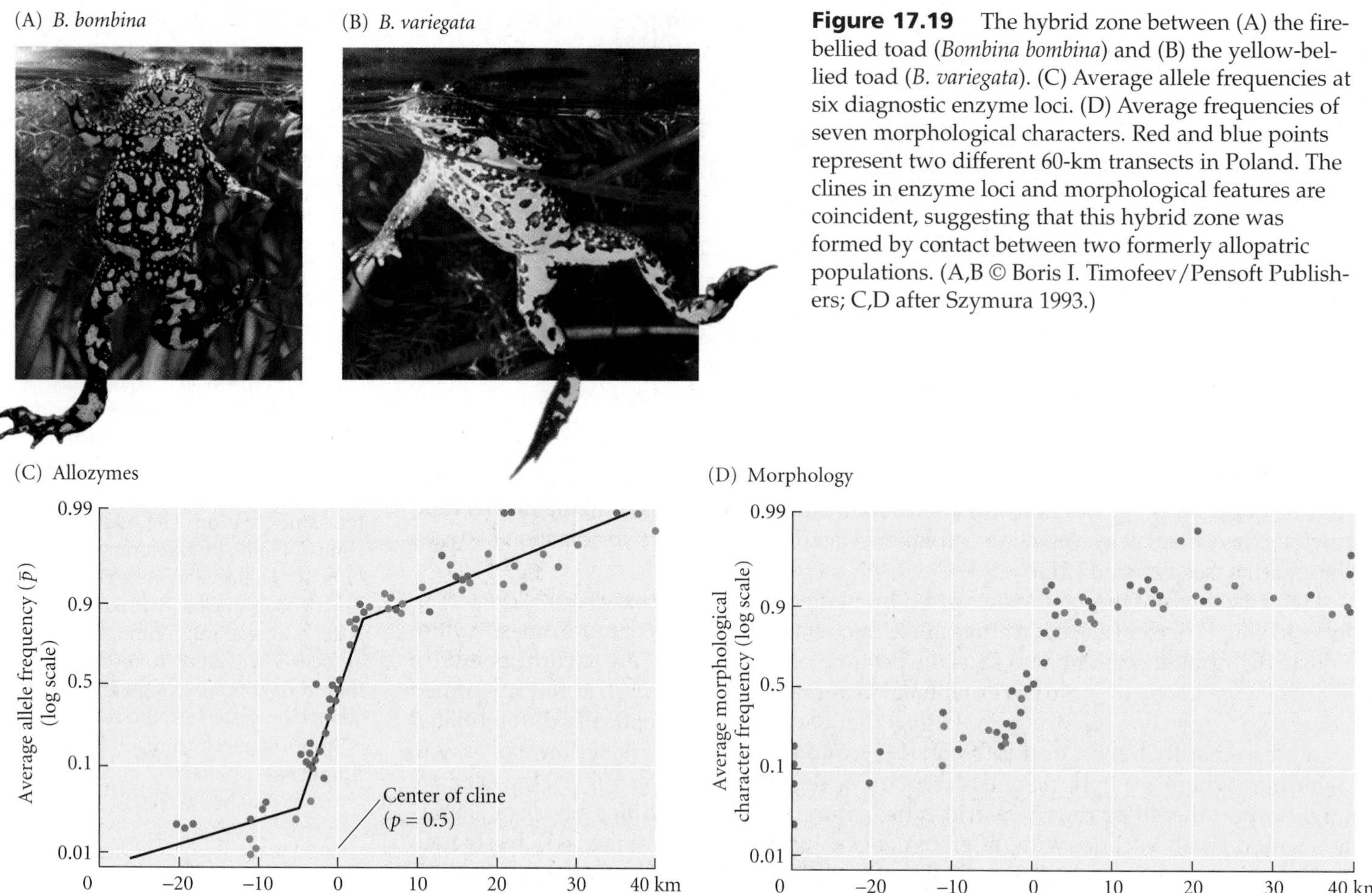

Figure 17.19 The hybrid zone between (A) the fire-bellied toad (*Bombina bombina*) and (B) the yellow-bellied toad (*B. variegata*). (C) Average allele frequencies at six diagnostic enzyme loci. (D) Average frequencies of seven morphological characters. Red and blue points represent two different 60-km transects in Poland. The clines in enzyme loci and morphological features are coincident, suggesting that this hybrid zone was formed by contact between two formerly allopatric populations. (A,B © Boris I. Timofeev/Pensoft Publishers; C,D after Szymura 1993.)

stitutes gene flow that tends to make the cline in allele frequency broader and shallower. Gene flow continues if F_1 hybrids backcross with the parental (nonhybrid) genotypes. However, allele A_1 cannot increase in frequency within population 2, nor can A_2 within population 1, because of the heterozygote's disadvantage (see Chapter 12). The term "tension zone" is based on this "standoff": neither population's allele can invade the other population. The lower the fitness of the F_1 hybrid, the fewer A_1 or A_2 allele copies will be introduced into the parent populations. Thus selection against hybrids acts as a barrier to gene flow, and the steepness of the cline at the *A* locus depends on the strength of selection against hybrids relative to the magnitude of dispersal. Unless the rate of dispersal or interbreeding changes, this cline persists indefinitely.

Now consider another locus, with alleles B_1 and B_2 fixed in populations 1 and 2. If there is selection against B_1B_2 heterozygotes, the cline at this locus will parallel the cline at the *A* locus; in fact, they will reinforce each other, since the two loci both reduce the F_1 hybrid's fitness. Suppose, though, that the B_1 and B_2 alleles are selectively neutral. If they are very closely linked to the *A* locus, then they will hitchhike with the *A* alleles and form a cline coincident with that of the *A* locus in position and steepness. In fact, they will mark the existence and location of the locus (*A*) that contributes to low hybrid fitness. However, if the *B* locus is on another chromosome, lacking genes that reduce hybrid fitness, the B_1 and B_2 alleles will diffuse through the hybrid zone and spread into the other semispecies, forming a cline that becomes more shallow over time (see Figure 17.17B). The diffusion occurs because if F_1 hybrids reproduce at all, some of the offspring of the backcross $A_1A_2B_1B_2 \times A_1A_1B_1B_1$ (i.e., $F_1 \times$ population 1) will be $A_1A_1B_1B_2$. Thus some copies of the B_2 allele, dissociated by recombination from the deleterious heterozygous allele combination A_1A_2, will be introduced into population 1. Backcrosses to population 2 will similarly spread B_1 allele copies into that population. Thus some alleles or characters can spread from each semispecies into the other (a process called **introgression**), whereas others cannot.

A lowered fitness of hybrids at one locus therefore reduces the flow of neutral (or advantageous) alleles between populations, but the reduction is greater for alleles at closely linked loci than at loosely linked or unlinked loci. If a chromosome rearrangement that includes a selected locus reduces crossing over, then neutral alleles at all loci within the rearrangement, as well as the rearrangement itself, will display the same steep cline as the selected locus. This may account for the pattern seen in the sunflower hybrid zone mentioned on page 462, in which rearranged chromosomes have steeper clines than structurally similar chromosomes (Rieseberg 2001). In fact, Loren Rieseberg and collaborators (1999) showed that genes in these chromosome rearrangements reduce hybrid fertility.

Neutral alleles that are tightly linked to selected loci will have a cline that is steep in the middle of the hybrid zone but shallow farther from the zone. Moreover, since some neutral genetic markers will be tightly linked to a locus that reduces hybrid fitness, whereas others may be unlinked, the number of markers with steep clines provides an estimate of the number of loci contributing to the low fitness of hybrids (i.e., to postzygotic reproductive isolation). For example, allozyme clines in the *Bombina* hybrid zone have been used to estimate that about 55 loci contribute to the reduction in the fitness of hybrids between fire-bellied and yellow-bellied toads (Barton and Gale 1993; Szymura 1993).

The fate of hybrid zones

As the last glaciers receded about 8000 years ago, many organisms rapidly colonized the newly exposed landscape. Thus many postglacial hybrid zones, such as that between the European carrion and hooded crows, are undoubtedly several thousand years old (Mayr 1963). Such apparent stability raises the question of whether or not hybridization will continue indefinitely.

Hybrid zones may have several fates:

1. A hybrid zone may persist indefinitely, with selection maintaining steep clines at some loci even while the clines in neutral alleles dissipate because of introgression. If the hybrid zone is a tension zone, it may move. It may become lodged in a region of low population density, or may eventually move to the far edge of the range of one of the semispecies, resulting in its extinction. If, however, some of the character differences are favored by different environments, the positions of the clines in those characters will be stable.
2. Natural selection may favor alleles that enhance prezygotic isolation, resulting ultimately in full reproductive isolation.
3. Alleles that improve the fitness of hybrids may increase in frequency. In the extreme case, the postzygotic barrier to gene exchange may break down, and the semispecies may merge into one species.
4. Some hybrids may become reproductively isolated from the parent forms and become a third species.

These possibilities will be discussed in the following chapter as we examine the processes of speciation.

Summary

1. Many definitions of "species" have been proposed. The biological species concept (or variants of it) is the one most widely used by evolutionary biologists. It defines species by reproductive discontinuity based on differences between populations, not by phenotypic differences (although phenotypic differences may be indicators of reproductive discontinuity). Among other species concepts, the most widely adopted is the phylogenetic species concept, according to which species are sets of populations with character states that distinguish them.
2. The biological species concept (BSC) has a restricted domain; moreover, some populations cannot be readily classified as species because the evolution of reproductive discontinuity is a gradual process. Although the BSC applies validly to allopatric populations, it is often difficult to determine whether or not such populations are distinct species.

3. The biological differences that constitute barriers to gene exchange are many in kind, the chief distinction being between prezygotic barriers (e.g., ecological or sexual isolation) and postzygotic barriers (hybrid inviability or sterility). Some species are also isolated by postmating, prezygotic barriers (e.g., gametic isolation).
4. Among prezygotic barriers to gene exchange, sexual (ethological) isolation is important in animals. It entails a breakdown in communication between the courting and the courted sexes, and therefore, usually, genetic divergence in both the signal and the response. The differences in male and female components are usually genetically independent.
5. Postzygotic isolation is usually caused by differences in nuclear genes and/or structural differences in chromosomes. Genic differences that yield hybrid sterility or inviability consist of differences at two or more (usually considerably more) loci that interact disharmoniously in the hybrid. The role of chromosome differences in hybrid sterility is not well understood.
6. Reproductive barriers evolve gradually. Hybrid sterility or inviability of the heterogametic sex (often male) usually evolves before its manifestation in the homogametic sex (often female); this is known as Haldane's rule.
7. Levels of molecular divergence between closely related species vary greatly, and some recently formed species cannot be distinguished by molecular markers. Some species share ancestral molecular polymorphisms. In some cases, some gene copies in one species are more closely related to gene copies in another species than to other gene copies in the same species. In such cases, the phylogeny of genes may not match the phylogeny of the species that carry the genes.
8. Species (or semispecies) sometimes hybridize, often in hybrid zones, which in many cases are regions of contact between formerly allopatric populations. Alleles at some loci, but not others, may be introduced between hybridizing populations by gene flow, forming allele frequency clines of varying steepness. The steepness of such a cline depends on the rate of dispersal, the strength of selection, and linkage to selected loci.

Terms and Concepts

behavioral isolation
biological species concept (BSC)
coadapted gene pool
conspecific
Dobzhansky-Muller incompatibility
F_2 breakdown
Haldane's rule
hybrid zone (primary, secondary)
introgression
isolating barriers
lineage sorting
phylogenetic species concept (PSC)
postzygotic barriers
premating barriers
prezygotic barriers
secondary contact
semispecies
sibling species
sister species
specific mate recognition system
superspecies
tension zone

Suggestions for Further Reading

Speciation by J. A. Coyne and H. A. Orr (Sinauer Associates, Sunderland, MA, 2004) is the most comprehensive book on speciation in more than 40 years. The authors analyze hypotheses and data about speciation carefully and summarize a great amount of relevant literature. They provide an extensive discussion of species concepts and a justification of the biological species concept in particular.

Three historically important books by Ernst Mayr—*Animal Species and Evolution* (Harvard University Press, Cambridge, MA, 1963), its abridged successor, *Populations, Species, and Evolution* (Harvard University Press, Cambridge, MA, 1970), and its predecessor, *Systematics and the Origin of Species* (Columbia University Press, New York, 1942)—are the classic works on the nature of animal species and speciation. For plants, an equally foundational work is *Variation and Evolution in Plants* by G. L. Stebbins (Columbia University Press, New York, 1950). This topic was also treated by V. Grant in *Plant Speciation* (Columbia University Press, New York, 1981).

Endless Forms: Species and Speciation, edited by D. J. Howard and S. H. Berlocher (Oxford University Press, New York, 1998), is a stimulating collection of essays on species concepts and speciation. *Hybrid Zones and the Evolutionary Process*, edited by R. G. Harrison (Oxford University Press,

New York, 1993), compiles essays, including case studies of animals and plants, on this complex topic.

Problems and Discussion Topics

1. Some degree of genetic exchange occurs in bacteria, which reproduce mostly asexually. What evolutionary factors should be considered in debating whether or not the biological species concept can be applied to bacteria?
2. Suppose the phylogenetic species concept were adopted in place of the biological species concept. What would be the implications for (a) evolutionary discourse on the mechanisms of speciation; (b) studies of species diversity in ecological communities; (c) estimates of species diversity on a worldwide basis; (d) conservation practices under such legal frameworks as the U.S. Endangered Species Act?
3. How might the fate of two hybridizing populations, i.e., whether or not they persist as distinct populations, depend on the kind of isolating barriers that reduce gene exchange between them?
4. Identify two species of plants in the same genus that grow in your area. Propose a research program that would enable you to (a) judge whether or not there is any gene flow between the species and (b) determine what the reproductive barriers are that maintain the differences between them.
5. Studies of hybrid zones have shown that mitochondrial and chloroplast DNA markers frequently introgress farther, and have higher frequencies, than nuclear gene markers (see Avise 1994, 284–290). Thus, far from the hybrid zone, individuals that have no phenotypic indications of hybrid ancestry may have mitochondrial or chloroplast genomes of the other species (or semispecies). How would you account for this pattern?
6. Suppose you are interested in a poorly studied genus of leaf beetles that occurs in a distant land you can't afford to visit. A biologist friend who will be doing research at a field station in that region volunteers to collect some specimens and preserve them in alcohol for you. You can then study both their morphology and DNA sequences. How will you decide how many species are in the collection your friend brings back? Suppose your friend extracts mitochondrial DNA from each specimen and brings it back safely, but loses the specimens en route. How can you decide how many species are represented—or can you?
7. Suppose that two or more related taxa are polyphyletic for gene lineages, as in the grasshoppers in Figure 2.21. What evidence might enable you to decide whether this is due to incomplete sorting (i.e., lack of coalescence) of ancestral polymorphism or to introgressive hybridization?
8. Postzygotic isolation between populations or species is sometimes based on interactions between nuclear genes and cytoplasmic genetic elements such as mitochondria or endosymbiotic bacteria. How might such incompatibility act and how might it arise? Refer to work on *Wolbachia*, an endosymbiotic bacterium in insects, as a particularly well-studied example (Werren 1998).

CHAPTER 18

Speciation

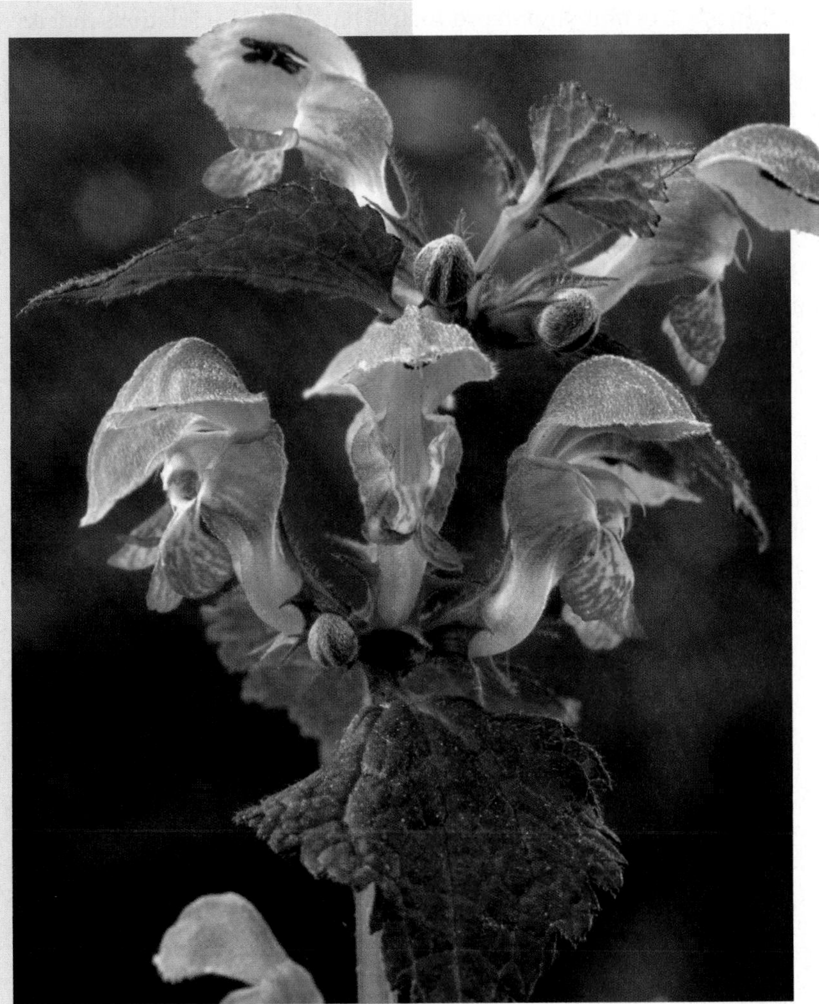

Speciation reenacted. The common hemp-nettle (*Galeopsis tetrahit*) of Europe is a wild species that was successfully "recreated" in 1930 by crossing its suspected progenitors. See p. 491. (Photo © Arco Images GmbH/Alamy.)

In this chapter, we turn to the question of how species arise. If we considered species to be merely populations with distinguishing characteristics, the question of how they originate would be easily answered: natural selection or genetic drift can fix novel alleles or characteristics (see Chapters 10–13). But if the permanence of these distinctions depends on reproductive isolation, and if we consider reproductive isolation a defining feature of species, then the central question about speciation must be how genetically based barriers to gene exchange arise.

The difficulty this question poses is most readily seen if we consider postzygotic reproductive barriers, such as hybrid inviability or sterility. If two populations are fixed for genotypes A_1A_1 and A_2A_2, but the heterozygote A_1A_2 has low reproductive success, how could these populations have diverged? Whatever allele the ancestral population may have carried (say, A_1), the low fitness of A_1A_2 would have prevented the alternative allele (A_2) from increasing in frequency and thus forming a reproductively incompatible population. Suppose, instead, that reproductive isolation between the populations is based on more than one locus. The problem with a polygenic trait as a cause of reproductive isolation is that recombination generates intermediates. If several loci govern, for example, time of breeding, $A_1A_1B_1B_1C_1C_1$ and $A_2A_2B_2B_2C_2C_2$ might breed early and late in the season, respectively, and so be reproductively isolated, but many other genotypes with intermediate breeding seasons, such as

$A_1A_2B_1B_1C_1C_2$, would constitute a "bridge" for the flow of genes between the two extreme genotypes. *The problem of speciation, then, is how two different populations can be formed without intermediates.* This problem holds, whatever the character that confers prezygotic or postzygotic isolation may be.

Modes of Speciation

The many conceivable solutions to this problem are the MODES OF SPECIATION. The modes of speciation that have been hypothesized can be classified by several criteria (Table 18.1), including the *geographic origin* of the barriers to gene exchange, the *genetic bases* of the barriers, and the *causes of evolution* of the barriers.

Speciation may occur in three kinds of geographic settings that blend one into another. **Allopatric speciation** is the evolution of reproductive barriers in populations that are prevented by a geographic barrier from exchanging genes at more than a negligible rate. A distinction is often made between allopatric speciation by **vicariance** (divergence of two large populations; Figure 18.1A) and **peripatric speciation** (divergence of a small population from a widely distributed ancestral form; Figure 18.1B). In **parapatric speciation**, spatially distinct populations, between which there is some gene flow, diverge and become reproductively isolated (Figure 18.1C). **Sympatric speciation** is the evolution of reproductive barriers within a single, initially randomly mating (panmictic) population (Figure 18.1D). Allopatric, parapatric, and sympatric speciation form a continuum, differing in the initial level of gene flow (m) between diverging populations. Strictly defined, $m = 0$ if speciation is allopatric, and is maximal ($m = 0.5$) if speciation is sympatric; intermediate cases ($0 < m < 0.5$) would represent parapatric speciation (Fitzpatrick et al. 2008). As usually envisioned, the initial reduction of gene exchange is accomplished by a physical barrier extrinsic to the organisms in allopatric speciation, but by evolutionary change in the biological characteristics of the organisms themselves in sympatric speciation. Allopatric or "nearly allopatric" (with low initial m) speciation is widely acknowledged to be a common mode of speciation; the incidence of sympatric or "nearly sympatric" (with high m) speciation is debated.

Table 18.1 *Modes of speciation*

- I. Classified by geographic origin of reproductive barriers
  - A. Allopatric speciation
    - 1. Vicariance
    - 2. Peripatric speciation
  - B. Parapatric speciation
  - C. Sympatric speciation
- II. Classified by genetic and causal bases[a]
  - A. Genetic divergence (allele substitutions)
    - 1. Genetic drift
    - 2. Peak shift (peripatric speciation)
    - 3. Natural selection
      - a. Ecological selection
        - i. Reproductive isolation
        - ii. Reproductive barriers
        - iii. Pleiotropic genes
      - b. Sexual selection
  - B. Cytoplasmic incompatibility
  - C. Cytological divergence
    - 1. Polyploidy
    - 2. Chromosome rearrangement
  - D. Recombinational speciation

[a]Most of the genetic and causal bases might act in an allopatric, parapatric, or sympatric geographic context, and some of the causal bases listed under "Genetic divergence" apply also to cytoplasmic incompatibility, cytological divergence, and/or recombinational speciation.

From a genetic point of view, the reproductive barriers that arise may be based on genetic divergence (allele differences at, usually, several or many loci), cytoplasmic incompatibility, or cytological divergence (polyploidy or structural rearrangement of chromosomes). We will devote most of this chapter to speciation by genetic divergence.

The causes of evolution of reproductive barriers, as of any characters, are genetic drift and natural selection of genetic alterations that have arisen by mutation. Peripatric speciation, a hypothetical form of speciation that is also referred to as TRANSILIENCE or SPECIATION BY PEAK SHIFT, requires both genetic drift and natural selection. Both sexual selection and ecological causes of natural selection may result in speciation. In some cases, there may be selection *for* reproductive isolation—that is, to prevent hybridization. (Recall the distinction between *selection for* and *selection of* characters, discussed in Chapter 11.) Alternatively, reproductive isolation may arise as a *by-product* of genetic changes that occur for other reasons (Muller 1940; Mayr 1963). In this case, there may be ADAPTIVE DIVERGENCE of the isolating character itself (e.g., climatic factors may favor breeding at two different seasons, with the effect that the populations do not interbreed), or the reproductive barrier may arise as a pleiotropic by-product of genes that are selected for their other functions.

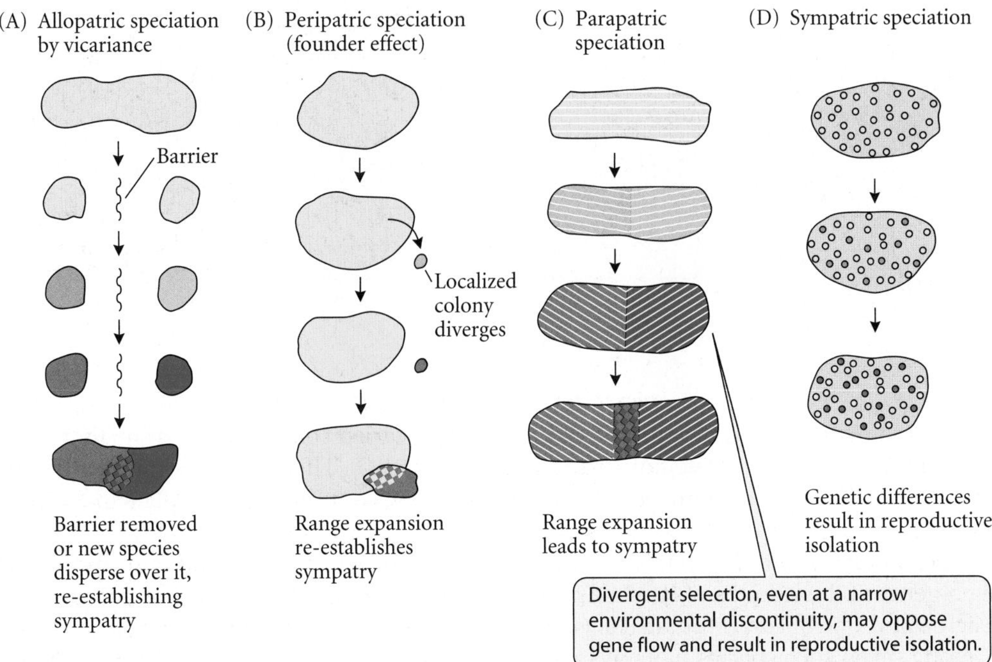

(A) Allopatric speciation by vicariance
(B) Peripatric speciation (founder effect)
(C) Parapatric speciation
(D) Sympatric speciation
Barrier
Localized colony diverges
Barrier removed or new species disperse over it, re-establishing sympatry
Range expansion re-establishes sympatry
Range expansion leads to sympatry
Genetic differences result in reproductive isolation
Divergent selection, even at a narrow environmental discontinuity, may oppose gene flow and result in reproductive isolation.

Figure 18.1 Schematic diagrams show the successive stages in models of speciation that differ in geographic setting. (A) The vicariance model of allopatric speciation. (B) The peripatric, or founder effect, model of allopatric speciation. (C) Parapatric speciation. (D) Sympatric speciation.

Allopatric Speciation

Allopatric speciation is the evolution of genetic reproductive barriers between populations that are geographically separated by a physical barrier such as topography, water (or land), or unfavorable habitat. The physical barrier reduces gene flow sufficiently for genetic differences between the populations to evolve that prevent gene exchange if the populations should later come into contact (see Figure 18.1A). Although some authors' definition of allopatric speciation requires zero gene flow ($m = 0$) between the populations, we will assume for this discussion only that m is so low that divergence by very weak selection or even genetic drift is possible. Allopatry is defined by a severe reduction of movement of individuals or their gametes, not by geographic distance. Thus, in species that disperse little or are faithful to a particular habitat, populations may be "microgeographically" isolated (e.g., among patches of a favored habitat along a lakeshore). All evolutionary biologists agree that allopatric speciation occurs, and many hold that it is the prevalent mode of speciation, at least in animals (Mayr 1963; Coyne and Orr 2004).

Allopatric populations may expand their range and come into contact. (Consider, for example, the postglacial expansion portrayed in Figure 6.15.) If sufficiently strong isolating barriers have evolved during the period of allopatry, the populations may become sympatric without exchanging genes. If incomplete reproductive isolation has evolved, they will form a hybrid zone (see Chapter 17, where we described the possible fates of hybrid zones). Because range expansion may have occurred in the past, sympatric sister species that we observe today may well have speciated allopatrically, not sympatrically.

Evidence for allopatric speciation

Because both natural selection and genetic drift cause populations to diverge in genetic composition, it is probably inevitable that if separated long enough, geographically separated populations will become different species. Many species show incipient prezygotic and/or postzygotic reproductive isolation among geographic populations. For example, Stephen Tilley and colleagues (1990) examined sexual isolation among dusky salamanders (*Desmognathus ochrophaeus*) from various localities in the southern Appalachian Mountains of the eastern United States. They brought males and females from different populations (heterotypic pairs) and from the same population (homotypic pairs) together and scored the proportion of these pairs that mated. Among the various pairs of pop-

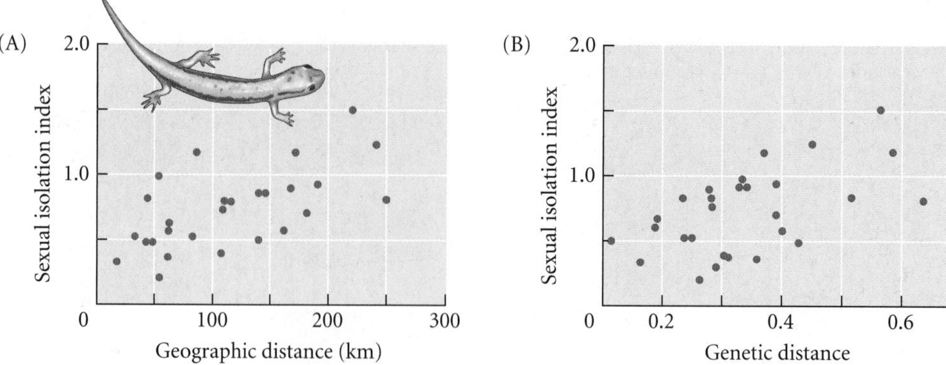

Figure 18.2 The degree of sexual isolation between populations of the salamander *Desmognathus ochrophaeus* is correlated with (A) the geographic distance between the populations as well as (B) their genetic distance (Nei's *D*, which measures the difference in allozyme frequencies at several loci). (After Tilley et al. 1990.)

ulations, an index of the strength of sexual isolation varied continuously, from almost no isolation to almost complete failure to mate (Figure 18.2). The more geographically distant the populations, the more genetically different they were, and the less likely they were to mate.

Speciation can often be related to the geological history of barriers. For example, the emergence of the Isthmus of Panama in the Pliocene divided many marine organisms into Pacific and Caribbean populations, some of which have diverged into distinct species. Among seven such species pairs of snapping shrimp, only about 1 percent of interspecific matings in the laboratory produced viable offspring (Knowlton et al. 1993). In some cases, CONTACT ZONES between differentiated forms mark the meeting of formerly allopatric populations. For example, Eldredge Bermingham and John Avise (1986; Avise 1994) analyzed the genealogy of mitochondrial DNA in samples of six fish species from rivers throughout the coastal plain of the southeastern United States. In all six species, DNA sequences form two distinct clades characterizing eastern and western populations, and the two clades make contact in the same region of western Florida (Figure 18.3). This pattern implies that gene flow between east and west was reduced at some time in the past. The amount of sequence divergence between the two DNA clades suggests that isolation occurred 3 to 4 Mya. At that time, sea level was much higher than at present, forming a barrier to dispersal by freshwater fishes.

In many cases, related species replace each other geographically, reflecting their presumably allopatric origins. For example, the leopard frogs of North America, most of which were at one time considered geographic races of *Rana pipiens*, consist of about 27 species in two clades (Hillis 1988). The geographic ranges of the species in each clade overlap only slightly or not at all (Figure 18.4). These species are actually or potentially isolated by habitat, breeding season, mating call, and/or hybrid inviability.

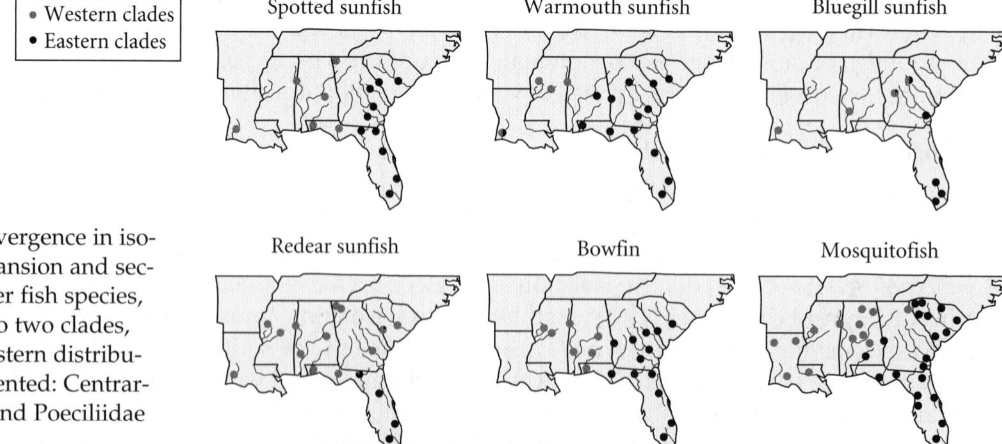

Figure 18.3 Evidence for genetic divergence in isolated "refuges," followed by range expansion and secondary contact. In each of six freshwater fish species, mitochondrial DNA haplotypes fall into two clades, one with a western and one with an eastern distribution. Three families of fishes are represented: Centrarchidae (sunfishes), Amiidae (bowfin), and Poeciliidae (mosquitofish). (After Avise 1994.)

(A)

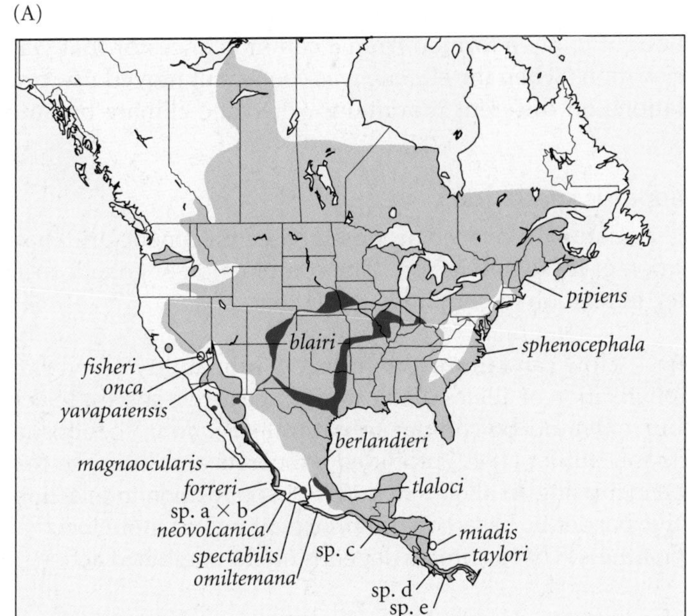

(B)

(C)

Figure 18.4 (A) The distribution of one clade of species in the *Rana pipiens* (leopard frog) complex. Studies of the overlap zones (dark) between some of these forms have shown that they are reproductively isolated species. The largely allopatric or parapatric distributions of these species suggest that spatial separation has enabled them to become differentiated. (B) *Rana pipiens*, a northern species. (C) The southern leopard frog, *R. sphenocephala*. (A after Hillis 1988; B © Rod Planck/Photo Researchers Inc.; C © Rex Lisman/istockphoto.)

Allopatric speciation is also supported by negative evidence. No pairs of sister species of birds occur together on any isolated island, implying that speciation does not occur on land masses that are too small to provide geographic isolation between populations (Coyne and Price 2000). Where two or more closely related species of birds do occur on an island, other islands or a continent can be identified as a source of invading species, and in all cases there is evidence that the several species invaded the island at separate times. For example, many of the islands in the Galápagos archipelago harbor two or more species of Darwin's finches, which evolved on different islands and later became sympatric. But Cocos Island, isolated far to the northeast of the Galápagos, has only one species of finch, which occupies several of the ecological niches that its relatives in the Galápagos Islands fill (Werner and Sherry 1987; see Figure 3.23).

Tim Barraclough and Alfried Vogler (2000) reasoned that over time, the amount of overlap between the geographic ranges of species that have formed by allopatric speciation can only increase from zero, whereas overlap between species that originated by sympatric speciation should stay the same or decrease. For several clades of closely related birds, insects, and fishes, they plotted degree of range overlap against degree of molecular difference between species, which they used as an index of time since gene exchange was curtailed. Several groups showed increasing overlap with time, as expected from allopatric speciation, whereas two groups of insects displayed a pattern consistent with the possibility of sympatric speciation (Figure 18.5).

The role of barriers such as the Isthmus of Panama in curtailing gene flow between populations is obvious, but what kinds of barriers could have produced the great numbers of species, in many taxa, that are found on continents? An important consideration is phylogenetic niche conservatism (see Chapter 6). Geographic distributions may be fragmented if populations maintain dependence on specific environmental conditions, such as climate regimes or habitats. Kenneth Kozak and John Wiens (2006) addressed the hypothesis that niche conservatism promotes allopatric speciation by using distribution data on plethodontid salamanders to estimate the "climatic niche space" of species. They found that, as predicted, allopatric sister species are found in locations with similar climatic conditions, and are absent from inter-

Figure 18.5 The degree of overlap in the geographic ranges of pairs of closely related species, plotted against the genetic divergence between them, which is an index of time since speciation. Overlap increases with time in fairy wrens and swordtail fish, as expected if speciation is allopatric in these groups. There is no correlation between overlap and time since divergence in tiger beetles or *Rhagoletis* fruit flies, a pattern consistent with sympatric speciation. (After Barraclough and Vogler 2000.)

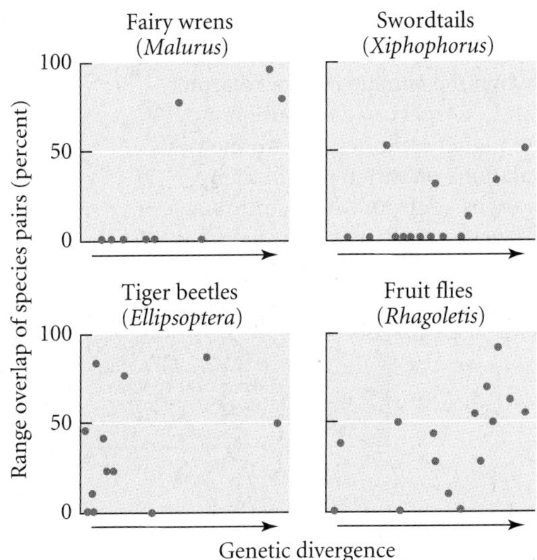

vening regions with different climate (Figure 18.6A). The sister-species pairs occur at fairly high elevations and are thought to have formed from a common ancestor that was widely distributed at lower elevations when the climate was cooler, but moved upward and formed separate populations on different mountains when the climate became warmer (Figure 18.6B).

Mechanisms of vicariant allopatric speciation

Models of vicariant allopatric speciation have been proposed based on genetic drift, natural selection, and a combination of these two factors. The combination of genetic drift and selection is discussed later, in relation to peripatric speciation.

THE ORIGIN OF INCOMPATIBILITY. How can failure to interbreed, or inability of hybrids to reproduce, arise if they imply fixation of alleles that lower reproductive success? Such an increase to fixation, of course, would be counter to natural selection. Theodosius Dobzhansky (1936) and Hermann Muller (1940) provided a theoretical solution to this problem that does not envision increasing an allele's frequency in opposition to selection. It requires that the reproductive barrier be based on differences at two or more loci that have complementary effects on fitness. That is, fitness depends on the combined action of

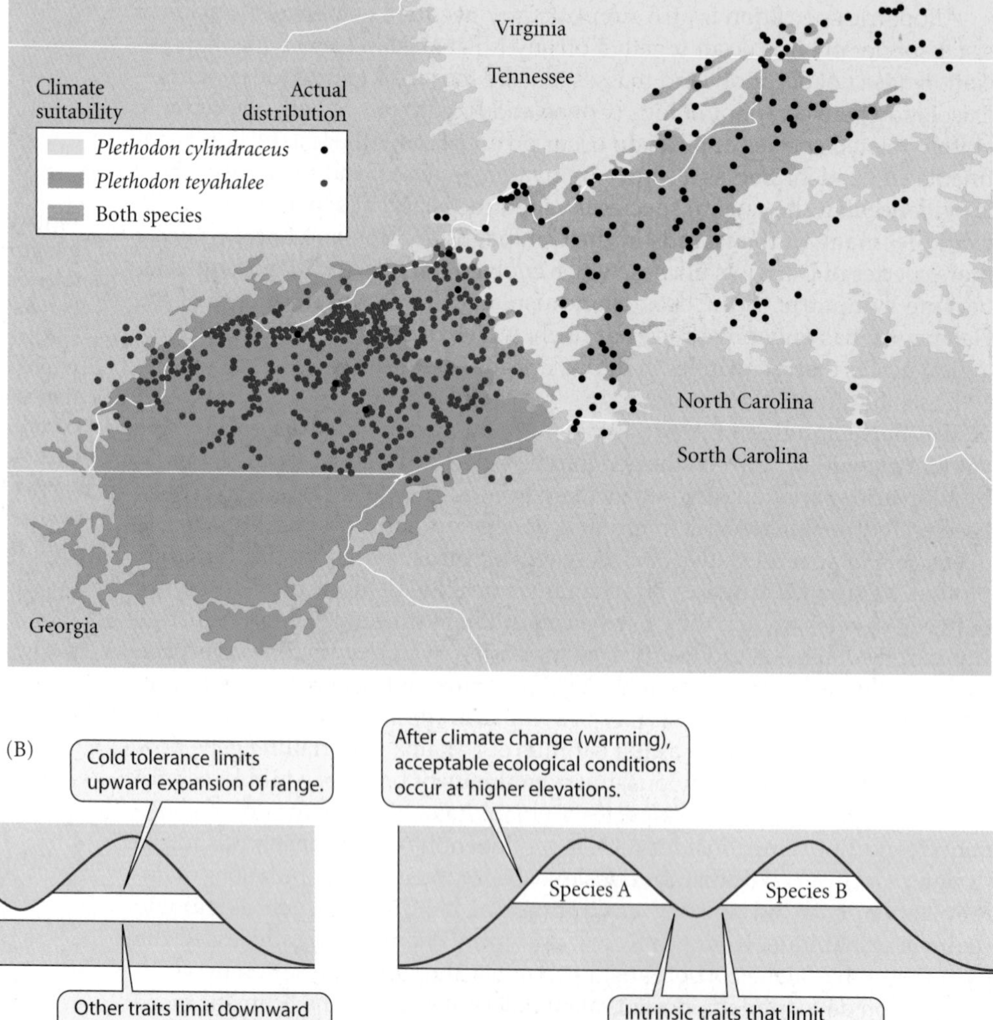

Figure 18.6 (A) Distribution data on plethodontid salamanders in the Appalachian Mountains, used to estimate their "climatic niche space." Climate data from localities where each of two sister species (*Plethodon cylindraceus* and *P. teyahalee*) has been recorded was used to predict the distribution of the other species, assuming that they share similar climate tolerances. The actual distribution of each species falls within the predicted distribution. (B) Sister species with similar climate tolerances, such as these salamanders, may have formed from a widespread ancestral species (left diagram) when the climate became warmer and the species' distribution moved to higher altitudes, separating populations on different mountain ranges. (A from Kozak and Wiens 2006; B after Wiens 2004.)

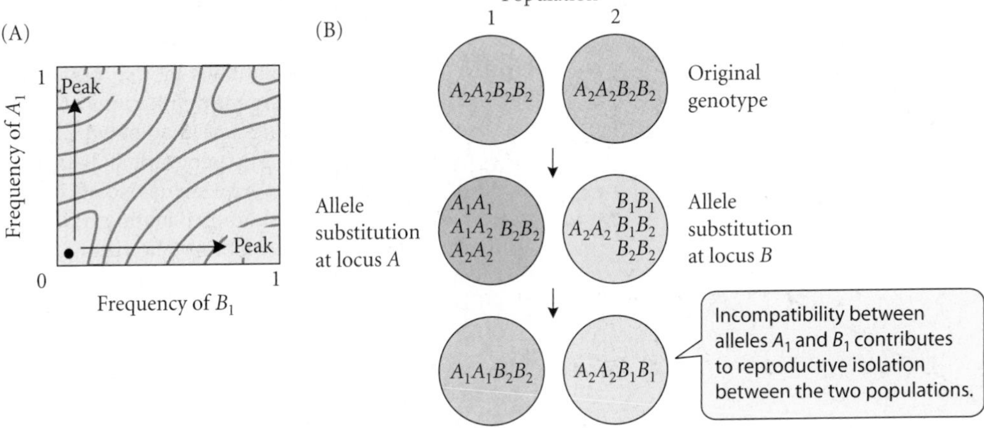

Figure 18.7 The Dobzhansky-Muller theory of allele substitution leading to reproductive isolation between two populations, each initially composed of genotype $A_2A_2B_2B_2$. (A) The adaptive landscape, in which contour lines represent mean fitness as a function of allele frequencies at both loci, shows how the two populations may move uphill toward different adaptive peaks. (B) Each population undergoes an allele substitution at a different locus (substituting either A_1 or B_1). The hybrid combination $A_1A_2B_1B_2$ has low fitness (as indicated by the "valley" in the center of the landscape) because of prezygotic or postzygotic incompatibility between A_1 and B_1.

the "right" alleles at both loci. The "wrong" combinations of alleles result in "Dobzhansky-Muller incompatibility," as illustrated by the *Drosophila* genes described in Figure 17.13.

Suppose the ancestral genotype in two allopatric populations is $A_2A_2B_2B_2$ (Figure 18.7). For some reason, A_1 replaces A_2 in population 1 and B_1 replaces B_2 in population 2, yielding populations monomorphic for $A_1A_1B_2B_2$ and $A_2A_2B_1B_1$, respectively. Both A_1A_2 and A_1A_1 have fitness equal to or greater than A_2A_2 in population 1, *as long as the genetic background is* B_2B_2; likewise, B_1B_2 and B_1B_1 are equal or superior to B_2B_2, as long as the genetic background is A_2A_2. Therefore these allele substitutions can occur by natural selection (if the fitnesses differ) or by genetic drift (if they do not). However, an epistatic interaction between A_1 and B_1 causes incompatibility, so that either the hybrid $A_1A_2B_1B_2$ has lowered viability or fertility, or $A_1A_1B_2B_2$ and $A_2A_2B_1B_1$ are isolated by a prezygotic barrier, such as different sexual behavior. The important feature of this model is that *neither population has passed through a stage in which inferior heterozygotes existed*. Neither of the incompatible alleles has ever been "tested" against the other within the same population.

A mutation at a third locus, say C_1, may be fixed in the $A_1A_1B_1B_1$ population but be incompatible with either A_2 or B_2, further lowering the hybrid's fitness. Likewise, mutation D_1 might be fixed, and be incompatible with A_2, B_2, or C_2. As more allelic differences accumulate, later mutations have more opportunities to cause incompatibility and lower the fitness of hybrids. Thus the degree of hybrid sterility or inviability may increase exponentially with the passage of time (Orr 1995).

This model is supported by genetic data showing that reproductive isolation is based on epistatic interactions among several or many loci (see Chapter 17). The allele substitutions could be caused by either genetic drift or natural selection. For the moment, we will leave open the possibility of speciation by random fixation of alleles, and consider the ways in which natural selection may contribute to the origin of species.

THE ROLE OF NATURAL SELECTION. The most widely held view of vicariant allopatric speciation is that it is caused by *natural selection, which causes the evolution of genetic differences that create prezygotic and/or postzygotic incompatibility*. Some—perhaps most—of the reproductive isolation evolves while the populations are allopatric, so that a substantial or complete barrier to gene exchange exists when the populations meet again if their ranges expand (Mayr 1963). Thus speciation is usually an *effect*—a by-product—of the divergent selection that occurred during allopatry. The divergent selection may be ecological selection or sexual selection.

A contrasting possibility is that natural selection favors prezygotic (e.g., sexual) reproductive barriers because of their isolating function—because they prevent their bearers from having unfit hybrid progeny. Selection would then result in *reinforcement of reproductive isolation*.

Ecological selection and speciation

Allopatric populations and species undergo both adaptive divergence and evolution of reproductive isolation, but showing that reproductive isolation is a *result* of adaptive

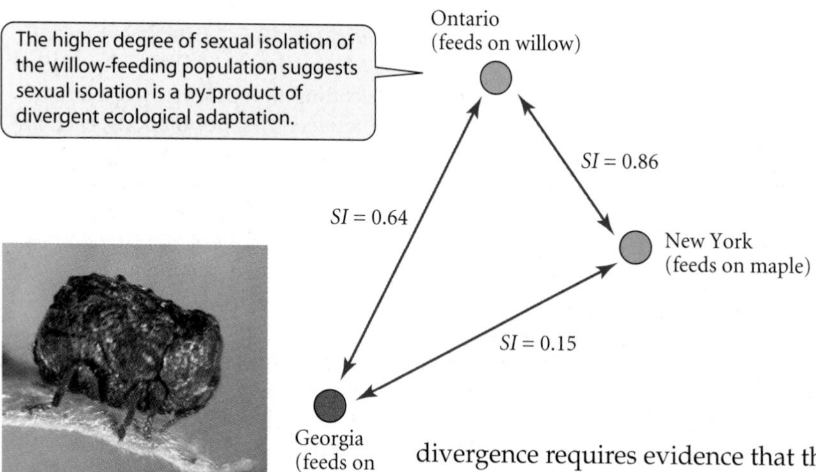

Figure 18.8 Sexual isolation is more pronounced between ecologically divergent populations of the leaf beetle *Neochlamisus bebbianae* than between populations that are separated by geographic distance alone. The numbers represent a sexual isolation index (*SI*) derived from laboratory mating trials. (After Funk 1998; photo by Christopher Brown, courtesy of Daniel Funk.)

divergence requires evidence that the two processes are genetically and causally related to each other. Molecular data provide some evidence that natural selection can contribute to speciation. The few genes that contribute to reproductive isolation and that have been sequenced, such as *Nup96* in *Drosophila*, show the signature of directional selection (Chapter 17). The most direct evidence comes from laboratory studies of *Drosophila* and houseflies, in which investigators have tested for reproductive isolation among subpopulations drawn from a single base population and subjected to divergent selection for various morphological, behavioral, or physiological characteristics (Rice and Hostert 1993). In many of these studies, partial sexual isolation or postzygotic isolation developed, demonstrating that substantial progress toward speciation can be observed in the laboratory, and that it can arise as a correlated response to divergent selection. That is, reproductive isolation was due to pleiotropic effects of genes for the divergently selected character, or to closely linked genes.

There is now considerable evidence for **ecological speciation**—the evolution of barriers to gene flow caused by divergent ecologically based selection (Rundle and Nosil 2005; Funk et al. 2006). Ecologically divergent characters may play a direct role in reproductive isolation. A good example is the case of the two monkeyflowers (*Mimulus*) described in Chapter 17, which have become adapted to different pollinators (see Figure 17.7). But reproductive isolation can be a by-product of ecological adaptation, presumably because of pleiotropic effects of alleles that affect adaptation. For example, Daniel Funk (1998) predicted that if this hypothesis were true, allopatric populations of an insect that are adapted to different host plant species should display stronger incipient isolation from each other than allopatric populations that feed on the same host plant. As predicted, a willow-adapted population of the leaf beetle *Neochlamisus bebbianae* was more strongly sexually isolated from two maple-feeding populations than the latter were from each other (Figure 18.8). In a similar vein, three-spined sticklebacks (*Gasterosteus aculeatus*) have undergone PARALLEL SPECIATION in several Canadian lakes, where a limnetic (open-water) ECOMORPH coexists with a benthic (bottom-feeding) ecomorph that is smaller and differs

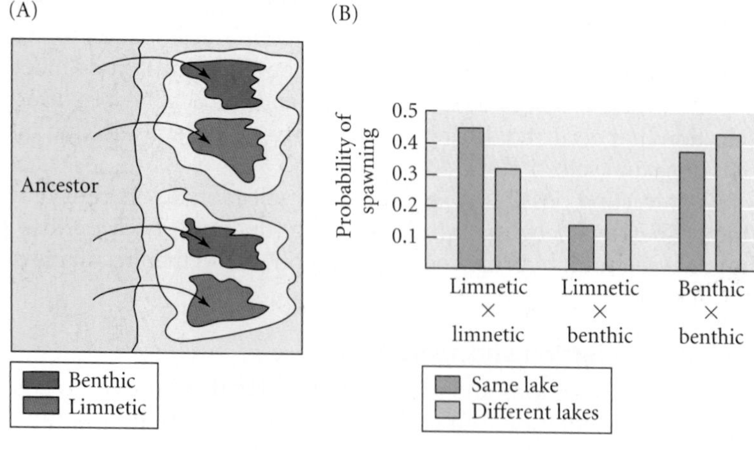

Figure 18.9 Parallel speciation in the three-spined stickleback (*Gasterosteus*). (A) Pairs of open-water (limnetic) and bottom-feeding (benthic) ecomorphs have arisen independently in different lakes. (B) Females mate preferentially with males on the basis of their morphology, whether they are from the same or different lakes. This isolating character is evidently adaptive, since it has repeatedly evolved in the same way. (A after Schluter and Nagel 1995; B after Rundle et al. 2000.)

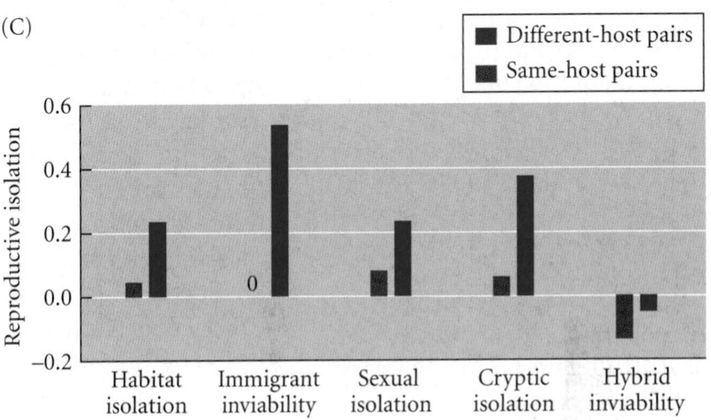

Figure 18.10 Ecomorphs of the stick insect *Timema cristinae* found on the shrubs *Adenostoma fasciculatum* (A) and *Ceanothus spinosus* (B) differ in body form and color pattern, closely matching the foliage on which they feed. (C) Pairs of populations from different hosts are more strongly reproductively isolated in several respects than pairs from the same host. The negative values for "hybrid inviability" indicate enhanced viability of hybrids, but here these values are not statistically significant. (Photos courtesy of Patrik Nosil; C after Nosil 2007.)

in shape. These ecomorphs are sexually isolated and have evolved independently in each lake; that is, speciation has occurred in parallel (Figure 18.9A). Parallel ecological divergence implies that ecological selection has shaped the differences between the ecomorphs. In laboratory trials, fish of the same ecomorph from different lakes mate almost as readily as those from the same lake, but different ecomorphs mate much less frequently (Figure 18.9B). Thus features associated with ecological divergence affect reproductive isolation (Rundle et al. 2000). Allopatric lake populations of sticklebacks also show sexual isolation between morphologically different forms adapted to different environments (Vines and Schluter 2006).

Two ecomorphs of the stick insect *Timema cristinae* are associated with different host plants in California, and differ in several morphological features that make each ecomorph more camouflaged on its own host plant than on the other. Patrik Nosil (2007) has studied several components of reproductive isolation between multiple pairs of different-host populations (Figure 18.10). Ecological divergence directly reduces gene flow in two ways. Habitat isolation reduces gene flow because large patches of chaparral vegetation are dominated by one or the other plant. "Immigrant inviability" results from the high mortality the insects suffer if they disperse to the "wrong" host and are easy targets for birds. Indirect effects of ecological divergence include sexual isolation, which is greater between different-host than same-host populations, and reduced fertility in cross-matings: females that mate with the other ecomorph lay fewer eggs.

Funk, Nosil, and their colleague William Etges (2006) provided evidence that divergent ecological adaptation commonly contributes to speciation, by compiling data from the literature on reproductive isolation, indicators of ecological divergence among species, and genetic distance among species in several groups of plants, insects, fishes, frogs, and birds. Genetic distance was used as an index of time since pairs of species had diverged from their common ancestor. By statistically controlling for time, they showed that the level of reproductive isolation is correlated with the degree of ecological divergence between species (Figure 18.11).

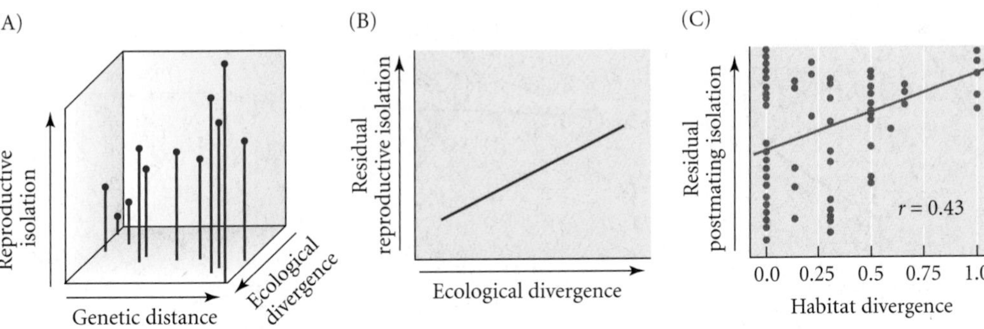

Figure 18.11 A method for testing the hypothesis that reproductive isolation evolves as a by-product of ecological divergence. (A) For several pairs of species, experimental estimates of reproductive isolation are plotted against both the time since common ancestry (estimated by genetic distance, as in Figure 17.10) and a measure of ecological difference. (B) The amount of "residual" reproductive isolation that is not accounted for by time since common ancestry is plotted against ecological divergence and tested for correlation. (C) An example of real data, showing that postzygotic isolation between pairs of flowering plant species is significantly correlated with difference in habitat use. (After Funk et al. 2006.)

Sexual selection and speciation

Closely related species of animals are often sexually isolated by female preferences for features of conspecific males. One hypothesis for the differences between species in these characteristics is that they enable individuals to recognize conspecific mates and avoid hybridization, which would be disadvantageous if hybrid offspring have low fitness (see below). Another hypothesis is that divergent sexual selection in different geographic populations of a species results in different male display traits and female preferences (Fisher 1930; West-Eberhard 1983). This hypothesis has been supported by mathematical models (Lande 1981; Pomiankowski and Iwasa 1998; Turelli et al. 2001).

It is very likely that sexual selection has been an important cause of speciation, especially in highly diverse groups such as Hawaiian *Drosophila*, birds of paradise, and hummingbirds (see Figure 17.6), in which males are often highly (and diversely) colored or ornamented (Panhuis et al. 2001). For example, the male color pattern of some closely related African lake cichlids acts as a reproductive barrier between species, and it is likely that sexual selection has contributed to the extraordinarily high species diversity of these fishes (Seehausen et al. 1999). Some comparisons of the species diversity of sister groups of birds suggest that sexual selection has enhanced diversity (Figure 18.12). Groups of birds with promiscuous mating systems have higher diversity than sister clades in which pair-bonds are formed and the variance in male mating success is presumably lower—resulting in weaker sexual selection (Mitra et al. 1996). Because sister clades, by definition, are equal in age, the difference in diversity implies a higher rate of speciation (or possibly a lower extinction rate) in clades that experience strong sexual selection. However, some studies have failed to find evidence that sexual dimorphism or other possible indicators of sexual selection are associated with higher rates of bird diversification (Price 2008).

Speciation by sexual selection is strongly suggested by cases such as two sister species of crickets (*Gryllus texensis* and *G. rubens*) in the southern United States (Gray and Cade

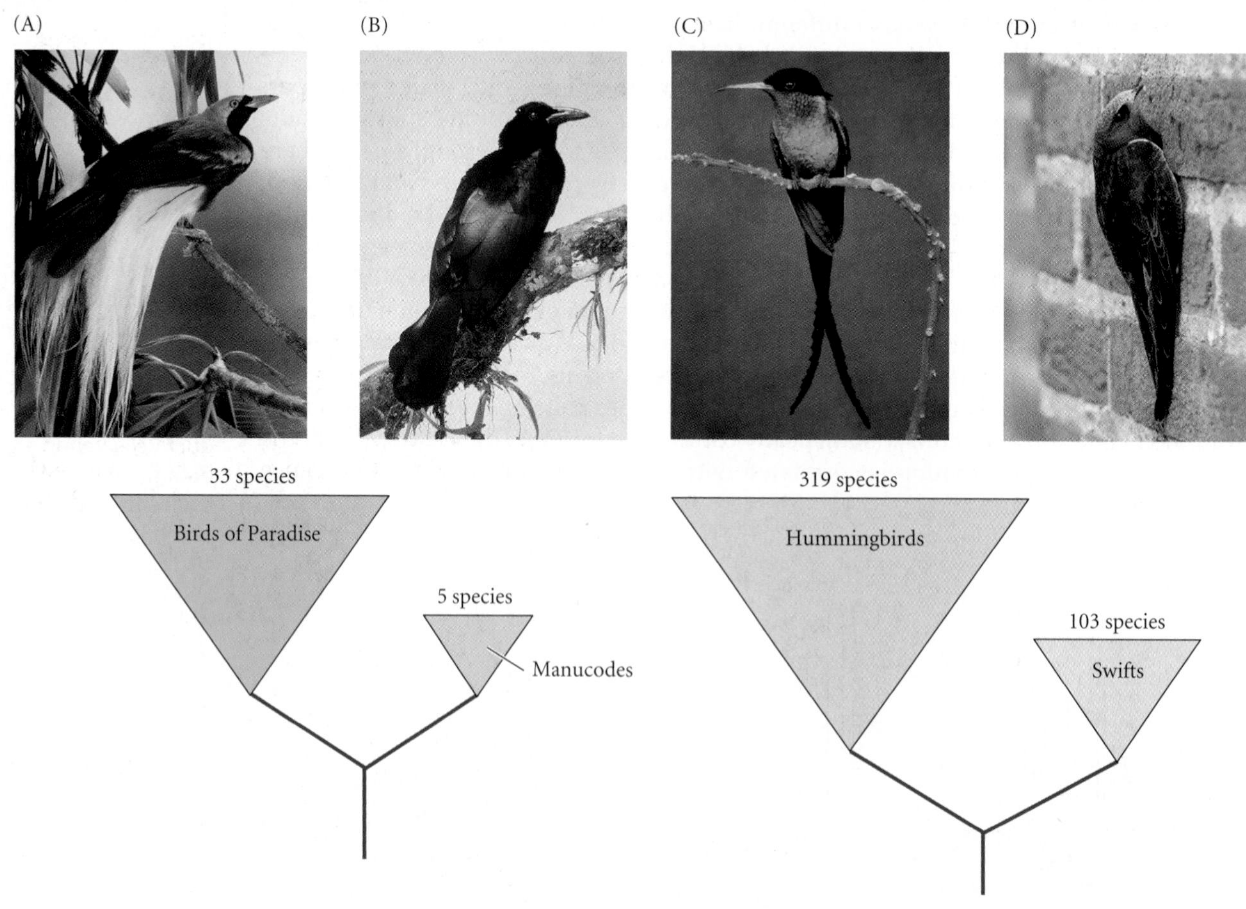

Figure 18.12 In sister clades of birds that differ in their mating system, those clades that mate promiscuously and do not form a pair-bond (A, C) tend to have more species than nonpromiscuous clades (B, D) that do form pair-bonds. The promiscuously mating clades are thought to experience stronger sexual selection. (A) A promiscuous male lesser bird of paradise (*Paradisaea minor*) and (B) a nonpromiscuous manucode (*Manucodia comrii*). (C) A male streamertail hummingbird (*Trochilus polytmus*) and (D) a common swift (*Apus apus*), member of a clade that forms pair-bonds. (A © Tony Tilford/photolibrary.com; B © W. Peckover, VIREO; C © RobertTyrrell/OSF/photolibrary.com; D © National Trust Photolibrary/Alamy Images.)

2000). The species differ in the pulse rate of the male song, and females of both species show a much stronger response to songs with pulse rates characteristic of their own species. Hybrids produced in the laboratory are viable and fully fertile, and there is no evidence that the differences in male song or female preference have evolved to reduce hybridization, because distant allopatric populations of the two species, which have no opportunity to hybridize, are just as different as sympatric populations.

Causes of speciation are often best studied by analyzing divergence among populations that have not yet become fully reproductively isolated. Populations of the South American frog *Physalaemus petersi*, studied in Ecuador by Kathryn Boul et al. (2007), provide strong evidence for incipient speciation by sexual selection. Populations only 20 to 30 km apart differ in the dominant frequency of the male's "whine" call and in whether the call is simple or more complex, with a second component (a "squawk"). Females in both a "simple-call" population (La Selva) and a "complex-call" population (Yasuní) were tested for their response to recordings of the whine call of both populations, with or without an appended squawk. Yasuní females preferred calls with a squawk, but La Selva females did not show any such discrimination. However, both females strongly preferred their own population's whine call. The authors suggest that divergence in the whine frequency is a by-product of evolution of a larger larynx in populations that include a squawk in the male call. Molecular markers show that gene flow between populations with different call types is 30 times lower than between equally distant populations with the same call type, suggesting that the divergence confers reproductive isolation and that the populations are incipient species.

These studies do not include analysis of sexual selection within the divergent populations or species, but Michael Ritchie (2000) performed such an analysis in the bushcricket *Ephippiger ephippiger*. The male song has a single syllable in Mediterranean populations but multiple syllables in the Pyrenees Mountains. Ritchie provided females with a choice between tape recordings of synthetic songs that varied in syllable number and observed which speaker the females approached. Mediterranean females showed increasing preference, the lower the syllable number, and should therefore exert directional selection on males for monosyllabic songs (Figure 18.13). In contrast, Pyrenees females responded most to songs with five syllables, which is actually more than most males in this polysyllabic population emit. Thus females exert directional sexual selection for greater syllable number in this population.

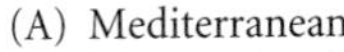

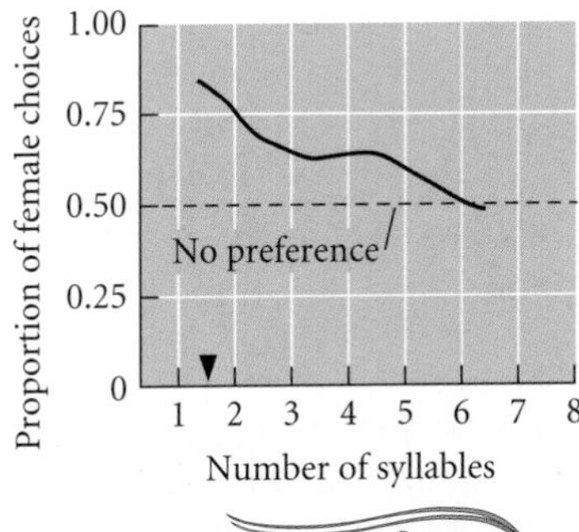

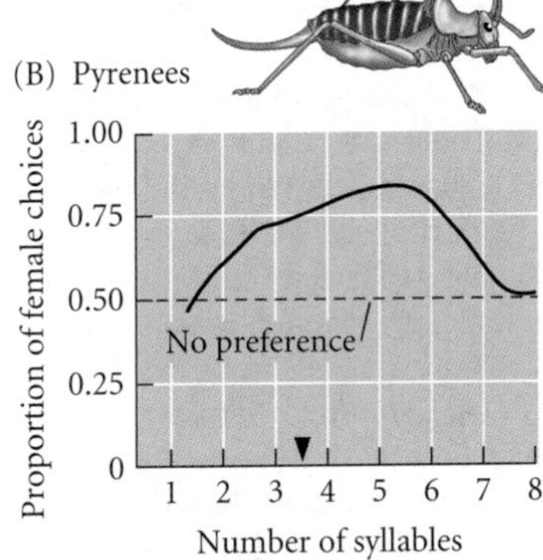

Figure 18.13 Divergent sexual selection for male song in the bushcricket *Ephippiger ephippiger*. The plots show the proportion of females that chose male songs with different numbers of syllables. Black triangles indicate the mean number of syllables in male song in each region. Females of (A) a Mediterranean population impose directional selection for few syllables, whereas (B) Pyrenees females impose selection for a greater number. Such differences in sexual selection can result in reproductive isolation. (After Ritchie 2000.)

Reinforcement of reproductive isolation

We have seen that reproductive isolation can arise as a *side effect* of genetic divergence due to natural selection. However, many persons have supposed that reproductive isolation evolves, at least in part, as an *adaptation to prevent the production of unfit hybrids*. The champion of this viewpoint was Theodosius Dobzhansky, who expressed the hypothesis this way:

> Assume that incipient species, A and B, are in contact in a certain territory. Mutations arise in either or in both species which make their carriers less likely to mate with the other species. The nonmutant individuals of A which cross to B will produce a progeny which is adaptively inferior to the pure species. Since the mutants breed only or mostly within the species, their progeny will be adaptively superior to that of the nonmutants. Consequently, natural selection will favor the spread and establishment of the mutant condition. (Dobzhansky 1951, 208)

Dobzhansky introduced the term "isolating mechanisms" to designate reproductive barriers, which he believed were indeed mechanisms designed to isolate. In contrast, Ernst Mayr (1963), among others, held that although natural selection might enhance reproductive isolation, reproductive barriers arise mostly as side effects of allopatric divergence, whatever its cause may be. Mayr cited several lines of evidence: sexual isolation exists among fully allopatric forms; it has not evolved in several hybrid zones that are thought to be thousands of years old; features that promote sexual isolation between species are usually not limited to regions where they are sympatric and face the "threat" of hybridization. It is now generally agreed that natural selection can enhance prezygotic reproductive isolation between hybridizing populations, but how often this process plays a role in speciation is not known (Howard 1993; Servedio and Noor 2003).

In most organisms, natural selection cannot strengthen *postzygotic* isolation between hybridizing populations because such a process would require that alleles that reduce fertility or survival increase in frequency, which would be precisely antithetical to the meaning of natural selection! (See Grant 1966 and Coyne 1974 for possible exceptions.) Postzygotic isolation arises, instead, by fixation of alleles at different loci in separate populations that happen to reduce fertility or viability when they are combined in hybrids. *Within* a population of hybrids, we would expect mutations that *improve* fitness to increase in frequency. Such an increase would reduce the advantage of genes that accentuate prezygotic isolation, and the likely result would be fusion of the populations. For example, populations of the common shrew (*Sorex araneus*) in England differ in complex chromosome rearrangements that should greatly reduce the fertility of heterozygotes. In contact zones between populations, however, the complex rearrangements have been replaced by simple chromosome arrangements that do not reduce hybrid fertility very much (Searle 1993).

The enhancement of prezygotic barriers that Dobzhansky envisioned is often called **reinforcement** of prezygotic isolation. This process has been cited as a cause of **character displacement**, meaning a *pattern* whereby characters differ more where two taxa are sympatric than where they are allopatric (Brown and Wilson 1956).

Figure 18.14 diagrams a simple model of reinforcement, in a highly simplified attempt to convey the basic idea. Suppose two allopatric populations have become fixed for different alleles at some loci, such as locus *A*, and that hybrids, denoted as genotype *Aa*, have lower fitness (1 – *s*) because they have reduced survival or fertility. Another locus, *M*, governs ASSORTATIVE MATING, such that *M* carriers have a strong preference for other *M* carriers as mates, and *m* carriers likewise prefer *m* mates. (The allele *m* might alter the season of mating or a phenotypic character that mates use to recognize each other.) The populations expand and meet in a hybrid zone in which, for the sake of argument, the *AA* and *aa* populations are equally abundant, so a female is equally likely to encounter an *AA* or an *aa* male. If both populations are fixed for the same mate preference allele (say, *M*), many individuals of both the *AAMM* and *aaMM* genotypes have reduced fitness because they produce unfit hybrid offspring (see the yellow mating table in Figure 18.14). But if some *aa* individuals have another mate preference genotype (*mm*) and mate almost exclusively with each other, they produce mostly nonhybrid offspring (*aamm*) with high fitness (see the red-bordered matings in the figure). Because the average fitness of *aamm* is greater than that of *aaMM*, the *m* allele will increase in frequency, so an increasing proportion of the *aa* population (which is increasingly *aamm* rather than *aaMM*) is sexually isolated from the *AAMM* population, and the incidence of matings that produce hybrid offspring declines.

Notice that if the mating allele *m* were randomly associated with both the *AA* and *aa* genotypes (as shown by the green-bordered matings in the figure), mating between *AAmm* and *aamm* would continue to produce unfit hybrids, so the *m* allele would not be associated with fitter offspring, and it would not increase in frequency: reinforcement of sexual isolation would not occur. Thus the evolution of assortative mating depends on an association (linkage disequilibrium) between the alleles at the mating locus (*M*, *m*) and those at a locus that determines fitness (*A*, *a*). Recombination between these loci will hinder the process of reinforcement.

This is perhaps the simplest model of reinforcement, and others may be more realistic (e.g., Liou and Price 1994; Cain et al. 1999; Kirkpatrick and Servedio 1999). For example, instead of assuming a single locus (*M*) for assortative mating, assume that one locus (*P*) governs female mate preference and another a male phenotypic trait (*T*) that differs between the populations. Then divergent sexual selection for different mate preference alleles (P_1, P_2) in the two populations may be reinforced by the low fitness of hybrids. Because sexual selection generally creates an association (linkage disequilibrium) between

Figure 18.14 A highly simplified illustration of how selection may favor assortative mating and reinforcement of sexual isolation in a hybrid zone in which the hybrid (heterozygote *Aa*) has reduced fitness. There is assortative mating with respect to the *M* locus, so *MM* and *mm* progeny (solid-bordered cells) are produced more often than *Mm* progeny (broken-bordered cells) Yellow cells show that if there is random mating because all individuals are *MM*, hybrids with low fitness are produced. Red cells show that the *m* allele can have high fitness if it is associated with the *a* allele, because *m* carriers mate mostly with each other, yielding *aamm* progeny with high fitness (solid-bordered red cell). Because few *Mm* heterozygotes are produced (broken-bordered red cells), *m* is seldom associated with the unfit *Aa* heterozygote. Green cells indicate that, in contrast, the *m* allele is associated with both high-fitness *AAmm* and low-fitness *Aamm* progeny if it is randomly associated with both *A* and *a* alleles. Therefore, the *m* allele can increase—causing an increase of assortative mating and sexual isolation—only if it is in linkage disequilibrium with an allele at the *A* locus.

♂ \ ♀	*AAMM*	*aaMM*	*aamm*	*AAmm*
AAMM	*AAMM* $w = 1$	*AaMM* $w = 1-s$	*AaMm* $w = 1-s$	*AAMm* $w = 1$
aaMM	*AaMM* $w = 1-s$	*aaMM* $w = 1$	*aaMm* $w = 1$	*AaMm* $w = 1-s$
aamm	*AaMm* $w = 1-s$	*aaMm* $w = 1$	*aamm* $w = 1$	*Aamm* $w = 1-s$
AAmm	*AAMm* $w = 1$	*AaMm* $w = 1-s$	*Aamm* $w = 1-s$	*AAmm* $w = 1$

Figure 18.15 Reinforcement in the pied flycatcher (*Ficedula hypoleuca*). Where this species is sympatric with the collared flycatcher (*F. albicollis*), males of the two species are similarly patterned in black and white; where the two species are sympatric, a brown form of the male pied flycatcher is prevalent. In regions of sympatry, female pied flycatchers greatly prefer brown over black-and-while pied males. In allopatric populations, female pied flycatchers prefer black-and-white males. (After Sætre et al. 1997.)

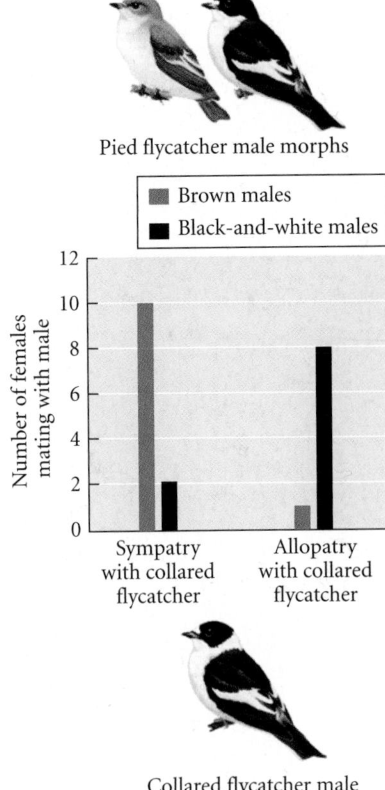

P-locus and *T*-locus alleles (see Chapter 15), and because the populations are assumed to differ already in the frequencies of the male trait alleles T_1 and T_2 because of sexual selection, associations between genes reducing hybrid fitness and genes affecting the mating system are less likely to be broken down by recombination than in the previous model.

Reinforcement of prezygotic isolation appears to occur fairly often (Howard 1993; Noor 1999). For instance, in allopatric populations of the European collared and pied flycatchers (*Ficedula albicollis* and *F. hypoleuca*), males are very similarly patterned in black and white, but a brown form of the male pied flycatcher is prevalent where the species are sympatric (Sætre et al. 1997). Female pied flycatchers prefer black-and-white males in allopatric populations, but in regions of sympatry they prefer brown males (Figure 18.15). This pattern of character displacement is the expected consequence of reinforcement. *Drosophila serrata* and its close relative *D. birchii* are sympatric in northern Australia, but *D. serrata* extends much farther south than *D. birchii* does. Females of both species choose males based on their relative proportions of several hydrocarbons in the cuticle (CHCs). There is an abrupt difference in male CHCs between allopatric and sympatric populations of *D. serrata* (Figure 18.16A). When Megan Higgie and Mark Blows (2007) confined allopatric *D. serrata* with *D. birchii* in laboratory populations, the composition of CHCs evolved within nine generations toward the composition found in sympatric *D. serrata* (Figure 18.16B), showing that reinforcement was not only predictable but could occur very rapidly. In another experiment, Higgie and Blows (2008) found that in laboratory populations formed by crossing sympatric and allopatric populations of *D. serrata*, male CHCs evolved toward the natural allopatric composition and female preference evolved toward that found in allopatric populations.

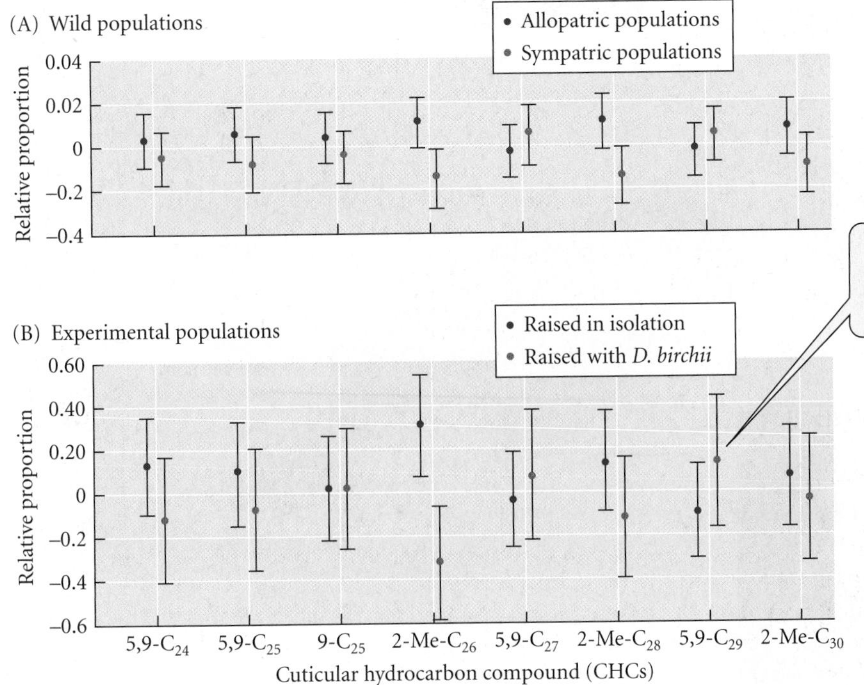

Figure 18.16 Evidence of reinforcement of differences in hydrocarbon composition of the male cuticle (CHC) in two species of *Drosophila*. Female mate choice is mediated by the relative proportions of these chemicals. (A) The relative proportions of eight CHCs in allopatric and sympatric populations of *D. serrata*. (B) The proportions of these compounds in allopatric *D. serrata* differed between experimental stocks kept in the same cages as *D. birchii* for nine generations and those kept isolated from *D. birchii*. (After Higgie and Blows 2007.)

Peripatric speciation

THE HYPOTHESIS. One of Ernst Mayr's most influential and controversial hypotheses was **founder effect speciation** (1954), which he later termed peripatric speciation (1982b). He based this hypothesis on the observation, in many birds and other animals, that isolated populations with restricted distributions, in locations peripheral to the distribution of a probable "parent" species, often are highly divergent, to the point of being classified as different species or even genera. For example, the paradise-kingfisher *Tanysiptera galatea* varies little throughout the large island of New Guinea but has differentiated into several distinctly different forms on small islands along the coast (Figure 18.17).

Mayr proposed that genetic change could be very rapid in localized populations founded by a few individuals and cut off from gene exchange with the main body of the species. He reasoned that allele frequencies at some loci would differ from those in the parent population because of accidents of sampling—i.e., genetic drift—simply because a small number of colonists would carry only some of the alleles from the source population, and at altered frequencies. (He termed this initial alteration of allele frequencies the FOUNDER EFFECT; see Chapter 10.) *Because epistatic interactions among genes affect fitness, this initial change in allele frequencies at some loci would alter the selective value of genotypes at other, interacting loci.* Hence selection would alter allele frequencies at these loci, and this in turn might select for changes at still other epistatically interacting loci. The "snowballing" genetic change that might result would incidentally yield reproductive isolation. As Mayr (1954) pointed out, this hypothesis implies that substantial evolution would occur so rapidly, and on so localized a geographic scale, that it would probably not be documented in the fossil record. If such a new species expanded its range, it would appear suddenly in the fossil record, without evidence of the intermediate phenotypic changes that had

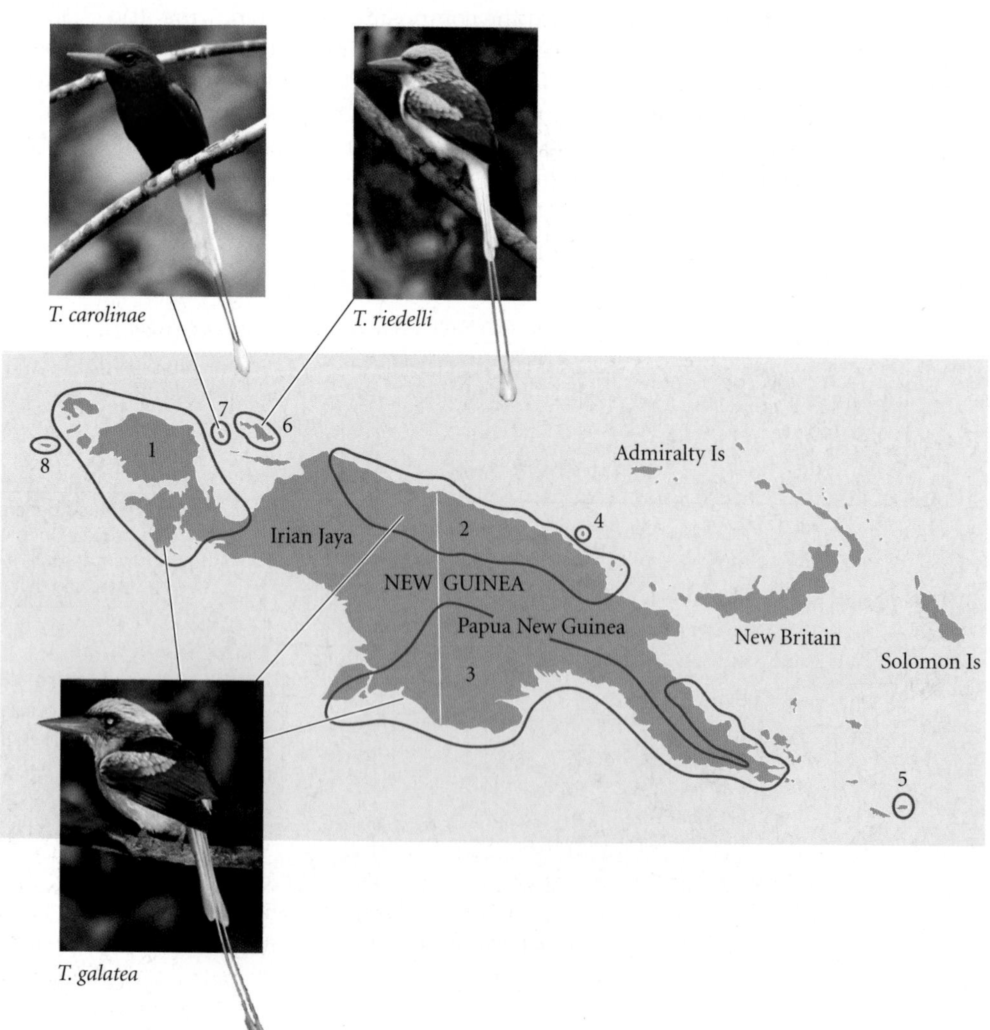

Figure 18.17 Variation among paradise-kingfishers in New Guinea. *Tanysiptera galatea* is distributed throughout the New Guinea lowlands (regions 1, 2, 3), whereas the very localized forms *T. riedelii* on Biak Island (6) and *T. carolinae* on Numfor Island (7) are now recognized as distinct species. (After Mayr 1954; *T. galatea* photo courtesy of Rob Hutchinson/Birdtour Asia; *T. riedelii* and*T. carolinae* courtesy of Mehd Halaouate.)

occurred. Thus this hypothesis, he said, might help to explain the rarity of fossilized transitional forms among species and genera. Mayr thus anticipated, and provided the theoretical foundation for, the idea of PUNCTUATED EQUILIBRIUM (see Chapters 4 and 22) advanced by Niles Eldredge and Stephen Jay Gould (1972). Similar hypotheses were later advanced by other authors (Lewis 1962; Carson 1975; Templeton 1980).

The usual interpretation of these hypotheses employs the metaphor of an adaptive landscape (see Figure 12.20). The colony undergoes a PEAK SHIFT from one adaptive genetic constitution (that of the parent population) through a less adaptive constitution (an "adaptive valley") to a new adaptive equilibrium (Figure 18.18A). The process begins when genetic drift in the small, newly founded population shifts allele frequencies from the vicinity of one adaptive peak to the slope of the other. This stage cannot be accomplished by natural selection, since selection cannot reduce mean fitness. However, selection can move the allele frequencies up the slope away from the valley toward the new peak. The most elementary peak shift would be substitution of one allele (or chromosome) for another when the heterozygote has lower fitness than either homozygote. Mayr, Carson, and Templeton envisioned a more complex shift between peaks represented by different "coadapted" gene combinations (such as $A_1A_1B_1B_1$ and $A_2A_2B_2B_2$). Most population geneticists, however, consider speciation by peak shift unlikely, because genetic drift is unlikely to oppose strong natural selection (Charlesworth and Rouhani 1988; Turelli et al. 2001). However, Sergei Gavrilets (2004) showed that speciation can occur by genetic drift if an isolated population moves along an "adaptive ridge" to the other side of an adaptive valley from the parent population (Figure 18.18B). Hybrids between the two populations would have low fitness, even though mean fitness would not decrease within either population (see also Figure 18.7). However, speciation by this process is likely to take much longer if the populations move along the ridges by genetic drift than by natural selection.

(A) Peak shift across an adaptive valley

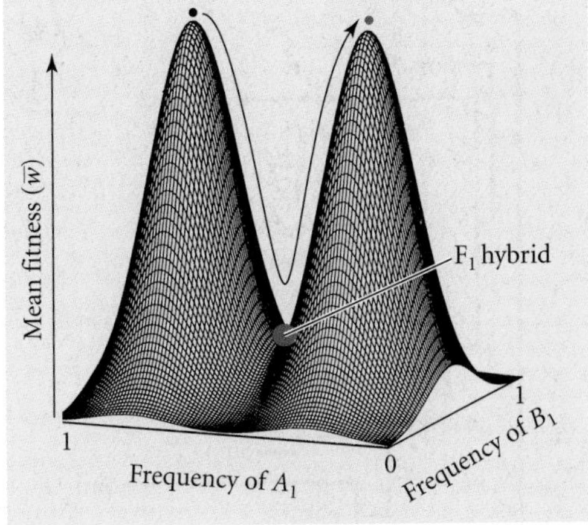

(B) Genetic drift along an adaptive ridge

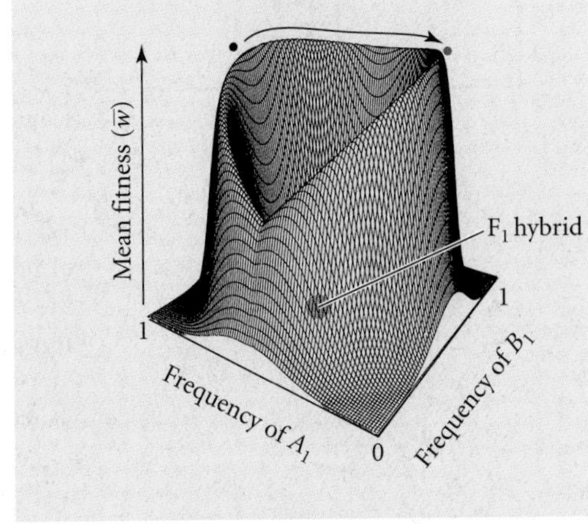

Figure 18.18 Two adaptive landscapes show how peripatric speciation might occur. The height of a point on the three-dimensional landscape represents the mean fitness of a population ($\overline{w}$). The mean fitness is a function of allele frequencies at loci *A* and *B*. (A) In a peak shift, a population evolves from one adaptive peak to another by moving downhill (lowering fitness) and then uphill. The F_1 hybrid of a cross between populations on the two peaks lies in the valley; i.e., it has low fitness, which causes some reproductive isolation between the populations. (B) Genetic drift along an adaptive ridge, in which genetic constitutions with the same fitness connect the beginning and end states of a population. The F_1 hybrid between populations with these genetic constitutions lies inside the crater. (After Gavrilets and Hastings 1996.)

EVIDENCE. Several investigators have passed laboratory populations through repeated bottlenecks to see whether reproductive isolation can evolve in this way (summarized by Rice and Hostert 1993; Rundle 2003; Coyne and Orr 2004). On the whole, these experiments have provided little evidence of drift-induced reproductive isolation.

Many species do originate, as Mayr said, as localized "buds" from a widespread parent species. This is shown by phylogenies of DNA sequences (gene genealogies), such as the *Greya* moths in Figure 17.4, in which a geographically localized species is more closely related to certain populations of the widespread species than the populations of the widespread species are to one another (Avise 1994). But this pattern, in itself, does not tell us if the evolution of reproductive isolation was initiated by genetic drift, or even if the species originated from a bottlenecked population. Evidence on this last point can be provided by the pattern of DNA sequence variation. The hypothesis that a species arose from a population that was not only peripheral, but also small, would be supported by evidence that the population had lost most of the genetic variation present in its more populous ancestor. A possible example is *Drosophila sechellia*, which is restricted to the Seychelles Islands, a group of small islands in the Indian Ocean east of Africa (Kliman et al. 2000). This species is closely related to *D. mauritiana*, found on the island of Mauritius in the Indian Ocean, and to *D. simulans*, which is native to Africa and has become cosmopolitan as a result of inadvertent human transport. At many loci, the gene genealogy of *D. sechellia* is nested within that of *D. simulans*, as expected if it arose by colonization from Africa, and the level of DNA sequence diversity within *D. sechellia* is much lower than in *D. mauritiana* or *D. simulans* (Figure 18.19).

Clearly, the effective population size of *D. sechellia* has been much smaller than those of the other species, but whether or not the reduced population size initiated the evolution of reproductive isolation is not known. In any case, the pattern seen in *D. sechellia*

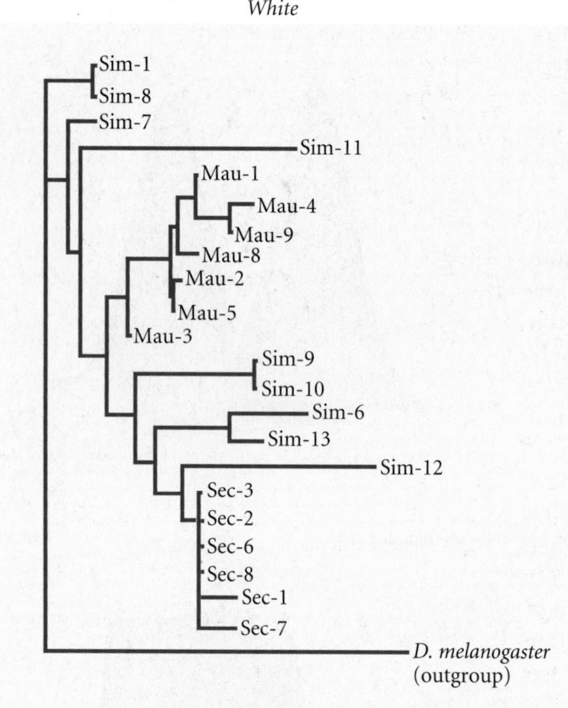

Figure 18.19 A gene tree for the *white* locus suggests that *Drosophila mauritiana* (Mau) and *D. sechellia* (Sec), which are restricted to small islands in the Indian Ocean, originated as small populations. Each numbered "twig" is a different haplotype. Since their common ancestry with the widespread species *D. simulans* (Sim), the smaller populations of the island species have coalesced faster, so that Mau and Sec sequences are nested within the Sim gene tree. Probably a *simulans*-like ancestor colonized each island and evolved into a new species. (After Kliman et al. 2000.)

is rather unusual. In most such analyses of recently derived sister species, such as the crested and least auklets portrayed in Figure 17.17, incomplete sorting of shared gene lineages provides evidence that both species have had large effective population sizes, and have not experienced the severe bottlenecks that are inherent in the peak-shift models of speciation.

Alternatives to Allopatric Speciation

Allopatric, parapatric, and sympatric speciation form a continuum, from little to more to much gene exchange between the diverging groups that eventually evolve biological barriers to gene exchange. Even in allopatric speciation, there may be some gene flow between populations, but it is very low compared with the divergent action of natural selection and/or genetic drift. Parapatric speciation is the same process, but since the rate of gene flow is higher, the force of selection must be correspondingly stronger to engender genetic differences that create reproductive isolation.

As we have seen, a parapatric or sympatric distribution of sister species does not necessarily provide evidence that they arose by parapatric or sympatric speciation, because species' distributions change over time. Because sympatric species may have originated by allopatric speciation, distributional evidence must be cautiously interpreted (McCune and Lovejoy 1998).

Parapatric speciation

Parapatric speciation can theoretically occur if gene flow between populations that occupy adjacent regions with different selective pressures is much weaker than divergent selection for different gene combinations (Endler 1977). Strong selection at a sharp border between different habitats poses a barrier to gene exchange, caused by the reproductive failure of individuals with the "wrong" genotype or phenotype that migrate across the border (Nosil et al. 2005). Consequently, clines at various loci may tend to develop at the same location, resulting in a primary hybrid zone that has developed in situ, but might look like a secondary hybrid zone (Endler 1977; Barton and Hewitt 1985). Steady genetic divergence may eventually result in complete reproductive isolation.

Another possibility is that populations isolated by distance can evolve reproductive incompatibility. Divergent features that arise at widely separated sites in the species' distribution may spread, supplanting ancestral features as they travel, and preventing gene exchange when they eventually meet. Russell Lande (1982) has theorized that prezygotic isolation could arise in this way as a result of divergent sexual selection.

Parapatric speciation undoubtedly occurs and may even be common, but it is difficult to demonstrate that it provides a better explanation than allopatric speciation for real cases (Coyne and Orr 2004). Possibly the best-documented example of the parapatric origin of reproductive isolation is attributable not to these theories but to selection for isolation—i.e., reinforcement. *Anthoxanthum odoratum* is one of several grasses that have evolved tolerance to heavy metals in the vicinity of mines within the last several centuries (see Chapter 13). Several populations, under strong selection for metal tolerance, have diverged from neighboring nontolerant populations (on uncontaminated soil) not only in

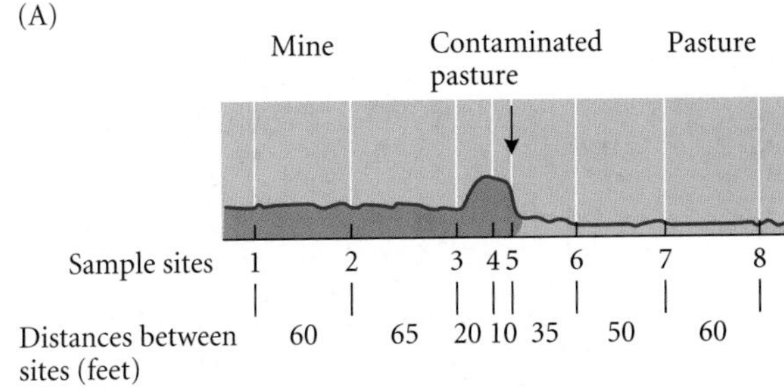

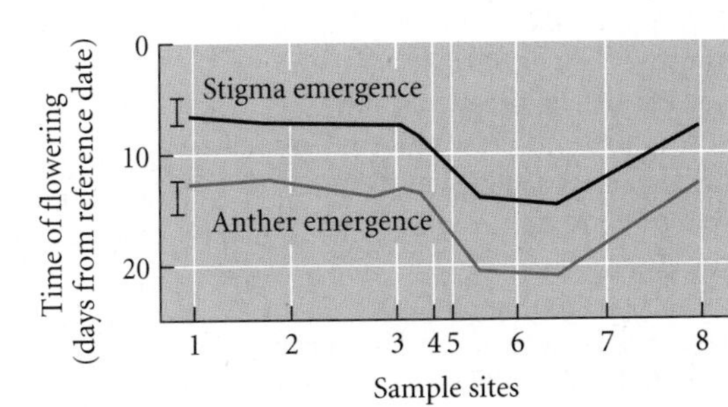

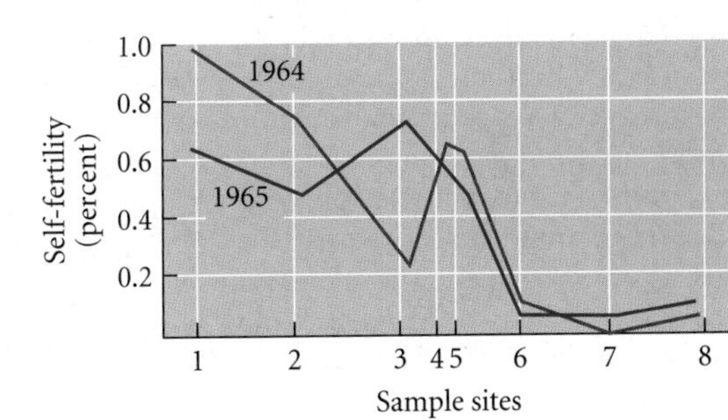

Figure 18.20 Parapatric evolution of reproductive isolation over a very short distance in the grass species *Anthoxanthum odoratum*. (A) Sample sites and distances along a transect from an area contaminated with lead and zinc (blue) into an area with uncontaminated soil. (B) Flowering time, in a common garden, of plants derived from different sample sites along the transect. (C) Seed set by plants with flowers bagged to prevent cross-pollination. The differences in both flowering time and capacity for self-fertilization are likely to reduce gene flow between plants from contaminated and uncontaminated environments. (After Antonovics 1968; McNeilly and Antonovics 1968.)

tolerance but in flowering time; moreover, they self-pollinate more frequently, having become more self-compatible (Figure 18.20). Both characteristics provide considerable reproductive isolation from adjacent nontolerant genotypes. In this case, the historical evidence indicates that the divergence is very recent, and the species is so ubiquitously distributed that parapatric divergence cannot be doubted.

Sympatric speciation

THE CONTROVERSY. Sympatric speciation is a highly controversial subject. Speciation would be sympatric if a biological barrier to gene exchange arose *within an initially randomly mating population without any spatial segregation of the incipient species*—that is, if speciation occurred despite high initial gene flow. The difficulty any model of sympatric speciation must overcome is how to reduce the frequency of the intermediate genotypes that would act as a conduit of gene exchange between the incipient species. For example, it is unlikely that complete sexual isolation could evolve by a single mutation, because most such characteristics are polygenic (Chapter 13), and some gene combinations will result in varying degrees of preference for one or the other kind of mate. (Besides, if there were such a single-gene mutant, it would have a hard time finding a mate.)

Ernst Mayr (1942, 1963) was the most vigorous and influential critic of the sympatric speciation hypothesis, demonstrating that many supposed cases are unconvincing and that the hypothesis must overcome severe theoretical difficulties. Under certain special circumstances, however, these difficulties are not all that severe (Turelli et al. 2001; Gavrilets 2004; Bolnick and Fitzpatrick 2007).

MODELS OF SYMPATRIC SPECIATION. Most models of sympatric speciation postulate disruptive (diversifying) selection (see Chapter 12) based on resource use. Simpler models postulate divergent adaptation to two distinct resources, based on one or on several loci ("type A" loci). For example, certain homozygous genotypes (say, A_1A_1 and A_2A_2) have high fitness on one or the other of two resources, and intermediate (heterozygous A_1A_2) phenotypes have lower fitness because they are not as well adapted to either resource. Selection may favor alleles at other loci ("type M" loci) that tend to make their carriers mate nonrandomly, if this would result in their having fewer poorly adapted heterozygous offspring. If such alleles increased in frequency, the result would be two partially reproductively isolated populations (say, M_1M_1 and M_2M_2) that are adapted to different resources (by their genotypic composition, A_1A_1 and A_2A_2). This would be a step toward speciation. (A "step" because the single M locus provides only partial premating isolation, and more such loci would be required to complete the process.) Two models of mating have been considered: (1) assortative mating, in which individuals prefer mates that match their own phenotype, and (2) trait-preference, in which different genes control female preference and male trait (as in most sexual selection models; Chapter 15). In either model, the trait in question may be either an "arbitrary" marker such as coloration, or a character [called a "magic trait" by Gavrilets (2004)] that enhances fitness on one of the resources. (For instance, there is evidence that Darwin's finches may choose mates partly on the basis of beak size, which is adapted to their typical food.)

The greatest obstacle to sympatric speciation is that recombination breaks down the association between the type A loci that affect fitness on the two resources and the type M loci that govern mating (Felsenstein 1981). Unless these are associated, there is no selection for a rare mating allele that would establish a partly reproductively isolated subpopulation. Among the several models, recombination most strongly opposes speciation in trait-preference models with arbitrary marker traits (Figure 18.21A). Suppose the mating (type M) loci are T, which affects male color (T_1 for pink, T_2 for orange), and P, which affects female preference (P_1 prefers pink, P_2 prefers orange), and that the population is initially nearly monomorphic for A_1, T_1, and P_1. A partly reproductively isolated subpopulation adapted to another resource would be mostly homozygous for A_2, T_2, and P_2. This subpopulation would develop only if these three alleles increase in frequency. In this model, the only factor that would cause rare alleles T_2 and P_2 to increase in frequency is their association with A_2, but recombination breaks down the necessary association. If full speciation requires different alleles at several loci in each category (A, T, and P), recombination is even more of an obstacle.

Progress toward sympatric speciation is theoretically somewhat more likely if mating is affected by a "magic trait" that also affects ecological divergence (Figure 18.21B). Suppose, for example, that a population of an herbivorous insect is divergently selected for adaptation both to its ancestral host plant (1) and another plant (2). At the A locus, alleles A_1 and A_2 improve survival on hosts 1 and 2, respectively. Locus B affects an individual's behavioral preference for host 1 or 2: carriers of the allele B_1 prefer host 1, and carriers of B_2 prefer host 2. The optimal gene combinations are $A_1A_1B_1B_1$ (adapted to and attracted to host 1) and $A_2A_2B_2B_2$ (adapted to and attracted to host 2); individuals with other gene combinations (e.g., $A_2A_2B_1B_1$) are attracted to a host plant on which their offspring will survive poorly. Thus selection favors the divergent gene combinations and promotes linkage disequilibrium, so that the antagonism between selection and recombination is lower than in the model described previously. If the sexes meet and mate on the host plant, as is true of many species of insects, divergence in behavior would result in

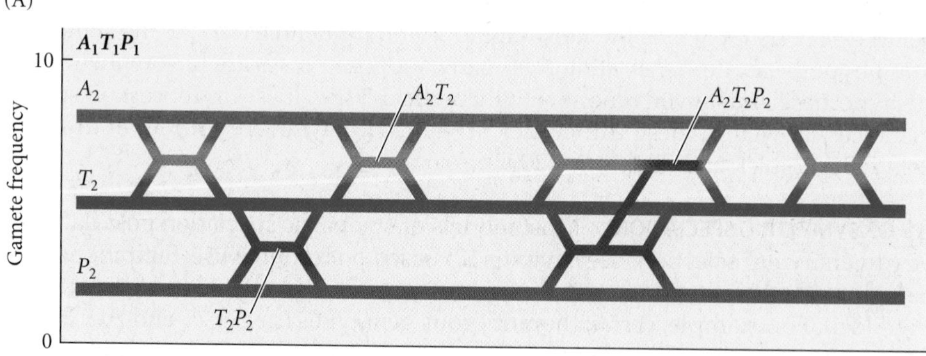

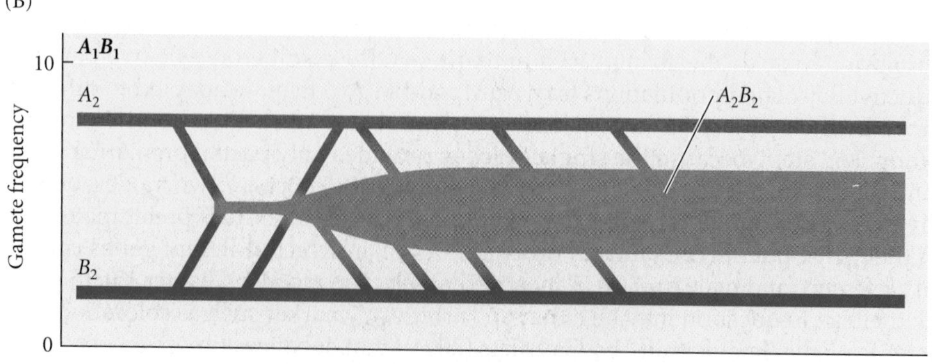

Figure 18.21 The problem of sympatric speciation by disruptive selection. The diagrams show the formation and dissolution of associations among alleles at different loci that are required to establish incipient sexual isolation between genotypes that are adapted to different resources. In both diagrams, the alleles A_1 and A_2 confer adaptation to two different resources, and we suppose that heterozygotes (not shown) have lower fitness on both resources. (A) Alleles T_1 and T_2 govern a difference in male color, and P_1 and P_2 govern females' partial preference for those colors. A_1, T_1, and P_1 are the predominant alleles. The low frequency of alleles A_2, T_2, and P_2 is shown by the narrow, colored "roadways." Allele combinations such as A_2T_2 and T_2P_2, which arise by recombination, are soon destroyed by recombination. A partially reproductively isolated population would form if the combination $A_2T_2P_2$ were to be established, but this too is soon destroyed by recombination. (B) Conditions are more favorable for two partly reproductively isolated subpopulations if reproductive isolation is a side effect of one of the genes that is disruptively selected. Here, the B_1 allele governs a behavioral preference for the resource to which A_1 is adapted. The rare B_2 allele (blue) causes a preference for the resource to which the rare A_2 allele (red) is adapted. The A_2B_2 combination (purple) is favored by natural selection. If mating occurs mostly on the preferred resource, individuals carrying the A_1B_1 and A_2B_2 combinations will tend not to interbreed, the allele combinations will not be as readily destroyed by recombination, and the frequency of A_2B_2 can increase, forming a partly isolated subpopulation.

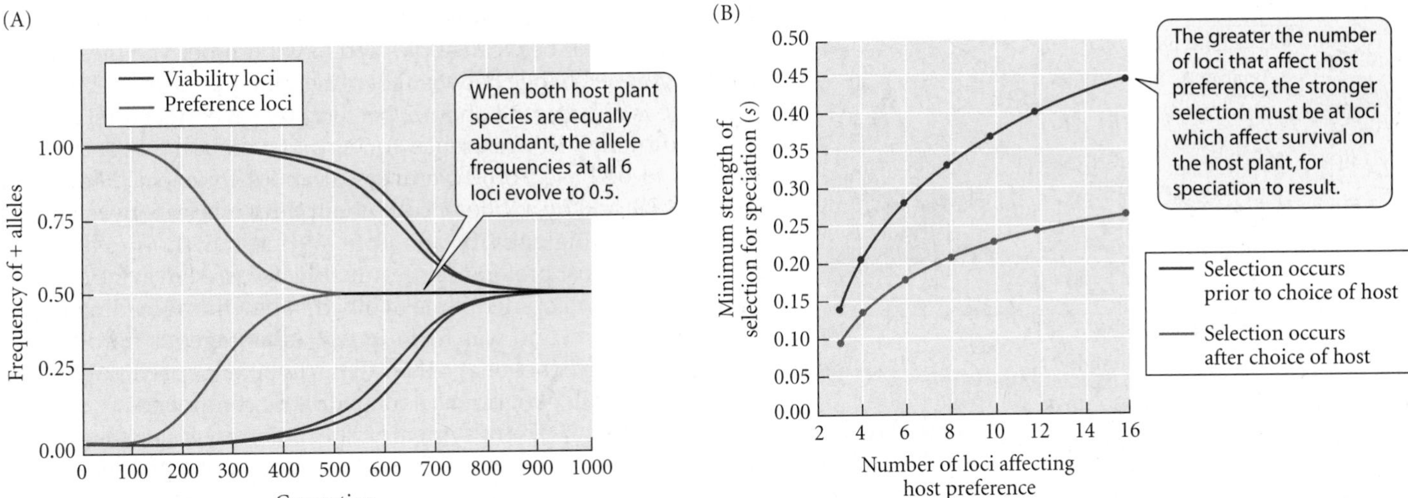

Figure 18.22 Some results of a computer simulation of sympatric speciation in an insect that mates on its host plant. (A) Alleles that enhance survival on or preference for one host plant species are referred to as "+ alleles"; those that have the complementary effect are called "– alleles." The simulation shows changes in the frequency of the + allele at four loci that affect survival (viability loci) and two loci that affect host preference (preference loci) when the + allele at each locus begins with a frequency near 0 or 1. Eventually, half the population prefers and survives better on one host and half on the other, representing progress toward reproductive isolation. (B) These curves show how strong selection at viability loci has to be to result in speciation when preference is controlled by multiple loci. The strength of selection at each viability locus is s (the coefficient of selection; see Chapter 12). The two curves model life histories in which selection occurs before and after the choice of host occurs. (After Fry 2003.)

partial isolation by habitat: host preference is a "magic trait." In computer simulations such as those by James Fry (2003), the frequencies of alleles such as A_2 and B_2 that confer adaptation to and preference for a new host may rapidly increase (Figure 18.22A). Gene flow may be strongly reduced, so that the population divides into two host-associated, ecologically isolated incipient species. However, if host preference is a continuous, polygenic trait, reproductive isolation will not evolve unless selection is strong (Figure 18.22B). Somewhat similar models describe sympatric speciation by adaptation to a continuously distributed resource, such as prey size (Dieckmann and Doebeli 1999; Kondrashov and Kondrashov 1999). Some authors have questioned how realistic these models are (Coyne and Orr 2004; Gavrilets 2004).

EVIDENCE FOR SYMPATRIC SPECIATION. The conditions required for sympatric speciation to occur are theoretically more limited than those for allopatric speciation, and sympatric speciation does not occur easily (Gavrilets 2004). Ecological and genetic research can help to determine whether or not conditions are favorable for sympatric speciation in specific groups of organisms (Bolnick and Fitzpatrick 2007). However, because there is so much evidence for allopatric speciation, sympatric speciation must be demonstrated rather than assumed for most groups of organisms. Such demonstration may be quite difficult, because evidence should show that a past allopatric phase of genetic differentiation is very unlikely (Coyne and Orr 2004). Nevertheless, many possible examples, supported by varying degrees of evidence, have been proposed.

Many experiments have been done in which laboratory populations of *Drosophila* have been subjected to disruptive selection and then tested for prezygotic isolation (Rice and Hostert 1993). In most, no sexual isolation developed. The exception was in experiments in which the disruptively selected character is a "magic trait" that automatically causes assortative mating as a correlated effect. In one experiment, for example, flies were disruptively selected for habitat preference and length of development. Within fewer than 30 generations, two subpopulations of flies had developed, strongly isolated by development time and to a lesser extent by habitat preference (Rice and Salt 1990).

"Host races" of specialized herbivorous insects—partially reproductively isolated subpopulations that feed on different host plants—have often been proposed to represent sympatric speciation in progress, although the evidence for this is very limited (Futuyma 2008). The most renowned case, studied first by Guy Bush (1969) and later by Jeffrey Feder and colleagues (2005), is the apple maggot fly (*Rhagoletis pomonella*). The larvae develop in ripe fruits and overwinter in the ground as pupae; adult flies emerge in July and August and mate on the host plant. The major ancestral host plants throughout eastern North America were hawthorns (*Crataegus*). About 150 years ago, *R. pomonella* was first recorded in the northeastern United States as a pest of cultivated apples (*Malus*), and infes-

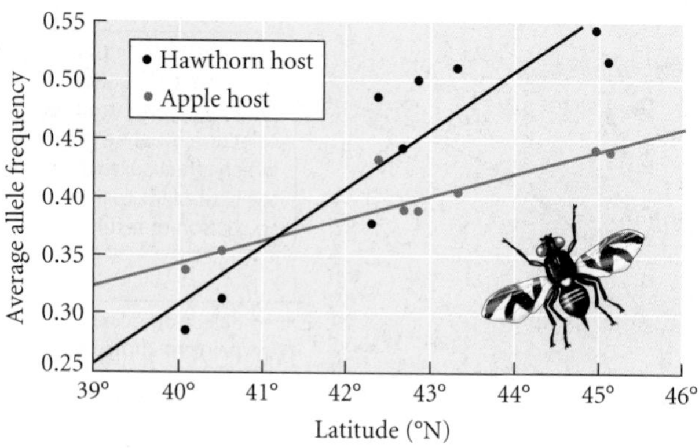

Figure 18.23 Genetic differentiation between *Rhagoletis pomonella* flies collected from hawthorn and apple hosts at several sites over a latitudinal transect. The average allele frequencies at several allozyme loci are plotted. The differentiation between these forms during the past century is associated with sympatric divergence in host use. (After Feder et al. 1990.)

tation of apples spread from there. Allele frequencies at several loci differ significantly between apple- and hawthorn-derived flies, showing that gene exchange between them is limited (Figure 18.23). Gene exchange is reduced (to about 2 percent) by several factors, including a difference in preference for apples versus hawthorns that appears to be based on differences at about four loci (Feder and Forbes 2008) and by a difference of about three weeks between emergence and mating activity on apple and on hawthorn. Although divergence in host preference presumably occurred in sympatry, the difference in development time evolved in hawthorn-feeding populations in Mexico, and was fortuitously advantageous for developing on apples (Michel et al. 2007). Thus incipient speciation in this case has had both geographic and sympatric components.

Probably the best-documented examples of sympatric speciation are sister species that inhabit small isolated islands where there has been no opportunity for spatial separation. The palms *Howea forsteriana* and *H. belmoreana*, the only members of the genus, are restricted to Lord Howe Island (east of Australia), which is so small that the pollen is dispersed by wind across the entire island (Figure 18.24). Reproductive isolation is based on partial separation by soil type and by different flowering seasons (Savolainen et al. 2006). A similar example has been described in two groups of cichlid fishes that are confined to two small crater lakes (Schliewen et al. 1994). Mitochondrial DNA sequence data indicate that the cichlids in each lake are monophyletic, suggesting that speciation has occurred within the crater lakes. The lakes lie in simple conical basins that lack habitat heterogeneity and opportunity for spatial isolation. It has often been suggested that the enormous diversity of cichlid fishes in the African Great Lakes (see Figure 3.25) arose by sympatric speciation, but there are plentiful opportunities for allopatric speciation within each lake, because these sedentary species are restricted to discontinuously distributed distinct habitats along the lake periphery. Conspecific populations of these cichlids differ genetically, even over short distances (Rico et al. 2003).

Polyploidy and Recombinational Speciation

Polyploidy

A POLYPLOID is an organism with more than two complements of chromosomes (see Chapter 8). A tetraploid, for example, has four chromosome complements in its somatic cells. Polyploid populations are reproductively isolated by postzygotic barriers from their

Figure 18.24 Sister species of palms on Lord Howe Island, less than 12 square km in area. (A) *Howea forsteriana* (kentia palm) has straight leaves with drooping leaflets. (B) *H. belmoreana* (curly palm) has curved leaves with ascending leaflets. (Photos courtesy of W. J. Baker, Royal Botanic Gardens, Kew.)

diploid (or other) progenitors and are therefore distinct biological species. Speciation by polyploidy is the only known mode of *instantaneous speciation* by a *single genetic event*.

For reasons that are not well understood, polyploid species are rare among sexually reproducing animals, although many parthenogenetic polyploid animals have been described. Polyploidy is very common in plants. Perhaps 30 to 70 percent of plant species are descended from polyploid ancestors, although changes in ploidy may accompany only 2 to 4 percent of speciation events in flowering plants (Otto and Whitton 2000).

Natural polyploids span a continuum between two extremes, called autopolyploidy and allopolyploidy. An AUTOPOLYPLOID is formed by the union of unreduced gametes from genetically and chromosomally compatible individuals that may be thought of as belonging to the *same species*. The cultivated potato (*Solanum tuberosum*), for example, is an autotetraploid of a South American diploid species. An ALLOPOLYPLOID is a polyploid derivative of a diploid hybrid between two species—or to be more exact, between two ancestors that form hybrids that are partly sterile because of genetic or chromosomal incompatibility. Ideally, auto- and allopolyploids can be distinguished by the behavior of their chromosomes in meiosis (see Figure 8.22). An autotetraploid, for example, has four homologues of each chromosome. These may not segregate two by two, so autopolyploids often form many aneuploid gametes and have reduced fertility. In contrast, the chromosomes of allopolyploids form BIVALENTS (synapsed pairs) in meiosis, form balanced, viable gametes, and generally have nearly normal fertility. However, there are many intermediate cases between auto- and allopolyploids. Moreover, chromosomes of autopolyploids may form bivalents because of natural selection for genes that enable chromosomes to pair normally. For example, a tetraploid strain of corn (*Zea mays*) in which multivalents were initially predominant evolved in just ten generations to a high level of bivalent formation (Gilles and Randolph 1951).

SPECIATION BY POLYPLOIDY. Polyploidy usually occurs because of a failure of the reduction division in meiosis (Ramsey and Schemske 1998). For example, the union of an unreduced ($2n$) gamete with a haploid (n) gamete yields a triploid ($3n$) individual; a tetraploid is then formed if an unreduced $3n$ gamete unites with a reduced (n) gamete. Plants with odd-numbered ploidy (e.g., triploids, $3n$, and pentaploids, $5n$) are generally nearly sterile. Because the hybrid between a tetraploid and its diploid ancestor is triploid, and is therefore largely sterile, the tetraploid is reproductively isolated and is therefore a distinct biological species (and the same is true at higher ploidy levels).

A milestone in the study of speciation was the experimental production of a natural polyploid species by Arne Müntzing in 1930. Müntzing suspected that the mint *Galeopsis tetrahit*, with $2n = 32$ chromosomes, might be an allotetraploid derived from the diploid ($2n = 16$) ancestors *G. pubescens* and *G. speciosa*. He crossed the two diploid species, and from the F_1 diploid hybrids he obtained one triploid offspring. The triploid, backcrossed to *G. pubescens*, yielded a single tetraploid offspring, which Müntzing propagated. The tetraploids closely resembled *G. tetrahit* in morphology, were highly fertile, and were reproductively isolated from the diploid species, but yielded fertile progeny when crossed with wild *G. tetrahit* (see the figure that opens this chapter).

In this and some other experimental studies of allopolyploids, the diploid hybrids between species are mostly sterile and form few bivalents in meiosis, whereas tetraploid offspring of these hybrids are highly fertile and have normal, bivalent chromosome pairing. In these cases, *the sterility of the diploid hybrid is not due to functional interactions between the genes of the two parent species, but instead to whatever mechanisms inhibit chromosome pairing*. The diploid and tetraploid hybrids have the same genes in the same proportions, so genic differences cannot account for the sterility of the one but not the other (Darlington 1939; Stebbins 1950).

Molecular studies have cast additional light on the origin of polyploid species. For example, three diploid European species of goatsbeards (Asteraceae), *Tragopogon dubius*, *T. porrifolius*, and *T. pratensis*, have become broadly distributed in North America. F_1 hybrids between them have low fertility. In 1950, Marion Ownbey described two fertile tetraploid species, *T. mirus* and *T. miscellus*, and postulated that *T. mirus* was a recent tetraploid

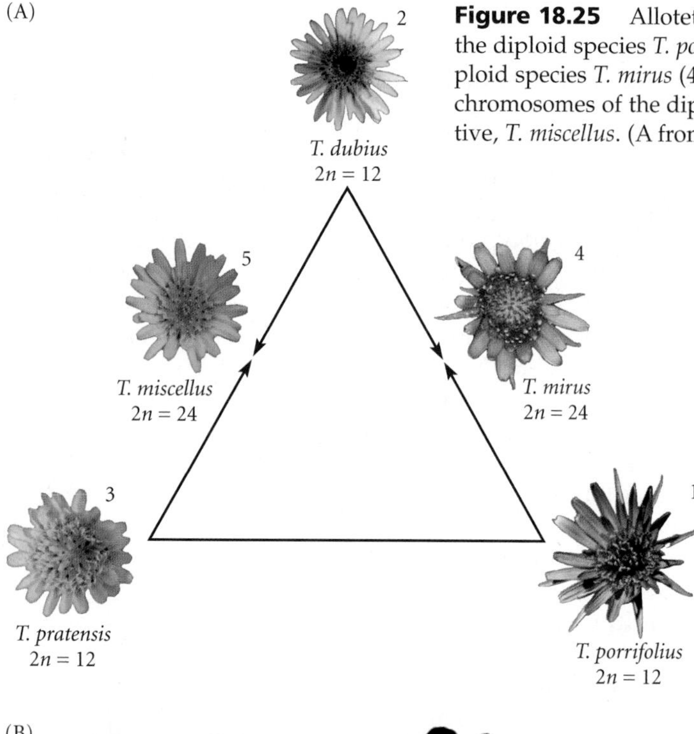

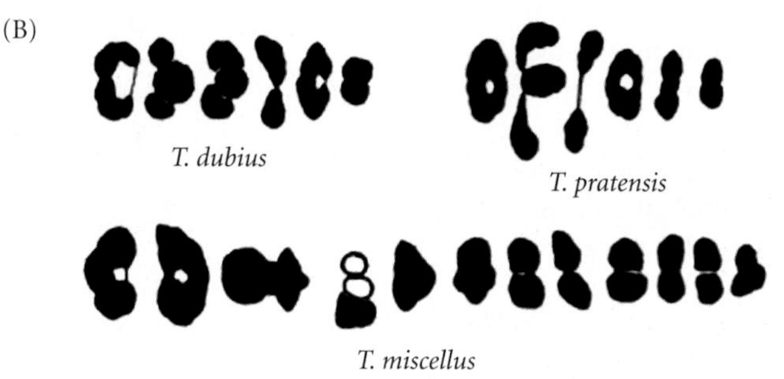

Figure 18.25 Allotetraploid species of goatsbeards (*Tragopogon*). (A) The flower heads of the diploid species *T. porrifolius* (1), *T. dubius* (2), and *T. pratensis* (3), and of the fertile tetraploid species *T. mirus* (4, from 1 × 2) and *T. miscellus* (5, from 2 × 3). (B) Drawings of the chromosomes of the diploids *T. dubius* and *T. pratensis*, and of their tetraploid hybrid derivative, *T. miscellus*. (A from Pires et al. 2004; B from Ownbey 1950.)

hybrid of *T. dubius* and *T. porrifolius*, and that *T. miscellus* was likewise derived from *T. dubius* × *T. pratensis* (Figure 18.25). Decades later, Douglas Soltis, Pamela Soltis, and their collaborators (2004) found that the tetraploid species have exactly the combinations of DNA markers from the diploid species that are predicted by Ownbey's hypothesis. In *Tragopogon* and other plants, allopolyploid species have typically arisen independently several times by hybridization between their diploid parents (Pires et al. 2004).

ESTABLISHMENT AND FATE OF POLYPLOID POPULATIONS. How polyploid species become established is not fully understood. If a newly arisen tetraploid within a diploid population crosses at random, its reproductive success should be lower than that of the diploids because many of its offspring will be inviable or sterile triploids, formed by backcrossing with the surrounding diploids. A study of mixed experimental populations of diploid and tetraploid fireweeds (*Chamerion angustifolium*) showed the lower the frequency of tetraploids in the population, the lower their seed production, as a result of increased pollination from the diploids (Husband 2000).

Several conditions—self-fertilization, vegetative propagation, higher fitness than the diploid, or niche separation from the diploid—might enable a new polyploid to increase and form a viable population (Fowler and Levin 1984; Rodríguez 1996). Indeed, many polyploid taxa reproduce by selfing or vegetative propagation, and most differ from their diploid progenitors in habitat and distribution. Increases in ploidy alter cell size, water content, rate of development, and many other physiological properties (Levin 1983; Otto and Whitton 2000), so many polyploids may be able to occupy new ecological niches immediately upon their origin.

Because it apparently confers new physiological and ecological capabilities, polyploidy may play an important role in plant evolution; moreover, their increased number of genes may enhance their adaptability (Otto and Whitton 2000). However, polyploidy does not confer major new morphological characteristics, such as differences in the structure of flowers or fruits, and seems unlikely to cause the evolution of new genera or other higher taxa (Stebbins 1950).

Recombinational speciation

Hybridization sometimes gives rise not only to polyploid species but also to distinct species with the same ploidy as their parents. Among the great variety of recombinant offspring produced by F_1 hybrids between two species, certain genotypes may be fertile but reproductively isolated from the parent species. These genotypes may then increase in frequency, forming a distinct population (Rieseberg 1997; Gross and Rieseberg 2005). This process has been called **recombinational speciation** or HYBRID SPECIATION (Grant 1981).

Although hybrid speciation has been described in animals (Mallet 2007), it seems to be more common in plants (Rieseberg 1997). In a molecular phylogenetic analysis of part of the sunflower genus, *Helianthus*, Loren Rieseberg and coworkers found that hybridization between *Helianthus annuus* and *H. petiolaris* has given rise to three other distinct

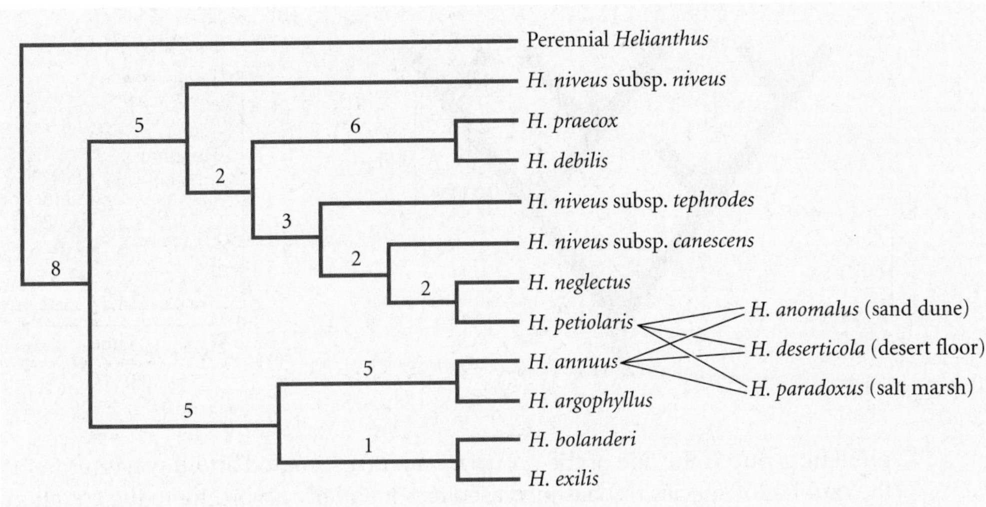

Figure 18.26 The hybrid origin of diploid species of sunflowers. The phylogeny, based on sequences of chloroplast DNA and nuclear ribosomal DNA, shows that *Helianthus anomalus*, *H. paradoxus*, and *H. deserticola* have arisen from hybrids between *H. annuus* and *H. petiolaris*. The numbers of synapomorphic base pair substitutions are shown along the branches of the phylogenetic tree. Hybridization results in a netlike rather than strictly branching phylogenetic tree. (After Gross and Rieseberg 2005.)

species (*H. anomalus*, *H. paradoxus*, and *H. deserticola*; Figure 18.26). Although F_1 hybrids between the parent species have low fertility, the derivative species are fully fertile and are genetically isolated from the parent species by postzygotic incompatibility. Because recombination breaks down the initial associations among genetic markers derived from the two parent species, the sizes of chromosomal blocks derived from each parent can be used to estimate how long it took for speciation to occur. On this basis, one of the hybrid species, *H. anomalus*, is estimated to have arisen within about 60 generations (Ungerer et al. 1998).

The recombinant species grow in very different (drier or saltier) habitats than either parent species, flower later, and have unique morphological and chemical features. *H. anomalus*, for example, has thicker, more succulent leaves and smaller flower heads than either parent species. Such "extreme" traits transgress the range of variation between the two parent species. Rieseberg and coworkers (2003) crossed *H. annuus* and *H. petiolaris*, the parent species, and grew the backcross progeny in a greenhouse for several generations. Using genetic markers on all the chromosomes, Rieseberg et al. found that the experimental hybrids had combinations of *annuus* and *petiolaris* chromosome segments that matched those found in the three hybrid species—confirming that these species indeed arose from hybridization. Almost all the extreme, "transgressive" traits of *H. anomalus* and the other two hybrid species, such as small flower heads, occurred among the experimentally produced hybrids. To a considerable degree, then, the hybridization experiment replayed the origin of these species. Thus hybridization, by generating diverse gene combinations on which selection can act, can be a source of new species with novel morphological and ecological features.

How Fast Is Speciation?

The phrase "rate of speciation" has several meanings (Coyne and Orr 2004). One is the TRANSITION TIME or TIME FOR SPECIATION (TFS)—the time required for (nearly) complete reproductive isolation to evolve once the process has started (Figure 18.27A). Another is the BIOLOGICAL SPECIATION INTERVAL (BSI), the average time between the origin of a new species and when that species branches (speciates) again. The BSI includes not only the TFS but also the "waiting time" before the process of speciation begins again. For example, in a clade that speciates by polyploidy, a new polyploid species may originate rarely (i.e., the waiting time is long), but when it does, reproductive isolation is achieved within one or two generations.

The diversification rate, *R*, or increase in species number per unit time, equals the difference between the rates of speciation (*S*) and extinction (*E*). *R* can be estimated for a mono-

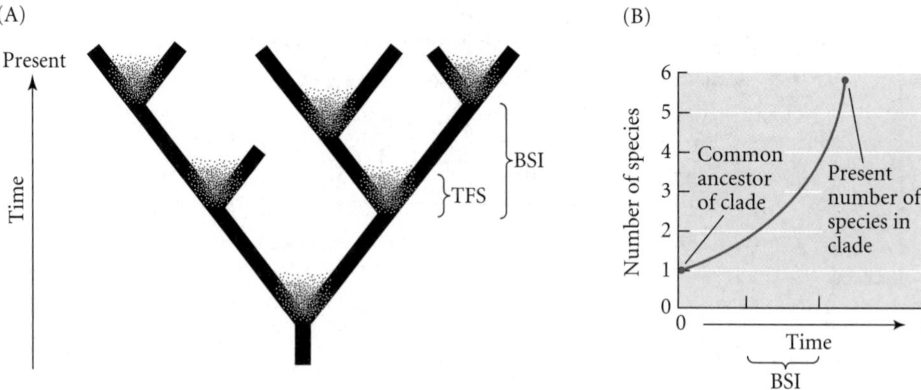

Figure 18.27 Two meanings of the "rate of speciation." (A) The time for speciation (TFS) is the time between the beginning and end of the evolution of reproductive isolation. The biological speciation interval (BSI) is the average time that has elapsed between two sequential forks in the phylogeny. (B) The BSI can be estimated from the amount of time between the present number of species and the common ancestor of the clade, assuming that the number has grown exponentially and that no extinctions have occurred. Extinctions (as in panel A) will cause us to overestimate BSI.

phyletic group if the age of the group (t) can be estimated and if we assume that the number of species (N) has increased exponentially according to the equation

$$N_t = e^{Rt}$$

(We encountered this approach in Chapter 7 when we considered long-term rates of diversification in the fossil record.) The average time between branching events on the phylogeny is $1/R$, the reciprocal of the diversification rate (Figure 18.27B). This number estimates BSI, the average time between speciation events, if we assume there has been no extinction ($E = 0$). According to estimates made using this approach, BSI in animals ranges from less than 0.3 Myr (in the phenomenal adaptive radiation of cichlid fishes in the Great Lakes of Africa) to more than 10 Myr in various groups of molluscs. When estimates of E from the fossil record are taken into account, BSI is about 3 Myr for horses and is still very long (6–11 Myr) for bivalve molluscs.

An upper bound on TFS can be estimated when geological evidence or calibrated DNA sequence divergence enables us to judge when young pairs of sister species were formed. For example, endemic species of *Drosophila* have evolved on the "big island" of Hawaii, which is less than 800,000 years old. Based on sequence divergence of mitochondrial DNA, several research groups disagree on the average time since sister species of North American birds diverged, but agree that some originated since the Pleistocene epoch began, 1.8 Mya (Figure 18.28). By correlating the degree of prezygotic or postzygotic reproductive isolation with estimated divergence time (see Figures 17.10 and 17.11), Coyne and Orr (1997) estimated that complete reproductive isolation takes 1.1 to 2.7 Myr for allopatric species of *Drosophila*, but only 0.08 to 0.20 Myr for sympatric species. (They attributed this difference to reinforcement of prezygotic isolation between sympatric forms.) A similar approach suggested that frogs take 1.5 Myr, on average, to complete speciation (Sasa et al. 1998).

Clearly, speciation rates vary greatly—as we would expect from theories of speciation. We expect the process of speciation (TFS) to be excruciatingly slow if

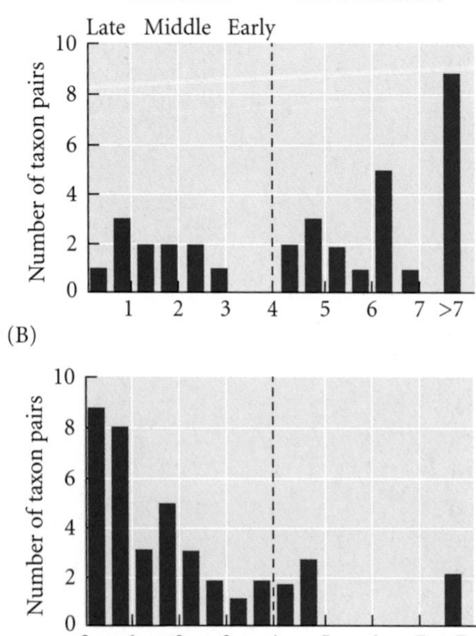

Figure 18.28 Two estimates of the divergence of mitochondrial DNA between pairs of closely related North American songbirds. Assuming that sequence divergence of mtDNA is 2 percent per Myr, less than 4 percent sequence divergence suggests that speciation occurred during the Pleistocene, and that TFS is less than 2 Myr. (A) Data from a study by Klicka and Zink (1997), who concluded that most speciation occurred before the Pleistocene. (B) Data from a study by Johnson and Cicero (2004), who came to the opposite conclusion. Johnson and Cicero argued that their data are based only on sister species and that some of Klicka and Zink's species pairs were not sister taxa. Johnson and Cicero also classified some forms as species that were classified as subspecies by other authors. (After Lovette 2005.)

it proceeds by mutation and drift of neutral alleles; we expect it to be faster if it is driven by ecological or sexual selection, and to be accelerated if reinforcement plays a role. Allopatric speciation could be slow or very rapid, depending on the strength of divergent selection and on genetic variation in relevant traits. Some possible modes of speciation, such as polyploidy, recombinational speciation, and sympatric speciation, should be very rapid when they occur—although they may occur rarely, resulting in long intervals (BSI) between speciation events. As we have already seen, ecological speciation can be rapid (Hendry et al. 2007): substantial reproductive isolation apparently evolved within about a century in the apple maggot fly (*Rhagoletis pomonella*) and the hybrid sunflower species *Helianthus anomalus*. On the other hand, some sister taxa of snapping shrimps (*Alpheus*) on opposite sides of the Isthmus of Panama have not achieved full reproductive incompatibility in the 3.5 Myr since the isthmus arose (Knowlton et al. 1993).

What characteristics favor high rates of speciation? The best way to approach this question is to compare the species diversity of replicated sister groups that differ in the characteristic of interest, although it is often hard to tell whether they enhance the speciation rate or diminish the extinction rate (see Figure 18.12). Among characteristics studied so far, those that seem most likely to have increased speciation rate as such seem to be animal (rather than wind) pollination in plants and features that indicate intense sexual selection in animals (Coyne and Orr 2004). These observations suggest that diversification in some groups of animals owes more to the simple evolution of reproductive isolation (due to sexual selection) than to ecological diversification. This conclusion may call into question the belief that ecological divergence is the main engine of evolutionary radiation (Schluter 2000).

Consequences of Speciation

The most important consequence of speciation is that it is the sine qua non of diversity. For sexually reproducing organisms, every branch in the great phylogenetic tree of life represents a speciation event, in which populations became reproductively isolated and therefore capable of independent, divergent evolution, including, eventually, the acquisition of those differences that mark genera, families, and still higher taxa. Speciation, then, stands at the border between MICROEVOLUTION—the genetic changes within and among populations—and MACROEVOLUTION—the evolution of the higher taxa in all their glorious diversity.

In their hypothesis of punctuated equilibrium, Eldredge and Gould (1972; see also Stanley 1979; Gould and Eldredge 1993) proposed that speciation may be required for morphological evolution to occur at all. They suggested, based on Mayr's (1954) proposal that founder events trigger rapid evolution from one genetic equilibrium to another, that most evolutionary changes in morphology are triggered by and associated with peripatric speciation.

Population geneticists generally reject this hypothesis (Charlesworth et al. 1982). Morphological characters vary among populations of a species, just as they do among reproductively isolated species, and as we have seen, there has been little evidence for the model of founder effect speciation. Nevertheless, there is evidence that speciation contributes to rates of both molecular and morphological evolution, based on the expected difference between patterns of divergence expected under the models of phyletic gradualism versus punctuated equilibrium (Figure 18.29; see Figure 4.19). In the gradual model, the rate of evolution of a lineage, and therefore the amount of evolutionary change from the root of a phylogenetic tree to any extant species (path length), is unaffected by the number of speciation events. In contrast, evolutionary change requires speciation, and increases with the number of speciation events, in the pure punctuated equilibrium model. By the same token, the variance among extant species increases with the number of speciation events in the punctuated equilibrium model, but not in the gradual model. Mark Pagel and colleagues (2006) found that in 27 percent of the phylogenies of animals, fungi, and plants, path lengths as measured by numbers of nucleotide substitutions were significantly correlated with the number of species, as predicted if speciation

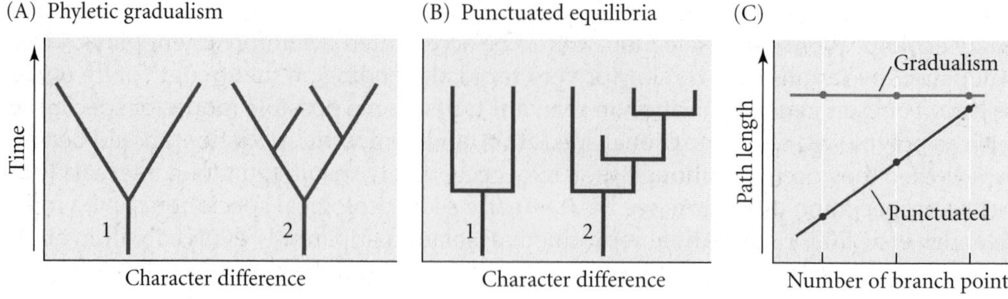

Figure 18.29 Models of phyletic gradualism (A) and punctuated equilibria (B) suggest how phylogenetic data might be used to determine if speciation is associated with enhanced evolution of molecular or morphological characters. In diagrams (A) and (B), lineages 1 and 2 differ in the number of speciation events (branch points). With phyletic gradualism, the variation among living species (tips) and the amount of evolution from the root to any living species (path length) are not affected by the number of speciation events. The correlation between path length and number of branch points in the phylogeny is expected to be zero, indicated by the horizontal line in (C). In the punctuated model, character evolution occurs only at speciation, so the variation among extant species and the path length is expected to be correlated with the number of branch points. Note that if some species have become extinct, as illustrated, the number of branch points in the phylogeny of extant species will underestimate the number of speciation events. (C after Pagel et al. 2006.)

accelerates evolution. In a more complex approach that takes into account the likelihood that a phylogeny of living species misses some speciation events because of extinction, Tiina Mattila and Folmer Bokma (2008) concluded that speciation explains more than two-thirds of the variance in body mass among species of mammals, and that gradual evolution accounts for little variation.

What might cause this pattern? As population geneticists have argued, and as Gould (2002) himself conceded, there is no reason to think that speciation (acquisition of reproductive isolation) triggers morphological evolution. Nevertheless, morphological change might be associated with speciation because *reproductive isolation enables morphological differences between populations to persist in the long term* (Futuyma 1987; Eldredge et al. 2005). Although different local populations may diverge rapidly as a result of selection, local populations are ephemeral: as climate and other ecological circumstances change, divergent populations move about and come into contact sooner or later. Much of the divergence that has occurred will then be lost by interbreeding—unless reproductive isolation has evolved (Figure 18.30). A succession of speciation events, each "capturing" further change in a character, may result in a long-term trend.

This hypothesis predicts, perhaps counterintuitively, that directional evolutionary change should be more pronounced during times of environmental stability (when the geographic isolation of populations lasts long enough for reproductive isolation to evolve)

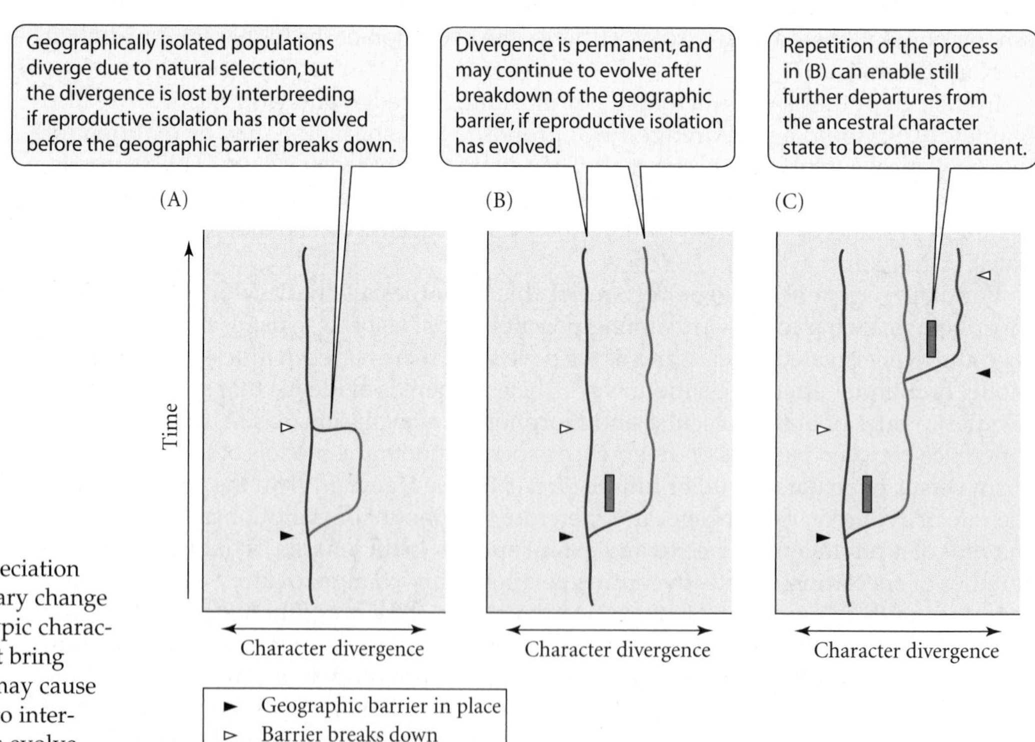

Figure 18.30 A model of how speciation might facilitate long-term evolutionary change in morphological and other phenotypic characters. Shifts in geographic ranges that bring divergent populations into contact may cause loss of their divergent features due to interbreeding (A), unless the populations evolve reproductive isolation while allopatric (B, C).

than during times of environmental change (when the distribution of habitats and populations changes, promoting gene flow). Paleontologists have described just such a pattern (Jansson and Dynesius 2002). Despite the great environmental changes during the Pleistocene glaciations, for example, early Pleistocene fossil beetles are identical to living species. Their geographic distributions have changed radically, however, and it is to these incessant changes in distribution, resulting in gene flow, that Coope (1979) attributed their morphological stability. Perhaps, as Ernst Mayr (1963, 621) said, "Speciation…is the method by which evolution advances. Without speciation, there would be no diversification of the organic world, no adaptive radiation, and very little evolutionary progress. The species, then, is the keystone of evolution."

Summary

1. Probably the most common mode of speciation is allopatric speciation, in which gene flow between populations is reduced by geographic or habitat barriers, allowing genetic divergence by natural selection and/or genetic drift.
2. In vicariant allopatric speciation, a widespread species becomes sundered by a geographic barrier, and one or both populations diverge from the ancestral state.
3. In a simple model of the evolution of reproductive isolation, complementary allele substitutions that do not reduce the fitness of heterozygotes occur at several loci in one or both populations. Epistatic interactions between alleles fixed in the two populations reduce the fitness of hybrids (postzygotic isolation). Likewise, genetic divergence may result in prezygotic isolation.
4. Reproductive isolation in allopatric populations appears to evolve as a side effect of divergent ecological or sexual selection. Both processes require further study before their relative importance can be assessed.
5. Prezygotic isolation evolves mostly while populations are allopatric, but may be reinforced when the populations become parapatric or sympatric.
6. Peripatric speciation, or founder effect speciation, is a hypothetical form of allopatric speciation in which genetic drift in a small peripheral population initiates rapid evolution, and reproductive isolation is a by-product. The likelihood of this form of speciation differs greatly depending on the mathematical model used. Although the geographic pattern of speciation predicted by this hypothesis may be common, there is little evidence for the process of drift-induced speciation.
7. Sympatric speciation, the origin of reproductive isolation within an initially randomly mating population, can occur as a result of disruptive selection. However, the sympatric evolution of sexual isolation is unlikely, due to recombination among loci affecting mating and loci affecting the disruptively selected character. Sympatric speciation may occur, however, if recombination does not oppose selection. For example, if disruptive selection favors preference for different habitats and if mating occurs within those habitats, prezygotic isolation may result. How often this occurs is debated.
8. Instantaneous speciation by polyploidy is common in plants. Allopolyploid species arise from hybrids between genetically divergent populations. Establishment of a polyploid population probably requires ecological or spatial segregation from the diploid ancestors because backcross offspring have low reproductive success. Polyploid species can have multiple origins.
9. In recombinational (hybrid) speciation, some genotypes of diploid hybrids are fertile and are reproductively isolated from the parent species, and so give rise to new species. This process appears to be uncommon, and has been documented more often in plants than in animals.
10. The time required for speciation to proceed to completion is highly variable. It is shorter for some modes of speciation (polyploidy, recombinational speciation) than others (especially speciation by mutation and drift of neutral alleles that confer incompatibility). The process of speciation may require 2 to 3 Myr, on average, for some groups of organisms; it is much longer in some cases and very much shorter in others.
11. Speciation is the source of the diversity of sexually reproducing organisms, and it is the event responsible for every branch in their phylogeny. It probably does not stimulate evolutionary change in morphological characters, as suggested by the hypothesis of punctuated equilibria, but rates of evolutionary change may nevertheless be correlated with speciation. This is probably because speciation prevents interbreeding between populations from undoing the changes wrought by natural selection or genetic drift.

GO TO THE
evolution
Companion Web Site at
www.sinauer.com/evolution
for quizzes, data analysis and simulation exercises, and other study aids.

Terms and Concepts

allopatric speciation
character displacement
ecological speciation
founder effect speciation
parapatric speciation
peripatric speciation
recombinational speciation
reinforcement
sympatric speciation
vicariance

Suggestions for Further Reading

As noted in Chapter 17, *Speciation*, by J. A. Coyne and H. A. Orr (Sinauer Associates, Sunderland, MA, 2004), is the most comprehensive recent work on the subject. Those with a mathematical bent will enjoy the wide-ranging treatment of models of speciation by S. Gavrilets in *Fitness Landscapes and the Origin of Species* (Princeton University Press, Princeton, NJ, 2004).

Problems and Discussion Topics

1. Why is it difficult to demonstrate that speciation has occurred parapatrically or sympatrically?
2. Coyne and Orr (1997) found that sexual isolation is more pronounced between sympatric populations than between allopatric populations of the same apparent age, and took this finding as evidence for reinforcement of sexual isolation. It might be argued, though, that any pairs of sympatric populations that were not strongly sexually isolated would have merged, and would have been unavailable for study. Thus the degree of sexual isolation in sympatric compared with allopatric populations might be biased. How might one rule out this possible bias? (Read Coyne and Orr after suggesting an answer.)
3. Suppose that full reproductive isolation between two populations has evolved. Can speciation in this case be reversed, so that the two forms merge into a single species? Under what conditions is this probable or improbable?
4. Referring to the discussion of parallel speciation in sticklebacks, can a single biological species arise more than once (i.e., polyphyletically)? How might this possibility depend on the nature of the reproductive barrier between such a species and its closest relative?
5. The heritability of an animal's preference for different habitats or host plants might be high or low. How might heritability affect the likelihood of sympatric speciation by divergence in habitat or host preference?
6. Biological species of sexually reproducing organisms usually differ in morphological or other phenotypic traits. The same is often true of taxonomic species of asexual organisms such as bacteria and apomictic plants. What factors might cause discrete phenotypic "clusters" of organisms in each case?
7. In many groups of plants, low levels of hybridization between related species are not uncommon, yet only a few cases of the origin of "hybrid species" by recombinational speciation have been documented. What factors make recombinational speciation likely versus unlikely?
8. If speciation occurs by divergent pathways of sexual selection in different populations, what might cause the nature of sexual selection to differ?
9. Genetic drift and natural selection give rise to geographic variation among populations of a species. How do we account, then, for the features that are uniform among all populations of a species? (See Morjan and Rieseberg 2004.)
10. Allochronic speciation is a hypothetical form of sympatric speciation, whereby a population becomes divided into subpopulations that breed at different seasons or in different years. Is this form of speciation theoretically "easy" or "difficult"? Consider some postulated examples in birds (Friesen et al. 2007) and insects (Marshall and Cooley 2000; Simon et al. 2000).
11. Choose a topic from this chapter and discuss how its treatment would be altered if one adopted a phylogenetic species concept rather than the biological species concept.

CHAPTER 19

Coevolution: Evolving Interactions among Species

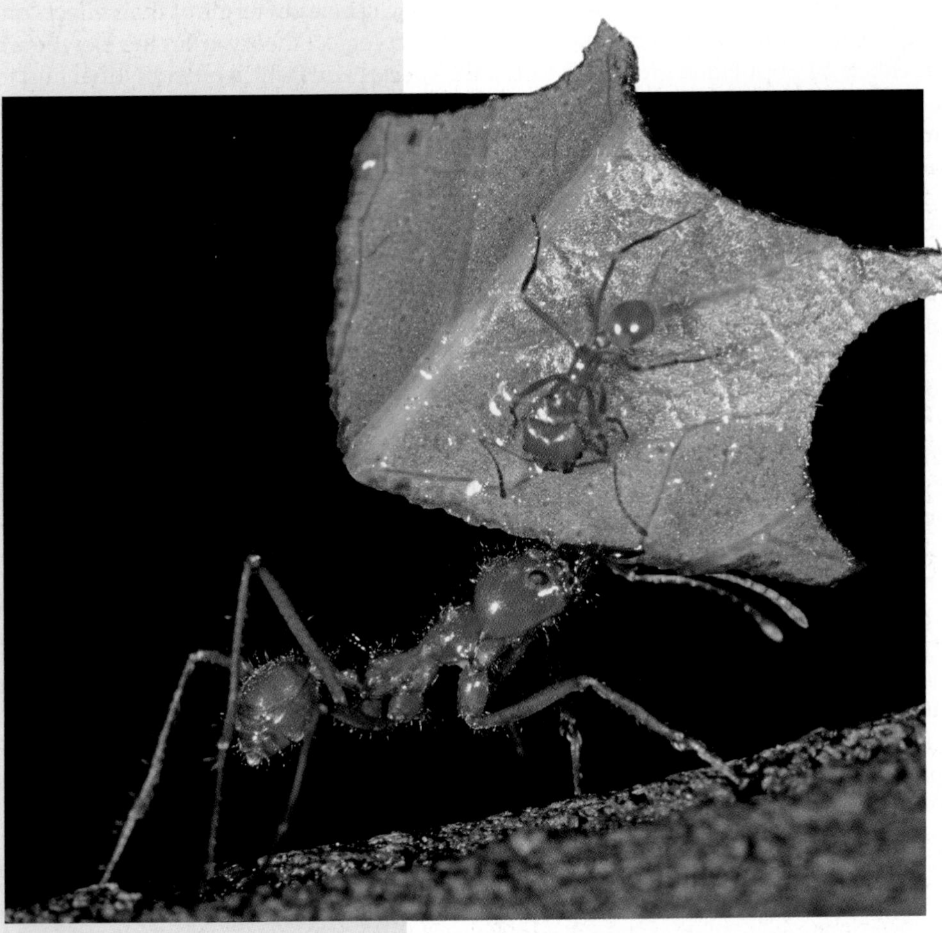

Interspecies interactions. *Atta colombica* and other leaf-cutter ants feed only on fungi that they cultivate on fragments of leaves (see Figure 2.4B). The mutually adapted ants and fungi are part of a web of interactions with other fungi and bacteria. (Photo © Gavriel Jecan/AGE Fotostock.)

If you walk through a forest in many parts of tropical America, you are likely to encounter large, reddish ants, each carrying a fragment of a leaf, rushing by the hundreds along their highway, a track worn into the soil by the tramp of many thousands of little ant feet in days and months past. They are leaf-cutter ants, genus *Atta* or *Acromyrmex*, transporting massive amounts of fresh leaf material to their nest, a series of underground caverns as much as 3 meters deep that may house 5 to 10 million worker ants, thousands of larvae, and their mother, the single queen who may live for 15 years and lay many millions of eggs. The ants do not eat leaves. Rather, small workers in the caverns chew the leaf material into tiny fragments and insert these into gardens of fungi, the sole food of both the larval and adult ants. This species of fungus (see Figure 2.4B) is found nowhere else and is carried by young queens from their natal nest in a special cavity in the pharynx. The fungus, in the family Lepiotaceae, is related to other fungi that digest leaf litter, and the phylogenetically most basal of the fungus-growing ants (tribe Attini) provide their lepiotaceous fungi with bits of dead and decaying leaves, instead of live leaves (Hölldobler and Wilson 1990; Mueller et al. 2005). This exquisite mutualistic symbiosis, however, is threatened by other fungi, in the genus *Escovopsis*, that attack and devastate the fungal garden, and are likewise found only in the nests of attine ants. The ants counter this threat partly by removing *Escovopsis* spores, but mostly by harboring actinomycete bacteria, *Pseudonocardia*, which produce an antibiotic that inhibits *Escovopsis* growth. The mutualistic bacteria are housed in many small pits on the ants' exoskeletons and are apparently nourished by secretions of glands, unique to attine ants, that open into these pits. These four organisms are thus bound to each other in a web of antagonistic and mutualistic interactions, and have been adapting to each other since the origin of the tribe Attini, about 50 Mya (Currie et al. 2003, 2006).

This is a rather extreme example of reciprocal adaptation of species to each other, but almost all species have adaptations for interacting with other species. Such adaptations, some of which are quite extraordinary, have enhanced the diversity of life and have had profound effects on the structure of ecological communities.

In this chapter, we will consider interactions among species in terms of their effects on the fitness of individual organisms (not, as in some ecological theory, from the viewpoint of their effects on population growth). Most of the species with which an individual might interact can be classified as RESOURCES (used as nutrition or habitat), COMPETITORS (for resources such as food, space, or habitat), ENEMIES (species for which the focal species is a consumable resource), or COMMENSALS (species that profit from but have no effect on the focal species). In MUTUALISTIC interactions (such as the relation between attine ants and their *Pseudonocardia* bacteria), each species obtains a benefit from the other. Some interactions are more complex, often because they are mediated by a third species. For example, different unpalatable species of butterflies that resemble one another may profit from their resemblance because predators that have learned to avoid one may avoid the other as well (see Figures 12.19 and 19.25). Moreover, the nature and strength of an interaction may vary depending on environmental conditions, genotype, age, and other factors. There is genetic variation, for example, in virulence within species of parasites and in resistance within species of hosts. Some mycorrhizal fungi, associated with plant roots, enhance plant growth in infertile soil but depress it in fertile soil. Thus the selection that species exert on each other may differ among populations, resulting in a "geographic mosaic" of coevolution that differs from one place to another (Thompson 1999).

The Nature of Coevolution

The possibility that an evolutionary change in one species evokes a reciprocal change in another species distinguishes selection in interspecific interactions from selection stemming from conditions in the physical environment. Reciprocal genetic change in interacting species, owing to natural selection imposed by each on the other, is **coevolution** in the narrow sense.

The term "coevolution" includes several concepts (Futuyma and Slatkin 1983; Thompson 1994). In its simplest form—**specific coevolution**—two species evolve in response to each other (Figure 19.1A). For example, Darwin envisioned predatory mammals, such as wolves, and their prey, such as deer, evolving ever greater fleetness, each improvement in one causing selection for compensating improvement in the other, in an "evolutionary arms race" between prey and predator. **Guild coevolution**, sometimes called **diffuse**

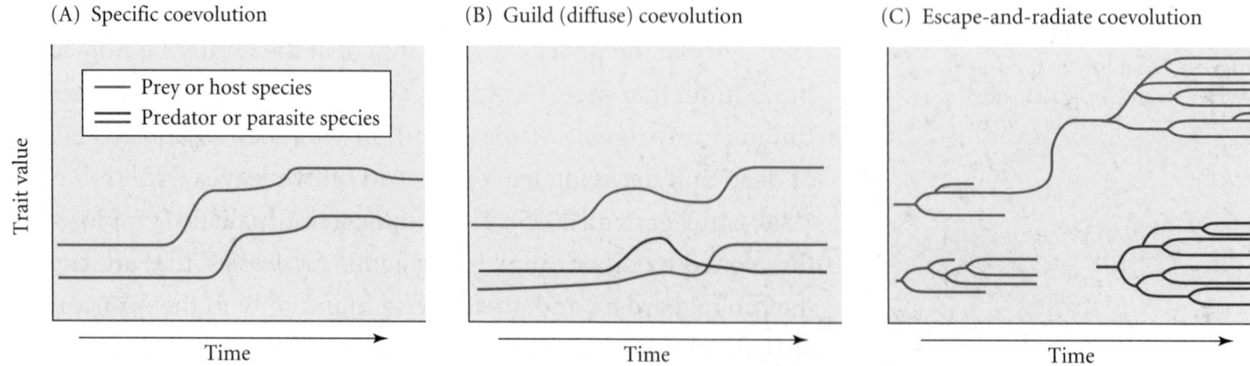

Figure 19.1 Three kinds of coevolution. In each graph, the horizontal axis represents evolutionary time and the vertical axis shows the state of a character in a species of prey or host and one or more species of predators or parasites. (A) Specific coevolution. (B) Guild, or diffuse, coevolution, in which a prey species interacts with two or more predators. (C) Escape-and-radiate coevolution. One of several prey or host species evolves a major new defense, escapes association with a predator or parasite, and diversifies. Later, a different predator or parasite adapts to the host clade and diversifies.

coevolution, occurs when several species are involved and their effects are not independent (Figure 19.1B). For example, genetic variation in the resistance of a host to two different species of parasites might be correlated (Hougen-Eitzman and Rausher 1994). In **escape-and-radiate coevolution**, a species evolves a defense against enemies and is thereby enabled to proliferate into a diverse clade (Figure 19.1C). For example, Paul Ehrlich and Peter Raven (1964) proposed that species of plants that evolved effective chemical defenses were freed from predation by most herbivorous insects, and thus diversified, evolving into a chemically diverse array of food sources to which different insects later adapted and then diversified in turn.

Phylogenetic Aspects of Species Associations

The term "coevolution" has also been applied to a history of parallel diversification, as revealed by concordant phylogenies, of associated organisms such as hosts and their parasites or endosymbionts. For example, aphids have endosymbiotic bacteria (*Buchnera*) that live in special cells and supply the essential amino acid tryptophan to their hosts (see Figure 16.22C). The phylogeny of these bacteria is completely concordant with that of their aphid hosts (Figure 19.2). The simplest interpretation of this pattern is that the association between *Buchnera* and aphids dates from the origin of this insect family, that there has been little if any cross-infection between aphid lineages, and that the bacteria have diverged in concert with speciation of their hosts. Differences among host species may reinforce host-specific associations and prevent symbionts from switching to distantly related species. For example, the phylogeny of feather lice in the genus *Columbicola*, which are parasites of pigeons and doves (Columbidae), is not a perfect match with that of their hosts, but it is similar enough to indicate that the lice have cospeciated with their hosts rather frequently, rather than switching from one species of columbid to another (Figure 19.3A). Dale Clayton and colleagues (2003) found that the body size of lice species is correlated with the body size of their host species, which have wider intervals between the adjacent units (barbs) that make up a feather. Therefore, lice fit better into the feathers of their normal host than into the feathers of larger or smaller birds. Lice transferred to smaller species of columbids had reduced survival because the birds could easily remove them by preening their feathers. On birds that could not preen effectively because their bills were fitted with bits, the lice survived quite well (Figure 19.3B). The investigators concluded that lice have seldom switched to bird species that differ in size because colonists would have had low fitness.

Associations between hosts and their parasites or symbionts rarely show much evidence of cospeciation. Instead, there are often mismatches, caused by horizontal transfer,

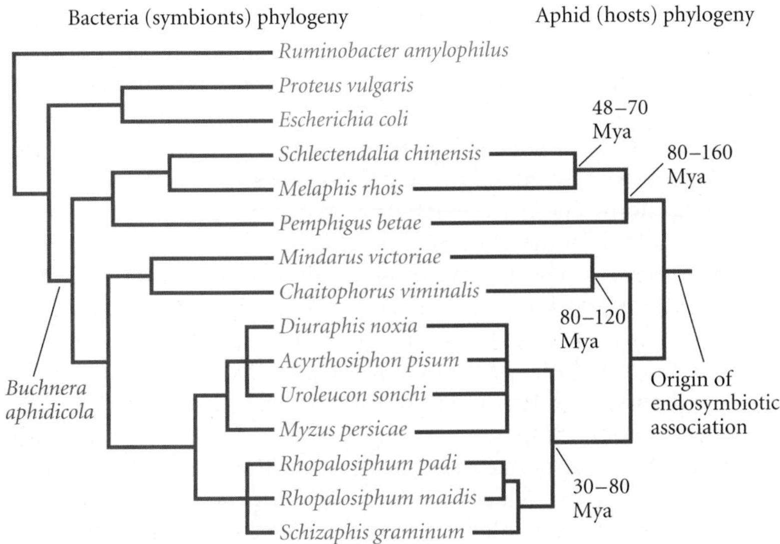

Figure 19.2 The phylogeny of endosymbiotic bacteria included under the name *Buchnera aphidicola* is perfectly congruent with that of their aphid hosts. Several related bacteria (names in red) were included as outgroups in this analysis. Names of the aphid hosts of the *Buchnera* lineages are given in blue. The estimated ages of the aphid lineages are based on fossils and/or biogeography. (After Moran and Baumann 1994.)

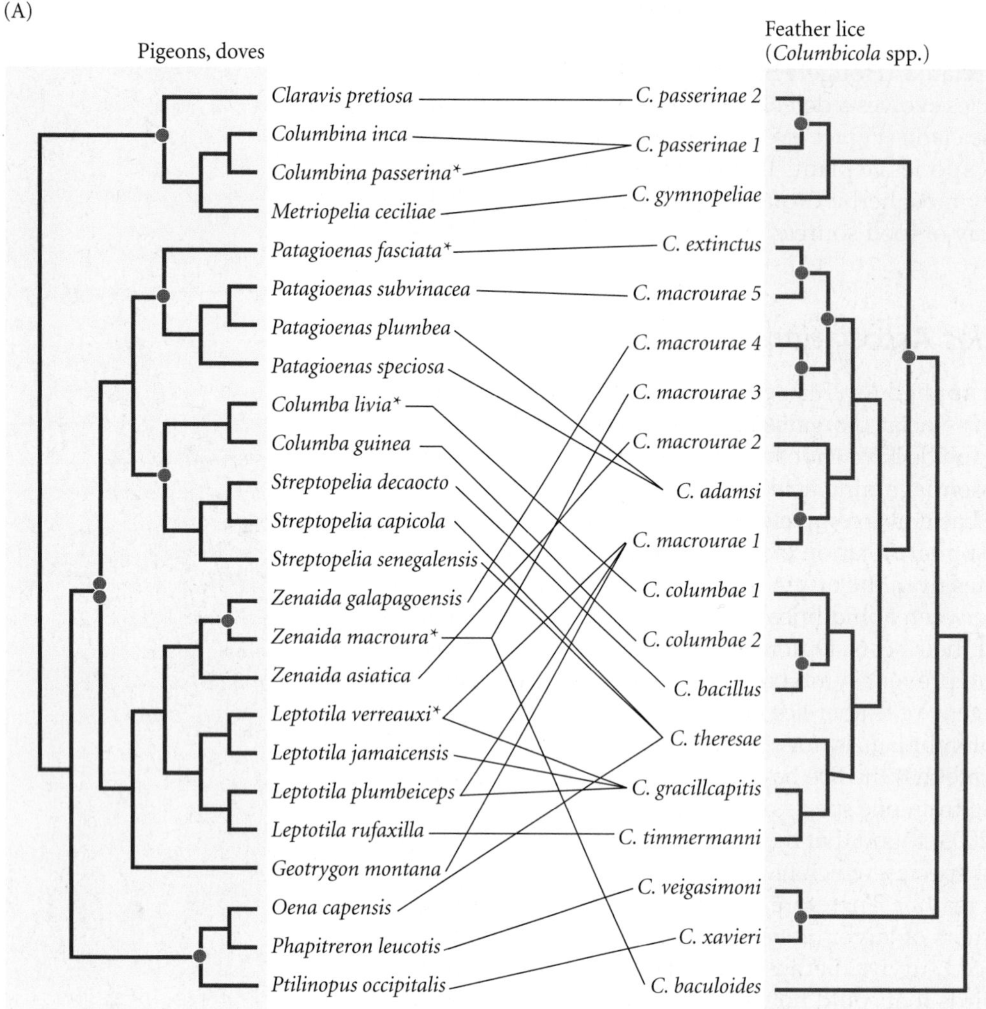

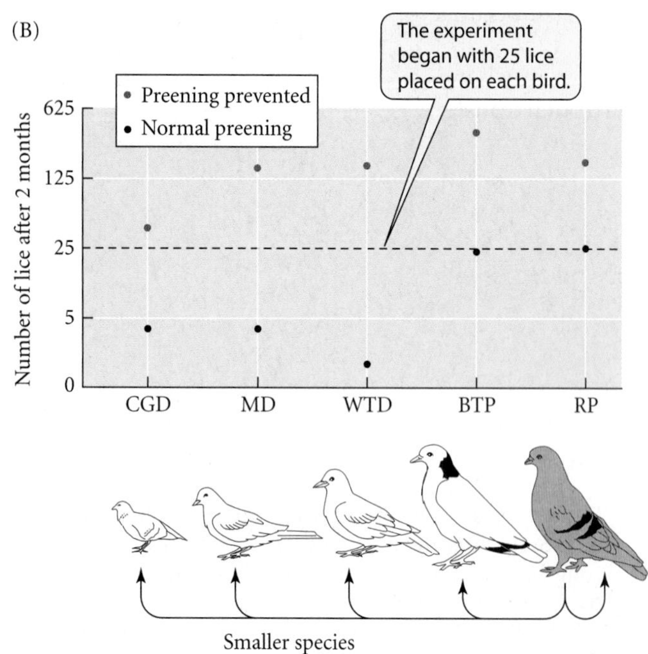

Figure 19.3 (A) A phylogeny of specialized feather lice (right) is mostly congruent with that of their dove and pigeon hosts (left), suggesting a history of cospeciation (with a few exceptions, shown by the crossing of lines that connect parasites to their hosts). Species marked with an asterisk were subjects in the experiment shown in (B). (B) A starting population of 25 lice transferred from rock pigeons (shaded) to other individuals of the same and four different species increased on all of these species when a bit placed on the birds' bills prevented them from preening. Louse populations on birds that could preen normally decreased on the three smaller species. RP, rock pigeon (*C. livia*); BTP, band-tailed pigeon (*P. fasciata*); WTD, white-tipped dove (*L. verreauxi*)l; MD, mourning dove (*Z. macroura*); CGD, common ground dove (*C. passerina*). (After Clayton et al. 2003).

or HOST SWITCHING, and by several other factors, such as extinction of parasite lineages (Page 2002). Host switching may be likely if parasites disperse from one host to another through the environment, as do plant-feeding insects. Among host-specific herbivorous insects, related species often feed on plants in the same family, and these associations may be very old. For example, phylogenetically basal lineages of leaf beetles, long-horned beetles, and weevils all feed mostly on cycads or conifers—plant lineages that evolved before the angiosperms, with which the more phylogenetically "advanced" beetles are associated (Farrell 1998). The fossil record also attests to the great age of some such associations (Labandeira 2002). Nevertheless, the phylogeny of the insect species seldom matches that of the host plants very closely, and the species of insects have often originated long after their host plant lineages diverged (Futuyma et al. 1995; Winkler and Mitter 2008). Flea beetles in the genus *Blepharida* are associated with the plant family Burseraceae in both Mexico and Africa, an association that is more than 100 Myr old. The beetles are adapted to the terpene-containing resins that appear to be the plants' main defense against nonadapted insects. Judith Becerra (1997) found that in Mexico, closely related host-specialized species of *Blepharida* are more likely to feed on chemically similar species of

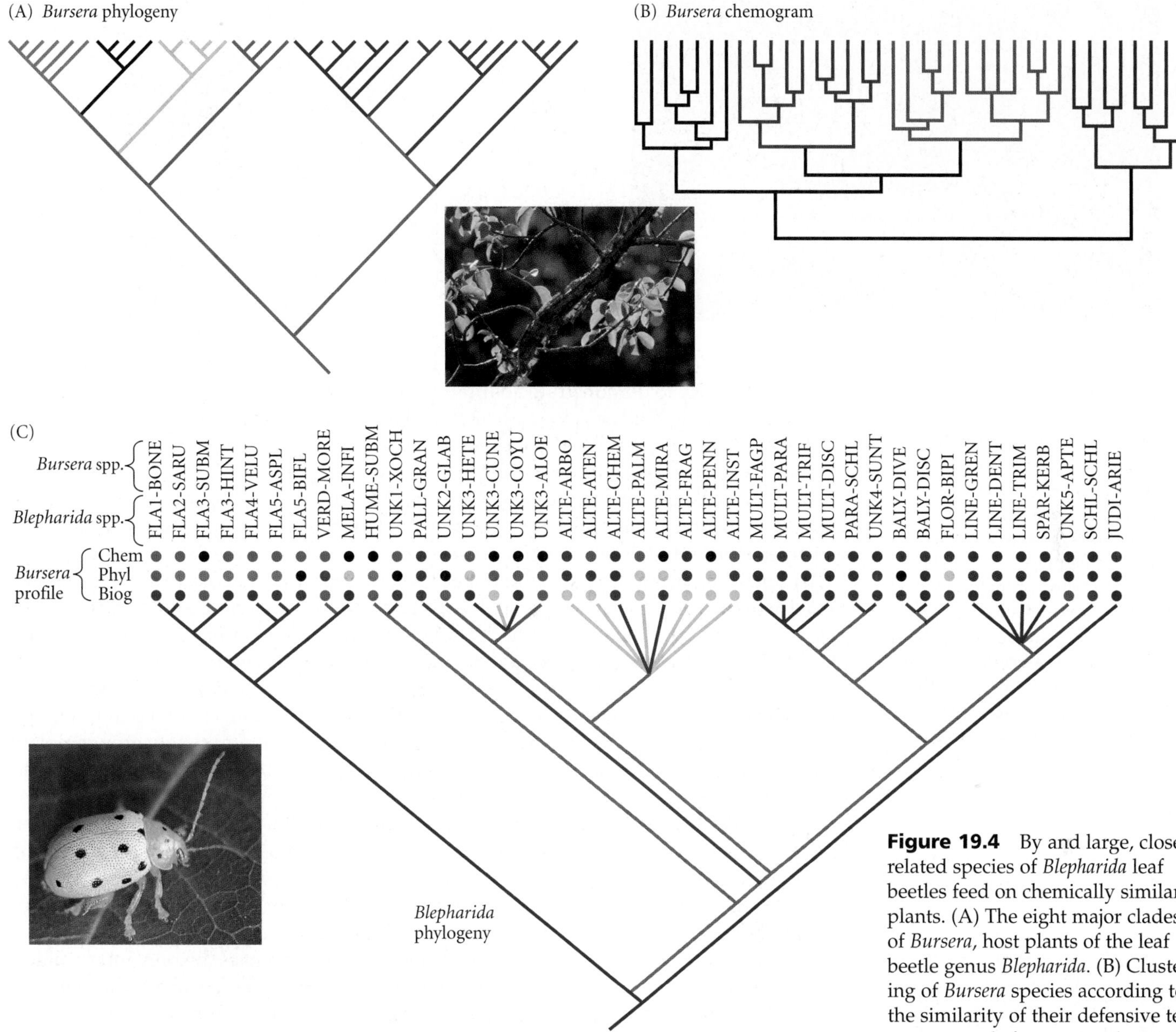

Figure 19.4 By and large, closely related species of *Blepharida* leaf beetles feed on chemically similar plants. (A) The eight major clades of *Bursera*, host plants of the leaf beetle genus *Blepharida*. (B) Clustering of *Bursera* species according to the similarity of their defensive terpenes reveals four major chemically coherent groups of species. (C) Phylogenetic relationships among species of *Blepharida* match the chemical similarity of their host plants more closely than they do the phylogeny or geographic distribution of their host. Colored circles above the tips of the beetle phylogeny profile the chemical group (top), major clade (middle), and major geographic region (bottom) of the beetle's host plant species. (From Becerra and Venable 1999; B, photo © tbkmedia.de/Alamy; C, photo courtesy of Judith Becerra).

Bursera than on phylogenetically closely related species (Figure 19.4). This pattern suggests that newly arisen species of beetles have adapted to plant lineages that had already evolved, and furthermore, that these insects adapt more readily to new hosts that are chemically similar to the beetles' ancestral host than to chemically different plants.

Coevolution of Enemies and Victims

In considering the *processes* of evolutionary change in interacting species, we will begin with interactions between enemies and victims: predators and their prey, parasites and their hosts, herbivores and their host plants. Predators and parasites have evolved some extraordinary adaptations for capturing, subduing, or infecting their victims (Figure 19.5). Defenses against predation and parasitism can be equally impressive, ranging from cryptic patterning (see Figure 12.5), to the highly toxic chemical defenses of both plants and animals (see Figure 19.8), to the most versatile of all defenses: the vertebrate immune system, which can generate antibodies against thousands of foreign compounds. Many such adaptations appear to be directed at a variety of different enemies or prey species, so

(A)

(B)

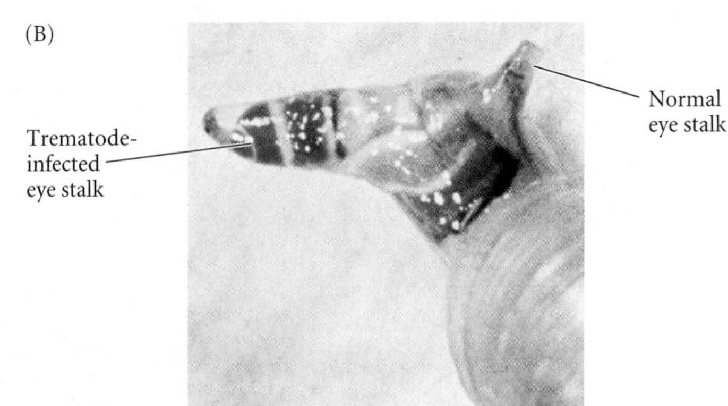

Figure 19.5 Predators and parasites have evolved many extraordinary adaptations to capture prey or infect hosts. (A) The dorsal fin spine of a deep-sea anglerfish (*Himantolophus*) is situated above the mouth and modified into a luminescent fishing lure. (B) The larva of a parasitic trematode (*Leucochloridium*) migrates to the eye stalk of its intermediate host, a land snail, and turns it a bright color to make the snail more visible to the next host in the parasite's life cycle, a snail-eating bird such as a thrush. (A © David Shale/naturepl.com; B photo by P. Lewis, courtesy of J. Moore.)

although it is easy to demonstrate adaptations in a predator or a prey species, it is usually difficult to show how any one species has coevolved with another.

Theoretically, the coevolution of predator and prey might take any of several courses (Abrams 2000): it might continue indefinitely in an unending escalation of an evolutionary arms race (Dawkins and Krebs 1979); it might result in a stable genetic equilibrium; it might cause continual cycles (or irregular fluctuations) in the genetic composition of both species; or it might even lead to the extinction of one or both species.

An unending arms race is unlikely because adaptations that increase the offensive capacity of the predator or the defensive capacity of the prey entail allocations of energy and other costs that at some point outweigh their benefits. Consequently, a stable equilibrium may occur when costs equal benefits. For example, the toxic SECONDARY COMPOUNDS that plants use as defenses against herbivores, such as the tannins of oaks and the terpenes of pines and of *Bursera*, can account for more than 10 percent of a plant's energy budget. Such high levels of chemical defense are especially typical of slowly growing plant species that inhabit nutrient-poor environments (Coley et al. 1985). Genetic lines of wild parsnip (*Pastinaca sativa*) containing high levels of toxic furanocoumarins suffered less attack from webworms, and matured more seeds, than lines with lower levels when grown outdoors; in the greenhouse, however, where they were free from insect attack, the lines with higher levels of furanocoumarins had lower seed production (Berenbaum and Zangerl 1988). Costs of this kind may explain why plants are not more strongly defended than they are, and thus why they are still subject to insect attack.

Another kind of cost arises if a defense against one enemy makes the prey more vulnerable to another enemy. For example, terpenoid compounds called cucurbitacins enhance the resistance of cucumber plants (*Cucumis sativus*) to spider mites, but they attract certain cucumber-feeding leaf beetles (Dacosta and Jones 1971).

Table 19.1 *Gene-for-gene interactions between a parasite and its host*

Parasite genotype	Host genotype			
	R_1— R_2—	R_1— r_2r_2	$r_1r_1R_2$—	$r_1r_1\ r_2r_2$
V_1—V_2—	–	–	–	+
V_1—v_2v_2	–	–	+	+
$v_1v_1\ V_2$—	–	+	–	+
$v_1v_1\ v_2v_2$	+	+	+	+

Source: After Frank 1992.

Note: In each species, two loci, with dominant and recessive alleles at each locus, control resistance (of the host) and infectivity (of the parasite). A + sign indicates that the parasite genotype can grow on a host of a given genotype (i.e., the parasite is infective and the host is susceptible); the – signs indicate that the host genotype is resistant to the parasite genotype.

Models of enemy-victim coevolution

GENE-FOR-GENE MODELS. Coevolution of enemies and victims has been modeled in several ways, appropriate to different kinds of characters. For example, models of evolution at one or a few loci are appropriate for **gene-for-gene interactions**, which were first described in cultivated flax (*Linum usitatissimum*) and in flax rust (*Melampsora lini*), a basidiomycete fungus. Similar systems have been described or inferred in several dozen other pairs of plants and fungi, and in the interaction between cultivated wheat (*Triticum*) and a major pest, the Hessian fly (*Mayetiola destructor*). In each such system, the host has several loci at which a dominant allele (R) confers resistance to the parasite. At each of several corresponding loci in the parasite, a recessive allele (v) confers infectivity—the ability to infect and grow in a host with a particular R allele (Table 19.1). If resistance has a cost, any particular resistance allele (R_i) will

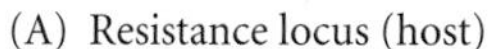

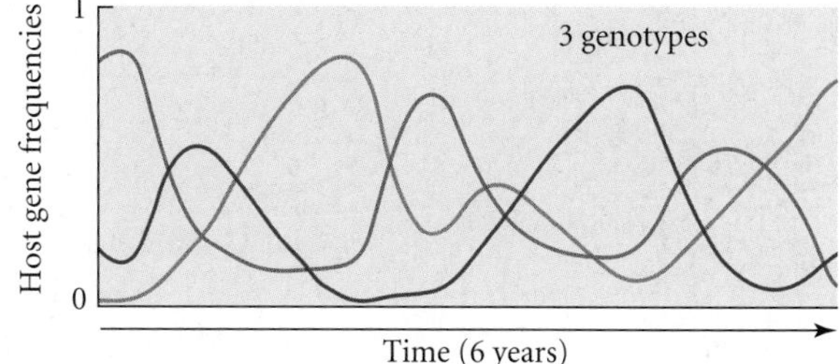

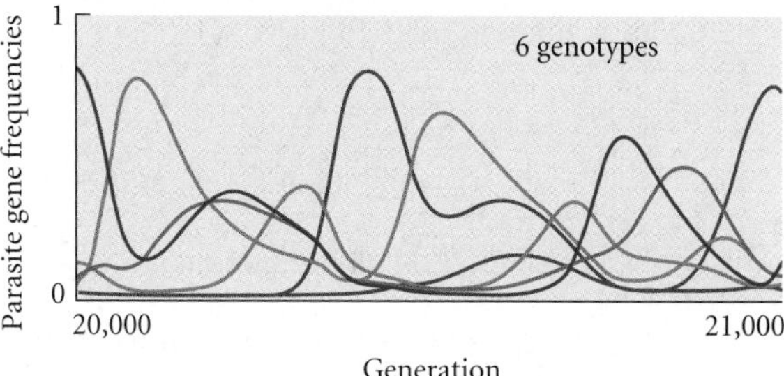

Figure 19.6 A computer simulation of genetic changes at (A) a resistance locus in a host and (B) an infectivity locus in a parasite. The host is diploid and has three resistance alleles; the parasite is haploid and has six infectivity alleles. Each parasite genotype can overcome the defenses of one of the six host genotypes (e.g., parasite P_1 can attack host H_1H_1). Both populations remain polymorphic and fluctuate irregularly in genetic composition. (After Seger 1992.)

decline in frequency when the parasite's corresponding infectivity allele (v_i) has high frequency, because R_i is then ineffective. As a different R allele (R_j) increases in frequency in the host population, the corresponding infectivity allele v_j increases in the parasite population.

According to computer simulations, such frequency-dependent selection can cause cycles or irregular fluctuations in allele frequencies (Figure 19.6). In populations of Australian flax (*Linum marginale*), the frequencies of different rust genotypes fluctuated from year to year. On the whole, highly infective genotypes—those that could attack the greatest number of flax genotypes—occurred in highly resistant flax populations, and less infective rusts were found in less resistant flax populations (Thrall and Burdon 2003). Resistance genes are now the subject of intense study at the molecular level (Stahl and Bishop 2000). The pattern of sequence variation in the resistance genes of *Arabidopsis*, a flowering plant in the mustard family (Brassicaceae), suggests that polymorphism has been maintained by balancing selection, as the frequency-dependent gene-for-gene model predicts (Bakker et al. 2006).

QUANTITATIVE TRAITS. Coevolutionary models of a defensive polygenic character (y) in a prey species and a corresponding polygenic character (x) in a predator are mathematically complex and include many variables that can affect the outcome (Abrams 2000). An important distinction is whether the capture rate of the prey by the predator increases with the difference $x - y$ (e.g., when the predator's speed is greater than the prey's), or decreases (e.g., if it depends on a close match between the size of the prey and the size of the predator's mouth). In the first case, mathematical analyses suggest that both species will often evolve in the same direction (e.g., toward greater speed), arriving at an equilibrium point that is determined by physiological limits or excessive investment costs. However, suppose the capture rate depends on a close match between x and y, that deviation too great in either direction increases the cost of x (or y), and that $\bar{x} = \bar{y}$. Then either increasing or decreasing y will improve prey survival. In this case, y will evolve in one or the other direction, and x will evolve to track y. Eventually y may evolve in the opposite direction as its cost becomes too great, and x will evolve likewise. The result might be continuing cycles of change in the characteristics of both species, and these genetic changes may contribute to cycles in population density (Figure 19.7).

Figure 19.7 Computer simulation of coevolution between prey and predator in which the optimal predator phenotype (e.g., mouth size) matches a prey phenotype (e.g., size). (A) Evolution of character state means. As a character state diverges from a reference value, its fitness cost prevents it from evolving indefinitely in either direction. The evolution of the predator's character state lags behind the prey's. (B) Changes in character state means may be paralleled by cycles in population density, arising partly from changes in the match between the predator's character and the prey's. (After Abrams and Matsuda 1997.)

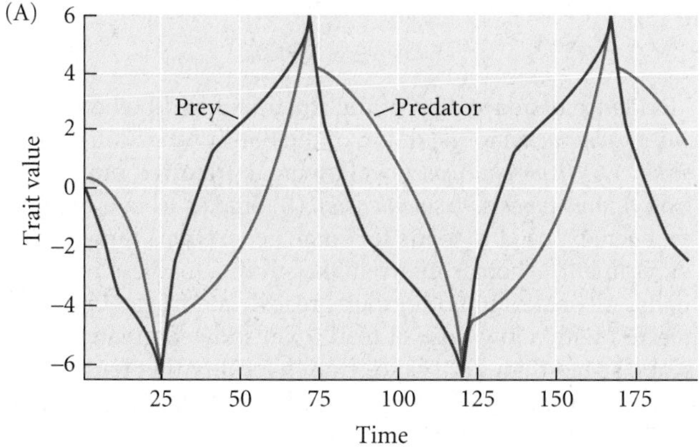

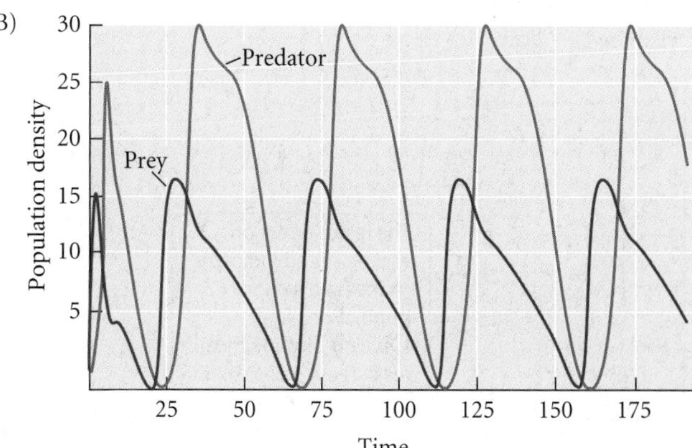

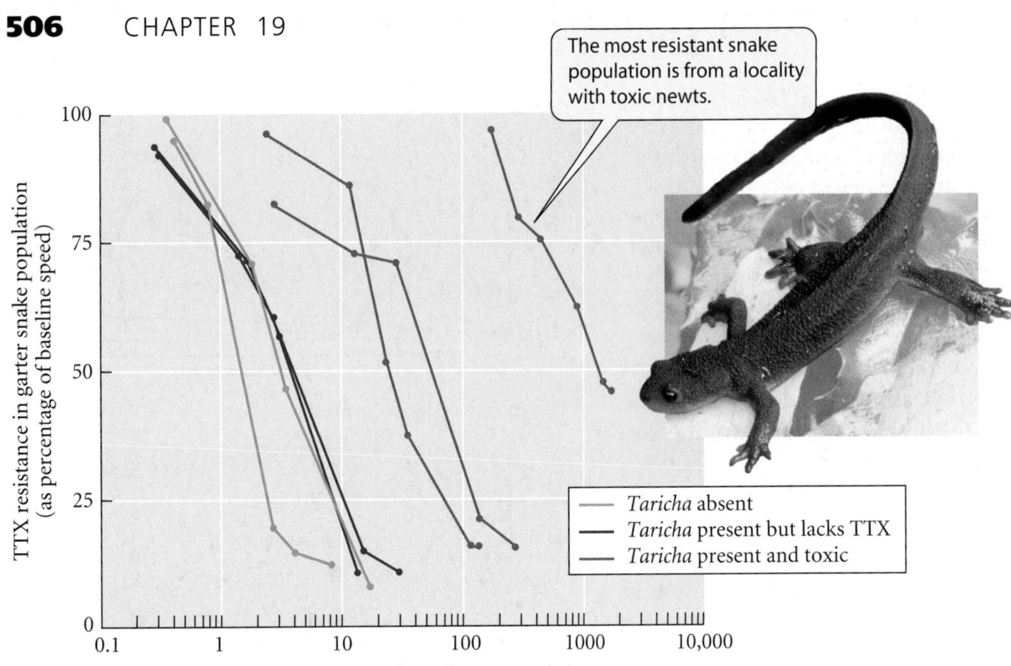

Figure 19.8 Variation in TTX resistance, measured by crawling speed after injection in relation to dose, in garter snakes (*Thamnophis sirtalis*) from several localities. The least resistant population is from Maine, where the toxic rough-skinned newt (*Taricha granulosa*) does not occur. Two of the other nonresistant populations coexist with newt populations that lack TTX. The three most resistant populations are sympatric with toxic newt populations. (After Brodie and Brodie 1999; photo © Mike Anich/Photolibrary.com.)

Examples of predator-prey coevolution

The rough-skinned newt (*Taricha granulosa*) of northwestern North America has one of the most potent known defenses against predation: the neurotoxin tetrodotoxin (TTX). Most populations have high levels of TTX in the skin (one newt has enough to kill 25,000 laboratory mice), but a few populations, such as the one on Vancouver Island, British Columbia, have almost none (Brodie and Brodie 1999; Brodie et al. 2002). Populations of the garter snake *Thamnophis sirtalis* from outside the range of this newt have almost no resistance to TTX. But populations that are sympatric with toxic newts feed on them, and can be as much as a hundred times more resistant to TTX than are allopatric populations (Figure 19.8). Resistance to toxic prey has evolved rapidly in the red-bellied black snake (*Pseudechis porphyriacus*), one of many Australian snake species that have been endangered by feeding on the South American cane toad (*Bufo marinus*), a highly toxic amphibian that was introduced into Australia in 1935 on the misguided presumption that it would control insect pests of sugar cane. (It did not.) Populations of several species of Australian snakes, which have never before been exposed to toxic prey, have declined greatly in abundance because they eat cane toads. Ben Phillips and Richard Shine (2004, 2006) determined that black snake populations that have been exposed to cane toads for about 23 generations have evolved several counteradaptations (Figure 19.9): they are physiolog-

(A)

(B)

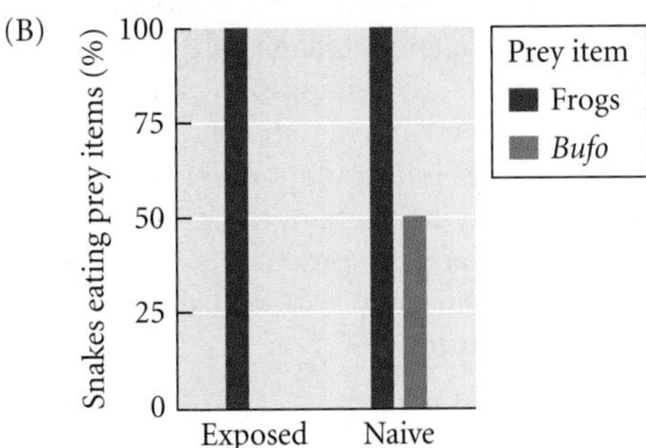

(C)

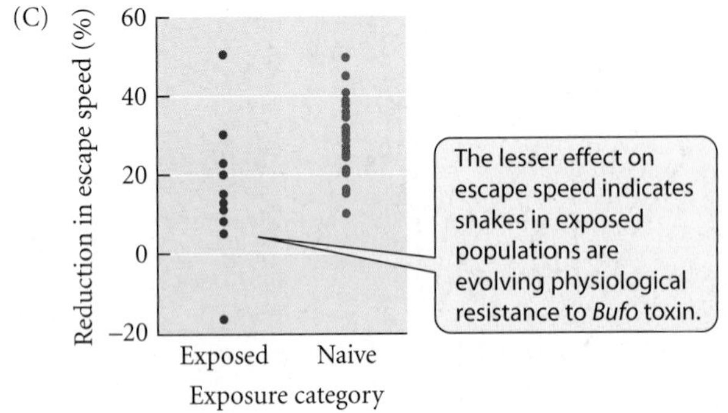

Figure 19.9 Evidence of adaptation of the Australian red-bellied black snake (*Pseudechis porphyriacus*) to incursion of the South American cane toad (*Bufo marinus*). (A) *P. porphyriacus* swallowing a preferred prey, the spotted grass frog (*Limnodynastes tasmaniensis*). (B) Snakes from a region where they have been exposed to toads for some generations refused to eat *Bufo* in an experiment, whereas many snakes from a toadless region ("naive") ate toads. All snakes in both groups readily ate frogs. (C) Toad-naive snakes injected with a low dose of toad toxin showed greater reduction in escape speed than did snakes from a locality that had been exposed to toads. (A © Robert Valentic/Naturepl.com; B, C after Phillips and Shine 2006.)

Figure 19.10 (A) A fledgling common cuckoo (*Cuculus canorus*) being fed by its foster parent, a much smaller reed warbler (*Acrocephalus scirpaceus*). (B) Mimetic egg polymorphism in the European cuckoo. The left column shows eggs of six species parasitized by the cuckoo (from top: robin, pied wagtail, dunnock, reed warbler, meadow pipit, great reed warbler). The middle column shows a cuckoo egg laid in the corresponding host's nest. The match is quite close except in the dunnock nest. The right column shows artificial eggs used by researchers to test rejection responses. (A © David Kjaer/naturepl.com; B photo by M. Brooke, courtesy of N. B. Davies.)

ically more resistant to toad toxin, they avoid eating toads, and they have evolved a narrower mouth, so they are less capable of swallowing large toads than toad-naive populations of black snakes far from the range of the cane toad.

Brood-parasitic birds, such as cowbirds and some species of cuckoos, lay eggs only in the nests of certain other bird species. Cuckoo nestlings hatch first and eject their host's eggs from the nest, so the host ends up rearing only the parasite (Figure 19.10A). Adults of host species do not treat parasite nestlings any differently from their own young, but some host species do recognize parasite eggs and either eject them or desert the nest and start a new nest and clutch.

The most striking counteradaptation among brood parasites is egg mimicry (Rothstein and Robinson 1998). Each population of the common cuckoo (*Cuculus canorus*) contains several different genotypes that prefer different hosts and lay eggs closely resembling those of their preferred hosts (Figure 19.10B). Some other individuals lay non-mimetic eggs. Some host species accept cuckoo eggs, some frequently eject them, and others desert parasitized nests. By tracing the fate of artificial cuckoo eggs placed in the nests of various bird species, Nicholas Davies and Michael Brooke (1998) found that species that are not parasitized by cuckoos (because of unsuitable nest sites or feeding habits) tend not to eject cuckoo eggs, whereas among the cuckoos' preferred hosts, those species whose eggs are mimicked by cuckoos rejected artificial eggs more often than those whose eggs are not mimicked. These species have evidently adapted to brood parasitism. Moreover, populations of two host species in Iceland, where cuckoos are absent, accepted artificial cuckoo eggs, whereas in Britain, where those two species are favored hosts, they rejected such eggs. Surprisingly, among suitable host species, those that are rarely parasitized by cuckoos did not differ in discriminatory behavior from those commonly parasitized. Davies and colleagues suspect that the rarely parasitized species were more commonly parasitized in the past, and that their ability to reject cuckoo eggs has selected against the cuckoo genotypes that parasitized these species.

Plants and herbivores

Almost all plants synthesize a variety of so-called secondary compounds that play little or no role in basic metabolism. Thousands of such compounds have been described, including many that humans have found useful as drugs (e.g., salicylic acid, the ingredient of aspirin), stimulants (caffeine), condiments (capsicin, the "hot" element in chili peppers), and in other applications (e.g., cannabinol, in marijuana). Higher taxa of plants are often characterized by particular groups of similar compounds, such as cardiac glycosides

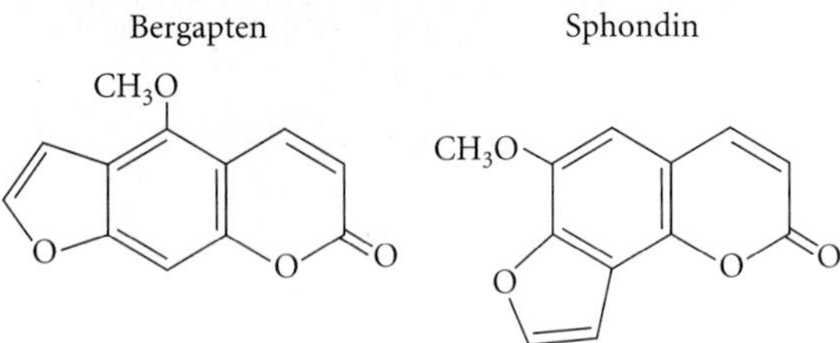

Figure 19.11 The furanocoumarins bergapten and sphondin are among the defensive compounds of wild parsnip (*Pastinaca sativa*), the host plant of a moth larva, the parsnip webworm (*Depressaria pastinacella*). (After Berenbaum et al. 1986; photo courtesy of May Berenbaum.)

in milkweeds (Apocynaceae) and glucosinolates in mustards (Brassicaceae). Some compounds are known to be toxic to animals, and some to be repellant, and so the hypothesis was posed that these features, as well as physical features such as spines, evolved as defenses against herbivores, especially insects (e.g., Fraenkel 1959). Paul Ehrlich and Peter Raven (1964) proposed a scenario of "escape and radiate" coevolution, in which a plant species that evolves a new and highly effective chemical defense may escape many of its associated herbivores, and give rise to a clade of species that share the novel defense. Eventually, though, some insect species from other hosts shift to these plants, adapt to them, and give rise to a clade of adapted herbivores. The association of the diverse species of *Blepharida* leaf beetles with the many species of *Bursera* (see Figure 19.4) is one of many examples of adaptive radiations of plants and associated insects that may have arisen in this way.

These hypotheses have been largely supported by subsequent research (Rosenthal and Berenbaum 1992). Insects and other herbivores certainly impose selection for chemical and other defenses in plants. For example, May Berenbaum and her colleagues (1986) found that genetic variation in resistance of wild parsnip (*Pastinaca sativa*) to its major herbivore, the seed-eating parsnip webworm (*Depressaria pastinacella*), is mostly attributable to the level of two furanocoumarin compounds in the seeds (Figure 19.11). Anurag Agrawal (2005) planted multiple families of the milkweed *Asclepias syriaca* in a common garden and measured a fitness component—seed production—in relation to the density of herbivorous insects on the plants, the damage they inflicted, and several putative defenses such as cardenolide content and production of the gummy white fluid (latex) for which the plant is named. He found protective effects of cardenolides, which reduced insect growth (Figure 19.12A), and latex, which reduced the abundance and impact of insects on a plant (Figure

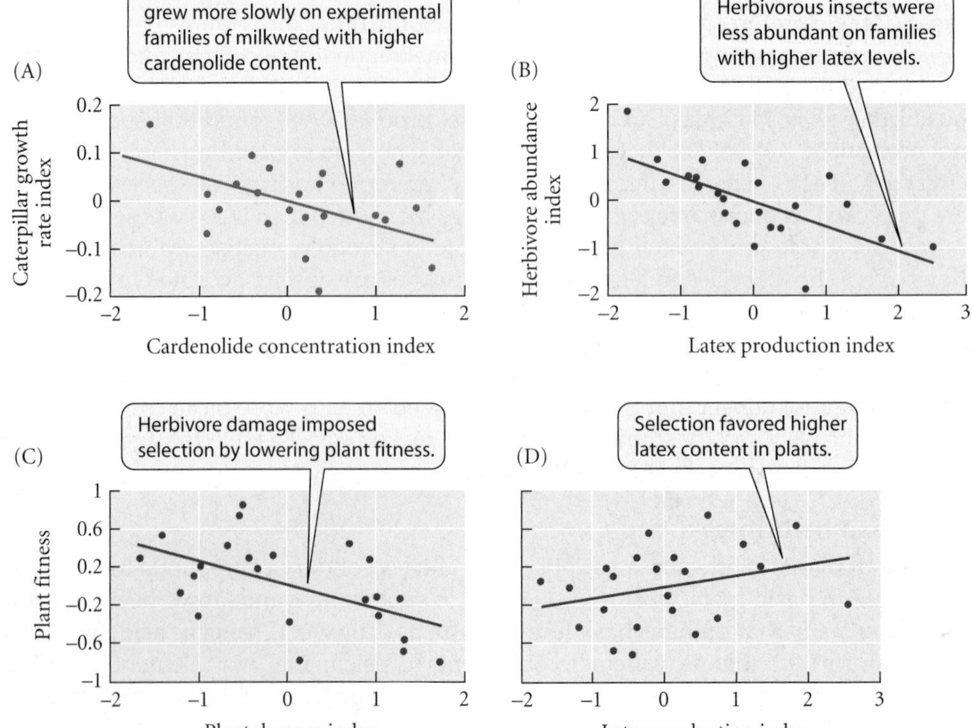

Figure 19.12 Evidence of selection for defensive traits in the common milkweed (*Asclepias syriaca*). (After Agrawal 2005; photo © Nancy Nehring/istockphoto.com.)

19.12B). Because plant fitness is significantly lowered by insect damage (Figure 19.12C), genetically higher levels of latex and other defenses improve fitness (Figure 19.12D). Thus natural selection by herbivores favors the evolution of these defenses.

Not surprisingly, herbivores often adapt to plant defenses (Bernays and Chapman 1994; Tilmon 2008). For example, among populations of parsnip webworms, resistance to the toxic effects of furamocoumarins is correlated with differences in the level of these compounds among populations of the host plant (Zangerl and Berenbaum 2003). A dramatic example of local adaptation involves a physical rather than chemical plant defense: the thick, woody fruit wall (pericarp) that encloses the seeds of the Japanese camellia (*Camellia japonica*; Figure 19.13A). A high proportion of seeds are consumed by larvae of the camellia weevil (*Curculio camelliae*) which feeds exclusively on this plant. The adult weevil inserts eggs into the seed chamber through a hole she bores with her mandibles, located at the end of her long snout, or rostrum (Figure 19.13B). Hirokazu Toju and Teiji Sota (2006) determined weevils' success in boring through to the seed chamber as a function of their rostrum length, relative to the thickness of a fruit's pericarp (Figure 19.13C), and determined the relation between these that would result in a 50 percent success rate for the weevil (or 50 percent seed mortality for the plant). They related these calculations to wild populations. Southern populations of the camellia have much thicker pericarps than do northern populations, and the rostrum length of the weevil likewise shows a strong cline. However, the clines differ in slope, so that northern weevil populations are "ahead" in this conflict—their rostra are long enough to ensure a success rate well over 50 percent—whereas in the south the plant population is "ahead," with pericarps thick enough to reduce the weevils' success (Figure 19.13D). These species may be engaged in an "arms race," although the reason for the thinner pericarp in northern camellia populations is not clear.

Figure 19.13 Imbalance in a coevolutionary conflict. (A) Fruits of the Japanese camellia (*Camellia japonica*) have a much thicker pericarp in the south than in the north. (B) The rostrum of the camellia weevil (*Curculio camelliae*) is much longer in southern than in northern populations. (C) The average boring success of the weevils as a function of their rostrum length and the pericarp thickness of a fruit, based on experimental data. The success of the plant in defense of its seeds is the opposite of the weevil's success in reaching the seeds. (D) Average rostrum length and pericarp thickness of associated populations of weevil and camellia in various localities in Japan are plotted as points relative to the line representing combinations of these variables that enable 50 percent success of the weevil. In populations to the right of the dashed line, the weevil is favored because its rostrum is long relative to the plant's pericarp thickness. In the population left of the line, the plant has the advantage. Points at the lower right represent more northern localities, in which the pericarp is thin and the rostrum short. (After Toju and Sota 2006; photos courtesy of Hiro Toju.)

(A)

(B)

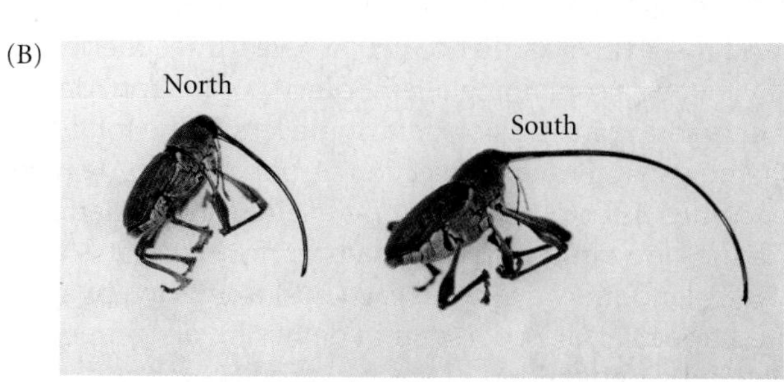

(C)

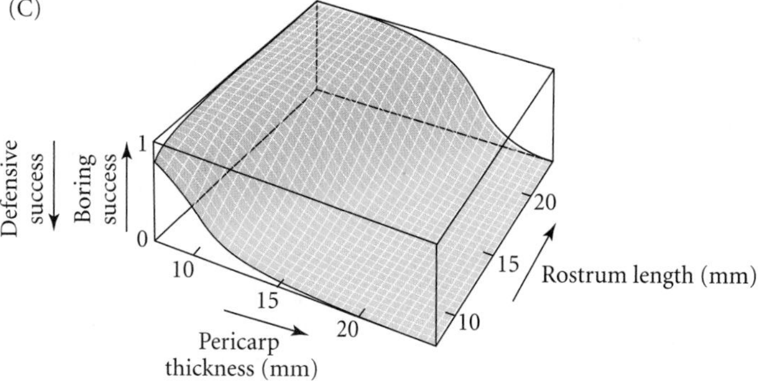

(D)

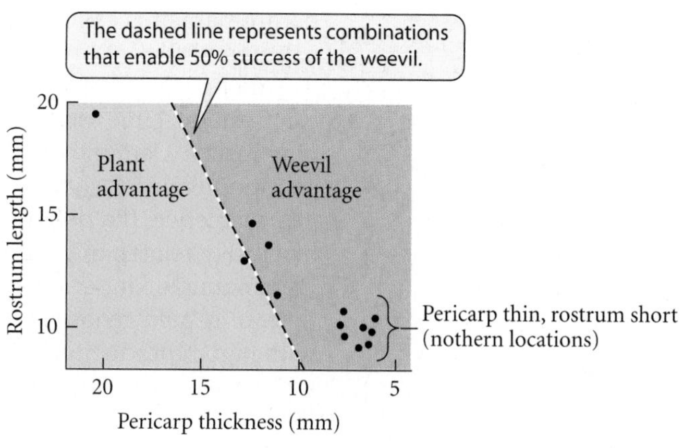

Infectious disease and the evolution of parasite virulence

The two greatest challenges a parasite faces are moving itself or its progeny from one host to another (transmission) and overcoming the host's defenses. Some parasites are transmitted **vertically**, from a host parent to her offspring, as in the case of *Wolbachia* bacteria, which are transmitted in insects' eggs. Other parasites are transmitted **horizontally** among hosts via the external environment (e.g., human rhinoviruses, the cause of the common cold, which are discharged by sneezing), via contact between hosts (e.g., the gonorrhea bacterium and other venereal disease agents), or via carriers (VECTORS, such as the mosquitoes that transmit the yellow fever virus and the malaria-causing protist *Plasmodium*).

The effects of parasites on their hosts vary greatly. Those that reduce the survival or reproduction of their hosts are considered VIRULENT. We are concerned here with understanding the evolutionary factors that affect the degree of virulence. This topic has immense medical implications because the evolution of virulence can be rapid in "microparasites" such as viruses and bacteria (Bull 1994; Ewald 1994). The level of virulence depends on the evolution of both host and parasite. For example (Fenner and Ratcliffe 1965), after the European rabbit (*Oryctolagus cuniculus*) became a severe rangeland pest in Australia, the myxoma virus, from a South American rabbit, was introduced to control it. Periodically after the introduction, wild rabbits were tested for resistance to a standard strain of the virus, and virus samples from wild rabbits were tested for virulence in a standard laboratory strain of rabbits. Over time, the rabbits evolved greater resistance to the virus, and the virus evolved a lower level of virulence. Although some almost avirulent strains were detected, the virus population did not become completely avirulent.

THEORY OF THE EVOLUTION OF VIRULENCE. Many people imagine that parasites generally evolve to be benign (avirulent) because the parasite's survival depends on that of the host population. However, a parasite may evolve to be more benign or more virulent depending on many factors (May and Anderson 1983; Bull 1994; Frank 1996).

The fitness of a parasite genotype is proportional to the number of hosts its progeny infect and may be measured by R_0, the number of new infections produced by an infected host:

$$R_0 = \frac{bN}{v + d + r}$$

where N is the number of hosts available for infection by the parasite progeny, b is the probability that the progeny will infect each such host, v is the mortality rate of hosts that is due to parasitism (and is a function of the parasite's virulence), d is the mortality rate of hosts that is due to other causes, and r is the rate at which infected hosts recover and become immune to further infection. Thus the denominator is the rate at which hosts move out of the infected class (and thus are not a source of new infections). In many cases, b depends on the number of parasite offspring produced within the host—but this variable is often proportional to host mortality v, since the parasite uses the host's resources (energy, protein, etc.) to reproduce. If b and v are correlated, the parasite may evolve greater virulence. If v becomes too great, however, the host may die before parasites can be transmitted to new hosts, so there is counterselection against extreme virulence.

Among many factors that may affect the level of virulence that evolves, three merit special mention. First, each host may be viewed as containing a temporary population (deme) of parasites. Demes that kill their host before transmission contribute less to the total parasite population than more benign demes, so interdemic selection (group selection) favors low virulence. If a host typically becomes infected by only one individual parasite, or by closely related individuals, the demes are kin groups, so interdemic selection is then tantamount to kin selection, and low virulence may evolve. If each host is infected by multiple, unrelated genotypes of parasites, however, selection within demes favors genotypes with high reproductive rates, which will be transmitted in greater numbers. Thus greater virulence is expected to evolve in parasite species in which multiple infection is frequent (Frank 1996).

Second, if hosts rapidly become immune to the parasite (i.e., if r in the above equation is large), selection favors rapid reproduction—that is, outrunning the host's immune system—by the parasite. Because this may entail greater virulence, an effective immune system (or a drug that rapidly kills the parasite) may sometimes induce the evolution of higher virulence.

A third factor affecting the level of virulence is whether parasites are transmitted horizontally or vertically. The transmission (and thus fitness) of horizontally transmitted parasites does not depend on the reproduction of their host (or, therefore, on its long-term survival). In contrast, the progeny of a vertically transmitted parasite are "inherited" directly, so b depends on the host's reproductive success. Hence we may expect evolution toward a relatively less virulent state in vertically transmitted parasites. This hypothesis was supported by an experiment with bacteriophage, in which a phage genotype that reduces its host's growth declined in frequency, and a more "benevolent" genotype increased, when horizontal transmission was prevented (Bull et al. 1991).

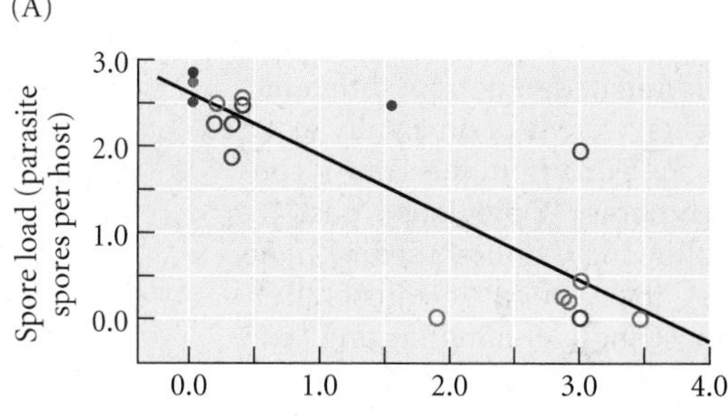

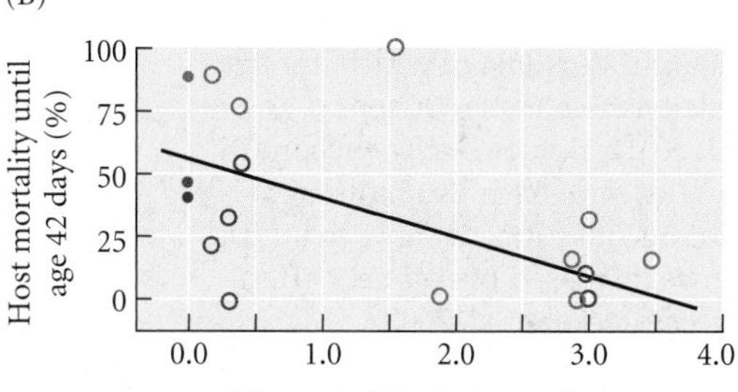

Figure 19.14 The fitnesses of three strains of a microsporidian parasite and their effects on various populations of the host species, the water flea *Daphnia magna*. Each strain, represented by a different color, was tested in hosts from its locality of origin (solid symbols) and from localities at various distances away (open symbols). (A) The number of parasite spores produced per host (spore load) was greatest when the parasite infected individuals from its own location, showing that parasites are best adapted to local host populations. (B) Host mortality was greatest in the parasite's own or nearby host populations, showing that the parasite is most virulent in the host population with which it has coevolved. (After Ebert 1994.)

VIRULENCE AND RESISTANCE IN NATURAL POPULATIONS. *Daphnia magna*, a planktonic crustacean, is parasitized by a microsporidian protist (*Pleistophora intestinalis*) that reproduces in the gut epithelium and releases daughter spores in the host's feces. In experimental pairs of infected and uninfected *Daphnia*, the greater the number of parasites in the infected individual, the more likely the other was to become infected. Moreover, the parasites produced more spores, and caused greater mortality, when they infected *Daphnia* from their own or nearby populations than when they infected hosts from distant populations (Figure 19.14). Thus populations of this parasite are best adapted to their local host population, and their more virulent effect on sympatric than on allopatric host populations contradicts the naive hypothesis that parasites necessarily evolve to be benign.

Daphnia magna can produce eggs that remain dormant in pond sediments for many years. The eggs may harbor resting spores of another parasite, the bacterium *Pasteuria ramosa*. Ellen Decaestecker and colleagues (2007) revived eggs and bacteria from layers of sediment deposited over a number of years, then experimentally cross-infected *Daphnia* from several different years with bacteria from the same ("contemporary") year, the preceding ("past") year, and the subsequent ("future") year. They discovered that hosts were more frequently infected by "contemporary" than by "past" or "future" bacteria (Figure 19.15A), and interpreted this to indicate that the *Daphnia* population underwent genetic change from year to year and that the bacteria evolved in concert, as in gene-for-gene models of coevolution (see Figure 19.6). The same research group had previously found that genotypes of *Daphnia* vary in their resistance to different genotypes of *Pasteuria*, as this hypothesis requires (Decaestecker et al. 2003). Even though both host and parasite underwent continual coevolution, the average virulence of the parasite increased over time (Figure 19.15B).

Populations of the freshwater snail *Potamopyrgus antipodarum* in New Zealand include both sexual and parthenogenetic genotypes (see Figure 15.7). Mark Dybdahl and Curtis Lively (1998) found that in a lake populated mostly by asexual snails, the frequencies of different clonal genotypes (distinguished by allozyme markers) changed over the course of 5 years, and the rate of infection of most of the genotypes peaked about a year after the genotype peaked in frequency. This observation suggests that rare snail genotypes have a selective advantage because they are resistant to the most prevalent parasite genotypes. This hypothesis was confirmed by exposing 40 rare clones and 4 clones that had recently been common to infection by parasites from the same lake: the rare clones were much less susceptible to infection. These results, like those with *Daphnia*, are consistent with gene-for-gene coevolution.

EVOLUTION AND EPIDEMICS. The genetics and evolution of parasite-host interactions are highly relevant to human health, as well as that of other species of concern. First, the

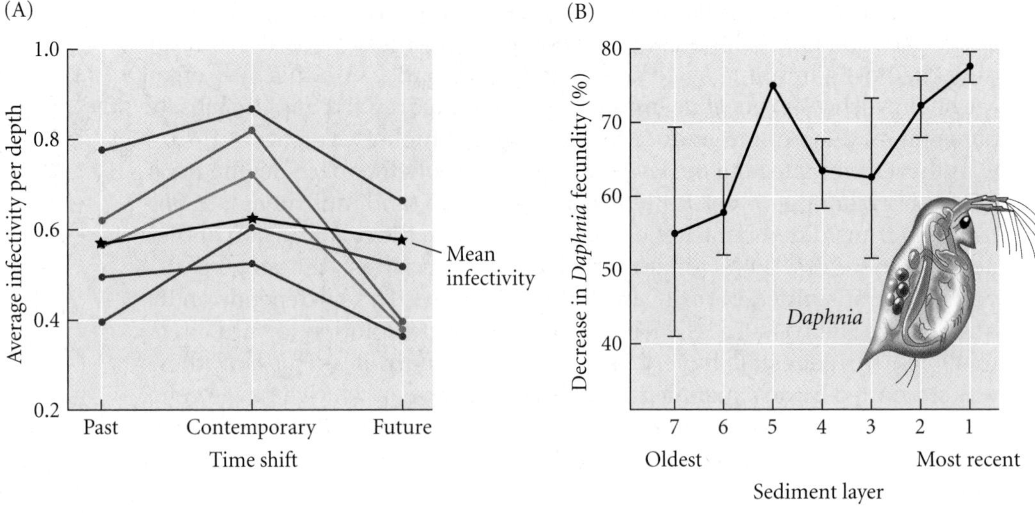

Figure 19.15 (A) When *Daphnia* eggs from several depths in pond sediment, dating from different years, were experimentally exposed to *Pasteuria* from the same ("contemporary"), previous ("past"), or following ("future") sediment layers, the bacteria were generally most successful in infecting "contemporary" *Daphnia*. Each line presents results from *Daphnia* from a particular sediment layer; the starred line is the mean infectivity of all trials, some of which, for simplicity, are not shown here. (B) Reduction of *Daphnia* fecundity by *Pasteuria* taken from different depths of sediment show that virulence increased over time. (After Decaestecker et al. 2007.)

examples we have considered suggest that genetic diversity in a host population may be important for maintaining resistance to pathogens. Conversely, populations that are inbred or have low genetic diversity may be at risk. For example, in 1970, 85 percent of the hybrid seed corn planted in the United States carried a cytoplasmic genetic factor for male sterility that was considered useful for preventing unintended cross-pollination. Unfortunately, this genetic factor also caused hypersusceptibility to the southern corn leaf blight (*Helminthosporium maydis*), and about 30 percent of the country's corn crop—and up to 100 percent in some places—was lost to this fungus (Ullstrup 1972). Widely planting a genetically uniform crop is a prescription for disaster.

Among the greatest threats to human health are "emerging pathogens," many of which enter the human population from other species. Some such events may occur entirely because of environmental changes that, for example, increase the likelihood of transmission from an animal "reservoir" to a human population, but in other cases, evolutionary change of the pathogen plays a role (Woolhouse et al. 2005). Phylogenetic analyses are routinely used to trace the origins of new pathogens, such as the evolution of human immunodeficiency viruses (HIV-1 and HIV-2) from chimpanzees and African monkeys (see Figure 3.4). Comparisons of the phylogeny of clades of RNA viruses with the phylogeny of their hosts show that some families of viruses have shifted to new host species much more frequently than others have (Jackson and Charleston 2004).

When the origins of a new pathogen can be discovered, it may be possible to determine the genetic basis by which the pathogen adapted to its new host. For instance, canine parvovirus arose and became pandemic in dogs throughout the world in 1978. Phylogenetic analysis showed that it arose from a feline panleukopenia virus that infects cats and several other carnivores. Six amino acid changes in the capsid protein of the virus enable it to infect dog cells, by specifically binding the canine transferrin receptor. After the virus first entered the dog population, several additional evolutionary changes made it more effective at binding the dog receptor, and unable to bind that of its original feline host (Hueffer et al. 2003).

Recombination plays a role in host shifts by influenza A viruses, which are major threats to human health. In 1918, the "Spanish flu" killed about 675,000 people in the United States (of whom about half were young adults) and 20 to 40 million worldwide. This was the most devastating of many influenza A pandemics. Influenza A viruses are widespread in birds, especially ducks and other aquatic species, and are thought to enter the human population usually through intermediate vehicles such as pigs, but sometimes directly from birds (Hay et al. 2001). Once the virus has established a population in humans, new antigenic variants continually evolve in response to immune suppression by human hosts. However, the shift from birds to mammals seems usually to be based, at least in part, on major genetic changes caused by recombination among fairly different strains. There is some evidence that recombination played a role in the emergence of the Spanish flu (Gibbs et al. 2001), and of the avian flu (H5N1) that arose in China and other eastern Asian countries in 2003 and 2004 and which caused disease in domestic poultry and several human deaths (Li et al. 2004). There is concern that H5N1 or a related strain could evolve further and become another pandemic (Longini et al. 2005).

Mutualisms

Mutualisms are interactions between species that benefit individuals of both species. In SYMBIOTIC MUTUALISMS, individuals are intimately associated for much of their lifetimes. Some mutualisms have promoted the evolution of extreme adaptations. Flowers that are pollinated by long-tongued moths usually have a long, tubular, white corolla and are fragrant at dusk or at night. Darwin, having seen the Madagascan orchid *Angraecum sesquipedale* in a London greenhouse, with a nectar spur up to 30 cm long, predicted that somewhere in Madagascar there must exist a moth with a similarly long proboscis, capable of pollinating it. More than a century later, such sphinx moths, with tongues more than 24 cm long, were found, and they do indeed pollinate this orchid and its relatives (Figure 19.16; Nilsson et al. 1985).

(A)

(B)

Figure 19.16 Mutualisms may result in extreme adaptations. (A) The orchid *Angraecum sesquipedale* bears nectar in an exceedingly long spur. (B) The long-tongued sphinx moth *Xanthopan morganii praedicta* pollinates *A. sesquipedale*. Darwin predicted the existence of such a pollinator based on seeing the orchid in a London greenhouse. (A © Pete Oxford / Naturepl.com; B © The Natural History Museum, London.)

In *The Origin of Species*, Darwin challenged his readers to find an instance of a species having been modified solely for the benefit of another species, "for such could not have been produced through natural selection." No one has met Darwin's challenge. Mutualisms exemplify not altruism but reciprocal exploitation, in which each species obtains something from the other. Some mutualisms, in fact, have arisen from parasitic or other exploitative relationships. Yuccas (*Yucca*, Agavaceae), for example, are pollinated only by female yucca moths (*Tegeticula* and *Parategeticula*), which carefully pollinate a yucca flower and then lay eggs in it (Figure 19.17A). The larvae consume some of the many seeds that develop. Some of the closest relatives of *Tegeticula* simply feed on developing seeds, and one of these species incidentally pollinates the flowers in which it lays its eggs, illustrating what may have been a transitional step from seed predation to mutualism (Figure 19.17B).

The theoretical bases for the evolution of mutualism between species are much the same as those governing the evolution of reciprocity between conspecific individuals, as decribed in Chapter 16. As in the case of intraspecific cooperation, there is always the potential for conflict within mutualisms because a genotype that "cheats" by exploiting its partner without paying the cost of providing a benefit in exchange is likely to have a selective advantage. Thus selection will always favor protective mechanisms, including punishing cheaters, to prevent overexploitation (Bull and Rice 1991). Moreover, selection will favor "honest" genotypes if the individual's genetic self-interest depends on the fitness of its host or partner (Herre et al. 1999). Thus the factors that should favor evolutionary stability of mutualisms include vertical transmission of endosymbionts from parents to offspring, repeated or lifelong association with the same individual host or partner, and restricted opportunities to switch to other partners or to use other resources. Some mutualisms indeed appear to conform to these principles. For example, the *Buchnera* bacteria that live in the cells of aphids and are vertically transmitted are all mutualistic, as far as is known. However, this is not an invariable rule, and some vertically transmitted symbionts are harmful to their hosts.

An example of how evolutionary stability can be achieved by punishment is provided by the interaction between yucca species and the moths that are their sole pollinators (Pellmyr and Huth 1994). Typically, the moth lays only a few eggs in each flower, so that only a few of the flower's many developing seeds are consumed by the larvae. The moth could lay more eggs per flower—indeed, she distributes eggs among many flowers—so

(A)

(B)

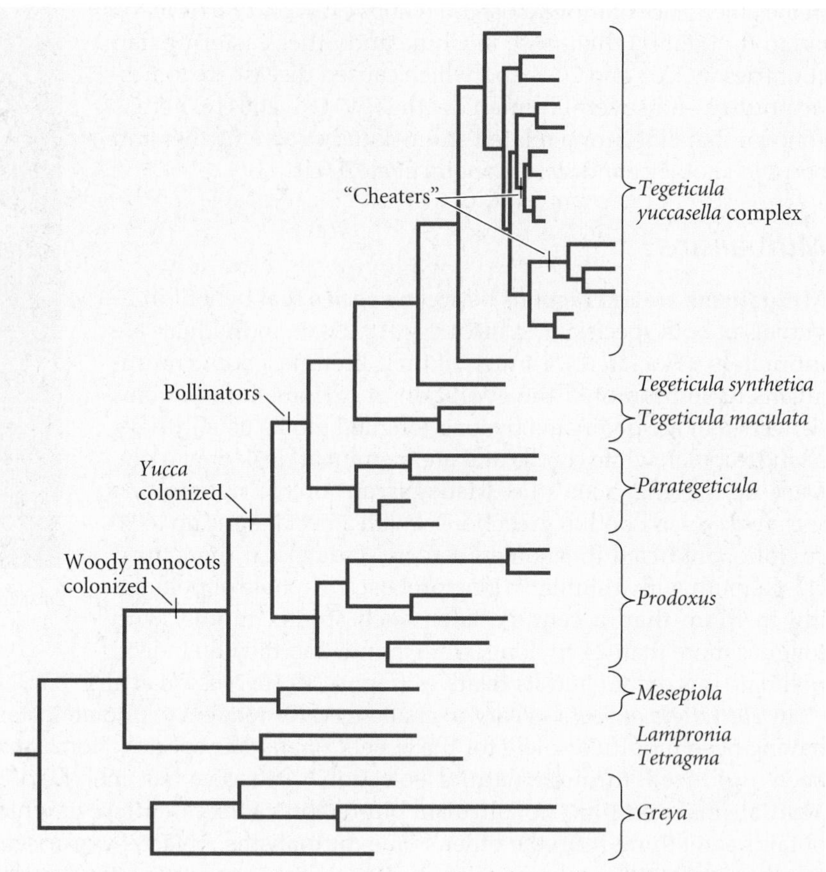

Figure 19.17 Yucca moths and their evolutionary history. (A) Yucca moths of the genus *Tegeticula* not only lay eggs in yucca flowers but use specialized mouthparts to actively pollinate the flowers in which they oviposit. (B) A phylogeny of the yucca moth family, showing major evolutionary changes. Some species in basal genera such as *Greya* incidentally pollinate the flowers in which they lay eggs. Intimate mutualism evolved in the ancestor of *Tegeticula* and *Parategeticula*, and "cheating" later evolved twice in *Tegeticula*. (A courtesy of O. Pellmyr; B after Pellmyr and Leebens-Mack 1999.)

why does she lay so few in each? The answer lies, in part, in the fact that the plant does not have enough resources to mature all of its many (often 500–1500) flowers into fruits. Pellmyr and Huth hand-pollinated all the flowers on some plants and found that only about 15 percent of the flowers yielded mature seed-bearing fruits; the rest were aborted and dropped from the plant. In the field, Pellmyr and Huth found more moth eggs, on average, in aborted than in maturing fruits, suggesting that the plant is more likely to abort a fruit if many eggs have been laid in it. Fruit abortion imposes strong selection on moths that lay too many eggs in a flower because the larvae in an aborted flower or fruit perish. Thus the moth has evolved restraint by individual selection and self-interest.

Mutualisms are not always stable over evolutionary time: many species cheat. For instance, many orchids secrete no nectar for their pollinators, and some practice downright deceit: they release a scent that mimics a female insect's sex pheromone, attracting male insects that accomplish pollination while "copulating" with the flower (see Figure 11.2). Two lineages of yucca moths that have evolved from mutualistic ancestors do not pollinate, and they lay so many eggs that the larvae consume most or all of the yucca seeds (see Figure 19.17B). These "cheaters" circumvent the plant's abortion response to high numbers of eggs by laying their eggs after the critical period in which fruit abortion occurs (Pellmyr and Leebens-Mack 1999).

Evolutionary shifts between mutualism and antagonism can sometimes occur rapidly. Many insects carry a maternally transmitted intracellular bacterium, *Wolbachia*, that in various insects causes parthenogenesis, cytoplasmic incompatibility, or death of male offspring. In the 1980s, *Wolbachia* invaded and swept through California populations of *Drosophila simulans*, reducing female fecundity by 15 to 20 percent. Recent tests show that the fecundity of *Wolbachia*-infected females from these populations is now 10 percent greater than that of uninfected females, suggesting that *Wolbachia* has evolved into a mutualist (Weeks et al. 2007). In the parasitoid wasp *Asobara tabida*, *Wolbachia* is actually required for the formation of eggs, because it regulates the apoptosis (cell death) of nurse cells in the ovary (Pannebakker et al. 2007).

MUTUALISM AND ADAPTATION. Partly because of genome studies, mutualism is increasingly recognized as an important basis for adaptation and the evolution of biochemical complexity (Moran 2007). The best-known examples are the evolution of mitochondria from purple bacteria and chloroplasts from cyanobacteria (see Figure 5.4). When a new, "compound" organism is formed from an intimate symbiosis, the subsequent evolution of both genomes is affected. For example, chloroplasts have fewer than 10 percent as many genes as free-living cyanobacteria, but many of the original cyanobacterial genes have been transferred to the plant nuclear genome. As many as 4500 of the protein-coding genes of *Arabidopsis*, or about 18 percent of the total, may have been acquired by transfer from the cyanobacterial ancestor early in plant evolution (Martin et al. 2002).

Mutualistic symbiosis often provides one or both partners with new capabilities. For example, many of the features of bacteria—including characteristics that make some bacteria pathogenic—are encoded by phage-borne genes. Some prokaryotes form mutualistic consortia, such as an association between methanogenic archaea and bacteria capable of fermentation. Bacteria and other microbes have formed intimate mutualisms with diverse multicellular organisms, especially animals, which are exceptional among eukaryotes in the extent to which they have lost the ability to synthesize essential amino acids and vitamins—probably because animals can often obtain these compounds from their food. Among the many animals that harbor symbiotic bacteria, the most extreme associations have been found in sap-sucking homopteran insects (aphids, leafhoppers, cicadas, and relatives), which have been extensively studied by Nancy Moran and her collaborators (Moran 2007). Plant sap lacks many nutrients, including essential amino acids, which in all homopterans are supplied by bacteria that are harbored in specialized cells (bacteriocytes). The phylogeny of these vertically transmitted bacteria closely matches the phylogeny of their hosts, denoting ancient associations. However, diverse groups of bacteria have formed such associations, and new endosymbionts have replaced ancestral ones in some homopteran lineages. Some individual insects carry up to six types of symbionts.

The phylogeny of a major group of homopterans that includes leafhoppers and cicadas matches that of a bacterial symbiont, *Sulcia*, indicating (together with the insects' fossil record) that this association has persisted since the late Permian, 270 Mya. In one subfamily of leafhoppers, known as sharpshooters, *Sulcia* was joined by another bacterium, *Baumannia*, in the Eocene (Figure 19.18). Like most endosymbionts, both mutualistic and parasitic, these bacteria have lost great numbers of genes, especially those whose function is supplied by their host's genome. The remarkable feature of *Sulcia* and *Baumannia* in the sharpshooters is that they have retained complementary functions: *Sulcia* can synthesize most of the essential amino acids required by its host, whereas *Baumannia* has lost

(A)

(B)

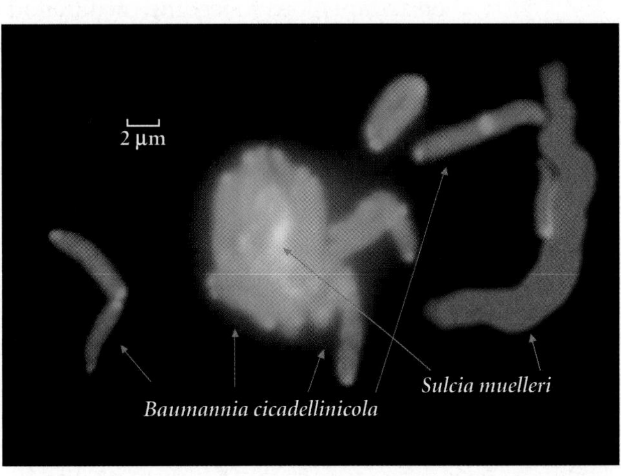

Figure 19.18 The members of an extraordinary mutualism. (A) A sharpshooter (*Cuerna sayi*), one of a large clade of sap-feeding homopteran insects that are nutritionally dependent on endosymbiotic bacteria. The bacteriocytes, revealed by dissecting away the exoskeleton, are the brightly colored masses on both sides of the abdomen. (B) Sharpshooters contain two endosymbionts, *Sulcia* (red) and *Baumannia* (green), here visualized by fluorescent probes for taxon-specific 16S rRNA sequences. The two bacteria and their host insect are codependent, and together can subsist on a diet of xylem sap, perhaps the least nutritionally adequate diet of any animal. (A courtesy of D. M. Takiya and R. Rakitov; B courtesy of N. A. Moran.)

most of these biosynthetic pathways but has retained a feature that *Sulcia* has lost: biosynthesis of vitamins (Wu et al. 2006). More remarkable still, *Baumannia* has retained the ability to synthesize a single essential amino acid, histidine—precisely the one that *Sulcia* cannot synthesize. The two endosymbionts and their host are codependent, and together can subsist on a diet of xylem sap, perhaps the least nutritionally adequate diet of any animal.

The Evolution of Competitive Interactions

The population densities of many species are limited, at least at times, by resources such as food, space, or nesting sites. Consequently, competition for resources occurs within many species (intraspecific competition) and between different species if they use some or all of the same resources (interspecific competition). Darwin postulated that competition would impose selection for divergence in resource use and viewed it as a major reason for the origin and divergence of species. There is now considerable evidence that evolution in response to competition is one of the major causes of adaptive radiation (Schluter 2000).

Ecologists have shown that sympatric animal species characteristically differ in resource use; for instance, species of Darwin's finches in the Galápagos Islands differ in diet and, accordingly, in bill shape (see Figure 3.23). It is plausible that such differences have evolved, at least in part, to avoid competition. Suppose individuals that differ in a phenotypic trait (e.g., bill depth in seed-eating birds) differ in the resource they use (e.g., seed size), that two species are both variable in this character, and that the frequency distributions of the two species overlap greatly, so that most individuals compete both with members of their own species and with the other species (Figure 19.19). Then, as long as there is a broad range of resource types, the individuals with the most extreme phenotypes (e.g., extremely small or large bills) will experience less intraspecific competition than more "central" phenotypes because they are less abundant, and they will experience less interspecific competition because they tend not to use the same resources as the other species. Therefore, the most extreme genotypes will have higher fitness. Such density-dependent diversifying selection can result in the two species evolving less overlap in their use of resources and in a shift of their phenotype distributions away from each other (Slatkin 1980; Taper and Case 1992). Divergence in response to competition between species is often called ecological **character displacement**, a term coined by William L. Brown and Edward O. Wilson (1956) to describe a pattern of geographic variation wherein sympatric populations of two species differ more greatly in a characteristic than do allopatric populations. The term is also used to mean the process of divergence that is due to competition.

The kind of geographic pattern that Brown and Wilson described has provided some of the best evidence for evolutionary divergence in response to competition (Taper and Case 1992; Schluter 2000). In northwestern North America, for example, several lakes have reproductively isolated open-water and bottom-dwelling forms of the three-spined stickleback fish (*Gasterosteus aculeatus*; see Figure 18.9). They differ in body shape (an adaptation to their habitat) and in mouth morphology and the number and length of the gill rakers (adaptations to feeding on different prey). Other lakes have only a single form of this

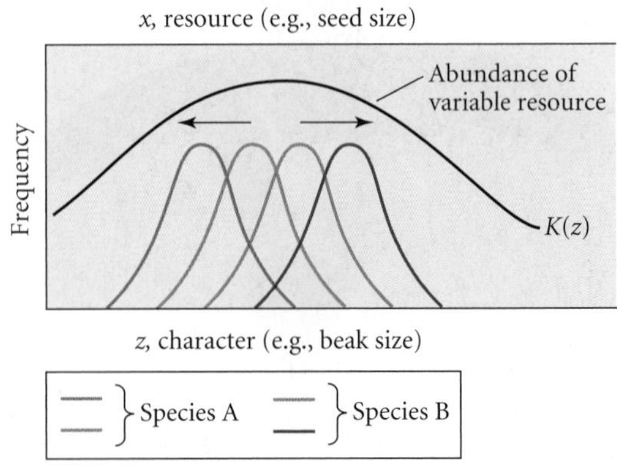

Figure 19.19 A model of evolutionary divergence in response to competition. The x-axis represents a quantitative phenotypic character (z), such as bill size, that is closely correlated with some quality of a resource (x), such as the average size of the food items eaten by that phenotype. The curve $K(z)$ represents the frequency distribution of food items that vary in size, and consequently the density that pure populations of various phenotypes, z, could attain. Two variable species (orange and green) initially overlap greatly in z, and therefore in the food items they depend on. Those phenotypes in each species that overlap with the fewest members of the other species experience less competition, and so may have higher fitness. Divergent selection on the two species is expected to shift their character distributions (red, dark green) so that they overlap less.

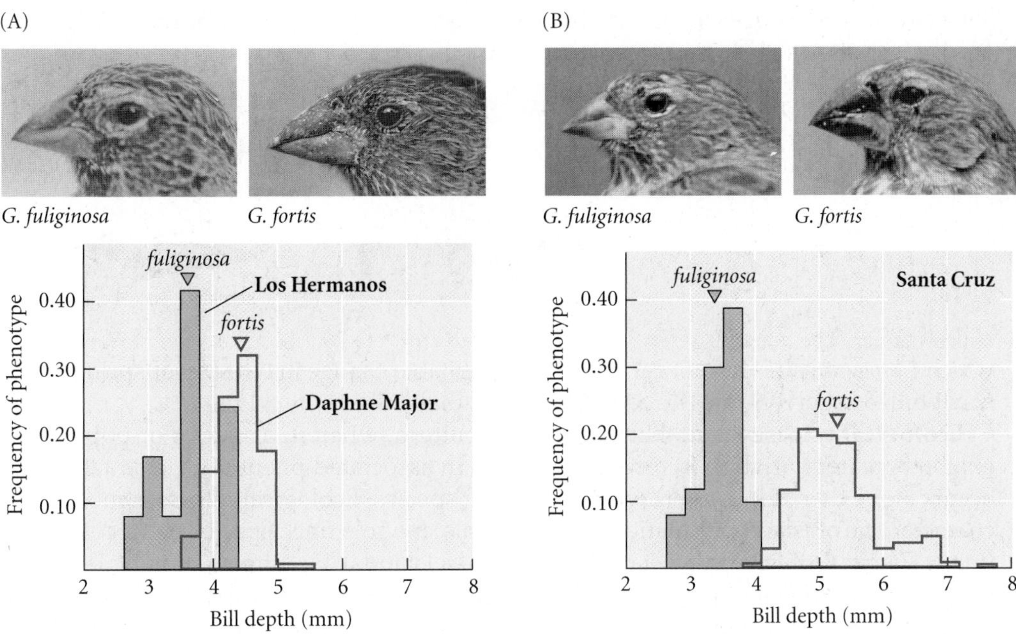

Figure 19.20 Character displacement in bill size in seed-eating ground finches of the Galápagos Islands. Bill depth is correlated with the size and hardness of seeds most used by each population; arrows show average bill depths. (A) Only *Geospiza fuliginosa* occurs on Los Hermanos, and only *G. fortis* occurs on Daphne Major. (B) The two species coexist on Santa Cruz, where they differ more in bill depth. (After Grant 1986; photos courtesy of Peter R. Grant.)

stickleback, with intermediate morphology (Schluter and McPhail 1992). Illustrating a famous case of character displacement, the Galápagos ground finches *Geospiza fortis* and *G. fuliginosa* differ more in bill size where they coexist than where they occur singly (Figure 19.20). Differences in bill size are correlated with the efficiency with which the birds process seeds that differ in size and hardness, and the population size of these finch species is often food-limited, resulting in competition (Grant 1986). Peter Grant and Rosemary Grant (2006) observed the process of character displacement in the population of *G. fortis* on the island of Daphne Major, where this species faced no competitors until *G. magnirostris*, a larger-billed species, colonized and built up a large population (Figure 19.21; see Figure 2.3B). Before *G. magnirostris* became abundant, a severe drought in 1977 that reduced seed availability caused the *G. fortis* population to evolve larger bills, because only larger-billed birds could eat the large, hard seeds of *Tribulus* plants after most smaller seeds had been consumed. When normal conditions returned, average bill size declined to its previous value. But by the time another drought occurred, in 2004, *G. magnirostris* had become abundant; it consumed most of the *Tribulus* seeds, and the *G. fortis* population evolved a smaller bill, for survival depended on eating the smallest available seeds,

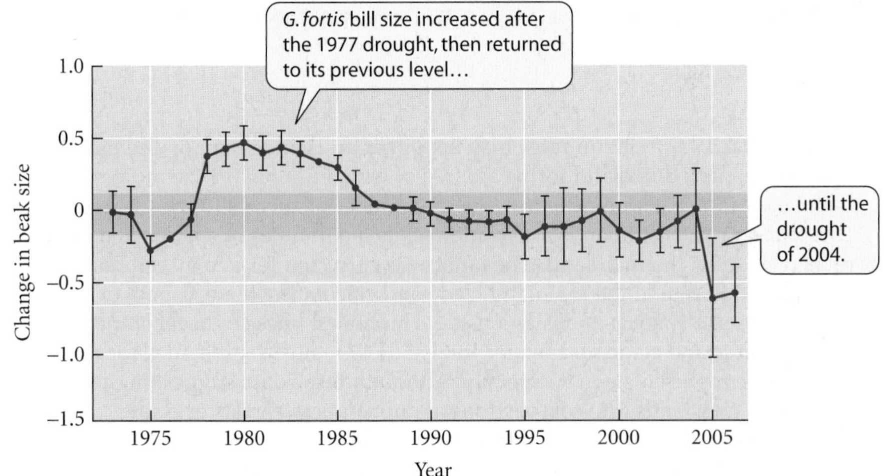

Figure 19.21 A history of change in mean beak size in the ground finch *Geospiza fortis* on the island of Daphne Major. Bill size increased after the 1977 drought and slowly returned to its previous level, until the drought of 2004, when bill size decreased because of character displacement from *G. magnirostris*, which has a larger bill and depleted the supply of large seeds. The darker shaded bar marks the range of mean beak size in 1973, with significant deviation showing up above and below that range. (After Grant and Grant 2006.)

Figure 19.22 Ecological release. The difference in bill size between the sexes is greater in *Melanerpes striatus* (A), the only species of woodpecker on the island of Hispaniola, than in continental species such as *M. aurifrons* (B), which is sympatric with other woodpecker species. Bill size is correlated with differences in feeding behavior, so greater sexual dimorphism results in broader resource use in *M. striatus.* (After Selander 1966.)

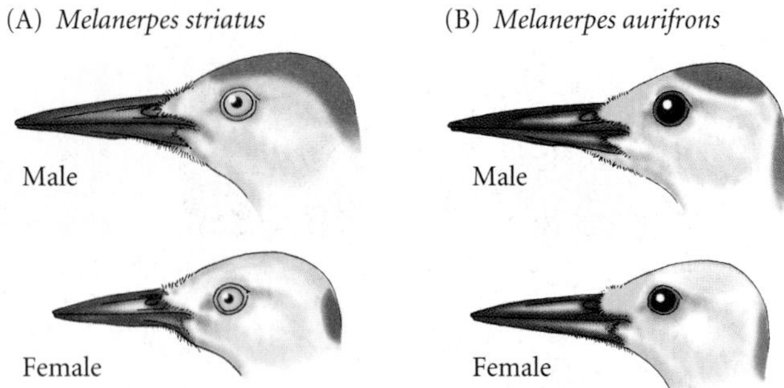

which *G. magnirostris* did not eat. This was a heritable change in bill size: offspring of the small-billed survivors, measured in 2006, had small bills.

Ecological release is another geographic pattern, wherein a species or population exhibits greater variation in resource use and in associated phenotypic characters if it occurs alone than if it coexists with competing species. Ecological release is most often characteristic of island populations. For example, the sole finch species on Cocos Island (in the Pacific Ocean, northeast of the Galápagos Islands) has a much broader diet, and forages in more different ways, than do any of its relatives in the Galápagos Islands, where there are many more species (Werner and Sherry 1987). Similarly, the only species of woodpecker on the Caribbean island of Hispaniola exhibits greater sexual dimorphism in the length of the bill and tongue than do related continental species that coexist with other woodpeckers, and the sexes differ in where and how they forage (Figure 19.22).

Competition among species is a cause of ecological diversification, as character displacement shows. On the other hand, alleviation of competition, resulting in ecological release, may enhance rates of speciation and diversification (Figure 19.23). The fossil record of diversity suggests that when incumbent taxa have become extinct, ecological opportunity has sometimes allowed diversification of other clades on a macroevolutionary scale (see Figure 7.19).

Multispecies interactions and community structure

The concept of character displacement between two competing species can readily be extended to "community-wide character displacement" among greater numbers of species that might evolve to partition resources (NICHE PARTITIONING). Using mathematical models, the pioneering evolutionary ecologists Robert MacArthur and Richard Levins (1967) predicted that coexisting species might be expected to be evenly spaced with respect to their average resources, such as the size of seeds or prey items. Morphological adaptations might reflect these ecological differences. Many examples fit this prediction (Schluter 2000). For example, among sympatric cats in Israel, differences between sexes and species in the size of the canine teeth, which are used to kill prey, are more nearly equal than expected by chance, and are thought to reflect differences in the average size of prey taken by these cats (Figure 19.24; Dayan et al. 1990).

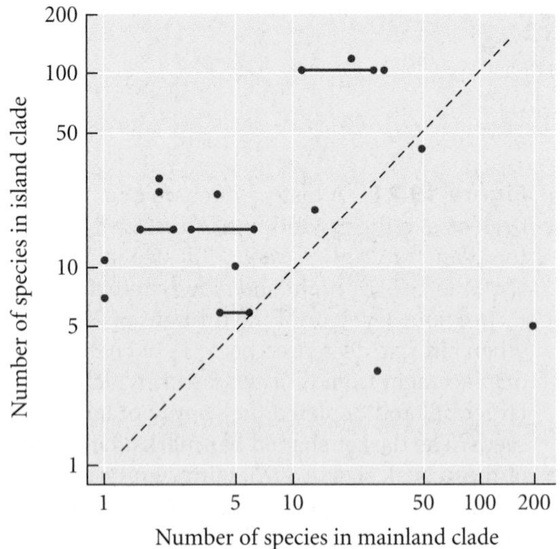

Figure 19.23 Speciation rates may be higher on islands than on the mainland, a pattern expected if island forms are free of competition with the more diverse mainland biota. Each point represents the number of species (of plants, birds, *Drosophila*, or *Anolis*) in an island clade versus the number of species in a closely related clade on the mainland. The latter was corrected for age by estimating the diversification rate in the mainland clade and calculating the diversity expected if it had the same age as the island clade. Horizontal lines connect comparisons between an island clade and two or three closely related mainland clades. The dashed line represents equal diversity. Although these data suggest higher speciation rates on islands, the comparisons are not all between sister clades, and the sample of island taxa may be biased toward those with high diversity. Thus more research is needed. (After Schluter 2000.)

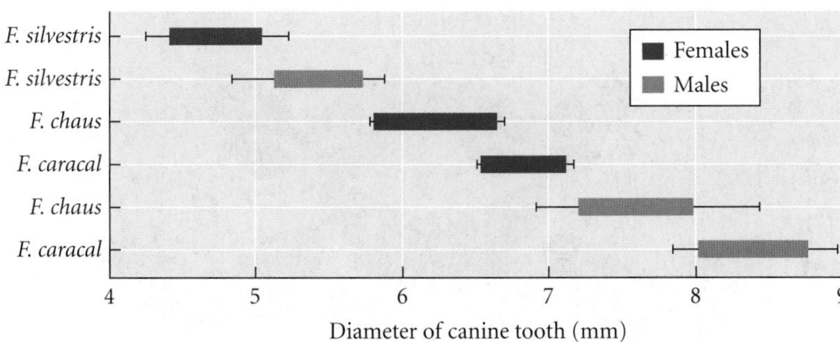

Figure 19.24 A nonrandom pattern of "equal spacing" among related predators may have evolved to minimize competition for food. The size of the canine teeth differs among three sympatric species of cats in Israel, and between the sexes in each species. These differences are thought to be adaptations to feeding on prey that differ in average size. The species are the wildcat (*Felis silvestris*), jungle cat (*F. chaus*), and caracal (*F. caracal*). (After Dayan et al. 1990.)

Several kinds of evolutionary outcomes may result from interactions between a predator species and multiple species of prey. If a predator's abundance increases with the abundance of one prey, a second species of prey may suffer heavier predation and so decline—a phenomenon called "apparent competition" (Holt 1977). One or both prey species may then be selected to diverge from the other in characteristics that affect susceptibility to the predator, such as defense features. No examples of this possibility have yet been well documented, but many lineages have diverged because of trade-offs in adaptation to different predators. For example, *Enallagma* damselfly larvae with large tail fans that enable them to swim rapidly are adapted to escape dragonfly larvae, but they cannot escape fish, which move more rapidly and cue in on moving prey. *Enallagma* species that inhabit ponds with fish are cryptic, have small tail fans, and move slowly—but they are easier prey for dragonfly larvae (McPeek 1995).

Models of such interactions show that prey species subjected to a common predator are likely to diverge in some cases, but to converge if that lowers the risk of predation (Abrams 2000). The most dramatic examples of convergence are cases of defensive mimicry (Turner 1977; Joron and Mallet 1998; Mallet and Joron 1999). Traditionally, two forms of defensive mimicry have been recognized. In BATESIAN MIMICRY, a palatable species (a mimic) resembles an unpalatable species (a model). Mimicry works because predators learn, from unpleasant experience, to avoid potential prey that look like the unpalatable species. Often, although not always, the models and mimics display conspicuous aposematic (warning) patterns. Selection on a mimetic phenotype can depend on both its density, relative to that of a model species, and the degree of unpalatability of the model. A predator is more likely to avoid eating a butterfly that looks like an unpalatable model if it has had a recent reinforcing experience (e.g., swallowing a butterfly with that pattern and then vomiting). If it has recently swallowed a tasty butterfly, however, it will be more, not less, inclined to eat the next butterfly with the same phenotype. Thus the rarer a palatable Batesian mimic is relative to an unpalatable model, the more likely predators are to associate its color pattern with unpalatability, and so the greater the advantage of resembling the model will be. If a rare new phenotype arises that mimics a different model species, it will have higher fitness, and so a mimetic polymorphism can be maintained by frequency-dependent selection, as is seen in the African swallowtail *Papilio dardanus* (see Figure 9.1B).

The other major form of mimicry is MÜLLERIAN MIMICRY, in which two or more unpalatable species are co-mimics (or co-models) that jointly reinforce aversion learning by predators. Groups of species that benefit from defensive mimicry are known as MIMICRY RINGS. In many cases, mimicry rings include strongly unpalatable species, mildly unpalatable species that may be "quasi-Batesian" mimics of the more unpalatable species, and fully palatable Batesian mimics (Figure 19.25). Müllerian mimics are almost never polymorphic within populations, but different geographic races of certain species may have different aposematic color patterns and may belong to different mimicry rings (e.g., the geographic races of *Heliconius erato* and *H. melpomene*; see Figures 12.19 and 19.25).

Convergent evolution of similar adaptations to similar environmental conditions is a common phenomenon (Chapter 3). Ecologists have speculated that the structure of ecological communities—the number of species, their niche partitioning, the pattern of links between trophic levels—might likewise converge, and become similar in similar abiotic

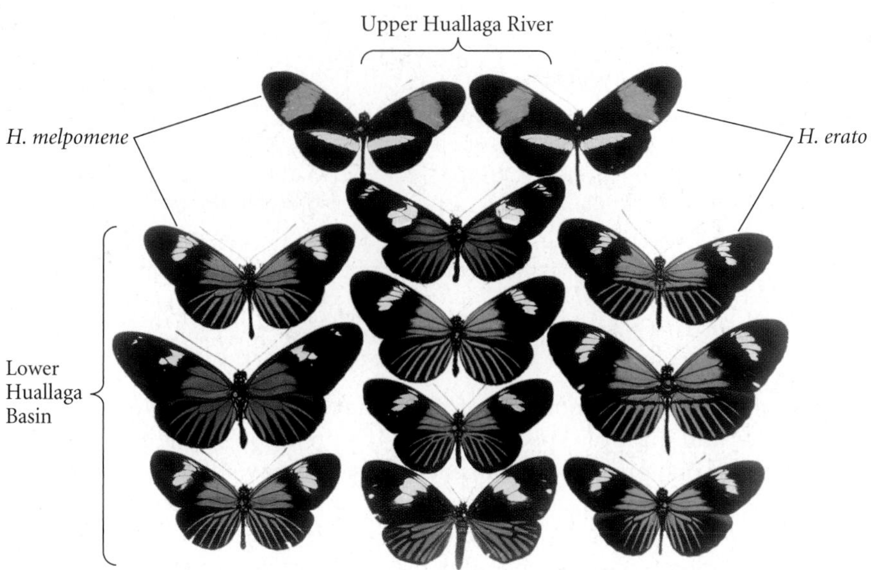

Figure 19.25 A mimicry ring. *Heliconius melpomene* and *H. erato* have a very different color pattern in the Mayo and upper Huallaga rivers, in eastern Peru, than in the lower Huallaga basin, where they join a mimicry ring with a "rayed" pattern. This ring of unpalatable species includes four other species of *Heliconius*, three other genera of butterflies (the top three species in the center column), and a moth (center column, bottom). (Courtesy of J. Mallet.)

environments. If species could usually adapt rapidly to the other species they encounter, the structure of local assemblages of species might be predictable. Phylogenetic niche conservatism, the retention of ancestral physiological and ecological properties, however, suggests limitations on the degree to which species are free to coevolve (Chapter 6). In that case, local communities are formed largely by processes of ECOLOGICAL SORTING: species with evolved features that enable them to coexist, without adaptive adjustment, do so, while others are excluded by competition or predation.

As we saw in Chapter 6, some examples of convergent assemblages arising from parallel multispecies evolution have been described, such as the ecomorphs of *Anolis* lizards that have independently evolved to use similar microhabitats on the several islands of the Greater Antilles (see Figures 6.22 and 6.23). However, the history of evolution has seldom been so predictable, and convergence of community structure can be hard to discern. Nevertheless, evolutionary history has left its imprint on ecological communities (Ricklefs and Schluter 1993; Webb et al. 2002). For example, Campbell Webb (2000) found that locally co-occurring species of trees in a Bornean rainforest are phylogenetically more closely related than if they had been randomly sampled from all the available species, evidently because they have similar (phylogenetically conservative) habitat requirements. Conversely, Jeannine Cavender-Bares and collaborators (2004) found that co-occurring oaks (*Quercus*) consistently are members of two different clades that have convergently adapted to each of several habitats. These two examples show the effects of ecological sorting versus local evolution on local species assemblages. Similarly, each of the small islands of the Lesser Antilles has either one or two species of *Anolis*. On each two-species island, the species differ in size, but the large and small species form two different clades; thus the two-species communities have been formed by ecological sorting of species that could coexist without coevolutionary adjustment (Losos 1992). This contrasts with the parallel adaptive radiations of *Anolis* that have occurred on the large islands of the Greater Antilles.

Summary

1. Coevolution is reciprocal evolutionary change in two or more species resulting from the interaction among them. Species also display many adaptations to interspecific interactions that appear one-sided rather than reciprocal.
2. Phylogenetic studies can provide information on the age of associations among species and on whether or not they have codiversified or acquired adaptations to each other. The phylogenies of certain symbionts and parasites are congruent with the phylogenies of their hosts, implying cospeciation, but in other cases such phylogenies are incongruent and imply shifts between host lineages.

3. Coevolution in predator-prey and parasite-host interactions can theoretically result in an ongoing evolutionary arms race, a stable genetic equilibrium, indefinite fluctuations in genetic composition, or even extinction.
4. Parasites (including pathogenic microorganisms) may evolve to be more or less virulent, depending on the correlation between virulence and the parasite's reproductive rate, vertical versus horizontal transmission between hosts, infection of hosts by single versus multiple parasite genotypes, and other factors. Parasites do not necessarily evolve to be benign.
5. Mutualism is best viewed as reciprocal exploitation. Selection favors genotypes that provide benefits to another species if this action yields benefits to the individual in return. Thus the conditions that favor low virulence in parasites, such as vertical transmission, can also favor the evolution of mutualisms. Mutualisms may be unstable if "cheating" is advantageous, or stable if it is individually advantageous for each partner to provide a benefit to the other.
6. Evolutionary responses to competition among species may lead to divergence in resource use and sometimes in morphology (character displacement). Competition has caused ecological diversification, whereas alleviation of competition can enhance the rate of increase in the number of species.
7. Character displacement between species that are subjected to a common predator may occur, but not many examples have as yet been reported. Convergence of prey species, as illustrated by defensive mimicry, occurs if it reduces predation.
8. The structure of ecological communities is affected by evolutionary adjustment of coexisting species to each other and by ecological sorting of species on the basis of characteristics that affect the likelihood of coexistence. Convergence of the structure of independently formed ecological communities can occur by both processes, but differences among communities may exceed similarities.

Terms and Concepts

character displacement
coevolution
ecological release
escape-and-radiate coevolution
gene-for-gene interactions
guild coevolution (= diffuse coevolution)
horizontal transmission
mutualism
specific coevolution
vertical transmission

Suggestions for Further Reading

J. N. Thompson, in *The Coevolutionary Process* (University of Chicago Press, 1994), discusses the evolution and ecology of many interactions, especially among plants and their herbivores and pollinators. He develops one of that book's themes further in *The Geographic Mosaic of Coevolution* (University of Chicago Press, 2004).

Plant-animal interactions are the focus of essays by prominent researchers in *Plant-Animal Interactions: An Evolutionary Approach*, edited by C. M. Herrera and O. Pellmyr (Blackwell Science, Oxford, 2002). "Models of parasite virulence" by S. A. Frank (*Quarterly Review of Biology* 71: 37–78, 1996) is an excellent entry into this subject. *The Ecology of Adaptive Radiation* by D. Schluter (Oxford University Press, Oxford, 2000) includes extensive treatment of the evolution of ecological interactions and their role in diversification. "Supplement on phylogenetic approaches to community ecology" appeared in the journal *Ecology* (vol. 87, July 2006).

Problems and Discussion Topics

1. How might coevolution between a specialized parasite and a host be affected by the occurrence of other species of parasites?
2. How might phylogenetic analyses of predators and prey, or of parasites and hosts, help to determine whether or not there has been a coevolutionary arms race?
3. The generation time of a tree species is likely to be 50 to 100 times longer than that of many species of herbivorous insects and parasitic fungi, so its potential rate of evolution should be slower. Why have trees, or other organisms with long generation times, not become extinct as a result of the potentially more rapid evolution of their natural enemies?
4. Design experiments to determine whether greater virulence is advantageous in a horizontally transmitted parasite and in a vertically transmitted parasite.

5. Some authors have suggested that selection by predators may have favored host specialization in herbivorous insects (e.g., Bernays and Graham 1988). How might this occur? Compare the pattern of niche differences among species that might diverge as a result of predation with the pattern that might evolve because of competition for resources.
6. Provide a hypothesis to account for the extremely long nectar spur of the orchid *Angraecum sesquipedale* (see Figure 19.16) and the long proboscis of its pollinator. How would you test your hypothesis?
7. In simple ecological models, two resource-limited species cannot coexist stably if they use the same resources. Hence coexisting species are expected to differ in resource use because of the extinction, by competition, of species that are too similar. Therefore, coexisting species could differ either because of this purely ecological process of "sorting" or because of evolutionary divergence in response to competition. How might one distinguish which process has caused an observed pattern? (See Losos 1992.)
8. Suppose that among related host species that carry related symbionts, the relationship is mutualistic in some pairs and parasitic in others. How would you (a) tell which is which, (b) determine what the direction of evolutionary change has been, and (c) determine whether the change from one to the other kind of interaction was a result of evolutionary change in the symbiont, in the host, or both?

CHAPTER 20

Evolution of Genes and Genomes

Immune complexity in an invertebrate. When the genome of the purple sea urchin (*Strongylocentrotus purpuratus*) was sequenced in its entirety, scientists found a surprising number of genes with immune functions (see Figure 20.5). In addition to this unexpected complexity in the innate immune system, the genome sequence also revealed signs of a rudimentary adaptive immune system, previously thought to be present only in vertebrates. (Photo © Georgie Holland/Photolibrary.com.)

From the time of Aristotle in the fourth century BC to Darwin, Huxley, and Owen in the nineteenth century, the study of the diversity and history of life focused on morphology, and to a lesser extent behavior. Only in the second half of the twentieth century did it become possible to compare the genes and molecules of different species in order to understand evolutionary relationships among species in the context of population processes such as gene flow and genetic drift.

The impact of molecular biology on evolutionary biology has been so profound that it is hard to imagine that evolutionary biology will experience further methodological and conceptual shifts of similar magnitude. Yet the tools of genomics—a suite of biotechnologies that can be described as molecular biology writ large—are causing just such an impact. Genomics is to twenty-first century evolutionary biology what protein electrophoresis and DNA sequencing were to the field in the twentieth century, and it likewise promises to open up as many questions as answers.

Today it is routine for the complete nucleotide sequence of all the genes of an organism to be determined, whether it be an archaean with a 2 megabase (million base pair, Mb) genome or a vertebrate with 2 or 3 gigabases (1 gb = 1 billion base pairs). As this chapter is being written, the list of microbial species whose genomes are completely sequenced is well over 700, and complete genome sequences exist for taxa sprinkled across the tree of life: several eukaryotic pathogens, yeasts, the nematode *C. elegans*, a sea urchin, a tunicate, several insects including a beetle and a bee, plant species including rice and the model organism *Arabidopsis*, several mammals and fishes, as well as the chicken and a frog. What a bonanza! Biologists can now study in detail the population genetics of hundreds of genes simultaneously, and compare the evolutionary trajectories of genes on the same or different chromosomes. Now that complete genomes can be characterized, we can fully catalogue the diversity of genes along a chromosome, and have a basis for studying the interactions among them and the ways in which these interactions influence their evolutionary fates. We can also get closer to the myriad links between genic and organismal evolution in exciting detail. For instance, multigene families, including the immunoglobulin superfamily and the family of calcium-dependent ion channels involved in brain function, provide the molecular foundation for many of the profound organismal adaptations that have occurred in the history of life.

BOX 20A Emerging Technologies in Evolutionary Genomics

When it was first conceived in the mid-1980s, whole-genome sequencing of a genome the size of a human's was believed to be an impossibly unrealistic task. Today, on average one new vertebrate genome is sequenced each year, and a new microbial genome (at about 2 Mb) is sequenced every week! What key technological and logistical elements have allowed this remarkable increase in sequencing capability? Technology for large-scale DNA sequencing is one of the most rapidly advancing fields influencing the study of genome evolution, and has become inexpensive enough that the sequencing of small or moderately sized genomes has come within the grasp of individual research groups rather than just large sequencing centers, and can now be extended to non-model species rather than only those with biomedical importance.

The first major advance was automation of nearly every step in the process of DNA sequencing. Today robots perform many routine tasks, including isolation of DNA from whole tissue or blood, amplification of specific DNA segments via the polymerase chain reaction (PCR), and the process of sequencing itself. The role of a typical human technician in a large genome center has shifted from hands-on manipulation of DNA to the programming of robots to perform myriad tasks rapidly. A single robot can do in an hour what would take a technician an entire day or more.

Up until about 2006, many of the chemistries used to sequence an entire genome were the same as those that had been used for the previous several decades to sequence DNA. The tried-and-true method of dideoxy chain termination sequencing, invented by Frederick Sanger in 1977, had not yet been modified for whole-genome sequencing (except that instead of using radioisotopes to visualize DNA fragments, brightly colored fluorescent dyes that could be read by lasers in automatic sequencers were used). Today, DNA sequencing has moved beyond the dideoxy method of sequencing PCR products, each containing thousands of DNA molecules, to approaches that determine the sequences of individual DNA molecules in picoliter wells. These individual DNA molecules are then copied in such a way that a different color light is emitted as each new base (A, G, C, or T) is added to the nascent strand and is detected with a camera. The result is a record of sequences on a scale several orders of magnitude greater than what the dideoxy approach could generate.

Like dideoxy sequencing, the new sequencing approaches (called PYROSEQUENCING) begin with the isolation of high-molecular-weight DNA from the target species and the shearing of large chromosome-sized DNA fragments into small fragments, usually around 2 kb. Unlike the dideoxy approach, pyrosequencing does not require laborious and costly cloning of these fragments, since each molecule is isolated from others in an individual picoliter well. Again, as in traditional sequencing, many short DNA sequences (as short as 32 base pairs or as long as 500 base pairs, depending on the technology) are generated and stitched together into larger "contigs" using bioinformatics tools. This computational phase generates a contiguous sequence spanning a particular region determined by the DNA starting material (this region could be a whole genome if total genomic DNA is used to start with, or it could be an individual genomic clone, such as a BAC, or "bacterial artificial chromosome," clone, which is usually about 100 to 200 kb). The method of stitching together these short reads is called "shotgun sequencing" and is enabled by relatively minor modifications of the same bioinformatics tools used to analyze sequences generated by the dideoxy method.

The "next-generation" sequencing approaches thus rely on a combination of new chemistry as well as modifications of older molecular and computational approaches. And researchers are already moving beyond these technologies to envision even vaster scales on which new genomes can be sequenced. A major, if somewhat arbitrary, goal is to be able to devise a set of approaches that will enable a single human genome to be sequenced for $1000 dollars or less. Whereas the first human genome to be sequenced cost about $15 billion, human genomes are now sequenced routinely for a few hundred thousand dollars each. Major initiatives and international consortia of scientists are aiming to sequence 1000 human genomes over the next few years, and those scientists working with non-model organisms and natural populations are now able to contemplate sequencing genomes of their focal species so as to enable the many tools and resources that a genome sequence can provide.

Comparative genomics—the comparative study of whole genomes—began only about 15 years ago, and for complex eukaryotes only in the last 5 years. The reach of comparative genomics has already extended beyond model species such as mice and *Drosophila*—or and species of immediate health concern such as the malaria parasite *Plasmodium falciparum* and its mosquito host *Anopheles gambiae*—to species that are better suited for other evolutionary questions and factors occurring in natural populations. Not only does comparative genomics promise to refine in ever greater detail the tree of life; it also enables us to trace the travels of genes as they evolve within species and occasionally move between species. Technologies such as microarrays, in which the expression status of thousands of genes can be monitored for an organism in different physiological or behavioral states, enable us to study the evolution of gene expression (see Figure 8.14).

New Molecules and Processes in Genomes

The study of the evolution of genes and genomes is a field that mixes new technologies with older evolutionary principles. Despite the advent of rapidly changing technologies and the discovery of novel molecular processes, the principles by which genes evolve in populations and by which species diverge over time are the same as those that have been understood since the modern synthesis. For example, RNA editing is a recently discovered process by which the sequence of a gene's mRNA transcript can be altered, often in ways that compensate for premature stop codons or other defects in the gene. Lateral gene transfer is now known to be much more frequent among microbial species than previously thought. Whole new classes of molecules, such as the many types of short RNA molecules that we now know are crucial to development and phenotypic variation, have been discovered and are completely changing traditional views of the functioning of gene regulation and expression. New molecules such as microRNAs (short—about 22 bp—RNA sequences that bind to RNA transcripts, repressing their translation into proteins) were initially of primary interest to molecular and developmental biologists but are now drawing the attention of evolutionary biologists (Figure 20.1). Wheeler and colleagues (2009), for example, used a "next-generation" sequencing approach called **pyrosequencing** (Box 20A) to study the diversity of microRNAs across the metazoans. By collecting sequences of microRNAs from many lineages of metazoans, they were able to confirm a pattern that was only dimly evident through the comparison of genome sequences from model organisms: that there is a strong signal of progressive accumulation of microRNAs from ancestral to derived lineages of metazoans, and that microRNAs tend to persist in metazoan genomes once they have evolved and become incorporated into the gene regulatory network.

With the discovery of incredible diversity of gene structures, patterns of duplication, gene splicing patterns, and regulatory interactions, even the definition of the gene itself has undergone radical revision. Some biologists now view the gene as a "computational module" rather than favoring the traditional definition emphasizing information content or a template for an eventual amino acid sequence. This juxtaposition of old and new presents a tension in the field that may not be resolved for several decades. A reasonable assessment, however, is that the genome appears increasingly fluid and dynamic, and that the categories of gene, exon, protein, and RNA transcript are becoming less distinct.

Genome Diversity and Evolution

Diversity of genome structure

The structures of genomes across the major branches of life differ widely. Whereas viral and bacterial genomes are models of efficiency, maximizing speed of genome replication and minimizing unnecessary genes, eukaryotic genomes—particularly those of mammals, amphibians, and some plants—are by comparison large and lumbering, harboring vast regions of noncoding and repeated DNA with unknown functions. Most of this noncoding DNA is no longer considered likely to be the "junk" it was postulated to be in the early 1970s, and a typical mammalian genome is by most measures vastly more complex than a bacterial genome.

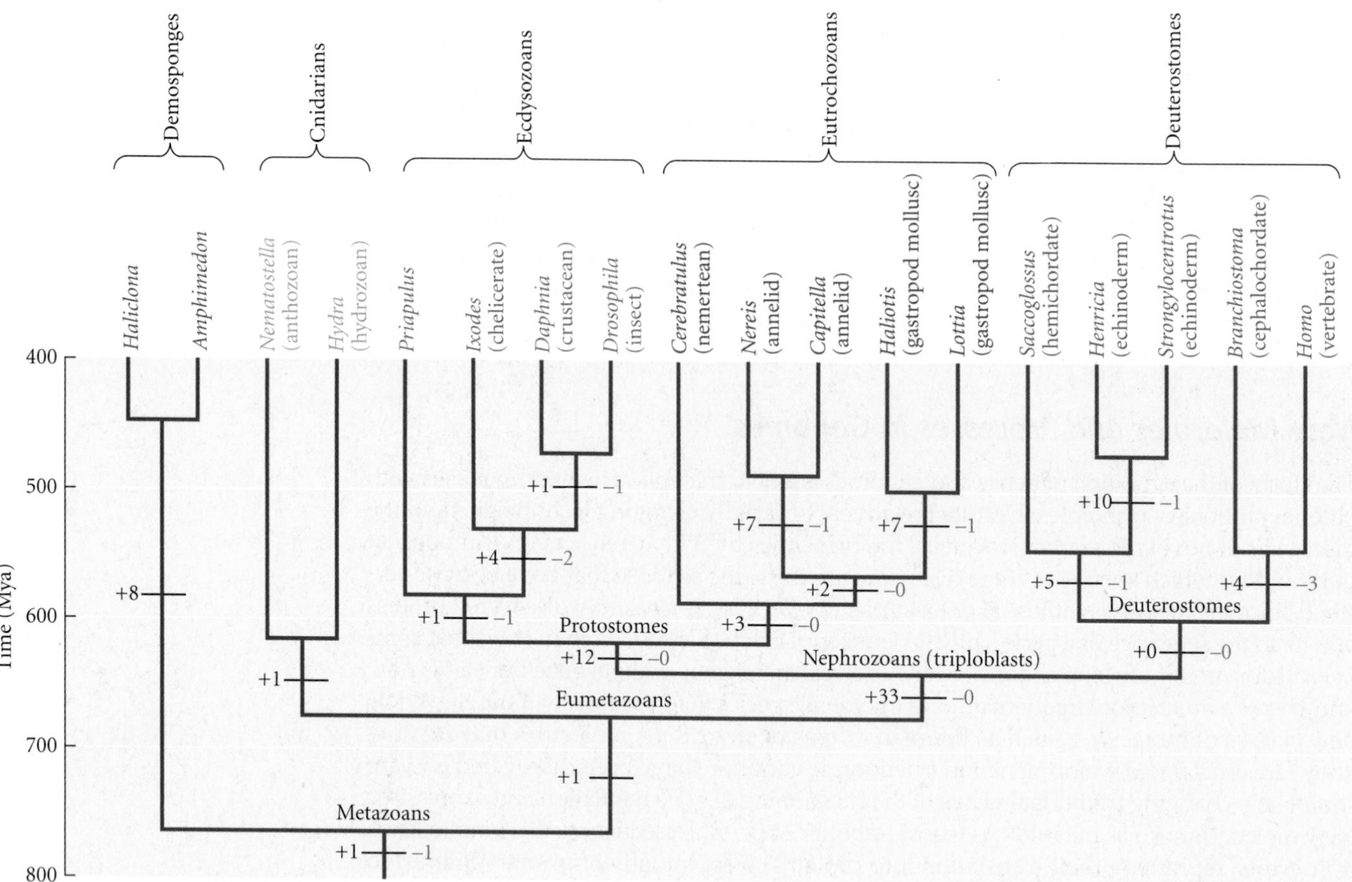

Figure 20.1 Continual acquisition of microRNA families through metazoan evolution. The numbers beside the branches indicate the addition (black) or loss (red) of identifiable microRNA families. MicroRNA families tend to persist in genomes once they are acquired and integrated into gene regulatory pathways. Thus losses of microRNA families are less common than gains. (From Wheeler et al. 2009.)

Michael Lynch and John Conery (2003) have pointed out that a variety of genomic features that initially may have had little fitness advantage for organisms—introns, transposable elements, large tracts of noncoding DNA—may be more prevalent in species with small effective population sizes. They have suggested that viruses and bacteria have extremely large population sizes that facilitate the sweep of advantageous mutations that enable genomic streamlining. By contrast, eukaryotes have smaller population sizes that facilitate the fixation of nonadaptive traits (Chapter 10). In this scenario, genomic features such as introns may originally have been acquired nonadaptively but were eventually co-opted to serve numerous functions in gene regulation, since it is known that introns routinely possess regulatory sequences such as microRNAs (Ghildiyal and Zamore 2009). Although population size alone cannot explain all the details of genome structure and variation among the clades of life, Lynch's hypothesis raises the important point that population size can strongly influence the fixation of nonadaptive traits in genomes (Lynch 2007).

Whereas many eukaryotic genes are interrupted by so-called **spliceosomal introns** (introns that require a complex group of proteins called a spliceosome to be removed from the mRNA), the genomes of bacteria and archaea have few introns, and what introns they do have are self-splicing and do not require a spliceosome. Because of some similarities between these two types of introns, Walter Gilbert, who discovered introns in the 1970s, suggested in 1987 that introns have been present since the common ancestor of all extant life, and have simply been lost in those genes and genomes that do not possess them today. This "introns early" hypothesis is still entertained (Poole et al. 1998), but the opposite ("introns late") hypothesis has gained further support because of the lack of introns in many phylogenetically basal eukaryotes. Also, most introns are restricted to specific clades of plants and animals, suggesting they arose relatively late in evolution and were not present in the common ancestor of eukaryotes, or of eukaryotes and bacteria (Palmer and Logsdon 1991).

A recent analysis of genome-scale data from 19 eukaryotes (Carmel et al. 2007) suggested a complex history of splicoeosomal introns in which a fairly high incidence of introns—over 2 introns per kilobase (kb)—is predicted for the common ancestor of all eukaryotes, and a variety of rates of intron gain or loss in different lineages has produced the diversity of intron numbers observed in eukaryotes today (Figure 20.2). This scenario posits that although some lineages, such as humans and other mammals, show a balanced pattern in which the rate of intron gain and loss is roughly equal, other lineages show higher rates of intron gain or loss; overall there is little correlation between rates of intron gain and loss across all 19 lineages. This scenario directly addresses Lynch's hypothesis for nonadaptive origin of introns by drift in small populations, which predicts an inverse correlation between intron gain and loss: in small populations, introns should on average be gained more often than lost, because of the increased fixation rates of deleterious mutations in small populations, whereas in large populations, introns should be more readily lost than gained because natural selection operates more efficiently in large populations (see Chapter 12). Regardless, the study by Carmel and colleagues shows that the picture of intron evolution is becoming vastly clearer with whole-genome data for many species.

Eukaryotic genomes contain far more noncoding DNA than prokaryotic genomes do. Only about 1.5 percent of the human genome, for example, is composed of protein-encoding sequences. Much of a typical human gene (up to 95 percent) consists of introns. Moreover, there are vast regions of noncoding DNA, whose function is still the subject of much debate. Traditionally, most of this noncoding DNA was deemed "**selfish DNA**," because it consists largely of transposable elements that have the capacity to replicate themselves and spread throughout the genome, sometimes debilitating genes as they jump to a new location. In 2001, however, scientists discovered that more than 10 percent of noncoding DNA is highly conserved between long-diverged species, such as humans and mice, suggesting a function maintained by purifying selection (Shabalina et al. 2001). Moreover, many noncoding regions, including introns, are transcribed into RNA sequences yet not translated into a protein. Other noncoding regions are composed of microRNAs, which can be identified in new genomes by computational algorithms as well as experiments on function. These diverse sequences, some represented by as many as 1,000 different types and up to 50,000 copies per genome, are now known to perform important functions in gene regulation and development (Bartel 2004).

Many eukaryotic genes, unlike prokaryotic genes, are subject to **alternative splicing** (**AS**), wherein many mRNAs, rather than just one, are encoded by a single gene. For example, the *CD44* gene, a cell-surface glycoprotein that modulates interactions between cells, contains 21 exons, at least 12 of which can undergo AS, potentially yielding thousands of splice variants (Roberts and Smith 2002). The protein DS-CAM is involved with controlling synaptic connections between neurons in the developing brain of *Drosophila*, and the gene encoding it has 95 exons (Zipursky et al. 2006). Remarkably, this gene is predicted to produce more than 38,000 isoforms based on AS and different rearrangements of its exons. This diversity is thought to underlie specificity of connections among the thousands of neurons in the developing fly brain (Hattori et al. 2007). AS appears to be a major mechanism by which metazoans can increase the functional diversity of a limited set of genes. Almost nothing is known yet about how AS patterns and control vary among related taxa, or how they evolve.

Viral and microbial genomes—the smallest genomes

According to life history theory (see Chapter 14), rapid growth and early reproduction are advantageous in organisms that frequently experience rapid population growth. Thus, in organisms such as viruses and bacteria, small genomes are advantageous because they can be copied faster—a key adaptation in situations where interspecific or host-parasite competition for resources is intense. Indeed, many viruses and bacteria have streamlined their genomes by doing away with many genes. Some of them, by exploiting the genomes of their hosts, make do with extremely small and focused genomes. For example, many viruses carry genes only for replication of their own genome, for construction of their

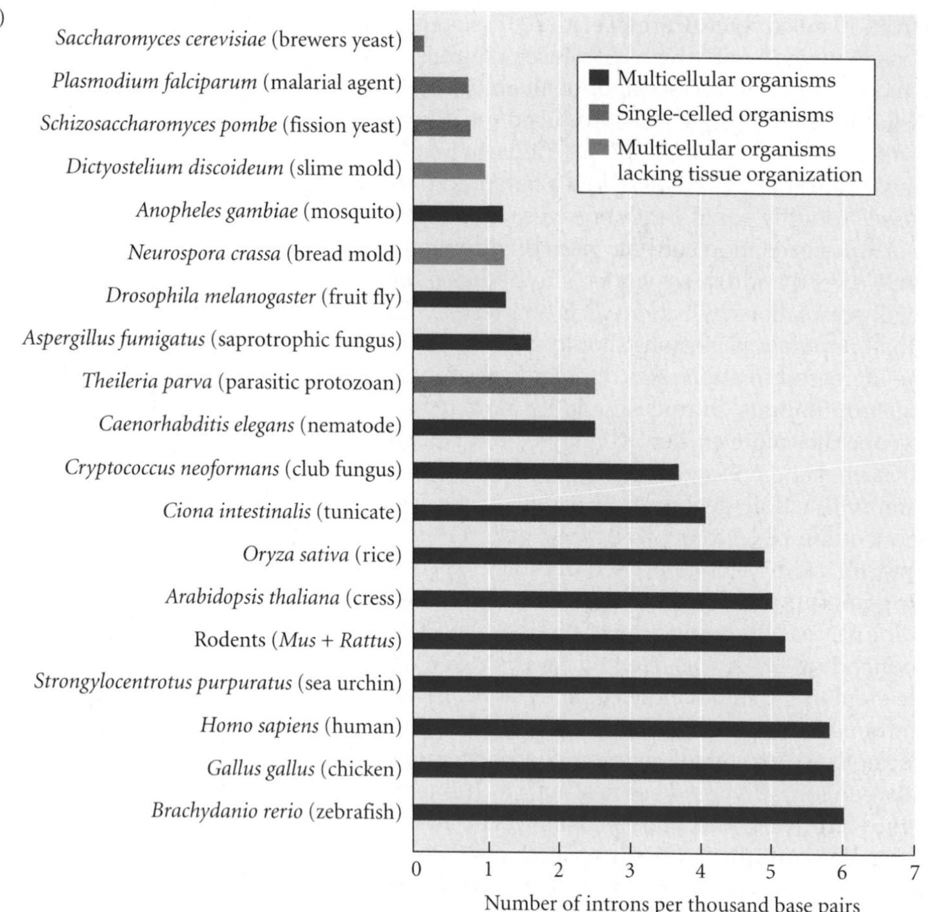

Figure 20.2 Phylogenetic distribution of introns. (A) Distribution of the number of introns per 1000 base pairs of genomic DNA in 19 eukaryotes. Note the increasing number of introns in moving from unicellular eukaryotes such as *Saccharomyces cerevisiae* to more complex ones. (Compare with the data for gene number shown in Figure 3.27.) (B) Rates of intron gain and loss for the species in (A) compared on a phylogenetic tree. The widths of the circles are proportional to the number of introns per 1000 base pairs. The dashed line (Ascomycetes) indicates that gains and losses in this lineage were significantly greater than the mean. (After Carmel et al. 2007.)

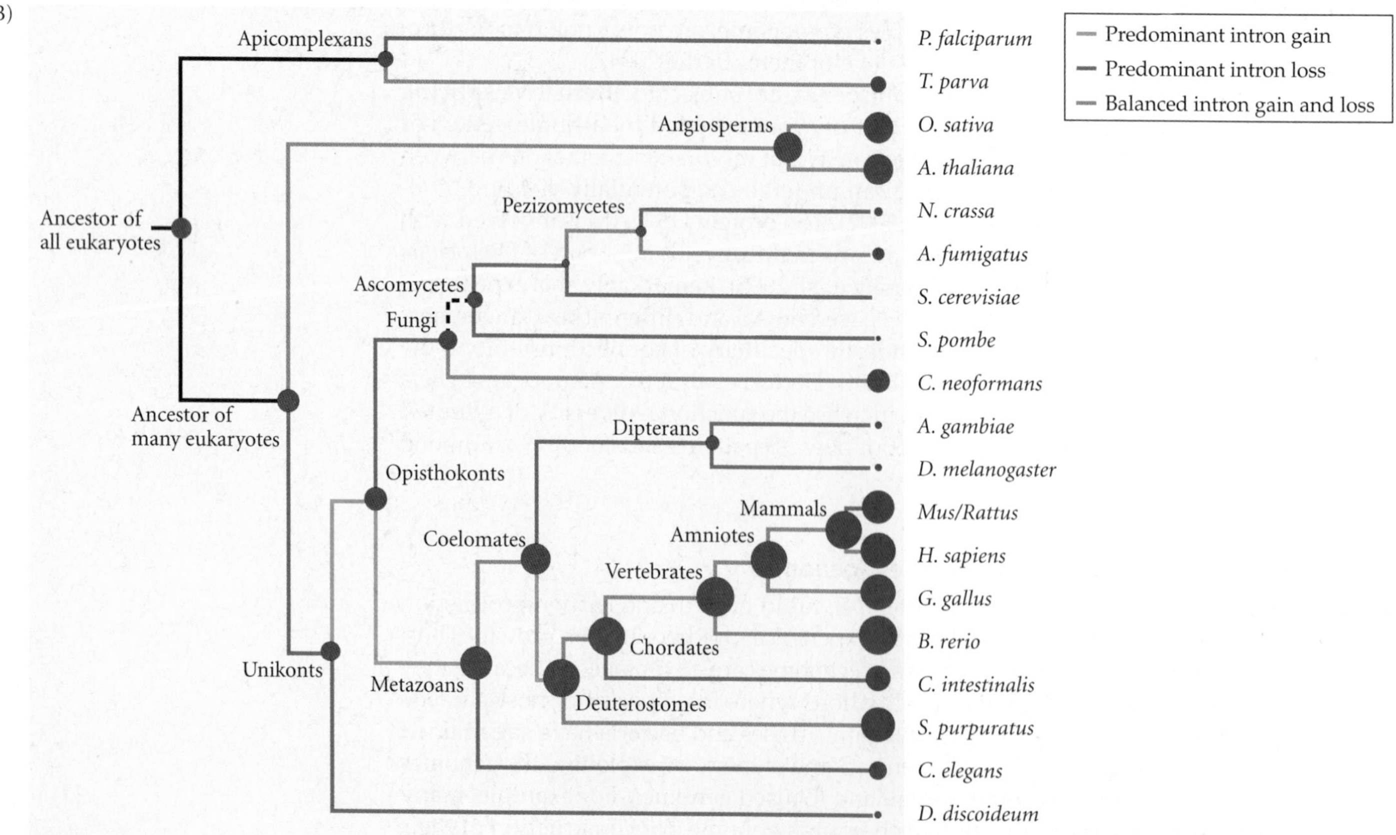

outer core and for integration into the host genome. The RNA genome of HIV, for instance, has only 9.8 kb, encoding 9 open reading frames (genes), and the bacterium *Mycoplasma genitalium*, a parasite of the genitalia and respiratory tracts of primates, including humans, has a genome that is only about 580 kb, containing 468 protein-coding genes that govern some basic molecular and metabolic functions as well as adaptations for parasitic life, such as variable surface proteins that evade the host's immune system (Razin 1997).

Nancy Moran and colleagues have extensively studied the evolution of reduced genomes in *Buchnera*, a bacterial clade that has been an intracellular symbiont of aphids for the past 200 Myr (Moran 2003; van Ham et al. 2003). *Buchnera* is a relative of better-known bacteria such as *E. coli*. Genome comparisons have revealed that the common ancestor of *Buchnera* species lost more than 2000 genes than *E. coli*. Because the host aphids provide many essential nutrients and metabolic functions for the symbionts, selection for retaining many genes has been relaxed. Intriguingly, the gene order has remained constant among *Buchnera* strains even though they have been diverging from one another for about 150 million years. This constancy may be a consequence of the loss of many ribosomal RNA genes and transposable elements that in other bacterial lineages can facilitate illegitimate recombination and rearrangement of gene order (see Figure 8.7). A recent comparison of seven closely related strains of *Buchnera* revealed a high rate of spontaneous point mutations and deletions, many of which are driven to fixation by genetic drift in the small population sizes of this symbiont, and go on to disrupt formerly functional genes (Moran et al. 2009).

Repetitive sequences and transposable elements

A substantial fraction of the genome—nearly half the human genome and about 34 percent of the *Drosophila* genome—consists of repeated DNA sequences that are called low repetitive, middle repetitive, and highly repetitive DNA, depending on copy number. They are also termed SATELLITE DNA because of their location in a chemical gradient in a centrifuge. A major source of repeated DNA in human and other mammalian genomes is transposable elements (TEs), sequences that can copy and transpose themselves, or "jump," to other regions of the genome (see Chapter 8). Most TEs are retroelements that are transposed by producing an RNA transcript followed by reintegration of new copies into the genome as DNA. The organism in which retroelements reside is often referred to as the host. The genes of many TEs do not contribute to development or function of the host organism; rather, they encode only proteins essential for replication and transposition of the retroelement itself, and are thus an example of a SELFISH GENETIC ELEMENT, or "selfish gene." An increasing number of such elements, however, are now known to have resulted in novel functions for the host and have proved a potent source of genomic innovation. For example, David Haussler and colleagues recently discovered a class of short interspersed nuclear element (SINE) that was active and proliferating in the ancient lineage of lobe-finned fishes represented by the coelocanth (Bejerano et al. 2006). Copies of this SINE were retained throughout vertebrate evolution and have been co-opted as a functional enhancer regulating the expression of the mammalian transcription factor *ISL1* (Figure 20.3), which controls the differentiation of motor neurons. This SINE is also conserved in a group of human exons known as *PCB2*, and is alternatively spliced in some cases. This example shows that some anciently proliferated TEs, many of which lay quiescent in the genome, can later be co-opted to perform important functions in the genome.

Variation in the number of repetitive elements in genomes goes a long way toward explaining the **C-value paradox**: the finding that organismal complexity and total genome size do not display a tight correlation, at least within eukaryotes (see Figure 3.26). Eukaryotes display a vast array of TEs, many of which are not shared between different major clades, such as insects and invertebrates. The accumulation of TEs in particular lineages appears to occur frequently once a particular progenitor TE is established in a genome, and scientists can use the number of DNA sequence differences accumulated among TE members in a particular genome to trace the rise and fall of particular TE families. How a TE first becomes established is not understood. But the number of TEs, and total genome size, for a species such as corn (*Zea mays*) vastly exceeds that for humans,

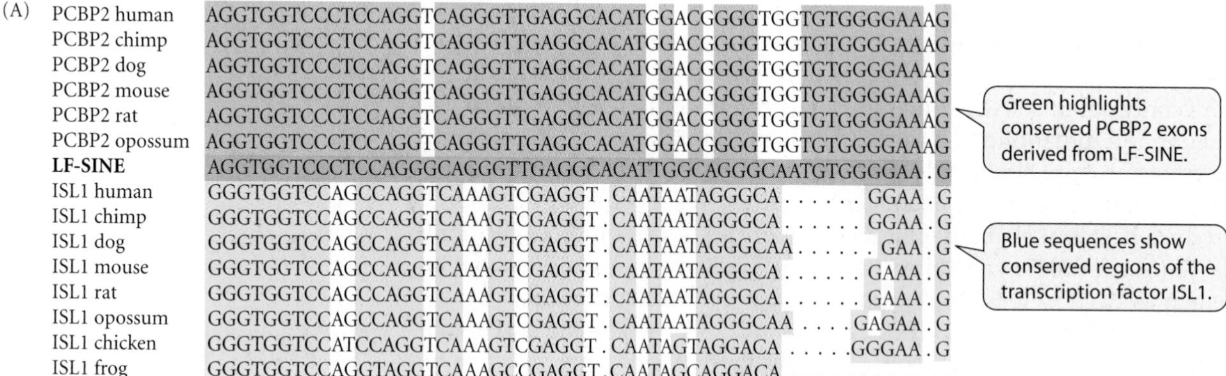

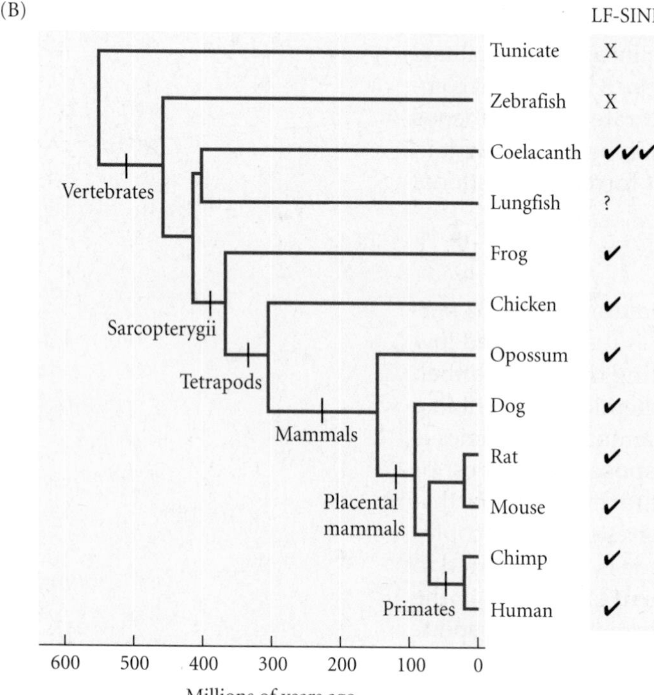

Figure 20.3 Conservation and evolution of a novel SINE in vertebrates. The LF-SINE (lobe-fin SINE) is found in many copies in the genome of the Indonesian coelacanth *Latimeria menadoensis*, a lobe-fin fish. (A) Conservation of the LF-SINE sequence (gray) in the genomes of various vertebrates as the exon PCBP2 (found in diverse proteins) and as the gene for the transcription factor ISL1. (B) Phylogenetic distribution of LF-SINE. Checkmarks indicate the relative abundance of the element in the various genomes; "X" indicates the element is not present. (From Bejerano et al. 2006.)

even though humans are generally considered more complex organisms than corn. It is patterns like this that contribute to the C-value paradox.

The copies of retroelements that are transposed to new sites in the genome undergo point mutations, just as host genes do. They either continue to produce daughter elements or, more frequently, degenerate by mutation and become inactive elements that no longer transpose. Mutations in these elements can be used to determine relationships among the copies in a genome, and the age of a family of TEs can be estimated in the same way that divergence times of species can be estimated. This requires an estimate of the absolute rate of point substitution, usually obtained from rates estimated for the host genome itself. For example, *Alu* elements, which are a type of SINE, are found throughout the primates generally and are abundant in hominoid primates. *Alu* elements are thought to have originated about 65 Mya and to have undergone a proliferation in the primate lineage 40–50 Mya. There are more than 500,000 copies in the human genome, making up over 10 percent of human DNA. The number of point mutations observed among hundreds of *Alu* elements scattered across the human genome implies that they underwent an ancient proliferation about 40 to 50 Mya and have since slowed down in their rate of transposition (Figure 20.4).

The effect of a new transposition event on the fitness of the individual bearing it depends largely on where it occurs. The production and insertion of new copies of TEs must often have little effect on the fitness of the host organism, but many transposition events are known to be deleterious, because they can affect the function of genes near which, or within which, they are inserted. TEs tend to occur in regions between genes and in introns, probably because those that occur in coding regions often cause deleterious mutations and are eliminated by purifying selection.

Transpositions can have at least two kinds of genetic effects. First, they cause mutations; 10 percent of all genetic mutations in mice are thought to result from retroelement transpositions, many of them into coding regions or control regions. Second, the repeated copies of a TE in different parts of the genome can provide templates for illegitimate

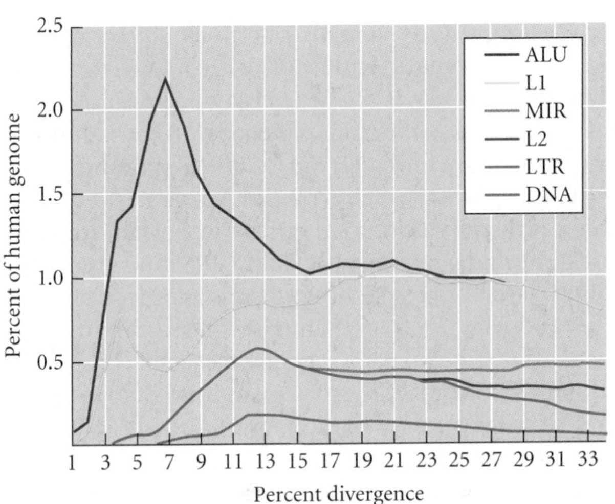

Figure 20.4 Age distribution of retroelements in the human genome. The x-axis shows the percent divergence between two sampled copies of a retroelement, whereas the y-axis shows a percent count of copies with that degree of divergence. Six different retroelements are indicated: *Alu* elements, L1 and L2 long interspersed nuclear elements (LINEs), mammalian interspersed repeats (MIRs), long-terminal repeat (LTR) elements, and DNA transposons. If percent substitution can be used as a rough proxy for time, the figure suggests that many *Alu* elements underwent a burst of proliferation about 40 to 50 Mya (7 percent divergence), with decreased rates of proliferation at earlier and later dates. By contrast, L1 LINEs seem to have had a protracted proliferative phase a very long time ago, with a second spike in numbers corresponding to roughly 4 percent divergence (about 25 Mya). MIRs and L2 LINEs are uniformly ancient and on their way to degradation and loss from the genome. (After Deininger and Batzer 2002.)

recombination, resulting in chromosome or gene rearrangements that often carry deletions of some genetic material (see Figure 8.7). Approximately 0.3 percent of all human genetic disorders, including many hereditary leukemias, are thought to arise in this way (Deininger and Batzer 2002). The reduction of host fitness that is due to deleterious mutations and chromosome rearrangements is thought to be the chief reason that TEs are not even more abundant in the host genome than they are (Charlesworth and Langley 1989).

New genomes reveal major events in the history of life

With each new genome that is sequenced, particularly those representing clades that have not yet been studied at the genomic level, scientists learn more about the major environmental and evolutionary events that have shaped that lineage, and often the genomic complexity of a particular species is unexpected. For example, the genome sequence of the purple sea urchin (*Strongylocentrotus purpuratus*), a model organism for the study of embryogenesis and gene regulation, was determined in 2006 (Sodergren et al. 2006). A major surprise from this genome was the unexpected complexity of its immune system as revealed by the number of genes with immune functions (Figure 20.5; Rast et al. 2006). It had been thought that the immune systems of vertebrates differed from those of echinoderms (such as sea urchins) and other basal deuterostomes primarily in possessing genes

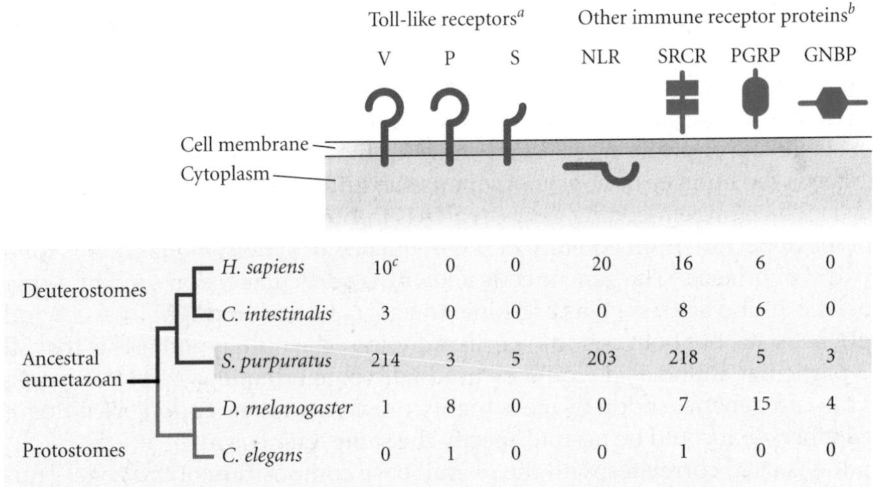

Toll-like receptors[a]
Other immune receptor proteins[b]
V P S NLR SRCR PGRP GNBP
Cell membrane
Cytoplasm
Deuterostomes
Ancestral eumetazoan
Protostomes
H. sapiens 10[c] 0 0 20 16 6 0
C. intestinalis 3 0 0 0 8 6 0
S. purpuratus 214 3 5 203 218 5 3
D. melanogaster 1 8 0 0 7 15 4
C. elegans 0 1 0 0 1 0 0
[a]Toll-like receptors include vertebrate type (V), protostome type (P), and short type (S).
[b]NLR = NACHT domain/LRR proteins; SRCR = scavenger receptor cysteine-rich proteins; PGRP = peptidoglycan (bacterial cell wall) recognition proteins; GNBP = Gram-negative bacterial binding proteins.
[c]Plus 1 pseudogene.

Figure 20.5 Immune receptor molecules in the genome of the purple sea urchin (*Strongylocentrotus purpuratus*). The diversity of these molecules rivals or exceeds that in the genomes of other metazoans, indicating unexpected complexity of innate immune defenses in sea urchins. (After Rast et al. 2006.)

for adaptive immunity—the arm of the immune system that is molecularly diverse, evolves rapidly, and sometimes undergoes somatic mutation in response to novel pathogens. The innate immune system, by contrast, is shared between vertebrates and other deuterostomes. The genome of the purple sea urchin was found to contain unexpected complexity in the innate immune system, and also the telltale beginnings of an adaptive immune system. For example, purple sea urchins posses 222 genes known as Toll-like receptors (TLRs), which exhibit little polymorphism but are important for combating pathogenic bacteria. By contrast, humans posses only about 10. In addition, the purple sea urchin possesses recombination-activating genes *rag1* and *rag2*, which were hitherto unknown from invertebrates and play an important role in coordinating the recombination and somatic diversification of T-cell receptor genes, a crucial component of the adaptive immune system. Although T-cell receptors have not been discovered in sea urchins, the *rag* genes suggest that sea urchins may harbor some capacity to engage in adaptive immunity like their vertebrate relatives.

The large number of new genomes and the new experiments that such genomes enable have resulted in several important conceptual paradigm shifts in the study of genome evolution. First, our view of the genome has become much more dynamic as a result of our ability to study it in detail at the level of whole genomes and nucleotide sequences. For example, as we will see with the *Drosophila* gene *jingwei* and with the antifreeze genes of arctic notothenioid fish, it is now routine for new and often cryptic genes to be discovered and for their often complex origin to be traced back to progenitor genes and genome fragments. Second, because of the many genes and other markers that are discovered with every new genome, whole genomes now allow a "systems-level" approach to the study of population genetics, in which a given compartment of the genome is studied not in isolation but in the context of all other genes, duplicated regions, and regulatory structures. This allows scientists to study the *distributions* of evolutionary phenomena and events, across thousands of genes, rather than be content to estimate evolutionary forces for a single gene. For example, as mentioned in Chapter 12, the distribution of selection coefficients and synonymous and nonsynonymous changes can now be determined for thousands of genes simultaneously, whereas only 10 years ago many evolutionary biologists were content to estimate such dynamics for just a single gene. Despite such advances, these processes do not alter the fact that the genome as a whole is subject to the same principles of mutation, drift, and adaptive evolution that have been studied for decades—although we can add a wealth of recent insights into chromosome evolution and genome structure to the evolutionary principles covered earlier in this book

Protein Evolution and Translational Robustness

Codon bias

Protein-coding genes are diverse, and their evolution has been modulated in important ways by the structure of the genetic code. For example, we have seen how many proteins evolve under stabilizing or purifying selection, in which the number of synonymous point mutations exceeds the number of nonsynonymous substitutions (d_n/d_s; see Chapter 12). Another way in which protein-coding genes evolve is through **codon bias**. Codon bias is a measure of the departure from equality of the frequency of synonymous codons for a given amino acid in proteins. The genetic code allows the particular codon used by a protein for a specific amino acid, such as arginine, to vary: one codon might be AGA and another might be AGG, but both codons specify arginine. The null hypothesis is that all codons for a particular amino acid would be used with equal frequency, but this usually is not the case. In general, codon usage is highly uneven, and one codon often dominates over all others that could be used to specify the same amino acid.

Which codon is used correlates with the overall base composition of the gene. Thus genes that are generally high in nucleotides G or C tend to have codons that end in G or C. Such correlations between overall base composition and the particular codons used have been found for genes both in mitochondrial DNA and in the nuclear genome (Li 1997). Correlations between transfer RNA abundance and codon usage have been found

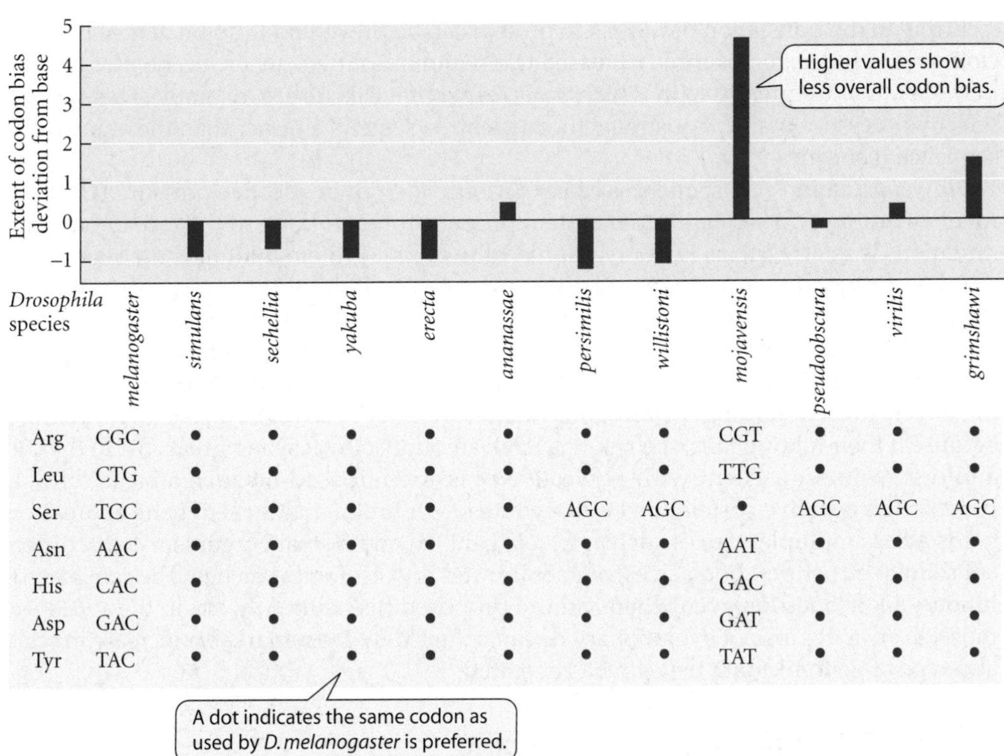

Figure 20.6 Extent of codon bias in 12 *Drosophila* species. The histogram shows the deviation in extent of codon bias of each species from *D. melanogaster*, which has a value set at 0. The particular measure of deviation used means that the higher values show less overall codon bias. The chart below shows the most numerous codons ("preferred codons") for the amino acids given at the left. (After Clark et al. 2007.)

in several organisms and are thought to underlie some cases of codon bias. Each tRNA is specific for a particular codon. If the tRNA for AGA is more abundant than that for AGG, the expression of a gene will be faster and more efficient if it uses AGA codons for arginine rather than AGG. Typically, genes with very skewed codon bias tend to be highly expressed and are thought to use those codons that correspond to the anticodon of the most abundant tRNAs. Surprisingly, codon bias can vary substantially among species, across hundreds of genes in the genome, as recent comparisons among whole genomes of *Drosophila* show (Figure 20.6; Clark et al. 2007).

Gene expression and selection on translation errors

We saw in Chapter 10 how the neutral theory predicts slower rates of evolution for genes under stronger functional constraint. With the availability of multiple types of genomic data for certain organisms, scientists have been able to study the many diverse factors influencing the evolutionary rate of proteins. For example, for baker's yeast (*Saccharomyces cerevisiae*)—the first eukaryote to have its genome sequenced—we now have not only the entire genome sequence, but also the complete inventory of expression levels for all genes in many physiological states, as well as multiple measurements of protein abundances for virtually all genes. In addition, for every gene in the yeast genome, we have a measure of **gene dispensability**—that is, the fitness of yeast cells in which each gene is physically deleted from the genome. These data sets allow biologists to compare the effects of multiple aspects of gene function and expression on protein evolution with a precision found in few if any other species.

These analyses try to find trends in the rate of protein evolution across thousands of genes, and as a result rely primarily on correlations of evolutionary rate to infer its determinants.Surprisingly, the correlation between an organism's ability to survive without a particular gene and the rate of evolution of that gene (predicted to be low for indispensable genes) is relatively weak compared to the analogous correlations with expression level or protein abundance.By contrast, recent studies by Drummond and Wilke (2008) suggest that the rate of protein evolution can largely be explained by selection to prevent and withstand missense errors during translation, which should impose the strongest selective pressure on highly expressed genes encoding abundant proteins.

Errors in the translation of mRNA to protein can be prevented to some degree by judicious codon usage; for example increased translational accuracy when a gene uses codons in the relative frequency with which their respective tRNAs are available.This type of selective pressure mainly constrains the nucleotide sequence rather than the amino acid sequence it encodes.

However, amino acid sequences can experience their own selective pressure. The ability of an amino acid sequence to maintain proper protein folding in the face of ongoing mutations is referred to as its **translational robustness**. Protein folding is the process by which most protein sequences attain a three-dimensional shape that possesses biological activity, such as the capacity of properly folded hemoglobin proteins to bind oxygen molecules. Because attaining functional structure depends critically on interactions between residues within a protein, even a single missense error in translation can have devastating effects protein folding. Abundant proteins are under the highest selection pressure because if their folding fails, the result is large amounts of waste and great cost to the cell and organism. Consistent with a selective pressure imposed on such a basic cellular process, the negative correlation between rate of evolution and level of gene expression holds across multiple species such as *E. coli*, fruit fly, and human, organisms with different cellular structures, body sizes, metabolic rates and ecological niches. The rate of evolution—a key question even when Kimura proposed his neutral theory in the 1960s—is thus still an active area of evolutionary research, fueled by the remarkable increase in scale and type of genomic data that can be generated.

Natural Selection across the Genome

As described in Chapter 12, the rapid rate of fixation of advantageous mutations is the key to an important approach to detecting the action of positive selection on DNA sequences. We can think of positive selection as accelerating the recruitment of advantageous mutations into a population. For example, whereas synonymous substitutions will accumulate at a rate governed by the mutation rate (d_s), nonsynonymous mutations that are advantageous can accumulate much faster (d_n; see Chapter 12). We can define the ratio of d_n/d_s as ω. If ω for a protein-coding gene exceeds 1, we can conclude that the number of advantageous substitutions has exceeded the number of neutral substitutions, resulting in a net trajectory marked by positive Darwinian selection.

As recently as a few years ago, it was feasible to measure positive selection only in one or, at best, a few genes at a time. Now, with the ability to compare multiple genomes, scientists can measure the rates of synonymous and nonsynonymous substitutions across thousands of genes and across multiple species. For example, researchers recently obtained the complete genome sequences from 12 species of *Drosophila*, allowing them to study patterns of adaptive evolution of genes in the context of a phylogeny of species as well as across genes classes with different biological functions. The *Drosophila* 12 Genomes Consortium (Clark et al. 2007) showed that most gene classes among 12 species of *Drosophila* were under stabilizing selection, with values of ω of less than 1. Several gene classes, however, including immune defense, proteolysis, and a class containing genes of unknown function, had higher average values of ω and contained some genes with values greater than 1.

The sequencing of 12 *Drosophila* genomes allowed researchers to study the evolution of coding regions in exquisite detail. For example, the consortium was able to measure the extent of both positive and purifying selection on genes on the X chromosome and on autosomes of these species in various ways. Theory predicts that, given the hemizygosity of the X chromosome in male fruit flies (the situation in which there are no homologues on the Y chromosome to shield novel mutations on the X), natural selection (both purifying and positive) should be more efficient on the X chromosome than on the autosomes when mutations are recessive. This is because on the autosomes of both males and females, novel recessive mutations will be masked by the phenotypic effects on the other allele on the corresponding chromosome. By contrast, novel adaptive or deleterious mutations on the X chromosome in males will not be masked by another allele because often the Y chromosome does not have a corresponding ortholog (see Figure 3.28). As a result, substitution rates of positively selected mutations should be higher, and rates of deleterious substitutions lower, on the X chromosome than on the autosomes.

Members of the *Drosophila* Consortium tested this hypothesis by estimating the rate of evolution of amino acid sequences (i.e., of nonsynonymous substitutions) in a sample of several thousand genes along each branch of the 12-species phylogeny. They expected to find a higher rate in X-linked than in autosomal genes throughout the phylogeny. Instead, in most lineages, X-linked and autosomal rates did not differ, and in two cases X-linked genes actually evolved more slowly (Figure 20.7; Singh et al. 2008). Thus, positive selection of adaptive mutations varied substantially among lineages, perhaps due to differences in the selective effects of mutations. Codon bias was consistently greater in X-linked than autosomal genes throughout the phylogeny (Clark et al. 2007). Codon bias is a sensitive indicator of weak but pervasive purifying selection, because it indicates that genes have been under selective pressure to maximize their translational efficiency by tRNAs (see the previous section). As with other mutations, variation in codon use, if at all recessive, is exposed to greater selection in X-linked than in autosomal genes. The greater codon bias in X-linked genes suggests that purifying selection has been efficient throughout the history of the genus *Drosophila*.

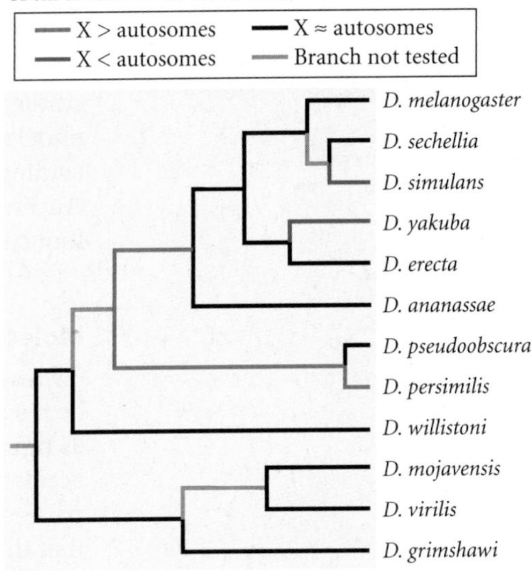

Figure 20.7 Relative rates of nonsynonymous substitution in protein-coding genes of the 12 *Drosophila* species with fully sequenced genomes. The substitution rate resulting from natural selection was expected to be higher in X-linked than in autosomal genes (as in the two lineages shown in red) because advantageous X-linked mutations are exposed to natural selection in males, which have only one X chromosome. However, the rate was more often found to be equal or lower (black, blue), suggesting that the efficiency of positive selection varies among lineages. All lineages displayed equal levels of codon bias throughout the genome, however, indicating that purifying selection was pervasive. (After Singh et al. 2008.)

Adaptive molecular evolution in primates

Molecular evolutionary processes are better understood in primates than in any other mammalian lineage. The first good example of adaptive convergent evolution at the molecular level was in the enzyme lysozyme (Figure 20.8), which breaks down bacterial cell walls. Both ruminants (such as cows, deer, sheep, and giraffes) and colobine monkeys (such as langurs) have evolved a modified foregut (the rumen) in which bacteria digest the cellulose from grasses and other plant material that the animals eat. Lysozyme in the rumen enables the animals to digest bacterial cell contents—a major source of protein. Caro-Beth Stewart and colleagues (Stewart et al. 1987) found that langurs, like cows, express high levels of lysozyme

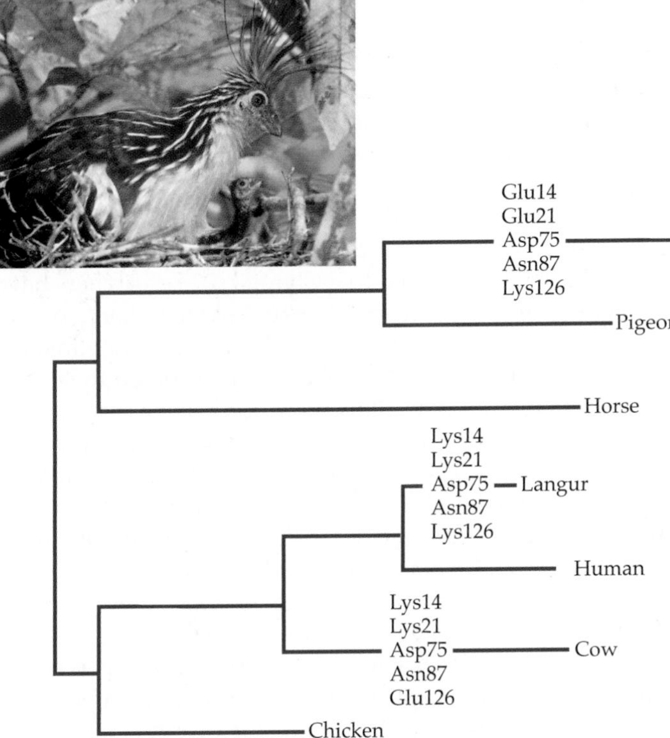

Figure 20.8 A gene tree for lysozyme provides phylogenetic evidence for molecular convergence in primate, ruminant, and avian lysozymes. Branch lengths are proportional to the total number of amino acid replacements along them. Convergent replacements occur at positions 14, 21, 75, 87, and 126 in the lineages leading to the lysozymes of three species that digest bacteria in the foregut: the cow, langur, and hoatzin (*Opisthocomus hoazin*; from Amazonian South America, it is the only bird known to possess this trait). For example, position 75 changed to aspartic acid independently in all three foregut-fermenting lineages. The convergent amino acid replacements presumably represent adaptive biochemical changes in the enzyme. (After Kornegay et al. 1994; photo © Photolibrary.com.)

in the foregut. More remarkably, the researchers found that the langur's protein has five amino acid substitutions (compared with other primates) that are also found in the lysozyme of the cow. Consequently, at specific amino acid positions, the lysozyme of the langur is more similar to that of the cow than to that of other primates. Messier and Stewart (1997) have since studied the DNA sequence evolution for the primate lysozyme gene and found that it underwent an episode of accelerated nonsynonymous substitution in the ancestral lineage leading to colobines, associated with the evolutionary change in diet from fruit to foliage. Adding to the history of convergence in this enzyme, Kornegay and colleagues (1994) found similar amino acid substitutions in the lysozyme of the hoatzin (*Opisthocomus hoazin*), a leaf-eating bird that also has high levels of cellulose-digesting bacteria.

Molecular evolution in the human lineage

Genes that exhibit adaptive evolution on the branch leading exclusively to humans could be responsible for some of the traits that distinguish humans from other primates—such as human speech. We saw in Chapter 8 how the rate of evolution of the gene *FOXP2* has accelerated in the human lineage during the 4 to 6 Myr since it separated from chimpanzees, representing a significant acceleration in the rate of amino acid substitution given that this gene is virtually invariant in another 28 orders of mammals over a time span of more than 100 Myr. Another gene, *sarcomeric myosin heavy chain* (*MYH*), is highly expressed in the chewing muscles of chimpanzees and is responsible for the large size of these muscles in this and other nonhuman hominins. In humans, this gene has been inactivated by a frameshift mutation that occurred about 2.4 Mya, after humans separated from chimpanzees (Stedman et al. 2004). The loss of function of *MYH* is thought to be responsible for the reduced size of these muscles in humans. In this example, ω increased from less than 0.1 to approximately 1, as expected if constraints were relaxed following loss of function.

Olfactory perception plays an important role in many mammalian behaviors, but primates, especially humans, possess a much less sensitive sense of smell than most other mammals. The olfactory receptor (OR) genes present a molecular example of the tendency in organisms for "structures of little use" to degrade and degenerate, as Darwin noted in *The Origin of Species*. There are approximately 1200 functional OR genes in the mouse genome but only about 550 in humans. Apparently, selective pressures to maintain OR gene functionality are relaxed in humans, and mutations that reduce function and cause pseudogene formation have been fixed (Gilad et al. 2003). Zhang and Webb (2003) found the same tendency in five genes that are normally expressed in the vomeronasal organ in primates. Mammals use this organ to perceive pheromones, which are used extensively in social interactions. Two distinct families of these genes are functional and conserved in several lineages of Old World primates (catarrhines), but they have degenerated dramatically in hominoids, having accumulated numerous stop codons since their separation from ancestral catarrhines some 23 Mya.

A major advance in our understanding of molecular evolution in humans will come with the completion of the genome sequence for a Neanderthal, a lineage of hominin that diverged from the human line about 650,000 years ago (see Chapter 4). Researchers have used the new "next-generation" sequencing approaches (see Box 20A) to obtain a partial genome sequence of a Neanderthal specimen approximately 30,000 years old (Green et al. 2006; Noonan et al. 2006). Although some of this genome was determined to be contaminated by modern human DNA, techniques now allow more confident assignment of sequence data to human or Neanderthal, despite their high level of sequence similarity. Complete sequences of mitochondrial and nuclear DNA for various extinct species, such as mastodons and moas, are now contemplated. For the first time, biologists will be able to directly compare the genomes of ancestral species with those that have survived to the present on the scale of whole genomes.

Scaling up: From gene to genome

Now that the genomes of more and more organisms are being completely sequenced, it is possible to move beyond descriptions of the evolutionary dynamics of single genes or gene families and to examine the distribution of selective histories and evolutionary rates across the entire genome (see Chapter 12, pp. 329–333). For example, human population genetic studies now include surveys of polymorphism in literally hundreds of genes, using a variety of new molecular approaches (e.g., Stephens et al. 2001). Andrew Clark and collaborators (Clark et al. 2003) estimated ω values (the ratio d_n/d_s) for the coding regions of 7645 genes from humans and chimpanzees, using mouse genes as an outgroup that enabled them to determine which substitutions occurred in the human and which in the chimpanzee lineage. Using a statistical method that permitted identification of specific codons exhibiting ω of greater than 1, they found evidence of adaptive evolution in 873 genes along the human lineage. Certain functional groups of genes were especially prone to show adaptive evolution, such as those encoding olfactory receptors and amino acid catabolism. The authors suggested that these changes reflect the behavioral and dietary changes in the human lineage. Several adaptively evolving genes play key roles in early development, pregnancy, and hearing.

Origin of New Genes

It is obvious that the approximately 24,000 different functional genes in mammalian genomes must have evolved from a much smaller number in the earliest ancestor of living organisms. Presumably, all genes in the human genome ultimately descend from a single gene or set of genes that provided the first programs for life on earth. Moreover, the number of functional genes differs among major groups of organisms. How do new genes arise?

Evolutionary biologists have described several mechanisms by which the genes in a species' genome have originated, either from pre-existing genes in the same genome or from the genome of a different species. These mechanisms include lateral gene transfer, exon shuffling, gene chimerism, retrotransposition, motif multiplication, and gene duplication (Long et al. 2003).

Lateral gene transfer

In the 1970s, Carl Woese analyzed 16S rRNA sequences and found that the tree of life is divided into three major "empires" or "domains": the Bacteria, Archaea, and Eukarya (see Figure 2.1). The Archaea and Eukarya appear to be sister clades. This phylogeny is supported by many gene sequences, although some gene sequences seem to indicate a sister-group relationship between Eukarya and Bacteria instead. As we saw in Chapter 2 (see Figure 2.24), when different genes provide strong support for different phylogenies, **lateral gene transfer** (LGT, also called **horizontal gene transfer**) between lineages is a likely hypothesis. It is now thought that genetic material was often transferred across quite different lineages early in the history of life, and some authors even favor a tree in which eukaryotes represent a fusion between Archaea and Bacteria (Doolittle 1999).

Lateral gene transfer has also occurred more recently; in fact, LGT among microbial lineages is thought to be an important source of gene functions that are novel for the recipient organism. For example, the eukaryotic protist *Entamoeba histolytica,* which causes more than 50 million cases of dysentery annually, can live anaerobically in the human colon and in tissue abscesses, as a result of fermentation enzymes that most other eukaryotes lack. A phylogenetic analysis showed that several of these fermentation genes were obtained by LGT from Archaea (Figure 20.9; Field et al. 2000). Perhaps 40 to 50 human genes have their origins in bacteria (Salzberg et al. 2001). New genome sequence data suggest that LGT may be quite frequent among prokaryotes, and that novel adaptive mechanisms, often borne on chromosomally distinct plasmids, are especially likely to spread phylogenetically by LGT (Ochman et al. 2000). In addition, intracellular symbiotic associations, such as the *Wolbachia* symbiont in many insect species, provide a context conducive to LGT. Researchers recently discovered that massive LGTs, involving segments ranging from fewer than 500 base pairs to nearly the entire 1 Mb *Wolbachia* genome, have occurred between *Wolbachia* and the nuclear genomes of four insect and four nematode hosts (Figure 20.10; Hotopp et al. 2007). Given the appropriate genomic environment, extensive LGT can occur between bacteria and multicellular eukaryotes.

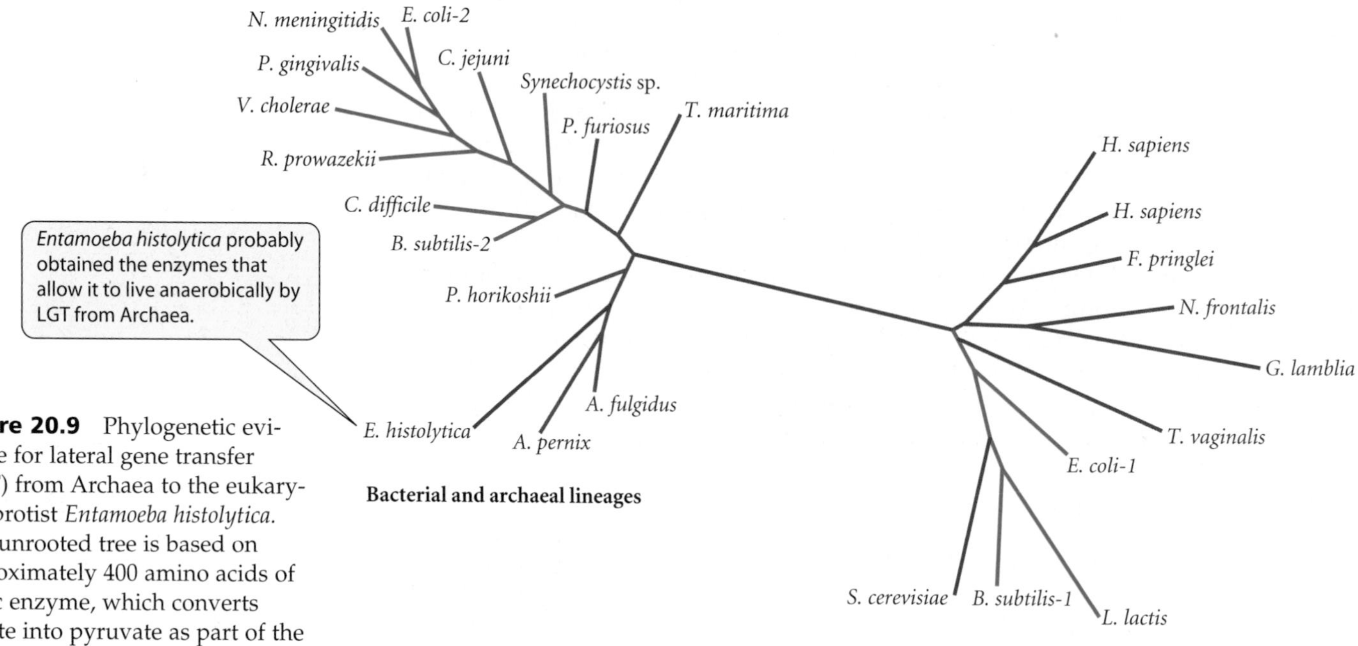

Figure 20.9 Phylogenetic evidence for lateral gene transfer (LGT) from Archaea to the eukaryotic protist *Entamoeba histolytica*. This unrooted tree is based on approximately 400 amino acids of malic enzyme, which converts malate into pyruvate as part of the fermentation pathway. Archaeal lineages are red, bacterial green, and eukaryotic blue. A tree in which the *E. histolytica* enzyme clusters with other protist lineages is 46 steps longer than the most parsimonious tree, and therefore a worse explanation of the amino acid data. Cell structure and other gene sequences, however, clearly show that *Entamoeba* is a typical eukaryote. (After Field et al. 2000.)

Origin of new genes from noncoding regions

Biologists are finding an increasing number of cases in which novel genes arise de novo from regions of the genome that are noncoding in related species and were presumably noncoding in recent ancestors. For example, Levine and colleagues (2006) documented five genes in *Drosophila melanogaster* that are not present in related species of *Drosophila*. These genes have no ortholog in other *Drosophila* species, and similar sequences are absent from their genomes. Nonetheless, the regions to which these novel genes map in *D. melanogaster* are in some cases transcribed in related species, leading to the suggestion that aberrant transcription in noncoding regions may facilitate the origin of new genes. A similar pattern was found in a novel gene called *BSC4* in the yeast *Saccharomyces cerevisiae*, in a region of the genome that contained no genes in related yeast species yet was transcribed into RNA and contained sequences that were distantly related to *BSC4* (Figure 20.11; Cai et al. 2008). These studies suggest that anonymous regions of the genome that are transcribed may generate novel genes in some lineages. Intriguingly, four of the five genes in the *Drosophila* study show evidence of adaptive evolution and are expressed in the testis, a tissue that is known to have frequent aberrant transcription of noncoding regions. Although the function of these genes is unknown, they may likely be involved in male reproduction. The origin of genes involved with novel male functions may frequently be driven by natural selection, the result of sperm competition or sexual conflict at the molecular level (see Chapter 15).

Exon shuffling

Lateral gene transfer adds genes to the genome of a particular lineage, but the genes already exist in the source species. Several other mechanisms produce truly new genes. One of these mechanisms is **exon shuffling**.

Eukaryotic genes include introns and exons. There is often a close correspondence between the division of a gene into exons and the division of the protein into **domains**.

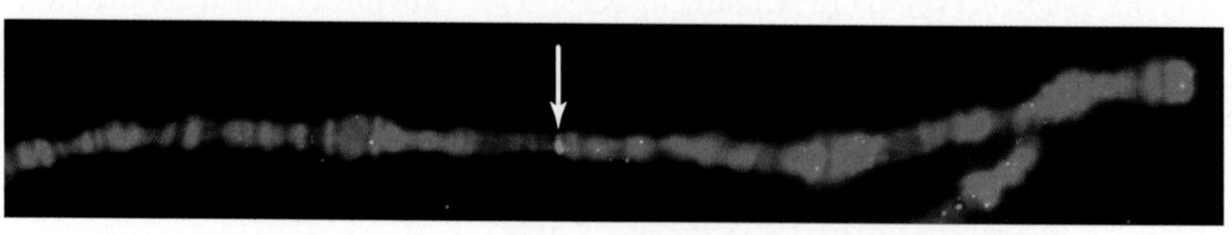

Figure 20.10 A polytene chromosome of *Drosophila ananassae* (red) with evidence of integration of a laterally transferred gene from the intracellular symbiont *Wolbachia* (green). In situ hybridization was used to label a segment of the *Wolbachia* genome with a fluorescent dye and hybridize this probe to the chromosomes of *D. ananassae*. The presence of *Wolbachia* sequences in the *Drosophila* chromosome is direct evidence of lateral transfer of DNA from the symbiont to the host genome. (After Hotopp et al. 2007.)

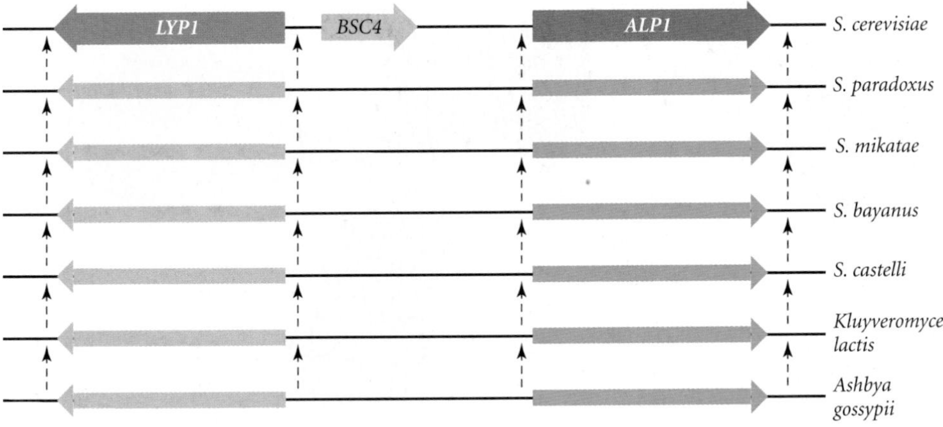

Figure 20.11 Origin of a new yeast gene from noncoding DNA. Each row depicts a segment of a homologous region of the genome of seven species of yeast. Species designations are at right. Colored arrows indicate known genes and the direction of transcription. *LYP1* and *ALP1* are paralogous genes occurring in inverted orientation in each of the seven species. The only species that exhibits an open reading frame, and produces a detectable protein product, between these two genes is *S. cerevisiae*, although the region between the paralogs in the other six species exhibits very low (~35 percent) similarity to *BSC4* and is transcribed in all seven species. (After Cai et al. 2008.)

A domain (or "module") is a small (~100 amino acids) segment that can fold into a specific three-dimensional structure independently of other domains. Domains frequently have specific functions, although they usually cannot perform these functions completely in the absence of other domains that would together make up a mature protein. For example, antibodies function primarily through their immunoglobulin domains, two of which together form the major cleft into which foreign proteins fit during an immune response.

Protein domains are frequently identifiable through bioinformatics and often contain specific amino acid motifs that are characteristic and distinguish them from other domains. Hundreds of different domains are known, and most proteins are mosaics that consist of several different domains (Figure 20.12). For example, various collagens have different sequences of up to five different types of domains, some of which may be repeated. Some domains, such as the nucleotide-binding domain, the immunoglobulin domain (Figure 20.13), and the heme-binding domain, occur in many different proteins.

Comparing gene structures in the human genome and in genomes from *Drosophila* and yeast has shown that many genes evolve by **domain accretion**, whereby new genes are produced by the addition of domains to the beginnings or ends of ancestral genes (Graur and Li 2000). For example, many chromatin-associated genes have evolved by addition of a variety of domains, such as chromodomains, zinc fingers, and helicase/ATPase domains. These domains, however, are found in other proteins as well.

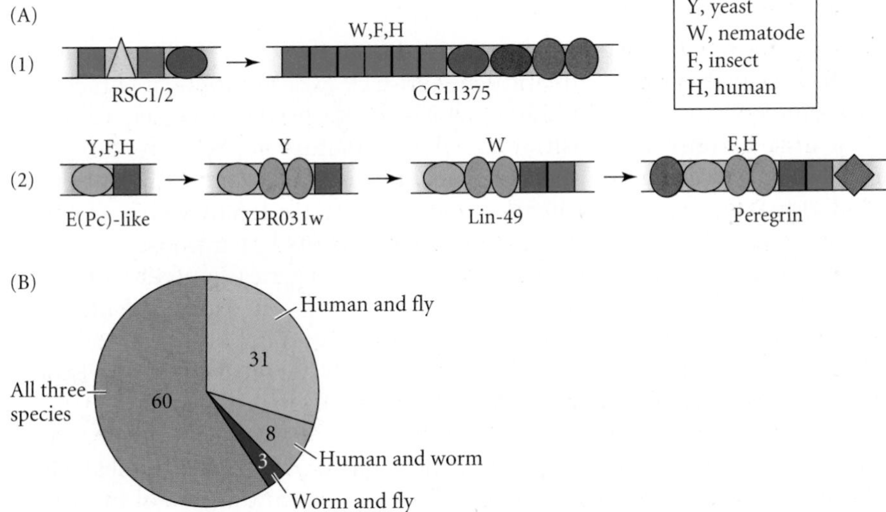

Figure 20.12 Evolution and conservation of domains in diverse proteins. (A) Diversification of chromatin proteins via exon shuffling and domain accretion. Each shape indicates a different protein domain (a bright red oval, for example, indicates a zinc finger domain). Arrows show how new proteins may have evolved by the terminal addition, intercalation, and shuffling of domains. In Group (2), for example, the protein YPR031w differs from the E(Pc)-like protein by the intercalation of two PHD domains (orange) between two enhancers (Ep1 and Ep2; green). The intercalation may have occurred after the origin of separate genes by gene duplication. Next in this series, the Lin-49 protein can be derived from the YPR031w protein by the terminal addition of a bromo domain (blue square). The further addition of an initial zinc finger domain (red) and a terminal BMB domain (brown) distinguishes the protein Peregrin from Lin-49. [These domains are combined into proteins observed in yeast (Y), the nematode *C. elegans* (W), fruit fly (F), and human (H)]. (B) Similarities in domain structure of different proteins such as those in (A) are called "conserved domain architecture." This diagram shows the proportion of chromatin proteins in which domain architecture is shared among three species with fully sequenced genomes. (After International Human Genome Consortium 2001.)

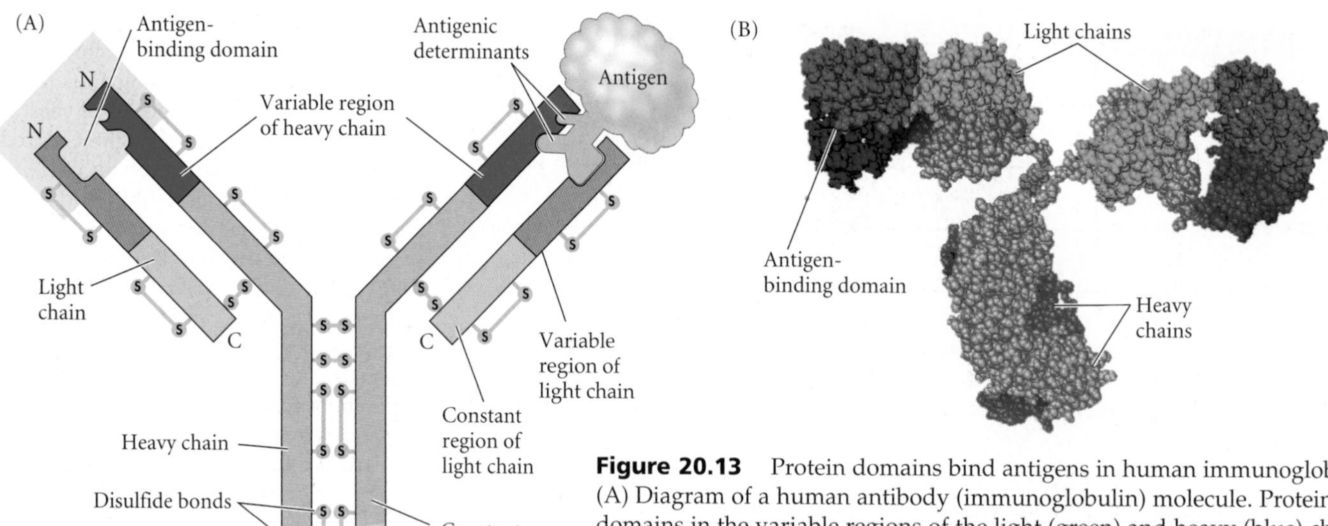

Figure 20.13 Protein domains bind antigens in human immunoglobulin. (A) Diagram of a human antibody (immunoglobulin) molecule. Protein domains in the variable regions of the light (green) and heavy (blue) chains form the cleft (antigen-binding site) that binds foreign proteins during the immune response. The immense variability of these domains is crucial to the antibody function of the adaptive immune sytem of vertebrates. (B) Three-dimensional molecular model in an orientation similar to that of (A).

The approximate correspondence between gene exons and protein domains in some proteins has led to the hypothesis of exon shuffling, which states that much of the diversity of genes has evolved as new combinations of exons have been produced by illegitimate (nonhomologous) recombination that occurs in the intervening introns. Manyuan Long and colleagues (2003) estimate that 19 percent of the exons in eukaryotic genomes have arisen from pre-existing exons via exon duplication and subsequent shuffling. One of the first well-documented examples of such an exon shuffling event in the potato plant (*Solanum tuberosum*) also illustrates how exon shuffling can confer new functionalities on genes. In the potato, cytochrome *c1* has nine exons, the first three of which are responsible for situating the protein in mitochondria, the site of enzyme catalysis. In an extensive search of the protein sequence data bases, Long et al. (1996) found that the first three exons of cytochrome *c1* of potato show significant sequence similarity to the first three exons of an otherwise unrelated gene (*GAPDH*) that encodes glyceraldehyde-3-phosphate dehydrogenase in maize, *Arabidopsis*, and other plants. Thus, the organelle-targeting function of a leader sequence of cytochrome *c1* in most other plants has been taken over by a sequence acquired from *GAPDH* in the origin of potato.

Gene chimerism

A **chimeric gene** is one that consists of pieces derived from two or more different ancestral genes. We have seen that some chimeric genes arise by exon shuffling, in which case the exons are separated by introns, the sites at which genetic exchange occurs. Chimeric genes may also arise by retrotransposition, in which a mature mRNA is reverse-transcribed into cDNA and inserted into another gene. The result is a gene with contiguous sequence—individual exons or entire genes—that is uninterrupted by introns, yet corresponds closely to sequences that in other genes are interrupted by introns.

The first discovered example of this process was the *jingwei* gene (Figure 20.14), which is found only in *Drosophila teissieri* and *D. yakuba* and consists of four exons. The first three exons, which are homologous to the *yellow-emperor* gene found in both these and other *Drosophila* species, have a length typical for *Drosophila* exons, whereas the fourth exon is as long as an entire typical gene. The sequence of this fourth exon is more than 90 percent similar to the entire coding sequence of a well-studied gene, *alcohol dehydrogenase* (*Adh*). Although *Adh* contains multiple introns, the fourth "exon" of *jingwei* is intronless: clearly, a retrotransposed copy of *Adh* landed in the middle of the third intron of the *yellow-*

emperor gene in the ancestor of *D. teissieri* and *D. yakuba*, producing a chimeric gene. The first three exons of *yellow-emperor* modified the expression pattern of the retrotransposed *Adh* gene, because *jingwei* exhibits the precise testis-specific expression pattern shown by *yellow-emperor*.

Even without corroborating evidence of a related gene possessing introns, any intronless gene is likely to have originated by retrotransposition. For example, most of the large and diverse family of G-protein coupled receptors (GPCRs) found in humans and other mammals are intronless, whereas many invertebrate GPCRs have introns (Gentles and Karlin 1999). It is likely that the mammalian gene family originated by retrotransposition of an ancestral gene that had introns. Approximately 1 percent of the genes in the human and mouse genomes show signatures of retrotransposition, such as a lack of introns compared with their progenitor genes. These genes are actively transcribed, often in novel tissues and contexts, suggesting that retrotransposition may play an important role in the origin of new genes.

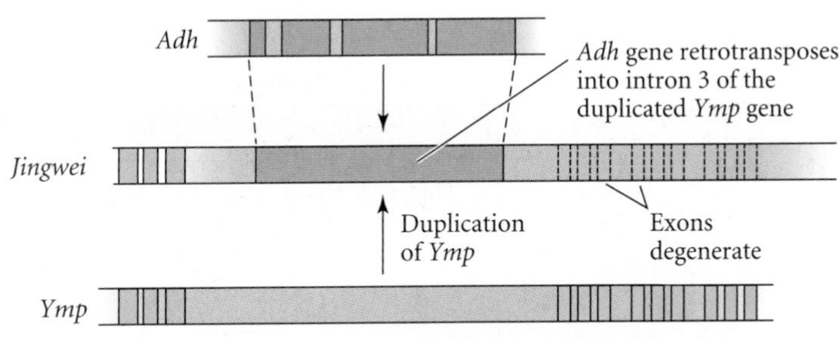

Figure 20.14 Origin of a new *Drosophila* gene, *jingwei*, via retrotransposition of a pre-existing gene into an intron of *Ymp* (*yellow-emperor*) to recruit new exons. Two copies of *Ymp* arose by gene duplication. An ancestral *Adh* gene retrotransposed into intron 3 of one of these copies of *Ymp* approximately 2.5 Mya. After the retrotransposition event, the exons downstream of the novel *Adh* exon of *Ymp* degenerated because the *Adh* transcript provided a new stop codon. Although *Adh* functions in regulation of alcohol tolerance and the function of *Ymp* is largely unknown, the function of *jingwei* is now known to be related to the metabolism of hormones and pheromones (Zhang et al. 2004), a new function that has evolved in the past 2 to 3 Myr. (After Long et al. 2003.)

Many such retrotransposition events result in new but nonfunctional sequences called **processed pseudogenes**. Although their DNA sequence may resemble that of related genes (from which they were originally copied), processed pseudogenes commonly have deletions that destroy the reading frame of the gene and stop codons that occur before the correct termination point. Processed pseudogenes are common in the human and other eukaryotic genomes; in humans there are at least 8000 processed pseudogenes. Ultimately, the ancestral genes from which processed pseudogenes arise, and even their identity as such, becomes unrecognizable because they accumulate mutations that erode sequence signatures of their origin.

Motif multiplication and exon loss

The multiplication of specific motifs within genes can give rise to new genes with new functions. The notothenioid fishes of the Southern Ocean, near Antarctica, are famous for their ability to thrive at ocean temperatures at which normal blood would freeze. They have evolved a variety of antifreeze glycoproteins (AFGPs)—short polypeptides that serve to break up ice crystals and prevent the fishes' blood from freezing. *AFGP* genes encode short three-amino-acid monomers (threonine-alanine-alanine, or ThrAlaAla) repeated over and over, comprising 234 amino acids in all. They share several features in common with the notothenioid gene for trypsinogen, which has a completely different function as a protease.

The entire beginning of the trypsinogen gene, extending as far as the beginning of exon 2, resembles structures in the beginning of the *AFGP* gene, including a single threonine-alanine-alanine motif at the start of trypsinogen exon 2. Additionally, the sequence of exon 6 and the 3′ end of the *AFGP* gene resembles that of trypsinogen. However, the *AFGP* gene lacks any equivalent of the central exons of the trypsinogen gene. The evidence suggests that the single triplet-amino-acid motif in exon 2 of an ancestor of the trypsinogen gene expanded to produce exon 2 of the *AFGP* gene. At some point during or after this event, exons 3 to 5 of the ancestral trypsinogen gene were lost. In fact, a chimeric *AFGP* gene has been found in one species, the giant Antarctic toothfish (*Dissostichus mawsoni*), which seems to represent an evolutionary intermediate in the transition from a trypsinogen-like ancestor and a modern *AFGP* gene (Figure 20.15). Such genes are believed to have played a crucial function permitting the notothenioids to colonize and diversify in the extremely cold ocean waters surrounding Antarctica.

A striking example of domain multiplication has occurred in the lineage leading to modern humans, and is implicated in the origin of higher cognitive functions that distinguish humans from other great apes. A bioinformatic survey conducted by Magdalena Popesco and colleagues (2006) revealed 134 genes that appear to have undergone extensive duplication in humans after their divergence from chimpanzees and gorillas. Further bioinformatics analysis, which was confirmed by direct quantification of cDNA abun-

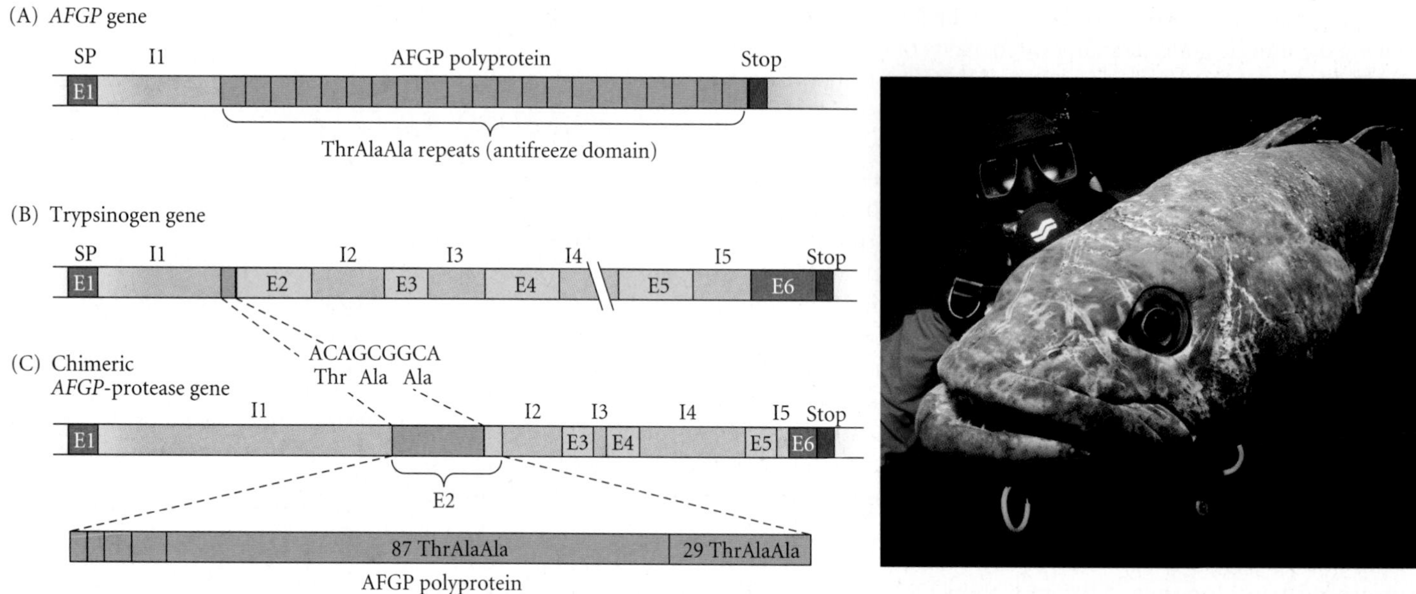

Figure 20.15 The evolution of *AFGP* genes of Antarctic notothenioid fishes. (A) A typical functional *AFGP* gene from an Antarctic notothenioid, showing the repeated ThrAlaAla motifs, encoded in exon 2 (E2), that comprise the antifreeze domain. (B) A typical trypsinogen gene showing the single ThrAlaAla motif at the beginning of exon 2, as well as exons 1 and 6 (in red) that are also found in the *AFGP* gene. (C) A chimeric *AFGP* gene from a giant Antarctic toothfish (*Dissostichus mawsoni*). The exon in blue, encoding the ThrAlaAla repeats, is a manyfold expansion of the single motif located just upstream of exon 2 in the trypsinogen gene. The *Dissostichus AFGP* gene possesses many of the exons of a typical trypsinogen gene but has a hybrid second exon (E2) consisting mostly of an expanded triplet motif antifreeze domain. (After Cheng and Chen 1999; photo © Rob Robbins.)

dances in tissues of humans and other primates, showed that a conspicuous protein domain of unknown function, called *DUF1220,* was present in 34 different human genes and totaled around 50 copies in the human genome, whereas fewer than 12 copies of it were present in other primate species (Figure 20.16). Statistical analysis showed that the proliferation of *DUF1220* domains in the human genome is marked by high rates of nonsynonymous substitution, implicating adaptive evolution. Furthermore, expression experiments showed that RNA transcripts containing the *DUF1220* domain were expressed extensively throughout regions of the human brain associated with higher cognitive functions, such as the hippocampus and neocortex. Although the function of *DUF1220* is at present unknown, these evolutionary clues suggest that *DUF1220* represents a domain amplification that occurred in the approximately 6 Myr period after the divergence of humans from great apes and has conferred novel, derived cognitive functions in humans.

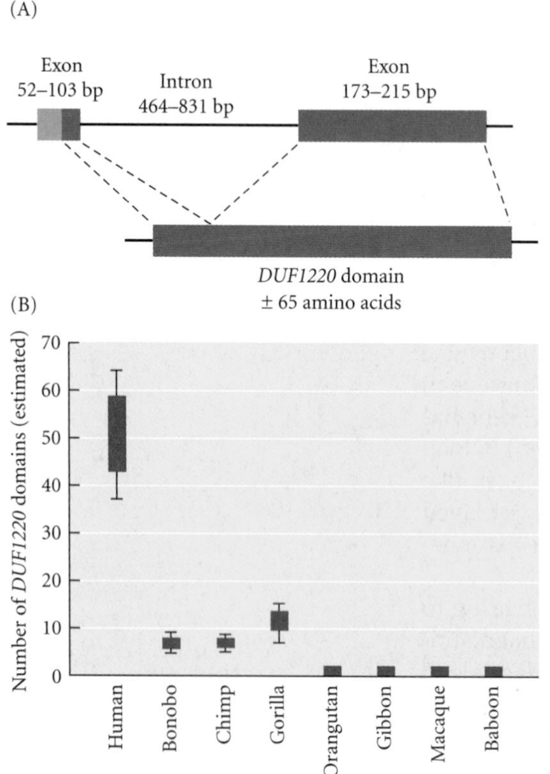

Figure 20.16 Amplification of the *DUF1220* domain in the human lineage. (A) The relationship of exons to domain structure of proteins containing *DUF1220*. The domain is encoded by two exons in this example. (B) The number of *DUF1220* domains in various primates estimated by direct analysis of levels of *DUF1220* transcript. The domain is estimated to be 4 to 5 times more numerous in the human genome than in the genomes of the great apes and other Old World primates. (After Popesco et al. 2006.)

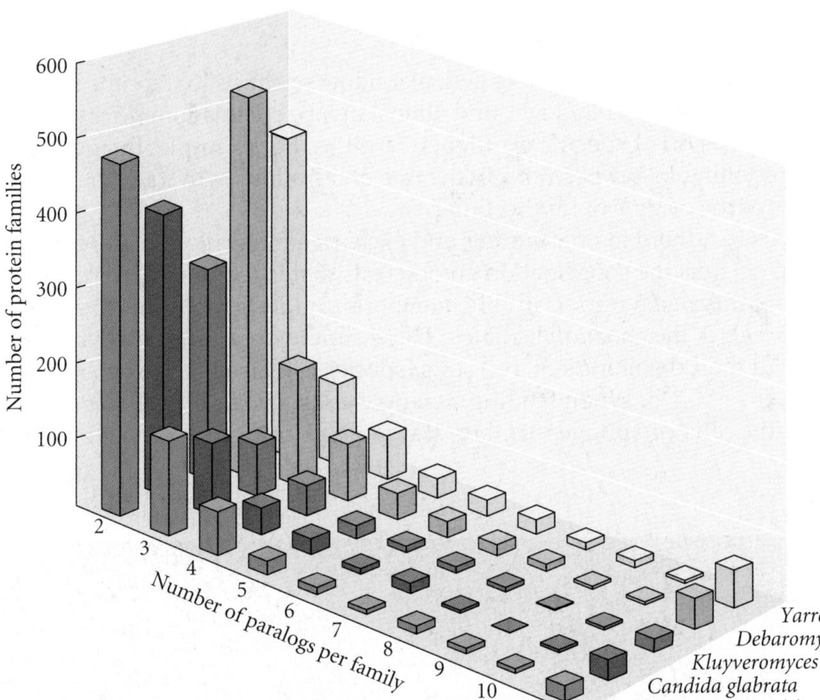

Figure 20.17 Distribution of the number of paralogs in the complete genomes of five species of yeast. This distribution has a similar shape in many organisms in which it has been tested, except that the largest size classes of paralogs are larger in complex eukaryotes such as humans than in simpler eukaryotes such as yeast. The diagram suggests that differential amplification of genes can take place in different species, often as a result of adaptation to different environments. (From Dujon et al. 2004.)

The Evolution of Multigene Families

Gene duplication

One of the most common ways in which new genes originate is by **gene duplication**, in which new genes arise as copies of pre-existing genes. Many genes are members of larger groups of genes, called **multigene families**, that are related to one another by clear ancestry and descent, and that often have diverse functions that nonetheless have a common theme. For example, globin genes (see Figure 8.3) all encode proteins that have a heme-binding domain and can bind oxygen. In mammals, the ε-, ζ-, and γ-globin chains have higher oxygen affinities and are expressed in embryonic tissues, whereas the α- and β-globins function in the adult. The molecular mechanisms of gene duplication are still poorly understood, although it occurs at the DNA level (since the members of gene families usually have similar intron-exon structure), and unequal crossing over (see Figure 8.6) is known to change copy number in multigene families.

The relationships among members of a multigene family can be analyzed phylogenetically, both among and within species. These two kinds of relationships among genes represent two forms of homology: *orthology* and *paralogy* (see Chapter 3). Orthologous genes found in different species have diverged from a common ancestral gene by phylogenetic splitting at the organismal level. In contrast, the members of multigene families are paralogous, and originated from a common ancestral gene by gene duplication.

Multigene families vary greatly in size (Figure 20.17), from just two members (the most common number in both yeast and human genomes; Gu et al. 2002; Dujon et al. 2004) to as many as 800 (e.g., the immunoglobulin superfamily) or 1000 or more (e.g., human ribosomal RNA genes). The size of multigene families often coincides with specific adaptations, or loss of adaptations. For example, mammals have hundreds of genes that encode olfactory receptor proteins, each of which binds one or a few odorant chemicals. In contrast, birds, most of which are less sensitive to smell than mammals, have relatively few olfactory receptor genes (Hillier et al. 2004).

Multigene families and the origin of key innovations

Genomes record in remarkable detail the molecular underpinnings of some of the major transitions of life, and also allow biologists to immediately grasp the molecular basis of novel adaptations. Biologists have now sequenced the genomes of several species falling in important positions in the tree of life; these genomes allow scientists to trace the increasing complexity of life at the molecular level, and almost invariably this complexity has its foundation in the expansion of specific multigene families. For example, the genomes of several organisms falling in key positions in the tree of life allow us to track the molecular underpinnings of the origins of multicellularity.

The ability of cells to adhere to one another and exchange molecules critical for conveying information permits the coordination among cells that is required for a functioning multicellular organism. An important lineage that provides insight into the origin of multicellular animals is the choanoflagellates. These simple organisms always have a single-celled stage in their development, but some species can also occur as colonies in the adult stage (Figure 20.18). When feeding as single cells, choanoflagellates closely resemble the feeding cells of sponges, among the most primitive of metazoans. But

(A)

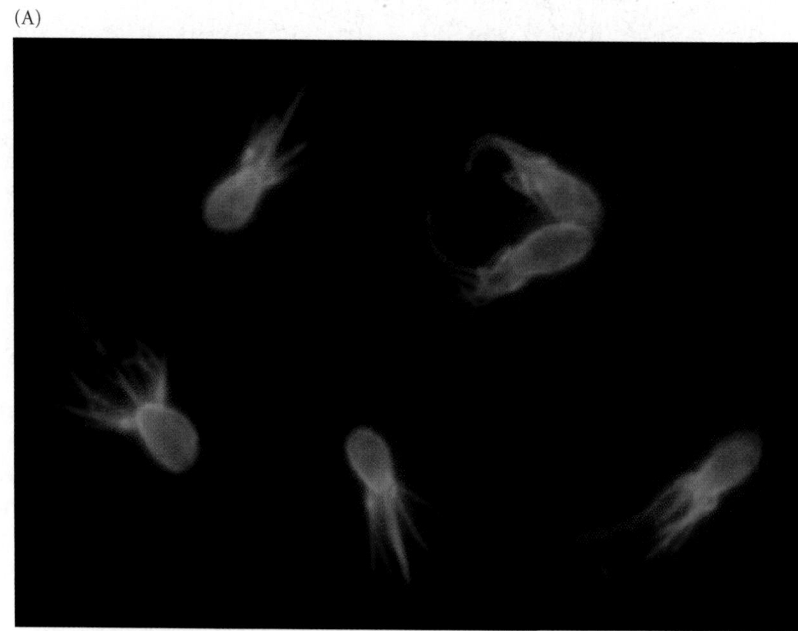

(B)

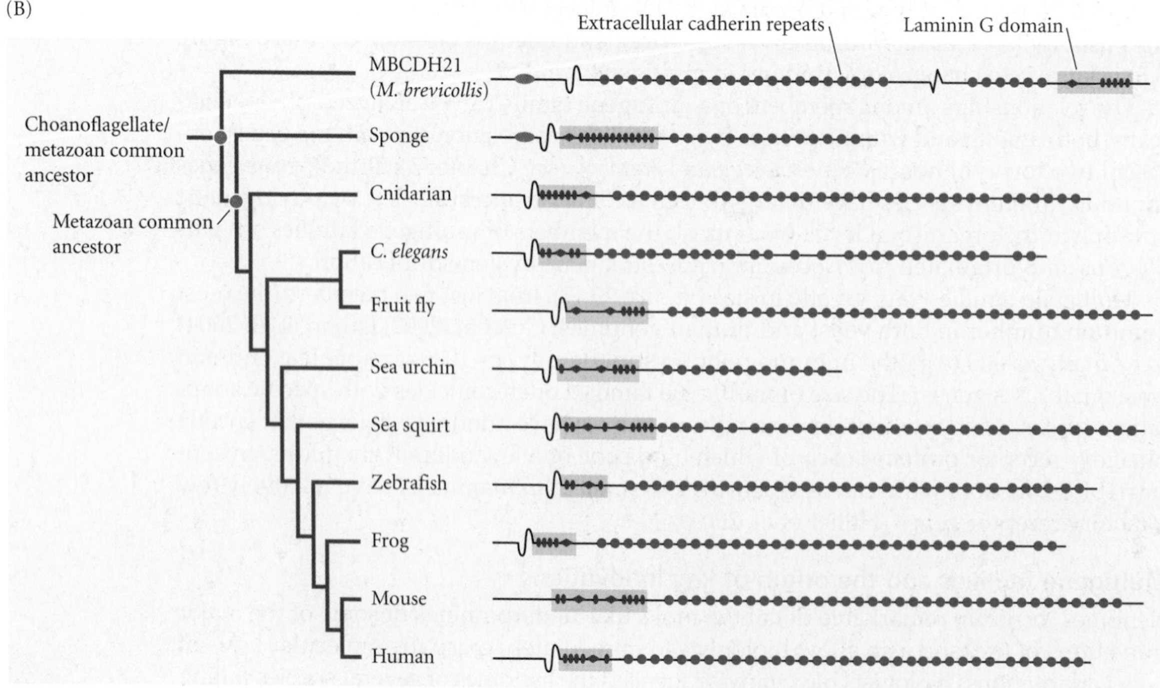

Figure 20.18 Ancient origin of cadherin genes as revealed in the genome of the choanoflagellate *Monosiga brevicollis*. (A) Like other choanoflagellates, *M. brevicollis* has a single-cell stage but also forms multicellular colonies. (B) Sister-group relationship of choanoflagellates to metazoans and structure of a fat-related cadherin protein, a type of cadherin that contains epidermal growth factor, laminin G domains (green boxes), as well as domains linked to extracellular cadherin repeats (purple). MBCDH21 refers to a specific transcript discovered in the *M. brevicollis* genome. (A photo courtesy of N. King; B after Abedin and King 2008.)

choanoflagellates are not metazoans; they fall outside the sponges and cnidarians at the base of the metazoan tree. Thus the genome of choanoflagellates was predicted to shed light on the earliest stages of multicellularity.

Cadherins are a class of molecules that play important roles in cell adhesion, movement, and communication. Prior to the sequencing of the first choanoflagellate genome, cadherins were known only from metazoans. Nicole King and colleagues published the first genome from a choanoflagellate (*Monosiga brevicollis*) in 2008 (King et al. 2008). To their surprise, they discovered 23 cadherin genes, constituting an important component of a genome that contains only a little over 9,000 genes (Abedin and King 2008). The multifunctional nature of the cadherin multigene family was evident in the many types of domains found in cadherin genes, including domains for growth factors and immunoglobulins. These diverse types of cadherin genes encoded a variety of proteins that transmit changes in the environment to intracellular processes by signaling between molecules. *M. brevicollis* is not known to engage in any contacts between cells, indicating that the diversity of cadherins in this species must have functions other than cell-cell communication. Thus the molecules enabling communication between cells in complex metazoans must have originally had different functions, with later functionality added to an initial, simpler toolkit. Thus genomes reveal a pattern in which a trait with a set of current functions originally had different functions—a pattern frequently encountered in the history of life (see Chapter 22).

Gene and Genome Duplication

Duplication of whole genomes and chromosomal segments

Many genes get duplicated as parts of large chromosomal blocks, or even as parts of whole genomes (polyploidy; see Chapter 18). Evidence for such events comes from finding that many duplicated genes seem to have diverged at approximately the same time, based on comparisons of DNA sequences. Whether an entire genome was duplicated in a "big bang" or much of it was duplicated piecemeal over a long time (the "continuous mode" model) can be determined by examining the age distribution of divergences between many duplicated paralogs, making judicious use of molecular clocks. For example, it has been postulated that the entire genome was duplicated in the common ancestor of jawed vertebrates and then again in the ancestor of the fishes (Prince and Pickett 2002). Gu and colleagues (2002) estimated the time of duplication for 1739 gene duplication events inferred from a phylogenetic analysis of 749 different gene families in the human genome. The distribution of divergences between paralogs provided evidence for both the "big-bang" hypothesis (a major peak in the frequency distribution at about 500 Myr ago, before the origin of fishes) and the "continuous mode" hypothesis (the continuous distribution of divergence times, including a peak near the base radiation of mammals; Figure 20.19). Thus the large-scale structure of mammalian genomes may have been produced by both a large- and many small-scale duplications.

Genome duplication and reduction have played important roles in the evolution of genes and genomes of plants. The genome of the black cottonwood (*Populus trichocarpa*), an important model in the genetic study of woody plants, was recently sequenced and found to contain more than 45,000 genes, about twice that estimated for humans, despite the fact that its genome is only about 500 Mb, which is only one-sixth the size of the human genome. About 8000 pairs of genes produced as a result of a whole-genome duplication continue to survive in the cottonwood genome, for reasons that are still unclear. The cottonwood genome contains on average one and a half putative paralogs for every one ortholog in *Arabidopsis*, a weedy mustard whose genome has also been sequenced. In contrast to the cottonwood genome, *Arabidopsis* has jettisoned many genes after its genome underwent a duplication in an ancestral species (see Figure 22.21). Clearly, whole-genome duplication and widespread gene deletion can have important effects on the gene complement and the number of genes available for adaptive evolution.

In addition to whole-genome duplication events, specific segments of chromosomes can duplicate and carry genes with them. Large duplications (paralogous regions) often

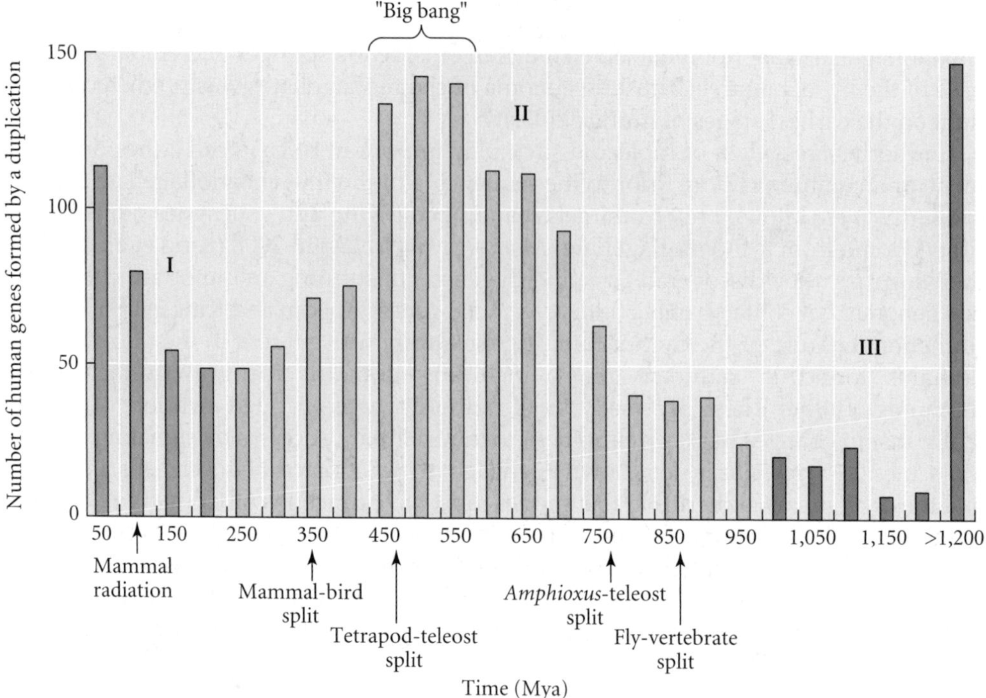

Figure 20.19 Use of age distribution of gene duplication events to infer whole-genome duplications. The *x*-axis shows the estimated age of divergence of gene duplicates in the human genome based on analysis of sequence differences. The *y*-axis shows the number of gene duplicates in each age class. The figure shows three waves, I, II, and III. Wave I corresponds to gene duplications that occurred during the radiation of mammals and consists of certain large gene families, such as the immunoglobulin family in mammals. Wave II corresponds to duplications that took place during early vertebrate evolution and consists of tissue-specific isoforms and other developmental loci. Wave III corresponds to duplications that took place very early in metazoan evolution, when several novelties in signal-transduction pathways are thought to have originated. (After Gu et al. 2002.)

contain hundreds of genes. For example, interspersed among the many pathogen-resistance genes of the major histocompatibility complex (MHC) on human chromosome 6 are many housekeeping genes such as collagen genes. These housekeeping genes were present in the same order on the ancestral chromosomal segment that gave rise to the MHC region. Additional sets of genes, in the same order, have been found on chromosomes 1, 9, and 19 as well (Figure 20.20). For several of the housekeeping genes, the existence of such paralogous regions has been detected in invertebrates such as the nematode *C. elegans*, which suggests that such paralogous regions are ancient components of metazoan genomes. What distinguishes the MHC from these paralogous regions is the presence of numerous genes involved in pathogen resistence that apparently were inserted and duplicated in one of the paralogous regions. MHCs are only found in jawed vertebrates, but the chromosomal region containing them may have a very ancient origin.

Copy number variants are emerging as an important source of variation in the human genome. They contribute to variation among individuals and other primate species 8and are also implicated in human cancers and disease. Recent genome-wide surveys in humans have revealed that somewhere on the order of 1500 regions spanning about 360 Mb, or about 12 percent of the genome, consist of chromosomal segments that have duplicated and diverged from paralagous ancestral segments.

Chromosome arm			
6p (MHC)	19p	1q	9q
RXRB	–	*RXRG*	*RXRA*
COL11A2	–	*COL11A1*	*COL5A1*
RING3	–	–	*RING3*-like
LMP2/LMP7	–	–	*PSMB7*
TAP1/TAP2	–	–	*ABC2*
NOTCH4 (INT3)	*NOTCH3*	*(NOTCH2)*	*NOTCH1*
PBX2	–	*PBX1*	*PBX3*
TNX	–	*TNR*	*HXB*
CYP21	*CYP2*	–	–
C4A/C4B	*C3*	–	*C5*
HSPA1	–	*(HSPA6/7)*	*GRP78*
HLA-A,-B,-C	–	*CD1*	–
–	*VAV1*	–	*VAV2*
–	*LMNB2*	*LMNA*	–
–	–	*SPTA*	*SPTAN1*
–	–	*ABL2*	*ABL1*

Paralogous *Notch* genes are present on all four chromosomes and occur upstream of a gene family known as PBX.

Figure 20.20 Block duplication. Paralogous genes belonging to a number of gene families are listed for four human chromosomes. Many of these paralogs occur in the same relative positions to one another on these chromosomes, suggesting duplication of an ancestral block. (After Kasahara 1997.)

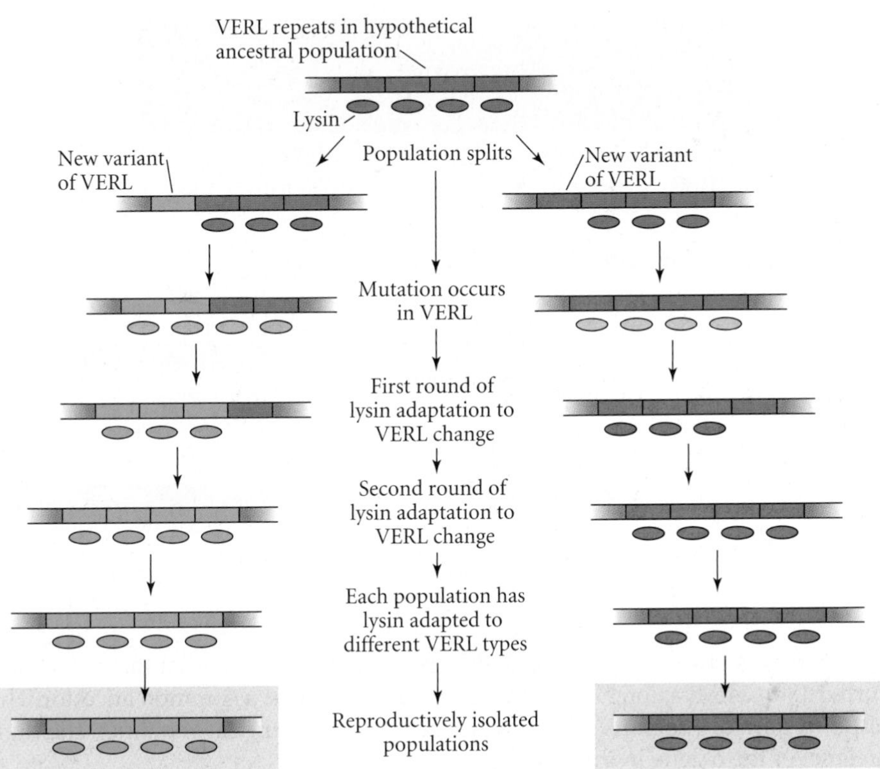

Figure 20.21 Evolution of species-specific differences in coevolving lysin and VERL proteins. Lysin, which is initially expressed on the acrosome of sperm, is released and then binds to VERL protein, in the vitelline envelope of the egg. Binding of VERL by lysin occurs in a species-specific manner, necessitating sequence changes in lysin when the conspecific VERL evolves. The several VERL repeats acquire the same new sequence, a process known as concerted evolution that occurs as a result of gene conversion. Lysin molecules adapt to VERL repeats until they match their respective VERL array. If enough divergence has occurred, lysin of one population will not interact properly with VERL of the other population, resulting in reproductive isolation and consequent speciation. (After Swanson and Vacquier 2002.)

Possible fates of duplicate genes

Duplicate genes can have several possible fates: (1) They may diverge in sequence and (usually) in function, in several different ways that we will describe in the next section. (2) One copy may remain functional, while the other becomes a nonfunctional pseudogene. It is also possible for a locus to be deleted. Thus genes may undergo "birth" and "death," resulting in a turnover of the membership of a gene family. (3) They may undergo **gene conversion**, which occurs when sequence information from one locus is transferred unidirectionally to other members of the gene family, so that all acquire essentially the same sequence. The molecular process of gene conversion is poorly understood in most organisms, but its consequence, **concerted evolution** of the gene family, is well known. It results in production of the same gene product from multiple loci, which can be adaptive if large quantities of the product are needed. Ribosomal RNA, a major component of ribosomes, is an example and is produced by a large multigene family. It would be deleterious if different transcripts of rRNA had different sequences. Abalone VERL protein, which consists of a series of amino acid motifs that have undergone extensive duplication but are still very similar in sequence to one another, is another example of concerted evolution (Figure 20.21).

The globin gene family of many animals is known to undergo concerted evolution and frequent gene conversion, sometimes with observable consequences for gene expression patterns. For example, the δ-globin gene of many primates is typically expressed at a very low level, whereas the β-globin gene is highly expressed. However, in galagos (a phylogenetically basal branch of African lemuroid primates), the δ-globin gene is highly expressed. To understand the basis of this pattern, Morris Goodman's group (Tagle et al. 1991) sequenced the δ- and β-globin genes of a galago and compared these with each other and with the pattern of divergence observed between these genes in humans. Whereas in humans the coding regions, upstream regulatory regions, and intron 1 are highly divergent between the δ- and β-globin genes, these same regions are virtually identical for the galago genes (Figure 20.22). By scrutinizing the sequence differences between the galago δ and β genes, it was determined that a conversion tract spanning the first third of the gene had been transferred from the β to the δ gene. By applying a molecular clock to the

Figure 20.22 Evidence for localized gene conversion and concerted evolution in the primate globin gene family. (A) Percent sequence divergence between various regions of the δ- and β-globin genes in humans and in a galago. Note the high sequence similarity between the 5′ upstream region, the coding regions, and intron 1 of the galago genes. (B) Conversion of the first third of the galago δ gene by the galago β gene, in effect converting it into another β gene. The bracket above the genes before gene conversion indicates the region participating in the conversion event. Colors of the exons and gray and white denoting introns indicate the origin of various regions. The hybrid origin of the galago δ gene (bottom left) is indicated by the multiple colors. (After Tagle et al. 1991.)

(A)

	Percent sequence divergence			
Globin pair	5′ region	Coding regions	Intron 1	Intron 2
Human δ - Human β	14.8	23.0	12.4	—
Galago δ - Galago β	0.8	5.0	0.0	18.0

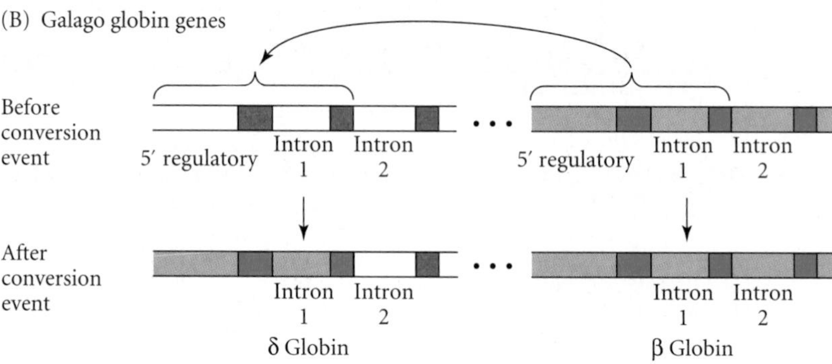

noncoding differences observed in intron 2, the researchers estimated that the conversion event occurred 18 to 24 Mya, long after galagos and humans had a common ancestor. The fact that the 5′ regulatory regions of the galago genes were identical explained the high expression level of the δ gene in this species.

When duplication occurs in the common ancestor of two or more species, these several fates of duplicate genes result in different phylogenetic patterns:

1. When gene duplication precedes speciation and the duplicates diverge in sequence both within and between species, the major clusters in the resulting phylogenetic tree will correspond to the different paralogs. Within each paralog, the phylogenetic relationships of the species sampled will be reflected (Figure 20.23A).
2. If loci have undergone concerted evolution, synapomorphic mutations that occurred in any ancestral species will be shared by all paralogs in all the species derived from that ancestor. Therefore, a tree based on these sequences will display the species' phylogeny, and the paralogs will cluster within each species (Figure 20.23B).
3. If some paralogs are lost in one or more species, or have not been characterized because their sequence has diverged so much that they have not been recognized as paralogs, the phylogenetic tree may display an intermingling of functional and nonfunctional genes (Figure 20.23C).

Figure 20.23 Phylogenetic consequences of duplication, speciation, and gene conversion in gene families. In each diagram, individual paralogs are indicated by Greek letters and a distinct color, and different species in which a given paralog is found are indicated by a number. This α1 is the copy of gene α in species 1; β2 is the copy of gene β in species 2. (A) In the divergent evolution model, gene duplication takes place before speciation. The result is a tree in which the paralogs from different species form separate clusters, within which the genes share the same phylogenetic relationships as the species do. (B) In the concerted evolution model, frequent interparalog gene conversion results in very closely related genes occurring in the same genome. Converted genes are colored purple because sometimes only portions of genes are converted, as in Figure 20.22. Such genes cluster closely with one another in a phylogenetic analysis and result in clusters of genes that are largely species-specific. (C) In the birth-and-death model, pseudogene formation balances gene duplication, the result being an equilibrium number of functional genes in a given genome. Pseudogene formation results in a patchwork of functional genes (colored circles) and pseudogenes (white circles) in the phylogeny.

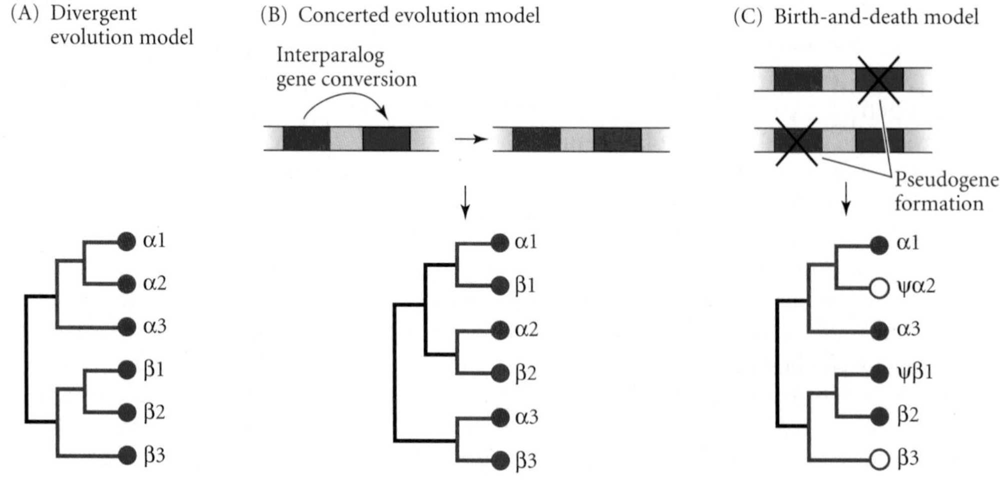

Selective fates of recently duplicated loci

Paralogous genes are initially redundant: when gene duplication occurs, two identical copies of a gene suddenly exist in a genome, opening the possibility for functional diversification if one of the copies does not become a pseudogene. This may occur according to two models.

Susumo Ohno (1970) first articulated the classical model of **neofuntionalization**, whereby one of the duplicates retains its original function and the other acquires a new function, because of fixation of certain new mutations. Evidence of this process is the rapid accumulation of nonsynonymous substitutions in only one of the recently duplicated copies. This process will occur only if advantageous new mutations occur before one of the duplicates loses function, and becomes a pseudogene, because of fixation of disabling mutations. Two genes in the ribonuclease gene family in primates, eosinophil cationic protein (ECP) and eosinophil-derived neurotoxin (EDN), provide an example. Zhang and colleagues (1998) found that the *ECP* gene experienced a large number of nonsynonymous substitutions after duplication from the *EDN* gene. Moreover, ECP possesses an antipathogen function not found in EDN, suggesting that functional diversity was acquired after duplication by rapid accumulation of amino acid substitutions.

Force and coworkers (1999) have proposed another mode of adaptive divergence, called **subfunctionalization**, whereby each gene duplicate becomes specialized for a subset of the functions originally performed by the ancestral single-copy gene. This model, called the DDC model (for duplication-degeneration-complementation), hypothesizes that an ancestral gene had two or more functions, and that in each paralog, complementary mutations are fixed that reduce or eliminate a different function. The paralogs are therefore no longer redundant, so both are preserved by natural selection, and may later undergo further functional specialization and evolutionary change. The DDC model differs from the classical model in that both duplicate loci are expected to undergo changes in sequence and function compared with the ancestral gene. It also predicts a higher rate of retention of gene duplicates. In polyploid plants, more than 15 percent of gene duplicates tend to be functionally retained, a much higher rate than predicted by the classical model (Prince and Pickett 2002).

In the zebrafish, the single *Hoxb1* gene of other vertebrates such as mice has been duplicated into *Hoxb1a* and *Hoxb1b*. Whereas mouse *Hoxb1* is expressed continuously in the developing hindbrain, zebrafish *Hoxb1a* and *b1b* are expressed sequentially, with *b1a* terminating about 10 hours after fertilization, at which point *b1b* takes over (Prince 2002; Prince and Pickett 2002). *Hoxb1a* and *b1b* have each lost single regulatory sequences upstream and downstream of the gene, both of which are present and functional in mouse *Hoxb1*. The complementary expression profiles and complementary degenerate mutations of the zebrafish genes exemplify subfunctionalization (Figure 20.24).

Rates of gene duplication

The great diversity of large multigene families indicates that gene duplication is a common and ubiquitous process. Michael Lynch and colleagues (2000) estimated that the rate of gene duplication in the nematode *C. elegans*, humans, and other recently sequenced genomes is about 0.01 duplication per gene per Myr, much higher than previously thought. However, Innan and colleagues (Gao and Innan 2004; Osada and Innan 2008) have found unexpectedly high rates of gene conversion between duplicated paralogs in yeast and *Drosophila*. These conversion events make the paralogs appear much more sim-

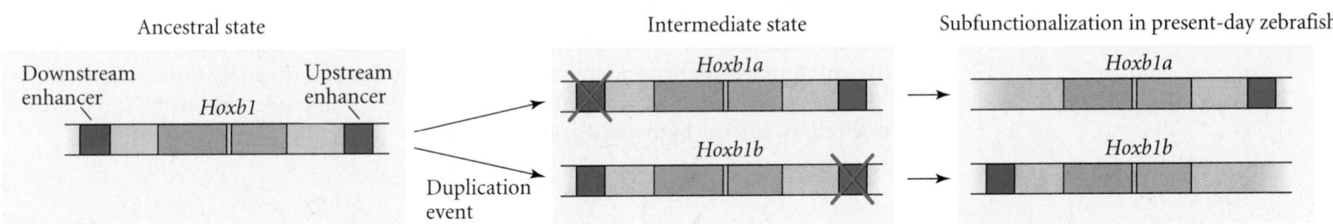

Figure 20.24 The DDC model of gene duplication as illustrated by *Hox* genes. In an ancestral vertebrate, the *Hoxb1* gene had dual functions and was regulated by dual upstream and downstream promoter motifs. In an intermediate derived species, this gene underwent a duplication event, followed by loss of one of these functions and change of expression pattern in the two novel paralogs (center). This loss is indicated by the crossing out of either the blue or green promoter motif, respectively. The derived state of the duplicated genes (right) consists of two paralogs, each with a single function—a process known as subfunctionalization. (After Prince and Pickett 2002.)

ilar in sequence—and hence much younger—than Lynch and colleagues initially thought, suggesting a much lower overall rate of gene duplication in these organisms.

Regardless of the challenges in estimating rates of gene duplication, and even though the vast majority of new duplicates simply degenerate and do not contribute to the functional genome, adaptive divergence of duplicate genes is clearly one of the keystones of adaptive diversification at the molecular level. The study of gene duplication illustrates the rapid pace of comparative genomics, and how multiple types of data and inference are required to refine our picture of genome evolution.

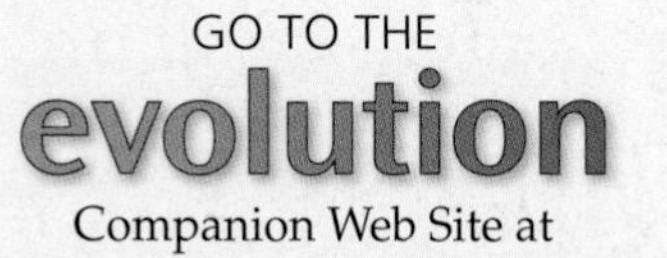

Companion Web Site at
www.sinauer.com/evolution
for quizzes, data analysis and simulation exercises, and other study aids.

Summary

1. New technologies that drive forward the field of genomics enable biologists to study genomes on hitherto unprecedented scales, and have revealed several new elements of genomes, such as diverse classes of untranslated RNA sequences. The structures of thousands of genes and entire genomes can be compared with one another to search for patterns of ancestry and evolution.
2. A true "theory of genomes" is still in its infancy. A major hypothesis that has emerged in recent years is that apparently deleterious trends, such as the accumulation of transposable elements, have been driven by genetic drift in lineages with small population size. Although not a universal explanation for genome characteristics, this idea makes specific predictions that can be tested with further data.
3. Proteins are composed of a diversity of building blocks called domains and exhibit widely varying rates of evolution. Level of gene expression (a proxy measure of protein abundance) seems to correlate negatively with a protein's evolutionary rate. This implies that highly expressed proteins may be under strong purifying selection to maximize translational robustness, that is, to minimize the occurrence of missense mutations that could alter a protein's folded three-dimensional structure.
4. Genome size varies by several orders of magnitude across life forms. The C-value paradox refers to the discrepancy between genome size and organismal complexity in eukaryotes. It was resolved by noting that the coding portion of genomes may increase with organismal complexity, whereas the noncoding portion, comprising highly repetitive DNA, transposable elements, and other types of "selfish DNA," varies with features other than complexity, such as population size.
5. New genes arise in genomes through a variety of mechanisms. Lateral gene transfer occurs when a gene is transferred between completely unrelated genomes, presumably by viruses or other genomic vehicles. New genes can also arise from pre-existing genes by exon shuffling of domains. Such shuffling can create chimeric genes with novel functions.
6. Genes can also arise by retrotransposition, a process that can generate new intronless genes and often produces processed pseudogenes. Such retrotransposed genes can sometimes produce new exons and regulatory regions up- and downstream, thereby gaining functions and expression patterns that differ from those of the progenitor gene.
7. Genes can duplicate individually or as parts of large chromosomal regions and sometimes as part of whole-genome duplications. Gene duplication is frequent in eukaryotic and prokaryotic genomes and provides an opportunity for the generation of novel genes with derived functions. Gene duplication is the major mode of growth of multigene families, which are the most complex expression of coding region diversity in genomes. Variation in the copy number of chromosomal segments is also a major source of genomic variation in species.
8. Duplicate genes sometimes undergo concerted evolution, wherein all or part of the DNA sequences of a gene are transferred unidirectionally to other members of the multigene family. Like gene duplication, concerted evolution has important consequences for the phylogenetic relationships of genes in gene families and can be inferred by phylogenetic analysis of orthologs and paralogs from multiple species.
9. Multigene families have been responsible for enabling some of the major transitions of life, such as the origin of multicellularity and immunity. Multigene families diversify through gene duplication, and the acquisition of new functions is sometimes accompanied by strong signatures of positive selection, such as high values of $\omega = d_n/d_s$.

10. Neofunctionalization is the process whereby a newly duplicated gene acquires a new function relative to its ancestral gene. In subfunctionalization, by contrast, new gene duplicates undergo complementary degenerative mutations that knock out one of several functions present in the ancestral gene; thus both duplicates retain one of the ancestral gene's functions and are preserved. Subfunctionalization is thought to be common among duplicated genes.

Terms and Concepts

alternative splicing
chimeric gene
codon bias
comparative genomics
concerted evolution
copy number variants
domain
domain accretion
exon shuffling
gene conversion
gene dispensability
gene duplication
lateral gene transfer
microRNAs
multigene family
neofunctionalization
processed pseudogenes
pyrosequencing
"selfish DNA"
spliceosomal intron
subfunctionalization
translational robustness

Suggestions for Further Reading

The Origins of Genome Architecture by Michael Lynch (Sinauer Associates, Sunderland, MA, 2007) provides a forceful argument for the importance of genetic drift and nonadaptive processes in the evolution of genomes. *Fundamentals of Molecular Evolution* by D. Graur and W.-H. Li (Sinauer Associates, Sunderland, MA, 2000) is a highly readable introduction to many aspects of sequence evolution. Many of these topics are treated in greater depth by A. L. Hughes in *Adaptive Evolution of Genes and Genomes* (Oxford University Press, Oxford, 2000). In *A Primer of Genome Science* (third edition, Sinauer Associates, Sunderland, MA, 2009), G. Gibson and S. V. Muse provide an introduction to the techniques and promise of this important new field.

Problems and Discussion Topics

1. What hypotheses could account for differences in the abundance of spliceosomal introns among species?
2. What are the major differences between viral, bacterial, and eukaryotic genomes? What demographic differences among species might account for some of the differences we observe at the level of whole genomes?
3. Why are highly expressed genes often characterized by a low rate of evolution?
4. Give an example of an exaptation—a trait whose current function is different from its original function, or a trait that had no original function but has been coopted to serve a functional role—at the level of the genome.
5. Give some examples of newly sequenced genomes that have revealed surprising complexity in species that were assumed to be simpler than mammals, for example.
6. Codon bias is a phenomenon whereby certain codons are substituted more often in phylogenetic lineages than other (synonymous) codons that encode the same amino acid. What might account for codon bias?
7. Distinguish the phylogenetic consequences of different scenarios involving gene duplication, speciation, and concerted evolution. How do these three processes interact to produce different types of phylogenies of multigene families?
8. Describe the difference between neofunctionalization and subfunctionalization of a newly duplicated gene.
9. Suppose you are discussing evidence for evolution with someone who does not believe in evolution. Describe three kinds of molecular data that provide evidence that different lineages (e.g., humans and chimpanzees) have evolved from a common ancestor.

CHAPTER 21

Evolution and Development

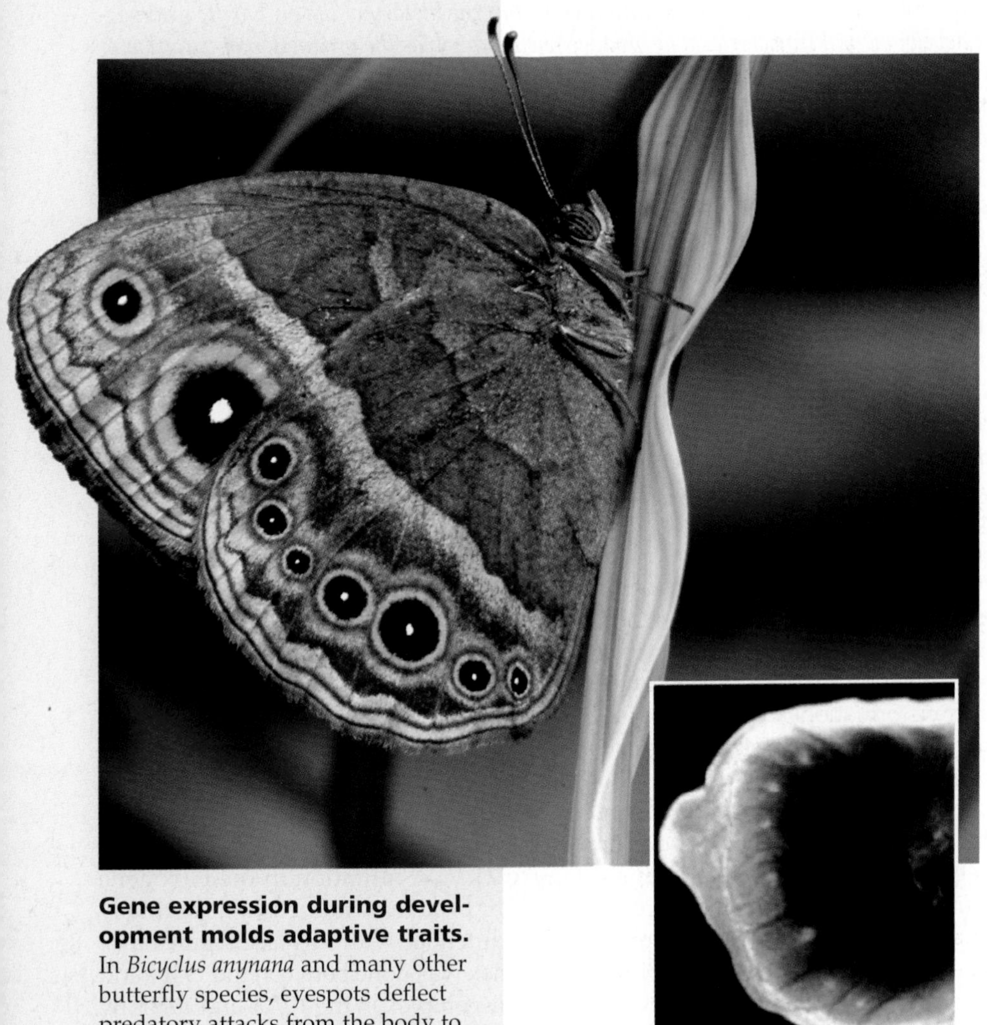

Gene expression during development molds adaptive traits. In *Bicyclus anynana* and many other butterfly species, eyespots deflect predatory attacks from the body to the wings. The inset shows fluorescent staining for the protein Distal-less in the developing eyespots of larval *B. anynana*. Previously known to be required for insect appendage development, the discovery that Distal-less is also used in butterfly color pattern development came as a surprise (see Brakefield et al. 1996). (Photos © W. H. Piel and A. Monteiro.)

The great morphological complexity and diversity that we see in multicellular organisms is produced by developmental processes that have evolved in response to natural selection. But how do these developmental processes evolve? Direct development in animals (see Figure 3.11) illustrates many of the issues involved in addressing this question. DIRECT DEVELOPMENT occurs when embryos develop directly into adultlike forms instead of progressing through a larval stage (INDIRECT DEVELOPMENT). This striking divergence in developmental mode has evolved independently in many animal lineages, including sea urchins, ascidians, frogs, and salamanders. The evolutionary forces and genetic mechanisms promoting such radical, and sometimes rapid, changes in development and life history have mystified biologists for more than a century. Comparisons of embryogenesis and larval morphogenesis, especially among marine invertebrates, are central topics in both classical developmental biology and modern evolutionary developmental biology.

These examples suggest several questions: What are the selective pressures that favor such a novel evolutionary trajectory? How could such a profound alteration of early development evolve so many times? And, perhaps most challenging, what genetic and developmental processes are involved in these evolutionary alterations? It is likely that selection for rapid development promotes the evolution of direct development. But even though some of the genes that underlie these alternative developmental trajectories are beginning to be uncovered, the developmental mechanisms involved—and more importantly, the reasons why these mechanisms are apparently more flexible in some groups of organisms than others—are still mysteries.

The field of **evolutionary developmental biology**, or **EDB** (often called "evo-devo"), seeks to understand the mechanisms by which development has evolved, both in terms of developmental processes (for example, what novel cell or tissue interactions are responsible for novel morphologies in certain taxa) and in terms of evolutionary processes (for example, what selection pressures promoted the evolution of these novel morphologies). Two of the main questions or themes that concern evolutionary developmental biologists are, first, *what role has developmental evolution played in the history of life on Earth?* and second, *do the developmental trajectories that produce phenotypes bias the production of variation or constrain trajectories of evolutionary change?* Natural selection acts on phenotypes produced by development, but ultimately we want to understand how the modes by which development produces those phenotypes affect evolutionary potentials and trajectories.

Hox Genes and the Dawn of Modern EDB

Biologists dating back to Geoffroy Saint-Hilaire (1772–1844), Karl Ernst von Baer (1792–1876), and Darwin himself were fascinated by the patterns of similarity and divergence in development among species. Until quite recently, the fields of evolutionary biology and developmental biology proceeded along mostly separate paths, with seemingly distinct research programs and methodologies (Gould 1977; Depew and Weber 1994; Wilkins 2002). In the past three decades, however, burgeoning information about the genetic mechanisms of morphogenesis in model organisms, as well as the molecular genetic techniques developed to obtain that information, have been integrated with many strands of evolutionary research to form the highly interdisciplinary field of EDB.

The discovery and characterization of the Hox cluster of **homeobox genes** in animals in the 1970s and 1980s marked the dawn of modern EDB (Wilkins 2002). The **Hox genes** are the best-known class of HOMEOTIC SELECTOR GENES, genes that control the patterning of specific body structures, as we saw in Chapter 8. Hox genes control the identity of segments along the anterior-posterior body axis of all metazoans. Mutations in Hox genes often cause transformations of one type of segment into another. In *Drosophila melanogaster*, for example, the Hox genes occur in two complexes (clusters) of genes on chromosome 3, termed the Antennapedia complex and the Bithorax complex. A mutation of the *Ultrabithorax* (*Ubx*) gene transforms the third thoracic segment (T3), which normally bears the halteres (the *Drosophila* homologue of the hindwing of four-winged insects), into a second thoracic segment (T2), which bears wings (Figure 21.1). A mutation in the Hox gene *Antennapedia* (*Antp*) causes the misexpression of Antp protein in the cells that normally give rise to antennae, resulting in the replacement of antennae with legs (see Figure 8.13).

The pioneering genetic work on the Bithorax complex was done between the 1940s and the 1970s by E. B. Lewis, and that on the Antennapedia complex in the 1970s and 1980s by Thomas Kaufman and his colleagues. These investigators found that the genes in both complexes control the anterior-posterior identity of segments corresponding to their order on the chromosome (Figure 21.2). They also discovered that the eight *Drosophila* Hox genes are members of a single gene family, and that the proteins these genes encode share a particular amino acid sequence that binds DNA. This sequence was subsequently named the **homeobox** (in the gene) or the **homeodomain** (in the protein). These findings supported

(A)

(B)

Figure 21.1 Effects of homeotic mutations. (A) A wild-type *Drosophila melanogaster* has a single pair of wings and a pair of small winglike structures called halteres. (B) This mutant fly was experimentally produced by combining several mutations in the regulatory region of the *Ultrabithorax* (*Ubx*) gene. The third thoracic segment has been transformed into another second thoracic segment, bearing wings instead of halteres. (Photos courtesy of E. B. Lewis.)

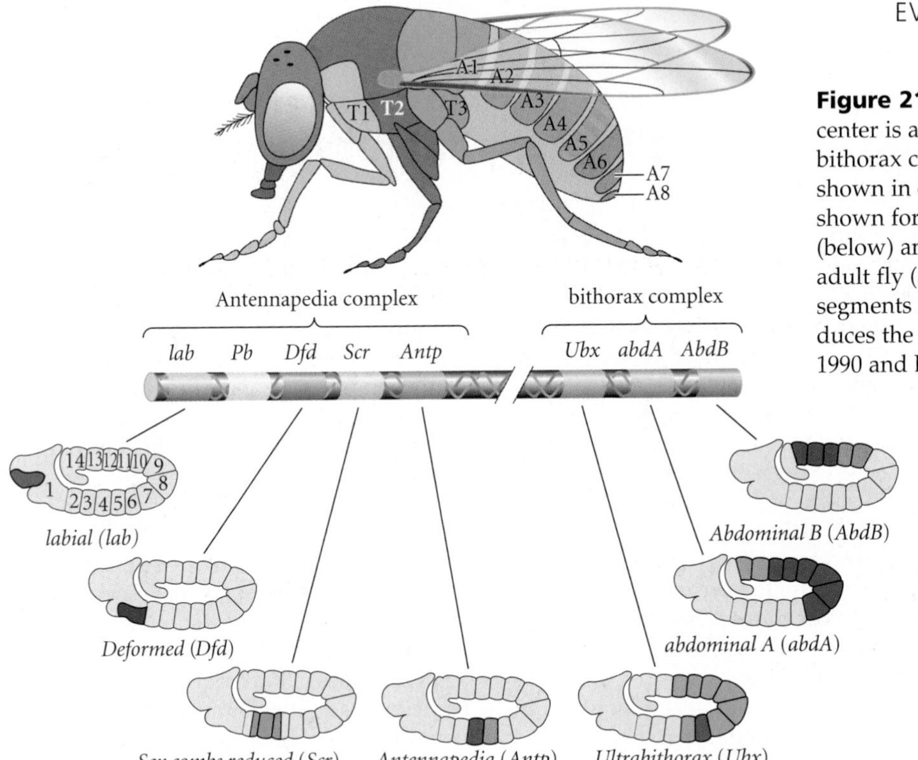

Figure 21.2 Hox gene expression in *Drosophila.* In the center is a map of the genes of the Antennapedia and bithorax complexes, with their functional domains shown in color. The regions of Hox gene expression are shown for the blastoderm of the *Drosophila* embryo (below) and for the regions that form from them in the adult fly (above). Darker shaded areas represent those segments in which gene expression is highest (i.e., produces the most protein product). (After Kaufman et al. 1990 and Dessain et al. 1992.)

Lewis's idea, proposed in the 1960s, that the Hox genes regulate the transcription of other genes. Other researchers were stunned to discover that *all* animal phyla appear to possess Hox genes. These genes have homeodomain sequences similar to those of their homologues in *Drosophila*; in addition, they have the same gene order and orientation as in *Drosophila.* Hox genes form a single complex in most animals. Mammals, however, have four Hox gene complexes (denoted *Hoxa*, *Hoxb*, *Hoxc*, and *Hoxd*) that arose by gene duplication (see Chapter 20) in different parts of the genome, and a total of 13 different Hox genes (as opposed to only 8 in *Drosophila*).

In *Drosophila*, staining for Hox proteins or mRNA (Box 21A) showed that the anterior-posterior expression of the Hox genes corresponds to their mutant phenotypes. For example, as predicted, *Ubx* is expressed in the T3 segment (as well as the anterior abdomen), where it was long known to be required for normal segment identity (see Figure 21.2). Vertebrate Hox expression patterns, although more complex, are also generally expressed in specific anterior-posterior patterns (Figure 21.3).

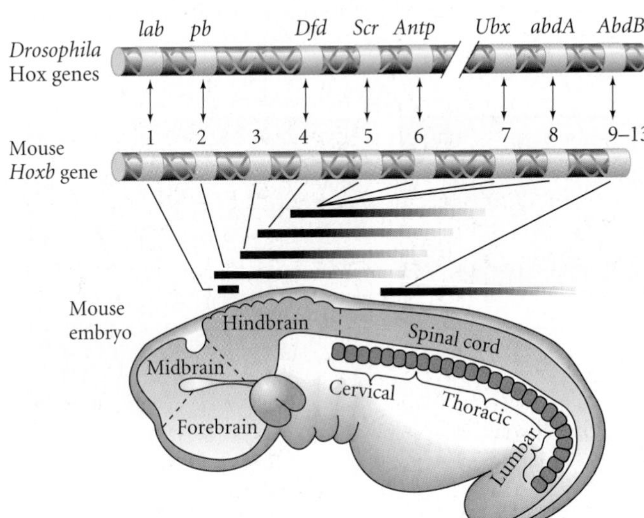

Figure 21.3 Segment-specific patterning functions of Hox genes in the vertebrate hindbrain. This schematic diagram of a mouse embryo shows the hindbrain, consisting of a series of segments. The horizontal bars indicate segmental patterns of *Hoxb* gene expression in the mouse hindbrain and spinal cord, with darker color corresponding to areas of relatively high gene expression. The double-headed arrows connect the genes in the *Hoxb* cluster to the homologous Hox genes in *Drosophila.* (After McGinnis and Krumlauf 1992.)

BOX 21A Characterizing Gene and Protein Expression during Development

Perhaps the most important type of data sought in developmental genetics and evolutionary developmental biology are the expression patterns of specific genes, and the proteins they encode, during development. These patterns are SPATIO-TEMPORAL, with a spatial component (referring to specific cells, tissues, segments, or structures) and a temporal component (referring to specific developmental stages). Gene expression patterns can be visualized by three methods, each requiring different tools and hence currently usable for certain species, but not others.

IN SITU HYBRIDIZATION subjects tissues or whole specimens to a chemical process designed to stabilize mRNA molecules in the cells in which they are produced. Then a species-specific, single-stranded RNA or DNA "probe" corresponding to the gene of interest is applied. It hybridizes by base-pairing with the mRNA of interest. The probe is either chemically modified so that it can be detected by a staining procedure or labeled with a radioisotope so that it can be detected by autoradiography (Figure 1A). Alternatively, extractions of mRNA from different tissues or developmental stages are run in separate lanes on an electrophoretic gel. The mRNA from the gel is then blotted onto a membrane, which is exposed to the same sort of radioactively or chemically labeled probe used in the in situ approach. This procedure is known as a NORTHERN BLOT.

Gene expression patterns can be analyzed at the protein level using antibodies. (The mRNA and protein expression patterns of a given gene may not be identical, due to translational regulation.) Antibodies are produced by injecting a mammal (e.g., a rat) with the protein of interest (the antigen). The animal produces antibodies (immunoglobulin molecules) that bind specifically to that protein. These "primary" antibodies are collected by passing the animal's blood serum over a resin column containing the antigen and then eluting the antibodies from the column in concentrated form. Tissues or embryos are prepared in a similar way as for in situ staining and incubated with the primary antibody. A secondary antibody, an immunoglobulin that specifically binds to the primary antibody, is then applied to the specimen. The secondary antibody is modified so that it can be detected either by an enzymatic reaction producing a colored product or by fluorescence (Figure 1B). An alternative to staining fixed tissue is to prepare protein extracts from different tissues or developmental stages and run each extract as a separate lane on an electrophoretic gel. The protein from the gel is then blotted onto a membrane, which is then incubated with primary and secondary antibodies as described above. This procedure is known as a WESTERN BLOT.

A third method is to study the transcription patterns of particular fragments of putative *cis*-regulatory DNA by using REPORTER CONSTRUCTS in cultured cells or in transgenic (genetically engineered) individuals. Reporter constructs consist of the regulatory DNA of interest, spliced upstream of a "reporter gene" that encodes a protein whose expression can be easily visualized under the microscope. One such protein is β-galactosidase, a bacterial enzyme that processes a particular sugar into a blue product. Another is a protein from jellyfish (GFP) that fluoresces bright green when irradiated with light of a particular wavelength. Because reporter construct analysis requires the use of gene transfer technology, it can be undertaken only in certain well-studied model species, such as *Drosophila*, *Caenorhabditis elegans*, *Arabidopsis*, and mice. Figure 1C shows the nematode *C. briggsae* expressing a GFP reporter construct containing *cis*-regulatory DNA from the *myo-2* gene, which directs the reporter gene's expression in the pharynx.

In genetic model species, the integration of genomic with genetic and developmental data provides a powerful base of knowledge for studies in evolutionary genetics and evolutionary developmental biology. Public online databases such as FlyBase (http://flybase.org) house many of these data, which are frequently

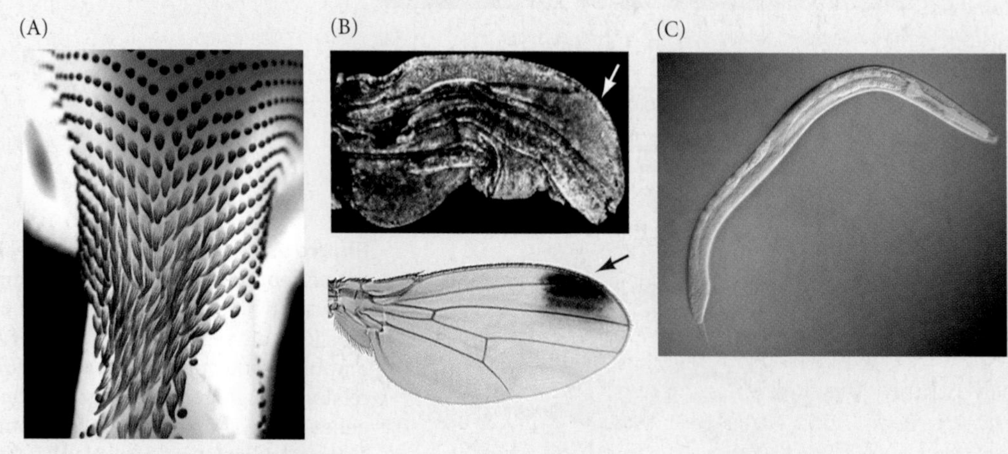

Figure 1 Methods of visualizing gene expression patterns in developing animal tissues. (A) In situ hybridization of *sonic hedgehog* mRNA in developing feathers on the neck region of a chicken. (B) Fluorescent antibody staining of the proteins Ebony (in green) and Yellow (in pink) in the pupal wing of male *Drosophila biarmipes*. The *yellow* gene is expressed where the pigmented spot is located in the fully developed wing (below). (C) Green fluorescent protein (GFP) reporter gene expression of a transgenic construct containing *cis*-regulatory DNA from the gene *myo-2*, which directs expression in the pharynx of the nematode *Caenorhabditis briggsae* (visualized in bright green). (A photo by Matthew Harris; B photos by John True; C photo by Eric Haag, courtesy of Takao Inoue and Eric Haag.)

BOX 21A (Continued)

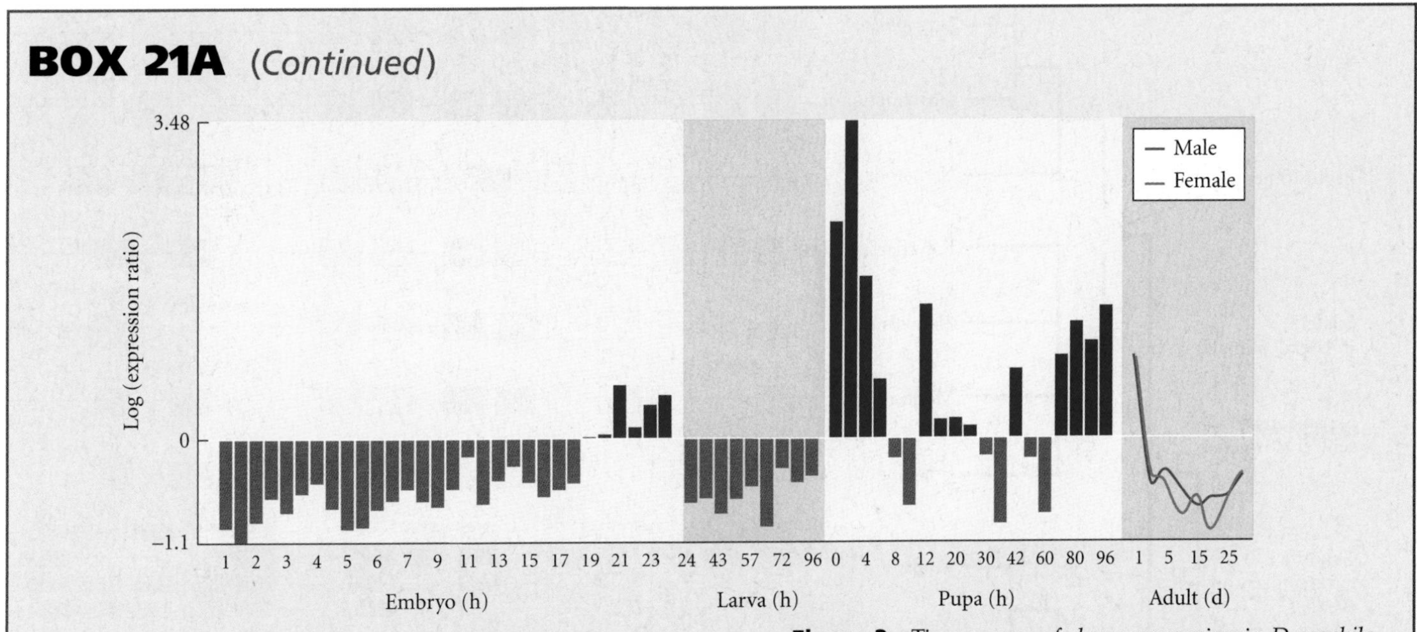

Figure 2 Time course of *ebony* expression in *Drosophila melanogaster*, determined by microarray analysis. The developmental stage is indicated by the *x*-axis, while levels of transcription are indicated in the *y*-axis. Red bars indicate low and green bars high expression, measured as the logarithm of the ratio of *ebony* expression to the average expression of a pool of all mRNAs in all stages of development. In adult insects (far right), levels of expression are different for males and females. (Redrawn from http://flybase.org/; data from Arbeitman et al. 2002.)

updated and augmented. Among the many forms of data available on FlyBase for most of the genes in the *Drosophila melanogaster* genome are compendiums of quantitative gene expression data across various tissues and developmental stages. An example of a temporal expression profile for the *ebony* gene, which is required for both pigment development and eye function, is shown in Figure 2. As data from more species are added, many investigations connecting gene expression data with the evolution of species differences will become possible.

Mapping the presence and absence of Hox genes on the metazoan phylogenetic tree shows their evolutionary history (Figure 21.4). At least four Hox genes are present in radially symmetrical Cnidaria (jellyfishes and corals; Figure 21.4 shows the two best known of these), which are the sister group of the bilaterally symmetrical animals, the Bilateria. Several novel Hox genes arose in the lineage leading to all Bilateria, representing new Hox classes (as evidenced by their homeodomain sequences) that presumably can define increasing degrees of anterior-posterior axis identity.

We can hardly overstate the importance of the Hox gene discoveries for our understanding of how animal diversity evolved. For the first time, a common developmental genetic framework unified the ontogeny of all metazoans; before then, few biologists imagined that vertebrates and invertebrates would share such fundamental developmental genetic underpinnings. These discoveries set the stage for further investigations into the potential commonalities among animals in other aspects of development (e.g., in dorsal-ventral patterning and limb development), which have turned out to be plentiful (Carroll et al. 2005; Wilkins 2002). Ironically, such interest in morphological evolution was precipitated by the discovery of *conserved*, as opposed to *evolving*, features. The discovery of such a fundamental underlying similarity raised new questions: What is the basis of body plan *differences* among animal taxa, and how can seemingly conserved genetic factors play a role in these differences?

These questions are among the most actively investigated in current EDB. For example, the *Ubx* gene was used to test an early hypothesis that major structural differences, such as hindwing differences between dipteran flies and butterflies, might result from simply turning transcription of a particular Hox gene on or off. Sean Carroll and his colleagues showed that differences between the tiny *Drosophila* haltere and the large butter-

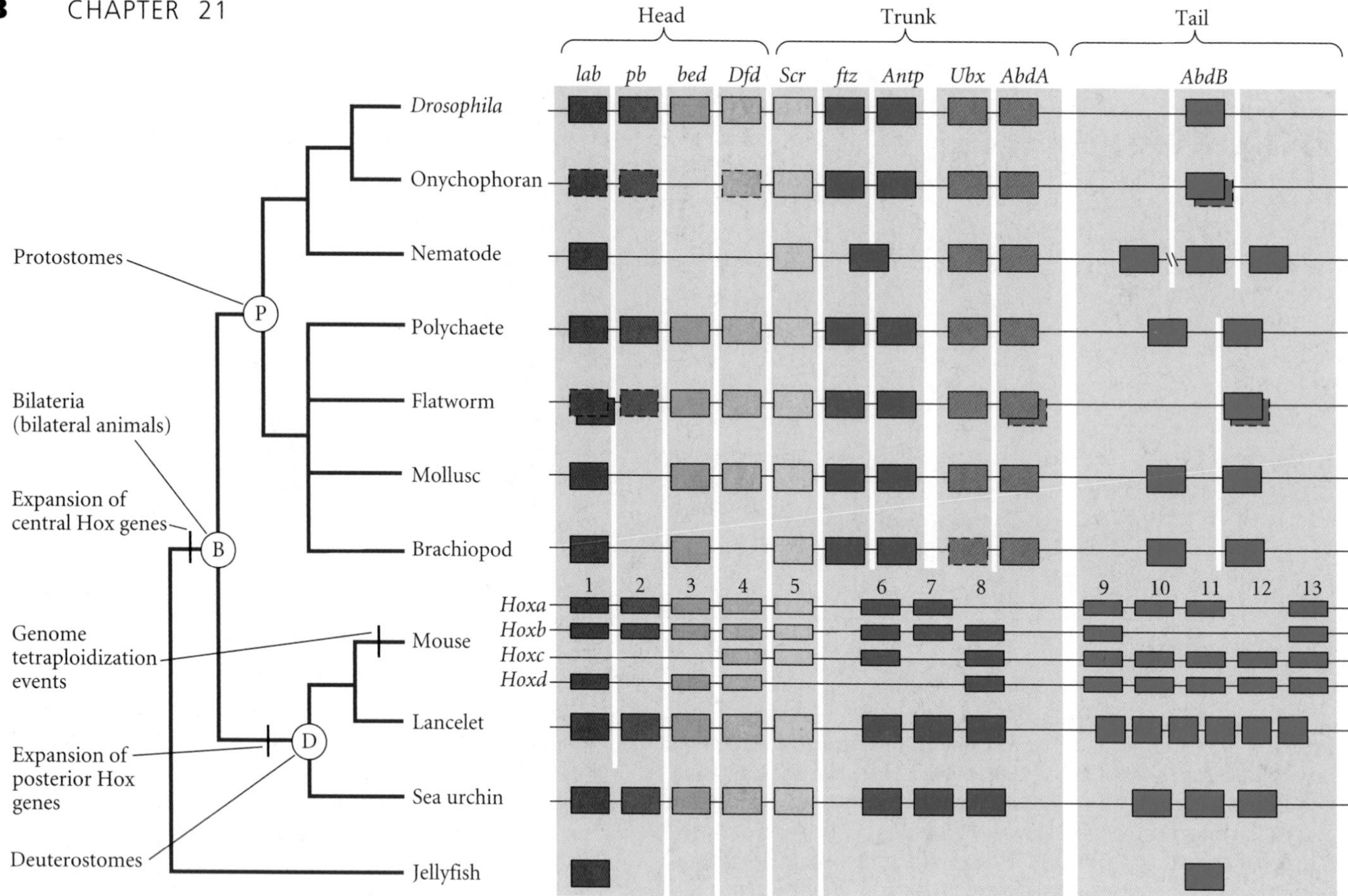

Figure 21.4 Probable evolution of the metazoan Hox gene complex. Vertical white lines delineate currently accepted groups of orthologous Hox genes. Important gene duplication events are indicated by the labeled tick marks. Genes with solid outlines indicate that complete homeobox sequences are known; dashed outlines indicate that only partial homeobox sequences are known. (After Carroll et al. 2005.)

fly hindwing were, in fact, *not* due to differences in Hox gene expression; *Ubx* is expressed in both (Warren et al. 1994). Therefore, divergence among taxa in hindwing morphology must be due to differences in the expression of other genes. The Hox genes encode proteins (**transcription factors**) that regulate transcription by binding to DNA control regions (called **promoters**, **enhancers**, or ***cis*-regulatory elements**) of "downstream" **target genes** (see Box 21B). Therefore, it is likely that morphological divergence is caused by changes in the expression of genes that the Hox genes regulate. In fact, *Drosophila* and butterfly hindwings differ in the expression of several *Ubx* target genes (Carroll et al. 1995; Weatherbee et al. 1998).

A second type of investigation has shown that differences in segmental Hox expression domains are strikingly correlated with the evolution of animal body plans. For example, in all the groups of crustaceans except the basal branchiopods, which have only one type of thoracic appendage, the anterior margin of expression of *Ubx* and *abdA* corresponds to the boundary between thoracic segments bearing maxillipeds (small, leglike appendages specialized for feeding) and those bearing thoracic limbs (see Figure 3.7). This observation suggests that the change in spatial expression of these Hox genes has enabled the segments and their appendages to become different (to become INDIVIDUALIZED; see Chapter 3). Many correlations of this kind have been found throughout the Bilateria (Carroll et al. 2005), implying that evolutionary change in the expression patterns of Hox genes may underlie key body plan adaptations both within and among phyla.

Investigations of differences among more closely related species have also implicated genetic differences at Hox loci. Using interspecies hybridization and gene mapping, David Stern (1998) demonstrated that differences between *Drosophila melanogaster* and *D. simulans* in the pattern of epidermal cell hairs on the legs of the T3 segment map to the *Ubx* locus. Furthermore, he found that the *Ubx* gene is expressed in patterns that precisely correlate with the hair patterns (Figure 21.5A). Because the two species do not differ in the amino acid sequence of the Ubx protein, Stern concluded that regulatory changes at the *Ubx* gene must be responsible for the differences between them. Another Hox protein,

Sex-combs reduced (Scr), controls the identity of the first thoracic segment (T1) in insects. In some species, Scr is involved in determining differences in a T1-specific bristle structure called the sex comb, which is found only in males and is thought to be used to stimulate females during copulation (Ng and Kopp 2008). From the early embryo stage in all flies, Scr is expressed throughout the developing T1 segment. But toward the end of development in the pupal stage in a lineage of the *Drosophila* genus, high levels of Scr are found only in a small region of the developing T1 leg that is destined to give rise to the sex comb (Barmina and Kopp 2007; Randsholt and Santamaria 2008). The pattern of this expression in each species reflects the size of the sex comb (Figure 21.5B). Since sex comb development depends on the presence of Scr, this indicates that regulation of this gene has evolved quickly, resulting in morphological diversity among closely related species (Barmina and Kopp 2007). We will say much more about gene regulation below.

Studies of the role of Hox genes in animal diversification established a new framework in which to think about morphological evolution, based on the idea that changes in the spatio-temporal regulation of a shared set of genes (sometimes referred to as a "toolkit"; see Carroll et al. 2005) are the primary causes of morphological evolution. In the rest of this chapter, we will see how this framework of gene regulatory evolution can help us understand how organismal form evolves.

Types of Evidence in Contemporary EDB

In EDB, patterns of gene expression are frequently used together with morphological, comparative embryological, and phylogenetic data to infer the developmental genetic origins and histories of morphological characters. Developmental genetic data, such as phenotypic information from mutants or individuals that have been genetically manipulated to under- or overexpress a gene or protein of interest, can definitively demonstrate that a particular gene is required for the development of a tissue or structure. Recent work on the evolution of flowers among the angiosperms illustrates how information from mutant phenotypes can be used to form evolutionary hypotheses. In the flowers of dicots, three classes of transcription factors—designated A, B, and C—are required to pattern the four whorls (concentric rings) of distinct floral structures: sepals, petals, stamens, and carpels (Figure 21.6A; Ma and DePamphilis 2000). The sepals are determined by A gene expression alone, the petals by overlapping A and B gene expression, the stamens by overlapping B and C gene expression, and the carpels by C gene expression alone. Loss of expression of a particular gene class during flower development results in conversion of

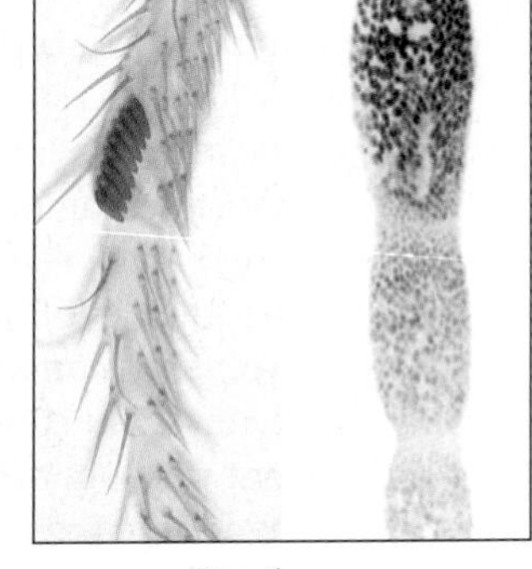

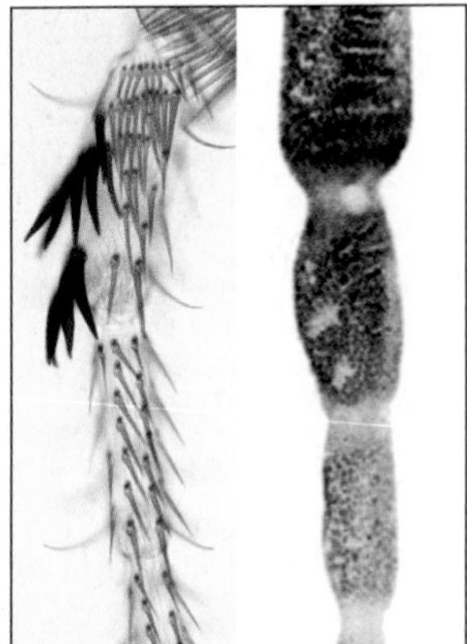

Figure 21.5 (A) The role of *Ubx* in the evolution of the pattern of epidermal cell hairs in the third thoracic leg of *Drosophila*. Areas that express *Ubx* during late leg development correspond with areas lacking epidermal cell hairs. This species difference was demonstrated to map genetically to the *Ubx* locus. (B) *Sex combs reduced* (*Scr*) expression patterns correlate with species differences in sex comb size in *Drosophila*. Antibody staining for Scr protein in *D. ficusphila*, *D. nikananu*, and *D. biarmipes*. (A after Carroll et al. 2005; B photos courtesy of Artyom Kopp.)

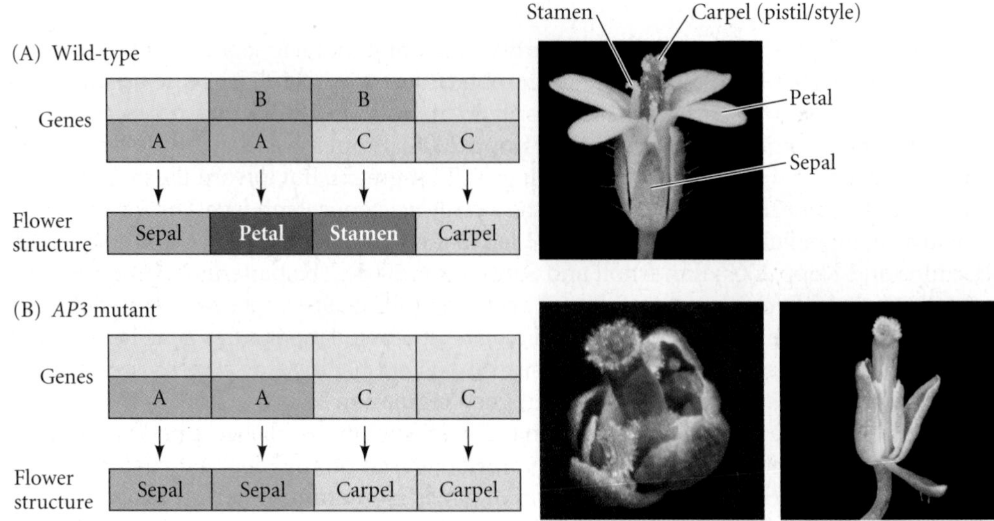

Figure 21.6 The ABC model of flower development. (A) Schematic of the four whorls of angiosperm floral organs with the expression patterns of the A-, B-, and C-class genes. A wild-type *Arabidopsis* flower (right) illustrates the phenotypic result of normal gene expression. (B) In *Arabidosis* mutants lacking expression of the B-class gene *AP3*, petals are transformed into sepals and stamens are transformed into carpels (pistils). (Photos courtesy of J. Bowman.)

one structure into another. For example, in *Arabidopsis thaliana* (a dicot), mutations in the B-class gene *APETALA3* (*AP3*) convert petals to sepals and stamens to pistils (Figure 21.6B). Similar mutant phenotypes and expression patterns of homologues of the ABC genes in a monocot, maize (corn), indicate that the floral patterning system is ancient and have helped to confirm long-supposed homologies between floral structures in different plant taxa (Ambrose et al. 2000). Nucleotide sequence data suggest that the ABC system originated long before the appearance of angiosperms (Purugganan 1997), and presumptive homologues of B- and C-class genes have been shown to be expressed in similar patterns in nonflowering seed plants ("gymnosperms") (Rutledge et al. 1998; Winter et al. 1999), which diverged from the angiosperms at least 300 Mya. The ancestral function of these genes may have been to pattern the male and female reproductive organs.

The Evolving Concept of Homology

Under the phylogenetic concept of homology, which is fundamental to all of comparative biology and systematics, homologous features are those that have been inherited, with more or less modification, from a common ancestor in which the feature first evolved. That is, homologous structures are synapomorphies (G. P. Wagner 1989b; Donoghue 1992). Homology may be suggested by a combination of similarity in position relative to other body structures, similarity in at least some structural features, and the presence of intermediate forms, either among species (fossil or extant) or during ontogeny. Once such observations are made, hypotheses of homology are evaluated by the congruence of the character with a phylogeny of the taxon derived from other characters (see Chapter 3).

We would therefore expect genetically and developmentally similar characters to be homologous, and phylogenetically homologous structures to have similar genetic and developmental bases. However, many observations conflict with these expectations, leading several evolutionary biologists to propose an additional concept, the **biological homology concept** (Roth 1988, 1991; G. P. Wagner 1989a; P. J. Wagner 1996).

Serially homologous structures, such as arthropod legs or vertebrate teeth, share much of the same developmental genetic machinery in their ontogeny, but they are clearly not historically homologous within a species. A more profound conflict is posed by evolutionary reversals. A long-standing notion in biology is Dollo's law (Simpson 1953), which states that evolutionary loss of complex characters is virtually always irreversible. However, recent comparative and developmental studies have found many cases that appear to run counter to this generalization (Collin and Miglietta 2008). For example, a recent phylogenetic analysis of insects (Whiting et al. 2003) found evidence that the winged lineages of the insect order Phasmatodea, the stick insects, are derived from a wingless ancestor and have thus regained the development of wings. Although many partial or complete losses of wings are known in various groups of insects, this is the first known case of apparent reversal of wing loss. The morphology and development of wings and the associated musculature were thought to strongly constrain the possibilities of evolving functional wings from a wingless state. It will be interesting to further characterize wing development in this order. One possibility is that in wingless lineages of stick insects, the tissues giving rise to adult wings go through some of the patterning events of ancestral wing development but that the process is arrested, as has been shown in the development of wingless castes of ants (Abouheif and Wray 2002). Thus, novel genetic variants preventing this arrest might promote the full development of wings such that stick insect wings could still be considered "developmental homologues" of the wings in the other insect groups with which they share a common winged ancestor. A case where there is evidence of the developmental potential to "re-evolve" a character is that of modern birds, which lack teeth but are descended from species such as *Archaeopteryx*, which possessed teeth. Matthew Harris and colleagues (2006) recently characterized a mutant strain of chicken, called *talpid*, that in early development forms toothlike structures indistinguishable from early-stage crocodilian teeth, which presumably represent the teeth of the common ancestor of birds, reptiles, and mammals. Vertebrate tooth development is well studied (Figure 21.7A); teeth arise from a ridge of oral epithelial cells that overlies mesenchyme tissue derived from embryonic neural crest cells. Invagination of the mesenchyme then leads to the formation of a tooth bud, which eventually develops into a tooth. These events require specific sets of genes to be expressed in the oral epithelium and mesenchyme. Very early in chicken development, the beginning of this process occurs but then stops. Earlier experiments demonstrated that the chicken tissue is capable of continuing further in tooth development when two proteins, BMP4 and FGF4, are added (Figure 21.7B; Collin and Miglietta 2008). In *talpid* mutants, these additions are not needed because the genes of the tooth pathway are expressed, resulting in the formation of early-stage tooth structures (Figure 21.7C; Harris et al. 2006).

Conflicts between phylogenetic and biological homology can also occur when phylogenetically homologous traits have different developmental and genetic foundations. For example, digits differentiate sequentially from back (postaxial) to front (preaxial) in all tetrapods except salamanders, whose digits differentiate in the reverse order. As another example, animal eye lenses all contain various crystallin proteins, but the lens crystallins

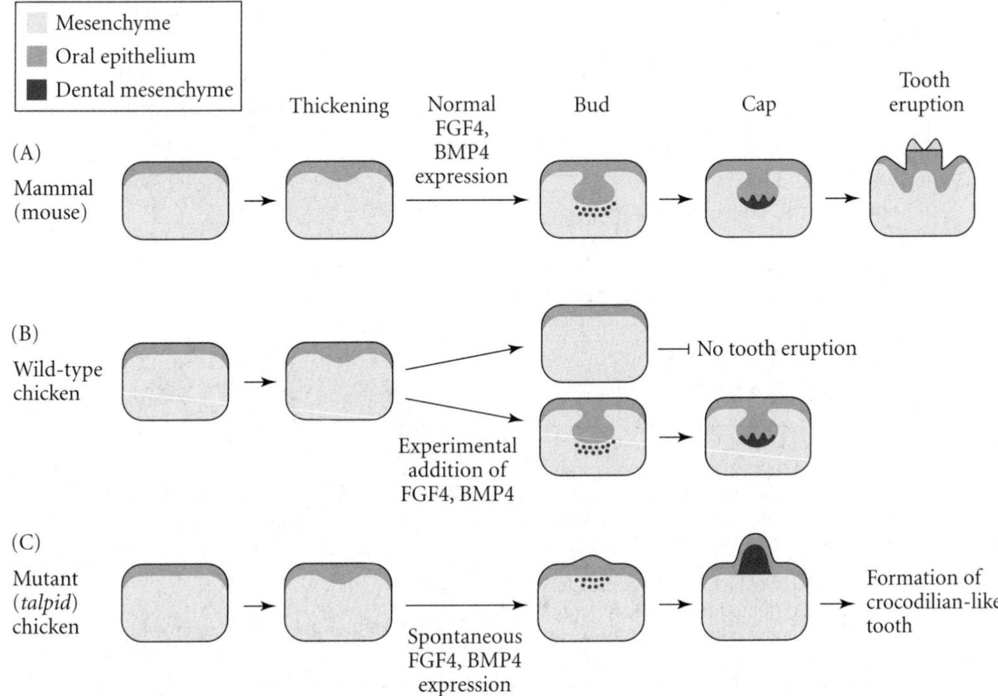

Figure 21.7 Early tooth development in a mouse, wild-type chicken, and mutant *talpid* chicken. (A) At the thickening stage, a battery of genes—two of which are the signaling proteins BMP4 and FGF4—are expressed in the mouse (and other mammals). (B) In wild-type chickens these genes are not expressed, resulting in an arrest after the thickening stage and no tooth formation. However, when BMP4 and FGF4 proteins are added in experimental (in vitro) cultures, dental mesenchyme forms and development proceeds to the cap stage. (C) In the mutant *talpid* chicken, BMP4 and FGF4 expression occurs spontaneously and dental mesenchyme forms, resulting in an outgrowth that eventually forms a toothlike structure similar to a crocodilian tooth. The process by which this structure forms lacks the later stages typical of mammalian tooth development. (After Collin and Miglietta 2008.)

BOX 21B Components of Developmental Pathways

Hox genes are examples of homeotic selector genes, which control cascades of gene expression (i.e., transcription) during the patterning and development of particular tissues, organs, or regions of the body. If a selector gene such as *Ultrabithorax* is not expressed properly, specific tissues or organs may not develop at all, or may be transformed into inappropriate structures (see Figure 21.1).

Transcription factors, including most homeotic selector genes, control the expression of many other genes, including the "structural" genes that encode the proteins, such as enzymes and cell structural components, that actually do the work of morphogenesis. The actions of transcription factors are often regulated in part by cell signaling pathways (Figure 1). These pathways rely on receptor proteins in the cell membrane that respond to extracellular signals such as hormones and short-range signaling proteins called MORPHOGENS. Receptor proteins relay these signals to the genes encoding transcription factors. Seven such cell signaling pathways (each named for a constituent protein, such as Hedgehog or Notch) have been found in animals; others are known in plants. All of these pathways are conserved between *Drosophila* and mammals, and all are involved in many aspects of morphogenesis and pattern formation throughout the developing body, having evolved many specific functions in particular animal lineages, often through gene duplication. Most cell signaling pathways are used multiple times during development, suggesting that morphogenetic novelty may often evolve by re-deploying these pathways in different tissues and at different developmental stages.

Cell signaling pathways and transcription factors are linked into developmental pathways (also called developmental circuits). Such circuits are involved, for example, in patterning the *Drosophila* wing, which takes place in the wing imaginal disc (Figure 2). The end result of developmental circuits are patterns of gene expression that guide the development of an adult structure such as the *Drosophila* wing. A recent striking example of evolutionary change

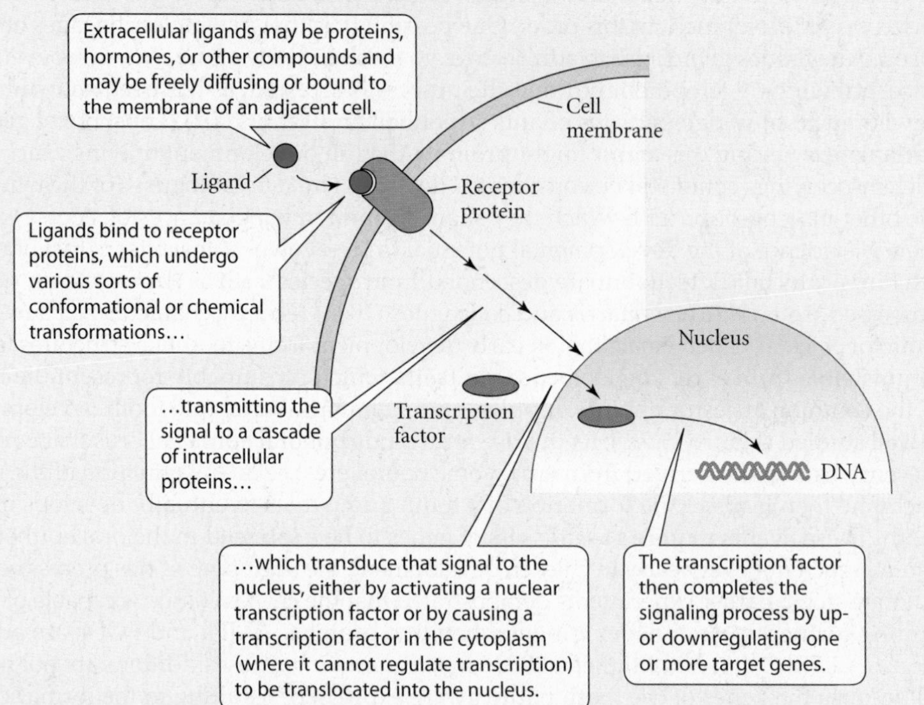
Extracellular ligands may be proteins, hormones, or other compounds and may be freely diffusing or bound to the membrane of an adjacent cell.
Cell membrane
Ligand
Receptor protein
Ligands bind to receptor proteins, which undergo various sorts of conformational or chemical transformations…
Nucleus
…transmitting the signal to a cascade of intracellular proteins…
Transcription factor
DNA
…which transduce that signal to the nucleus, either by activating a nuclear transcription factor or by causing a transcription factor in the cytoplasm (where it cannot regulate transcription) to be translocated into the nucleus.
The transcription factor then completes the signaling process by up- or downregulating one or more target genes.

Figure 1 Transduction of intercellular signals through a cell signaling pathway.

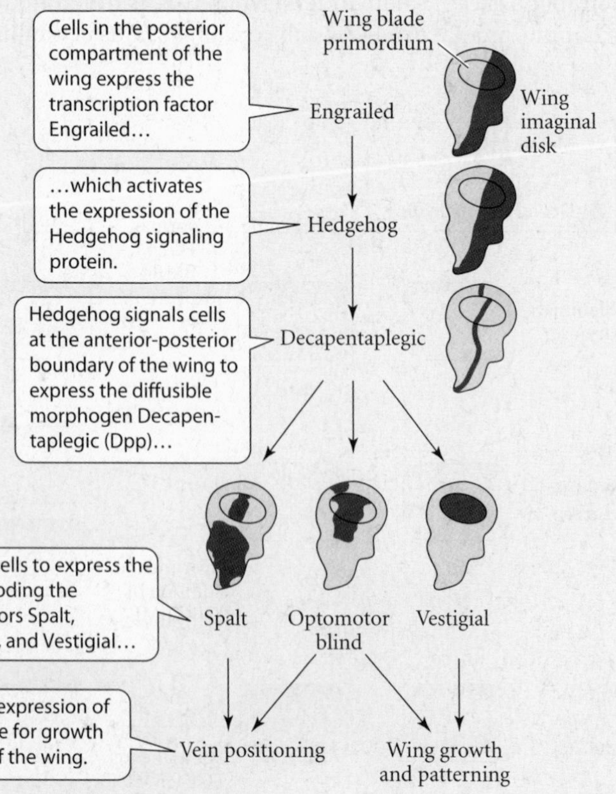
Cells in the posterior compartment of the wing express the transcription factor Engrailed…
Wing blade primordium
Engrailed
Wing imaginal disk
…which activates the expression of the Hedgehog signaling protein.
Hedgehog
Hedgehog signals cells at the anterior-posterior boundary of the wing to express the diffusible morphogen Decapentaplegic (Dpp)…
Decapentaplegic
…which signals cells to express the target genes encoding the transcription factors Spalt, Optomotor blind, and Vestigial…
Spalt
Optomotor blind
Vestigial
…which control expression of genes responsible for growth and patterning of the wing.
Vein positioning
Wing growth and patterning

Figure 2 A developmental pathway consisting of cell signaling and transcriptional activation events in the developing *Drosophila* wing imaginal disc. The gene expression patterns are shown in dark brown. Imaginal discs are sacs of epidermal cells that are set aside early in larval development of some insects. These discs undergo growth and patterning throughout the larval stages; during the pupal stage, they develop into external adult structures, such as wings and genitalia.

of different animal lineages have evolved from a wide variety of different ancestral proteins (see Figure 11.19). Conversely, developmentally and functionally similar structures in different taxa may not be phylogenetically homologous. In perhaps the best example, animal eyes evolved independently in very diverse taxa, but in all of these taxa, a highly conserved transcription factor, Pax6, controls eye development. We will examine these examples in more detail later in this chapter.

The concept of biological homology suggests that a feature may be homologous among species at one level of organization (e.g., phenotypic), but not at another level (e.g., genetic or developmental). It emphasizes the idea that multicellular organisms are constructed from a set of more or less conserved tools. This "genetic toolkit" consists of individual genes and proteins as well as multiprotein circuits. Morphological evolution has consisted, in large part, of "tinkering" with this toolkit (Jacob 1977). One of the greatest challenges of current EDB studies is to understand how biologically homologous structural units are assembled from these tools.

Evolutionarily Conserved Developmental Pathways

The genes that regulate morphogenesis function in hierarchies or networks termed **developmental pathways** or **developmental circuits** (Box 21B). These genes encode signal-

BOX 21B *(Continued)*

associated with differential expression of a cell signaling protein was uncovered in Darwin's finches. Arhat Abzhanov and colleagues (2004) showed that variation in beak size and shape correlated with the expression pattern of bone morphogenetic protein-4 (BMP4), a member of a protein family known to be required for bone formation and development. When the *BMP4* gene was misexpressed in chickens, the resulting beak morphology was similar to that seen in the finches (Abzhanov et al. 2004; Wu et al. 2004).

Whether transcription of a gene is increased (upregulated) or decreased (downregulated) depends on the binding of transcription factors to that gene's enhancer sequences. Enhancers therefore act as genetic switches by controlling transcription of the gene in specific spatio-temporal patterns determined by the transcription factors. Some transcription factors act exclusively to increase or to decrease gene expression, whereas others may have either effect in particular contexts, depending on the other proteins they interact with and on the specific enhancer sequence they bind to. Figure 3 shows two enhancers in the *Drosophila vestigial* gene that affect *vestigial* expression in the developing wing. Transcription factors in the Notch signaling pathway bind to an enhancer that directs *vestigial* expression at the anterior-posterior and dorsal-ventral boundaries of the wing field, and transcription factors in the Dpp pathway bind to an enhancer that directs expression in the four quadrants of the wing field that complement the boundary pattern. Normal wing development requires this pattern of expression of *vestigial*.

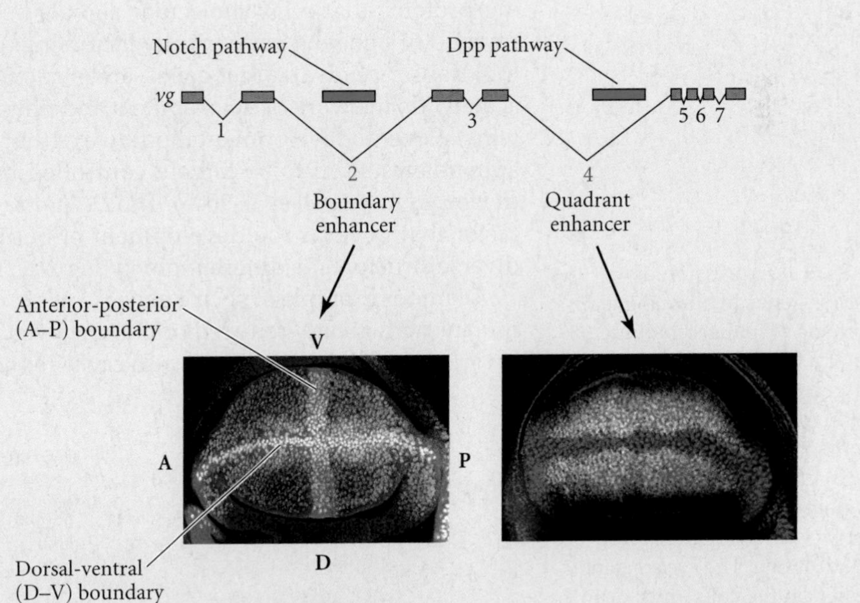

Figure 3 Enhancers in the *Drosophila vestigial* (*vg*) gene. Exons of *vestigial* are shown in gray; the introns are numbered. Two different enhancers have been characterized in *vestigial*. One, the "boundary" enhancer" (in intron 2), is controlled in part by the Notch cell signaling pathway and activates *vestigial* expression in a cross-like pattern at the anterior-posterior and dorsal-ventral boundaries of the wing imaginal disc. The other, the "quadrant" enhancer (in intron 4), is controlled in part by the Dpp signaling pathway and activates *vestigial* expression in a complementary four-quadrant pattern. (Photos courtesy of Sean Carroll.)

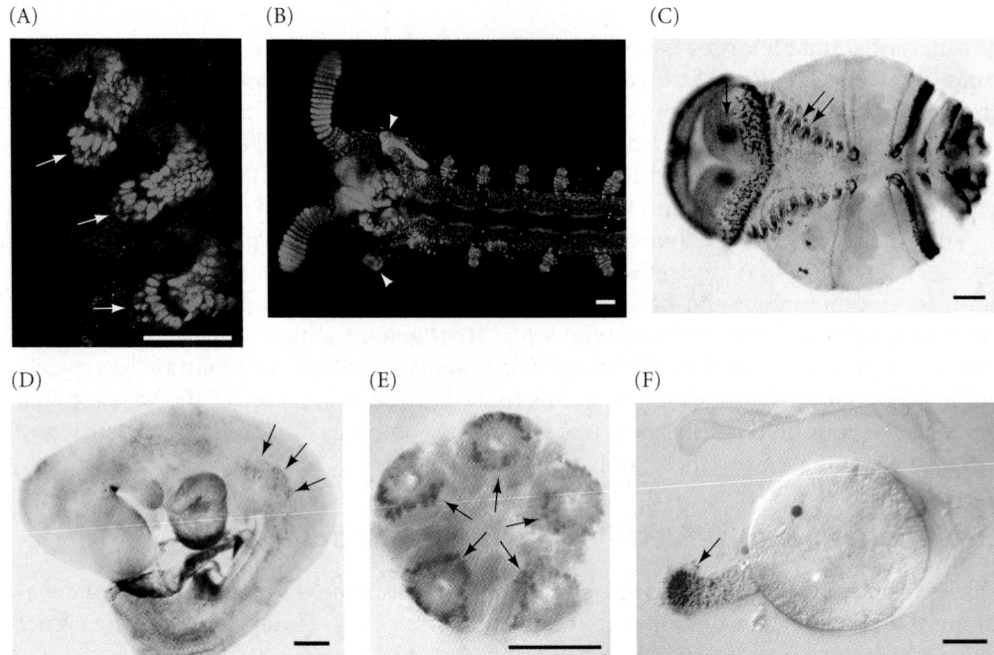

Figure 21.8 Expression of *Distal-less* family genes in the primordia of various animal appendages. (A) Abdominal prolegs (arrows) of a *Precis coenia* butterfly larva. (B) Antennae and oral papillae (arrowheads), and lobopods of the onychophoran *Peripatopsis capensis*. (C) Parapodial rudiments (arrows) of an embryonic polychaete annelid (*Chaetopterus variopedatus*). (D) Mouse forelimb bud (arrows). (E) Tube feet (arrows) of a metamorphosing larva of the sea urchin *Strongylocentrotus droebachiensis*. (F) Ampulla (arrow) of a larval ascidian (*Molgula occidentalis*). Scale bars = 0.1 mm. (Photos courtesy of Grace Boekhoff-Falk.)

ing proteins that relay molecular signals between cells, transcription factors, which respond to signaling pathways by increasing (upregulating) or decreasing (downregulating) transcription at target genes, and structural genes, which encode the proteins that actually do the work of development and physiology (e.g., enzymes and cytoskeletal proteins). Several developmental pathways that control the formation of major organs or appendages seem to be largely controlled by highly conserved transcription factors (reviewed in Carroll et al. 2005). The *Distal-less* gene, for example, encodes a transcription factor that governs the development of body outgrowths that differentiate into very diverse structures in different phyla (Figure 21.8; Panganiban et al. 1997).

A famous example of such a gene is *eyeless*, originally discovered in a classic *Drosophila* mutant with greatly reduced or missing eyes. Mutations in the mammalian homologue of *eyeless*, which is called *Pax6*, also cause reduction of the eyes. *Pax6/eyeless* activates the

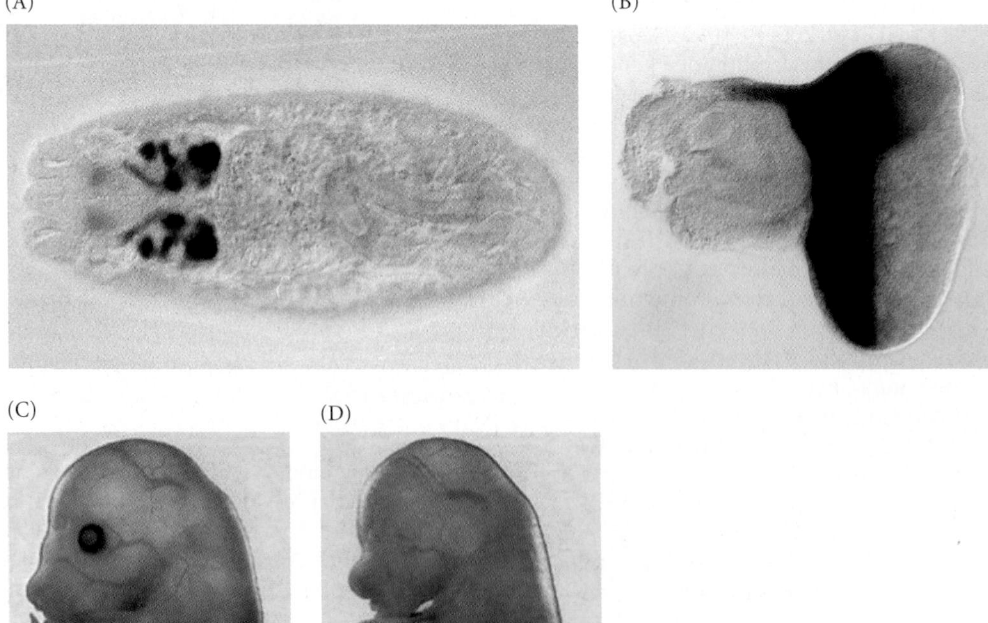

Figure 21.9 Expression of *Pax6/eyeless* genes in animal eye development, visualized by blue staining of a β-galactosidase reporter gene. (A,B) *Drosophila*. (A) Expression of *eyeless* in the larval eye precursors and other nearby tissues. (B) Close-up of eye imaginal disc (large lobe on right), a group of larval cells that develop into the adult eye. The *eyeless* gene is expressed in the cells anterior to a boundary-like feature called the morphogenetic furrow, which moves from posterior to anterior during eye development and delineates the boundary between differentiated and undifferentiated cells. The smaller lobe on the left is the antennal imaginal disc, which gives rise to the adult antenna. (C,D) Mouse. (C) The developing mouse eye can be seen as a circle of pigment. (D) The mouse mutant Small eye (Sey) lacks *Pax6* expression and has small or missing eyes. (A,B courtesy of Georg Halder; C,D courtesy of Robert Hill.)

transcription of a hierarchy of regulatory proteins that control the development and differentiation of the eye. Expression of *Pax6/eyeless* is localized to the developing eye in embryos (Figure 21.9). Amazingly, when researchers genetically engineered *Drosophila* to express *eyeless* in various regions of the body where it is not normally expressed (**ectopic expression**), they discovered that this gene was sufficient to induce the development of ectopic eyes at these positions (Figure 21.10A,B). Even more astounding is the functional conservation of the *Pax6/eyeless* gene between vertebrates and invertebrates: *Pax6* genes from other animal species, including humans, can induce ectopic eyes when expressed in *Drosophila* (Figure 21.10C). These ectopic eyes, no matter which species' *Pax6* was expressed, exhibit many aspects of eye morphology and function, including innervation and responses to light (Clements et al. 2008). Subsequent studies found that at least two genes regulated by *Pax6* have conserved functions in *Drosophila* and mammalian eye development (Oliver et al. 1995; Xu et al. 1997), which helps to explain how *Pax6* homologues can function when placed in the genomes of such divergent animal species.

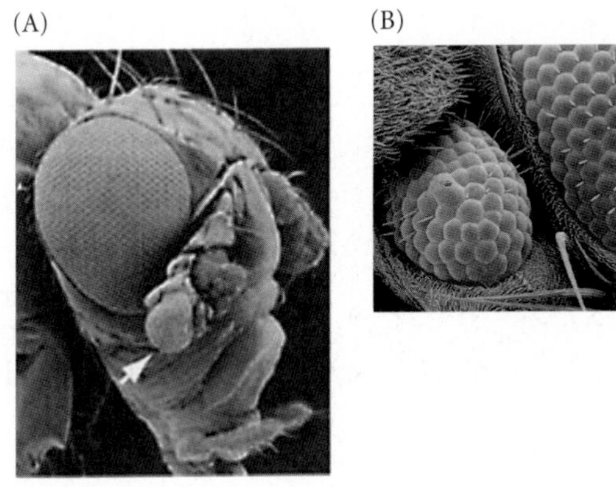

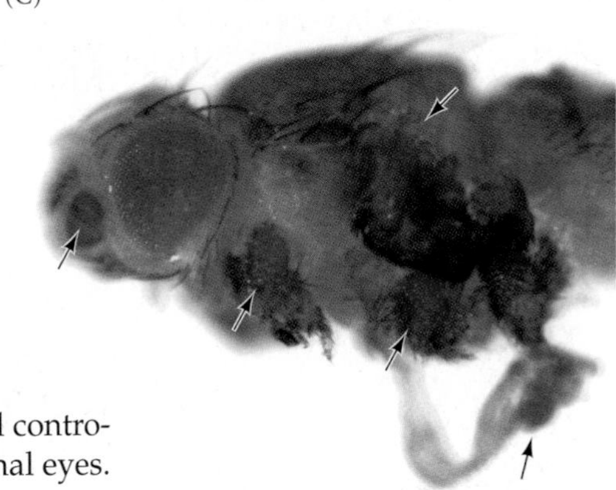

Figure 21.10 Ectopic eye formation in *Drosophila.* (A) A small ectopic eye on the base of the antenna (arrow), formed by ectopic expression of the *eyeless* gene. (B) A close-up of an ectopic eye, showing the same morphology as the large, normal compound eye nearby. (C) Ectopic expression of the human *PAX6* gene (driven by enhancers for the gene *dpp*) causes eyes to form in many locations (arrows). (A, B courtesy of W. G. Gehring; C courtesy of Nadean Brown.)

The question of how an organ as complex as the human eye evolved is one in which Darwin himself, and many theorists after him, have invested a great deal of thought. Darwin suggested that various intermediate stages, capable of photoreception, may have had adaptive value, leading to the evolution of more complex eyes. Consistent with this idea, basal animal phyla such as Cnidaria have very simple "eyespots," consisting of a few cells containing photoreceptive pigments (rhodopsins), and many types of eyes, varying in complexity, are found among other invertebrates (see Figure 22.14). There is still controversy over whether the common ancestor of all the Bilateria possessed functional eyes. Nevertheless, an explanation for the diversity of animal eyes that has gained wide acceptance among evolutionary developmental biologists is that *Pax6* has a very ancient function in regulating the expression of universal components of photoreceptor organs, such as rhodopsin pigments. Thus in various lineages of animals, different morphological features evolved independently, possibly because of the actions of different genes that became regulated by the *Pax6* pathway (see Wilkins 2002, pp. 148–155). Therefore, the capacity for photoreception may be homologous, but the structural features that carry out this function—eyes—may not be. Such an explanation may also apply to other cases of highly conserved regulators of organogenesis, such as the transcription factor NKX2-5/tinman in bilaterian heart development.

Gene Regulation: A Keystone of Developmental Evolution

The regulation of gene expression, whereby genes are activated or repressed at particular times and in particular tissues, is achieved by discrete ENHANCERS for each gene. These DNA sequences bind particular sets of transcription factors in specific cells or at specific developmental stages. For example, several genes expressed in the developing *Drosophila* wing are regulated by the transcription factors Scalloped and Vestigial, which activate genes required for wing development (Guss et al. 2001). The noncoding DNA (introns) of these genes contains one to several binding sites for the proteins Scalloped and Vestigial, each of which is 8 to 9 nucleotides long (see Box 21B, Figure 3). An enhancer consists of one or more such sites, which bind one or more transcription factors. Enhancers can be studied using reporter constructs in cultured cells or transgenic (genetically engineered) individuals (see Box 21A).

A particular gene often has a number of different enhancers. This **regulatory modularity** is thought to enable evolutionary changes in the development of specific tissues and body structures. That is, changes in the enhancers, rather than changes in the amino acid sequences of proteins, are thought to be responsible for many phenotypic adaptations. However, demonstrating the validity of this attractive idea is difficult. The best examples thus far have been found in *Drosophila*.

The *Drosophila* gene *yellow*, which encodes an extracellular protein required for dark melanin pigmentation, is expressed in flies wherever there is dark pigmentation on the body. This pigmentation is often expressed in rapidly evolving sexually dimorphic patterns, such as in the abdomen (see below). This protein is also expressed in a spot pattern in the pupae in species with male-specific wing spots. Benjamin Prud'homme and colleagues (2006) demonstrated that in spot-bearing species such as *D. elegans* (Figure 21.11A), the *yellow* wing spot expression pattern is directed by an enhancer 5′ of the *yellow* coding region (Figure 21.11B). This enhancer resides within a larger enhancer that was previously demonstrated to direct expression in the wing blade (all of the wing cells except the veins) in *D. melanogaster* and other *Drosophila* species. Interestingly, *D. tristis*,

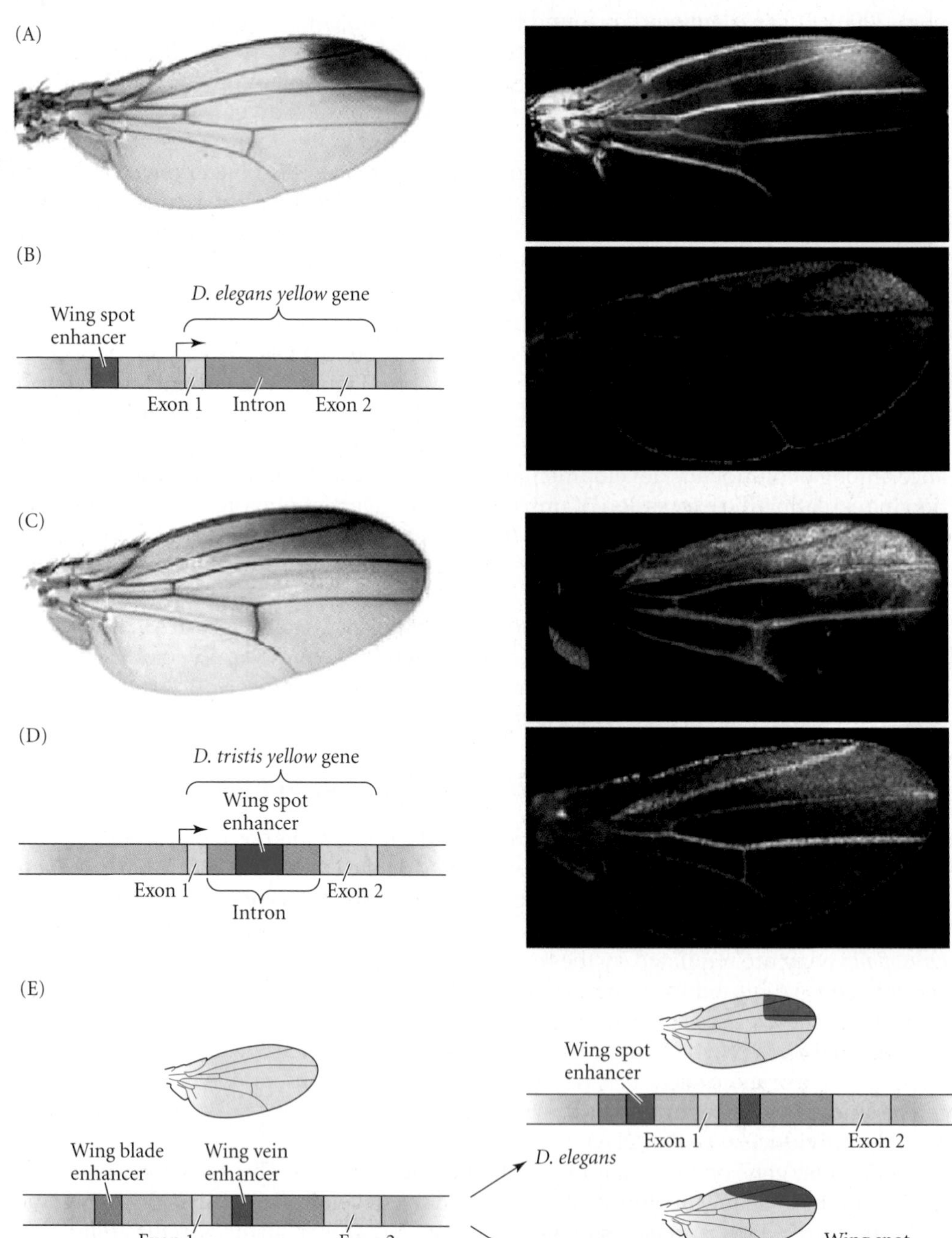

Figure 21.11 Independent evolution of *Drosophila* wing spot enhancers in the *yellow* gene. (A) *D. elegans* male wing spot (left) and Yellow protein expression as detected using fluorescent antibody staining (right). (B) Location of wing spot enhancer in the *D. elegans yellow* gene (left) and wing spot reporter expression (right). Arrow indicates transcription start site and direction. (C) *D. tristis* male wing spot (left) and Yellow protein expression as detected using fluorescent antibody staining (right). (D) Location of wing spot enhancer in the *D. tristis yellow* gene (left) and wing spot reporter expression (right). (E) Model of convergent *yellow* wing spot evolution in *Drosophila*. (From Prud'homme et al. 2006; images courtesy of Sean Carroll.)

which evolved similar male wing spots independently, also expresses *yellow* in the developing spot region (Figure 21.11C). *D. tristis*, also possesses a "spot enhancer" in the *yellow* gene, but this enhancer does not occur in its 5′ "wing blade" enhancer. Instead it occurs in a different, pre-existing enhancer found in the intron of the *yellow* gene that directs expression in the wing veins (Figure 21.11D). Thus convergent evolution has involved surprisingly different regulatory components of the same gene (Figure 21.11E).

Genes such as *yellow* are often referred to as "candidate genes" for phenotypic evolution because their known functions are consistent with causation of a difference between species. Often the first type of observation for hypothesizing the involvement of a candidate gene is its mutant phenotype. Another example in *Drosophila* in which a candidate gene has been implicated in a species difference is in *D. santomea*, which has a very light abdominal melanin pigmentation compared with its sibling species *D. yakuba*. The light pigmentation of *D. santomea* resembles that of several *D. melanogaster* pigmentation mutants, including a mutant called *tan*. Mary Anna Carbone and collaborators (2005) used quantitative trait locus (QTL) mapping (see Chapter 13) to determine that the region around the *tan* locus did contribute strongly to this species difference. Sangyun Jeong and coworkers (2008) further determined that differences in *tan* expression explained most of the abdominal pigmentation divergence between *D. yakuba* and *D. santomea*. The expression was shown to be due to just two nucleotide substitutions in a small region of *cis*-regulatory DNA containing an enhancer for abdominal expression (Figure 21.12). In *D. santomea*, expression of *tan* in the abdomen is essentially turned off. Surprisingly, when Jeong

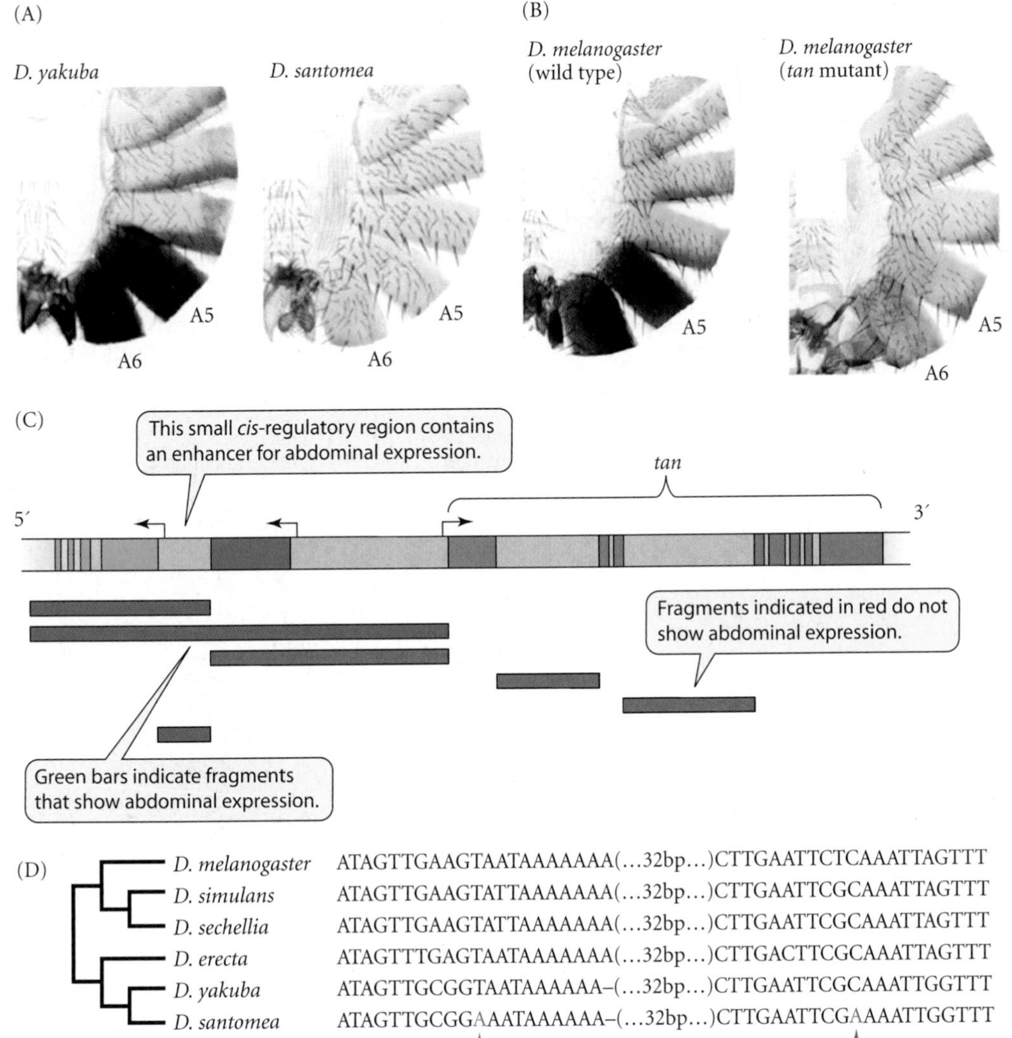

Figure 21.12 Abdominal pigmentation differences between *Drosophila yakuba* and *D. santomea* are due in large part to *cis*-regulatory changes in the *tan* gene, which is required for melanin synthesis. (A) Abdominal cuticle of *D. yakuba* male, which has dark abdominal pigmentation in the fifth and sixth abdominal segments, and *D. santomea* male which has little abdominal pigmentation. (B) This pigmentation difference resembles that between wild-type and *tan* mutant *D. melanogaster*, suggesting *tan* as a candidate gene for the distinction. (C) Abdominal expression of *tan* is directed by a *cis*-regulatory element residing between the two genes (pink exons) 5′ of *tan* as determined by reporter assays in transgenic *D. melanogaster*. Black arrows indicate start and direction of gene transcription; note that the 5′ genes are transcribed in the opposite direction from *tan*. (D) The loss of abdominal expression in one naturally occurring *D. santomea tan* allele is due primarily to two derived nucleotide substitutions (red arrows), as shown in this molecular phylogeny. This expression difference is presumably the result of changes in transcription factor binding sites, although these factors have not yet been identified. (A,B courtesy of Sean Carroll; C,D after Jeong et al. 2008.)

et al. sequenced several *tan* alleles from *D. santomea*, they found that three independent alleles, which appear to have evolved hundreds of thousands of years ago, are associated with reduced expression of *tan*. After arising independently, each of these alleles must have been selectively favored to have been maintained in the population for so long. Normally, we would have expected to find only one evolutionary origin of a novel, selectively favored phenotype within a particular lineage.

As we have just seen, an evolutionary change in the level of expression of a gene can be caused by an alteration of a closely linked (*cis*) regulatory sequence (such as an enhancer) that is affected differently by the same transcription factor—produced by another locus—in both species. Alternatively, the difference in gene expression between the species might be the result of a ***trans*-regulatory** difference: a difference in the transcription factor itself. Patricia Wittkopp and colleagues (2004) reasoned that if a gene expression difference between two related species is caused by the divergence of a *cis*-regulatory sequence, the same expression difference between their genes should also occur within an F_1 hybrid, because each species' gene would remain associated with its linked regulatory sequence (Figure 21.13; compare cases 1 and 2). However, if the expression difference between the species is caused by divergent transcription factors (*trans*-regulators), both transcription factors would be produced in the F_1 hybrid, and both would diffuse through the nucleoplasm to the target genes of both species—so they should cause both target genes to be expressed at about the same level (see Figure 21.13, case 3). To test this, Wittkopp et al. studied the expression of 29 genes in hybrids between *D. melanogaster* and *D. simulans.* They found that about half the genes displayed only *cis*-regulatory divergence, while the other half showed signs of both *cis*- and *trans*-regulatory divergence. This seminal study suggests that changes in enhancers are likely to account for many evolutionary changes in gene expression.

An important pursuit in recent EDB studies is the comparison of gene regulatory changes that contribute to morphological differences both between and within species. As described in Chapter 13, the three-spined stickleback (*Gasterosteus aculeatus*) has invaded freshwater habitats from its ancestral marine habitat many times in the last 10,000 years. These invasions have involved essentially the same morphological adaptations for living either at the bottom of lakes and streams (benthic habitats) or farther up in the water column (limnetic forms). The best characterized of these adaptations are reductions in the pelvic spines and armor plates, which are needed for predator defense in the ocean, but are lost in many freshwater habitats in which predation is not a strong selection force. Since all of these populations of *G. aculeatus* remain interfertile, genetic methods such as

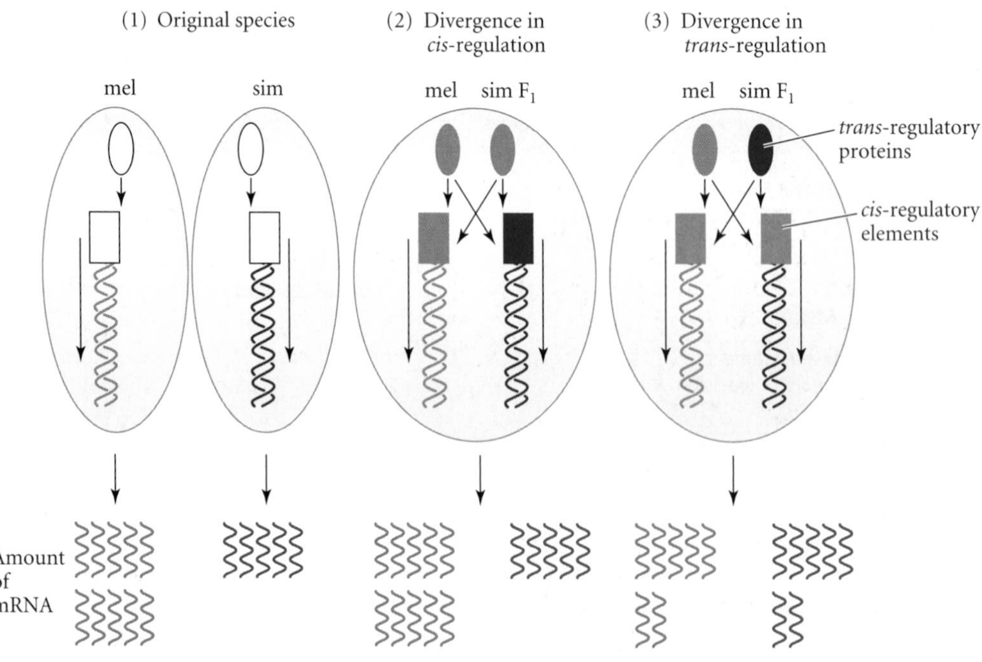

Figure 21.13 A test for divergence of gene expression resulting from divergence in *cis*-regulatory elements or *trans*-regulatory proteins. The *cis*-regulatory elements are represented by rectangles; *trans*-regulatory proteins (which bind to the *cis*-regulatory elements) are shown as ovals. (1) The expression of the gene of interest (measured by the production of mRNA) differs between *D. melanogaster* (mel) and *D. simulans* (sim); the ratio of mel/sim mRNA is 2:1. Whether this difference is due to differences in *cis*-regulatory elements or *trans*-regulatory proteins is unknown (open symbols). (2) In an F_1 hybrid, the same relative difference in gene expression is expected if the two species' *cis*-regulatory elements have diverged (different shades of blue), since neither element alters the expression of the homologous gene from the other species. (The *trans*-regulatory proteins are presumed to be the same in both species, indicated by gray.) (3) If the species difference is due to divergence in *trans*-regulatory proteins, equal expression of the two species' genes (1:1 mRNA ratio) is expected, since the homologous genes from both species are equally exposed to the *trans*-regulatory proteins of both parental species.

QTL mapping can readily determine the genetic basis of these adaptations. Developmental genetic data from zebrafish and other vertebrate model species have also been useful in identifying candidate genes.

In the first such study, Michael Shapiro and colleagues (2004) found that among five QTL for the pelvic reduction phenotypes, the one with the predominant effect maps to the *Pitx1* gene, a transcriptional regulatory protein that in mice is known to be required for pelvic limb development. *Pitx1* is expressed in many places during development, but most importantly, in marine fish it is expressed in the developing pelvis. This pattern has been lost in the benthic freshwater study population of sticklebacks, which has pelvic reduction. Since no coding region differences occur between marine and benthic genotypes, the authors concluded that a regulatory change reducing *Pitx1* expression is responsible for the freshwater adaptation. Shapiro et al. demonstrated that other populations of pelvic-reduced *G. aculeatus* also carry loss-of-function *Pitx1* alleles. These results suggest that the ancestral marine population likely harbors this recessive allele at a low frequency. Further work (Shapiro et al. 2006) demonstrated that another pelvic-reduced species, the nine-spined stickleback (*Pungitius pungitius*), carries similar alleles of *Pitx1*, suggesting that this gene may contribute to the evolution of pelvic morphology in many species.

Pamela Colosimo and coworkers (2005) found that parallel reductions in armor plates in freshwater populations of *G. aculeatus* are caused by regulatory changes in the gene encoding ectodysplasin (Eda; see Figure 13.24), a secreted signaling protein required for development of teeth, hair, and other ectodermal structures in mammals. Anne Knecht and colleagues (2007) went on to ask whether naturally occurring variation in other genes in the Eda signaling pathway might be involved in stickleback armor- plate reductions but found that variation in only one of five genes tested, the Eda receptor (EdaR) affects this trait. This suggests that only particular genes in the armor developmental pathway contribute to adaptation in nature.

Evolution of protein-coding sequences is also an important contributor to phenotypic evolution

Changes in the regulation of genes are thought by many researchers to be more likely causes of phenotypic trait evolution than changes in the amino acid sequences. This is because genes often have more than one function during development (that is, they are pleiotropic). Changing the protein coding sequence of a gene is expected to alter most or all of the functions of that gene, whereas changing the regulation at particular stages or in particular cells or tissues will only affect the gene's function in those specific stages or tissues. However, the current evidence is rather equivocal on the relative importance of regulatory versus protein-coding changes in adaptive evolution (Hoekstra and Coyne 2007; see also Wray 2007). A well-known and surprisingly recurrent example of the latter type of evolution occurs in the gene for the melanocortin-1 receptor (*Mc1r*), which was illustrated in Figure 3.8. The *Mc1r* gene is expressed on the surfaces of vertebrate melanocytes (pigment-producing cells), where it receives signals from the signaling peptides (hormones) melanocortin and Agouti and controls whether dark pigmentation (eumelanin) or lighter pigmentation (pheomelanin) is produced. Signaling by melanocortin tells the melanocyte to produce eumelanin, whereas signaling by Agouti tells the vertebrate melanocyte to switch to pheomelanin production. In birds and mammals, some amino acid substitutions in *Mc1r* result in genetically dominant melanism because of the inability of melanocytes to sense the Agouti signal to turn off eumelanin production during development of hair, fur, or feathers. Such mutations have been found to cause melanic polymorphisms in a large number of both domesticated strains and natural populations of birds and mammals (Majerus and Mundy 2003).

Modularity in morphological evolution

The body plans of multicellular organisms are composed of discernible units, such as body segments, appendages, or the petals and sepals of angiosperm flowers. In animals, many serially homologous structures have quite different morphologies (e.g., arthropod appendages). Similarly, elements such as teeth that repeat within a segment may vary greatly in form. Much of the spectacular diversity that we see in nature is the result of such changes in individual segments or structures. Thus the individual parts of the body plan show dissociation, or individualization—that is, they can develop and evolve in an independent manner. This phenomenon is an example of mosaic evolution (see Chapter 3).

The degree to which the development of different body structures is independent is referred to as MODULARITY, and the individual structures or units can be thought of as MODULES. How is modularity achieved by developmental pathways? This central question in EDB is only beginning to be explored. It is clear that the genes that produce the final form of each morphological module or unit, such as enzymes and structural components of cells and organs (sometimes called "effector genes"), are expressed differentially by the action of regulatory genes such as those encoding transcription factors. The Hox genes discussed above are the best-known example of how animals achieve their modular body plans. Because different segments express distinct Hox proteins, the "effector genes" are able to be expressed in very different ways in the different segments. The mechanisms by which Hox genes regulate this segment specificity are likely to be different in each segment. For example, Takuya Tsubota and colleagues (2008) found that development of the sternopleural (Sp) bristles in *Drosophila*, which normally are only expressed in the second thoracic segment (T2), is repressed differently between the first (T1) and third (T3) segments (Figure 21.14). In T1, the Hox protein Scr represses expression of the Hox gene *Antp* and the gene *spineless* (*ss*), which are jointly required for Sp bristle development in T2. Ectopic *ss* expression (in genetically engineered flies) is sufficient to cause the development of ectopic Sp bristles in T1. However, in T3, ectopic *ss* expression does not cause the formation of Sp bristles. Instead, the Hox protein Ubx represses development of these bristles by acting at a different stage, perhaps downstream of *ss*. It is not yet clear whether the Sp bristles first evolved only in T2 or in multiple thoracic segments and then disappeared from T1 and/or T3. In any case, the observed differences in how Hox genes repress expression of the same genes and structures in different segments seem to reflect the mosaic evolutionary history of these segments.

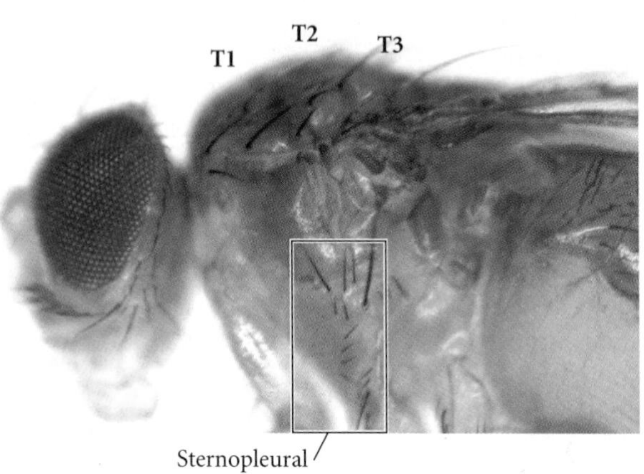

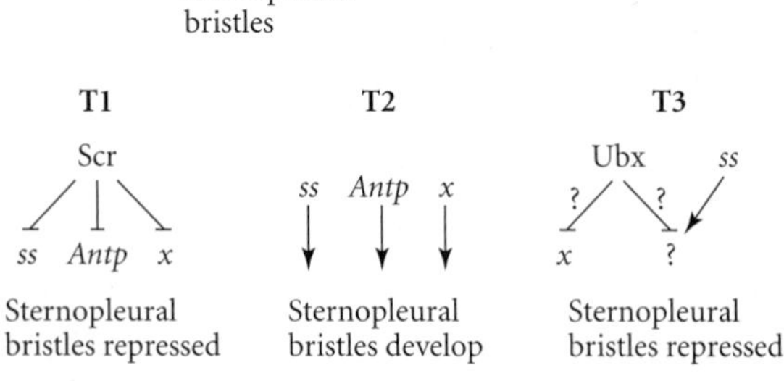

Figure 21.14 Hox proteins repress sternopleural (Sp) bristle development differently in the first (T1) and third (T3) thoracic segments of *Drosophila melanogaster*. Top: Sp bristles normally develop only in the second thoracic segment (T2). Bottom center: In T2, the *spineless* gene (*ss*) along with the Hox gene *Antennapedia* (*Antp*) and possible unknown genes (*x*) promote development of Sp bristles. Bottom left: Experiments by Tsubota et al. (2008) showed that the Hox protein Scr represses *ss* and *Antp* expression in T1. Bottom right: Tsubota et al. also showed that in T3, Sp bristle development is repressed through a different mechanism than in T1 because ectopic *ss* expression in T3 does not promote Sp bristle development. Instead, the Hox protein Ubx may act by inhibiting expression of one or more genes downstream of *ss* (question marks), or on genes required in parallel to *ss* (indicated by *x*). (Photograph courtesy of J. True.)

Co-option and the evolution of novel characters

Many genes and signaling pathways have multiple developmental roles. For example, the transcription factor Distal-less (see Figure 21.8) is required to organize the development of legs, wings, and antennae of all insects, but in some butterflies, it is also expressed later in specific positions on the developing wing, where it is involved in setting up eyespot color patterns (see the chapter-opening figure). Such cases suggest that, over the course of evolution, genes and pathways have been redeployed to serve new functions.

Change in the function of pre-existing features in adaptive evolution has been known ever since Darwin. Stephen Jay Gould and Elisabeth Vrba (1982) coined the term **exaptations** to refer to novel uses of pre-existing morphological traits (see Chapter 11). Developmental biologists have used the terms **recruitment** (Wilkins 2002) and **co-option** (reviewed by True and Carroll 2002) to refer to the evolution of novel functions for pre-existing genes and developmental pathways.

There are two ways in which a new structure might evolve by cooption of a developmental pathway. First, expression of a developmental regulatory protein may persist after the developmental stage in which it is needed. If additional target genes evolve the ability to be activated by the protein (e.g., by evolving a novel enhancer), then a new morphogenetic process may arise later in development. Alternatively, a developmental pathway originally expressed in one region of the embryo may become expressed in a different region (referred to as heterotopy; see p. 62 and Figure 3.20), leading to a duplication of that structure in the new region. For example, Stanislav Gorb and colleagues (2006) recently discovered protein secretions from tarantula feet which enable these spiders to walk on vertical and inverted surfaces. These proteins are the same ones that compose the silk secreted by the spinnerets during web production. It is unclear whether the foot or spinneret secretions represent the ancestral condition; spinnerets are believed to be serially homologous to feet. In either case, the genes encoding silk proteins have been adaptively co-opted in one of the two structures.

Similar redeployments of genes controlling developmental pathways are also known. The eyespots on the wings of nymphalid butterflies, for example (see the chapter-opening photo), which are among the last morphological features to develop, express several regulatory proteins during their development that are also required earlier for segmentation and early wing patterning (Figure 21.15A). Patterning of the tetrapod limb involves expression of Hox genes in nested anterior-posterior patterns, similar to the way in which they are initially expressed during patterning of the anterior-posterior body axis (Figure 21.15B).

In plants, a developmental circuit has been recruited in the evolution of divided, or compound, leaves, in which more or less separate leaflets develop along the main axis of the leaf (reviewed by Bharathan and Sinha 2001; Byrne et al. 2001). *KNOX1*, a member of a homeobox protein family, is expressed in the apical meristem of most plants, in association with the maintenance of growth of the stem. *KNOX1* expression must be deactivated in the leaf primordium to enable normal leaf development. However, in the developing compound leaves of the tomato plant, leaflet primordia form as bulges on the main leaf primordium, and *KNOX1* expression reappears in these regions. A growth control pathway has apparently been co-opted in the evolution of compound leaves, which has occurred many times in angiosperms (Goliber et al. 1999; Bharathan and Sinha 2001).

The developmental genetics of heterochrony

Much of morphological evolution has entailed HETEROCHRONY—evolutionary changes in the timing of development (see Chapter 3). The developmental genetic basis of hete-

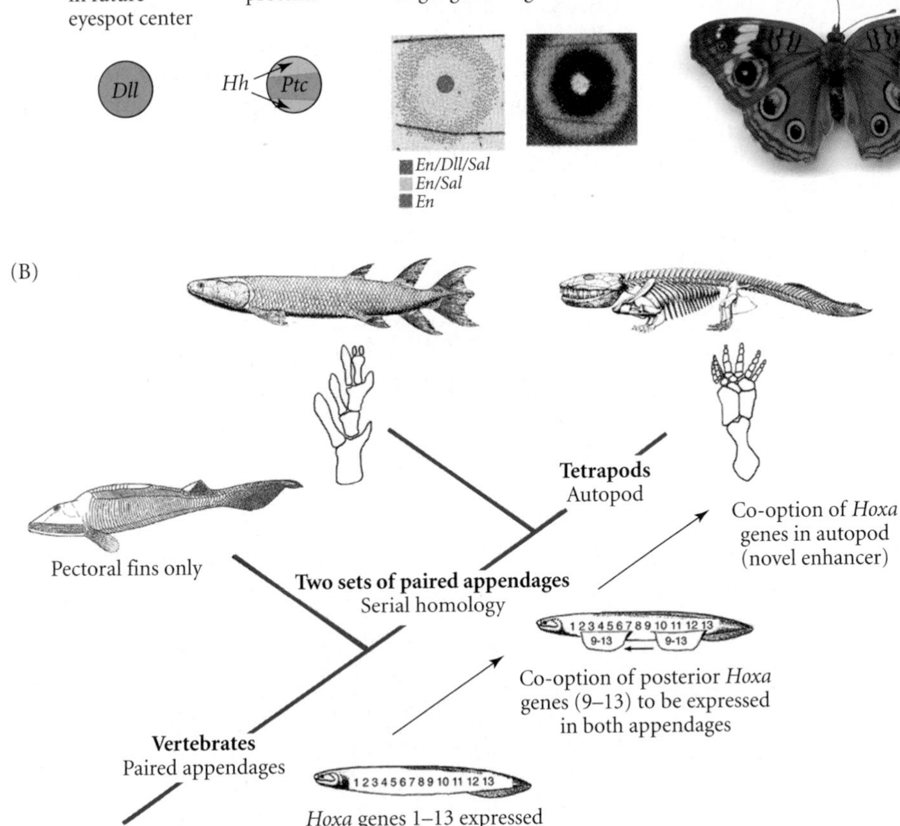

Figure 21.15 Co-option of developmental circuits in the evolution of novelties. (A) Butterfly eyespots are the developmental products not only of genes for pigmentation but also of many co-opted genes and pathways that play important roles in establishing the body plan. These include the signaling proteins Hedgehog (Hh) and Patched (Ptc) and the transcription factors Distal-less (Dll), Spalt (Sal), and Engrailed (En). (B) Co-option of the vertebrate *Hoxa* genes during evolution of the tetrapod appendage. Ancestrally, Hox genes were expressed only along the anterior-posterior axis of the developing body. The evolution of paired fore- and hindlimbs involved novel gene expression, presumably using novel enhancer sequences of *Hoxa* genes 9–13. The evolution of hands and feet (autopods) involved further novel expression patterns of the *Hoxa* genes. (From True and Carroll 2002.)

rochrony has been little studied in model organisms, but genetic approaches have been used to study neoteny in salamanders (Voss and Shaffer 2000; Voss et al. 2003). In tiger salamanders (*Ambystoma tigrinum*), the standard ontogenetic sequence involves loss of the tail fin and external gills in terrestrial adults. Several related species, including the axolotl (*A. mexicanum*), attain reproductive maturity while remaining fully aquatic, retaining the gills and other larval features (see Figure 3.16). Some species are capable of facultatively transforming into the typical terrestrial adult form if their habitats dry up, whereas others have lost this ability completely.

In salamanders, thyrotropin-releasing hormone (TRH) stimulates the pituitary gland to release another hormone, thyroid-stimulating hormone (TSH), which stimulates the thyroid gland to release a third hormone, thyroxin. Thyroxin triggers metamorphosis by inducing morphogenetic events in several different tissues. In the neotenic salamander *Ambystoma mexicanum*, the TRH cascade does not take place, but metamorphosis can be induced by injecting TRH into the animal (Shaffer and Voss 1996). This observation suggests that the evolution of neoteny in *A. mexicanum* involved inactivation of the TRH cascade. Further genetic analysis on hybrids between a metamorphic strain (referred to as *Att*) and two different nonmetamorphic strains (called *lab* and *wild*) revealed that the *Att* allele of a single locus, called *met*, is necessary for metamorphosis. However, the existence of another gene was also implicated because metamorphic individuals were 4.5 times more likely to occur in the backcross of *Att/lab* F_1 individuals to the *lab/lab* parental strain than in the backcross of *Att/wild* F_1 individuals to the *wild/wild* parental strain (Voss and Smith 2005). This second locus differs between the *lab* and *wild* paedomorphic strains and affects whether the *met* allele from the *Att* strain is able to cause metamorphosis. It is not yet clear whether these loci correspond to known genes in the TRH cascade.

Interestingly, neoteny appears to have a different genetic basis in each of three species in which it has evolved independently (Voss et al. 2003). Recently, genomic analysis has been used to understand the gene expression changes associated with changes in the timing of metamorphosis in the axolotl (Page et al. 2008). The hormone thyroxine was used to induce early metamorphosis in *A. mexicanum* and genome-level analysis revealed that more than 500 genes showed expression changes in head tissue from the early-metamorphosing animals. The current challenge is to understand the regulatory connections between the TRH cascade and this huge battery of responding genes and to identify which points in this gene network have evolved, leading to variation in heterochrony in natural populations and between species.

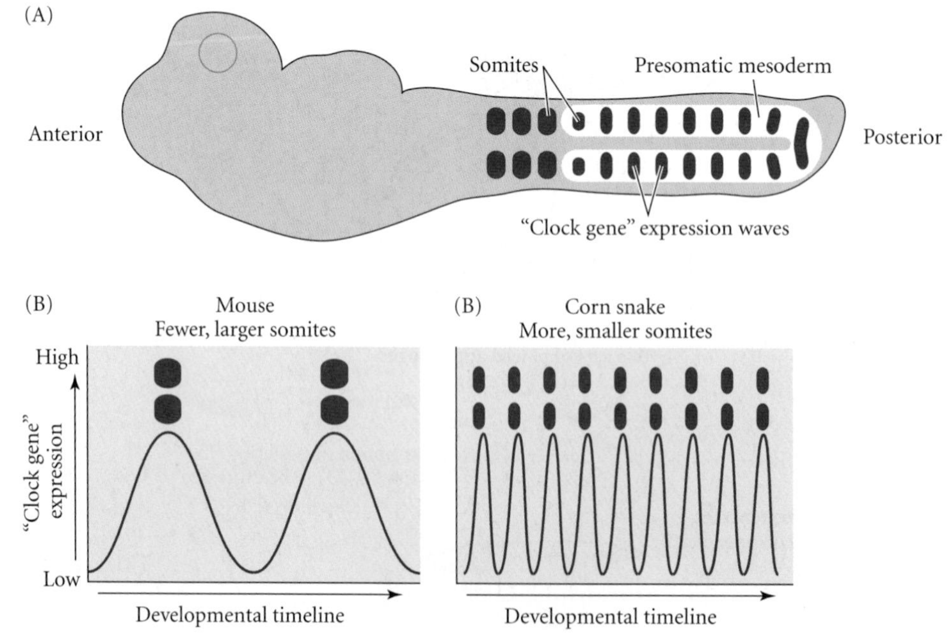

Figure 21.16 The segmentation clock in vertebrates. (A) New somites, which develop into individual vertebral segments and their associated structures, develop sequentially from the anterior presomitic mesoderm as a result of regular waves of expression of a battery of patterning genes called "clock genes." (B) In the mouse and other vertebrates with relatively few vertebral segments, the "clock" runs slowly. (C) The "clock" runs approximately four times faster in the corn snake, resulting in the formation of many more vertebral segments. (After Vonk and Richardson 2008.)

Evolutionary changes in the timing of developmental events in certain tissues of the embryo can produce a huge diversity of final forms. An excellent example was uncovered recently by Céline Gomez and colleagues (2008), who were interested in explaining how snakes evolved so many vertebrae compared with other vertebrates. The repeated segments of the vertebrate body develop from segments of the developing embryo called somites. Gomez et al. compared somite differentiation in zebrafish, chickens, mice, and corn snakes (*Elaphe guttata*) and found that all of these species employ a "clock-and-wavefront" mechanism whereby the mesoderm that eventually gets divided into somites (the presomitic mesoderm, or PSM) first proliferates and then buds off somite segments from its anterior end, which form the vertebral segments. Somites form at regular time intervals, like a ticking clock. Since there is a limited amount of starting tissue in the PSM, there is a trade-off between somite number and somite size. When there is no more PSM to bud off, somite differentiation ceases. Snakes are able to develop many more segments than most other vertebrates because their somite differentiation clock ticks much faster (Figure 21.16; Gomez et al. 2008). However, snake PSM produces small segments compared with other species. It is clear from the evolution of this timing mechanism that a continuous and wide range of final vertebral numbers is possible.

The evolution of allometry

ALLOMETRY refers to the differential growth rates of different parts of the body (see Chapter 3). Morphological evolution, including changes in shape, often involves alterations in such allometries, achieved by heterochronic changes in the growth of individual body parts. For example, in the horse lineage from *Hyracotherium* to *Equus* (see Chapter 4), the length of the face and depth of the lower jaw (mandible) are allometrically related to body size, and both increased along with body size during the evolution of browsing forms in the Eocene and Oligocene (Figure 21.17A). In the Miocene, *Merychippus* became adapted to grazing on grasses, and evolved a relatively deeper mandible, as well as more anteriorly situated molars relative to the eye socket (Figure 21.17B). These evolutionary changes are thought to have been necessary to accommodate the bases of the molars, which extend deep into the jaw bones. These teeth descend throughout life to compensate for abrasion of the grinding surfaces by grasses, which wear down teeth more rapidly than other plant foods.

Allometric relationships between different body parts can evolve quite rapidly, as in the dung beetle *Onthophagus taurus*. Males of this beetle exhibit a conspicuous POLYPHENISM, which is a non-genetically based polymorphism in which environmental differences, such as seasonal or nutritional factors, cause development to produce a small set (typically two) of alternative states. In *O. taurus*, small males do not develop horns, whereas those that attain a threshold body size develop horns that are used in male-male combat. In the mid-twentieth century, this beetle was introduced from Europe to North America and Australia in order to reduce dung accumulation in pastures. The threshold size for horn development has diverged from

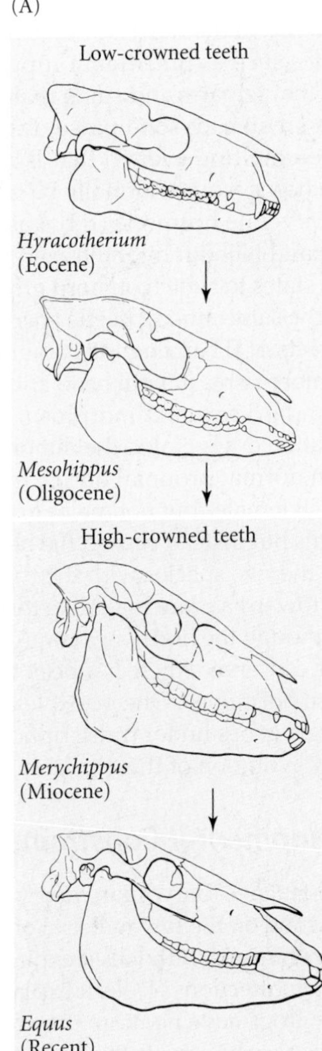

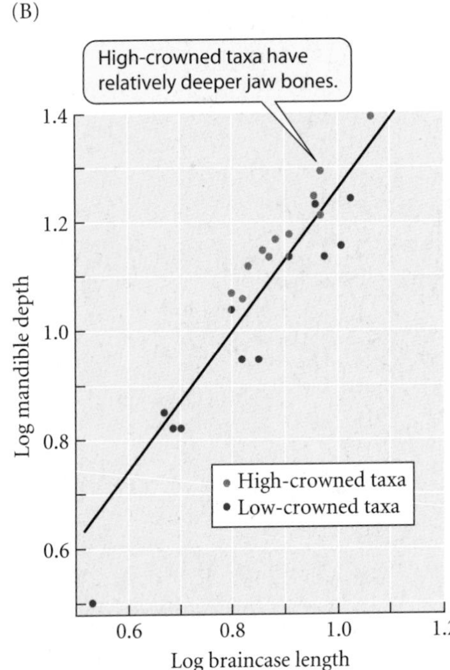

Figure 21.17 Evolution by allometry in the horse lineage. (A) Skulls of four equid taxa in the lineage from *Hyracotherium* to the modern horse, *Equus*. The face is relatively longer, and the lower jaw is relatively deeper, in the more recent taxa. The posterior molars are shifted forward of the eye socket in *Merychippus* and *Equus*, which have high-crowned teeth that extend deep into the skull. (B) A log-log plot of two measurements in various fossil horses shows that lower jaw depth increased disproportionately in relation to body size (for which length of the braincase was used as an index). (After Radinsky 1984.)

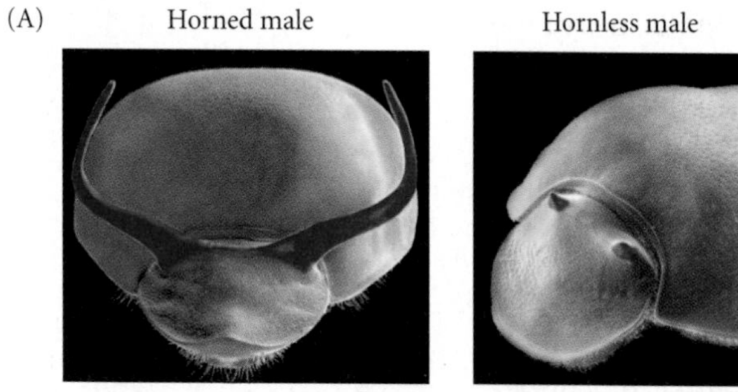

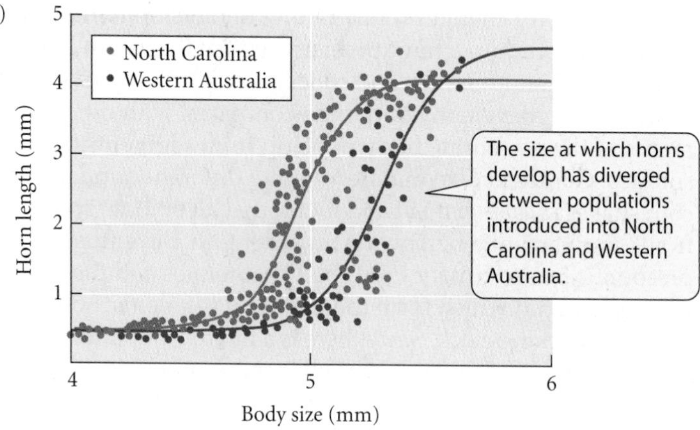

Figure 21.18 Rapid evolution of an allometric threshold in the dung beetle *Onthophagus taurus*. (A) Morphology of horned and hornless males at the same developmental stage; horns have been artificially emphasized with blue. (B) Horn length shows a steep allometric relationship to body size, and has diverged between populations introduced into Western Australia and North Carolina. (A photos courtesy of Doug Emlen; B after Moczek et al. 2002.)

the ancestral condition in both introduced populations, even though variation of this magnitude was not evident in the founding populations (Figure 21.18).

As in the cases of heterochrony discussed above, these evolutionary changes are believed to be caused by changes in the response of specific tissues (e.g., the cells destined to give rise to the horns) to global hormonal signals that appear to detect environmental states such as nutritional supplies. Douglas Emlen and H. Frederick Nijhout (1999) found that females and small males of *O. taurus*, which normally do not produce horns, have a small ecdysone pulse during the second of the five larval instars (size stages between molting events) that is not seen in large, horn-producing males. Also, males whose horns would normally be small could be induced to develop large horns if an analog of juvenile hormone (JH) was injected at a critical stage of the fourth larval instar. Emlen and Nijhout hypothesized that these two hormones thus read the nutritional and sexual states to switch on horn growth only in well-fed males. Surprisingly, the role of JH is not the same among beetle species. J. Andrew Shelby and coworkers (2007) compared the effects of JH analog injection in females among three *Onthophagus* species with sexual dimorphisms in both head and pronotal (anterior thoracic) horns. In two species with the standard sexual dimorphism in which males but not females grow large horns, JH application exaggerated the dimorphism by causing injected females to grow even smaller than normal pronotal horns. In a third species with the reverse type of dimorphism in which females but not males grow large horns, JH injection had no effect on the pronotal horns but instead caused the head horns to increase in size in injected females. Finally, one of the two species with the typical sexual dimorphism showed an alteration of horn shape toward a more male-like form in response to the JH treatment. These results show that although the hormones controlling development of sexually dimorphic horns in beetles are conserved, there has been considerable evolutionary change in the specific actions of these hormones. One possible explanation would be if these species had diverged in the target genes under transcriptional regulation by the JH pathway as a result of *cis*-regulatory evolution of these targets.

Developmental Constraints and Morphological Evolution

Traditional neo-Darwinian theory explains how natural selection, genetic drift, and gene flow, acting on the raw material of genetic variation, have produced the astonishing variety of organisms. But does it explain why organisms have *not* evolved certain features, or in certain directions? Does it explain why no turtle species bears live young, for instance, or why frogs have no more than four digits on their forelimbs? Such questions have led evolutionary biologists to ask what the **constraints** on evolution might be.

Several kinds of constraints on evolution have been distinguished. Some are universal, in that they affect all organisms; an example is the constant presence of gravity during morphogenesis. Others, referred to as PHYLOGENETIC CONSTRAINTS, are more local, affecting only a group of related organisms. There are several types of constraints on evolution.

1. *Physical constraints.* Some structures do not evolve because the properties of biological materials (e.g., bones, epidermis, DNA, RNA, etc.) do not permit them. Physical constraints can be phylogenetically local. Insects, for example, conduct oxygen and carbon dioxide by means of diffusion in narrow tubes, or tracheae, which branch throughout the body. Limits on diffusion rates are thought to set an upper limit on insect body size.
2. *Selective (or functional) constraints.* Some features do not appear in particular lineages because they are always disadvantageous, or because they might interfere with the function of an existing trait.
3. *Genetic constraints.* Genetic variation in a particular phenotype may not be present, as discussed in Chapter 13. Developmental pathways are expected to have varying degrees of tolerance for variation in their components, for example in the levels of expression of a transcriptional regulatory protein; too much or too little expression may result in misexpression of responding genes. Also, if two traits share a common pathway of morphogenesis, the genes underlying that pathway will have strong pleiotropic effects, resulting in genetic correlations that limit the freedom with which those traits can vary relative to each other (see Chapter 13). Thus genetic constraints, such as paucity of variation and genetic correlation, are closely related to developmental constraints. Butterfly eyespots provide excellent evidence of the different strengths of genetic correlations. In species bearing more than one eyespot on their wings, there are often strong correlations between the sizes and color patterns of the different eyespots. This is to be expected, because the different eyespots are likely to use similar sets of genes for their development. Paul Brakefield's research group has used several artificial selection experiments to test whether correlations on the ventral hindwing of *Bicyclus anynana* could be uncoupled. Patricia Beldade and coworkers (2002) showed that the sizes of two phenotypically correlated eyespots were readily uncoupled after only 11 generations of selection. Cerisse Allen and coworkers (2008) confirmed this finding (Figure 21.19A) but also found that, unlike eyespot size, different color composition of eyespots could not be uncoupled (Figure 21.19B). This indicates an absence of genetic variation in the regulation of pigmentation genes that would allow them to respond differently to selection on different regions of the wing.
4. *Developmental constraints.* John Maynard Smith and colleagues (1985) defined a developmental constraint as "a bias on the production of various phenotypes caused by the structure, character, composition, or dynamics of the developmental system." The two most common phenomena attributed to developmental constraint are absence or paucity of variation, including the absence of morphogenetic capacity (i.e., lack of cells, proteins, or genes required for the development of a structure), and strong correlations among characters, which may result from interaction between tissues during development or the involvement of the same genes or developmental pathways in multiple morphogenetic processes.

Developmental constraints can be revealed by embryological manipulations in the laboratory. In a classic experiment, Pere Alberch and Emily Gale (1985) used the mitosis-inhibiting chemical colchicine to inhibit digit development in the limb buds of frogs (*Xenopus*) and salamanders (*Ambystoma*; Figure 21.20). The treatment consistently caused specific digits to be missing in each species, and the missing digits were the preaxial ones in frogs and the postaxial ones in salamanders. These results reflected the different order of digit differentiation in the two taxa; the last digits to form tended to be the most sensitive to the colchicine treatment. Furthermore, the results strongly reflected evolutionary trends: salamanders have often lost postaxial digits, and frogs have repeatedly experi-

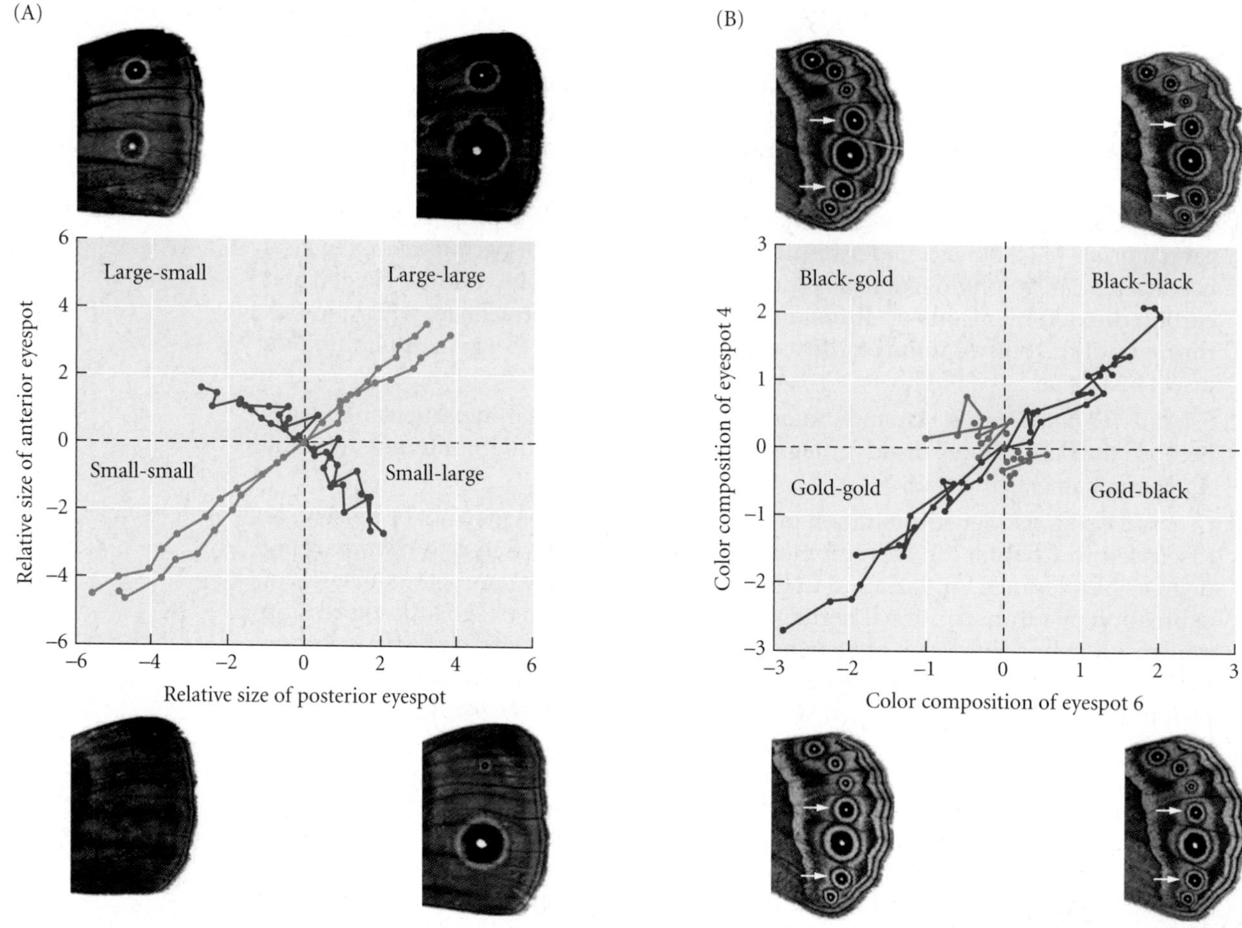

Figure 21.19 Artificial selection experiments showing different levels of constraint on different eyespot traits in *Bicyclus anynana*. (A) Trajectories of selection, starting from center of plot, on dorsal forewing eyespots, with axes in units of phenotypic standard deviations. The graph shows selection for both eyespots to be large or small (green trajectories) or for one to be large and the other to be small (purple trajectories). Each point represents one generation. (B) Results of similar experiments on the fourth and sixth eyespots of the ventral hindwing (arrows in wing photos). "Black" and "gold" refer to selection for the eyespots to exhibit more of these two respective colors. Note that the red trajectories do not extend far from the starting point at the center, indicating that selection for decoupling the colors of the two eyespots was not sucessful. This lack of success is reflected by the lack of distinctiveness in the color of these eyespots (compare the upper left and lower right wing photos). (Courtesy of Cerisse Allen and Paul Brakefield.)

enced preaxial digit reduction, during evolution. Although the digit number variation in this study was produced artificially, the results suggest that naturally occurring variation in developmental systems may be constrained by intrinsic, species-specific developmental programs.

Although in practice it is very difficult to rule out selective constraints, developmental or genetic constraints might explain some common evolutionary patterns:

1. *The absence of features in certain lineages.* For example, although viviparity (giving birth to live offspring rather than laying eggs) has evolved in some lizards and snakes, it has not evolved in turtles, even though it might well be advantageous in sea turtles by freeing them from the need to come to land to lay eggs (Williams 1992b).
2. *Directional trends* (see Chapter 3), such as cumulative elaboration of a particular structure. The developmental system may impose a bias such that certain kinds of variation are produced and not others, enabling particular evolutionary trajectories to be taken. For example, if a morphogenetic novelty arises in the ancestor of a lineage, the descendant species may gradually evolve structural modifications that could not have evolved otherwise. Bony frontal horns have evolved in only a few groups of mammals, such as the Bovidae (antelopes, goats, sheep, and cattle), among which horn sizes and shapes vary greatly. In contrast to the unbranched horns of bovids, the evolution of branching in the antlers of the Cervidae (deer) has enabled a very different variety of forms to evolve (Figure 21.21).
3. *Parallel evolution of traits in independent lineages.* For example, similar patterns of abdominal pigmentation and cuticular cell hair morphology have evolved in several lineages of *Drosophila*. Nicolas Gompel and Sean Carroll (2003)

(A)

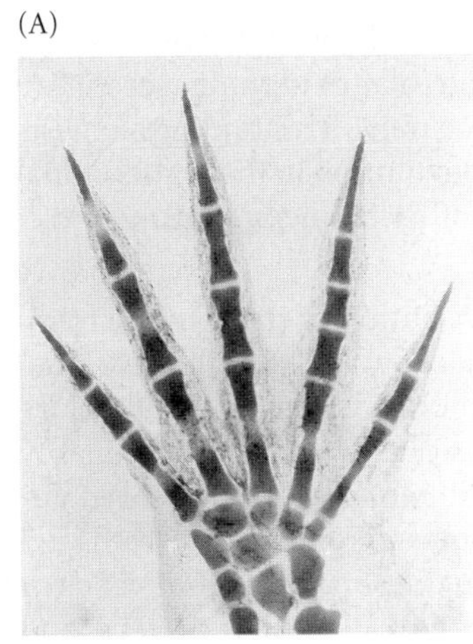

(B)

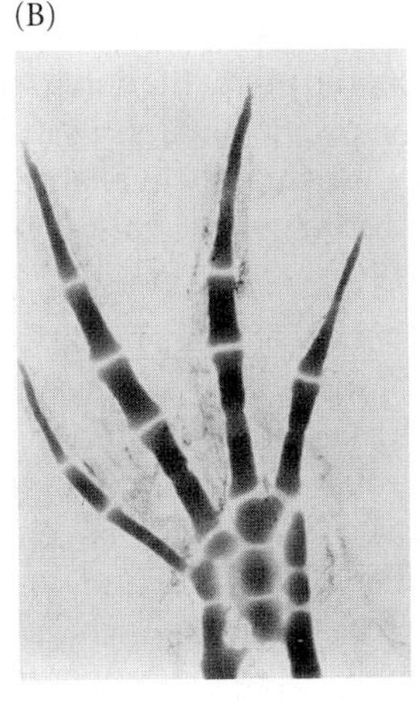

(C)

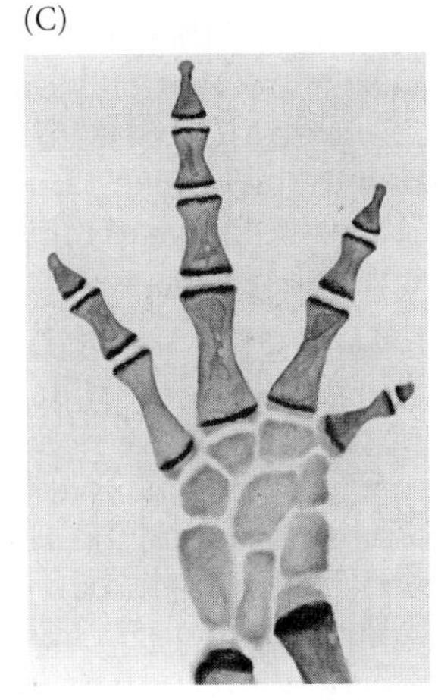

Figure 21.20 Evidence for developmental constraints. (A) X-ray of the right hind foot of an axolotl salamander (*Ambystoma mexicanum*), showing the normal five-toed condition. (B) The left hind foot of the same individual, which was treated by an inhibitor of mitosis during the limb bud stage. The foot lacks the postaxial toe and some toe segments, and is smaller than the control foot. (C) A normal left hind foot of the four-toed salamander (*Hemidactylium scutatum*) has the same features as the experimentally treated foot of the axolotl. (From Alberch and Gale 1985; photos courtesy of P. Alberch.)

(A) Bovidae (unbranched horns)

(B) Cervidae (branched antlers)

Figure 21.21 Bony horns are a novelty that have evolved in only a few mammalian lineages. (A) The frontal horns of the Bovidae (antelope, cattle, sheep, and goats) have evolved a stunning variety of forms, all of which are unbranched. Left to right: Cape buffalo (*Syncerus caffer*), gemsbok (*Oryx gazella*), and Nubian ibex (*Capra nubiana*). (B) The antlers borne by males of many deer species (Cervidae) also display wide diversity, but on a "branched template." Left to right: white-tailed deer (*Odocoileus virginianus*), Indian muntjac (*Muntiacus muntiacus*), and moose (*Alces alces*). (Buffalo, gemsbok courtesy of Andrew Sinauer; ibex and deer courtesy of David McIntyre; muntjac © OSF/photolibrary.com; moose courtesy of Donna Dewhurst/U.S. Fish & Wildlife Service.)

demonstrated that changes in the expression of the transcription factor bric-a-brac, which controls sexually dimorphic pigmentation and cuticular patterning (see Figure 21.22), are correlated with interspecific differences in these traits. This finding suggests that only one genetic pathway was readily available for evolution in these abdominal traits, possibly due to limitations in genetic variation at other pigmentation and cuticular development genes.

4. *"Standardization,"* or the reduction of morphological variation of traits in the fossil record following an initial phase of high variation. One of several possible explanations for this pattern is **canalization**: the evolution of modifications of the developmental system such that the most highly advantageous phenotype is more reliably produced (Waddington 1942; see Chapter 13). Several proteins that are present in virtually all eukaryotic cells, called "heat shock proteins" or chaperonins, have been suggested to play roles in this process, although what those roles may be is not yet clear (see Figure 13.19). These proteins aid in the proper folding of all cellular proteins and are expressed at high levels when the organism experiences various environmental stresses, such as thermal shock or infection.
5. *Morphological stasis over long periods of evolutionary time.* The absence of evolutionary change has many possible explanations, of which developmental constraint is one (see Chapter 22).
6. *Similarities in embryological stages among higher taxa.* Such similarities might result from the need to conserve early developmental processes so as not to disturb the later events that depend on them (Riedl 1978). von Baer's law (see Figure 3.14) may be a consequence of developmental constraints. For example, the notochord, which persists throughout life in the most "primitive" vertebrates, is almost completely lost in the postembryonic forms of "advanced" vertebrates, but is needed in the embryo in order to induce the differentiation of central nervous system tissues.

The Molecular Genetic Basis of Gene Regulatory Evolution

Morphological novelty and diversity originate in intraspecific variation and in differences between closely related species. Most of this variation is polygenic and can be studied by QTL mapping. The kinds of genes and the kinds of alleles involved in these differences (i.e., *cis*-regulatory or amino acid-coding sequences) are particularly interesting to EDB researchers. The few QTL studies that implicate specific genes suggest that developmental regulatory loci are commonly involved in morphological differences between species, and that these loci can include major developmental regulatory genes such as *Ubx*, as we saw above. Thus the many crucial developmental functions of these genes do not preclude their involvement in short-term evolutionary change. Changes in the regulatory regions of these genes have been implicated more often than changes in their amino acid-coding sequences, confirming the importance of variation in gene regulation in morphological evolution.

In *Drosophila melanogaster*, naturally occurring variation in bristle number is caused partly by nucleotide substitutions at regulatory loci encoding both cell signaling proteins and transcription factors (Lai et al. 1994; Long et al. 1998, 2000). An example of regulatory differences among closely related species involves the gene *bric-a-brac* (*bab*), which encodes a transcription factor required for sexually dimorphic pigmentation and cuticle morphology in *Drosophila* and is associated with a QTL for abdominal pigmentation in *D. melanogaster* (Kopp et al. 2003). Sexually dimorphic abdominal pigmentation evolved once, in the lineage leading to the *melanogaster* species group after its divergence from the *obscura* species group. In the *melanogaster* group, *bab* is expressed in the posterior abdomen of females, in which it is required to prevent the development of male abdominal pigmentation, and it is repressed in males, enabling the development of pigmentation (Figure 21.22A–C). However, the species in the *montium* lineage, a subgroup of the *melanogaster* group, generally lack sex-specific pigmentation patterns, even though *bab* expression is downregulated in males—in which it may regulate sexual dimorphism in abdominal cuti-

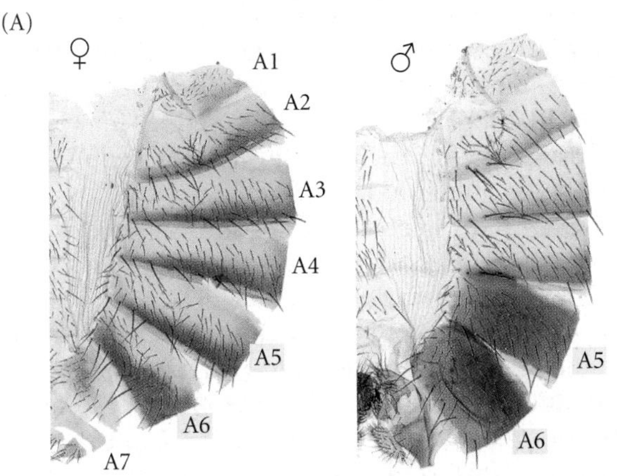

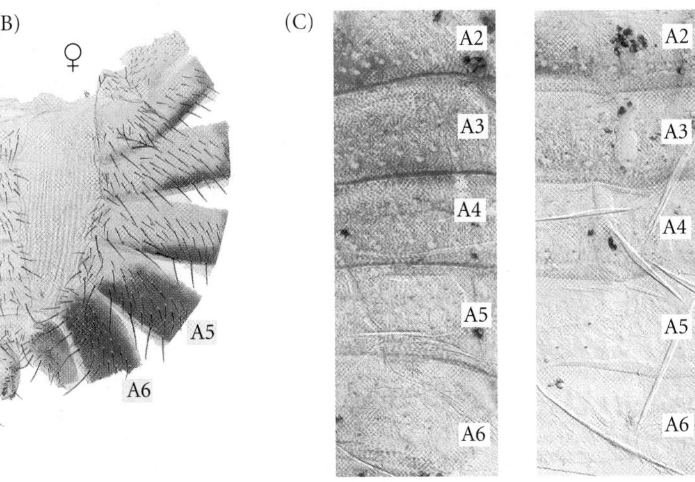

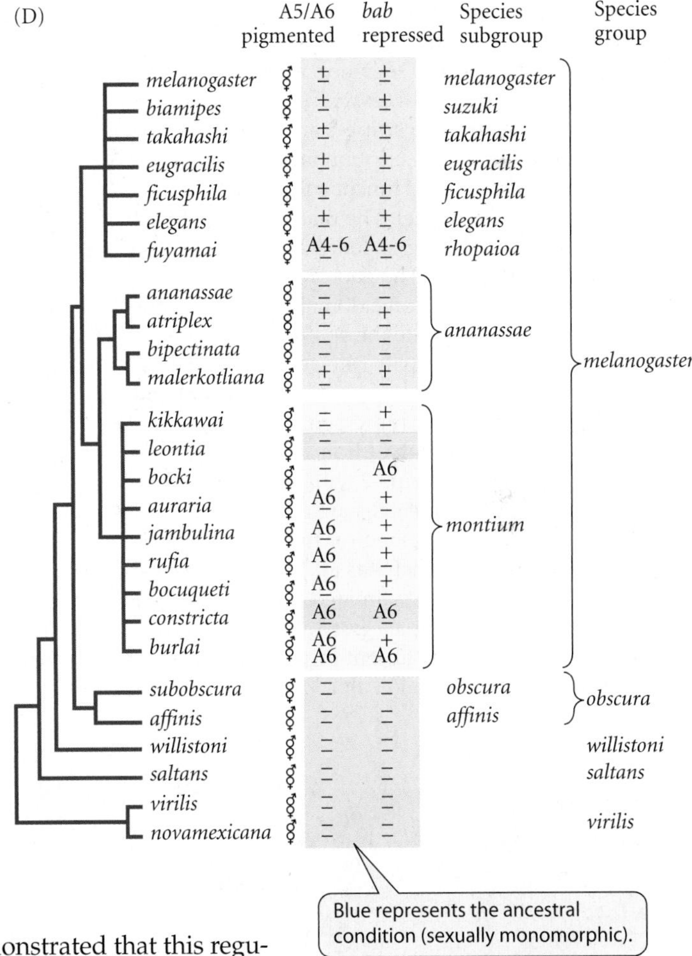

(D)
A5/A6 pigmented
bab repressed
Species subgroup
Species group
melanogaster ⚥ +/− +/− melanogaster
biamipes ⚥ +/− +/− suzuki
takahashi ⚥ +/− +/− takahashi
eugracilis ⚥ +/− +/− eugracilis
ficusphila ⚥ +/− +/− ficusphila
elegans ⚥ +/− +/− elegans
fuyamai ⚥ A4-6/− A4-6/− rhopaioa
ananassae ⚥ −/− −/−
atriplex ⚥ +/− +/−
bipectinata ⚥ −/− −/−
malerkotliana ⚥ +/− +/−
ananassae
melanogaster
kikkawai ⚥ −/− +/−
leontia ⚥ −/− −/−
bocki ⚥ −/− A6/−
auraria ⚥ A6/− +/−
jambulina ⚥ A6/− +/−
rufia ⚥ A6/− +/−
bocuqueti ⚥ A6/− +/−
constricta ⚥ A6/− A6/−
burlai ⚥ A6/A6 +/A6
montium
subobscura ⚥ −/− −/− obscura
affinis ⚥ −/− −/− affinis
obscura
willistoni ⚥ −/− −/− willistoni
saltans ⚥ −/− −/− saltans
virilis ⚥ −/− −/−
novamexicana ⚥ −/− −/− virilis
Blue represents the ancestral condition (sexually monomorphic).

Figure 21.22 Role of *bric-a-brac* (*bab*) expression in sexually dimorphic abdominal pigmentation in *Drosophila.* (A) Abdominal cuticle of wild-type female and male *D. melanogaster*, showing differences in the pigmentation of the fifth (A5) and sixth (A6) abdominal segments. (B) A *bab* mutant female, showing male-like pigmentation in A5 and A6. (C) A reporter construct shows *bab* expression (blue) in developing female (left) and male (right) abdomens. Note the lack of *bab* expression in A5 and A6 of the male. (D) Sexually dimorphic *bab* expression is associated with sexually dimorphic abdominal pigmentation. The + and – signs indicate presence or absence of sexually dimorphic A5 and A6 pigmentation. In the right column, a + sign indicates the repression of *bab* in A5 and A6 segments. Orange indicates species with correlated sexually dimorphic abdominal pigmentation and *bab* expression. Yellow indicates species with male-specific downregulation of *bab* but no male-specific pigmentation. Blue indicates species that are sexually monomorphic in both abdominal pigmentation and abdominal *bab* expression (the ancestral condition). (After Kopp et al. 2000. Photos courtesy of A. Kopp and S. Carroll.)

cle patterns rather than in pigmentation (Figure 21.22D). Developmental genetic data from *D. melanogaster* indicate that *bab*'s role is to produce sexual dimorphism in specific abdominal segments by integrating information about anterior-posterior position in the abdomen, conferred by expression of the Hox gene *Abdominal B* (*AbdB*), and about sexual identity, conferred by expression of the gene *doublesex* (*dsx*). Artyom Kopp and colleagues (2000) have proposed that, following the divergence of the *melanogaster* and *obscura* lineages about 25 Mya, transcriptional regulation of *bab* evolved to integrate the *AbdB* and *dsx* signals, presumably through changes in its *cis*-regulatory DNA. Thomas Williams and colleagues (2008) recently demonstrated that this regulation occurs by the direct binding of AbdB and Dsx to a *cis*-regulatory enhancer of *bab*. AbdB directly activates the expression of the gene *yellow*, which is required for dark pigmentation, in species with dark male abdomens. Jeong et al. (2006) showed that that evolution of regulation of the *yellow* gene by AbdB in the common ancestor of the sexually dimorphic lineages occurred via nucleotide substitutions in the *yellow cis*-regulatory DNA that created novel AbdB binding sites. This work is among the first demonstrations of a direct regulatory interaction between a Hox protein and a target gene that contributes to morphological evolution.

Molecular evolutionary studies in closely related species show that even when enhancer function is conserved, the binding sites within an enhancer can change in surprising ways. In a groundbreaking research program, Michael Ludwig, Martin Kreitman, and colleagues (Ludwig and Kreitman 1995; Ludwig et al. 1998, 2000) examined sequence and functional differences among *Drosophila* species in a specific *cis*-regulatory sequence from the gene *even-skipped*, a transcription factor required for segmentation. In *Drosophila melanogaster*, this enhancer, called *eve stripe-2*, is a 670-base-pair segment of DNA that is necessary and sufficient to direct expression of *even-skipped* in the second of seven stripes in which the protein is expressed in wild-type embryos (Figure 21.23A). This enhancer contains 17 binding sites for transcription factors encoded by four different regulatory genes that are expressed in anterior-posterior gradients and that, by activation and repression, combinatorially determine the precise limits of the *even-skipped* stripes (Figure 21.23B). Kreitman's group sequenced *eve stripe-2* from five *Drosophila* species and examined the number and positions of these binding sites, as well as the function of these species' enhancers when placed in transgenic *D. melanogaster*. Surprisingly, even though the sequences from all five species directed the correct stripe of expression, the nucleotide sequence of only 3 of the 17 binding sites was perfectly conserved across all five species (Figure 21.23C). In fact, two of the binding sites were not found at all in two of the more distantly related species.

Ludwig and Kreitman wondered whether the function of the *eve stripe-2* enhancer was conserved despite these sequence differences because each individual part of the enhancer maintained its independent role, or because changes in some parts were accompanied by compensatory changes in other parts (i.e., coevolution of the entire element). They constructed enhancers made up of DNA from two different *Drosophila* species. When these chimeric enhancers were placed in transgenic *D. melanogaster*, the discrete stripe of *even-skipped* expression was usually not seen, showing that the chimeric enhancers could not promote the normal pattern of transcriptional activation. Therefore, it appears that within each species, the entire enhancer has evolved in a concerted fashion to maintain its function. If the fixation (by selection or drift) of a nucleotide substitution in one position in the enhancer caused a minor alteration in stripe expression, that change might provide selection pressure for a compensatory substitution elsewhere in the enhancer in order to maintain the stripe expression.

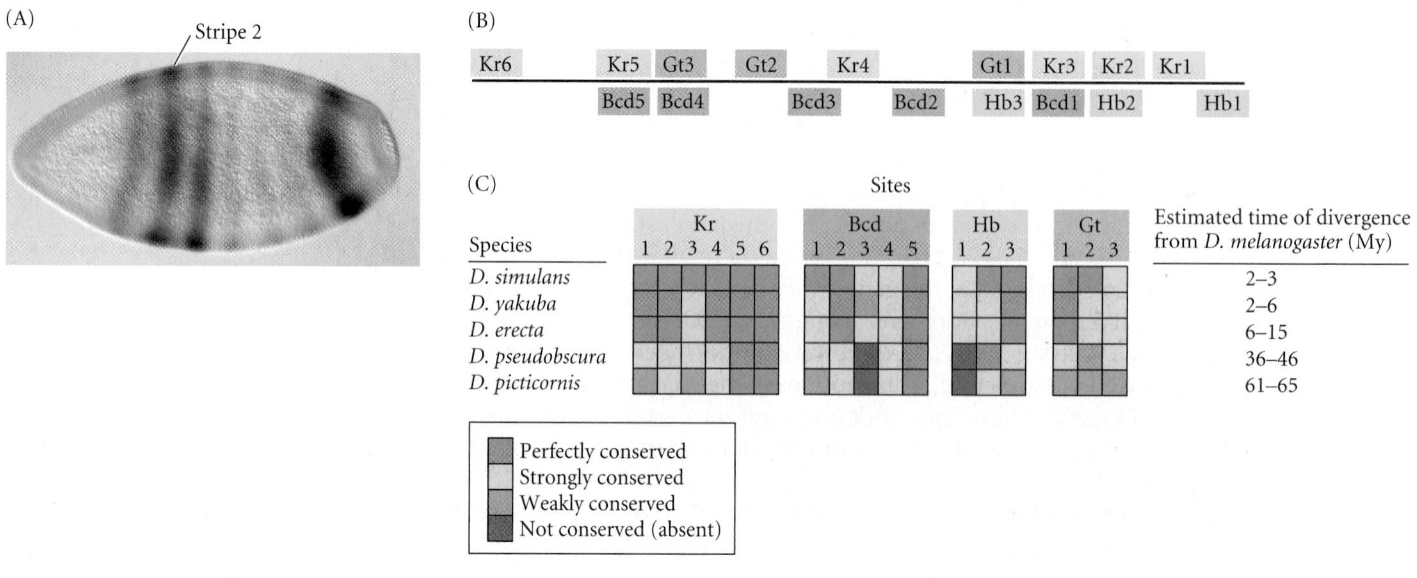

Figure 21.23 Evolution of the even-skipped stripe 2 (eve stripe-2) enhancer in *Drosophila*. (A) Embryo in the blastoderm cellularization stage stained for expression of an eve stripe-2 reporter construct (blue) and the Even-skipped protein (brown). The *even-skipped* (*eve*) gene is a "pair-rule" gene that is expressed in seven stripes, one in every other segment, in the embryo. The eve stripe-2 construct is also expressed in stripes 3 and 7. (B) Schematic of the eve stripe-2 enhancer, showing binding sites for four transcription factors: the maternal effect proteins Bicoid and Hunchback and the "gap" proteins Giant and Krüppel. The maternal effect proteins are laid down during oogenesis in anterior-posterior gradients in the egg. The gap proteins are expressed in broad embryonic domains in the zygote and are required to regulate the spatial expression patterns of *eve* and the other pair-rule genes. Bicoid and Hunchback upregulate transcription of *eve;* Giant and Krüppel downregulate, or repress, its transcription. (C) Conservation and evolution of transcription factor binding sites in the eve stripe-2 enhancer among *Drosophila* species. Binding sites are numbered according to a convention established for the *D. melanogaster* eve stripe-2 sequence, which has all 17 sites. The column at right indicates the estimated time of divergence of each species from *D. melanogaster*. More distantly related species have more absent or weakly conserved binding sites. (After Carroll et al. 2005.)

Toward the EDB of Homo sapiens

As the field of evolutionary developmental biology unfolds, one of its most important and fascinating endeavors will be to elucidate the developmental genetic and evolutionary mechanisms involved in the appearance of traits unique to humans, such as our large brain size; craniofacial morphology; vertebral, limb, and digit innovations; reduced hair cover; and, of course, our complex behavioral and cultural traits (reviewed by Carroll 2003). Based on studies in model organisms, we can expect that many of the innovations that evolved in the human lineage involved several or many genes. Comparative genomic data suggest that many or most of the DNA-level changes responsible for unique human traits caused alterations in the regulation of developmental and structural proteins that we share with our primate and mammalian relatives. Identifying the potential functional roles of nucleotide differences in noncoding DNA is much more difficult than in protein-coding regions, but these potential regulatory regions have long been thought to be of primary importance in differences between humans and chimpanzees (King and Wilson 1975). Bioinformatics is providing powerful new approaches to this question.

One clue that a particular regulatory region might have an important function is if its sequence is highly conserved among species. Recently Shyam Prabhakar and colleagues (2008) scanned for such regions by comparing the genome sequences of humans and other primates. They assayed the development of one such region, called *human-accelerated conserved noncoding sequence-1*, or *HACNS1*, which showed an intriguing cluster of human-specific nucleotide substitutions at positions that were otherwise conserved among non-human vertebrates (Figure 21.24A). The expression assays were done by engineering reporter constructs (see Box 21A) of this sequence from humans and from two other primates, chimpanzees and rhesus monkeys, and inserting these into transgenic mice. Constructs from all three species were expressed in the developing eye, ear, and pharyngeal arch, but interestingly, only the human construct was expressed in the anterior of the forelimb and hindlimb buds (Figure 21.24B). In further experiments, these expression differences were attributed to a cluster of 13 nucleotide substitutions in humans, found in an 81 base pair region that is otherwise highly conserved among terrestrial vertebrates. It is still uncertain which gene the *HACNS1* protein

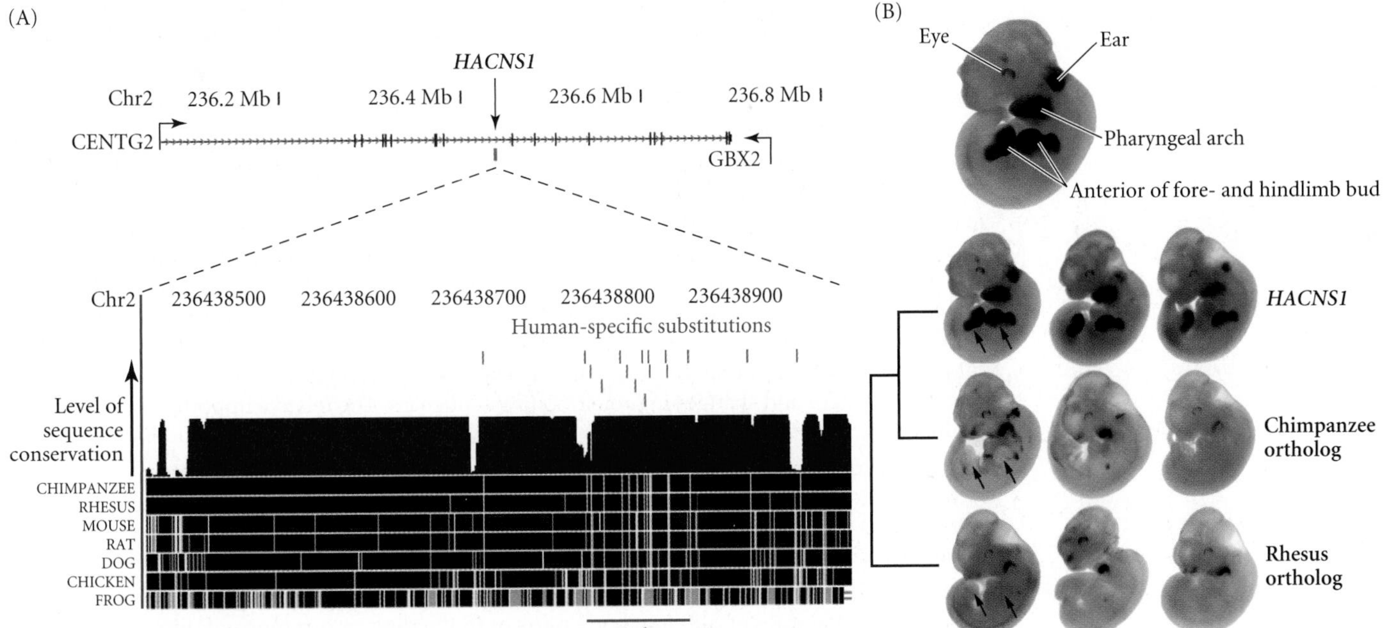

Figure 21.24 Gain of function in a developmentally expressed enhancer in the human lineage. (A) The human-accelerated conserved noncoding sequence-1 (*HACNS1*) region is highly conserved among vertebrates. The top indicates the human genome region surrounding *HACNS1* with vertical bars representing known transcribed genes. The bottom indicates conservation of the *HACNS1* region among vertebrates. Blue bars indicate the degree of conservation. Black vertical bars below indicate conserved regions among taxa. Red vertical bars indicate human-specific nucleotide substitions. Yellow vertical bars indicate conservation of these positions (which have different nucleotides in humans than in other vertebrates) among nonhuman vertebrates. (B) Evolution of *HACNS1* expression in humans as assayed by reporter construct expression (dark blue) in transgenic mice during embryonic development. The top figure and top row show human *HACNS1* expression with primordia of structures indicated. The second and third rows show expression of chimpanzee and Rhesus monkey orthologs, respectively. Arrows indicate that expression in the anterior of fore- and hindlimb buds is absent in these two species, which is the ancestral condition. (From Prabhakar et al. 2008; © AAAS, used by permission.)

regulates, but a nearby gene, *GBX2*, encodes a transcriptional regulatory protein that is expressed in developing limbs.

The exciting possibility that this work suggests is that key aspects of human limb development, such as the modifications allowing tool use or bipedalism, may have resulted in part from selection on this regulatory element. As the work of Prabhakar et al. illustrates, large-scale identification of the regulatory and coding region substitutions involved in human evolution will be much more complex than simply sequencing and comparing the human and chimpanzee genomes. It will also require extensive experimental work on gene expression using microarrays and other new technologies. Many clues to the genetic bases of human traits will also come from studying human variation, including genetic disorders, and development in many different model species. The examples cited in this chapter provide new opportunities to study the history and dynamics of molecular genetic variation contributing to phenotypic evolution at scales ranging from within populations to among higher taxonomic groups. Insights at all of these levels will be invaluable to "human EDB," for example in shedding light on whether the genes underlying trait variation within the human species are the same as those involved in divergence from our cousins, the great apes. Advances in genomics enabling insights on evolutionary genetics throughout the tree of life are expected to accelerate our knowledge of how our own branch of the tree emerged and adapted.

Companion Web Site at
www.sinauer.com/evolution
for quizzes, data analysis and simulation exercises, and other study aids.

Summary

1. Evolutionary developmental biology (EDB) seeks to integrate data from comparative embryology and developmental genetics with theory and data on morphological evolution and population genetics.
2. Phylogenetic homology can differ from biological homology, a concept based on the realization that homology must take into account information on the developmental genetics of morphological traits.
3. Many of the genes and developmental pathways underlying morphogenesis in multicellular organisms are conserved across wide phyletic ranges, showing that the vast diversity of multicellular eukaryotes is largely due to diverse uses of a highly conserved "toolkit" of genes and developmental pathways.
4. Modularity among body parts is achieved by segment-specific patterning mechanisms that result from modularity in gene regulation, allowing genes to be independently regulated in different parts of the body and at different developmental stages. Modularity has been important in enabling different parts of the body to develop divergent morphologies. Differences in morphology among segments of bilaterian animals, regulated by Hox proteins, provide many classic examples of mosaic evolution enabled by modular developmental pathways that can be deployed differently in each segment.
5. Noncoding DNA sequences called enhancers or *cis*-regulatory elements independently control expression of each gene by binding different sets of transcription factors that are present in different areas of the developing body. The evolution of differences in gene expression is largely due to evolution of these enhancers.
6. The genes responsible for developmental pathways include those that encode signaling proteins, transcription factors, and structural genes. Evolutionary change in the regulatory connections among signaling pathways and transcription factors, and between transcription factors and their targets, is believed to underlie much of the phenotypic diversity seen in nature. Evolution of protein-coding sequences has also given rise to many novel adaptive traits. Morphological variation within and among species has been found to be caused by both regulatory changes and changes in protein coding sequences. The relative importance of these two fundamental types of genetic changes to phenotypic evolution is currently under debate.
7. During evolution, genes and developmental pathways have often been co-opted, or recruited, for new functions that probably are responsible for the evolution of many novel morphological traits. This process results from evolutionary changes in gene regulation and diversification in the function and expression patterns of duplicated genes.
8. Many differences among species are due to heterochronic or allometric changes in the relative developmental rates of different body parts or in the rates or durations of different life histo-

ry stages. The modularity of morphogenesis in different body parts and in different developmental stages facilitates such changes.

9. Several kinds of constraints on evolution may determine that certain evolutionary trajectories are followed and not others. Some correlations of traits in natural populations can be easily broken using artificial selection, whereas others cannot. This suggests that some pairs of traits are constrained to vary in limited ways due to limitations in the function and regulation of the underlying genes.
10. Some genes may be involved in phenotypic evolution more often than others in the same pathway (e.g., *yellow* in *Drosophila* species and *Mc1r* in vertebrates), suggesting that not all components of developmental genetic pathways are readily available for natural selection to act upon. This could be because of differential pleiotropy among genes.
11. "Loss-of-function" alleles of genes involved in the development of morphological traits can segregate at appreciable frequencies in natural populations and be key components of morphological evolution when environmental changes occur or when small subpopulations colonize new habitats, as seen in stickleback fish.

Terms and Concepts

biological homology concept
canalization
***cis*-regulatory elements**
constraints
co-option
developmental circuits (= developmental pathways)
ectopic expression
enhancers
evolutionary developmental biology (EDB)
exaptations
homeobox
homeobox genes
homeodomain
Hox genes
promoters
recruitment
regulatory modularity
target genes
transcription factors
***trans*-regulation**

Suggestions for Further Reading

Evolutionary biologists' interest in development was rekindled in the 1970s largely by Stephen Jay Gould, who also portrayed the early history of the subject in *Ontogeny and Phylogeny* (Harvard University Press, Cambridge, MA, 1977). An excellent, very readable introduction to contemporary evolutionary developmental biology, emphasizing regulation of gene expression, is *From DNA to Diversity: Molecular Genetics and the Evolution of Animal Design* by S. B. Carroll, J. K. Grenier, and S. D. Weatherbee (2nd edition, Blackwell Science, Malden, MA, 2005). A more extended treatment is *The Evolution of Developmental Pathways* by A. S. Wilkins (Sinauer Associates, Sunderland, MA, 2002).

For the more general science reader, Sean B. Carroll describes the development of EDB in *Endless Forms Most Beautiful: The New Science of Evo-Devo* (W.W. Norton, New York, 2006) and connects EDB to both the fossil record and the fundamentals of natural selection in *The Making of the Fittest: DNA and the Ultimate Forensic Record of Evolution* (W.W. Norton, New York, 2007).

Rudolph A. Raff, one of the founders of EDB, portrayed the field somewhat earlier in his influential book *The Shape of Life: Genes, Development, and the Evolution of Animal Form* (University of Chicago Press, 1996). E. H. Davidson, a developmental biologist whose insights helped to shape the field, does much the same in *Genomic Regulatory Systems: Development and Evolution* (Academic Press, San Diego, 2001).

For a view of EDB from a plant perspective, see W. E. Friedman, R. C., Moore, and M. D. Purugganan, "The evolution of plant development" in *American Journal of Botany* 91:1726–1741, 2004.

A popular article by Carl Zimmer ("A fin is a limb is a wing," *National Geographic*, November 2006) discusses how morphological novelty and diversity have evolved from a common set of developmental genetic tools from the perspective of vertebrate limbs.

A thought-provoking synthesis of EDB and the emerging field of ecological developmental biology can be found in *Ecological Developmental Biology: Integrating Epigenetics, Medicine, and Evolution* by Scott Gilbert and David Epel (Sinauer Associates, Sunderland, MA, 2009). In addition, Gilbert's textbook *Developmental Biology* (8th edition, Sinauer Associates, 2006) is a thorough exploration of the rapidly moving field of modern developmental biology.

Problems and Discussion Topics

1. If two allometrically related traits show a strong correlation both within and among species, what kinds of experiments would you use to test whether these correlations are due to natural selection or to developmental genetic constraints? When some correlations but not others are found to be constrained, what does this suggest about the underlying genes?
2. How might differential expression and regulation by Hox genes contribute to mosaic evolution in which different segments of an animal body plan evolve different morphology?
3. If mutations such as those of the *Ubx* gene can drastically change morphology in a single step, why should most evolutionary biologists maintain that evolution has generally proceeded by successive small steps?
4. Describe how convergent evolution of similar morphology in two different lineages might involve DNA sequence evolution in different parts of the same developmental gene.
5. Would you predict that novel structures that require complex morphogenetic processes are more often gained or lost in evolution? How might you address this question given a group of organisms with known phylogenetic relationships that vary with respect to presence or absence of a complex structure, such as an eyespot or a pair of appendages modified for feeding?
6. How might the regulatory DNA sequences underlying a heterochronic change (e.g., an increase in the developmental rate of the larval stage of an insect, resulting in a shorter larval period) differ from those responsible for an evolutionary novelty in a particular body segment (e.g., a novel wing pigmentation spot)? What spatio-temporal components of the developmental system would you expect to be acting in these two cases, and what sorts of genes (encoding transcription factors, signaling proteins, hormones) would you look for as candidate genes underlying these two types of evolutionary change?
7. Development of a morphological structure involves many different types of gene products, including transcription factors, signaling proteins, and "effector" genes such as enzymes. When a morphological change occurs in a single mutational step, which of these types might be more or less likely to be involved? Within a gene, would such single-step events be more likely to involve coding or non-coding sequences, and what characteristics of the gene's function might affect this likelihood?
8. When a lineage loses an ancestral character (such as the loss of wings in the ancestor of the Phasmatodea), how do you think a derived lineage, such as the stick insect, might regain the lost trait? What ecological or population-level factors might act to cause the disappearance of the trait, and what factors might be involved in regaining or maintaining the developmental pathway needed for morphogenesis of that trait?

CHAPTER 22

Macroevolution: Evolution above the Species Level

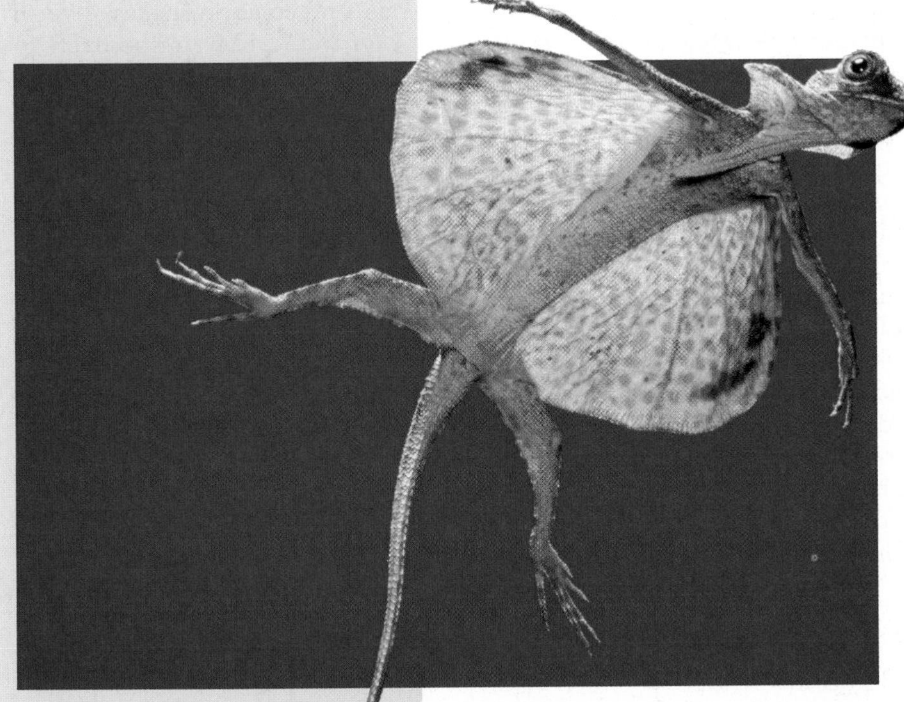

Structural changes accompany changes in function. The ribs of the arboreal gliding lizards (*Draco* spp.) of Asia are greatly elongated and support a membrane that is opened to form a "wing" when the lizard glides between trees. The flattened hind limbs and neck flaps also increase the surface area. (Photo © Satoshi Kuribayashi/OSF/Photolibrary.com.)

The phenomena of evolution are often divided into **microevolution** (meaning, mostly, processes that occur within species) and **macroevolution**, which is often defined as "evolution above the species level." "Macroevolution" has slightly different meanings to different authors. To Stephen Jay Gould (2002, p. 38), it meant "evolutionary phenomenology from the origin of species on up." These phenomena include patterns of origination, extinction, and diversification of higher taxa, the subject of Chapter 7. To other authors, macroevolution is restricted to the evolution of great phenotypic changes, or the origin of characteristics that diagnose higher taxa (e.g., Levinton 2001). The subject matter of macroevolutionary studies, however defined, includes patterns that have developed over great periods of evolutionary time—patterns that are usually revealed by paleontological or comparative phylogenetic studies, even if their explanation lies in genetic and ecological processes that can be studied in living organisms. Thus we want to know how fast evolution happens and what determines its pace, whether the great differences that distinguish higher taxa have arisen gradually or discontinuously, what the mechanisms are by which novel features have come into existence, and whether or not there are grand trends, or progress of any kind, in the history of life.

Much of the modern study of macroevolution stems from themes and principles developed by the paleontologist George Gaylord Simpson (1947, 1953), who focused on rates and directions of evolution perceived in the fossil record, and Bernhard Rensch (1959), a zoologist who inferred patterns of evolution from comparative morphology. Contemporary macroevolutionary studies draw on the fossil record, on phylogenetic patterns of evolutionary change, on evolutionary developmental biology, and on our understanding of genetic and ecological processes.

Rates of Evolution

As we noted in Chapters 4 and 7, rates of evolution vary greatly. Simpson (1953), who pioneered the study of evolutionary rates, distinguished rates at which single characters or complexes of characters evolve (which he called PHYLOGENETIC RATES) from TAXONOMIC RATES, the rates at which taxa with different characteristics originate, become extinct, and replace one another. For example, Steven Stanley (1979) analyzed several clades that have increased exponentially in species diversity during the Cenozoic. Among the most rapidly radiating groups are murid rodents (mice and rats) and colubrid snakes, diverse groups that arose in the Miocene and took about 1.98 Myr and 1.24 Myr, respectively, to double in species number. These rates mean that, without extinction, each rodent species would speciate, on average, within about 2 Myr (assuming that each species bifurcates into two "daughter" species). This interval is roughly the same as, or even greater than, the time required for speciation that has been estimated from genetic differences between sister species of living organisms (see Chapter 18). Thus a duration of 1 or 2 million years per speciation event is more than enough to account for the evolution of great diversity, even in the most species-rich groups.

Individual characters evolve at rates that differ greatly (see Chapters 4 and 13). Many features in fossilized lineages, such as the tooth dimensions of the early horse *Hyracotherium* (see Figure 4.25A), show the pattern that Niles Eldredge and Gould (1972) called **punctuated equilibrium**: long periods of little change (which they called **stasis**) interrupted by brief episodes of much more rapid change (Figure 22.1A, left panel). During these brief periods (of hundreds of thousands of years), the rate of change per generation is about the same as or less than the evolutionary rates of characteristics that have been altered by novel selection pressures within the last few centuries (see Figure 4.25B). Over time intervals of several million years, the average rate of evolution of most charac-

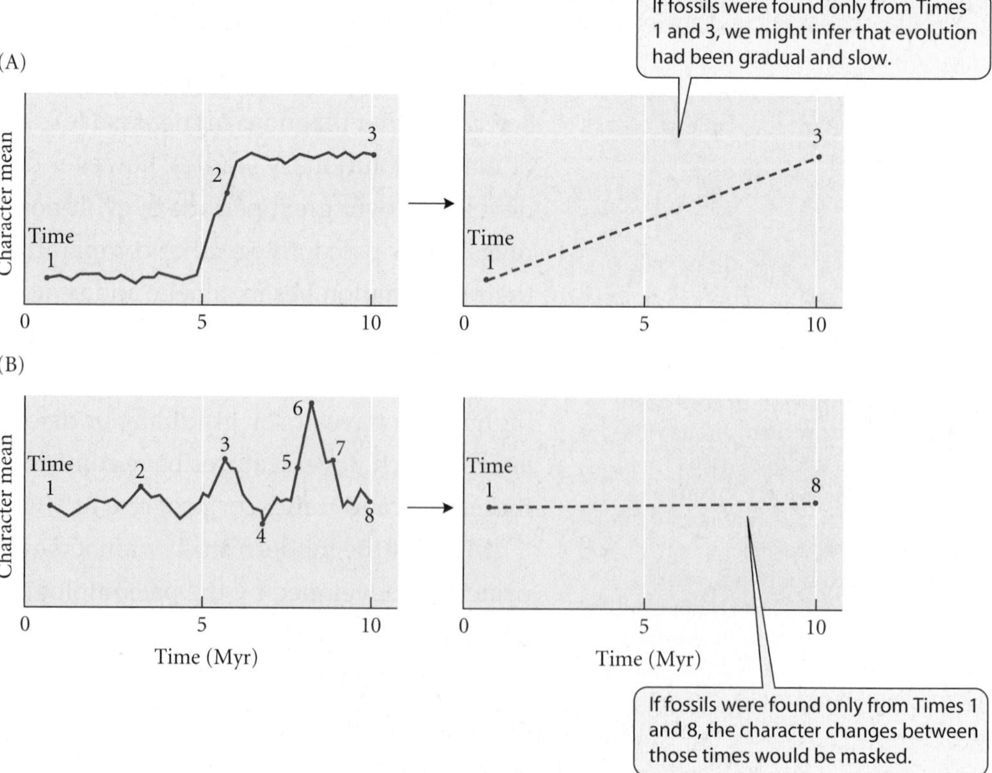

Figure 22.1 Averaged over long periods, the rate of evolution may be low, even though there are episodes of rapid evolution. (A) A punctuated pattern of shift from one rather static character mean to another. If we had fossils only from Times 1 and 3, we would not know there had been periods of slow and rapid evolution. (B) A pattern of rapid fluctuations in the character mean, but with little net change over a long period. Given only fossils from Times 1 and 8, we might think that little evolution had occurred.

ters is much lower, because the long-term average rate masks not only "punctuational" episodes of rapid evolution (Figure 22.1A, right panel) but also rapid fluctuations that may transpire even during periods of apparent stasis (Figure 22.1B).

Punctuated equilibrium and stasis

In their theoretical *model* of punctuated equilibrium, Eldredge and Gould (1972) proposed that stasis may be caused by genetic constraints that prevent species from adapting to changes in the selective environment, and that rapid changes (the punctuational events) represent evolution in newly formed species that originate as small, local populations. That is, Eldredge and Gould applied Ernst Mayr's model of founder effect speciation (peripatric speciation, Chapter 18) to macroevolutionary change. Paleontologists have described examples in which morphological change has been associated with true speciation (e.g., the bryozoan genus *Metrarhabdotos*; see Figure 4.21), and of course there must be some such instances because fossils of different species cannot be distinguished except by phenotype. But the punctuated equilibrium hypothesis requires that morphological evolution be almost inevitably accompanied by speciation, and it is not clear that the evidence supports this expectation. Moreover, geographic variation within species, as well as the rapid adaptive evolution of populations exposed to new selection pressures, show that speciation is not required for adaptive phenotypic change. Furthermore, Mayr's model of speciation, which requires that genetic drift (the founder effect) oppose the action of natural selection and move a small population from one adaptive peak to another across an adaptive valley, is theoretically unlikely to account for many speciation events, and indeed is not supported by genetic data on speciation (Chapter 18). Eldredge (1989) and Gould (2002, p. 796) themselves came to agree that speciation is not a necessary trigger of adaptive, directional morphological evolution.

The controversy about punctuated equilibrium had the healthy effect of drawing attention to many interesting questions about macroevolution. Perhaps most importantly, it established "stasis" as an important and puzzling phenomenon (Figure 22.2). Rapid evolution is not a problem for evolutionary biology to explain, because rapid rates in the fossil record are fully consistent with information on mutation, genetic variation, short-term rates of evolution in the recent past (Chapter 13), and divergence among populations and among closely related species (Chapter 17). The problem, rather, is to explain why evolution is often so slow. For example, Stanley and Yang (1987) measured 24 shell characters in large samples of 19 lineages of bivalves, in which they compared early Pliocene (4 Mya) fossils with their nearest living relatives (most of which bear the same species name as the fossils; Figure 22.3B). They compared these differences, in turn, with variation among geographic populations of eight of these species (Figure 22.3A). With few exceptions, the difference over the span of 4 Myr was no greater than the difference among contemporary conspecific populations.

Three major hypotheses have been proposed to account for stasis within species lineages.

First, Eldredge and Gould (1972) proposed that stasis is caused by internal *genetic or developmental constraints*, which would be manifested by lack of genetic variation or by genetic correlations too strong to permit characters to evolve independently to new optima. But although such constraints may indeed play a role in evolution, they cannot explain the constancy of size and shape of many quantitative characters, which are almost always genetically variable and only imperfectly correlated with one another (see Chapter 13).

The second, and most commonly suggested, explanation of stasis is *stabilizing selection* for a constant optimum phenotype (Charlesworth et al. 1982). Susan Estes and Stevan Arnold (2007) calculated the theoretical rates of evolution of a polygenic quantitative character (Chapter 13) under a wide range of values of effective population size, intensity of

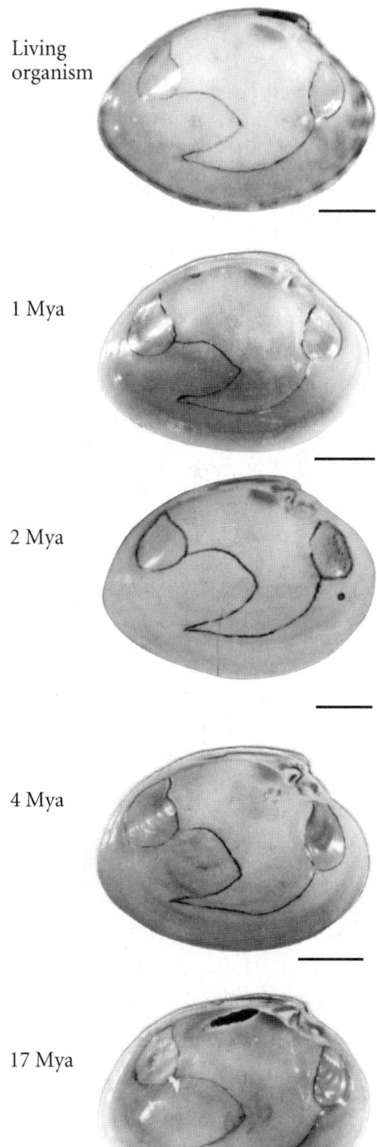

Figure 22.2 An example of stasis: specimens of the bivalve *Macrocallista maculata* from a living population and from fossil deposits dated at 1, 2, 4, and 17 Mya. All are from Florida. Scale bars = 1 cm. (Photos courtesy of Steven M. Stanley.)

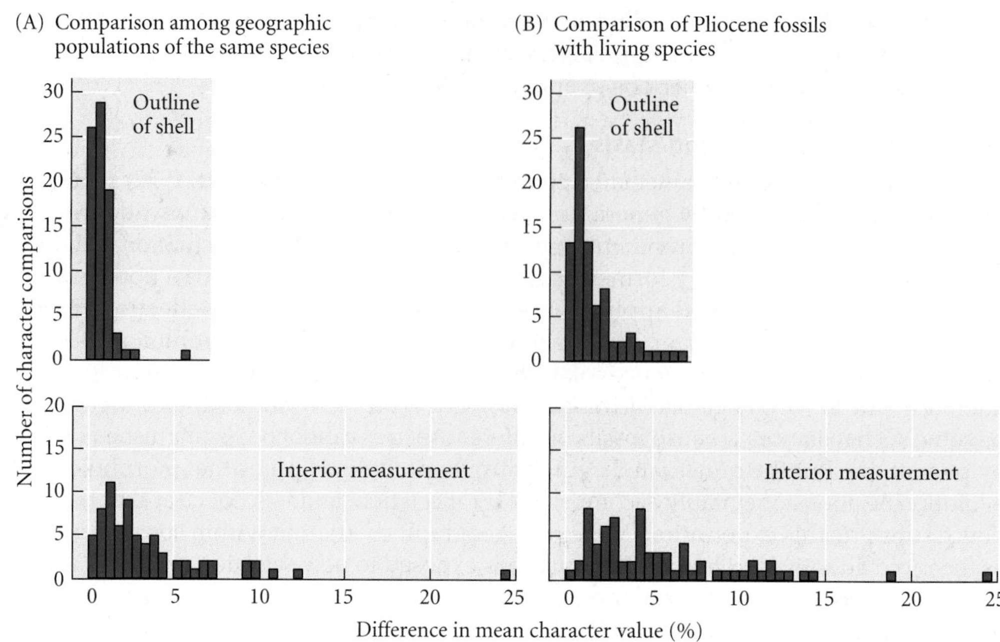

Figure 22.3 A quantitative expression of stasis in shell characters of bivalves in the fossil record. Each graph plots the number of characters, in several species, that show a certain percentage difference in the mean between samples of (A) different living geographic populations and (B) living and Pliocene populations. The two upper graphs show measurements pertaining to the outline of the shell; the lower graphs show interior measurements. The variation between Pliocene and living populations is about the same, overall, as that between different geographic populations of living species. (After Stanley and Yang 1987.)

stabilizing selection, and heritability, including heritabilities so low (0.001) as to imply genetic constraint. They compared these theoretical rates of evolution against a database of 2639 rates of evolution of diverse characters of diverse species (ranging from foraminiferans to dinosaurs) that had been compiled by paleontologist Philip Gingerich (2001; Figure 22.4A). Estes and Arnold simulated evolution under several different models. For example, a model of neutral evolution (no selection on the trait) consistently predicted less trait evolution over short time intervals, and much greater evolutionary divergence over long intervals, than are seen in the data (Figure 22.4B). A model of selection for an optimum phenotype that varies randomly (by "Brownian movement") over time failed to match the data in exactly the same way. The only model that matched the data was one of a "displaced optimum," in which a single episode of directional selection (as might be caused by an environmental change) brings the character mean to a new optimum, at which stabilizing selection maintains it thereafter (Figure 22.4C). This model fit the data for a fairly wide range of character shifts, as long as heritability was high enough ($h^2 \geq 0.01$).

It may seem unlikely that natural selection could favor the same character state over millions of years, during which both physical and biotic environmental factors would almost inevitably change. A dramatic example of such change is the succession of the many glacial and interglacial episodes of the Pleistocene, during which climates, geographic ranges of species, and associations among species changed drastically and repeatedly (see Chapters 5 and 6). However, a species' "effective environment" may be far more constant over time than we might expect, because of **habitat tracking** (Eldredge 1989): the shifting of the geographic distributions of species in concert with the distribution of their typical habitat. The distribution of cold-climate plants such as spruce, for example, has shifted southward during glacial times and northward during interglacial times; similarly, many groups of aquatic and semiaquatic insects are found in desert regions, but they live in such water as is available and have not adapted to dry habitats. Of course, the optimum might not remain exactly the same throughout long periods, but it might fluctuate only within narrow limits if habitat tracking keeps the species within narrow environmental bounds.

A related hypothesis is that *most evolutionary changes in local populations do not contribute to long-term change* because they remain spatially localized and are temporally evanescent. Niles Eldredge, John Thompson, and coauthors (2005) point out that an adaptive change arises in a geographically localized population of a species, and will become a feature of the entire species (and perhaps of descendant species as well), only if it is spread to other populations by gene flow and is advantageous in those populations. They note that many, perhaps most, advantageous variants are advantageous in only part of a species' geographic range, because of geographic variation in selection. For example, interactions with other species, such as predators, parasites, and competitors, vary from place to place, in a "geographic mosaic of coevolution," and so will select for different and perhaps ever-changing characteristics in different populations of a species (Thompson 2004; see Chapter 19).

A related hypothesis (Futuyma 1987) proposed that, although most adaptations are only locally advantageous and so would not contribute to long-term evolution, a microhabitat or a new food item (e.g., a different host plant of an insect) may have a wide, albeit patchy, distribution. An adaptation to such a new niche might spread over a broad geographic range—except that interbreeding of migrants with intervening ancestral-type populations would result in recombination and loss of the distinctive phenotype (especially if it were polygenic). Moreover, if the geographic distribution of patches with the new resource shifted (perhaps because of climate change), the divergent character state would be lost as a result of interbreeding with the (probably widespread) ancestral phenotype as individuals dispersed and colonized newly favorable sites, and as the populations from which they dispersed became extinct. Therefore, the existence of the new phenotype might be as brief as the intervals between climate-induced shifts in the geographic distribution of resources and species, and for this reason might not be registered in the fossil record. However, speciation could preserve the new adaptation. If the phenotypically divergent local population became reproductively isolated from the ancestral phenotype, it could track the geographic distribution of its habitat or resource without interbreeding with the ancestral phenotype, and so maintain a long-lasting, divergent identity (see Figure 18.30). Thus, although speciation may not cause anagenetic adaptive change, it may confer long life

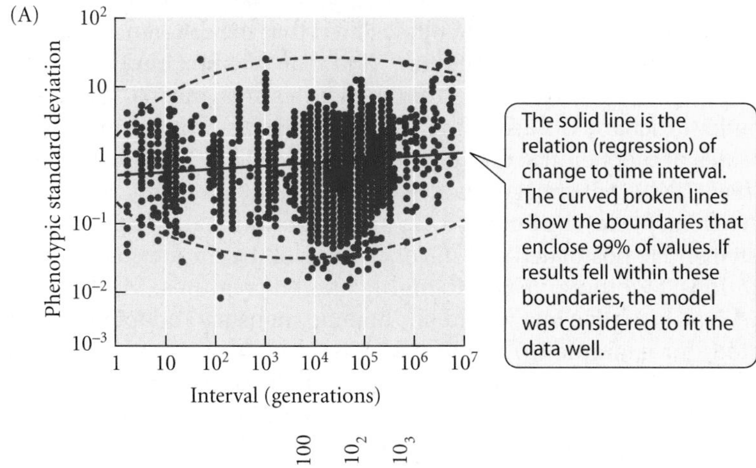

(A)
Phenotypic standard deviation
10^2
10
1
10^{-1}
10^{-2}
10^{-3}
1 10 10^2 10^3 10^4 10^5 10^6 10^7
Interval (generations)
The solid line is the relation (regression) of change to time interval. The curved broken lines show the boundaries that enclose 99% of values. If results fell within these boundaries, the model was considered to fit the data well.

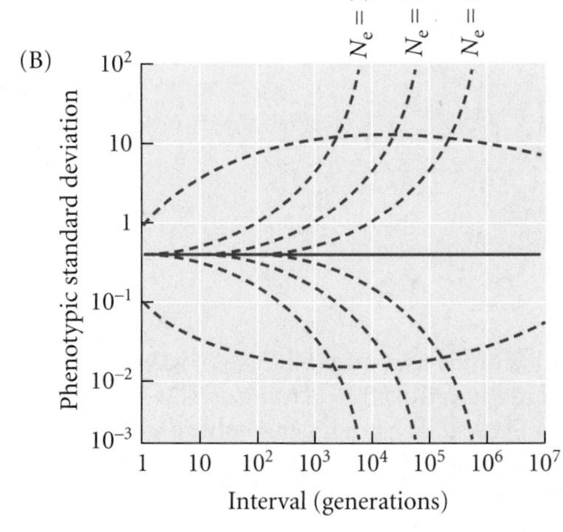

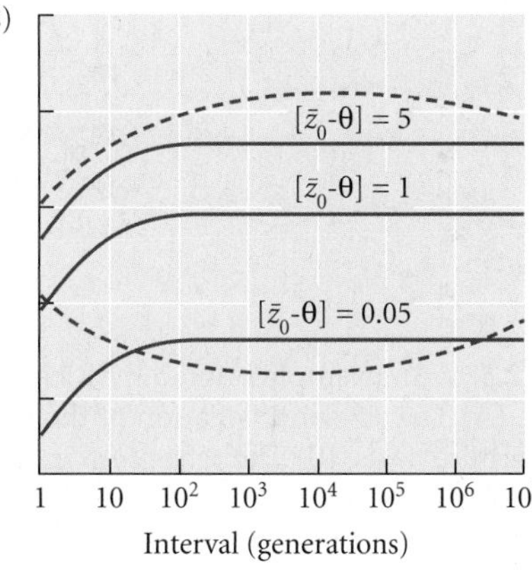

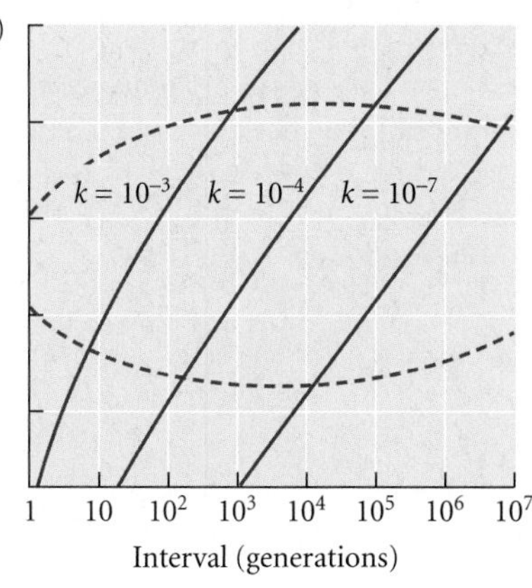

Figure 22.4 A comparison of empirically measured rates of character evolution with results from several models. (A) Evolutionary change (divergence) of various characters in relation to the time interval over which change occurred. Divergence is expressed in phenotypic standard deviations, a measure of within-population variation. Both variables are on a log scale. (B) Results of a neutral model, in which a character evolves by genetic drift. The average divergence (solid line) was zero, because increases and decreases occurred equally frequently in multiple simulations. The dotted lines show the range of values within which 95 percent of random changes occurred, in simulations of populations with three different effective sizes (N_e). These curves show that the character frequently evolved far outside the boundaries that represent data. (C) Solid lines show the average character change in a model in which the character evolves from an initial mean, $\bar{z}_0$, to a new optimum, θ. Results are shown for three such differences. This model fits the data fairly well. (D) Solid lines show the change in a character mean if the optimum changes slowly and steadily at three different rates, k (expressed in fractional standard deviations per generation). This model of steady, gradual evolution does not fit the data at all well. (After Estes and Arnold 2007.)

on such changes, leading to a possible association between speciation and morphological evolution (i.e., the *pattern* of punctuated equilibrium).

This last scenario leads us to the perhaps counterintuitive conclusion that long-term evolutionary change is more likely to occur in relatively stable than in frequently changing environments. Evidence of sustained evolution in stable environments has led some paleontologists to the same conclusion, and there is considerable evidence that the drastic climatic fluctuations in the Pleistocene inhibited both speciation and persistent adaptive phenotypic change (Jansson and Dynesius 2002). For example, even though many groups of beetles had undergone enormous diversification before the Pleistocene, G. R. Coope (1995) has found that species of Quaternary fossil beetles can be matched in precise morphological detail with living species. They "have remained constant both in their morphology and their environmental requirements throughout the whole of the Quaternary period. … From the British Isles alone there are now over two thousand fossil species known that are the precise match of their present-day equivalents." Coope attributes this constancy to habitat tracking, by which "populations continuously split up and reform as they progress to and fro across a complex landscape," during which "the gene pools were kept well stirred." Adaptive changes that are advantageous among most of the geographic populations of a species may be favored by widespread environmental changes, but otherwise such instability is likely to dampen evolutionary change.

Gradualism and Saltation

Darwin proposed that evolution proceeds gradually, by small steps. In *The Origin of Species*, he wrote that "if it could be demonstrated that any complex organ existed, which could not possibly have been formed by numerous, successive, slight modifications, my theory would absolutely break down." His ardent supporter Thomas Henry Huxley, however, cautioned that Darwin's theory of evolution would be just as valid even if evolution proceeded by leaps. Some later evolutionary biologists, such as the paleontologist Otto Schindewolf (1950), proposed just this. Schindewolf declared that the differences among higher taxa have "arisen discontinuously, by saltation," and that "the first bird [*Archaeopteryx*] hatched from a reptilian egg." (**Saltation**, from the Latin *saltus*, means "a jump.") The accomplished geneticist Richard Goldschmidt, one of the first to propose that genes act by controlling the rates of biochemical and developmental reactions, argued in *The Material Basis of Evolution* (1940) that species and higher taxa arise not from the genetic variation that resides within species, but instead "in single evolutionary steps as completely new genetic systems." He postulated that major changes of the chromosomal material, or "systemic mutations," would give rise to highly altered creatures. Most would have little chance of survival, but some few would be "hopeful monsters" adapted to new ways of life. He pointed, for example, to a mutation that transforms the halteres of *Drosophila* into more winglike structures that resemble, he said, the rudimentary wings of a dipteran fly (*Termitoxenia*) that inhabits termite nests (Figure 22.5).

If, between even the most extremely different organisms, there existed a full panoply of all possible intermediate forms, each differing from similar forms ever so slightly, we

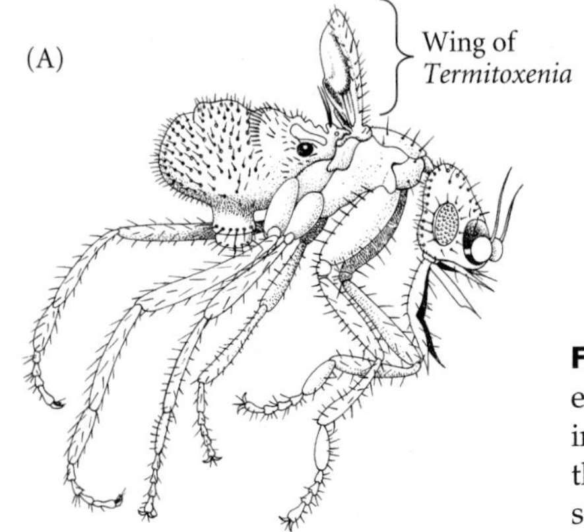

Figure 22.5 An example offered by Richard Goldschmidt as a possible case of saltational evolution. (A) The size and venation of the wing is greatly reduced in *Termitoxenia*, a fly that inhabits termite nests. (B) The tetraptera mutation in *Drosophila* changes the haltere into a wing that resembles the wing of *Termitoxenia*. Goldschmidt suggested that a similar mutation caused saltational evolution of the wing of *Termitoxenia*. (After Goldschmidt 1940.)

would have little doubt that evolution is a history of very slight changes. Such is often the case when we examine differences among individuals in a population, among populations of a species, and among closely related species, such as those in the same genus. Moreover, quite different species are often connected by intermediate forms, so that it becomes arbitrary whether the complex is classified as two genera (or subfamilies, or families) or as one (see Chapter 3). Nonetheless, there exist many conspicuous gaps among phenotypically similar clusters of species, especially among those classified as higher taxa such as orders and classes. No living species bridge the gap between cetaceans (dolphins and whales) and other mammals, for example.

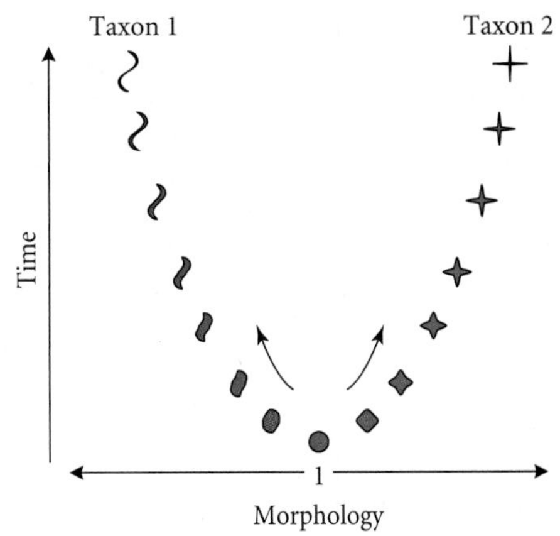

Figure 22.6 Two very different taxa may have evolved gradually from a common ancestor, even though no form precisely intermediate between them ever existed.

The most obvious explanation of phenotypic gaps among living species is extinction of intermediate forms that once existed—as the cetaceans themselves illustrate. Of course, the common ancestor of two quite different forms need not have appeared precisely intermediate between them, because the two phyletic lines may have undergone quite different modifications (Figure 22.6). For instance, DNA sequences imply that, among living animals, whales are most closely related to hippopotamuses, but early fossil cetaceans do not have a hippo-like appearance in the slightest (see Figure 4.11). Certainly, the fossil record provides many examples of the gradual evolution of higher taxa (see Chapter 4). Nevertheless, many higher taxa appear in the fossil record without intermediate antecedents.

One of the most enduring controversies in evolutionary biology has been whether such phenotypic gaps simply represent an inadequate fossil record—that is, evolution was gradual, but we simply lack the data to prove it—or whether evolution really proceeded by saltation. A saltation would result from the fixation of a single mutation of large effect. This concept is entirely different from the hypothesis of punctuated equilibrium, which allows that evolutionary changes in morphology might (or might not) have been continuous, passing through many intermediate stages, but were so rapid and so geographically localized that the fossil record presents the appearance of a discontinuous change. The saltation hypothesis, in contrast, holds that intermediates never existed—that mutant individuals differed drastically from their parents.

In judging gradualism and saltationism, we must distinguish between the evolution of *taxa* and of their *characters*. Higher taxa often differ in many characters: for example, a great many features distinguish modern birds from Cretaceous dinosaurs. Gradualists hold that many characters of higher taxa evolved independently and sequentially (MOSAIC EVOLUTION). Both comparisons of living species and the fossil record provide abundant evidence of mosaic evolution (see Chapters 3 and 4), although developmentally integrated features can sometimes evolve as a unit (see Chapter 13). There is some disagreement, however, on whether each of the distinguishing *characters* of a higher taxon—the reduction and the fusion of birds' tail vertebrae, for example—might have evolved discontinuously.

Certainly many mutations arise that have discontinuous, large, even drastic effects on the phenotype. Many of these, however, have such important pleiotropic effects that they greatly reduce viability. A mutation of the *Ultrabithorax* gene in *Drosophila*, for example, converts the halteres into wings. It may be tempting to think that this mutation reverses evolution, and that a mutation in this gene caused the evolutionary transformation of the second pair of wings into halteres in the ancestor of the Diptera, but the *Ultrabithorax* mutation is lethal in homozygous condition. Moreover, we understand that the normal form of the *Ultrabithorax* gene regulates the many genes that determine the distinctive development of the third thoracic segment, including the form of the halteres (see Chapter 21). Thus mutations that reduce the function of this master gene interfere with a complex developmental pathway, and development is routed into a "default" pathway that produces the features of the second thoracic segment (including wings). The whole system can be shut down in a single step by turning a master switch, but that does not mean the system came into existence by a single step.

Neo-Darwinians have always accepted, however, that characters can evolve by minor jumps—that is, by mutations that have fairly large (but not huge) effects. For example, variation both within and among species in characters such as bristle number in *Drosophila*

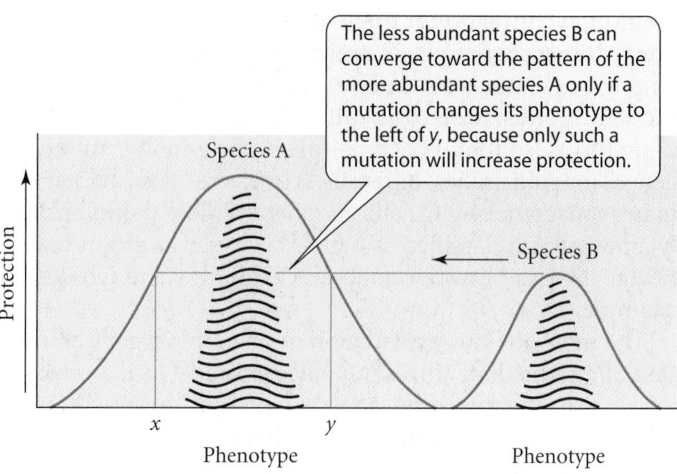

Figure 22.7 A model of the evolution of color pattern in Müllerian mimics such as *Heliconius* butterflies. The red curves represent the degree of protection against predators for two species, A and B, that differ in phenotype and abundance. The wavy lines indicate the distribution of phenotypes within each species. Selection favors convergence of the less abundant species B toward the pattern of the more abundant species A because predators more often learn to avoid the more abundant species. A mutation of small effect that slightly alters the phenotype of species B will be selectively disadvantageous. However, a mutation of large effect that causes members of species B to acquire a phenotype with a modest resemblance to species A (e.g., just left of phenotype *y*) will be selectively advantageous. Subsequently, allele substitutions with small effects that bring B closer to the peak for A will be advantageous. (After Charlesworth 1990.)

is often caused by a mixture of quantitative trait loci with both small and large effects (Orr and Coyne 1992; see Chapters 13 and 18). Alleles with large effects contribute importantly to mimetic polymorphisms in butterflies such as the African swallowtail *Papilio dardanus* (see Figure 9.1B), in which each of several very different forms is a mimic of a different unpalatable species (or model). Were phenotypes to arise that deviated only slightly from one mimetic pattern toward another, they would lack protective resemblance to either unpalatable model and presumably would suffer a disadvantage. Thus it is likely that the evolution of one mimetic pattern from another was initiated by a mutation of large enough effect to provide substantial resemblance to a different model species, followed by selection of alleles with smaller effects that "fine-tuned" the phenotype (Figure 22.7). Genetic analysis of the color patterns of *P. dardanus* supports this hypothesis (Ford 1971). Perhaps the most dramatic single-gene differences yet described are those with strong heterochronic effects, such as the few genetic changes that determine the difference between metamorphosing tiger salamanders (*Ambystoma tigrinum*) and their paedomorphic relative the axolotl (*A. mexicanum*; see Chapter 21). These genes do not engender an entirely new complex morphology, but merely truncate a complex, integrated pathway of development that presumably evolved by many small steps.

Phylogenetic Conservatism and Change

Given the great diversity of phenotypes among even closely related species, and given the rapidity with which evolution can occur, biologists are challenged to understand the existence of "living fossils"—organisms such as the ginkgo (see Figure 5.19B), the tadpole shrimp *Triops cancriformis* (Figure 22.8A), and the coelacanths (Figure 22.8B) that have changed so little over many millions of years that they closely resemble their Mesozoic or even Paleozoic relatives. The synapomorphies of large clades also represent conservatism: almost all mammals, no matter how long or short their necks, have seven neck vertebrae, and almost no tetrapods have had more than five digits per limb. (The earliest amphibians had more but soon settled on five.) The problem of stasis in individual lineages is magnified in such instances, in which characters have remained little changed despite extensive speciation, and despite diversification in other characteristics. The hypotheses for phylogenetic conservatism include stabilizing selection and internal constraints. These may be related to each other, and perhaps are much the same thing.

One important reason that the optimal condition of a character may remain unchanged is **niche conservatism**: long-continued dependence of related species on much the same resources and environmental conditions (Holt 1996; Wiens and Graham 2005). For example, closely related species of plants have climatically similar geographic ranges in Asia and North America (Ricklefs and Latham 1992); the larvae of all species of the butterfly tribe Heliconiini feed on passionflower plants (Passifloraceae), and apparently have done so since the tribe originated in the Oligocene. By occupying one niche (e.g., host plant, cli-

Figure 22.8 Two "living fossils." (A) The tadpole shrimp *Triops cancriformis*, found in temporary pools in arid regions of Eurasia and northern Africa, has undergone no evident morphological evolution since the Triassic. (B) Coelacanths are lobe-finned fishes that originated in the Devonian and were thought to have become extinct in the Cretaceous, until this living species, *Latimeria chalumnae*, was discovered in 1938. (A © OSF/photolibrary.com; B © Peter Scoones/PhotoResearchers Inc.)

matic zone) rather than another, a species subjects itself to some selective pressures and screens off others; it may even be said to "construct" or determine its own niche, and therefore many aspects of its potential evolutionary future (Lewontin 2000; Odling-Smee et al. 2003). Niches remain conservative for two major, sometimes interacting, reasons: First, other species, often by acting as competitors, may prevent a species from shifting or expanding its niche; conversely, lineages may evolve and radiate when competition is alleviated (see Chapters 3, 4, and 7). Second, and more generally, if there is gene exchange among individuals that inhabit the ancestral niche (e.g., microhabitat) and those that inhabit a novel niche, and if there is a fitness trade-off between character states that improve fitness in the two environments, then selection will generally favor the ancestral character state (i.e., stabilizing selection will prevail) simply because most of the population occupies the ancestral environment (Holt 1996).

As the degree of adaptation to any one environment increases, the differential in fitness between that environment and a new environment also increases, so that adaptation to an alternative environment may become steadily less likely. In some cases, a species may lose the ability to vary in features that would be necessary for a substantial ecological shift. For example, blue and purple pigments, products of a biosynthetic pathway that uses the substrate cyanidin, are ancestral in the morning glory genus *Ipomoea* but are replaced in a bird-pollinated clade by red pigments that are synthesized from pelargonidin (Figure 22.9). Rebecca Zufall and Mark Rausher (2004) showed that in a red-flowered species, the F3′H enzyme that initiates the cyanidin pathway has undergone both regu-

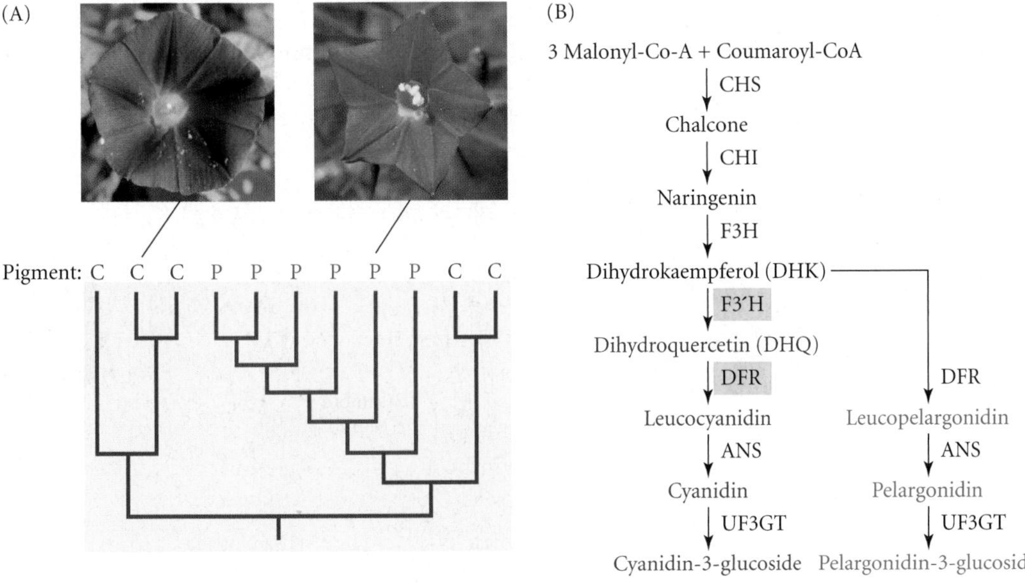

Figure 22.9 Adaptive genetic changes may restrict subsequent evolutionary potential. (A) A molecular phylogeny of species of *Ipomoea* shows that a red-flowered clade, with pelargonidin-derived pigment (P), is derived from blue-flowered ancestors with cyanidin-derived pigment (C). (B) The anthocyanin biosynthetic pathway, with branches to cyanidin (blue) and pelargonidin (red) pigments. Evolutionary changes in enzymes F3′H and DFR (highlighted) are discussed in the text. (After Zufall and Rausher 2004; photos courtesy of Mark D. Rausher.)

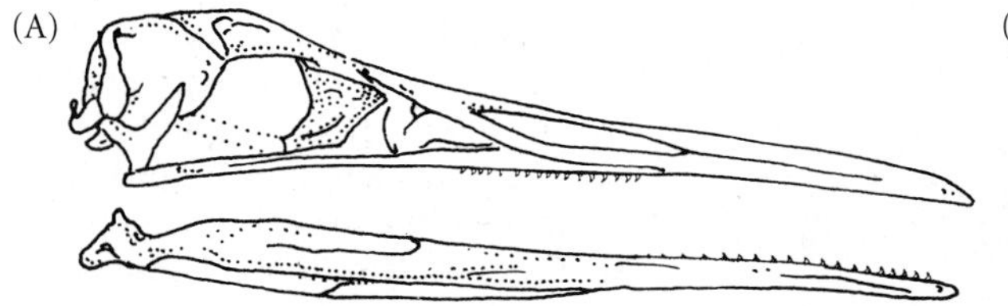

Figure 22.10 Complex structures, if lost, are generally not regained, but their function may be. (A) *Hesperornis*, a marine bird of the Late Cretaceous, had teeth that enabled it to grip fish. (B) The bill of a typical living fish-eating bird, the anhinga (darter), as drawn by John James Audubon. No living bird has teeth. (C) The bill of a merganser, a fish-eating duck, has a substitute for teeth: serrations that enable it to grip fish. (A courtesy of Larry Martin in Feduccia 1999; C art by Nancy Haver.)

latory and functional inactivation, so that only the pelargonidin pathway is active. Perhaps as a consequence of the inactivation of the cyanidin pathway, another enzyme, DFR, has lost the ability to metabolize its substrate in the cyanidin pathway (although it is still active in the pelargonidin pathway). The authors point out that it would require at least two or three restorative changes to evolve blue pigments anew. In general, unused genes acquire disabling mutations such as stop signals and become pseudogenes, as shown by such examples as the reduced number of functional olfactory receptors in primates, and the degeneration of many genes and phenotypic functions in parasites.

The loss of gene function is one likely basis of DOLLO'S LAW, the generalization that a complex character, if lost, is seldom regained in its original form. Although exceptions to Dollo's law are known (see Figure 3.9; Chapter 21), the loss of ancestral developmental and biosynthetic pathways is highlighted by cases in which other features are modified to serve the function of a lost character. For example, tooth development is initiated in birds but stops at a very early stage unless two proteins are added that are not normally expressed. A mutation is known that enables tooth development to proceed further (Collin and Miglietta 2008); nevertheless, birds have not had teeth since the end of the Cretaceous. "Substitute teeth," serrations of the bill margin, have evolved in mergansers, enabling these ducks to catch fish (Figure 22.10). Such examples suggest that development can impose constraints on the rate or direction of evolution of a character. The consequences of such constraints are made clear when an adaptive function is performed not by the structure we might expect, but by another structure that has been modified instead. The giant panda, for example, has six apparent fingers, evidently useful for manipulating the bamboo on which it feeds. The outermost "finger" (or "thumb"), however, is not a true digit but a sesamoid bone that develops from cartilage (Figure 22.11). We have noted that differences among clades in developmental pathways affect the kinds of variation that can arise: salamanders are more likely to lose

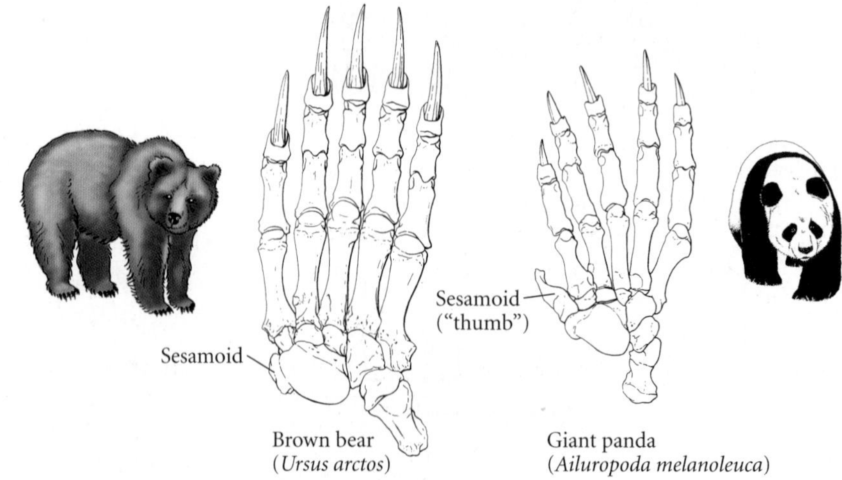

Figure 22.11 The right hand (in dorsal view) of two members of the bear family, a brown bear (left) and the giant panda (right). A small sesamoid bone in bears has been modified into a false finger ("thumb") in the panda, which uses it to help manipulate the bamboo on which it feeds. Developmental constraints probably prevented the evolution of a sixth true digit. The panda's "thumb" is an example of what has been called "tinkering" by natural selection: evolving adaptations from whatever variable characters a lineage happens to already have. (After Davis 1964.)

preaxial toes, and frogs postaxial toes, corresponding to the order in which the digits develop in the two groups (see Chapter 21).

From a genetic point of view, there are many possible kinds of constraints. Because of genetic correlations, there can exist genetic variation for certain combinations of characters but not others, even if each character individually is variable (Blows and Hoffmann 2005). Some such patterns of character correlation may be very long lasting because mutations may have consistent patterns of pleiotropy. This is probably the most important determinant, in the long run, of the pattern of genetic variation that is available for evolutionary change (Jones et al. 2007). Perhaps for this reason, researchers are increasingly finding that the directions of evolutionary differences among species are correlated with the "genetic lines of least resistance" that are estimated from genetic or phenotypic correlations in contemporary populations (see Figure 13.16). For example, evolution of carapace shape characteristics in multiple species lineages of the early Tertiary ostracode *Poseidonamicus* proceeded largely along character combinations that showed the greatest variation within populations (Hunt 2007; Figure 22.12). Similarly, the pattern of divergence in wing shape among various clades of *Drosophila* is broadly similar to the genetic variance-covariance matrix within *D. melanogaster*, suggesting that the pattern of genetic variation has been fairly consistent for more than 50 Myr (Hansen and Houle 2008).

The evolution of a character (or reacquisition, as in the case of blue pigments in the morning glory described earlier) may be unlikely if a number of specific mutations are required to assemble it. In this context, it is useful to bear in mind that many events in evolutionary history have evidently been improbable, since they have occurred only once among many milllions of lineages and countless numbers of mutations over the course of hundreds of millions of years. Only one invertebrate lineage ever evolved wings; only one animal lineage evolved vertebrae; only one evolved the human brain.

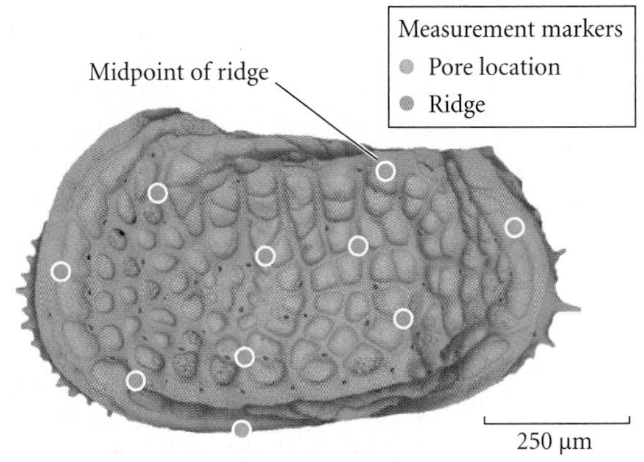

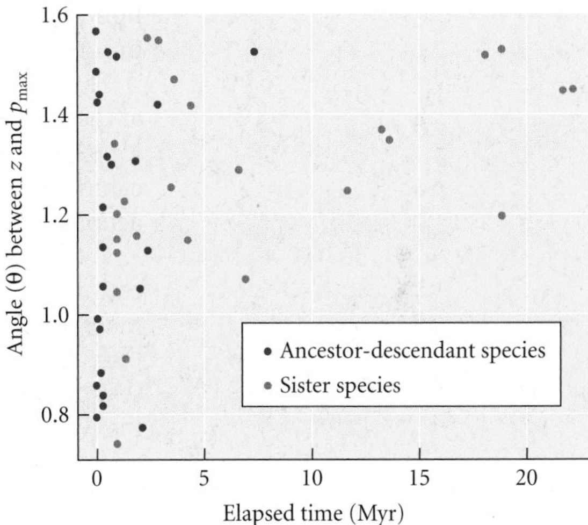

Figure 22.12 The relationship between elapsed time and the propensity for evolutionary changes to be in directions close to the axis of greatest phenotypic variation, in fossilized lineages of the marine ostracode crustacean *Poseidonamicus*. For multiple measurement characters of the carapace (above), p_{max} expresses the character combination with maximal variation within populations, and z is the direction of change between ancestor-descendant transitions or between sister species derived from a common ancestor. Smaller angles represent a direction of evolution closer to p_{max} (see Figure 13.16). Although there is much variation in the direction of evolution over short intervals, it tends to depart more over long time spans from the direction that the variation within populations would predict. (After Hunt 2007; photo courtesy of Gene Hunt.)

The Evolution of Novelty

How do major changes in characters evolve, and how do new features originate? These questions have two distinct meanings. First, we can ask what the genetic and developmental bases of such changes are—the subject of Chapter 21. Second, we can ask what role natural selection plays in their evolution. For instance, we may well ask whether each step, from the slightest initial alteration of a feature to the full complexity of form displayed by later descendants, could have been guided by selection. What functional advantage can there be, skeptics ask, in an incompletely developed eye? And we can ask how complex characters could have evolved if their proper function depends on the mutually adjusted form of each of their many components.

Accounting for incipient and novel features

Several pathways of evolutionary change account for the macroevolution of phenotypic characters (see Mayr 1960; Nitecki 1990; Müller and Wagner 1991; Galis 1996).

A feature may originate as a new structure or as a novel modification of an existing structure. For example, sesamoid bones often develop in connective tissue in response to embryonic movement. Such bones are the origin of novel skeletal elements, such as the extra "finger" of the giant panda (see Figure 22.11) and the patella (kneecap) in the leg of mammals, which is lacking in reptiles (Müller and Wagner 1991).

A feature may be a developmental by-product of other adaptive features, perhaps initially nonadaptive but later recruited or modified to serve an adaptive function. For instance, by excreting nitrogenous wastes as crystalline uric acid, insects lose less water than if they excreted ammonia or urea. Excreting uric acid is surely an adaptation, but the white color of uric acid is not. However, pierine butterflies such as the cabbage white butterfly (*Pieris rapae*) sequester uric acid in their wing scales, imparting to the wings a white color that plays a role in thermoregulation and probably in other functions.

Decoupling the multiple functions of an ancestral feature frees it from functional constraints and may lead to its elaboration. For example, the locomotory muscles of many "reptiles" insert on the ribs, so that these animals cannot breathe effectively while running. In birds and mammals, the muscle insertions have shifted to processes on the vertebrae, so that breathing and running (or flying) are decoupled, and many lineages have evolved features associated with rapid locomotion (Galis 1996). David Wake (1982) has proposed that the loss of lungs in the largest family of salamanders (Plethodontidae) has relieved a functional constraint on the evolution of the tongue. In other salamanders, the bones that support the tongue are also used for moving air in and out of the lungs. In plethodontids, this hyobranchial skeleton, no longer used for ventilating the lungs, has been modified into a set of long elements that can be greatly extended from a folded configuration. This modification enables plethodontids to catch prey by projecting the tongue, in some species to extraordinary lengths at extraordinary speed (Figure 22.13).

Duplication with divergence gives rise to diversity at the morphological level, as it does at the level of genes and proteins (see Chapter 20). The diversity of teeth in a mammal, for example, provides enormous functional possibilities: molars can slice or grind, while canines stab. Some teeth—such as the tusks of elephants—are used more for social interactions than for eating.

A change in the function of a feature alters the selective regime, leading to its modification. This principle, already recognized by Darwin, is one of the most important in macroevolution (Mayr 1960), and every group of organisms presents numerous examples. A bee's sting is a modified ovipositor, or egg-laying device. The wings of auks and several other aquatic birds are used in the same way in both air and water; in penguins, the wings have become entirely modified for underwater flight (see Figure 11.18). Many genes and proteins have been co-opted or modified for new functions, as Joram Piatigorsky (2007) has emphasized. He and his coworkers have found that two types of crystallins, α and β, that are common to all vertebrates are derived from stress proteins, and that various animal lineages have also derived taxon-specific crystallins from various enzymes, such as lactate dehydrogenase in reptiles and alcohol dehydrogenase is some mammals (see Figure 11.19). Some crystallins have undergone amino acid substitutions since they were co-opted from their ancestral function, but they share extensive sequence homology with the unmodified enzyme. These crystallins are derived from duplicated genes, but Piatigorsky emphasizes that the same gene often encodes both the enzyme and the crystallin, a condition termed GENE SHARING.

Figure 22.13 A lungless bolitoglossine salamander (*Hydromantes supramontis*) captures prey with its extraordinarily long tongue. The rapid tongue extension is accomplished with a modified hyobranchial apparatus, which in other families of salamanders plays an important role in ventilating the lungs. (From Deban et al. 1997, courtesy of S. Deban.)

Complex characteristics

FUNCTIONAL INTERMEDIATES. A common argument against Darwinian evolution is based on what is sometimes termed "irreducible complexity": the proposition that a complex organismal feature cannot function effectively except by the coordinated action of all its components, so that an organism's fitness would be reduced by elimination or alteration of any one of the interacting components. From this proposition, opponents of evolution argue that the feature must have required all of its components from the beginning, for the entire functional complex could not have arisen in a single mutational step.

Needless to say, the first person to recognize this potential problem was Darwin himself, in *The Origin of Species:* "That the eye, with all its inimitable contrivances for adjusting the focus to different distances, for admitting different amounts of light, and for the correction of spherical and chromatic aberration, could have been formed by natural selection seems, I freely confess, absurd in the highest possible degree." But he then proceeded to supply examples of animals' eyes as evidence that "if numerous gradations from a perfect and complex eye to one very imperfect and simple, each grade being useful to its possessor, can be shown to exist; if further, the eye does vary ever so slightly, and the variations be inherited, which is certainly the case; and if any variation or modification in the organ be ever useful to an animal under changing conditions of life, then the difficulty of believing that a perfect and complex eye could be formed by natural selection, though insuperable by our imagination, can hardly be considered real."

Darwin's claim has been fully supported by later research (e.g., Nilsson and Pelger 1994; Osorio 1994). The eyes of various animals range from small groups of merely light-sensitive cells (in some flatworms, annelid worms, and others), to cuplike or "pinhole camera" eyes (in cnidarians, molluscs, cephalochordates, and others), to the "closed" eyes, capable of registering precise images, that have evolved independently in cnidarians, snails, bivalves, polychaete worms, arthropods, and vertebrates (Figure 22.14). The evolution of eyes is apparently not so improbable! Each of the many grades of photoreceptors, from the simplest to the most complex, serves an adaptive function. Simple epidermal photoreceptors and cups are most common in slowly moving or burrowing animals; highly elaborated structures are typical of more mobile animals. Not only eye anatomy, but the molecular basis of vision, has evolved by comprehensible steps. We have already seen that lens crystallin proteins have evolved from a variety of other proteins, and the visula pigments, or opsins, evolved from a G-protein-coupled receptor protein of the kind that is ubiquitous in animals and fungi (Oakley and Pankey 2008). The mystery of how a

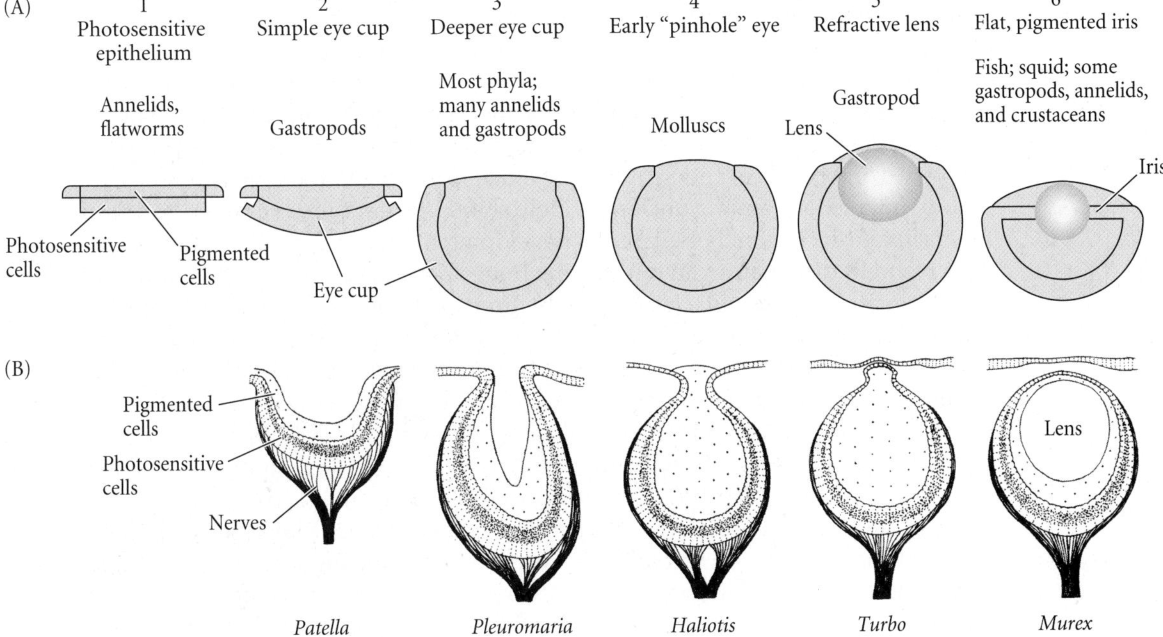

Figure 22.14 Intermediate stages in the evolution of complex eyes. (A) Schematic diagrams of stages of eye development in various animals, from a simple photosensitive epithelium, through the deepening of the eye cup (providing progressively more information on the direction of the light source). There is gradual evolution toward a "pinhole camera" eye, eventually including a refractive lens and a pigmented iris for more perfect focusing. (B) Most of these stages can be found among various gastropod species, as shown in these drawings. (A after Osorio 1994; B after Salvini-Plawen and Mayr 1977.)

simple eye could be adaptive is no great mystery after all. [Volume 1, issue 4 (October 2008) of the journal *Evolution: Education and Outreach* is a special issue on the evolution of eyes.] Neither at the morphological nor the molecular level is the notion of "irreducible complexity" a barrier to evolution.

The antievolutionary argument, then, ignores evidence that intermediate stages in the evolution of complex systems exist and have adaptive value. It also fails to recognize that a component of a functional complex that was initially merely superior can become indispensable because other characters evolve to become functionally integrated with it. Although the eyes of many animals do not have a lens, these animals do quite well without the visual acuity that a lens can provide. But a lens is indispensable for eagles, since their way of hunting prey has been acquired and made possible only by such acuity. Eagles and mammals have *acquired* dependence on the elements of a complex eye. Such dependence, indeed, is often lost: many burrowing and cave-dwelling vertebrates have degenerate eyes.

DEVELOPMENTAL BIOLOGY AND COMPLEXITY. Throughout the sciences, understanding of complex phenomena has been advanced by "reductionist" investigation: dissecting the system into low-level components, analyzing them, and then reconstructing elements of the larger system. Today, biologists' understanding of how complex organismal phenotypes evolve is being greatly enhanced by advances in molecular, cell, and developmental biology. For example, we now know that animal development is heavily based on a GENETIC TOOLKIT, consisting largely of genes that encode transcription factors and components of signaling pathways, and that many of them have been conserved since the origin of animals or even earlier (Carroll et al. 2001). No fewer than 79 percent of mouse genes are homologous across phyla, and thus date from Precambrian ages (Gerhardt and Kirschner 2007). Many of these genes have retained similar functions, such as *Pax6*, which initiates the development of photoreceptors, and *Distal-less*, which controls development of outgrowths such as limbs. But we also know that the location and time of gene expression can be altered by changes in the cis-regulatory elements (CREs) of the genes that transcription factors regulate. Furthermore, changes in CREs can alter functional linkages between controlling and controlled genes (Prud'homme et al. 2007; Chapter 21). *Distal-less* controls not only limb development but also spot patterns in butterfly wings. *Ultrabithorax* (*Ubx*), a Hox gene instrumental in specifying the identity of body regions, represses the development of hindwings in *Drosophila*, so that they develop into halteres, whereas in the beetle *Tribolium*, *Ubx* is necessary for normal development of the hindwings, and activates different genes than in *Drosophila* (Tomoyasu et al. 2005; Prud'homme et al. 2007). In such cases, changes in transcription factors evoke different expression of entire downstream pathways: a single genetic change may trigger coherent activity of many genes.

These revelations of evolutionary developmental biology contribute to a theory of "facilitated variation," espoused by Marc Kirschner and John Gerhardt (2005). These authors argue that the "core processes" of protein activity and cell and organ development have properties of robustness and adaptability that, together with the establishment of new regulatory interactions among ancient, widely shared genes, cause variation to arise in ways that facilitate evolution. For example, developing muscles, nerves, and blood vessels in a limb respond to signals from developing bone and dermis, and so grow into their proper positions. Thus genetic changes in the limb skeleton result in altered but functional limbs, without the necessity for independent genetic changes in musculature and vasculature. When investigators administered Bmp protein to the beak primordia of a chicken embryo, the animals developed a deeper, broad, and well-formed beak: the development of its parts was coordinated. The *Bmp* gene is largely responsible for the more massive bill of the large ground finch (*Geospiza magnirostris*; see Figure 2.3B) compared with other finches in the Galápagos Islands (Abzhanov et al. 2004). Kirschner and Gerhardt suggest that because developmental and other organismal processes have been selected to be coordinated and resistant to perturbations—because they have been canalized (Chapter 13)—genetic and environmental variation in the resulting characters is more likely to be viable than we might otherwise suppose, facilitating the evolution of complex characters.

EVOLVABILITY. The idea that the production of genetic and phenotypic variation may be structured in a way that makes adaptive evolution more likely is one of several concepts that have been termed **evolvability**. Among the definitions of this term (Pigliucci 2008) is David Houle's (1992): evolvability is the ability of a population to respond to natural or artificial selection, measured by the "additive coefficient of genetic variation," CV_a, that relates the additive genetic variance of a character to the mean ($CV_a = [100\sqrt{V_A}]/\bar{x}$). Günter Wagner and Lee Altenberg (1996), however, define evolvability as "the genome's ability to produce adaptive variants when acted upon by the genetic system." The difference between these concepts of evolvability parallels Wagner and Altenberg's distinction between genetic *variation*, which is a directly observed, ephemeral property of a population that can change rapidly, and genetic *variability*, which is the *ability to vary*, and which is a property of features of the genome, such as the complement of genes and their mutation rate. In their terms, Houle's CV_a is a measure of variation, not the ability to produce variants.

Wagner and his colleagues are among those who have argued that variability, and therefore evolvability, can evolve (Wagner 1996; Hansen et al. 2006; Draghi and Wagner 2008). Some aspects of the evolution of evolvability are uncontroverisal. For example, in asexual populations, mutation rates can theoretically evolve to an optimum, based on "mutator" alleles that affect mutation; recombination rates and sexual versus asexual reproduction certainly affect the production of genetic variation, and they clearly evolve (see Chapter 15). The more recent (and more controversial) emphasis is on whether evolvability can evolve by changes in the relationship between genotype and phenotype, as expressed in developmental pathways. For example, the traits of an organism are bundled into MODULES to a greater or lesser extent, such as different organs or the different segments of an arthropod. Modularity exists to the extent that sets of genes contribute to different pathways of development of different complex characters such as eyes or legs (Figure 22.15). To the extent that genes pleiotropically affect different pathways, the resulting characteristics will be genetically correlated. Genetic correlations among traits can enhance adaptation if the characters are subject to correlational selection, that is, if selection favors specific combinations of characteristics (see Figures 13.13 and 13.16), or they can constrain evolution if selection on the several traits is antagonistic (e.g., Figure 13.14). We can envision that functionally related features might evolve to be more highly correlated (INTEGRATION) so that they are functionally matched even if they vary. They would then be less capable of independent evolution (i.e., less "evolvable"). However, selection might favor PARCELLATION of a highly integrated network of characters into different modules, each of a few functionally related characters, and each able to evolve independently of other modules (Figure 22.16).

Simulation models show that the degree of modularity (hence the pattern of genetic correlation) can evolve by natural selection within populations. That is, the developmental pathways expressed by the "genotype-phenotype map" can evolve so that the expression of mutations, the effect of mutations on genetic variances and genetic correlations among characters (the "**M** matrix"), is altered. Selection in a changing environment can

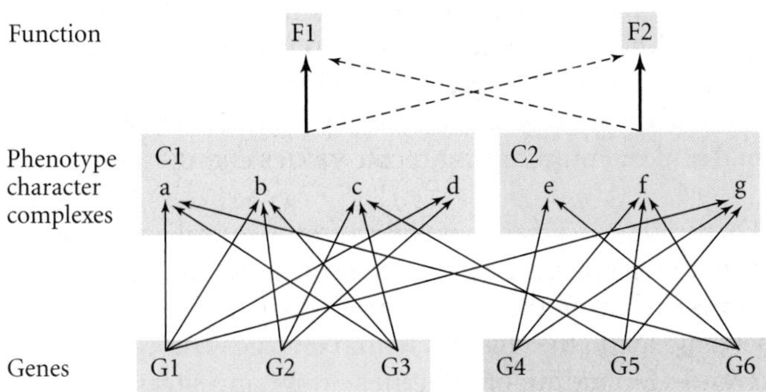

Figure 22.15 A schematic representation of a modular organization of the phenotype and its genetic basis. The phenotype is organized into complex characters, such as C1 and C2, each of which consists of component characters (a–d in C1, e–g in C2). Genes (G1–G6) have pleiotropic effects, indicated by arrows from each gene to multiple characters. The phenotype is more modular if each gene's pleiotropic effects are mostly on component characters within each complex (C1 or C2), as shown. The more modular the character, the stronger the genetic correlations among the component characters, compared with genetic correlations among characters that are components of different complexes. The diagram suggests that each complex character has a primary function (F1 for C1, F2 for C2), and weakly affects the other function. (After Wagner and Altenberg 1996.)

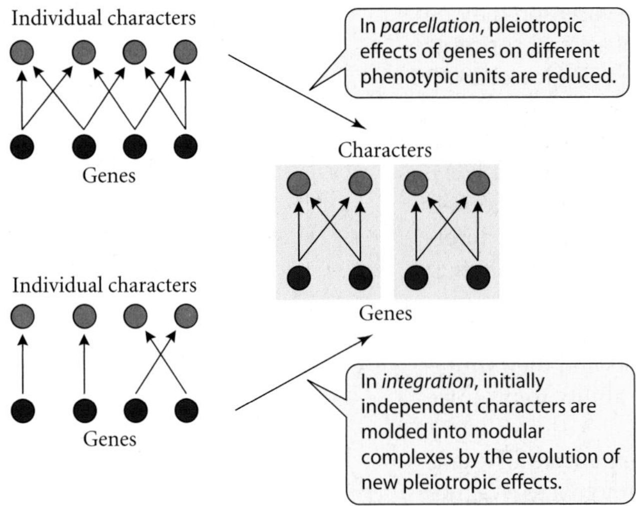
Individual characters
Genes
In *parcellation*, pleiotropic effects of genes on different phenotypic units are reduced.
Characters
Genes
Individual characters
Genes
In *integration*, initially independent characters are molded into modular complexes by the evolution of new pleiotropic effects.

Figure 22.16 Two ways in which interactions among suites of genes and characters can evolve by changes in pleiotropic effects. In parcellation, the pleiotropic effects of genes on different phenotypic characters are reduced. Such changes increase the possibility of independent character evolution. In integration, the pleiotropic effects of genes are increased, molding initially independent characters into modular complexes. Such changes may act as constraints because independent evolution becomes less likely. (After G. P. Wagner 1996.)

theoretically cause the evolution of reduced pleiotropy and greater modularity, whereas stabilizing selection in a stable environment can align the **M** matrix with the adaptive landscape (Jones et al. 2007; Draghi and Wagner 2008). For example, mutations might produce a greater correlation between the relative length of a plant's stamens and pistils if the developmental pathways have been evolving in a stable environment, but might produce more independent variation if selection has fluctuated in such a way that independent variation has enhanced fitness. It is important to recognize that this process does not involve group selection (survival versus extinction of populations or species that differ in mutational pattern), much less any anticipation of the "need" for evolutionary flexibility.

Theoretical analyses of evolvability are quite recent, and there has been little experimental research on the subject so far. Whether the apparent evolvability of organisms is a by-product of developmental processes or is itself the target of selection is not yet known, but research in this area will illuminate one of the least-explored areas of evolutionary biology, the origin of the hereditary variation on which all of evolution depends.

Trends and Progress

For many decades after the publication of *The Origin of Species*, many of those who accepted the historical reality of evolution viewed it as a cosmic history of progress. As humanity had been the highest earthly link in the pre-evolutionary Great Chain of Being, just below the angels (see Chapter 1), so humans were seen as the supreme achievement of the evolutionary process (and Western Europeans as the pinnacle of human evolution). Darwin distinguished himself from his contemporaries by denying the necessity of progress or improvement in evolution (Fisher 1986), but almost everyone else viewed progress as an intrinsic, even defining, property of evolution. In this section, we will examine the nature and possible causes of trends in evolution and ask whether the concept of evolutionary progress is meaningful.

A **trend** may be described objectively as a directional shift over time. **Progress**, in contrast, implies betterment, which requires a value judgment of what "better" might mean. We will come to terms with evolutionary "progress" after discussing trends.

Trends: Kinds and causes

A trend is a persistent, directional change in the average value of a feature, or perhaps its maximal (or minimal) value, in a clade over the course of time. A phylogenetically LOCAL TREND applies to an individual clade, whereas a GLOBAL TREND characterizes all of life. Trends can also be classified as passive or driven (McShea 1994). In a **passive trend**, lineages in the clade evolve in both directions with equal probability, but if there is an impassable boundary on one side (e.g., a minimal possible body size), the variation among lineages can expand only in the other direction. Because the variance expands, so do the mean and the maximum (Figure 22.17A). Although the mean increases, some lineages may remain near the ancestral character value. In a **driven**, or **active**, **trend**, changes within lineages in one direction are more likely than changes in the other (i.e., there is a "bias" in direction), so both the maximal and the minimal character values change along with the mean (Figure 22.17B).

Both driven and passive trends could have any of several causes. Neutral evolution by *mutation and genetic drift* results in increasing variance among lineages (see Chapter 12) and could produce a passive trend if variation were bounded as in Figure 22.17A. *Individual selection* could be responsible for all the changes within lineages and could result in either a passive or a driven trend, depending on whether or not ancestral character

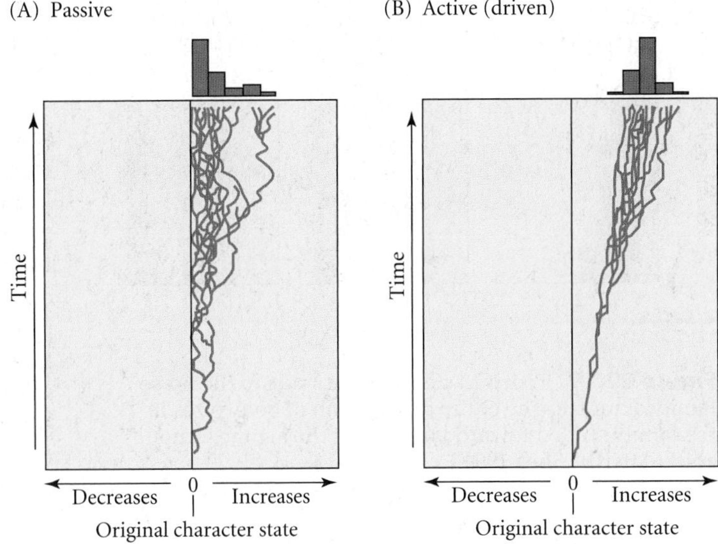

Figure 22.17 Computer simulations of the diversification of a clade. (A) A passive trend. A character shift in either direction is equally likely, but the character value cannot go beyond the boundary at left. The mean increases, but many lineages retain the original character value. (B) A driven trend. The entire distribution of character values is shifted by a bias in the direction of change, caused by a factor such as natural selection. (After McShea 1994.)

states remained advantageous for some lineages. The mean character state among species in a clade could also change as a result of a correlation with speciation or extinction rates (see Figure 11.16). The character might cause a rate difference (SPECIES SELECTION in the broad sense; see Chapter 11) or might simply be correlated with another character that causes a rate difference. This process has been called **species hitchhiking** (Levinton 2001) by analogy with the hitchhiking of linked genes in populations.

Examples of trends

Increases in the body size of mammals may represent a passive trend. Paleontologists noticed long ago that the maximal body size of species in many animal groups has tended to increase over time, a trend dubbed COPE'S RULE. A plot of the body sizes of 1534 species of North American late Cretaceous and Cenozoic mammals (Figure 22.18A) against their dates in the fossil record shows such a passive trend (Alroy 1998). Mammals were small before the K/T mass extinction, and the lower size limit has remained nearly the same ever since. However, mean and maximal sizes have increased, especially after the K/T extinction, when the explosive diversification of mammals began. Changes in body size between 779 matched pairs of older and younger species in the same genera (likely ancestor-descendant pairs) occurred in both directions but were significantly biased toward increases (Figure 22.18B). This nonrandomness strongly suggests that the trend was caused by natural selection rather than genetic drift.

Whereas data for all mammals taken together suggest a passive trend, body size in the horse family (Equidae) conforms to a driven trend (MacFadden 1986; McShea 1994). Not

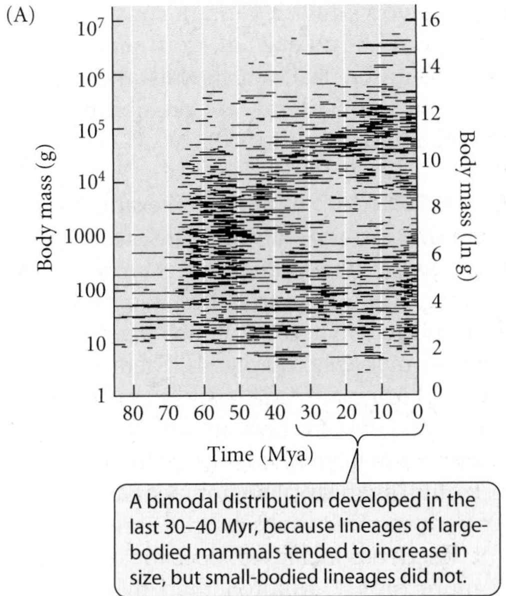

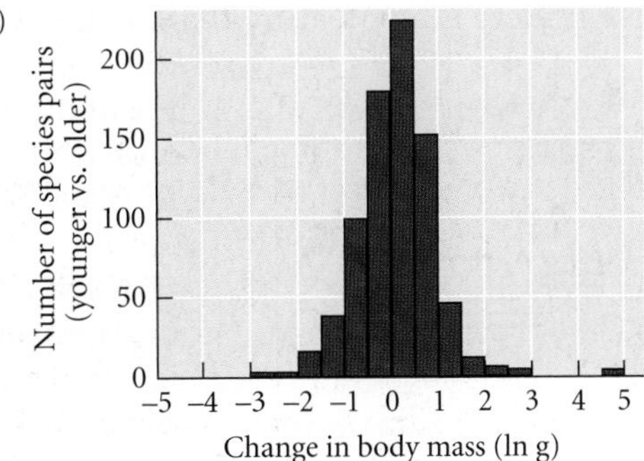

Figure 22.18 A passive trend: Cope's rule in Late Cretaceous North American mammals. (A) Each of 1534 species is plotted as a line showing its temporal extension and its body mass (estimated from tooth size). Although small mammals persist throughout the Cenozoic, there is an increasing number of large species over the course of time. (B) Change in body mass (negative or positive) is plotted for 779 pairs of older and younger species in the same genera (likely ancestor-descendant pairs). There are significantly more positive than negative changes, implicating natural selection as the cause of the trend toward increased size. (After Alroy 1998.)

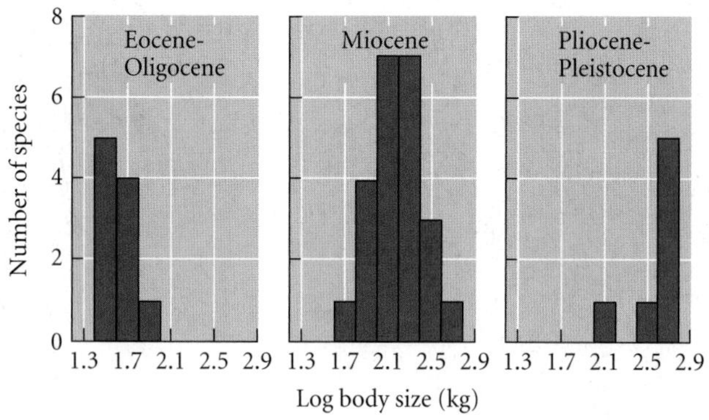

Figure 22.19 A driven trend: Cope's rule in the horse family, Equidae. The entire distribution of body sizes in the family shifted toward larger sizes during the Cenozoic. (After McShea 1994.)

only the maximal and the mean, but also the minimal, size increased during the Cenozoic (Figure 22.19). Ancestor-descendant pairs showed an increase in size much more often than a decrease.

These trends can be attributed to individual selection, and studies of selection in contemporary populations, especially of birds, insects, and plants, show that directional selection generally favors larger size (Kingsolver and Pfennig 2004). A good example of a trend that is due to species selection, in contrast, is the increase in the ratio of nonplanktotrophic to planktotrophic species in several clades of Cenozoic gastropods (Figure 22.20). Species that lack a planktotrophic dispersal stage are more susceptible to extinction than are planktotrophic species (species that feed as planktonic larvae). However, the nonplanktotrophic species more than compensate by their higher rate of speciation, probably because their lower rate of dispersal reduces the rate of gene flow among populations (Hansen 1980; Jablonski and Lutz 1983).

Trends that are due to lineage sorting by species hitchhiking are probably very common because if any one character causes one clade to become richer in species than other clades as a result of its effect on the rate of speciation or extinction, then all the other features of that clade will also tend toward greater frequencies. For example, coiled, sucking adult mouthparts may have become more prevalent among insects because they are a feature of the extremely diverse Lepidoptera (moths and butterflies). Lepidoptera larvae are herbivorous, and the herbivorous habit has consistently been associated with a high diversification rate in insects (Mitter et al. 1988; see Chapter 7). So the increased frequency of sucking mouthparts may be a result of hitchhiking with herbivory.

The boundaries that enforce passive trends may be due to either functional or developmental genetic constraints. For example, the smallest birds and mammals may have reached the lower functional limit of body size because a smaller animal might be unable to maintain a high body temperature because of its greater surface/volume ratio. In addition, developmental pathways may evolve that act as RATCHETS—mechanisms that make reversal unlikely, such as evolutionary losses of complex features, which in some cases may be irreversible (Bull and Charnov 1985).

Are there major trends in the history of life?

Do any trends or directions characterize the entire evolutionary history of life? Although many have been postulated, it is probably safe to say that no uniform driven trends can be discerned, because exceptions can be cited for every proposed trend. Still, one might ask if there is any feature that, on the whole, has evolved with enough consistency of direction that one would be able to tell, from snapshots of life at different times in the past, which was taken earlier and which later. Let us consider a few possibilities (see McShea 1998; Knoll and Bambach 2000).

Figure 22.20 A trend caused by species selection. The bars show the stratigraphic distributions of fossil species of volutid snails. Although nonplanktotrophic species had shorter durations, they arose by speciation at a higher rate, so the ratio of nonplanktotrophic to planktotrophic species increased over time. (After Hansen 1980.)

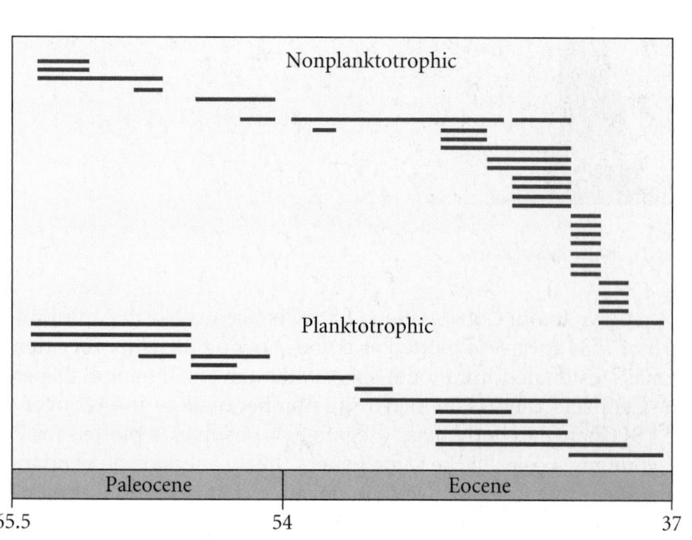

EFFICIENCY AND ADAPTEDNESS. Innumerable examples exist of improvements in the form of features that serve a specific function. The mammal-like reptiles, for example, show trends in feeding and locomotory structures associated with higher metabolism and activity levels, culminating in the typical body plan of mammals (see Chapter 4). There may well be a global trend toward greater efficiency (Ghiselin 1995). However, efficiency and effectiveness are difficult to measure, and they must always be defined relative to the task set by the context—by the organism's environment and way of life, which differ from species to species.

If efficiency of design has increased, does that mean that organisms are more highly adapted than in the past?

Darwin thought this likely; he imagined that if long-extinct species were revived and had to compete with today's species, they would lose the competition badly. We might find evidence of the evolution of competitive superiority if the fossil record provided many examples of competitive *displacement* of early by later taxa—but as we have seen (in Chapter 7), this pattern is much less common than *replacement* by later taxa well after the earlier ones became extinct. And even though natural selection within populations increases mean fitness (specifically, *relative* fitness), fitness values are always context-dependent. We cannot meaningfully compare the level of adaptedness of a shark and a falcon, or even of a bird-hunting falcon and a bat-hunting falcon, since they are as adapted to different tasks as are flat-head and Phillips-head screwdrivers. And it may be difficult, even in principle, to compare the adaptedness of a species with that of its long-extinct ancestor, since they may have experienced quite different selective regimes.

We might suppose that species longevity would be a measure of increase in adaptedness, but this need not be so. The consequence of natural selection is the adaptation of a population to the currently prevailing environment, not to future environments, so selection does not imbue a species with insurance against environmental change. We have seen that within many clades, the age of a genus or family does not influence its probability of extinction, implying that a lineage does not become more extinction-resistant over time (see Figure 7.9). It is true that the rate of "background" extinction of taxa has declined through the Phanerozoic (see Figure 7.7). This decline might indicate increasing adaptedness, but it might also be explained by the greater numbers of species in later taxa, or by the demise of major extinction-prone clades in the Paleozoic. Neither of these hypotheses requires us to postulate that extinction resistance evolves within lineages.

COMPLEXITY. It is difficult to define, measure, and compare complexity among very different organisms, although anatomical complexity may be considered to be proportional to the number of different kinds of parts that comprise an organism and to the irregularity of their arrangement (McShea 1991).

One of the most striking aspects of the history of life has been an increase in the maximal level of HIERARCHICAL ORGANIZATION, whereby entities have emerged that consist of functionally integrated associations of lower-level individuals (Maynard Smith and Szathmáry 1995; McShea 2001). The first cells arose from compartments of replicating molecules; the eukaryotic cell evolved from an association of prokaryotic cells; multicellular organisms with differentiated cell types evolved from aggregations of unicellular ancestors; and aggregates of a few kinds of multicellular organisms (e.g., social insects, humans) form highly integrated colonies. The difficulty to be overcome in all these transitions was that selection at the level of the component units (e.g., individual cells) could threaten the integrity of the larger unit (e.g., multicellular organism; see Chapter 16). In general, such conflict has been suppressed by development through a stage (e.g., the unicellular egg) that establishes high relationship (and thus the power of kin selection) among the component units, such as the genetic identity of the cells of a multicellular organism (Maynard Smith and Szathmáry 1995; Michod 1999). Richard Grosberg and Richard Strathmann (2007) describe the many other defenses that multicellular organisms have against disruption by rogue cells, and argue that the evolution of multicellularity was rather easy because many of the prerequisites, such as cell adhesion mechanisms, had already evolved in unicellular forms. (See Newman et al. 2006.)

The major changes in hierarchical organization represent only a few evolutionary events, in which the great majority of lineages did not participate. In contrast, the anatomical complexity of Cambrian animals was arguably as great as that of living forms, and many characteristics have evolved toward simplification or loss in innumerable clades (see Chapter 3). In fact, phylogenetic evidence seems to indicate that less pronounced changes in hierarchical complexity, such as the difference between green algae with one cell type versus two, have been almost equally divided between increases and decreases, with a bias toward decreased complexity in several clades (Marcot and McShea 2007). This is true of both morphology and behavior. For example, eusociality has evolved several times in the Hymenoptera but has frequently been lost (Danforth et al. 2003). The advantage of euso-

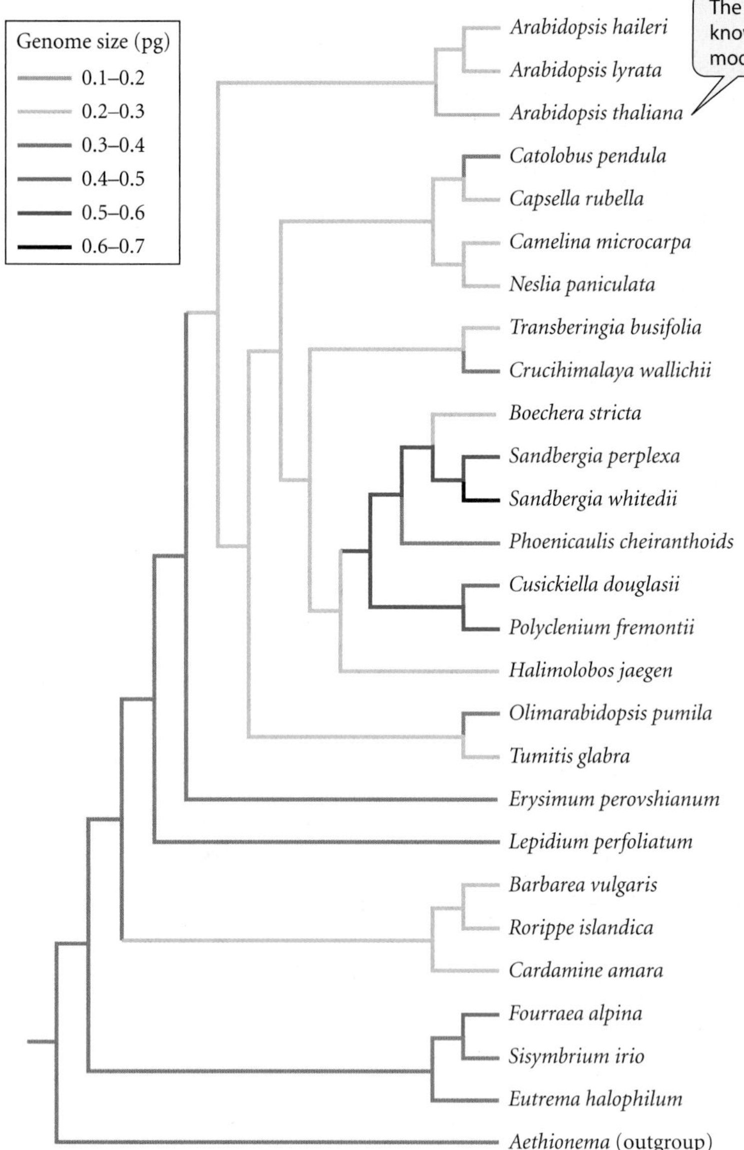

Figure 22.21 A phylogeny of some genera in the mustard family (Brassicaceae) reveals that DNA content (C-value) has both increased and decreased. (After Oyama et al. 2008.)

ciality, or probably the capacity for any other complex behavior, must depend on the environment, and there is no guarantee that it will always increase.

Genome size, meaning the amount of DNA in the cell nucleus (C-value; see Chapter 3), varies by five orders of magnitude among eukaryotes (Oliver et al. 2007). Although genome size has decreased in some lineages, such as *Arabidopsis* (Figure 22.21), increases have been more prevalent because of mechanisms such as polyploidy and the proliferation of transposable elements. However, the number of functional genes recognized in sequenced eukaryotic genomes, which might be a more meaningful measure of organismal complexity, varies by less than one order of magnitude; the variation in genome size is largely due to introns and noncoding DNA, most of which is thought not to encode genetic information (Oliver et al. 2007). The number of coding sequences does not appear to be correlated with our traditional (although superficial and perhaps erroneous) impressions of phenotypic complexity; we like to think that mammals such as humans, with 24,000 genes, are more complex than rice, with 60,000 (see Figure 3.27). The number of functional genes has not yet been determined for enough organisms to permit a phylogenetic analysis of changes in their number, but we may expect that just as for DNA content, both increases and decreases will ultimately be revealed. We know that gene number has declined in endosymbiotic bacteria (Chapter 19), and the same will surely be found in eukaryotic parasites. So it is not yet possible to say if there has been a trend in the information content of genomes throughout evolution. In any case, it is not clear that the number of genes is the best measure of organismal complexity, because the variety of functions that reside in a genome may be greatly amplified by alternative splicing of genes, multiple binding sites for different transcription factors, and other processes.

DIVERSITY AND DISPARITY. It appears, then, that no active trends have been found for any characteristic of individual organisms over the entire tree of life. There are, however, collective properties of organisms taken together that might have increased. Although many higher taxa have dwindled to extinction, and although several mass extinctions have temporarily reduced diversity, the total number of taxa (species, genera, families) on Earth has certainly increased over time. The likely causes of this increase are discussed in Chapter 7. A trend toward greater diversity is qualitatively different, of course, from trends in the properties of individual organisms, such as structural complexity.

In many clades, and in life as a whole, not only the number of species and higher taxa, but also the amount of phenotypic variation among species, has increased (Foote 1997). Increases in the phenotypic diversity (often referred to as the **disparity**) of a taxon or clade are often strongly correlated with increases in the number of taxa within that clade, which indeed may be used as an indicator of disparity. Morphological disparity, in turn, reflects ecological disparity—the variety of adaptive zones that have been occupied over evolutionary time (Knoll and Bambach 2000). Disparity often increases more rapidly early in a clade's history than later. It is not known if developmental integration increases and prevents large phenotypic changes from evolving later, or if increasing numbers of species occupy major resources, so that there is less room for ecological and morphological divergence of newly arisen species (Foote 1997).

The question of progress

Many people who accept evolution still conceive it as a purposive, progressive process, culminating in the emergence of consciousness and intellect. Even some evolutionary biologists have taken some of the qualities we most prize in ourselves, such as intellect or empathy, as criteria of progress, and quite ignoring the innumerable lineages that have not evolved at all in these directions, have seen in evolution a history of progress toward the emergence of humankind (see Ruse 1996).

The word "progress" usually implies movement toward a goal, as well as improvement or betterment. The processes of evolution, such as mutation and natural selection, cannot imbue evolution with a goal. We cannot suppose that humans, or any creatures with our mental faculties, were destined to evolve. Every step in the long evolutionary history of humans—or of any other species—was contingent on genetic and environmental events that could well have been otherwise (Gould 1989). For this reason, reflective evolutionary biologists have often concluded that humanity's existence has never been predetermined in any sense, and moreover, that it is very unlikely that any other "humanoid" species exists elsewhere in the near reaches of the universe—especially any with which we have the faintest hope of communicating (Simpson 1964; Mayr 1988b).

Progress in the sense of betterment implies a value judgment, and here we must guard against the parochial, anthropocentric view that human features are better than those of other species. Were a rattlesnake or a knifefish capable of conscious reflection, it doubtless would measure evolutionary progress by the elegance of an animal's venom delivery system or its ability to communicate by electrical signals. The great majority of animal lineages—to say nothing of plants and fungi—show no evolutionary trend toward greater "intelligence" (however it might be defined and measured), which must be seen as a special adaptation appropriate to some ways of life but not others. It is difficult, if not impossible, to specify a universal criterion by which to measure "improvement" that is not laden with our human-centered values.

Most evolutionary biologists have therefore concluded that we cannot objectively find progress in evolutionary history (Ruse 1996). As we have seen, it is hard to document even objective trends, especially active trends, in any feature such as complexity or adaptedness. The most characteristic feature of the history of evolution, rather, is the unceasing proliferation of new forms of life, of new ways of living, of seemingly boundless, exquisite diversity. The majesty of this history inspired Darwin to end *The Origin of Species* by reflecting on the "grandeur in this view of life," that "whilst this planet has gone cycling on according to the fixed law of gravity, from so simple a beginning endless forms most beautiful and most wonderful have been, and are being, evolved."

Summary

1. The average rate of evolution of most characters is very low because long periods of little change (stasis) are averaged with short periods of rapid evolution, or the character mean fluctuates without long-term directional change. The highest rates of character evolution in the fossil record are comparable to rates observed in contemporary populations and can readily be explained by known processes such as mutation, genetic drift, natural selection, and speciation. The duration of speciation is short enough to account for the observed rates of increase in species diversity.
2. The fossil record provides examples of both gradual change and the pattern called punctuated equilibrium: a rapid shift from one static phenotype to another. The hypothesis that such shifts require the occurrence of peripatric speciation is not widely accepted because responses to selection do not depend on speciation.

3. Stasis can be explained by genetic constraints, stabilizing selection (owing largely to habitat tracking), or the erasure of divergence by episodic massive gene flow among populations with ancestral and derived character states.
4. Higher taxa arise not in single steps, by macromutational jumps (saltation), but by multiple changes in genetically independent characters (mosaic evolution). Most such characters evolve gradually, through intermediate stages, but some characters evolve discontinuously as a result of mutations with moderately large effects.
5. Characters may be phylogenetically conservative because of limits on the origin of variation (genetic and developmental constraints) or because of niche conservatism (resulting in stabilizing selection).
6. New features often are advantageous even at their inception. They often evolve by modification of pre-existing characters to serve accentuated or new functions, or sometimes as by-products of the development of other structures. Evolutionary novelties often result when two or more functions of a structure are decoupled, or when structures are duplicated and diverge in structure and function.
7. Complex structures such as eyes evolve by rather small, individually advantageous steps. They may acquire functional integration with other structures so that they become indispensable.
8. Some fundamental characteristics of developmental processes and organismal integration, such as modularity, may enhance "evolvability," the capacity of a genome to produce variants that are potentially adaptive. In theory, some aspects of evolvability can evolve by natural selection, but whether or not this has commonly occurred is not known.
9. Long-term trends may result from individual selection, species selection, or species hitchhiking, the phylogenetic association of a character with other characters that affect speciation or extinction rates. Active trends, whereby the entire frequency distribution of a character among species in a clade shifts in a consistent direction over time, are less common than passive trends, in which variation among species (and therefore the mean of the clade) expands from an ancestral state that is located near a boundary (e.g., the clade may begin near a minimal body size).
10. Probably no feature exhibits a trend common to all living things. Features such as genome size and structural complexity display passive trends, in that the maximum has increased since very early in evolutionary history, but such changes have been inconsistent among lineages. There is no clear evidence of trends in measures of adaptedness such as the longevity of species or higher taxa in geological time. The most conspicuous directional change in evolutionary history is an increase (with setbacks due to major extinctions) in the species diversity and the phenotypic and ecological disparity of organisms taken as a whole.
11. If "progress" implies a goal, then there can be no progress in evolution. If it implies betterment or improvement, we still cannot identify objective criteria by which the history of evolution can be shown to be one of "improvement." Characters improve in their capacity to serve certain functions, but these functions are specific to the ecological context of each species.

Terms and Concepts

disparity
driven trend (= active trend)
evolvability
habitat tracking
macroevolution
microevolution
niche conservatism
passive trend
progress
punctuated equilibrium
saltation
species hitchhiking
stasis
trend

Suggestions for Further Reading

Tempo and Mode in Evolution by G. G. Simpson (Columbia University Press, New York, 1944) and *Evolution above the Species Level* by B. Rensch (Columbia University Press, New York, 1959) are classic works of the evolutionary synthesis, in which the authors reconcile macroevolutionary phenomena with neo-Darwinian theory. *Punctuated Equilibrium* by Stephen Jay Gould (Harvard University Press, Cambridge, MA, 2007) is the posthumously published central chapter of his magnum opus, *The Structure of Evolutionary Theory.*

J. S. Levinton, in *Genetics, Paleontology, and Macroevolution* (second edition, Cambridge University Press, Cambridge, 2001), discusses many of the topics of this chapter, taking a strong stand on controversial issues. D. W. McShea provides a good introduction to contemporary research on evolutionary trends in "Possible largest-scale trends in organismal evolution: Eight "live hypotheses" (*Annual Review of Ecology and Systematics* 29: 293–318, 1998). Evolvability is treated by A. Wagner in *Robustness and Evolvability in Living Systems* (Princeton University Press, Princeton, NJ, 2005).

Newman, S. A., G. Forgacs, and G. B. Müller. 2006. Before programs: The physical origination of multicellular forms. Internat. J. Devel. Biol. 50 (S1):289-299.

Problems and Discussion Topics

1. Snapdragons (*Antirrhinum*) and their relatives in the traditionally recognized family Scrophulariaceae have bilaterally symmetrical flowers, derived from the radially symmetrical condition of their ancestors. A mutation in the *cycloidea* gene makes snapdragon flowers radially symmetrical. The *cycloidea* gene has been mapped and sequenced and was found to be asymmetrically expressed in the developing snapdragon flower, where it probably regulates other genes that contribute to flower development (Luo et al. 1996). Is it possible to determine whether this gene was the primary basis of the original evolution of bilateral symmetry in the Scrophulariaceae? How might one determine whether bilateral symmetry evolved continuously through intermediate stages or by a single discrete ("saltational") change at this locus?
2. Despite the close relationship between humans and chimpanzees, and despite the genetic variation in most of their morphological characteristics, there are no records of humans giving birth to babies with chimpanzee morphology. Would you expect this to occur if humans have evolved gradually from apelike ancestors? If not, why not?
3. Find and evaluate the evolutionary literature that describes how wings could have evolved gradually, through advantageous intermediate steps. See Dudley et al. 2007.
4. How might one test the three hypotheses proposed to explain stasis that are mentioned in this chapter?
5. How can we tell whether a trend was caused by species selection or individual selection? How might we tell whether a passive trend was due to natural selection or genetic drift?
6. Paleontologists and some biologists commonly infer function, and even behavior, from anatomical details. Skeletal features, for example, are often used to infer that an extinct mammal (such as an early hominin) was highly, somewhat, or not at all arboreal. This inference assumes a good fit of form to function (i.e., optimal form). Can this assumption be justified?
7. Would you expect "living fossils," such as a tadpole shrimp, to differ from other species in amount of genetic variation, genetic correlations among characters, canalization, or any other feature that might affect "evolvability"?
8. Are you surprised that the number of distinct genes in the human genome seems not to differ greatly from that in most other animals? Why or why not?
9. When the stress protein hsp90 is rendered less functional by mutation in *Drosophila* or *Arabidopsis*, individuals display various morphological abnormalities (Chapter 13). Does this suggest that the *hsp90* gene acts as a regulator of variation and is therefore a mechanism of evolvability?
10. Explain how the relationship of phenotype to genotype might evolve so that genetic variation might be more likely to include potentially adaptive variants. How is this theory different from Lamarckian inheritance or from supposing that natural selection has foresight?
11. Stephen Jay Gould (1989) and others have argued that the evolution of a self-conscious, intelligent species (i.e., humans) was historically contingent: it would not have occurred had any of a great many historical events been different. The philosopher Daniel Dennett (1995) and others have disagreed, arguing that convergent evolution is so common that if humans had not evolved, some other lineage would probably have given rise to a species with similar mental abilities. What do you think, and why? If Gould's position is right, what are its philosophical implications, if any?

CHAPTER 23

Evolutionary Science, Creationism, and Society

Too complex? The eye of a spectacled caiman (*Caiman crocodylus*). Creationists have argued that many features of organisms, ranging from subcellular structures to the eyes of vertebrates, are too complex to have evolved in incremental stages. Overwhelming evidence and the principles of evolution decisively refute this claim. (Photo © Mark MacEwen/OSF/Photolibrary.com.)

We cannot pretend to know more than a small fraction of the evolutionary history of life, or to understand fully all the mechanisms of evolution, and there are many differences of opinion among evolutionary biologists about details of both history and mechanisms. However, the historical reality of evolution—the descent, with modification, of all organisms from common ancestors—has not been in question among scientists for well over a century. It is as much a scientific fact as the atomic constitution of matter or the revolution of the Earth around the Sun.

Nevertheless, many people do not believe in evolution. More than 40 percent of people in the United States believe that the human species was created directly by God, rather than having evolved from a common ancestor shared with other primates (and from a much more remote common ancestor shared with all other species). People who hold this belief are often referred to as **creationists**. In contrast, most people in Europe do not question the reality of evolution (even in countries such as Italy that have an officially established religion), and are often astonished that antiscientific attitudes on evolution flourish in the technologically and scientifically most prominent country in the world. However, creationists are becoming more vocal in western Europe, and an Islamic creationist movement is very active in Turkey. Its propaganda includes a huge, lavishly illustrated "Atlas of Creation" that has been shipped to prominent scientists throughout the world.

In this chapter, we will look at the beliefs of creationists and their arguments against evolution, and we will review the evidence for evolution presented throughout this book. We will also look at the many ways in which human society depends on an understanding of evolutionary science.

Creationists and Other Skeptics

Most disbelievers in evolution reject the idea because they think it conflicts with their religious beliefs. Many, perhaps most, are Christian, Muslim, or (rarely) Jewish fundamentalists who adhere to a literal or almost literal reading of sacred texts. For Christian and Jewish fundamentalists, evolution conflicts with their interpretation of the Bible, especially the first chapters of Genesis, which portray God's creation of the heavens, the Earth, plants, animals, and humans in 6 days. However, many Western religions understand these biblical descriptions to contain symbolic truths, not literal or scientific ones. Many deeply religious people believe in evolution, viewing it as the natural mechanism by which God has enabled creation to proceed. Many scientists subscribe to this view, including some researchers in evolutionary biology and some of the most impassioned opponents of creationism (see Miller 1999, 2008). Some religious leaders have made clear their acceptance of the reality of evolution. (See an array of such statements under "Voices for Evolution" at www.ncseweb.org, the Web site of the National Center for Science Education.) For example, Pope John Paul II affirmed the validity of evolution in 1996 and emphasized that there is no conflict between evolution and the Catholic Church's theological doctrines. [The text of his letter was reprinted in *The Quarterly Review of Biology* 72: 381–396 (December 1997)]. The Pope's position was close to the argument generally known as **theistic evolution**, which holds that God established natural laws (such as natural selection) and then let the universe run on its own, without further supernatural intervention.

The beliefs of creationists vary considerably (Figure 23.1; Numbers 2006). The most extreme interpret every statement in the Bible literally. They include "young Earth" creationists who believe in **special creation** (the doctrine that each species, living and extinct, was created independently by God, essentially in its present form) and in a young universe and Earth (less than 10,000 years old), a deluge that drowned the Earth, and an ark in which Noah preserved a pair of every living species. They must therefore deny not only evolution but also most of geology and physics (including radioactive dating and astronomical evidence of the great age of the universe). Some creationists share many or most

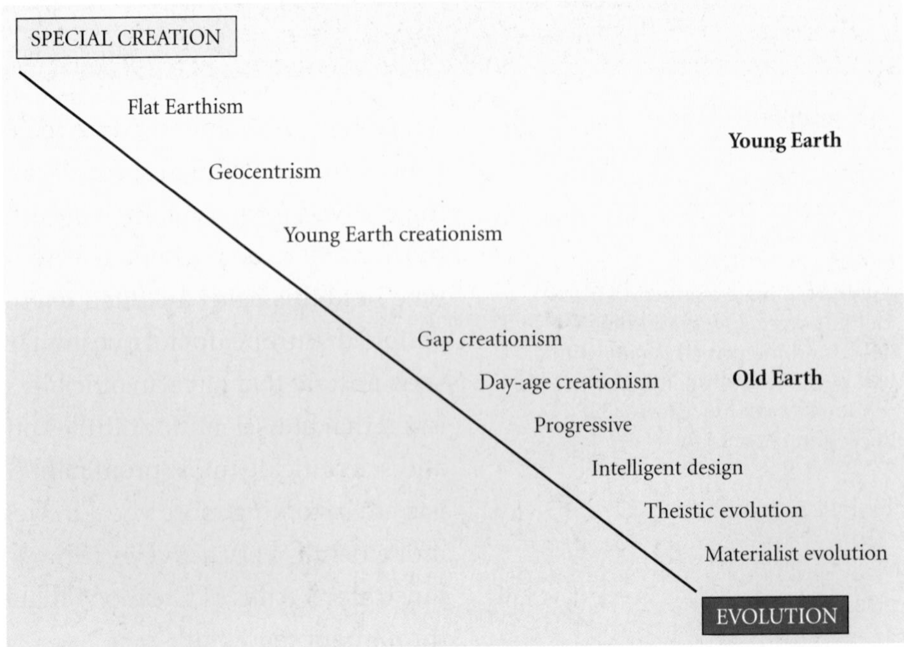

Figure 23.1 The special creation–evolution continuum. "Young Earth" proponents affirm that the Earth is young (thousands of years old), not old (billions of years old). Adherents of special creation tend to consider the Bible as literally true. (After Scott 1997.)

of the biblical literalists' beliefs, including special creation, but grant the antiquity of the Earth and the rest of the universe. (The "day-age" position holds that each of the days of creation cited in Genesis was millions of years long.) Other creationists allow that mutation and natural selection can occur, and even that very similar species can arise from a common ancestor. However, they deny that higher taxa (genera, families, etc.) have evolved from common ancestors, and most of them vigorously assert that the human species was specially created by God, in His image—which is their major reason for caring about evolution at all.

Most of the efforts of activist creationists are devoted to suppressing the teaching of evolution in schools, or at least insisting on "equal time" for their views. In the United States, however, the Constitutional prohibition of state sponsorship of religion has been interpreted by the courts to mean that the biblical version, or any other explicitly religious version, of the origin of life's diversity cannot be promulgated in public schools. The activists have therefore adopted several forms of camouflage. One is **scientific creationism** or **creation science**, which consists of attacks on, and supposedly scientific disproofs of, evolution. These arguments have not succeeded, chiefly because they do not have, and cannot have, any scientific content. The more recent camouflage is **intelligent design (ID)** theory. ID proponents generally do not publicly invoke special creation by God. Some of them even seem to accept certain aspects of evolution, such as development of different species from common ancestors. However, they argue that many biological phenomena are too complicated to have arisen by natural processes, and can only be explained by an intelligent designer. They claim that intelligent design is a scientific, not a religious, concept. The designer they envision, however, is a supernatural rather than a material being (i.e., a being equivalent to God). Some ID advocates have explicitly stated that their aim is much broader than evolutionary biology: that they propose to defeat "scientific materialism" and to replace materialistic explanations with the understanding that nature and human beings are created by God (Forrest and Gross 2004). Creationists' targets include geology, physics, and cosmology (Kaiser 2007).

Bills requiring public schools to give "creation science" equal time have been introduced in state legislatures, but the U.S. Supreme Court in 1987 found one such law unconstitutional because it "endorses religion by advocating the religious belief that a supernatural being created humankind," and because it was written "to restructure the science curriculum to conform with a particular religious viewpoint." As a result, local and state legislators and school officials use a variety of subterfuges to get creationism into school science curricula. In the most prominent recent case, parents in the town of Dover, Pennsylvania, filed a suit against local school board officials who read a statement to science classes, warning them that evolution is a theory rather than fact and suggesting that students read a particular book in the school library that advocates intelligent design. In December 2005, the judge ruled for the parents in a decision that is well worth reading as a model statement of what science is and why intelligent design is not science, but thinly guised religious doctrine. (See www.ncseweb.org or www.pamd.uscourts.gov/kitzmiller/kitzmiller_342.pdf.)

Since the Dover decision, the creationist strategy in the United States has been to advocate that students be encouraged to examine the evidence for both sides of "controversial scientific issues"—which sounds marvelously enlightened, except that evolution and one or two other topics, such as global warming, are usually singled out, with the clear implication that these are particularly doubtful. Bear in mind that the fundamental principles of evolution are not controversial within science.

Science, Belief, and Education

Biologists and others who argue against including any form of creationism in science curricula are not against free speech, and they are not trying to extinguish religious belief. They simply hold that although it might be acceptable to teach about creation stories in classes on, say, history or contemporary society, such beliefs are not valid *scientific* hypotheses and have no place in *science* classes. Unfortunately, most people have little understanding of what science is and how it works, even if they have had science courses—and this understanding is critically important to the evolution-versus-creationism debate.

Science is not a collection of facts, contrary to popular belief, but rather a *process* of acquiring understanding of natural phenomena. This process consists largely of posing hypotheses and testing them with observational or experimental evidence. Despite loose talk about "proving" hypotheses, most scientists agree with philosophers of science that scientists cannot absolutely prove hypotheses; that is, they cannot attain absolute, guaranteed proof of the kind that can be obtained in mathematics. Rather, the hypothesis that currently best explains the data is *provisionally* accepted, with the understanding that it may be altered, expanded, or rejected if subsequent evidence warrants doing so, or if a better hypothesis, not yet imagined, is devised. Sometimes, indeed, a radically new "paradigm" replaces an old one; for example, plate tectonics revolutionized geology in the 1950s, replacing the conviction that continents are fixed in position. More often, old hypotheses are incrementally modified and expanded over time. For instance, Mendel's laws of assortment and independent segregation, which initiated modern genetics, were modified when phenomena such as linkage and meiotic drive were discovered, but his underlying principle of inheritance based on "particles" (genes) holds true today.

This process reflects one of the most important and valuable features of science: even if individual scientists may be committed to a hypothesis, scientists as a group are not irrevocably committed to any belief, and do not maintain it in the face of convincing contrary evidence. They must, and do, change their minds if the evidence so warrants. Indeed, much of science consists of seeking chinks in the armor of established ideas, and few successes will burnish a scientist's reputation more than showing that an important orthodox hypothesis is inadequate or flawed. Thus science, as a social process, is tentative; it questions belief and authority; it continuously tests its views against evidence. Scientific claims, in fact, are the outcome of a process of natural selection, for ideas (and scientists) compete with one another, so that the body of ideas in a scientific field grows in explanatory content and power (Hull 1988). Science differs in this way from creationism, which does not use evidence to test its claims, does not allow evidence to shake its a priori commitment to certain beliefs, and does not grow in its capacity to explain the natural world.

How could it? Suppose an advocate of "intelligent design" says that multicellular organisms are so complex, compared with single-celled organisms, that they must have originated by the intervention of an intelligent designer. Unless the ID advocate proposes that extraterrestrial creatures are responsible, this designer must be a supernatural rather than a material being. So what was this designer, how did it equip organisms with new features, how long did it take, and why did it do this? Natural science can at least imagine ways to address such questions. (For example, we can look for phylogenetic intermediates, analyze sequence differences in genes that encode the relevant character differences, look for fossils, do experiments on the selective advantage of multicellularity). But the ID hypothesis generates no research ideas. It stops science dead in its tracks.

Scientific research requires that we have some way of testing hypotheses based on experimental or observational data. *The most important feature of scientific hypotheses is that they are testable*, at least in principle. Sometimes we can test a hypothesis by direct observation, but more often we do not see processes or causes directly. (For example, electrons, atoms, hydrogen bonds, molecules, and genes are not directly visible, and we cannot watch the occurrence of mutation during DNA replication. We cannot see DNA replication itself, for that matter.) Rather, we infer such processes by comparing the outcome of observations or experiments with predictions made from competing hypotheses. In order to make such inferences, we must assume that the processes obey **natural laws**: statements that certain patterns of events will always occur if certain conditions hold. That is, science depends on the consistency, or predictability (at least in a statistical sense), of natural phenomena, as exemplified by the laws of physics and chemistry. Because supernatural events or agents are supposed to suspend or violate natural laws, science cannot infer anything about them, and indeed, cannot judge the validity of any hypotheses that involve them.

It is important to understand that just as religion does not provide scientific, mechanistic explanations for natural phenomena, science cannot provide answers to any questions that are not about natural phenomena: it cannot tell us what is beautiful or ugly, good or bad, moral or immoral. It cannot tell us what the meaning of life is, and it cannot tell us whether or not supernatural beings exist (see Gould 1999; Pigliucci 2002).

Scientists can test and falsify some specific creationist claims, such as the occurrence of a worldwide flood or the claim that the Earth and all organisms are less than 10,000 years old, but scientists cannot test the hypothesis that God exists, or that He created anything, because we do not know what consistent patterns these hypotheses might predict. (Try to think of any observation at all that would definitively rule out these supernatural possibilities.) Science must therefore adopt the position that natural causes are responsible for whatever we wish to explain about the natural world. This is not necessarily a commitment to METAPHYSICAL NATURALISM (the assumption that everything *truly* has natural rather than supernatural causes), but it is a commitment to METHODOLOGICAL NATURALISM (the *working principle* that we can entertain only natural causes when we seek scientific explanations). The fundamental claim of creationism—that biological diversity is the result of supernatural powers—is not testable. This is equally true of "intelligent design" theory; it cannot be evaluated by the methods of science.

We have been using the words "hypothesis," "theory," and "fact," and it is imperative that we understand what they mean. A **hypothesis** is a proposition, a supposition. Before 1944, the idea that the genetic material is DNA was a plausible hypothesis that had little evidence to support it. After 1944, this hypothesis became stronger and stronger as evidence grew. Now we consider it a fact. A **fact** is, simply, a hypothesis that has become so well supported by evidence that we feel safe in acting as if it were true. To use a courtroom analogy, it has been "proven" beyond reasonable doubt. Not beyond any conceivable doubt, but reasonable doubt.

A **theory**, as the word is used in science, doesn't mean an unsupported speculation or hypothesis (the popular use of the word). A theory is, instead, a big idea that encompasses other ideas and hypotheses and weaves them into a coherent fabric. It is a mature, interconnected body of statements, based on reasoning and evidence, that explains a wide variety of observations. It is, in one of the definitions offered by the *Oxford English Dictionary*, "a scheme or system of ideas and statements held as an explanation or account of a group of ideas or phenomena; … a statement of what are known to be the general laws, principles, or causes of something known or observed." Thus atomic theory, quantum theory, and plate tectonic theory are not mere speculations or opinions, but strongly supported ideas that explain a great variety of phenomena. There are few theories in biology, and among them evolution is surely the most important.

So is evolution a fact or a theory? In light of these definitions, evolution is a scientific fact. That is, the descent of all species, with modification, from common ancestors is a hypothesis that in the last 150 years or so has been supported by so much evidence, and has so successfully resisted all challenges, that is has become a fact. But this history of evolutionary change is explained by evolutionary theory, the body of statements (about mutation, selection, genetic drift, developmental constraints, and so forth) that together account for the various changes that organisms have undergone.

Because creationist explanations for the diversity and characteristics of living things cannot be evaluated by the methods of science, they should not be given "equal time" in a science classroom. Neither should other hypotheses that are either unscientific or demonstrably wrong. Chemistry teachers do not and should not teach about alchemy (the old idea that one element, such as lead, can be magically transformed into another, such as gold); earth science classes should not even mention the hypothesis that Earth is flat; history and psychology teachers should not consider astrological explanations for historical events or personality traits—even though there are people who believe in all of these pseudoscientific ideas. The ideal of democracy doesn't extend to ideas—some are simply wrong, and as a purely practical matter it is imperative that we recognize them as such. In everyday life, we assume and depend on natural, not supernatural, explanations. Unlike the Puritans of Salem, Massachusetts, who in 1692 condemned people for witchcraft, we no longer seriously entertain the notion that someone can be victimized by a witch's spell or possessed by devils, and we would be outraged if a criminal successfully avoided conviction because he claimed "the Devil made me do it." Even those who most devoutly believe that God sustains them in the palm of His hand will panic if the airplane's wing flaps malfunction. We depend on scientific explanations, and we know that science has proven its ability—because it works.

The consequences of rejecting science in favor of ideological dogma or folk belief or religious belief can be tragic. The day before I wrote this passage, news reports described the arrest of a couple on charges of child abuse. They had allowed their two children to die because they rejected medical help in favor of "faith healing." It has been estimated that the South African government could have prevented the premature death by AIDS of 330,000 adults and 35,000 babies, had it not been for President Thabo Mbeki's denial that AIDS is caused by HIV, and his appointment of a like-minded health minister who prescribed garlic, lemon juice, and beetroot as remedies (*New York Times*, November 26, 2008). The perils of ignoring or distorting fundamental science became tragically evident in the Soviet Union between the 1930s and the late 1960s. Before that time, Russia and the early Soviet Union had led the world in genetics, including evolutionary genetics. Under Stalin, however, Mendelism and Darwinism came to be viewed as unwelcome or dangerous Western ideologies. Trofim Lysenko, an agronomist with no scientific training, gained Stalin's favor by declaring genetics a capitalist, bourgeois threat to the state, and led a campaign in which geneticists were persecuted and imprisoned and genetics research and training were extinguished. Rejecting the Mendelian-Darwinian view of adaptation, Lysenko preached a Lamarckian doctrine that exposing organisms to an environmental factor, such as low temperature, would induce inherited adaptive changes in their offspring. Lysenko completely altered the practice of Soviet agriculture to fit this notion. The result was a disaster for Soviet food production and the Soviet people. Moreover, the Soviet Union was left out of major advances in food production that were developed by Western research (Soyfer 1994). Rejection or ideological control of science, the most reliable basis yet developed for knowing about the natural, material world, is profoundly dangerous to the welfare of society.

The Evidence for Evolution

The evidence for evolution has been presented throughout the preceding chapters of this book. The examples provided represent only a very small percentage of the studies that might be cited for each particular line of evidence. In this section, we will simply review the sources of evidence for evolution and refer back to earlier chapters for detailed examples.

The fossil record

The fossil record is extremely incomplete, for reasons that geologists understand well (see Chapter 4). Consequently, the transitional stages that we postulate in the origin of many higher taxa have not (yet) been found. But there is absolutely no truth to the claim, made by many creationists, that the fossil record does not provide any intermediate forms. There are many examples of such forms, both at low and high taxonomic levels; Chapter 4 provides several examples in the evolution of the classes of tetrapod vertebrates. Critically important intermediates are still being found: just in the last few years, several Chinese fossils, including feathered dinosaurs, have greatly expanded the record of the origin of birds, and the discovery of *Tiktaalik* has enlarged our understanding of the origin of tetrapods (Chapter 4). The fossil record, moreover, documents two important aspects of character evolution: mosaic evolution (e.g., the more or less independent evolution of different features in the evolution of mammals) and gradual change of individual features (e.g., cranial capacity and other features of hominins).

Many discoveries in the fossil record fit predictions made based on phylogenetic or other evidence. The earliest fossil ants, for instance, have the wasplike features that had been predicted by entomologists, and the discovery of feathered dinosaurs was to be expected, given the consensus that birds are modified dinosaurs. Likewise, phylogenetic analyses of living organisms imply a sequence of branching events, as well as a sequence of origin of the diagnostic characters of those branches. The fossil record often matches the predicted sequences (as we saw in Chapters 4 and 5): for example, prokaryotes precede eukaryotes in the fossil record, wingless insects precede winged insects, fishes precede tetrapods, amphibians precede amniotes, algae precede vascular plants, ferns and "gymnosperms" precede flowering plants.

Phylogenetic and comparative studies

Although many uncertainties about phylogenetic relationships persist (e.g., the branching order of the major groups of birds), phylogenies that are well supported by one class of characters usually match the relationships implied by other evidence quite well (see Chapter 2). For example, molecular phylogenies support many of the relationships that have long been postulated from morphological data. These two data sets are entirely independent (the molecular phylogenies are often based on sequences that have no biological function), so their correspondence justifies confidence that the relationships are real: that the lineages form monophyletic groups and have indeed descended from common ancestors.

The largest monophyletic group encompasses all organisms. Although Darwin allowed that life might have originated from a few original ancestors, we can be confident today that all known living things stem from a single ancestor because of the many features that are universally shared. These features include most of the codons in the genetic code, the machinery of nucleic acid replication, the mechanisms of transcription and translation, proteins composed only of "left-handed" (L-isomer) amino acids, and many aspects of fundamental biochemistry. Many genes are shared among all organisms, including the three major "empires" (Bacteria, Archaea, and Eukaryota; see Chapters 4 and 20), and these genes have been successfully used to infer the deepest branches in the tree of life.

Systematists have shown that the differences among related species often form gradual series, ranging from slight differences to great differences with stepwise intermediates (e.g., see Figures 3.12 and 3.22). Such intermediates often make it difficult to establish clear-cut families or other higher taxa, so that classification often becomes a somewhat arbitrary choice between "splitting" species among many taxa and "lumping" them into few. Systematic studies have also demonstrated the common origin, or homology, of characteristics that may differ greatly among taxa—the most familiar examples being the radically different forms of limbs among tetrapod vertebrates. Homology of structures is often more evident in early developmental stages than in adult organisms, and contemporary developmental biology demonstrates that Hox genes and other developmental mechanisms are shared among animal phyla that diverged from common ancestors a billion or more years ago (see Chapter 21).

Genes and genomes

The revolution in molecular biology and genomics is yielding data about evolution on a larger scale than ever before. These data increasingly show the extraordinary commonality of all living things. Because of this commonality, the structure and function of genes and genomes can be understood through comparisons among species and evolutionary models. Indeed, it is only because of this common ancestry that there has ever been any reason to think that human biochemistry, physiology, or brain function, much less genome function, could be understood by studying yeast, flies, rats, or monkeys (Figure 23.2).

Molecular studies show that the genomes of most organisms have similar elements, such as a great abundance of noncoding pseudogenes and satellite DNA and a plethora of "selfish" transposable elements that generally provide no advantage to the organism. These features are readily understandable under evolutionary theory, but would hardly be expected of an intelligent, omnipotent designer. Molecular evolutionary analyses have shown in great detail how new genes arise by processes such as unequal crossing over, and how duplicate genes diverge in function, increasing the genetic repertoire (see Chapter 20). Some DNA polymorphisms are shared between species, so that, for example, some major histocompatibility sequences of humans are more similar and more close-

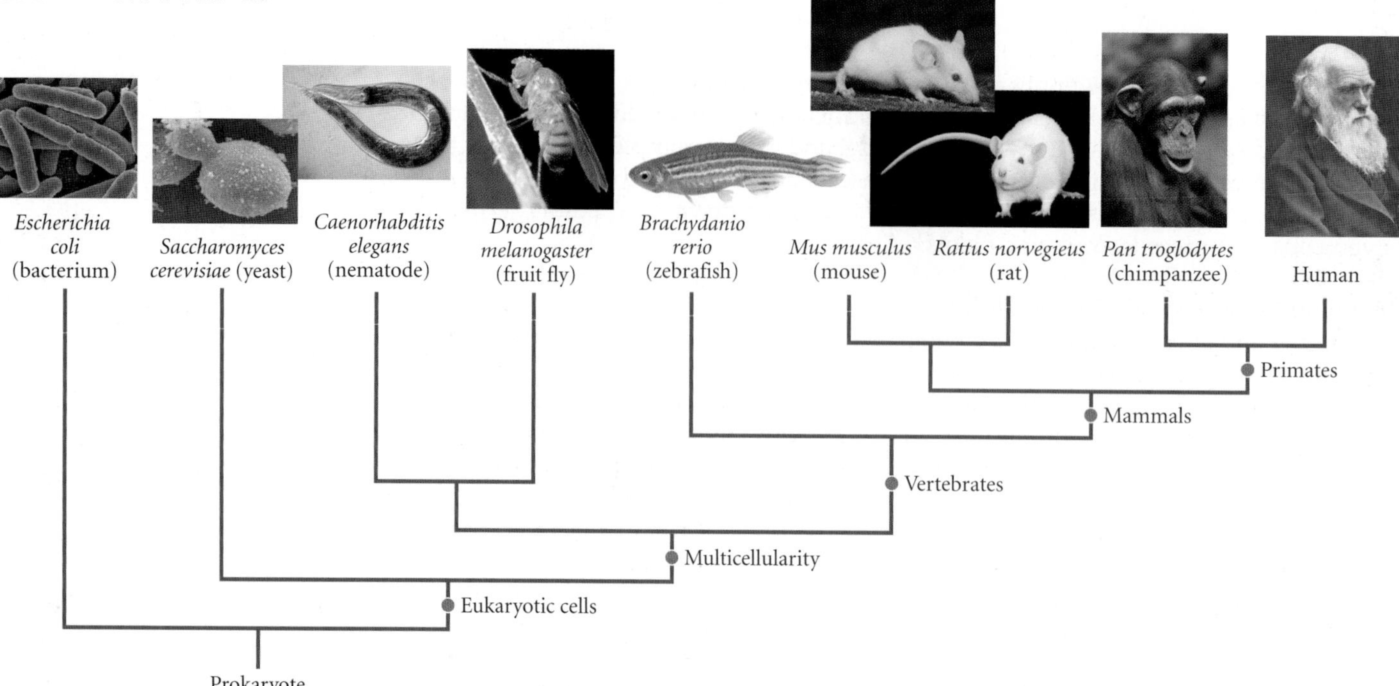

Figure 23.2 Only common ancestry can provide a scientific explanation for why the National Institutes of Health should support basic biological research on organisms ranging from bacteria to chimpanzees, on the supposition that it will contribute to understanding human health and disease. (Photo credits from left to right: courtesy of Rocky Mountain Laboratories, NIAID, NIH; courtesy of NASA; courtesy of Amy Pasquinelli/NIH; courtesy of David McIntyre; © David Dohnal/ShutterStock; © Kurt G./ShutterStock; © Podfoto/ShutterStock; © Holger Ehlers/ShutterStock.)

ly related to chimpanzee sequences than to other human sequences (see Figure 12.27B). What more striking evidence of common ancestry could there be?

Among the many other ways in which molecular studies affirm the reality of evolution, consider just one more: molecular clocks. They are far from perfectly accurate, but sequence differences between species nevertheless are roughly correlated with the time since common ancestry, judged from other evidence such as biogeography or the fossil record (see Chapters 2 and 4). The sequence differences do not simply encode the phenotypic differences that the organisms manifest, and the phenotypic differences themselves are much less correlated with time since common ancestry. No theory but evolution makes sense of these patterns of DNA differences among species.

Biogeography

We noted in Chapter 6 that the geographic distributions of organisms provided Darwin with abundant evidence of evolution, and they have continued to do so. For example, the distributions of many taxa correspond to geological events such as the movement of land masses and the formation and dissolution of connections between them. We saw that the phylogeny of Hawaiian species matches the sequence by which the islands came into existence. We saw that diverse ecological niches in a region are typically filled not by the same taxa that occupy similar niches in other parts of the world, but by different monophyletic groups that have undergone independent adaptive radiation (such as the anoles of the Greater Antilles). We saw, as did Darwin, that an isolated region such as an island is not populated by all the kinds of organisms that could thrive there, as we might suppose a thoughtful designer could arrange. Instead, whole groups are commonly missing, and human-introduced species often come to dominate.

Failures of the argument from design

Since God cannot be known directly, theologians such as Thomas Aquinas have long attempted to infer His characteristics from His works. Theologians have argued, for instance, that order in the universe, such as the predictable movement of celestial bodies, implied that God must be orderly and rational, and that He created according to a plan. From the observation that organisms have characteristics that serve their survival, it could similarly be inferred that God is a rational, intelligent designer who, furthermore, is beneficent: He not only conferred on living things the boon of existence but equipped them for all their needs. Such a beneficent God would not create an imperfect world; so as the philosopher Leibniz said, this must be "the best of all possible worlds." (Leibniz's position was actually more complicated, but his phrase was mercilessly ridiculed by Voltaire in his marvelous satire *Candide*.) The adaptive design of organisms, in fact, has long been cited as evidence of an intelligent designer. This was the thrust of William Paley's (1831) famous example: as the design evident in a watch implies a watchmaker, so the design evident in organisms implies a designer of life. This "argument from design" has been renewed in the "intelligent design" version of creationism, and it is apparently the most frequently cited reason people give for believing in God (Pigliucci 2002).

Of course, Darwin made this particular theological argument passé by providing a natural mechanism of design: natural selection. Moreover, Darwin and subsequent evolutionary biologists have described innumerable examples of biological phenomena that are hard to reconcile with beneficent intelligent design. Just as Voltaire showed (in *Candide*) that cruelties and disasters make a mockery of the idea that this is "the best of all possible worlds," biology has shown that organisms have imperfections and anomalies that can be explained only by the contingencies of history, and characteristics that make sense only if natural selection has produced them. If "good design" were evidence of a kindly, omnipotent designer, would "inferior design" be evidence of an unkind, incompetent, or handicapped designer?

Only evolutionary history can explain vestigial organs—the rudiments of once-functional features, such as the tiny, useless pelvis and femur of whales, the reduced wings under the fused wing covers of some flightless beetles, and the nonfunctional stamens or pistils of plants that have evolved separate-sexed flowers from an ancestral hermaphroditic condition. Likewise, only history can explain why the genome is full of "fossil" genes: pseudogenes that have lost their function. Only the contingencies of history can explain the arbitrary nature of some adaptations. For instance, whereas ectothermic ("cold-blooded") tetrapods have two aortic arches, "warm-blooded" endotherms have only one. This difference is probably adaptive, but can anything except historical chance explain why birds have retained the right arch and mammals the left?

Because characteristics evolve from pre-existing features, often undergoing changes in function, many features are poorly engineered, as anyone who has suffered lower back pain or wisdom teeth can testify. Once the pentadactyl limb became developmentally canalized, tetrapods could not evolve more than five digits even if they would be useful: the extra "finger" of the giant panda's hand is not a true digit at all, and lacks the flexibility of true fingers because it is not jointed (see Figure 22.11). Similarly, animals would certainly be better off if they could synthesize their own food, and corals do so by harboring endosymbiotic algae—but no animal is capable of photosynthesis.

If a designer were to equip species with a way to survive environmental change, it might make sense to devise a Lamarckian mechanism, whereby genetic changes would occur in response to need. Instead, adaptation is based on a combination of a random process (mutation) that cannot be trusted to produce the needed variation (and often does not) and a process that is the very epitome of waste and seeming cruelty (natural selection, which requires that great numbers of organisms fail to survive or reproduce). It would be hard to imagine a crueler instance of natural selection than sickle-cell anemia, whereby part of the human population is protected against malaria at the expense of hundreds of thousands of other people, who are condemned to die because they are homozygous for a gene that happens to be worse for the malarial parasite than for heterozygous carriers (see Chapter 12). Indeed, Darwin's theory of the cause of evolution was widely rejected just because people found it so distasteful, even horrifying, to contemplate. And, of course, this process often does not preserve species in the face of change: more than 99 percent of all species that have ever lived are extinct. Were they the products of an incompetent designer? Or one that couldn't foresee that species would have to adapt to changing circumstances?

Many species become extinct because of competition, predation, and parasitism. Some of these interactions are so appalling that Darwin was moved to write, "What a book a devil's chaplain might write on the clumsy, wasteful, blundering, low, and horribly cruel works of Nature!" Darwin knew of maggots that work their way up the nasal passages into the brains of sheep, and wasp larvae that, having consumed the internal organs of a living caterpillar, burst out like the monsters in the movie *Alien*. The life histories of parasites, whether parasitic wasp or human immunodeficiency virus (HIV), ill fit our concept of an intelligent, kindly designer, but they are easily explained by natural selection (see Chapter 19).

No one has yet demonstrated a characteristic of any species that serves only to benefit a different species, or only to enhance the so-called balance of nature—for as Darwin saw, "such could not have been produced through natural selection." Because natural selection consists only of differential reproductive success, it results in "selfish" genes and genotypes, some of which have results that are inexplicable by intelligent design (see Chapter 16). We have seen that genomes are brimming with sequences such as transposable elements that increase their own numbers without benefiting the organism. We have seen maternally transmitted cytoplasmic genes that cause male sterility in many plants, and nuclear genes that have evolved to override them and restore male fertility. Such conflicts among genes in a genome are widespread. Are they predicted by intelligent design theory? Likewise, no theory of design can predict or explain features that we ascribe to sexual selection, such as males that remove the sperm of other males from the female's reproductive tract, or chemicals that enhance a male's reproductive success but shorten his mate's life span. Nor can we rationalize why a beneficent designer would shape the many other selfish behaviors that natural selection explains, such as cannibalism, siblicide, and infanticide.

Evolution and its mechanisms, observed

Anyone can observe erosion, and geologists can measure the movement of continental plates, which travel at up to 10 centimeters per year. No geologist doubts that these mechanisms, even if they accomplish only slight changes on the scale of human generations, have shaped the Grand Canyon and have separated South America from Africa over the course of millions of years. Likewise, biologists do not expect to see anything like the origin of mammals played out on a human time scale, but they have documented the processes that will yield such grand changes, given enough time.

Evolution requires genetic variation, which originates by mutation. From decades of genetic study of initially homozygous laboratory populations, we know that mutations arise that have effects, ranging from very slight to drastic, on all kinds of phenotypic characters (see Chapter 8). These mutations can provide new variation in quantitative characters, seemingly without limit. This variation has been used for millennia to develop strains of domesticated plants and animals that differ in morphology more than whole families of natural organisms do. In experimental studies of laboratory populations of microorganisms, we have seen new advantageous mutations arise and enable rapid adaptation to temperature changes, toxins, or other environmental stresses. Laboratory studies have documented the occurrence of the same kinds of mutations, at the molecular level, that are found in natural populations and distinguish species. No geneticist or molecular biologist doubts that the differences among species in their genes and genomes originated by natural mutational processes that, by and large, are well understood.

We know also that most natural populations carry a great deal of genetic variation that can yield rapid responses to artificial or natural selection (see Chapters 9, 11, and 12). We have seen allele frequency differences among recently established populations that can be confidently attributed to genetic drift (see Chapter 10). Evolutionary biologists have documented literally hundreds of examples of natural selection acting on genetic and phenotypic variation (see Chapters 12 and 13). They have described hundreds of cases in which populations have responded to directional selection and have adapted to new environmental factors, ranging from the evolution of resistance to insecticides, herbicides, and antibiotics to the evolution of different diets (see Chapter 13).

Speciation generally takes a very long time, but some processes of speciation can also be observed. Substantial reproductive isolation has evolved in laboratory populations, and species of plants that apparently originated by polyploidy and by hybridization have been "re-created" de novo by crossing their suspected parent forms and selecting for the species' diagnostic characters (see Chapter 18).

In summary, the major causes of evolution are known, and they have been extensively documented. The two major aspects of long-term evolution, anagenesis (changes of characters within lineages) and cladogenesis (origin of two or more lineages from common ancestors), are abundantly supported by evidence from every possible source, ranging from molecular biology to paleontology. Over the past century we have certainly learned of evolutionary processes that were formerly unknown; we now know, for example, that some species may arise from hybridization, and that some DNA sequences are mobile and can cause mutations in other genes. But no scientific observations have ever cast serious doubt on the reality of the basic mechanisms of evolution, such as natural selection, or on the reality of the basic historical patterns, such as transformation of characters and the origin of all known forms of life from common ancestors. Contrast this mountain of evidence with the evidence for supernatural creation or intelligent design: *there is no such evidence whatever.*

Refuting Creationist Arguments

Creationists attribute the existence of diverse organisms and their characteristics to miracles: direct supernatural intervention. As we have seen, it is impossible to predict miracles or to do experiments on supernatural processes, so creationists do not do original research in support of their theory. Thus, rather than providing positive evidence of divine creation, "creation science" consists entirely of attempts to demonstrate the falsehood or inadequacy of evolutionary science and to show that biological phenomena must, by default, be the products of intelligent design. Here are some of the most commonly encountered creationist arguments, together with capsule counterarguments.

1. ***Evolution is outside the realm of science because it cannot be observed.***

   Evolutionary changes have indeed been observed, as we saw earlier in this chapter. In any case, most of science depends not on direct observation but on testing hypotheses against the predictions they make about what we should observe. Observation of the postulated processes or entities is not required in science, and most of the fundamental processes and objects in physics, chemistry, and cell biology cannot be seen or observed.

2. ***Evolution cannot be proved.***

   Nothing in science is ever absolutely proved. "Facts" are hypotheses in which we can have very high confidence because of massive evidence in their favor and the absence of contradictory evidence. Abundant evidence from every area of biology and paleontology supports evolution, and there exists no contradictory evidence.

3. ***Evolution is not a scientific hypothesis because it is not testable: no possible observations could refute it.***

   Many conceivable observations could refute or cast serious doubt on evolution, such as finding incontrovertibly mammalian fossils in incontrovertibly Precambrian rocks. In contrast, any puzzling quirk of nature could be attributed to the inscrutable will and infinite power of a supernatural intelligence, so creationism is untestable.

4. ***The orderliness of the universe, including the order manifested in organisms' adaptations, is evidence of intelligent design.***

   Order in nature, such as the structure of crystals, arises from natural causes and is not evidence of intelligent design. The order displayed by the correspondence between organisms' structures and their functions is the consequence of natural selection acting on genetic variation, as has been observed in many experimental and natural populations; we have described numerous cases of both throughout this book. Darwin's realization that the combination of a random process (the origin of genetic variation) and a nonrandom process (natural selection) can account for adaptations provided a natural explanation for the apparent design and purpose in the living world and made a supernatural account unnecessary and obsolete.

5. ***Evolution of greater complexity violates the second law of thermodynamics, which holds that entropy (disorder) increases.***

   This is a common misrepresentation of one of the most important laws of physics. The second law applies only to closed systems, such as the universe as a whole. Order and complexity can increase in local, open systems as a result of an influx of energy. This is evident in the development of individual organisms, in which biochemical reactions are powered by energy derived ultimately from the Sun.

6. ***It is almost infinitely improbable that even the simplest life could arise from nonliving matter. The probability of random assembly of a functional nucleotide sequence only 100 bases long is $1/4^{100}$, an exceedingly small number. And scientists have never synthesized life from nonliving matter.***

   It is true that a fully self-replicating system of nucleic acids and replicase enzymes has not yet arisen from simple organic constituents in the laboratory, but the history of scientific progress shows that it would be foolish and arrogant to assert that what science has not accomplished in a few decades cannot be accomplished. (And even if, given our human limitations, we should never succeed in this endeavor, why should that require us to invoke the supernatural?) Critical steps in the probable origin of life, such as abiotic synthesis of purines, pyrimidines, and amino acids and self-replication of short RNAs, have been demonstrated in the laboratory (see Chapter 5). And there is no reason to think that the first self-replicating or polypeptide-encoding nucleic acids had to have had any particular sequence. If there are many possible sequences with such properties, the probability of their formation rises steeply. Moreover, the origin of life is an entirely different problem from the modification and diversification of life once it has arisen. We do not need to know anything about the origin of life in order to understand and document the evolution of different life forms from their common ancestor.

7. ***Mutations are harmful and do not give rise to complex new adaptive characteristics.***

   Most mutations are indeed harmful and are purged from populations by natural selection. Some, however, are beneficial, as shown in many experiments (see Chapters 8 and 13). Complex adaptations usually are based not on single mutations but on combinations of mutations that jointly or successively increase in frequency as a result of natural selection.

8. ***Natural selection merely eliminates unfit mutants, rather than creating new characters.***

   "New" characters, in most cases, are modifications of pre-existing characters, which are altered in size, shape, developmental timing, or organization (see Chapters 3, 4, and 21). This is true at the molecular level as well (see Chapter 20). Natural selection "creates" such modifications by increasing the frequencies of alleles at several or many loci so that combinations of alleles, initially improbable because of their rarity, become probable. Observations and experiments on both laboratory and natural populations have demonstrated the efficacy of natural selection.

9. ***Chance could not produce complex structures.***

   This is true, but natural selection is a deterministic, not a random, process. The random processes of evolution—mutation and genetic drift—do not result in the evolution of complexity, as far as we know. Indeed, when natural selection is relaxed, complex structures, such as the eyes of cave-dwelling animals, slowly degenerate, due in part to the fixation of neutral mutations by genetic drift.

10. ***Complex adaptations such as wings, eyes, and biochemical pathways could not have evolved gradually because the first stages would not have been adaptive. The full complexity of such an adaptation is necessary, and this could not arise in a single step by evolution.***

This was one of the first objections that greeted *The Origin of Species*, and it has recently been christened "irreducible complexity" by advocates of intelligent design. Our answer has two parts. First, many such features, such as hemoglobins and eyes, do show various stages of increasing complexity among different organisms (see Chapters 3, 4, 5, and 22). "Half an eye"—an eye capable of discriminating light from dark, but incapable of forming a focused image—is indeed better than no eye at all (see Figure 22.14). Second, many structures have been modified for a new function after being elaborated to serve a different function (see Chapters 3 and 22). The "finished version" of an adaptation that we see today may indeed require precise coordination of many components in order to perform its current function, but the earlier stages, performing different or less demanding functions, and performing them less efficiently, are likely to have been an improvement on the ancestral feature. The evolution of the mammalian skull and jaw (see Chapter 4) provides a good example.

11. ***If an altered structure, such as the long neck of the giraffe, is advantageous, why don't all species have that structure?***

This naive question ignores the fact that different species and populations have different ecological niches and environments, for which different features are adaptive. This principle holds for all features, including "intelligence."

12. ***If gradual evolution had occurred, there would be no phenotypic gaps among species, and classification would be impossible.***

Many disparate organisms are connected by intermediate species, and in such cases, classification into higher taxa is indeed rather arbitrary (see Chapter 3). In other cases, gaps exist because of the extinction of intermediate forms (see Chapter 4). Moreover, although much of evolution is gradual, some advantageous mutations with large, discrete effects on the phenotype have probably played a role. Whether or not evolution has been entirely gradual is an empirical question, not a theoretical necessity.

13. ***The fossil record does not contain any transitional forms representing the origin of major new forms of life.***

This very common claim is flatly false, for there are many such intermediates (see Chapter 4). Creationists sometimes use rhetorical subterfuge in presenting this argument, such as defining *Archaeopteryx* as a bird because of its feathers and then claiming that there are no known intermediates between reptiles and birds.

14. ***The fossil record does not objectively represent a time series because strata are ordered by their fossil contents, and then are assigned different times on the assumption that evolution has occurred.***

Even before *The Origin of Species* was published, geologists who did not believe in evolution recognized the temporal order of fossils that are characteristic of different periods, and named most of the geological periods. Since then, radioactive dating and other methods have established the absolute dates of geological strata.

15. ***Vestigial structures are not vestigial but functional.***

According to creationist thought, an intelligent Creator must have had a purpose, or design, in each element of His creation. Thus all features of organisms must be functional. For this reason, creationists view adaptations as support for their position. However, nonfunctional, imperfect, and even maladaptive structures are expected if evolution is true, especially if a change in an organism's environment or way of life has rendered them superfluous or harmful. As noted earlier, organisms display many such features at both the morphological and molecular levels.

16. ***The classic examples of evolution are false.***

Some creationists have charged that some of the best-known studies of evolution, cited in almost every textbook, are flawed and that evolutionary biologists have dishonestly perpetuated these supposed falsehoods. For example, H. B. D. Kettlewell, who performed the classic study of industrial melanism in the peppered moth, is accused of having obtained spurious evidence for natural selection by predatory birds by pinning moths to unnatural resting sites (tree trunks).

Suppose this and the other cited studies were indeed flawed. First, it does not follow that textbook authors and other contemporary biologists have dishonestly perpetuated falsehood; they simply might not have checked and analyzed the original studies, but instead borrowed from other books and "secondary" sources, since no textbook author can check every study in depth. In any case, there is no reason to suspect intellectual dishonesty. Second, whether or not Kettlewell is guilty as charged is completely irrelevant to the validity of the basic claims involved. Both natural selection and rapid evolutionary changes have been demonstrated in so many species that these principles would stand firmly even if the peppered moth story were completely false. (Kettlewell's evidence that birds differentially attack dark and light peppered moths was based on a variety of experiments, and other investigators have added to this evidence since then. They have shown that other selective factors also affect the evolution and loss of melanism in this species.) Whatever flaws there may be in a few old studies, they do not invalidate decades of subsequent research by hundreds of scientists.

Of course, the creationists who cite these examples of supposed flaws and frauds realize that the strength of evolutionary biology does not rest on these studies. After all, most creationists accept natural selection and "microevolution," such as changes in moth coloration. Rather, these attacks enable their readers to doubt the truthfulness of evolutionary scientists and to justify their disbelief in evolution. But remember that even if individual scientists are stupid (which a few are) or dishonest (which almost none are), the social process of science uncovers errors and justifies our confidence in its major claims.

17. ***Disagreements among evolutionary biologists show that Darwin was wrong.***

Disagreements among scientists exist in every field of inquiry, and are in fact the fuel of scientific progress. They stimulate research and are a sign of vitality. Creationists misunderstand or misinterpret evolutionary biologists who have argued (a) that the fossil record displays abrupt shifts rather than gradual change (punctuated equilibrium); (b) that many characteristics of species may not be adaptations; (c) that evolution may involve mutations with large effects as well as those with small effects; and (d) that natural selection does not explain certain major events and trends in the history of life. In fact, none of the evolutionary biologists who hold these positions deny the central proposition that adaptive characteristics evolve by the action of natural selection on random mutations. All these debates arise from differing opinions on the relative frequency and importance of factors known to influence evolution: large-effect versus small-effect mutations, genetic drift versus natural selection, individual selection versus species selection, adaptation versus constraint, and so forth (see Chapters 11, 21, and 22). These arguments about the relative importance of different processes do not at all undermine the strength of the evidence for the historical fact of evolution—i.e., descent, with modification, from common ancestors. On this point, there is no disagreement among evolutionary biologists.

18. ***There are no fossil intermediates between apes and humans; australopithecines were merely apes. And there exists an unbridgeable gap between humans and all other animals in cognitive abilities.***

This is a claim about one specific detail in evolutionary history, but it is the issue about which creationists care most. This claim is simply false. The array of fossil hominids shows numerous stages in the evolution of posture, hands and feet, teeth, facial structure, brain size, and other features. Both functional and nonfunctional DNA sequences of humans and African apes are extremely similar and clearly demonstrate that they share a recent common ancestor. The mental abilities of humans are indeed developed to a far greater degree than those of other species, but many of our mental faculties seem to be present in more rudimentary form in other primates and mammals.

19. ***As a matter of fairness, alternative theories, such as supernatural creation and intelligent design, should be taught, so that students can make their own decisions.***

This train of thought, if followed to its logical conclusion, would have teachers presenting hundreds of different creation myths, in fairness to the peoples who hold them, and it would compel teachers to entertain supernatural explanations of everything in earth science and astronomy, in chemistry and physics, because anything explained by these sciences, too, could be argued to have a supernatural cause. It would imply teaching students that to do a proper job of investigating an airplane crash, federal agencies should consider the possibility of mechanical failure, a terrorist bomb, a missile impact—and supernatural intervention (Alters and Alters 2001).

Science teachers should be expected to teach the content of contemporary science—which means the hypotheses that have been strongly supported and the ideas that are subjects of ongoing research. That is, they should teach what scientists do. Several scientists have searched the scientific literature for research reports on intelligent design and "creation science" and have found no such reports. Nor is there any evidence that so-called "creation scientists" have carried out scientific research that a biased community of scientists has refused to publish. As noted earlier in this chapter, there is no way of testing hypotheses about the supernatural, so there cannot be any scientific research on the subject. And that means that the subject should not be taught in a science course.

On arguing for evolution

How, and whether or not, to convince people to accept evolution is not a subject for this book, which is devoted to the content of evolutionary science. Brian and Sandra Alters have addressed this topic in *Defending Evolution in the Classroom: A Guide to the Creation/Evolution Controversy* (2001). As these authors point out, for many people, religious beliefs take precedence over scientific evidence, especially among people who believe that their fate for all eternity depends on adhering to their belief system. The points made in this chapter will do little to persuade such people. Presenting the nature of science and the evidence for evolution will, however, have an impact on people who genuinely question what the truth is; but the way in which it is presented is important, as Alters and Alters describe. In this context, anyone who is considering holding a public debate on evolution with a creationist should first become thoroughly familiar with creationist arguments and strategies, and especially with the intelligent design movement, which aggressively markets itself as science when it is not. Jerry Coyne's book *Why Evolution Is True* (2008) provides the best overall summary of the argument. Other Internet resources and books are listed at the end of this chapter.

Why Should We Teach Evolution?

If evolution is so controversial, why invite trouble? Why not just drop it from the science curriculum? After all, it's an academic subject that doesn't really affect people's lives, right?

Wrong. Evolution is a foundation of all of biology. Understanding evolution is as relevant to everyday life as understanding physics or chemistry, and research in evolutionary biology affects our lives both directly and indirectly (Figure 23.3).

Educating students about evolution should mean teaching them not only the basic principles and facts of evolutionary processes and evolutionary history, but also the concepts and ways of testing hypotheses that are used in evolutionary science. These

Figure 23.3 Cartoonist Gary Trudeau affirms the importance of evolutionary science. (Doonesbury © 2005 by G. B. Trudeau. Reprinted with permission of Universal Press Syndicate.)

approaches have many applications. For example, evolutionary biologists specialize in studying variation both within and among species, and the conceptual approaches and methods they employ are broadly useful. Simply being aware of the importance of genetic and nongenetic variation is a useful lesson; for example, everyone, whether patient or doctor, should be aware that people may vary in their reaction to a drug (or to a disease, for that matter). Understanding the difference between genetic and nongenetic variation is profoundly important for interpreting claims that are made about differences among ethnic or "racial" groups (see Chapter 9). Evolutionary biology shows, moreover, that a trait may have high heritability and nevertheless be readily altered by environmental change (e.g., medical or educational intervention).

Learning about evolution would be important even if it had no practical utility at all, simply because of the insights and understanding it provides. Nonetheless, evolutionary science does bear on applied research and does have actual applications. We can mention only a few of the many practical applications of evolutionary biology, which are treated more fully by David Mindell (2006) and are the subject of a recently founded journal, *Evolutionary Applications*.

Health and medicine

The direct and indirect uses of evolutionary biology are probably more numerous, and more important, in medicine and public health than in any other area. Depending on the topic, evolutionary theory may provide a new conceptual approach to medically relevant research (e.g., the evolution of senescence), principles that medical research and practice should take into account (e.g., natural selection for drug resistance in pathogens), or methods for making inferences and discoveries (e.g., phylogenetic methods for tracing the spread of pathogens). Randolph Nesse and Stephen Stearns (2008) explain why some education in evolutionary biology is essential for every medical researcher, because it bears directly on almost every field of biological study, and useful for every clinical practitioner, because it deepens one's understanding both of the human body and its ills, and of the organisms that cause harm. The following paragraphs are based largely on their authoritative review of this topic.

Two major evolutionary questions can be asked of any trait: what is its evolutionary history, and what has been its evolutionary cause? Often this means, what is its adaptive function? Both questions can lead to deeper understanding of the traits and the underlying genes of both humans and pathogens.

For instance, a phylogenetic approach enables us to understand why some populations have a higher prevalence of lactase persistence, malaria resistance, and other traits (Chapter 12). Together with population genetics, phylogeny helps to explain sickle-cell hemoglobin and the *ApoE4* subtype of apolipoprotein, which increases the risk of developing atherosclerosis and Alzheimer's disease, and which has increased in populations that inhabit cold climates, perhaps as an adaptation to a diet high in meat. Population genetic theory enables researchers to identify alleles that have recently increased due to strong selection (Chapter 12), some of which can increase the risk of infection (Dean et al. 2002). Genomic comparisons of human and other genomes indicate which genes are under strong purifying selection and are therefore likely to cause genetic disease when mutated (Miller and Kumar 2001; see Figure 12.30).

Evolutionary genetics is helpful for interpreting human genetic variation, but widespread traits also call for explanation. An appreciation of natural selection can help us view human characteristics in new ways. For example, the theory of genetic conflict can explain the remarkable tug of war between a pregnant woman and her fetus, based on the conflict of interest between maternal and imprinted paternal genes (Chapter 16). Because of the theory of the evolution of life histories (Chapter 14), senescence is no longer viewed as an inevitable result of wear and tear, but as a genetically controlled process based on pleiotropic effects of alleles that are advantageous at younger ages. This theory suggests that genes that reduce oxidative metabolism and senescence may lower reproduction—for which there is some evidence. If we posit that universal human traits are effects of natural selection, we may wonder if the annoying symptoms of infectious disease, such as fever and coughing, might be defenses against pathogens. Research based on this hypothesis might result in changes in medical practice. The adaptationist approach leads naturally to ask what the effects are of our modern environment, so unlike the environments we evolved in. It is now almost trite to say that we are not adapted to a life of abundant calories and little exercise, the joint cause of the epidemic of obesity and diabetes in the United States and some other countries. But what about our exposure to light for 18 hours per day? Light reduces the synthesis of melatonin, and there are suggestions that melatonin reduces the incidence of breast cancer. In psychiatry, the recognition is growing that severe emotional disorders are often the result of gene × environment interactions, not context-independent conditions.

The evolution of antibiotic resistance by pathogens is one of the greatest threats to public health: infections caused by resistant *Staphylococcus aureus* cause more than 18,000 deaths annually in the United States, where the total annual cost of antibiotic resistance is estimated to be about $80 billion. It is frustrating that the popular press and even medical journals tend not to call this phenomenon by its rightful name—*the evolution of resistance*—but to use euphemisms such as "develop" instead (Antonovics et al. 2007). Population genetic modeling shows that some strategies are better than others in delaying the evolution of drug resistance, especially applying a mixture of drugs simultaneously, such as the "cocktail" prescribed for people with HIV. Pathogens can also evolve to be more or less virulent, and recognizing that natural selection will increase virulence under some conditions will probably have significant implications for devising public

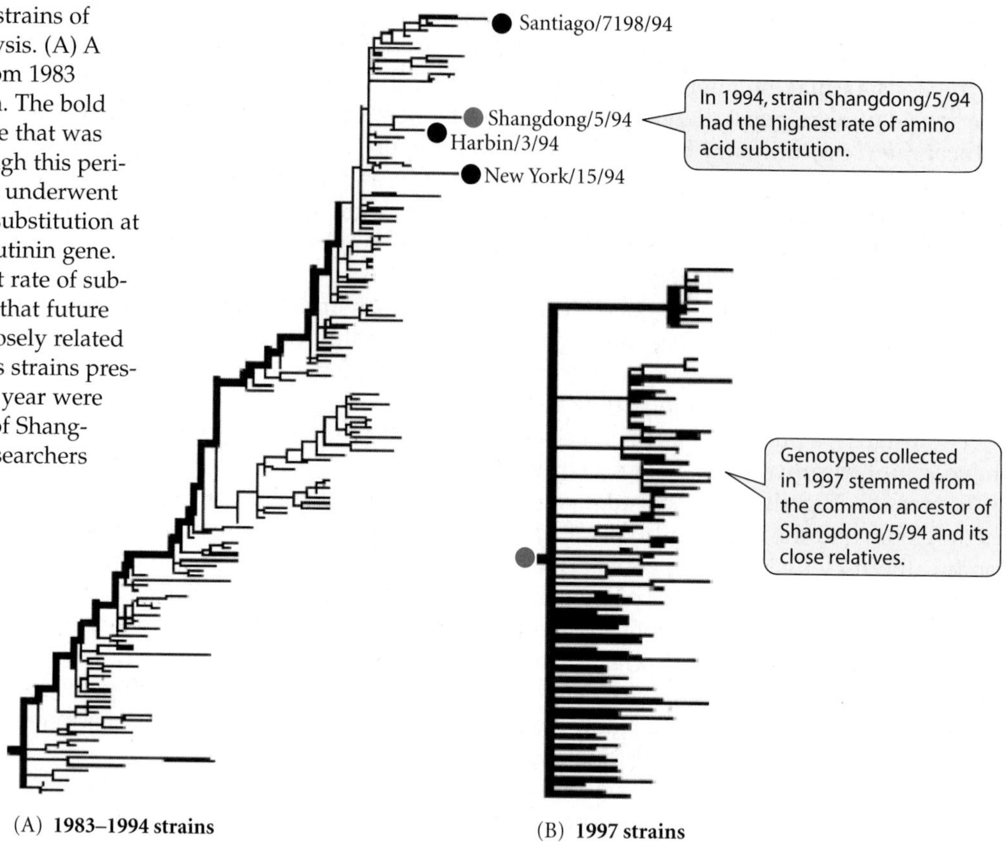

Figure 23.4 Prediction of epidemic strains of influenza A virus by phylogenetic analysis. (A) A phylogeny of influenza virus strains from 1983 through the 1993–1994 influenza season. The bold line, or "trunk," traces the single lineage that was successful in leaving descendants through this period. The labeled dots mark lineages that underwent exceptionally high rates of amino acid substitution at positively selected sites in the hemagglutinin gene. Strain Shangdong/5/94 had the highest rate of substitution, and the researchers predicted that future epidemics would stem from this or a closely related lineage. (B) A partial phylogeny of virus strains present in 1997. Genotypes collected in that year were descended from the common ancestor of Shangdong/5/94 and its close relatives, as researchers had predicted. (After Bush et al. 1999.)

Santiago/7198/94
Shangdong/5/94
Harbin/3/94
New York/15/94
In 1994, strain Shangdong/5/94 had the highest rate of amino acid substitution.
Genotypes collected in 1997 stemmed from the common ancestor of Shangdong/5/94 and its close relatives.
(A) 1983–1994 strains
(B) 1997 strains

health measures in the future. At present, the greatest application of evolutionary biology in public health is using phylogenetic methods to trace the origin and spread of disease agents, such as bacterial contaminants of food and the pathways of transmission among human hosts. (Phylogenetic relationships among samples of HIV were used to convict a doctor of attempting to murder his former girlfriend by injecting her with blood from an HIV-infected patient; see Metzker et al. 2002.)

Evolutionary studies may also prove useful in vaccine development, as new strains of pathogens continually arise that are resistant to existing vaccines. Robin Bush and colleagues (1999) tested the hypothesis that new strains of influenza A virus spread because their fitness is increased by mutations at codons in the hemagglutinin gene (the main target of human antibodies) that have a history of positive selection. New vaccines are required to provide protection against these strains. Bush et al. determined phylogenetic relationships among strains that had been preserved between 1983 and 1994 and identified those strains (e.g., Shangdong/5/94) that had undergone the most amino acid substitutions at positively selected sites in the hemagglutinin gene (Figure 23.4A). They predicted that Shangdong/5/94 or its close relatives would be ancestral to strains that later spread widely. Their prediction was correct (Figure 23.4B). This method may provide a way of estimating which currently rare strains are likely to give rise to epidemic strains, so that vaccines can be developed before those strains spread.

We have seen that phylogenetic methods are a major way of determining the origin of new and emerging infectious diseases (Chapter 19). The several forms of HIV evolved from viruses carried by chimpanzees and mangabey monkeys (see Figure 3.4), and biologists have recently determined that gorillas may be a reservoir for one form, the HIV-1, strain O (Van Heuverswyn et al. 2006). The pathogens that cause infectious diseases in humans range from those that are mostly endemic to other animals and only occasionally infect humans (e.g., rabies), through those that are animal-borne but can be transmitted among humans to a greater or lesser extent, to those that are specific to humans, such as the agents that cause smallpox, syphilis, and falciparum malaria (Figure 23.5). Jared

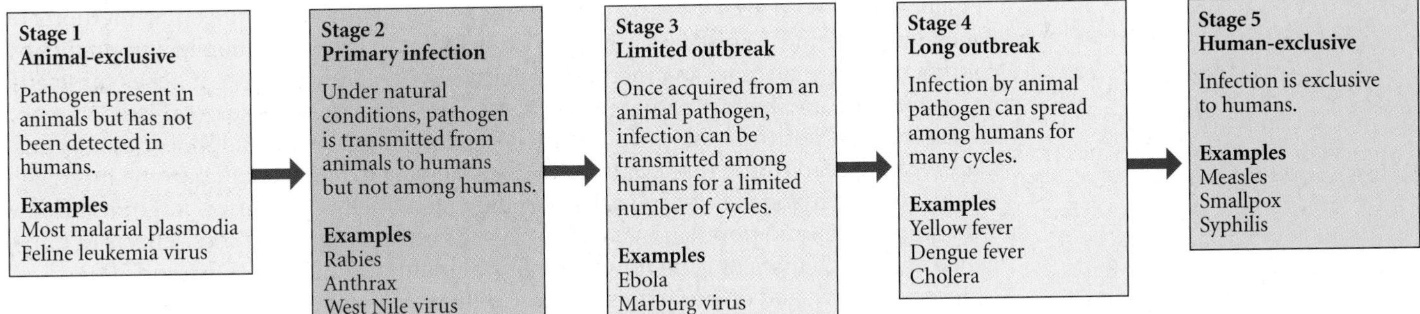

Stage 1
Animal-exclusive
Pathogen present in animals but has not been detected in humans.
Examples
Most malarial plasmodia
Feline leukemia virus
Stage 2
Primary infection
Under natural conditions, pathogen is transmitted from animals to humans but not among humans.
Examples
Rabies
Anthrax
West Nile virus
Stage 3
Limited outbreak
Once acquired from an animal pathogen, infection can be transmitted among humans for a limited number of cycles.
Examples
Ebola
Marburg virus
Monkeypox
Stage 4
Long outbreak
Infection by animal pathogen can spread among humans for many cycles.
Examples
Yellow fever
Dengue fever
Cholera
Stage 5
Human-exclusive
Infection is exclusive to humans.
Examples
Measles
Smallpox
Syphilis

Figure 23.5 The ecological relationships between pathogens and humans vis-à-vis nonhuman hosts can be expressed as five stages of increasing cycling within human populations. Whether or not all pathogens in the higher stages have evolved through the lower stages is not yet known. Some exemplar pathogens are given for each stage. (Data from Wolfe et al. 2007.)

Diamond and colleagues (Diamond 1997; Wolfe et al. 2007) note that most human diseases originated in tropical animals, probably because pathogens shift most readily among closely related hosts (Chapter 19), and most primates have tropical distributions. To an overwhelming degree, the major human pathogens originated in Asia or Africa rather than in the Western Hemisphere, for several probable reasons: there was much more time for pathogens to transfer and become adapted to humans in the Old World because hominids evolved there; humans are much more closely related to other Old World (catarrhine) primates than to New World (platyrrhine) primates (Chapter 6), so pathogen shift was more likely; and most domesticated animals, which have been reservoirs for some pathogens such as influenza, are Old World species.

One form of selection at the level of the gene is meiotic drive, wherein more than half the gametes of a heterozygote carry one of the two alleles (Chapter 11). A "driver" allele can attain high frequency, even fixation if its pleiotropic costs are not too great. One of the most exciting potential applications of evolutionary biology is GENETIC PEST MANAGEMENT, in which genotypes of pest insects are released that carry a driver allele coupled with a gene that has a desirable effect. For instance, falciparum malaria is carried by the mosquito *Aedes aegypti*. Fred Gould and colleagues (Huang et al 2007; Gould 2008) are studying the feasibility of releasing mosquitoes that carry a driver allele and a gene that prevents the pathogen from surviving and multiplying in the mosquito. Similar applications might be used to control insect pests of crops.

Agriculture and natural resources

The development of improved varieties of domesticated plants and animals is evolution by artificial selection. Evolutionary genetics and plant and animal breeding have had an intimate, mutually beneficial relationship for more than a century. Both theoretical evolutionary methods and experimental studies of *Drosophila* and other model organisms have contributed both to traditional breeding and to modern QTL (quantitative trait locus) analysis, which is used to locate and characterize genes that contribute to traits of interest (see Chapter 13).

Agronomists have learned by bitter experience what evolutionary biologists have long known: that genetic diversity is essential for a population's long-term success. For example, a costly epidemic of the southern corn leaf blight fungus in 1970 was a consequence of widely planting genetically uniform corn with a specific genetic factor that increased susceptibility (Chapter 19). This example shows why it is important to build "germ plasm" banks of different crop strains, containing genes for various characteristics such as disease resistance. Useful genes can also be found in wild species that are related to crop plants. Genetic and evolutionary studies of the tomato, for example, led to study of the many related species in South America. At least 40 genes for resistance to various diseases have been found in these species, of which about 20 have been crossed into commercial tomato stocks. Studies of wild species' genes for other traits, such as tolerance to salinity and drought, are under way. This approach depends on phylogenetics, evolutionary genetics, and evolutionary ecology.

Genes from wild species can also be transferred to crops by molecular methods of genetic engineering. Evolutionary biology contributes to this revolution in agronomics by contributing to gene mapping methods, by identifying likely sources of genes for useful characteristics, and by evaluating possible risks posed by transgenic organisms. Genetic engineering is controversial for several reasons; for example, there is concern that transgenes may spread from crop plants to wild species, which then could become more vigorous weeds. Phylogenetic studies help to indicate which wild species might hybridize with crop plants, and population genetic methods can estimate the fitness effects of transgenes and the chances of gene flow into natural plant populations (Ellstrand 2003).

Insects, weeds, and other organisms cause billions of dollars worth of crop losses. Much of this loss is caused by the evolution of resistance to chemical insecticides and herbicides by crop pests (see Chapter 12). This resistance not only increases the costs of agriculture but also results in a steady increase in the amount of toxic chemicals sprayed on the landscape (some of which find their way up the food chain, affecting humans and other consumers). For this reason, many crop pests are now managed by INTEGRATED PEST MANAGEMENT (IPM), which uses a combination of chemical pesticides and nonchemical methods, such as release of natural enemies of the pest (BIOLOGICAL CONTROL). In some places, regulations on the use of pesticides follow recommendations made by evolutionary biologists about how to manage pest populations in order to keep them susceptible to pesticides (see Gould 1998; Gould et al. 2002). The development of biological control also benefits from evolutionary analysis. When a new insect species suddenly appears on a crop, phylogenetic systematics is the first approach to identifying the pest and determining where in the world it has come from. That is where entomologists will search for natural enemies, scrutinizing in particular those that are related to known enemies of species that are related to the new pest. Likewise, herbivorous insects used to control weeds or invasive plants must be screened to be sure they will not also attack crops or native plants. A good approach is to see whether they have the potential to feed on or adapt to plants that are related to the target plant species (Futuyma 2000).

Evolutionary biology bears on the management and development of other biological resources as well. The journal *Evolutionary Applications* devoted an entire issue (May 2008) to evolutionary aspects of managing and conserving salmon and trout species. For example, fisheries biologists use genetic markers and population genetic methods to distinguish stocks of fish that migrate from different spawning locations (an issue with great political and economic implications). Genetic analyses are similarly important in forestry, for example, in developing commercial stocks of conifers. Hunting and fishing have caused evolutionary changes in natural populations, especially earlier reproduction at a smaller size (see Figure 13.12).

Environment and conservation

In a few instances, evolutionary biology has contributed to environmental remediation and restoration of degraded land. For example, populations of grasses that have evolved tolerance to nickel and other heavy metals (see Chapter 13) can be used for revegetating land made barren by mining activities, and some such plants may even be used to remove contaminants from soil and water.

There is little doubt that a major extinction event has been initiated by the huge and accelerating impact of human activities on every aspect of the environment. The most important means of conservation are obvious and require mostly political, legal, and economic expertise: save natural habitats in preserves, establish and enforce limits on the exploitation of fish populations and other biological resources, reduce pollution and global warming. But biologists, including evolutionary biologists, also make indispensable contributions to conservation efforts. They use phylogenetic information to determine where potential nature reserves should be located to protect the greatest variety of biologically different species; they use evolutionary biogeography to identify regions with many endemic species (e.g., Madagascar); they use genetic methods and theory to prevent inbreeding depression in rare species and to distinguish genetically unique populations (Frankham et al. 2002; Allendorf and Luikart 2007); and they use genetic markers to identify illegal traffic in endangered species (see Baker et al. 2000).

Human behavior

No topic in evolutionary biology is more intriguing or more controversial than the evolutionary foundations of human behavior, especially social behavior. Resistance to hypotheses about human behavioral evolution, or even research on the subject, is widespread for several reasons. Many people are emotionally reluctant to see human abilities as extensions of those of other species, and justify making a sharp distinction by pointing to the immense difference between human mental capacities and those of any other mammal. Almost every aspect of our behavior varies greatly among individuals and among populations because of learning and cultural differences, so the very hypothesis that there are species-specific, genetically based behaviors is viewed with skepticism, especially by many social scientists. Finally, any intimation of biological determinism summons memories of a long, sordid history of political and ideological abuse of evolutionary ideas. Just as the Bible has been used to justify crusades, inquisitions, and witch hunts, scientific ideas have been used to justify social inequity and even persecution. From its earliest days, evolutionary science, under the banner of "social Darwinism," was misappropriated to justify racism and imperialist domination, to exclude women from political and economic power, and to ascribe poverty, illiteracy, and crime not to the social conditions that exclude much of society from education and economic self-sufficiency, but to genetic inferiority (Hofstadter 1955). Evolution was used to justify the American eugenics movement, which advocated encouraging "superior" people to reproduce and discouraging or preventing "inferior" people from doing so; to justify discriminatory quotas in United States immigration policy; and worst of all, to justify the racism that found its most horrifying expression in Nazi Germany. All these abuses were based on misunderstanding or twisting of the data and theory of evolution and genetics (and, to their credit, some evolutionary biologists and geneticists said so). A proper understanding of evolutionary biology, as of any science, is necessary to prevent it from being misused. Research on the evolution of human behavior may as well be used for good as for ill.

Darwin devoted a book, *The Expression of the Emotions in Man and Animals*, to the thesis that rudimentary homologues of human mental abilities and emotions can be seen in other species. It is now clear that many animals are capable of learning, and that some, especially primates, can learn by watching other individuals. Thus, cultural differences exist among troops and populations of chimpanzees and some other primates (Figure 23.6). Tools are made and used by chimpanzees and, surprisingly, New Caledonian crows (Hunt and Gray 2004). Chimpanzees and other apes can learn to use sign language or other nonverbal symbolic representation, at levels equivalent to those of a young human child. Social interactions, including reciprocal aid and alliances, are quite elaborate in some primates, leading researchers to describe "baboon metaphysics" (Cheney and Seyfarth 2007) and "chimpanzee politics" (de Waal 2007).

It would be contrary to everything we know about evolution to suppose that the hominid lineage has diverged so far from other apes that there should remain no trace of homology in the inherited components of brain organization and behavior, or that the human brain should be a complete tabula rasa, or blank slate, without genetically determined predispositions. Evolutionary psychologists have compiled a long list of supposedly universal human behavioral tendencies and capacities, such as language, while recognizing that the expression of most such behaviors is culturally highly variable. (All humans speak a grammatical language, but grammar and vocabulary vary greatly.)

Several schools of evolutionary research on human behavior have developed in the last few decades (Laland and Brown 2002). HUMAN SOCIOBIOLOGY, announced by E. O. Wilson in 1975, proposed to interpret a wide range of behaviors (such as conflicts between offspring and parents) as adaptations that had evolved by natural selection, especially kin selection. This approach, which emphasized genetic over cultural explanations of particular behaviors, was hugely controversial. It has been succeeded by several movements,

Figure 23.6 Chimpanzees use tools, and cultural traditions of tool use develop, based on imitation. As a female chimpanzee (*Pan troglodytes*) and her infant watch, a male cracks nuts using a rock as a hammer. (Photo © Clive Bromhall/OSF/Photolibrary.com.)

including HUMAN BEHAVIORAL ECOLOGY, which uses adaptive models to predict and explain a variety of behaviors, including cultural norms. For instance, we may ask why, in polygynous societies, a woman would choose to marry a man who already has one or more wives. Adopting a model that was first developed to explain polygyny in birds, behavioral ecologists have proposed that women decide on the basis of the resources they may expect a man to provide.

Complementing this approach, Leda Cosmides, John Tooby, and others have developed the active (and controversial) field of EVOLUTIONARY PSYCHOLOGY (Barkow et al. 1992; Pinker 1997, 2002; Buss 1999). Evolutionary psychologists propose that over the course of human evolution, especially during the Pleistocene, specific mental "organs" or modules evolved to solve different classes of problems, especially social challenges such as choosing mates and detecting cheaters. This approach has been criticized on the grounds that it is difficult to separate innate from learned behaviors, to identify discrete behavioral traits, and to demonstrate that they are adaptations; however, it is credited with developing insights into language, some aspects of sexual behavior, and many other topics. For example, David Buss (1994) reasoned that because reproduction entails a far greater commitment and investment of resources by women than by men, women should have evolved to seek mates who are likely to provide resources, while men, as a consequence of sexual selection, might be expected to place more value on young, physically attractive mates who are likely to be fertile. This sounds like the epitome of the sexism that pervades American culture, yet Buss claimed, based on more than 10,000 interviews in 37 broadly different cultures, that the predicted differences between the sexes are cross-culturally consistent. This study does not conclusively demonstrate that mate preferences are based on evolved psychological mechanisms, but it suggests that the subject may reward further study.

Some of the elements of evolutionary theory, such as natural selection and genetic drift, have been adopted, *mutatis mutandi*, to the evolution of culture, and some methods developed for evolutionary analysis are being applied to cultural evolution as well (Mace and Holden 2005). Phylogenetic methods are used to trace the evolution of languages and have revealed interesting patterns; for example, frequently used words evolve at slower rates than infrequently used words (Pagel et al. 2007). Cultural evolution and genetic evolution of behavior have been joined in models of "gene-culture evolution" (Cavalli-Sforza and Feldman 1981) or "dual inheritance" (Boyd and Richerson 1985; Richerson and Boyd 2005). Surely the most interesting problem addressed by these models is what accounts for the uniquely hypertrophied organization and cooperation in human societies—far greater than in any other species. Complex social relationships required that an individual be able to recognize many other individuals and remember their past interactions; they may have required the ability to interpret the state of mind and intentions of others; they were probably facilitated by punishing cheaters (Fehr and Fischbacher 2003). Many authors have proposed that multilevel selection—selection among groups or tribes, stemming from competition and warfare—played a major role in the evolution of the extraordinary human capacity for cooperation and altruism. Richerson and Boyd (2005) propose that such group selection was facilitated by processes of cultural evolution, such as pressure to conform to the group's norms, that reduce behavioral variation within groups and increase the variance among groups. This would make multilevel selection more effective than it usually is in purely genetic models (see Chapters 11 and 16).

Understanding nature and humanity

"All art," said Oscar Wilde, "is perfectly useless." He meant that as high praise: art is a human creation that needs no utilitarian justification, a creation that is justified simply by being an expression—indeed, one of the defining characteristics—of humanity.

Much of what is most meaningful to us is "perfectly useless": music, sunsets, walking on a clean beach, baseball, soccer, movies, gardening, spiritual inspiration—and understanding. Whether the subject be mathematics, the natural world, philosophy, or human nature, attempting to understand is rewarding in itself, quite aside from whatever practical consequences it may yield.

To know about the extraordinary diversity of organisms, about the complexities of the cell, of development, or of our brains, and about how these marvels came to be is deeply rewarding to anyone with a sense of curiosity and wonder. To have achieved such knowledge is, like other advances in science and technology, among humanity's great accomplishments. Likewise, to have some understanding, however imperfect, of what we humans are and how we came into existence is richly rewarding. It is fascinating and ennobling to learn of our 3.5-billion-year-old pedigree, of when and how and possibly why our ancestors evolved the characteristics that led to our present condition, of how and when modern humans emerged from Africa and colonized the rest of the Earth, of how genetically unified all humans are with one another, and yet how genetically diverse we are. It is both challenging and important to try to understand "human nature"—to understand how our behavior is shaped by our genes and therefore by our evolutionary past, and how it is shaped by culture, social forces, and our unique individual history of learning and experience.

Evolution is the unifying theme of the biological sciences and an important foundation for the "human sciences" of medicine, psychology, and sociology. Although psychologists and anthropologists may differ among themselves on the role of evolution in determining "human nature," most will agree that some knowledge of evolutionary principles is essential. And although evolutionary biologists and social scientists do not set social policy, they can speak out against abuses of their science. They can point out misunderstandings of evolutionary theory, such as racist interpretations of differences among human populations, or the "naturalistic fallacy" that what is natural is good and can therefore guide human conduct. Social Darwinism was based on this fallacy; so is the belief that homosexuality is morally wrong because it does not lead to reproduction, and the belief that women should be subservient to men because of their "natural," evolved sex roles.

Evolution has neither moral nor immoral content, and evolutionary biology provides no philosophical basis for aesthetics or ethics. But evolutionary biology, like other knowledge, can serve the cause of human freedom and dignity by helping to relieve disease and hunger and by helping us to understand and appreciate both the unity and diversity of humankind. And it can enhance our appreciation of life in all its diversity: in Darwin's phrase, "endless forms most beautiful and most wonderful."

Summary

1. Evolution is a fact—a hypothesis that is so thoroughly supported that it is extremely unlikely to be false. The theory of evolution is not a speculation, but rather a complex set of well-supported hypotheses that explain how evolution happens.
2. Although many people do not think there is any necessary incompatibility between evolution and religion, many others reject evolution and instead accept divine creation, because they think evolution conflicts with their religious beliefs. The positions taken by creationists on issues such as the age of the Earth and of life vary.
3. Science is tentative, it accepts hypotheses provisionally and changes in the face of convincing new evidence, and it is concerned only with testable hypotheses; it depends on empirical studies that are subject to peer scrutiny and that can be verified and repeated by others. Supernatural hypotheses, in contrast, cannot be tested. Creationism has none of the features of science, so it has no claim to be taught in science classes.
4. The evidence for evolution comes from all realms of biology and geology, including comparative studies of morphology, development, life histories, and other features, molecular biology, genomics, paleontology, and biogeography. Evolutionary principles can explain features of organisms that would not be expected of a beneficent intelligent designer, such as imperfect adaptation, useless or vestigial features, extinction, selfish DNA, sexually selected characteristics, conflicts among genes within the genome, and infanticide. Furthermore, all the proposed mechanisms of evolution have been thoroughly documented, and evolution has been observed.
5. The arguments used by creationists are all logically refutable in scientific terms and are contradicted by data.
6. It is important to understand evolution because it has broad implications for how we think about nature and humanity, and because it has many practical ramifications. Evolutionary science contributes to many aspects of medicine and public health, agriculture and natural-resource management, pest management, and conservation.
7. One of the most difficult and controversial challenges is to join biological and social science in order to understand how the distinctively human cognitive and behavioral characteristics evolved, the extent to which human behaviors have an evolved genetic foundation, how that foundation interacts with cultural and other environmental factors to shape individual behavior, and how genes and culture have coevolved.

Terms and Concepts

creationists
fact
hypothesis
intelligent design (ID)
natural laws
scientific creationism (= creation science)
special creation
theistic evolution
theory

Suggestions for Further Reading

On creationism, ID, and defending evolution and science, books include:

Why Evolution Is True by J. A. Coyne (Oxford University Press, New York, 2008) is an outstanding, well-written description of the evidence for evolution and its mechanisms by a leading evolutionary biologist.

Evolution vs. Creationism: An Introduction (second edition, Greenwood Press, Westport, CT, 2009) by Eugenie C. Scott, director of the National Center for Science Education, who probably knows more about the evolution-creationism controversy than anyone else.

Defending Evolution in the Classroom: A Guide to the Creation/Evolution Controversy by B. J. Alters and S. M. Alters (Jones and Bartlett, Sudbury, MA, 2001) is a perceptive analysis of how to present and teach evolution.

Denying Evolution: Creationism, Scientism, and the Nature of Science by M. Pigliucci (Sinauer Associates, Sunderland, MA, 2002) shows why creationism fails as science, but treats the limits of science as well.

Science, Evolution, and Creationism (National Academy Press, Washington, D.C., 2008) is a 70-page booklet issued by the most prestigious scientific organization in the United States, the National Academy of Sciences and the Institute of Medicine of the National Academies. It is available at www.nap.edu.

Several books address the relationship between religion and evolution, including:

Finding Darwin's God: A Scientist's Search for Common Ground between God and Evolution (HarperCollins, New York, 1999) by K. Miller, a cell biologist at a leading university, who argues that one can fully accept evolution and reconcile it with religion. See also his later book, *Only a Theory: Evolution and the Battle for America's Soul* (Viking, New York, 2008).

Living with Darwin: Evolution, Design, and the Future of Faith by P. Kitcher (Oxford University Press, New York, 2007) is a short reflection by an eminent philosopher on reconciling evolution with our need for meaning in life, whether this is provided by religion or other sources of fulfilment.

Several excellent Web sites provide information about evolution and can serve as valuable teaching aids. These include the following:

"Understanding Evolution" (http://evolution.berkeley.edu) is an outstanding site, developed by the University of California Museum of Paleontology to provide content and resources for teachers at all grade levels.

www.ncseweb.org is the Web site of the National Center for Science Education (P.O. Box 9477, Berkeley, CA 94709-0477), which actively supports the teaching of evolution and combats creationism. This is the most comprehensive site on the conflict, and provides links to a great range of resources.

The Talk Origins Archive (www.talkorigins.org) has a wealth of material on many aspects of evolution and the social controversy. It includes a comprehensive list of creationist claims and rebuttals to them, by Mark Isaak, that is also available in book form (M. Isaak, *The Counter-Creationism Handbook*, University of California Press, Berkeley, 2007).

Ken Miller's Evolution Resources (www.millerandlevine.com/km/evol/) includes text, video clips of interviews, and other material, especially on debunking creationist claims and discussing Miller's position that religion and evolution are compatible.

Darwin Day Celebration (www.darwinday.org) describes annual educational events held around the world on or near the anniversary of Darwin's birth (February 12) and includes a link to a free online course on evolution and science.

Applications of evolutionary biology are described by D. P. Mindell in *The Evolving World: Evolution in Everyday Life* (Harvard University Press, Cambridge, MA, 2006). Other sources of information on this topic are "Applied evolution" by J. J. Bull and H. A. Wichman (*Annual Review of Ecology and Systematics* 32: 183–217, 2001) and "Evolution, science, and society" at www.amnat.org or in *The American Naturalist* 158 (supplement): S1–S47, 2001. Slightly dated but still interesting is *Why We Get Sick: The New Science of Darwinian Medicine* (Times Books, New York, 1994) by R. M. Nesse and G. C. Williams, who launched the movement to integrate evolutionary biology and medicine.

A huge literature, of variable quality, concerns the evolution of human behavior. A very good introduction to the major contemporary approaches is *Sense and Nonsense: Evolutionary Perspectives on Human Behaviour* (Oxford University Press, Oxford, 2002) by K. N. Laland and G. R. Brown, who are most inclined to the gene-culture coevolution approach but provide a fair and lively description of the other schools of evolutionary thought. A very readable, convincing treatment of culture and human evolution is *Not by Genes Alone: How Culture Transformed Human Evolution* (University of Chicago Press, 2005) by P. J. Richerson and R. Boyd, leading figures in this field.

Glossary

Most of the terms in this glossary appear at several or many places in the text of this book. Most terms that are used broadly in biology or are used in this book only near their definition in the text are not included here.

A

adaptation A process of genetic change in a population whereby, as a result of natural selection, the average state of a character becomes improved with reference to a specific function, or whereby a population is thought to have become better suited to some feature of its environment. Also, *an* adaptation: a feature that has become prevalent in a population because of a selective advantage conveyed by that feature in the improvement in some function. A complex concept; see Chapter 11.

adaptive peak That allele frequency, or combination of allele frequencies at two or more loci, at which the mean fitness of a population has a (local) maximum. Also, the mean phenotype (for one or more characters) that maximizes mean fitness. An **adaptive valley** is a set of allele frequencies at which mean fitness has a minimum.

adaptive radiation Evolutionary divergence of members of a single phylogenetic lineage into a variety of different adaptive forms; usually the taxa differ in the use of resources or habitats, and have diverged over a relatively short interval of geological time. The term **evolutionary radiation** describes a pattern of rapid diversification without assuming that the differences are adaptive.

adaptive zone A set of similar **ecological niches** occupied by a group of (usually) related species, often constituting a higher taxon.

additive effect The magnitude of the effect of an allele on a character, measured as half the phenotypic difference between homozygotes for that allele compared with homozygotes for a different allele.

additive genetic variance That component of the **genetic variance** in a character that is attributable to additive effects of alleles.

allele One of several forms of the same gene, presumably differing by mutation of the DNA sequence. Alleles are usually recognized by their phenotypic effects; DNA sequence variants, which may differ at several or many sites, are usually called **haplotypes**.

allele frequency The proportion of gene copies in a population that are a given allele; i.e., the probability of finding this allele when a gene is taken randomly from the population; also called **gene frequency**.

allometric growth Growth of a feature during ontogeny at a rate different from that of another feature with which it is compared.

allopatric Of a population or species, occupying a geographic region different from that of another population or species. *Cf.* **parapatric**, **sympatric**.

allopolyploid A **polyploid** in which the several chromosome sets are derived from more than one species.

allozyme One of several forms of an enzyme encoded by different alleles at a locus.

alternative splicing Splicing of different sets of exons from mRNA to form mature transcripts that are translated into different proteins (thus allowing the same gene to encode different proteins).

altruism Conferral of a benefit on other individuals at an apparent cost to the donor.

anagenesis Evolution of a feature within a lineage over an arbitrary period of time.

aneuploid Of a cell or organism, possessing too many or too few of one or more chromosomes.

antagonistic selection A source of natural selection that opposes another source of selection on a trait.

apomixis Parthenogenetic reproduction in which an individual develops from one or more mitotically produced cells that have not experienced recombination or syngamy.

apomorphic Having a **derived** character or state, with reference to another character or state. *See* **synapomorphy**.

aposematic Coloration or other features that advertise noxious properties; warning coloration.

artificial selection Selection by humans of a deliberately chosen trait or combination of traits in a (usually captive) population; differing from natural selection in that the criterion for survival and reproduction is the trait chosen, rather than fitness as determined by the entire genotype.

asexual Pertaining to reproduction that does not entail meiosis and syngamy.

assortative mating Nonrandom mating on the basis of phenotype; usually refers to positive assortative mating, the propensity to mate with others of like phenotype.

autopolyploid A **polyploid** in which the several chromosome sets are derived from the same species.

autosome A chromosome other than a sex chromosome.

B

background extinction A long-prevailing rate at which taxa become extinct, in contrast to the highly elevated rates that characterize **mass extinction**.

background selection Elimination of deleterious mutations in a region of the genome; may explain low levels of neutral sequence variation.

balancing selection A form of natural selection that maintains polymorphism at a locus within a population.

benthic Inhabiting the bottom, or substrate, of a body of water. *Cf.* **planktonic**.

biogeography The study of the geographic distribution of organisms.

biological species A population or group of populations within which genes are actually or potentially exchanged by interbreeding, and which are reproductively isolated from other such groups.

bottleneck A severe, temporary reduction in population size.

C

canalization The evolution of internal factors during development that reduce the effect of perturbing environmental and genetic influences, thereby constraining variation and consistently producing a particular (usually wild-type) phenotype.

candidate gene A gene thought to be involved in the evolution of a particular trait based on its mutant phenotype or the function of the protein it encodes.

carrying capacity The population density that can be sustained by limiting resources.

category In taxonomy, one of the ranks of classification (e.g., genus, family). *Cf.* **taxon**.

character A feature, or trait. *Cf.* **character state**.

character displacement Usually refers to a pattern of geographic variation in which a character differs more greatly between sympatric than between allopatric populations of two species; sometimes used for the evolutionary process of accentuation of differences between sympatric populations of two species as a result of the reproductive or ecological interactions between them.

character state One of the variant conditions of a character (e.g., yellow versus brown as state of the character "color of snail shell").

chronospecies A segment of an evolving lineage preserved in the fossil record that differs enough from earlier or later members of the lineage to be given a different binomial (name). Not equivalent to biological species.

***cis*-regulatory element** A noncoding DNA sequence in or near a gene required for proper spatiotemporal expression of that gene, often containing binding sites for transcription factors. Often used interchangeably with **enhancer**.

clade The set of species descended from a particular ancestral species.

cladistic Pertaining to branching patterns; a cladistic classification classifies organisms on the basis of the historical sequences by which they have diverged from common ancestors.

cladogenesis Branching of lineages during phylogeny.

cladogram A branching diagram depicting relationships among taxa; i.e., an estimated history of the relative sequence in which they have evolved from common ancestors. Used by some authors to mean a branching diagram that displays the hierarchical distribution of derived character states among taxa.

cline A gradual change in an allele frequency or in the mean of a character over a geographic transect.

clone A lineage of individuals reproduced asexually, by mitotic division.

coadapted gene pool A population or set of populations in which prevalent genotypes are composed of alleles at two or more loci that confer high fitness in combination with each other, but not with alleles that are prevalent in other such populations.

coalescence Derivation of the gene copies in one or more populations from a single ancestral copy, viewed retrospectively (from the present back into the past).

coevolution Strictly, the joint evolution of two (or more) ecologically interacting species, each of which evolves in response to selection imposed by the other. Sometimes used loosely to refer to evolution of one species caused by its interaction with another, or simply to a history of joint divergence of ecologically associated species.

commensalism An ecological relationship between species in which one is benefited but the other is little affected.

common garden A place in which (usually conspecific) organisms, perhaps from different geographic populations, are reared together, enabling the investigator to ascribe variation among them to genetic rather than environmental differences. Originally applied to plants, but now more generally used to describe any experiment of this design.

comparative method A procedure for inferring the adaptive function of a character by correlating its states in various taxa with one or more variables, such as ecological factors hypothesized to affect its evolution.

compartment A contiguous group of cells, descended from the same progenitor cell, that form a spatially discrete part of a developing organ or structure and often act as a discrete developmental unit. Cells from one compartment typically do not intermix with cells from other compartments.

competition An interaction between individuals of the same species or different species whereby resources used by one are made unavailable to others.

competitive exclusion Extinction of a population due to competition with another species.

concerted evolution Maintenance of a homogeneous nucleotide sequence among the members of a gene family, which evolves over time.

conspecific Belonging to the same species.

convergent evolution (**convergence**) Evolution of similar features independently in different evolutionary lineages, usually from different antecedent features or by different developmental pathways.

co-option The evolution of a function for a gene, tissue, or structure other than the one it was originally adapted for. At the gene level, used interchangeably with **recruitment** and, occasionally, **exaptation**.

cost A reduction in fitness caused by a correlated effect of a feature that provides an increment in fitness (i.e., a benefit).

correlation A statistical relationship that quantifies the degree to which two variables are associated. For *phenotypic correlation, genetic correlation, environmental correlation* as applied to the relationship between two traits, see Chapter 13.

creationism The doctrine that each species (or perhaps higher taxon) was created separately, essentially in its present form, by a supernatural Creator.

C-value paradox The lack of correlation between the DNA content of eukaryotic genomes and a given organism's phenotypic complexity (i.e., the genome of a less complex eukaryotic organism, such as a plant, may contain far more DNA than that of a more complex organism, such as a human being). The paradox is explained by the amount of noncoding repetitive DNA sequences in a genome.

D

deme A local population; usually, a small, panmictic population.

demographic Pertaining to processes that change the size of a population (i.e., birth, death, dispersal).

density-dependent Affected by population density.

derived character (**state**) A character (or character state) that has evolved from an antecedent (ancestral) character or state.

deterministic Causing a fixed outcome, given initial conditions. *Cf.* **stochastic**.

differential gene expression Differences in the time, location, and/or quantitative level at which a gene expresses the protein it encodes. Differential gene expression involves differences between species, developmental stages, or physiological states in the specific cells, tissues, structures, or body segments that express a given gene; it is believed to be a significant agent of morphological change over evolutionary time.

diploid Of a cell or organism, possessing two chromosome complements. *See also* **haploid, polyploid**.

direct development A life history in which the intermediate larval stage is omitted and development proceeds directly from an embryonic form to an adult-like form. *Cf.* **indirect development**.

directional selection Selection for a value of a character that is higher or lower than its current mean value.

dispersal In population biology, movement of individual organisms to different localities; in biogeography, extension of the geographic range of a species by movement of individuals.

disruptive selection Selection in favor of two or more modal phenotypes and against those intermediate between them; also called **diversifying selection**.

divergence The evolution of increasing difference between lineages in one or more characters.

diversification An evolutionary increase in the number of species in a clade, usually accompanied by divergence in phenotypic characters.

diversifying selection *See* **disruptive selection.**

domain A relatively small protein segment or module (100 amino acids or less) that can fold into a specific three-dimensional structure independently of other domains.

dominance Of an allele, the extent to which it produces when heterozygous the same phenotype as when homozygous. Of a species, the extent to which it is numerically (or otherwise) predominant in a community.

duplication The production of another copy of a locus (or other sequence) that is inherited as an addition to the genome.

E

ecological niche The range of combinations of all relevant environmental variables under which a species or population can persist; often more loosely used to describe the "role" of a species, or the resources it utilizes.

ecological release The expansion of a population's niche (e.g., range of habitats or resources used) where competition with other species is alleviated.

ecotype A genetically determined phenotype of a species that is found as a local variant associated with certain ecological conditions.

effective population size The effective size of a real population is equal to the number of individuals in an ideal population (i.e., a population in which all individuals reproduce equally) that produces the rate of genetic drift seen in the real population.

enhancer A DNA sequence that, when acted on by **transcription factors** controls transcription of an associated gene. *Cf.* ***cis*-regulatory element, promoter**.

endemic Of a species, restricted to a specified region or locality.

environment Usually, the complex of external physical, chemical, and biotic factors that may affect a population, an organism, or the expression of an organism's genes; more generally, anything external to the object of interest (e.g., a gene, an organism, a population) that may influence its function or activity. Thus, other genes within an organism may be part of a gene's environment, or other individuals in a population may be part of an organism's environment.

environmental variance Variation among individuals in a phenotypic trait that is caused by variation in the environment rather than by genetic differences.

epistasis An effect of the interaction between two or more gene loci on the phenotype or fitness whereby their joint effect differs from the sum of the loci taken separately.

equilibrium An unchanging condition, as of population size or genetic composition. Also, the value (e.g., of population size, allele frequency) at which this condition occurs. An equilibrium need not be stable. *See* **stability, unstable equilibrium**.

ESS *See* **evolutionarily stable strategy**.

essentialism The philosophical view that all members of a class of objects (such as a species) share certain invariant, unchanging properties that distinguish them from other classes.

evolution In a broad sense, the origin of entities possessing different states of one or more characteristics and changes in the proportions of those entities over time. *Organic evolution*, or *biological evolution*, is a change over time in the proportions of individual organisms differing genetically in one or more traits. Such changes transpire by the origin and subsequent alteration of the frequencies of genotypes from generation to generation within populations, by alteration of the proportions of genetically differentiated populations within a species, or by changes in the numbers of species with different characteristics, thereby altering the frequency of one or more traits within a higher taxon.

evolutionarily stable strategy (ESS) A phenotype such that, if almost all individuals in a population have that phenotype, no alternative phenotype can invade the population or replace it.

evolutionary radiation *See* **adaptive radiation**.

evolutionary reversal The evolution of a character from a derived state back toward a condition that resembles an earlier state.

evolutionary synthesis The reconciliation of Darwin's theory with the findings of modern genetics, which gave rise to a theory that emphasized the coaction of random mutation, selection, genetic drift, and gene flow; also called the **modern synthesis**.

exaptation The evolution of a function of a gene, tissue, or structure other than the one it was originally adapted for; can also refer to the adaptive use of a previously nonadaptive trait.

exon That part of a gene that is translated into a polypeptide (protein). *Cf.* **intron**.

exon shuffling The formation of new genes by assembly of exons from two or more preexisting genes. The classical model of exon shuffling generates new combinations of exons mediated via recombination of intervening introns; however, exon shuffling can also come about by retrotransposition of exons into pre-existing genes.

F

fecundity The quantity of gametes (usually eggs) produced by an individual.

fitness The success of an entity in reproducing; hence, the average contribution of an allele or genotype to the next generation or to succeeding generations. *See also* **relative fitness**.

fixation Attainment of a frequency of 1 (i.e., 100 percent) by an allele in a population, which thereby becomes **monomorphic** for the allele.

founder effect The principle that the founders of a new population carry only a fraction of the total genetic variation in the source population.

frameshift mutation An insertion or deletion of base pairs in a translated DNA sequence that alters the reading frame, resulting in multiple downstream changes in the gene product.

frequency In this book, usually used to mean *proportion* (e.g., the frequency of an allele is the proportion of gene copies having that allelic state).

frequency-dependent selection A mode of natural selection in which the fitness of each genotype varies as a function of its frequency in the population.

function The way in which a character contributes to the fitness of an organism.

G

gene The functional unit of heredity. A complex concept.

gene complex A group of two or more genes that are members of the same family and in most cases are located in close proximity to one another in the genome, often in tandem separated by various amounts of intergenic, noncoding DNA.

gene conversion A process involving the unidirectional transfer of DNA information from one gene to another. In a typical conversion event, a gene or part of a gene acquires the same sequence as the other allele at that locus (intralocus or intra-allelic conversion), or the same sequences as a different, usually paralogous, locus (interlocus conversion). One consequence of gene conversion may be the homogenization of sequences among members of a gene family.

gene duplication When new genes arise as copies of preexisting gene sequences. The result can be a **gene family**.

gene family Two or more loci with similar nucleotide sequences that have been derived from a common ancestral sequence.

gene flow The incorporation of genes into the gene pool of one population from one or more other populations.

gene frequency *See* **allele frequency**.

gene pool The totality of the genes of a given sexual population.

gene tree A diagram representing the history by which gene copies have been derived from ancestral gene copies in previous generations.

genetic conflict Antagonistic fitness relationships between alleles at different loci in a genome.

genetic correlation Correlated differences among genotypes in two or more phenotypic characters, due to pleiotropy or linkage disequilibrium.

genetic distance Any of several measures of the degree of genetic difference between populations, based on differences in allele frequencies.

genetic drift Random changes in the frequencies of two or more alleles or genotypes within a population.

genetic load Any reduction of the mean fitness of a population resulting from the existence of genotypes with a fitness lower than that of the most fit genotype.

genetic variance Variation in a trait within a population, as measured by the variance that is due to genetic differences among individuals.

genic selection A form of selection in which the single gene is the unit of selection, such that the outcome is determined by fitness values assigned to different alleles. *See* **individual selection, kin selection, natural selection**.

genome The entire complement of DNA sequences in a cell or organism. A distinction may be made between the nuclear genome and organelle genomes, such as those of mitochondria and plastids.

genotype The set of genes possessed by an individual organism; often, its genetic composition at a specific locus or set of loci singled out for discussion.

genotype × environment interaction Phenotypic variation arising from the difference in the effect of the environment on the expression of different genotypes.

geographic variation Differences among spatially distributed populations of a species.

grade A group of species that have evolved the same state in one or more characters and typically constitute a **paraphyletic** group relative to other species that have evolved further in the same direction.

gradualism The proposition that large differences in phenotypic characters have evolved through many slightly different intermediate states.

group selection The differential rate of origination or extinction of whole populations (or species, if the term is used broadly) on the basis of differences among them in one or more characteristics. *See also* **interdemic selection, species selection**.

H

habitat selection The capacity of an organism (usually an animal) to choose a habitat in which to perform its activities. Habitat selection is not a form of natural selection.

haploid Of a cell or organism, possessing a single chromosome complement, hence a single gene copy at each locus.

haplotype A DNA sequence that differs from homologous sequences at one or more base pair sites.

Hardy-Weinberg Pertaining to the genotype frequencies expected at a locus under ideal equilibrium conditions in a randomly mating population.

heritability The proportion of the **variance** in a trait among individuals that is attributable to differences in genotype. Heritability in the narrow sense is the ratio of **additive genetic variance** to phenotypic variance.

heterochrony An evolutionary change in phenotype caused by an alteration of timing of developmental events.

heterokaryotype A genome or individual that is heterozygous for a chromosomal rearrangement such as an inversion. *Cf.* **homokaryotype**.

heterozygosity In a population, the proportion of loci at which a randomly chosen individual is heterozygous, on average.

heterozygote An individual organism that possesses different alleles at a locus.

heterozygous advantage The manifestation of higher fitness by heterozygotes than by homozygotes at a specific locus.

hitchhiking Change in the frequency of an allele due to linkage with a selected allele at another locus.

homeobox genes A large family of eukaryotic genes that contain a DNA sequence known as the **homeobox**. The homeobox sequence encodes a protein **homeodomain** about 60 amino acids in length that binds DNA. Most homeobox genes are transcriptional regulators. *Cf.* **domain**; **Hox genes**.

homeostasis Maintenance of an equilibrium state by some self-regulating capacity of an individual.

homeotic mutation A mutation that causes a transformation of one structure into another of the organism's structures.

homokaryotype A genome or individual that is homozygous for a chromosomal rearrangement such as an inversion. *Cf.* **heterokaryotype**.

homology Possession by two or more species of a character state derived, with or without modification, from their common ancestor. **Homologous chromosomes** are those members of a chromosome complement that bear the same genes.

homonymous Pertaining to biological structures that occur repeatedly within one segment of the organism, such as teeth or bristles.

homoplasy Possession by two or more species of a similar or identical character state that has not been derived by both species from their common ancestor; embraces **convergence**, **parallel evolution**, and **evolutionary reversal**.

homozygote An individual organism that has the same allele at each of its copies of a genetic locus.

horizontal transmission Movement of genes or symbionts (such as parasites) between individual organisms other than by transmission from parents to their offspring (which is **vertical transmission**). Horizontal transmission of genes is also called **lateral gene transfer**.

Hox genes A subfamily of **homeobox genes**, conserved in all metazoan animals, that controls anterior-posterior segment identity by regulating the transcription of many genes during development.

hybrid An individual formed by mating between unlike forms, usually genetically differentiated populations or species.

hybrid zone A region in which genetically distinct populations come into contact and produce at least some offspring of mixed ancestry.

hypermorphosis An evolutionary increase in the duration of ontogenetic development, resulting in features that are exaggerated compared to those of the ancestor.

I

identical by descent Of two or more gene copies, being derived from a single gene copy in a specified common ancestor of the organisms that carry the copies.

inbreeding Mating between relatives that occurs more frequently than if mates were chosen at random from a population.

inbreeding depression Reduction, in inbred individuals, of the mean value of a character (usually one correlated with fitness).

inclusive fitness The fitness of a gene or genotype as measured by its effect on the survival or reproduction of both the organism bearing it and the genes, identical by descent, borne by the organism's relatives.

indirect development A life history consisting of a larval stage between embryo and adult stages. *Cf.* **direct development**.

individual selection A form of natural selection consisting of nonrandom differences among different genotypes (or phenotypes) within a population in their contribution to subsequent generations. *See also* **genic selection**, **natural selection**.

inter-, intra- Prefixes meaning, respectively, "between" and "within." For example, "interspecific" differences are differences between species and "intraspecific" differences are differences among individuals within a species.

interaction Strictly, the dependence of an outcome on a combination of causal factors, such that the outcome is not predictable from the average effects of the factors taken separately. More loosely, an interplay between entities that affects one or more of them (as in interactions between species). *See also* **genotype × environment interaction.**

interdemic selection **Group selection** of populations within a species.

intrinsic rate of natural increase The potential per capita rate of increase of a population with a stable age distribution whose growth is not depressed by the negative effects of density.

introgression Movement of genes from one species or population into another by hybridization and backcrossing; carries the implication that some genes in a genome undergo such movement, but others do not.

intron A part of a gene that is not translated into a polypeptide. *Cf.* **exon**.

inversion A 180° reversal of the orientation of a part of a chromosome, relative to some standard chromosome.

isolating barrier, isolating mechanism A genetically determined difference between populations that restricts or prevents gene flow between them. The term does not include spatial segregation by extrinsic geographic or topographic barriers.

iteroparous Pertaining to a life history in which individuals reproduce more than once. *Cf.* **semelparous**.

K

karyotype The chromosome complement of an individual.

key adaptation An adaptation that provides the basis for using a new, substantially different habitat or resource.

kin selection A form of selection whereby alleles differ in their rate of propagation by influencing the impact of their bearers on the reproductive success of individuals (kin) who carry the same alleles by common descent.

L

Lamarckism The theory that evolution is caused by inheritance of character changes acquired during the life of an individual due to its behavior or to environmental influences.

lateral gene transfer *See* **horizontal transmission**.

lethal allele An allele (usually recessive) that causes virtually complete mortality, usually early in development.

lineage A series of ancestral and descendant populations through time; usually refers to a single evolving species, but may include several species descended from a common ancestor.

lineage sorting The process by which each of several descendant species, carrying several gene lineages inherited from a common ancestral species, acquires a single gene lineage; hence, the derivation of a monophyletic gene tree, in each species, from the paraphyletic gene tree inherited from their common ancestor.

linkage Occurrence of two loci on the same chromosome: the loci are functionally linked only if they are so close together that they do not segregate independently in meiosis.

linkage disequilibrium The association of two alleles at two or more loci more frequently (or less frequently) than predicted by their individual frequencies.

linkage equilibrium The association of two alleles at two or more loci at the frequency predicted by their individual frequencies.

locus (plural: **loci**) A site on a chromosome occupied by a specific gene; more loosely, the gene itself, in all its allelic states.

logistic equation An equation describing the idealized growth of a population subject to a density-dependent limiting factor. As density increases, the rate of growth gradually declines until population growth stops.

M

macroevolution A vague term, usually meaning the evolution of substantial phenotypic changes, usually great enough to place the changed lineage and its descendants in a distinct genus or higher taxon. *Cf.* **microevolution**.

mass extinction A highly elevated rate of extinction of species, extending over an interval that is relatively short on a geological time scale (although still very long on a human time scale).

maternal effect A nongenetic effect of a mother on the phenotype of her offspring, stemming from factors such as cytoplasmic inheritance, transmission of symbionts from mother to offspring, or nutritional conditions.

maximum parsimony *See* **parsimony**.

mean Usually the arithmetic mean or average; the sum of n values, divided by n. The mean value of x, symbolized as $\bar{x}$, equals $(x_1 + x_2 + \ldots + x_n)/n$.

mean fitness The arithmetic average fitness of all individuals in a population, usually relative to some standard.

meiotic drive Used broadly to denote a preponderance (> 50 percent) of one allele among the gametes produced by a heterozygote; results in genic selection.

metapopulation A set of local populations, among which there may be gene flow and patterns of extinction and recolonization.

microevolution A vague term, usually referring to slight, short-term evolutionary changes within species. *Cf.* **macroevolution**.

microsatellite A short, highly repeated, untranslated DNA sequence.

migration Used in theoretical population genetics as a synonym for gene flow among populations; in other contexts, refers to directed large-scale movements of organisms that do not necessarily result in gene flow.

mimicry Similarity of certain characters in two or more species due to convergent evolution when there is an advantage conferred by the resemblance. Common types include *Batesian* mimicry, in which a palatable *mimic* experiences lower predation because of its resemblance to an unpalatable *model*; and *Müllerian* mimicry, in which two or more unpalatable species enjoy reduced predation due to their similarity.

modern synthesis *See* **evolutionary synthesis**.

modularity The ability of individual parts of an organism, such as segments or organs, to develop or evolve independently from one another; the ability of developmental regulatory genes and pathways to be regulated independently in different tissues and developmental stages.

molecular clock The concept of a steady rate of change in DNA sequences over time, providing a basis for dating the time of divergence of lineages if the rate of change can be estimated.

monomorphic Having one form; refers to a population in which virtually all individuals have the same genotype at a locus. *Cf.* **polymorphism**.

monophyletic Refers to a taxon, phylogenetic tree, or gene tree whose members are all derived from a common ancestral taxon. In cladistic taxonomy, the term describes a taxon consisting of all the known species descended from a single ancestral species. *Cf.* **paraphyletic**, **polyphyletic**.

mosaic evolution Evolution of different characters within a lineage or clade at different rates, hence more or less independently of one another.

mutation An error in the replication of a nucleotide sequence, or any other alteration of the genome that is not manifested as reciprocal recombination.

mutational variance The increment in the genetic variance of a phenotypic character caused by new mutations in each generation.

mutualism A symbiotic relation in which each of two species benefits by their interaction.

N

natural selection The differential survival and/or reproduction of classes of entities that differ in one or more characteristics. To constitute natural selection, the difference in survival and/or reproduction cannot be due to chance, and it must have the potential consequence of altering the proportions of the different entities. Thus natural selection is also definable as a deterministic difference in the contribution of different classes of entities to subsequent generations. Usually the differences are inherited. The entities may be alleles, genotypes or subsets of genotypes, populations, or, in the broadest sense, species. A complex concept; see Chapter 11. *See also* **genic selection**, **individual selection**, **kin selection**, **group selection**.

neo-Darwinism The modern belief that natural selection, acting on randomly generated genetic variation, is a major, but not the sole, cause of evolution.

neofunctionalization Divergence of duplicate genes whereby one acquires a new function. *Cf.* **subfunctionalization**.

neoteny Heterochronic evolution whereby development of some or all somatic features is retarded relative to sexual maturation, resulting in sexually mature individuals with juvenile features. *See also* **paedomorphosis**, **progenesis**.

neutral alleles Alleles that do not differ measurably in their effect on fitness.

nonsynonymous substitution A base pair substitution in DNA that results in an amino acid substitution in the protein product; also called **replacement substitution**. *Cf.* **synonymous substitution**.

norm of reaction The set of phenotypic expressions of a genotype under different environmental conditions. *See also* **phenotypic plasticity**.

normal distribution A bell-shaped frequency distribution of a variable; the expected distribution if many factors with independent, small effects determine the value of a variable; the basis for many statistical formulations.

nucleotide substitution The complete replacement of one nucleotide base pair by another within a lineage over evolutionary time.

O

ontogeny The development of an individual organism, from fertilized zygote until death.

organism Usually used in this book to refer to an individual member of a species.

orthologous Refers to corresponding (homologous) members of a gene family in two or more species. *Cf.* **paralogous**.

outcrossing Mating with another genetic individual. *Cf.* **selfing**.

outgroup A taxon that diverged from a group of other taxa before they diverged from one another.

overdominance The expression by two alleles in heterozygous condition of a phenotypic value for some character that lies outside the range of the two corresponding homozygotes.

P

paedomorphosis Possession in the adult stage of features typical of the juvenile stage of the organism's ancestor.

panmixia Random mating among members of a population.

parallel evolution (**parallelism**) The evolution of similar or identical features independently in related lineages, thought usually to be based on similar modifications of the same developmental pathways.

paralogous Refers to the homologous relationship between two different members of a gene family, within a species or in a comparison of different species. *Cf.* **orthologous**.

parapatric Of two species or populations, having contiguous but non-overlapping geographic distributions.

paraphyletic Refers to a taxon, phylogenetic tree, or gene tree whose members are all derived from a single ancestor, but which does not include all the descendants of that ancestor. *Cf.* **monophyletic**.

parental investment Parental activities or processes that enhance the survival of existing offspring but whose **costs** reduce the parent's subsequent reproductive success.

parsimony Economy in the use of means to an end (*Webster's New Collegiate Dictionary*); the principle of accounting for observations by that hypothesis requiring the fewest or simplest assumptions that lack evidence; in systematics, the principle of invoking the minimal number of evolutionary changes to infer phylogenetic relationships.

parthenogenesis Virgin birth; development from an egg to which there has been no paternal contribution of genes.

PCR (**polymerase chain reaction**) A laboratory technique by which the number of copies of a DNA sequence is increased by replication in vitro.

peak shift Change in allele frequencies within a population from one to another local maximum of mean fitness by passage through states of lower mean fitness.

peripatric Of a population, peripheral to most of the other populations of a species.

peripatric speciation Speciation by evolution of reproductive isolation in peripatric populations as a consequence of a combination of genetic drift and natural selection.

phenetic Pertaining to phenotypic similarity, as in a phenetic classification.

phenotype The morphological, physiological, biochemical, behavioral, and other properties of an organism manifested throughout its life; or any subset of such properties, especially those affected by a particular allele or other portion of the **genotype**.

phenotypic plasticity The capacity of an organism to develop any of several phenotypic states, depending on the environment; usually this capacity is assumed to be adaptive.

phylogenetic species concept Any of several related concepts of species as sets of populations that are diagnosably different from other populations.

phylogeny The history of descent of a group of taxa such as species from their common ancestors, including the order of branching and sometimes the absolute times of divergence; also applied to the genealogy of genes derived from a common ancestral gene.

planktonic Living in open water. *Cf.* **benthic**.

pleiotropy A phenotypic effect of a gene on more than one character.

ploidy The number of chromosome complements in an organism.

polygenic character A character whose variation is based wholly or in part on allelic variation at more than a few loci.

polymorphism The existence within a population of two or more genotypes, the rarest of which exceeds some arbitrarily low frequency (say, 1 percent); more rarely, the existence of phenotypic variation within a population, whether or not genetically based. *Cf.* **monomorphic**.

polyphenism The capacity of a species or genotype to develop two or more forms, with the specific form depending on specific environmental conditions or cues, such as temperature or day length. A polyphenism is distinct from a **polymorphism** in that the former is the property of a single genotype, whereas the latter refers to multiple forms encoded by two or more different genotypes.

polyphyletic Refers to a taxon, phylogenetic tree, or gene tree composed of members derived by evolution from ancestors in more than one ancestral taxon; hence, composed of members that do not share a unique common ancestor. *Cf.* **monophyletic**.

polyploid Of a cell or organism, possessing more than two chromosome complements.

population A group of conspecific organisms that occupy a more or less well defined geographic region and exhibit reproductive continuity from generation to generation; ecological and reproductive interactions are more frequent among these individuals than with members of other populations of the same species.

positive selection Selection for an allele that increases fitness. *Cf.* **purifying selection**.

postzygotic Occurring after union of the nuclei of uniting gametes; usually refers to inviability or sterility that confers reproductive isolation.

preadaptation Possession of the necessary properties to permit a shift to a new niche, habitat, or function. A structure is preadapted for a new function if it can assume that function without evolutionary modification.

prezygotic Occurring before union of the nuclei of uniting gametes; usually refers to events in the reproductive process that cause reproductive isolation.

primordium A group of embryonic or larval cells destined to give rise to a particular adult structure.

processed pseudogene A **pseudogene** that has arisen via the retrotransposition of mRNA into cDNA.

progenesis A decrease during evolution of the duration of ontogenetic development, resulting in retention of juvenile features in the sexually mature adult. *See also* **neoteny**, **paedomorphosis**.

promoter Usually refers to the DNA sequences immediately 5′ to (upstream of) a gene that are bound by the RNA polymerase and its cofactors and/or are required in order to transcribe the gene. Sometimes used interchangeably with **enhancer**.

provinciality The degree to which the taxonomic composition of a biota is differentiated among major geographic regions.

pseudogene A nonfunctional member of a gene family that has been derived from a functional gene. *Cf.* **processed pseudogene**.

punctuated equilibria A pattern of rapid evolutionary change in the phenotype of a lineage separated by long periods of little change; also, a hypothesis intended to explain such a pattern, whereby phenotypic change transpires rapidly in small populations, in concert with the evolution of reproductive isolation.

purifying selection Elimination of deleterious alleles from a population. *Cf.* **positive selection**.

Q

QTL Quantitative trait locus (or loci); a chromosome region containing at least one gene that contributes to variation in a quantitative trait. **QTL mapping** is a procedure for determining the map positions of QTL on chromosomes.

quantitative trait A phenotypic character that varies continuously rather than as discretely different character states.

R

race A poorly defined term for a set of populations occupying a particular region that differ in one or more characteristics from populations elsewhere; equivalent to **subspecies.** In some writings, a distinctive phenotype, whether or not allopatric from others.

radiation *See* **adaptive radiation**.

random genetic drift *See* **genetic drift**.

recruitment (1) In evolutionary genetics, the evolution of a new function for a gene other than the function for which that gene was originally adapted. (2) In population biology, refers to the addition of new adult (breeding) individuals to a population via reproduction (i.e., individuals born into the population that reach reproductive age).

recurrent mutation Repeated origin of mutations of a particular kind within a species.

refugia Locations in which species have persisted while becoming extinct elsewhere.

regression In geology, withdrawal of sea from land, accompanying lowering of sea level; in statistics, a function that best predicts a dependent from an independent variable.

reinforcement Evolution of enhanced reproductive isolation between populations due to natural selection for greater isolation.

relative fitness The fitness of a genotype relative to (as a proportion of) the fitness of a reference genotype, which is often set at 1.0.

relict A species that has been "left behind"; for example, the last survivor of an otherwise extinct group. Sometimes, a species or population left in a locality after extinction throughout most of the region.

replacement substitution *See* **nonsynonymous substitution**.

reporter construct A DNA segment in which a putative *cis*-regulatory sequence is spliced upstream of a gene whose expression can be easily assayed, such as β-galactosidase or green fluorescent protein.

reproductive effort The proportion of energy or materials that an organism allocates to reproduction rather than to growth and maintenance.

response to selection The change in the mean value of a character over one or more generations due to selection.

restriction enzyme An enzyme that cuts double-stranded DNA at specific short nucleotide sequences. Genetic variation within a population results in variation in DNA sequence lengths after treatment with a restriction enzyme, or **restriction fragment length polymorphism (RFLP)**.

reticulate evolution Union of different lineages of a clade by hybridization.

RFLP *See* **restriction enzyme**.

S

saltation A jump; a discontinuous mutational change in one or more phenotypic traits, usually of considerable magnitude.

scala naturae The "scale of nature," or Great Chain of Being: the pre-evolutionary concept that all living things were created in an orderly series of forms, from lower to higher.

selection Nonrandom differential survival or reproduction of classes of phenotypically different entities. *See* **natural selection, artificial selection**.

selection coefficient The difference between the mean relative fitness of individuals of a given genotype and that of a reference genotype.

selective advantage The increment in fitness (survival and/or reproduction) provided by an allele or a character state.

selective sweep Reduction or elimination of DNA sequence variation in the vicinity of a mutation that has been fixed by natural selection relatively recently.

selfing Self-fertilization; union of female and male gametes produced by the same genetic individual. *Cf.* **outcrossing**.

"selfish DNA" A DNA sequence that has the capacity for its own replication, or replication via other self-replicating elements, but has no immediate function (or is deleterious) for the organism in which it resides.

semelparous Pertaining to a life history in which individuals (especially females) reproduce only once. *Cf.* **iteroparous**.

semispecies One of several groups of populations that are partially but not entirely isolated from one another by biological factors (**isolating mechanisms**).

serial homology A relationship among repeated, often differentiated, structures of a single organism, defined by their similarity of developmental origin; for example, the several legs and other appendages of an arthropod.

sex-linked Of a gene, being carried by one of the sex chromosomes; it may be expressed phenotypically in both sexes.

sexual reproduction Production of offspring whose genetic constitution is a mixture of those of two potentially genetically different gametes.

sexual selection Differential reproduction as a result of variation in the ability to obtain mates.

sibling species Species that are difficult or impossible to distinguish by morphological characters, but may be discerned by differences in ecology, behavior, chromosomes, or other such characters.

silent substitution *See* **synonymous substitution**.

single nucleotide polymorphism (SNP) Variation in the identity of a nucleotide base pair at a single position in a DNA sequence, within or among populations of a species.

sister taxa Two species or higher taxa that are derived from an immediate common ancestor, and are therefore each other's closest relatives.

speciation Evolution of reproductive isolation within an ancestral species, resulting in two or more descendant species.

species In the sense of biological species, the members of a group of populations that interbreed or potentially interbreed with one another under natural conditions; a complex concept (see Chapter 15). Also, a fundamental taxonomic category to which individual specimens are assigned, which often but not always corresponds to the biological species. *See also* **biological species, phylogenetic species concept**.

species selection A form of **group selection** in which species with different characteristics increase (by speciation) or decrease (by extinction) in number at different rates because of a difference in their characteristics.

stability Often used to mean constancy; more often in this book, the propensity to return to a condition (a stable equilibrium) after displacement from that condition.

stabilizing selection Selection against phenotypes that deviate in either direction from an optimal value of a character.

standard deviation The square root of the **variance**.

stasis Absence of evolutionary change in one or more characters for some period of evolutionary time.

stochastic Random. *Cf.* **deterministic**.

strata Layers of sedimentary rock that were deposited at different times.

subfunctionalization Divergence of duplicate genes whereby each retains only a subset of the several functions of the ancestral gene. *Cf.* **neofunctionalization**.

subspecies A named geographic race; a set of populations of a species that share one or more distinctive features and occupy a different geographic area from other subspecies.

substitution The complete replacement of one allele by another within a population or species over evolutionary time. *Cf.* **fixation**.

superspecies A group of semispecies.

symbiosis An intimate, usually physical, association between two or more species.

sympatric Of two species or populations, occupying the same geographic locality so that the opportunity to interbreed is presented.

synapomorphy A derived character state that is shared by two or more taxa and is postulated to have evolved in their common ancestor.

synonymous substitution Fixation of a base pair change that does not alter the amino acid in the protein product of a gene; also called **silent substitution**. *Cf.* **nonsynonymous substitution**.

T

target gene In developmental genetics, a gene regulated by a transcription factor of interest. This regulation may be direct or indirect.

taxon (plural: **taxa**) The named taxonomic unit (e.g., *Homo sapiens*, Hominidae, or Mammalia) to which individuals, or sets of species, are assigned. **Higher taxa** are those above the species level. *Cf.* **category**.

teleology The belief that natural events and objects have purposes and can be explained by their purposes.

territory An area or volume of habitat defended by an organism or a group of organisms against other individuals, usually of the same species; **territorial behavior** is the behavior by which the territory is defended.

trade-off The existence of both a fitness benefit and a fitness cost of a mutation or character state, relative to another.

transcription factor A protein that interacts with a regulatory DNA sequence and affects the transcription of the associated gene.

transition A mutation that changes a nucleotide to another nucleotide in the same class (purine or pyrimidine). *Cf.* **transversion**.

translocation The transfer of a segment of a chromosome to another, nonhomologous, chromosome; the chromosome formed by the addition of such a segment.

transposable element A DNA sequence, copies of which become inserted into various sites in the genome.

***trans*-regulatory element** A nucleotide sequence, usually encoding a regulatory protein, that is not closely linked to the structural gene whose expression it regulates. *Cf.* ***cis*-regulatory element.**

transversion A mutation that changes a nucleotide to another nucleotide in the opposite class (purine or pyrimidine). *Cf.* **transition**.

U

unstable equilibrium An **equilibrium** to which a system does not return if disturbed.

V

variance (σ^2, s^2, V) The average squared deviation of an observation from the arithmetic mean; hence, a measure of variation. $s^2 = [\Sigma(x_i - \bar{x})^2]/(n - 1)$, where $\bar{x}$ is the mean and n the number of observations.

vertical transmission *See* **horizontal transmission**.

vestigial Occurring in a rudimentary condition as a result of evolutionary reduction from a more elaborated, functional character state in an ancestor.

viability Capacity for survival; often refers to the fraction of individuals surviving to a given age, and is contrasted with inviability due to deleterious genes.

vicariance Separation of a continuously distributed ancestral population or species into separate populations due to the development of a geographic or ecological barrier.

virulence Usually, the damage inflicted on a host by a pathogen or parasite; sometimes, the capacity of a pathogen or parasite to infect and develop in a host.

W

wild-type The allele, genotype, or phenotype that is most prevalent (if there is one) in wild populations; with reference to the wild-type allele, other alleles are often termed mutations.

Z

zygote A single-celled individual formed by the union of gametes. Occasionally used more loosely to refer to an offspring produced by sexual reproduction.

Literature Cited

A

Abbott, R. J. 1992. Plant invasions, interspecific hybridization and the evolution of new plant taxa. *Trends Ecol. Evol.* 7: 401–405. [17]

Abedin, M., and N. King. 2008. The premetazoan ancestry of cadherins. *Science* 319: 946–948. [5, 20]

Abouheif, E., and G. A. Wray. 2002. Evolution of the gene network underlying wing polyphenism in ants. *Science* 297: 249–252. [21]

Abrams, P. A. 2000a. The evolution of predator-prey interactions: Theory and evidence. *Annu. Rev. Ecol. Syst.* 31: 79–108. [19]

Abrams, P. A. 2000b. Character shifts of prey species that share predators. *Am. Nat.* 156 (Suppl.): S45–S61. [19]

Abrams, P. A., and H. Matsuda. 1997. Fitness minimization and dynamic instability as a consequence of predator-prey coevolution. *Evol. Ecol.* 11: 1–20. [19]

Abzhanov, A., M. Protas, B. R. Grant, P. R. Grant, and C. J. Tabin. 2004. Bmp4 and morphological variation of beaks in Darwin's finches. *Science* 305: 1462–1465. [21]

Adamowicz, S. J., and A. Purvis. 2006. From more to fewer? Testing an allegedly pervasive trend in the evolution of morphological structure. *Evolution* 60: 1402–1416. [3]

Adoutte, A., G. Balavoine, N. Lartillot, O. Lespinet, B. Prud'homme, and R. de Rosa. 2000. The new animal phylogeny: reliability and implications. *Proc. Natl. Acad. Sci. USA* 97: 4453–4456. [5]

Agrawal, A. A. 2005. Natural selection on common milkweed (*Asclepias syriaca*) by a community of specialized insect herbivores. *Evol. Ecol. Res.* 7: 651–667. [19]

Ahlberg, P. E., and J. A. Clack. 2006. A firm step from water to land. *Nature* 440: 747–749. [4]

Akey, J. M., M. A. Eberle, M. J. Rinder, C. S. Carlson, M. D. Shriver, D. A. Nickerson, and L. Kruglyak. 2004. Population history and natural selection shape patterns of genetic variation in 132 genes. *PLoS Biol.* 2: e286 (doi:10.1371/journal.pbio.0020286). [12]

Alberch, P., and E. A. Gale. 1985. A developmental analysis of an evolutionary trend: Digital reduction in amphibians. *Evolution* 39: 8–23. [21]

Alcock, J. 2009. *Animal Behavior: An Evolutionary Approach*, 9th ed. Sinauer, Sunderland, MA. [16]

Allen, C. E., P. Beldade, B. J. Zwaan, and P. M. Brakefield. 2008. Differences in the selection response of serially repeated color pattern characters: standing variation, development, and evolution. *BMC Evol. Biol.* 8: 94. [21]

Allendorf, F. W., and G. Luikart. 2007. *Conservation and the Genetics of Populations.* Blackwell, Oxford. [23]

Allison, A. C. 1955. Aspects of polymorphism in man. *Cold Spring Harbor Symp. Quant. Biol.* 20: 239–255. [9]

Alroy, J. 1998. Cope's rule and the dynamics of body mass evolution in North American fossil mammals. *Science* 280: 731–734. [7, 22]

Alroy, J. 1998. Equilibrial diversity dynamics in North American mammals. In M. L. McKinney and J. H. Drake (eds.), *Biodiversity Dynamics: Turnover of Populations, Taxa, and Communities,* pp. 233–287. Columbia University Press, New York. [7]

Alroy, J. 2001. A multispecies overkill simulation of the end-Pleistocene megafaunal mass extinction. *Science* 292: 1893–1896. [5]

Alroy, J., P. L. Koch, and J. C. Zachos. 2000. Global climate change and North American mammal evolution. In D. H. Erwin and S. L. Wing (eds.), *Deep Time: Paleobiology's Perspective*, pp. 259–288. *Paleobiology* 25 (4), supplement. [7]

Alroy, J., and 34 others. 2008. Phanerozoic trends in the global diversity of marine invertebrates. *Science* 321: 97–100. [7]

Alters, B. J., and S. M. Alters. 2001. *Defending Evolution: A Guide to the Creation/Evolution Controversy*. Jones and Bartlett, Sudbury, MA. [1, 22, 23]

Alvarez, L. W., W. Alvarez, F. Asaro, and H. V. Michel. 1980. Extraterrestrial cause for the Cretaceous-Tertiary extinction. *Science* 208: 1095–1108. [7]

Ambrose, B. A., D. R. Lerner, P. Ciceri, C. M. Padilla, M. F. Yanofsky, and R. J. Schmidt. 2000. Molecular and genetic analyses of the *silky1* gene reveal conservation in floral organ specification between eudicots and monocots. *Molec. Cell* 5: 569–579. [21]

Anderson, E. 1949. *Introgressive Hybridization*. Wiley, New York. [17]

Andersson, M. B. 1982. Female choice selects for extreme tail length in a widowbird. *Nature* 299: 818–820. [11]

Andersson, M. B. 1994. *Sexual Selection*. Princeton University Press, Princeton, NJ. [15]

Andersson, M. B., and Y. Iwasa. 1996. Sexual selection. *Trends Ecol. Evol.* 11: 53–58. [15]

Andersson, M. B., and L. W. Simmons. 2006. Sexual selection and mate choice. *Trends Ecol. Evol.* 21: 296–302. [15]

Andrade, M. C. B. 1996. Sexual selection for male sacrifice in the Australian redback spider. *Science* 271: 70–72. [11]

Antolin, M. F. 1993. Genetics of biased sex ratios in subdivided populations: Models, assumptions, and evidence. *Oxford Series in Evolutionary Biology* 9: 239–281. Oxford University Press, Oxford. [15]

Antonovics, J. 1968. Evolution in closely adjacent plant populations. V. Evolution of self-fertility. *Heredity* 23: 219–238. [18]

Antonovics, J., A. D. Bradshaw, and R. G. Turner. 1971. Heavy metal tolerance in plants. *Adv. Ecol. Res.* 7: 1–85. [13]

Antonovics, J., and 13 others. 2007. Evolution by any other name: Antibiotic resistance and avoidance of the e-word. *PLoS Biology* 5: e30 doi:10.1371/journal.pbio.0050030. [23]

Aquadro, C. F., D. J. Begun, and E. C. Kindahl. 1994. Selection, recombination, and DNA polymorphism in *Drosophila*. In B. Golding (ed.), *Non-neutral Evolution: Theories and Molecular Data*, pp. 46–56. Chapman & Hall, New York. [12]

Arbeitman, M. N., and 9 others. 2002. Gene expression during the life cycle of *Drosophila melanogaster*. *Science* 297: 2270–2275. [21]

Arbogast, B. S., S. V. Edwards, J. Wakeley, P. Beerli, and J. B. Slowinski. 2002. Estimating divergence times from molecular data on phylogenetic and population genetic timescales. *Annu. Rev. Ecol. Evol. Syst.* 33: 707–740. [2]

Arendt, J., and D. Reznick. 2008. Convergence and parallelism reconsidered: What have we learned about the genetics of adaptation? *Trend Ecol. Evol.* 23: 26–32. [3]

Armbruster, P., W. E. Bradshaw, and C. M. Holzapfel. 1998. Effects of postglacial range expansion on allozyme and quantitative genetic variation of the pitcher-plant mosquito, *Wyeomyia smithii*. *Evolution* 52: 1697–1704. [9]

Armbruster, W. S., V. S. Di Stilio, J. D. Tuxill, T. C. Flores, and J. L. Velasquez Runk. 2000. Covariance and decoupling of floral and vegetative traits in nine neotropical plants: A reevaluation of Berg's correlation-pleiades concept. *Am. J. Bot.* 86: 39–55. [13]

Armbruster, W. S., C. Pélabon, T. F. Hansen, and C. P. H. Mulder. 2004. Floral integration, modularity, and accuracy. In M. Pigliucci and K. Preston (eds.), *Phenotypic Integration: Studying the Ecology and Evolution of Complex Phenotypes*, pp. 23–49. Oxford University Press, Oxford. [13]

Arnold, M. J. 1997. *Natural Hybridization and Evolution*. Oxford University Press, Oxford. [17]

Arnold, S. J. 1981. Behavioral variation in natural populations. I. Phenotypic, genetic, and environmental correlations between chemoreceptive responses to prey in the garter snake, *Thamnophis elegans*. *Evolution* 35: 489–509. [13]

Arnqvist, G., and L. Rowe. 2002. Correlated evolution of male and female morphologies in water striders. *Evolution* 56: 936–947. [15]

Arnqvist, G., and L. Rowe. 2005. *Sexual Conflict*. Princeton University Press, Princeton, NJ. [15]

Atlan, A., H. Merçot, C. Landre, and C. Montchamp-Moreau. 1997. The sex-ratio trait in *Drosophila simulans*: Geographical distribution of distortion and resistance. *Evolution* 51: 1886–1895. [16]

Atwood, R. C., L. K. Schneider, and F. J. Ryan. 1951. Selective mechanisms in bacteria. *Cold Spring Harbor Symp. Quant. Biol.* 16: 345–355. [11]

Ausich, W. I., and N. G. Lane. 1999. *Life of the past*, 4th ed. Prentice-Hall, Upper Saddle River, NJ. [5]

Averoff, M., and N. H. Patel. 1997. Crustacean appendage evolution associated with changes in Hox gene expression. *Nature* 388: 682–686. [3]

Avise, J. C. 1994. *Molecular Markers, Natural History, and Evolution*. Chapman & Hall, New York. [17, 18]

Avise, J. C. 1998. *The Genetic Gods: Evolution and Belief in Human Affairs*. Harvard University Press, Cambridge, MA. [8]

Avise, J. C. 2000. *Phylogeography*. Harvard University Press, Cambridge, MA. [6]

Avise, J. C. 2004. *Molecular Markers, Natural History, and Evolution*, 2nd ed. Sinauer, Sunderland, MA. [9, 12]

Avise, J. C., and R. M. Ball, Jr. 1990. Principles of genealogical concordance in species concepts and biological taxonomy. *Oxford Surv. Evol. Biol.* 7: 45–67. [17]

Avise, J. C., B. C. Bowen, T. Lamb, A. B. Meylan, and E. Bermingham. 1992. Mitochondrial DNA evolution at a turtle's pace: Evidence for low genetic variability and reduced microevolutionary rate in Testudines. *Mol. Biol. Evol.* 9: 433–446. [10]

Axelrod, R., and W. D. Hamilton. 1981. The evolution of cooperation. *Science* 211: 1390–1396. [16]

Ayala, F. J. 1968. Genotype, environment, and population numbers. *Science* 162: 1453–1459. [14]

B

Backwell, P. R. Y., J. H. Christy, S. R. Telford, M. D. Jennions, and N. I. Passmore. 2000. Dishonest signaling in a fiddler crab. *Proc. R. Soc. Lond. B* 267: 719–724. [16]

Bailey, W. J., D. H. A. Fitch, D. A. Tagle, J. Czelusniak, J. L. Slightom, and M. Goodman. 1991. Molecular evolution of the ψη-globin gene locus: Gibbon phylogeny and the hominoid slowdown. *Mol. Biol. Evol.* 8: 155–184. [2]

Baker, C. S., et al. 2000. Predicted decline of protected whales based on molecular genetic monitoring of Japanese and Korean markets. *Proc. R. Soc. Lond. B* 267: 1191–1199. [23]

Bakker, E. G., C. Toomajian, M. Kreitman, and J. Bergelson. 2006. A genome-wide survey of *R* gene polymorphisms in *Arabidopsis*. *Plant Cell* 18: 1803–1818. [19]

Baldauf, S. L., D. Bhattacharya, J. Cockrill, P. Hugenholtz, J. Pawlowski, and A. G. B. Simpson. 2004. The tree of life: An overview. In J. Cracraft and M. J. Donoghue (eds.), *Assembling the Tree of Life*, pp. 44–75. Oxford University Press, New York. [2]

Baldwin, B. G., and M. J. Sanderson. 1998. Age and rate of diversification of the Hawaiian silversword alliance (Compositae). *Proc. Natl. Acad. Sci. USA* 95: 9402–9406. [7]

Bambach, R. K. 1985. Classes and adaptive variety: The ecology of diversification in marine faunas through the Phanerozoic. In J. W. Valentine (ed.), *Phanerozoic Diversity Patterns: Profiles in Macroevolution*, pp. 191–253. Princeton University Press, Princeton, NJ. [7]

Bambach, R. K. 2006. Phanerozoic biodiversity mass extinctions. *Annu. Rev. Earth Planet. Sci.* 34: 127–155. [7]

Bambach, R. K., A. H. Knoll, and J. J. Sepkoski, Jr. 2002. Anatomical and ecological constraints on Phanerozoic animal diversity in the marine realm. *Proc. Natl. Acad. Sci. USA* 99: 6854–6859. [7]

Barber, I., S. A. Arnott, V. A. Braithwaite, J. Andrew, and F. Huntingford. 2001. Indirect fitness consequences of mate choice in sticklebacks: offspring of brighter males grow slowly but resist parasitic infections. *Proc. R. Soc. Lond. B* 268: 71–76. [15]

Barkow, J. H., L. Cosmides, and J. Tooby (eds.). 1992. *The Adapted Mind: Evolutionary Psychology and the Generation of Culture*. Oxford University Press, New York. [16, 23]

Barmina, O. and A. Kopp. 2007. Sex-specific expression of a Hox gene associated with rapid morphological evolution. *Dev. Biol.* 311: 277–286. [21]

Barraclough, T. G., and A. P. Vogler. 2000. Detecting the geographical pattern of speciation from species–level phylogenies. *Am. Nat.* 155: 419–434. [18]

Barrett, R. D. H., and D. Schluter. 2008. Adaptation from standing genetic variation. *Trends Ecol. Evol.* 23: 38–44. [13]

Barrett, S. C. H., and D. Charlesworth. 1991. Effects of a change in the level of inbreeding on the genetic load. *Nature* 352: 522–524. [15]

Bartel, D. P. 2004. MicroRNAs: Genomics, biogenesis, mechanism, and function. *Cell* 116: 281–297. [20]

Bartolomé, C., and B. Charlesworth. 2006. Rates and patterns of chromosomal evolution in *Drosophila pseudoobscura* and *D. miranda*. *Genetics* 173: 779–791. [8]

Barton, N. H., and B. Charlesworth. 1984. Genetic revolutions, founder effects, and speciation. *Annu. Rev. Ecol. Syst.* 15: 133–164. [12]

Barton, N. H., and B. Charlesworth. 1998. Why sex and recombination? *Science* 281: 1986–1990. [15]

Barton, N. H., and K. S. Gale. 1993. Genetic analysis of hybrid zones. In R. G. Harrison (ed.), *Hybrid Zones and the Evolutionary Process*, pp. 13–45. Oxford University Press, New York. [17]

Barton, N. H., and G. M. Hewitt. 1985. Analysis of hybrid zones. *Annu. Rev. Ecol. Syst.* 16: 113–148. [18]

Barton, N. H., and M. Turelli. 1989. Evolutionary quantitative genetics: How little do we know? *Annu. Rev. Genet.* 23: 337–370. [13]

Barton, N., and L. Partridge. 2000. Limits to natural selection. *BioEssays* 22: 1075–1084. [9, 13]

Basolo, A. L. 1994. The dynamics of Fisherian sex-ratio evolution: Theoretical and experimental investigations. *Am. Nat.* 144: 473–490. [12]

Basolo, A. L. 1995. Phylogenetic evidence for the role of a preexisting bias in sexual selection. *Proc. R. Soc. Lond.* B 259: 307–311. [15]

Basolo, A. L. 1998. Evolutionary change in a receiver bias: A comparison of female preference functions. *Proc. R. Soc. Lond. B* 265: 2223–2228. [15]

Bateman, A. J. 1947. Contamination of seed crops. II. Wind pollination. *Heredity* 1: 235–246. [9]

Baum, D. A., and K. L. Shaw. 1995. Genealogical perspectives on the species problem. In P. C. Hoch and A. G. Stephenson (eds.), *Experimental and Molecular Approaches to Plant Biosystematics*, pp. 289–303. Monographs in Systematic Botany, Missouri Botanical Garden, St. Louis, MO. [17]

Becerra, A., L. Delaye, S. Islas, and A. Lazcano. 2007. The very early stages of biological evolution and the nature of the last common ancestor of the three major cell domains. *Annu. Rev. Ecol. Evol. Syst.* 38: 361–379. [5]

Becerra, J. X. 1997. Insects on plants: Macroevolutionary chemical trends in host use. *Science* 276: 253–256. [19]

Becerra, J. X., and D. L. Venable. 1999. Macroevolution of insect-plant associations: The relevance of host biogeography to host affiliation. *Proc. Natl. Acad. Sci. USA* 96: 12626–12631. [19]

Beck, N. R., M. C. Double, and A. Cockburn. 2003. Microsatellite evolution at two hypervariable loci revealed by extensive avian pedigrees. *Mol. Biol. Evol.* 20: 54–61. [8]

Bégin, M., and D. A. Roff. 2004. From micro- to macroevolution through quantitative genetic variation: Positive evidence from field crickets. *Evolution* 58: 2287–2304. [13]

Behrensmeyer, A. K., J. D. Damuth, W. A. DiMichele, R. Potts, H.-D. Sues, and S. L. Wing (eds.). 1992. *Terrestrial Ecosystems Through Time: Evolutionary Paleoecology of Terrestrial Plants and Animals.* University of Chicago Press, Chicago. [5]

Bejerano, G., M. Pheasant, I. Makurin, S. Stephen, W. J. Kent, J. S. Mattick, and D. Haussler. 2004. Ultraconserved elements in the human genome. *Science* 304: 1321–1325. [12]

Bejerano, G., and 8 others. 2006. A distal enhancer and an ultraconserved exon are derived from a novel retroposon. *Nature* 441: 87–90. [20]

Beldade, P., K. Koops, and P. M. Brakefield. 2002. Developmental constraints versus flexibility in morphological evolution. *Nature* 416: 844–847. [21]

Bell, G., and V. Koufopanou. 1986. The cost of reproduction. *Oxford Surv. Evol. Biol.* 3: 83–131. [14]

Bell, M. A., J. V. Baumgartner, and E. C. Olson. 1985. Patterns of temporal change in single morphological characters of a Miocene stickleback fish. *Paleobiology* 11: 258–271. [4]

Bennett, A. F., R. E. Lenski, and J. E. Mittler. 1992. Evolutionary adaptation to temperature. I. Fitness responses of *Escherichia coli* to changes in its thermal environment. *Evolution* 46: 16–30. [8]

Bennetzen, J. L. 2000. Transposable element contributions to plant gene and genome evolution. *Plant Mol. Biol.* 42: 251–269. [8]

Benton, M. J. (ed.). 1988. *The Phylogeny and Classification of the Tetrapods.* Clarendon Press, Oxford. [5]

Benton, M. J. 1990. The causes of the diversification of life. In P. D. Taylor and G. P. Larwood (eds.), *Major Evolutionary Radiations*, pp. 409–430. Clarendon Press, Oxford. [7]

Benton, M. J. 1996. On the nonprevalence of competitive replacement in the evolution of tetrapods. In D. Jablonski, D. H. Erwin, and J. Lipps (eds.), *Evolutionary Paleobiology*, pp. 185–210. University of Chicago Press, Chicago. [7]

Benton, M. J., and R. Hitchin. 1997. Congruence between phylogenetic and stratigraphic data on the history of life. *Proc. R. Soc. Lond. B* 264: 885–890. [4]

Benton, M. J., M. A. Wills, and R. Hitchin. 2000. Quality of the fossil record through time. *Nature* 403: 534–537. [4]

Berenbaum, M. R., and A. R. Zangerl. 1988. Stalemates in the coevolutionary arms race: Synthesis, synergisms, and sundry other sins. In K. C. Spencer (ed.), *Chemical Mediation of Coevolution*, pp. 113–132. Academic Press, San Diego, CA. [19]]

Berenbaum, M. R., A. R. Zangerl, and J. K. Nitao. 1986. Constraints on chemical coevolution: Wild parsnips and the parsnip webworm. *Evolution* 40: 1215–1228. [19]

Bermingham, E., and J. C. Avise. 1986. Molecular zoogeography of freshwater fishes in the southeastern United States. *Genetics* 113: 939–965. [18]

Bernasconi, G., and J. E. Strassmann. 1999. Cooperation among unrelated individuals: The ant foundress case. *Trends Ecol. Evol.* 14: 477–482. [16]

Bernays, E. A., and R. F. Chapman. 1994. *Host-Plant Selection by Phytophagous Insects.* Chapman & Hall, London. [19]

Bernays, E. A., and M. Graham. 1988. On the evolution of host specificity by phytophagous arthropods. *Ecology* 69: 886–892. [19]

Bersaglieri, T., and 8 others. 2004. Genetic signatures of strong recent positive selection at the lactase gene. *Am. J. Hum. Genet.* 74: 1111–1120. [12]

Berthold, P., A. J. Heibig, G. Mohr, and U. Querner. 1992. Rapid microevolution of migratory behavior in a wild bird species. *Nature* 360: 668–670. [13]

Bharathan, G., and R. Sinha, R. 2001. The regulation of compound leaf development. *Plant Physiol.* 127: 1–5. [21]

Bininda-Emonds, O. R. P., and 9 others. 2007. The delayed rise of present-day mammals. *Nature* 446: 507–512. [5]

Birch, L. C., T. Dobzhansky, P. D. Elliott, and R. C. Lewontin. 1963. Relative fitness of geographic races of *Drosophila serrata. Evolution* 17: 72–83. [14]

Birkhead, T. R. 2000. *Promiscuity: An Evolutionary History of Sperm Competition.* Harvard University Press, Cambridge, MA. [15]

Birkhead, T. R., and A. P. Møller. 1992. *Sperm Competition in Birds: Evolutionary Causes and Consequences.* Academic Press, London. [15]

Bishop, J. A. 1981. A neo-Darwinian approach to resistance: Examples from mammals. In J. A. Bishop and L. M. Cook (eds.), *Genetic Consequences of Man Made Change*, pp. 37–51. Academic Press, London. [12]

Blount, Z. D., C. Z. Borland, and R. E. Lenski. 2008. Historical contingency and the evolution of a key innovation in an experimental population of *Escherichia coli. Proc. Natl. Acad. Sci. USA* 105: 7899–7906. [8]

Blows, M. W., and A. A. Hoffmann. 2005. A reassessment of genetic limits to evolutionary change. *Ecology* 86: 1371–1384. [13]

Blows, M. W., S. Chenoweth, and E. Hine. 2004. Orientation of the genetic variance–covariance matrix and the fitness surface for multiple male sexually selected traits. *Am. Nat.* 163: 329–340. [13]

Boag, P. T. 1983. The heritability of external morphology in Darwin's ground finches (*Geospiza*) on Isla Daphne Major, Galápagos. *Evolution* 37: 877–894. [9]

Bodmer, W., and M. Ashburner. 1984. Conservation and change in the DNA sequences coding for alcohol dehydrogenase in sibling species of *Drosophila. Nature* 309: 425–540. [8]

Bolnick, D. I., and B. M. Fitzpatrick. 2007. Sympatric speciation: Models and empirical evidence. *Annu. Rev. Ecol. Evol. Syst.* 38: 459–487. [18]

Bonaparte, J. F. 1978. El Mesozoico de America del Sur y sus tetrapodos. *Opera Lilloana* 26: 1–596. [5]

Bonnell, M. L., and R. K. Selander. 1974. Elephant seals: Genetic variation and near extinction. *Science* 184: 908–909. [10]

Bonner, J. T. 1988. *The Evolution of Complexity.* Princeton University Press, Princeton, NJ. [21]

Borries, C., K. Launhardt, C. Epplen, J. T. Epplen, and P. Winkler. 1999. DNA analyses support the hypothesis that infanticide is adaptive in langur monkeys. *Proc. R. Soc. Lond. B* 266: 901–904. [16]

Bouchard, T. J. Jr., D. T. Lykken, M. McGue, N. L. Segal, and A. Tellegen. 1990. Sources of human psychological differences: The Minnesota study of twins reared apart. *Science* 250: 223–228. [9]

Boucher, Y., C. J. Douady, R. T. Papke, D. A. Walsh, M. E. R. Boudreau, C. L. Nesbo, R. J. Case, and W. F. Doolittle. 2003. Lateral gene transfer and the origins of prokaryotic groups. *Annu. Rev. Genet.* 37: 283–328. [5]

Boucot, A. J. 1975. *Evolution and Extinction Rate Controls.* Elsevier, Amsterdam. [7]

Boul, K. E., W. C. Funk, C. R. Darst, D. C. Cannatella, and M. J. Ryan. 2007. Sexual selection drives speciation in an Amazonian frog. *Proc. R. Soc. Lond. B* 274: 399–406. [18]

Boureau, E. 1964. *Traité de paléobotanique*, Vol. III. Masson, Paris. [5]

Bourke, A. F. G., and N. R. Franks. 1995. *Social Evolution in Ants.* Princeton University Press, Princeton, NJ. [16]

Bowler, P. J. 1989. *Evolution: The History of an Idea.* University of California Press, Berkeley. [1]

Bowler, P. J. 1996. *Life's Splendid Drama: Evolutionary Biology and the Reconstruction of Life's Ancestry 1860–1940.* University of Chicago Press. Chicago. [1]

Boyce, M. S., and C. J. Perrins. 1987. Optimizing great tit clutch size in a fluctuating environment. *Ecology* 68: 142–153. [17]

Boyd, R., and P. J. Richerson. 1985. *Culture and the Evolutionary Process.* University of Chicago Press, Chicago. [23]

Bradbury, J. W., and S. L. Vehrencamp. 1998. *Principles of Animal Communication.* Sinauer, Sunderland, MA. [16]

Bradshaw, A. D. 1991. Genostasis and the limits to evolution. The Croonian Lecture, 1991. *Phil. Trans. R. Soc. Lond.* B 333: 289–305. [8, 13]

Bradshaw, H. D., and D. W. Schemske. 2003. Allele substitution at a flower colour locus produces a pollinator shift in monkeyflowers. *Nature* 426: 176–178. [17]

Brakefield, P. M., J. Gates, D. Keys, F. Kesbeke, P. J. Wijngaarden, A. Montelro, V. French, and S. B. Carroll. 1996. Development, plasticity and evolution of butterfly eyespot patterns. *Nature* 384: 236–242. [21]

Brideau, N. J., H. A. Flores, J. Wang, S. Maheshwari, X. Wang, and D. A. Barbash. 2006. Two Dobzhansky-Muller genes interact to cause hybrid lethality in *Drosophila. Science* 314: 1292–1295. [17]

Bridle, J. R., and T. H. Vines. 2007. Limits to evolution at range margins: When and why does adaptation fail? *Trends Ecol. Evol.* 22: 140–147. [6]

Briggs, D. E. G., and P. R. Crowther (eds.). 1990. *Palaeobiology: A Synthesis.* Blackwell Scientific, Oxford. [4]

Briskie, J. V., and M. Mackintosh. 2004. Hatching failure increases with severity of population bottlenecks in birds. *Proc. Natl. Acad. Sci. USA* 101: 558–561. [10]

Bro-Jorgensen, J. 2007. The intensity of sexual selection predicts weapon size in male bovids. *Evolution* 61: 1316–1326. [15]

Broadhead, T. W., and J. A. Waters. 1980. *Echinoderms: Notes for a Short Course.* Studies in Geology 3. University of Tennessee Dept. of Geological Sciences, Knoxville. [5]

Brodie, E. D. III, and E. D. Brodie, Jr. 1999. Predator-prey arms races. *BioScience* 49: 557–568. [19]

Brodie, E. D. III. 1992. Correlational selection for color pattern and antipredator behavior in the garter snake *Thamnophis ordinoides. Evolution* 46: 1284–1298. [13]

Brodie, E. D. III. 1993. Homogeneity of the genetic variance-covariance matrix for antipredator traits in two natural populations of the garter snake *Thamnophis ordinoides. Evolution* 47: 844–854. [13, 16]

Brodie, E. D., Jr., B. J. Ridenhour, and E. D. Brodie III. 2002. The evolutionary response of predators to dangerous prey: Hotspots and coldspots in the geographic mosaic of coevolution between garter snakes and newts. *Evolution* 56: 2067–2082. [19]

Brooks, D. R. 1990. Parsimony analysis in historical biogeography and coevolution: methodological and theoretical update. *Syst. Zool.* 39: 14–30. [6]

Brooks, D. R., and D. A. McLennan. 2002. *The Nature of Diversity: An Evolutionary Voyage of Discovery.* University of Chicago Press, Chicago. [3]

Brooks, L. D. 1988. The evolution of recombination rates. In B. R. Levin and R. E. Michod (eds.), *The Evolution of Sex,* pp. 87–105. Sinauer, Sunderland, MA. [15]

Brown, J. H. 1995. *Macroecology.* University of Chicago Press, Chicago. [7]

Brown, J. H., and A. C. Gibson. 1983. *Biogeography.* Mosby, St. Louis. [6]

Brown, J. H., and M. V. Lomolino. 1998. *Biogeography,* 2nd ed. Sinauer, Sunderland, MA [5, 6]

Brown, W. L. Jr., and E. O. Wilson. 1956. Character displacement. *Syst. Zool.* 5: 49–64. [18, 19]

Brown, W. M., J. M. George, and A. C. Wilson. 1979. Rapid evolution of animal mitochondrial DNA. *Proc. Natl. Acad. Sci. USA* 76: 1967–1971. [10]

Brusca, R. C., and G. J. Brusca. 1990. *Invertebrates.* Sinauer, Sunderland, MA. [3]

Buchanan, K. L., and C. K. Catchpole. 2000. Song as an indicator of male parental effort in the sedge warbler. *Proc. R. Soc. Lond.* 267: 321–326. [16]

Bull, J. J. 1994. Perspective: Virulence. *Evolution* 48: 1423–1437. [19]

Bull, J. J., and E. L. Charnov. 1985. On irreversible evolution. *Evolution* 39: 1149–1155. [22]

Bull, J. J., and W. R. Rice. 1991. Distinguishing mechanisms for the evolution of cooperation. *J. Theor. Biol.* 149: 63–74. [19]

Bull, J. J., and H. A. Wichman. 2001. Applied evolution. *Annu. Rev. Ecol. Syst.* 32:183–217. [23]

Bull, J. J., I. J. Molineaux, and W. R. Rice. 1991. Selection of benevolence in a host–parasite system. *Evolution* 45:875–882. [19]

Burbrink, F. T., R. Lawson, and J. B. Slowinski. 2000. Mitochondrial DNA phylogeography of the polytypic North American rat snake (*Elaphe obsoleta*): a critique of the subspecies concept. *Evolution* 54:2107–2118. [9]

Buri, P. 1956. Gene frequency drift in small population of mutant *Drosophila. Evolution* 10: 367–402. [10]

Burney, D. A., and T. F. Flannery. 2005. Fifty millennia of catastrophic extinctions after human contact. *Trends Ecol. Evol.* 20: 395–401. [5]

Burt, A., and R. Trivers. 2006. *Genes in Conflict: The Biology of Selfish Genetic Elements.* Harvard University Press, Cambridge, MA. [11, 16]

Bush, G. L. 1969. Sympatric host race formation and speciation in frugivorous flies of the genus *Rhagoletis* (Diptera, Tephritidae). *Evolution* 23: 237–251. [18]

Bush, R. M., C. A. Bender, K. Subbarao, N. J. Cox, and W. M. Fitch. 1999. Predicting the evolution of human influenza A. *Science* 286: 1921–1925. [23]

Buss, D. M. 1994.*The Evolution of Desire: Strategies of Human Mating.* HarperCollins, New York. [23]

Buss, D. M. 1999. Sex differences in human mate preferences: Evolutionary hypotheses tested in 37 cultures. *Behavioral and Brain Sciences* 12: 1–49. [16]

Bustamante, C. D., and 13 others. 2005. Natural selection on protein-coding genes in the human genome. *Nature* 437: 1153–1157. [12]

Byrne, M., M. Timmermans, C. Kidner, and R. Martienssen. 2001. Development of leaf shape *Curr. Opin. Plant Biol.* 4: 38–43. [21]

C

Cai, J., R. Zhao, H. Jiang, and W. Wang. 2008. De novo origination of a new protein-coding gene in *Saccharomyces cerevisiae. Genetics* 179: 487–496. [20]

Cain, M. L., V. Andreasen, and D. J. Howard. 1999. Reinforcing selection is effective under a relatively broad set of conditions in a mosaic hybrid zone. *Evolution* 53: 1343–1353. [18]

Camperio-Ciani, A., F. Corna, and C. Capiluppi. 2004. Evidence for maternally inherited factors favouring male homosexuality and promoting female fecundity. *Proc. R. Soc. Lond. B* 271: 2217–2221. [16]

Camperio-Ciani, A., Paolo Cermelli, and G. Zanzotto. 2008. Sexually antagonistic selection in human male homosexuality. *PLoS One* 3: 1–8 (e2282). [16]

Cann, R. L., M. Stoneking, and A. C. Wilson. 1987. Mitochondrial DNA and human evolution. *Nature* 325: 31–36. [6]

Carbone, M. A., A. Llopart, M. deAngelis, J. A. Coyne, and T. F. Mackay. 2005. Quantitative trait loci affecting the difference in pigmentation between *Drosophila yakuba* and *D. santomea*. *Genetics* 171: 211–225. [21]

Carlquist, S., B. G. Baldwin, and G. E. Carr (eds.). 2003. *Tarweeds and Silverswords: Evolution of the Madiinae (Asteraceae)*. Missouri Botanical Garden Press, St. Louis, MO. [3]

Carmel, L., Y. I. Wolf, I. B. Rogozin, and E. V. Koonin. 2007. Three distinct modes of intron dynamics in the evolution of eukaryotes. *Genome Res.* 17: 1034–1044. [20]

Carroll, R. L. 1988. *Vertebrate Paleontology and Evolution*. W. H. Freeman, New York. [4, 5]

Carroll, S. B. 2003. Genetics and the making of *Homo sapiens*. *Nature* 422: 849–857. [21]

Carroll, S. B. 2006. *The Making of the Fittest: DNA as the Ultimate Forensic Record of Evolution*. W.W. Norton, New York. [3]

Carroll, S. B., and C. Boyd. 1992. Host race radiation in the soapberry bug: Natural history with the history. *Evolution* 46: 1052–1069. [13]

Carroll, S. B., J. K. Grenier, and S. D. Weatherbee. 2001. *From DNA to Diversity: Molecular Genetics and the Evolution of Animal Design*. Blackwell Science, Malden, MA. [21, 22]

Carroll, S. P., H. Dingle, and S. P. Klassen. 1997. Genetic differentiation of fitness-associated traits among rapidly evolving populations of the soapberry bug. *Evolution* 51: 1182–1188. [13]

Carson, H. L. 1967. Selection for parthenogenesis in *Drosophila mercatorum*. *Genetics* 55: 157–171. [15]

Carson, H. L. 1975. The genetics of speciation at the diploid level. *Am. Nat.* 109: 73–92. [18]

Carson, H. L., and B. A. Clague. 1995. Geology and biogeography of the Hawaiian Islands. In W. L. Wagner and V. A. Funk (eds.), *Hawaiian Biogeography*, pp. 14–29. Smithsonian Institution Press, Washington, DC. [6]

Carson, H. L., and K. Y. Kaneshiro. 1976. *Drosophila* of Hawaii: Systematics and ecological genetics. *Annu. Rev. Ecol. Syst.* 7: 311–345. [3]

Carstens, B. C., and L. L. Knowles. 2007. Estimating species phylogeny from gene-tree probabilities despite incomplete lineage sorting: an example from *Melanoplus* grasshoppers. *Syst. Biol.* 56: 400–411. [2]

Cartwright, J. 2000. *Evolution and Human Behavior: Darwinian Perspectives on Human Nature*. MIT Press, Cambridge, MA. [16]

Cavalier-Smith, T. 2006. Cell evolution and Earth history: Stasis and revolution. *Phil. Trans. R. Soc. Lond. B* 361: 969–1006. [5]

Cavalli-Sforza, L. L., and W. F. Bodmer. 1971. *The Genetics of Human Populations*. W. H. Freeman, San Francisco. [9, 12, 13]

Cavalli-Sforza, L. L., and M. W. Feldman. 1981. *Cultural Transmission and Evolution: A Quantitative Approach*. Princeton University Press, Princeton, NJ. [23]

Cavalli-Sforza, L. L., P. Menozzi, and A. Piazza. 1994. *The History and Geography of Human Genes*. Princeton University Press, Princeton, NJ. [6, 9]

Cavender-Bares, J., D. D. Ackerly, D. A. Baum, and F.A. Bazzaz. 2004. Phylogenetic overdispersion in Floridian oak communities. *Am. Nat.* 163: 823–843. [19]

Cela-Conde, C. J., and F. J. Ayala. 2007. *Human Evolution: Trails from the Past*. Oxford University Press. New York. [4]

Chaline, J., and B. Laurin. 1986. Phyletic gradualism in a European Plio–Pleistocene *Mimomys* lineage (Arvicolidae, Rodentia). *Paleobiology* 12: 203–216. [4]

Chang, B. S. W., K. Jönsson, M. A. Kazmi, M. J. Donoghue, and T. P. Sakmar. 2002. Recreating a functional ancestral archosaur visual pigment. *Mol. Biol. Evol.* 19: 1483–1489. [3]

Chapman, T., L. F. Liddle, J. Kalb, M. Wolfner, and L. Partridge. 1995. Cost of mating in *Drosophila melanogaster* females is mediated by male accessory gland products. *Nature* 373: 241–244. [15]

Charlesworth, B. 1989. The evolution of sex and recombination. *Trends Ecol. Evol.* 4: 264–267. [15]

Charlesworth, B. 1990. The evolutionary genetics of adaptation. In M. Nitecki (ed.), *Evolutionary Innovations*, pp. 47–70. University of Chicago Press, Chicago. [22]

Charlesworth, B. 1994a. The effect of background selection against deleterious mutations on weakly selected, linked variants. *Genet. Res.* 63: 213–227. [12, 14]

Charlesworth, B. 1994b. *Evolution in Age-Structured Populations*. Cambridge University Press, Cambridge. [14]

Charlesworth, B., and D. Charlesworth. 1978. A model for the evolution of dioecy and gynodioecy. *Am. Nat.* 112: 975–997. [15]

Charlesworth, B., and C. H. Langley. 1989. The population genetics of *Drosophila* transposable elements. *Annu. Rev. Genet.* 23: 251–287. [20]

Charlesworth, B., and S. Rouhani. 1988. The probability of peak shifts in a founder population. II. An additive polygenic trait. *Evolution* 42: 1129–1145. [18]

Charlesworth, B., R. Lande, and M. Slatkin. 1982. A neo-Darwinian commentary on macroevolution. *Evolution* 36: 474–498. [18, 22]

Charlesworth, B., M. T. Morgan, and D. Charlesworth. 1993. The effect of deleterious mutations on neutral molecular variation. *Genetics* 134: 1289–1303. [12]

Charlesworth, B., D. Charlesworth, and N. H. Barton. 2003. The effects of genetic and geographic structure on neutral variation. *Annu. Rev. Ecol. Evol. Syst.* 34: 99–125. [10]

Charnov, E. L. 1982. *The Theory of Sex Allocation*. Princeton University Press, Princeton, NJ. [12, 15]

Cheetham, A. H. 1987. Tempo of evolution in a Neogene bryozoan: Are trends in single morphological characters misleading? *Paleobiology* 13: 286–296. [4]

Cheney, D. L., and R. M. Seyfarth. 2007. *Baboon Metaphysics: The Evolution of a Social Mind*. University of Chicago Press, Chicago. [23]

Cheng, C.-H., and L. Chen. 1999. Evolution of an antifreeze glycoprotein. *Nature* 401: 443–444. [20]

Chiappe, L. M. 2007. *Glorified Dinosaurs: The Origin and Early Evolution of Birds*. John Wiley and Sons, New York. [4]

Chiappe, L. M., and G. J. Dyke. 2002. The Mesozoic radiation of birds. *Annu. Rev. Ecol. Syst.* 33: 91–124. [4]

Chippindale, P. T., R. M. Bonett, A. S. Baldwin, and J. J. Wiens. 2004. Phylogenetic evidence for a major reversal of life-history evolution in plethodontid salamanders. *Evolution* 58: 2809–2822. [3]

Christiansen, F. B. 1984. The definition and measurement of fitness. In B. Shorrocks (ed.), *Evolutionary Ecology*, pp. 65–79. Blackwell Scientific, Oxford. [12]

Civetta, A., and A. G. Clark. 2000. Correlated effects of sperm competition and postmating female mortality. *Proc. Natl. Acad. Sci. USA* 97: 13162–13165. [15]

Clack, J. A. 2002. *Gaining Ground: The Origin and Evolution of Tetrapods*. Indiana University Press, Bloomington. [4]

Clark, A. G., and D. J. Begun. 1998. Female genotypes affect sperm displacement in *Drosophila*. *Genetics* 149: 1487–1493. [15]

Clark, A. G., M. Aguadé, T. Prout, L. G. Harshman, and C. H. Langley. 1994. Variation in sperm displacement and its association with accessory gland protein loci in *Drosophila melanogaster*. *Genetics* 139: 189–201. [15]

Clark, A. G., and 16 others. 2003. Inferring non-neutral evolution from human-chimp-mouse orthologous gene trios. *Science* 302: 1960–1963. [20]

Clark, A. G. and the *Drosophila* 12 Genomes Consortium. 2007. Evolution of genes and genomes on the *Drosophila* phylogeny. *Nature* 450: 203–218. [20]

Clausen, J., D. D. Keck, and W. M. Hiesey. 1940. Experimental studies on the nature of species. I. Effect of varied environments on western North American plants. Carnegie Institution of Washington Publication no. 520: 1–452. [9]

Clausen, J., D. D. Keck, and W. M. Hiesey. 1947. Heredity of geographically and ecologically isolated races. *Am. Nat.* 81: 114–133. [9]

Clayton, D. H., S. E. Bush, B. M. Goates, and K. P. Johnson. 2003. Host defense reinforces host–parasite coevolution. *Proc. Natl. Acad. Sci. USA* 100: 15694–15699. [19]

Clegg, J. B., and D. J. Weatherall. 1999. Thalassaemia and malaria: new insights into an old problem. *Proc. Assoc. Am. Physicians* 111: 278–282. [12]

Clegg, S. M., S. M. Degnan, C. Moritz, A. Estoup, J. Kikkawa, and I. P. F. Owens. 2002. Microevolution in island forms: The role of drift and directional selection in morphological divergence of a passerine bird. *Evolution* 56: 2090–2099. [13]

Clemens, J., Z. Lu, W. J. Gehring, I. A. Meinertzhagen, and P. Callaerts. 2008. Central projections of photoreceptor axons originating from ectopic eyes in *Drosophila*. *Proc. Natl. Acad. Sci. USA* 105: 8968–8973. [21]

Clutton-Brock, T. H. 1991. *The Evolution of Parental Care*. Princeton University Press, Princeton, NJ. [16]

Clutton-Brock, T. H. 2002. Breeding together: Kin selection and mutualism in cooperative vertebrates. *Science* 296: 69–72. [16]

Cockburn, A. 1998. *Evolution* of helping behavior in cooperatively breeding birds. *Annu. Rev. Ecol. Syst.* 29: 141–177. [16]

Cohan, F. M. 1984. Can uniform selection retard random genetic divergence between isolated conspecific populations? *Evolution* 38: 495–504. [8]

Colbert, E. H. 1980. *Evolution of the Vertebrates*, 3rd ed. Wiley, New York. [4]

Cole, C. T. 2003. Genetic variation in rare and common plants. *Annu. Rev. Ecol. Evol. Syst.* 34: 213–237. [10]

Colegrave, N. 2002. Sex releases the speed limit on evolution. *Nature* 420: 664–666. [15]

Coley, P. D., J. P. Bryant, and F. S. Chapin III. 1985. Resource availability and plant antiherbivore defense. *Science* 230: 895–899. [19]

Collin, R., and M. P. Miglietta. 2008. Reversing opinions on Dollo's Law. *Trends Ecol. Evol.* 23: 602–609. [21]

Colosimo, P. F., and 9 others. 2005. Widespread parallel evolution in sticklebacks by repeated fixation of *Ectodysplasin* alleles. *Science* 307: 1928–1933. [13, 21]

Coltman, D. W., P. O'Donoghue, J. T. Jorgenson, J. T. Hogg, C. Strobeck, and M. Festa-Blanchet. 2003. Undesirable evolutionary consequences of trophy hunting. *Nature* 426: 655–658. [13]

Conant, R. 1958. *A Field Guide to Reptiles and Amphibians*. Houghton Mifflin, Boston, MA. [9]

Cook, C. D. K. 1968. Phenotypic plasticity with particular reference to three amphibious plant species. In V. Heywood (ed.), *Modern Methods in Plant Taxonomy*, pp. 97–111. Academic Press, London. [13]

Cook, L. G., and M. D. Crisp. 2005. Not so ancient: The extant crown group of *Nothofagus* represents a post-Gondwanan radiation. *Proc. R. Soc. Lond. B* 272: 2535–2544. [6]

Cook, L. M. 2003. The rise and fall of the *carbonaria* form of the peppered moth. *Q. Rev. Biol.* 78: 399–417. [12]

Coope, G. R. 1979. Late Cenozoic fossil Coleoptera: Evolution, biogeography, and ecology. *Annu. Rev. Ecol. Syst.* 10: 249–267. [5, 18]

Coope, G. R. 1995. Insect faunas in ice age environments: Why so little extinction? In J. H. Lawton and R. M. May (eds.), *Extinction Rates*, pp. 55–74. Oxford University Press, Oxford. [22]

Cooper, S. B. J., K. M. Ibrahim, and G. M. Hewitt. 1995. Postglacial expansion and genome subdivision in the European grasshopper *Chorthippus parallelus*. *Mol. Ecol.* 4: 49–60. [6]

Cooper, V. S., and R. E. Lenski. 2000. The population genetics of ecological specialization in evolving *Escherichia coli* populations. *Nature* 407: 736–739. [8]

Cosmides, L., and J. Tooby. 1992. Cognitive adaptations for social exchange. In J. H. Barkow, L. Cosmides, and J. Tooby (eds.), *The Adapted Mind*, pp. 163–228. Oxford University Press, New York. [16]

Coyne, J. A. 1974. The evolutionary origin of hybrid inviability. *Evolution* 28: 505–506. [18]

Coyne, J. A. 1984. Genetic basis of male sterility in hybrids between two closely related species of *Drosophila*. *Proc. Natl. Acad. Sci. USA* 81: 4444–4447. [17]

Coyne, J. A. 2008. *Why Evolution Is True*. Viking, New York. [1, 23]

Coyne, J. A., and H. A. Orr. 1989a. Patterns of speciation in *Drosophila*. *Evolution* 43: 362–381. [17]

Coyne, J. A., and H. A. Orr. 1989b. Two rules of speciation. In D. Otte and J. A. Endler (eds.), *Speciation and Its Consequences*, pp. 180–207. Sinauer, Sunderland, MA. [17]

Coyne, J. A., and H. A. Orr. 1997. "Patterns of speciation in *Drosophila*" revisited. *Evolution* 51: 295–303. [17, 18]

Coyne, J. A., and H. A. Orr. 2004. *Speciation*. Sinauer, Sunderland, MA. [17, 18]

Coyne, J. A., and T. D. Price. 2000. Little evidence for sympatric speciation in island birds. *Evolution* 54: 2166–2171. [18]

Cracraft, J. 1989. Speciation and its ontology: The empirical consequences of alternative species concepts for understanding patterns and processes of differentiation. In D. Otte and J. A. Endler (eds.), *Speciation and Its Consequences*, pp. 29–59. Sinauer, Sunderland, MA. [17]

Cracraft, J. 1991. Patterns of diversification within continental biotas: Hierarchical congruence among the areas of endemism of Australian vertebrates. *Austral. Syst. Bot.* 4: 211–227. [6]

Cracraft, J. 2001. Avian evolution, Gondwana biogeography, and the Cretaceous-Tertiary mass extinction event. *Proc. R. Soc. Lond. B* 268: 459–469. [6]

Crawford, D. C., D. T. Akey, and D. A. Nickerson. 2005. The patterns of natural variation in human genes. *Annu. Rev. Genomics Human Genet.* 6: 287–312. [9]

Creel, S., And P. M. Waser. 1991. Failures of reproductive suppression In dwarf mongooses: Accident or adaptation? *Behav. Ecol.* 2: 7–15. [16]

Crow, J. F. 1993. Mutation, mean fitness, and genetic load. *Oxford Surv. Evol. Biol.* 9: 3–42. [12]

Crow, J. F., and M. Kimura. 1965. Evolution in sexual and asexual populations. *Am. Nat.* 99: 439–450. [15]

Crow, J. F., and M. Kimura. 1970. *An Introduction to Population Genetics Theory*. Harper & Row, New York. [10]

Crozier, R. H., and P. Pamilo. 1996. *Evolution of Social Insect Colonies*. Oxford University Press, Oxford. [16]

Cruzan, M. B., and M. L. Arnold. 1993. Ecological and genetic associations in an *Iris* hybrid zone. *Evolution* 47: 1432–1445. [17]

Cubas, P., C. Vincent, and E. Coen. 1999. An epigenetic mutation responsible for natural variation in floral symmetry. *Nature* 401: 157–161. [9]

Culver, D. C. 1982. *Cave Life: Evolution and Ecology*. Harvard University Press, Cambridge, MA. [14]

Cunningham, C., W.-H. Zhu, and D. M. Hillis. 1998. Best-fit maximum-likelihood models for phylogenetic inference: empirical tests with known phylogenies. *Evolution* 52: 978–987. [2]

Currat, M., and L. Excoffier. 2004. Modern humans did not admix with Neanderthals during their range expansion into Europe. *PLoS Biology* 2:2264–2274. [6]

Currie, C. R., and 8 others. 2003. Ancient tripartite coevolution in the attine ant–microbe symbiosis. *Science* 299: 386–388. [19]

Currie, C. R., M. Poulsen, J. Mendenhall, J. J. Boomsma, and J. Billen. 2006. Coevolved crypts and exocrine glands support mutualistic bacteria in fungus-growing ants. *Science* 311: 81–83. [19]

Curtsinger, J. W., P. M. Service, and T. Prout. 1994. Antagonistic pleiotropy, reversal of dominance, and genetic polymorphism. *Am. Nat.* 144: 210–228. [12]

D

Daan, S., C. Jijkstra, and J. M. Tinbergen. 1990. Family planning in the kestrel (*Falco tinnunculus*): The ultimate control of covariation of laying date and clutch size. *Behaviour* 114: 83–116. [14]

Dacosta, C. P., and C. M. Jones. 1971. Cucumber beetle resistance and mite susceptibility controlled by the bitter gene in *Cucumis sativus*. *Science* 172: 1145–1146. [19]

Daeschler, E. B., N. H. Shubin, and F. A. Jenkins, Jr. 2006. A Devonian tetrapod-like fish and the evolution of the tetrapod body plan. *Nature* 440: 757–763. [4]

Daly, M., and M. Wilson. 1983. *Sex, Evolution, and Behavior*. Willard Grant Press, Boston, MA. [15, 16]

Damuth, J., and I. L. Heisler. 1988. Alternative formulations of multilevel selection. *Biol. Phil.* 3: 407–430. [11, 16]

Danforth, B. N., L. Conway, and S. Ji. 2003. Phylogeny of eusocial *Lasioglossum* reveals multiple losses of eusociality within a primitively eusocial clade of bees (Hymenoptera: Halictidae). *Syst. Biol.* 52: 23–36. [22]

Dao, D. N., R. H. Kessin, and H. L. Ennis. 2000. Developmental cheating and the evolutionary biology of *Dictyostelium* and *Myxococcus*. *Microbiology* 146: 1505–1512. [16]

Darlington, C. D. 1939. *The Evolution of Genetic Systems*. Cambridge University Press, Cambridge. [18]

Darwin, C. 1854. *A Monograph of the Sub-class Cirripedia, with Figures of All the Species.* The Ray Society, London. [4]

Darwin, C. 1859. *The Origin of Species by Means of Natural Selection, or the Preservation of Favored Races in the Struggle for Life.* Modern Library, New York. [1, 2, 22]

Davies, N. B. 1991. Mating systems. In J. R. Krebs and N. B. Davies (eds.), *Behavioural Ecology: An Ecological Approach*, third edition, pp. 263–294. Blackwell Scientific, Oxford. [16]

Davies, N. B. 1992. *Dunnock Behaviour and Social Evolution*. Oxford University Press, Oxford. [16]

Davies, N. B., and M. de L. Brooke. 1998. Cuckoos versus hosts: Experimental evidence for coevolution. In S. I. Rothstein and S. K. Robinson (eds.), *Parasitic Birds and Their Hosts: Studies in Coevolution*, pp. 59–79. Oxford University Press, New York. [19]

Davis, C. C., and K. J. Wurdack. 2004. Host-to-parasite gene transfer in flowering plants: phylogenetic evidence from Malpighiales. *Science* 305: 676–678. [2]

Dawkins, R. 1986. *The Blind Watchmaker*. W. W. Norton, New York. [11]

Dawkins, R. 1989. *The Selfish Gene.* Revised edition. Oxford University Press, Oxford. [11, 16]

Dawkins, R., and J. R. Krebs. 1979. Arms races between and within species. *Proc. R. Soc. London B* 205: 489–511. [19]

Dayan, T., D. Simberloff, E. Tchernov, and Y. Yom-Tov. 1990. Feline canines: Community-wide character displacement among the small cats of Israel. *Am. Nat.* 136: 39–60. [19]

de Jong, G. 2005. Evolution of phenotypic plasticity: Patterns of plasticity and the emergence of ecotypes. *New Phytol.* 166: 101–118. [13]

de Muizon, C. 2001. Walking with whales. *Nature* 413: 259–161. [4]

de Pontbriand, A., X.-P. Wang, Y. Cavaloc, M.-G. Mattei, and F. Galibert. 2002. Synteny comparison between apes and human using fine-mapping of the genome. *Genomics* 80: 395–401. [8]

de Queiroz, A. 2005. The resurrection of oceanic dispersal in historical biogeography. *Trends Ecol. Evol.* 9: 68–73. [6]

de Queiroz, K., and M. J. Donoghue. 1990. Phylogenetic systematics or Nelson's version of cladistics? *Cladistics* 6: 61–75. [17]

de Waal, F. 2007. *Chimpanzee Politics: Power and Sex among Apes.* Johns Hopkins University Press, Baltimore. [23]

Dean, A. M., D. E. Dykhuizen, and D. L. Hartl. 1986. Fitness as a function of β-galactosidase activity in *Escherichia coli*. *Genet. Res.* 48: 1–8. [11]

Dean, M., M. Carrington, and S. J. O'Brien. 2002. Balanced polymorphism selected by genetic versus infectious human disease. *Annu. Rev. Genomics Hum. Genet.* 3: 263–292. [23]

Deban, S. M., D. B. Wake, and G. Roth. 1997. Salamander with a ballistic tongue. *Nature* 389: 27–28. [22]

Decaestecker, E., A. Vergote, D. Ebert, and L. De Meester. 2003. Evidence for strong host clone–parasite species interactions in the *Daphnia* microparasite system. *Evolution* 57: 784–792. [19]

Decaestecker, E., S. Gaba, J. A. M. Raeymaekers, R. Stoks, L. Van Kerckhoven, D. Ebert, and L. De Meester. 2007. Host–parasite 'Red Queen' dynamics archived in pond sediment. *Nature* 450: 870–873. [19]

Deininger, P. L., and M. A. Batzer. 2002. Mammalian retroelements. *Genome Res* 12: 1455–65. [20]

Delson, E., I. Tattersall, J. A. Van Couvering, and A. S. Brooks (eds.) 2000. *Encyclopedia of Human Evolution and Prehistory*, 2nd ed. Garland Press, New York. [2]

Dennett, D. C. 1995. *Darwin's Dangerous Idea: Evolution and the Meanings of Life.* Simon & Schuster, New York. [11, 22]

Dennett, D. C. 2006. *Breaking the Spell: Religion as a Natural Phenomenon.* Viking, New York. [16]

Depew, D. J. and B. H. Weber. 1994. *Darwinism Evolving: Systems Dynamics and the Genealogy of Natural Selection.* MIT Press, Cambridge, MA. [21]

Desmond, A., and J. Moore. 1991. *Darwin*. Warner Books, New York. [1]

Dessain, S., C. T. Gross, M. A. Kuziora, and W. McGinnis. 1992. *Antp*-type homeodomains have distinct DNA-binding specificities that correlate with their different regulatory functions in embryos. *EMBO J.* 11: 991–1002. [21]

Diamond, J. M. 1975. Assembly of species communities. In M. L. Cody and J. M. Diamond (eds.), *Ecology and Evolution of Communities*, pp. 342–444. Harvard University Press, Cambridge, MA. [6]

Diamond, J. M. 1997. *Guns, Germs, and Steel: The Fates of Human Societies.* Norton, New York. [23]

Dickinson, W. J. 2008. Synergistic fitness interactions and a high frequency of beneficial changes among mutations accumulated under relaxed selection in *Saccharomyces cerevisiae*. *Genetics* 178: 1571–1578. [8]

Dieckmann, U., and M. Doebeli. 1999. On the origin of species by sympatric speciation. *Nature* 400: 354–357. [18]

Diego, F., E. A. Herniou, C. Boschetti, M. Caprioli, G. Melone, C. Ricci, and T. G. Barraclough. 2007. Independently evolving species in asexual bdelloid rotifers. *PLoS Biology* 5(4): e87 [15]

Dilda, C. L., and T. F. C. Mackay. 2002. The genetic architecture of *Drosophila* sensory bristle number. *Genetics* 162: 1655–1674. [13]

Dobzhansky, Th. 1934. Studies on hybrid sterility. I. Spermatogenesis in pure and hybrid *Drosophila pseudoobscura*. *Z. Zellforsch. Mikrosk. Anat.* 21: 169–221. [17]

Dobzhansky, Th. 1936. Studies on hybrid sterility. II. Localization of sterility factors in *Drosophila pseudoobscura* hybrids. *Genetics* 21: 113–135. [17, 18]

Dobzhansky, Th. 1937. *Genetics and the Origin of Species*. Columbia University Press, New York. [1, 17]

Dobzhansky, Th. 1948. Genetics of natural populations. XVIII. Experiments on chromosomes of *Drosophila pseudoobscura* from different geographic regions. *Genetics* 33: 588–602. [11]

Dobzhansky, Th. 1951. *Genetics and the Origin of Species*, third edition. Columbia University Press, New York. [17, 18]

Dobzhansky, Th. 1955. A review of some fundamental concepts and problems of population genetics. *Cold Spring Harbor Symp. Quant. Biol.* 20: 1–15. [17]

Dobzhansky, Th. 1956. What is an adaptive trait? *Am. Nat.* 90: 337–347. [13]

Dobzhansky, Th. 1970. *Genetics of the Evolutionary Process*. Columbia University Press, New York. [8, 9]

Dobzhansky, Th., and O. Pavlovsky. 1953. Indeterminate outcome of certain experiments on *Drosophila* populations. *Evolution* 7: 198–210. [11]

Dobzhansky, Th., and B. Spassky. 1969. Artificial and natural selection for two behavioral traits in *Drosophila pseudoobscura*. *Proc. Natl. Acad. Sci. USA* 62: 75–80. [9, 13]

Donoghue, M. J. 1992. Homology. In E. F. Keller and E. A. Lloyd (eds.), *Keywords in Evolutionary Biology*, pp. 170–179. Harvard University Press, Cambridge, MA. [21]

Donoghue, M. J., and S. A. Smith. 2004. Patterns in the assembly of temperate forests around the Northern Hemisphere. *Phil. Trans. R. Soc. Lond. B* 359: 1633–1644. [6]

Donoghue, M. J., J. A. Doyle, J. Gauthier, and A. G. Kluge. 1989. The importance of fossils in phylogeny reconstruction. *Annu. Rev. Ecol. Syst.* 20: 431–460. [4]

Donoghue, P. C. J., and M. J. Benton. 2007. Rocks and clocks: calibrating the tree of life using fossils and molecules. *Trends Ecol. Evol.* 22: 424–431. [2]

Doolittle, W. F. 1999. Phylogenetic classification and the universal tree. *Science* 284: 2124–9. [20]

Draghi, J., and G. P. Wagner. 2008. Evolution of evolvability in a developmental model. *Evolution* 62: 301–315. [22]

Drake, J. W., B. Charlesworth, D. Charlesworth, and J. F. Crow. 1998. Rates of spontaneous mutation. *Genetics* 148: 1667–1686. [8]

Drummond, D. A., and C. O. Wilke. 2008. Mistranslation-induced protein misfolding as a dominant constraint on coding-sequence evolution. *Cell* 134: 341-352. [20]

Duberman, M. B., M. Vicinus, and G. Chauncey (eds.). 1989. *Hidden from History: Reclaiming the Gay and Lesbian Past*. Penguin, New York. [16]

Dubrova, Y. E., V. N. Nesterov, N. G. Krouchinsky, V. A. Ostapenko, R. Neumann, D. L. Neil, and A. J. Jeffreys. 1996. Human minisatellite mutation rate after the Chernobyl accident. *Nature* 380: 683–686. [8]

Dudley, R., G. Byrnes, S. P. Yanoviak, B. Borrell, R. M. Brown, and J. A. McGuire. 2007. Gliding and the functional origins of flight: Biomechanical novelty or necessity? *Annu. Rev. Ecol. Evol. Syst.* 38: 179–201. [22]

Dujon, B., and 66 others. 2004. Genome evolution in yeasts. *Nature* 430: 35–44. [20]

Dybdahl, M. F., and C. M. Lively. 1998. Host-parasite coevolution: Evidence for rare advantage and time-lagged selection in a natural population. *Evolution* 52: 1057–1066. [19]

Dyer, L. A., and 12 others. 2007. Host specificity of Lepidoptera in tropical and temperate forests. *Nature* 448: 696–699. [7]

Dykhuizen, D. E. 1990. Experimental studies of natural selection in bacteria. *Annu. Rev. Ecol. Syst.* 21: 393–398. [8]

E

Easteal, S., and C. Collett. 1994. Consistent variation in amino-acid substitution rate, despite uniformity of mutation rate: Protein evolution in mammals is not neutral. *Mol. Biol. Evol.* 11: 643–647. [10]

Eberhard, W. G. 1996. *Female Control: Sexual Selection by Cryptic Female Choice*. Princeton University Press, Princeton, NJ. [17]

Ebert, D. 1994. Virulence and local adaptation of a horizontally transmitted parasite. *Science* 265: 1084–1086. [19]

Edwards, S. V. 2009. Is a new and general theory of molecular systematics emerging? *Evolution* 63: 1–19 [2]

Ehrlich, P. R., and P. H. Raven. 1964. Butterflies and plants: A study in coevolution. *Evolution* 18: 586–608. [7, 19]

Eicher, D. L. 1976. *Geologic Time*. Prentice-Hall, Englewood Cliffs, NJ. [4]

Eldredge, N. 1989. *Macroevolutionary Patterns and Evolutionary Dynamics: Species, Niches and Adaptive Peaks*. McGraw-Hill, New York. [22]

Eldredge, N., and S. J. Gould. 1972. Punctuated equilibria: An alternative to phyletic gradualism. In T. J. M. Schopf (ed.), *Models in Paleobiology*, pp. 82–115. Freeman, Cooper and Co., San Francisco. [4, 11, 18, 22]

Eldredge, N., and 9 others. 2005. The dynamics of evolutionary stasis. *Paleobiology* 31: 133–145. [18, 22]

Elena, S. F., and R. E. Lenski. 2003. *Evolution* experiments with microorganisms: The dynamics and genetic bases of adaptation. *Nature Rev. Genetics* 4: 457–469. [8]

Elena, S. F., and R. Sanjuán. 2007. Virus evolution: Insights from an experimental approach. *Annu. Rev. Ecol. Evol. Syst.* 38: 27–52. [8, 15]

Ellegren, H., G. Lindgren, C. R. Primmer, and A. P. Møller. 1997. Fitness loss and germline mutations in barn swallows breeding in Chernobyl. *Nature* 389: 593–596. [8]

Ellstrand, N. C. 2003. Current knowledge of gene flow in plants: Implications for transgenic flow. *Phil. Trans. R. Soc. B* 358: 1163–1170 [23]

Ellstrand, N. C., R. Whitkus, and L. H. Rieseberg. 1996. Distribution of spontaneous plant hybrids. *Proc. Natl. Acad. Sci.* 93: 5090–5093. [17]

Ellstrand, N., and J. Antonovics. 1984. Experimental studies of the evolutionary significance of sexual reproduction. I. A test of the frequency-dependent selection hypothesis. *Evolution* 38: 103–115. [12]

Emlen, D. J. 2000. Integrating development with evolution: a case study with beetle horns. *BioScience* 50: 403–418. [9]

Emlen, D. J. 2008. The evolution of animal weapons. *Annu. Rev. Ecol. Evol. Syst.* 39: 387–413. [15]

Emlen, D. J., and H. F. Nijhout. 1999. Hormonal control of male horn length dimorphism in the dung beetle *Onthophagus taurus* (Coleoptera: Scarabaeidae). *J. Insect Physiol.* 45: 45–53. [21]

Emlen, S. T. 1991. Evolution of cooperative breeding in birds and mammals. In J. R. Krebs and N. B. Davies (eds.), *Behavioural Ecology*, third edition, pp. 301–337. Blackwell Scientific, Oxford. [16]

Enard, W., M. Przeworski, S. E. Fisher, C. S. Lai, V. Wiebe, T. Kitano, A. P. Monaco, and S. Pääbo. 2002. Molecular evolution of *FOXP2*, a gene involved in speech and language. *Nature* 418: 869–872. [8]

Endler, J. A. 1973. Gene flow and population differentiation. *Science* 179: 243–250. [12]

Endler, J. A. 1977. *Geographic Variation, Speciation, and Clines*. Princeton University Press, Princeton, NJ. [17, 18]

Endler, J. A. 1980. Natural selection on color patterns in *Poecilia reticulata*. *Evolution* 34: 76–91. [11]

Endler, J. A. 1983. Testing causal hypotheses in the study of geographic variation. In J. Felsenstein (ed.), *Numerical Taxonomy*, pp. 424–443. Springer-Verlag, Berlin. [6]

Endler, J. A. 1986. *Natural Selection in the Wild*. Princeton University Press, Princeton, NJ. [11, 12, 13]

Erwin, D. H. 1993. *The Great Paleozoic Crisis: Life and Death in the Early Permian*. Columbia University Press, New York. [7]

Erwin, D. H. 2006. *Extinction: How Life on Earth Nearly Ended 250 Million Years Ago*. Princeton University Press, Princeton, NJ. [5, 7]

Estes, S., and S. J. Arnold. 2007. Resolving the paradox of stasis: Models with stabilizing selection explain evolutionary divergence on all timescales. *Am. Nat.* 169: 227–244. [22]

Estes, S., B. C. Ajie, M. Lynch, and P. C. Phillips. 2005. Spontaneous mutational correlations for life-history, morphological and behavioral characters in *Caenorhabditis elegans*. *Genetics* 170:645–653. [8, 13]

Etterson, J. R. 2004. Evolutionary potential of *Chamaecrista fasciculata* in relation to climate change. II. Genetic architecture of three populations reciprocally planted along an environmental gradient in the Great Plains. *Evolution* 58: 1549–1471. [13]

Etterson, J. R., and R. G. Shaw. 2001. Constraint to adaptive evolution in response to global warming. *Science* 294: 151–154. [13]

Ewald, P. W. 1994. *Evolution of Infectious Disease*. Oxford University Press, Oxford. [19]

Eyre-Walker, A., and P. D. Keightley. 1999. High genomic deleterious mutation rates in hominids. *Nature* 397: 344–347. [8]

F

Falconer, D. S., and T. F. C. Mackay. 1996. *Introduction to Quantitative Genetics*, 4th ed. Longman, Harlow, U.K. [13]

Farrell, B. D. 1998. "Inordinate fondness" explained: Why are there so many beetles? *Science* 281: 555–559. [7, 19]

Farrell, B. D., and A. S. Sequeira. 2004. Evolutionary rates in the adaptive radiation of beetles on plants. *Evolution* 58: 1984–2001. [7]

Farrell, B., D. Dussourd, and C. Mitter. 1991. Escalation of plant defenses: Do latex and resin canals spur plant diversification? *Am. Nat.* 138: 881–900. [7]

Fay, J. C., and C.-I. Wu. 2003. Sequence divergence, functional constraint, and selection in protein evolution. *Annu. Rev. Genomics Hum. Genet.* 4: 213–235. [12]

Feder, J. L., C. A. Chilcote, and G. L. Bush. 1990. Geographic pattern of genetic differentiation between host-associated populations of *Rhagoletis pomonella* (Diptera: Tephritidae) in the eastern United States and Canada. *Evolution* 44: 570–594. [18]

Feder, J. L., and 8 others. 2005. Mayr, Dobzhansky and Bush and the complexities of sympatric speciation in *Rhagoletis*. *Proc. Natl. Acad. Sci. USA* 102 (Suppl. 1): 6573–6580. [18]

Feder, J. L., and A. A. Forbes. 2008. Host fruit-odor discrimination and sympatric host-race formation. pp. 101–116. In K. J. Tilmon (ed.), *Specialization, Speciation, and Radiation: The Evolutionary Biology of Herbivorous Insects*. University of California Press, Berkeley. [18]

Fedigan, L. M. 1986. The changing role of women in models of human evolution. *Annu. Rev. Anthropol.* 15: 25–66. [4]

Feduccia, A. 1999. *The Origin and Evolution of Birds*, 2nd ed. Yale University Press, New Haven, CT. [22]

Fehr, E., and U. Fischbacher. 2003. The nature of human altruism. *Nature* 425: 785–791. [23]

Felsenstein, J. 1976. The theoretical population genetics of variable selection and migration. *Annu. Rev. Genet.* 10: 253–280. [12]

Felsenstein, J. 1981. Skepticism towards Santa Rosalia, or why are there so few kinds of animals? *Evolution* 35: 124–138. [18]

Felsenstein, J. 1985. Phylogenies and the comparative method. *Am. Nat.* 125: 1–15. [11]

Felsenstein, J. 2004. *Inferring Phylogenies*. Sinauer, Sunderland, MA. [2]

Fenner, F., and F. N. Ratcliffe. 1965. *Myxomatosis*. Cambridge University Press, Cambridge. [19]

ffrench-Constant, R. H., R. T. Roush, D. Mortlock, and G. P. Dively. 1990. Isolation of dieldrin resistance from field populations of *Drosophila melanogaster* (Diptera: Drosophilidae). *J. Econ. Entomol.* 83: 1733–1737. [8]

Field, J., B. Rosenthal, and J. Samuelson J. 2000. Early lateral transfer of genes encoding malic enzyme, acetyl-CoA synthetase and alcohol dehydrogenases from anaerobic prokaryotes to *Entamoeba histolytica*. *Molec. Microbiol.* 38: 446–455. [20]

Field, J., G. Shreeves, S. Summer, and M. Casiraghi. 2000. Insurance-based advantage to helpers in a tropical hover wasp. *Nature* 404: 869–871. [16]

Fine, P. V. A., and R. H. Ree. 2006. Evidence for a time-integrated species-area effect on the latitudinal gradient in species diversity. *Am. Nat.* 168: 796–804. [6]

Finlayson, C. 2005. Biogeography and evolution of the genus *Homo*. *Trends Ecol. Evol.* 20: 457–463. [6]

Fischer, C. S., M. Hout, M. S. Jankowski, S. R. Lucas, A. Swidler, and K. Voss. 1996. *Inequality by Design: Cracking the Bell Curve Myth*. Princeton University Press, Princeton, NJ. [9]

Fisher, D. C. 1986. Progress in organismal design. In D. M. Raup and D. Jablonski (eds.), *Patterns and Processes in the History of Life*, pp. 99–117. Springer-Verlag, Berlin. [22]

Fisher, R. A. 1930. *The genetical theory of natural selection*. Clarendon Press, Oxford. [12, 15, 18]

Fitzpatrick, B. M., J. A. Fordycce, and S. Gavrilets. 2008. What, if anything, is sympatric speciation? *J. Evol. Biol.* 21: 1452–1459. [18]

Fleischer, R. C., C. E. McIntosh, and C. L. Tarr. 1998. Evolution on a volcanic conveyor belt: Using phylogeographic reconstructions and K-Ar-based ages of the Hawaiian Islands to estimate molecular evolutionary rates. *Mol. Ecol.* 7: 533–545. [2]

Foote, M. 1988. Survivorship analysis of Cambrian and Ordovician trilobites. *Paleobiology* 14: 258–271. [7]

Foote, M. 1997. The evolution of morphological diversity. *Annu. Rev. Ecol. Syst.* 28: 129–152. [22]

Foote, M. 2000a. Origination and extinction components of diversity: general problems. In D. H. Erwin and S. L. Wing (eds.), *Deep Time: Paleobiology's Perspective*, pp. 74–102. *Paleobiology* 26 (4), supplement. [7]

Foote, M. 2000b. Origination and extinction components of taxonomic diversity: Paleozoic and post-Paleozoic dynamics. *Paleobiology* 26: 578–605. [7]

Foote, M., and A. I. Miller. 2007. *Principles of Paleontology*, 3rd ed. W. H. Freeman, New York. [7]

Force, A., M. Lynch, F. B. Pickett, A. Amores A., Yan, and J.Postlethwait. 1999. Preservation of duplicate genes by complementary, degenerative mutations. *Genetics* 151: 1531–1545. [20]

Ford, E. B. 1971. *Ecological Genetics*. Chapman & Hall, London. [9, 22]

Fornoni, J., P. L. Valverde, and J. Núñez-Farfán. 2003. Quantitative genetics of plant tolerance and resistance against natural enemies of two natural populations of *Datura stramonium*. *Evol. Ecol. Res.* 5: 1049–1065. [13]

Forrest, B., and P. R. Gross. 2004. *Creationism's Trojan Horse: The Wedge of Intelligent Design*. Oxford University Press, New York. [23]

Forster, L. M. 1992. The stereotyped behaviour of sexual cannibalism in *Latrodectus hasselti* Thorell (Araneae: Theridiidae), the Australian redback spider. *Aust. J. Zool.* 40: 1–11. [11]

Foster, K. R., T. Wenseleers, and F. L. W. Ratnieks. 2006. Kin selection is the key to altruism. *Trends Ecol. Evol.* 21: 57–60. [16]

Foster, P. L. 2000. Adaptive mutation: implications for evolution. *BioEssays* 22: 1067–1074. [8]

Fowler, K., and L. Partridge. 1989. A cost of mating in female fruitflies. *Nature* 338: 760–761. [14]

Fowler, N. L., and D. A. Levin. 1984. Ecological constraints on the establishment of a novel polyploid in competition with its diploid progenitor. *Am. Nat.* 124: 703–711. [18]

Fox, C. W. and J. B. Wolf (eds.). *Evolutionary Genetics*. Oxford University Press, New York. [8]

Fraenkel, G. 1959. The *raison d'être* of secondary plant substances. *Science* 129: 1466–1470. [19]

Frank, S. A. 1990. Sex allocation theory for birds and mammals. *Annu. Rev. Ecol. Syst.* 21: 13–55. [15, 17]

Frank, S. A. 1992. Models of plant-pathogen coevolution. *Trends Genet.* 8: 213–219. [19]

Frank, S. A. 1996. Models of parasite virulence. *Q. Rev. Biol.* 71: 37–78. [19]

Frank, S. A. 1997. Models of symbiosis. *Am. Nat.* 150: S80–S99. [16]

Frank, S. A. 1998. *Foundations of Social Evolution*. Princeton University Press, Princeton, NJ. [16]

Frank, S. A. 2003. Perspective: Repression of competition and the evolution of cooperation. *Evolution* 57: 693–705. [16]

Frankham, R. 1995. Effective population-size : adult population-size ratios in wildlife—A review. *Genet. Res.* 66: 95–107. [10]

Frankham, R., J. D. Ballou, and D. A. Briscoe. 2002. *Introduction to Conservation Genetics*. Cambridge University Press, Cambridge. [9, 10, 23]

Fraser, S. (ed.). 1995. *The Bell Curve Wars: Race, Intelligence, and the Future of America*. Basic Books, New York. [9]

Friesen, V. L., A. L. Smith, E. Gómez-Díaz, M. Bolton, R. W. Furness, J. González-Solís, and L. R. Monteiro. 2007. Sympatric speciation by allochrony in a seabird. *Proc. Natl. Acad. Sci. USA* 104: 18589–18594. [18]

Fry, J. D. 2003. Multilocus models of sympatric speciation: Bush vs. Rice vs. Felsenstein. *Evolution* 57: 1735–1746. [18]

Fryer, G., and T. D. Iles. 1972. *The Cichlid Fishes of the Great Lakes of Africa*. T.F.H. Publications, Neptune City, NJ. [3]

Funk, D. J. 1998. Isolating a role for natural selection in speciation: Host adaptation and sexual isolation in *Neochlamisus bebbianae* leaf beetles. *Evolution* 52: 1744–1759. [18]

Funk, D. J., and K. E. Omland. 2003. Species-level paraphyly and polyphyly: Frequency, causes, and consequences, with insights from mitochondrial DNA. *Annu. Rev. Ecol. Syst.* 34: 397–423. [17]

Funk, D. J., P. Nosil, and W. J. Etges. 2006. Ecological divergence exhibits consistently positive associations with reproductive isolation across disparate taxa. *Proc. Natl. Acad. Sci. USA* 103: 3209–3213. [18]

Fürsich, F. T., and D. Jablonski. 1984. Lake Triassic naticid drillholes: Carnivorous gastropods gain a major adaptation but fail to radiate. *Science* 224: 78–80. [7]

Futuyma, D. J. 1987. On the role of species in anagenesis. *Am. Nat.* 130: 465–473. [18, 22]

Futuyma, D. J. 1991. A new species of *Ophraella* Wilcox (Coleoptera: Chrysomelidae) from the southeastern United States. *J. NY Entomol. Soc.* 99: 643–653. [17]

Futuyma, D. J. 1995. *Science on Trial: The Case for Evolution*. Sinauer, Sunderland, MA. [3, 4]

Futuyma, D. J. 2000. Some current approaches to the evolution of plant-herbivore interactions. *Plant Species Biol.* 15: 1–9. [23]

Futuyma, D. J. 2004. The fruit of the tree of life: Insights into evolution and ecology. In J. Cracraft and M. J. Donoghue (eds.), *Assembling the Tree of Life*, pp. 25–39. Oxford University Press, New York. [3]

Futuyma, D. J. 2008. Sympatric speciation: Norm or exception? pp. 1136–148. In K. J. Tilmon (ed.), *Specialization, Speciation, and Radiation: The Evolutionary Biology of Herbivorous Insects*. University of California Press, Berkeley. [18]

Futuyma, D. J., and M. Slatkin (eds.). 1983. *Coevolution*. Sinauer, Sunderland, MA. [19]

Futuyma, D. J., M. C. Keese, and D. J. Funk. 1995. Genetic constraints on macroevolution: The evolution of host affiliation in the leaf beetle genus *Ophraella*. *Evolution* 49: 797–809. [13]

G

Galindo, B. E., V. D. Vacquier, and W. J. Swanson. 2003. Positive selection in the egg receptor for abalone sperm lysin. *Proc. Natl. Acad. Sci. USA* 100: 4639–4643. [17]

Galis, F. 1996. The application of functional morphology to evolutionary studies. *Trends Ecol. Evol.* 11: 124–129. [22]

Gao, F., E. Bailes, D. L. Robertson et al. 1999. Origin of HIV-1 in the chimpanzee *Pan troglodytes troglodytes*. *Nature* 397: 436–441. [1]

Gao, L. Z., and H. Innan. 2004. Very low gene duplication rate in the yeast genome. *Science* 306: 1367–1370. [20]

Garland, T., Jr., R. B. Huey, and A. F. Bennett. 1991. Phylogeny and coadaptation of themal physiology in lizards: A reanalysis. *Evolution* 45: 1969–1975. [11]

Garrigan, D., and M. F. Hammer. 2006. Reconstructing human origins in the genomic era. *Nature Rev. Genet.* 7: 669–680. [6]

Garrigan, D., and 11 others. 2007. Inferring human population sizes, divergence times and rates of gene flow from mitochondrial, X and Y chromosome resequencing data. *Genetics* 177: 2195–2207. [10]

Gaston, K. J., and T. M. Blackburn. 2000. *Pattern and Process in Macroecology*. Blackwell Science, Oxford. [7]

Gatesy, J., M. Milinkovitch, V. Waddell, and M. Stanhope. 1999. Stability of cladistic relationships between Cetacea and higher-level artiodactyls taxa. *Syst. Biol.* 48: 6–20. [4]

Gavrilets, S. 2004. *Fitness Landscapes and the Origin of Species*. Princeton University Press, Princeton, NJ. [18]

Gavrilets, S., and A. Hastings. 1996. Founder effect speciation: A theoretical reassessment. *Am. Nat.* 147: 466–491. [18]

Gavrilets, S., and W. R. Rice. 2006. Genetic models of homosexuality: Generating testable predictions. *Proc. R. Soc. Lond. B* 273: 3031–3038. [16]

Gehring, W. J., and K. Ikeo. 1999. *Pax6* mastering eye morphogenesis and evolution. *Trends Genet.* 15: 371–377. [8]

Gentles, A. J., and S. Karlin. 1999. Why are human G-protein-coupled receptors predominantly intronless? *Trends Genet.* 15: 47–49. [20]

Gerhart, J., and M. Kirschner. 2007. The theory of facilitated variation. *Proc. Natl. Acad. Sci. USA* 104 (Suppl. 1): 8582–8589. [22]

Gerrish, P. J., and R. E. Lenski. 1998. The fate of competing beneficial mutations in an asexual population. *Genetica* 102/103: 127–144. [15]

Ghalambor, C. K., J. K. McKay, S. P. Carroll, and D. N. Reznick. 2007. Adaptive versus non-adaptive phenotypic plasticity and the potential for contemporary adaptation in new environments. *Functional Ecol.* 21: 394–407. [13]

Ghildiyal, M., and P. D. Zamore. 2009. Small silencing RNAs: An expanding universe. *Nature Rev. Genetics* 10: 94–108. [20]

Ghiselin, M. T. 1969. *The Triumph of the Darwinian Method*. University of California Press, Berkeley. [11]

Ghiselin, M. T. 1995. Perspective: Darwin, progress, and economic principles. *Evolution* 49: 1029–1037. [22]

Gibbs, M. J., J. S. Armstrong, and A. J. Gibbs. 2001. The haemagglutinin gene, but not the neuraminidase gene, of 'Spanish flu' was a recombinant. *Phil. Trans. R. Soc. Lond. B* 356: 1845–1855. [19]

Gilad, Y., O. Man, S. Pääbo, and D. Lancet. 2003. Human-specific loss of olfactory receptor genes. *Proc. Natl. Acad. Sci. USA* 100: 3324–3327. [20]

Gilbert, W. 1987. The exon theory of genes. *Cold Spring Harbor Symp. Quant. Biol.* 52: 901–905. [20]

Gilinsky, N. L. 1994. Volatility and the Phanerozoic decline of background extinction. *Paleobiology* 20: 424–444. [7]

Gilinsky, N. L., and R. K. Bambach. 1987. Asymmetrical patterns of origination and extinction in higher taxa. *Paleobiology* 13: 427–445. [7]

Gill, F. B. 1995. *Ornithology*, 2nd ed. W. H. Freeman, New York. [11]

Gilles, A., and L. F. Randolph. 1951. Reduction in quadrivalent frequency in autotetraploid maize during a period of 10 years. *Am. J. Bot.* 38: 12–16. [18]

Gingerich, P. D. 1983. Rates of evolution: Effects of time and temporal scaling. *Science* 222:159–161. [4]

Gingerich, P. D. 1993. Quantification and comparison of evolutionary rates. *Am. J. Sci.* 293A: 453–478. [4]

Gingerich, P. D. 2001. Rates of evolution on the time scale of the evolutionary process. *Genetica* 112/113: 127–144. [4, 22]

Gingerich, P. D. 2003. Land-to-sea transition of early whales: Evolution of Eocene Archaeoceti (Cetacea) in relation to skeletal proportions and locomotion of living semiaquatic mammals. *Paleobiology* 29: 429–454. [4]

Gingerich, P. D., M. ul Haq, I. S. Zalmout, I. H. Khan, and M. S, Malkani. 2001. Origin of whales from early artiodactyls: Hands and feet of Eocene Protocetidae from Pakistan. *Science* 293: 2239–2242. [4]

Godfray, H. C. J. 1999. Parent–offspring conflict. In L. Keller (ed.), *Levels of Selection in Evolution*, pp. 100–120. Princeton University Press, Princeton, NJ. [16]

Goldblatt, P. (ed.). 1993. *Biological Relationships Between Africa and South America*. Yale University Press, New Haven, CT. [6]

Goldschmidt, R. B. 1940. *The Material Basis of Evolution*. Yale University Press, New Haven, CT. [17, 22]

Goldstein, D. B., A. Ruiz-Linares, L. L. Cavalli–Sforza, and M. W. Feldman. 1995. Genetic absolute dating based on microsatellites and the origin of modern humans. *Proc. Natl. Acad. Sci. USA* 92: 6723–6727. [10]

Goldstein, D. B., and K. E. Holsinger. 1992. Maintenance of polygenic variation in spatially structured populations: Roles for local mating and genetic redundancy. *Evolution* 46: 412–429. [13]

Goliber, T., S. Kessler, J.-J. Chen, G. Bharathan, and N. Sinha. 1999. Genetic, molecular, and morphological analysis of compound leaf development. *Curr. Topics Dev. Biol.* 43: 260–290. [21]

Gómez-Zurita, J., T. Hunt, F. Kopliku, and A. P. Vogler. 2007. Recalibrated tree of leaf beetles (Chrysomelidae) indicates independent diversification of angiosperms and their insect herbivores. *PLoS ONE* 2(4): e360. doi:10.1371/journal.pone.0000360.

Gomez, C., E. M. Özbudak, J. Wunderlich, D. Baumann, J. Lewis, and O. Pourquié. 2008. Control of segment number in vertebrate embryos. *Nature* 454: 335–339. [21]

Gompel, N., and S. B. Carroll. 2003. Genetic mechanisms and constraints governing the evolution of correlated traits in drosophilid flies. *Nature* 424: 931–935. [21]

Goodman, M., B. F. Koop, J. Czelusniak, D. H. A. Fitch, D. A. Tagle, and J. L. Slightom. 1989. Molecular phylogeny of the family of apes and humans. *Genome* 31: 316–335. [2]

Goodwin, D. 1986. *Crows of the World*. British Museum (Natural History), London. [17]

Gorb, S. N., S. Niederegger, C. Y. Hayashi, A. P. Summers, W. Votsch, and P. Walther. 2006. Biomaterials: Silk-like secretion from tarantula feet. *Nature* 443: 407. [21]

Gould, F. 1998. Sustainability of transgenic insecticidal cultivars: Integrating pest genetics and ecology. *Annu. Rev. Entomol.* 443: 701–726. [23]

Gould, F. 2008. Broadening the application of evolutionarily based genetic pest management. *Evolution* 62: 500–510. [23]

Gould, F., N. Blair, M. Reid, T. L. Rennie, J. Lopez, and S. Micinski. 2002. *Bacillus thuringiensis*-toxin resistance management: Stable isotope assessment of alternate host use by *Helicoverpa zea*. *Proc. Natl. Acad. Sci. USA* 99: 16581–16586. [23]

Gould, J. and C. G. Gould. 1989. *Sexual Selection*. Scientific American Library, New York. [15]

Gould, S. J. 1974. The origin and function of "bizarre" structures: Antler size and skull size in the "Irish elk," *Megaloceros giganteus*. *Evolution* 28: 191–220. [3]

Gould, S. J. 1977. *Ontogeny and Phylogeny*. Harvard University Press, Cambridge, MA. [3, 21]

Gould, S. J. 1981. *The Mismeasure of Man*. Norton, New York. [9]

Gould, S. J. 1982. The meaning of punctuated equilibrium and its role in validating a hierarchical approach to macroevolution. In R. Milkman (ed.), *Perspectives on Evolution*, pp. 83–104. Sinauer, Sunderland, MA. [11]

Gould, S. J. 1985. The paradox of the first tier: An agenda for paleobiology. *Paleobiology* 11: 2–12. [7]

Gould, S. J. 1989. *Wonderful Life: The Burgess Shale and the Nature of History*. W. W. Norton, New York. [5, 22]

Gould, S. J. 1999. *Rocks of Ages*. Ballantine, New York. [23]

Gould, S. J. 2002. *The Structure of Evolutionary Theory*. Belknap Press of Harvard University Press, Cambridge, MA. [4, 18, 22, 23]

Gould, S. J. 2007. *Punctuated Equilibrium*. Belknap Press of Harvard University Press, Cambridge, MA. [4, 22]

Gould, S. J., and E. S. Vrba. 1982. Exaptation: A missing term in the science of form. *Paleobiology* 8: 4–15. [11, 21]

Gould, S. J., and N. Eldredge. 1993. Punctuated equilibrium comes of age. *Nature* 366: 223–227. [18]

Gould, S. J., and R. C. Lewontin. 1979. The spandrels of San Marco and the Panglossian paradigm. *Proc. R. Soc. Lond. B* 205: 581–598. [14]

Gradstein, F., J. Ogg, and A. Smith. 2004. *A Geologic Time Scale 2004*. Cambridge University Press, Cambridge. [4]

Grafen, A. 1991. Modelling in behavioural ecology. In J. R. Krebs and N. B. Davies (eds.), *Behavioural Ecology*, 3rd ed., pp. 5–31. Blackwell Scientific, Oxford. [16]

Grafen, A. 2006. Various remarks on Lehmann and Keller's article. *J. Evol. Biol.* 19: 1397–1399. [16]

Grant, B. R., and P. R. Grant. 1989. *Evolutionary Dynamics of a Natural Population: The Large Cactus Finch of the Galápagos*. University of Chicago Press, Chicago. [13]

Grant, B. S. 2002. Sour grapes of wrath. *Science* 297: 940–941. [12]

Grant, B. S., and L. L. Wiseman. 2002. Recent history of melanism in the American peppered moth. *J. Hered.* 93: 86–90. [12]

Grant, P. R. 1986. *Ecology and Evolution of Darwin's Finches*. Princeton University Press, Princeton, NJ. [9, 13, 19]

Grant, P. R., and B. R. Grant. 2006. Evolution of character displacement in Darwin's finches. *Science* 313: 224–226. [19]

Grant, P. R., and B. R. Grant. 2008. *How and Why Species Multiply: The Radiation of Darwin's Finches*. Princeton University Press, Princeton. [3]

Grant, V. 1966. The selective origin of incompatibility barriers in the plant genus *Gilia*. *Am. Nat.* 100: 99–118. [18]

Grant, V. 1981. *Plant Speciation*. Columbia University Press, New York. [17, 18]

Graur, D., and W.-H. Li. 2000. *Fundamentals of Molecular Evolution*, 2nd ed. Sinauer, Sunderland, MA. [10, 20]

Gray, D. A., and W. H. Cade. 2000. Sexual selection and speciation in field crickets. *Proc. Natl. Acad. Sci. USA* 97: 14449–14454. [18]

Green, P. M., J. A. Naylor, and F. Giannelli. 1995. The hemophilias. *Adv. Genet.* 32: 99–139. [8]

Green, R. E., and 10 others. Analysis of one million base pairs of Neanderthal DNA. *Nature* 444: 330–336. [6, 20]

Griffin, A. S., S. A. West, and A. Buckling. 2004. Cooperation and competition in pathogenic bacteria. *Nature* 430: 1024–1027. [16]

Grimaldi, D. A. 1987. Phylogenetics and taxonomy of *Zygothrica* (Diptera: Drosophilidae). *Bull. Am. Mus. Nat. Hist.* 186: 103–268. [3]

Grimaldi, D. A., and M. S. Engel. 2005. *Evolution of the Insects*. Cambridge University Press, New York. [5]

Grosberg, R. K., and R. R. Strathmann. 2007. The evolution of multicellularity: A minor major transition? *Annu. Rev. Ecol. Evol. Syst.* 38: 621–654. [22]

Gross, B. L., and L. H. Rieseberg. 2005. The ecological genetics of homoploid hybrid speciation. *J. Hered.* 96: 241–252. [18]

Gross, M. 1984. Sunfish, salmon, and the evolution of alternative reproductive strategies and tactics in fishes. In G. W. Potts and R. J. Wootton (eds.), *Fish Reproduction: Strategies and Tactics*, pp. 55–75. Academic Press, London. [15]

Gu, X., Y. Wang, and J. Gu. 2002. Age distribution of human gene families shows significant roles of both large- and small-scale duplications in vertebrate evolution. *Nature Genet.* 31: 205–209. [20]

Gupta, A. P., and R. C. Lewontin. 1982. A study of reaction norms in natural populations of *Drosophila pseudoobscura*. *Evolution* 36: 934–948. [9]

Guss, K. A., C. E. Nelson, A. Hudson, M. E. Kraus, and S. B. Carroll. 2001. Control of a genetic regulatory network by a selector gene. *Science* 292: 1164–1167. [21]

Gwynne, D. T. 1984. Sexual selection and sexual differences in Mormon crickets (*Anabrus simplex*). *Evolution* 38: 1011–1022. [15]

Gwynne, D. T. 1993. Food quality controls sexual selection in Mormon crickets by altering male mating investment. *Ecology* 74: 1406–1413. [15]

Gwynne, D. T. 2008. Sexual conflict over nuptial gifts in insects. *Annu. Rev. Entomol.* 53: 83–101. [15]

H

Haag-Liautard, C., M. Dorris, X. Maside, S. Macaskill, D. L. Halligan, B. Charlesworth, and P. D. Keightley. 2007. Direct estimation of *per* nucleotide and genomic deleterious mutation rates in *Drosophila*. *Nature* 445: 82–85. [8]

Haddrath, O., and A. J. Baker. 2001. Complete mitochondrial DNA genome sequences of extinct birds: ratite phylogenetics and the vicariance biogeography hypothesis. *Proc. R. Soc. Lond. B* 268: 939–945. [6]

Hahn, B. H., G. M. Shaw, K. M. De Cock, and P. M. Sharp. 2000. AIDS as a zoonosis: Scientific and public health implications. *Science* 287: 607–614. [3]

Haig, D. 2002. *Genomic Imprinting and Kinship*. Rutgers University Press, New Brunswick, NJ. [16]

Haldane, J. B. S. 1932. *The Causes of Evolution*. Longmans, Green, New York. [12]

Hall, B. G. 1982. Evolution on a petri dish: The evolved β-galactosidase system as a model for studying acquisitive evolution in the laboratory. *Evol. Biol.* 15: 85–150. [8]

Hall, B. G. 2007. *Phylogenetic Trees Made Easy. A How-To Manual for Molecular Biologists*, 3rd ed. Sinauer, Sunderland, MA. [2]

Hall, M. C., C. J. Basten, and J. H. Willis. 2006. Pleiotropic quantitative trait loci contribute to population divergence in traits associated with life history variation in *Mimulus guttatus*. *Genetics* 172: 1829–1844. [8]

Hallström, B. M., M. Kullberg, M. A. Nilsson, and A. Janke. 2007. Phylogenomic data analyses provide evidence that Xenarthra and Afrotheria are sister groups. *Mol. Biol. Evol.* 24: 2059–2068. [2]

Hamilton, W. D. 1964. The genetical evolution of social behavior, I and II. *J. Theor. Biol.* 7: 1–52. [11, 16]

Hamilton, W. D. 1967. Extraordinary sex ratios. *Science* 156: 477–488. [15]

Hamilton, W. D., and M. Zuk. 1982. Heritable true fitness and bright birds: A role for parasites? *Science* 218: 384–387. [15]

Hammer, M. F. 1995. A recent common ancestry for human Y chromosomes. *Nature* 378: 376–378. [10]

Hammerstein, P. 2003. Why is reciprocity so rare in social animals? A protestant appeal. In P. Hammerstein (ed.), *Genetic and Cultural Evolution of Cooperation*, pp. 481–496. MIT Press, Cambridge, MA. [16]

Hanken, J. 1984. Miniaturization and its effects on cranial morphology in plethodontid salamanders, genus *Thorius* (Amphibia: Plethodontidae). I. Osteological variation. *Biol. J. Linn. Soc.* 23: 55–75. [3]

Hansen, T. A. 1980. Influence of larval dispersal and geographic distribution on species longevity in neogastropods. *Paleobiology* 6: 193–207. [22]

Hansen, T. F. 2006. The evolution of genetic architecture. *Annu. Rev. Ecol. Evol. Syst.* 37: 123–157. [13]

Hansen, T. F., and D. Houle. 2004. Evolvability, stabilizing selection, and the problem of stasis. In M. Pigliucci and K. Preston (eds.), *Phenotypic Integration: Studying the Ecology and Evolution of Complex Phenotypes*, pp 131–150. Oxford University Press, Oxford. [13]

Hansen, T. F., and D. Houle. 2008. Measuring and comparing evolvability and constraint in multivariate characters. *J. Evol. Biol.* 21: 1201–1219. [22]

Hansen, T. F., J. M. Álvarez-Castro, A. J. R. Carter, J. Hermisson, and G. P. Wagner. 2006. Evolution of genetic architecture under directional selection. *Evolution* 60: 1523–1536. [22]

Harcourt, A. H., P. H. Harvey, S. G. Larson, and R. V. Short. 1981. Testis weight, body weight and breeding system in primates. *Nature* 293: 55–57. [16]

Harris, H., and D. A. Hopkinson. 1972. Average heterozygosity in man. *J. Hum. Genet.* 36: 9–20. [9]

Harris, M. P., S. M. Hasso, M. W. J. Ferguson, and J. F. Fallon. 2006. The development of archosaurian first-generation teeth in a chicken mutant. *Current Biology* 16: 371–377. [3, 21]

Harrison, R. G. 1979. Speciation in North American field crickets: Evidence from electrophoretic comparisons. *Evolution* 33: 1009–1023. [17]

Harrison, R. G. 1990. Hybrid zones: Windows on evolutionary process. *Oxford Surv. Evol. Biol.* 7: 69–128. Oxford University Press, New York. [17]

Harrison, R. G. (ed.). 1993. *Hybrid Zones and the Evolutionary Process*. Oxford University Press, New York. [17]

Harrison, R. G. 1998. Linking evolutionary pattern and process: the relevance of species concepts for the study of speciation. In D. J. Howard and S. H. Berlocher (eds.), *Endless Forms: Species and Speciation*, pp. 19–31. Oxford University Press, New York. [17]

Hartl, D. L., and A. G. Clark. 1989. *Principles of Population Genetics*, 2nd ed. Sinauer, Sunderland, MA. [10]

Hartl, D. L., and A. G. Clark. 1997. *Principles of Population Genetics*, 3rd ed. Sinauer, Sunderland, MA. [9, 12]

Hartl, D. L., and E. W. Jones. 2001. *Genetics: Analysis of Genes and Genomes*. Jones and Bartlett, Sudbury, MA. [8]

Hartwell, L. H., L. Hood, M. L. Goldberg, A. E. Reynolds, L. M. Silver, and R. C. Veres. 2000. *Genetics: From Genes to Genomes*. McGraw-Hill Higher Education, Boston. [3, 8]

Harvey, P. H., and M. D. Pagel. 1991. *The Comparative Method in Evolutionary Biology*. Oxford University Press, Oxford. [11]

Harvey, P. H., and R. M. May. 1989. Out for the sperm count. *Nature* 337: 508–509. [16]

Hattori, D., E. Demir, H. W. Kim, E. Viragh, S. L. Zipursky, and B. J. Dickson. 2007. *Dscam* diversity is essential for neuronal wiring and self-recognition. *Nature* 449: 223–227. [20]

Hausfater, G., and S. Hrdy (eds.). 1984. *Infanticide: Comparative and Evolutionary Perspectives*. Aldine Publishing Co., New York. [16]

Hausmann, M. F., D. W. Winkler, K. M. O'Reilly, C. E. Huntington, I. C. T. Nisbet, and C. M. Vleck. 2003. Telomeres shorten more slowly in long-lived birds and mammals than in short-lived ones. *Proc. Natl. Acad. Sci. USA* 270: 1387–1392. [14]

Hawks, J., E. T. Wang, G. M. Cochran, H. C. Harpending, and R. K. Moyzis. 2007. Recent acceleration of human adaptive evolution. *Proc. Natl. Acad. Sci. USA* 104: 20753–20758. [12]

Hay, A. J., V. Gregory, A. R. Douglas, and Y. P. Lin. 2001. The evolution of human influenza viruses. *Phil. Trans. R. Soc. Lond. B* 356: 1861–1870. [19]

Hayman, P., J. Marchant, and T. Prater. 1986. *Shorebirds: An Identification Guide to the Waders of the World*. Houghton Mifflin, Boston, MA. [3]

Hazen, R. M. 2005. *Genesis: The Scientific Quest for Life's Origin*. Joseph Henry Press, Washington, D.C. [5]

Head, M. L., J. Hunt, M. D. Jennions, and R. Brooks. 2005. The indirect benefits of mating with attractive males outweigh the direct costs. *PLoS Biology* 3: 0289–0294 (doi: 10.1371/journal.pbio.0030033). [15]

Hedrick, P. W. 1986. Genetic polymorphisms in heterogeneous environments: A decade later. *Annu. Rev. Ecol. Syst.* 17: 535–566. [12]

Hendry, A. P., and M. T. Kinnison. 1999. Perspective: The pace of modern life: Measuring rates of contemporary microevolution. *Evolution* 53: 1637–1653. [4, 13]

Hendry, A. P., P. Nosil, and L. H. Rieseberg. 2007. The speed of ecological speciation. *Func. Ecol.* 21: 455-464. [18]

Hennig, W. 1966. *Phylogenetic Systematics*. University of Illinois Press, Urbana. [2]

Herre, E. A., N. Knowlton, U. G. Mueller, and S. A. Rehner. 1999. The evolution of mutualism: Exploring the paths between conflict and cooperation. *Trends Ecol. Evol.* 14: 49–53. [19]

Herrera, C. M., and O. Pellmyr (eds.). 2002. *Plant–Animal Interactions: An Evolutionary Approach*. Blackwell Science, Malden, MA. [19]

Herrera, C. M., X. Cerdá, M. B. García, J. Guitán, M. Medrano, P. J. Rey and A. M. Sánche-Lafuente. 2002. Floral integration, phenotypic covariance structure and pollinator variation in bumblebee-pollinated *Helleborus foetidus*. *J. Evol. Biol.* 15: 108–121. [13]

Herrnstein, R. J., and C. Murray. 1994. *The Bell Curve: Intelligence and Class Structure in American Life*. The Free Press, New York. [9]

Hewitt, G. M. 2000. The genetic legacy of the Quaternary ice ages. *Nature* 405: 907–913. [6]

Higgie, M., and M. W. Blows. 2007. Are traits that experience reinforcement also under sexual selection? *Am. Nat.* 170: 409–420. [18]

Higgie, M., and M. W. Blows. 2008. The evolution of reproductive character displacement conflicts with how sexual selection operates within a species. *Evolution* 62: 1192–1203. [18]

Hill, W. G. 1981. Estimate of effective population size from data on linkage disequilibrium. *Genet. Res.* 38: 209–216. [10]

Hill, W. G., and A. Caballero. 1992. Artificial selection experiments. *Annu. Rev. Ecol. Syst.* 23: 287–310. [13]

Hillier, L. W. and the International Chicken Genome Sequencing Consortium. 2004. Sequence and comparative analysis of the chicken genome provide unique perspectives on vertebrate evolution. *Nature* 432: 695–716. [20]

Hillis, D. M. 1988. Systematics of the *Rana pipiens* complex: Puzzle and paradigm. *Annu. Rev. Ecol. Syst.* 19: 39–63. [18]

Hillis, D. M., J. J. Bull, M. E. White, M. R. Badgett, and I. J. Molineaux. 1992. Experimental phylogenetics: Generation of a known phylogeny. *Science* 255: 589–592. [2]

Hobbs, H. H., M. S. Brown, J. L. Goldstein, and D. W. Russell. 1986. Deletion of exon encoding cysteine-rich repeat of low density lipoprotein receptor alters its binding specificity in a subject with familial hypercholesterolemia. *J. Biol. Chem.* 261: 13114–13120. [8]

Hoekstra, H. E., and J. A. Coyne. 2007. The locus of evolution: Evo devo and the genetics of adaptation. *Evolution* 61: 995–1016. [21]

Hoekstra, H. E., R. J. Hirschmann, R. A. Bundey, P. A. Insel, and J. P. Crossland. 2006. A single amino acid mutation contributes to adaptive beach mouse color pattern. *Science* 313:101–104. [3]

Hoffmann, A. A., R. Hallas, J. Dean, and M. Schiffer. 2003. Low potential for climatic stress adaptation in a rainforest *Drosophila* species. *Science* 301: 100–102. [13]

Hofstadter, R. 1955. *Social Darwinism in American Thought*. Beacon Press, Boston, MA. [1, 23]

Holden, C., and R. A. Mace. 1997. A phylogenetic analysis of the evolution of lactose digestion. *Hum. Biol.* 69: 605–628. [11]

Holland, B. and W. R. Rice. 1998. Chase-away selection: Antagonistic seduction versus resistance. *Evolution* 52: 1–7. [15]

Holland, B. and W. R. Rice. 1999. Experimental removal of sexual selection reverses intersexual antagonistic coevolution and removes a reproductive load. *Proc. Natl. Acad. Sci. USA* 96: 5083–5088. [15]

Hölldobler, B., and E. O. Wilson. 1983. The evolution of communal nest-weaving in ants. *Am. Sci.* 71: 490–499. [11]

Hölldobler, B., and E. O. Wilson. 1990. *The Ants*. Harvard University Press, Cambridge, MA. [19]

Holsinger, K. E. 1991. Inbreeding depression and the evolution of plant mating systems. *Trends Ecol. Evol.* 6: 307–308. [15]

Holt, R. D. 1977. Predation, apparent competition, and the structure of prey communities. *Theor. Pop. Biol.* 12: 197–229. [19]

Holt, R. D. 1996. Demographic constraints in evolution: Towards unifying the evolutionary theories of senescence and niche conservatism. *Evol. Ecol.* 10: 1–11. [22]

Holt, R. D. 2003. On the evolutionary ecology of species' ranges. *Evol. Ecol. Res.* 5: 159–178. [6]

Holt, R. D., and M. S. Gaines. 1992. Analysis of adaptation in heterogeneous landscapes: Implications for the evolution of fundamental niches. *Evol. Ecol.* 6: 433–447. [6]

Holt, S. B. 1955. Genetics of dermal ridges: Frequency distribution of total finger ridge count. *Ann. Hum. Genet.* 20: 270–281. [9]

Hooper, J. 2002. *Of Moths and Men: Intrigue, Tragedy and the Peppered Moth*. Fourth Estate, London. [12]

Horai, S., K. Hayasaka, R. Kondo, K. Tsugane, and N. Takahata. 1995. Recent African origin of humans revealed by complete sequences of hominoid mitochondrial DNAs. *Proc. Natl. Acad. Sci. USA* 92: 532–536. [10]

Hori, M. 1993. Frequency-dependent natural selection in the handedness of scale-eating cichlid fish. *Science* 260: 216–219. [12]

Hotopp, J. C. D., and 19 others. 2007. Widespread lateral gene transfer from intracellular bacteria to multicellular eukaryotes. *Science* 317: 1753–1756. [20]

Hougen-Eitzman, D., and M. D. Rausher. 1994. Interactions between herbivorous insects and plant-insect coevolution. *Am. Nat.* 143: 677–697. [19]

Houle, D. 1989. The maintenance of polygenic variation in finite populations. *Evolution* 43: 1767–1780.

Houle, D. 1992. Comparing evolvability and variability of quantitative traits. *Genetics* 130: 195–204. [13, 22]

Houle, D., and A. S. Kondrashov. 2006. Mutation. In C. W. Fox and J. B. Wolf (eds.), *Evolutionary Genetics*, pp. 32–48. Oxford University Press, New York. [8]

Houle, D., B. Morikawa, and M. Lynch. 1996. Comparing mutational variabilities. *Genetics* 143: 1467–1483. [13]

Houston, A. I., T. Székely, and J. M. McNamara. 2005. Conflict between parents over care. *Trends Ecol. Evol.* 20: 33–38. [16]

Howard, D. J. 1993. Reinforcement: Origin, dynamics, and fate of an evolutionary hypothesis. In R. G. Harrison (ed.), *Hybrid Zones and the Evolutionary Process*, pp. 46–69. Oxford University Press, New York. [18]

Howard, D. J. 1999. Conspecific sperm and pollen precedence and speciation. *Annu. Rev. Ecol. Syst.* 30: 109–132. [17]

Howard, D. J., and S. H. Berlocher (eds.). 1998. *Endless Forms: Species and Speciation*. Oxford University Press, New York. [17]

Howell, F. C. 1978. Hominidae. In V. J. Maglio and H. B. S. Cooke (eds.), *Evolution of African Mammals*, pp. 154–248. Harvard University Press, Cambridge, MA. [4]

Huang, Y.-X., K. Magori, A. L. Lloyd, and F. Gould. 2007. Introducing desirable transgenes into insect populations using Y-linked meiotic drive – A theoretical assessment. *Evolution* 61: 717–726. [23]

Hudson, R. R. 1990. Gene genealogies and the coalescent process. In D. J. Futuyma and J. Antonovics (eds.), *Oxford Surv. Evol. Biol.*, pp. 1–44. Oxford University Press, New York. [10]

Hueffer, K., J. S. L. Parker, W. S. Weichert, R. E. Geisel, J.-Y. Sgro, and C. R. Parrish. 2003. The natural host range shift and subsequent evolution of canine parvovirus resulted from virus-specific binding to the canine transferrin receptor. *J. Virol.* 77: 1718–1726. [19]

Hull, D. L. 1973. *Darwin and his Critics: The Reception of Darwin's Theory of Evolution by the Scientific Community*. Harvard University Press, Cambridge, MA. [1]

Hull, D. L. 1988. *Science as a Process: An Evolutionary Account of the Social and Conceptual Development of Science.* University of Chicago Press, Chicago. [23]

Humphries, C. J., and L. R. Parenti. 1986. *Cladistic biogeography.* Clarendon Press, Oxford. [6]

Hunt, G. 2007a. The relative importance of directional change, random walks, and stasis in the evolution of fossil lineages. *Proc. Natl. Acad. Sci. USA* 104: 18404–18408. [4]

Hunt, G. 2007b. Evolutionary divergence in directions of high phenotypic variance in the ostracode genus *Poseidonamicus. Evolution* 61: 1560–1576. [22]

Hunt, G. R., and R. D. Gray. 2004. The crafting of hook tools by wild New Caledonian crows. *Proc. R. Soc. Lond. B* 271 (Suppl. 3): S88–S90. [23]

Huntley, J. W., and M. Kowalewski. 2007. Strong coupling of predation intensity and diversity in the Phanerozoic fossil record. *Proc. Natl. Acad. Sci. USA* 104: 15006–15010. [5]

Hurst, G. D. D., and J. H. Werren. 2001. The role of selfish genetic elements in eukaryotic evolution. *Nature Rev. Genet.* 2: 597–606. [11]

Hurst, L. D., A. Atlan, and B. D. Bengtsson. 1996. Genetic conflicts. *Q. Rev. Biol.* 71: 317–364. [16]

Husband, B. C. 2000. Constraints on polyploid evolution: A test of the minority cytotype exclusion principle. *Proc. R. Soc. Lond. B* 267: 217–223. [18]

Huston, M. 1994. Biological diversity: The coexistence of species on changing landscapes. Cambridge University Press, New York. [6]

Hutchings, J. A., and D. J. Fraser. 2008. The nature of fisheries- and farming-induced evolution. *Mol. Ecol.* 17: 294–313. [13]

Hutchinson, G. E. 1957. Concluding remarks. *Cold Spring Harbor Symp. Quant. Biol.* 22: 415–427. [6]

Hutchinson, J. 1969. *Evolution and Phylogeny of Flowering Plants.* Academic Press, New York. [3]

I

Ingman, M., H. Kaessmann, S. Pääbo, and U. Gyllensten. 2000. Mitochondrial genome variation and the origin of modern humans. *Nature* 408: 708–713. [6, 10]

International Human Genome Sequencing Consortium. 2001. Initial sequencing and analysis of the human genome. *Nature* 409: 860–921. [3, 8, 20]

Ioerger, T. R., A. G. Clark, and T.-H. Kao. 1990. Polymorphism at the self-incompatibility locus in Solanaceae predates speciation. *Proc. Natl. Acad. Sci. USA* 87: 9732–9735. [12]

Islam, M. S., P. Roessingh, S. J. Simpson, and A. R. McCaffery. 1994. Effects of population density experienced by parents during mating and oviposition on the phase of hatchling desert locusts, *Schistocerca gregaria. Proc. R. Soc. Lond. B* 257: 93–98. [9]

Iyengar, V. K., and T. Eisner. 1999. Female choice increases offspring fitness in an arctiid moth (*Utetheisa ornatrix*). *Proc. Natl. Acad. Sci. USA* 96: 15013–15016. [15]

J

Jablonka, E., and M. J. Lamb. 2005. *Evolution in Four Dimensions: Genetic, Epigenetic, Behavioral, and Symbolic Variations in the History of Life.* MIT Press, Cambridge, MA. [9]

Jablonka, E., and M. J. Lamb. 2006. Evolutionary epigenetics. In C. W. Fox and J. B. Wolf (eds.), *Evolutionary Genetics: Concepts and Case Studies*, pp. 252–264. Oxford University Press, New York [9]

Jablonski, D. 1987. Heritability at the species level: Analysis of geographic ranges of Cretaceous molluscs. *Science* 238: 360–363. [7, 11]

Jablonski, D. 1995. Extinctions in the fossil record. In J. H. Lawton and R. M. May (eds.), *Extinction Rates*, pp. 25–44. Oxford University Press, Oxford. [5, 7]

Jablonski, D. 2002. Survival without recovery after mass extinctions. *Proc. Natl. Acad. Sci. USA* 99: 8139–8144. [7]

Jablonski, D. 2005. Mass extinctions and macroevolution. *Paleobiology* 31 (Suppl.): 192–210. [7]

Jablonski, D. 2008. Biotic interactions and macroevolution: Extensions and mismatches across scales and levels. *Evolution* 62: 715–739. [7]

Jablonski, D., and D. J. Bottjer. 1990. Onshore-offshore trends in marine invertebrate evolution. In R. M. Ross and W. D. Allmon (eds.), *Causes of Evolution: A Paleontological Perspective*, pp 21–75. University of Chicago Press, Chicago. [7]

Jablonski, D., and G. Hunt. 2006. Larval ecology, geographic range, and species survivorship in Cretaceous molluscs. *Am. Nat.* 168: 556–564. [11]

Jablonski, D., and R. A. Lutz. 1983. Larval ecology of marine benthic invertebrates: Paleobiological implications. *Biol. Rev.* 58: 21–89. [22]

Jablonski, D., and K. Roy. 2003. Geographical range and speciation in fossil and living molluscs. *Proc. R. Soc. Lond. B* 270: 401–406. [7]

Jablonski, D., S. J. Gould, and D. M. Raup. 1986. The nature of the fossil record: A biological perspective. In D. M. Raup and D. Jablonski (eds.), *Patterns and Processes in the History of Life*, pp. 7–22. Springer-Verlag, Berlin. [4]

Jablonski, D., D. H. Erwin, and J. H. Lipps (eds.). 1996. *Evolutionary Paleobiology: Essays in Honor of James W. Valentine.* University of Chicago Press, Chicago. [4, 7]

Jablonski, D., K. Roy, and J. W. Valentine. 2006. Out of the tropics: Evolutionary dynamics of the latitudinal diversity gradient. *Science* 314: 102–106. [6]

Jackson, A. P., and M. A. Charleston. 2004. A cophylogenetic perspective of RNA-virus evolution. *Mol. Biol. Evol.* 21: 45–57. [19]

Jackson, J. B. C. 1974. Biogeographic consequences of eurytopy and stenotopy among marine bivalves and their biogeographic significance. *Am. Nat.* 104: 541–560. [7]

Jackson, J. B. C. 1995. Constancy and change of life in the sea. In J. H. Lawton and R. M. May (eds.), *Extinction Rates*, pp. 45-54. Oxford University Press, Oxford. [5]

Jackson, J. B. C., and K. G. Johnson. 2000. Life in the last few million years. *Paleobiology* 26 (Suppl.): 221–235. [7]

Jackson, S. T., and J. T. Overpeck. 2000. Responses of plant populations and communities to environmental changes of the late Quaternary. *Paleobiology* 26 (Suppl.): 194–220. [6]

Jackson, S. T., and J. W. Williams. 2004. Modern analogs in Quaternary paleoecology: Here today, gone yesterday, gone tomorrow? *Annu. Rev. Earth Planet. Sci.* 32: 495–537. [5]

Jacob, F. 1977. Evolution and tinkering. *Science* 196: 1161–1166. [21]

Jain, S. K., and D. R. Marshall. 1967. Population studies on predominantly self-pollinating species. X. Variation in natural populations of *Avena fatua* and *A. barbata. Am. Nat.* 101: 19–33. [9]

Jakobsson, M., and 23 others. Genotype, haplotype and copy-number variation in worldwide human populations. *Nature* 451: 998–1005. [10]

Janis, C. M. 1993. Tertiary mammal evolution in the context of changing climates, vegetation, and tectonic events. *Annu. Rev. Ecol. Syst.* 24: 467–500. [7]

Jansson, R., and M. Dynesius. 2002. The fate of clades in a world of recurrent climatic change: Milankovitch oscillations and evolution. *Annu. Rev. Ecol. Syst.* 33: 741–777. [18, 22]

Janz, N., S. Nylin, and N. Wahlberg. 2006. Diversity begets diversity: Host expansions and the diversification of plant-feeding insects. *BMC Evolutionary Biology* 6: 4, doi:10.1186/1471-2148/6/4. (http:www.biomedcentral.com/1471-2148/6/4) [7]

Jarne, P., and D. Charlesworth. 1993. The evolution of the selfing rate in functionally hermaphroditic plants and animals. *Annu. Rev. Ecol. Syst.* 24: 441–466. [15]

Jensen, A. R. 1973. *Educability and Group Differences.* Harper & Row, New York. [9]

Jeong, S., A. Rokas, and S. B. Carroll. 2006. Regulation of body pigmentation by the Abdominal-B Hox protein and its gain and loss in *Drosophila* evolution. *Cell* 125: 1387–1399. [21]

Jeong, S., M. Rebeiz, P. Andolfatto, T. Werner, J. True, and S. B. Carroll. 2008. The evolution of gene regulation underlies a morphological difference between two *Drosophila* sister species. *Cell* 132: 783–793. [21]

Johnson, J. A., M. R. Bellinger, J. E. Toepfer, and P. Dunn. 2004. Temporal changes in allele frequencies and low effective population size in greater prairie-chickens. *Mol. Ecol.* 13: 2617–2630. [10]

Johnson, N. K., and C. Cicero. 2004. New mitochondrial DNA data affirm the importance of Pleistocene speciation in North American birds. *Evolution* 58: 1122–1130. [18]

Johnstone, R. A. 1995. Sexual selection, honest advertisement and the handicap principle: Reviewing the evidence. *Biol. Rev.* 70: 1–65. [15]

Jokela, J., and C. M. Lively. 1995. Parasites, sex, and early reproduction in a mixed population of freshwater snails. *Evolution* 49: 1268–1271. [15]

Jones, A. G., S. J. Arnold, and R. Bürger. 2007. The mutation matrix and the evolution of evolvability. *Evolution* 61: 727–745. [22]

Jones, I. L., and F. M. Hunter. 1999. Experimental evidence for mutual inter- and intrasexual selection favouring a crested auklet ornament. *Animal Behaviour* 57: 521–528. [15]

Jones, S., R. Martin, and D. Pilbeam (eds.). 1992. *The Cambridge Encyclopedia of Human Evolution*. Cambridge University Press, Cambridge. [4]

Jones, T. M., R. J. Quinnell, and A. Balmford. 1998. Fisherian flies: Benefits of female choice in a lekking sandfly. *Proc. R. Soc. Lond. B* 265: 1651–1657. [15]

Joron, M., and J. L. B. Mallet. 1998. Diversity in mimicry: Paradox or paradigm? *Trends Ecol. Evol.* 13: 461–466. [19]

Joyce, G. F. 2002. The antiquity of RNA-based evolution. *Nature* 418: 214–221. [5]

Judd, W. S., C. S. Campbell, E. A. Kellogg, P. F. Stevens, and M. J. Donoghue. 2002. *Plant Systematics: A Phylogenetic Approach*, 2nd ed. Sinauer, Sunderland, MA. [5]

K

Kaiser, D. 2007. The other evolution wars. *Am. Sci.* 95: 518–525. [23]

Kareiva, P. M., J. G. Kingsolver, and R. B. Huey (eds.). 1993. *Biotic Interactions and Global Change*. Sinauer, Sunderland, MA. [5]

Karn, M. N., and L. S. Penrose. 1951. Birth weight and gestation time in relation to maternal age, parity, and infant survival. *Ann. Eugenics* 16: 147–164. [13]

Kasahara, M. 1997. New insights into the genome organization and origin of the major histocompatibility complex: role of chromosomal (genome) duplication and emergence of the adaptive immune system. *Hereditas* 127: 59–66. [20]

Kassen, R., and T. Bataillon. 2006. Distribution of fitness effects among beneficial mutations before selection in experimental populations of bacteria. *Nature Genetics* 38: 484–488. [8]

Katakura, H., and T. Hosogai. 1994. Performance of hybrid ladybird beetles (*Epilachna*) on the host plants of parental species. *Entomol. Exp. Appl.* 71: 81–85. [17]

Kaufman, T. C., M. A. Seeger, and G. Olsen. 1990. Molecular and genetic organization of the *Antennapedia* gene complex of *Drosophila melanogaster. Adv. Genet.* 27: 309–362. [21]

Kawecki, T. J. 2000. The evolution of genetic canalization under fluctuating selection. *Evolution* 54: 1–12. [13]

Kazazian, H. H., Jr. 2004. Mobile elements: Drivers of genome evolution. *Science* 303: 1626–1632. [8]

Kearsey, M. J., and B. W. Barnes. 1970. Variation for metrical characters in *Drosophila* populations. II. Natural selection. *Heredity* 25: 11–21. [13]

Keeling, P. J. 2004. Diversity and evolutionary history of plastids and their hosts. *Am. J. Bot.* 91: 1481–1493. [5]

Keeling, P. J. 2007. Deep questions in the tree of life. *Science* 317: 1875–1876. [5]

Keightley, P. D., and S. P. Otto. 2006. Interference among deleterious mutations favours sex and recombination in finite populations. *Nature* 443: 89–92. [15]

Keller, L. (ed.) 1993. *Queen Number and Sociality in Insects*. Oxford University Press, Oxford. [16]

Keller, L. (ed.) 1999. *Levels of Selection in Evolution*. Princeton University Press, Princeton, NJ. [11, 16]

Keller, L., and H. K. Reeve. 1994. Partitioning of reproduction in animal societies. *Trends Ecol. Evol.* 9: 98–102. [16]

Keller, L., and H. K. Reeve. 1999. Dynamics of conflicts within insect societies. In L. Keller (ed.), *Levels of Selection in Evolution*, pp. 153–175. Princeton University Press, Princeton, NJ. [16]

Kellermann, V. M., B. van Heerwaarden, A. A. Hoffmann, and C. M. Sgrò. 2006. Very low additive genetic variance and evolutionary potential in multiple populations of two rainforest *Drosophila* species. *Evolution* 60: 1104–1108. [6]

Kelley, S. T., and B. D. Farrell. 1998. Is specialization a dead end? The phylogeny of host use in *Dendroctonus* bark beetles (Scolytidae). *Evolution* 52: 1731–1743. [6]

Kemp, T. S. 1982. *Mammal-like Reptiles and the Origin of Mammals*. Academic Press, London. [4]

Kenrick, P., and P. R. Crane. 1997a. The origin and early evolution of plants on land. *Nature* 389: 33–39. [5]

Kenrick, P., and P. R. Crane. 1997b. *The Origin and Early Diversification of Land Plants*. Smithsonian Institution Press, Washington. [5]

Kidston, R., and W. H. Lang. 1921. On Old Red Sandstone plants showing structure from the Rhynie chert bed, Aberdeenshire, Part IV. Restorations of the vascular cryptogams, and discussion of their bearing on the general morphology of Pteridophyta and the origin of the organization of land plants. *Trans. R. Soc. Edinburgh* 32: 477–487. [5]

Kiers, E. T., R. A. Rousseau, S. A. West, and R. F. Denison. 2003. Host sanctions and the legume-rhizobium mutualism. *Nature* 425: 78–81. [16]

Kimura, M. 1955. Solution of a process of random genetic drift with a continuous model. *Proc. Natl. Acad. Sci. USA* 41: 144–150. [10]

Kimura, M. 1968. Evolutionary rate at the molecular level. *Nature* 217: 624–626. [10]

Kimura, M. 1983. *The Neutral Theory of Molecular Evolution*. Cambridge University Press, Cambridge. [1, 10]

King, J. L., and T. H. Jukes. 1969. Non-Darwinian evolution. *Science* 164: 788–798. [10]

King, M. 1993. *Species Evolution: The Role of Chromosome Change*. Cambridge University Press, Cambridge. [17]

King, M.-C., and A. C. Wilson. 1975. Evolution at two levels in humans and chimpanzees. *Science* 188: 107–116. [21]

King, N., C. T. Hittinger, and S. B. Carroll. 2003. Evolution of key cell signaling and adhesion protein families predates animal origins. *Science* 301: 361–363. [5]

King, N. and 35 others. 2008. The genome of the choanoflagellate *Monosiga brevicollis* and the origin of metazoans. *Nature* 451: 783–788. [20]

Kingsolver, J. G., and D. W. Pfennig. 2004. Individual-level selection as a cause of Cope's rule of phyletic size increase. *Evolution* 58: 1608–1612. [22]

Kirkpatrick, M. 1982. Sexual selection and the evolution of female choice. *Evolution* 3: 1–12. [15]

Kirkpatrick, M., and N. H. Barton. 1997a. Evolution of a species' range. *Am. Nat.* 150: 1–23. [6]

Kirkpatrick, M., and N. H. Barton. 1997b. The strength of indirect selection on female mating preferences. *Proc. Natl. Acad. Sci. USA* 94: 1282–1286. [15]

Kirkpatrick, M., and M. R. Servedio. 1999. The reinforcement of mating preferences on an island. *Genetics* 151: 865–884. [18]

Kirkpatrick, R. C. 2000. The evolution of human homosexual behavior. *Curr. Anthropol.* 41: 385–414. [16]

Kirschner, M., and J. Gerhart. 2005. *The Plausibility of Life*. Yale University Press, New Haven, CT. [22]

Kitcher, P. 1985. *Vaulting Ambition*. MIT Press, Cambridge, MA. [15]

Kittles, R. A., and K. M. Weiss. 2003. Race, ancestry, and genes: Implications for defining disease risk. *Annu. Rev. Genomics Human Genet.* 4: 33-67. [9]

Klein, J., and N. Takahata. 2002. *Where Do We Come From? The Molecular Evidence for Human Descent*. Springer-Verlag, Berlin. [6]

Klicka, J., and R. M. Zink. 1997. The importance of recent ice ages in speciation: A failed paradigm. *Science* 277: 1666–1669. [18]

Kliman, R. M., P. Andolfatto, J. A. Coyne, F. Depaulis, M. Kreitman, A. J. Berry, M. McCarter, J. Wakeley, and J. Hey. 2000. The population genetics of the origin and divergence of the *Drosophila simulans* complex species. *Genetics* 156: 1913–1931. [18]

Knecht, A. K., K. E. Hosemann, and D. M. Kingsley. 2007. Constraints on utilization of the EDA signaling pathway in threespine stickleback evolution. *Evol. Dev.* 9: 141–154. [21]

Knoll, A. H., and R. K. Bambach. 2000. Directionality in the history of life: Diffusion from the left wall or repeated scaling of the right? In D. H. Erwin and S. L. Wing (eds.), *Deep Time: Paleobiology's Perspective*, pp. 1–14. The Paleontological Society, Allen Press, Lawrence, Kan. [22]

Knoll, A. H., R. K. Bambach, J. L. Payne, S. Pruss, and W. Fischer. 2007. Paleophysiology and end-Permian mass extinction. *Earth and Planetary Sci. Lett.* 256: 295–313. [5]

Knowles, L. L. 2004. The burgeoning field of statistical phylogeography. *J. Evol. Biol.* 17: 1–10. [6]

Knowlton, N., L. A. Weigt, L. A. Solórzano, D. K. Mills, and E. Bermingham. 1993. Divergence in proteins, mitochondrial DNA, and reproductive compatibility across the Isthmus of Panama. *Science* 260: 1629–1632. [18]

Koch, P. L., and A. D. Barnosky. 2006. Late Quaternary extinctions: State of the debate. *Annu. Rev. Ecol. Evol. Syst.* 37: 215–250. [5]

Kocher, T. D. 2004. Adaptive evolution and explosive speciation: the cichlid fish model. *Nature Rev. Genet.* 5: 288–298. [3]

Koehn, R. K., and J. J. Hilbish. 1987. The adaptive importance of genetic variation. *Am. Sci.* 75: 134–141. [12]

Kojak, K. H., and J. J. Wiens. 2006. Does niche conservatism promote speciation? A case study in North American salamanders. *Evolution* 60: 2604–2621. [18]

Kokko, H. 1998. Should advertising parental care be honest? *Proc. R. Soc. Lond. B* 265: 1871–1878. [16]

Kokko, H. 2007. *Modelling for Field Biologists*. Cambridge University Press, Cambridge. [14]

Kokko, H., M. D. Jennions, and R. Brooks. 2006. Unifying and testing models of sexual selection. *Annu. Rev. Ecol. Evol. Syst.* 37: 43–66. [15]

Kokko, H., R. Brooks, J. M. McNamara, and A. Houston. 2002. The sexual selection continuum. *Proc. R. Soc. Lond. B* 269: 1331–1340. [15]

Kondrashov, A. S. 1988. Deleterious mutations and the evolution of sexual reproduction. *Nature* 336: 435–441. [15]

Kondrashov, A. S. 1993. Classification of hypotheses on the advantage of amphimixis. *J. Hered.* 84: 372–387. [15]

Kondrashov, A. S., and F. A. Kondrashov. 1999. Interactions among quantitative traits in the course of sympatric speciation. *Nature* 400: 351–354. [18]

Kopp, A., I. Duncan, and S. B. Carroll. 2000. Genetic control and evolution of sexually dimorphic characters in *Drosophila*. *Nature* 408: 553–559. [21]

Kopp, A., R. M. Graze, S. Xu, S. B. Carroll, and S. V. Nuzhdin. 2003. Quantitative trait loci responsible for variation in sexually dimorphic traits in *Drosophila melanogaster*. *Genetics* 163: 771–787. [21]

Korber, B., M. Muldoon, J. Theiler, F. Gao, R. Gupta, A. Lapedes, B. H. Hahn, S. Wolinksy, and T. Bhattacharya. 2000. Timing the ancestor of the HIV-1 pandemic strains. *Science* 288: 1789–1796. [1, 10]

Kornegay, J. R., J. W. Schilling, and A. C. Wilson. 1994. Molecular adaptation of a leaf-eating bird: Stomach lysozyme of the hoatzin. *Mol. Biol. Evol.* 11: 921–928. [20]

Kornfield, I., and P. F. Smith. 2000. African cichlid fishes: Model systems for evolutionary biology. *Annu. Rev. Ecol. Syst.* 31: 163–196. [3]

Kotiaho, J. S., L. W. Simmons, and J. L. Tomkins. 2001. Towards a resolution of the lek paradox. *Nature* 410: 684–686. [15]

Koufopanou, V., and G. Bell. 1991. Developmental mutants of *Volvox*: Does mutation recreate patterns of phylogenetic diversity? *Evolution* 45: 1806–1822. [8]

Krebs, J. R., and R. H. McCleery. 1984. Optimization in behavioural ecology. In J. R. Krebs and N. B. Davies (eds.), *Behavioural Ecology: An Evolutionary Approach*, 2nd ed., pp. 91–121. Sinauer, Sunderland, MA. [14]

Kreitman, M. 1983. Nucleotide polymorphism at the alcohol dehydrogenase locus of *Drosophila melanogaster*. *Nature* 304: 412–417. [9]

Kucera, M., and B. A. Malmgren. 1998. Differences between evolution of mean form and evolution of new morphotypes: An example from Late Cretaceous planktonic foraminifera. *Paleobiology* 24: 49–63. [4]

Kumar, S., and S. Subramanian. 2002. Mutation rates in mammalian genomes. *Proc. Natl. Acad. Sci. USA* 99: 803–808. [8]

Kuparinen, A., and J. Merilä. 2007. Detecting and managing fisheries-induced evolution. *Trends Ecol. Evol.* 22: 652–659. [13]

L

Labandeira, C. C. 2002. The history of associations between plants and animals. In C. M. Herrera and O. Pellmyr (eds.), *Plant-Animal Interactions: An Evolutionary Approach*, pp. 26–74. Blackwell Science, London. [19]

Labandeira, C. C., and J. J. Sepkoski Jr. 1993. Insect diversity in the fossil record. *Science* 261: 310–315. [7]

Lack, D. 1954. *The Natural Regulation of Animal Numbers*. Oxford University Press, Oxford. [14]

Lai, C., R. F. Lyman, A. D. Long, C. H. Langley, and T. F. C. Mackay. 1994. Naturally occurring variation in bristle number and DNA polymorphisms at the *scabrous* locus of *Drosophila melanogaster*. *Science* 266: 1697–1702. [13, 21]

Laland, K. N., and G. R. Brown. 2002. *Sense and Nonsense: Evolutionary Perspectives on Human Behaviour*. Oxford University Press, Oxford. [23]

Lande, R. 1976a. The maintenance of genetic variability by mutation in a polygenic character with linked loci. *Genet. Res.* 26: 221–235. [13]

Lande, R. 1976b. Natural selection and random genetic drift in phenotypic evolution. *Evolution* 30: 314–334. [13]

Lande, R. 1979. Effective deme sizes during long-term evolution estimated from rates of chromosome rearrangement. *Evolution* 33: 234–251. [8]

Lande, R. 1980. Sexual dimorphism, sexual selection and adaptation in polygenic characters. *Evolution* 34: 292–305. [13]

Lande, R. 1981. Models of speciation by sexual selection on polygenic traits. *Proc. Natl. Acad. Sci. USA* 78: 3721–3725. [15, 18]

Lande, R. 1982. Rapid origin of sexual isolation and character divergence in a cline. *Evolution* 36: 213–223. [18]

Lande, R., and S. J. Arnold. 1983. The measurement of selection on correlated characters. *Evolution* 37: 1210–1226. [11, 13]

Lande, R., and D. W. Schemske. 1985. The evolution of self-fertilization and inbreeding depression. I. Genetic models. *Evolution* 39: 24–40. [15]

Landry, C. R., B. Lemos, S. A. Rifkin, W. J. Dickinson, and D. L. Hartl. 2007. Genetic properties influencing the evolvability of gene expression. *Science* 317: 118–121. [8]

Law, R., A. D. Bradshaw, and P. D. Putwain. 1979. The cost of reproduction in annual meadow grass. *Am. Nat.* 113: 3–16. [14]

Lawton, J. H., and R. M. May (eds.). 1995. *Extinction Rates*. Oxford University Press, Oxford. [7]

Lederberg, J., and E. M. Lederberg. 1952. Replica plating and indirect selection of bacterial mutants. *J. Bacteriol.* 63: 399–406. [8]

Lee, M. S. Y., T. W. Reeder, J. B. Slowinski, and R. Lawson. 2004. Resolving reptile relationships: Molecular and morphological markers. In J. Cracraft and M. J. Donoghue (eds.), *Assembling the Tree of Life*, pp. 451–467. Oxford University Press, New York. [5]

Lehmann, L., and L. Keller. 2006. The evolution of cooperation and altruism – a general framework and a classification of models. *J. Evol. Biol.* 19: 1365-1376. [16]

Leigh, E. G. Jr. 1973. The evolution of mutation rates. *Genetics* Suppl. 73: 1–18. [15]

Lenormand, T. 2002. Gene flow and the limits to natural selection. *Trends Ecol. Evol.* 17: 183–189. [12]

Lessios, H. A. 1998. The first stage of speciation as seen in organisms separated by the Isthmus of Panama. In D. J. Howard and S. H. Berlocher (eds.), *Endless Forms: Species and Speciation*, pp. 186–201. Oxford University Press, Oxford. [6]

Levin, B. R., and R. M. Anderson. 1995. The population biology of anti-infective chemotherapy and the evolution of drug resistance: more questions than answers. In S. C. Stearns (ed.), *Evolution in Health and Disease*, pp. 125–137. Oxford University Press, Oxford. [1]

Levin, D. A. 1983. Polyploidy and novelty in flowering plants. *Am. Nat.* 122: 1–25. [18]

Levin, D. A. 1984. Immigration in plants: An exercise in the subjunctive. In R. Dirzo and J. Sarukhán (eds.), *Perspectives on Plant Population Ecology*, pp. 242–260. Sinauer, Sunderland, MA. [9]

Levine, M. T., C. D. Jones, A. D. Kern, H. A. Lindfors, and D. J. Begun. 2006. Novel genes derived from noncoding DNA in *Drosophila* melanogaster are frequently X-linked and exhibit testis-biased expression. *Proc. Natl. Acad. Sci. USA* 103: 9935–9939. [20]

Levinton, J. S. 2001. *Genetics, Paleontology, and Macroevolution*, 2nd ed. Cambridge University Press, Cambridge. [4, 22]

Lewin, B. 1985. *Genes II*. Wiley, New York. [8]

Lewin, R. 2005. *Human Evolution: An Illustrated Introduction*. Blackwell Publishing, Oxford. [4]

Lewis, H. 1962. Catastrophic speciation as a factor in evolution. *Evolution* 16: 257–271. [18]

Lewontin, R. C. 1974. *The Genetic Basis of Evolutionary Change*. Columbia University Press, New York. [9]

Lewontin, R. C. 2000. *The Triple Helix: Gene, Organism, and Environment*. Harvard University Press, Cambridge, MA. [22]

Lewontin, R. C., and J. L. Hubby. 1966. A molecular approach to the study of genic heterozygosity in natural populations. II. Amount of variation and degree of heterozygosity in natural populations of *Drosophila pseudoobscura*. *Genetics* 54: 595–609. [9, 10]

Lewontin, R. C., S. Rose, and L. Kamin. 1984. *Not in Our Genes: Biology, Ideology, and Human Nature*. Pantheon, New York. [9]

Li, J. Z., and 10 others. 2008. Worldwide human relationships inferred from genome-wide human variation. *Science* 319: 1100–1104. [9]

Li, K. S., and 21 others. 2004. Genesis of a highly pathogenic H5N1 influenza virus in eastern Asia. *Nature* 430: 209–213. [19]

Li, W.-H. 1997. *Molecular Evolution*. Sinauer, Sunderland, MA. [3, 8, 9, 10, 20]

Li, W.-H., and D. Graur. 1991. *Fundamentals of Molecular Evolution*. Sinauer, Sunderland, MA. [2]

Lieberman, B. S. 2003. Paleobiogeography: The relevance of fossils to biogeography. *Annu. Rev. Ecol. Evol. Syst.* 34: 51–69. [6]

Linville, S. U., R. Breitwitsch, and A. J. Schilling. 1998. Plumage brightness as an indicator of parental care in northern cardinals. *Anim. Behav.* 55: 119–127. [16]

Liou, L. W., and T. D. Price. 1994. Speciation by reinforcement of premating isolation. *Evolution* 48: 1451–1459. [18]

Liu, L., and D. K. Pearl. 2007. Species trees from gene trees: reconstructing Bayesian posterior distribution of a species phylogeny using estimated gene tree distributions. *Syst. Biol.* 50: 504–514. [2]

Lively, C. M., and M. F. Dybdahl. 2000. Parasite adaptation to locally common host genotypes. *Nature* 405: 679–681. [15]

Lloyd, D. G. 1992. Evolutionarily stable strategies of reproduction in plants: Who benefits and how? In R. Wyatt (ed.), *Ecology and Evolution of Plant Reproduction*, pp. 137–168. Chapman & Hall, New York. [15]

Lohmueller, K. E. and 11 others. 2008. Proportionally more deleterious genetic variation in European than in African populations. *Nature* 451: 994–997. [10]

Lomolino, M. V., B. R. Riddle, and J. H. Brown. 2006. *Biogeography*, 3rd ed. Sinauer, Sunderland, MA. [6]

Long, A. D., R. F. Lyman, A. H. Morgan, C. H. Langley, and T. F. C. Mackay. 2000. Both naturally occurring inserts of transposable elements and intermediate frequency polymorphisms at the *achaete-scute* complex are associated with variation in bristle number in *Drosophila melanogaster*. *Genetics* 154: 1255–1269. [21]

Long, A. D., R. F. Lyman, C. H. Langley, and T. F. C. Mackay. 1998. Two sites in the *Delta* gene region contribute to naturally occurring variation in bristle number in *Drosophila melanogaster*. *Genetics* 149: 999–1017. [21]

Long, A. D., S. L. Mullaney, L. A. Reid, J. D. Fry, C. H. Langley, and T. F. C. Mackay. 1995. High resolution mapping of genetic factors affecting abdominal bristle number in *Drosophila melanogaster*. *Genetics* 139: 1273–1291. [13]

Long, M., S. J. de Souza, C. Rosenberg, and W. Gilbert. 1996. Exon shuffling and the origin of the mitochondrial targeting function in plant cytochrome *c1* precursor. *Proc. Natl. Acad. Sci. USA* 93: 7727–7731. [20]

Long, M., E. Betran, K. Thornton, and W. Wang. 2003a. The origin of new genes: Glimpses of the young and old. *Nature Rev. Genet.* 4: 865–875. [20]

Long, M., M. Deutsch, W. Wang, E. Betran, F. G. Brunet, and J. Zhang. 2003b. Origin of new genes: Evidence from experimental and computational analyses. *Genetica* 118: 171–182. [20]

Longini, I. M., Jr., A, Nizam, S. Xu, K. Ungchusak, W. Hanshaoworakul, D. A. T. Cummings, and M. E. Halloran. 2005. Containing pandemic influenza at the source. *Science* 309: 1083–1087. [19]

López, M. A., and C. López-Fanjul. 1993. Spontaneous mutation for a quantitative trait in *Drosophila melanogaster*. I. Response to artificial selection. *Genet. Res.* 61: 107–116. [13]

Losos, J. B. 1992. The evolution of convergent structure in Caribbean *Anolis* communities. *Syst. Biol.* 41: 403–420. [19]

Losos, J. B., T. R. Jackman, A. Larson, K. de Queiroz, and L. Rodríguez-Schettino. 1998. Contingency and determinism in replicated adaptive radiations of island lizards. *Science* 279: 2115–2118. [6]

Losos, J. B., and 8 others. 2000. Evolutionary implications of phenotypic plasticity in the hindlimb of the lizard *Anolis sagrei*. *Evolution* 54: 301–305. [13]

Losos, J. B., R. E. Glor, J. J. Kolbe, and K. Nicholson. 2006. Adaptation, speciation, and convergence: a hierarchical analysis of adaptive radiation in Caribbean *Anolis* lizards. *Annals Missouri Bot. Gard.* 93: 24–33. [6]

Lovejoy, A. O. 1936. *The Great Chain of Being: A Study of the History of an Idea*. Harvard University Press, Cambridge, MA. [1]

Lovejoy, C. O. 1981. The origins of man. *Science* 211: 341–350. [4]

Lovette, I. J. 2005. Glacial cycles and the tempo of avian speciation. *Trends Ecol. Evol.* 20: 57–59. [18]

Ludwig, M. Z. and M. Kreitman. 1995. Evolutionary dynamics of the enhancer region of *even-skipped* in *Drosophila*. *Mol. Biol. Evol.* 12: 1002–1011. [21]

Ludwig, M. Z., C. Bergman, N. H. Patel, and M. Kreitman. 2000. Evidence for stabilizing selection in a eukaryotic enhancer element. *Nature* 403: 564–567. [21]

Ludwig, M. Z., N. H. Patel, and M. Kreitman. 1998. Functional analysis of *eve stripe 2* enhancer evolution in *Drosophila*: Rules governing conservation and change. *Development* 125: 949–958. [21]

Luo, D., R. Carpenter, C. Vincent, L. Copsey, and E. Coen. 1996. Origin of floral asymmetry in *Antirrhinum*. *Nature* 383: 794-799. [22]

Luo, Z.-X. 2007. Transformation and diversification in early mammal evolution. *Nature* 450: 1011–1019. [4]

Luo, Z.-X., A. W. Crompton, and A.-L. Sun. 2001. A new mammaliaform from the early Jurassic and evolution of mammalian characteristics. *Science* 292: 1535–1540. [4]

Lupia, R., S. Lidgard, and P. R. Crane. 1999. Comparing palynological abundance and diversity: implications for biotic replacement during the Cretaceous angiosperm radiation. *Paleobiology* 25: 305–340. [7]

Lyman, R. F., F. Lawrence, S. V. Nuzhdin and T. F. C. Mackay. 1996. Effects of single *P*-element insertions on bristle number and viability in *Drosophila melanogaster*. Genetics 143: 277–292. [8]

Lynch, M., and J. S. Conery. 2000. The evolutionary fate and consequences of duplicate genes. *Science* 290**:** 1151–1155. [20]

Lynch, M. 1988. The rate of polygenic mutation. *Genet. Res.* 51: 137–148. [8, 13]

Lynch, M. 1990. The rate of morphological evolution in mammals from the standpoint of the neutral expectation. *Am. Nat.* 136: 727–741. [13]

Lynch, M. 2007. *The Origins of Genome Architecture*. Sinauer, Sunderland, MA. [3, 20]

Lynch, M., and J. B. Walsh. 1998. *Genetics and Analysis of Quantitative Traits*. Sinauer, Sunderland, MA. [9, 13]

Lynch, M., and J. S. Conery. 2000. The evolutionary fate and consequences of duplicate genes. *Science* 290: 1151–1155. [20]

Lynch, M., and J. S. Conery. 2003. The origins of genome complexity. *Science* 302: 1401–1404. [20]

Lynch, M., and R. Lande. 1993. Evolution and extinction in response to environmental change. In P. M. Kareiva, J. G. Kingsolver, and R. B. Huey (eds.), *Biotic Interactions and Global Change*, pp. 234–250. Sinauer, Sunderland, MA. [7, 13, 15]

Lynch, M., J. Blanchard, D. Houle, T. Kibota, S. Schultz, L. Vassilieva, and J. Willis. 1999. Perspective: Spontaneous deleterious mutation. *Evolution* 53: 645–663. [8] Lynch, M., J. Blanchard, D. Houle, T. Kibota, S. Schultz, L. Vassilieva, and J. Willis. 1999. Perspective: Spontaneous deleterious mutation. *Evolution* 53: 645–663. [8]

Lynch, M., R. Bürger, D. Butcher, and W. Gabriel. 1993. The mutational meltdown in an asexual population. *J. Hered.* 84: 339–344. [15]

Lyon, B. E. 1993. Conspecific brood parasitism as a flexible reproductive tactic in American coots. *Anim. Behav.* 46: 911–928. [16]

Lyon, B. E. 2003. Egg recognition and counting reduce costs of avian conspecific brood parasitism. *Nature* 422: 495–499. [16]

Lyon, B. E., and J. McA. Eadie. 2008. Conspecific brood parasitism in birds: A life-history perspective. *Annu. Rev. Ecol. Evol. Syst.* 39: 343–363. [16]

Lyon, B. E., J. McA. Eadie, and L. D. Hamilton. 1994. Parental choice selects for ornamental plumage in American coot chicks. *Nature* 371: 240–243. [16]

M

Ma, H. and C. dePamphilis. 2000. The ABCs of floral evolution. *Cell* 101: 5–8. [21]

Mace, R., and C. J. Holden. 2005. A phylogenetic approach to cultural evolution. *Trends Ecol. Evol.* 20: 116–123. [23]

MacFadden, B. J. 1986. Fossil horses from "Eohippus" (*Hyracotherium*) to *Equus*: Scaling, Cope's law, and the evolution of body size. *Paleobiology* 12: 355–369. [4, 22]

Mackay, T. F. C. 2001a. Quantitative trait loci in *Drosophila*. *Nat. Rev. Genet.* 2: 11–20. [13]

Mackay, T. F. C. 2001b. The genetic architecture of quantitative traits. *Annu. Rev. Genet.* 35: 303–339. [13]

Mackay, T. F. C., and R. F. Lyman. 2005. *Drosophila* bristles and the nature of quantitative genetic variation. *Phil. Trans. R. Soc. Lond. B* 360: 1513–1527. [13]

Mackay, T. F. C., J. D. Fry, R. F. Lyman, and S. V. Nuzhdin. 1994. Polygenic mutation in *Drosophila melanogaster*: Estimates from response to selection in inbred strains. *Genetics* 136: 937–951. [13]

Mackay, T. F. C., R. F. Lyman, and M. S. Jackson. 1992. Effects of P-element insertion on quantitative traits in *Drosophila melanogaster*. *Genetics* 130: 315–332. [8, 13]

Macnair, M. R. 1981. Tolerance of higher plants to toxic materials. In J. A. Bishop and L. M. Cook (eds.), *Genetic Consequences of Man Made Change*, pp. 177–207. Academic Press, New York. [13]

Macnair, M. R. 1997. The evolution of plants in metal-contaminated environments. In R. Bijlsma and V. Loeschcke (eds.), *Environmental Stress, Adaptation and Evolution*, pp. 2–24. Birkhäuser, Basel, Switzerland. [13]

Maddison, W. 1995. Phylogenetic histories within and among species. In P. C. Hoch and A. G. Stevenson (eds.), *Experimental and Molecular Approaches to Plant Biosystematics*, pp. 273–287. Monographs in Systematic Botany 53. Missouri Botanical Garden, St. Louis. [2]

Maddison, W. P., and D. R. Maddison. 2000. *MacClade 4*. Sinauer, Sunderland, MA. [2]

Maddison, W. P., and L. L. Knowles. 2006. Inferring phylogeny despite incomplete lineage sorting. *Syst. Biol.* 55: 21–30. [2]

Madin, J. S., J. Alroy, M. Aberhan, F. T. Fürsich, W. Kiessling, M. A. Kosnik, and P. J. Wagner. 2006. Statistical independence of escalatory ecological trends in Phanerozoic marine invertebrates. *Science* 312: 897–900. [7]

Madsen, T., B. Stille, and R. Shine. 1995. Inbreeding depression in an isolated population of adders *Vipera berus*. *Biol. Conserv.* 75: 113–118. [9]

Madsen, T., R. Shine, M. Olsson, and H. Wittzell. 1999. Restoration of an inbred adder population. *Nature* 402: 34–35. [9]

Majerus, M. E. 1998. *Melanism: Evolution in Action*. Oxford University Press, Oxford. [12]

Majerus, M. E., and N. I. Mundy. 2003. Mammalian melanism: Natural selection in black and white. *Trends Genet.* 19: 585–588. [21]

Makova, K. D., and W.-H. Li. 2002. Strong male-driven evolution of DNA sequences in humans and apes. *Nature* 416: 624–626. [8]

Mallet, J. 2007. Hybrid speciation. *Nature* 446: 279–283. [17, 18]

Mallet, J., and M. Joron. 1999. Evolution of diversity in warning color and mimicry: polymorphisms, shifting balance, and speciation. *Annu. Rev. Ecol. Syst.* 30: 201–233. [3, 18, 19]

Mallet, J., and N. Barton. 1989. Strong natural selection in a warning-color hybrid zone. *Evolution* 43: 421–431. [12]

Malmgren, B. A., W. A. Berggren, and G. P. Lohmann. 1983. Evidence for punctuated gradualism in the Late Neogene *Globorotalia tumida* lineage of planktonic Foraminifera. *Paleobiology* 9: 377–389. [4]

Manica, A., W. Amos, F. Balloux, and T. Hanihara. 2007. The effect of ancient population bottlenecks on human phenotypic variation. *Nature* 448: 346–348. [10]

Manno, T. G., F. S. Dobson, J. L. Hoogland, and D. W. Foltz. 2007. Social group fission and gene dynamics among black-tailed prairie dogs. *J. Mammal.* 88: 448–456. [16]

Marcot, J. D., and D. W. McShea. 2007. Increasing hierarchical complexity throughout the history of life: Phylogenetic tests of trend mechanisms. *Paleobiology* 33: 182–200. [22]

Margulis, L. 1993. *Symbiosis in Cell Evolution*, 2nd ed. W. H. Freeman, San Francisco. [5]

Margulis, L., and R. Fester (eds.) 1991. *Symbiosis as a Source of Evolutionary Innovation*. MIT Press, Cambridge, MA. [5]

Marshall, C. R.2006. Explaining the Cambrian "explosion" of animals. *Annu. Rev. Earth Planet. Sci.* 34: 355–384. [5]

Marshall, D. C., and J. R. Cooley. 2000. Reproductive character displacement and speciation in periodical cicadas, with description of a new species, 13-year *Magicicada neotredecim*. *Evolution* 54: 1313–1325. [18]

Martin, P. S., and R. G. Klein (eds.). 1984. *Quaternary Extinctions: A Prehistoric Revolution*. University of Arizona Press, Tucson. [5]

Martin, W., and 9 others. 2002. Evolutionary analysis of *Arabidopsis*, cyanobacterial, and chloroplast genomes reveals plastid phylogeny and thousands of cyanobacterial genes in the nucleus. *Proc. Natl. Acad. Sci.* 99: 12246–12251. [19]

Martínez Wells, M., and C. S. Henry. 1992a. Behavioural responses of green lacewings (Neuroptera: Chrysopidae: *Chrysoperla*) to synthetic mating songs. *Anim. Behav.* 44: 641–652. [17]

Mather, K. 1949. *Biometrical Genetics: The Study of Continuous Variation*. Methuen, London. [9]

Mattila, T. M., and F. Bokma. 2008. Extant mammal body masses suggest punctuated equilibrium. *Proc. R. Soc. Lond. B* 275: 2195–2199. [18]

Matton, D. P., N. Nass, A. G. Clark, and E. Newbigin. 1994. Self-incompatibility: How plants avoid illegitimate offspring. *Proc. Natl. Acad. Sci. USA* 91: 1992–1997. [15]

May, R. M., and R. M. Anderson. 1983. Parasite-host coevolution. In D. J. Futuyma and M. Slatkin (eds.), *Coevolution*, pp. 186–206. Sinauer, Sunderland, MA. [19]

Maynard Smith, J. 1978. *The Evolution of Sex*. Cambridge University Press, Cambridge. [15]

Maynard Smith, J. 1982. *Evolution and the Theory of Games*. Cambridge University Press, Cambridge. [14, 16]

Maynard Smith, J. 1988. The evolution of recombination. In B. R. Levin and R. E. Michod (eds.), *The Evolution of Sex*, pp. 106–125. Sinauer, Sunderland, MA. [15]

Maynard Smith, J., and J. Haigh. 1974. The hitch-hiking effect of a favourable gene. *Genet. Res.* 23: 23–35. [12]

Maynard Smith, J., and E. Szathmáry. 1995. *The Major Transitions in Evolution*. W. H. Freeman, San Francisco. [5, 16, 22]

Maynard Smith, J., R. Burian, S. Kauffman, P. Alberch, et al. 1985. Developmental constraints and evolution. *Q. Rev. Biol.* 60: 265–287. [21]

Mayr, E. 1942. *Systematics and the Origin of Species*. Columbia University Press, New York. [1, 17, 18]

Mayr, E. 1954. Change of genetic environment and evolution. In J. Huxley, A. C. Hardy, and E. B. Ford (eds.), *Evolution as a Process*, pp. 157–180. Allen and Unwin, London. [4, 18]

Mayr, E. 1960. The emergence of evolutionary novelties. In S. Tax (ed.), *The Evolution of Life*, pp. 157–180. University of Chicago Press, Chicago. [22]

Mayr, E. 1963. *Animal Species and Evolution*. Harvard University Press, Cambridge, MA. [9, 17, 18]

Mayr, E. 1982a. *The Growth of Biological Thought: Diversity, Evolution, and Inheritance*. Harvard University Press, Cambridge, MA. [1]

Mayr, E. 1982b. Processes of speciation in animals. In C. Barigozzi (ed.), *Mechanisms of Speciation*, pp. 1–19. Alan R. Liss, New York. [18]

Mayr, E. 1988a. Cause and effect in biology. In E. Mayr (ed.), *Toward a New Philosophy of Biology*, pp. 24–37. Harvard University Press, Cambridge, MA. [11]

Mayr, E. 1988b. The probability of extraterrestrial intelligent life. In E. Mayr (ed.), *Toward a New Philosophy of Biology*, pp. 67–74. Harvard University Press, Cambridge, MA. [22]

Mayr, E. 2004. *What Makes Biology Unique? Considerations on the Autonomy of a Scientific Discipline*. Cambridge University Press, Cambridge. [3]

Mayr, E., and W. B. Provine (eds.). 1980. *The Evolutionary Synthesis: Perspectives on the Unification of Biology*. Harvard University Press, Cambridge, MA. [1]

McCarrey, J. R., M. Kumari, M. J. Aivaliotis, Z. Q. Wang, P. Zhang, F. Marshall, and J. L. Vandenberg. 1996. Analysis of the cDNA and encoded protein of the human testis-specific PGK-2 gene. *Devel. Genet.* 19: 321–332. [8]

McCauley, D. E. 1993. Evolution in metapopulations with frequent local extinction and recolonization. In D. J. Futuyma and J. Antonovics (eds.), *Oxford Surv. Evol. Biol.*, pp. 109–134. Oxford University Press, Oxford. [9]

McClearn, G. E., B. Johansson, S. Berg, N. L. Pedersen, F. Ahern, S. A. Petrill, and R. Plomin. 1997. Substantial genetic influence on cognitive abilities in twins 80 or more years old. *Science* 276: 1560–1563. [9]

McClintock, W. J., and G. W. Uetz. 1996. Female choice and pre-existing bias: visual cues during courtship in two *Schizocosa* wolf spiders (Araneae: Lycosidae). *Animal Behaviour* 52: 167–181. [15]

McCommas, S. A., and E. H. Bryant. 1990. Loss of electrophoretic variation in serially bottlenecked populations. *Heredity* 64: 315–321. [10]

McCune, A. R., and N. J. Lovejoy. 1998. The relative rate of sympatric and allopatric speciation in fishes: tests using DNA sequence divergence between sister species and among clades. In D. J. Howard and S. H. Berlocher (eds.), *Endless Forms: Species and Speciation*, pp. 172–185. Oxford University Press, New York. [17, 18]

McDonald, D. B., and W. K. Potts. 1994. Cooperative display and relatedness among males in a lek-mating bird. *Science* 266: 1030–1032. [16]

McDonald, J. H., and M. Kreitman. 1991. Adaptive protein evolution at the *Adh* locus in *Drosophila*. *Nature* 351: 652–654. [10]

McElwain, J. C., and S. W. Punyasena. 2007. Mass extinction events and the plant fossil record. *Trends Ecol. Evol.* 22: 548–557. [5]

McGinnis, W., and R. Krumlauf. 1992. Homeobox genes and axial patterning. *Cell* 68: 283–302. [21]

McGuigan, K., S. F. Chenoweth, and M. W. Blows. 2005. Phenotypic divergence along lines of genetic variance. *Am. Nat.* 165: 32–43. [13]

McKenzie, J. A., and G. M. Clarke. 1988. Diazinon resistance, fluctuating asymmetry and fitness in the Australian sheep blowfly, *Lucilia cuprina*. *Genetics* 120: 213–220. [13]

McKinney, F. K. 1992. Competitive interactions between related clades: Evolutionary implications of overgrowth interactions between encrusting cyclostome and cheilostome bryozoans. *Mar. Biol.* 114: 645–652. [7]

McKinney, M. L., and K. J. McNamara. 1991. *Heterochrony: The Evolution of Ontogeny*. Plenum, New York. [3]

McNamara, K. J. 1997. *Shapes of Time*. Johns Hopkins University Press, Baltimore. [3]

McNeilly, T., and J. Antonovics. 1968. Evolution in closely adjacent plant populations. IV. Barriers to gene flow. *Heredity* 23: 205–218. [18]

McPeek, M. A. 1995. Morphological evolution mediated by behavior in the damselflies of two communities. *Evolution* 49: 749–769. [19]

McPeek, M. A., and J. M. Brown. 2007. Clade age and not diversification rate explains species richness among animal taxa. *Am. Nat.*169: E97–E106. [7]

McShea, D. W. 1991. Complexity and evolution: What everybody knows. *Biol. Phil.* 6: 303–324. [22]

McShea, D. W. 1994. Mechanisms of large-scale evolutionary trends. *Evolution* 48: 1747–1763. [22]

McShea, D. W. 1998. Possible largest–scale trends in organismal evolution: eight "live hypotheses." *Annu. Rev. Ecol. Syst.* 29: 293–318. [22]

McShea, D. W. 2001. The hierarchical structure of organisms: a scale and documentation of a trend in the maximum. *Paleobiology* 27: 405–423. [22]

Medawar, P. B. 1952. *An Unsolved Problem of Biology*. H. K. Lewis, London. [14]

Mendelson, T. C., and K. L. Shaw. 2005. Rapid speciation in an arthropod. *Nature* 433: 375. [6]

Messier, W., and C.-B. Stewart. 1997. Episodic adaptive evolution of primate lysozymes. *Nature* 385: 151–154. [20]

Metcalf, R. L., and W. H. Luckmann (eds.). 1994. *Introduction to Insect Pest Management*, 3rd ed. Wiley, New York. [12]

Metzker, M. L., D. P. Mindell, X.-M. Liu, R. G. Ptak, R. A. Gibbs, and D. M. Hillis. 2002. Molecular evidence of HIV-1 transmission in a criminal case. *Proc. Natl. Acad. Sci. USA* 99: 14292–14297. [23]

Meyer, A., and R. Zardoya. 2003. Recent advances in the (molecular) phylogeny of vertebrates. *Annu. Rev. Ecol. Evol. Syst.* 34: 311–338. [2]

Mezey, J. G., and D. Houle. 2005. The dimensionality of genetic variation for wing shape in *Drosophila melanogaster*. Evolution 59: 1027–1038. [13]

Michel, A. P., J. Rull, M. Aluja, and J. L. Feder. 2007. The genetic structure of hawthorn-infesting *Rhagoletis pomonella* populations in Mexico: Implications for sympatric speciation. *Mol. Ecol.* 16: 2867–2878. [18]

Michod, R. E. 1999. *Darwinian Dynamics: Evolutionary Transitions in Fitness and Individuality*. Princeton University Press, Princeton, NJ. [5, 16, 22]

Michod, R. E. 2007. Evolution of individuality during the transition from unicellular to multicellular life. *Proc. Natl. Acad. Sci. USA* 104: 8613–8618. [5]

Michod, R. E., and B. R. Levin. 1988. *The Evolution of Sex*. Sinauer, Sunderland, MA. [15]

Michor, F., S. A. Frank, R. M. May, Y. Iwasa, and M. A. Nowak. 2003. Somatic selection for and against cancer. *J. Theor. Biol.* 225: 377–382. [11]

Milá, B., D. J. Girman, M. Kimura, and T. B. Smith. 2000. Genetic evidence for the effect of a postglacial population expansion on the phylogeography of a North American songbird. *Proc. R. Soc. Lond. B* 267: 1033–1040. [2, 9]

Millais, J. G. 1897. *British Deer and Their Horns*. Henry Sotheran and Co., London. [3]

Miller, A. I. 2000. Conversations about Phanerozoic global diversity. *Paleobiology* 26 (Suppl.): 53–73. [7]

Miller, K. R. 2008. *Only a Theory: Evolution and the Battle for America's Soul*. Viking, New York. [23]

Miller, M. P. and S. Kumar. 2001. Understanding human disease mutations through the use of interspecific genetic variation. *Hum. Molec. Genet.* 21: 2319–2328. [23]

Miller, S. L. 1953. Production of amino acids under possible primitive earth conditions. *Science* 117: 528–529. [5]

Mindell, D. F., and C. E. Thacker. 1996. Rates of molecular evolution: phylogenetic issues and applications. *Annu. Rev. Ecol. Syst.* 27: 279–303. [2]

Mindell, D. O. 2006. *The Evolving World: Evolution in Everyday Life*. Harvard University Press, Cambridge, MA. [23]

Miralles, R., P. J. Gerrish, A. Moya, and S. F. Elena. 1999. Clonal interference and the evolution of RNA viruses. *Science* 285: 1745–1747. [15]

Mitra, S., H. Landel, and S. Pruett-Jones. 1996. Species richness covaries with mating systems in birds. *Auk* 113: 544–551. [18]

Mittelbach, G. G. and 21 others. 2007. Evolution and the latitudinal diversity gradient: Speciation, extinction and biogeography. *Ecol. Lett.* 10: 315–331. [6]

Mitter, C., and B. D. Farrell. 1991. Macroevolutionary aspects of insect-plant relationships. In E. A. Bernays (ed.), *Insect/Plant Interactions*, vol. 3, pp. 35–78. CRC Press, Boca Raton, FL. [6]

Mitter, C., B. D. Farrell, and B. Wiegmann. 1988. The phylogenetic study of adaptive zones: Has phytophagy promoted insect diversification? *Am. Nat.* 132: 107–128. [7, 22]

Mitton, J. B., and M. C. Grant. 1984. Associations among protein heterozygosity, growth rate, and developmental homeostasis. *Annu. Rev. Ecol. Syst.* 15: 479–499. [12]

Mock, D. W. 2004. *More than Kin and Less than Kind: The Evolution of Family Conflict*. Harvard University Press, Cambridge, MA. [16]

Moczek, A. P., J. Hunt, D. J. Emlen, and L. W. Simmons. 2002. Threshold evolution in exotic populations of a polyphenic beetle. *Evol. Ecol. Res.* 4: 587–601. [21]

Møller, A. P., and R. V. Alatalo. 1999. Good-genes effects in sexual selection. *Proc. R. Soc. Lond. B* 266: 85–91. [15]

Montgomery, S. L. 1982. Biogeography of the moth genus *Eupithecia* in Oceania and the evolution of ambush predation in Hawaiian caterpillars (Lepidoptera: Geometridae). *Entomologia Generalis* 8: 27–34. [7]

Moore, F. B. G., D. E. Rozen, and R. E. Lenski. 2000. Pervasive compensatory adaptation in *Escherichia coli*. *Proc. R. Soc. Lond. B* 267: 515–522. [8]

Moore, R. C. (ed.) 1966. *Treatise on Invertebrate Paleontology, Part U, Echinodermata*. Geological Society of America and University of Kansas Press 3(2), Boulder, CO. [5]

Moore, W. S., and J. T. Price. 1993. Nature of selection in the northern flicker hybrid zone and its implications for speciation theory. In R. G. Harrison (ed.), *Hybrid Zones and the Evolutionary Process*, pp. 196–225. Oxford University Press, New York. [9]

Moran, N. 1996. Accelerated evolution and Muller's ratchet in endosymbiotic bacteria. *Proc. Natl. Acad. Sci. USA* 93: 2873–2878. [15]

Moran, N. 2003. Tracing the evolution of gene loss in obligate bacterial symbionts. *Curr. Opin. Microbiol.* 6: 512–518. [3, 20]

Moran, N. A. 2007. Symbiosis as an adaptive process and source of phenotypic complexity. *Proc. Natl. Acad. Sci. USA* 104: 8627–8633. [5, 19]

Moran, N. A., and P. Baumann. 1994. Phylogenetics of cytoplasmically inherited microorganisms of arthropods. *Trends Ecol. Evol.* 9: 15–20. [19]

Moran, N. A., H. J. McLaughlin, and R. Sorek. 2009. The dynamics and time scale of ongoing genomic erosion in symbiotic bacteria. *Science* 323: 379–382. [20]

Morjan, C. L., and L. H. Rieseberg. 2004. How species evolve collectively: Implications of gene flow and selection for the spread of advantageous alleles. *Mol. Ecol.* 13: 1341–1356. [18]

Morton, W. F., J. F. Crow, and H. J. Muller. 1956. An estimate of the mutational damage in man from data on consanguineous marriages. *Proc. Natl. Acad. Sci. USA* 42: 855–863. [9]

Mousseau, T. A., and D. A. Roff. 1987. Natural selection and the heritability of fitness components. *Heredity* 59: 181–197. [9, 13]

Moyle, P. B., and J. J. Cech Jr. 1983. *Fishes: An Introduction to Ichthyology*. Prentice-Hall, Englewood Cliffs, NJ. [6]

Mueller, L. D., P. Guo, and F. J. Ayala. 1991. Density-dependent natural selection and trade-offs in life history traits. *Science* 253: 433–435. [14]

Mueller, U. G., N. M. Gerardo, D. K. Aanen, L. Six, and T. R. Schultz. 2005. The evolution of agriculture in insects. *Annu. Rev. Ecol. Evol. Syst.* 36: 563–595. [19]

Mukai, T., S. I. Chigusa, L. E. Mettler, and J. F. Crow. 1972. Mutation rate and dominance of genes affecting viability in *Drosophila melanogaster*. *Genetics* 72: 335–355. [8]

Müller, G. B. 1990. Developmental mechanisms at the origin of morphological novelty: A side-effect hypothesis. In M. H. Nitecki (ed.), *Evolutionary Innovations*, pp. 99–130. University of Chicago Press, Chicago. [3, 18]

Müller, G. B., and G. P. Wagner. 1991. Novelty in evolution: Restructuring the concept. *Annu. Rev. Ecol. Syst.* 23: 229–256. [22]

Müller, G. B., and G. P. Wagner. 1996. Homology, *Hox* genes, and developmental integration. *Am. Zool.* 36: 4–13. [3]

Muller, H. J. 1940. Bearing of the *Drosophila* work on systematics. In J. S. Huxley (ed.), *The New Systematics*, pp. 185–268. Clarendon Press, Oxford. [17, 18]

Muller, H. J. 1964. The relation of recombination to mutational advance. *Mutat. Res.* 1: 2–9. [15]

Muñoz-Fuentes, V., A. J. Green, M. D. Sorenson, J. J. Negro, and C. Vila. 2006. The ruddy duck *Oxyura jamaicensis* in Europe: Natural colonization or human introduction? *Mol. Ecol.* 15: 1441–1453. [10]

Müntzing, A. 1930. Über Chromosomenvermehrung in Galeopsis-Kreuzungen und ihre phylogenetische Bedeutung. *Hereditas* 14: 153–172. [18]

Myers, A. A., and P. S. Giller (eds.). 1988. *Analytical Biogeography*. Chapman & Hall, London. [6] Ma, H., and C. dePamphilis. 2000. The ABCs of floral evolution. *Cell* 101: 5–8. [21]

N

Nachman, M. W., and S. L. Crowell. 2000. Estimate of the mutation rate per nucleotide in humans. *Genetics* 156: 297–304. [8]

Nachman, M. W., H. E. Hoekstra, and S. L. D'Agostino. 2003. The genetic basis of adaptive melanism in pocket mice. *Proc. Natl. Acad. Sci. U.S.A.* 100: 5268–5273. [3]

Nason, J. D., N. C. Ellstrand, and M. L. Arnold. 1992. Patterns of hybridization and introgression in populations of oaks, manzanitas, and irises. *Am. J. Bot.* 79: 101–111. [17]

Nee, S. 2006. Birth-death models in macroevolution. *Annu. Rev. Ecol. Evol. Syst.* 37: 1–17. [7]

Nei, M. 1987. *Molecular Evolutionary Genetics*. Columbia University Press, New York. [9]

Nei, M., and A. L. Hughes. 1991. Polymorphism and evolution of the major histocompatibility complex loci in mammals. In R. K. Selander, A. G. Clark, and T. S. Whittam (eds.), *Evolution at the Molecular Level*, pp. 222–247. Sinauer, Sunderland, MA. [12]

Nei, M., T. Maruyama, and R. Chakraborty. 1975. The bottleneck effect and genetic variability in populations. *Evolution* 29: 1–10. [10]

Neigel, J. E., and J. C. Avise. 1986. Phylogenetic relationship of mitochondrial DNA under various demographic models of speciation. In E. Nevo and S. Karlin (eds.), *Evolutionary Processes and Theory*, pp. 515–534. Academic Press, London. [17]

Nesse, R. M., and S. C. Stearns. 2008. The great opportunity: Evolutionary applications to medicine and public health. *Evol. Applications* 1: 28–48. [23]

Nesse, R., and G. C. Williams. 1994. *Why We Get Sick: The New Science of Darwinian Medicine*. Times Books, New York. [23]

Nestmann, E. R., and R. F. Hill. 1973. Population genetics in continuously growing mutator cultures of *Escherichia coli*. *Genetics* 73: 41–44. [11]

Nevo, E. 1991. Evolutionary theory and processes of active speciation and adaptive radiation in subterranean mole rats, *Spalax ehrenbergi* superspecies, in Israel. *Evol. Biol.* 25: 1–125. [17]

Newcomb, R. D., P. M. Campbell, D. L. Olles, E. Cheah, R. J. Russell, and J. G. Oakeshott. 1997. A single amino acid substitution converts a carboxylesterase to an organophosphate hydrolase and confers insecticide resistance on a blowfly. *Proc. Natl. Acad. Sci. USA* 94: 7464–7468. [8]

Ng, C. S., and A. Kopp. 2008. Sex combs are important for male mating success in *Drosophila melanogaster*. *Behav. Genet.* 38: 195–201. [21]

Nielsen, R. 2005. Molecular signatures of natural selection. *Annu. Rev. Genet.* 39: 197–218. [12]

Nielsen, R., I. Hellmann, M. Hubisz, C. Bustamante, and A. G. Clark. 2007. Recent and ongoing selection in the human genome. *Nature Rev. Genet.* 8: 857–868. [12]

Niklas, K. J., B. H. Tiffney, and A. H. Knoll. 1983. Patterns in vascular land plant diversification. *Nature* 303: 614–616. [7]

Nilsson, D. E., and S. Pelger. 1994. A pessimistic estimate of the time required for an eye to evolve. *Proc. R. Soc. Lond. B* 256: 59–65. [22]

Nilsson, L. A., L. Jonsson, L. Ralison, and E. Randrianjohany. 1985. Monophily and pollination mechanisms in *Angraecum arachnites* Schltr. (Orchidaceae) in a guild of long-tongued hawkmoths (Sphingidae). *Biol. J. Linn. Soc.* 26: 1–19. [19]

Nisbett, R. 1995. Race, IQ, and scientism. In S. Fraser (ed.), *The Bell Curve Wars*, pp. 36–57. BasicBooks, New York. [9]

Nitecki, M. H. (ed.). 1988. *Evolutionary Progress*. University of Chicago Press, Chicago. [11]

Nitecki, M. H. (ed.). 1990. *Evolutionary Innovations*. University of Chicago Press, Chicago. [22]

Noë, R., and P. Hammerstein. 1995. Biological markets. *Trends Ecol. Evol.* 10: 336–339. [16]

Noonan, B. P., and P. T. Chippindale. 2006. Vicariant origin of Malagasy reptiles supports Late Cretaceous Antarctic land bridge. *Am. Nat.* 168: 730–741. [6]

Noonan, J. P., and 10 others. 2006. Sequencing and analysis of Neanderthal genomic DNA. *Science* 314: 1113–1118. [6, 20]

Noor, M. A. F. 1995. Speciation driven by natural selection in *Drosophila*. *Nature* 375: 674–675. [18]

Noor, M. A. F. 1999. Reinforcement and other consequences of sympatry. *Heredity* 83: 503–508. [18]

Norell, M. A., and M. J. Novacek. 1992. The fossil record and evolution: Comparing cladistic and paleontologic evidence for vertebrate history. *Science* 255: 1690–1693. [4]

Norell, M. A., and X. Xu. 2005. Feathered dinosaurs. *Annu. Rev. Earth & Planetary Sci.* 33: 277–299. [4]

Normark, B. B., O. P. Judson, and N. A. Moran. 2003. Genomic signatures of ancient asexual lineages. *Biol. J. Linn. Soc.* 79: 69–84. [11, 15]

Nosil, P. 2007. Divergent host plant adaptation and reproductive isolation between ecotypes of *Timema cristinae* walking sticks. *Am. Nat.* 169: 151–162. [18]

Nosil, P., T. H. Vines, and D. J. Funk. 2005. Perspective: Reproductive isolation caused by natural selection against immigrants from divergent habitats. *Evolution* 59: 705–719. [18]

Novembre, J., and 11 others. 2008. Genes mirror geography within Europe. *Nature* (31 Aug), doi: 10.1038/nature07331. [9]

Novotny, V., and 15 others. 2007. Low beta diversity of herbivorous insects in tropical forests. *Nature* 448: 692–695. [7]

Novotny, V., P. Drozd, S. E. Miller, M. Kulfan, M. Janda, Y. Basset, and G. D. Weiblen. 2006. Why are there so many species of herbivorous insects in tropical forests? *Science* 313: 1115–1118. [7]

Nowak, M. A. 2006. Five rules for the evolution of cooperation. *Science* 314: 1560–1563. [16]

Numbers, R. L. 2006. *The Creationists: From Scientific Creationism to Intelligent Design.* Harvard University Press, Cambridge, MA. [22, 23]

Nummela, S., J. G. M. Thewissen, S. Bajpal, S. T. Hussain, and K. Kumar. 2004. Eocene evolution of whale hearing. *Nature* 430: 776–778. [4]

Nuzhdin, S. V., and T. F. C. Mackay. 1994. Direct determination of retrotransposon transposition rates in *Drosophila melanogaster*. *Genet. Res.* 63: 139–144. [8]

Nuzhdin, S. V., C. L. Dilda, and T. F. C. Mackay. 1999. The genetic architecture of selection response: inferences from fine-scale mapping of bristle number quantitative trait loci in *Drosophila melanogaster*. *Genetics* 153: 1317–1331. [13]

O

Oakeshott, J. G., J. B. Gibson, P. R. Anderson, W. R. Knib, D. G. Anderson, and G. K. Chambers. 1982. Alcohol dehydrogenase and glycerol-3-phosphate dehydrogenase clines in *Drosophila melanogaster* on different continents. *Evolution* 36: 86–96. [9]

Oakley, T. H., and M. S. Pankey. 2008. Opening the "black box": The genetic and biochemical basis of eye evolution. *Evolution: Education and Outreach* 1: 390–402. [22]

Ochman, H., J. G. Lawrence, and E. A. Groisman. 2000. Lateral gene transfer and the nature of bacterial innovation. *Nature* 405: 299–304. [2, 20]

Odling-Smee, F. J., K. N. Laland, and M. W. Feldman. 2003. *Niche Construction: The Neglected Process in Evolution.* Princeton University Press, Princeton, NJ. [22]

Ohno, S. 1970. *Evolution by Gene Duplication.* Springer-Verlag, Berlin. [20]

Okasha, S. 2006. *Evolution and the Levels of Selection.* Oxford University Press, Oxford. [11]

Oliver, G., A. Mailhos, R. Wehr, N. G. Copeland, N. A. Jenkins, and P. Gruss. 1995. *Six3*, a murine homolog of the *sine oculis* gene, demarcates the most anterior border of the developing neural plate and is expressed during eye development. *Development* 121: 4045–4055. [21]

Oliver, M. J., D. Petrov, D. Ackerley, P. Falkowski, and O. M. Schofield. 2007. The mode and tempo of genome size evolution in eukaryotes. *Genome Research* 17: 594–601. [22]

Olivieri, I., Y. Michalakis, and P.-H. Gouyon. 1995. Metapopulation genetics and the evolution of dispersal. *Am. Nat.* 146: 202–228. [14]

Olmstead, R. G., and J. A. Sweere. 1994. Combining data in phylogenetic systematics: An empirical approach using three molecular data sets in the Solanaceae. *Syst. Biol.* 43: 467–481. [2]

Olsen, E. M., M. Heino, G. R. Lilly, M. J. Morgan, J. Brattley, B. Ernande, and U. Dieckmann. 2004. Maturation trends indicative of rapid evolution preceded the collapse of northern cod. *Nature* 428: 932–935. [13]

Olson, E. C., and R. L. Miller. 1958. *Morphological Integration.* University of Chicago Press, Chicago. [13]

Orr, H. A. 1995. The population genetics of speciation the evolution of hybrid incompatibilities. *Genetics* 139: 1805–1813. [18]

Orr, H. A. 1998. The population genetics of adaptation: The distribution of factors fixed during adaptive evolution. *Evolution* 52: 935–949. [13]

Orr, H. A. 2000. Adaptation and the cost of complexity. *Evolution* 54: 13–20. [13]

Orr, H. A., and J. Coyne. 1992. The genetics of adaptation: A reassessment. *Am. Nat.* 140: 725–742. [22]

Orr, H. A., and S. Irving. 2001. Complex epistasis and the genetic basis of hybrid sterility in the *Drosophila pseudoobscura* Bogotá-USA hybridization. *Genetics* 158: 1089–1100. [17]

Osada, N., and H. Innan. 2008. Duplication and gene conversion in the *Drosophila* melanogaster genome. *PLoS Genetics* 4(12): e1000305. [20]

Osorio, D. 1994. Eye evolution: Darwin's shudder stilled. *Trends Ecol. Evol.* 9: 241–242. [22]

Ostrom, J. H. 1976. On a new specimen of the Lower Cretaceous theropod dinosaur *Deinonychus antirrhopus*. *Breviora* 439: 1–21. [4]

Otto, S. P., and J. Whitton. 2000. Polyploid incidence and evolution. *Annu. Rev. Genet.* 34: 401–437. [18]

Otto, S. P., and S. L. Nuismer. 2004. Species interactions and the evolution of sex. *Science* 304: 1018–1020. [15]

Ownbey, M. 1950. Natural hybridization and amphiploidy in the genus *Tragopogon*. *Am. J. Bot.* 37: 489–499. [18]

Oyama, R. K., and 8 others. 2008. The shrunken genome of *Arabidopsis thaliana*. *Plant Syst. Evol.* 273: 257–271. [22]

P

Page, R. B., S. R. Voss, A. K. Samuels, J. J. Smith, S. Putta, and C. K. Beachy. 2008. Effect of thyroid hormone concentration on the transcriptional response underlying induced metamorphosis in the Mexican axolotl (*Ambystoma*). *BMC Genomics* 9: 78. [21]

Page, R. D. M. 1994. Maps between trees and cladistic analyses of historical associations among genes, organisms, and areas. *Syst. Biol.* 43: 58–77. [6]

Page, R. D. M. 2002. Introduction. In R. D. M. Page (ed.), *Tangled Trees: Phylogeny, Cospeciation, and Coevolution*, pp. 1–21. University of Chicago Press, Chicago. [19]

Pagel, M., C. Venditti, and A. Meade. 2006. Large punctuational contribution of speciation to evolutionary divergence at the molecular level. *Science* 314: 119–121. [18]

Pagel, M., Q. D. Atkinson, and A. Meade. 2007. Frequency of word-use predicts rates of lexical evolution throughout Indo-European history. *Nature* 449: 717–720. [23]

Palmer, A. R. 1982. Predation and parallel evolution: Recurrent parietal plate reduction in balanomorph barnacles. *Paleobiology* 8: 31–44. [4]

Palmer, J. D., and J. M. Logsdon, Jr. 1991. The recent origin of introns. *Curr. Opin. Genet. Dev.* 1: 470–477. [20]

Palmer, J. D., D. E. Soltis, and M. W. Chase. 2004. The plant tree of life: An overview and some points of view. *Am. J. Bot.* 91: 1437–1445. [5]

Palstra, F. P., and D. E. Ruzzante. 2008. Genetic estimates of contemporary effective population size: What can they tell us about the importance of genetic stochasticity for wild population persistence? *Mol. Ecol.* 17: 3428–3447. [10]

Palumbi, S. R. 1998. Species formation and the evolution of gamete recognition loci. In D. J. Howard and S. H. Berlocher (eds.), *Endless Forms: Species and Speciation*, pp. 271–278. Oxford University Press, New York. [15, 17]

Palumbi, S. R. 2001. *The Evolution Explosion: How Humans Cause Rapid Evolutionary Change.* W. W. Norton, New York [1, 13]

Pamilo, P., and M. Nei. 1988. Relationships between gene trees and species trees. *Mol. Biol. Evol.* 5: 568–583. [17]

Panganiban, G., S. M. Irvine, C. Lowe, H. Roehl, L. S. Corley, B. Sherbon, J. K. Grenier, J. F. Fallon, J. Kimble, M. Walker, G. A. Wray, B. J. Swalla, M. Q. Martindale, and S. B. Carroll. 1997. The origin and evolution of animal appendages. *Proc. Natl. Acad. Sci. USA* 94: 5162–5166. [21]

Panhuis, T. M., R. Butlin, M. Zuk, and T. Tregenza. 2001. Sexual selection and speciation. *Trends Ecol. Evol.* 16: 364–371. [18]

Pannebakker, B. A., B. Loppin, C. P. H. Elemans, L. Humblot, and F. Vavre. 2007. Parasitic inhibition of cell death facilitates symbiosis. *Proc. Natl. Acad. Sci. USA* 104: 213–215. [19]

Panopoulou, G., and A. J. Poustka. 2005. Timing and mechanism of ancient vertebrate genome duplications: The adventure of a hypothesis. *Trends Genet.* 21: 559–567. [3]

Paradis, J., and G. C. Williams. 1989. *Evolution and Ethics: T. H. Huxley's Evolution & Ethics with New Essays on its Victorian and Sociobiological Context*. Princeton University Press, Princeton, NJ. [1]

Parker, G. A. 1970. Sperm competition and its evolutionary consequences in the insects. *Biol. Rev.* 45: 525–567. [15]

Parker, G. A. 1979. Sexual selection and sexual conflict. In M. S. Blum and N. A. Blum (eds.), *Sexual Selection and Reproduction*, pp. 123–166. Academic Press, New York. [15]

Parker, G. A., and J. Maynard Smith. 1990. Optimality theory in evolutionary biology. *Nature* 348: 27–33. [14]

Parmesan, C., N., and 12 others. 1999. Poleward shift of butterfly species' ranges associated with regional warming. *Nature* 399: 579–583. [6]

Parmesan, C., S. Gaines, L. Gonzalez, D. M. Kaufman, J. Kingsolver, A. T. Peterson, and R. Sagarin. 2005. Empirical perspectives on species borders: from traditional biogeography to global change. *Oikos* 108: 58–75. [6]

Partridge, L. 2001. Evolutionary theories of ageing applied to long-lived organisms. *Exper. Gerontol.* 36: 641–650. [14]

Partridge, L., and L. D. Hurst. 1998. Sex and conflict. *Science* 281: 2003–2008. [15]

Partridge, L., N. Prowse, and P. Pignatelli. 1999. Another set of responses and correlated responses to selection on age at reproduction in *Drosophila melanogaster*. *Proc. R. Soc. Lond. B* 266: 255–261. [14]

Paterson, H. E. H. 1985. The recognition concept of species. In E. S. Vrba (ed.), *Species and Speciation*, pp. 21–29. Transvaal Museum Monograph No. 4, Pretoria, South Africa. [17]

Patton, J. L., and S. Y. Yang. 1977. Genetic variation in *Thomomys bottate* pocket gophers: Macrogeographic patterns. *Evolution* 31: 697–720. [10]

Pellmyr, O., and C. J. Huth. 1994. Evolutionary stability of mutualism between yuccas and yucca moths. *Nature* 372: 257–260. [19]

Pellmyr, O., and J. Leebens-Mack. 1999. Forty million years of mutualism: Evidence for Eocene origin of the yucca-yucca moth association. *Proc. Natl. Acad. Sci. USA* 96: 9178–9183. [19]

Pennock, R. T. 1999. *Tower of Babel: The Evidence against the New Creationism.* MIT Press, Cambridge, MA. [1]

Perfeito, L., L. Fernandes, C Mota, and I. Gordo. 2007. Adaptive mutations in bacteria: High rate and small effects. *Science* 317: 813–815. [8]

Peterson, A. T., J. Soberón, and V. Sánchez-Cordero. 1999. Conservatism of ecological niches in evolutionary time. *Science* 285: 1265–1267. [6]

Peterson, K. J., J. A. Cotton, J. G. Kehling, and D. Pisani. 2008. The Ediacaran emergence of bilaterians: congruence between the genetic and the geological fossil records. *Phil. Trans. R. Soc. Lond. B* 363: 1435–1443. [5]

Petrov, D., and J. F. Wendel. 2006. Evolution of eukaryotic genome structure. In C. W. Fox and J. B. Wolf (eds.), *Evolutionary Genetics*, pp.144–156. Oxford University Press, New York. [8]

Pfennig, D. W., and W. A. Frankino. 1997. Kin-mediated morphogenesis in facultatively cannibalistic tadpoles. *Evolution* 51: 1993–1999. [16]

Pfennig, D. W., J. P. Collins, and R. E. Ziemba. 1999. A test of alternative hypotheses for kin recognition in cannibalistic tiger salamanders. *Behav. Ecol.* 10: 436–443. [11]

Phillips, B. L., and R. Shine. 2004. Adapting to an invasive species: Toxic cane toads induce morphological change in Australian snakes. *Proc. Natl. Acad. Sci. USA* 101: 17150–17155. [19]

Phillips, B. L., and R. Shine. 2006. An invasive species induces rapid adaptive change in a native predator: Cane toads and black snakes in Australia. *Proc. R. Soc. Lond. B* 273: 1545–1550. [19]

Piatigorsky, J. 2007. *Gene Sharing and Evolution: The Diversity of Protein Functions*. Harvard University Press, Cambridge, MA. [11, 22]

Pielou, E. C. 1991. *After the Ice Age: The Return of Life to Glaciated North America*. University of Chicago Press, Chicago. [5]

Pigliucci, M. 2001. *Phenotypic Plasticity: Beyond Nature and Nurture*. Johns Hopkins University Press, Baltimore. [13]

Pigliucci, M. 2002. *Denying Evolution: Creationism, Scientism, and the Nature of Science*. Sinauer, Sunderland, MA. [1, 23]

Pigliucci, M. 2008. Is evolvability evolvable? *Nature Rev. Genet.* 9: 75–82. [22]

Pigliucci, M., and K. Preston (eds.). 2004. *Phenotypic Integration: Studying the Ecology and Evolution of Complex Phenotypes*. Oxford University Press, Oxford. [13]

Pinker, S. 1997. *How the Mind Works.* Norton, New York. [23]

Pinker, S. 2002. *The Blank Slate: The Modern Denial of Human Nature*. Penguin Books, New York. [16]

Pires, J. C., and 9 others. Molecular cytogenetic analysis of recently evolved *Tragopogon* (Asteraceae) allopolyploids reveal a karyotype that is additive of the diploid progenitors. *Am. J. Bot.* 91: 1022–1035. [18]

Pitnick, S., and T. A. Markow. 1994. Large-male advantages associated with costs of sperm production in *Drosophila hydei*, a species with giant sperm. *Proc. Natl. Acad. Sci. USA* 91: 9277–9281. [15]

Plomin, R., J. C. DeFries, G. E. McClearn, and M. Rutter. 1997. *Behavioral Genetics*, 3rd ed. W. H. Freeman, New York. [9]

Pomiankowski, A. 1988. The evolution of female mate preferences for male genetic quality. *Oxford Surv. Evol. Biol.* 5: 136–184. [15]

Pomiankowski, A., and Y. Iwasa. 1998. Runaway ornament diversity caused by Fisherian sexual selection. *Proc. Natl. Acad. Sci. USA* 95: 5106–5111. [15, 18]

Pomiankowski, A., Y. Iwasa, and S. Nee. 1991. The evolution of costly mate preferences. I. Fisher and biased mutation. *Evolution* 45: 1422–1430. [15]

Poole, A. M., and D. Penny. 2006. Evaluating hypotheses for the origin of eukaryotes. *BioEssays* 29: 74–84. [5]

Poole, A. M., D. C. Jeffares, and D. Penny. 1998. The path from the RNA world. *J. Molec. Evol.* 46: 1–17. [20]

Poon, A., and L. Chao. 2004. Drift increases the advantage of sex in RNA bacteriophage ?6. *Genetics* 166: 19–24. [15]

Popesco, M. C., and 8 others. 2006. Human lineage-specific amplification, selection, and neuronal expression of DUF1220 domains. *Science* 313: 1304–1307. [20]

Porter, K. R. 1972. *Herpetology*. W. B. Saunders, Philadelphia, PA. [11]

Potts, R. 1988. *Early Hominid Activity at Olduvai*. Aldine de Gruyter, New York. [4]

Prabhakar, S., and 12 others. 2008. Human-specific gain of function in a developmental enhancer. *Science* 321: 1346–1350. [21]

Presgraves, D. C., and W. Stephan.2007. Pervasive adaptive evolution among interactors of the *Drosophila* hybrid inviability gene, *Nup96*. *Mol. Biol. Evol.* 24: 306–314. [17]

Presgraves, D. C., L. Balagopalan, S. M. Abmayr, and H. A. Orr. 2003. Adaptive evolution drives divergence of a hybrid inviability gene between two species of *Drosophila*. *Nature* 423: 715–719. [17]

Price, T. D. 2008. *Speciation in Birds*. Roberts and Co., Greenwood Village, Colo. [18]

Prince, V. E. 2002. The Hox paradox: More complex(es) than imagined. *Dev. Biol.* 249: 1–15. [20]

Prince, V. E., and F. B. Pickett. 2002. Splitting pairs: The diverging fates of duplicated genes. *Nature Rev. Genet.* 3: 827–837. [20]

Promislow, D. E. L., and P. H. Harvey. 1991. Mortality rates and the evolution of mammalian life histories. *Acta Oecologica* 220: 417–437. [14]

Prothero, D. R. 2006. *After the Dinosaurs: The Age of Mammals*. Indiana University Press, Bloomington. [5]

Prothero, D. R. 2007. *Evolution: What the Fossils Say and Why it Matters*. Columbia University Press, New York. [4]

Prothero, D. R., and C. Buell. 2007. *Evolution: What the Fossils Say and Why it Matters*. Columbia University Press, New York. [5]

Prout, T. 1980. Some relationships between density-independent selection and density-dependent population growth. *Evol. Biol.* 13: 1–68. [14]

Prud'homme, B., N. Gompel, A. Rokas, V. A. Kassner, T. M. Williams, S.-D. Yeh, J. R. True, and S. B. Carroll. 2006. Repeated morphological evolution through *cis*-regulatory changes in a pleiotropic gene. *Nature* 440: 1050–1053. [21]

Prud'homme, B., N. Gompel, and S. M. Carroll. 2007. Emerging principles of regulatory evolution. *Proc. Natl. Acad. Sci. USA 104* (Suppl. 1): 8605–8612. [22]

Przeworski, M., G. Coop, and J. D. Wall. 2005. The signature of positive selection on standing genetic variation. Evolution 59: 2312–2323. [13]

Purugganan, M. D. 1997. The MADS-box floral homeotic gene lineages predate the origin of seed plants: Phylogenetic and molecular clock estimates. *J. Mol. Evol.* 45: 392–396. [21]

Q

Quammen, D. 2006. *The Reluctant Mr. Darwin: An Intimate Portrait of Charles Darwin and Making of His Theory of Evolution*. W. W. Norton, New York. [1]

Quattrocchio, F., J. Wing, K. Van der Woude, E. Souer, N. van Netten, J. Mol, and R. Koes. 1999. Molecular analysis of the *ANTHOCYANIN2* gene of petunia and its role in the evolution of flower color. *Plant Cell* 11: 1433–1444. [8]

Queitsch, C., T. A. Sangster, and S. Lindquist. 2002. *Hsp90* as a capacitor of phenotypic variation. *Nature* 417: 618–624. [13]

Queller, D. C. 1994. Genetic relatedness in viscous populations. *Evol. Ecol.* 8: 70–73. [16]

Queller, D. C., F. Zacchi, R. Cervo, S. Turillazzi, M. T. Henshaw, L. A. Santorelli, and J. E. Strassmann. 2000. Unrelated helpers in a social insect. *Nature* 405: 784–787. [16]

Queller, D. C., E. Ponte, S.Bozzaro, and J. E. Strassmann. 2003. Single-gene greenbeard effects in the social amoeba *Dictyostelium discoideum*. *Science* 299: 105–106. [16]

R

Radinsky, L. B. 1984. Ontogeny and phylogeny in horse skull evolution. *Evolution* 38: 1–15. [21]

Raff, R. A. 1996. *The Shape of Life: Genes, Development, and the Evolution of Animal Form*. University of Chicago Press, Chicago. [3]

Ralls, K., P. H. Harvey, and A. M. Lyles. 1986. Inbreeding in natural populations of birds and mammals. In M. E. Soulé (ed.), *Conservation Biology: The Science of Scarcity and Diversity*, pp. 35–56. Sinauer, Sunderland, MA. [15]

Ramsey, J., and D. W. Schemske. 1998. Pathways, mechanisms, and rates of polyploid formation in flowering plants. *Annu. Rev. Ecol. Syst.* 29: 467–501. [8, 18]

Randsholt, N. B., and P. Santamaria. 2008. How *Drosophila* change their combs: the Hox gene *Sex combs reduced* and sex comb variation among *Sophophora* species. *Evol. Dev.* 10: 121–133. [21]

Ranz, J. M., F. Casals, and A. Ruiz. 2001. How malleable is the eukaryotic genome? Extreme rate of chromosomal rearrangement in the genus *Drosophila*. *Genome Res.* 11: 230–239. [8]

Rasmussen, S., L. Chen, D. Deamer, D. . Krakauer, N. H. Packard, P. F. Stadler, and M. A. Bedau. 2004. Transitions from nonliving to living matter. *Science* 303: 963–965. [5]

Rast, J. P., L. C. Smith, M. Loza-Coll, T. Hibino, and G. W. Litman. 2006. Review: Genomic insights into the immune system of the sea urchin. *Science* 314: 952–956. [20]

Ratnieks, F. L. W., and T. Wenseleers. 2007. Altruism in insect societies and beyond: Voluntary or enforced? *Trends Ecol. Evol.* 23: 45–52. [16]

Ratnieks, F. L. W., K. R. Foster, and T. Wenseleers. 2006. Conflict resolution in insect societies. *Annu. Rev. Entomol.* 51: 581–608. [16]

Raup, D. M. 1972. Taxonomic diversity during the Phanerozoic. *Science* 177: 1065–1071. [7]

Raup, D. M., and J. J. Sepkoski Jr. 1982. Mass extinctions in the marine fossil record. *Science* 215: 1501–1503. [7]

Raxworthy, C. J., M. R. J. Forstner, and R. A. Nussbaum. 2002. Chamaeleon radiation by oceanic dispersal. *Nature* 415: 784–787. [6]

Raymond, M., C. Merticat, M. Weill, N. Pasteur, and C. Chevillon. 2001. Insecticide resistance in the mosquito *Culex pipiens*: What have we learned about adaptation? *Genetica* 112-113: 287–296. [12]

Razin, S. 1997. The minimal cellular genome of mycoplasma. *Indian J. Biochem. Biophys.* 34: 124–130. [20]

Reeve, H. K., and P. W. Sherman. 1993. Adaptation and the goals of evolutionary research. *Q. Rev. Biol.* 68: 1–32. [11]

Relethford, J. H. 2008. Genetic evidence and the modern human origins debate. *Heredity* 100: 555–563. [6]

Rendel, J. M., B. L. Sheldon, and D. E. Finlay. 1966. Selection for canalization of the scute phenotype. II. *Am. Nat.* 100: 13–31. [13]

Rensch, B. 1959. *Evolution Above the Species Level*. Columbia University Press, New York. [1, 3, 22]

Reznick, D. 1985. Cost of reproduction: An evaluation of the empirical evidence. *Oikos* 44: 257–267. [14]

Reznick, D., and J. Travis. 2002. Adaptation. In C.W. Fox, D. A. Roff, and D. J. Fairbairn (eds.), *Evolutionary Ecology: Concepts and Case Studies*, pp. 44–57. Oxford University Press, New York. [14]

Reznick, D., H. Bryga, and J. A. Endler. 1990. Experimentally induced life-history evolution in a natural population. *Nature* 346: 357–359. [14]

Rice, W. R. 1992. Sexually antagonistic genes: Experimental evidence. *Science* 256: 1436–1439. [15]

Rice, W. R. 1996. Sexually antagonistic male adaptation triggered by experimental arrest of female evolution. *Nature* 381: 232–234. [15]

Rice, W. R., and B. Holland. 1997. The enemies within: Intergenomic conflict, interlocus contest evolution (ICE), and the intraspecific Red Queen. *Behav. Ecol. Sociobiol.* 41: 1–10. [15]

Rice, W. R., and E. E. Hostert. 1993. Laboratory experiments on speciation: What have we learned in forty years? *Evolution* 47: 1637–1653. [18]

Rice, W. R., and G. W. Salt. 1990. The evolution of reproductive isolation as a correlated character under sympatric conditions: Experimental evidence. *Evolution* 44: 1140–1152. [18]

Richardson, D. S., T. Burke, and J. Kondeur. 2002. Direct benefits and the evolution of female-biased cooperative breeding in Seychelles warblers. *Evolution* 56: 2313–2321. [16]

Richardson, M. K., J. Hanken, L. Selwood, G. M. Wright, R. J. Richards, C. Pieau, and A. Raynaud. 1998. Haeckel, embryos, and evolution. *Science* 280: 983–984. [3]

Richerson, P. J., and R. Boyd, 2005. *Not by Genes Alone: How Culture Transformed Human Evolution*. University of Chicago Press, Chicago. [11, 16, 23]

Ricklefs, R. E. 2004. A comprehensive framework for global patterns in biodiversity. *Ecol. Lett.* 7: 1–15. [6]

Ricklefs, R. E., and D. Schluter (eds.). 1993. *Species Diversity in Ecological Communities*. University of Chicago Press, Chicago. [6, 19]

Ricklefs, R. E., and R. E. Latham. 1992. Intercontinental correlation of geographical ranges suggests stasis in ecological traits of relict genera of temperate perennial herbs. *Am. Nat.* 139: 1305–1321. [22]

Rico, P., P. Bouteillon, M. J. H. van Oppen, M. E. Knight, G. M. Hewitt, and G. F. Turner. 2003. No evidence for parallel sympatric speciation in cichlid species of the genus *Pseudotropheus* from northwestern Lake Malawi. *J. Evol. Biol.* 16: 37–46. [18]

Ridley, M. 1983. *The Explanation of Organic Diversity: The Comparative Method and Adaptations for Mating*. Oxford University Press, Oxford. [11]

Riedl, R. 1978. *Order in Living Organisms: A Systems Analysis of Evolution*. Wiley, New York. [21]

Rieseberg, L. H. 1997. Hybrid origins of plant species. *Annu. Rev. Ecol. Syst.* 28: 359–389. [18]

Rieseberg, L. H. 2001. Chromosomal arrangements and speciation. *Trends Ecol. Evol.* 16: 351–358. [17]

Rieseberg, L. H., and J. H. Willis. 2007. Plant speciation. *Science* 317: 910–914. [17]

Rieseberg, L. H., D. M. Raymond, Z. Lai, K. Livingstone, J. L. Durphy, A. E. Schwarzbach, L. A. Donovan, and C. Lexer. 2003. Major ecological transitions in wild sunflowers facilitated by hybridization. *Science* 301: 1211–1216. [18]

Rieseberg, L. H., J. Whitten, and K. Gardner. 1999. Hybrid zones and the genetic architecture of a barrier to gene flow between two sunflower species. *Genetics* 152: 713–727. [17]

Rieseberg, L. H., T. E. Wood, and E. J. Baack. 2006. The nature of plant species. *Nature* 440: 524–527. [17]

Rifkin, S. A., D. Houle, J. Kim, and K. P. White. 2005. A mutation accumulation assay reveals a broad capacity for rapid evolution of gene expression. *Nature* 438: 220–223. [8]

Risch, N.H. Tang, H. Katzenstein, and J. Ekstein. 2003. Geographic distribution of disease mutations in the Ashkenazi Jewish population supports genetic drift over selection. *Am. J. Hum. Genet.* 72: 812–822. [10]

Ritchie, M. G. 2000. The inheritance of female preference functions in a mate recognition system. *Proc. R. Soc. Lond. B* 267: 327–332. [18]

Ritchie, M. G., and S. D. F. Phillips. 1998. The genetics of sexual isolation. In D. J. Howard and S. H. Berlocher (eds.), *Endless Forms: Species and Speciation*, pp. 291–308. Oxford University Press, New York. [17]

Roberts, G. C., and C. W. Smith. 2002. Alternative splicing: Combinatorial output from the genome. *Curr. Opin. Chem. Biol.* 6: 375–383. [20]

Rodríguez, D. J. 1996. A model for the establishment of polyploidy in plants. *Am. Nat.* 147: 33–46. [18]

Roff, D. A. 2002. *Life History Evolution*. Sinauer, Sunderland, MA. [14]

Rogers, A. R. 1995. Genetic evidence for a Pleistocene population explosion. *Evolution* 49: 608–615. [10]

Rogers, A. R., and H. Harpending. 1992. Population growth makes waves in the distribution of pairwise differences. *Mol. Biol. Evol.* 9: 552–569. [10]

Romer, A. S. 1966. *Vertebrate Paleontology*. University of Chicago Press, Chicago. [3, 5]

Romer, A. S., and T. S. Parsons. 1986. *The Vertebrate Body*. Saunders College Publishing, Philadelphia. [5]

Ronquist, F. 1997. Dispersal-vicariance analysis: A new approach to the quantification of historical biogeography. *Syst. Biol.* 46: 195–203. [6]

Rose, M. R. 1991. *The Evolutionary Biology of Aging*. Oxford University Press, Oxford. [14]

Rosenthal, G. A., and M. R. Berenbaum (eds.). 1992. *Ecological and Evolutionary Processes*, 2nd ed. Vol II: *Herbivores: Their Interaction with Secondary Plant Metabolites*. Academic Press, San Diego, CA. [19]

Rosenzweig, M. L., and R. D. McCord. 1991. Incumbent replacement: Evidence for long-term evolutionary progress. *Paleobiology* 17: 202–213. [7]

Rossiter, MC. 1996. Incidence and consequences of inherited environmental effects. *Annu. Rev. Ecol. Syst.* 27: 451–476. [9]

Roth, V. L. 1988. The biological basis of homology. In C. J. Humphries (ed.), *Ontogeny and Systematics*, pp. 1–26. British Museum (Natural History). [21]

Roth, V. L. 1991 Homology and hierarchies: Problems solved and unresolved. *J. Evol. Biol.* 4: 167–194. [21]

Rothschild, L. J. and A. M. Lister (eds.). 2003. *Evolution on Planet Earth: The Impact of the Physical Environment*. Academic Press, San Diego, CA. [7]

Rothstein, S. I., and S. K. Robinson (eds.). 1998. *Parasitic Birds and Their Hosts: Studies in Coevolution*. Oxford University Press, New York. [19]

Roughgarden, J. 1971. Density-dependent natural selection. *Ecology* 52: 453–468. [14]

Roughgarden, J. 2004. *Evolution's Rainbow: Diversity, Gender, and Sexuality in Nature and People*. University of California Press, Berkeley. [15, 16]

Roush, R. T., and J. A. McKenzie. 1987. Ecological genetics of insecticide and acaricide resistance. *Annu. Rev. Entomol.* 32: 361–380. [12]

Roush, R. T., and B. E. Tabashnik (eds.). 1990. *Pesticide Resistance in Arthropods*. Chapman and Hall, New York. [12]

Rowe, L. E., and D. Houle. 1996. The lek paradox and the capture of genetic variance by condition dependent traits. *Proc. R. Soc. Lond. B* 263: 1415–1421. [15]

Royo-Torres, R., A. Cobos, and L. Alcalá. 2006. A giant European dinosaur and a new sauropod clade. *Science* 314: 1925–1927. [4]

Rundle, H. D. 2003. Divergent environments and population bottlenecks fail to generate premating isolation in *Drosophila pseudoobscura*. *Evolution* 57: 2557–2565. [18]

Rundle, H. D., and P. Nosil. 2005. Ecological speciation. *Ecol. Lett.* 8: 336–352. [18

Rundle, H. D., L. Nagel, J. W. Boughman, and D. Schluter. 2000. Natural selection and parallel speciation in sympatric sticklebacks. *Science* 287: 306–308. [18]

Ruse, M. 1979. *The Darwinian Revolution*. University of Chicago Press, Chicago. [11]

Ruse, M. 1996. *Monad to Man: The Concept of Progress in Evolutionary Biology*. Harvard University Press, Cambridge, MA. [11, 22]

Rutherford, S. L., and S. Lindquist. 1998. Hsp90 as a capacitor for morphological evolution. *Nature* 396: 336–342. [13, 22]

Rutledge, R., and 9 others. Characterization of an *AGAMOUS* homologue from the conifer black spruce (*Picea mariana*) that produces floral homeotic conversions when expressed in *Arabidopsis*. *Plant J.* 15: 625–634. [20]

Ruvolo, M. 1997. Molecular phylogeny of the hominoids: Inferences from multiple independent DNA sequence data sets. *Mol. Biol. Evol.* 14: 248–265. [2]

Ryan, M. J. 1998. Sexual selection, receiver biases, and the evolution of sex differences. *Science* 281: 1999–2003. [15]

S

Sabeti, P. C., and 16 others. 2002. Detecting recent positive selection in the human genome from haplotype structure. *Nature* 419: 832–837. [12]

Sabeti, P. C., and 9 others. 2006. Positive natural selection in the human lineage. *Science* 312: 1614–1620. [12]

Sachs, J. L., U. G. Mueller, T. P. Wilcox, and J. J. Bull. 2004. The evolution of cooperation. *Q.. Rev. Biol.* 79: 135–160. [16]

Sætre, G.-P., T. Moum, S. Bureš, M. Král, M. Adamjan, and J. Moreno. 1997. A sexually selected character displacement in flycatchers reinforces premating isolation. *Nature* 387: 589–592. [18]

Sala, O. E., and 18 others. 2000. Global biodiversity scenarios for the year 2100. *Science* 287: 1770–1774. [7]

Salvini-Plawen, L. V., and E. Mayr. 1977. On the evolution of photoreceptors and eyes. *Evol. Biol.* 10: 207–263. [22]

Salzberg, S. L., O. White, J. Peterson, and J. A. Eisen. 2001. Microbial genes in the human genome: Lateral transfer or gene loss? *Science* 292: 1903–1906. [20]

Sang, T., and Y. Zhong. 2000. Testing hybridization hypotheses based on incongruent gene trees. *Syst. Biol.* 49: 422–434. [2]

Sanmartín, I., H. Enghoff, and F. Ronquist. 2001. Patterns of animal dispersal, vicariance and diversification in the Holarctic. *Biol. J. Linn. Soc.* 73: 345–390. [6]

Santos, F. R., and 8 others. 1999. The central Siberian origin for Native American Y chromosomes. *Am. J. Hum. Genet.* 64: 619–628. [6]

Sasa, M., P. T. Chippendale, and N. A. Johnson. 1998. Patterns of postzygotic isolation in frogs. *Evolution* 52: 1811–1820. [18]

Satta, Y., and N. Takahata. 2002. Out of Africa with regional interbreeding? Modern human origins. *BioEssays* 24: 871–875. [6]

Savolainen, V., and 9 others. 2006. Sympatric speciation in palms on an oceanic island. *Nature* 441: 210–213. [18]

Scharloo, W. 1991. Canalization: Genetic and developmental aspects. *Annu. Rev. Ecol. Syst.* 22: 65–93. [13]

Schemske, D. W., and H. D. Bradshaw, Jr. 1999. Pollinator preference and the evolution of floral traits in monkeyflkowers (*Mimulus*). *Proc. Natl. Acad. Sci. USA* 96: 11910–11915. [17]

Schindewolf, O. H. 1950. *Grundfrage der Paläontologie*. Schweitzerbart, Jena, Germany. [22]

Schlichting, C. D., and M. Pigliucci. 1998. *Phenotypic Evolution: A Reaction Norm Perspective*. Sinauer, Sunderland, MA. [13]

Schliewen, U. K., D. Tautz, and S. Pääbo. 1994. Sympatric speciation suggested by monophyly of crater lake cichlids. *Nature* 368: 629–632. [18]

Schluter, D. 1996. Adaptive radiation along genetic lines of least resistance. *Evolution* 50: 1766–1774. [13]

Schluter, D. 2000. *The Ecology of Adaptive Radiation*. Oxford University Press, Oxford. [3, 13, 18, 19]

Schluter, D., and J. D. McPhail. 1992. Ecological character displacement and speciation in sticklebacks. *Am. Nat.* 140: 85–108. [19]

Schluter, D., and L. M. Nagel. 1995. Parallel speciation by natural selection. *Am. Nat.* 146: 292–301. [18]

Schluter, D., and R. E. Ricklefs. 1993. Convergence and the regional component of species diversity. In R. E. Ricklefs and D. Schluter (eds.), *Species Diversity in Ecological Communities: Historical and Geographical Perspectives*, pp. 230-240. University of Chicago Press, Chicago. [6]

Scott, E. C. 1997. Antievolution and creationism in the United States. *Annu. Rev. Anthropol.* 26: 263–289. [23]

Scott, E. C. 2005. *Evolution versus Creationism: An Introduction*. University of California Press, Berkeley. [1]

Searle, J. B. 1993. Chromosomal hybrid zones in eutherian mammals. In R. G. Harrison (ed.), *Hybrid Zones and the Evolutionary Process*, pp. 309–353. Oxford University Press, New York. [18]

Seehausen, O., P. J. Mayhew, and J. J. van Alphen. 1999. *Evolution* of colour patterns in East African cichlid fish. *J. Evol. Biol.* 12: 514–534. [18]

Seger, J. 1992. Evolution of exploiter-victim relationships. In M. J. Crawley (ed.), *Natural Enemies: The Population Biology of Predators, Parasites, and Disease*, pp. 3–25. Blackwell Scientific, Oxford. [19]

Selander, R. K. 1966. Sexual dimorphism and differential niche utilization in birds. *Condor* 68: 113–151. [19]

Sepkoski, J. J. Jr. 1984. A kinetic model of Phanerozoic taxonomic diversity. III. Post-Paleozoic families and mass extinctions. *Paleobiology* 10: 246–267. [7]

Sepkoski, J. J. Jr. 1993. Ten years in the library: New data confirm paleontological patterns. *Paleobiology* 19: 43–51. [7]

Sepkoski, J. J. Jr. 1996a. Competition in macroevolution: The double wedge revisited. In D. Jablonski, D. H. Erwin, and J. Lipps (eds.), *Evolutionary Paleobiology*, pp. 211–255. University of Chicago Press, Chicago. [7]

Sepkoski, J. J. Jr. 1996b. Large-scale history of biodiversity. In V. H. Heywood (ed.), *Global Biodiversity Assessment*, pp. 202–212. United Nations Environmental Programme. Cambridge University Press, Cambridge. [7]

Sepkoski, J. J. Jr., F. K. McKinney, and S. Lidgard. 2000. Competitive displacement among post-Paleozoic cyclostome and cheilostome bryozoans. *Paleobiology* 26: 7–18. [7]

Sereno, P. C. 1999. The evolution of dinosaurs. *Science* 284: 2137–2147. [4, 5]

Servedio, M. R., and M. A. F. Noor. 2003. The role of reinforcement in speciation: Theory and data. *Annu. Rev. Ecol. Evol. Syst.* 34: 339–364. [18]

Sgrò, C. M., and L. Partridge. 1999. A delayed wave of death from reproduction in *Drosophila*. *Science* 286: 2521–2524. [14]

Shabalina S. A., A. Y. Ogurtsov, V. A. Kondrashov, and A. S. Kondrashov. 2001. Selective constraint in intergenic regions of human and mouse genomes. *Trends Genet.* 17**:** 373–376. [20]

Shabalina, S. A., L. Y. Yampolsky, and A. S. Kondrashov. 1997. Rapid decline of fitness in panmictic populations of *Drosophila melanogaster* maintained under relaxed natural selection. *Proc. Natl. Acad. Sci. USA* 94: 13034–13039. [8]

Shaffer, H. B. and S. R. Voss. 1996. Phylogenetic and mechanistic analysis of a developmentally integrated character complex: Alternate life history modes in ambystomatid salamanders. *Am. Zool.* 36: 24–35. [21]

Shapiro, M. D., M. A. Bell, and D. M. Kingsley. 2006. Parallel genetic origins of pelvic reduction in vertebrates. *Proc. Natl. Acad. Sci. USA* 103: 13753–13758. [21]

Shapiro, M. D., M. E. Marks, C. L. Peichel, B. K. Blackman, K. S. Nereng, B. Jonsson, D. Schluter, and D. M. Kingsley. 2004. Genetic and developmental basis of evolutionary pelvic reduction in threespine sticklebacks. *Nature* 428: 717–723. [21]

Sharp, A. J., Z. Cheng, and E. E. Eichler. 2006. Structural variation of the human genome. *Annu. Rev. Genomics Hum. Genet.* 7: 407–442. [8]

Shaver, A. C., P. G. Dombrowski, J. Y. Sweeney, T. Treis, R. M. Zappala, and P. D. Sniegowski. 2002. Fitness evolution and the rise of mutator alleles in experimental *Escherichia coli* populations. *Genetics* 162: 557–566. [15]

Shelby, J. A., R. Madewell, and A. P. Moczek. 2007. Juvenile hormone mediates sexual dimorphism in horned beetles. *J. Exp. Zoolog. B. Mol. Dev. Evol.* 308: 417–427. [21]

Sherman, P. W., H. K. Reeve, and D. W. Pfennig. 1997. Recognition systems. In J. R. Krebs and N. B. Davies (eds.), *Behavioural Ecology: An Evolutionary Approach*, 4th ed., pp. 69–96. Blackwell Scientific, Oxford. [16]

Shine, R., and E. L. Charnov. 1992. Patterns of survival, growth, and maturation in snakes and lizards. *Am. Nat.* 139: 1257–1269. [14]

Short, L. L. 1965. Hybridization in the flickers (*Colaptes*) of North America. *Bull. Am. Mus. Nat. Hist.* 129: 307–428. [9]

Shoshani, J., C. P. Groves, E. L. Simons, and G. F. Gunnell. 1996. Primate phylogeny: Morphological and molecular results. *Mol. Phyl. Evol.* 5: 102–154. [2]

Shu, D.-G., and 10 others. 2003. Head and backbone of the Early Cambrian vertebrate *Haikouichthys*. *Nature* 421: 526–529. [5]

Shubin, N. 2008. *Your Inner Fish: A Journey into the 3.5-Billion-Year History of the Human Body.* Allen Lane/Pantheon, New York. [4]

Shubin, N. H., E. B. Daeschler, and F. A. Jenkins, Jr. 2006. The pectoral fin of *Tiktaalik roseae* and the origin of the tetrapod limb. *Nature* 440: 764–771. [4]

Sidor, C. A. 2001. Simplification as a trend in synapsid cranial evolution. *Evolution* 55: 1419–1442. [4]

Sidor, C. A., and J. A. Hopson. 1998. Ghost lineages and "mammalness:" Assessing the temporal pattern of character acquisition in the Synapsida. *Paleobiology* 24: 254–273. [4]

Signor, P. W. III. 1985. Real and apparent trends in species richness through time. In J. W. Valentine (ed.), *Phanerozoic Diversity Patterns: Profiles in Macroevolution*, pp. 129–150. Princeton University Press, Princeton, NJ. [7]

Signor, P. W. III. 1990. The geological history of diversity. *Annu. Rev. Ecol. Syst.* 21: 509–539. [7]

Simmons, L. W. 2001. *Sperm Competition and its Evolutionary Consequences in the Insects.* Princeton University Press, Princeton, NJ. [15]

Simon, C. J. Tang, S. Dalwadi, G. Staley, J. Deniega, and T. R. Unnasch. 2000. Genetic evidence for assortative mating between 13-year cicadas and sympatric "17-year cicadas with 13-year life cycles" provides support for allochronic speciation. *Evolution* 54: 1326–1336. [18]

Simons, E. L. 1979. The early relatives of man. In G. Isaac and R. E. F Leakey (eds.), *Human Ancestors,* pp. 22–42. W. H Freeman, San Francisco. [5]

Simpson, G. G. 1944. *Tempo and Mode in Evolution.* Columbia University Press, New York. [1, 22]

Simpson, G. G. 1953. *The Major Features of Evolution.* Columbia University Press, New York. [1, 21, 22]

Simpson, G. G. 1964. *This View of Life: The World of an Evolutionist.* Harcourt, Brace and World, New York. [22]

Sinervo, B. 1990. The evolution of maternal investment in lizards: An experimental and comparative analysis of egg size and its effect on offspring performance. *Evolution* 44: 279–294. [14]

Singh, N. D., A. M. Larracuente, and A. G. Clark. 2008. Contrasting the efficacy of selection on the X and autosomes in *Drosophila. Mol. Biol. Evol.* 25: 454–467. [20]

Slatkin, M. 1980. Ecological character displacement. *Ecology* 61: 163–177. [19]

Slatkin, M. 1985. Gene flow in natural populations. *Annu. Rev. Ecol. Syst.* 16: 393–430. [10]

Slatkin, M. 2004. A population-genetic test of founder effects and implications for Ashkenazi Jewish disease. *Am. J. Hum. Genet.* 75: 282–293. [10]

Slatkin, M., and W. P. Maddison. 1989. A cladistic measure of gene flow inferred from the phylogenies of alleles. *Genetics* 123: 603–613. [10]

Smith, A. B., and K. J. Peterson. 2002. Dating the time of origin of major clades: Molecular clocks and the fossil record. *Annu. Rev. Earth Planet. Sci.* 30: 65–88. [5]

Smith, T. B. 1993. Disruptive selection and the genetic basis of bill size polymorphism in the African finch *Pyrenestes. Nature* 363: 618–620. [12]

Smith, T. B., and L. Bernatchez. 2008. Evolutionary change in human-altered environments. *Mol. Ecol.* 17: 1–8. [13]

Smocovitis, V. B. 1996. *Unifying Biology: The Evolutionary Synthesis and Evolutionary Biology.* Princeton University Press, Princeton, NJ. [1]

Smuts, B. 1995. The evolutionary origins of patriarchy. *Human Nature* 6: 1–32. [16]

Sniegowski, P. D., and R. E. Lenski. 1995. Mutation and adaptation: The directed mutation controversy in evolutionary perspective. *Annu. Rev. Ecol. Syst.* 26: 553–578. [8]

Sniegowski, P. D., P. J. Gerrish, T. Johnson, and A. Shaver. 2000. The evolution of mutation rates: Separating causes from consequences. *Bioessays* 22: 1057–1066. [15]

Sober, E. 1984. *The Nature of Selection: Evolutionary Theory in Philosophical Focus.* MIT Press, Cambridge, MA. [11]

Sodergren, E., and 232 others. 2006. The genome of the sea urchin *Strongylocentrotus purpuratus. Science* 314: 941–952. [2, 20]

Soltis, D. E., P. S. Soltis, J. C. Pires, A. Kovarik, J. A. Tate, and E. Mavrodiev. 2004. Recent and recurrent polyploidy in *Tragopogon* (Asteraceae): Cytogenetic, genomic and genetic comparisons. 2004. *Biol. J. Linn. Soc.* 82: 485–501. [18]

Somers, C. M., B. E. McCarry, F. Malek and J. S. Quinn. 2004. Reduction of particulate air pollution lowers the risk of heritable mutations in mice. *Science* 304: 1008–1010. [8]

Soulé, M. 1966. Trends in the insular radiation of a lizard. *Am. Nat.* 100: 47–64. [18]

Soulé, M. E. 1976. Allozyme variation, its determinants in space and time. In F. J. Ayala (ed.), *Molecular Evolution*, pp. 60–77. Sinauer, Sunderland, MA. [10]

Soyfer, V. 1994. *Lysenko and the Tragedy of Soviet Science.* Rutgers University Press, New Brunswick, NJ. [23]

Spiegelman, S. 1970. Extracellular evolution of replicating molecules. In F. O. Schmitt (ed.), *The Neuro Sciences: A Second Study Program*, pp. 927–945. Rockefeller University Press, New York. [5]

Stace, C. A. 1989. Hybridization and the plant species. In K. M. Urbanska (ed.), *Differentiation Patterns in Higher Plants*, pp. 115–127. Academic Press, New York. [8]

Stahl, E. A., and J. G. Bishop. 2000. Plant-pathogen arms races at the molecular level. *Current Opinion Plant Biol.* 3: 299–304. [19]

Stanley, S. M. 1979. *Macroevolution: Pattern and Process.* W. H. Freeman, San Francisco. [7, 11, 18, 22]

Stanley, S. M. 1986. Anatomy of a regional mass extinction: Plio-Pleistocene decimation of the western Atlantic bivalve fauna. *Palaios* 1: 17–36. [7]

Stanley, S. M. 1990. The general correlation between rate of speciation and rate of extinction: Fortuitous causal linkages. In R. M. Ross and W. D. Allmon (eds.), *Causes of Evolution: A Paleontological Perspective*, pp. 103–127. University of Chicago Press, Chicago. [7]

Stanley, S. M. 1993. *Earth and Life Through Time*, 2nd ed. W.H. Freeman, New York. [5]

Stanley, S. M. 2005. *Earth System History.* W.H. Freeman, New York. [4]

Stanley, S. M., and L. A. Campbell. 1981. Neogene mass extinction of western Atlantic mollusks. *Nature* 293: 457–459. [5]

Stanley, S. M., and X. Yang. 1987. Approximate evolutionary stasis for bivalve morphology over millions of years: A multivariate, multi-lineage study. *Paleobiology* 13: 113–139. [22]

Stearns, S. C. 1992. *The Evolution of Life Histories.* Oxford University Press, Oxford. [14]

Stebbins, G. L. 1950. *Variation and Evolution in Plants.* Columbia University Press, New York. [1, 17, 18]

Stebbins, G. L. 1974. *Flowering Plants: Evolution Above the Species Level.* Belknap Press of Harvard University Press, Cambridge, MA. [13]

Stebbins, R. C. 1954. *Amphibians and Reptiles of Western North America.* McGraw-Hill, New York. [17]

Stedman, H. H., and 9 others. 2004. Myosin gene mutation correlates with anatomical changes in the human lineage. *Nature* 428: 415–418. [20]

Stephens J. C., and 27 others. 2001. Haplotype variation and linkage disequilibrium in 313 human genes. *Science* 293: 489–493. [20]

Steppan, S. J., P. C. Phillips, and D. Houle. 2002. Comparative quantitative genetics: evolution of the G matrix. *Trends Ecol. Evol.* 17: 320–327. [13]

Stern, C. 1973. *Principles of Human Genetics.* W. H. Freeman, San Francisco. [9]

Stern, D. L. 1998 A role of Ultrabithorax in morphological differences between *Drosophila* species. *Nature* 396: 463–466. [21]

Stewart, C.-B., J. W. Schilling, and A. C. Wilson. 1987. Adaptive evolution in the stomach lysozymes of foregut fermenters. *Nature* 330: 401–404. [20]

Stewart, W. N. 1983. *Paleobotany and the Evolution of Plants*. Cambridge University Press, Cambridge. [5]

Storfer, A., and A. Sih. 1998. Gene flow and ineffective antipredator behavior in a stream-breeding salamander. *Evolution* 52: 558–565. [12]

Storfer, A., J. Cross, V. Rush, and J. Caruso. 1999. Adaptive coloration and gene flow as a constraint to local adaptation in the streamside salamander, *Ambystoma barbouri*. *Evolution* 53: 889–898. [12]

Strait, D. S., F. E. Grine,and M. A. Moniz. 1997. A reappraisal of early hominid phylogeny. *J. Hum. Evol.* 32: 17–82. [4]

Strassmann, J. E., and D. C. Queller. 2007. Insect societies as divided organisms: The complexities of purpose and cross-purpose. *Proc. Natl. Acad. Sci. USA* 104 (Suppl. 1): 8619–8626. [16]

Strickberger, M. W. 1968. *Genetics*. Macmillan, New York. [8]

Strobeck, C. 1983. Expected linkage disequilibrium for a neutral locus linked to a chromosomal arrangement. *Genetics* 103: 545–555. [12]

Strömberg, C. A. E. 2006. Evolution of hypsodonty in equids: testing a hypothesis of adaptation. *Paleobiology* 32: 236–258. [4]

Stutt, A. D., and M. T. Siva-Jothy. 2001. Traumatic insemination and sexual conflict in the bed bug *Cimex lectularius*. *Proc. Natl. Acad. Sci. USA* 98: 5683–5687. [15]

Sundström, L., M. Chapuisat, and L. Keller. 1996. Conditional manipulation of sex ratio by ant workers: A test of kin selection theory. *Science* 274: 993–995. [16]

Suzuki, Y., and H. F. Nijhout. 2006. Evolution of a polyphenism by genetic accommodation. *Science* 311: 650–652. [13]

Swanson, W. J., and V. D. Vacquier. 2002. The rapid evolution of reproductive proteins. *Nature Rev. Genet.* 3: 137–144. [12, 20]

Szathmáry, E. 1993. Coding coenzyme handles: A hypothesis for the origin of the genetic code. *Proc. Natl. Acad. Sci. USA* 90: 9916–9920. [5]

Szymura, J. M. 1993. Analysis of hybrid zones with *Bombina*. In R. G. Harrison (ed.), *Hybrid Zones and the Evolutionary Process*, pp. 261–289. Oxford University Press, New York. [17]

T

Taberlet, P., L. Fumagalli, A.-G. Wust-Saucy, and J.-F. Cosson. 1998. Comparative phylogeography and postglacial colonization routes in Europe. *Mol. Ecol.* 7: 453–464. [6]

Tagle, D. A., J. L. Slightom, R. T. Jones, and M. Goodman. 1991. Concerted evolution led to high expression of a prosimian primate δ-globin gene locus. *J. Biol. Chem.* 266: 7469–7480. [20]

Tajima, F. 1989. Statistical method for testing the neutral mutation hypothesis by DNA polymorphism. *Genetics* 123: 585–595. [12]

Taper, M. L., and T. J. Case. 1992. Coevolution among competitors. *Oxford Surv. Evol. Biol.* 8: 63–109. [19]

Tauber, M. J., C. A. Tauber, and S. Masaki. 1986. *Seasonal Adaptations of Insects*. Oxford University Press, New York. [13]

Templeton, A. R. 1980. The theory of speciation via the founder principle. *Genetics* 94: 1011–1038. [18]

Templeton, A. R. 1989. The meaning of species and speciation: A genetic perspective. In D. Otte and J. A. Endler (eds.), *Speciation and Its Consequences*, pp. 3–27. Sinauer, Sunderland, MA. [17]

Templeton, A. R. 1998. Nested clade analyses of phylogeographic data: testing hypotheses about gene flow and population history. *Mol. Ecol.* 7: 381–397. [6]

Templeton, A. R. 2007. Genetics and recent human evolution. *Evolution* 61: 1507–1519. [6]

Thewissen, J. G. M., and S. Bajpai. 2001. Whale origins as a poster child for macroevolution. *BioScience* 51: 1017–1029. [4]

Thewissen, J. G. M., and E. M. Williams. 2002. The early radiations of Cetacea (Mammalia): Evolutionary pattern and developmental correlations. *Annu. Rev. Ecol. Syst.* 33: 73–90. [4]

Thewissen, J. G. M., L. N. Cooper, M. T. Clementz, S. Bajpal, and B. N. Tiwari. 2007. Whales originated from aquatic artiodactyls in the Eocene epoch of India. *Nature* 450: 1190–1194. [4]

Thomas, C. D., A. Cameron, R. E. Green, and 16 others. 2004. Extinction risk from climate change. *Nature* 427: 145–148. [5, 7]

Thompson, J. N. 1994. *The Coevolutionary Process*. University of Chicago Press, Chicago. [19]

Thompson, J. N. 1999. Specific hypotheses on the geographic mosaic of coevolution. *Am. Nat.* 153: S1–S14. [19]

Thompson, J. N. 2004. *The Geographic Mosaic of Coevolution*. University of Chicago Press, Chicago. [22]

Thomson, J. D., and J. Brunet. 1990. Hypotheses for the evolution of dioecy in seed plants. *Trends Ecol. Evol.* 5: 11–16. [15]

Thornhill, N. W. (ed.). 1993. *The Natural History of Inbreeding and Outbreeding*. University of Chicago Press, Chicago. [15]

Thornhill, R., and J. Alcock. 1983. *The Evolution of Insect Mating Systems*. Harvard University Press, Cambridge, MA. [15]

Thornton, J. W. 2004. Resurrecting ancient genes: Experimental analysis of extinct molecules. *Nat. Rev. Genet.* 5: 366–375. [3]

Thrall, P. H., and J. J. Burdon. 2003. Evolution of virulence in a plant host-pathogen metapopulation. *Science* 299: 1735–1737. [19]

Tilley, S. G., P. A. Verrell, and S. J. Arnold. 1990. Correspondence between sexual isolation and allozyme differentiation: A test in the salamander *Desmognathus ochrophaeus*. *Proc. Natl. Acad. Sci. USA* 87: 2715–2719. [18]

Tilmon, K. J. (ed.) 2008. *Specialization, Speciation, and Radiation: The Evolutionary Biology of Herbivorous Insects*. University of California Press, Berkeley. [19]

Tinbergen, J. M., and S. Daan. 1990. Family planning in the great tit (*Parus major*): optimal clutch size as integration of parent and offspring fitness. *Behaviour* 114: 161–190. [14]

Tishkoff, S. A., and 18 others. 2007. Convergent adaptation of human lactase persistence in Africa and Europe. *Nature Genet.* 39: 31–40. [11, 12]

Tizard, B. 1973. IQ and race. *Nature* 247: 316. [9]

Toju, H., and T. Sota. 2006. Imbalance of predator and prey armament: Geographic clines in phenotypic interface and natural selection. *Am. Nat.* 167: 105–117. [19]

Tomoyasu, Y., S. R. Wheeler, and R. E. Denell. 2005. *Ultrabithorax* is required for membranous wing identity in the beetle *Tribolium castaneum*. *Nature* 433: 643–647. [22]

Tooby, J., and L. Cosmides. 1992. The psychological foundations of culture. In J. H. Barkow, L. Cosmides, and J. Tooby (eds.), *The Adapted Mind: Evolutionary Psychology and the Generation of Culture*, pp. 19-136. Oxford University Press, New York. [16]

Travis, J. 1989. The role of optimizing selection in natural populations. *Annu. Rev. Ecol. Syst.* 20: 279–296. [13]

Trivers, R. L. 1971. The evolution of reciprocal altruism. *Q. Rev. Biol.* 46: 35–57. [16]

Trivers, R. L. 1972. Parental investment and sexual selection. In B. Campbell (ed.), *Sexual Selection and the Descent of Man*, pp. 136–179. Heinemann, London. [15]

Trivers, R. L. 1974. Parent-offspring conflict. *Am. Zool.* 11: 249–264. [16]

True, J. R., and S. B. Carroll. 2002. Gene co-option in physiological and morphological evolution. *Annu. Rev. Cell. Dev. Biol.* 18: 53–80. [11, 21]

Tsubota, T., K. Saigo, and T. Kojima. 2008. Hox genes regulate the same character by different strategies in each segment. *Mech. Dev.* 125: 894–905. [21]

Turelli, M. 1984. Heritable genetic variation via mutation-selection balance: Lerch's zeta meets the abdominal bristle. *Theor. Pop. Biol.* 25: 138–193. [13]

Turelli, M. 1988. Phenotypic evolution, constant covariances, and the maintenance of additive variance. *Evolution* 42: 1342–1347. [13]

Turelli, M., J. H. Gillespie, and R. Lande. 1988. Rate tests for selection on quantitative characters during macroevolution and microevolution. *Evolution* 42: 1085–1089. [13]

Turelli, M., N. H. Barton, and J. A. Coyne. 2001. Theory and speciation. *Trends Ecol. Evol.* 16: 330–343. [18]

Turner, J. R. G. 1977. Butterfly mimicry: The genetical evolution of an adaptation. *Evol. Biol.* 10: 163–206. [19]

Turner, T. F., J. P. Wares, and J. R. Gold. 2002. Genetic effective size is three orders of magnitude smaller than adult census size in an abundant, estuarine-dependent marine fish (*Sciaenops ocellatus*). *Genetics* 162: 1329–1339. [10]

U

Ullstrup, A. J. 1972. The impacts of the southern corn leaf blight epidemic of 1970–1971. *Annu.* Rev. Phytopathol. 19: 37–50. [19]

Underhill, P. A., and T. Kivisild. 2007. Use of Y chromosome and mitochondrial DNA population structure in tracing human migrations. *Annu. Rev. Genetics* 41: 539–544. [6]

Underhill, P. A., G. Passarino, A. A. Lin, P. Shen, M. Mirazón Lahr, R. A. Foley, P. J. Oefner, and L. L. Cavalli-Sforza. 2001. The phylogeography of Y chromosome binary haplotypes and the origins of human populations. *Ann. Hum. Genet.* 65: 43–62. [6]

Ungerer, M. C., S. J. E. Baird, J. Pan, and L. H. Rieseberg. 1998. Rapid hybrid speciation in wild sunflowers. *Proc. Natl. Acad. Sci. USA* 95: 11757–11762. [18]

Uyenoyama, M. K., K. E. Holsinger, and D. M. Waller. 1993. Ecological and genetic factors directing the evolution of self-fertilization. *Oxford Surv. Evol. Biol.* 9: 327–381. [15]

V

Vacquier, V. D. 1998. Evolution of gamete recognition proteins. *Science* 281: 1995–1998. [15]

Valentine, J. W. 2004. *On the Origin of Phyla*. University of Chicago Press, Chicago. [5]

Valentine, J. W., T. C. Foin, and D. Peart. 1978. A provincial model of Phanerozoic marine diversity. *Paleobiology* 4: 55–66. [7]

van Ham, R. C., and 15 others. 2003. Reductive genome evolution in *Buchnera aphidicola*. *Proc. Natl. Acad. Sci. USA* 100: 581–586. [3, 20]

Van Heuverswyn, F., and 15 others. SIV infection in wild gorillas. Nature 444: 164. [23]

van Noordwijk, A. J., and G. deJong. 1986. Acquisition and allocation of resources: Their influence on variation in life history tactics. *Am. Nat.* 128: 137–142. [14]

van Rheede, T., T. Bastiaans, D. S. Boone, S. B. Hedges, W. W. de Jong, and O. Madsen. 2006. The platypus is in its place: nuclear genes and indels confirm the sister group relation of monotremes and therians. *Mol. Biol. Evol.* 23: 587–597. [2]

Van Valen, L. 1973. A new evolutionary law. *Evol. Theory* 1: 1–30. [7]

Vanhooydonck, B., A. Y. Herrel, R. van Damme, and D. J. Irschick. 2005. Does dewlap size predict male bite performance in Jamaican *Anolis* lizards? *Func. Ecol.* 19: 38–42. [16]

Vasey, P. L. 1995. Homosexual behavior in primates: A review of evidence and theory. *Internatl. J. Primatol.* 16: 173–204. [16]

Vaughan, T. A. 1986. *Mammalogy*, 3rd ed. Saunders College Publishing, Philadelphia, PA. [3]

Vehrencamp, S. L. 1983. Optimal degree of skew in cooperative societies. *Am. Zool.* 23: 327–335. [16]

Venter, J. C. et al. 2001. The sequence of the human genome. *Science* 291: 1304–1351. [3, 8]

Vermeij, G. J. 1987. *Evolution and Escalation: An Ecological History of Life*. Princeton University Press, Princeton, NJ. [5]

Vigilant, L., M. Stoneking, H. Harpending, K. Hawkes, and A. C. Wilson. 1991. African populations and the evolution of human mitochondrial DNA. *Science* 253: 1503–1507. [6, 10]

Vines, T. H., and D. Schluter. 2006. Strong assortative mating between allopatric sticklebacks as a by-product of adaptation to different environments. *Proc. R. Soc. Lond. B* 273: 911–916. [18]

Voelker, R. A., H. E. Schaffer, and T. Mukai. 1980. Spontaneous allozyme mutations in *Drosophila melanogaster*: Rate of occurrence and nature of the mutants. *Genetics* 94: 961–968. [10]

Voight, B. F., S. Kudaravalli, X. Wen, and J. K. Pritchard. 2006. A map of recent positive selection in the human genome. *PLoS Biol.* 4: e72 (doi: 10.1371/journal.pbio.0040072). [12]

von Baer, K. E. 1828. *Entwicklungsgeschichte der Thiere: Beobachtung und Reflexion*. Bornträger, Konigsberg. [3]

Vonk, F. J., and M. K. Richardson. 2008. Developmental biology: Serpent clocks tick faster. *Nature* 454: 282–283. [21]

Voss, S. R., and H. B. Shaffer. 2000. Evolutionary genetics of metamorphic failure using wild-caught vs. laboratory axolotls (*Ambystoma mexicanum*). *Molec. Ecol.* 9: 1401–1407. [21]

Voss, S. R., and J. J. Smith. 2005. Evolution of salamander life cycles: A major-effect quantitative trait locus contributes to discrete and continuous variation for metamorphic timing. *Genetics* 170: 275–281. [21]

Voss, S. R., K. L. Prudic, J. C. Oliver, and H. B. Shaffer. 2003. Candidate gene analysis of metamorphic timing in ambystomatid salamanders. *Molec. Ecol.* 12: 1217–1223. [21]

W

Waddington, C. H. 1942. Canalization of development and the inheritance of acquired characters. *Nature* 150: 563–565. [21]

Waddington, C. H. 1953. Genetic assimilation of an acquired character. *Evolution* 7: 118–126. [13]

Wade, M. J. 1977. An experimental study of group selection. *Evolution* 31: 134–153. [11]

Wade, M. J. 1979. The primary characteristics of *Tribolium* populations group selected for increased and decreased population size. *Evolution* 33: 749–764. [11]

Wade, M. J., and S. J. Arnold. 1980. The intensity of sexual selection in relation to male sexual behavior, female choice, and sperm precedence. *Anim. Behav.* 28: 446–461. [15]

Wagner, A. 2005. *Robustness and Evolvability in Living Systems*. Princeton University Press, Princeton, NJ. [22]

Wagner, A., G. P. Wagner, and P. Similion. 1994. Epistasis can facilitate the evolution of reproductive isolation by peak shifts: A two-locus two-allele model. *Genetics* 138: 533–545. [18]

Wagner, G. P. 1988. The influence of variation and of developmental constraints on the rate of multivariate phenotypic evolution. *J. Evol. Biol.* 1: 45–66. [13]

Wagner, G. P. 1989a. The biological homology concept. *Annu. Rev. Ecol. Syst.* 20: 51–69. [3, 21]

Wagner, G. P. 1989b. The origin of morphological characters and the biological basis of homology. *Evolution* 43: 1157–1171. [21]

Wagner, G. P. 1996. Homologues, natural kinds and the evolution of modularity. *Am. Zool.* 36: 36–43. [3, 22]

Wagner, G. P., and L. Altenberg. 1996. Perspective: Complex adaptations and the evolution of evolvability. *Evolution* 50: 967–976. [13, 22]

Wagner, G. P., G. Booth, and H. Bagheri-Chaichian. 1997. A population genetic theory of canalization. *Evolution* 51: 329–347. [13]

Wagner, G. P., J. P. Kenney-Hunt, M. Pavlicev, J. R. Peck, D. Waxman, and J. M. Cheverud. 2008. Pleiotropic scaling of gene effects and the 'cost of complexity'. *Nature* 452: 470–472. [13]

Wagner, P. J. 1996. Contrasting the underlying patterns of active trends in morphologic evolution. *Evolution* 50: 990–1007. [21, 22]

Wake, D. B. 1982. Functional and developmental constraints and opportunities in the evolution of feeding systems in urodeles. In D. Mossakowski and G. Roth (eds.), *Environmental Adaptation and Evolution*, pp. 51–66. G. Fischer, Stuttgart. [22]

Walsh, H. E., I. L. Jones, and V. L. Friesen. 2005. A test of founder effect speciation using multiple loci in the auklets (*Aethia* spp.) *Genetics* 171: 1885–1894. [17]

Wang, J., and M. C. Whitlock. 2003. Estimating effective population size and migration rates from genetic samples over space and time. *Genetics* 163: 429–446. [10]

Waples, R. S. 1989. A generalized approach for estimating effective population size from temporal changes in allele frequency. *Genetics* 121: 379–391. [10]

Warner, R. R. 1984. Mating behavior and hermaphroditism in coral reef fishes. *Am. Sci.* 72: 128–136. [15]

Warren, R. W., L. Nagy, J. Selegue, J. Gates, and S. B. Carroll. 1994. Evolution of homeotic gene function in flies and butterflies. *Nature* 372: 458–461. [21]

Waser, N. M., and M. V. Price. 1989. Optimal outcrossing in *Ipomopsis aggregata*: Seed set and offspring fitness. *Evolution* 43: 1097–1109. [15]

Weatherbee, S. D., G. Halder, J. Kim, A. Hudson, and S. B. Carroll. 1998 Ultrabithorax regulates genes at several levels of the wing-patterning hierarchy to shape the development of the *Drosophila* haltere. *Genes Devel.* 12: 1474–1482. [21]

Webb, C. O. 2000. Exploring the phylogenetic structure of ecological communities: An example for rain forest trees. *Am. Nat.* 156: 145–155. [19]

Webb, C. O., D. D. Ackerley, M. A. McPeek, and M. J. Donoghue. 2002. Phylogenies and community ecology. *Annu. Rev. Ecol. Evol. Syst.* 33: 475–505. [19]

Weber, K. E. 1996. Large genetic change at small fitness cost in large populations of *Drosophila melanogaster* selected for wind tunnel flight: rethinking fitness surfaces. *Genetics* 144: 205–213. [13]

Weber, K. E., and L. T. Diggins. 1990. Increased selection response in larger populations. II. Selection for ethanol vapor resistance in *Drosophila melanogaster* at two population sizes. *Genetics* 125: 585–597. [13]

Weeks, A. R., M. Turelli, W. R. Harcombe, K. T. Reynolds, and A. A. Hoffmann. 2007. From parasite to mutualist: Rapid evolution of *Wolbachia* in natural populations of *Drosophila*. *PLoS Biology* 5: 997–1005. [19]

Weis, A. E., W. G. Abrahamson, and M. C. Andersen. 1992. Variable selection on *Eurosta's* gall size. I. The extent and nature of variation in phenotypic selection. *Evolution* 46: 1674–1697. [13]

Weishampel, D. B., P. Dodson, and H. Osmólska. 2004. *The Dinosauria*, 2nd ed. University of California Press, Berkeley. [5]

Welch, J. J., and L. Bromham. 2005. Molecular dating when rates vary. *Trends Ecol. Evol.* 20: 320–327. [2]

Wellman, C. H., P. L. Osterloff, and U. Mohiuddin. 2003. Fragments of the earliest land plants. *Nature* 425: 282–290. [5]

Wells, J. 2000. *Icons of Evolution: Science or Myth?* Regnery, Washington, D.C. [12]

Wen, J. 1999. *Evolution* of eastern Asian and eastern North American disjunct distributions of flowering plants. *Annu. Rev. Ecol. Syst.* 30: 421–455. [6]

Wenseleers, T. and F. L. W. Ratnieks. 2006. Comparative analysis of worker reproduction and policing in eusocial Hymenoptera supports relatedness theory. *Am. Nat.* 168: E163–E179. [16]

Werner, P. A., and W. J. Platt. 1976. Ecological relationships of co-occurring goldenrods (*Solidago*: Compositae). *Am. Nat.* 110: 959–971. [14]

Werner, T. K., and T. W. Sherry. 1987. Behavioral feeding specialization in *Pinaroloxias inornata*, the "Darwin's Finch" of Cocos Island, Costa Rica. *Proc. Natl. Acad. Sci. USA* 84: 5506–5510. [18, 19]

Werren, J. H. 1980. Sex ratio adaptations to local mate competition in a parasitic wasp. *Science* 208: 1157–1159. [15]

Werren, J. H. 1998. *Wolbachia* and speciation. In D. J. Howard and S. H. Berlocher (eds.), *Endless Forms: Species and Speciation*, pp. 245–260. Oxford University Press, New York. [17]

West-Eberhard, M. J. 1983. Sexual selection, social competition, and speciation. *Q. Rev. Biol.* 58: 155–183. [15, 18]

West-Eberhard, M. J. 2003. *Developmental Plasticity and Evolution*. Oxford University Press, New York. [13]

West, S. A., E. T. Kiers, E. L. Simms, and R. F. Denison. 2002. Sanctions and mutualism stability: Why do rhizobia fix nitrogen? *Proc. R. Soc. Lond. B* 269: 685–694. [16]

Westneat, D. F., and I. R. K. Stewart. 2003. Extra-pair paternity in birds: causes, correlates, and conflict. *Annu. Rev. Ecol. Evol. Syst.* 34: 365–396. [11]

Westoby, M., M. R. Leishman, and J. M. Lord. 1997. On misinterpreting the "phylogenetic correlation." *J. Ecol.* 83: 531–534. [11]

Wheeler, B. M., A. M. Heimberg, V. N. Moy, E. A. Sperling, T. W. Holstein, S. Heber and K. J. Peterson. 2009. The deep evolution of metazoan microRNAs. *Evol. Dev.* 11: 50–68. [20]

White, M. J. D. 1978. *Modes of Speciation*. W. H. Freeman, San Francisco. [12]

Whitfield, J. B., and P. J. Lockhart. 2007. Deciphering ancient rapid radiations. *Trends Ecol. Evol.* 22: 258–265. [2]

Whiting, M. F., S. Bradler, and T. Maxwell. 2003. Loss and recovery of wings in stick insects. *Nature* 421: 264–267. [21]

Whittall, J. B., and S. A. Hodges. 2007. Pollinator shifts drive increasingly long nectar spurs in columbine flowers. *Nature* 447: 706–709. [3]

Wible, J. R., G. W. Rougier, M. J. Novacek, and R. J. Asher. 2007. Cretaceous eutherians and Laurasian origin for placental mammals near the K/T boundary. *Nature* 447: 1003–918. [5]

Wichman, H. A., L. A. Scott, C. D. Yarber, and J. J. Bull. 2000. Experimental evolution recapitulates natural evolution. *Phil. Trans. R. Soc. Lond. B* 355: 1677–1684. [8]

Wiens, J. J. 2001. Widespread loss of sexually selected traits: How the peacock lost its spots. *Trends Ecol. Evol.* 16: 517–523. [15]

Wiens, J. J. 2004. Speciation and ecology revisited: Phylogenetic niche conservatism and the origin of species. *Evolution* 58: 193–197. [18]

Wiens, J. J., and M. J. Donoghue. 2004. Historical biogeography, ecology and species richness. *Trends Ecol. Evol.* 19: 639–644. [6]

Wiens, J. J., and C. H. Graham. 2005. Niche conservatism: Integrating evolution, ecology, and conservation biology. *Annu. Rev. Ecol. Evol. Syst.* 36: 519–539. [6, 22]

Wiens, J. J., C. H. Graham, D. S. Moen, S. A. Smith, and T. W. Reeder. 2006. Evolutionary and ecological causes of the latitudinal diversity gradient in hylid frogs: Treefrog trees unearth the roots of high tropical diversity. *Am. Nat.* 168: 579–596. [6]

Wildman, D. E., and 9 others. 2007. Genomics, biogeography, and the diversification of placental mammals. *Proc. Natl. Acad. Sci. U.S.A.* 104: 14395–14400. [2]

Wiley, E. O. 1978. The evolutionary species concept reconsidered. *Syst. Zool.* 27: 17–26. [17]

Wilkins, A. S. 2002. *The Evolution of Developmental Pathways*. Sinauer, Sunderland, MA. [21]

Wilkinson, G. S. 1988. Reciprocal altruism in bats and other mammals. *Ethol. Sociobiol.* 9: 85–100. [16]

Williams, E. E. 1972. The origin of faunas: Evolution of lizard congeners in a complex island fauna—a trial analysis. *Evol. Biol.* 6: 47–89. [6]

Williams, G. C. 1957. Pleiotropy, natural selection, and the evolution of senescence. *Evolution* 11: 398–411. [14]

Williams, G. C. 1966. *Adaptation and Natural Selection*. Princeton University Press, Princeton, NJ. [11, 14]

Williams, G. C. 1992a. *Gaia*, nature worship and biocentric fallacies. *Q. Rev. Biol.* 67: 479–486. [11]

Williams, G. C. 1992b. *Natural Selection: Domains, Levels, and Challenges*. Oxford University Press, New York. [11, 21]

Williams, T. M., J. E. Selegue, T. Werner, N. Gompel, A. Kopp, and S. B. Carroll. 2008. The regulation and evolution of a genetic switch controlling sexually dimorphic traits in *Drosophila*. *Cell* 134: 610–623. [21]

Willig, M. R., D. M. Kaufman, and R. D. Stevens. 2003. Latitudinal gradients of biodiversity: pattern, process, scale, and synthesis. *Annu. Rev. Ecol. Evol. Syst.* 34: 273–309. [6]

Williston, S. W. 1925. *The Osteology of the Reptiles*. Harvard University Press, Cambridge, MA. [5]

Wilson, A. C., S. S. Carlson, and T. J. White. 1977. Biochemical evolution. *Annu. Rev. Biochem.* 46: 573–639. [2]

Wilson, A. M., and K. Thompson. 1989. A comparative study of reproductive allocation in 40 British grasses. *Functional Ecology* 3: 297–302. [14]

Wilson, D. S. 1975. A theory of group selection. *Proc. Natl. Acad. Sci. USA* 72: 143–146. [16]

Wilson, D. S. 1983. The group selection controversy: History and current status. *Annu. Rev. Ecol. Syst.* 14: 159–187. [11]

Wilson, D. S. 1997. Altruism and organism: Disentangling the themes of multilevel selection theory. *Am. Nat.* 150 (Suppl.): S122–S134. [16]

Wilson, D. S. 2002. *Darwin's Cathedral: Evolution, Religion, and the Nature of Society*. University of Chicago Press, Chicago. [16]

Wilson, D. S., and R. K. Colwell. 1981. The evolution of sex ratio in structured demes. *Evolution* 35: 882–897. [15]

Wilson, D. S., and E. O. Wilson. 2007. Rethinking the theoretical foundation of sociobiology. *Quart. Rev. Biol.* 82: 327–348. [16]

Wilson, E. O. 1971. *The Insect Societies*. Harvard University Press, Cambridge, MA. [16]

Wilson, E. O. 1975. *Sociobiology: The New Synthesis*. Harvard University Press, Cambridge, MA. [16, 23]

Wilson, E. O. 1992. *The Diversity of Life*. Harvard University Press, Cambridge, MA. [5, 7]

Wilson, E. O., F. M. Carpenter, and W. L. Brown Jr. 1967. The first Mesozoic ants. *Science* 157: 1038–1040. [4]

Wilson, G., and Q. Rahman. 2005. *Born Gay: The Psychobiology of Sexual Orientation*. Peter Owen Publishers, London. [16]

Wilson, R. S., M. J. Angilleta, Jr., R. S. James, C. Navos, and F. Seebacher. 2007. Dishonest signals of strength in male slender crayfish (*Cherax dispar*) during agonistic encounters. *Am. Nat.* 170: 284–291. [16]

Winkler, I. S., and C. Mitter. 2008. The phylogenetic dimension of insect–plant interactions: A review of recent evidence. pp. 240–263. In K. J. Tilmon (ed.), *Specialization, Speciation, and Radiation: The Evolutionary Biology of Herbivorous Insects*. University of California Press, Berkeley. [6. 7, 19]

Winter, K. U., A. Becker, T. Munster, J. T. Kim, H. Saedler, and G. Theissen. 1999. MADS-box genes reveal that gnetophytes are more closely related to conifers than to flowering plants. *Proc. Natl. Acad. Sci. USA* 96: 7342–7347. [21]

Wittkopp, P. J., B. K. Haerum, and A. G. Clark. 2004. Evolutionary changes in *cis* and *trans* gene regulation. *Nature* 430: 85–88. [21]

Woese, C. R. 2000. Interpreting the universal phylogenetic tree. *Proc. Natl. Acad. Sci. USA* 97: 8392–8396. [5]

Wolfe, N. D., C. P. Dunavan, and J. Diamond. 2007. Origins of major human infectious diseases. *Nature* 447: 279–283. [23]

Wood, B. A. 2002. Hominid revelations from Chad. *Nature* 418: 133–135. [4]

Wood, B. A., and M. C. Collard. 1999. The human genus. *Science* 284: 65–71. [4]

Woodruff, R. C., H. Huai and J. N. Thonpson, Jr. 1996. Clusters of identical new mutation in the evolutionary landscape. *Genetica* 98: 149–160. [8]

Woolhouse, M. E. J., D. T. Haydon, and R. Antia. 2005. Emerging pathogens: The epidemiology and evolution of species jumps. *Trends Ecol. Evol.* 20: 238–244. [19]

Wray, G. A. 2007. The evolutionary significance of *cis*-regulatory mutations. *Nat. Rev. Genet.* 8: 206–216. [21]

Wright, S. 1931. Evolution in Mendelian populations. *Genetics* 16: 97–159. [10]

Wright, S. 1935. The analysis of variance and the correlations between relatives with respect to deviations from an optimum. *J. Genet.* 30: 243–256. [10]

Wu, C.-I., and H. Hollocher. 1998. Subtle is nature: Differentiation and speciation. In D. J. Howard and S. H. Berlocher (eds.), *Endless Forms: Species and Speciation*, pp. 339–351. Oxford University Press, New York. [17]

Wu, D., and 12 others. 2006. Metabolic complementarity and genomics of the dual bacterial symbiosis of sharpshooters. *PLoS Biology* 4(6): 1079–1092. [19]

Wu, P., T. X. Jiang, S. Suksaweang, R. B. Widelitz, and C. M. Chuong. 2004. Molecular shaping of the beak. *Science* 305: 1465–1466. [21]

Wyatt, R. 1988. Phylogenetic aspects of the evolution of self-pollination. In L. D. Gottlieb and S. K. Jain (eds.), *Plant Evolutionary Biology*, pp. 109–131. Chapman & Hall, London. [15]

X

Xu, X., Z. Zhou, X. Wang, X. Kuang, F. Zhang, and X. Du. 2003. Four-winged dinosaurs from China. *Nature* 421: 335–340. [4]

Xu, Z. P., I. Woo, H. Her, D. R. Beier, and R. L. Maas. 1997. Mouse *Eya* homologues of the *Drosophila eyes absent* gene require *Pax6* for expression in lens and nasal placode. *Development* 124: 219–231. [21]

Y

Yoder, A. D., and M. D. Nowak. 2006. Has vicariance or dispersal been the predominant biogeographic force in Madagascar? Only time will tell. *Annu. Rev. Ecol. Evol. Syst.* 37: 405–431. [6]

Yoder, A. D., and Z. Yang. 2004. Divergence dates for Malagasy lemurs estimated from multiple gene loci: Geological and evolutionary context. *Mol. Ecol.* 13: 757–773. [6]

Yoo, B. H. 1980. Long-term selection for a quantitative character in large replicate populations of *Drosophila melanogaster*. I. Response to selection. II. Lethals and visible mutants with large effects. *Genet. Res.* 35: I. 1–17, II. 19–31. [13]

Yu, N., M. I. Jensen-Seaman, L. Chemnick, O. Ryder, and W.-H. Li. 2004. Nucleotide diversity in gorillas. *Genetics* 166: 1375–1383. [10]

Z

Zahavi, A. 1975. Mate selection: A selection for a handicap. *J. Theor. Biol.* 53: 205–214. [15]

Zangerl, A. R., and M. R. Berenbaum. 2003. Phenotype matching in wild parsnip and parsnip webworm: Causes and consequences. *Evolution* 57: 806–815. [19]

Zeh, J. A. 2004. Sexy sons: A dead end for cytoplasmic genes. *Proc. R. Soc. Lond. B* 271: S306–S309. [16]

Zeng, Z.-B., J. Liu, L. F. Stam, C.-H. Kao, J. M. Mercer, and C. C. Laurie. 2000. Genetic architecture of a morphological shape difference between two *Drosophila* species. *Genetics* 154: 299–310. [17]

Zhang, J., A. M. Dean, F. Brunet, and M. Long. 2004. Evolving protein functional diversity in new genes of *Drosophila. Proc. Natl. Acad. Sci. USA* 101: 16246–16250. [20]

Zhang, J., and D. M. Webb. 2003. Evolutionary deterioration of the vomeronasal pheromone transduction pathway in catarrhine primates. *Proc. Natl. Acad. Sci. USA* 100: 8337–8341. [20]

Zhang, J., D. M. Webb and O. Podlaha. 2002. Accelerated protein evolution and origins of human-specific features: FOXP2 as an example. *Genetics* 162: 1825–1835. [8]

Zhang, J., H. F. Rosenberg, and M. Nei. 1998. Positive Darwinian selection after gene duplication in primate ribonuclease genes. *Proc. Natl. Acad. Sci. USA* 95: 3708–3713. [20]

Zhang, X.-S., and W. G. Hill. 2005. Genetic variability under mutation selection balance. *Trends Ecol. Evol.* 20: 468–470. [13]

Zielenski, J., and L.-C. Tsui. 1995. Cystic fibrosis: Genotypic and phenotypic variations. *Annu. Rev. Genet.* 29: 777–807. [8]

Zipursky, S. L., W. M. Wojtowicz, and D. Hattori. 2006. Got diversity? Wiring the fly brain with Dscam. *Trends in Biochemical Sciences* 31: 581–588. [20]

Zouros, E. 1987. On the relation between heterozygosity and heterosis: An evaluation of the evidence from marine mollusks. In M. C. Rattazi, J. G. Scandalios, and G. S. Whitt (eds.), *Isozymes: Current Topics in Biological and Medical Research*, pp. 255–270. Alan R. Liss, New York. [12]

Zuckerkandl, E., and L. Pauling. 1965. Evolutionary divergence and convergence of proteins. In V. Bryson and H. J. Vogel (eds.), *Evolving Genes and Proteins*, pp. 97–166. Academic Press, New York. [2]

Zufall, R. A., and M. D. Rausher. 2004. Genetic changes associated with floral adaptation restrict future evolutionary potential. *Nature* 428: 847–850. [22]

Zuk, M., J. T. Rotenberry, and R. M. Tinghitella. 2006. Silent night: Adaptive disappearance of a sexual signal in a parasitized population of field crickets. *Biol. Lett.* 2: 521–524. [15]

Index

Numbers in *italic* refer to information in an illustration or illustration caption.

A

C

D

E

H

I

J

N

O

Q

R

About the Book

Editor: Andrew D. Sinauer

Project Editor: Carol J. Wigg

Copy Editor: Elizabeth Pierson

Photo Editor: David McIntyre

Production Manager: Christopher Small

Book and Cover Design: Jefferson Johnson

Book Layout: Jefferson Johnson

Illustration Program: Elizabeth Morales Illustration Service

Manufacturer: Courier Corporation, Inc.